MW01680250

AutoCAD LT 2002
A Problem Solving Approach

Sham Tickoo

Professor
Department of Manufacturing Engineering Technologies
Purdue University Calumet
Hammond, Indiana
U.S.A.

THOMSON LEARNING™

autodesk® Press

AutoCAD LT 2002®: A Problem-Solving Approach
Sham Tickoo

Business Unit Director:
Alar Elken

Executive Editor:
Sandy Clark

Acquisitions Editor:
James DeVoe

Development Editor:
John Fisher

Editorial Assistant:
Jasmine Hartman

Executive Marketing Manager:
Maura Theriault

Channel Manager:
Mary Johnson

Marketing Coordinator:
Karen Smith

Executive Production Manager:
Mary Ellen Black

Production Manager:
Larry Main

Production Editor:
Tom Stover

Art/Design Coordinator:
Mary Beth Vought

Printed in the United States
1 2 3 4 5 XXX 05 04 03 02 01

For more information contact Autodesk Press,
3 Columbia Circle, PO Box 15015,
Albany, NY 12212-5015.

Or find us on the World Wide Web at
www.autodeskpress.com

Library of Congress
Cataloging-in-Publication Data
Tucker, Irvin B.
Macroeconomics for today/Irvin B.
Tucker - 2nd ed.
p.000 00.00cm.
Includes bibliographical references and index.
ISBN 0-7668-38544
1. Macroeconomics, 2. United States — Economic conditions—1981-
1. Title.
HB 172.5 T782 1999
339-dc21 99-2972

Table of Contents

Preface xv
Dedication xix

AutoCAD LT Part I

Chapter 1: Getting Started-I

Starting AutoCAD LT 1-1
Starting a Drawing 1-2
Open Drawings 1-3
Create Drawings 1-3
AutoCAD LT Screen Components 1-7
Invoking Commands in AutoCAD LT 1-10
AutoCAD LT Dialog Boxes 1-13
Drawing Lines in AutoCAD LT 1-14
Coordinate Systems 1-18
Absolute Coordinate System 1-19
Relative Coordinate System 1-21
Direct Distance Entry 1-26
Erasing Objects 1-28
Canceling and Undoing a Command 1-29
Object Selection Methods 1-29
The Window Option 1-29
The Crossing Option 1-30
Drawing Circles 1-31
Saving Your Work 1-35
Automatic Timed Save 1-39
Creation of Backup Files 1-40
Closing a Drawing 1-40
Quitting AutoCAD LT 1-40
AutoCAD LT's Help 1-40
Additional Help Resources 1-44

Chapter 2: Getting Started-II

Opening an Existing File 2-1
Starting a New Drawing 2-5

Multiple Document Environment (MDE) 2-5
Basic Display Commands 2-7
ZOOM Command 2-7
Panning in Realtime 2-9
Setting Units 2-9
The Drawing Units Dialog Box 2-9
Forcing Default Angles 2-14
LIMITS Command 2-16
Introduction to Plotting Drawings 2-21
Options Dialog Box 2-26

Chapter 3: Draw Commands

Drawing Arcs 3-1
Drawing Rectangles 3-12
Drawing Ellipses 3-16
Drawing Elliptical Arcs 3-19
Drawing Regular Polygons 3-22
Drawing Polylines 3-25
Convert Command 3-31
Drawing Donuts 3-32
Drawing Points 3-34
Drawing Infinite Lines 3-37
XLINE Command 3-38
RAY Command 3-40
Creating Text 3-41

Chapter 4: Drawing Aids

Layers 4-1
Layer Properties Manager Dialog Box 4-3
LINETYPE Command 4-16
LWEIGHT Command 4-16
Object Properties 4-18
Object Properties Toolbar 4-18
Properties Window 4-19
Global and Current Linetype Scaling 4-20
Ltscale Factor for Plotting 4-21
AutoCAD DesignCenter 4-22
Drafting Settings Dialog Box 4-23
Setting Ortho Mode 4-29
Object Snaps 4-29
Running Object Snap Mode 4-40
Using Auto Tracking 4-43
Function and Control Keys 4-45

Chapter 5: Editing Commands-I

Creating a Selection Set 5-1
Editing Commands 5-7
MOVE Command 5-7
COPY Command 5-8
COPYBASE Command 5-9
PASTEBLOCK Command 5-10
PASTEORIG Command 5-10
OFFSET Command 5-11
ROTATE Command 5-12
SCALE Command 5-13
FILLET Command 5-15
CHAMFER Command 5-18
TRIM Command 5-21
EXTEND Command 5-23
STRETCH Command 5-26
LENGTHEN Command 5-27
ARRAY Command 5-29
Rectangular Array 5-29
Polar Array 5-31
MIRROR Command 5-35
BREAK Command 5-36
MEASURE Command 5-38
DIVIDE Command 5-39

Chapter 6: Editing Commands-II

Editing With Grips 6-1
Grip Types 6-2
Adjusting Grips Settings 6-3
Editing the Objects with Grips 6-4
Stretching Objects with Grips (Stretch Mode) 6-4
Moving Objects with Grips (Move Mode) 6-7
Rotating Objects with Grips (Rotate Mode) 6-8
Scaling Objects with Grips (Scale Mode) 6-9
Mirroring Objects with Grips (Mirror Mode) 6-10
Loading Hyperlinks 6-12
Editing Gripped Objects 6-12
Grip System Variables 6-13
PROPERTIES Command 6-13
Changing Properties Using Grips 6-14
Matching Properties 6-14
QSELECT Command 6-16
Managing Contents Using the AutoCAD DesignCenter 6-18
Making Inquiry about a Drawing (Inquiry Commands) 6-25

Calculating the Area of the Objects (AREA Command) 6-26
Calculating Distance between Two Points (DIST Command) 6-29
Identifying a Position on the Screen (ID Command) 6-30
Listing Information about Objects (LIST Command) 6-30
Checking the Time-related Information (TIME Command) 6-31
Displaying Drawing Properties (DWGPROPS Command) 6-33

Chapter 7: Controlling the Drawing Display and Creating Text

Basic Display Options 7-1
REDRAW Command 7-2
REGEN Command 7-2
ZOOM Command 7-3
PAN Command 7-10
Creating Views 7-12
Aerial View 7-14
Creating Text 7-16
Creating Single Line Text 7-16
Drawing Special Characters 7-22
Creating Paragraph Text 7-23
Editing Text 7-30
Editing Text Using DDEDIT Command 7-30
Editing Text Using Properties Window 7-31
Substituting Fonts 7-31
Specifying an Alternate Default Font 7-33
STYLE Command 7-33
Determining Text Height 7-34
Checking Spelling 7-35
Text Quality and Text Fill 7-36
Finding and Replacing Text 7-37

Chapter 8: Basic Dimensioning, Geometric Dimensioning and Tolerancing

Need for Dimensioning 8-1
Dimensioning in AutoCAD LT 8-2
Fundamental Dimensioning Terms 8-2
Associative Dimensions 8-6
Definition Points 8-7
Selecting Dimensioning Commands 8-8
Linear Dimensioning 8-11
Aligned Dimensioning 8-14
Rotated Dimensioning 8-15
Baseline Dimensioning 8-16
Continue Dimensioning 8-17

Angular Dimensioning 8-19
Diameter Dimensioning 8-22
Radius Dimensioning 8-23
Generating Center Marks and Centerlines 8-24
Ordinate Dimensioning 8-25
Creating True Associative Dimensions* 8-27
DIMREASSOCIATE Command* 8-27
Removing Dimension Associativity (DIMDISASSOCIATE Command*) 8-28
Drawing Leaders 8-28
Using Leader with the DIM Command 8-33
Geometric Dimensioning and Tolerancing 8-33
Geometric Characteristics and Symbols 8-34
Adding Geometric Tolerance (TOLERANCE Command) 8-35
Complex Feature Control Frames 8-37
Combining Geometric Characteristics 8-37
Composite Position Tolerancing 8-39
Using Feature Control Frames with Leaders 8-40
Projected Tolerance Zone 8-40

Chapter 9: Editing Dimensions

Editing Dimensions 9-1
Editing the Dimensions by Stretching 9-1
Editing Dimensions by Trimming and Extending 9-3
Editing the Dimensions 9-4
Editing Dimension Text 9-6
Updating Dimensions 9-7
Editing Dimensions with GRIPS 9-8
Editing Dimensions Using the Properties Window 9-8
Properties Window (Dimension) 9-8
Properties Window (Leader) 9-10
Model Space and Paper Space Dimensioning 9-11

Chapter 10: Dimension Styles and Dimensioning System Variables

Using Styles and Variables to Control Dimensions 10-1
Creating and Restoring Dimension Styles 10-2
New Dimension Style Dialog Box 10-3
Controlling Dimension Text Format 10-9
Fitting Dimension Text and Arrowheads 10-13
Formatting Primary Dimension Units 10-16
Formatting Alternate Dimension Units 10-20
Formatting the Tolerances 10-22
Other Dimensioning Variables 10-26
Dimension Style Families 10-27

Using Dimension Style Overrides 10-30
Comparing and Listing Dimension Styles 10-31
Using Externally Referenced Dimension Styles 10-32

Chapter 11: Model Space Viewports, Paper Space Viewports, and Layouts

Model Space and Paper Space/Layouts 11-1
Model Space Viewports (Tiled Viewports) 11-2
Creating Tiled Viewports 11-3
Making a Viewport Current 11-4
Joining Two Adjacent Viewports 11-5
Obtaining Plan Views Automatically (UCSFOLLOW System Variable) 11-7
Paper Space Viewports (Floating Viewports) 11-7
Creating Floating Viewports (VPORTS Command) 11-7
Temporary Model Space 11-9
Editing the Viewports 11-10
Controlling the Display of the Objects in the Viewports 11-11
Locking the Display in the Viewports 11-11
Controlling the Display of the Hidden Lines in the Viewports 11-11
Manipulating the Visibility of Viewport Layers 11-12
Controlling the Layers in the Viewports Using the Layer Properties Manager Dialog box 11-13
Paper Space Linetype Scaling (PSLTSCALE System Variable) 11-14
Creating and Working with the Layouts 11-15
PAGESETUP Command 11-18
LAYOUTWIZARD Command 11-19

Chapter 12: Plotting Drawings

Plotting Drawings in AutoCAD LT 12-1
PLOT Command 12-1
Adding Plotters 12-13
PLOTTERMANAGER Command 12-13
Editing Plotter Configuration 12-16
Importing PCP/PC2 Configuration Files 12-18
Setting the Plot Parameters 12-18
PAGESETUP Command 12-18
PSETUPIN Command 12-21
Using Plot Styles 12-21
STYLESMANAGER Command 12-22
Applying Plot Styles 12-28
PLOTSTYLE Command 12-29

Chapter 13: Hatching

Hatching 13-1
BHATCH Command 13-3
Boundary Hatch Options 13-4
Ray Casting 13-17
    Using the BHATCH Command 13-17
    Using the -BHATCH Command 13-18
Hatching around Text, Attributes, Shapes, and Solids 13-21
Editing Hatch Patterns 13-21
    Using the HATCHEDIT Command 13-21
    Using the PROPERTIES Command 13-23
Editing Hatch Boundary 13-24
    Using Grips 13-24
    Using AutoCAD LT Editing Commands 13-25
Hatching Blocks and Xref Drawings 13-26
Pattern Alignment During Hatching 13-26
BOUNDARY Command 13-27
Other Features of Hatching 13-30
Hatching by Using the HATCH Command 13-30

Chapter 14: Blocks

The Concept of Blocks 14-1
Formation of Blocks 14-2
Converting Objects into a Block 14-4
    BLOCK Command 14-4
    -BLOCK Command 14-7
Inserting Blocks (INSERT Command) 14-8
    -INSERT Command 14-15
Using AutoCAD DesignCenter to Insert Blocks 14-16
Layers, Colors, Linetypes and Lineweights for Blocks 14-17
Nesting of Blocks 14-18
Creating Drawing Files (WBLOCK Command) 14-20
    Using the -WBLOCK Command 14-23
Defining the Insertion Base Point (BASE Command) 14-24
Breaking up a Block 14-25
    Selecting the Explode check box in the Insert dialog box 14-27
    Using the EXPLODE Command 14-26
    Using the XPLODE Command 14-26
Renaming Blocks 14-28
Deleting Unused Blocks (PURGE Command) 14-29

AutoCAD LT Part II

Chapter 15: Defining Block Attributes

Attributes 15-1
Defining Attributes 15-2
Editing Attribute Definitions 15-7
Inserting Blocks with Attributes 15-8
Extracting the Attributes (ATTEXT Command) 15-11
Controlling Attribute Visibility 15-17
Editing Block Attribute Values 15-18
    Editing Attributes Using the ATTEDIT Command 15-18
    Global Editing of Attributes 15-19
    Individual Editing of Attributes 15-25
Inserting Text Files in the Drawing 15-28

Chapter 16: External References

External References 16-1
Dependent Symbols 16-2
Managing External References in a Drawing 16-4
Overlay Option 16-13
XATTACH Command 16-16
Adding Dependent Symbols to a Drawing 16-18
Displaying Clipping Frames 16-20
Demand Loading 16-20
Spatial and Layer Indexes 16-21

Chapter 17: More Draw Commands

Creating Double Lines (Using the DLINE Command) 17-1
Creating Revision Clouds (REVCLOUD Command) 17-5
Creating NURBS 17-6
Editing Splines 17-7
Digitizing Drawings 17-11
    TABLET Command 17-12

Chapter 18: Object Grouping and Editing Commands

Group Manager (GROUP Command) 18-1
Group Manager Dialog Box Options 18-2
Selecting Groups 18-4
Cycling through Groups 18-5
Changing Properties of an Object 18-5
    Using the Properties Window 18-5
    CHANGE Command 18-10
    CHPROP Command 18-13

EXPLODE Command 18-13
Polyline Editing 18-14
PEDIT Command 18-15
Undoing Commands 18-30
REDO Command 18-34
Renaming Named Objects 18-35
Removing Unused Named Objects 18-36
Object Selection Modes 18-38

Chapter 19: Data Exchange, and Object Linking and Embedding

Data Exchange in AutoCAD LT 19-1
DXF File Format (Data Interchange File) 19-2
Other Data Exchange Formats 19-5
Data Exchange through Raster Images 19-5
Raster Images 19-5
Managing Raster Images (Image Manager dialog box) 19-6
Editing Raster Image Files 19-8
Postscript Files 19-9
Creating the PostScript Files 19-9
Object Linking and Embedding 19-12

Chapter 20: The User Coordinate System

Conventions in AutoCAD LT 20-1
World Coordinate System (WCS) 20-3
UCSICON Command 20-4
UCS Command 20-6
Managing UCS through Dialog Box (UCSMAN Command) 20-19
System Variables 20-23

Chapter 21: Drawing and Viewing 3D Objects

Changing the Viewpoint to View the 3D Models 21-1
DDVPOINT Command 21-2
VPOINT Command 21-4
View Toolbar 21-8
3D Coordinate Systems 21-8
Absolute Coordinate System 21-8
Relative Coordinate System 21-12
Trim, Extend, and Fillet Commands in 3D 21-16
Setting Thickness and Elevation for the New Objects 21-16
ELEV Command 21-17
Suppressing the Hidden Edges 21-18
HIDE Command 21-18
Drawing 3D Polylines (3DPOLY Command) 21-19

Creating Regions (REGION Command) 21-20
Creating Complex Regions by Applying the Boolean Operations 21-20
UNION Command 21-20
SUBTRACT Command 21-20
INTERSECT Command 21-21
Dynamic Viewing of 3D Objects (DVIEW Command) 21-22
Creating Shaded Images 21-38
SHADE Command 21-38
SHADEMODE Command 21-39
Analyzing the Regions (MASSPROP Command) 21-40

AutoCAD LT Part III (Customizing)

Chapter 22: Template Drawings

Creating Template Drawings 22-1
Standard Template Drawings 22-2
Loading a Template Drawing 22-7
Customizing Drawings with Layers and Dimensioning Specifications 22-7
Customizing Drawings with Layout 22-12
Customizing Drawings with Viewports 22-15
Customizing Drawings According to Plot Size and Drawing Scale 22-17

Chapter 23: Script Files and Slide Shows

What are Script Files? 23-1
SCRIPT Command 23-4
RSCRIPT Command 23-9
DELAY Command 23-10
RESUME Command 23-10
Command Line Switches 23-11
Invoking a Script File When Loading AutoCAD LT 23-11
What Is a Slide Show? 23-20
What Are Slides? 23-20
MSLIDE Command 23-21
VSLIDE Command 23-22
Preloading Slides 23-24
Slide Libraries 23-26

Chapter 24: Creating Linetypes and Hatch Patterns

Standard Linetypes 24-1
Linetype Definition 24-2
Elements of Linetype Specification 24-3
Creating Linetypes 24-3

Alignment Specification 24-9
LTSCALE Command 24-9
LTSCALE Factor for Plotting 24-12
Alternate Linetypes 24-12
Modifying Linetypes 24-13
Current Linetype Scaling (CELTSCALE) 24-16
Complex Linetypes 24-17
Hatch Pattern Definition 24-22
How Hatch Works 24-24
Simple Hatch Pattern 24-24
Effect of Angle and Scale Factor on Hatch 24-26
Hatch Pattern with Dashes and Dots 24-26
Hatch with Multiple Descriptors 24-30
Saving Hatch Patterns in a Separate File 24-34
Custom Hatch Pattern File 24-34

Chapter 25: Pull-down, Shortcut, Partial Menus, and Customizing Toolbars

AutoCAD LT Menu 25-1
Standard Menus 25-3
Writing a Menu 25-3
Loading Menus 25-10
Restrictions 25-12
Cascading Submenus in Menus 25-12
Shortcut and Context Menus 25-18
Submenus 25-22
Loading Menus 25-23
Partial Menus 25-27
Accelerator Keys 25-31
Special Virtual Keys 25-33
Toolbars 25-33
Menu-Specific Help 25-38
Customizing the Toolbars 25-38

Chapter 26: AutoCAD LT on the Internet

Introduction 26-1
Changes from AutoCAD LT Release 97 26-2
WHIP! 3.1 Limitations 26-2
Changed Internet Commands 26-3
Understanding URLs 26-4
Drawings on the Internet 26-6
Inserting a Block from the Internet 26-11
Accessing Other Files on the Internet 26-11

Saving the Drawing to the Internet 26-12
AutoCAD LT Today Window 26-12
Using Hyperlinks with AutoCAD LT 26-15
Pasting as Hyperlinks 26-21
Highlighting Objects with URLs 26-22
Editing Hyperlinks 26-22
Removing Hyperlinks from Objects 26-22
Drawing Web Format 26-23
    Drag and Drop from Browser to Drawing 26-31
Embedding a DWF File 26-32

Index

Preface

AutoCAD LT, developed by Autodesk Inc., is the most popular PC-CAD system available in the market. AutoCAD LT 2002 is a Windows-based application used to generate various kinds of drawings. AutoCAD LT's drafting system is a two-dimensional CAD application with many graphical user interface tools. This package is very useful for architects, engineers, and draftspersons for creating drawings, site plans, and fine tuning design details accurately and quickly. AutoCAD LT has also provided facilities that allow the users to customize AutoCAD LT to make it more efficient and therefore, increase productivity.

This book contains a detailed explanation of AutoCAD LT 2002 commands and how to use them in solving drafting and design problems. It also unravels the customizing power of AutoCAD LT. Every AutoCAD LT command and customizing technique is thoroughly explained with examples and illustrations that make it easy to understand their function and application. At the end of each topic, there are examples that illustrate the function of the command and how it can be used in the drawing. When you are done reading this book, you will be able to use AutoCAD LT commands to make a drawing, create text, make and insert symbols, dimension a drawing, create 3D objects, write script files, define linetypes and hatch patterns, and write your own menus. This book also unleashes the Internet power of AutoCAD LT 2002. You will learn how to access various Internet related utilities of AutoCAD LT in this book.

While you are learning to use this book, you will discover some new applications of AutoCAD LT that are unique and might have a significant effect on your drawings. You will also get a better idea of why AutoCAD LT has become such a popular software package and an international standard in PC-CAD. Please refer to the following table for conventions used in this text.

Convention	**Example**
• Command names are capitalized and bold.	The **MOVE** command
• A key icon appears when you should respond by pressing the ENTER or RETURN key.	Enter
• Command sequences are indented. The responses are indicated by boldface. The directions are indicated by italics and the comments are enclosed in parentheses.	Command: **MOVE** Select object: **G** Enter group name: *Enter a group name (the group name is group1)*

Convention	**Examples**
• The command selection from the toolbars, menus, and Command prompt are enclosed in a shaded box.	**Toolbar:** Draw > Arc **Menu:** Draw > Arc **Command:** ARC
• AutoCAD LT 2002 features are indicated by an asterisk symbol at the end of the feature.	DIMREASSOCIATE Command*

Author's Web Sites

For Faculty: Please contact the author at **stickoo@calumet.purdue.edu** to access the Web Site that contains the following:

1. PowerPoint presentations, program listings, and drawings used in this textbook.
2. Syllabus, chapter objectives and hints, and questions with answers for every chapter.

For Students: You can download drawing exercises, tutorials, programs, and special topics by accessing the author's Web Site at **www.cadcim.com** or **www.calumet.purdue.edu/public/mets/tickoo/index.htm**.

Chapters on the Web Site

The following chapters are posted on the author's Web Sites.

1. Technical Drawing with AutoCAD LT
2. Isometric Drawing
3. Customizing the ACLT.PGP File, DIESEL: A String Expression Language
4. Tablet Menus
5. Image Tile Menus
6. Button and Auxiliary Menus

Users can download these chapters by accessing the following Web Sites:

1. www.calumet.purdue.edu/public/mets/tickoo/index.htm
2. www.cadcim.com

ACKNOWLEDGMENTS

Peter Lucasiewicz, Glenmont, NY
Ed Wheeler, University of Tennessee, Martin, TN

DEDICATION

*To teachers, who make it possible to disseminate knowledge
to enlighten the young and curious minds
of our future generations*

*To students, who are dedicated to learning new technologies
and making the world a better place to live*

Thanks

*To the faculty and students of the MET department of
Purdue University Calumet for their cooperation*

*To Deepak Maini, Abhinav Rai, Puja Bahl, and Santosh Tickoo,
CADCIM/CADSoft Technologies
for their valuable help*

AutoCAD LT Part I

AutoCAD LT Part I

Author's Web Sites

For Faculty: Please contact the author at **stickoo@calumet.purdue.edu** or **tickoo@cadcim.com** to access the Web site that contains the following:

1. PowerPoint presentations, programs, and drawings used in this textbook.
2. Syllabus, chapter objectives and hints, and questions with answers for every chapter.

For Students: You can download drawing-exercises, tutorials, programs, and special topics by accessing the author's web site at **www.cadcim.com** or **www.calumet.purdue.edu/public/mets/tickoo/index.htm**.

Chapter 1

Getting Started-I

Learning Objectives

After completing this chapter, you will be able to:

- *Start AutoCAD LT and start a drawing in AutoCAD LT.*
- *Understand the different components of the initial AutoCAD LT screen.*
- *Invoke AutoCAD LT commands from the keyboard, menu, toolbar, shortcut menu, and digitizer.*
- *Understand the functioning of dialog boxes in AutoCAD LT.*
- *Draw lines using the **LINE** command and its options.*
- *Understand different coordinate systems used in AutoCAD LT.*
- *Use the **ERASE** commands to clear the drawing area.*
- *Understand the two basic object selection methods: Window and Crossing options.*
- *Draw circles using different options of the **CIRCLE** command.*
- *Save the work using different file-saving commands.*
- *Use the different options of AutoCAD LT's help.*
- *Understand the use Active Assistance, Learning Assistance, and other interactive help topics provided in AutoCAD LT 2002.*

STARTING AUTOCAD LT

When you turn on your computer, the operating system (Windows 95, Windows NT, or Windows 98) is automatically loaded. This will display the Windows screen with various application icons (Figure 1-1). You can load AutoCAD LT by double-clicking on the AutoCAD LT 2002 icon. You can also load AutoCAD LT from the Windows taskbar by choosing the **Start** button at the bottom left corner of the screen (default position) to display the menu. Choose **Programs** to display the program folders. Now, choose the **AutoCAD LT 2002** folder to display AutoCAD LT programs and then choose **AutoCAD LT 2002** to start AutoCAD LT.

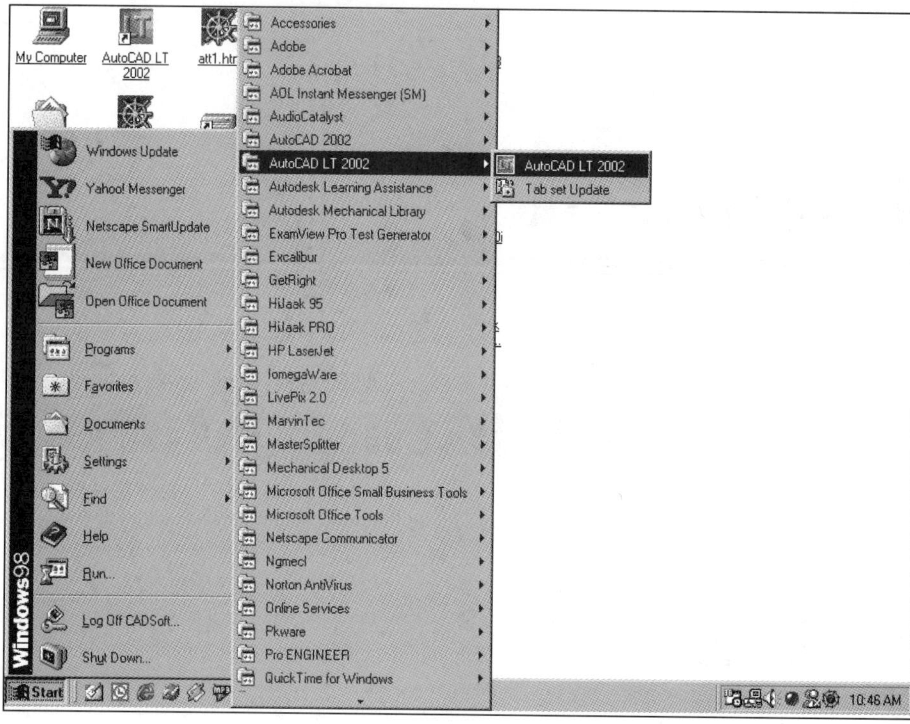

Figure 1-1 Windows screen with taskbar and application icons

STARTING A DRAWING

When you start AutoCAD LT, the **AutoCAD LT Today** window is displayed on the screen (Figure 1-2). The **AutoCAD LT Today** window has two main areas of work labeled as **My Workplace** and **The Web**. The **My Workplace** section has the **My Drawings** area that gives you the various options for starting a new drawing, opening a previously saved drawing, and loading the system libraries. It also has the **Bulletin Board** area that gives you space to display information and create a customized bulletin board. **The Web** section connects you to the Internet and provides you with the different links and resources. The **AutoCAD LT Today** window can be disabled for the subsequent sessions of AutoCAD LT by choosing a different option from the **Startup** drop-down list in the **System** tab of the **Options** dialog box. The **Options** dialog box can be accessed by choosing **Options** from the **Tools** menu. Depending on the Startup option you chose, an initial view will be displayed while starting AutoCAD LT. The other options apart from the **AutoCAD LT Today** window include the display of the traditional **Startup** dialog box, or bypassing the **AutoCAD LT Today** window and the **Startup** dialog box.

To start a new drawing you can simply minimize the **AutoCAD LT Today** window using the minus (-) button on the top right corner of the window. You have an initial drawing open with the name **Drawing1.dwg**. Instead of working on a drawing, you can first explore the different options in the **Today** window. Maximize the **AutoCAD LT Today** window by clicking on the **AutoCAD LT Today** button at the bottom of the screen.

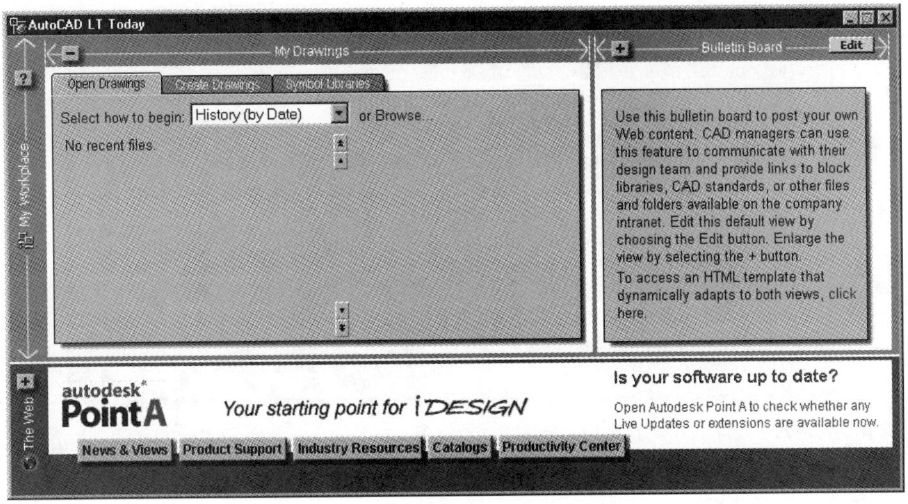

*Figure 1-2 **AutoCAD LT Today** window*

Open Drawings

The **Open Drawings** tab in the **My Drawings** area (Figure 1-2) is open by default. A list of the recently opened drawings is displayed for you to select from. If you are opening AutoCAD LT for the first time, the list box will be empty. You can select a category for opening the drawings from the **Select how to begin** drop-down list. Depending on these categories the drawings are grouped and then displayed in the list box. This option is explained in detail in Chapter 2.

Create Drawings

The **Create Drawings** tab in the **My Drawings** area (Figure 1-3) allows you to start a new drawing with the specifications you specify. The different options for starting a new drawing are listed in the **Select how to begin** drop-down list.

Template

When you select **Template** from the **Select how to begin** drop-down list, AutoCAD LT displays a list of templates supplied with AutoCAD LT (Figure 1-3). The default template file is aclt.dwt or acltiso.dwt, depending on the installation. If you use a template file, the new drawing will have the same settings as specified in the template file. All the drawing parameters of the new drawing such as the units, limits, and other settings are already set according to the template file used. The preview of the template file selected is displayed in the dialog box. You can also define your own template files that are customized to your requirements (See Chapter 22, Template Drawings). To differentiate the template files from the drawing files, the template files have **.dwt** extension whereas the drawing files have **.dwg** extension. Any drawing file can be saved as a template file. You can use the **Browse** button to select other template files. When you choose the **Browse** button, the **Select File** dialog box is displayed with the **Template** folder open displaying all the template files.

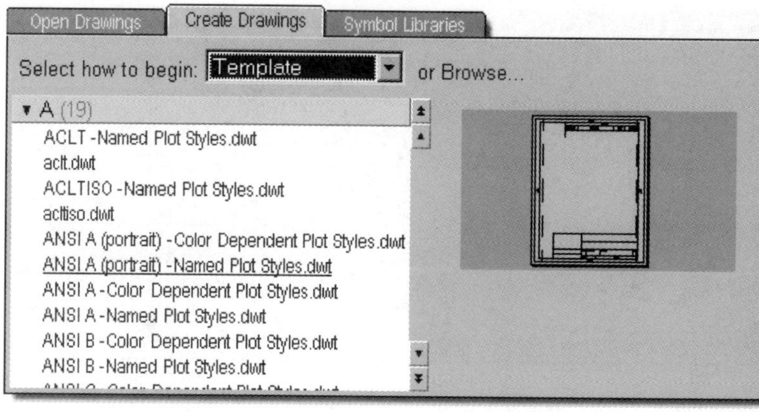

*Figure 1-3 Selecting **Template** in the **Create Drawings** tab of the **Today** window*

Start from Scratch

When you select **Start from Scratch** from the **Select how to begin** drop-down list (Figure 1-4), AutoCAD LT lets you start a new drawing that contains the default setup for **English** (aclt.dwt) or **Metric** drawings units (acltiso.dwt). For example, if you select the English default setting, the limits are 12 x 9, text height is 0.20, and dimensions and linetype scale factors are 1. If you start a drawing using the **Start from Scratch** settings, the units and limits for the drawing can be changed later.

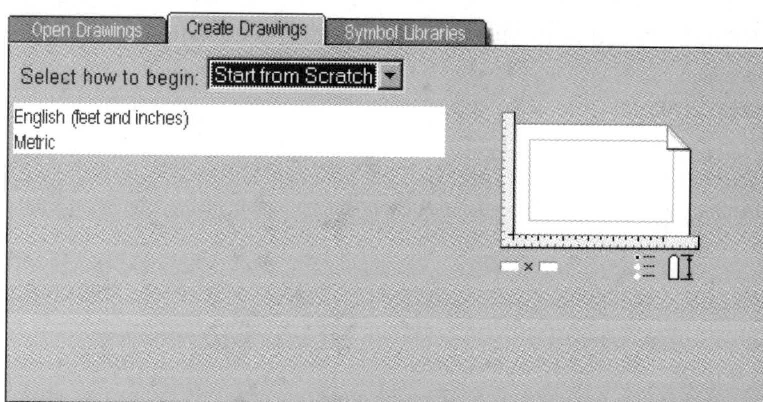

*Figure 1-4 When using **Start from Scratch**, the new drawing contains the default settings*

Wizard

The **Wizard** allows you to set the initial drawing settings. When you select **Wizard** from the **Select how to begin** drop-down list, you have the choice of using **Quick Setup** or **Advanced Setup** (Figure 1-5) options. In the **Quick Setup**, you can specify the units and the limits of the work area. In the **Advanced Setup** you can set the units, limits, and the different angle settings for a drawing.

Figure 1-5 *Use the* ***Wizard*** *option in the* ***Create Drawings*** *tab*

Quick Setup. When you choose **Quick Setup**, the **Quick Setup** dialog box is displayed with the **Units** page (Figure 1-6) open by default. You can select **Next** to display the **Area** page. The **Units** page is used to set the units and is displayed as default. A small arrow is displayed along its left side to indicate that this page is open. The units of measurements can be set by selecting the radio button for it and the respective preview image is displayed along with it. The different units of measurement you can choose from are Decimal, Engineering, Architectural, Fractional, and Scientific.

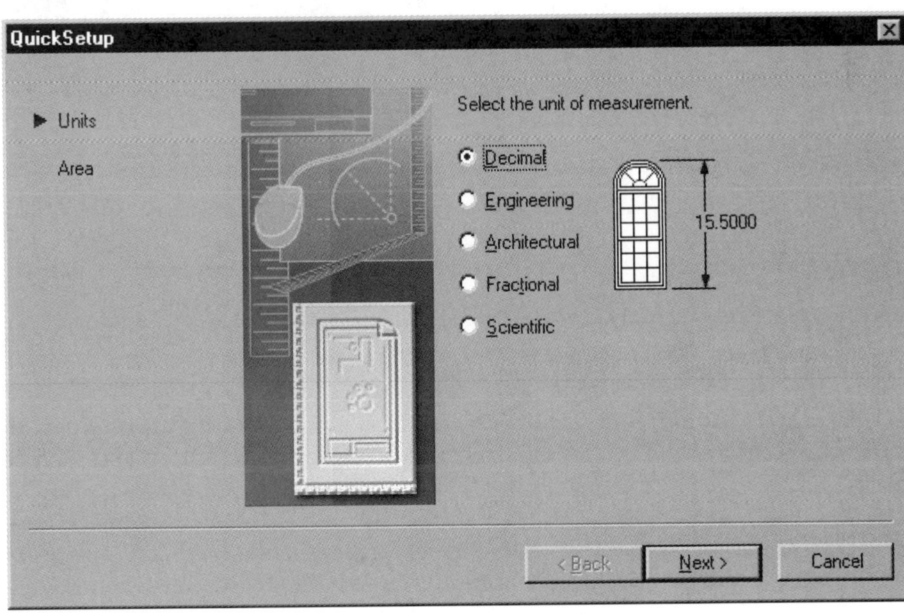

Figure 1-6 *The* ***Units*** *page of the* ***Quick Setup*** *dialog box*

To set the limits of the drawing choose the **Next** button in the dialog box to display the **Area** page of the **Quick Setup** dialog box. You can enter the width and length of the drawing area in the respective edit boxes.

Advanced Setup. When you choose **Advanced Setup**, the **Advanced Setup** dialog box is displayed with the **Units** page open by default. The **Units** page has the same options as in the **Quick Setup** with the difference being you can also set the number of decimal places using the **Precision** drop-down list. The **Area** page also is similar to the one in the **Quick Setup** where you can set the drawing limits.

Choose the **Next** button to open the **Angle** page of the dialog box (Figure 1-7). In this page the different angle measuring systems and a **Precision** drop-down list are displayed. The angle of measurement can be set by selecting the radio button for it and the preview of that particular angular unit is displayed along with it. The format of precision also changes in the **Precision** edit box depending on the angle measuring system selected. You can then select the precision from the drop-down list. The angle measuring systems are Decimal Degrees, Deg/Min/Sec, Grads, Radians, and Surveyor.

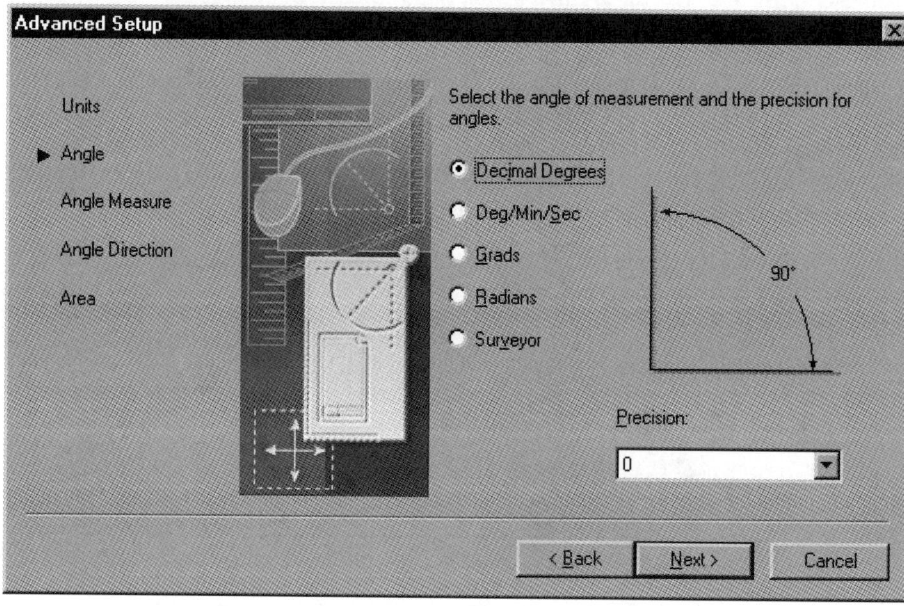

Figure 1-7 *The **Angle** page of the **Advanced Setup** dialog box*

The next page is the **Angle Measure** page where you can choose the direction of the base angle. You can set your own direction by selecting the **Other** radio button and then entering the value in its edit box, which becomes available when you select **Other**.

Choose **Next** to display the **Angle Direction** page to set the orientation for angle measurement. By default the angles are positive if measured in a counterclockwise direction.

Note

*It is possible to change the drawing units and area later while working on the drawing using the **UNITS** and **LIMITS** commands. These commands are discussed in detail in Chapter 2, Getting Started-II.*

Symbol Libraries

The **Symbol Libraries** tab in the **My Drawings** area (Figure 1-8) allows you to load the different symbol libraries into the **DesignCenter**. A list of the symbol libraries are displayed when you choose the **Symbol Libraries** tab. You can modify the symbol libraries by using the **Edit DesignCenter Symbol Libraries** dialog box, which can be accessed by choosing the **Edit** button. The DesignCenter is explained in detail in Chapter 6, Editing Commands-II.

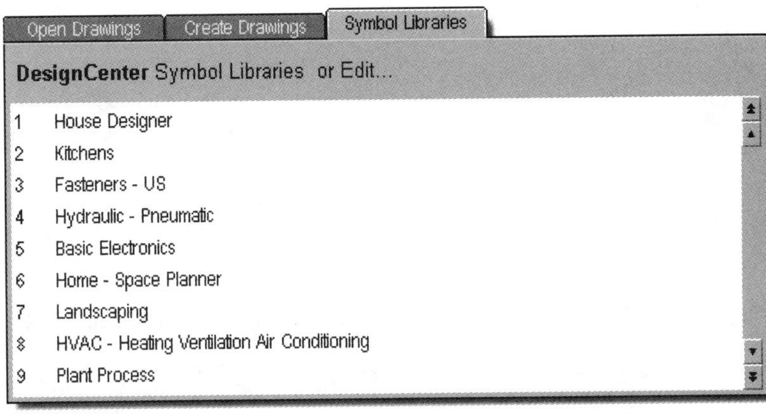

Figure 1-8 ***Symbol Libraries*** *tab in the* ***AutoCAD LT Today*** *window*

Note

The ***Bulletin Board*** *and the area for* ***The Web*** *that includes* ***Autodesk Point A*** *are discussed in Chapter 26, AutoCAD LT on the Internet.*

AUTOCAD LT SCREEN COMPONENTS

After minimizing the **AutoCAD LT Today** window, or starting a new drawing using any of the start options in the **AutoCAD LT Today** window, the initial AutoCAD LT screen is displayed. The different components of the initial AutoCAD LT screen are the drawing area, the command window, menu bar, several toolbars, model and layout tabs, and the status bar (Figure 1-9). A title bar is displayed on top of the screen that has the AutoCAD LT symbol and the current drawing name.

Drawing Area

The drawing area covers the major portion of the screen. Here you can draw the different objects and use the different commands. To draw the objects, you need to define the coordinate points which can be selected by using your pointing device. The position of the pointing device is represented on the screen by the cursor. There is a coordinate system icon at the lower left corner of the drawing area. The window has the standard windows buttons such as close, minimize, scroll bar, and so on, which have the same functions as for any other standard window.

Command Window

The command window is present at the bottom of the drawing area and has the Command

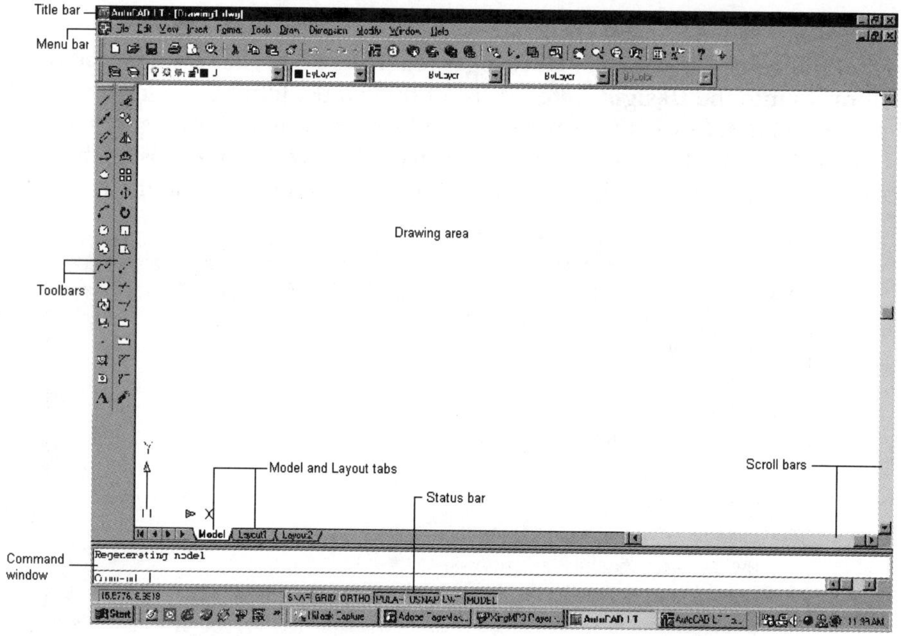

Figure 1-9 AutoCAD LT screen components

prompt where you can enter the commands. It also displays the subsequent prompt sequences and the messages. You can change the size of the window by placing the cursor on the top edge (double line bar known as the grab bar) and then dragging it. This way you can increase its size to see all the previous commands you have used. By default the command window displays only three lines. You can also press the **F2** key to display the **Text window** that displays the previous commands and prompts.

Status Bar

The status bar is displayed at the bottom of the screen (Figure 1-10). This bar contains some useful information and buttons that will make it easy to change the status of some AutoCAD LT functions. To change the status, you must choose the buttons that toggle between on and off.

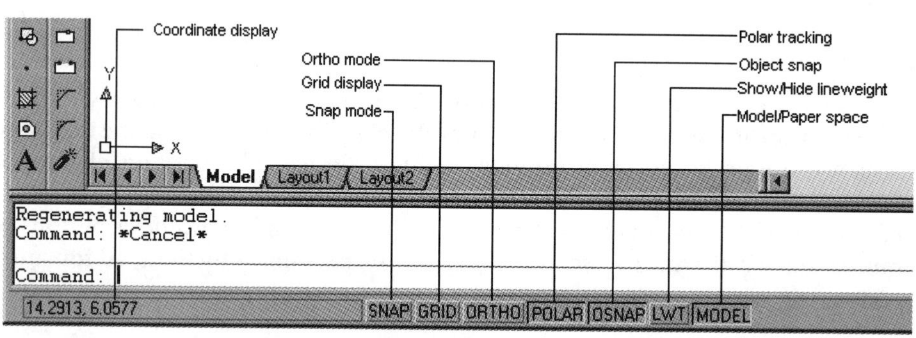

Figure 1-10 Default status bar display

Coordinate display. The **Coordinates** information is displayed in the left corner of the status bar. You can select this coordinate button to toggle between on and off. The **COORDS** system variable controls the type of display of the coordinates. If the value of the **COORDS** variable is set to 0, the coordinate display is static, that is, the coordinate values displayed in the status bar change only when you specify a point. If the value of the **COORDS** variable is set to 1 or 2, coordinate display is dynamic. When the variable is set to 1, AutoCAD LT constantly displays the absolute coordinates of the graphics cursor with respect to the UCS origin. The polar coordinates (length<angle) are displayed if you are in an AutoCAD LT command and the **COORDS** variable is set to 2. You can use the key F6 to turn the coordinate display on or off.

Snap. If **Snap Mode** is on, the **Snap** button is displayed as pressed in the status bar; otherwise, it is not displayed. This allows you to move the cursor in fixed increments. You can also use the function key F9 as a toggle key to turn Snap off or on.

Grid. If **Grid Display** is on, grid lines are displayed on the screen. These lines are used as reference lines in a drawing. The function key F7 can be used to turn the grid display on or off.

Ortho. If **Ortho Mode** is on, the **Ortho** button is pressed in the status bar. This allows you to draw lines at right angles only. You can use the F8 key to turn Ortho on or off.

Polar. Choosing the **Polar** button in the status bar turns on the **Polar Tracking**. The movement of the cursor is restricted along a path based on the angle set as the polar angle settings. You can also use the function key, F10. Turning the Polar Tracking on, automatically turns off the Ortho mode.

Osnap. If **Object Snap** is on, the **Osnap** button is displayed as pressed in the status bar. You can also use the F3 key to turn the Object Snap on or off. When Object Snap is on, you can use the running object snaps. If Osnap is off, the running Object Snaps are temporarily disabled. The status of Osnap (Off or On) does not prevent you from using immediate mode object snaps.

LWT. Choosing this button in the status bar, allows you to **Show/Hide Lineweight**. If the **LWT** button is not chosen, the display of Lineweight is turned off.

Model. The Model button is displayed in the status bar when you are working in the model space to create drawings. You can select this button to shift to the layouts (paper space) where you can create drawing views. Once you are working with the layouts, you can choose the **Model** tab to shift back to the model space.

Note

All the buttons in the status bar are discussed in detail in Chapter 4, Drawing Aids.

The menu bar and toolbar are discussed in the following section. The model and layout tab is discussed in Chapter 11, Model Space Viewports, Paper Space Viewports, and Layouts.

INVOKING COMMANDS IN AUTOCAD LT

When you start AutoCAD LT and you are in the drawing area, you need to invoke AutoCAD LT commands to perform any operation. For example, if you want to draw a line, first you have to invoke the **LINE** command, and then you define the start point and endpoint of the line. Similarly, if you want to erase objects, you must enter the **ERASE** command, and then select the objects for erasing. AutoCAD LT has provided the following methods to invoke commands:

Keyboard	**Menu**	**Toolbar**
Shortcut menu	**Digitizing tablet**	

Keyboard

You can invoke any AutoCAD LT command at the keyboard by typing the command name at the Command prompt, and then pressing ENTER or the SPACEBAR. Before you enter a command, make sure the Command prompt is displayed as the last line in the command window area. If the Command prompt is not displayed, you must cancel the existing command by pressing ESC. KEY (escape) on the keyboard. The following example shows how to invoke the **LINE** command from the keyboard:

Command: **LINE or L** Enter (L is command alias)

Menu

You can also select commands from the menu. The **menu bar** that displays the menu bar titles is at the top of the screen. As you move the cursor over the menu bar, different titles are highlighted. You can choose the desired item by pressing the pick button of your pointing device. Once the item is selected, the corresponding menu is displayed directly under the title. You can invoke a command from the menu by pressing the pick button of your pointing device. Some of the menu items in the menu display an arrow on the right side, which indicates that the menu item has a cascading menu. You can display the cascading menu by selecting the menu item or by just moving the arrow pointer to the right of that item. You can then choose any item in the cascading menu by highlighting the item or command and pressing the pick button of your pointing device. For example, if you want to draw an ellipse using the Center option, choose **Draw** from the menu bar, choose **Ellipse** from the **Draw** menu, then choose **Center** from the cascading menu (Figure 1-11). In this text, this command selection sequence will be referenced as **Draw > Ellipse > Center. The menus and toolbars tend to vary with every release of AutoCAD LT.**

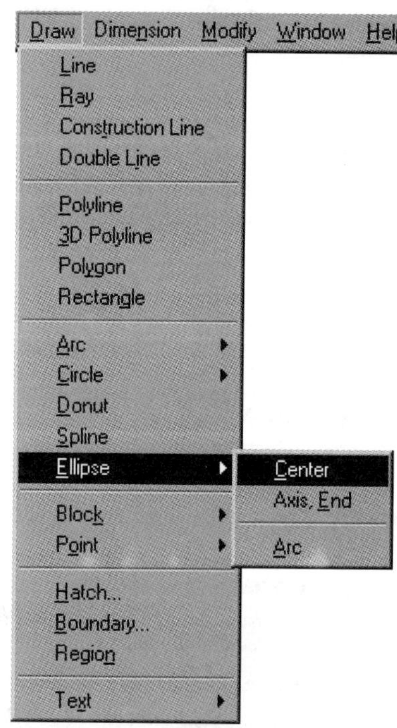

***Figure 1-11** Invoking the **ELLIPSE** command from the **Draw** menu*

Toolbar

In Windows, the toolbar is an easy and convenient way to invoke a command. For example, you can invoke the **LINE** command by choosing the **Line** button (the upper left button, Figure 1-12) in the **Draw** toolbar. When you choose a command from the toolbar, the command prompts are displayed in the command window. By default the **Standard Toolbar**, **Object Properties**, **Draw**, and **Modify** toolbars are displayed on the screen and are docked to the top and the left side edges of the drawing area.

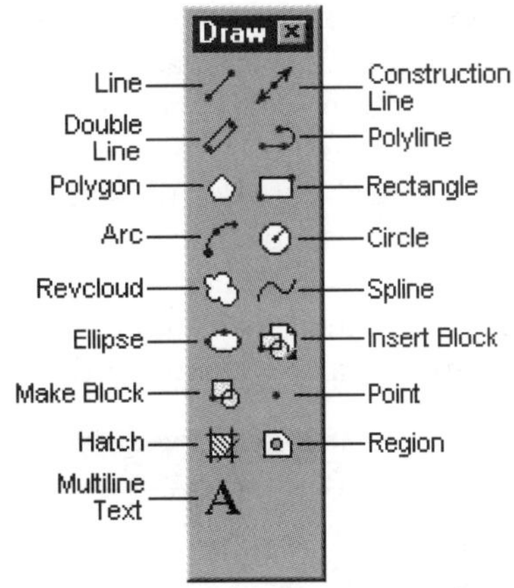

*Figure 1-12 The **Draw** toolbar*

Displaying toolbars. The different toolbars can be displayed by selecting their respective check boxes in the **Toolbars** tab of the **Customize** dialog box (Figure 1-13). The **Customize** dialog box can be invoked by selecting **Toolbars** in the **View** menu. You can also display a toolbar from the shortcut menu by selecting the name of the toolbar. The shortcut menu can be displayed by right-clicking anywhere on any toolbar on the screen. Each toolbar contains a group of buttons representing different AutoCAD LT commands. When you move the cursor over the buttons of a toolbar, the button gets lifted and a 3D box encloses the button on which the cursor is resting. The tooltip (the name of the command) is also displayed below the button. Once you locate the desired button, the command associated with that button can be invoked by choosing the button. Some of the buttons in a toolbar have a small triangular mark at the lower right corner. This indicates that the button has a flyout attached to it. If you hold the button down, the flyout is displayed. The flyout contains the different options for the command.

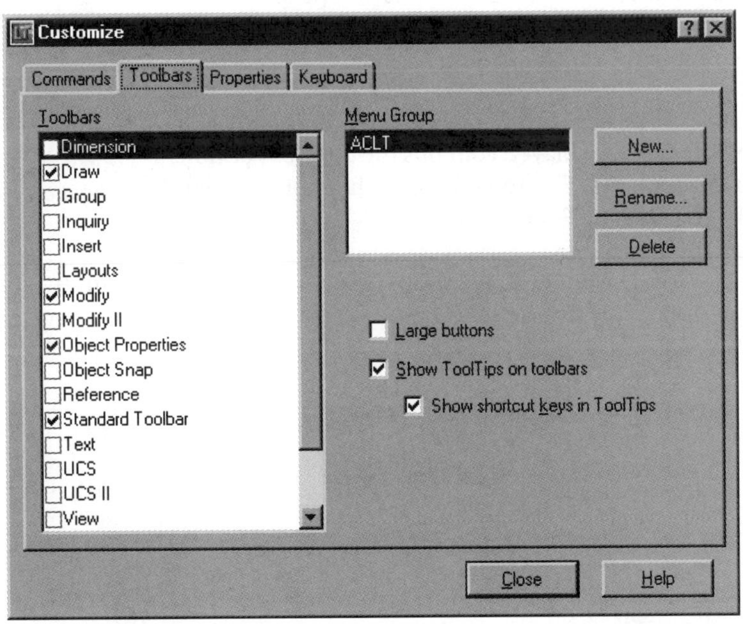

*Figure 1-13 List of toolbars in the **Toolbars** tab of the **Customize** dialog box*

Moving toolbars. The toolbars can be moved anywhere on the screen by placing the cursor on the title bar area and then dragging it to the desired location. You must hold the pick button down while dragging. While moving the toolbars you can dock them to the top or sides of the screen by dropping them in the docking area. You may also prevent docking by holding the **Control** key when moving the toolbar to a desired location. You can also change the shape of the toolbars by placing the cursor anywhere on the border of the toolbar where it takes the shape of a double arrow (Figure 1-14), and then pulling it in the desired direction (Figure 1-15). You can also customize toolbars to meet your requirements (See Chapter 25, Pull-down, Shortcut, Partial Menus, and Customizing Toolbars).

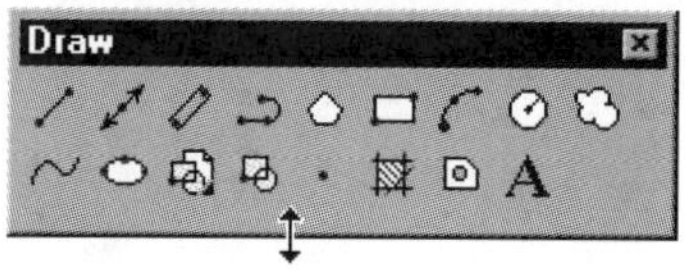

***Figure 1-14** Reshaping the **Draw** toolbar*

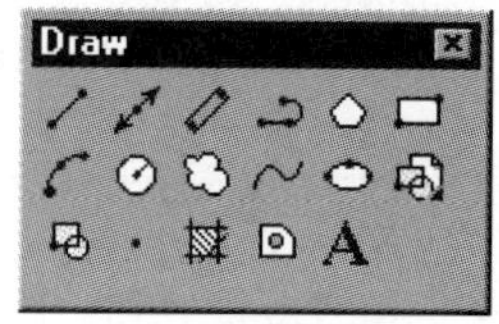

***Figure 1-15 Draw** toolbar reshaped*

Shortcut Menu

AutoCAD LT has provided shortcut menus as an easy and convenient way of invoking commands. These menus are context sensitive which means that the commands present in them are dependent on the place/object for which they are displayed. These menus are invoked by right-clicking and are displayed at the cursor location.

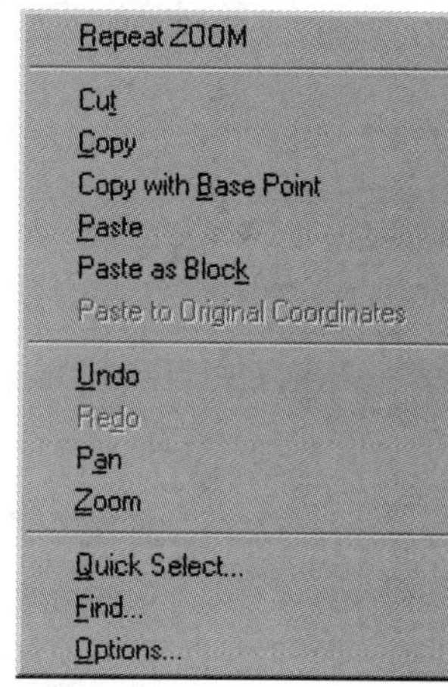

***Figure 1-16** Shortcut menu without an active command*

You can right-click anywhere on the drawing area to display the general shortcut menu. It generally contains an option to select the previously invoked command again (Figure 1-16) apart from the common commands for Windows.

If you right-click on the drawing area while a command is in effect, the shortcut menu displayed contains the different options of that particular command. Figure 1-17 shows the shortcut menu when the **PLINE** command is active.

If you right-click on the layout tabs, the shortcut menu displayed contains the different options for the layouts (Figure 1-18).

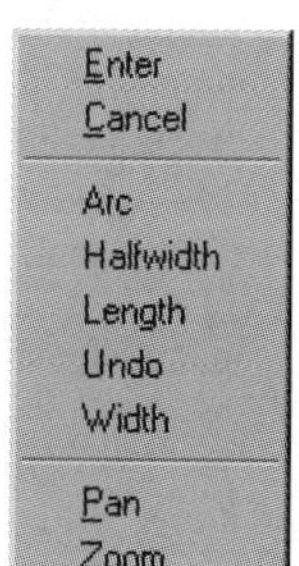

***Figure 1-17** Shortcut menu for the **PLINE** command*

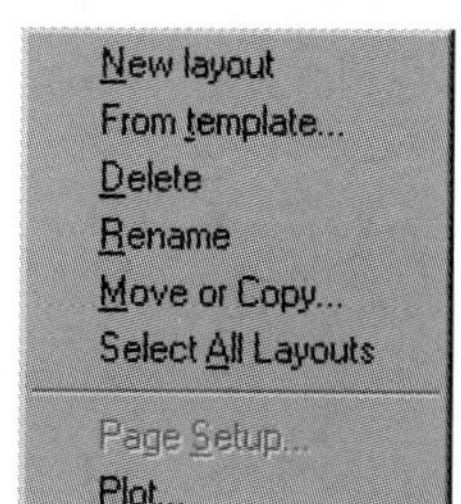

***Figure 1-18** Shortcut menu for the Layout tab*

You can also right-click on the command window to display the shortcut menu. This menu displays the six most recently used commands and some of the Window's options like Copy and Paste (Figure 1-19). The commands and their prompt entries are displayed in the History window (previous command lines not visible) and can be selected, copied, and pasted in the command line using the shortcut menu. As you press the up arrow key, the previously entered commands are displayed in the command window. Once the desired command is displayed at the Command prompt, you can execute the command by simply pressing the ENTER key. You can also copy and edit any previously invoked command by locating it in the History window and then selecting the lines. Right-click in the command window to display the shortcut menu (Figure 1-19), select copy, and then paste the selected lines in the command line. After the lines are pasted, you can edit them.

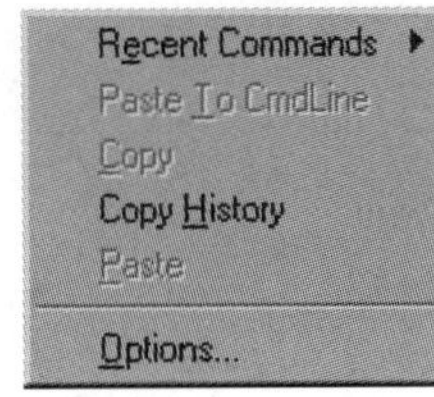

***Figure 1-19** Command line window shortcut menu*

You can right-click on the status bar to display the shortcut menu. This menu contains the options to change the settings of drawing tools (Figure 1-20).

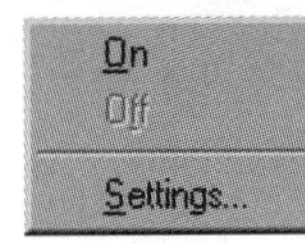

***Figure 1-20** Status bar shortcut menu*

You can also right-click on any of the toolbars to display the shortcut menu from where you can choose any toolbar to be displayed.

Tip

A shortcut menu is available for any situation while working in AutoCAD LT. You should try to make use of it frequently by right-clicking at different positions.

Digitizing Tablet

In the digitizing tablet, the commands are selected from the template that is secured on the surface of the tablet. To use the tablet menu, you need a digitizing tablet and a pointing device. You also need a tablet template that contains AutoCAD LT commands arranged in various groups for easy identification. To invoke a command from the digitizer, move your pointing device so that the crosshairs of the pointing device (or the tip of the stylus) are directly over the cell that contains the command you want to invoke. Now press the pick button of the pointing device, or simply press down the stylus, and AutoCAD LT will automatically invoke that command.

AUTOCAD LT DIALOG BOXES

There are certain commands which when invoked display a dialog box. A dialog box is a convenient method of user interface. In the menus, the menu item with the ellipses [...] displays the dialog box when you choose that item. For example, by default, **New...** in the **File** menu displays the **Create Drawings** tab in the **AutoCAD LT Today** window. Any dialog box contains a number of parts like the dialog label, radio buttons, text or edit boxes, check boxes, slider bars, image boxes, and command buttons. These components are also referred to as TILES. Some of the components of a dialog box are shown in Figure 1-21.

You can select the desired tile using the pointing device which is represented by an arrow when

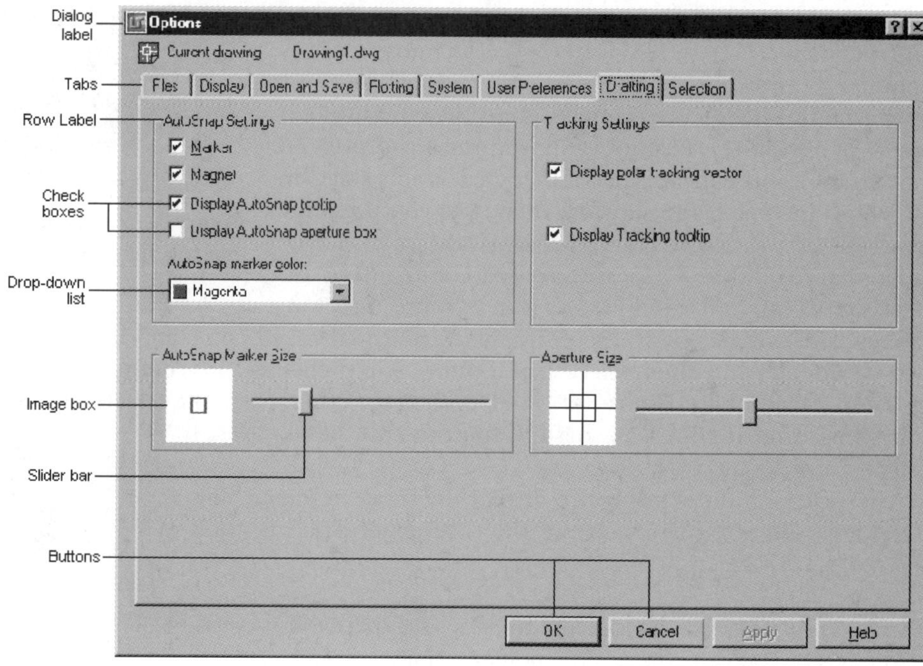

Figure 1-21 Components of a dialog box

a dialog box is invoked. The title bar displays the name of the dialog box. The **tabs** specify the different sections with a group of related options under them. The **check boxes** are toggle buttons for making the particular option available or unavailable. The **drop-down** list displays an item and an arrow on the right which when selected displays a list of items to choose from. You can make a selection in the **radio buttons.** Only one can be selected at a time. The **image box** displays the preview image of the item selected. The **text box** is an area where you can enter a text like a file name. It is also called an **edit box** because you can make any change to the text entered. In some dialog boxes there is the **[...] button** which displays another related dialog box. There are certain **command buttons** (OK, Cancel, Help) at the bottom of the dialog box. The name implies their functions. The button with a dark border is the default button. The dialog box has a **Help** button for getting help on the different features of the dialog box. There is also a **question mark** (?) button near the top right corner of the dialog box meant for feature-specific help. If you want help on a particular feature of a dialog box, select the ? button. The ? gets attached with the cursor. You can then select the feature to display its help.

DRAWING LINES IN AUTOCAD LT

Toolbar:	Draw > Line
Menu:	Draw > Line
Command:	LINE or L

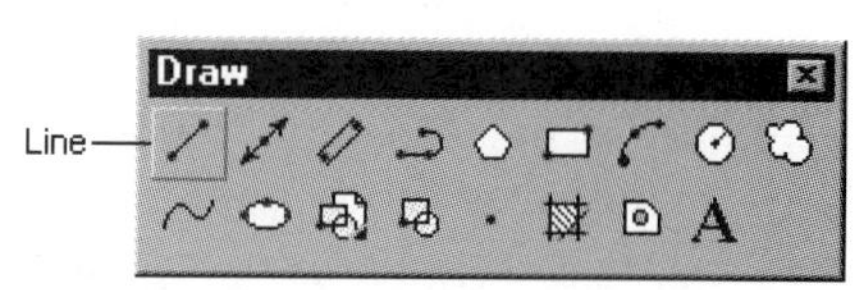

*Figure 1-22 Invoking the **LINE** command from the **Draw** toolbar*

The most fundamental object in a drawing is the line. A line can be drawn between any two points by using AutoCAD LT's **LINE** command. You can invoke the **LINE** command by choosing the **Line** button in the **Draw** toolbar(Figure 1-22), or by choosing **Line** from the **Draw** menu (Figure1-23), or by entering **LINE** at the Command prompt. Once you have invoked the **LINE** command, the next prompt, **Specify first point,** requires you to specify the starting point of the line. You can either select a point using the pointing device or you can enter its coordinates. After the first point is selected, AutoCAD LT will prompt you to enter the second point at the **Specify next point or [Close/Undo]** prompt. When you select the second point of the line, AutoCAD LT will again display the prompt **Specify next point or [Close/Undo]**. At this point you may continue to select points or terminate the **LINE** command by pressing ENTER, ESC, or the SPACEBAR. You can also right-click to display the shortcut menu from where you can choose the **Enter** or **Cancel** options to exit from the **LINE** command. After terminating the **LINE** command, AutoCAD LT will again display the Command prompt. The prompt sequence when you choose the **Line** button for the drawing Figure 1-24 is as follows:

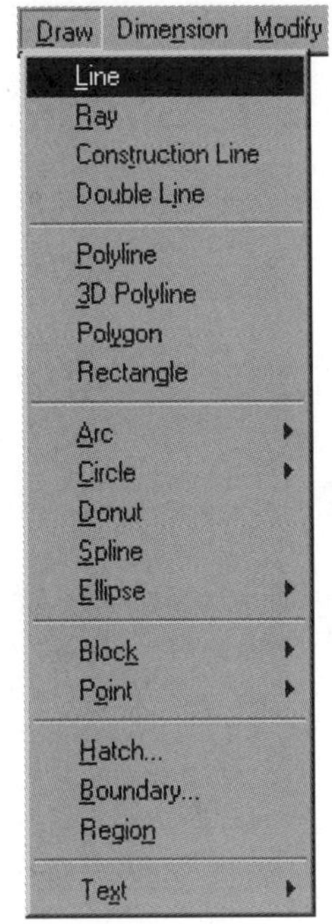

***Figure 1-23** Invoking the **LINE** command from the **Draw** menu*

Specify first point: *Move the cursor (mouse) and left-click to specify the first point.*
Specify next point or [Undo]: *Move the cursor and left-click to specify the second point.*
Specify next point or [Undo]: *Specify the third point.*
Specify next point or [Closc/Undo]: Enter *(Press ENTER to exit the **LINE** command.)*

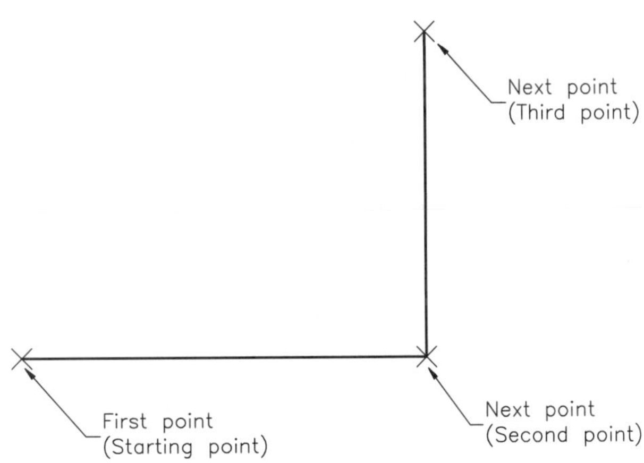

***Figure 1-24** Drawing lines using the **LINE** command*

The **LINE** command has the following three options:

Continue **Close** **Undo**

Tip

When you select the points with the cursor, a rubber-band line appears that stretches between the previous point selected and the current position of the cursor. This line is sensitive to the movement of the cursor and helps you to select the direction and the placement of the next point for the line.

Note

To clear the drawing area to gain space to work on the exercises and the examples, choose the ***Erase*** *button from the* ***Modify*** *toolbar or type* ***ERASE*** *at the Command prompt and press ENTER. The screen crosshairs will change into a box called a pick box and AutoCAD LT will prompt you to select objects. You can select the object by positioning the pick box anywhere on the object and pressing the pick button of the pointing device. Once you have finished selecting the objects, press ENTER to terminate the* ***ERASE*** *command and the objects you selected will be erased. If you enter* ***All*** *at the* ***Select objects:*** *prompt, AutoCAD LT will erase all objects from the screen. (See "Erasing Objects" discussed later in this chapter). You can use the* ***U*** (undo) *command to undo the last command by choosing the* ***Undo*** *button from the* ***Standard*** *toolbar.*

Command: **ERASE or E** [Enter] *(**E** is the command alias of **the ERASE** command.)*
Select objects: *Select objects.* *(Select objects using the pick box.)*
Select objects: [Enter]

Command: **ERASE** [Enter]
Select objects: **ALL** [Enter]
Select objects: [Enter]
Command: **U** [Enter] *(**The U** command will undo the last command.)*

Continue Option

After exiting from the **LINE** command, you may want to draw another line starting from the point where the previous line ended. In such cases you can use the **Continue** option. This option enables you to grab the endpoint of the previous line and to continue drawing the line from that point (Figure 1-25). The following is the prompt sequence for the **Continue** option:

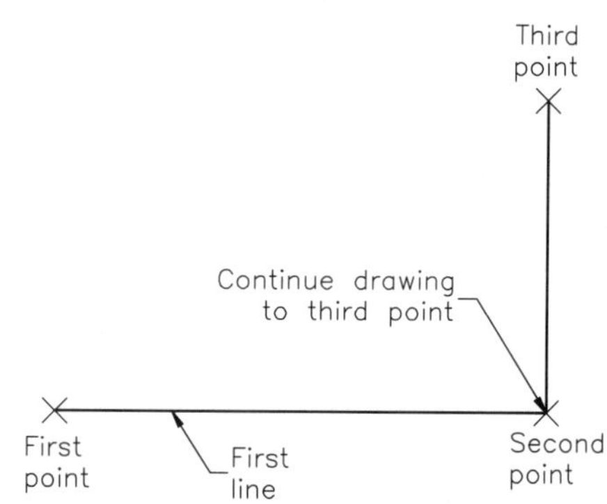

Figure 1-25 *Using the **Continue** option with the **LINE** command*

Command: **LINE or L** [Enter] *(**L** is the command alias of **the LINE** command.)*
Specify first point: *Specify the first point of the line.*
Specify next point or [Undo]: *Specify the second point.*
Specify next point or [Undo]: [Enter]

Command: **LINE** [Enter] *(Or select **Repeat Line** from the shortcut menu.)*
Specify first point: [Enter] *(Press ENTER or right-click to continue the line from the last line.)*

Specify next point or [Undo]: *Pick second point of second line (third point in Figure 1-25).*
Specify next point or [Undo]: Enter

You can also type the @ symbol to start the line from the **last point**. For example, if you draw a circle and then immediately start the **LINE** command, the @ will snap to the center point of the circle. The Continue option snaps to the endpoint of the last line or arc, even if other points have been defined after the line was drawn.

Command: **LINE** Enter
Specify first point: *Pick first point of the line.*
Specify next point or [Undo]: *Pick second point.*
Specify next point or [Undo]: Enter

Command: **LINE or L** Enter *(**L** is the command alias for **LINE** command.)*
Specify first point: **@** Enter *(Continues drawing the line from the last point.)*
Specify next point or [Undo]: *Pick second point of the second line.*
Specify next point or [Undo]: Enter

Close Option

The **Close** option can be used to join the current point with the initial point of the first line when two or more lines are **drawn in continuation**. For example, this option can be used when an open figure needs one more line to close it and make a polygon (a polygon is a closed figure with at least three sides, for example, a triangle or rectangle). The following is the prompt sequence for the **Close** option (Figure 1-26):

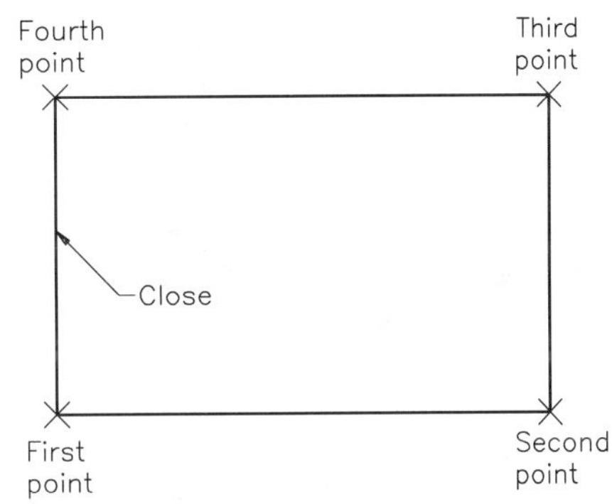

Figure 1-26 *Using the **Close** option with the **LINE** command*

Command: **LINE** Enter
Specify first point: *Pick first point.*
Specify next point or [Undo]: *Pick second point.*
Specify next point or [Undo]: *Pick third point.*
Specify next point or [Close/Undo]: *Pick fourth point.*
Specify next point or [Close/Undo]: **C** Enter *(Joins the fourth point with the first point.)*

You can also choose the **Close** option from the shortcut menu, which appears when you right-click in the drawing area.

The Undo Option

If you draw a line, and then realize that you made an error, you can remove the line using the **Undo** option. If you need to remove more than one line, you can use this option multiple times and go as far back as you want. In this option, you can type **Undo** (or just **U**) at the **Specify next point or [Undo]** prompt. You can also right-click to display the shortcut menu which gives you the **Undo** option. The following example illustrates the use of the **Undo** option (Figure 1-27):

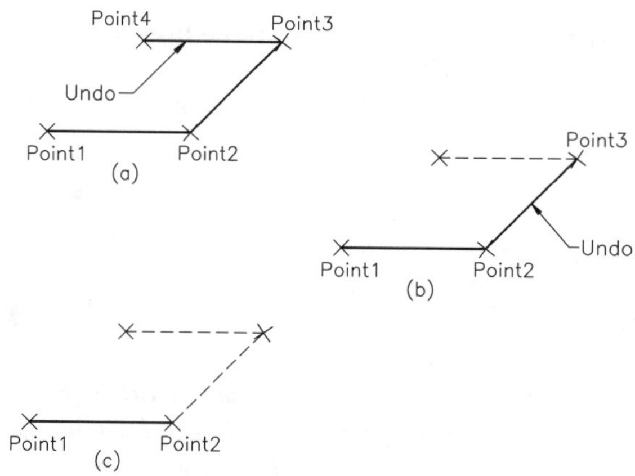

Figure 1-27 *Removing lines using the* ***Undo*** *option of the* ***LINE*** *command*

Command: **LINE or L** [Enter] *(**L** is the command alias of **LINE** command.)*
Specify first point: *Pick first point (Point 1 in Figure 1-27).*
Specify next point or [Undo]: *Pick second point (Point 2).*
Specify next point or [Undo]: *Pick third point.*
Specify next point or [Close/Undo]: *Pick fourth point.*
Specify next point or [Close/Undo]: **U** [Enter] *(Removesthe last line from Point 3 to Point 4.)*
Specify next point or [Close/Undo]: **U** [Enter] *(Removes the next line from Point 2 to Point 3.)*
Specify next point or [Close/Undo]: [Enter]

Tip
*AutoCAD LT allows you to enter the command aliases in place of the complete command name. For example, you can enter **L** instead of **LINE** at the Command prompt to invoke the **LINE** command.*

COORDINATE SYSTEMS

To specify a point in a plane, two mutually perpendicular lines are taken as references. The horizontal line is called the **X axis**, and the vertical line is called the **Y axis.** The point of intersection of these two axes is called the **origin**. The X and Y axes divide the XY plane into four parts, generally known as quadrants. The X coordinate measures the horizontal distance from the origin (how far left or right) on the X axis. The Y coordinate measures the vertical distance from the origin (how far up or down) on the Y axis. The origin has the coordinate values of X = 0, Y = 0. The origin is taken as the reference for locating any point in the XY plane. The X coordinate is positive if measured to the right of the origin and negative if measured to the left of the origin. The Y coordinate is positive if measured above the origin and negative if measured below the origin. This method of specifying points is called the **Cartesian coordinate system** (Figure 1-28). In AutoCAD LT, the default origin is located at the lower left corner of the graphics area of the screen. AutoCAD LT uses the following coordinate systems to locate a point in an XY plane:

1. **Absolute coordinates**
2. **Relative coordinates**
   a. **Relative rectangular coordinates**
   b. **Relative polar coordinates**
3. **Direct distance entry**

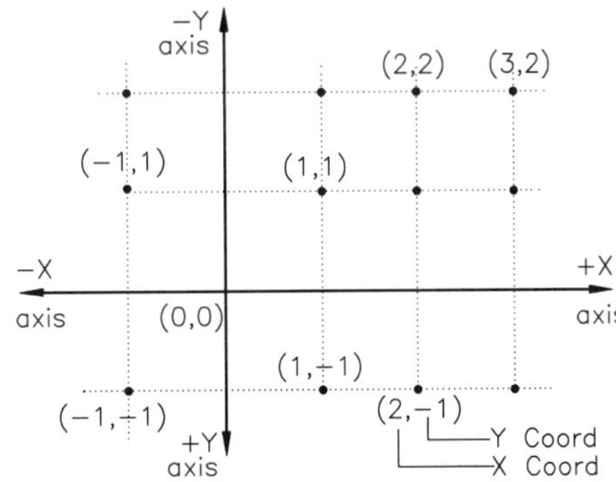

Figure 1-28 *Cartesian coordinate system*

Absolute Coordinate System

In the absolute coordinate system the points are located with respect to the origin (0,0), Figure 1-29. For example, a point with X = 4 and Y = 3 is measured 4 units horizontally (displacement along the X axis) and 3 units vertically (displacement along the Y axis) from the origin (Figure 1-30).

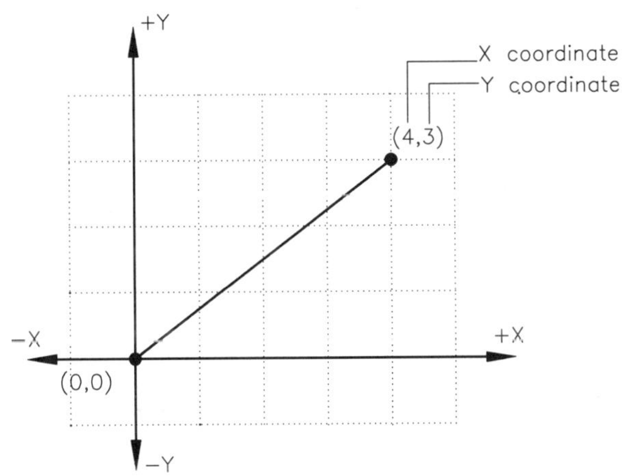

Figure 1-29 *Absolute coordinate system*

In AutoCAD LT, the absolute coordinates are specified by entering X and Y coordinates, separated by a comma. The following example illustrates the use of absolute coordinates (Figure 1-30):

Command: **LINE** Enter
Specify first point: **1,1** Enter
(X = 1 and Y = 1.)
Specify next point or [Undo]: **4,1** Enter
(X = 4 and Y = 1.)
Specify next point or [Undo]: **4,3** Enter
Specify next point or [Close /Undo]: **1,3** Enter
Specify next point or [Close/Undo]: **1,1** Enter

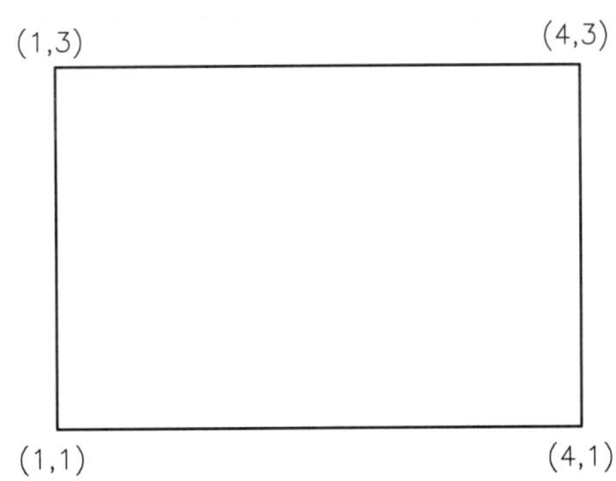

Figure 1-30 *Drawing lines using absolute coordinates*

Specify next point or [Close/Undo]: Enter

Example 1 *General*

For Figure 1-31, enter the absolute coordinates of the points in the following table. Then draw the figure using absolute coordinates. Save the drawing under the name Exam1.dwg.

Point	Coordinates	Point	Coordinates
1	3,1	5	5,2
2	3,6	6	6,3
3	4,6	7	7,3
4	4,2	8	7,1

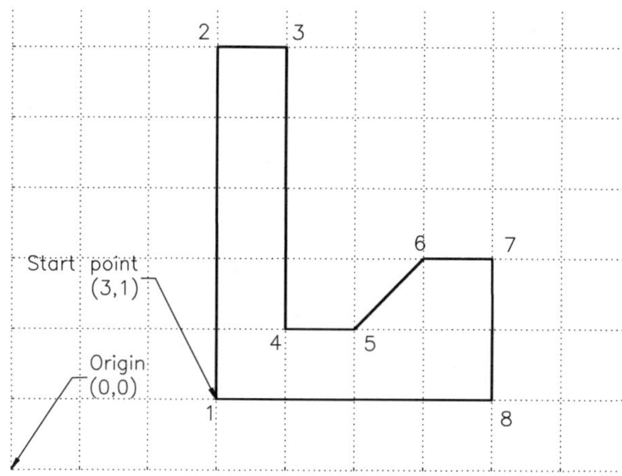

***Figure 1-31** Drawing a figure using absolute coordinates*

Once the coordinates of the points are known, you can draw the figure by using AutoCAD LT's **LINE** command. The prompt sequence is:

Command: **LINE** Enter
Specify first point: **3,1** Enter *(Start point.)*
Specify next point or [Undo]: **3,6** Enter
Specify next point or [Undo]: **4,6** Enter
Specify next point or [Close/Undo]: **4,2** Enter
Specify next point or [Close/Undo]: **5,2** Enter
Specify next point or [Close/Undo]: **6,3** Enter
Specify next point or [Close/Undo]: **7,3** Enter
Specify next point or [Close/Undo]: **7,1** Enter
Specify next point or [Close/Undo]: **3,1** Enter
Specify next point or [Close/Undo]: Enter

Save this drawing. Enter **SAVE** at the Command prompt and then press ENTER. The **Save Drawing As** dialog box is displayed. Enter the name **Exam2** in the **File name** edit box to replace Drawing1.dwg and then choose the **Save** button. The drawing will be saved with the given name in the **AutoCAD LT 2002** default directory and you can continue using the original drawing file (Drawing1.dwg) for further practice.

Exercise 1 *General*

For Figure 1-32, enter the absolute coordinates of the points in the following table, then use these coordinates to draw the same figure. Distance between the dotted lines is 1 unit.

Point	Coordinates	Point	Coordinates
1	2, 1	6	__________
2	__________	7	__________
3	__________	8	__________
4	__________	9	__________
5	__________		

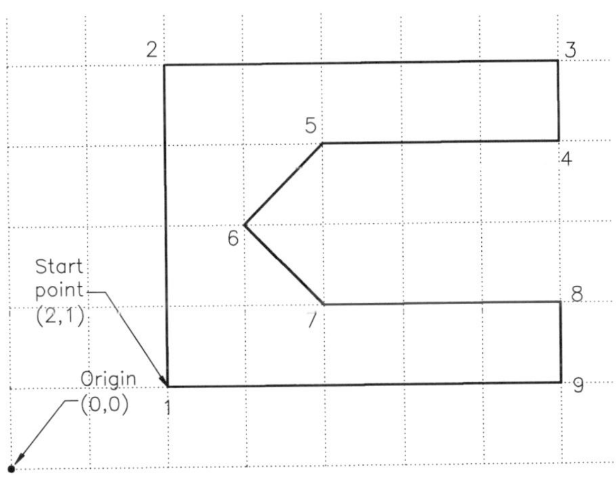

Figure 1-32 Drawing for Exercise 1

Relative Coordinate System

There are two types of relative coordinates: the relative rectangular and relative polar coordinates.

Relative Rectangular Coordinates

In the relative rectangular coordinate system, the displacements along the X and Y axes (DX and DY) are measured with reference to the previous point rather than to the origin (Figure 1-33). In AutoCAD LT, the relative coordinate system is designated by the symbol @ and it should precede any relative entry. The syntax of the Relative Rectangular coordinate system is @x displacement,y displacement. The following prompt sequence illustrates the use of the relative rectangular coordinate system to draw a rectangle with the lower left corner at point (1,1). The length of the rectangle is 4 units and the width is 3 units (Figure 1-34).

Command: **LINE** [Enter]
Specify first point: **1,1** [Enter] *(Start point.)*
Specify next point or [Undo]: **@4,0** [Enter]
(Second point DX = 4, DY = 0.)
Specify next point or [Undo]: **@0,3** [Enter]
(Third point DX = 0, DY = 3.)
Specify next point or [Close/Undo]:
@-4,0 [Enter] *(Fourth point DX = -4, DY = 0.)*
Specify next point or [Close/Undo]:
@0,-3 [Enter] *(Start point DX = 0, DY = -3.)*
Specify next point or [Close/Undo]: [Enter]

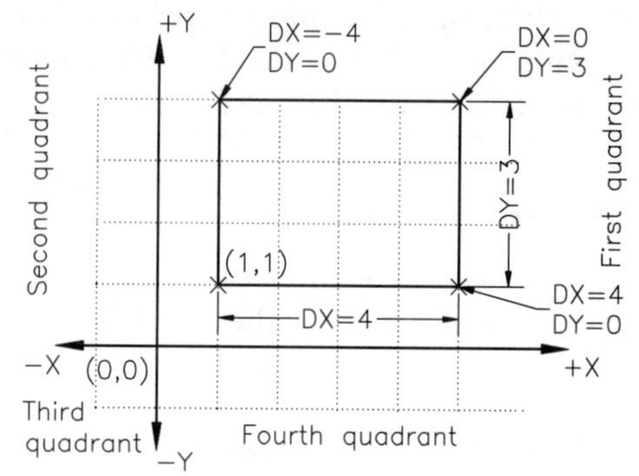

Figure 1-33 Drawing lines using relative rectangular coordinates

Sign Convention. As just mentioned, in the relative rectangular coordinate system the displacements along the X and Y axes are measured with respect to the previous point. Imagine a horizontal line and a vertical line passing through the previous point so that you get four quadrants. If the new point is located in the first quadrant, the displacements DX and DY are both positive. If the new point is located in the third quadrant, the displacements DX and DY are both negative. In other words, up or right are positive and down or left are negative.

Example 2 — *General*

Draw Figure 1-34 using relative rectangular coordinates of the points given in the table.

Point	Coordinates	Point	Coordinates
1	3,1	8	@-1,-1
2	@4,0	9	@-1,1
3	@0,1	10	@-1,0
4	@-1,0	11	@0,-2
5	@1,1	12	@1,-1
6	@0,2	13	@-1,0
7	@-1,0	14	@0,-1

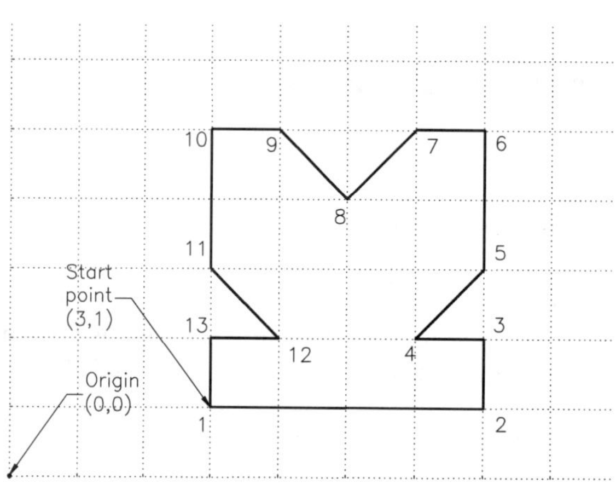

*Figure 1-34 Using relative rectangular coordinates with the **LINE** command*

Once you know the coordinates of the points, you can draw the figure by using AutoCAD LT's **LINE** command and entering the coordinates of the points.

Command: **LINE** [Enter]
Specify first point: **3,1** [Enter] *(Start point.)*
Specify next point or [Undo]: **@4,0** [Enter]
Specify next point or [Undo]: **@0,1** [Enter]
Specify next point or [Close/Undo]: **@-1,0** [Enter]
Specify next point or [Close/Undo]: **@1,1** [Enter]
Specify next point or [Close/Undo]: **@0,2** [Enter]
Specify next point or [Close/Undo]: **@-1,0** [Enter]
Specify next point or [Close/Undo]: **@-1,-1** [Enter]
Specify next point or [Close/Undo]: **@-1,1** [Enter]
Specify next point or [Close/Undo]: **@-1,0** [Enter]
Specify next point or [Close/Undo]: **@0,-2** [Enter]
Specify next point or [Close/Undo]: **@1,-1** [Enter]
Specify next point or [Close/Undo]: **@-1,0** [Enter]
Specify next point or [Close/Undo]: **@0,-1** [Enter]
Specify next point or [Close/Undo]: [Enter]

Exercise 2 *General*

For Figure 1-35, enter the relative rectangular coordinates of the points in the following table, then use these coordinates to draw the figure. The distance between the dotted lines is 1 unit.

Point	Coordinates	Point	Coordinates
1	2, 1	12	________
2	________	13	________
3	________	14	________
4	________	15	________
5	________	16	________
6	________	17	________
7	________	18	________
8	________	19	________
9	________	20	________
10	________	21	________
11	________	22	________

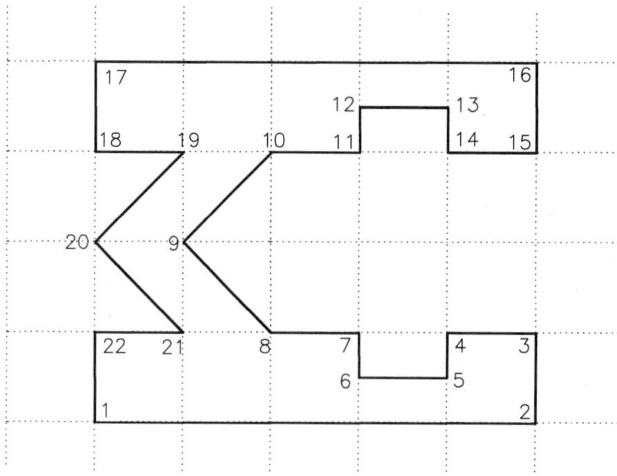

Figure 1-35 Drawing for Exercise 2

Relative Polar Coordinates

In the relative polar coordinate system, a point can be located by defining both the distance of the point from the current point and the angle that the line between the two points makes with the positive X axis. The prompt sequence to draw a line from a point at 1,1 to a point at a distance of 5 units from the point (1,1), and at an angle of 30 degrees to the X axis (Figure 1-36) is:

Command: **LINE** [Enter]
Specify first point: **1,1** [Enter]
Specify next point or [Undo]: **@5<30** [Enter]

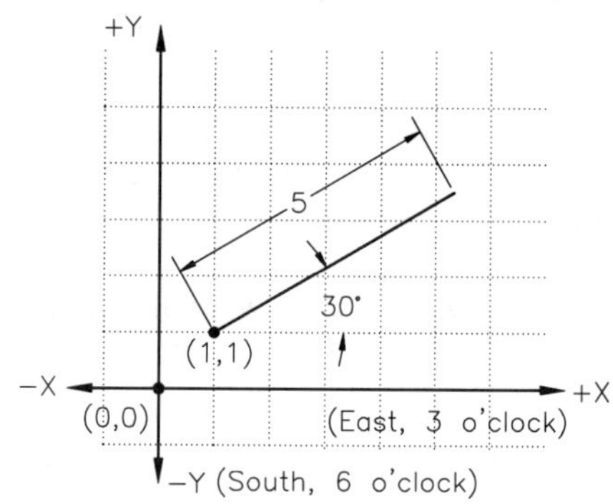

***Figure 1-36** Drawing a line using relative polar coordinates*

Sign Convention. In the relative polar coordinate system, the angle is measured from the horizontal axis (3 o' clock) as the zero degree baseline. Also, the angle is positive if measured in a counterclockwise direction and negative if measured in a clockwise direction. Here we assume that the default setup of angle measurement has not been changed.

Example 3 — *General*

For Figure 1-37, enter the relative polar coordinates of each point in the table, then generate the drawing. Use absolute coordinates for the start point (1.5, 1.75). The dimensions are shown in the drawing. Also, save this drawing as **Exam3.dwg**.

Point	Coordinates	Point	Coordinates
1	1.5,1.75	7	@1.0<180
2	@1.0<90	8	@0.5<270

3	@2.0<0	9	@1.0<0
4	@2.0<30	10	@1.25<270
5	@0.75<0	11	@0.75<180
6	@1.25<-90 (or <270)	12	@2.0<150

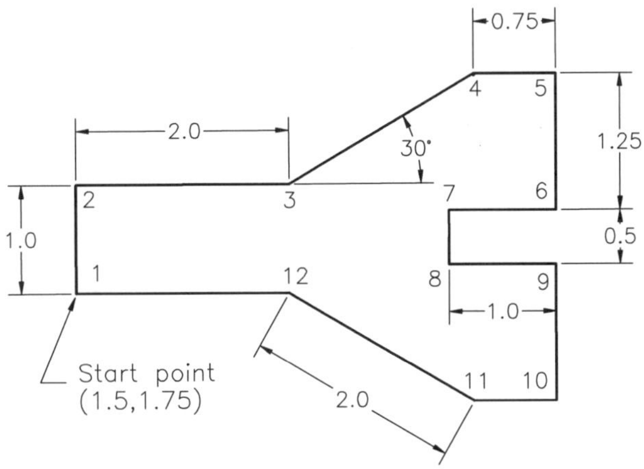

Figure 1-37 *Drawing for Example 3*

Once you know the coordinates of the points, you can generate the drawing by using AutoCAD LT's **LINE** command and entering the coordinates of the points.

Command: **LINE** Enter
Specify first point: **1.5,1.75** Enter *(Start point.)*
Specify next point or [Undo]: **@1<90** Enter
Specify next point or [Undo]: **@2.0<0** Enter
Specify next point or [Close/Undo]: **@2<30** Enter
Specify next point or [Close/Undo]: **@0.75<0** Enter
Specify next point or [Close/Undo]: **@1.25<-90** Enter
Specify next point or [Close/Undo]: **@1.0<180** Enter
Specify next point or [Close/Undo]: **@0.5<270** Enter
Specify next point or [Close/Undo]: **@1.0<0** Enter
Specify next point or [Close/Undo]: **@1.25<270** Enter
Specify next point or [Close/Undo]: **@0.75<180** Enter
Specify next point or [Close/Undo]: **@2.0<150** Enter
Specify next point or [Close/Undo]: **C** Enter *(Joins the last point with the first point.)*

Save this drawing by entering **SAVE** at the Command prompt and then press ENTER. The **Save Drawing As** dialog box is displayed. Enter the name **Exam3** in the **File name** edit box to replace Drawing1.dwg and then choose the **Save** button. The drawing will be saved with the given name in the **AutoCAD LT 2002** default directory and you can continue using the original drawing file (Drawing1.dwg) for further practice.

Exercise 3 *General*

Draw the object shown in Figure 1-38 using the absolute, relative rectangular, and relative polar coordinate systems to locate the points. Do not draw the dimensions; they are for reference only.

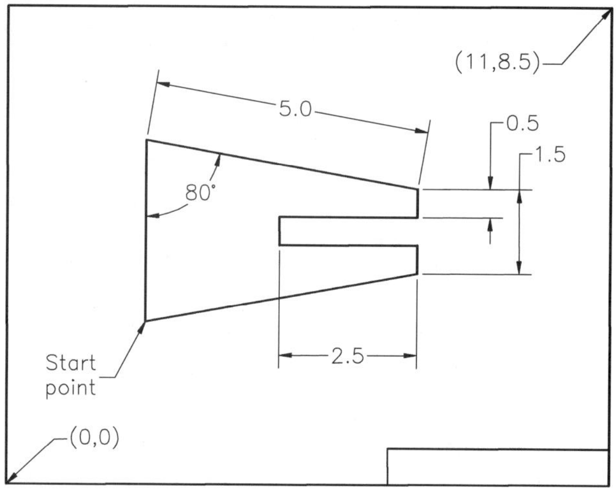

Figure 1-38 *Drawing for Exercise 3*

Direct Distance Entry

You can draw a line by specifying the length of the line and its direction, using the Direct Distance Entry (Figure 1-39). The direction is determined by the position of the cursor, and the length of the line is entered from the keyboard. If Ortho is on, you can draw lines along the X or Y axis by specifying the length of line and positioning the cursor along ortho direction. You can also use it with other draw commands like **RECTANGLE**. You can also use the Direct Distance Entry with polar tracking and **SNAPANG.** For example, if **SNAPANG** is 45 degrees and ortho is off, you can draw a line at 45 or 135 degrees direction by positioning the cursor and entering the distance from the keyboard. Similarly, if the polar tracking is on, you can position the cursor at the predefined angles and then enter the length of the line from the keyboard.

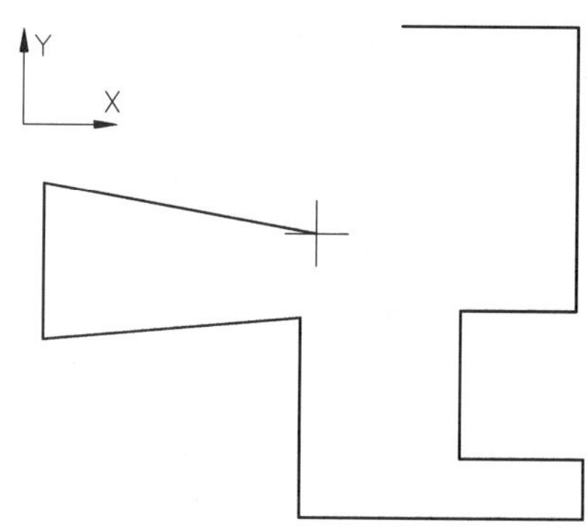

Figure 1-39 *Using Direct Distance Entry to draw lines*

Command: **LINE** [Enter]
Specify first point: *Start point.*
Specify next point or [Undo]: *Position the cursor and then enter distance.*
Specify next point or [Undo]: *Position the cursor and then enter distance.*

Example 4 *General*

In this example you will draw the object as shown in Figure 1-40, using Direct Distance Entry. The starting point is 2,2.

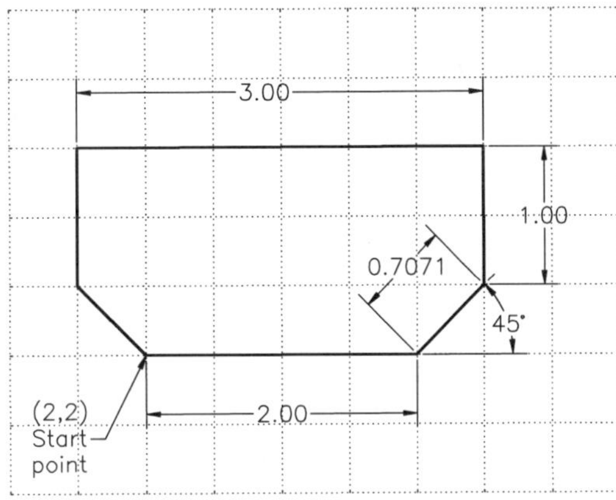

Figure 1-40 *Drawing for Example 4*

Before you invoke the **LINE** command, you should turn on polar tracking. This will make it easier to specify the direction of lines. To turn polar tracking on, choose the **POLAR** button in the status bar. You can also turn polar tracking on or off while you are in a command. As you move the cursor, AutoCAD LT displays a dotted line when the position of the cursor matches one of the predefined angles for polar tracking. The following is the Command prompt sequence for drawing the object in Figure 1-40:

Command: **LINE**
Specify first point: **2,2**
Specify next point or [Close/Undo]: **2** Enter *(Move the cursor horizontally and enter the length of the line, 2, from the keyboard.)*
Specify next point or [Close/Undo]: **@0.7071<45** *(Select* ***POLAR*** *in the status bar and position the cursor in a 45-degree direction and enter .7071.)*
Specify next point or [Close/Undo]: **1** *Move the cursor up vertically, then enter 1.*
Specify next point or [Close/Undo]: **3** *Move the cursor left horizontally, then enter 3.*
Specify next point or [Close/Undo]: **1** *Move the cursor down vertically, then enter 1.*
Specify next point or [Close/Undo]: **C**

Tip

It is possible to set the angle for the polar tracking. Right-click on the ***POLAR*** *button in the status bar to display the shortcut menu and choose* ***Settings*** *to display the* ***Drafting Settings*** *dialog box. Set the Increment angle to 45.*

Exercise 4 *General*

Use the direct distance entry method to draw a parallelogram. The base of the parallelogram equals 4 units, the side equals 2.25 units, and the angle equals 45 degrees. Draw the same parallelogram using absolute, relative, and polar coordinates. Note the differences and the advantage of using direct distance entry.

ERASING OBJECTS

Toolbar:	Modify > Erase
Menu:	Modify > Erase
Command:	ERASE

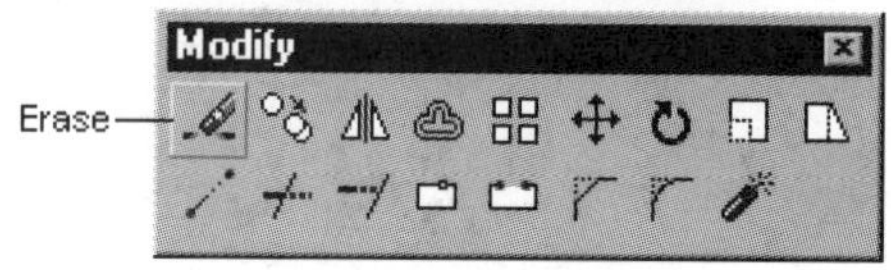

Figure 1-41 *Invoking the* ***ERASE*** *command from the* ***Modify*** *toolbar*

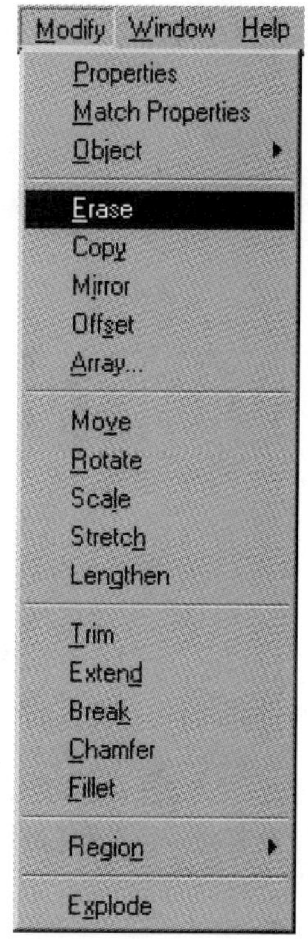

Figure 1-42 *Invoking the* ***ERASE*** *command from the* ***Modify*** *menu*

After drawing some objects you may want to erase some of them from the screen. To erase you can use AutoCAD LT's **ERASE** command (Figure 1-41 and Figure 1-43). This command is used exactly the same way as an eraser is used in manual drafting to remove unwanted information. When you invoke the **ERASE** command, a small box known as the pick box replaces the screen cursor. To erase an object, move the **pick box** so it touches the object. You can select the object by pressing the pick button of your pointing device, Figure 1-43. AutoCAD LT confirms the selection by changing the selected objects into dashed lines, and the **Select objects** prompt returns. You can either continue selecting objects or press ENTER to terminate the object selection process and erase the selected objects. If you are entering the command from the keyboard, you can type **E** or **ERASE**. The following is the prompt sequence:

Command: **ERASE** [Enter]
Select objects: *Select first object.*
Select objects: *Select second object.*
Select objects: [Enter]

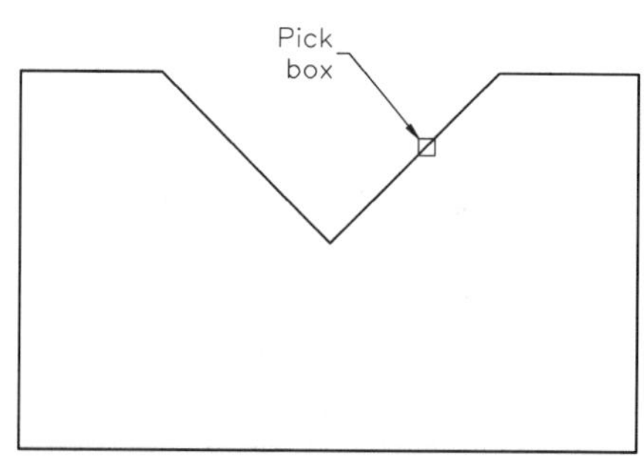

Figure 1-43 *Selecting objects by positioning the pick box at the top of the object and then pressing the pick button on the pointing device*

If you enter All at the **Select objects:** prompt, AutoCAD LT will erase all objects in the drawing, even if the objects are outside the screen display area.

Command: **ERASE** [Enter]
Select objects: **All**

You can also first select the objects to be erased from the drawing and then right-click on the drawing area to display the shortcut menu. From here, you can choose the **Erase** option.

CANCELING AND UNDOING A COMMAND

If you are in a command and you want to cancel or get out of that command, press the ESC (Escape) key on the keyboard.

Command: **ERASE** Enter
Select objects: *Press ESC (Escape) to cancel the command.*

Similarly, sometimes you unintentionally erase some object from the screen. When you discover such an error, you can correct it by restoring the erased object by means of the **OOPS** command. The **OOPS** command restores objects that have been accidentally erased by the previous **ERASE** command, Figure 1-44. You can also use the **U** (Undo) command to undo the last command.

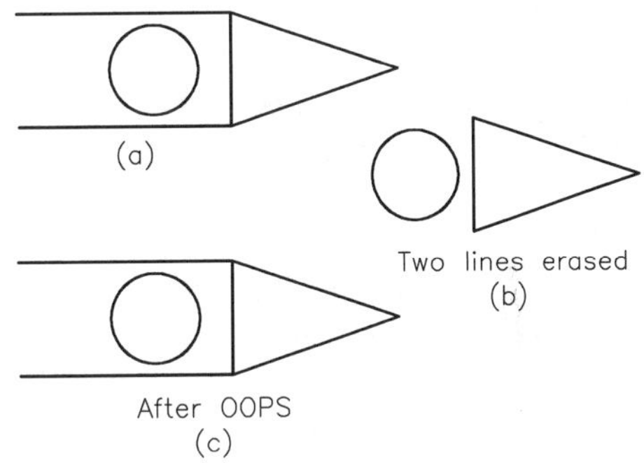

Figure 1-44 *Use of the **OOPS** command*

Command: **OOPS** Enter *(Restores erased objects.)*
Command: **U** Enter *(Undoes the last command.)*

OBJECT SELECTION METHODS

One of the ways to select objects is to select them individually, which can be time consuming if you have a number of objects to edit. This problem can be solved by creating a selection set that enables you to select several objects at a time. The selection set options can be used with those commands that require object selection, such as **ERASE** and **MOVE**. There are many object selection methods, such as All, Last, and Add. At this point we will explore the two options: **Window** and **Crossing.** The remaining options are discussed in the Chapter 5.

The Window Option

This option is used to select an object or group of objects by enclosing them by a box or window. The objects to be selected should be completely enclosed within the window; those objects that lie partially inside the boundaries of the window are not selected. You can select the **Window** option by typing W at the **Select objects** prompt. You are prompted to select the two opposite corners of the window. After selecting the first corner, you can select the other corner by dragging the cursor to the desired position and specifying the particular point. As you move the cursor, a window is displayed that changes in size as you move the cursor. The objects selected by the **Window** option are displayed as dashed objects (Figure 1-45).

The prompt sequence for using the **Window** option with the **ERASE** command is:

Command: **ERASE** Enter
Select objects: **W** Enter
Specify first corner: *Select the first corner.*

Specify opposite corner: *Select the second corner.*
Select objects: Enter

You can also select the **Window** option by selecting a blank point on the screen at the **Select objects** prompt. This is automatically taken as the first corner of the window. Dragging the cursor to the right will display a window. After getting all the objects to be selected inside this window, you can specify the other corner with your pointing device. The objects that are completely enclosed within the window will be selected and highlighted. The following is the prompt sequence for automatic window selection with the **ERASE** command:

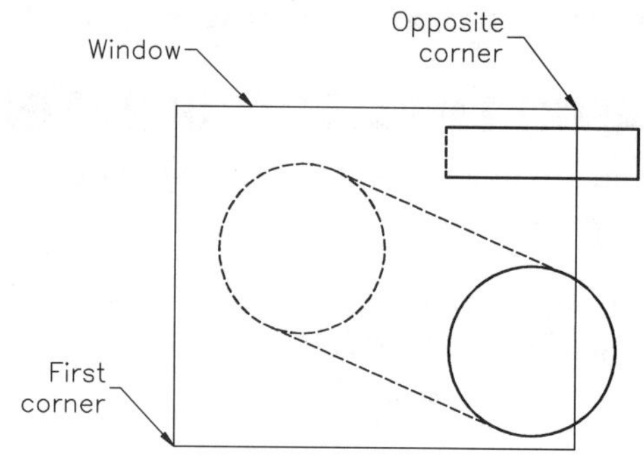

Figure 1-45 *Selecting objects using the Window option*

Command: **ERASE** Enter
Select objects: *Select a blank point as the first corner of the window.*
Specify opposite corner: *Drag the cursor to the right to select the other corner of the window.*
Select objects: Enter

The Crossing Option

This option is used to select an object or group of objects by creating a box or window around them. The objects to be selected should be touching the window boundaries or completely enclosed within the window. You can invoke the **Crossing** option by entering **C** at the **Select objects** prompt. After you choose the Crossing option, AutoCAD LT prompts you to select the first corner at the **Specify first corner** prompt. Once you have selected the first corner, a box or window made of dashed lines is displayed. By moving the cursor you can change the size of the crossing box, hence putting the objects to be selected within (or touching) the box. Here you can select the other corner. The objects selected by the Crossing option are highlighted by displaying them as dashed objects Figure 1-46. The following prompt sequence illustrates the use of the Crossing option when you choose the **Erase** button is:

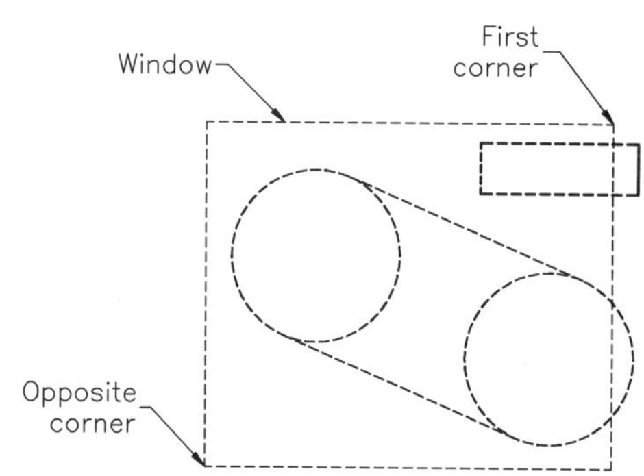

Figure 1-46 *Selecting objects using the Crossing option*

Select objects: **C** Enter
Specify first corner: *Select the first corner of the crossing window.*
Specify opposite corner: *Select the other corner of the crossing window.*
Select objects: Enter

You can also select the **Crossing** option automatically by selecting a blank point on the screen at the **Select objects** prompt and dragging the cursor to the left. The blank point you selected becomes the first corner of the crossing window and AutoCAD LT will then prompt you to select the other corner. As you move the cursor, a box or window made of dashed lines is displayed. The objects that are touching or completely enclosed within the window will be selected. The objects selected by the Crossing option are highlighted by being displayed as dashed objects. The prompt sequence for automatic crossing selection when you choose the **Erase** button is:

Select objects: *Select a blank point as the first corner of the crossing window.*
Specify opposite corner: *Drag the cursor to the left to select the other corner of the crossing window.*
Select objects: [Enter]

DRAWING CIRCLES

Toolbar:	Draw > Circle
Menu:	Draw > Circle
Command:	CIRCLE

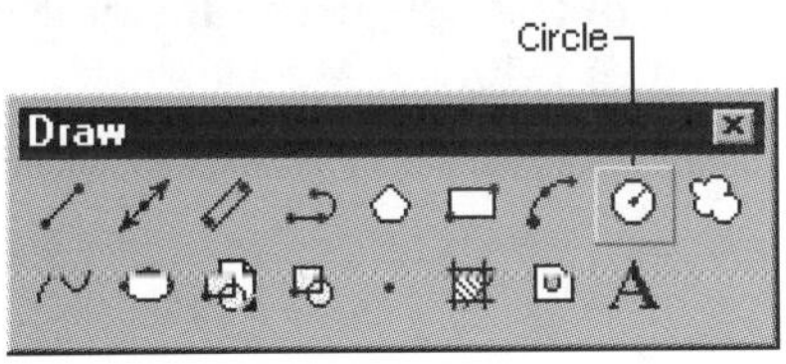

*Figure 1-47 Invoking the **CIRCLE** command from the **Draw** toolbar*

To draw the circles you can use the AutoCAD LT's **CIRCLE** command (Figure 1-47 and Figure 1-48). The following is the prompt sequence for the **CIRCLE** command:

Command: **CIRCLE** [Enter]
Specify center point for circle or [3P/2P/Ttr (tan tan radius)]:

The different options of the **CIRCLE** command are explained in the following sections.

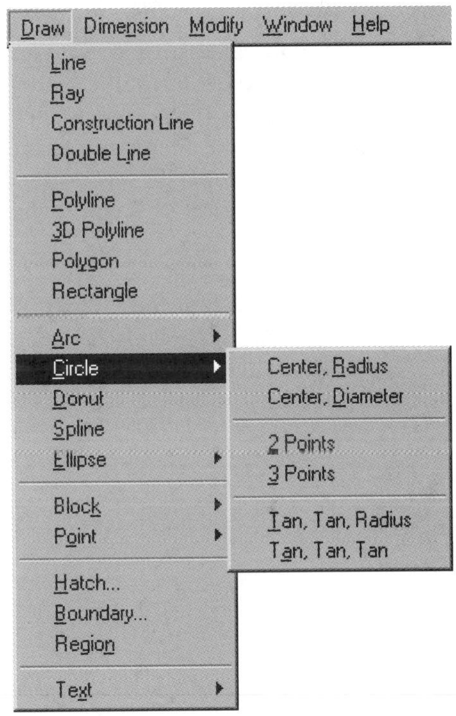

*Figure 1-48 Invoking the **CIRCLE** command from the **Draw** menu*

The Center and Radius Option

In this option you can draw a circle by defining the center and the radius of the circle, Figure 1-49. After entering the **CIRCLE** command, AutoCAD LT will prompt you to enter the center of the circle, which can be selected by specifying a point on the screen or by entering the coordinates of the center point. Next, you will be prompted to enter the radius of the circle. Here you can accept the default value, enter a new value, or select a point on the circumference of the circle to specify the radius. The following is the prompt sequence for drawing a circle with a center at 3,2 and a radius of 1 unit:

Command: **CIRCLE** Enter
Specify center point for circle or [3P/2P/Ttr(tan tan radius): **3,2** Enter
Specify radius of circle or [Diameter]<current>: **1** Enter

Note
*You can also set the radius by assigning a value to the **CIRCLERAD** system variable. The value you assign becomes the default value for radius.*

The Center and Diameter Option

In this option you can draw a circle by defining the center and diameter of the circle. After invoking the **CIRCLE** command, AutoCAD LT prompts you to enter the center of the circle, which can be selected by specifying a point on the screen or by entering the coordinates of the center point. Next, you will be prompted to enter the radius of the circle. At this prompt enter **D**. After this you will be prompted to enter the diameter of the circle. For entering the diameter you can accept the default value, enter a new value, or drag the circle to the desired diameter and select a point. If you use a menu option to select the **CIRCLE** command with the **Diameter** option, the menu automatically enters the **Diameter** option and prompts for the diameter after you specify the center. The following is the prompt sequence for drawing a circle with the center at (2,3) and a diameter of 2 units, Figure 1-50:

Command: **CIRCLE** Enter
Specify center point for circle or [3P/2P/Ttr(tan tan radius): **2,3** Enter
Specify radius of circle or [Diameter]<current>: **D** Enter
Specify diameter of circle <current>: **2** Enter

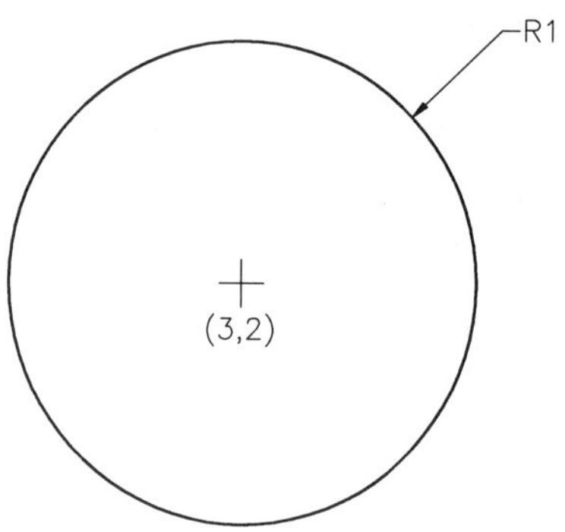

Figure 1-49 *Drawing a circle using the Center and Radius option*

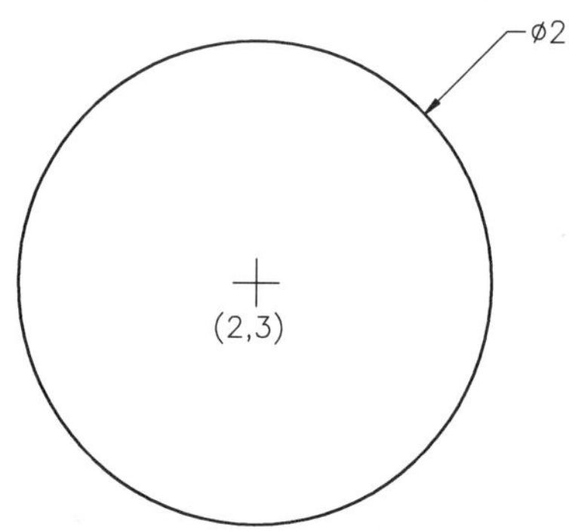

Figure 1-50 *Drawing a circle using the Center and Diameter option*

The Two-Point Option

You can also draw a circle using the **Two-Point** option. In this option AutoCAD LT lets you draw the circle by specifying the two endpoints of the circle's diameter. For example, if you want to draw a circle that passes through the points (1,1) and (2,1), you can use the **CIRCLE** command with 2P option, as shown in the following example (Figure 1-51):

Command: **CIRCLE** Enter

Specify center point for circle or [3P/2P/Ttr(tan tan radius): **2P** Enter
Specify first endpoint of circle's diameter: **1,1** Enter
Second endpoint of circle's diameter: **2,1** Enter *(You can also use the relative coordinates.)*

The Three-Point Option

For drawing a circle, you can also use the **Three-Point** option by defining three points on the circumference of the circle. The three points may be entered in any order. To draw a circle that passes through the points 3,3, 3,1 and 4,2, Figure 1-52, the prompt sequence is:

Command: **CIRCLE** Enter
Specify center point for circle or [3P/2P/Ttr(tan tan radius)]: 3P Enter
Specify first point on circle: **3,3** Enter
Specify second point on circle: **3,1** Enter
Specify third point on circle: **4,2** Enter

You can also use **relative rectangular coordinates** to define the points:

Command: **CIRCLE** Enter
Specify center point for circle or [3P/2P/Ttr(tan tan radius)]: **3P** Enter
Specify first point on circle: **3,3** Enter
Specify second point on circle: **@0,-2** Enter
Specify third point on circle: **@1,1** Enter

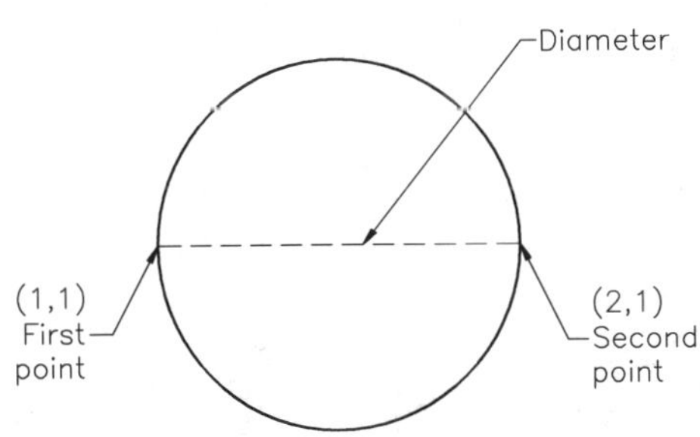

Figure 1-51 *Drawing a Circle using the Two-Point option*

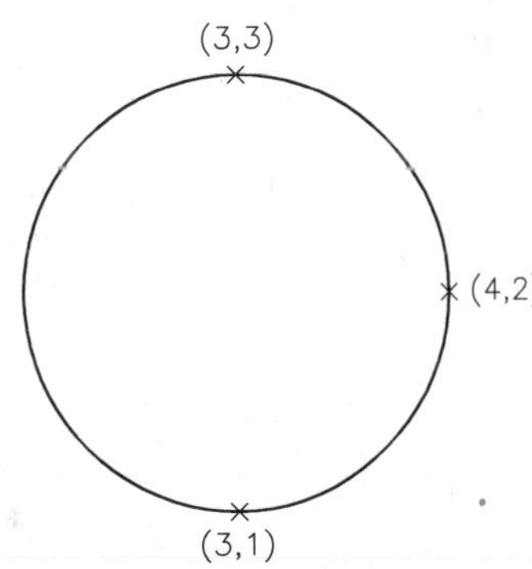

Figure 1-52 *Drawing a Circle using the Three-Point option*

The Tangent Tangent Radius Option

A tangent is an object (line, circle, or arc) that contacts the circumference of a circle at only one point. In this option AutoCAD LT uses the Tangent object snap to locate two tangent points on the selected objects that are to be tangents to the circle. Then you have to specify the radius of the circle. The prompt sequence for drawing a circle using the **Ttr** option is:

Command: **CIRCLE** Enter
Specify center point for circle or [3P/2P/Ttr(tan tan radius)]: **T** Enter
Specify point on object for first tangent of circle: *Select first line, circle, or arc.*

Specify point on object for second tangent of circle: *Select second line, circle, or arc.*
Specify radius of circle <current>: **0.75** Enter

In Figures 1-53 through 1-56, the dotted circles represent the circles that are drawn by using the **Ttr** option. The circle actually drawn depends on how you select the objects that are to be tangent to the new circle. The figures show the effect of selecting different points on the objects. The dashed circles are the circles that are drawn using the **Ttr** option. If you specify too small or large radius, you may get unexpected results or the "Circle does not exist" prompt.

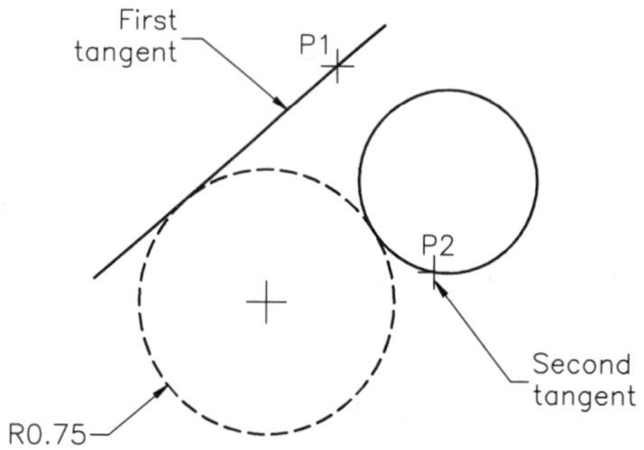

Figure 1-53 *Tangent, tangent, radius (Ttr) option*

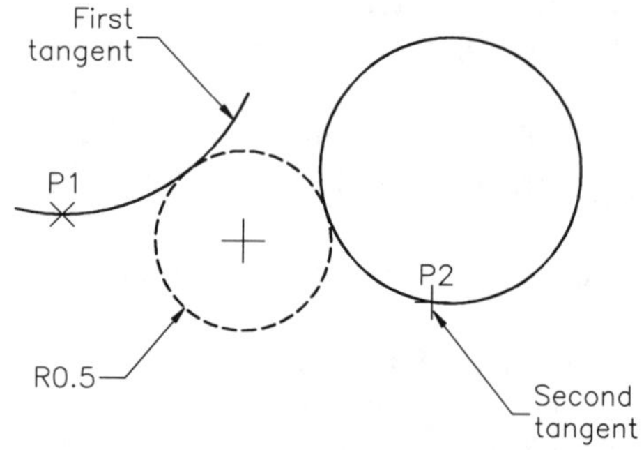

Figure 1-54 *Drawing a circle using the tangent, tangent, radius (Ttr) option*

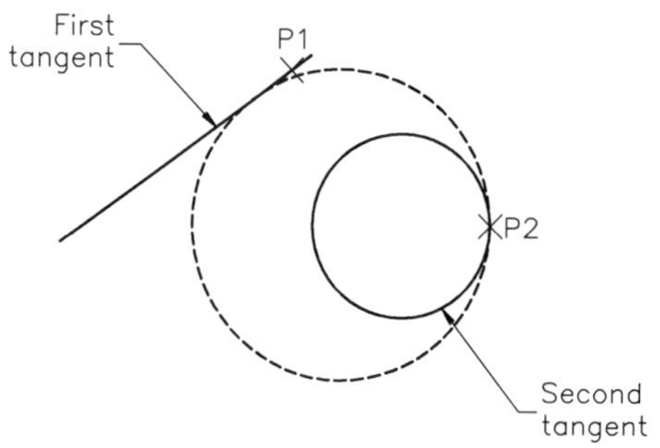

Figure 1-55 *Tangent, tangent, radius option*

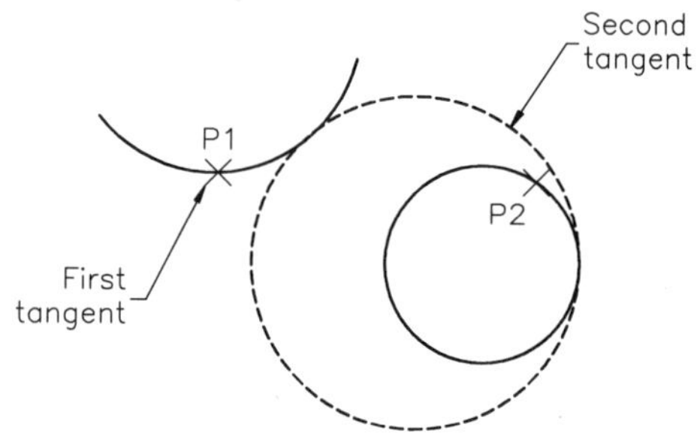

Figure 1-56 *Tangent, tangent, radius (Ttr) option*

The Tangent, Tangent, Tangent Option

You can invoke this option from the menu bar. This option is a modification of the Three-Point option. In this option AutoCAD LT uses the Tangent Osnap to locate three points on three selected objects to which the circle is drawn tangent. The following is the prompt sequence for drawing a circle using the **Tan, Tan, Tan** option (Figure 1-57):

Command: Choose **Circle > Tan, Tan, Tan** from the **Draw** menu.
Specify center point for circle or [3P/2P/Ttr(tan tan radius): *Select Tan, Tan, Tan option from the **Draw** menu.*

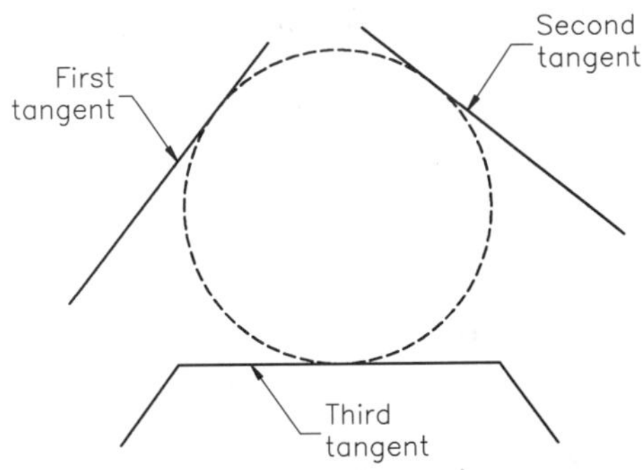

Figure 1-57 *Drawing a circle using the Tan, Tan, Tan option*

_ 3P Specify first point on circle: _tan to *Select the first object.*
Specify second point on circle: _tan to *Select the second object.*
Specify third point on circle: _tan to *Select the third object.*

Exercise 5 — *Mechanical*

Draw Figure 1-58 using different options of the **LINE** and **CIRCLE** commands. Use absolute, relative rectangular, or relative polar coordinates for drawing the triangle. The vertices of the triangle will be used as the center of the circles. The circles can be drawn using the Center and Radius, Center and Diameter, or Tan, Tan, Tan option. (Height of triangle = 4.5 x sin 60 = 3.897.) Do not draw the dimensions; they are for reference only.

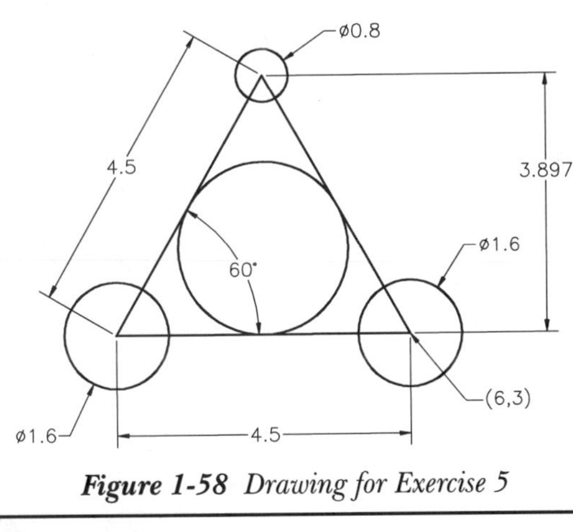

Figure 1-58 *Drawing for Exercise 5*

SAVING YOUR WORK

Toolbar:	Standard > Save
Menu:	File >Save or Save As
Command:	QSAVE , SAVEAS , SAVE

In AutoCAD LT or any computer system, you must save your work before you exit from the drawing editor or turn the system off. Also, it is recommended that you save your drawings after regular time intervals. In case of a power failure, an editing error, or other problems, all work saved before the problem started will be retained.

AutoCAD LT has provided the following commands that let you save your work on the hard disk of the computer or on the floppy diskette:

QSAVE **SAVEAS** **SAVE**

The **QSAVE**, **SAVEAS**, and **SAVE** commands allow you to save your drawing by writing it to a permanent storage device, such as a hard drive, or on a diskette in the A drive.

When you choose **Save** from the **File** menu, or choose the **Save** button in the **Standard** toolbar, the **QSAVE** command is invoked. If the current drawing is unnamed and you are saving the drawing for the first time in the present session, the **QSAVE** command will prompt you to enter the file name in the **Save Drawing As** dialog box (Figure 1-60). You can enter the name for the drawing and then choose the **Save** button in the dialog box. Once the drawing is saved and you make some changes to it, you can

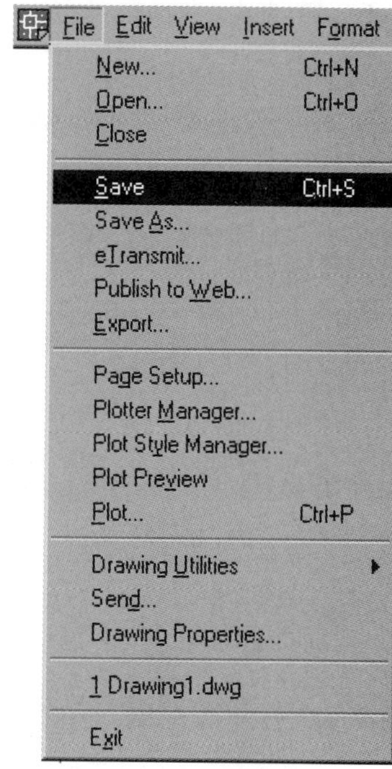

***Figure 1-59** Different Save options in the **File** menu*

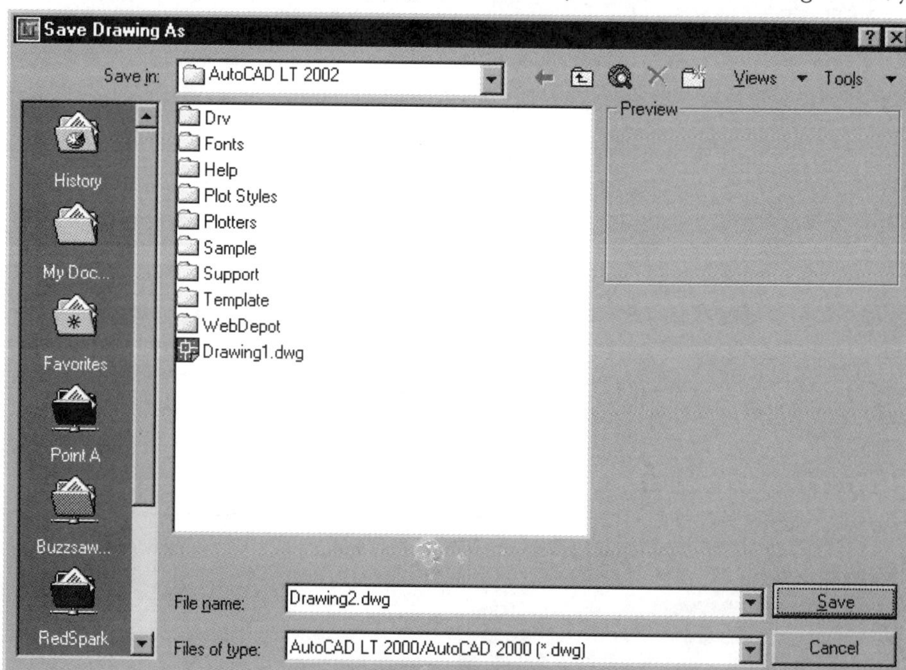

***Figure 1-60** **Save Drawing As** dialog box*

use the **QSAVE** command to save the drawing with the current name without prompting you to enter a file name. This allows you to do a quick save.

When you invoke the **SAVEAS** command, the **Save Drawing As** dialog box (Figure 1-60) is displayed always. Even if the drawing has been saved with a file name, this command gives you an option to save it with a different file name. In addition to saving the drawing, it sets the name of the current drawing to the file name you specify, which is displayed in the title bar. This command is used when you want to save a previously saved drawing under a different file name. You can also use this command when you make certain changes to a template and want to save the changed template drawing but leave the original template unchanged.

The **SAVE** command is the most rarely used command and can be invoked only from the command line by entering **SAVE** at the Command prompt. It is similar to the **SAVEAS** command and displays the **Save Drawing As** dialog box always. With this command you can save a previously saved drawing under a different file name, but this command does not set it as the current drawing.

Save Drawing As Dialog Box

The **Save Drawing As** dialog box displays the information related to the drawing files on your system. The different components of the dialog box are described next.

Places list

A column of icons is displayed on the left side of the dialog box. These icons contain the shortcuts to the folders that are frequently used. You can quickly save your drawings in one of these folders. The **History** folder displays the list of most recently saved drawings. You can save your personal drawings in the **My documents** or the **Favorites** folder. The **Point A**, **RedSpark**, and **Buzzsaw** icons connect you to their respective pages on the Web. You can add any new folder in this list for easy access by simply dragging this folder on to the Places list area and then leaving it. You can rearrange all these folders by dragging them and then placing them at the desired locations. It is also possible to remove the folders when not in frequent use. Right-click on the particular folder and then select **Remove** from the shortcut menu. The Point A, RedSpark, and the Buzzsaw shortcuts are permanent and cannot be removed.

File name edit box

To save your work, enter the name of the drawing in the **File Name** edit box. This can be done by typing the file name or selecting it from the drop-down list. If you select the file name you want from the drop-down list, the name you select automatically appears in the **File name** edit box. If you have already assigned a name to the drawing, the current drawing name is taken as default. If the drawing is unnamed, the default name **Drawing1** is displayed in the **File Name** edit box. You can also choose the down arrow at the right of the edit box to display the names of the previously saved drawings and choose a name here.

Files of type drop-down list

The **Files of type** drop-down list (Figure 1-61) is used to specify the drawing format in which you want to save the file. For example, to save the file as AutoCAD LT 2000 drawing file, select **AutoCAD LT 2000 Drawing** from the drop-down list.

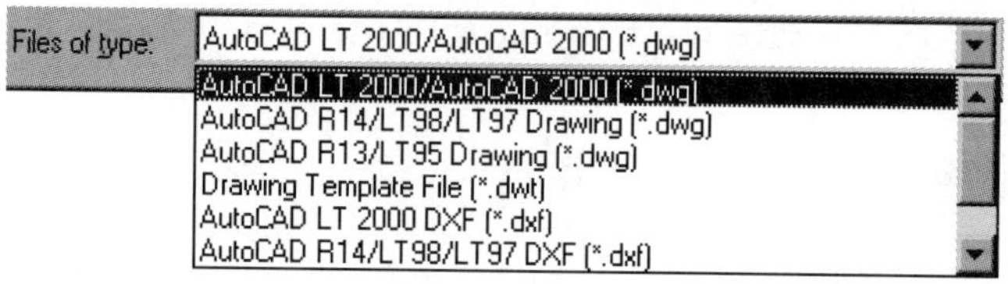

***Figure 1-61 Files of type** drop-down list*

Save in drop-down list

The current drive and path information is listed in the **Save in** drop-down list (Figure 1-62). AutoCAD LT will initially save the drawing in the default directory, but if you want to save the drawing in a different directory, you have to specify the path. For example, if you want to save the present drawing under the file name **HOUSE** in the c1 subdirectory of the Edited 2002 directory, choose the arrow button in the **Save in** drop-down list to display the drop-down list and select D:. When you select D:, all directories in the D drive will be listed in the **File** list box. Double-click on **Edited 2002** or select **Edited 2002** and choose the **Open** button to display its directories. Again double-click on **C1** or select it and choose the **Open** button to display drawing names in the **File** list box. Select HOUSE from the list, if it is already listed there, or enter it in the **File name** edit box and then choose the **Save** button. Your drawing (House) will be saved in the folder (D:\Edited 2002\C1\HOUSE.dwg). If you want to save the drawing on the A drive, select A: in the **Save in** drop-down list.

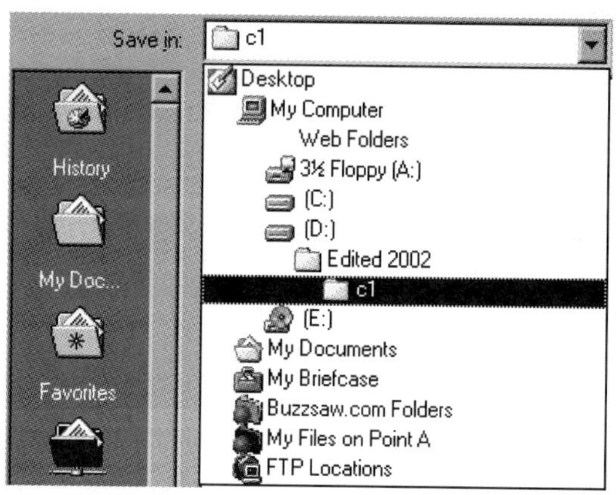

***Figure 1-62 Save in** drop-down list*

Tip

The file name you enter to save a drawing should match the contents of the drawing. This helps you to remember the drawing details and makes it easier to refer to them later. Also the file name can be 255 characters long and can contain spaces and punctuation marks.

Note

If you want to save a drawing on the A or B drive, make sure the diskette you are using to save the drawings is formatted.

Views List

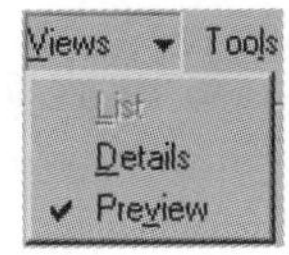

The **Views** drop-down list has options for the type of listing of files and displaying the preview images (Figure 1-63).

Figure 1-63 Views list

List, Detail, and Preview options

If you choose the **Detail** option, it will display detailed information about the files (size, type, date, and time of modification) in the **Files** list box. In the detailed information if you click on the **Name** label, the files are listed with the names in alphabetical order. If you double-click on the Name label, the files will be listed in reverse order. Similarly if you click on the **Size** label, the files are listed according to the size in ascending order. Double-clicking on the **Size** label will list the files descending by order of size. Similarly you can click on the **Type** label or the **Modified** label to list the files accordingly. If you choose the **List** option, all files present in the current directory will be listed in the **File** list box. If you select the **Preview** option, the list box displays the Preview image box wherein the bitmap image of the file chosen is displayed. If cleared, the Preview box is not displayed.

Folder Button

If you choose the **Create New Folder** button, AutoCAD LT creates a new directory under the name **New Folder**. The new folder is displayed in the **File** list box. You can accept the name or change it to your requirement.

Up One Level Button

The **Up One Level** button displays the directories that are up by one level. For example, if you are in the **Samples** subdirectory of the **AutoCAD LT 2002** directory, then choosing the **Up one level** button will take you to the **AutoCAD LT 2002** directory.

Search the Web

It displays the **Browse the Web** dialog box that enables you to access and store AutoCAD LT files on the Internet.

Tools List

The **Tools** drop-down list (Figure 1-64) has an option for adding or modifying the FTP sites. These sites can then be browsed from the FTP shortcut in the **Places** list. The **Prompt for PointA Login** option can be used to reenter your login and password if you have asked the password to be remembered while opening the PointA portal from the Places list which bypasses the login prompt. The **Add Current Folder to Places** and **Add to Favorites** options add the folder displayed in the **Save in** edit box to the Places list or to the Favorites folder. The **Options** button displays the **Saveas Options** dialog box where you can save the proxy images of custom objects. It has the **DWG** and **DXF** options tabs.

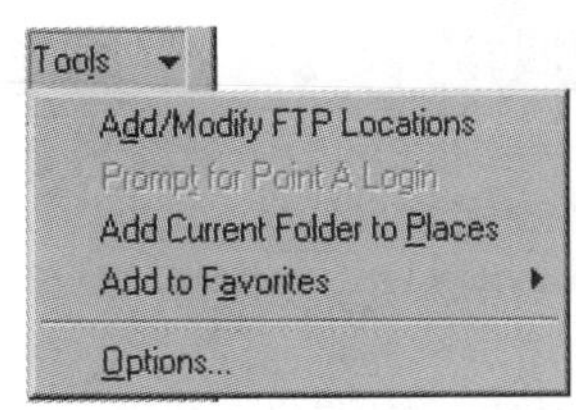

Figure 1-64 ***Tools*** *drop-down list*

AUTOMATIC TIMED SAVE

AutoCAD LT allows you to save your work automatically at specific intervals. To change the time intervals you can enter the time intervals in minutes in the **Minutes between saves** text box in the **Options** dialog box (**Open and Save** tab). This dialog box can be invoked from the **Tools** menu. Depending on the power supply, hardware, and type of drawings, you should decide on an appropriate time and assign that time to this variable. AutoCAD LT saves the drawing under the file name **AUTO.SV$.** The extension of the automatic save file is **.SV$.** You can also change the time interval by using the **SAVETIME** system variable.

CREATION OF BACKUP FILES

If the drawing file already exists and you use **SAVE** or **SAVEAS** commands to update the current drawing, AutoCAD LT creates a backup file. AutoCAD LT takes the previous copy of the drawing and changes it from a file type **.DWG** to **.BAK**, and the updated drawing is saved as a drawing file with the **.DWG** extension. For example, if the name of the drawing is **MYPROJ.DWG**, AutoCAD LT will change it to **MYPROJ.BAK** and save the current drawing as **MYPROJ.DWG.**

Tip

Although the Automatic save saves your drawing after a certain time interval, you should not completely depend on it because the procedure for converting the SV$ file into a drawing file is cumbersome. Therefore it is recommended that you save your files regularly using the QSAVE or SAVEAS commands.

CLOSING A DRAWING

You can use the **CLOSE** command to close the current drawing file without actually quitting AutoCAD LT. If you choose **Close** from the **File** menu or enter **CLOSE** at the Command prompt, the current drawing file is closed. If you have not saved the drawing after making the last changes to it, and you invoke the **CLOSE** command, AutoCAD LT displays a dialog box that allows you to save the drawing before closing. This box gives you an option to discard the current drawing or changes made to it. It also gives you an option to cancel the command. After closing the drawing you are still in AutoCAD LT from where you can open a new or an already saved drawing file. You can also use the close button (**X**) of the drawing window to close the drawing.

QUITTING AUTOCAD LT

You can exit from the AutoCAD LT program by using the **EXIT** or **QUIT** command. Even if you have a drawing file open you can use the EXIT command to close it as well as quit the AutoCAD LT program. In case the drawing has not been saved, it allows you to save the work first through a dialog box. You an also use the close button (**X**) of the main AutoCAD LT window (present in the title bar) to end the AutoCAD LT session.

AUTOCAD LT'S HELP

Toolbar:	Standard > Help
Menu:	Help > Help
Command:	HELP

You can get the online help and documentation on the working of AutoCAD LT 2002 commands from the **Help** menu (Figure 1-65).

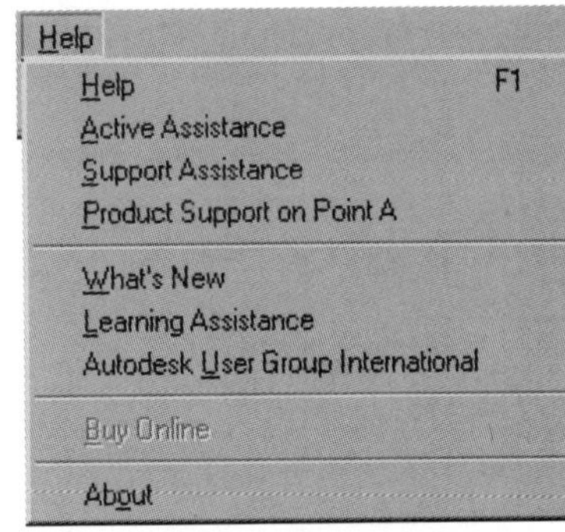

Figure 1-65 ***Help*** *menu*

Help

Invoking **HELP** displays the **AutoCAD LT 2002 Help** dialog box (Figure 1-66). You can use this dialog box to access help on different topics and commands. It has five tabs: **Contents**, **Index**, **Search**, **Favorites**, and **Ask me**, which display the corresponding help topics. If you are in the middle of a command and need help for using it, choose the **Help** button. The information about that particular command will be displayed in the dialog box.

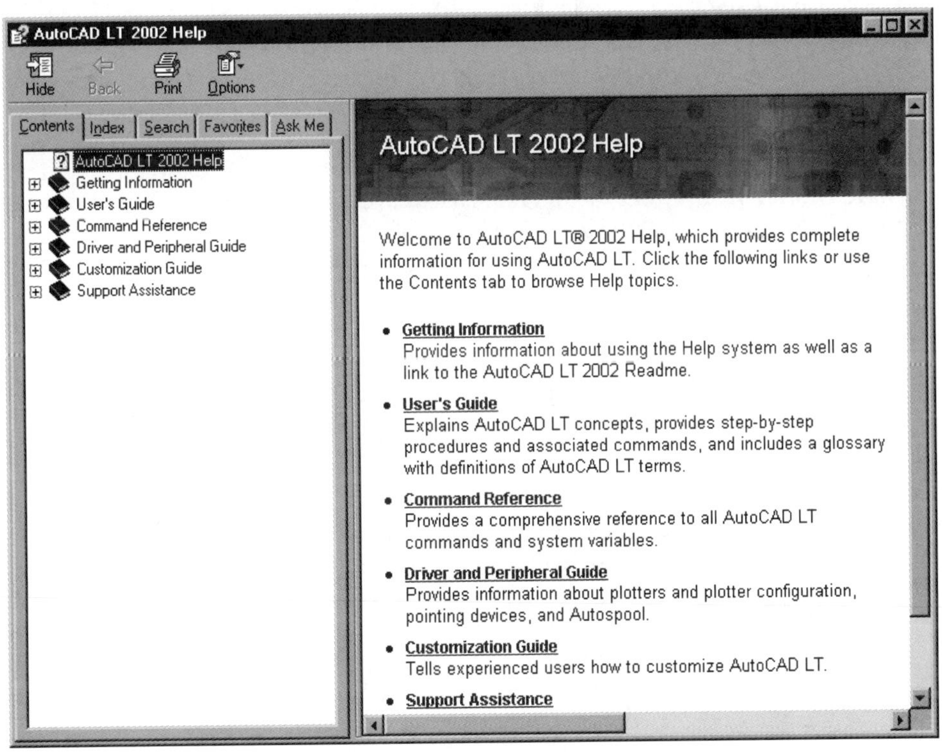

Figure 1-66 ***AutoCAD LT 2002 Help*** *dialog box (**Contents** tab)*

Contents

This tab displays the help topics that are organized by categories pertaining to different sections of AutoCAD LT such as the User's Guide, Command Reference, and so on. To select a category, double-click on the corresponding book icon or choose the plus sign on the left. The icon becomes an open book with a minus (–) sign and a list of headings associated with that category are displayed. Use the plus sign (+) to further open the headings until you reach the help topic that has a question mark (?) displayed with it. Choose the topic to display information about the selected topic or command in the window present on the right side of the dialog box.

Index

This tab displays the complete index (search keywords) in alphabetical order. To display information about an item or command, type the item (word) or command name in the edit box. With each letter entered the listing keeps on changing in the list area, displaying the possible topics. When you enter the word and if AutoCAD LT finds that word, it is automatically highlighted in the list area. Choose the **Display** button to display information about it.

Search

This tab creates a word list based on all the keywords present in the online help files. When you type a word you are looking for and then choose the **List Topics** button, a list of matching words appears in a window below to narrow down your search. This search is dependent on the option you have selected at the bottom of the dialog box. Use the scroll bar to scroll through the list, select the desired topic, and then choose the **Display** button to display its help.

Favorites

This tab lets you create a list of your own topics that you need to access regularly. The topic that you have chosen in any other tab of Help is displayed in the **Current Topic** box when you choose the **Favorites** tab. Use the **Add** button to add it to your own list. You can use the **Remove** button to remove a topic from your list and the **Display** button to display its help.

Ask Me

When you choose this tab, you are allowed to enter a query in the edit box and then press ENTER. A list of topics related to the question follow. It also shows the book (category) from which it has been selected.

Active Assistance

This option gives you an access to context sensitive help. This help can be started by choosing the **Active assistance** button in the **Standard** toolbar, or the **Help** menu, or by entering **ASSIST** at the Command prompt. This mode of help is on by default when you start AutoCAD LT. Whenever you enter a command, an Active Assistance window with information regarding that command is displayed. Its icon is displayed in the main Task bar of the Windows at the bottom of the screen. To disable this help, right-click on the Active Assistance icon in the window's bar and choose Exit.

What's New

This option gives you an interactive list of all the new features in AutoCAD LT 2002. You can choose this option from the **Help** menu that displays the **What's New in AutoCAD LT 2002** screen with a list of new features and enhanced features. Open the **New Features** folder to display a list of topics (Figure 1-67). When you choose a topic, a description of the feature improvement is displayed. It has a file that displays an overview of the features added since AutoCAD LT R98 and AutoCAD LT 2000.

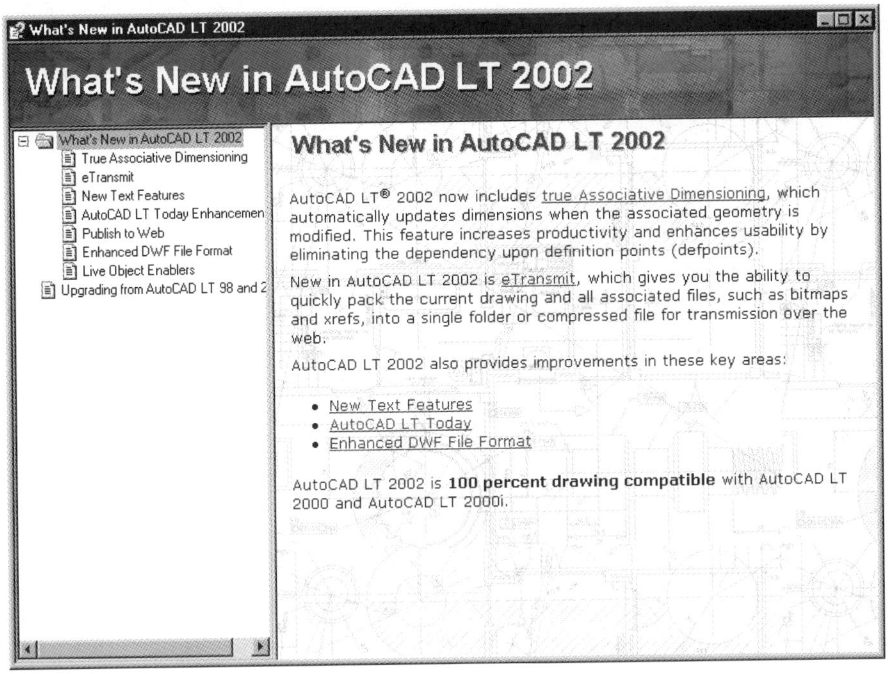

Figure 1-67 What's New in AutoCAD LT 2002 screen

Learning Assistance

This is a multimedia learning option that provides you an online environment where you can learn the AutoCAD LT software application interactively. You can choose this option from the **Help** menu. Choosing this option opens the **AutoCAD LT 2000 Help: User Documentation** dialog box with the **Learn AutoCAD LT** screen displayed. You can move the cursor on **Go to Autodesk e-Learning** where the cursor changes to a hand. Choosing this option connects you to the Uniform Resource Locator (URL) **http://www.autodesk.com/AutoCAD LT-elearning**. The AutoCAD LT page is opened using the Internet Explorer and displays information regarding the AutoCAD LT products. You can select Home to display the Autodesk home page. Here you can register with Headlight.com which is the Autodesk vendor. You can also select other pages such as My Profile, Team Profile to get specific help.

Support Assistance

This help option displays the Autodesk Support Assistance screen with five main headings: **Search for solution**, **Download**, **What's hot**, **More Resources,** and **Phone Support**. Choosing any one of these topics displays further subtopics to choose from with their brief explanations. You can choose any of these to look for their respective descriptions and related topics. To go to Search or Up one level, you choose the **Search** or **Up one level** buttons, provided at the end of the screen. This option gives you all the Technical Support information in a question-and-answer format. It also provides information about other help resources.

Autodesk User Group International

This utility connects you to the **Autodesk User Group International** Web pages and sites through the Microsoft Internet Explorer.

About

This option gives you information about the Release, Serial number, Licensed To, and also the legal description about AutoCAD LT.

ADDITIONAL HELP RESOURCES

1. You can get help for a command while working by pressing the F1 key. The **Help** dialog box containing information about the command is displayed. You can exit the dialog box and continue with the command.

2. You can get help about a certain dialog box by choosing the **Help** button in that dialog box.

3. Some of the dialog boxes have a **question mark** (?) button at the top right corner just adjacent to the **close** button. When you choose this button, the ? gets attached to the cursor. You can then drop it on any item in the dialog box to display information about that particular item.

4. Autodesk has provided several resources that you can use to get assistance with your AutoCAD LT questions. You can also use these resources to get information on Autodesk products, product updates, and other services provided by Autodesk. The following is a list of some of the resources:

   a. Autodesk Web Site **http://www.autodesk.com**
   b. AutoCAD LT 2002 Web Site: **http://www.autodesk.com/autocadlt**
   c. Autodesk Fax Information System: **(415) 446-1919**
   d. AutoCAD LT Technical Assistance Web site **autodesk.com/support**.
   e. AutoCAD LT Discussion Groups Web site **autodesk.com/support/discsgrp/acad.htm**.

5. **The Web** section in the **AutoCAD LT Today** window connects you to the Internet and provides you with the different links and resources.

6. You can also get help by contacting the author, Sham Tickoo, at **stickoo@calumet.purdue.edu** or **tickoo@cadcim.com**.

7. You can download AutoCAD LT drawings, programs, and special topics by accessing the author's Web site at **www.cadcim.com** or **www.calumet.purdue.edu/public/mets/tickoo/index.html**.

Self-Evaluation Test

Answer the following questions and then compare your answers to the correct answers given at the end of this chapter.

1. It is possible to change the initial (start-up) view displayed while starting AutoCAD LT. (T/F)

2. If the value of the **COORDS** variable is set to 1, the coordinate display is static, that is, the coordinate values displayed in the status bar change only when you specify a point. (T/F)

3. Using the **Crossing** method of object selection, the objects to be selected should be completely enclosed within the boundaries of the crossing box. (T/F)

4. If the current drawing is unnamed and you are saving the drawing for the first time in the present session, the **QSAVE** command will prompt you to enter the file name in the **Save Drawing As** dialog box. (T/F)

5. You can draw a line by specifying the length of the line and its direction, using the _______________.

6. In the _______________ coordinate system, the displacements along the X and Y axes (DX and DY) are measured with reference to the previous point rather than to the origin.

7. You can use the _______________ command to close the current drawing file without actually quitting AutoCAD LT.

8. The __________ option of the **CIRCLE** command can be used to draw a circle, if you want the circle to be tangent to two previously drawn objects.

9. You can erase a previously drawn line using the __________ option of the **LINE** command.

10. The __________ system variable can be used to change the time interval for automatic save.

Review Questions

Answer the following questions.

1. The **My Drawings** area of the **AutoCAD LT Today** window includes the **Bulletin Board** that gives you space to display information and create a customized bulletin board. (T/F)

2. The shortcut menu invoked by right-clicking in the command window displays the six most recently used commands and some of the Window's options such as Copy, Paste. (T/F)

3. The **Three-Point** option of the **CIRCLE** command lets you draw the circle by specifying the two endpoints of the circle's diameter. (T/F)

4. The file name you enter to save a drawing in the **Save Drawing As** dialog box file name can be 255 characters long but cannot contain spaces and punctuation marks. (T/F)

5. The **My Drawings** area in the **AutoCAD LT Today** window does not give you the option for:

   (a) Opening a drawing (b) Creating a new drawing
   (c) Saving a drawing (d) Opening Symbol libraries

6. You cannot terminate the **LINE** command by pressing which of the following keys on the keyboard at the **Specify next point or [Close/Undo]:** prompt?

   (a) SPACEBAR (b) BACKSPACE
   (c) ENTER (d) ESC

7. Which of the following commands is used to close the current drawing file without actually quitting AutoCAD LT?

   (a) **EXIT** (b) **END**
   (c) **CLOSE** (d) **QUIT**

8. Which of the following shortcuts can be removed from the **Places** list in the **Save Drawing As** dialog box?

   (a) Favorites (b) Point A
   (c) RedSpark (d) Buzzsaw

9. When you choose **Save** from the **File** menu, or choose the **Save** button in the **Standard** toolbar, which of the following commands is invoked?

   (a) **SAVE** (b) **LSAVE**
   (c) **QSAVE** (d) **SAVEAS**

10. Using the **Advanced Setup** option of the **Wizard** you can set the units, limits, and the different __________ settings for the new drawing.

11. A __________ is displayed on top of the AutoCAD LT screen that has the AutoCAD LT symbol and the current drawing name.

12. You can change the shape of the toolbars by placing the cursor anywhere on the __________

of the toolbar where it takes the shape of a double-sided arrow.

13. The __________ option of the **LINE** command can be used to join the current point with the initial point of the first line when two or more lines are drawn in continuation.

14. In the __________ coordinate system, the points are located with respect to origin 0,0.

15. To draw a circle, you can use the **Three-Point** option of the **CIRCLE** command by defining three points on the __________ of the circle.

16. The __________ tab of the **AutoCAD LT 2002 Help** dialog box displays the help topics that are organized by categories pertaining to different sections of AutoCAD LT.

Exercises

Exercise 6 *General*

Use the following relative rectangular and absolute coordinate values in the **LINE** command to draw the object.

Point	Coordinates	Point	Coordinates
1	3.0, 3.0	5	@3.0,5.0
2	@3,0	6	@3,0
3	@-1.5,3.0	7	@-1.5,-3
4	@-1.5,-3.0	8	@-1.5,3

Exercise 7 *General*

For Figure 1-68, enter the relative rectangular and relative polar coordinates of the points in the following table, then use these coordinates to draw the figure. The distance between the dotted lines is 1 unit. Save this drawing as Exer7.dwg.

1	3.0, 1.0	9	__________
2	__________	10	__________
3	__________	11	__________
4	__________	12	__________
5	__________	13	__________
6	__________	14	__________
7	__________	15	__________
8	__________	16	__________

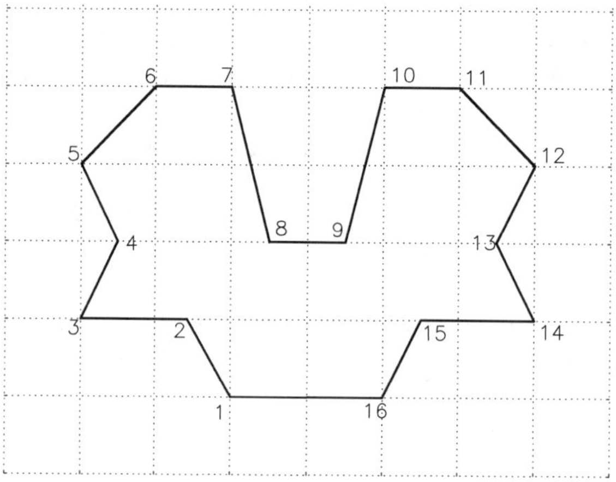

Figure 1-68 *Drawing for Exercise 7*

Exercise 8

General

For Figure 1-69, enter the relative polar coordinates of the points in the following table. Then use these coordinates to draw the figure. Do not draw the dimensions.

Point	Coordinates	Point	Coordinates
1	1.0, 1.0	6	________
2	________	7	________
3	________	8	________
4	________	9	________
5	________		

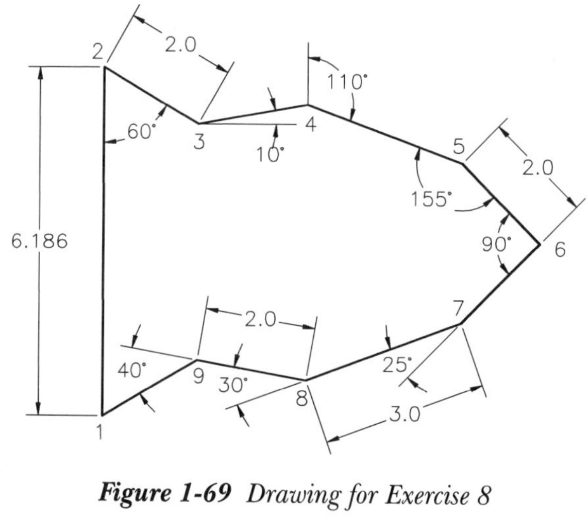

Figure 1-69 *Drawing for Exercise 8*

Exercise 9 *Mechanical*

Draw Figure 1-70, using the **LINE** and **CIRCLE** commands. The distance between the dotted lines is 1.0 units.

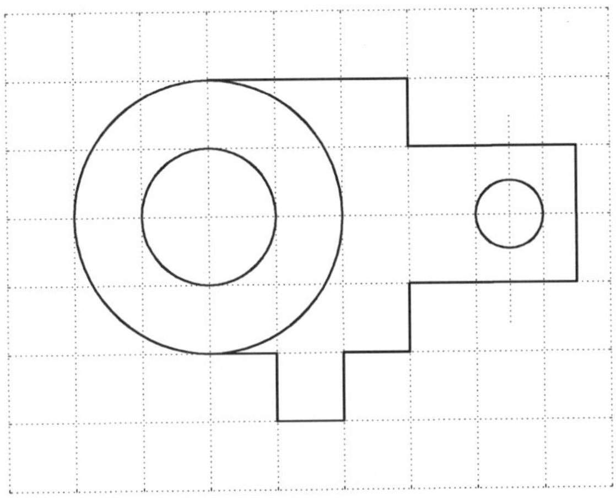

Figure 1-70 *Drawing for Exercise 9*

Exercise 10 *General*

Draw Figure 1-71, using the **LINE** command and the **Ttr** option of the **CIRCLE** command.

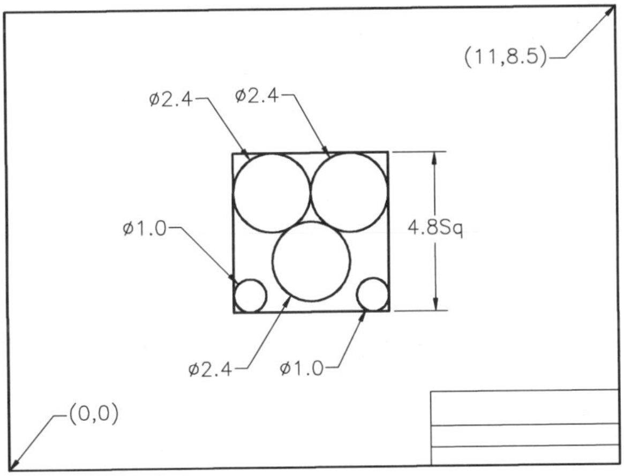

Figure 1-71 *Drawing for Exercise 10*

Problem Solving Exercise 1 *General*

Draw the object shown in Figure 1-72, using the **LINE** and **CIRCLE** commands. In this exercise only the diameters of the circles are given. To draw the lines and small circles (Dia 0.6), you need to find the coordinate points for the lines and the center points of the circles. For example, if the center of concentric circles is at 5,3.5, then the X coordinate of the lower left corner of the rectangle is 5.0 - 2.4 = 2.6.

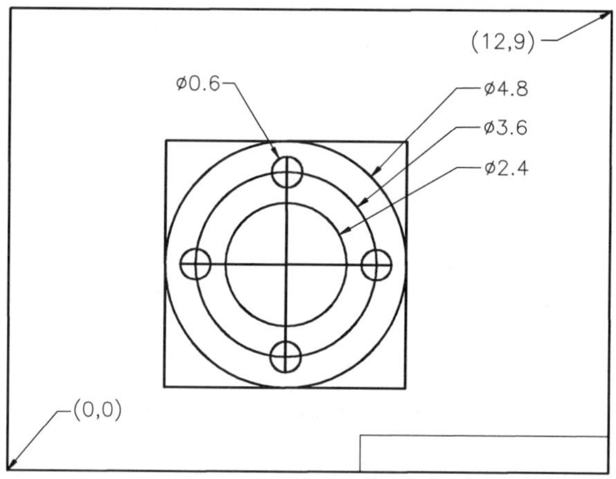

***Figure 1-72** Drawing for Problem Solving Exercise 1*

Problem Solving Exercise 2 *General*

Draw Figure 1-73, using various options of the **CIRCLE** and **LINE** commands. In this exercise you have to find the coordinate points for drawing the lines and circles. Also, you need to determine the best and easiest method to draw the 0.85 diameter circles along the outermost circle.

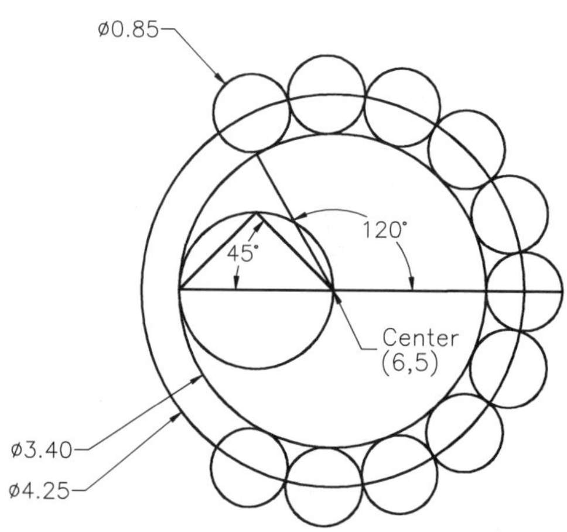

***Figure 1-73** Drawing for Problem Solving Exercise 2*

Problem Solving Exercise 3

General

Draw the drawing in Figure 1-74, using the absolute, relative rectangular, or relative polar coordinate system. Draw according to the dimensions shown in the figure, but do not draw the dimensions.

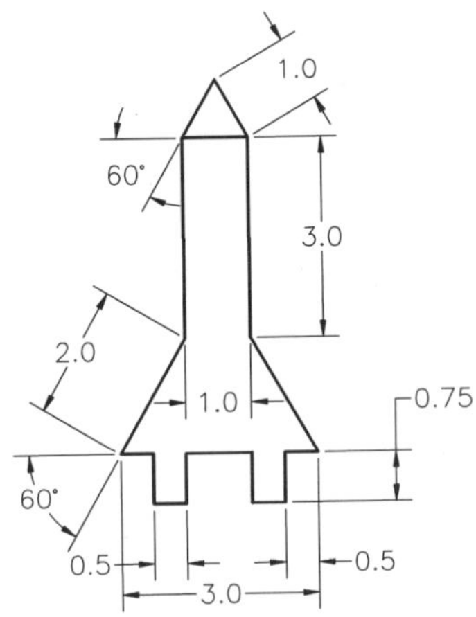

***Figure 1-74** Drawing for Problem Solving Exercise 3*

Answers to Self-Evaluation Test

1 - T, **2** - F, **3** - F, **4** - T, **5** - Direct distance entry, **6** - relative rectangular, **7** - **CLOSE**, **8** - Tangent tangent radius, **9** - Undo, **10** - **SAVETIME**

Chapter 2

Getting Started-II

Learning Objectives

After completing this chapter, you will be able to:

- *Open an existing file and start a new drawing.*
- *Use the MDE feature to open more than one drawing in one AutoCAD LT session.*
- *Use the* ***ZOOM*** *and* ***PAN*** *display commands.*
- *Set up units using the* ***UNITS*** *command.*
- *Set up and determine limits for a given drawing.*
- *Plot drawings using the basic plotting options.*
- *Use the* ***Options*** *dialog box to specify the settings.*

OPENING AN EXISTING FILE

You can open an existing drawing file that has been saved previously. There are two methods you can use to open a drawing file: the **AutoCAD LT Today** window and the **OPEN** command.

Using the AutoCAD LT Today Window

You can open a drawing using the **AutoCAD LT Today** window whenever you are starting a new session of AutoCAD LT. When you start AutoCAD LT, the **AutoCAD LT Today** window is displayed (Figure 2-1). Select the **Open Drawings** tab. A list of existing drawings are displayed depending upon the category chosen in the **Select how to begin** list box. You can choose a drawing from the list to open it. The **History (by Date)** option is the default category and the saved drawings are listed by day for the current week and by week for the previous weeks. The latest date when a drawing is saved is displayed on the top and the total number of drawings saved on that day are given in parentheses. You can click on the date to list the drawings. If you place the cursor on a particular drawing name, it gets underlined and a preview is displayed (Figure 2-1). Also a tooltip showing the full path is displayed.

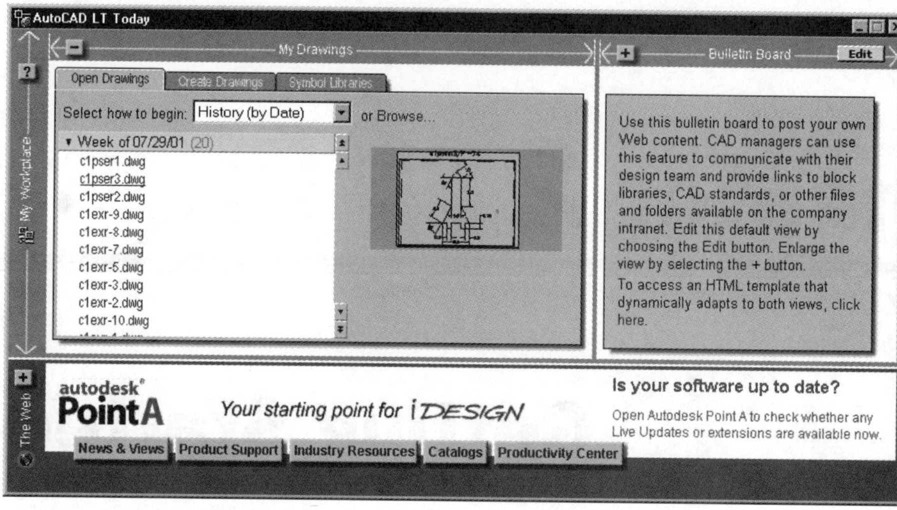

Figure 2-1 ***Open Drawings*** *option of the* ***AutoCAD LT Today*** *window,* ***History (by Date)*** *category*

If you select the **Most Recently Used** option (Figure 2-2) from the **Select how to begin** drop-down list, a list of four most recently opened drawings is displayed. A maximum of nine drawings can be displayed depending upon the number entered in the **Options** dialog box (**Open and Save** tab).

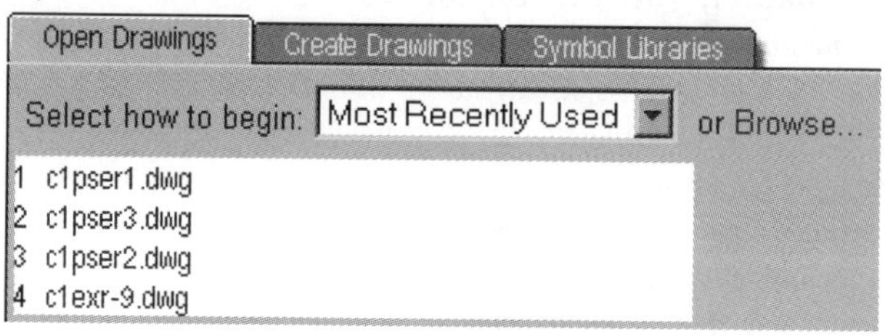

Figure 2-2 *Listing drawings under* ***Most Recently Used*** *category*

If you select the **History (by Filename)** option (Figure 2-3) from the **Select how to begin** drop-down list, a list of alphabetically sorted drawings is displayed. The drawings are sorted alphabetically by their file names.

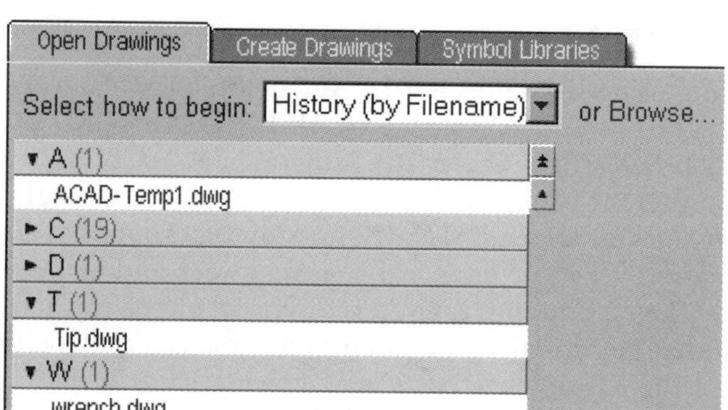

Figure 2-3 *Listing drawings under* ***History (by Filename)*** *category*

If you select the **History (by Location)** option (Figure 2-4) from the **Select how to begin** drop-down list, a list of drawings is sorted and displayed according to their location along with the path in which they are saved.

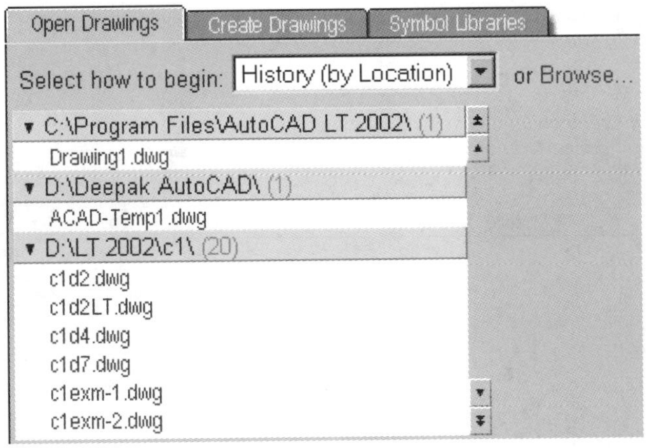

Figure 2-4 Listing drawings under ***History (by Location)*** *category*

If the drawing you want to open is not displayed in the list, choose the **Browse** button to display the **Select File** dialog box (Figure 2-5), where you can select the drawing file to open. The **Select File** dialog box has been discussed in the following section.

Using the OPEN Command

Toolbar:	Standard Toolbar > Open
Menu:	File > Open
Command:	OPEN

If you are already in the drawing editor, and you want to open a drawing file, you can use the **OPEN** command. The **OPEN** command also displays the **Select File** dialog box (Figure 2-5) from which you can select the drawing. This dialog box is similar to the standard dialog boxes. You can choose the particular file you want to open from the particular directory. The default directory displayed in the dialog box is AutoCAD LT 2002. You can change the directory from the **Look in** drop-down list. You can then select the name of the drawing from the list box or you can enter the name of the drawing file you want to open in the **File name** edit box. After selecting the drawing file you can choose the **Open** button to open the file. Here you can choose **Drawing1.dwg** from the list and then choose the **Open** button to open the drawing.

When you select a file name, its image is displayed in the **Preview** box. If you are not sure about the file name of a particular drawing but know the contents, you can select the file names and look for the particular drawing in the **Preview** box. You can also change the file type by selecting it in the **Files of type** drop-down list. Apart from the dwg files, you can open the dwt (template) files or the dxf files. You have all the standard icons in the **Places** list that can be used to open the drawing files from different locations. The **Open** button has a drop-down list (Figure 2-6) where you can choose a method for opening the file. These methods

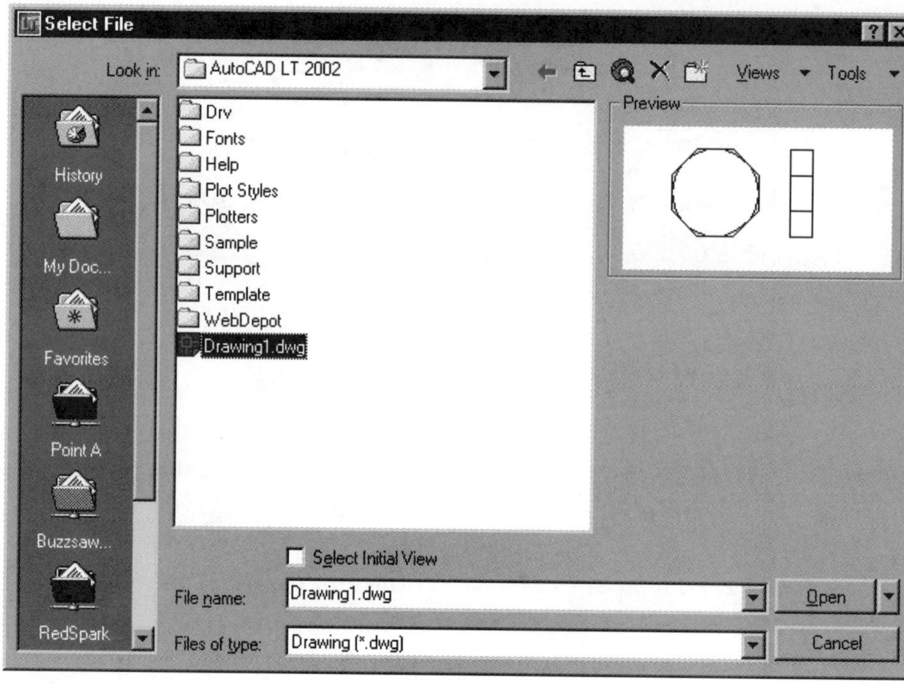

Figure 2-5 ***Select File*** *dialog box*

are described next.

Open Read-Only

If you want to view a drawing without altering it, you must select the **Open Read-Only** option from the **Open** drop-down list. In other words, read only protects the drawing file from changes. AutoCAD LT does not prevent you from editing the drawing, but if you try to save the opened drawing to the original file name, AutoCAD LT warns you that the drawing file is **write protected**. However, you can save the edited drawing to a file with a different file name using the **SAVEAS** command. This way you can preserve your drawing.

Figure 2-6 ***Open*** *drop-down list*

Select Initial View

A view is defined as the way you look at an object. The **Select Initial View** option allows you to specify the view you want to load initially when AutoCAD LT loads the drawing. This option will work if the drawing has saved views. You can save a desired view by using AutoCAD LT's **VIEW** command (see "Creating Views", Chapter 7). If the drawing has no saved views, selecting this option will load the last view. If you select the **Select Initial View** check box and then choose the **Open** button, AutoCAD LT will display the **Select Initial View** dialog box. You can select the view name from this dialog box, and AutoCAD LT will load the drawing with the selected view displayed.

Tip

Apart from opening a drawing from the ***AutoCAD LT Today*** *window or the* ***Select File*** *dialog box, you can also open a drawing from the* ***File*** *menu, which displays the four most recently opened drawings, by simply choosing the desired file name.*

It is possible to open a Release 97 or 98 drawing in AutoCAD LT 2002. When you save this drawing it is automatically converted and saved as an AutoCAD LT 2000 drawing file.

Note

If you are in the drawing editor and want to open a drawing file, instead of using the ***OPEN*** *command you can use the* ***TODAY*** *command to open the* ***AutoCAD LT Today*** *window. You can then open the drawing from this dialog box. This command can be invoked by selecting the* ***Today*** *button in the* ***Standard Toolbar****, or by choosing* ***Today*** *from the* ***Tools*** *menu, or by entering* ***TODAY*** *at the Command prompt.*

STARTING A NEW DRAWING

Toolbar:	Standard Toolbar > New
Menu:	File > New
Command:	NEW

When you are in AutoCAD LT you can use the **NEW** command to open a new drawing file. After you invoke the **NEW** command, AutoCAD LT will display the **AutoCAD LT Today** window with the **Create Drawings** tab open. The different options such as **Start from Scratch**, **Template**, and **Wizards** for starting a new drawing have already been discussed in Chapter 1, Getting Started-I.

MULTIPLE DOCUMENT ENVIRONMENT (MDE)

The MDE is a Windows feature that allows you to open more than one drawing at a time in one AutoCAD LT session. This feature is very helpful when you want to work in different drawings simultaneously and make changes to them with reference to each other. For example, if you are working on an architectural plan of a building and you want to make changes to it while referring to its elevations, this feature allows you to open all these drawings simultaneously and work on them. Sometimes you may want to incorporate certain features from different old drawings into a new drawing. With the help of MDE you can open all these drawings and the new drawing and then copy the features from the old drawings into the new drawing by simply dragging and dropping. This way it is very convenient to create an assembly by copying the components created in separate drawings. You can also use the **Cut/Copy** and **Paste** instead of drag and drop. These commands are invoked from the shortcut menu, or the **Edit** menu, or you can use the keyboard shortcuts such as CTRL + C, CTRL + V, and so on. It is possible for you to shift from one drawing to another leaving a command incomplete and then start working in another drawing. When you come back to that particular drawing, the command resumes from the point you had left it.

Tip

If you want to place an object at a particular location in a drawing, it is better to use the ***Copy/Cut*** *and* ***Paste*** *options rather than drag and drop. AutoCAD LT allows you to define the coordinates for the insertion point while pasting the object. This makes the placement of objects accurate.*

When you open the first drawing, by default its file name merges with the AutoCAD LT title bar. When you want to work on more than one drawing simultaneously, choose the **Restore** button present in the right corner of the menu bar to make it floating. This drawing will be placed as a floating window inside the main AutoCAD LT window. Now if you open another drawing, this window will be placed over the first drawing window but the title bar of the previous window will remain visible. Similarly the rest of the drawings opened are placed over the previous ones with the latest drawing opened being active.

The active drawing is specified by the dark blue band in the title bar. You can make any drawing active by simply clicking in it. This type of arrangement is the cascading arrangement (Figure 2-7) and can be changed as per your requirements. If the title bar of a particular drawing is not visible and you want it active, choose the drawing name in the **Window** menu (Figure 2-7). A maximum of nine drawing names are listed in the **Window** menu. If more than nine drawings are opened at a time, **More Windows** is displayed in the list. Choosing this option opens the **Select Window** dialog box where all the drawing names are listed. You can arrange the drawing windows vertically and horizontally by choosing the desired option from the **Window** menu. You can also minimize the drawing files using the **Minimize** button at the top right corner of any window. Here only a small portion of the title bar is visible at the bottom of the window area. You can get the drawings to fill the drawing area using the **Restore** button, which replaces the Minimize button after minimizing. The **Arrange Icons** option in the **Window** menu arranges these minimized title bars at the bottom.

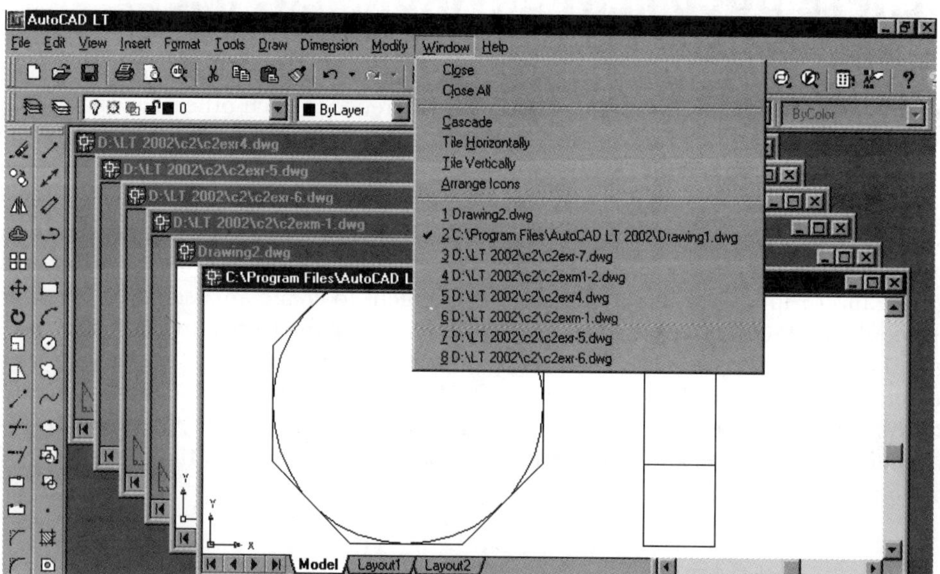

Figure 2-7 Cascading arrangement in MDE and the ***Window*** *menu*

Tip

*When you invoke the **OPEN** command and the **Select File** dialog box is displayed, you can select multiple drawings together using the CTRL or SHIFT keys and then choose the **Open** button. All the drawings selected will be opened one after the other. This way you don't have to use the **OPEN** command repeatedly to open multiple drawings.*

BASIC DISPLAY COMMANDS

Drawing in AutoCAD LT is much simpler than manual drafting in many ways. Sometimes while drawing, it is very difficult to see and alter minute details. In AutoCAD LT, you can overcome this problem by viewing only a specific portion of the drawing. This is done using the **ZOOM** command as shown in Figure 2-8 and Figure 2-9. This command lets you enlarge or reduce the size of the drawing displayed on the screen. Some of the drawing display commands, such as **ZOOM** and **PAN** will be introduced here. A detailed explanation of these commands and other display options appears in Chapter 7, (Controlling the Drawing Display and Creating Text).

ZOOM Command

Toolbar:	Zoom, Standard Toolbar > Zoom flyout
Menu:	View > Zoom
Command:	ZOOM

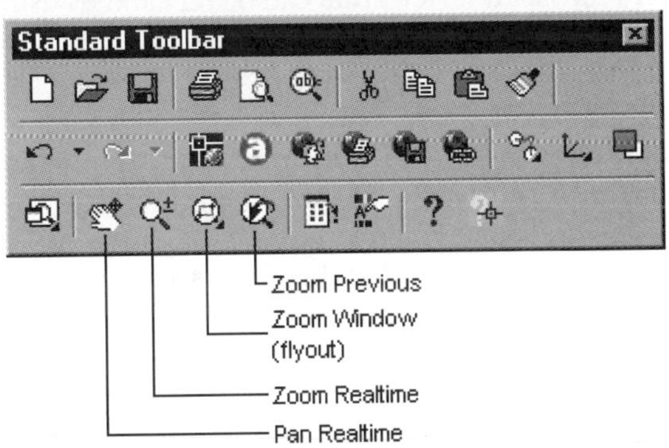

*Figure 2-8 Selecting **Zoom** options from the **Standard Toolbar***

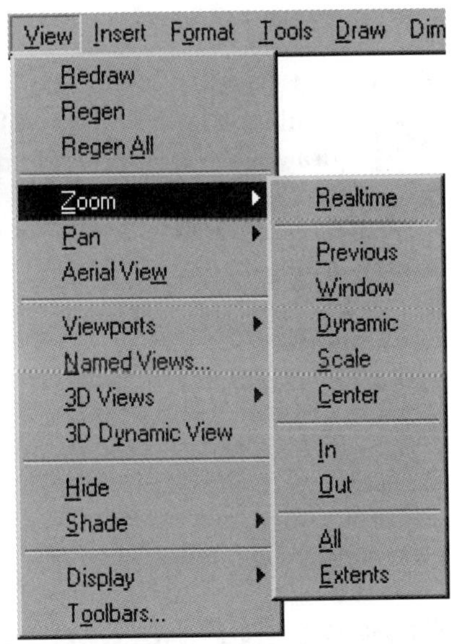

*Figure 2-9 Invoking the **ZOOM** command from the **View** menu*

The **ZOOM** command enlarges or reduces the view of the drawing on the screen, but it does not affect the actual size of the entities. After the **ZOOM** command has been invoked, different options can be used to obtain the desired display. If you use a menu, it issues the appropriate option at the initial **ZOOM** prompt. The following is the prompt sequence of the **ZOOM** command:

Command: **ZOOM** or **Z** [Enter]
Specify corner of window, enter a scale factor (nX or nXP), or [All/Center/Dynamic/Extents/Previous/Scale/Window]<real time>:

Realtime Zooming

You can use the **Realtime Zoom** to zoom in and zoom out interactively. To zoom in, invoke the command, then hold the pick button down and move the cursor up. If you want to zoom in further, bring the cursor down, specify a point, and move the cursor up. Similarly, to zoom out, hold the pick button down and move the cursor down. Realtime zoom is the default setting for the **ZOOM** command. At the Command prompt, pressing ENTER after invoking the **ZOOM** command automatically invokes the realtime zoom. To exit the Realtime Zoom, right-click to display the shortcut menu and choose **Exit**. You can also press ESC or the ENTER key to exit the command.

Window Option

This is the most commonly used option of the **ZOOM** command. It lets you specify the area you want to zoom in on by letting you specify two opposite corners of a rectangular window. The center of the specified window becomes the center of the new display screen. The area inside the window is magnified in size to fill the drawing area as completely as possible. The points can be specified by selecting them with the help of the pointing device or by entering their coordinates.

Previous Option

While working on a complex drawing, you may need to zoom in on a portion of the drawing to edit some minute details. When you have completed the editing you may want to return to the previous view. This can be done using the **Previous** option of the **ZOOM** command. AutoCAD LT remembers the last ten views that can be recalled by using the **Previous** option.

All Option

This option zooms to the drawing limits or the extents, whichever is greater. Whenever you increase the limits (**Quick Setup** in the **Use a Wizard** dialog box) the current display is not affected and hence does not show. You need to use the **Zoom All** option to display the limits of the drawing. Sometimes it is possible that the objects are drawn beyond the limits. In such a case the **Zoom All** option zooms to fill the drawn objects in the drawing area irrespective of its limits.

Panning in Realtime

You can use the **Realtime Pan** to pan the drawing interactively, by sliding the drawing and placing it at the required position. To pan a drawing, invoke the command and then hold the pick button down and move the cursor in any direction. When you select the realtime pan, AutoCAD LT displays an image of a hand indicating that you are in PAN mode. You can drag the hand anywhere on the screen to move the drawing. To exit the Realtime Pan, right-click to display the shortcut menu and choose **Exit**. You can also press ESC or the ENTER key to exit the command.

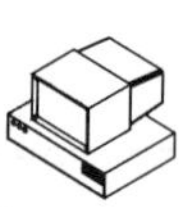

Tip

*You can right-click to display a shortcut menu while the realtime zoom or realtime pan options are active. The **Realtime Zoom, Realtime Pan, Exit,** and other **ZOOM** command options are available in this shortcut menu.*

SETTING UNITS

In Chapter 1 you have already learned to set units while starting a drawing from the **AutoCAD LT Today** window using the **Wizards** option. If you want to change the units while you are already working on a drawing, you can use the **UNITS** command.

The Drawing Units Dialog Box

Menu:	Format > Units
Command:	UNITS

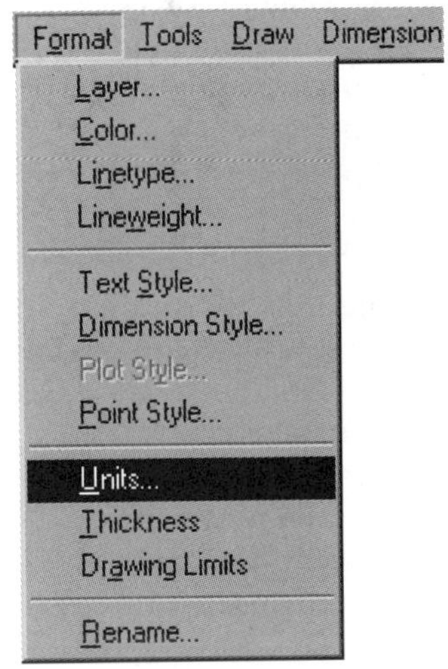

***Figure 2-10** Invoking the **UNITS** command from the **Format** menu*

The **UNITS** (Figure 2-10) command is used to select a format for the units of distance and angle measurement. The **UNITS** command displays the **Drawing Units** dialog box (Figure 2-11) which can be used to set the units and angles. You can then specify the precision for the units and angles from the corresponding **Precision** drop-down list (Figure 2-12). You can also set the units from the command line by entering **-UNITS** at the Command prompt.

Specifying Units

In the **Drawing Units** dialog box (Figure 2-11), you can select a desired format of units from the drop-down list displayed when you choose the down arrow to the right of the **Type** edit box. You can select one of the following five formats:

1. Architectural
2. Decimal
3. Engineering
4. Fractional
5. Scientific

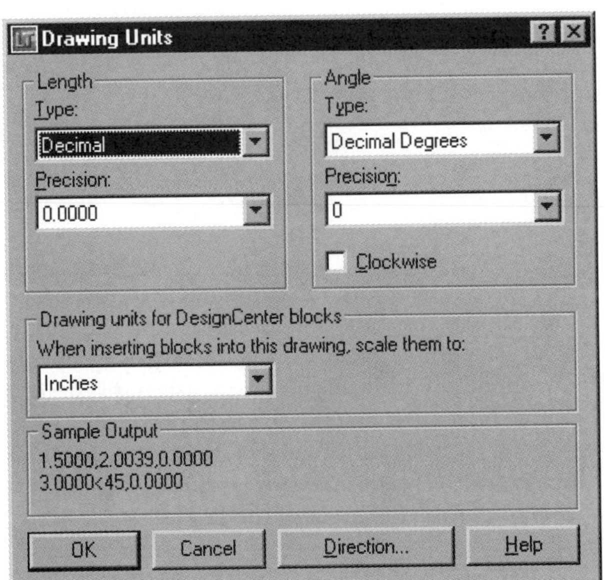

***Figure 2-11** **Drawing Units** dialog box*

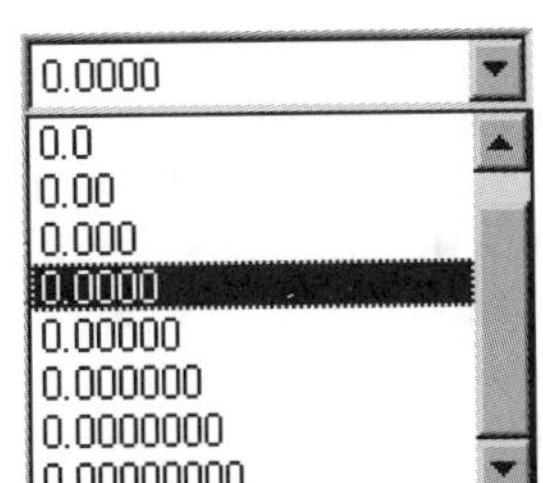

***Figure 2-12** Specifying **Precision** from the **Drawing Units** dialog box*

If you choose the Scientific, Decimal, or Fractional format, you can enter the distances or coordinates in any of these three formats, but not in engineering or architectural units. In the following example, the units are set as decimal, scientific, fractional, and decimal and fractional to enter the coordinates of different points:

Command: **LINE** [Enter]
Specify from point: **1.75,0.75** [Enter] *(Decimal.)*
Specify next point or [Undo]: **1.75E+01, 3.5E+00** [Enter] *(Scientific.)*
Specify next point or [Undo]: **10-3/8,8-3/4** [Enter] *(Fractional.)*
Specify next point or [Close/Undo]: **0.5,17/4** [Enter] *(Decimal and fractional.)*

If you choose the Engineering or Architectural format, you can enter the distances or coordinates in any of the five formats. In the following example, the units are set as architectural; hence, different formats are used to enter the coordinates of points:

Command: **LINE** [Enter]
Specify first point: **1-3/4,3/4** [Enter] *(Fractional.)*
Specify next point or [Undo]: **1'1-3/4",3-1/4** [Enter] *(Architectural.)*
Specify next point or [Undo]: **0'10.375,0'8.75** [Enter] *(Engineering.)*
Specify next point or [Close/Undo]: **0.5,4-1/4"** [Enter] *(Decimal and engineering.)*

Note

The inch symbol (") is optional. For example, 1'1-3/4" is the same as 1'1-3/4, and 3/4" is the same as 3/4.

Remember that you cannot use the feet (') or inch (") symbols if you have selected Scientific, Decimal, or Fractional unit formats.

Specifying Angle. You can select one of the following five angle measuring systems:

1. Decimal Degrees
2. Deg/min/sec
3. Grads
4. Radians
5. Surveyor's Units

If you select any of the first four measuring systems, you can enter the angle in the Decimal, Degrees/minutes/seconds, Grads, or Radians system, but you cannot enter the angle in Surveyor's units. However, if you select Surveyor's units, you can enter the angles in any of the five systems. If you enter an angle value without any indication of a measuring system, it is taken in the current system. To enter the value in another system, use the appropriate suffixes and symbols such as r (Radians), d (Degrees), g (Grads), or the others shown in the following examples. In the following example, the system of angle measure is Surveyor's units and different systems of angle measure are used to define the angle of the line:

Command: **LINE** [Enter]
Specify first point: **3,3** [Enter]

Specify next point or [Undo]: **@3<45.5** Enter *(Decimal degrees.)*
Specify next point or [Undo]: **@3<90d30'45"** Enter *(Degrees/min/sec.)*
Specify next point or [Close/Undo]: **@3<75'0g** Enter *(Grads.)*
Specify next point or [Close/Undo]: **@3<N45d30'E** Enter *(Surveyor's units.)*

In **Surveyor's units** you must specify the bearing angle that the line makes with the north-south direction (Figure 2-13). For example, if you want to define an angle of 60 degrees with north, in the Surveyor's units the angle will be specified as N60dE. Similarly, you can specify angles such as S50dE, S50dW, N75dW, as shown in Figure 2-13. You cannot specify an angle that exceeds 90 degrees (N120E). The angles can also be specified in **radians** or **grads**, for example, 180 degrees is equal to **PI** (3.14159) radians. You can convert degrees into radians or radians into degrees using the following equations:

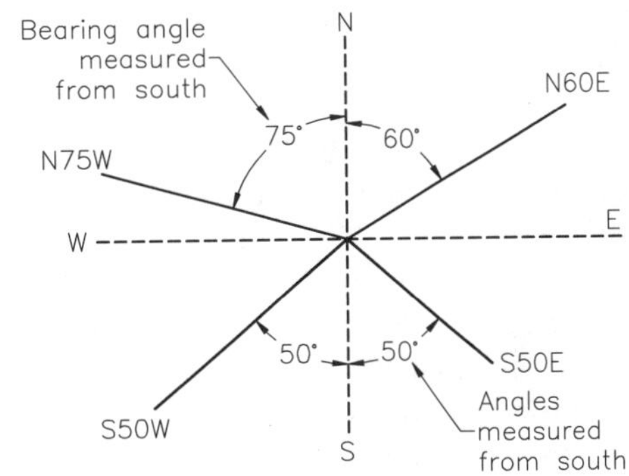

Figure 2-13 *Specifying angles in Surveyor's units*

radians = degrees x 3.14159/180; degrees = radians x 180/3.14159

Grads are generally used in land surveys. There are 400 grads or 360 degrees in a circle. A 90-degree angle is equal to 100 grads.

Tip
*An example corresponding to the type of unit and angle selected from the **Length** or **Angle** area of the dialog box can be seen in the **Sample Output** area of the dialog box.*

In AutoCAD LT, by default the angles are positive if measured in the counterclockwise direction (Figure 2-14) and the angles are measured from the positive X axis see East, Figure 2-15. The angles are negative if measured clockwise. If you want the angles measured as positive in the clockwise direction, select the **Clockwise** check box. The positive angles will be measured in the clockwise direction and the negative angles in the counterclockwise direction.

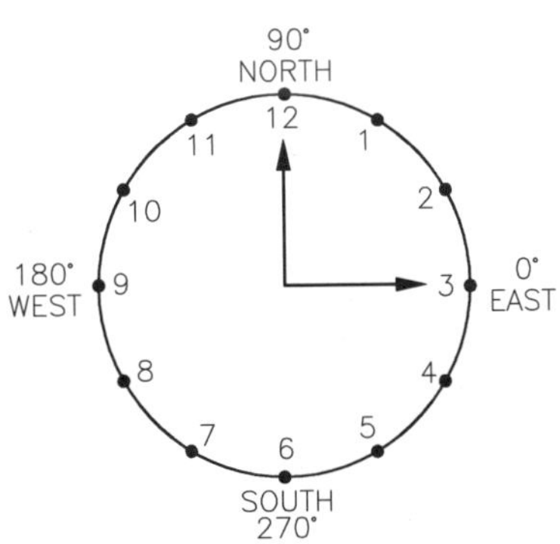

Figure 2-14 *N,S,E,W directions*

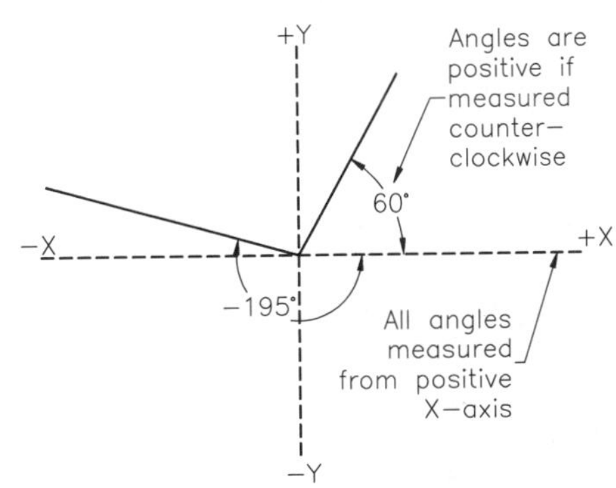

Figure 2-15 *Measuring angles counterclockwise from the positive X axis (default)*

When you choose the **Direction** button in the **Drawing Units** dialog box, the **Direction Control** dialog box appears. It provides the options of selecting the setting for direction of the Base Angle (see Figure 2-16). If you select the **Other** option, you can set your own direction for the Base Angle by either entering a value in the **Angle** edit box or choosing the **Pick an Angle** button to specify the angle on the screen. After selecting an angle, choose the **OK** button to apply the settings. This will get you back to the **Drawing Units** dialog box.

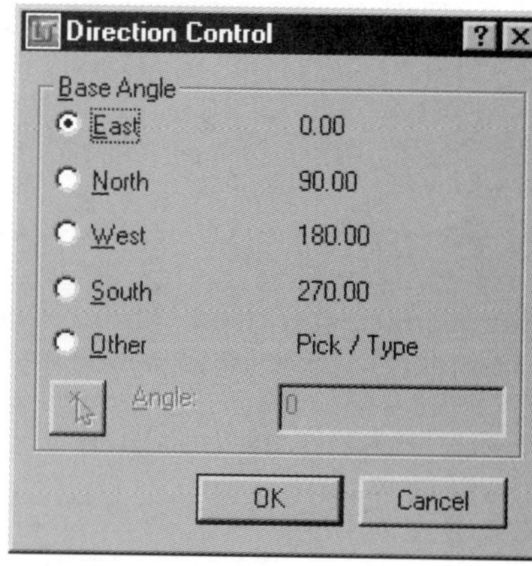

Figure 2-16 *Setting direction from the **Direction Control** dialog box*

You can also set the units of measure while inserting a block from the AutoCAD LT DesignCenter. In the **Drawing Units** dialog box choose any measuring unit from the **When inserting blocks in this drawing, scale them to** drop-down list (Figure 2-17). Now while inserting a block from the DesignCenter, AutoCAD LT inserts the block with the chosen unit. Even if the block was created using a different measuring unit, AutoCAD LT scales it and inserts it using the specified measuring unit. If you want to insert the block with the original units, then choose **Unitless** from the drop-down list.

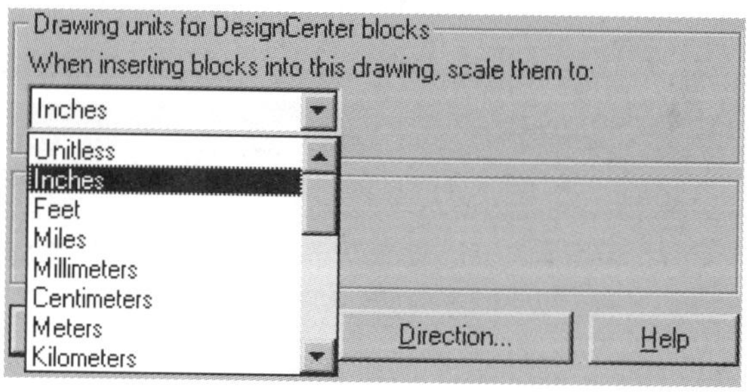

Figure 2-17 *Choosing measuring units for inserting the DesignCenter blocks*

Note

The insertion of blocks from the DesignCenter into a drawing is discussed in detail in Chapter 6, Editing Commands-II.

The **Sample Output** area in the **Drawing Units** dialog box shows an example of the current format of the units and angles. When you change the type of length and angle measure in the Length and Angle areas of the **Drawing Units** dialog box, the corresponding example is displayed in the **Sample Output** area.

Example 1 *General*

In this example you will set the units for a drawing according to the following specifications and then draw Figure 2-18.

1. Set **UNITS** to fractional, with the denominator of the smallest fraction equal to 32.

2. Set the angular measurement to Surveyor's units, with the number of fractional places for display of angles equal to zero.

3. Set the direction to 90 degrees (north) and the direction of measurement of angles to clockwise (angles measured positive in clockwise direction), Figure 2-18.

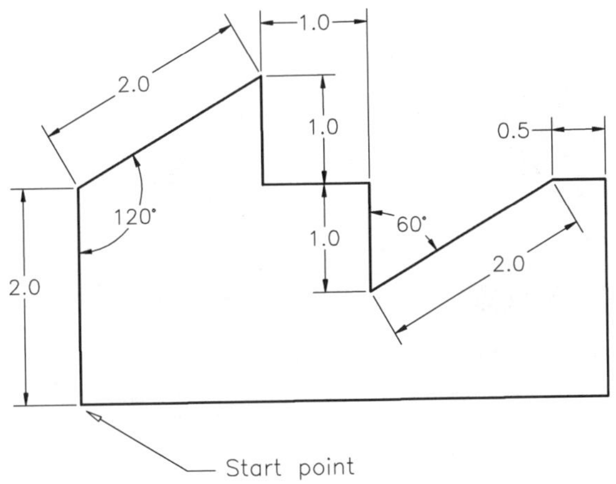

***Figure 2-18** Drawing for Example 1*

1. Invoke the **Drawing Units** dialog box (Figure 2-19) by choosing **Units** from the **Format** menu. You can also invoke the dialog box by entering **UNITS** at the Command prompt.

2. In the **Length** area of the dialog box, select **Fractional** from the **Type** drop-down list. From the **Precision** drop-down list select **0 1/32**.

3. In the **Angle** area of the dialog box, select **Surveyor's Units** from the **Type** drop-down list. From the **Precision** drop-down list select **N 0d E** if it is not already selected. Also, select the **Clockwise** check box to set the clockwise angle measurement as positive.

4. Choose the **Direction** button to display the **Direction Control** dialog box. Select the **North** radio button. Choose the **OK** button to exit the **Direction Control** dialog box.

5. Choose the **OK** button to exit the **Drawing Units** dialog box.

6. With the units set, draw Figure 2-18 using relative polar coordinates. Here the units are fractional and the **angles are measured from north** (90-degree axis). Also, the angles are measured as positive in a clockwise direction and **negative** in a counterclockwise direction (Figure 2-20). Invoke the **LINE** command and specify the points as follows:

Command: **LINE** [Enter]
Specify first point: **2,2** [Enter]
Specify next point or [Undo]: **@2.0<0** [Enter]
Specify next point or [Undo]: **@2.0<60** [Enter]
Specify next point or [Close/Undo]: **@1<180** [Enter]

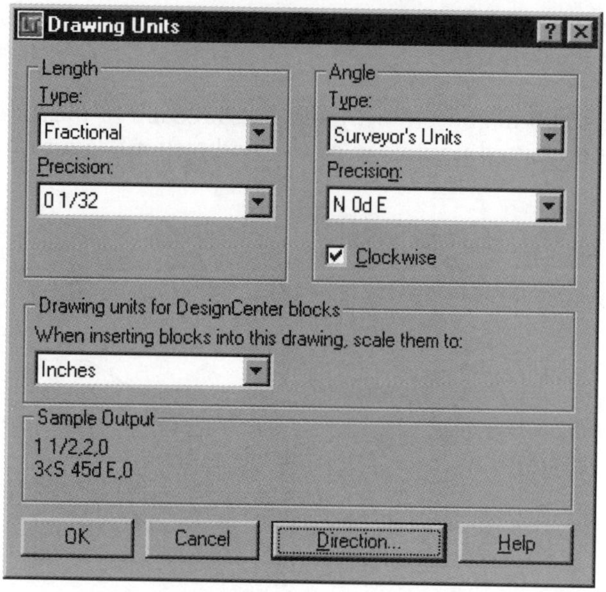

Figure 2-19 *Setting units for Example 1 in the* ***Drawing Units*** *dialog box*

Specify next point or [Close/Undo]: **@1<90** Enter
Specify next point or [Close/Undo]: **@1<180** Enter
Specify next point or [Close/Undo]: **@2.0<60** Enter
Specify next point or [Close/Undo]: **@0.5<90** Enter
Specify next point or [Close/Undo]: **@2.0<180** Enter
Specify next point or [Close/Undo]: **C** Enter

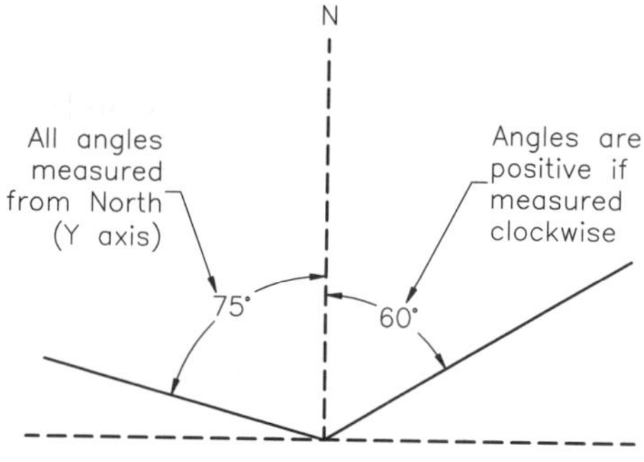

Figure 2-20 *Angles measured from north (Y axis)*

Forcing Default Angles

When you define the direction by specifying the angle, the output of the angle depends on the following (Figure 2-21):

Angular units
Angle direction
Angle base

For example, if you are using the AutoCAD LT default setting, <70 represents an angle of 70 decimal degrees from the positive X axis, measured counterclockwise. The decimal degrees represent angular units, the X axis represents the angle base, and the counterclockwise represents the angle direction. If you have changed the default settings for measuring angles, it might be confusing to enter the angles. AutoCAD LT lets you bypass the current settings by entering << or <<< before the angle. If you enter << before the angle, AutoCAD LT will bypass the current angle settings and use the angle as decimal degrees from the default angle base and default angle direction (angles are referenced from the positive X axis and are positive if measured counterclockwise). If you enter <<< in front of the angle, AutoCAD LT will use current angular units, but it will bypass the current settings for angle base and angle direction and use the default angle setting (angles are referenced from the positive X axis and are positive if measured counterclockwise).

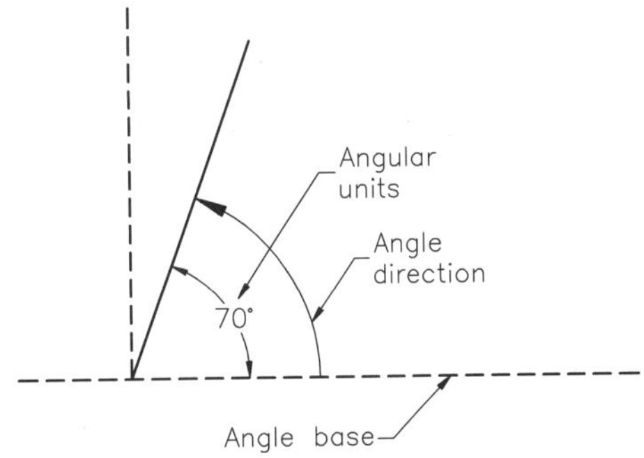

Figure 2-21 *Default angular units, direction, and base*

Assume, for example, that you have changed the current settings and made the system of angle measure radians with two places of precision: angle base north (the Y axis), and the direction clockwise. Now, if you enter <1.04 or <1.04r, all the current settings will be taken into consideration and you will get an angle of 1.04 radians, measured in a clockwise direction from the positive Y axis, Figure 2-22(a). If you enter <<60, AutoCAD LT will bypass the current settings and reference the angle in degrees from the positive X axis, measuring 60 degrees in a counterclockwise direction, Figure 2-22(b). If you enter <<<1.04r, AutoCAD LT will use the current angular units, but will bypass the current angle base and angle direction. Hence, the angle will be referenced from the positive X axis, and will be measured 1.04 radians in a counterclockwise direction, Figure 2-22(c).

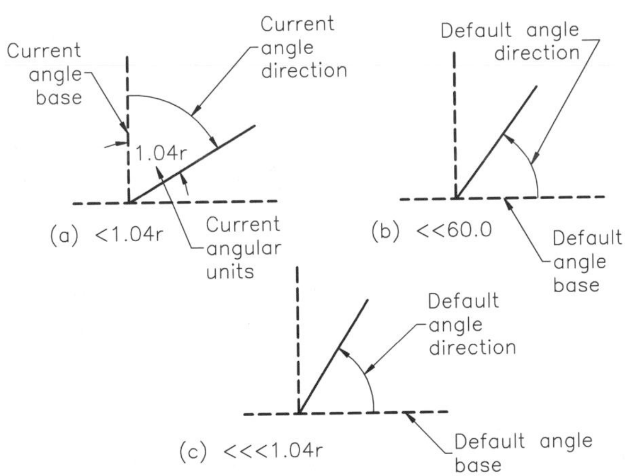

Figure 2-22 *Forcing default angles*

The effect of using the open angle brackets is summarized in the following table.

Angle prefix	Angular units	Angle direction	Angle base	Example
<	Current	Current	Current	<1.04r
<<	Degrees	Counterclockwise	Default	<<60.0
<<<	Current	Counterclockwise	Default	<<<1.04r

You can also override the current angle units by entering the appropriate suffix such as 60d for degrees.

LIMITS COMMAND

Menu:	Format > Drawing Limits
Command:	LIMITS

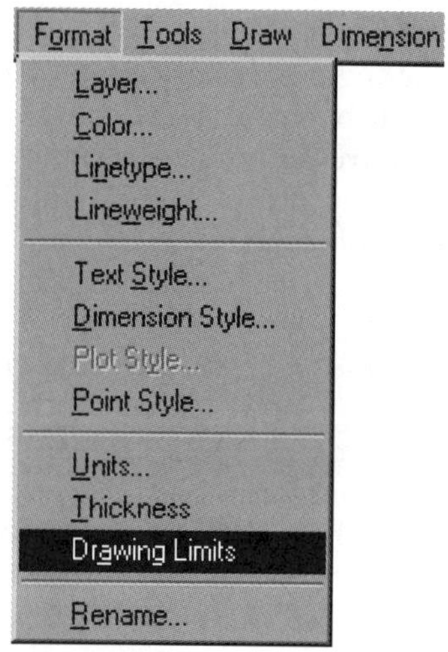

***Figure 2-23** Choosing **Drawing Limits** from the **Format** menu*

In Chapter 1 you have already learned to set the limits while starting a drawing from the **AutoCAD LT Today** window using the **Wizards** option. If you want to change the limits while you are already working in a drawing, you can use the **LIMITS** command. When you start AutoCAD LT, the default limits are 12.00,9.00. You can use the **LIMITS** command to set up new limits. The following is the prompt sequence of the **LIMITS** command for setting the limits of 24,18:

Command: **LIMITS** [Enter]
Specify lower left corner or [ON/OFF]<current>: **0,0** [Enter]
Specify upper right corner <current>: **24,18** [Enter]

At the preceding two prompts you are required to specify the lower left corner and the upper right corner of the sheet. Normally you choose (0,0) as the lower left corner, but you can enter any other point. If the sheet size is 24 x 18, enter 24,18 as the coordinates of the upper right corner.

Tip

*Whenever you increase the drawing limits, the display area does not change. You need to use the **ZOOM All** command to display the complete area inside the drawing window.*

Setting Limits

In AutoCAD LT, the drawings must be drawn full scale and, therefore, limits are needed to size up a drawing area. The limits of the drawing area are usually determined by the following factors:

1. The actual size of the drawing.
2. The space needed for putting down the dimensions, notes, bill of materials, and other necessary details.

3. The space between different views so that the drawing does not look cluttered.
4. The space for the border and a title block, if any.

To get a good idea of how to set up limits, it is always better to draw a rough sketch of the drawing to help calculate the area needed. For example, if an object has a front view size of 5 x 5, a side view size of 3 x 5, and a top view size of 5 x 3, the limits should be set so that they can accommodate the drawing and everything associated with it. In Figure 2-24, the space between the front and side views is 4 units and between the front and top views is 3 units. Also, the space between the border and the drawing is 5 units on the left, 5 units on the right, 3 units at the bottom, and 2 units at the top. (The space between the views and between the borderline and the drawing depends on the drawing.)

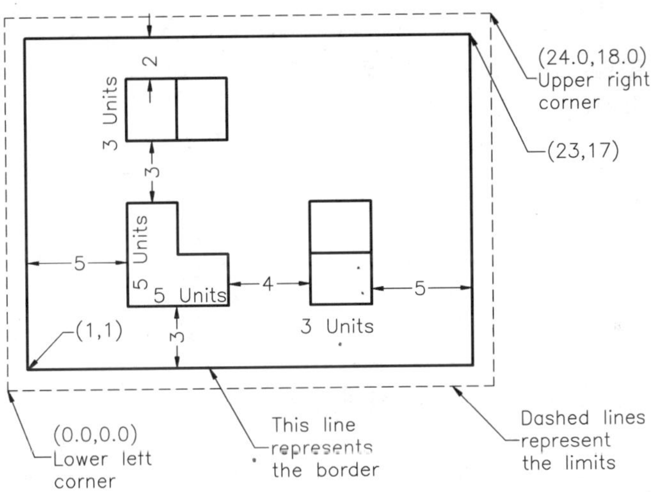

***Figure 2-24** Setting limits in a drawing*

After you know the sizes of different views and have determined the space required between views, between the border and the drawing, and between the borderline and the edges of the paper, you can calculate the space you need as follows:

Space along (X axis) = 1 + 5 + 5 + 4 + 3 + 5 + 1 = 24
Space along (Y axis) = 1 + 3 + 5 + 3 + 3 + 2 + 1 = 18

Thus, the space or the work area you need for the drawing is 24 x 18. Once you have determined the space you need, select the sheet size that can accommodate your drawing. In the case just explained, you will select a D size (34 x 22) sheet. Therefore, the actual drawing limits are 34,22.

Standard Sheet Sizes

When you make a drawing, you might want to plot the drawing to get a hard copy. Several standard sheet sizes are available to plot your drawing. Although in AutoCAD LT you can select any work area, it is recommended that you select the work area based on the sheet size you will be using to plot the drawing. The sheet size is the deciding factor for determining the limits (work area), text size **(TEXTSIZE)**, dimensioning scale factor **(DIMSCALE)**, linetype scale factor **(LTSCALE)** and other drawing-related parameters. The following tables list standard sheet sizes and the corresponding drawing limits for different scale factors:

Standard U.S. Size

Letter size	Sheet size	Limits (1:1)	Limits (1:4)	Limits (1/4"=1')
A	8.5 x 11	8.5,11	34,44	34',44'
B	11 x 17	11,17	44,68	44',68'
C	17 x 22	17,22	68,88	68',88'
D	22 x 34	22,34	88,136	88',136'
E	34 x 44	34,44	136,176	136'x176'

International Size

Letter size	Sheet size	Limits (1:1)	Limits (1:20)
A4	210 x 297	210,297	4200,5940
A3	297 x 420	297,420	5940,8400
A2	420 x 594	420,597	8400,11940
A1	594 x 841	595,841	11940,16820
A0	841 x 1189	841,1189	16820,23780

Limits for Architectural Drawings

Most architectural drawings are drawn at a scale of 1/4" = 1', 1/8" = 1', or 1/16" = 1'. You must set the limits accordingly. The following example illustrates how to calculate the limits in architectural drawings:

Given
Sheet size = 24 x 18
Scale is 1/4" = 1'

Calculate limits
Scale is 1/4" = 1'
or 1/4" = 12"
or 1" = 48"
X limit = 24 x 48
= 1152" or 1152 Units
= 96'
Y limit = 18 x 48
= 864" or 864 Units
= 72'

Thus, the scale factor is 48 and the limits are 1152", 864", or 96',72'.

Example 2 *Architectural*

In this example you will calculate limits and determine an appropriate drawing scale factor for Figure 2-25. The drawing is to be plotted on a 12" x 9" sheet.

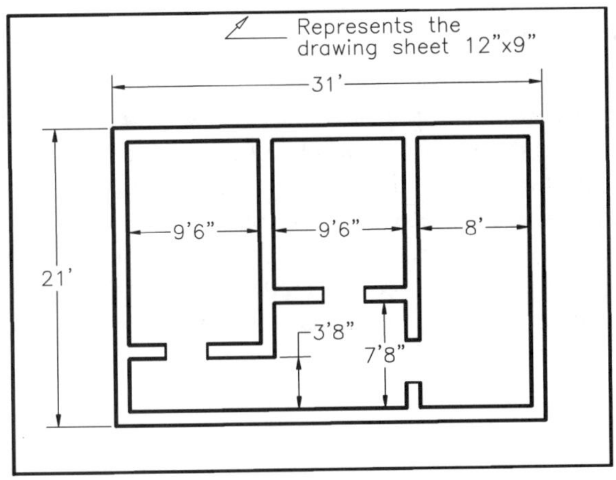

Figure 2-25 *Drawing for Example 2*

The scale factor can be calculated as follows:

Given or known
Overall length of the drawing = 31'
Length of the sheet = 12"
Approximate space between the drawing and the edges of the paper = 2"

Calculate scale factor
To calculate the scale factor, you have to try different scales until you find one that satisfies the given conditions. After some experience you will find this fairly easy to do. For this example, assume a scale factor of 1/4" = 1'.

Scale factor 1/4" = 1'
or 1" = 4'
Thus, a line 31' long will be = 31'/4' = 7.75" on paper. Similarly, a line 21' long = 21'/4' = 5.25".

Approximate space between the drawing and the edges of paper = 2"
Therefore, total length of the sheet = 7.75 + 2 + 2 = 11.75"
Similarly, total width of the sheet = 5.25 + 2 + 2 = 9.25"

Because you selected the scale 1/4" = 1', the drawing will definitely fit on the given sheet of paper (12" x 9"). Therefore, the scale for this drawing is 1/4" = 1'.

Calculate limits
Scale factor = 1" = 48" or 1" = 4'
The length of the sheet is 12"
Therefore, X limit = 12 x 4' = 48'
Also, Y limit = 9 x 4' = 36'

Limits for Metric Drawings

When the drawing units are metric, you must use **standard metric size sheets** or calculate the limits in millimeters (mm). For example, if the sheet size you decide to use is 24 x 18, the limits after conversion to the metric system will be 609.6,457.2 (multiply length and width by 25.4). You can round these numbers to the nearest whole numbers 610,457. Note that metric drawings do not require any special setup, except for the limits. Metric drawings are like any other drawings that use decimal units. As with architectural drawings, you can draw metric drawings to a scale. For example, if the scale is 1:20 you must calculate the limits accordingly. The following example illustrates how to calculate the limits for metric drawings.

Given
Sheet size = 24" x 18"
Scale = 1 : 20

Calculate limits
Scale is 1 : 20
Therefore, scale factor = 20
X limit = 24 x 25.4 x 20 = 12192 units
Y limits = 18 x 25.4 x 20 = 9144 units

Thus, the limits are 12192 and 9144.

Exercise 1 *Mechanical*

Set the units of the drawing according to the following specifications and then make the drawing shown in Figure 2-26 (leave a space of 3 to 5 units around the drawing for dimensioning and title block). The space between the dotted lines is 1 unit.

1. Set **UNITS** to decimal units, with two digits to the right of the decimal point.

2. Set the angular measurement to decimal degrees, with the number of fractional places for display of angles equal to 1.

3. Set the direction to 0 degrees (east) and the direction of measurement of angles to counterclockwise (angles measured positive in a counterclockwise direction).

4. Set the limits leaving a space of 3 to 5 units around the drawing for dimensioning and title block.

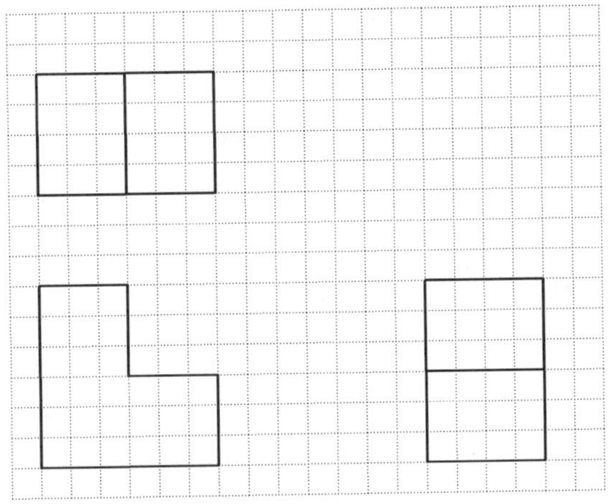

Figure 2-26 Drawing for Exercise 1

INTRODUCTION TO PLOTTING DRAWINGS

Toolbar:	Standard toolbar > Plot
Menu:	File > Plot
Command:	PLOT or PRINT

Once you have created a drawing in the current session of AutoCAD LT, you may need to have its hard copy. This hard copy is very useful in the industry and can be created by plotting and printing it on a sheet of paper. Suppose you have drawn an architectural plan on the computer, and you need to send its hard copy to the site for implementation. Similarly, if you have created a mechanical component, you may need to send its hard copy to the shop floor for manufacturing. Drawings can be plotted by using the **PLOT** command (Figure 2-27), and AutoCAD LT will display the **Plot** dialog box when you invoke the **PLOT** command.

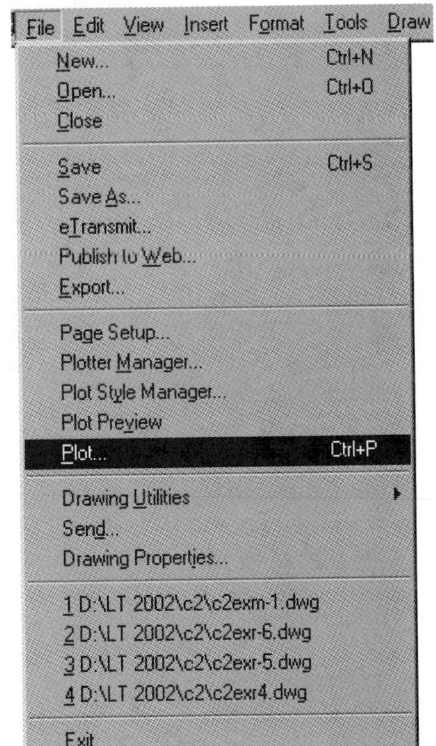

*Figure 2-27 Invoking the **PLOT** command from the **File***

The values in this dialog box are the ones that were set during the configuring of AutoCAD LT. If the displayed values conform to your requirements, you can start plotting without making any changes. If necessary, you can make changes in the default values according to your plotting requirements.

Basic Plotting

In this section you will learn how to set up the basic plotting parameters. Later, you will learn about the advance options that allow you to plot according to your plot drawing specifications. Basic plotting involves selecting the correct output device (plotter), specifying the area to plot, selecting paper size, specifying the plot origin, orientation, and the plot scale.

Example 3 *General*

You will plot the drawing in Figure 2-28 using the **Window** option to select the area to plot. The drawing was drawn in **Example 3** of Chapter 1 and here it is assumed to be open on the screen. Assume that AutoCAD LT is configured for two output devices: Default System Printer and **HP Laserjet 2100 Series PS.**

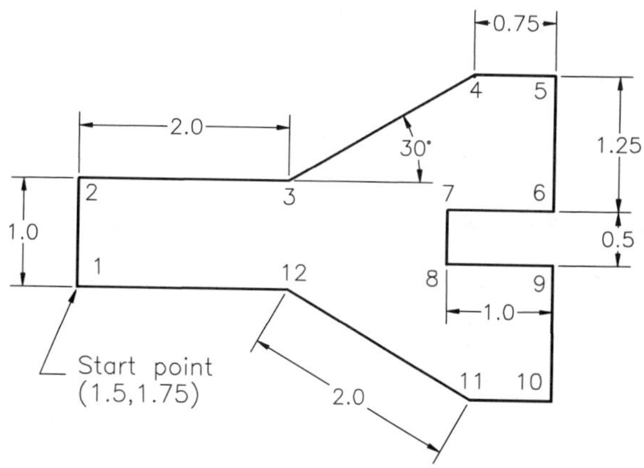

Figure 2-28 *Drawing for Example 3*

1. Invoke the **Plot** dialog box (Figure 2-29) from the **Standard Toolbar**, or the **File** menu (choose **Plot**), or by entering **PLOT** at the Command prompt. You can also invoke it by choosing **Plot** from the shortcut menu which is displayed by right-clicking on the **Model/Layout** tabs.

2. Choose the **Plot Device** tab if it is not already open, Figure 2-29. Under the **Plotter Configuration** area, the name of the default system printer will be displayed (in this example it is **HP Laserjet 2100 Series PS**). You can use any other printer by selecting the name of the device from the drop-down list displayed when you select the arrow button in the **Name** edit box. This displays the list of the currently configured devices. Related information about the selected device is displayed below this box.

Note

The default system printer varies from system to system. It is the printer chosen as the default printer while configuring your system for the output device.

3. Now choose the **Plot Settings** tab and here, choose the **Window** button in the **Plot area**. You can select the object for plotting by creating a window around it on the screen. When you choose the **Window** button, the dialog box will temporarily disappear and the drawing

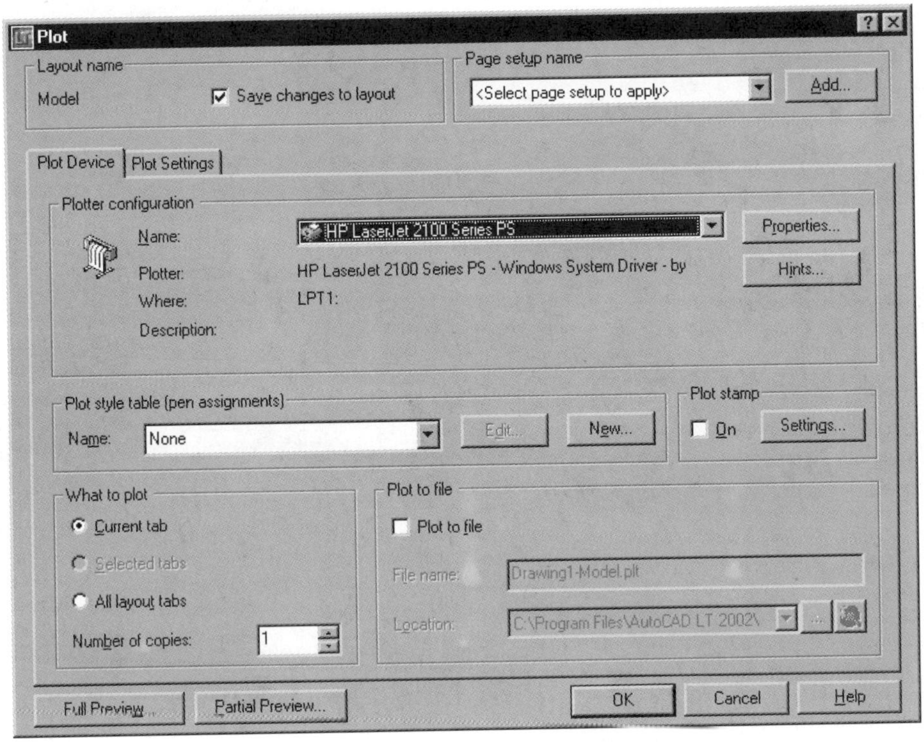

Figure 2-29 ***Plot*** *dialog box with the* ***Plot Device*** *tab*

area will appear. Now, select the first and second corners (Points **P1** and **P2**) and specify the plot area (the area you want to plot). Once you have defined the two corners, the **Plot** dialog box will be redisplayed.

4. To set the size for the plot, you can select a size from the **Paper size** drop-down list in the **Paper size and paper units** area (Figure 2-30), which lists all the plotting sizes that the present plotter can support. You can select any one of the sizes listed in the dialog box or specify a size (width and height) of your own through the **Plotter Manager**. (This option is discussed later in Chapter 12, Plotting Drawings.) Once you select a size, you can also select the orientation by choosing either the **Landscape** or **Portrait** radio buttons under the **Drawing orientation** area. The sections in the **Plot** dialog box pertaining to paper size and orientation are automatically revised to reflect the new paper size and orientation. In this example you will specify Paper size **A4** and Orientation **Portrait**.

5. You can also modify values for **Plot offset**; the default values for X and Y are 0. For this example you can select the **Center the plot** check box to get the drawing in the center of the paper. Similarly, you can enter values for **Plot scale**. Open the **Scale** drop-down list to display the various scale factors. From this list you can select a scale factor you want to use. For example if you select the scale factor 1/4"=1'-0", the **Custom** edit boxes show 1 inches = 48 drawing units. If you want the drawing to be plotted so that it fits on the specified sheet of paper, select the **Scaled to Fit** option. When you select this option, AutoCAD LT will determine the scale factor and display the scale factor in the **Custom**

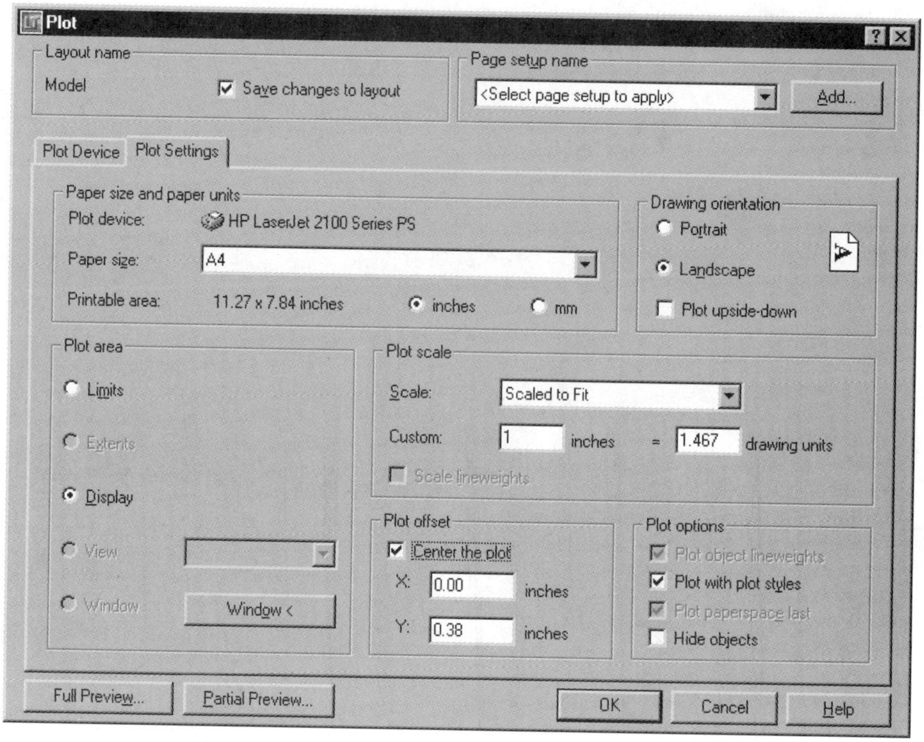

***Figure 2-30 Plot** dialog box with **Plot Settings** tab*

edit boxes. In this example, you will plot the drawing so that it scales to fit the paper. Therefore, select the **Scaled to Fit** option and notice the change in the **Custom** (Inches = Drawing Units) edit boxes. You can also enter your own values in the **Custom** edit boxes.

6. You can view the plot on the specified paper size before actually plotting it by choosing the **Preview** buttons in the **Plot** dialog box. This way you can save time and stationery. AutoCAD LT provides two types of Plot Previews, partial and full. To generate a partial preview of a plot, choose the **Partial Preview** button. The **Partial Plot Preview** dialog box is displayed (Figure 2-31). To display the drawing just as it would be plotted on the paper, choose the **Full Preview** button. **Full Preview** takes more time than **Partial Preview** because the drawing is regenerated. Once regeneration is complete, the preview image is displayed on the screen, Figure 2-32. Here, in place of the cursor, a realtime zoom icon is displayed. You can hold the pick button of your pointing device and then move it up to zoom into the preview image and move the cursor down to zoom out of the preview image.

7. If the plot preview is satisfactory, you can directly plot your drawing by choosing **Plot** from the shortcut menu, which is displayed by right-clicking. If you want to make some changes in the settings, choose **Exit** in the shortcut menu or press the ESC or the ENTER key to get back to the dialog box. You can choose the **OK** button in the dialog box to plot the drawing.

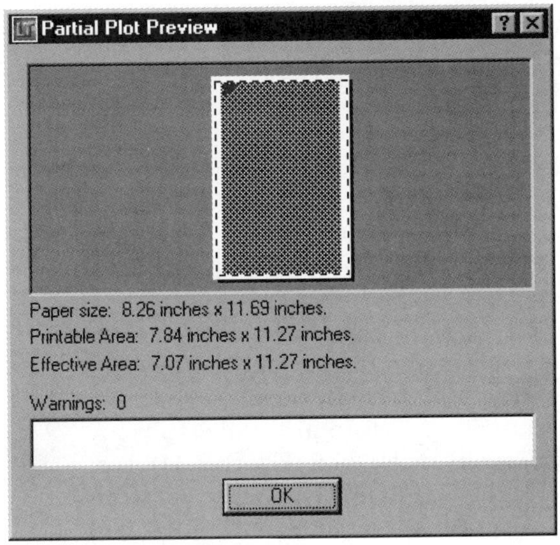

***Figure 2-31 Partial Plot Preview** dialog box*

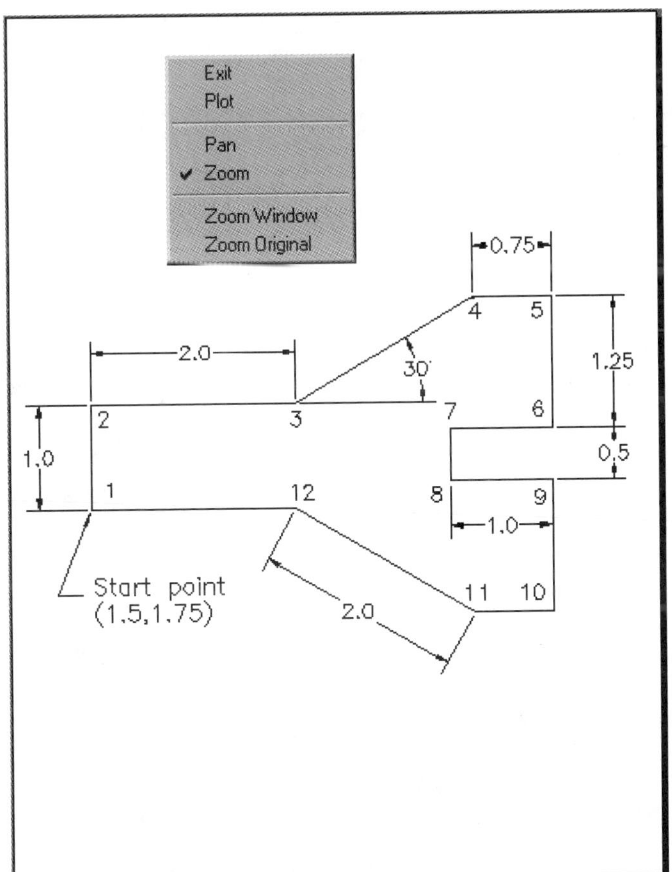

***Figure 2-32** Full plot preview with the shortcut menu*

OPTIONS DIALOG BOX

Menu:	Tools > Options
Command:	OPTIONS

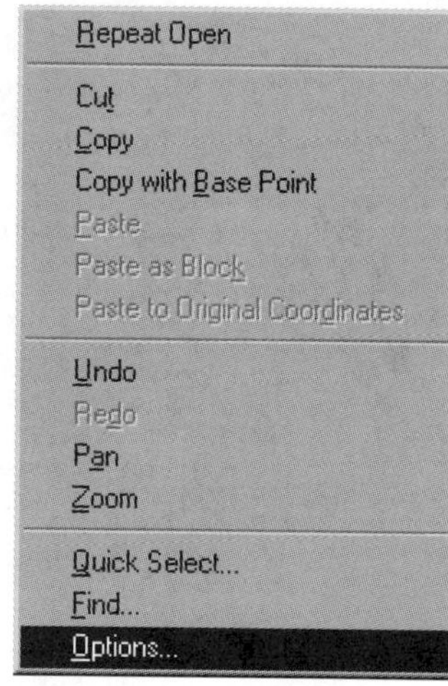

Figure 2-33 *Options in the drawing area shortcut menu*

You can use the **Options** dialog box to change the settings that affect the drawing environment or the AutoCAD LT interface. This dialog box is displayed by choosing **Options** from the **Tools** menu, or by entering **OPTIONS** at the Command prompt. You can also invoke this command by choosing **Options** from the shortcut menu (Figure 2-33), which is displayed by right-clicking in the command window or in the drawing area when no command is active or no object is selected. You can change the AutoCAD LT default settings and customize them to your requirements using the **Options** dialog box (Figure 2-34). For example, you can use this dialog box to remove the shortcut menu by right-clicking or specify the support directories that contain the files you need.

The dialog box contains eight tabs that display the sections to change the different environment aspects. The Current drawing: name are displayed on the top, above the tabs.

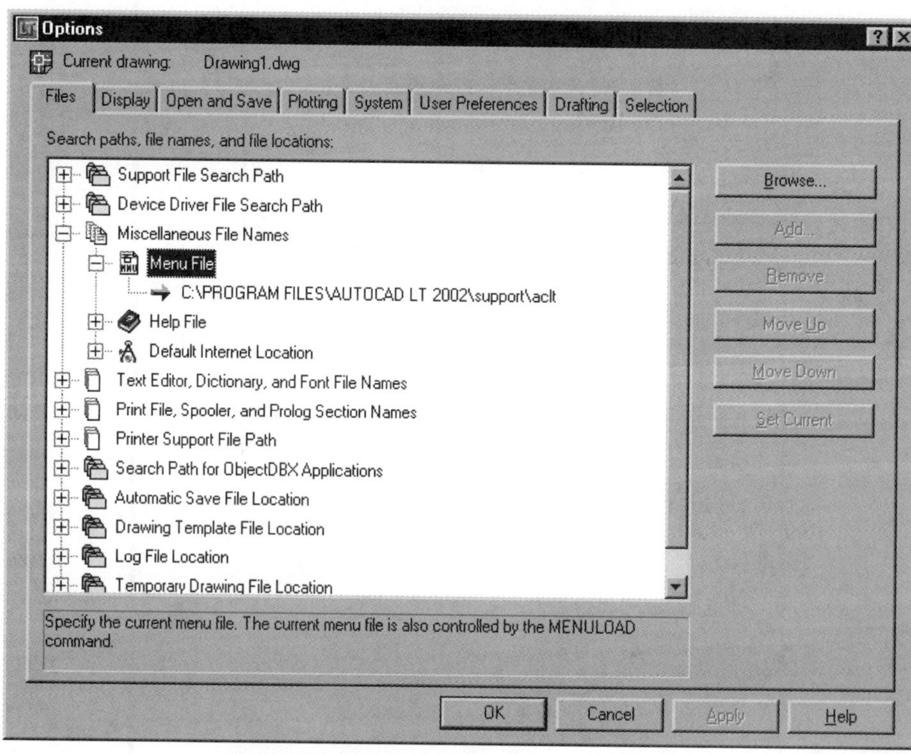

Figure 2-34 ***Options*** *dialog box (**Files** tab)*

The Options Dialog Box Tabs

Files

This tab stores the directories in which AutoCAD LT looks for driver, support, menu, temporary files, template files, and other files. It uses three icons: folder, paper stack, and file cabinet. The folder icon is for search path, the paper stack icon is for files, and the file cabinet icon is for a specific folder. Suppose you want to know the path of the menu file. You can select the **Miscellaneous File Names** folder and then select the **Menu File** icon to display the path (Figure 2-34). Similarly you can define a custom hatch pattern file and then add its search path. This way AutoCAD LT can locate the custom hatch pattern.

Display

This tab controls drawing and window settings like screen menu display and scroll bar. For example, if you want to display the scroll bars, select the **Display scroll bars in drawing window** check box in the **Window Elements** area. You can also change the color of the graphics window background, layout window background, command line background and also the color of the command line text using the **Colors** button in the **Window Elements** area. It also allows you to modify the Display resolution and Display performance. In this tab you can also set the smoothness and resolutions of certain objects such as the circle, arc, and polyline curve. Here you can toggle on and off the different layout elements such as the layout tabs on the screen, margins, paper background, and so on. You can also toggle on and off the display of a box instead of the text, apply the solid fills, and so on.

Open and Save

This tab controls the parameters of opening and saving files in AutoCAD LT. You can specify the file type for saving while using the **SAVEAS** command. The different formats are AutoCAD LT 2000.dwg, R14/LT98/LT97.dwg, .dxf, and so on. You can also set the different file safety precautions such as the **Automatic Save** feature, or the creation of a backup copy. You can change the number of recently saved files to be displayed in the **File** menu for opening. You can also set the different parameters for External references.

Plotting

The **Plotting** tab controls the parameters related to the plotting of drawings in AutoCAD LT. You can set the default output device and also add a new plotter. You can set the general parameters such as the layout or plot device paper size. It is possible to select the spool alert for the system printer and also the OLE plot quality. You can also set the parameters for the plot style such as using the color dependent plot styles or the named plot styles.

System

This tab contains AutoCAD LT system settings options such as the Pointing device settings options where you can choose the pointing device driver. Here you can also set the different system parameters such as the single drawing mode instead of MDE, the display of the **Startup** option while opening a new session of AutoCAD LT and the **OLE Properties** dialog box, and beep for wrong user input. You can also set the **Live Enabler Options** using the **System** tab of the **Options** dialog box.

User Preferences

This tab controls settings that depend on the way the user prefers working on AutoCAD LT, such as the Right-click Customization where you can change the shortcut menus. You can set the units parameters for the AutoCAD LT DesignCenter as well as the priorities for different coordinates' data entry methods. Here it is possible to set the order of object sorting methods and also set the lineweight options.

Drafting

This tab controls settings such as the Autosnap Settings and the Aperture Size. Here you can also set the toggles on and off for the different tracking settings.

Selection

This tab controls settings related to the methods of Object Selection such as the Grips, that enables you to change the various grip colors and the grip size. You can also set the toggles on or off for the different selection modes.

Note

*The various options in the different tabs of the **Options** dialog box have been discussed throughout the book wherever applicable.*

Tip

*Some options in the different tabs of the **Options** dialog box have a drawing file icon in front of them. For example, the options in the **Display resolution** area of the **Display** tab have the drawing file icons. This specifies that these parameters are saved with the current drawing only and therefore affects it. The rest of the options (without the drawing file icon) are saved with the current profile and also affect all the drawings present in that AutoCAD LT session.*

Self-Evaluation Test

Answer the following questions and then compare your answers to the correct answers given at the end of this chapter.

1. In the **Open Drawings** tab of the **AutoCAD LT Today** window, the already existing drawings are displayed randomly. (T/F)

2. In the **Open Drawings** tab of the **AutoCAD LT Today** window, if you choose the **Browse** button, the **Select File** dialog box is displayed where you can save the drawing file. (T/F)

3. If you choose the Engineering or Architectural format for units in the **Drawing Units** dialog box, you can enter the distances or coordinates in any of the five formats. (T/F)

4. For plotting a drawing, **Partial Preview** takes more time than **Full Preview** because the drawing is regenerated. (T/F)

5. If you want to work on a drawing without altering the original, you must select the __________ option from the **Open** drop-down list in the **Select File** dialog box.

6. The __________ tab of the **Options** dialog box stores the path in which AutoCAD LT looks for the driver support, menu, and other files.

7. The __________ command enlarges or reduces the view of the drawing on the screen, but it does not affect the actual size of the entities.

8. In __________ units you must specify the bearing angle that the line makes with the north-south direction.

9. After you increase the drawing limits, you need to use the __________ option of the **ZOOM** command to display the complete area inside the drawing area.

10. You can view the plot on the specified paper size before actually plotting it by choosing thc __________ buttons in the **Plot** dialog box.

Review Questions

Answer the following questions.

1. It is not possible for you to open the views of the drawing earlier saved. (T/F)

2. In AutoCAD LT, by default the angles are positive if measured in the counterclockwise direction and the angles are measured from the positive X axis. (T/F)

3. You can also invoke the **PLOT** command by choosing **Plot** from the shortcut menu, which is displayed by right-clicking on the Command window. (T/F)

4. The **Files** tab of the **Options** dialog box stores the directories in which AutoCAD LT looks for driver, support, menu, template and other files. (T/F)

5. What is the maximum number of most recently opened drawings that can be displayed when you choose the **Most Recently Used** option from the **Select how to begin** list box in the **AutoCAD LT Today** window?

   (a) Four (b) Six
   (c) Seven (d) Nine

6. Which of the following options of the **ZOOM** command zooms to the drawing limits or the extents, whichever is greater?

   (a) **Previous** (b) **Window**
   (c) **All** (d) **Realtime**

7. How many formats of units can you choose from in the **Drawing Units** dialog box?

   (a) Three (b) Five
   (c) Six (d) Seven

8. Which of the following input methods cannot be used to invoke the **OPTIONS** command, which displays the **Options** dialog box?

   (a) Menu (b) Toolbar
   (c) Shortcut menu (d) Command prompt

9. When you define the direction by specifying the angle, the output of the angle does not depend on which one of the following factors?

   (a) Angular units (b) Angle value
   (c) Angle direction (d) Angle base

10. The __________ is the Windows feature that allows you to open more then one drawing at a time in one AutoCAD LT session.

11. You can also use the __________ and __________ instead of dragging and dropping the objects from one drawing to another while multiple drawings are opened.

12. When you select any type of unit and angle in the **Length** or **Angle** area of the **Drawing Units** dialog box, the corresponding example is displayed in the __________ area of the dialog box.

13. You can also invoke the **PLOT** command by choosing **Plot** from the shortcut menu, which is displayed by right-clicking on the __________ tabs.

14. If you want the drawing to be plotted so that it fits on the specified sheet of paper, select the __________ option in the **Plot** dialog box.

15. The __________ tab in the **Options** dialog box stores the information about the format in which the drawings will be saved when the **SAVEAS** command is invoked.

16. You can use the __________ command to change the settings that affect the drawing environment or the AutoCAD LT interface.

Exercises

Exercise 2 *General*

Open five drawings in the same session of AutoCAD LT. You can open any two drawings you have saved in the previous chapter and create three new drawings. Use the options in the **Window** menu to tile the drawings differently. Also copy and move the objects from one drawing into another by dragging/dropping and by using the Copy/Cut and Paste options.

Exercise 3 *Architectural*

Set the units for a drawing according to the following specifications:

1. Set the **UNITS** to architectural, with the denominator of the smallest fraction equal to 16.
2. Set the angular measurement to degrees/minutes/seconds, with the number of fractional places for display of angles equal to 0d00'.
3. Set the direction to 0 degrees (east) and the direction of measurement of angles to counterclockwise (angles measured positive in a counterclockwise direction).

Based on Figure 2-35, determine and set the limits of the drawing. The scale for this drawing is 1/4" = 1'. Leave enough space around the drawing for dimensioning and title block. (HINT: Scale factor = 48; sheet size required is 12 x 9; therefore, the limits are 12 x 48, 9 x 48 = 576, 432. Use the **ZOOM** command and then select the **All** option to display the new limits.)

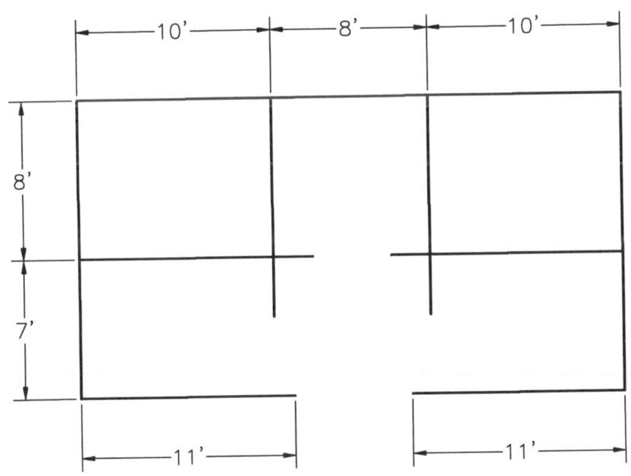

Figure 2-35 *Drawing for Exercise 3*

Exercise 4 *Mechanical*

Make the drawing as shown in Figure 2-36. The distance between the dotted lines is 10 feet. Determine the limits for this drawing and use the Architectural units with 0'-01/32" precision.

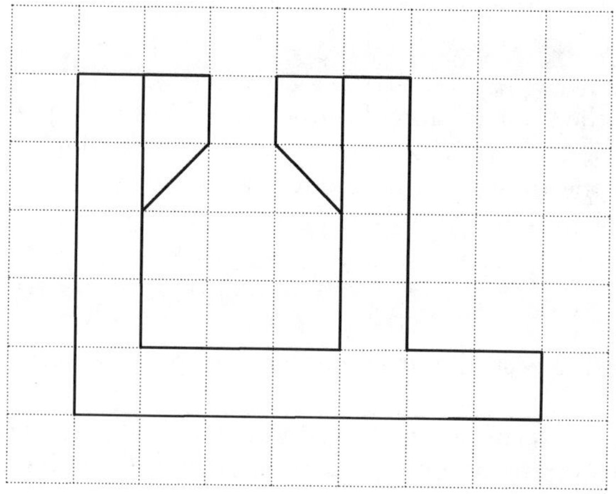

Figure 2-36 Drawing for Exercise 4

Exercise 5

Mechanical

Make the drawing as shown in Figure 2-37. The distance between the dotted lines is 5 units. Determine the limits for this drawing and use the Fractional units with 0 1/16 precision.

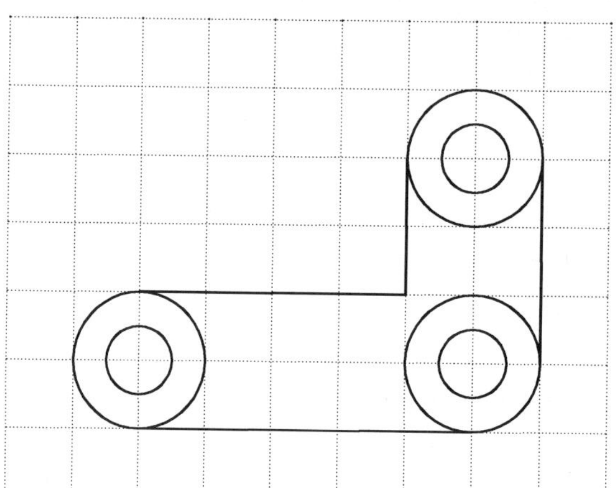

Figure 2-37 Drawing for Exercise 5

Exercise 6 — *Mechanical*

Make the drawing as shown in Figure 2-38. The distance between the dotted lines is 1 unit. Determine the limits for this drawing and use the Decimal units with 0.00 precision.

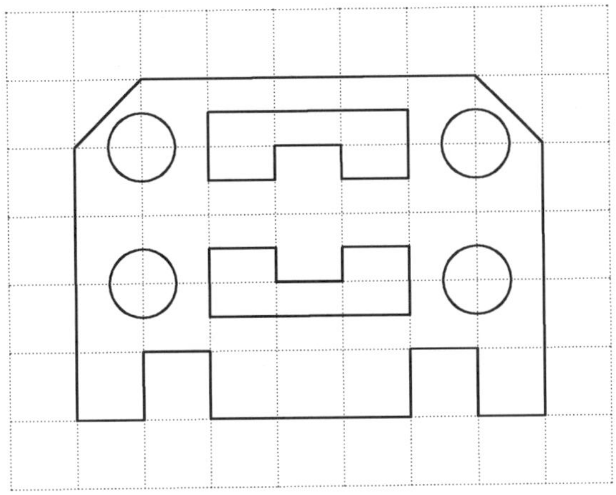

Figure 2-38 *Drawing for Exercise 6*

Exercise 7 — *Mechanical*

Make the drawing as shown in Figure 2-39. The distance between the dotted lines is 10 units. Determine the limits for this drawing and use the Decimal units with 0.00 precision.

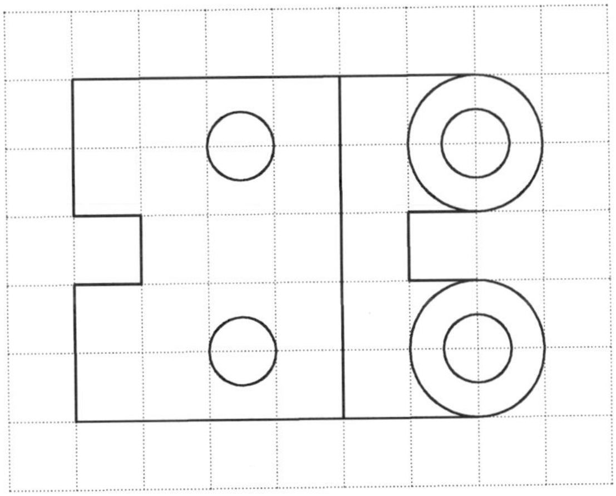

Figure 2-39 *Drawing for Exercise 7*

Answers to Self-Evaluation Test

1 - F, **2** - F, **3** - T, **4** - F, **5** - **Open Read-Only**, **6** - **Files**, **7** - ZOOM, **8** - Surveyor's, **9** - All, **10** - **Preview**

Chapter 3

Draw Commands

Learning Objectives

After completing this chapter, you will be able to:

- *Draw arcs using different options.*
- *Draw rectangles, ellipses, and elliptical arcs.*
- *Draw polygons such as hexagons and pentagons.*
- *Draw polylines and donuts.*
- *Draw points and change point style and point size.*
- *Draw infinite lines and create simple text.*

DRAWING ARCS

Toolbar:	Draw > Arc
Menu:	Draw > Arc
Command:	ARC or A

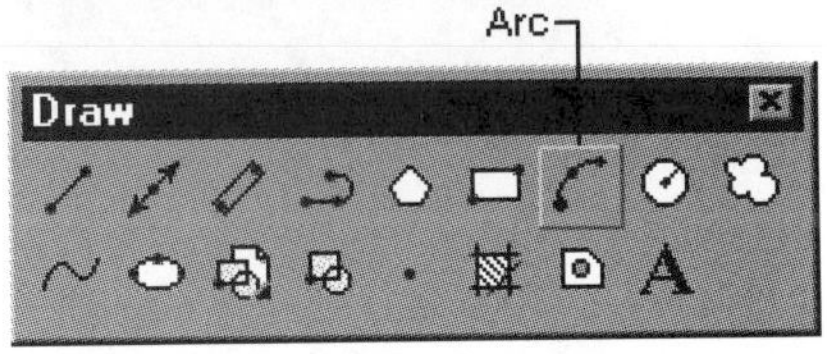

*Figure 3-1 Invoking the **ARC** command from the **Draw** toolbar*

An arc is defined as a part of a circle; it can be drawn using the **ARC** command (Figure 3-1). An arc can be drawn in eleven distinct ways using the options listed under the **ARC** command (Figure 3-2). The default method for drawing an

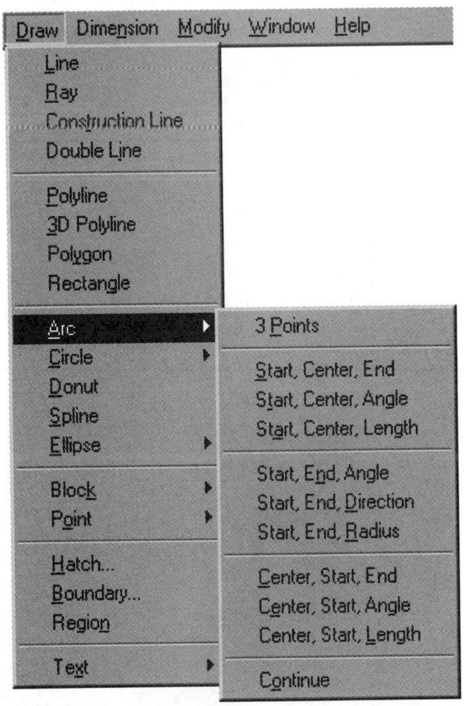

*Figure 3-2 Invoking the **ARC** command from the **Draw** menu*

arc is the **3 Points** option. Other options can be invoked by entering the appropriate letter to select an option. The last parameter to be specified in any arc generation is automatically dragged into the relevant location. This is dependent on the **DRAGMODE** variable which should be set to **Auto** (default).

The 3 Points Option

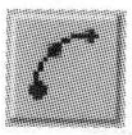

When you choose the **Arc** button from the **Draw** toolbar, or enter **ARC** at the Command prompt, you automatically enter the **3 Points** option. The 3 Points option requires the start point, the second point, and the endpoint of the arc (Figure 3-3 and Figure 3-4). The arc can be drawn in a clockwise or counterclockwise direction by dragging the arc with the cursor. The following is the prompt sequence to draw an arc with a start point at (2,2), second point at (3,3), and an endpoint at (3,4). (You can also specify the points on the screen by moving the cursor.)

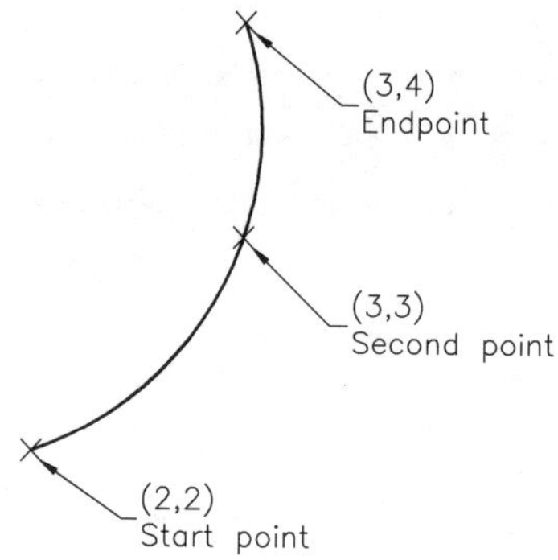

***Figure 3-3** Drawing an arc using the 3 Points option*

Command: *Choose the **Arc** button from the **Draw** toolbar.*
Specify start point of arc or [Center]: **2,2** [Enter]
Specify second point of arc or [Center/End]: **3,3** [Enter]
Specify end point of arc: **3,4** [Enter]

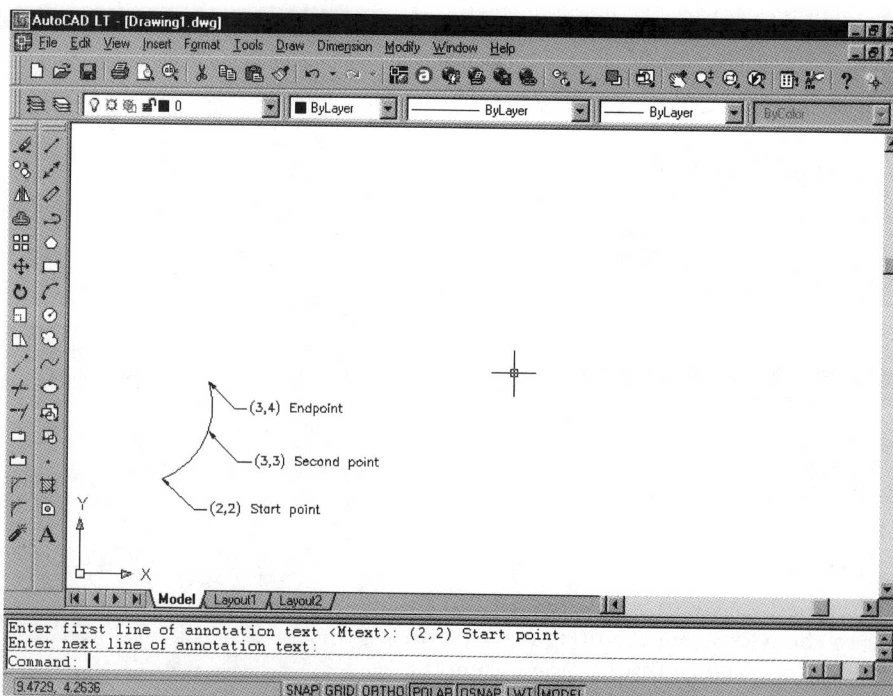

***Figure 3-4** Drawing an arc using the **3 Points** option*

Exercise 1 *General*

Draw several arcs using the **3 Points** option. The points can be selected by entering coordinates or by specifying points on the screen. Also, try to create a circle by drawing two separate arcs and by drawing a single arc. Notice the limitations of the **ARC** command.

The Start, Center, End Option

This option is slightly different from the 3 Points option. In this option, instead of entering the second point, you enter the center of the arc. Choose this option when you know the start point, endpoint, and center point of the arc. The arc is drawn in a counterclockwise direction from the start point to the endpoint around the specified center. The endpoint specified need not be on the arc and is used only to calculate the angle at which the arc ends. The radius of the arc is determined by the distance between the center point and the start point. The prompt sequence for drawing an arc with a start point of (3,2), center point at (2,2), and endpoint of (2,3.5) (Figure 3-5) is as follows:

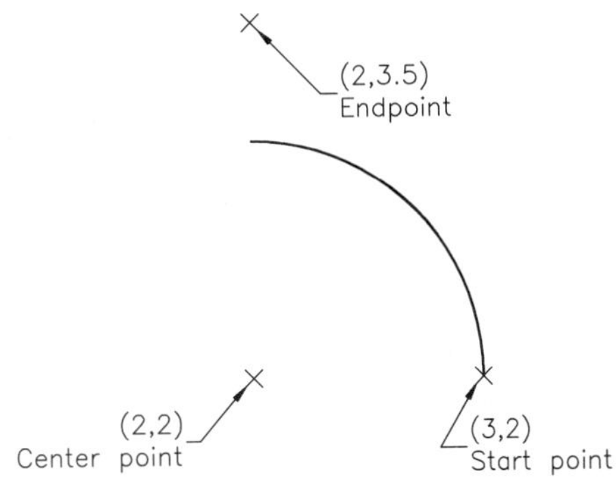

Figure 3-5 *Drawing an arc using the* ***Start, Center, End*** *option*

Command: **ARC** [Enter]
Specify start point of arc or [Center]: **3,2** [Enter]
Specify second point of arc or [Center/End]: **C** [Enter]
Specify center point of arc: **2,2** [Enter]
Specify end point of the arc or [Angle/chord Length]: **2,3.5** [Enter]

Note

If you choose the ***Start, Center, End*** *option from the* ***Draw > Arc*** *menu, after entering the start point, the center option is automatically invoked. This way you bypass the prompt* ***Specify second point of arc or [Center/End]****. You need to simply specify the center point.*

The Start, Center, Angle Option

This option is the best choice if you know the **included angle** of the arc. The included angle is the angle formed by the start point and the endpoint of the arc with the specified center. This option draws an arc in a counterclockwise direction with the specified center and start point spanning the indicated angle (Figure 3-6). If the specified angle is negative, the arc is drawn in a clockwise direction (Figure 3-7).

The prompt sequence for drawing an arc with center at (2,2), a start point of (3,2), and an included angle of 60 degrees (Figure 3-6) is:

Command: **ARC** [Enter]
Specify start point of arc or [Center]: **3,2** [Enter]

Chapter 3

Specify second point of arc or [Center/End]: **C** Enter
Specify center point of arc: **2,2** Enter
Specify end point of the arc or [Angle/chord Length]: **A** Enter
Specify included angle: **60** Enter

You can draw arcs with negative angle values in the Start, Center, Included Angle (St,C,Ang) option by entering "-" (a negative sign) followed by the angle values of your requirement at the **Specify included angle** prompt (Figure 3-7). The following is the prompt sequence:

Command: **ARC** Enter
Specify start point of arc or [Center]: **4,3** Enter
Specify second point of arc or [Center/End]: **C** Enter
Specify center point of arc: **3,3** Enter
Specify end point of the arc or [Angle/ chord Length]: **A** Enter
Specify included angle: **-180** Enter

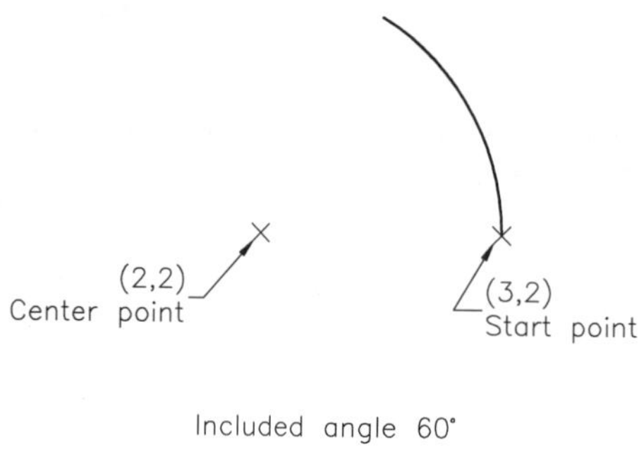

*Figure 3-6 Drawing an arc using the **Start, Center, Angle** option*

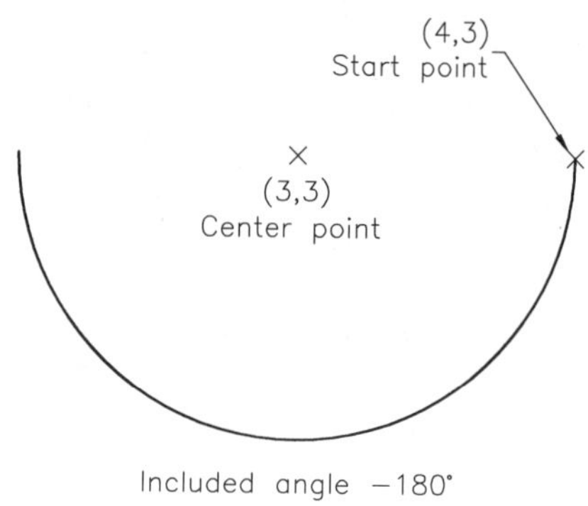

*Figure 3-7 Drawing an arc using a negative angle in the **Start, Center, Angle** option*

Note

*If you choose this option from the **Draw** menu, the center and the angle options are invoked and you simply have to specify the start point, center point, and the angle.*

*To invoke the rest of the options, you will use the **Draw** menu.*

Exercise 2 *Mechanical*

a. Draw an arc using the **St,C,Ang** option. The start point is (6,3), the center point is (3,3), and the angle is 240 degrees.

b. Make the drawing shown in Figure 3-8. The distance between the dotted lines is 1.0 unit. Create the radii by using the **ARC** command options as indicated in the drawing.

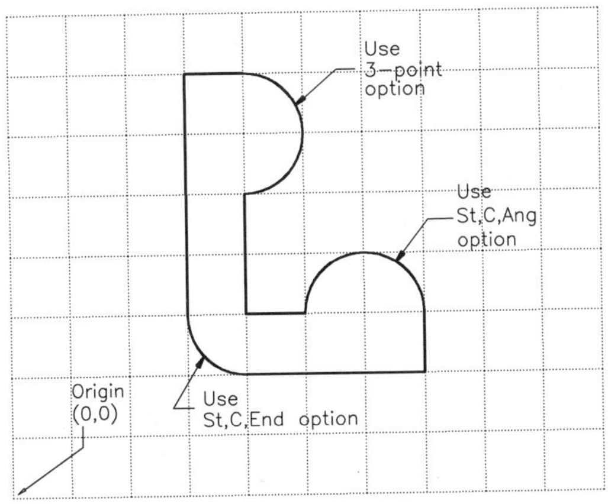

***Figure 3-8** Drawing for Exercise 2(b)*

The Start, Center, Length Option

In this option you are required to specify the start point, center point, and length of chord. A **chord** is defined as the straight line connecting the start point and the endpoint of an arc. The chord length needs to be specified so that AutoCAD LT can calculate the ending angle. Identical start, center, and chord length specifications can be used to define four different arcs. AutoCAD LT always draws this type of arc counterclockwise from the start point. Therefore, a positive chord length gives the smallest possible arc with that length. This is known as the minor arc. The minor arc is less than 180 degrees. A negative value for chord length results in the largest possible arc, also known as the major arc. The chord length can be determined by using the standard chord length tables or using the mathematical relation (L = 2*Sqrt [h(2r-h)]). For example, an arc of radius 1 unit, with an included angle of 30 degrees, has a chord length of 0.51764 units. The prompt sequence for drawing an arc that has a start point of (3,1), center of (2,2) and the chord length of (2) (Figure 3-9) is:

Command: *Choose the **Start, Center, Length** option from the **Draw** > **Arc** menu.*
Specify start point of arc or [Center]: **3,1** Enter
Specify second point of arc or [Center/End]: _c Specify center point of arc: **2,2** Enter
Specify end point of the arc or [Angle/ chord Length]: _l Length of chord: **2** Enter

You can draw the major arc by defining the length of the chord as negative (Figure 3-10). In this case the arc with a start point of (3,1), a center point of (2,2), and a negative chord length of (-2) is drawn with the following prompt sequence:

Command: *Choose the **Start, Center, Length** option from the **Draw** > **Arc** menu.*
Specify start point of arc or [Center]: **3,3** Enter
Specify second point of arc or [Center/End]: _c Specify center point of arc: **2,2** Enter
Specify end point of the arc or [Angle/ chord Length]: _l Length of chord: **-2** Enter

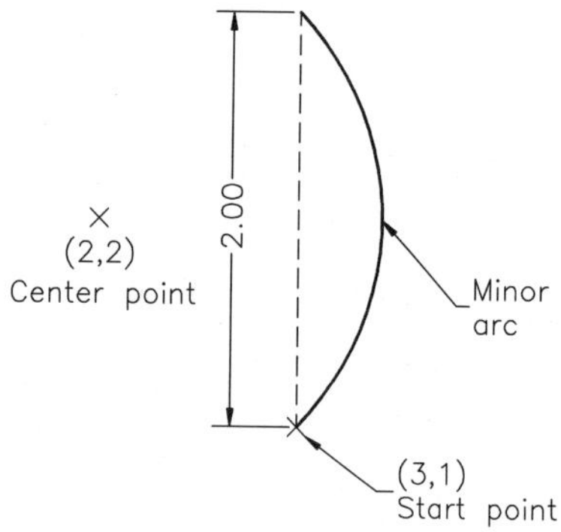

Figure 3-9 *Drawing an arc using the* ***Start, Center, Length*** *option*

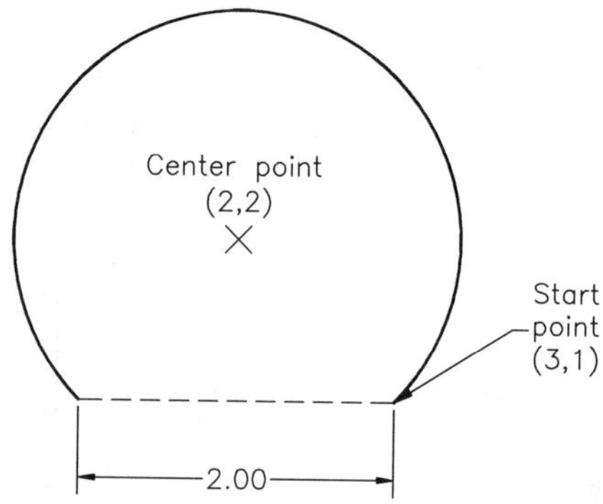

Figure 3-10 *Drawing an arc using a negative chord length in the* ***Start, Center, Length*** *option*

Note

The points that you have specified for drawing the arcs using different options are quite close to each other and therefore the arcs created might overlap each other. You can use the ***ERASE*** *command to erase some of the previously drawn arcs and gain space to draw new arcs.*

Exercise 3 *General*

Draw a minor arc with the center point at (3,4), start point at (4,2), and chord length of 4 units.

The Start, End, Angle Option

With this option you can draw an arc by specifying the start point of the arc, the endpoint, and the included angle. A positive included angle value draws an arc in a counterclockwise direction from the start point to the endpoint, spanning the included angle; a negative included angle value draws the arc in a clockwise direction. The prompt sequence for drawing an arc with a start point of (3,2), endpoint of (2,4), and included angle of 120 degrees (Figure 3-11) is:

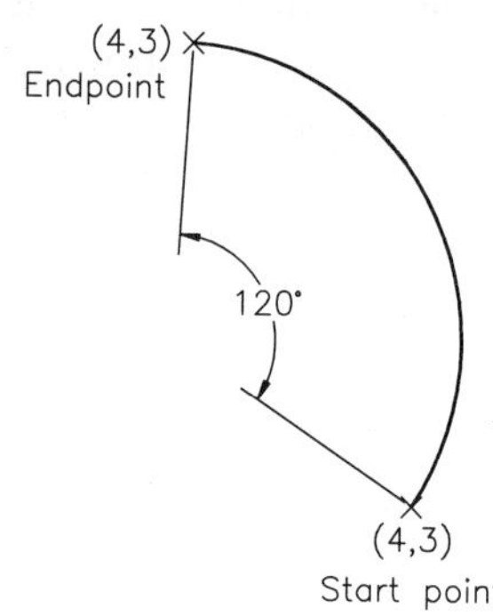

Figure 3-11 *Drawing an arc using the* ***Start, End, Angle*** *option*

Command: *Choose the* ***Start, End, Angle*** *option from the* ***Draw*** **>** ***Arc*** *menu.*
Specify start point of arc or [Center]: **3,2** [Enter]
Specify second point of arc or [Center/End]: _e Specify end point of arc: **2,4** [Enter]
Specify center point of arc or [Angle/Direction/Radius]: _a Specify included angle: **120** [Enter]

The Start, End, Direction Option

In this option you can draw an arc by specifying the start point, endpoint, and starting direction of the arc, in degrees. In other words, the arc starts in the direction you specify (the start of the arc is established tangent to the direction you specify). This option can be used to draw a major or minor arc, in a clockwise or counterclockwise direction. The size and position of the arc are determined by the distance between the start point and endpoint and the direction specified. To illustrate the positive direction option (Figure 3-12), the prompt sequence for an arc having a start point of (4,3), endpoint of (3,5), and direction of 90 degrees is:

> Command: *Choose the* ***Start, End, Direction*** *option from the* ***Draw > Arc*** *menu.*
> Specify start point of arc or [Center]: **4,3** Enter
> Specify second point of arc or [Center/End]: _e Specify end point of arc: **3,5** Enter
> Specify center point of arc or [Angle/Direction/Radius]: _d Specify tangent direction for the start point of arc: **90** Enter

To illustrate the option using a negative direction degree specification (Figure 3-13), the prompt sequence for an arc having a start point of (4,3), endpoint of (3,4), and direction of -90 degrees is:

> Command: *Choose the* ***Start, End, Direction*** *option from the* ***Draw > Arc*** *menu.*
> Specify start point of arc or [Center]: **4,3** Enter
> Specify second point of arc or [Center/End]: _e Specify end point of arc: **3,5** Enter
> Specify center point of arc or [Angle/Direction/Radius]: _d Specify tangent direction for the start point of arc: **-90** Enter

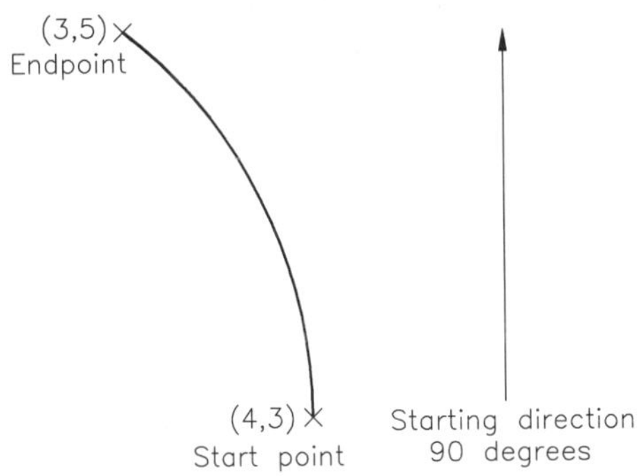

Figure 3-12 *Drawing an arc using the* ***Start, End, Direction*** *option*

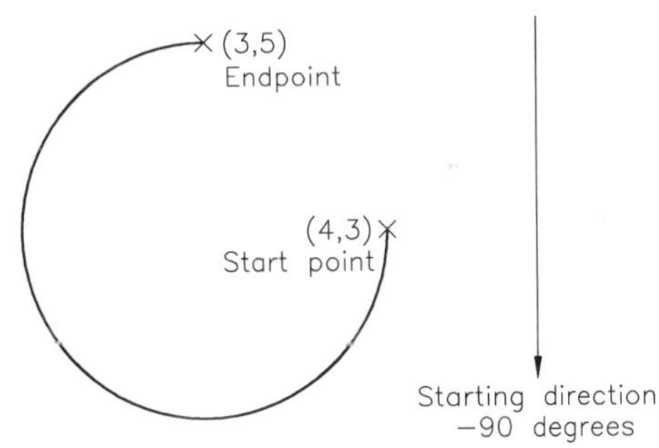

Figure 3-13 *Drawing an arc using a negative direction in the* ***Start, End, Direction*** *option*

Note

With the ***Start, End, Direction*** *option, if you do not specify a start point but just press ENTER at the* ***Specify start point of arc or [Center]*** *prompt, the start point and direction of the arc will be taken from the* ***endpoint and ending direction*** *of the previous line or arc drawn on the current screen. You are then required to specify only the endpoint of the arc.*

Exercise 4 — *Mechanical*

a. Specify the directions and the coordinates of two arcs in such a way that they form a circular figure.

b. Make the drawing shown in Figure 3-14. Create the radii by using the **ARC** command options. The distance between the dotted lines is 1.0 units and the diameter of the circles is 1 unit.

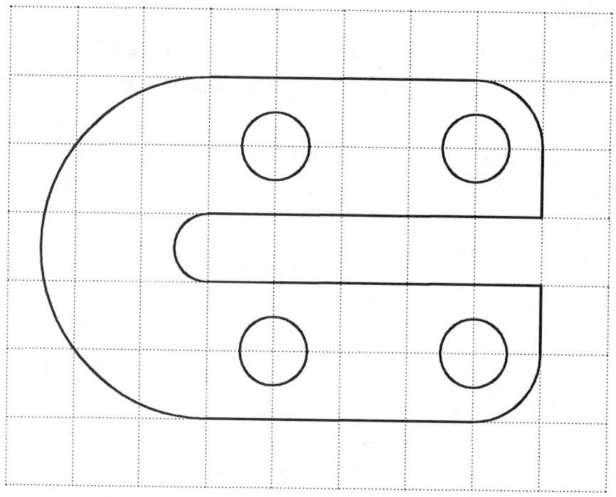

Figure 3-14 Drawing for Exercise 4(b)

The Start, End, Radius Option

This option is used when you know the start point, endpoint, and radius of the arc. The same values for the three variables (start point, endpoint, and radius) can result in four different arcs. AutoCAD LT resolves this by always drawing this type of arc in a counterclockwise direction from the start point. Therefore, a positive radius value results in a **minor arc** (the smallest arc between the start point and the endpoint), Figure 3-15(b), while a negative radius value results in a **major arc** (the largest arc between two endpoints), Figure 3-15(a). The prompt sequence to draw a major arc having a start point of (3,3), endpoint of (2,5), and radius of -2, Figure 3-15(a), is:

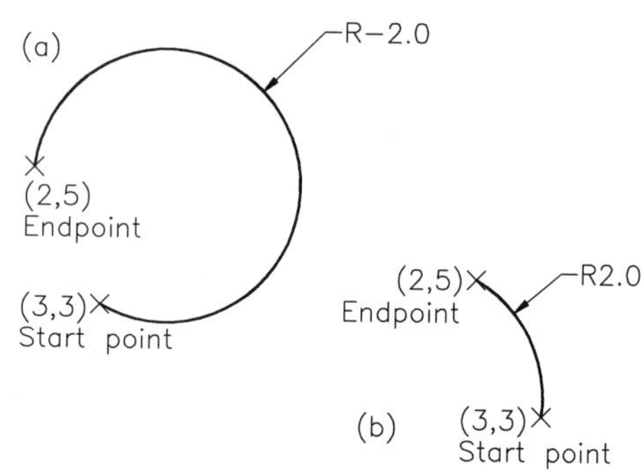

*Figure 3-15 Drawing an arc using the **Start, End, Radius** option*

Command: *Choose the **Start, End, Radius** option from the **Draw > Arc** menu.*
Specify start point of arc or [Center]: **3,3** Enter
Specify second point of arc or [Center/End]: _e Specify end point of arc: **2,5** Enter
Specify center point of arc or [Angle/Direction/Radius]: _r Specify radius of arc: **-2** Enter

The prompt sequence to draw a minor arc having its start point at (3,3), endpoint at (2,5), and radius as 2 as shown in Figure 3-15(b) is:

Command: *Choose the **Start, End, Radius** option from the **Draw/Arc** menu.*
Specify start point of arc or [Center]: **3,3** Enter
Specify second point of arc or [Center/End]: _e Specify end point of arc: **2,5** Enter
Specify center point of arc or [Angle/Direction/Radius]: _r Specify radius of arc: **2** Enter

The Center, Start, End Option

The **Center, Start, End** option is a modification of the Start, Center, End option. Use this option whenever it is easier to start drawing an arc by establishing the center first. Here the arc is always drawn in a counterclockwise direction from the start point to the endpoint, around the specified center. The prompt sequence for drawing an arc that has a center point at (3,3), start point at (5,3), and endpoint at (3,5) (Figure 3-16) is:

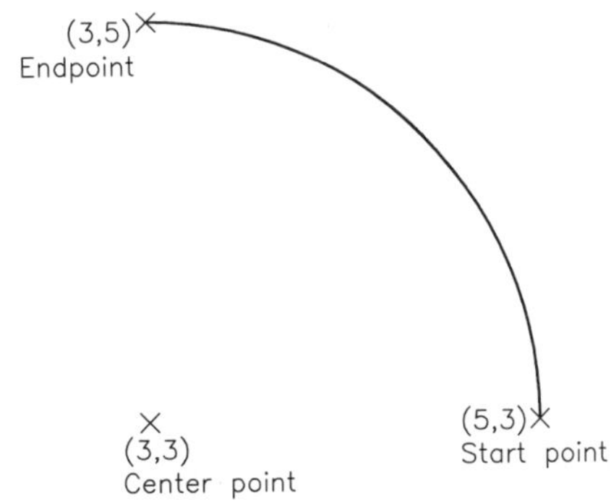

Figure 3-16 Drawing an arc using the ***Center, Start, End*** *option*

Command: *Choose the **Center, Start, End** option from the **Draw > Arc** menu.*
Specify start point of arc or [Center]: _c
Specify center point of arc: **3,3** [Enter]
Specify start point of arc: **5,3** [Enter]
Specify end point of arc or [Angle/chord Length]: **3,5** [Enter]

The Center, Start, Angle Option

This option is a variation of the **Start, Center, Angle** option. Use this option whenever it is easier to draw an arc by giving the center first. The prompt sequence for drawing an arc that has a center point at (4,5), start point at (5,4), and included angle of 120 degrees (Figure 3-17) is:

Command: *Choose the **Center, Start, Angle** option from the **Draw > Arc** menu.*
Specify start point of arc or [Center]: _c Specify center point of arc: **4,5** [Enter]
Specify start point of arc: **5,4** [Enter]
Specify end point of arc or [Angle/chord Length]: _a Specify included angle: **120** [Enter]

The Center, Start, Length Option

The **Center, Start, Length** option is a modification of the Start, Center, Length option. This option is used whenever it is easier to draw an arc by establishing the center first. The prompt sequence for drawing an arc that has a center point at (2,2), start point at (4,3), and length of chord of 3 (Figure 3-18) is:

Command: *Choose the **Center, Start, Length** option from the **Draw > Arc** menu.*
Specify start point of arc or [Center]: _c Specify center point of arc: **2,2** [Enter]
Specify start point of arc: **4,3** [Enter]
Specify end point of arc or [Angle/chord Length]: _l Specify length of chord: **3** [Enter]

Continue Option

With this option you can continue drawing an arc from a previously drawn arc or line. When you select the **Continue** option (**Draw** menu), the start point and direction of the arc will be taken from the **endpoint and ending direction** of the previous line or arc drawn on the current AutoCAD session. When this option is used to draw arcs, each successive arc is tangent to the

Chapter 3

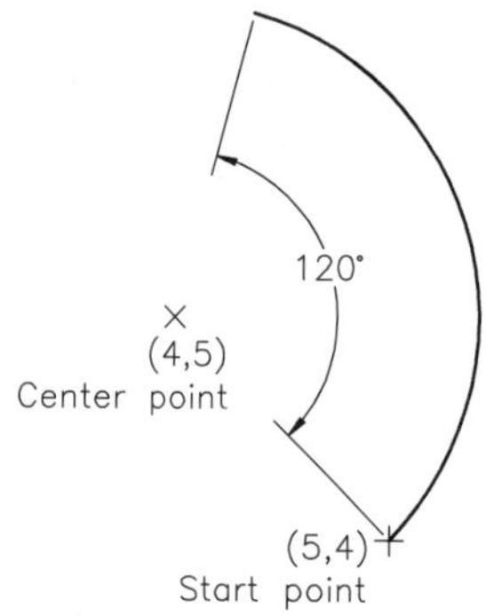

Figure 3-17 *Drawing an arc using the* ***Center, Start, Angle*** *option*

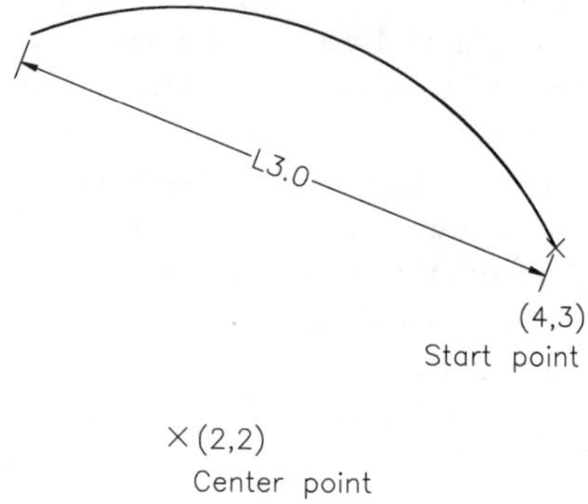

Figure 3-18 *Drawing an arc using the* ***Center, Start, Length*** *option*

previous one. Most often this option is used to draw arcs tangent to a previously drawn line. The prompt sequence to draw an arc tangent to an earlier drawn line using the Continue option (Figure 3-19) is:

Command: **LINE** Enter
Specify first point: **2,2** Enter
Specify next point or [Undo]: **4,3** Enter
Specify next point or [Undo]: Enter

Command: *Choose the* **Continue** *option from the* ***Draw > Arc*** *menu.*
Specify endpoint of arc: **4,5** Enter

The prompt sequence to draw an arc continued from a previously drawn arc (Figure 3-20) is:

Command: **ARC** Enter
Specify start point of arc or [Center]: **2,2** Enter
Specify second point of arc or [Center/End]: **E** Enter
Specify endpoint of arc : **3,4** Enter
Specify center point of arc or [Angle/Direction/Radius]: **R** Enter
Specify radius of arc: **2** Enter

Command: *Choose the* **Continue** *option from the* ***Draw > Arc*** *menu.*
Specify end point of arc: **5,4** Enter

Tip

The ***Continue*** *option can also be invoked by pressing the ENTER key at the* ***Specify start point of arc or [Center]*** *prompt after entering* ***ARC*** *at the Command prompt. Even when you select any option of the ARC command from the* ***Draw > Arc*** *menu, pressing ENTER at the* ***Specify start point of arc or [Center]*** *prompt invokes the* ***Continue*** *option by terminating the selected option. Here you are prompted to specify the endpoint to complete the arc.*

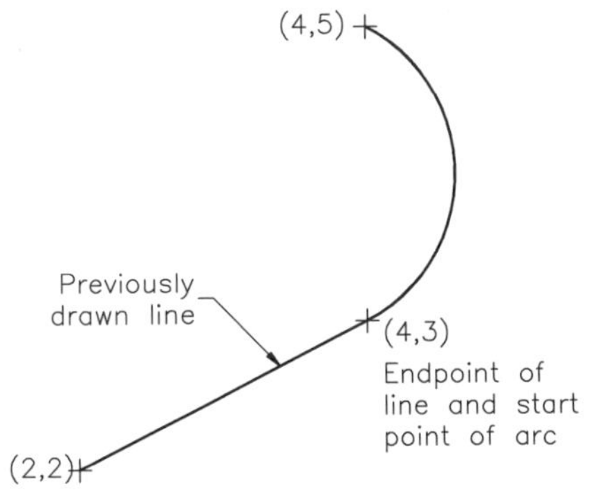

***Figure 3-19** Drawing an arc using the **Continue** option*

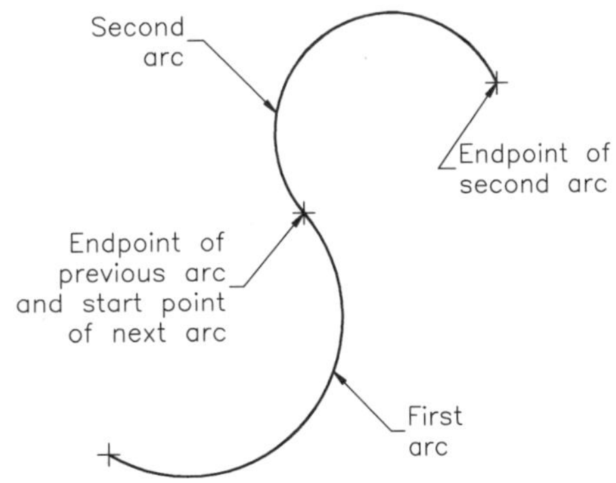

***Figure 3-20** Drawing an arc using the **Continue** option*

Continue (LineCont:) Option

This option is used when you want to continue drawing a line from the endpoint of a previously drawn arc. When you use this option, the start point and direction of the line will be taken from the **endpoint and ending direction** of the previous arc. In other words, the line will be tangent to the arc drawn on the current screen. This option is invoked when you press ENTER at the **Specify first point:** prompt of the LINE command. The prompt sequence to draw a line tangent to an earlier drawn arc (Figure 3-21) is:

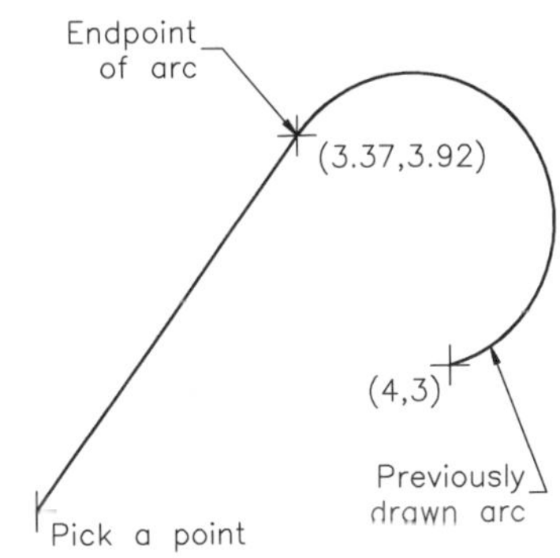

***Figure 3-21** Drawing a line from the endpoint of an arc*

Command: **ARC** [Enter]
Specify start point of arc or [Center]: **4,3** [Enter]
Specify second point of arc or [Center/End]: **E** [Enter]
Specify end point of arc : **3.37,3.92** [Enter]
Specify center point of arc or [Angle/Direction/Radius]: *Specify the center point.*

Command: **LINE** [Enter]
Specify first point: [Enter]
Length of line: *Enter a value or pick a point.*

Exercise 5 — *General*

a. Use the **Center, Start, Angle** and the **Continue** options to draw the figures shown in Figure 3-22.

b. Make the drawing shown in Figure 3-23. The distance between the dotted lines is 1.0 unit. Create the radii as indicated in the drawing by using the **ARC** command options.

Figure 3-22 *Drawing for Exercise 5(a)*

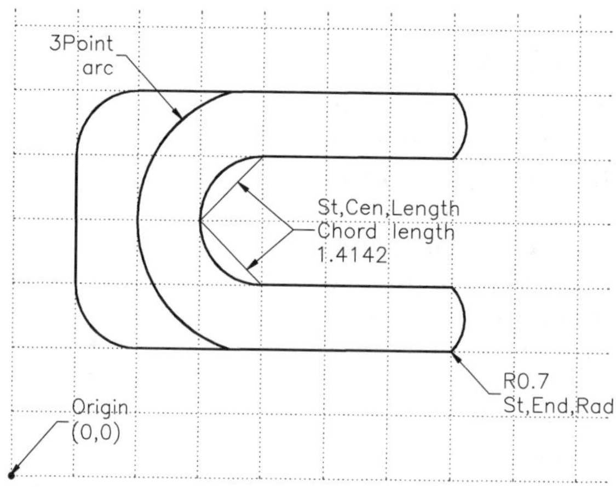

Figure 3-23 *Drawing for Exercise 5(b)*

DRAWING RECTANGLES

Toolbar:	Draw > Rectangle
Menu:	Draw > Rectangle
Command:	RECTANG

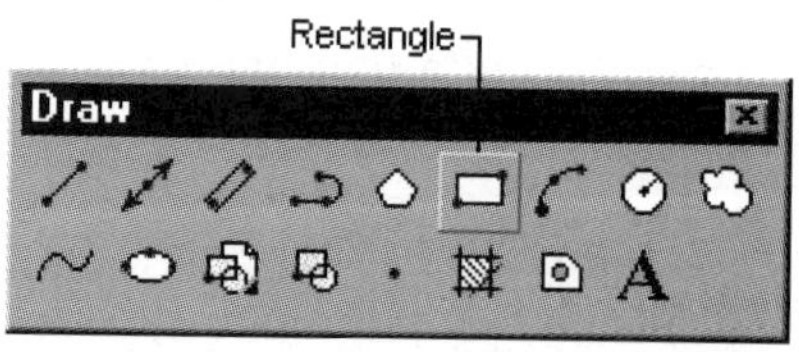

Figure 3-24 Invoking the ***RECTANG*** *command from the* ***Draw*** *toolbar*

A rectangle can be drawn using the **RECTANG** command that can be invoked using the **Draw** toolbar (Figure 3-24) or from the **Draw** menu (Figure 3-25). After invoking the **RECTANG** command, you are prompted to specify the first corner of the rectangle at the **Specify first corner point or [Chamfer/Elevation/Fillet/Thickness/Width]** prompt. Here you can enter the coordinates of the first corner or specify the desired point with the pointing device. The first corner can be any one of the four corners. Then you are prompted to enter the coordinates or specify the other corner at the **Specify other corner point or [Dimensions]** prompt. This corner is taken as the corner diagonally opposite the first corner. You can specify the coordinates for the other corner or simply drag the cursor to specify it. The prompt sequence for drawing a rectangle with (3,3) as its lower left corner coordinate and (6,5) as its upper right corner (Figure 3-26) is:

> Command: **RECTANG**
> Specify first corner point or [Chamfer/Elevation/Fillet/Thickness/Width]: **3,3** *(Lower left corner location.)*
> Specify other corner point or [Dimensions]: **6,5** *(Upper right corner location.)*

After specifying the first corner instead of specifying the diagonally opposite corner you can give the dimensions of the rectangle. This is possible by entering **D** at the **Specify other corner point or [Dimensions]** prompt which then allows you to enter the length and the width of the rectangle. The prompt sequence for drawing a rectangle with a length of **5** units and a width of **3** units is:

> Command: **RECTANG**
> Specify first corner point or [Chamfer/Elevation/Fillet/Thickness/Width]: **3,3**
> Specify other corner point or [Dimensions]: **D**
> Specify length for rectangles <0.0000>: **5**
> Specify width for rectangles <0.0000>: **3**

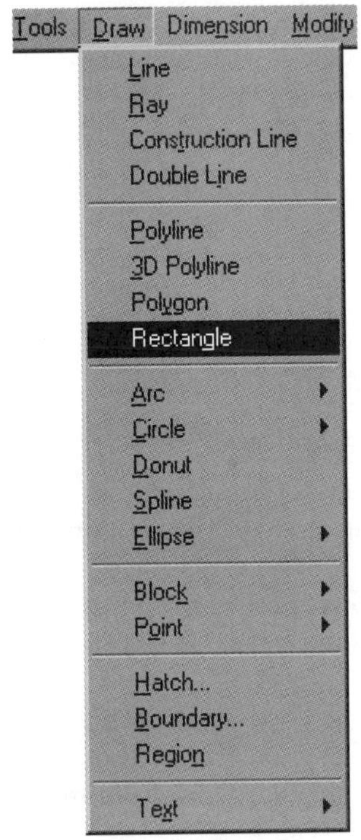

*Figure 3-25 Invoking the **RECTANG** command from the **Draw** menu*

Here you are allowed to choose any one of the four locations for placing the rectangle. You can move the cursor to see the four locations of the rectangle. Depending upon the location of the cursor, the specified first corner point holds the position of either the lower left corner, or the lower right corner, or the upper right corner, or the upper left corner. After deciding the position you can then click to place the rectangle.

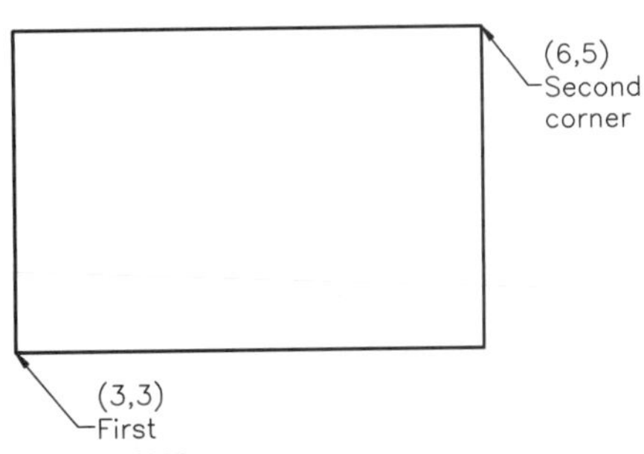

*Figure 3-26 Drawing a rectangle using the **RECTANG** command*

Chamfer

The **Chamfer** option creates a chamfer, which is an angled corner, by specifying the chamfer distances (Figure 3-27). The chamfer is created at all the four corners. You can give two different chamfer values to create an unequal chamfer.

> Command: **RECTANG**
> Specify first corner point or [Chamfer/Elevation/Fillet/Thickness/Width]: **C**

Specify first chamfer distance for rectangles <0.0000>: *Enter a value.*
Specify second chamfer distance for rectangles <0.0000>: *Enter a value.*
Specify first corner point or [Chamfer/Elevation/Fillet/Thickness/Width]: *Select a point as lower left corner location.*
Specify other corner point or [Dimensions]: *Select a point as upper right corner location.*

Note

The first corner point that you specify need not be the lower left corner location. While selecting the other corner you can select a location such that the first corner point becomes the lower right corner or the upper left corner or the upper right corner.

Fillet

The **Fillet** option allows you to create a filleted rectangle by specifying the fillet radius (Figure 3-28). A fillet is a rounded corner which is created at all the four corners of the rectangle.

Specify first corner point or [Chamfer/Elevation/Fillet/Thickness/Width]: **F** Enter
Specify fillet radius for rectangles <0.0000>: *Enter a value.*

Now, if you draw a rectangle it will be filleted provided the length and width of the rectangle is equal to or greater than twice the value of the specified fillet. Otherwise, AutoCAD LT will draw a rectangle without fillet.

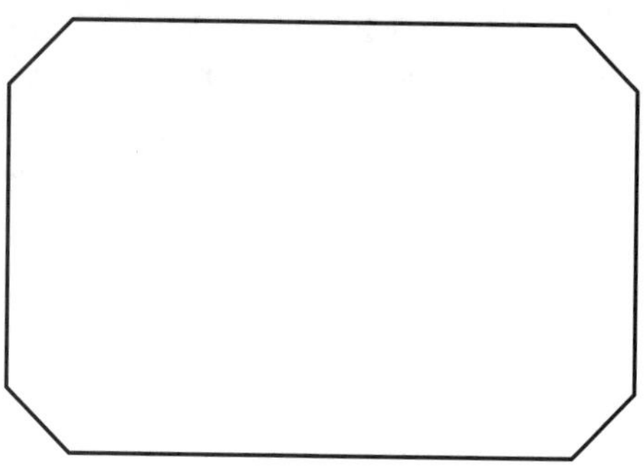

Figure 3-27 *Drawing a rectangle with chamfer*

Figure 3-28 *Drawing a rectangle with fillet*

Note

You can draw a rectangle either with chamfers or with fillets. If you specify the chamfer distances first and then specify the fillet radius in the same ***RECTANG*** *command, the rectangle will be drawn with fillets only.*

Width

The **Width** option allows you to control the line width of the rectangle by specifying the width (Figure 3-29).

Specify first corner point or [Chamfer/Elevation/Fillet/Thickness/Width]: **W** Enter

Specify lines width for rectangles <0.0000>: *Enter a value.*

Thickness

The **Thickness** option allows you to draw a rectangle that is extruded in the Z direction by the specified value of thickness. For example, if you draw a rectangle with thickness of 2 units, you will get a rectangular box whose height is 2 units (Figure 3-30). To view the box, use the **SE Isometric View** button in the **Named View** flyout of the **Standard** toolbar.

Specify first corner point or [Chamfer/Elevation/Fillet/Thickness/Width]: **T** Enter
Specify thickness for rectangles <0.0000>: *Enter a value.*

Elevation

The **Elevation** option allows you to draw a rectangle at a specified distance from the XY plane along the Z axis. For example, if the elevation is 2 units, the rectangle will be drawn two units above the XY plane. If the thickness of the rectangle is 1 unit, you will get a rectangular box of 1 unit height located 2 units above the XY plane (Figure 3-30).

Chamfer/Elevation/Fillet/Thickness/Width/<First corner>: **E** Enter
Specify elevation for rectangles <0.0000>: *Enter a value.*

To view the objects in 3D space, change the viewpoint (Select **View, 3D Views > SE Isometric**).

Figure 3-29 *Drawing a rectangle with specified width*

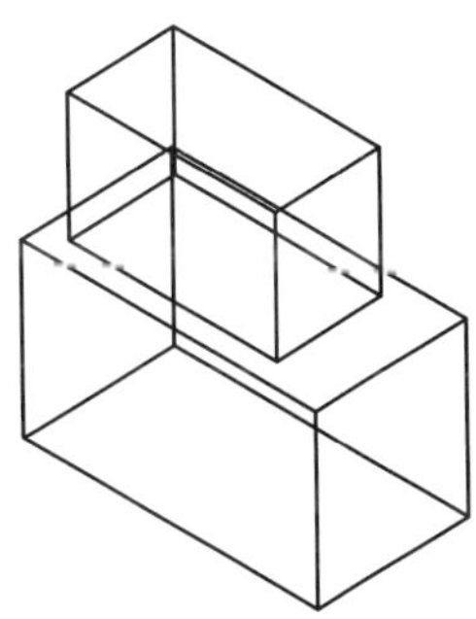

Figure 3-30 *Drawing rectangles with specified thickness and elevation*

Note

*You can change back your viewpoint from SE Isometric by choosing the **Top View** button in the **Named View** flyout of the **Standard Toolbar** or by selecting **Top** in the **View > 3D Views** menu.*

The value you enter for Fillet, Width, Elevation, and Thickness becomes the current value for the subsequent rectangle command. Therefore, you must first reset the values if they are different from the current values. The thickness of rectangle is always controlled by its thickness settings.

*The rectangle generated on the screen is treated as a single object. Therefore, the individual sides can be edited only after the rectangle has been exploded using the **EXPLODE** command.*

Tip

You can combine the different options in one ***RECTANG*** *command and then draw the rectangle with the specified characteristics. When you invoke the* ***RECTANG*** *command again the previously set options and their values are displayed before the first prompt. This allows you to change the settings according to the new specifications.*

Exercise 6 *General*

Draw a rectangle 4 units long, 3 units wide, and with its first corner at (1,1). Draw another rectangle of length 2 units and width 1 unit, and with first corner at 1.5,1.5.

DRAWING ELLIPSES

Toolbar:	Draw > Ellipse
Menu:	Draw > Ellipse
Command:	ELLIPSE

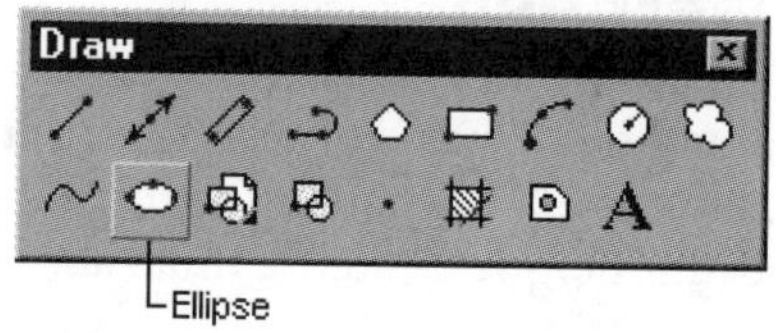

Figure 3-31 Invoking the ***ELLIPSE*** *command from the* ***Draw*** *toolbar*

If a circle is observed from an angle, the shape seen is called an **Ellipse**. An ellipse can be created using various options listed within the **ELLIPSE** command. AutoCAD LT creates a true ellipse, also known as a NURBS-based (Non-Uniform Rational Bezier Spline) ellipse. The true ellipse has a center and quadrant points. If you select it, the grips (small blue squares) will be displayed at the center and the quadrant points of the ellipse. If you move one of the grips located on the perimeter of the ellipse, the major or minor axis will change, which changes the size of the ellipse, as shown in Figure 3-32(d). The creation of a true ellipse is dependent on the **PELLIPSE** system variable that has a value 0 by default.

Note

Up to AutoCAD LT 2.0 the ellipses were based on polylines. They were made of multiple polyarcs and as a result, it was difficult to edit an ellipse. For example, if you select a polyline-based ellipse, the grips will be displayed at the endpoints of each polyarc. If you move a vertex point, you get the shape shown in Figure 3-32(b). Also, you cannot snap to the center or the quadrant points of a polyline-based ellipse. In AutoCAD LT 2002, you can still draw the polyline-based ellipse by setting the value of the ***PELLIPSE*** *system variable to* ***1*** *which is* ***0*** *(true ellipse) by default.*

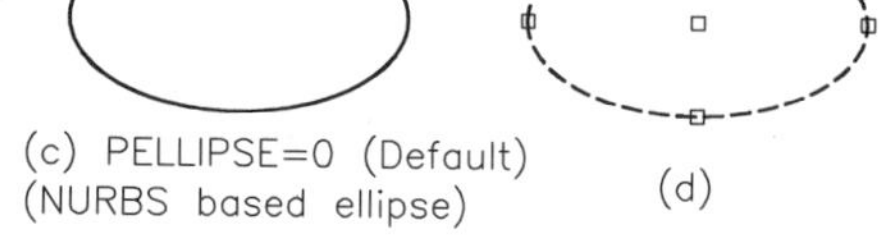

Figure 3-32 Drawing polyline and NURBS-based ellipses

Once you invoke the **ELLIPSE** command, AutoCAD LT will acknowledge with the prompt **Specify axis endpoint of ellipse or [Arc/Center]** or **Specify axis endpoint of ellipse or [Arc/Center/Isocircle]** (if Isometric snap is on). The response to this prompt depends on the option you want to choose. The different options are explained next.

Note

*The **Isocircle** option is not available by default in the **ELLIPSE** command. To display this option, you have to select the **Isometric snap** radio button in the **Snap and Grid** tab of the **Drafting Settings** dialog box.*

*The **Arc** option is not available if you set the **PELLIPSE** system variable to 1 for drawing the polyline-based ellipse.*

Drawing an Ellipse Using the Axis and Endpoint Option

In this option you draw an ellipse by specifying one of its axes and the endpoint of the other axis. To use this option, acknowledge the **Specify axis endpoint of ellipse or [Arc/Center]** prompt by specifying a point, either by specifying a point using a pointing device or by entering its coordinates. This is the first endpoint of one axis of the ellipse. AutoCAD LT will then respond with the prompt **Specify other endpoint of axis**. Here, specify the other endpoint of the axis. The angle at which the ellipse is drawn depends on the angle made by these two axis endpoints. Your response to the next prompt determines whether the axis is the **major axis** or the **minor axis**.

The next prompt is **Specify distance to other axis or [Rotation]**. If you specify a distance, it is presumed as half the length of the second axis. You can also specify a point. The distance from this point to the midpoint of the first axis is again taken as half the length of this axis. The ellipse will pass through the selected point only if it is perpendicular to the midpoint of the first axis. To visually analyze the distance between the selected point and the midpoint of the first axis, AutoCAD LT appends an elastic line to the crosshairs, with one end fixed at the midpoint of the first axis. You can also drag the point, dynamically specifying half of the other axis distance. This helps you to visualize the ellipse. The prompt sequence for drawing an ellipse with one axis endpoint located at (3,3), the other at (6,3), and the distance of the other axis at 1 (Figure 3-33) is:

Command: **ELLIPSE** Enter
Specify axis endpoint of ellipse or [Arc/Center]: **3,3** Enter
Specify other endpoint of axis : **6,3** Enter
Specify distance to other axis or [Rotation]: **1** Enter

Another example for drawing an ellipse (Figure 3-34) using this option is illustrated by the following prompt sequence:

Command: **ELLIPSE** Enter
Specify axis endpoint of ellipse or [Arc/Center]: **3,3** Enter
Specify other endpoint of axis : **4,2** Enter
Specify distance to other axis or [Rotation]: **2** Enter

If you enter Rotation or R at the **Specify distance to other axis or [Rotation]** prompt, the first axis specified is automatically taken as the major axis of the ellipse. The next prompt is **Specify rotation around major axis**. The major axis is taken as the diameter line of the circle, and the rotation takes place around this diameter line into the third dimension. The ellipse is formed when AutoCAD LT projects this rotated circle into the drawing plane. You can enter

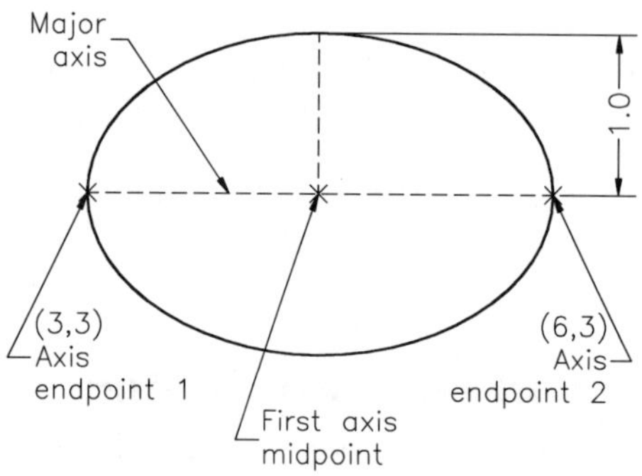

Figure 3-33 *Drawing an ellipse using the Axis and Endpoint option*

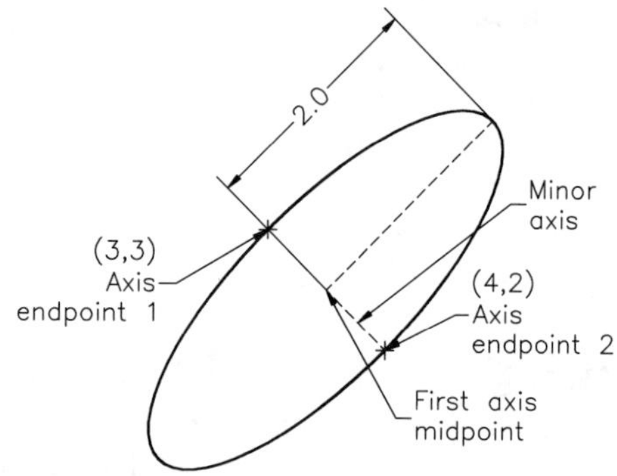

Figure 3-34 *Drawing an ellipse using the Axis and Endpoint option*

the rotation angle value in the range of 0 to 89.4 degrees only, because an angle value greater than 89.4 degrees changes the circle into a line. Instead of entering a definite angle value at the **Specify rotation around major axis** prompt, you can specify a point relative to the midpoint of the major axis. This point can be dragged to specify the ellipse dynamically. The following is the prompt sequence for a rotation of 0 degrees around the major axis, as shown in Figure 3-35(a).

Command: **ELLIPSE** [Enter]
Specify axis endpoint of ellipse or [Arc/Center]: *Select point (P1).*
Specify other endpoint of axis: *Select another point (P2).*
Specify distance to other axis or [Rotation]: **R** [Enter]
Specify rotation around major axis: **0** [Enter]

Figure 3-35 also shows rotations of 45 degrees, 60 degrees, and 89.4 degrees.

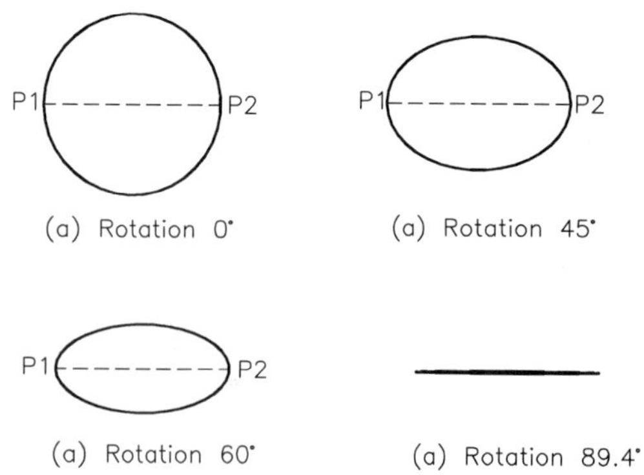

Figure 3-35 *Rotation about the major axis*

Exercise 7 *General*

Draw an ellipse whose major axis is 4 units and whose rotation around this axis is 60 degrees. Draw another ellipse whose rotation around the major axis is 15 degrees.

Drawing Ellipse Using the Center and Two Axes Option

In this option you can construct an ellipse by specifying the center point, the endpoint of one axis, and the length of the other axis. The only difference between this method and the ellipse by axis and endpoint method is that instead of specifying the second endpoint of the first axis, the center of the ellipse is specified. The center of an ellipse is defined as the point of intersection of the major and minor axes. In this option, the first axis need not be the major axis. For example, to draw an ellipse with center at (4,4), axis endpoint at (6,4), and length of the other axis as 2 units (Figure 3-36), the prompt sequence is:

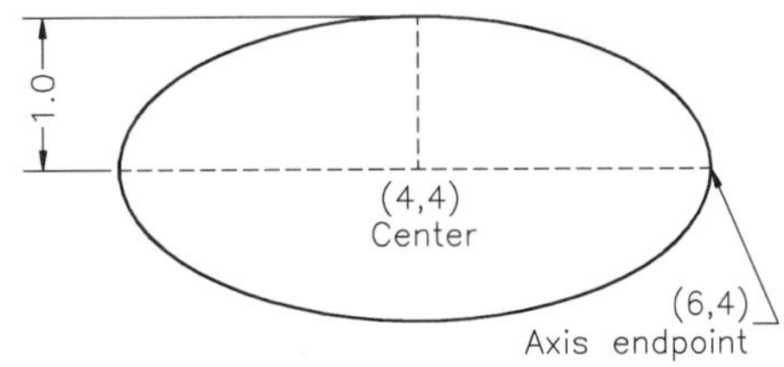

Figure 3-36** Drawing an ellipse using the* ***Center *option*

Command: **ELLIPSE** [Enter]
Specify axis endpoint of ellipse or [Arc/Center]: **C** [Enter]
Specify center of ellipse: **4,4** [Enter]
Specify endpoint of axis: **6,4** [Enter]
Specify distance to other axis or [Rotation]: **1** [Enter]

Instead of entering the distance, you can enter Rotation or R at the **Specify distance to other axis [Rotation]** prompt. This takes the first axis specified as the major axis. The next prompt, **Specify rotation around major axis**, prompts you to enter the rotation angle value. The rotation takes place around the major axis, which is taken as the diameter line of the circle. The rotation angle values should range from 0 to 89.4 degrees.

Drawing Elliptical Arcs

Menu:	Draw > Ellipse > Arc
Command:	ELLIPSE > Arc

You can use the **Arc** option of the **ELLIPSE** command to draw an elliptical arc. When you choose **Ellipse > Arc** from the **Draw** menu, the **ELLIPSE** command is invoked with the **Arc** option selected. AutoCAD LT will prompt you to enter information about the geometry of the ellipse and the arc limits. You can define the arc limits by using the following options:

1. Start and End angle of the arc.

2. Start and Included angle of the arc.

3. Start and End parameters.

The angles are measured from the first point and in a counterclockwise direction if AutoCAD LT's default setup has not been changed. The following example illustrates the use of these three options.

Chapter 3

Example 1

General

Draw the following elliptical arcs as shown in Figure 3-37 and Figure 3-38.

a. Start angle = -45, end angle = 135
b. Start angle = -45, included angle = 225
c. Start parameter = @1,0, end parameter = @1<225

Specifying Start and End Angle of the Arc [Figure 3-37(a)]

Command: *Choose* ***Ellipse > Arc*** *from the* ***Draw*** *menu.*
Specify axis endpoint of ellipse or [Arc/Center]: _a
Specify axis endpoint of elliptical arc or [Center]: *Select the first endpoint.*
Specify other endpoint of axis: *Select the second point.*
Specify distance to other axis or [Rotation]: *Select a point or enter a distance.*
Specify start angle or [Parameter]: **-45** [Enter]
Specify end angle or [Parameter/Included angle]: **135** [Enter] *(Angle where arc ends.)*

Specifying Start and Included Angle of the Arc [Figure 3-37(b)]

Command: *Choose* ***Ellipse > Arc*** *from the* ***Draw*** *menu.*
Specify axis endpoint of ellipse or [Arc/Center]: _a
Specify axis endpoint of elliptical arc or [Center]: *Select the first endpoint.*
Specify other endpoint of axis : *Select the second point.*
Specify distance to other axis or [Rotation]: *Select a point or enter a distance.*
Specify start angle or [Parameter]: **-45** [Enter]
Specify end angle or [Parameter/Included angle]: **I** [Enter]
Specify included angle for arc<current>: 225 [Enter] *(Included angle.)*

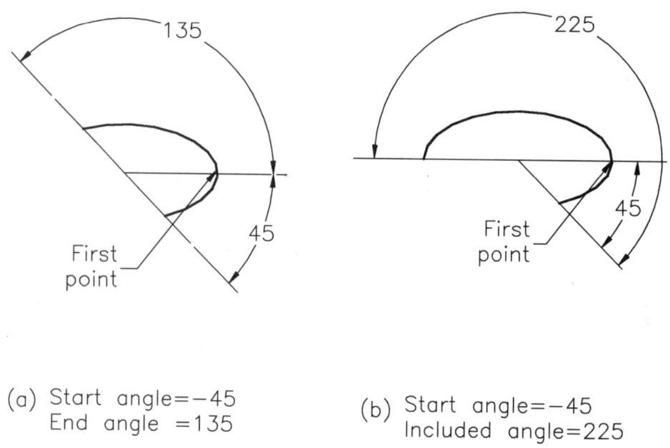

Figure 3-37 *Drawing elliptical arcs*

Specifying Start and End Parameters (Figure 3-38):

Command: *Choose **Ellipse > Arc** from the **Draw** menu.*
Specify axis endpoint of ellipse or [Arc/Center]: _a
Specify axis endpoint of elliptical arc or [Center]: *Select the first endpoint.*
Specify other endpoint of axis: *Select the second endpoint.*
Specify distance to other axis or [Rotation]: *Select a point or enter a distance.*
Specify start angle or [Parameter]: **P**
Specify start parameter or [Angle]: **@1,0**
Specify end parameter or [Angle/ Included angle]: **@1<225**

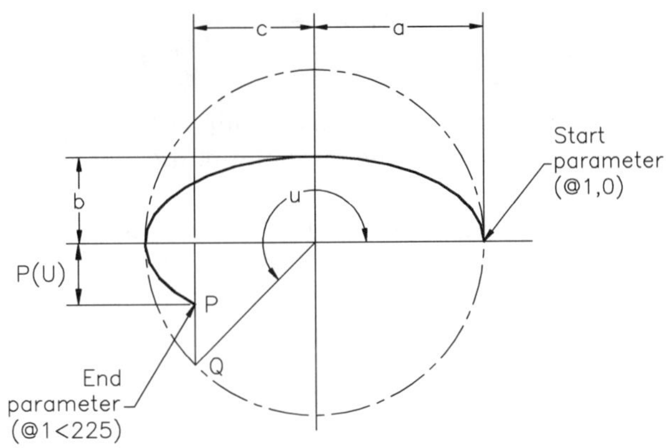

***Figure 3-38** Drawing an elliptical arc by specifying the start and end parameters*

Calculating Parameters for an Elliptical Arc

The start and end parameters of an elliptical arc are determined by specifying a point on the circle whose diameter is equal to the major diameter of the ellipse as shown in Figure 3-38. In this drawing, the major axis of the ellipse is 2.0 and the minor axis is 1.0. The diameter of the circle is 2.0. To determine the start and end parameters of the elliptical arc, you must specify the points on the circle. In the example, the start parameter is @1,0 and the end parameter is @1<225. Once you specify the points on the circle, AutoCAD LT will project these points on the major axis and determine the endpoint of the elliptical arc. In the Figure 3-38, Q is the end parameter of the elliptical arc. AutoCAD LT projects point Q on the major axis and locates intersection point P, which is the endpoint of the elliptical arc. The coordinates of point P can be calculated by using the following equations:

The equation of an ellipse with center as origin is
x^2/a^2 + y^2/b^2 = 1
In parametric form x = a * cos(u)
y = b * sin(u)
For the example a = 1
b = 0.5
Therefore x = 1 * cos(225) = -0.707

y = 0.5 * sin(225) = -0.353

The coordinates of point P are (-0.707, -0.353) with respect to the center of the ellipse.

Note: v = atan(b/a*tan(u)) = end angle

v = atan(0.5/1*tan(225)) = 206.56o

Also e = 1-b^2/a^2)^.5 = eccentricity

e = 1-.5^2/1^2)^.5 = .866

r = x^2 + y^2)^.5

r = .707^2 + .353^2)^.5 = 0.790

or using the polar equation r = b/(1 - e^2 * cos(v)^2)^.5

r = .5/(1 - .866^2 * cos(206.56)^2)^.5

r = 0.790

Exercise 8 — *General*

a. Construct an ellipse with a center at (2,3), axis endpoint at (4,6), and the other axis endpoint a distance of 0.75 units from the midpoint of the first axis.

b. Make the drawing as shown in Figure 3-39. The distance between the dotted lines is 1.0 unit. Create the elliptical arcs using the **ELLIPSE** command options.

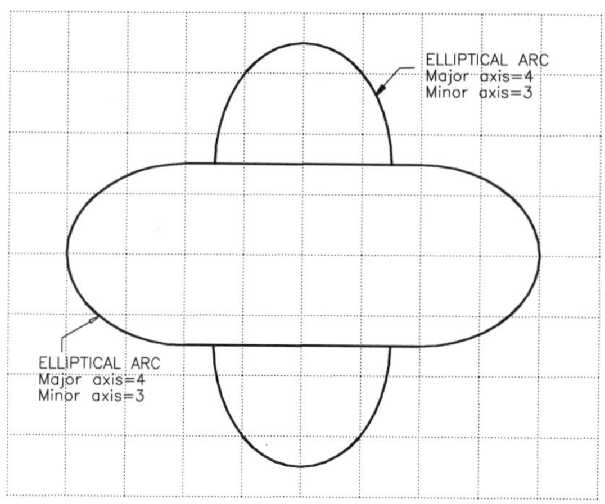

Figure 3-39 *Drawing for Exercise 8(b)*

DRAWING REGULAR POLYGONS

Toolbar:	Draw > Polygon
Menu:	Draw > Polygon
Command:	POLYGON

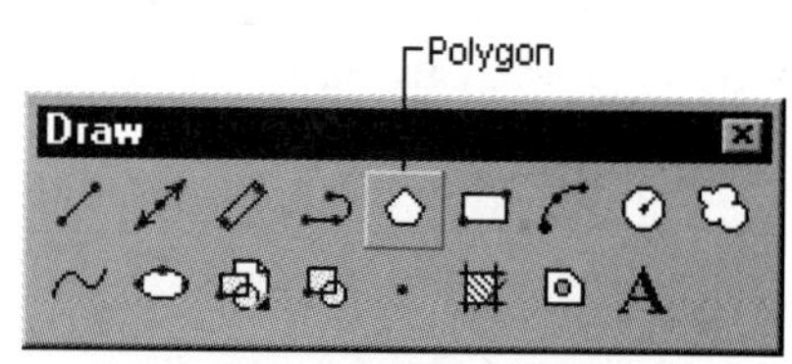

Figure 3-40 *Invoking the* ***POLYGON*** *command from the* ***Draw*** *toolbar*

A **regular polygon** is a closed geometric figure with equal sides and equal angles. The number of sides varies from **3** to **1024**. For example, a triangle is a three-sided polygon and a pentagon is a five-sided polygon. In AutoCAD LT, the **POLYGON** command (Figure 3-40) is used to draw regular two-dimensional polygons. The characteristics of a polygon drawn in AutoCAD LT are those of a closed polyline having 0 width. You can change the width of the polyline forming the polygon. The prompt sequence is:

Command: **POLYGON** Enter
Enter number of sides <4>:

Once you invoke the **POLYGON** command, it prompts you to enter the number of sides. The number of sides determines the type of polygon (for example, six sides define a hexagon). The default value for the number of sides is **4**. You can change the number of sides to your requirement and then the new value becomes the default. You can also set a different default value for the number of sides by using the **POLYSIDES** system variable.

The Center of Polygon Option

After you specify the number of sides, the next prompt is **Specify center of polygon or [Edge]**, where the default option prompts you to select a point that is taken as the center point of the polygon. The next prompt is **Enter an option [Inscribed in circle/Circumscribed about circle]<I>**. A polygon is said to be **inscribed** when it is drawn inside an imaginary circle and its vertices (corners) touch the circle (Figure 3-41). Likewise, a polygon is **circumscribed** when it is drawn outside the imaginary circle and the sides of the polygon are tangent to the circle (the midpoint of each side of the polygon will lie on the circle) (Figure 3-42). If you want to have an inscribed polygon, enter **I** at the prompt. The next prompt issued is **Specify radius of circle**. Here you are required to specify the radius of the circle on which all the vertices of the polygon will lie. Once you specify the radius, a polygon will be generated. If you want to select the circumscribed option, enter **C** at the prompt **Enter an option[Inscribed in circle/ Circumscribed about circle]<I>**. After this, enter the radius of the circle. The inscribed or circumscribed circle is not drawn on the screen. The radius of the circle can be dynamically dragged instead of a numerical value entered. The prompt sequence for drawing an inscribed octagon with a center at (4,4), and a radius of 1.5 units (Figure 3-41) is:

Command: **POLYGON** Enter
Enter number of sides<4>: **8** Enter
Specify center of polygon or [Edge]: **4,4** Enter
Enter an option[Inscribed in circle/Circumscribed about circle]<I>: **I** Enter
Specify radius of circle: **1.5** Enter

The prompt sequence for drawing a circumscribed pentagon with a center at (4,4) and a radius of 1.5 units (Figure 3-42) is:

Command: **POLYGON** Enter
Enter number of sides<4>: **5** Enter
Specify center of polygon or [Edge]: **4,4** Enter
Enter an option[Inscribed in circle/Circumscribed about circle]<I>: **C** Enter
Specify radius of circle: **1.5** Enter

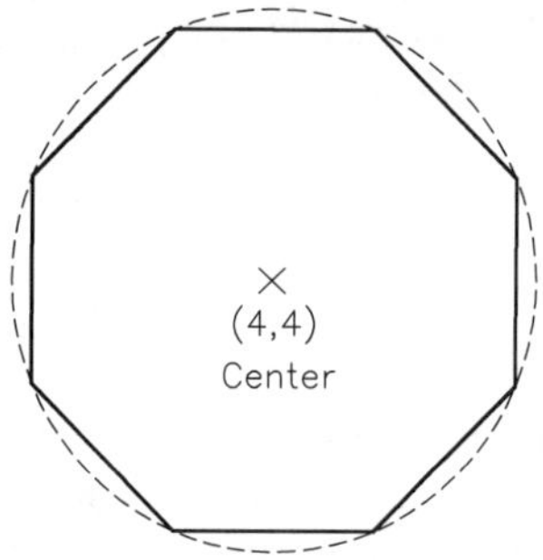

*Figure 3-41 Drawing an inscribed polygon using the **Center of Polygon** option*

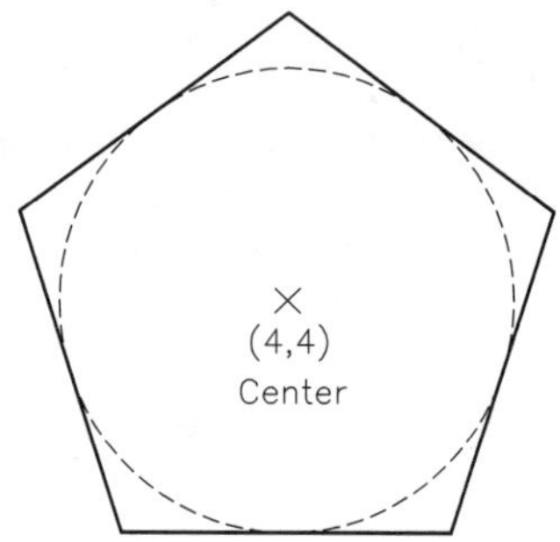

*Figure 3-42 Drawing a circumscribed polygon using the **Center of Polygon** option*

Note

If you select a point to specify the radius of an inscribed polygon, one of the vertices is positioned on the selected point. In the case of circumscribed polygons, the midpoint of an edge is placed on the point you have specified. In this manner you can specify the size and rotation of the polygon.

In the case of numerical specification of the radius, the bottom edge of the polygon is rotated by the prevalent snap rotation angle.

Exercise 9 — *General*

Draw a circumscribed polygon of eight sides. Use the **Center of Polygon** method.

The Edge Option

The other method for drawing a polygon is to select the **Edge** option. This can be done by entering E at the **Specify center of polygon or [Edge]** prompt. The next two prompts issued are **Specify first endpoint of edge** and **Specify second endpoint of edge**. Here you need to specify the two endpoints of an edge of the polygon. The polygon is drawn in a counterclockwise direction, with the two points entered defining its first edge. To draw a hexagon (six-sided polygon) using the Edge option, with the first endpoint of the edge at (2,4) and the second endpoint of the edge at (2,2.5) (Figure 3-43), the prompt sequence is:

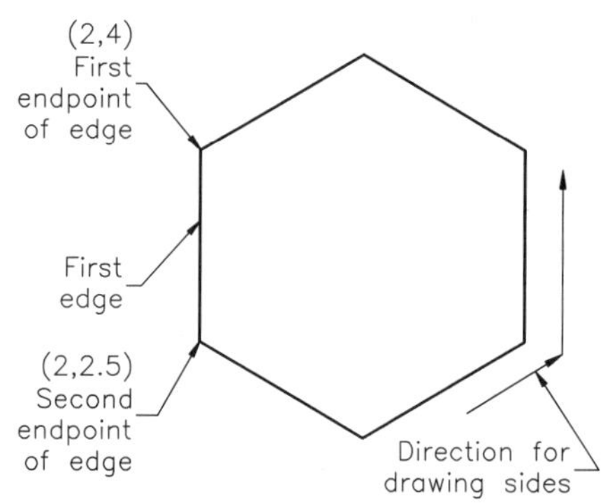

*Figure 3-43 Drawing a polygon (hexagon) using the **Edge** option*

Command: **POLYGON** [Enter]
Enter number of sides<4>: **6** [Enter]
Specify center of polygon or [Edge]: **E** [Enter]
Specify first endpoint of edge: **2,4** [Enter]
Specify second endpoint of edge: **2,2.5** [Enter]

Exercise 10 General

Draw a polygon with ten sides using the **Edge** option and an elliptical arc as shown in Figure 3-44. Let the first endpoint of the edge be at (7,1) and the second endpoint be at (8,2).

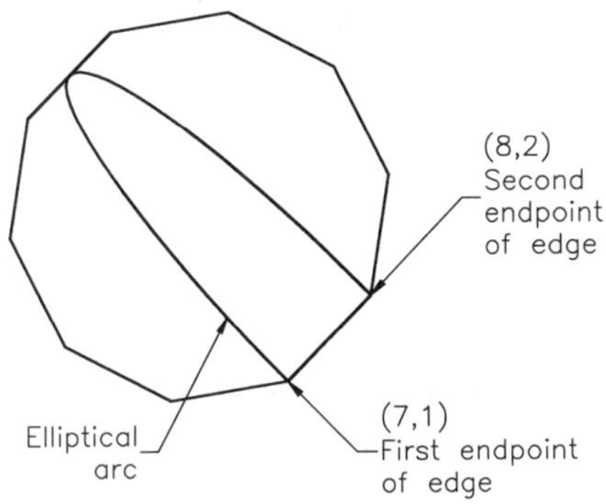

***Figure 3-44** Polygon and elliptical arc for Exercise 10*

DRAWING POLYLINES

Toolbar:	Draw > Polyline
Menu:	Draw > Polyline
Command:	PLINE

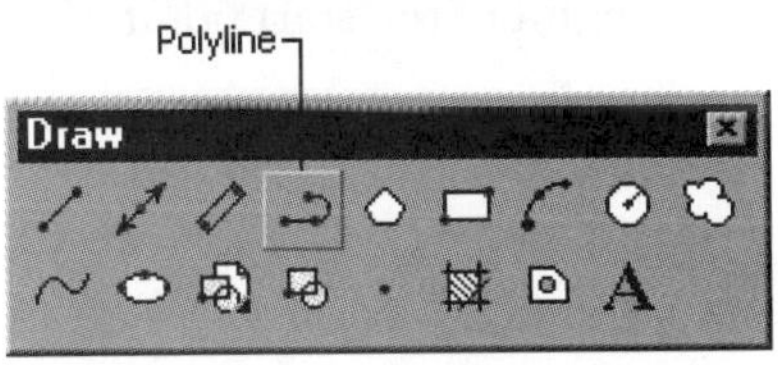

***Figure 3-45** Invoking the **Polyline** command from the **Draw** toolbar*

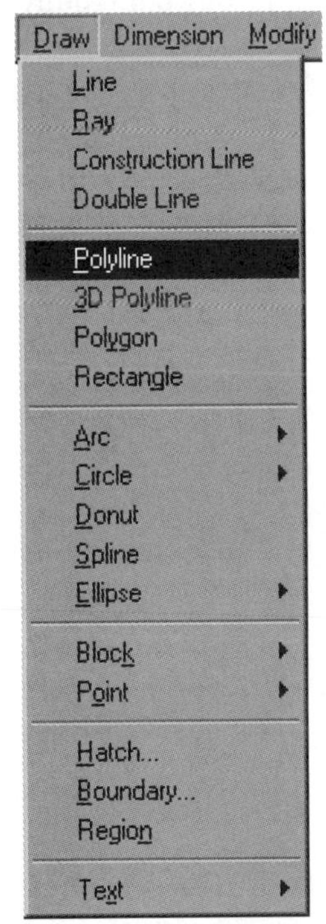

***Figure 3-46** Invoking the **Polyline** command from the **Draw** menu*

A **polyline** (Figure 3-45 and Figure 3-46) is a line that can have different characteristics. The term POLYLINE can be broken into two parts: POLY and LINE. POLY means "many". This signifies that a polyline can have many features. Some of the features of polylines are:

1. Polylines, like traces, are thick lines having a desired width.
2. Polylines are very flexible and can be used to draw any shape, such as a filled circle or a donut.
3. Polylines can be used to draw objects in any linetype (for example, hidden linetype).
4. Advanced editing commands can be used to edit polylines (for example, the **PEDIT** command).

5. A single polyline object can be formed by joining polylines and polyarcs of different thicknesses.
6. It is easy to determine the area or perimeter of a polyline feature. Also, it is easy to offset when drawing walls.

The **PLINE** command functions fundamentally like the **LINE** command, except that additional options are provided and all the segments of the polyline form a single object. After invoking the **PLINE** command, the next prompt is:

Specify start point: *Specify the starting point or enter its coordinates.*
Current line width is nn.n.

Current line width is nn.n is displayed automatically, which indicates that the polyline drawn will have nn.n width. If you want the polyline to have a different width, invoke the Width option at the next prompt and then set the polyline width. The next prompt is:

Specify next point or [Arc/Halfwidth/Length/Undo/Width]: *Specify next point or enter an option.*

Depending on your requirements, the options that can be invoked at this prompt are as follows.

Next Point of Line

This option is maintained as the default and is used to specify the next point of the current polyline segment. If additional polyline segments are added to the first polyline, AutoCAD LT automatically makes the endpoint of the previous polyline segment the start point of the next polyline segment. The prompt sequence is:

Command: **PLINE** [Enter]
Specify start point: *Specify the starting point of the polyline.*
Current line width is 0.0000.
Specify next point or [Arc/Halfwidth/Length/Undo/Width]: *Specify the endpoint of the first polyline segment.*
Specify next point or [Arc/Halfwidth/Length/Undo/Width]: *Specify the endpoint of the second polyline segment, or press ENTER to exit the command.*

Width

You can change the current polyline width by entering W (width option) at the last prompt. You can also right-click and choose the Width option from the shortcut menu. Then you are prompted for the starting width and the ending width of the polyline.

Specify next point or [Arc/Close/Halfwidth/Length/Undo/Width]: **W** [Enter]
Specify starting width <current>: *Specify the starting width.*
Specify ending width <starting width>: *Specify the ending width.*

The starting width value is taken as the default value for the ending width. Hence, to have a uniform polyline you need to press ENTER at the **Specify ending width <>** prompt. As in the case of traces, the start point and endpoint of the polyline are located at the center of the line width. To draw a uniform polyline (Figure 3-47) with a width of 0.25 units, start point at (4,5), endpoint at (5,5), and the next endpoint at (3,3), the following is the prompt sequence:

Command: **PLINE** Enter
Specify start point: **4,5** Enter
Current line-width is 0.0000
Specify next point or [Arc/Halfwidth/Length/Undo/Width]: **W** Enter
Specify starting width <current>: **0.25** Enter
Specify ending width <0.25>: Enter
Specify next point or [Arc/Halfwidth/Length/Undo/Width]: **5,5** Enter
Specify next point or [Arc/Close/Halfwidth/Length/Undo/Width]: **3,3** Enter
Specify next point or [Arc/Close/Halfwidth/Length/Undo/Width]: Enter

You can get a tapered polyline by entering two different values at the starting width and the ending width prompts. To draw a tapered polyline (Figure 3-48) with a starting width of 0.5 units and an ending width of 0.15 units, a start point at (2,4), and an endpoint at (5,4), the prompt sequence is:

Command: **PLINE** Enter
Specify start point: **2,4** Enter
Current line-width is 0.0000
Specify next point or [Arc/Halfwidth/Length/Undo/Width]: **W** Enter
Specify starting width <0.0000>: **0.50** Enter
Specify ending width <0.50>: **0.15** Enter
Specify next point or [Arc/Halfwidth/Length/Undo/Width]: **5,4** Enter
Specify next point or [Arc/Close/Halfwidth/Length/Undo/Width]: Enter

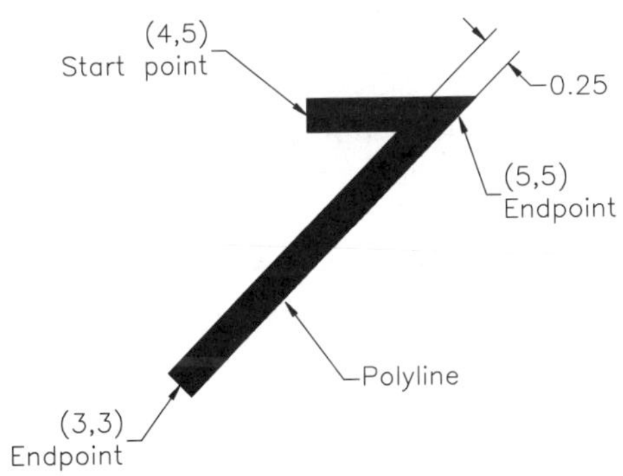

***Figure 3-47** Drawing a uniform polyline using the **PLINE** command*

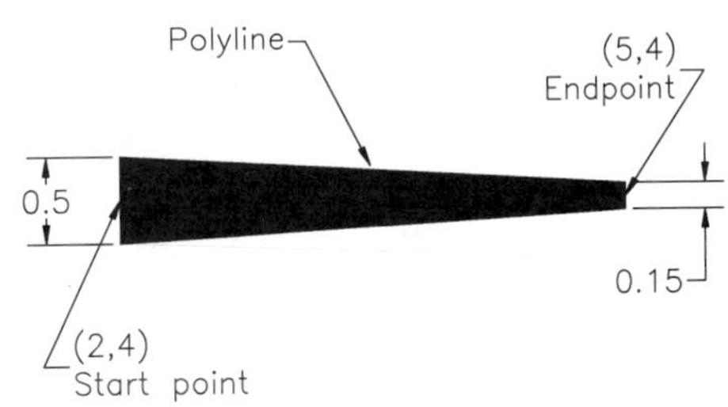

***Figure 3-48** Drawing a tapered polyline using the **PLINE** command*

Halfwidth

With this option you can specify the starting and ending halfwidth of a polyline. This halfwidth distance is equal to half of the actual width of the polyline. This option can be invoked by entering H or choosing **Halfwidth** from the shortcut menu at the following prompt:

Specify next point or [Arc/Close/Halfwidth/Length/Undo/Width]: **H**
Specify starting half-width <0.0000>: **0.12** *(Specify desired starting halfwidth.)*
Specify ending half-width <0.1200>: **0.05** *(Specify desired ending halfwidth.)*

Length

This option prompts you to enter the length of a new polyline segment. The new polyline segment will be the length you have entered. It will be drawn at the same angle as the last polyline segment or tangent to the previous polyarc segment. This option can be invoked by entering L at the following prompt or by choosing **Length** from the shortcut menu:

Specify next point or [Arc/Close/Halfwidth/Length/Undo/Width]: **L**
Specify length of line: *Specify the desired length of the Pline.*

Undo

This option erases the most recently drawn polyline segment. This option can be invoked by entering U at the following prompt:

Specify next point or [Arc/Close/Halfwidth/Length/Undo/Width]: **U**

You can use this option repeatedly until you reach the start point of the first polyline segment. Further use of the **Undo** option evokes the message **All segments already undone**.

Close

This option is available when two segments of the polyline are drawn. It closes the polyline by drawing a polyline segment from the most recent endpoint to the initial start point, and when doing so exits from the **PLINE** command. The width of the closing segment can be changed by using the Width/Halfwidth option before invoking the Close option.

Arc

This option is used to switch from drawing polylines to drawing polyarcs, and provides you the options associated with drawing polyarcs. The prompt sequence is:

Specify next point or [Arc/Close/Halfwidth/Length/Undo/Width]: **A**
Specify endpoint of arc or [Angle/CEnter/Direction/Halfwidth/Line/Radius/Second pt/Undo/Width]: *Enter an option.*

By default the arc segment is drawn tangent to the previous segment of the polyline. The direction of the previous line, arc, or polyline segment is the default for the polyarc. The preceding prompt contains options associated with the PLINE Arc. The detailed explanation of each of these options follows.

Angle

This option prompts you to enter the included angle for the arc. If you enter a positive angle, the arc is drawn in a counterclockwise direction from the start point to the endpoint. If the angle specified is negative, the arc is drawn in a clockwise direction. The prompts are:

Specify included angle: *Specify the included angle.*
Specify endpoint of arc or [Center/Radius]:

Center refers to the center of the arc segment, Radius refers to the radius of the arc, and Endpoint draws the arc.

CEnter

This option prompts you to specify the center of the arc to be drawn. As mentioned before, usually the arc segment is drawn so that it is tangent to the previous polyline segment; in such cases AutoCAD LT determines the center of the arc automatically. Hence, the **CEnter** option provides the freedom to choose the center of the arc segment. The CEnter option can be invoked by entering **CE** at the **Specify end point of arc or [Angle/CEnter/Direction/Halfwidth/Line/Radius/Second pt/Undo/Width]** prompt. Once you specify the center point, AutoCAD LT issues the following prompt:

Specify endpoint of arc or [Angle/Length]:

Angle refers to the included angle, Length refers to the length of the chord, and Endpoint refers to the endpoint of the arc.

Direction

Usually, the arc drawn with the **PLINE** command is tangent to the previous polyline segment. In other words, the starting direction of the arc is the ending direction of the previous segment. The Direction option allows you to specify the tangent direction of your choice for the arc segment to be drawn. You can specify the direction by specifying a point. The prompts are:

Specify tangent direction for the start point of arc: *Specify the direction.*
Specify endpoint of arc: *Specify the endpoint of arc.*

Halfwidth

This option is the same as for the PLine and prompts you to specify the starting and ending halfwidth of the arc segment.

Line

This option takes you back to the **Line mode**. You can draw polylines only in Line mode.

Radius

This option prompts you to specify the **radius** of the arc segment. The prompt sequence is:

Specify radius of arc: *Specify the radius of the arc segment.*
Specify endpoint of arc or [Angle]:

If you specify a point, the arc segment is drawn. If you enter an angle, you will have to specify the angle and the direction of the chord at the **Specify included angle** and **Specify direction of chord for arc<current>** prompts, respectively.

Second pt

This option selects the second point of an arc in the three-point arc option. The prompt sequence is:

Specify second point of arc: *Specify the second point on the arc.*
Specify endpoint of arc: *Specify the third point on the arc.*

Undo

This option reverses the changes made in the previously drawn segment.

Width

This option prompts you to enter the width of the arc segment. To draw a tapered arc segment you can enter different values at the starting width and ending width prompts. The prompt sequence is identical to that for the polyline. Also, a specified point on a polyline refers to the midpoint on its width.

Endpoint of arc

This option is maintained as the default and prompts you to specify the endpoint of the current arc segment. The following is the prompt sequence for drawing an arc with a start point at (3,3), endpoint at (3,5), starting width of 0.50 units, and ending width of 0.15 units (Figure 3-49):

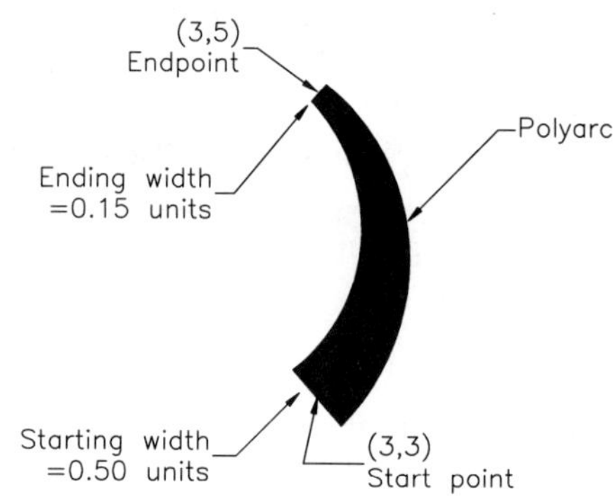

Figure 3-49 *Drawing a polyarc*

Command: **PLINE** **Enter**
Specify start point: **3,3** **Enter**
Current line-width is 0.0000
Specify nextpoint or [Arc/Close/
Halfwidth/Length/Undo/Width]: **A** **Enter**
Specify endpoint of arc or [Angle/
CEnter/CLose/Direction/Halfwidth/Line/Radius/Second pt/Undo/Width]: **W** **Enter**
Specify starting width <current>: **0.50** **Enter**
Specify ending width <0.50>: **0.15** **Enter**
Specify endpoint of arc or [Angle/CEnter/CLose/Direction/Halfwidth/Line/Radius/
Second pt/Undo/Width]: **3,5** **Enter**
Specify endpoint of arc or [Angle/CEnter/CLose/Direction/Halfwidth/Line/Radius/
Second pt/Undo/Width]: **Enter**

Tip

*After invoking the **PLINE** command and specifying the start point, you can right-click to display the shortcut menu. You can choose any Pline option directly from the shortcut menu instead of entering the appropriate letters at the Command prompt. Similarly after invoking the **Arc** option of the PLINE command, you can right-click to display the shortcut menu and choose any polyarc option.*

Note

*If FILL is On or if **FILLMODE** is 1, the polylines are drawn filled. If you change FILL to Off or **FILLMODE** to 0, only the outlines are drawn for the new polylines and previously drawn polylines are also changed from filled to empty, but the change is effective on regeneration. Similarly it works in reverse also.*

*Also, **PLINEGEN** system variable controls the linetype pattern between the vertex points of a 2D polyline. A value of 0 centers the linetype for each polyline segment and 1 makes them continuous.*

Optimized Polylines

The optimized polylines also known as **lightweight polylines** are created when you use the **PLINE** command (an AutoCAD LT feature since AutoCAD LT 97). The optimized polylines are similar to regular 2D polylines in functionality, but the data base format of an optimized polyline is different from a 2D polyline. In case of an optimized polyline, the vertices are not stored as separate entities, but as a single object with an array of information. This feature results in reduced object and file size.

If you use the **PEDIT** command (discussed in Chapter 18, Object Grouping and Editing Commands) on an optimized polyline and use the Spline or Fit options, the polyline automatically becomes a regular 2D polyline. Also, when you load an AutoCAD LT 2.0 or earlier drawing, AutoCAD LT automatically converts the polylines into optimized polylines. This is dependent on the **PLINETYPE** system variable. The default value of **2** for this variable automatically converts a 2D polyline into an optimized polyline while opening any old drawing.

Note

*Commands such as **POLYGON**, **DONUT**, **ELLIPSE**, **PEDIT**, **BOUNDARY**, and **RECTANG** create optimized polylines. To see the difference between these two polylines, draw a polyline consisting of several segments and then use the **LIST** command to list the information about the object. Now, use the **PEDIT** command to change this polyline into a splined polyline and then use the **LIST** command again. Notice the difference in the database associated with these polylines.*

CONVERT Command

You can use the **CONVERT** command to manually change a 2D polyline to an optimized (lightweight) polyline. You can also use this command to change an AutoCAD LT 2.0 associative hatch pattern to an AutoCAD LT 2002 hatch object, with the exception of the solid-fill hatches.

> Command: **CONVERT** [Enter]
> Enter type of objects to convert [Hatch/Polyline/All]: *Enter an option.*
> Enter object selection preference [Select/All] <All>:

AutoCAD LT will display a dialog box that warns you that hatch objects may change their appearance when converted. Choose the **Yes** button if you want to proceed with the conversion and choose the **No** button if you do not want to convert the polyline.

Tip

*The **CONVERT** command is used very rarely to change the polylines because the 2D polylines in any old drawing are automatically converted to lightweight polylines when you open the drawings in AutoCAD LT 2002. This is dependent on the **PLINETYPE** system variable that has a default value of 2. If you want the old 2D polylines to not get converted to lightweight polylines when you open the drawings, set the value of the **PLINETYPE** variable to 0. This also lets you create 2D polylines in AutoCAD LT 2002.*

Exercise 11 — *General*

Draw the objects shown in Figure 3-50 and Figure 3-51. Approximate the width of different polylines.

Figure 3-50 *Drawing for Exercise 11*

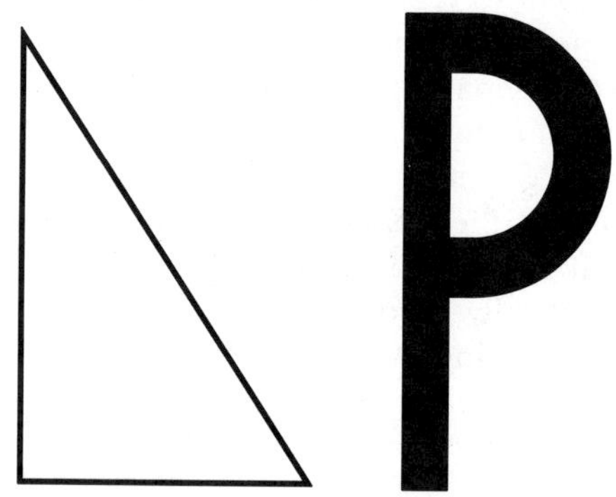

Figure 3-51 *Drawing for Exercise 11*

DRAWING DONUTS

Menu:	Draw > Donut
Command:	DONUT

In AutoCAD LT the **DONUT** command is issued to draw an object that looks like a filled circle ring called a donut. Actually, AutoCAD LT's donuts are made of two semicircular polyarcs having a certain width. Hence the **DONUT** command allows you to draw a thick circle. The donuts can have any inside and outside diameters. If the FILLMODE is off, the donuts look like circles (if the inside diameter is zero) or concentric circles (if the inside diameter is not zero). After specifying the two diameters, the donut gets attached to the crosshair. You can select a point for the center of the donut anywhere on the screen with the help of a pointing device, and then place the donut. You can place the donuts by clicking your pointing device. The prompt sequence for drawing donuts is:

Command: **DONUT** Enter
Specify inside diameter of donut <current>: *Specify the inner diameter of the donut.*
Specify outside diameter of donut <current>: *Specify the outer diameter of the donut.*
Specify center of donut or <exit>: *Specify the center of the donut.*
Specify center of donut or <exit>: *Specify the center of the donut to draw more donuts of previous specifications or give a null response to exit.*

The defaults for the inside and outside diameters are the respective diameters of the most recent donut drawn. The values for the inside and outside diameters are saved in the **DONUTID** and **DONUTOD** system variables. A solid-filled circle is drawn by specifying the inside diameter as zero (FILLMODE is on). Once the diameter specification is completed, the donuts are formed at the crosshairs and can be placed anywhere on the screen. For the location, you can enter the coordinates of the point or specify the point by dragging the center point. Once you have specified the center of the donut, AutoCAD LT repeats the **Specify center of donut or <exit>** prompt. As you go on specifying the locations for the center point, donuts with the specified diameters at specified locations are drawn. To end the **DONUT** command, give a null response to this prompt by pressing ENTER. Since donuts are circular polylines, the donut can be edited with the **PEDIT** command or any other editing command that can be used to edit polylines.

Example 2 *General*

You will draw an unfilled donut with an inside diameter of 0.75 units, an outside diameter of 2.0 units, and centered at (2,2) (Figure 3-52). You will also draw a filled donut and a solid-filled donut with the specifications given.

The following is the prompt sequence to draw an unfilled donut (Figure 3-52):

Command: **FILLMODE** [Enter]
New value for FILLMODE <1>: **0** [Enter]

Command: **DONUT** [Enter]
Specify inside diameter of donut<0.5000>: **0.75** [Enter]
Specify outside diameter of donut <1.000>: **2** [Enter]
Specify center of donut or <exit>: **2,2** [Enter]
Specify center of donut or <exit>: [Enter]

The following is the prompt sequence for drawing a filled donut with an inside diameter of 0.5 units, outside diameter of 2.0 units, centered at a specified point (Figure 3-53).

Command: **FILLMODE** [Enter]
Enter new value for FILLMODE <0>: **1** [Enter]

Command: **DONUT** [Enter]
Specify inside diameter of donut<0.5000>: **0.50** [Enter]
Specify outside diameter of donut <1.000>: **2** [Enter]
Specify center of donut or <exit>: *Specify a point.*
Specify center of donut or <exit>: [Enter]

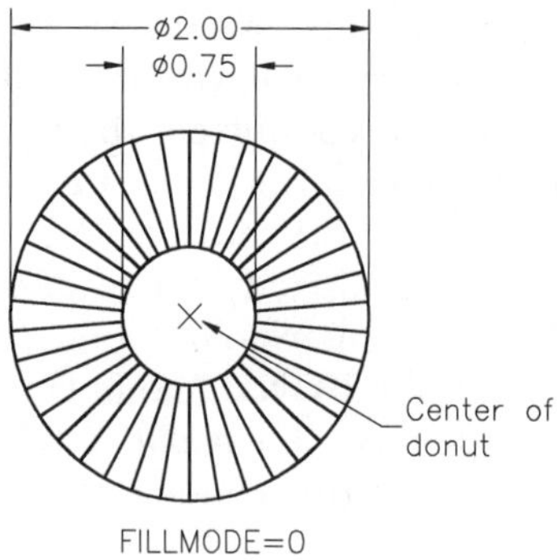

Figure 3-52 *Drawing an unfilled donut using the **DONUT** command*

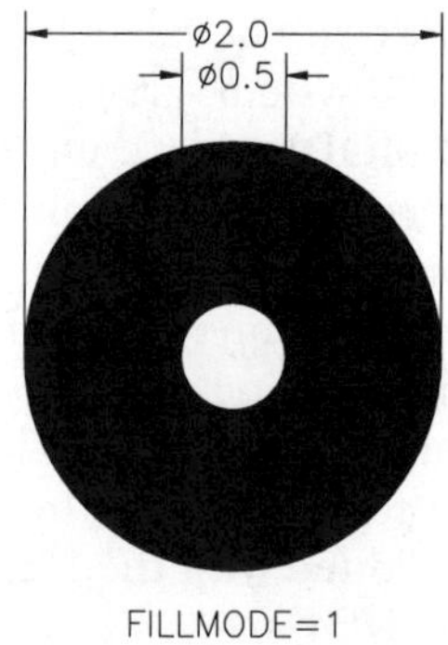

Figure 3-53 *Drawing a filled doughnut using the **DONUT** command*

To draw a solid-filled donut with an outside diameter of 2.0 units (Figure 3-54), the following is the prompt sequence:

Command: **DONUT** [Enter]
Specify inside diameter of donut <0.50>: **0** [Enter]
Specify outside diameter of donut <1.0>: **2** [Enter]
Specify center of donut or <exit>: *Specify a point.*
Specify center of donut or <exit>: [Enter]

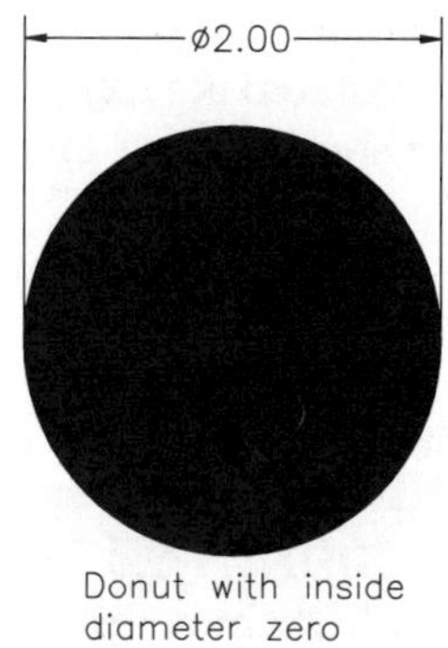

Figure 3-54 *Solid-filled donut*

DRAWING POINTS

Toolbar:	Draw > Point
Menu:	Draw > Point
Command:	POINT

The **point** is the basic drawing object. Points are invaluable in building a drawing file. To draw a point anywhere on the screen, AutoCAD LT provides the **POINT** command (Figure 3-55).

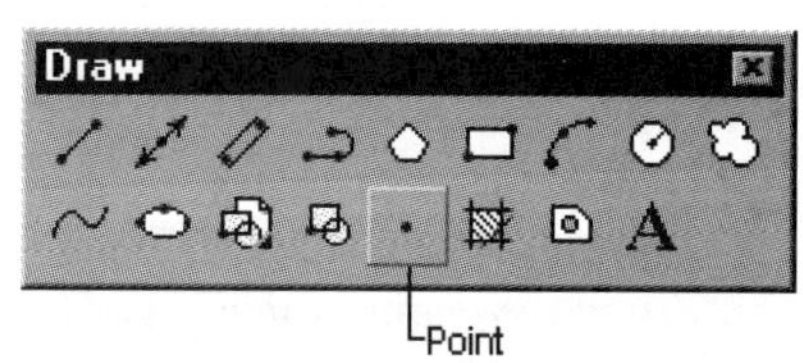

Figure 3-55 *Invoking the **POINT** command from the **Draw** toolbar*

Command: **POINT** [Enter]
Current pointmodes: PDMODE=n PDSIZE=n.nn
Specify a point: *Specify the location where you want to place the point.*

If you invoke the **POINT** command from the toolbar or the menu (**Multiple Point** option), you can draw as many points as you desire in a single command (Figure 3-56). In this case you can exit from the **POINT** command by pressing ESC. If you invoke this command by entering **POINT** at the Command prompt or use the **Single Point** option from the menu, you can draw only one point in a single point command.

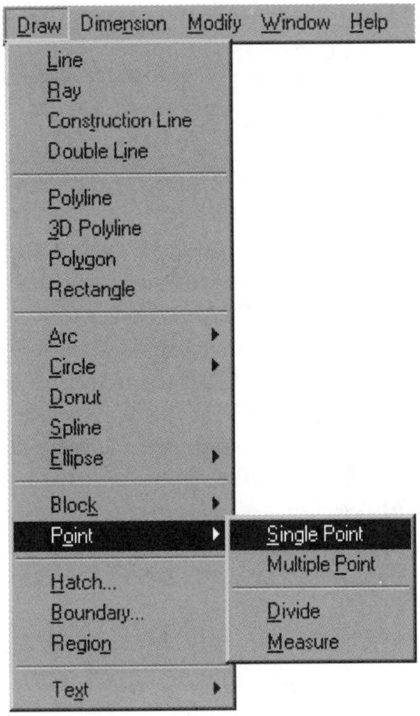

*Figure 3-56 Invoking the **POINT** command from the **Draw** menu*

Note
*It is possible to have a temporary construction marker for the point known as blip. A mark appears on the screen where you place the point. This blip mark can then be cleared once the screen is redrawn using the **REDRAW** command and the point is left on the screen. The visibility of blips can be controlled using the **BLIPMODE** system variable.*

Changing the Point Type

Menu:	Format > Point Style
Command:	DDPTYPE

The point type can be set either from the **Point Style** dialog box (Figure 3-57). There are twenty combinations of point types. The **Point Style** dialog box can be accessed from the **Format** menu (Select **Format > Point Style**) as shown in Figure 3-58. You can choose a point style in this dialog box which is indicated by highlighting that particular point style. Next, choose the **OK** button. Now all the points will be drawn in the selected style until you change it to a new style. The type of point drawn is stored in the **PDMODE** (Point Display MODE) system variable. You can change the point style by entering a numeric value in the **PDMODE** variable.

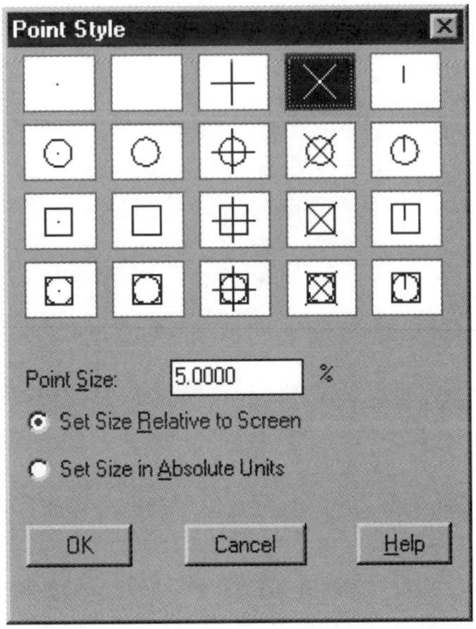

*Figure 3-57 **Point Style** dialog box*

The PDMODE values for different point types are:

1. A value of 0 is the default for the **PDMODE** variable and generates a dot at the specified point.
2. A value of 1 for the **PDMODE** variable generates nothing at the specified point.
3. A value of 2 for the **PDMODE** variable generates a plus sign (+) through the specified point.

4. A value of 3 for the **PDMODE** variable generates a cross mark (X) through the specified point.
5. A value of 4 for the **PDMODE** variable generates a vertical line in the upward direction from the specified point.
6. When you add **32** to the **PDMODE** values of 0 to 4, a circle is generated around the symbol obtained from the original **PDMODE** value. For example, to draw a point having a cross mark and a circle around it, the value of the **PDMODE** variable will be **3 + 32 = 35**. Similarly, you can generate a square around the symbol with the **PDMODE** value of 0 to 4 by adding **64** to the original **PDMODE** value. For example, to draw a point having a plus sign and a square around it, the value of the **PDMODE** variable will be **2 + 64 = 66**. You can also generate a square and a circle around the symbol with the **PDMODE** of value 0 to 4 by adding **96** to the original **PDMODE** value. For example, to draw a point having a dot mark and a circle and a square around it, the value of the **PDMODE** variable will be **0 + 96 = 96** (See Figure 3-59).

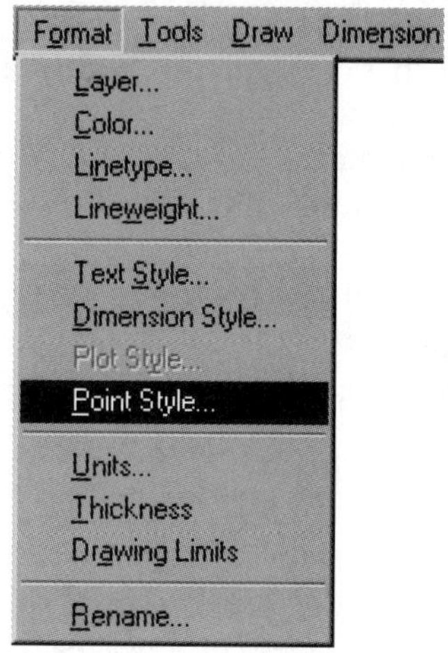

***Figure 3-58** Choosing **Point Style** from the **Format** menu*

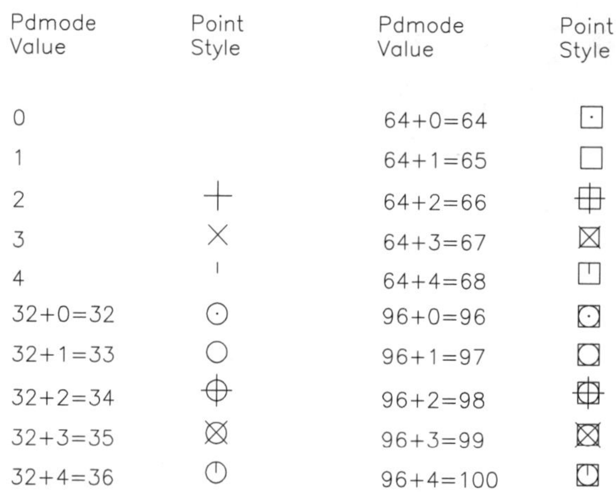

***Figure 3-59** Different point style for **PDMODE** values*

Exercise 12 — *General*

Check what types of points are drawn for each value of the PDMODE variable.

Changing the Point Size

Menu:	Format > Point Style
Command:	DDPTYPE

The size of a point can be set from the **Point Style** dialog box (Figure 3-57) by entering the desired point size in the **Point Size** edit box. You can generate the point at a specified percentage of the graphics area height or define an absolute size for the point. An absolute size for the point can be specified by selecting the **Set Size in Absolute Units** radio button in the **Point Style** dialog box and then enter a value in the **Point Size** edit box. The point size can also be set by changing the value of **PDSIZE** (Figure 3-60). The variable **PDSIZE** governs the size of the point (except for the **PDMODE** values of 0 and 1). You can set the size in absolute units by specifying a positive value for the **PDSIZE** variable. If the **Set Size Relative to Screen** radio button is selected in the **Point Style** dialog box, the size is taken as a percentage of the viewport size. This can also be set by entering a negative value for the **PDSIZE** variable. For example, a setting of 5 makes the point 5 units high; a setting of -5 makes the point 5 percent of the current drawing area.

Figure 3-60 *Changing point size using the* ***PDSIZE*** *variable*

Exercise 13 *General*

a. Try various combinations of the **PDMODE** and **PDSIZE** variables.
b. Check the difference between the points generated from negative values of **PDSIZE** and points generated from positive values of **PDSIZE**.

DRAWING INFINITE LINES

The **XLINE** and **RAY** commands can be used to draw construction or projection lines (Figure 3-61 and Figure 3-62). These are lines that aid in construction or projection and are drawn very lightly when drafting manually. An **xline** (construction line) is a 3D line that extends to infinity on both ends (Figure 3-63). Since the line is infinite in length, it does not have any endpoints. A **ray** is a 3D line that extends to infinity on only one end. The other end of the ray has a finite endpoint. The xlines and rays have zero extents. This means that the extents of the drawing will not change if you use the commands that change the drawing extents, such as the **ZOOM** command with the **All** option. Most of the object snap modes work with both xlines and rays, with some limitations: You cannot use the **Endpoint** object snap with the xline because by definition an xline does not have any endpoints. However, for rays you can use the **Endpoint** snap on one end only. Also, xlines and rays take the properties of the layer in which they are drawn.

Tip

Xlines and rays plot like any other objects in a drawing and may create confusion. It is therefore a good idea to create the construction lines in a different layer altogether, so that you can recognize them easily.

XLINE Command

Toolbar:	Draw > Construction Line
Menu:	Draw > Construction Line
Command:	XLINE

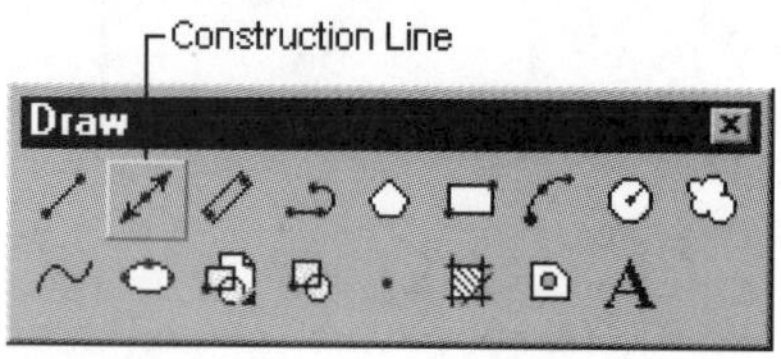

Figure 3-61 *Choosing the* ***Construction Line*** *button from the* ***Draw*** *toolbar*

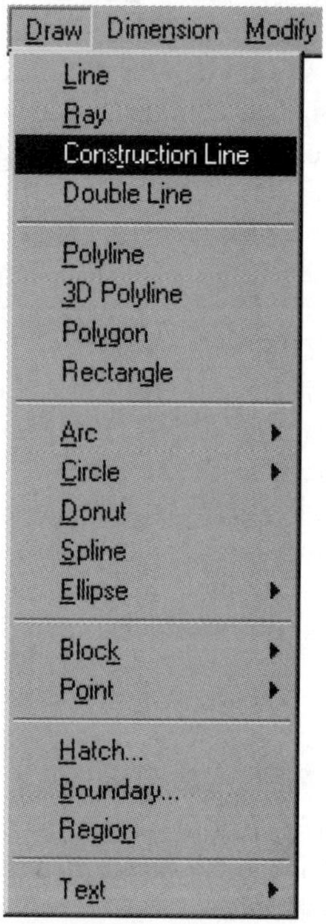

Figure 3-62 *Choosing* ***Construction Line*** *from the* ***Draw*** *menu*

When you invoke the **XLINE** command, you will follow this prompt sequence:

Command: **XLINE** [Enter]
Specify a point or [Hor/Ver/Ang/Bisect/Offset]: *Specify an option or select a point through which the xline will pass.*

The various options of the command are discussed next.

Point

If you use the default option, AutoCAD LT will prompt you to select two points through which the xline shall pass at the **Specify a point** and the **Specify through point** prompts. After you select the first point, AutoCAD LT will dynamically rotate the xline through the specified point as you move the cursor. When you select the second point, an xline will be created that passes through the first and second points (Figure 3-63).

Command: **XLINE** [Enter]
Specify a point or [Hor/Ver/Ang/Bisect/Offset]: *Specify a point.*
Specify through point: *Specify the second point.*

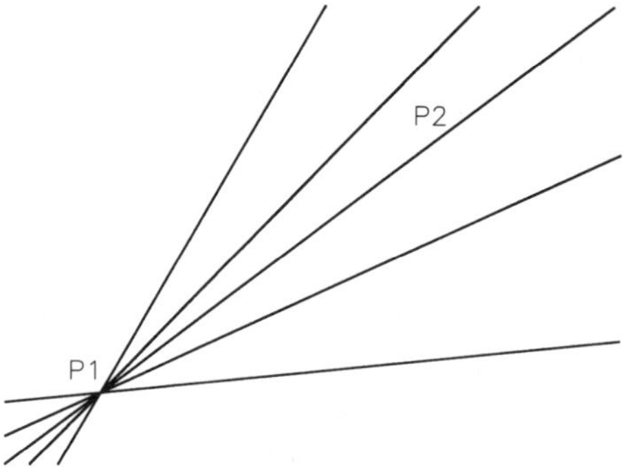

Figure 3-63 *Drawing xlines*

You can continue to select more points to create more xlines. All these xlines will pass through the first point you had selected at the **Specify a point** prompt. This point is also called the root point. Right-click or press ENTER to end the command.

Horizontal

This option will create horizontal xlines of infinite length that pass through the selected points. The xlines will be parallel to the X axis of the current UCS (Figure 3-64). As you invoke this option the horizontal xline gets attached to the cursor. You are prompted to select only one point through which the horizontal xline passes. You can continue selecting points to draw horizontal xlines and right-click or press ENTER to end the command.

Vertical

This option will create vertical xlines of infinite length that pass through the selected points. The xlines will be parallel to the Y axis of the current UCS (Figure 3-64). As you invoke this option the vertical xline gets attached to the cursor. You are prompted to select only one point through which the vertical xline passes. You can continue selecting points to draw vertical xlines and right-click or press ENTER to end the command.

Angular

This option will create xlines of infinite length that pass through the selected point at a specified angle (in Figure 3-65, the angle specified is 38 degrees). The angle can be specified by entering a value at the keyboard. You can also use the reference option by selecting an object and then specifying an angle relative to it. The **Reference** option is useful when the actual angle is not known but the angle relative to an existing object can be specified.

Command: **XLINE** [Enter]
Specify a point or [Hor/Ver/Ang/Bisect/Offset]: **A** [Enter]
Enter angle of xline (0) or [Reference]: **R** [Enter] *(Here we shall use the **Reference** method for specifying the angle.)*
Select a line object: *Select a line.*
Enter angle of xline <0>: *Enter angle (the angle will be measured counterclockwise with respect to the selected line.)*
Specify through point: *Specify the second point.*

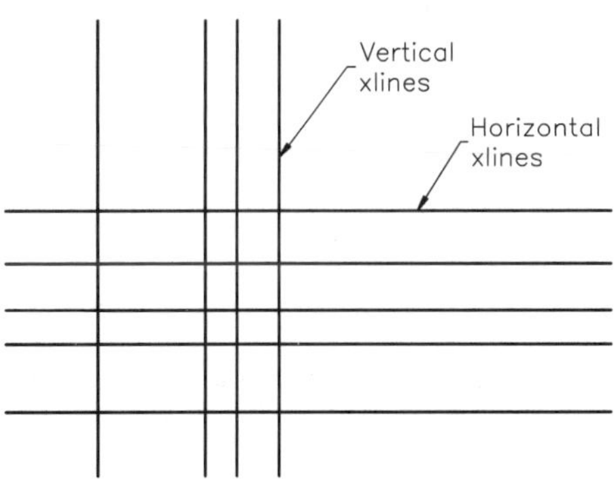

Figure 3-64 *Horizontal and vertical xlines*

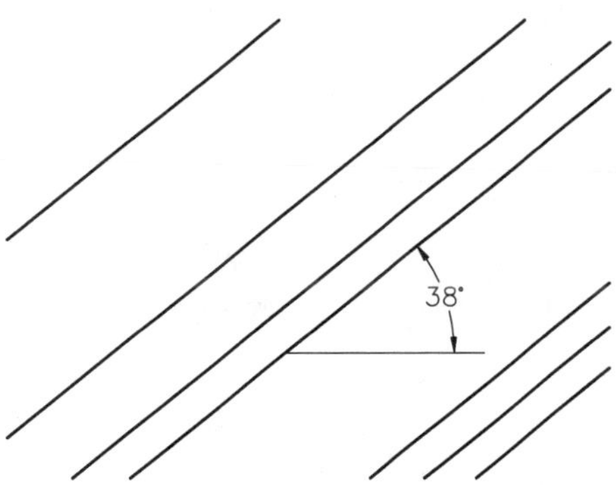

Figure 3-65 *Angular X lines*

Bisect

This option will create an xline that passes through the angle vertex and bisects the angle you specify by selecting two points. The xline created using this option will lie in the plane defined by the selected points. You can use the object snaps to select the points on the existing objects. The following is the prompt sequence for this option (Figure 3-66):

Command: **XLINE** [Enter]
Specify a point or [Hor/Ver/Ang/Bisect/Offset]: **B** [Enter]
Specify angle vertex point: *Enter a point (P1).*
Specify angle start point: *Enter a point (P2).*
Specify angle end point: *Enter a point (P3).*
Specify angle end point: *Select more points or press ENTER or right-click to end the command.*

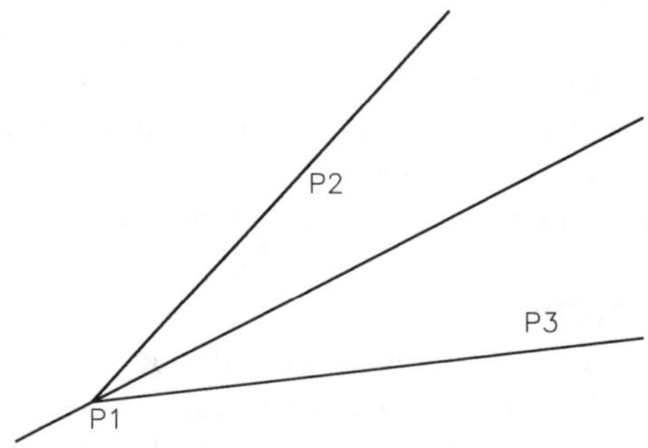

Figure 3-66 *Using the **Bisect** option to draw xlines*

Offset

The **Offset** option creates xlines that are parallel to the selected line/xline at a specified offset distance. You can specify the offset distance by entering a numerical value or by selecting two points on the screen. If you select the **Through** option, the offset line will pass through the selected point. This option works like the **OFFSET** editing command. The prompts at the command line are as follows:

Command: **XLINE** [Enter]
Specify a point or [Hor/Ver/Ang/Bisect/Offset]: **O** [Enter]
Specify offset distance or [Through] <Through>: *Press ENTER to accept the **Through** option or specify a distance from the selected line object at which the xline shall be drawn.*
Select a line object: *Select the object to which the xline is drawn parallel at a specified distance.*
Specify through point: *Select a point through which the xline should pass.*

If you specify the offset distance, and after you have selected a line object, you are prompted to specify the direction in which the xline is to be offset. You can continue drawing xlines or right-click or press ENTER to end the command.

RAY Command

Menu:	Draw > Ray
Command:	RAY

A ray is a 3D line similar to the xline construction line with the difference being that it extends to infinity only in one direction. It starts from a point you specify and extends to infinity through the specified point. The prompt sequence is:

Command: **RAY** [Enter]
Specify start point: *Select the starting point for the ray.*
Specify through point: *Specify the second point.*

Press ENTER or right-click to exit the command.

Note

When you trim an xline, it gets converted into a ray and when a ray is trimmed at the end that is infinite, it gets converted into a line object.

CREATING TEXT

Toolbar:	Text > Single Line Text
Menu:	Draw > Text > Single Line Text
Command:	TEXT

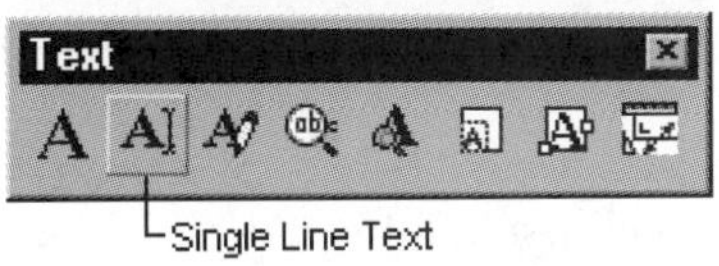

***Figure 3-67** Invoking the **TEXT** command from the **Draw** toolbar*

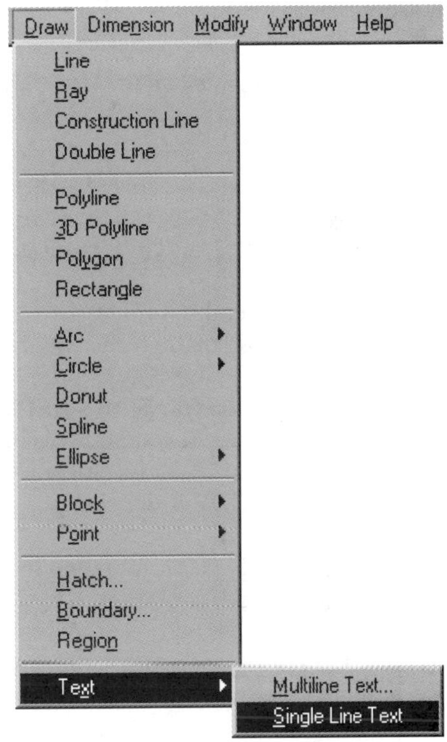

***Figure 3-68** Invoking the **TEXT** command from the **Draw** menu*

The **TEXT** command lets you write several lines of text on the drawing (Figure 3-67 and Figure 3-68). After invoking this command you have to specify the start point for the text. Then you need to specify the height of the text and also the rotation angle. The characters appear on the screen as you enter them. When you press ENTER after typing a line, the cursor automatically places itself at the start of the next line and repeats the prompt for entering another line. You can end the command by pressing the ENTER key. You can use the BACKSPACE key to edit the text on the screen while you are writing it. The prompt sequence is:

Command: **TEXT** Enter
Current text style: "current" Text height: current
Specify start point of text or [Justify/Style]: *Specify the starting point of the text.*
Specify height<current>: *Enter the text height.*
Specify rotation angle of text <0>: Enter
Enter text: *Enter first line of text.*
Enter text: *Enter the second line of text.*
Enter text: Enter

Note

*The **TEXT** commands are discussed in detail in Chapter 7, Controlling the Drawing Display and Creating Text.*

*If you want to move the objects, use the **MOVE** command and then select the objects, specify the base point, and the second point of displacement. The **MOVE** command is discussed in detail in Chapter 5, Editing Commands-I.*

Self-Evaluation Test

Answer the following questions and then compare your answers to the correct answers given at the end of this chapter.

1. A negative value for chord length in the **Start, Center, Length** option of the **ARC** command results in the largest possible arc, also known as the major arc. (T/F)
2. In the **ARC** command, if you do not specify a start point but just press ENTER or choose the **Continue** option, the start point and direction of the arc will be taken from the endpoint and ending direction of the previous line or arc drawn on the current screen. (T/F)
3. If the **PELLIPSE** is set to 1, AutoCAD LT creates a true ellipse, also known as a NURBS-based (Non-Uniform Rational Bezier Spline) ellipse. (T/F)
4. The start and end parameters of an elliptical arc are determined by specifying a point on the circle whose diameter is equal to the minor diameter of the ellipse. (T/F)
5. The _______ option of the **RECTANG** command allows you to draw a rectangle at a specified distance from the XY plane along the Z axis.
6. If the **FILLMODE** is off, the _________ donuts are drawn.
7. You can get a _______ polyline by entering two different values at the starting width and the ending width prompts.
8. In case of a(n) __________ polyline the vertices are not stored as separate entities, but as a single object with an array of information.
9. If you invoke the **POINT** command from the __________, you can draw as many points as you desire in a single command.
10. The size of a point is taken as a percentage of the viewport size if you enter a __________ value for the **PDSIZE** variable.

Review Questions

Answer the following questions.

1. Using the **Start, End, Angle** option of the **ARC** command, a negative included angle value draws the arc in a clockwise direction. (T/F)

2. When the **Continue** option of the **ARC** command is used to draw arcs, each successive arc is perpendicular to the previous one. (T/F)

3. If you specify the chamfer distances first and then specify the fillet radius in the same **RECTANG** command, the rectangle will be drawn with chamfers only. (T/F)

4. Using the **RECTANG** command, the rectangle generated on the screen is treated as a combination of different objects; therefore, individual sides can be edited immediately. (T/F)

5. Using the **Start, Center, Length** option of the **ARC** command a positive chord length generates the smallest possible arc (minor arc) with this length and this arc is always less than:

   (a) 90 degrees (b) 180 degrees
   (c) 270 degrees (d) 360 degrees

6. Which one of the following options of the **RECTANG** command allows you to draw a rectangle that is extruded in the Z direction by the specified value?

   (a) **Elevation** (b) **Thickness**
   (c) **Extrude** (d) **Width**

7. Which of the following commands draws a line anywhere in three-dimensional spacc which starts from a point that you specify with the other end extending to infinity?

   (a) **PLINE** (b) **RAY**
   (c) **XLINE** (d) **MLINE**

8. If the old 2D polylines should not get converted to lightweight polylines when opening the drawings in AutoCAD LT 2002, and also the new polylines drawn should be 2D polylines, the **PLINETYPE** variable should be set to which of the following values?

   (a) 0 (b) 1
   (c) 2 (d) 3

9. Which of the following values should be assigned to the **PDMODE** variable such that a cross mark (X) is generated through the specified point.

   (a) 1 (b) 3
   (c) 5 (d) 7

10. A polygon is said to be __________ when it is drawn inside an imaginary circle and its vertices (corners) touch the circle.

11. If additional polyline segments are added to the first polyline, AutoCAD LT automatically makes the __________ of the first polyline segment the start point of the next polyline segment.

12. With the **DONUT** command, a solid-filled circle is drawn by specifying the inside diameter as __________ and FILLMODE is on.

13. The visibility of blips can be controlled using the __________ system variable.

14. An absolute size for the point can be specified by entering a __________ value for **PDSIZE** variable.

15. The __________ option of the **XLINE** creates xlines of infinite length that are parallel to the Y axis of the current UCS.

16. You can use the __________ key to edit the text on the screen while you are writing it using the **TEXT** command.

Exercises

Exercise 14 — *Mechanical*

Make the drawing shown in Figure 3-69. The distance between the dotted lines is 1.0 unit. Create the radii using appropriate **ARC** command options.

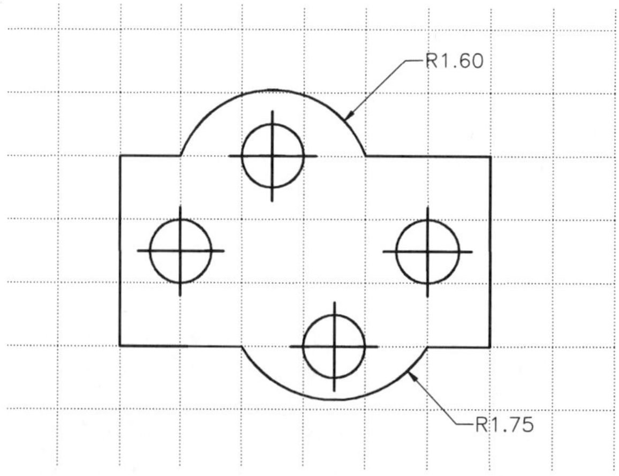

Figure 3-69 *Drawing for Exercise 14*

Exercise 15 *Graphics*

Make the drawing shown in Figure 3-70. The distance between the dotted lines is 1.0 unit. Create the radii using appropriate **ARC** command options.

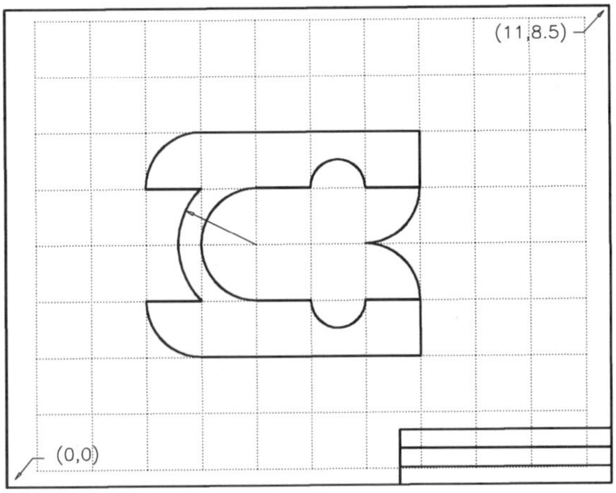

Figure 3-70 *Drawing for Exercise 15*

Exercise 16 *Mechanical*

Make the drawing shown in Figure 3-71. The distance between the dotted lines is 0.5 unit. Create the ellipses using the **ELLIPSE** command.

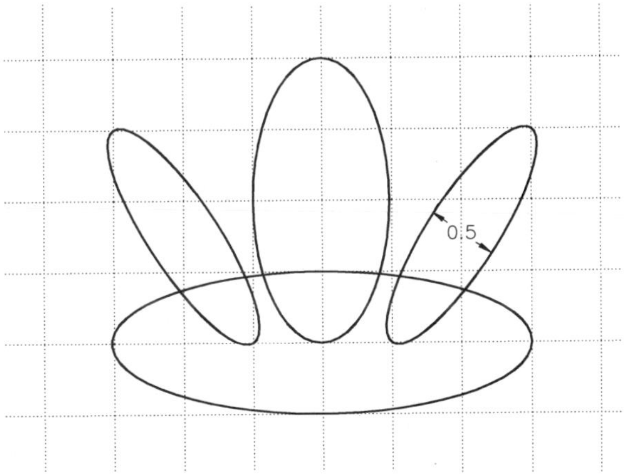

Figure 3-71 *Drawing for Exercise 16*

Exercise 17 *Mechanical*

Draw the following Figure 3-72 using the **LINE**, **CIRCLE**, and **ARC** commands. The distance between the dotted lines is 1.0 unit and the diameter of the circles is 1.0 unit.

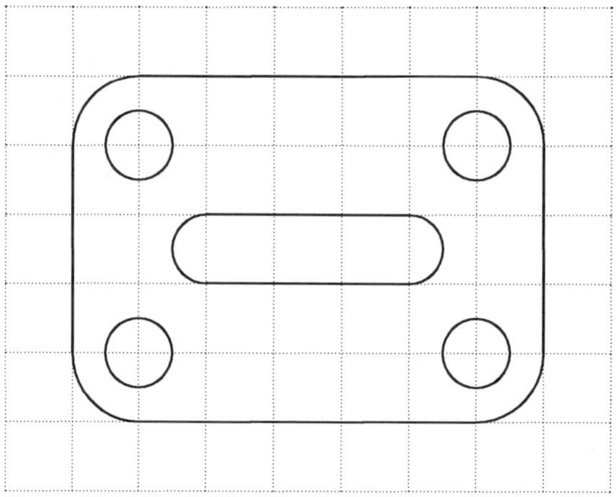

Figure 3-72 *Drawing for Exercise 17*

Exercise 18 *Mechanical*

Draw Figure 3-73 using the **LINE**, **CIRCLE**, and **ARC** commands or their options. The distance between the grid lines is 1.0 unit and the diameter of the circle is 1.0 unit.

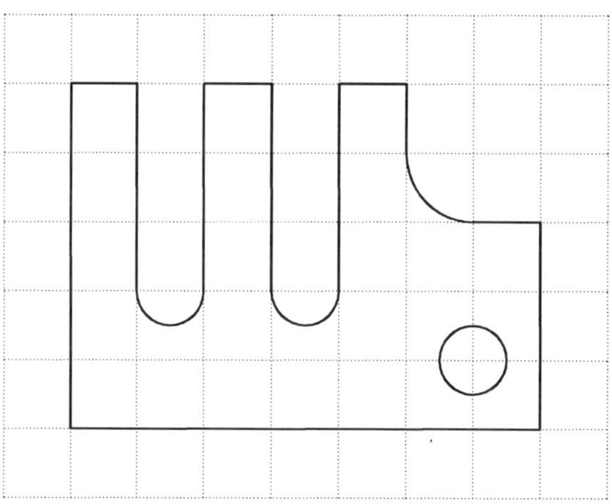

Figure 3-73 *Drawing for Exercise 18*

Problem Solving Exercise 1

Mechanical

Make the drawing shown in Figure 3-74. Create the radii by using the **ARC** command options indicated in the drawing. (Use the @ symbol to snap to the previous point. Example: Specify start point of arc or [Center]: @)

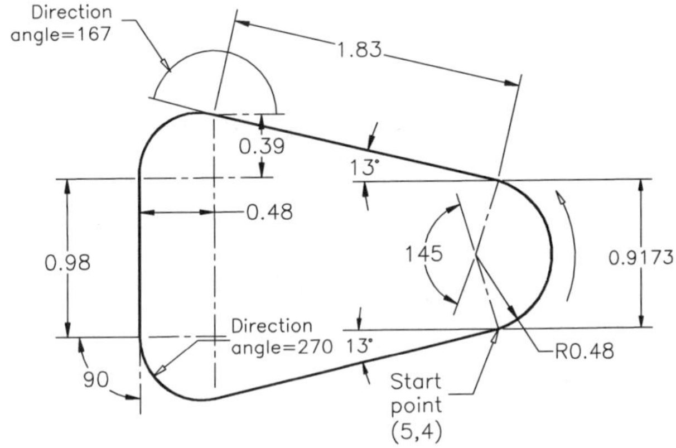

***Figure 3-74** Drawing for Problem Solving Exercise 1*

Problem Solving Exercise 2

Mechanical

Make the drawing shown in Figure 3-75. Create the radii by using the **ARC** command options. The distance between the dotted lines is 0.5 units.

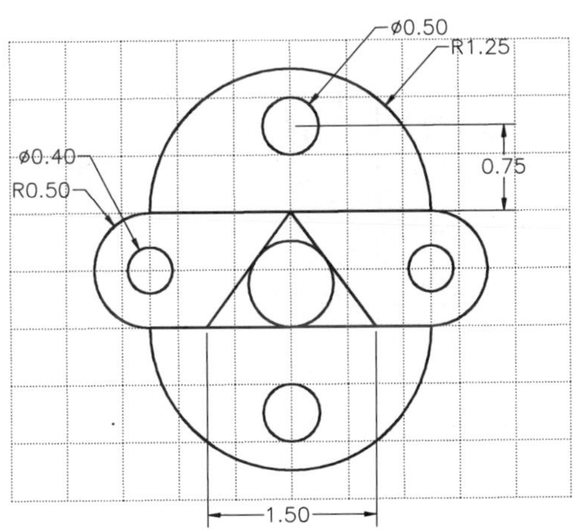

***Figure 3-75** Drawing for Problem Solving Exercise 2*

Problem Solving Exercise 3

Mechanical

Make the drawing shown in Figure 3-76. Create the radii by using the **ARC** command options. The distance between the dotted lines is 1.0 unit.

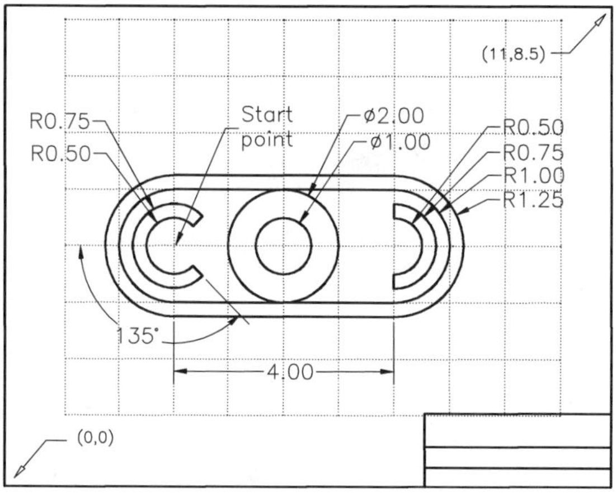

Figure 3-76 *Drawing for Problem Solving Exercise 3*

Problem Solving Exercise 4

Mechanical

Make the drawing shown in Figure 3-77 using the **POLYGON**, **CIRCLE**, and **LINE** commands.

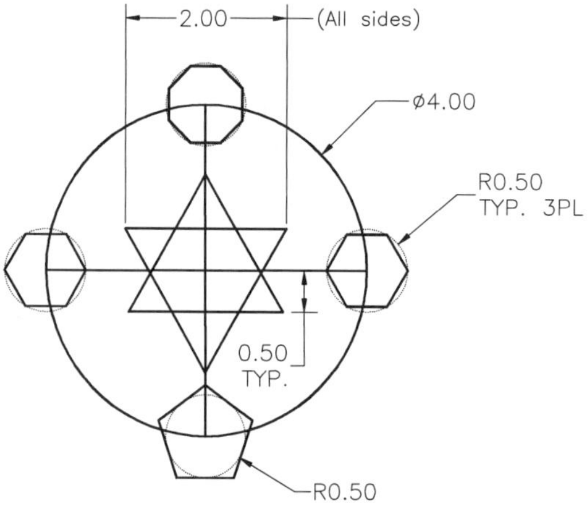

Figure 3-77 *Drawing for Problem Solving Exercise 4*

Problem Solving Exercise 5 — *Mechanical*

Make the drawing shown in Figure 3-78 using the **DRAW** commands. Note, Sin 30=0.5, Sin 60=0.866. The distance between the dotted lines is 1 unit.

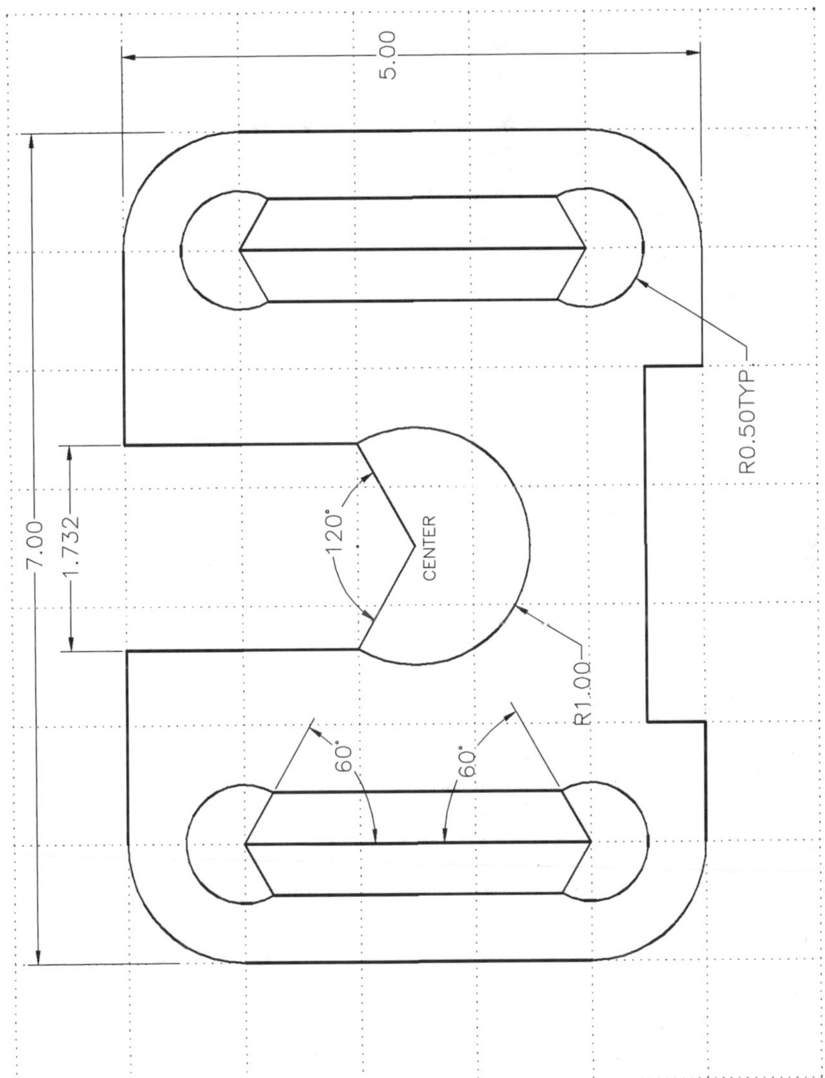

Figure 3-78 Drawing for Problem Solving Exercise 5

Answers to Self-Evaluation Test

1 - T, **2** - T, **3** - F, **4** - F, **5** - **Elevation**, **6** - unfilled, **7** - tapered, 8 - optimized, **9** - toolbar, **10** - negative.

Chapter 4

Drawing Aids

Learning Objectives

After completing this chapter, you will be able to:

- *Set up layers and assign colors and linetypes to them.*
- *Use the **Object Properties** toolbar to directly change the general object properties.*
- *Change the properties of objects using the **PROPERTIES** command.*
- *Determine current and global linetype scaling and **LTSCALE** factor for plotting.*
- *Set up Grid, Snap, and Ortho modes based on the drawing requirements.*
- *Use object snaps and understand their applications.*
- *Combine Object Snap modes and set up running Object Snap modes.*
- *Use Polar Tracking to locate key points in a drawing.*

In this chapter you will learn about the drawing setup and what factors affect the quality and accuracy of your drawing. This chapter contains a detailed description of how to set up layers. You will also learn about some of the drawing aids such as Grid, Snap, and Ortho. These aids will help you to draw accurately and quickly.

LAYERS

The concept of layers can be best explained by using the concept of overlays in manual drafting. In manual drafting, different details of the drawing can be drawn on different sheets of paper, or overlays. Each overlay is perfectly aligned with the others, and when all of them are placed on top of each other you can reproduce the entire drawing. As shown in Figure 4-1, the object lines have been drawn in the first overlay and the dimensions in the second overlay. You can place these overlays on top of each other and get a combined look at the drawing.

In AutoCAD LT, instead of using overlays you use layers. Each layer is assigned a name. You can also assign a color and linetype to these layers. For example, in Figure 4-2 the object lines have been drawn in the OBJECT layer and the dimensions have been drawn in the DIM layer.

The object lines will be red because the red color has been assigned to the OBJECT layer. Similarly, the dimension lines will be green because the DIM layer has been assigned the green color. You can display all of the layers or you can display the layers individually or in any combination.

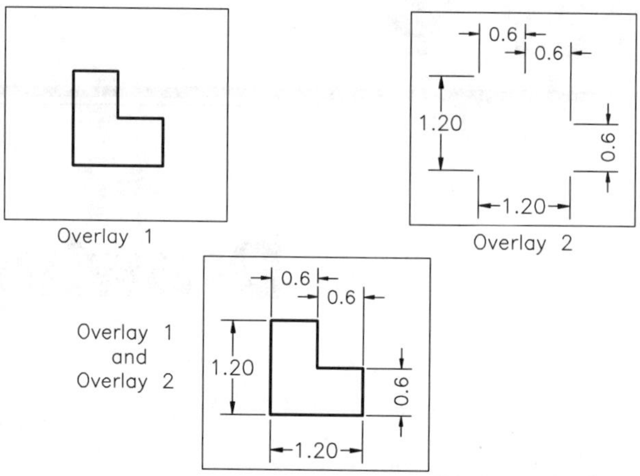

Figure 4-1 *Drawing lines and dimensions in different overlays*

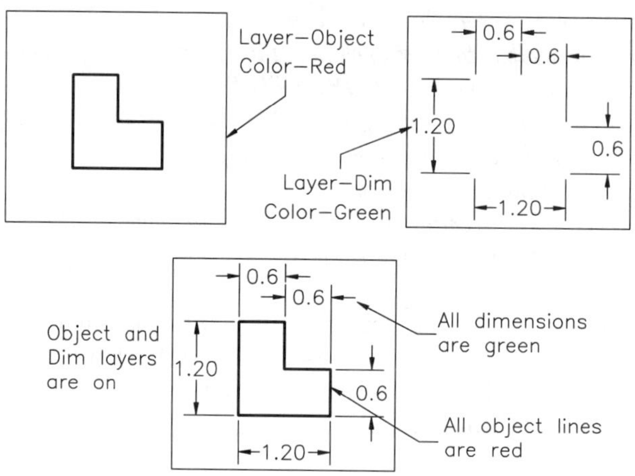

Figure 4-2 *Drawing lines and dimensions in different layers*

Advantages of Layers

1. Each layer can be assigned a different color. Assigning a particular color to a group of objects is very important for plotting. For example, if all object lines are red, at the time of plotting you can assign the red color to a slot (pen) that has the desired tip width (e.g., medium). Similarly, if the dimensions are green, you can assign the green color to another slot (pen) that has a thin tip. By assigning different colors to different layers you can control the width of the lines when the drawing is plotted. You can also make a layer plottable or nonplottable.

2. The layers are useful for some editing operations. For example, if you want to erase all dimensions in a drawing, you can freeze all layers except the dimension layer and then erase all dimensions by using the **Crossing** option to select objects.
3. You can turn a layer off or freeze a layer that you do not want to be displayed or plotted.
4. You can lock a layer, which will prevent the user from accidentally editing the objects in that layer.
5. The colors also help you to distinguish different groups of objects. For example, in architectural drafting, the plans for foundation, floors, plumbing, electrical, and heating systems may all be made in different layers. In electronics drafting and in PCB (printed circuit board), the design of each level of a multilevel circuit board can be drawn on separate layers. Similarly in mechanical engineering the main components of an assembly can be made in one layer, other components such as the nuts, bolts, keys, and washers can be made in another layer, and the annotations such as datum symbols and identifiers, texture symbols, Balloons, and Bill of Materials can be made in another layer.

LAYER PROPERTIES MANAGER DIALOG BOX

Toolbar:	Object Properties > Layers
Menu:	Format > Layer
Command:	LAYER

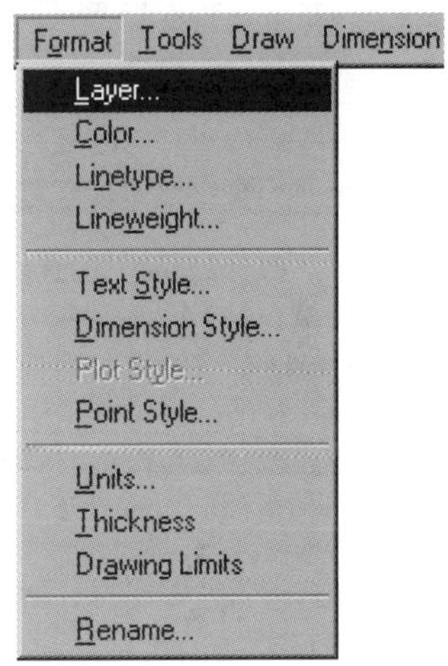

*Figure 4-3 Invoking the **LAYER** command from the **Format** menu*

You can use the **Layer Properties Manager** dialog box to perform the functions associated with the **LAYER** command (Figure 4-3). For example, you can create new layers, assign colors, assign linetypes, or perform any operation that is shown in the dialog box. Using this dialog box for layers is an efficient and convenient way to use the various options. You can also perform some layer functions such as freeze, thaw, lock, unlock, and so on, directly from the **Object Properties** toolbar (Figure 4-4). When you invoke the **Layer Properties Manager** dialog box a default layer with the name **0** is displayed. It is the current layer and any object you draw is created in this layer. There are certain features associated with each layer such as the color, linetype, and lineweight. The layer 0 has a default color of white, linetype continuous, and lineweight default.

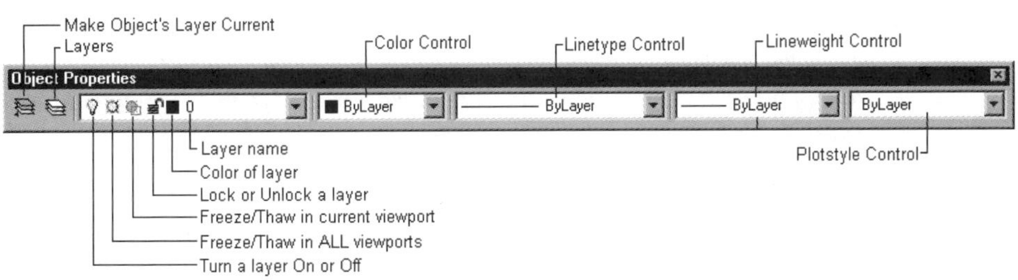

***Figure 4-4 Object Properties** toolbar*

Chapter 4

Creating New Layers

If you want to create new layers, choose the **New** button in the **Layer Properties Manager** dialog box (Figure 4-5). A new layer with name **Layer1** and having the properties of layer 0 is created and listed in the dialog box just below layer 0. If you have more layers apart from the layer 0, the new layer has the properties of the selected layer and is placed below the same. You can change or edit the name by selecting it and then entering a new name. If no layer is selected, the new layer is placed at the end of the layers list and has the properties of the default layer 0. If more than one layer is selected, the new layer is placed below the last highlighted layer and has its properties. Also, right-clicking anywhere in the Layers list area of the **Layer Properties Manager** dialog box displays a shortcut menu that gives you an option to create a new layer. You can right-click a layer whose properties you want to use in the new layer and then select **New Layer** from the shortcut menu.

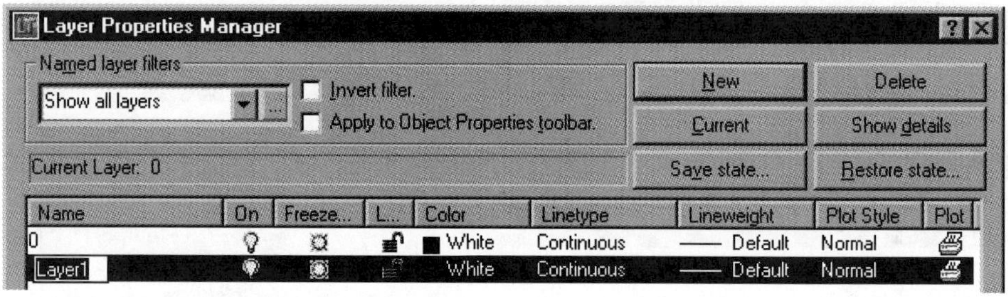

***Figure 4-5 Layer Properties Manager** dialog box with a new layer created*

Layer names

1. A layer name can be up to 255 characters long, including letters (a – z), numbers (0 – 9), special characters ($ _ -) and spaces. Any combination of lower and uppercase letters can be used while naming a layer.

2. The layers should be named to help the user identify the contents of the layer. For example, if the layer name is HATCH, a user can easily recognize the layer and its contents. On the other hand, if the layer name is X261, it is hard to identify the contents of the layer.

3. Layer names should be short, but should also convey the meaning.

Note

All the previous releases of AutoCAD LT, before AutoCAD LT 2000, allowed the layer name to be thirty-one characters long. If you give a long name to the layer in AutoCAD LT 2002 and save the drawing in the previous release, the long name is shortened to thirty-one characters and the illegal characters are replaced by underscores.

*The length of the layer name is controlled by the **EXTNAMES** system variable that has a default value 1. If you change it to 0, the layer name is allowed to be up to thirty-one characters long and cannot include the special characters and spaces.*

Tip

If you exchange drawings with or provide drawings to consultants or others, you must standardize and coordinate layer names and other layer settings.

Making a Layer Current

To draw an object in a particular layer, you need to make it the current layer. Only one layer can be made current and it is the layer where new objects will be drawn. To make a layer current, select the name of the desired layer and then choose the **Current** button in the dialog box. AutoCAD LT will display the name of the selected layer next to **Current Layer** above the list of layers in the dialog box. Choose **OK** to exit the dialog box. You can also make a layer current by double-clicking on the particular layer from the layer list box of the **Layer Properties Manager** dialog box. Right-clicking a layer in the layer list box displays a shortcut menu which gives you an option (**Make Current**) to make the selected layer current (Figure 4-6). The name and properties of the current layer is also displayed in the **Object Properties** toolbar. You can also make a layer current by selecting the layer from the **Layer Control** drop-down list in the **Object Properties** toolbar. You can use the **CLAYER** system variable to make the layer current from the Command prompt. Choosing the **Make Object's Layer Current** button from the **Object Properties** toolbar prompts you to select the object whose layer you wish to make current. After selecting an object, the layer associated with that object will be made current.

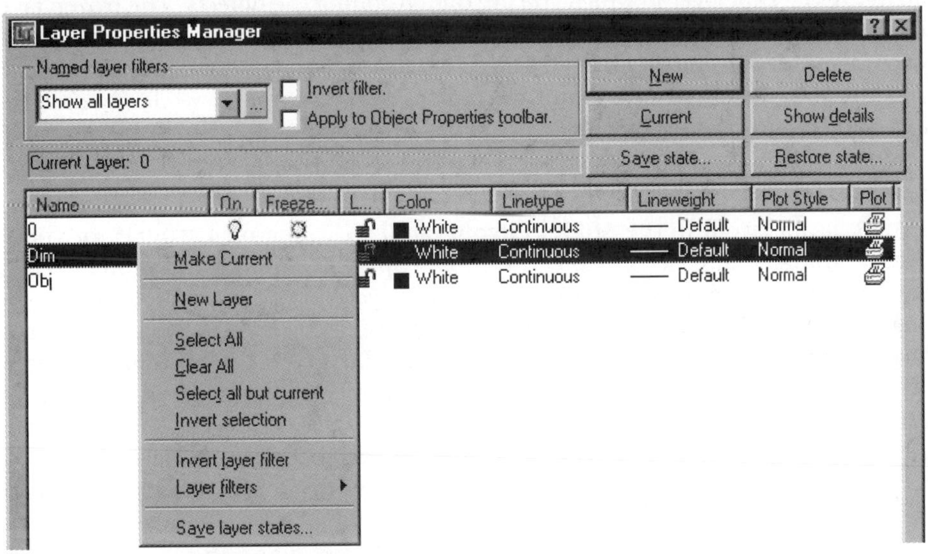

Figure 4-6 ***Layer Properties Manager*** *dialog box with the Layer shortcut menu*

Note

When you select more than one layer at a time using the SHIFT key, the ***Make Current*** *option is not displayed in the shortcut menu. Only one layer can be made current at one time.*

Controlling the Display of Layers

You can control the display of the layers by selecting the **Turn a layer On or Off**, **Freeze or thaw in ALL viewports**, or **Lock or Unlock a layer** toggle buttons in the list box of any layer.

Turn a Layer On or Off

With the **Turn a layer On or Off** toggle icon (light bulb) you can turn the layers on or off. The layers that are turned on are displayed and can be plotted. The layers that are turned off are not displayed and cannot be plotted. You can perform all the operations such as drawing and editing in the layer that has been turned off. You can turn the current layer off, but AutoCAD LT will display a warning box informing you that the current drawing layer has been turned off. You can also turn the layer on or off by clicking on the On/Off toggle icon from the **Layer** drop-down list in the **Object Properties** toolbar (Figure 4-7).

***Figure 4-7** Turning layer **Dim** Off from the **Object Properties** toolbar*

Freeze or Thaw in ALL Viewports

While working on a drawing, if you do not want to see certain layers you can also use the **Freeze or thaw in ALL viewports** toggle icon (sun/snowflakes) to freeze the layers. You can use the **Object Properties** toolbar or the **Layer Properties Manager** dialog box to freeze or thaw a layer. No modifications can be done in the Frozen layer. For example, while editing a drawing you may not want the dimensions to be changed and displayed on the screen. To avoid this, you can freeze the Dim layer in which you are dimensioning the objects. The frozen layers are invisible and cannot be plotted. The Thaw option negates the effect of the Freeze option, and the frozen layers are restored to normal. The difference between the **Off** option and the **Freeze** option is that the frozen layers are not calculated by the computer while regenerating the drawing, and this saves time. The current layer cannot be frozen.

Current or New VP Freeze

When you select a layout in the **Model/Layout** tab (by clicking on Layout1), or you set the **TILEMODE** variable to 0 (see Chapter 11, Model Space Viewports, Paper Space Viewports, and Layouts), you can freeze or thaw the selected layers in the active floating viewport by selecting the **Current VP Freeze** icon for the selected layers. Once you are in a floating viewport, the Current and New VP Freeze icons are added in the **Layer Properties Manager** dialog box towards the right side (Figure 4-8). If the icon is not visible you have a scroll bar at the bottom of the layer list box that you can move to display the icons. Also the **Freeze or Thaw in Current viewport** icon in the **Object Properties** toolbar becomes available once you have viewports. Selecting the **Current VP Freeze** icon makes the selected layers invisible in the

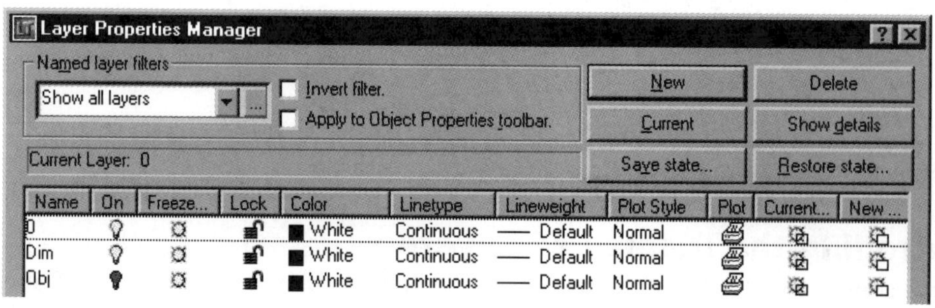

***Figure 4-8** **Layer Properties Manager** dialog box with **Current and New VP Freeze** icons*

active floating viewport only. The frozen layers will still be visible in other viewports. If you want to freeze some layers in the new floating viewports, then select the **New VP Freeze** toggle icon for the selected layers. AutoCAD LT will freeze the layers in subsequently created new viewports without affecting the viewports that already exist. (Paper space is discussed in Chapter 11.) Also, check the **VPLAYER** command for selectively freezing the layers in viewports.

Tip

*The widths of the column headings in the **Layer Properties Manager** dialog box can be decreased or increased, by positioning the cursor between the column headings on the separator (the cursor turns into a plus sign) and then holding down the pick button of your pointing device and dragging the cursor to the right or the left. This way you can make the widths smaller to display all the headings.*

Lock or Unlock a Layer

While working on a drawing, if you do not want to accidentally edit some objects on a particular layer but still need to have them visible, you can use the **Lock/Unlock** toggle icon to lock the layers. When a layer is locked you can still use the objects in the locked layer for Object Snaps and inquiry commands such as **LIST**. You can also make the locked layer the current layer and draw objects on it. The locked layers are plotted. The **Unlock** option negates the **Lock** option and allows you to edit objects on the layers previously locked.

Plot/Don't Plot

If you do not wish to plot a particular layer, for example, construction lines, you can use the **Plot/Don't Plot** toggle icon (printer) to make the layer nonplottable. This icon is available in the **Layer Properties Manager** dialog box. The construction lines will not be plotted if its layer is made nonplottable.

Tip

*It is faster and convenient to use the layer drop-down list in the **Object Properties** toolbar to make a layer current and control the display features of the layer (**On/Off**, **Freeze/Thaw**, **Lock/Unlock**).*

Assigning Linetype to Layer

To assign a new linetype to a layer, click on the current linetype displayed with a particular layer in the **Layer Properties Manager** dialog box. When you click on the linetype, AutoCAD LT will display the **Select Linetype** dialog box (Figure 4-9), which displays the linetypes that are defined and loaded on your system. Select the new linetype and then choose the **OK** button. The linetype that you select is assigned to the layer you selected initially. The layers are, by default, assigned continuous linetype and white color if no layer is selected at the time of creating a new layer. Otherwise the new layer created takes the properties of the selected layer.

If you have not loaded the linetypes and are opening the **Select Linetype** dialog box for the first time, only the **Continuous** linetype is displayed. You need to load the linetypes you want and then assign them to the layers. To load the linetypes, choose the **Load** button in the **Select Linetype** dialog box. This displays the **Load or Reload Linetypes** dialog box (Figure 4-10), which displays all linetypes in the **ACLT.LIN** file. In this dialog box you can select individual

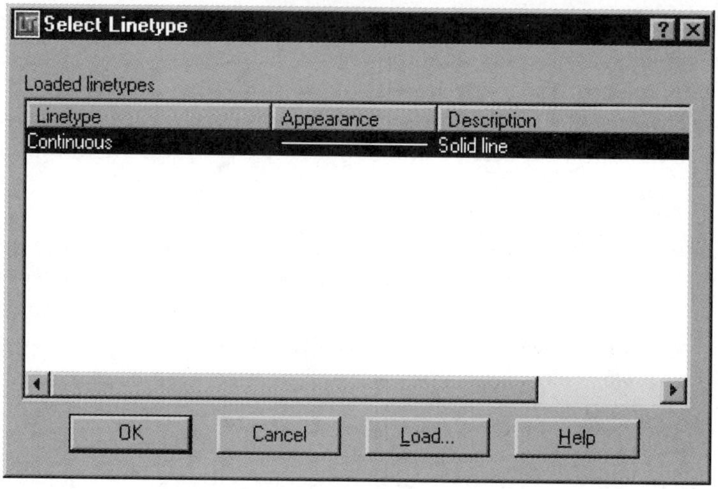

***Figure 4-9 Select Linetype** dialog box*

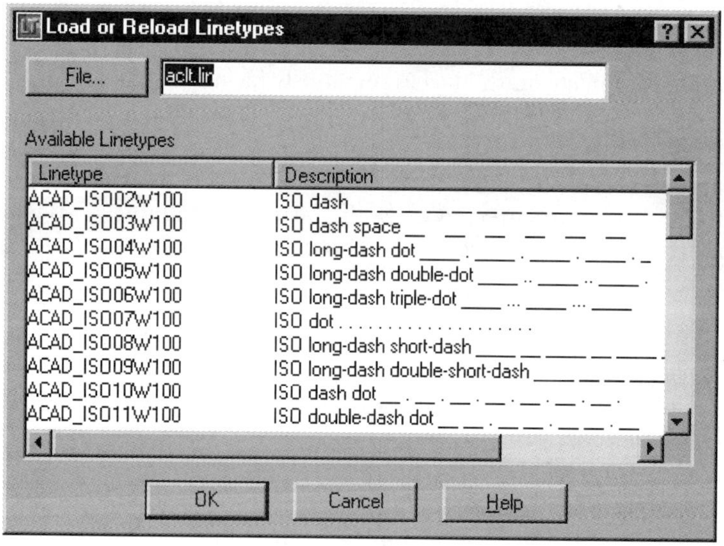

***Figure 4-10 Load or Reload Linetypes** dialog box*

linetypes or a number of linetypes by holding the SHIFT or CTRL key on the keyboard and then selecting the linetypes. If you right-click, AutoCAD LT displays the shortcut menu that you can use to select all linetypes. Then by choosing the **OK** button, the selected linetypes are loaded and therefore displayed in the **Select Linetype** dialog box. Here you can select the desired linetype and select **OK**. The selected linetype is assigned to the selected layer.

Note

*By default the linetypes in the ACLT.LIN file are displayed in the **Load or Reload Linetypes** dialog box. You can select the linetypes in the ACLTISO.LIN file by choosing the **File** button in the dialog box and then opening the ACLTISO.LIN file from the **Select Linetype Files** dialog box.*

Assigning Color to a Layer

To assign a color, select the color swatch in a particular layer in the **Layer Properties Manager** dialog box; AutoCAD LT will display a **Select Color** dialog box. Select a desired color and then choose the **OK** button. The color you selected will be assigned to the selected layer. The number of colors is determined by your graphics card and monitor. Most color systems support eight or more colors. If your system allows it, you may choose a color number between 0 and 255 (256 colors). The following are the first seven standard colors:

Color number	Color name	Color number	Color name
1	Red	5	Blue
2	Yellow	6	Magenta
3	Green	7	White
4	Cyan		

Note

Using nonstandard colors may cause compatibility problems if the drawings are used on other systems with different colors. On a light background, the color "white" appears black. Certain colors are hard to see on light backgrounds and others are hard to see on dark backgrounds, so you may have to use different colors than those specified in some examples and exercises.

Assigning Lineweight to Layer

Lineweight is used to give thickness to the objects in a given layer. This thickness is displayed on the screen if the display of the lineweight is on. The lineweight assigned to the objects can also be plotted. If you are making a sectional plan at a certain height, you can assign a layer with a larger value of lineweight to create the objects through which the section is made. Another layer with lesser lineweight can be used to show the objects through which the section does not pass. To assign a lineweight to a layer, select the layer and then click on the lineweight associated with it; the **Lineweight** dialog box (Figure 4-11) appears. Select a lineweight from the **Lineweights** list. Choose **OK** to return to the **Layer Properties Manager** dialog box.

Note

*The **LWEIGHT** command for editing and displaying the lineweight is discussed in detail later in this chapter.*

Assigning Plot Style to Layer

Plot Style is a group of property settings such as color, linetype, and lineweight that can be assigned to a layer. The applied plot style affects the drawings while plotting only. The drawing in which you are working should be in named plot style mode (.stb) to make the plot style available in the **Layer Properties Manager** dialog box. If the **Plot Style** icon in the dialog box is not available, then you are in color dependent mode (.ctb). To make it available, open the **Options** dialog box from the **Tools** menu and select the **Use named plot style** radio button in the **Plotting** tab. Choose **OK**. After changing the plot style to named plot style dependent, you have to open a new drawing to apply this setting. The default Plot Style is **Normal** in which the color, linetype, and lineweight are BYLAYER. To assign a Plot Style to a layer, select the layer and then click on its plot style; the **Select Plot Style** dialog box (Figure 4-12) appears where you can select a specific plot style from the **Plot Styles** list of available plot styles. The plot styles have to be created before you can use them (see Chapter 12, Plotting Drawings). Choose **OK** to return to the **Layer Properties Manager** dialog box.

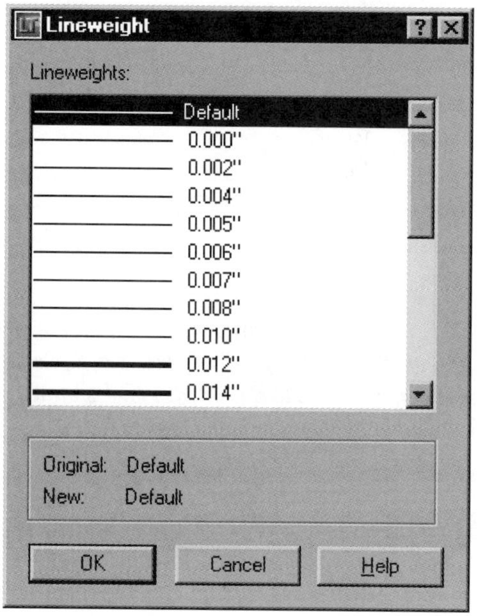

Figure 4-11 ***Lineweight*** *dialog box*

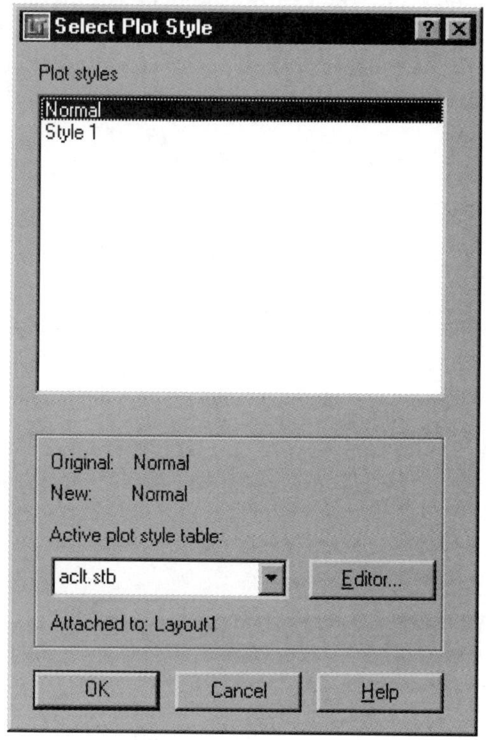

Figure 4-12 ***Select Plot Style*** *dialog box*

Note

*After changing the plot style mode (color dependent or named) in the **Plotting** tab of **Options** dialog box you need to open a new drawing to apply the settings.*

Tip

*You can also change the plot style mode from the command line by using the **PSTYLEPOLICY** system variable. A value of **0** sets it to color dependent mode and a value of **1** sets it to named mode.*

Deleting Layers

You can delete a layer by selecting the layer and then choosing the **Delete** button in the **Layer Properties Manager** dialog box. To delete a layer it is necessary that the layer should not contain any objects. You cannot delete the Layers 0 and Defpoints, a current layer and an xref dependent layer.

Show Details

You can also control the display of layers by first selecting a layer and then choosing the **Show**

details button in the **Layer Properties Manager** dialog box. This displays the layer details area in the lower portion of the **Layer Properties Manager** dialog box. The details area displays the properties associated with the selected layer. Using this box, you can change the name, color, linetype, lineweight, and other properties of the selected layer. Note that you cannot change the name of 0 layer. Once you have chosen the **Show details** button, the **Hide details** button becomes available. This button closes the details area.

Selective Display of Layers

If you have a limited number of layers it is easy to scan through them. However, if you have many layers it is sometimes difficult to search through the layers. To solve this problem you can use the drop-down list in the **Named layer filters** area of the **Layer Properties Manager** dialog box to display the layers selectively. If you want to display only **XRef dependent** layers, you can select this option in the drop-down list. You can also define your own filter specification by choosing the **[...]** button to display the **Named Layer Filters** dialog box (Figure 4-13). When loaded, initially this dialog box contains the default specifications. If you want to list only those layers that are red in color, enter red in the **Color** edit box, then give this filter a name in the **Filter name** edit box (Red color) and then choose the **Add** button. This filter gets added to the list of Named layer filters. Now, when you select this named filter from the drop-down list in the Named layer filters area, AutoCAD LT will display only the layers that

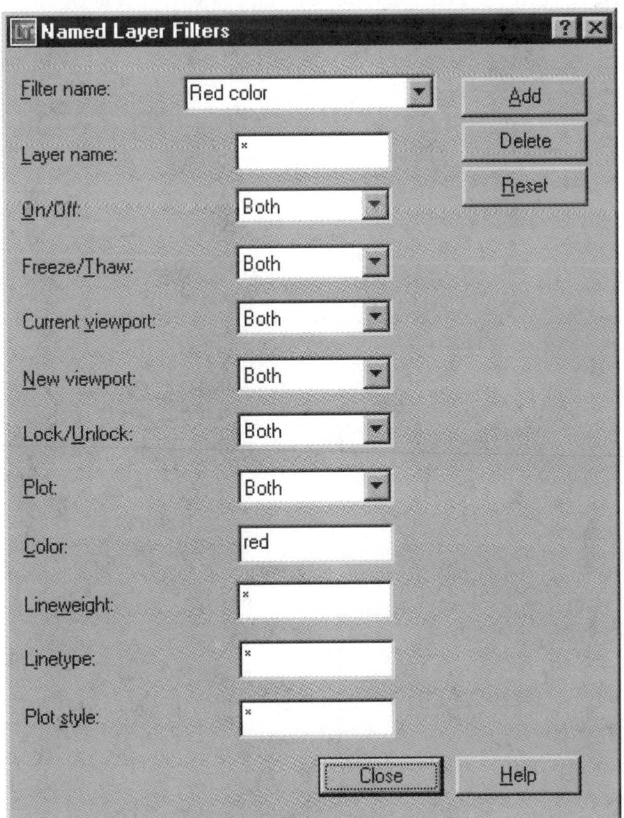

***Figure 4-13 Named Layer Filters** dialog box*

are red in color in the **Layer Properties Manager** dialog box. If you select **Show all layers** in the drop-down list, all layers will be displayed.

In the **Layer Properties Manager** dialog box, when you select the **Invert filter** check box, you invert the filter that you have selected. For example, if you have selected **Show all layers**, none of the layers will be displayed. You can right-click in the layers list box of the dialog box to display the shortcut menu. You can then use the **Layer Filter** and **Named Filter** options to selectively display layers. You can also apply the current layer filter to the **Layer Control** list in the **Object Properties** toolbar by selecting the **Apply to Object Properties toolbar** check box. Choose **OK** in the dialog box. You will notice that only the filtered layers are displayed in the **Layer Control** drop-down list of the **Object Properties** toolbar. Also if you place your cursor on the arrow of the drop-down list, the tooltip indicates that the filter has been applied. If the current layer is not one among the filtered layers, it will still be displayed in the **Layer Control** list box of the **Object Properties** toolbar. In the **Layer Properties Manager** dialog box, the current layer is not displayed in the list box if it is not among the filtered layers.

Layer States

You can save and then restore the properties of all the layers in a drawing using the **Save state** and **Restore state** buttons in the **Layer Properties Manager** dialog box. While working on a drawing, at any point in time you can save all the layers with their present properties settings under one name and then restore it anytime later. Choosing the **Save state** button displays the **Save Layer States** dialog box, where you can give the name for saving the layers and you can specify the different states and properties of the layers that you want to save. Choosing the **OK** button saves the checked states and properties of the layers together under the specified name with and extension **.las**.

You can restore the layer states later from the **Layer States Manager** dialog box that is displayed by choosing the **Restore state** button in the **Layer Properties Manager** dialog box. In this dialog box you are allowed to edit the states and properties of the saved state. You can rename and delete a state. You are also allowed to import and export the layer states from and to different other folders using the **Import** and **Export** buttons in the **Layer States Manager** dialog box.

Note

You can also set layers from the command line by entering ***-LAYER*** *at the Command prompt.*

Example 1 *Mechanical*

Set up four layers with the following linetypes and colors. Then make the drawing (without dimensions) as shown in Figure 4-14.

Layer name	Color	Linetype	Lineweight
Obj	Red	Continuous	0.031"
Hid	Yellow	Hidden	0.008"
Cen	Green	Center	0.006"

In this example, assume that the limits and units are already set. Before drawing the lines, you

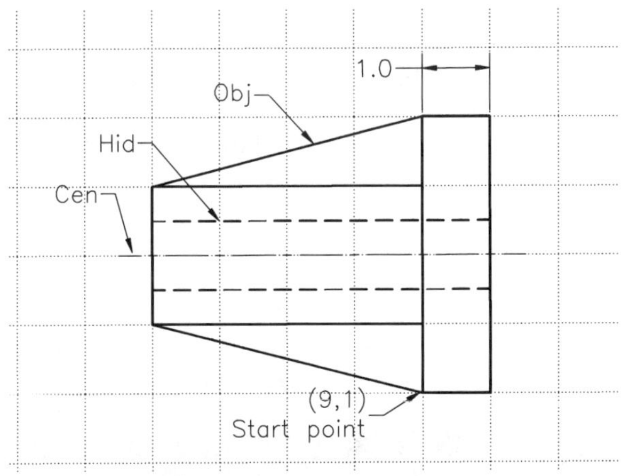

Figure 4-14 Drawing for Example 1

need to create layers and assign colors, linetypes, and lineweights to these layers. Also, depending on the objects that you want to draw, you need to set that layer as current. In this example you will create the layers using the **Layer Properties Manager** dialog box. You will use the **Object Properties** toolbar to set the layers current and then draw the figure.

1. As the lineweights specified are in inches, first change the units for lineweight if they are in millimeters. Choose **Lineweight** from the **Format** menu to display the **Lineweight Settings** dialog box. Select the **Inches [In]** radio button in the **Units for Listing** area of the dialog box and then choose the **OK** button.

2. Choose the **Layers** button in the **Object Properties** toolbar, or choose **Layer** in the **Format** menu, or enter **LAYER** at the Command prompt to display the **Layer Properties Manager** dialog box. The layer **0** with default properties is displayed in the list box.

3. Choose the **New** button. AutoCAD LT automatically creates a new layer (**Layer1**) having the default properties and displays it in the list box. Change its name by entering **Obj** in place of Layer1.

4. Now choose the color swatch to display the **Select Color** dialog box. Select color **Red** and then choose **OK**. Red color is assigned to layer **Obj**.

5. Choose the lineweight of the layer to display the **Lineweight** dialog box and choose 0.031" and then choose **OK**. The 0.031 lineweight is assigned to layer **Obj**.

6. Again choose the New button. AutoCAD LT automatically creates a new layer (Layer1) having the properties of layer Obj. Change its name by entering **Hid** in place of Layer1.

7. Choose the color swatch to display the **Select Color** dialog box. Select color **Yellow** and then choose **OK**.

8. Choose the linetype of the layer to display the **Select Linetype** dialog box. If the linetype

Hidden is not displayed in the dialog box, choose the **Load** button to display the **Load and Reload** dialog box. Choose **Hidden** from the list and choose the **OK** button. Choose **HIDDEN** in the **Select Linetype** dialog box and then choose **OK**.

9. Choose the lineweight of the layer to display the **Lineweight** dialog box and choose **0.008"** and then choose OK. The 0.008 lineweight is assigned to layer **Hid**.

10. Similarly create the new layer **Cen** and assign the color Green, linetype CENTER, and lineweight 0.006" to it.

11. Select **Obj** layer and then choose the **Current** button to make the **Obj** layer current (Figure 4-15). Choose the **OK** button to exit the dialog box.

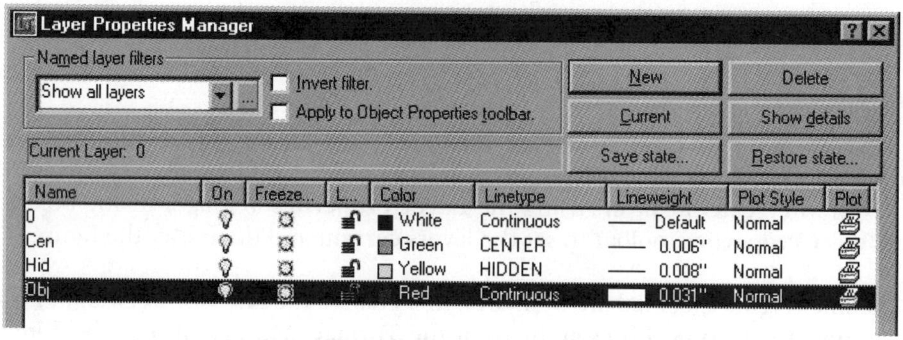

Figure 4-15 *Layers created for Example 1*

12. Using the **LINE** command, draw the object lines in layer **Obj**.

Command: **LINE** [Enter]
Specify first point: **9,1** [Enter]
Specify next point or [Undo]: **9,9** [Enter]
Specify next point or [Undo]: **11,9** [Enter]
Specify next point or [Close/Undo]: **11,1** [Enter]
Specify next point or [Close/Undo]: **9,1** [Enter]
Specify next point or [Close/Undo]: **1,3** [Enter]
Specify next point or [Close/Undo]: **1,7** [Enter]
Specify next point or [Close/Undo]: **9,9** [Enter]
Specify next point or [Close/Undo]: [Enter]

Command: [Enter] *(Invokes the LINE command.)*
Specify first point: **1,3** [Enter]
Specify next point or [Undo]: **9,3** [Enter]
Specify next point or [Undo]: [Enter]

Command: [Enter] *(Invokes the LINE command.)*
Specify first point: **1,7** [Enter]
Specify next point or [Undo]: **9,7** [Enter]
Specify next point or [Undo]: [Enter]

13. Now, make the **Hid** layer current from the **Object Properties** toolbar. Select the down arrow in the **Layer Control** list box to display the drop-down list (Figure 4-16). Choose the **Hid** layer from the list to make it current.

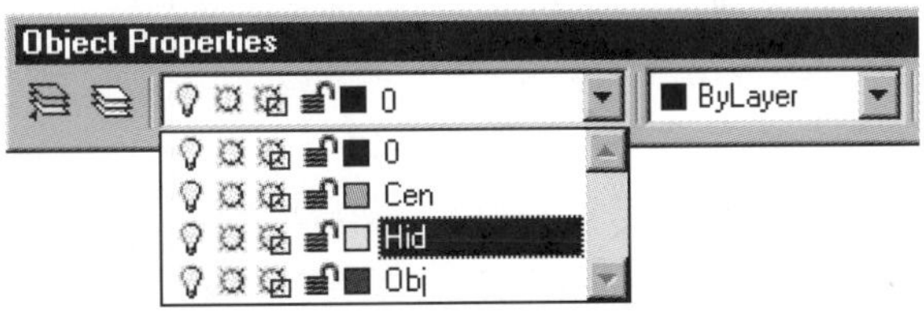

Figure 4-16 Making Hid layer current from the Object Properties toolbar

14. Now draw the following lines using the **LINE** command.

Command: **LINE** [Enter]
Specify first point: **1,4** [Enter]
Specify next point or [Undo]: **11,4** [Enter]
Specify next point or [Undo]: [Enter]

Command: [Enter] *(Invokes the LINE command.)*
Specify first point: **1,6** [Enter]
Specify next point or [Undo]: **11,6** [Enter]
Specify next point or [Undo]: [Enter]

15. Again choose the down arrow in the **Layer Control** list box to display the drop-down list Choose the **Cen** layer from the list to make it current.

16. Now draw the following lines using the **LINE** command.

Command: **LINE** [Enter]
Specify first point: **0,5** [Enter]
Specify next point or [Undo]: **12,5** [Enter]
Specify next point or [Undo]: [Enter]

Exercise 1 *Mechanical*

Set up layers with the following linetypes and colors. Then make the drawing (without dimensions) as shown in Figure 4-17. The distance between the dotted lines is 1 unit.

Layer name	Color	Linetype
Object	Red	Continuous
Hidden	Yellow	Hidden
Center	Green	Center
Dimension	Blue	Continuous

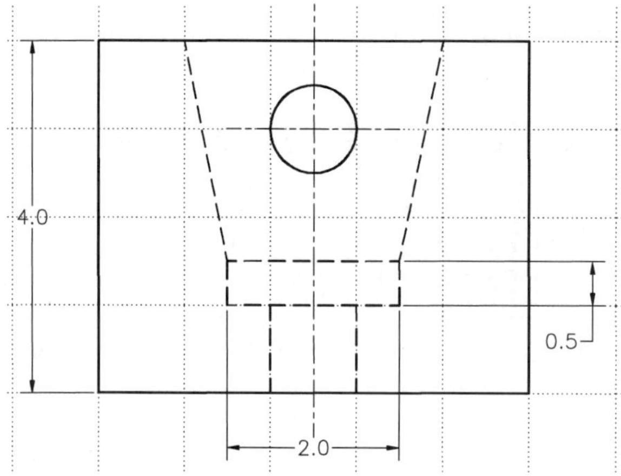

Figure 4-17 *Drawing for Exercise 1*

Tip

*You should note that the **Linetype Control**, **Lineweight Control**, **Color Control**, and the **Plot Style Control** drop-down lists in the **Object Properties** toolbar should display **ByLayer** as the current properties of the objects. This is to ensure that when you draw an object, it takes those properties from the current layer in which the object is drawn.*

LINETYPE COMMAND

The **LINETYPE** command can be used to load, delete, and make current the linetype which can be used to draw an object. You can invoke the LINETYPE command from the **Format** menu or by entering LINETYPE at the Command prompt. The **Linetype Manager** dialog box (Figure 4-18) is displayed when you invoke this command. By default the linetypes displayed are ByLayer, ByBlock, and Continuous. Choosing the **Load** button displays the **Load and Reload Linetype** dialog box, which is also displayed from the **Layer Properties Manager** dialog box. The linetypes that you load here will be displayed in the **Select Linetypes** dialog box also. Similarly, if you have loaded some linetypes using the **LAYER** command, they will be displayed in the **Linetype Manager** dialog box. You can make a linetype current by using the **Current** button. All the new objects will be drawn with the current linetype. If you have chosen the current linetype as **ByLayer** then the object will be drawn with the linetype of the layer. You can also make any linetype current other than ByLayer, and the object will be drawn with the particular linetype. The current linetype is also displayed in the **Linetype Control** field of the **Object Properties** toolbar.

LWEIGHT COMMAND

The **LWEIGHT** command can be used to assign a lineweight to an object. You can invoke the LWEIGHT command by selecting **Lineweight** from the **Format** menu or by entering **LWEIGHT** at the Command prompt to display the **Lineweight Settings** dialog box (Figure 4-19). You can also use the **Settings** option in the shortcut menu, displayed on right-clicking on the **LWT** button on the status bar, to invoke the dialog box. In this dialog box you can choose the current lineweight for the objects. You can change the units for lineweight and also the display

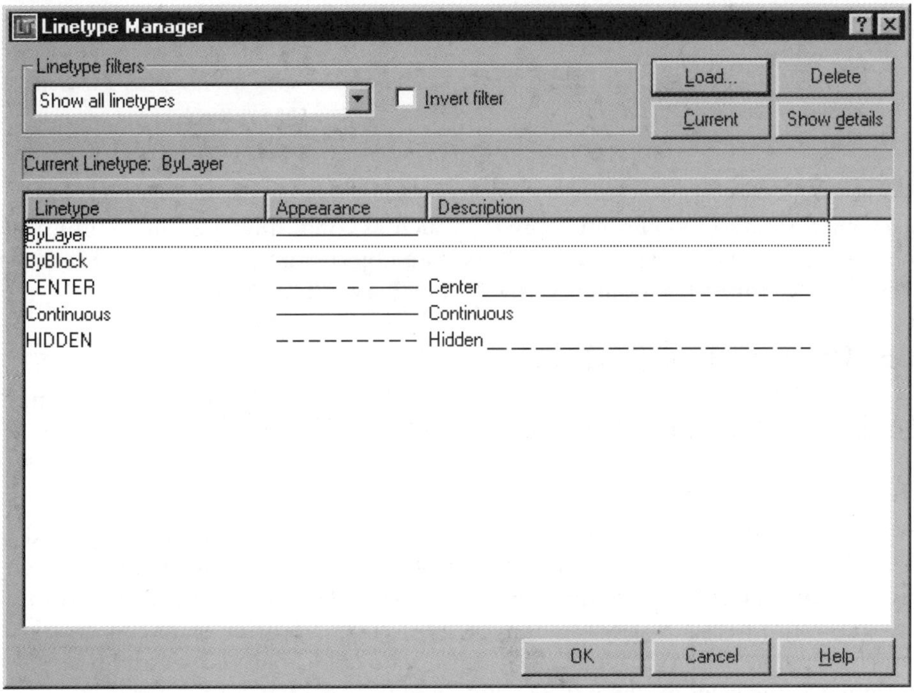

Figure 4-18 ***Linetype Manager*** *dialog box*

of lineweights for the current drawing. The lineweights are displayed in pixel widths and depending upon the lineweight value chosen, the lineweights are displayed if the **Display Lineweight** check box is checked. For a large drawing this increases the regeneration time and should be cleared. You can also turn the display of lineweights to on or off directly from the status bar by choosing the **Lwt** button. The slider bar for **Adjust Display Scale** affects the regeneration time as well. You can keep the slider bar at **Max** for getting a good display of different lineweights on the screen in the Model space; otherwise, keep it at **Min** for faster regeneration.

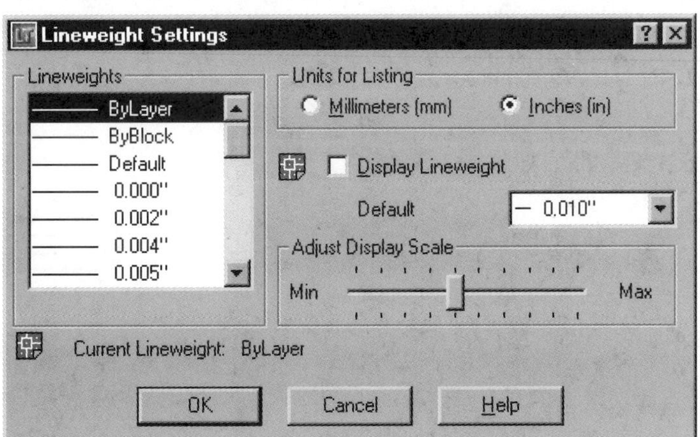

Figure 4-19 ***Lineweight Settings*** *dialog box*

Tip

For large drawings you should keep the display scale at the minimum value for increasing the performance of AutoCAD LT. It is also recommended to clear the ***Display Lineweight*** *check box or simply turn the display of lineweight to off from the status bar.*

OBJECT PROPERTIES

An object when created has certain properties such as color, linetype, lineweight, plot style, and layer associated with it. These properties of an object can be changed and made current using the **Object Properties** toolbar or the **Properties** window.

Object Properties Toolbar

You can use the **Object Properties** toolbar to directly change the general properties of the selected objects and also set those properties current.

Color

Select the object whose color you want to change. The current color of the object is displayed in the **Color Control** list box. Display the list of colors by selecting the drop-down arrow (Figure 4-20) and select a new color. The color of the selected object is changed to the new color. If you want to set a different color current, choose a color from the drop-down list without selecting any object. By default the color of an object is **ByLayer** and hence takes the color assigned to the layer in which it is created. From the **Color** drop-down list you can select any other color to assign the color explicitly to the objects drawn. This color is set as current and all the new objects will be drawn in this color. If you want to assign a color that is not displayed in the list, choose the **Other** option to display the **Select Color** dialog box. You can select the desired color in this dialog box, and then choose the **OK** button.

Linetype

From the **Linetype Control** drop-down list (Figure 4-21), you can select the new linetype and assign it to the selected object. You can also make a linetype current by selecting it from the list. By default the linetype of an object is **ByLayer**. The linetypes that are loaded from the **LAYER** or the **LINETYPE** command will be listed. If you want to assign a linetype that is not displayed in the list, choose the **Other** option to display the **Linetype Manager** dialog box. You can use the **Load** button to load other linetypes.

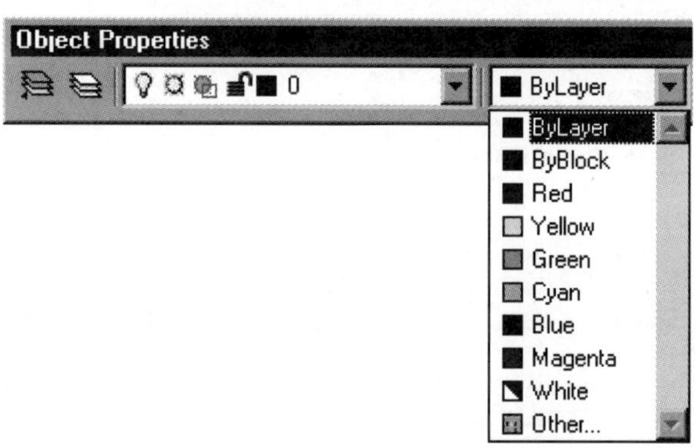

Figure 4-20 *Setting object color from the* ***Object Properties*** *toolbar*

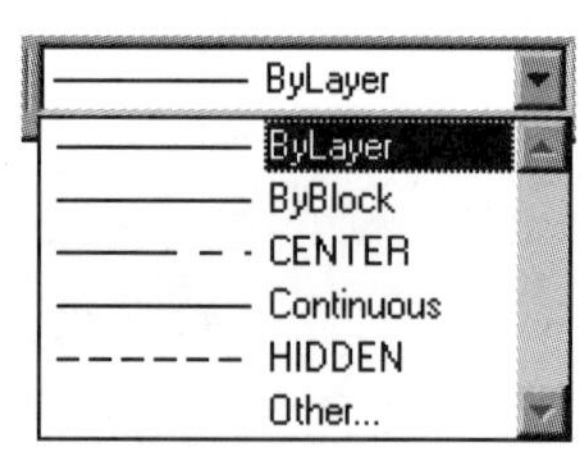

Figure 4-21 *Linetype Control list box*

Lineweight

Similarly, you can select a different lineweight value from the **Lineweight Control** drop-down list in the Object Properties toolbar to assign to the selected object. Also you can set a new lineweight current by selecting it from the list.

Plot Style

You can select a different plot style for the selected object from the **Plot Style Control** drop-down list. Also you can set a new plot style current by selecting it from the list.

Note

When you set a linetype, lineweight, or color current, all the objects drawn thereafter will have the current linetype, lineweight, and color. The properties of the current layer are not considered. You need to set current those properties to ***ByLayer*** *again so that the objects assume the linetype, lineweight, color, or plot style of the layer in which they are created.*

Properties Window

Toolbar:	Standard > Properties
Menu:	Tools > Properties
Command:	PROPERTIES, CH, MO

You can also use the **Properties** window to change the general properties of the selected objects and also set those properties current. The Properties window can be displayed by invoking the **PROPERTIES** command. It can also be invoked by selecting an object and then right-clicking to display a shortcut menu; choose **Properties** from the shortcut menu. The **Properties** window is displayed (Figure 4-22), from where you can change the different properties of the selected object. When you select the objects whose properties you want to change, the **Properties** window displays the properties of the selected object. Depending on the object selected, the properties differ. You can also make the different properties current by selecting no object and then selecting the particular properties in the window. Right-clicking in the **Properties** window displays a shortcut menu where you can choose to **Allow Docking** or **Hide** the window.

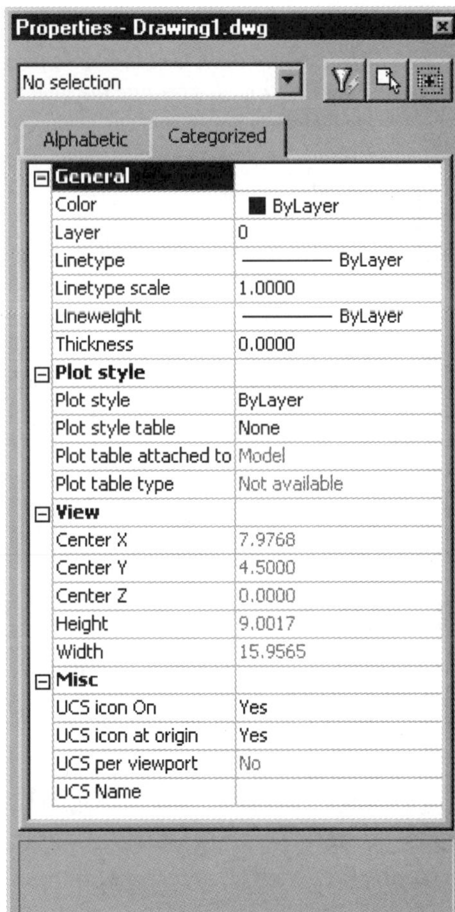

Figure 4-22 ***Properties*** *window*

When you select **Color** from the **General** list, the color of the selected object is displayed along with a drop-down arrow. By default the color of an object is **ByLayer** and hence takes the color assigned to the layer in which it is created. From the **Color** drop-down list you can select any other color you wish to assign to the selected object. If you want to assign a color that is not displayed in the list, choose the **Other** option to display the **Select Color** dialog box. You can select the desired color in this dialog box, and then choose the **OK** button.

Similarly, you can set current the linetype, lineweight, and linetype scale and other properties for the other objects individually. You can also assign a hyperlink to an object. The **Hyperlink** name and description assigned to the object is displayed. Choosing the **[...]** button, displays the **Insert Hyperlink** dialog box where you can enter the path of the URL or file to which you want to link the selected object.

Tip

*You can keep the **Properties** window open while working in a drawing. When no object is selected, the current settings of the drawing are listed. Whenever you need to change a certain property of an object, select the object so that its properties are listed in the **Properties** window and make the required changes.*

Note

*You can use the **CHPROP** command to change the properties of an object from the command line. You can also change the general properties of an object by using the **Properties** option of the **CHANGE** command. Both these command along with the **Geometry** portion of the **Properties** window to edit the objects will be discussed in Chapter 18, Object Grouping and Editing Commands.*

*The system variables **CECOLOR**, **CELTYPE**, and **CELWEIGHT** controls the current color, linetype, and lineweight of the object. You can set the properties of the objects current using these variables from the command line.*

When you set the property of the object current, it does not consider the property of the layer in which the object will be drawn.

Tip

*It is easier and faster to set the properties of the object current from the **Object Properties** toolbar. It is better to leave the different properties of objects at **ByLayer** to avoid confusion.*

Exercise 2 *General*

Draw a hexagon on layer OBJ in red color. Let the linetype be hidden. Now, use the **PROPERTIES** command to change the layer to some other existing layer, the color to yellow, and the linetype to continuous.

GLOBAL AND CURRENT LINETYPE SCALING

The LTSCALE system variable controls the global scale factor of the lines in a drawing. For example, if LTSCALE is set to 2, all lines in the drawing will be affected by a factor of 2. Like LTSCALE, the **CELTSCALE** system variable controls the linetype scaling. The difference is that CELTSCALE determines the current linetype scaling. For example, if you set CELTSCALE to 0.5, all lines drawn after setting the new value for CELTSCALE will have the linetype scaling factor of 0.5. The value is retained in the CELTSCALE system variable. Line (a) in Figure 4-23 is drawn with a CELTSCALE factor of 1; line (b) is drawn with a CELTSCALE factor of 0.5. The length of the dash is reduced by a factor of 0.5 when CELTSCALE is 0.5. The net scale factor is equal to the product of CELTSCALE and LTSCALE. Figure 4-23(c) shows a line that is drawn with LTSCALE of 2 and CELTSCALE of 0.25. The net scale factor = LTSCALE x CELTSCALE = 2 x 0.25 = 0.5. You can also change the global and current scale factors by entering a desired value in the **Linetype Manager** dialog box. If you choose the **Show details** button, the properties associated with the selected linetype are displayed (Figure 4-24). You can change the values according to your drawing requirements.

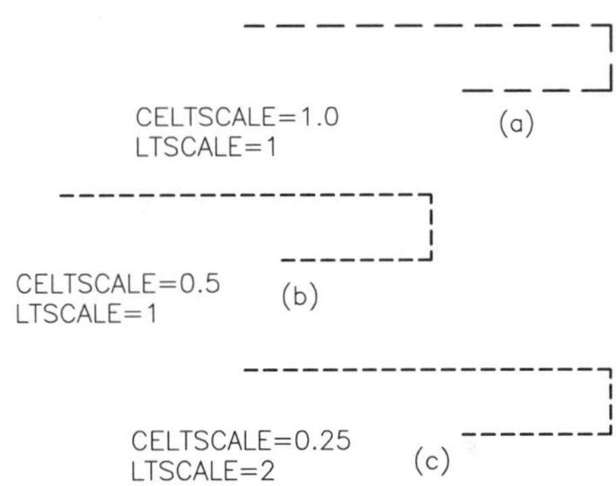

***Figure 4-23** Using **CELTSCALE** to control current linetype scaling*

***Figure 4-24 Details** area of the **Linetype Manager** dialog box*

LTSCALE FACTOR FOR PLOTTING

The **LTSCALE** factor for plotting depends on the size of the sheet you are using to plot the drawing. For example, if the limits are 48 by 36, the drawing scale is 1:1, and you want to plot the drawing on a 48" by 36" size sheet, then the **LTSCALE** factor is 1. If you check the specification of the Hidden linetype in the **ACLT.LIN** file, the length of each dash is 0.25. Hence, when you plot a drawing with 1:1 scale, the length of each dash in a hidden line is 0.25.

However, if the drawing scale is 1/8" = 1' and you want to plot the drawing on 48" by 36" paper, the **LTSCALE** factor must 8 x 12 = 96. The length of each dash in the hidden line will increase by a factor of 96, because the **LTSCALE** factor is 96. Therefore, the length of each dash will be (0.25 x 96 = 24) units. At the time of plotting, the scale factor for plotting must be 1:96 to plot the 384' by 288' drawing on 48" by 36" paper. Each dash of the hidden line that was 24" long on the drawing will be 24/96 = 0.25" long when plotted. Similarly, if the desired text size on the paper is 1/8", the text height in the drawing must be 1/8 x 96 = 12".

Ltscale Factor for PLOTTING = Drawing Scale

Sometimes your plotter may not be able to plot a 48" by 36" drawing, or you might like to decrease the size of the plot so that the drawing fits within a specified area. To get the correct dash lengths for hidden, center, or other lines, you must adjust the **LTSCALE** factor. For example, if you want to plot the previously mentioned drawing in a 45" by 34" area, the correction factor is:

Correction factor = 48/45
= 1.0666

New LTSCALE factor = **LTSCALE** factor x Correction factor
= 96 x 1.0666
= 102.4

New Ltscale Factor for PLOTTING = Drawing Scale x Correction Factor

Note
If you change the LTSCALE factor, all lines in the drawing are affected by the new ratio.

Changing Linetype Scale Using the PROPERTIES Command

You can also change the linetype scale of an object by using the **PROPERTIES** command. When you choose this command, AutoCAD LT will display the **Properties** window. All the properties of the selected objects are displayed in the **Properties** window. To change the current linetype scale, select the objects and then invoke the **PROPERTIES** command to display the **Properties** window (Figure 4-25). In this window, select the **Categorized** tab and locate the **Linetype scale** edit box. Here enter the new linetype scale. The linetype scale of the selected objects is changed to the new value you entered.

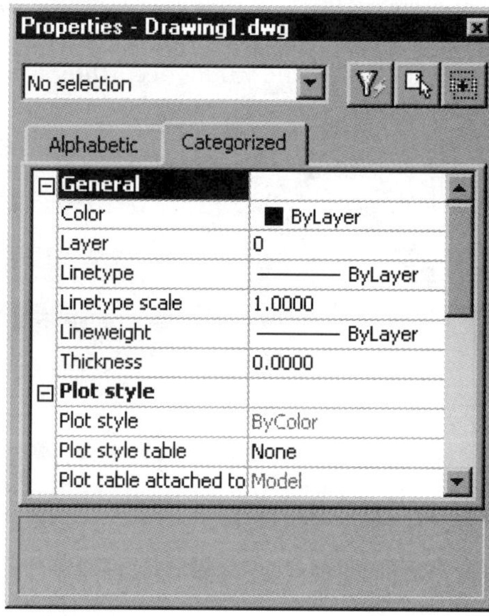

Figure 4-25 ***Properties*** *window*

AUTOCAD DESIGNCENTER

Toolbar:	Standard Toolbar > AutoCAD DesignCenter
Menu:	Tools > AutoCAD DesignCenter
Command:	ADCENTER

The AutoCAD DesignCenter allows you to reuse and share contents in different drawings. You can use this window to locate the drawing data with the help of search tools and then use it in your drawing. You can insert layers, linetypes, blocks, layouts, external references, and other drawing content in any number of drawings. Therefore, if a layer is created once, it can be used repeatedly any number of times.

Open the AutoCAD DesignCenter window and choose the **Tree View Toggle** button to display the **Tree pane** and the **Palette** side by side (if they are not already displayed). Open the folder in which the drawing is saved containing the layers and linetypes you want to insert. You can use the **Desktop** and the **Up** buttons to open the folder you want. For example, you want to insert some layers and linetypes from the drawing **c1d2** in the directory D:\LT 2002\c1. Open the desired folders and open c1d2 to display its contents. Choose Layers and all the layers created in **c1d2** are displayed in the Palette. Select **Box** and **Cen** using the CTRL key and right-click to display the shortcut menu (Figure 4-26). Choose **Add Layer[s]**. The two layers are added to the current drawing. You can also drag and drop the desired layers into the current drawing. If you open the **Layer control** drop-down list in the **Object Properties** toolbar, the two layers are listed. Similarly you can insert the linetypes.

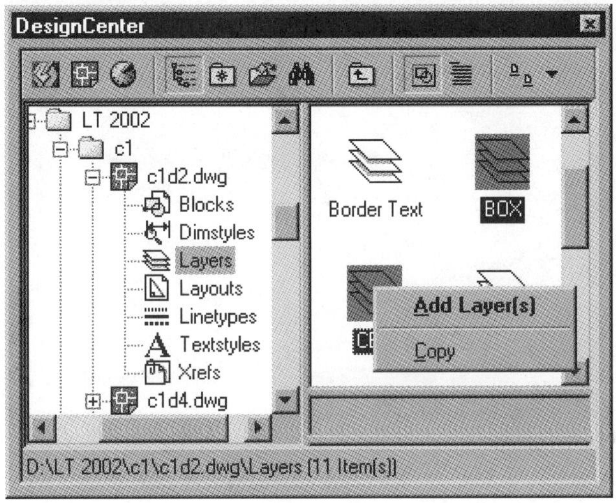

Figure 4-26 *AutoCAD DesignCenter with the shortcut menu*

Note
AutoCAD DesignCenter is discussed in detail in Chapter 6, Editing Commands-II.

DRAFTING SETTINGS DIALOG BOX

Menu:	Tools > Drafting Settings
Command:	DSETTINGS

You can use the **Drafting Settings** dialog box to set drawing modes such as Grid, Snap, Object Snap, and Polar Tracking. All these aids help you to draw accurately and also increase the drawing speed. You can right-click on the **Snap**, **Grid**, **Polar**, or **Osnap** buttons on the status bar to display a shortcut menu (Figure 4-27) and choose **Settings** to display the **Drafting Settings** dialog box. This dialog box operates under three main tabs: **Snap and Grid** (Figure 4-28)**, Object Snap** and **Polar Tracking**. When you start AutoCAD LT you are provided with the default settings of these aids. You can change them according to your requirements by using the **Drafting Settings** dialog box.

Figure 4-27 Settings *in the shortcut menu*

Setting Grid

The grid lines are lines of dots on the screen at predefined spacing (Figure 4-29). These

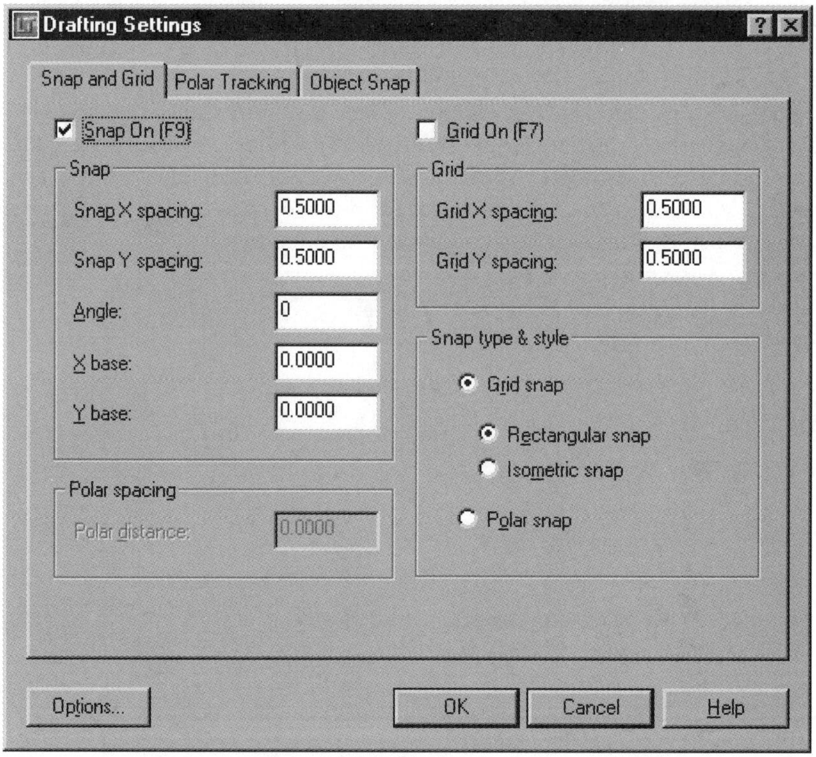

***Figure 4-28 Drafting Settings** dialog box (**Snap and Grid** tab)*

dotted lines act as a graph that can be used as reference lines in a drawing. You can change the distance between the grid dots as per your requirement. The grid pattern appears within the drawing limits, which helps to define the working area. The grid also gives you a sense of the size of the drawing objects.

***Figure 4-29** Grid lines*

Grid On (F7): Turning the Grid On or Off

You can turn the grid on/off by using the **Grid On** check box in the **Drafting Settings** dialog box (Figure 4-28). You can also turn grid on or off by choosing the **Grid** button in the status bar, or use the **On** or **Off** option in the shortcut menu when you right-click on the **Grid** button in the status bar, or use the **GRID** command. The function key **F7** acts as a toggle key for turning the grid on or off. When grid is turned on after it has been off, the grid is set to the previous grid spacing.

Grid X Spacing and Grid Y Spacing

The **Grid X spacing** and **Grid Y spacing** edit boxes in the **Drafting Settings** dialog box are used to define a desired grid spacing along the X and Y axis. For example, to set the grid spacing to 0.5 units, enter 0.5 in the **Grid X spacing** and **Grid Y spacing** edit boxes (Figure 4-30). You can also enter different values for horizontal and vertical grid spacing (Figure 4-31). If you enter only the grid X spacing and then choose the **OK** button in the dialog box, the corresponding Y Spacing value is automatically set to match the X spacing value. Therefore, if you want different X and Y spacing values, you need to set the X spacing first, then set the Y spacing.

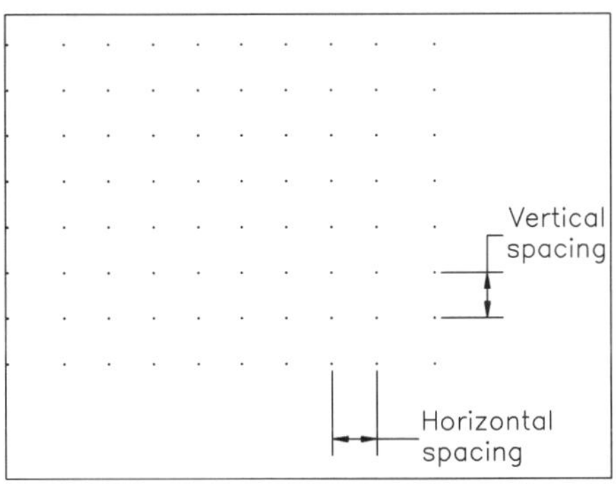

Figure 4-30 *Controlling grid spacing*

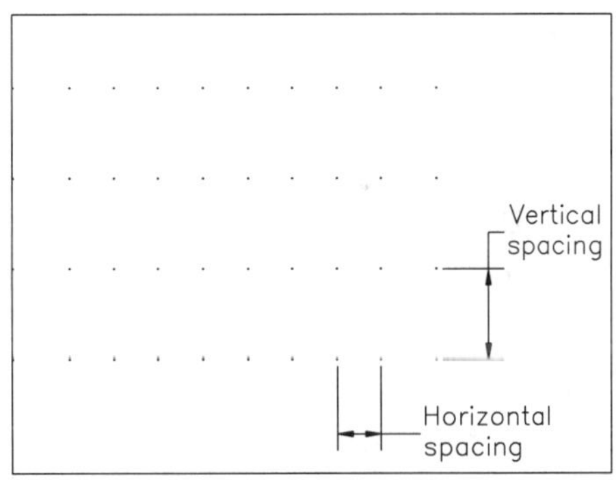

Figure 4-31 *Creating unequal grid spacing*

Note

*You can also use the **GRID** command to set the grid from the command line. At the prompt sequence for the **GRID** command you can use the **Aspect** option to assign a different value to the horizontal and vertical grid spacing.*

Tip

*The grid and the snap grid (discussed in the next section) are independent of each other. However, you can automatically display the grid lines at the same resolution as that of the Snap grid by using the **Snap** option of the **GRID** command. When you use this option, AutoCAD LT will automatically change the grid spacing to zero and display the grid lines at the same resolution as set for Snap. Therefore, in the **Drafting Settings** dialog box, if the grid spacing is specified as zero, it automatically adjusts to equal the Snap resolution.*

*In the **GRID** command, to specify the grid spacing as a multiple of the Snap spacing, enter X after the value (2X).*

Setting Snap

The snap is used to set increments for cursor movement. While moving the cursor it is sometimes difficult to position a point accurately. The **SNAP** command allows you to set up an invisible grid (Figure 4-32) that allows the cursor to move in fixed increments from one snap point to another. The snap points are the points where the invisible snap lines intersect. The snap spacing is independent of the grid spacing, so the two can have equal or different values. You generally set snap to an increment of the grid setting, for example, Snap=2 and Grid=10.

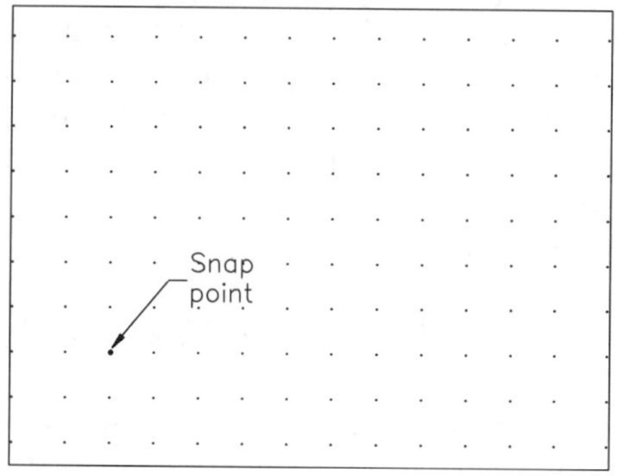

Figure 4-32 *Invisible Snap grid*

Snap On

You can turn the snap on/off and set the snap spacing from the **Drafting Settings** dialog box (Figure 4-28). As in the case of GRID, the Snap On/Off selection turns the invisible snap grid on or off and moves the cursor by increments. If the SNAP is off, the cursor will not move by increments. It will move freely as you move the pointing device. When you turn the snap off, AutoCAD LT remembers the value of the snap, and this value is restored when you turn the snap back on. You can also turn snap on or off by choosing the **Snap** button in the status bar, or from the shortcut menu displayed by right-clicking on the Snap button in the status bar, or by using the function key F9 as a toggle key for turning snap on or off.

Snap Angle

The **Angle** edit box in the **Drafting Settings** dialog box is used to rotate the snap grid through an angle (Figure 4-33). Normally, the snap grid has horizontal and vertical lines, but sometimes you need the snap grid at an angle. For example, when you are drawing an auxiliary view (a drawing view that is at an angle to other views of the drawing), it is more useful to have the snap grid at an angle. If the rotation angle is positive, the grid rotates in a counterclockwise direction; if the rotation angle is negative, it rotates in a clockwise direction.

Figure 4-33 *Snap grid rotated 30 degrees*

X Base and Y Base

The **X base** and **Y base** edit boxes can be used to define the snap and grid origin point in the current viewport relative to current UCS. When you specify the snap angle, the grid is rotated around the base point whose coordinates are specified as X base and Y base. Hence, you will always have a grid point at this base point. The value is stored in the **SNAPBASE** system variable.

Tip

It is generally preferable to rotate the UCS rather than rotating the snap grid. See Chapter 20, The User Coordinate System, for UCS.

Example 2 *Mechanical*

Draw the auxiliary view (Figure 4-35) of the object whose front view is shown in Figure 4-34. The upper edge of the object is 2.5 units. The X and Y spacing of the grid is 0.5 units. The thickness of the plate is 2 units; the length of the incline's face is 5 units.

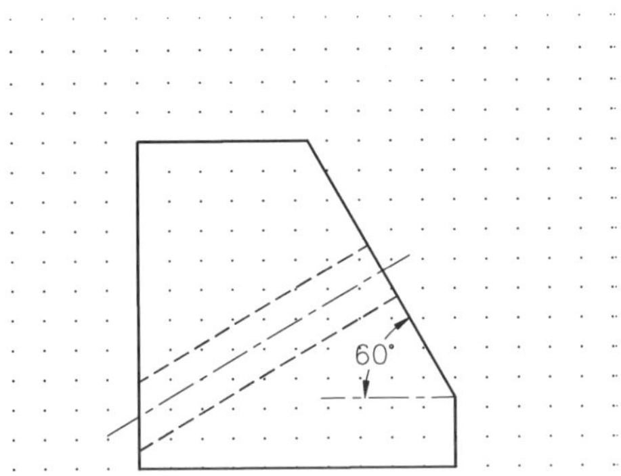

Figure 4-34 *Front view of the plate for Example 2*

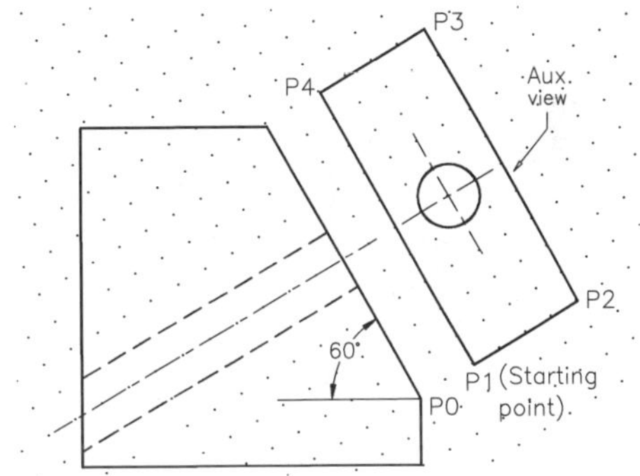

Figure 4-35 *Auxiliary view using rotated snap grid*

Rotating the Snap Grid

It is assumed that the front view of the plate is already drawn with the coordinates of its starting point as 1,1. Invoke the **Drafting Settings** dialog box and open the **Snap and Grid** tab. In the **Angle** edit box, enter **30**, and in the **X base** and **Y base** edit boxes, enter **6** and **2** respectively as the coordinates of point **P0**. This will rotate the snap grid at an angle of 30-degrees through point P0. Now you can draw the auxiliary view easily.

Command: **LINE** [Enter]
Specify first point: *Select point (P1).*
Specify next point or [Undo]: *Move the cursor 2 units (4 grid points) right and select point (P2).*
Specify next point or [Undo]: *Move the cursor 5 units up and select point (P3).*
Specify next point or [Close/Undo]: *Move the cursor 2 units left and select point (P4).*
Specify next point or [Close/Undo]: **C** [Enter]

Snap Type and Style

There are two types of Snap types, **Polar** snap and **Grid** snap. **Grid** snap snaps along the grid and is either a **Rectangular** style or **Isometric** style. Rectangular is the default standard style.

Isometric Snap/Grid

You can choose the **Isometric snap** radio button in **Snap type & style** area (Figure 4-36) of the **Drafting Settings** dialog box to set the snap grid to isometric mode. The default is off (standard).

The isometric mode is used to make isometric drawings. In isometric drawings, the isometric axes are at angles of 30, 90, and 150 degrees. The Isometric snap/grid enables you to display the grid lines along these axes (Figure 4-37). Once you select the **Isometric snap** radio button and choose OK in the dialog box, AutoCAD LT automatically changes the cursor to align with the isometric axis. You can adjust the cursor orientation when you are working on the left, top, or right plane of the drawing by using the **F5** key or holding down the CTRL key and then pressing the E key to cycle the cursor through different isometric planes (left, right, top). You can change the vertical snap and grid spacings by entering values in the **Snap Y spacing** and **Grid Y spacing** edit boxes. You will notice that the X spacing is not available for this option.

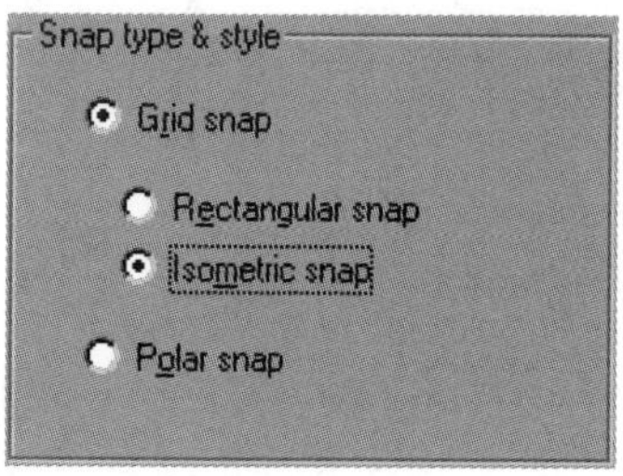

*Figure 4-36 Selecting **Isometric snap** in **Snap type & style** area*

Figure 4-37 Isometric snap grid

Polar Snap

You can use polar snap with polar tracking. Here the cursor snaps to points at a specified distance along the Polar alignment angles. You can choose the **Polar snap** radio button in the **Drafting Settings** dialog box to set the snap grid to the polar mode. When you choose this option the normal snap options are not available. You can enter a desired distance in the **Polar distance** edit box. If this value is zero, it assumes the same value as the Snap X spacing. The cursor snaps along an imaginary path based on polar tracking angles, relative to the last point selected or acquired. The angles can be set in the **Polar Tracking** tab of the **Drafting Settings** dialog box (discussed later in the chapter).

For example, choose the **Polar snap** radio button, enter 0.5 in the **Polar distance** edit box and then choose the **OK** button. Select the **Polar** button in the status bar for polar tracking. Invoke the **LINE** command and select the starting point anywhere on the screen. Now move the cursor; an imaginary path will be displayed at 0, 90, 180, and 270 angles (polar tracking) with small cross marks displayed on this imaginary line at a distance of 0.5 distance (polar snap).

Note

*The polar snap works along with polar tracking. Therefore, it is used when the polar tracking is on by selecting the **Polar** button on the status bar, or by selecting the **Polar Tracking On** check box in the **Polar Tracking** tab of the **Drafting Settings** dialog box.*

*The snap type is also controlled by the **SNAPTYPE** system variable.*

SETTING ORTHO MODE

You can turn the Ortho mode on or off by choosing the **ORTHO** button in the status bar, or by using the function key F8, or using the **ORTHO** command. The **ORTHO** mode allows you to draw lines at right angles only. Whenever you are using the pointing device to specify the next point, the movement of the rubber-band line connected to the cursor is either horizontal (parallel to the X axis) or vertical (parallel to the Y axis). If you want to draw a line in the Ortho mode, specify the starting point at the **Specify first point** prompt. To specify the second point, move the cursor with the pointing device and specify a desired point. The line drawn will be either vertical or horizontal, depending on the direction in which you moved the cursor (Figure 4-38 and Figure 4-39).

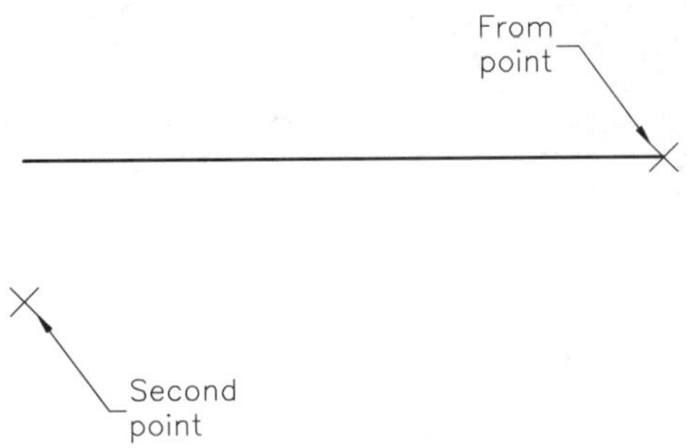

Figure 4-38 Drawing a horizontal line using the Ortho mode

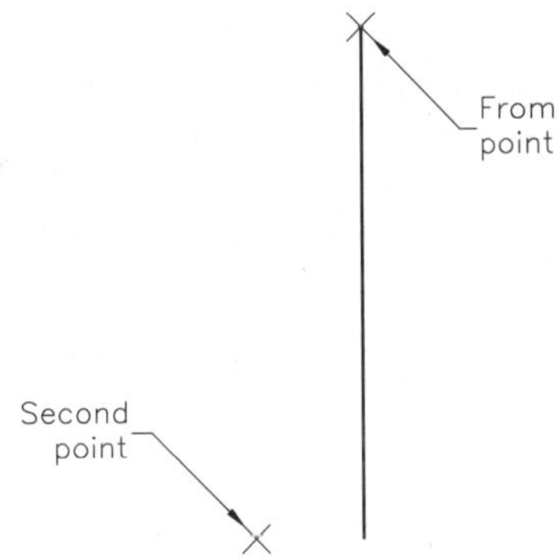

Figure 4-39 Drawing a vertical line using the Ortho mode

Tip

You can use the status bar at the bottom of the graphics area to easily and conveniently toggle between on or off for the different drafting functions like the Snap, Grid, and Ortho.

OBJECT SNAPS

Toolbar:	Object Snap or Standard Toolbar > Tracking flyout

Object snaps are one of the most useful features of AutoCAD LT. They improve your performance and the accuracy of your drawing and make drafting much simpler than it normally would be. Object snaps can also be invoked from the shortcut menu (Figure 4-40), which can be accessed by holding down the SHIFT key on the keyboard and then pressing the ENTER button (right-click) on your pointing device. You can also invoke the different object snaps from the **Standard** toolbar by choosing the **Object Snap** button and holding the Pick button of your pointing device to display the flyout with the object snap buttons. The term

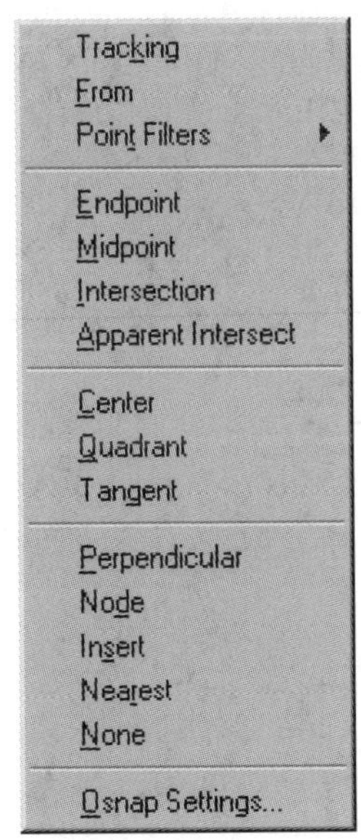

Figure 4-40 Selecting Object Snap modes from the shortcut menu

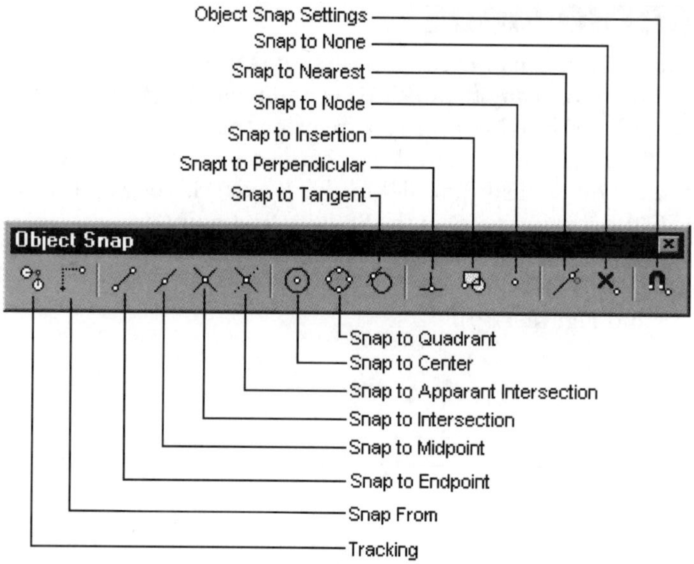

***Figure 4-41 Object Snap** toolbar*

OBJECT SNAP refers to the cursor's ability to snap exactly to a geometric point on an object. The advantage of using object snaps is that you do not have to specify an exact point. For example, if you want to place a point at the midpoint of a line, you may not be able to specify the exact point. Using the **MIDpoint** object snap, all you do is move the cursor somewhere on the object. You will notice a marker (in the form of a geometric shape, a triangle for Midpoint) is automatically displayed at the middle point (snap point). You can click to place a point at the position of the **marker**. You also have a **tooltip** with the object snap marker. When you place the cursor on the marker, a tooltip having the name of the object snap will be displayed. The object snaps recognize only the objects that are visible on the screen, which include the objects on locked layers. The objects on layers that are turned off or frozen are not visible, so they cannot be used for object snaps. The following are the object snap modes available in AutoCAD LT:

ENDpoint	CENter	PERpendicular	NEArest
MIDpoint	QUAdrant	INTersection	NONe
TANgent	INSert	From	NODe
APParant Intersection			

You can also invoke an object snap by entering an abbreviation in the command line.

AutoSnap

The AutoSnap feature controls the various characteristics for object snap. As you move the target box over the object, AutoCAD LT displays the geometric marker corresponding to the shapes shown in the **Object Snap** tab of the **Drafting Settings** dialog box. You can change the different AutoSnap settings such as attaching a target box to the cursor when you invoke any object snap, or change the size and color of the marker. These settings can be changed from the **Options** dialog box (**Drafting** tab). You can invoke the **Options** dialog box from the **Tools** menu (**Tools** > **Options**), or from the shortcut menu when you right-click when no command is active, or by entering **OPTIONS** at the Command prompt. You can also invoke it by selecting the **Options** button in the **Drafting Settings** dialog box. When you choose the **Drafting** tab in the **Options** dialog box, the **AutoSnap** options are displayed (Figure 4-42). You can select the **Marker** check box to toggle the display of marker. You can use the **Magnet** check box to toggle the magnet that snaps the crosshair to the particular point of the object for that object snap. You have check boxes to toggle the **AutoSnap Tooltip** and **Aperture box** display. The **AutoSnapTooltip** displays a flag that gives the name of the object snap that AutoCAD LT has detected. You can change the size of the marker and the aperture box by moving the **Autosnap Marker Size** and **Aperture Size** slider bars respectively. You can also change the color of the markers through the **AutoSnap Marker Color** list box. The size of the aperture is measured in PIXELS, short for picture elements. Picture elements are dots that make up the screen picture. The aperture size can also be changed using the **APERTURE** command. In AutoCAD LT the default value for the aperture size is 10 pixels.

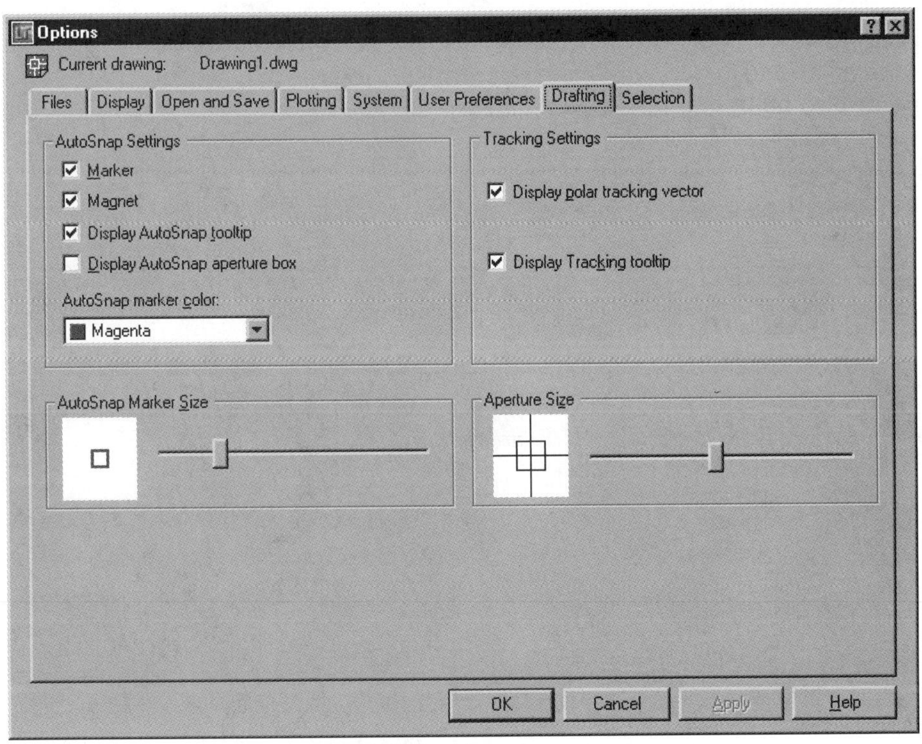

*Figure 4-42 **Drafting** tab of the **Options** dialog box*

The display of the Marker and the AutoSnapTooltip is controlled by the **AUTOSNAP** system variable. The following are the bit values for AUTOSNAP:

Bit Values	Function
0	Turns off the Marker, AutoSnap Tooltip, and Magnet
1	Turns on the Marker

2	Turns on the AutoSnapTooltip
4	Turns on the Magnet
8	Turns on polar tracking
16	Turns on object snap tracking
32	Turns on tooltips for polar tracking and object snap tracking

Endpoint

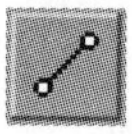

The **ENDpoint** Object Snap mode snaps to the closest endpoint of a line or an arc. To use this Object Snap mode, select the Endpoint button, and move the cursor (crosshairs) anywhere close to the endpoint of the object. The marker will be displayed at the endpoint, and then click to specify that point. AutoCAD LT will grab the endpoint of the object. If there are several objects near the cursor crosshairs, AutoCAD LT will grab the endpoint of the object that is closest to the crosshairs, or if the Magnet is on you can move to grab the desired endpoint. For Figure 4-43, invoke the **LINE** command from the **Draw** toolbar. The following is the prompt sequence:

Specify first point: *Select the **Endpoint** button from the **Object Snap** toolbar.*
_endp of *Move the crosshair and select the arc.*
Specify next point or [Undo]: *Select the endpoint of the line.*

Midpoint

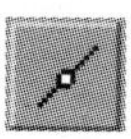

The **MIDpoint** Object Snap mode snaps to the midpoint of a line or an arc. To use this Object Snap mode, select Midpoint osnap and select the object anywhere. AutoCAD LT will grab the midpoint of the object. For Figure 4-44, invoke the **LINE** command from the **Draw** toolbar. The following is the prompt sequence:

Specify first point: *Select the starting point of the line.*
Specify next point or [Undo]: *Choose the **Midpoint** button from the **Object Snap** toolbar.*
_mid of *Move the cursor and select the original line.*

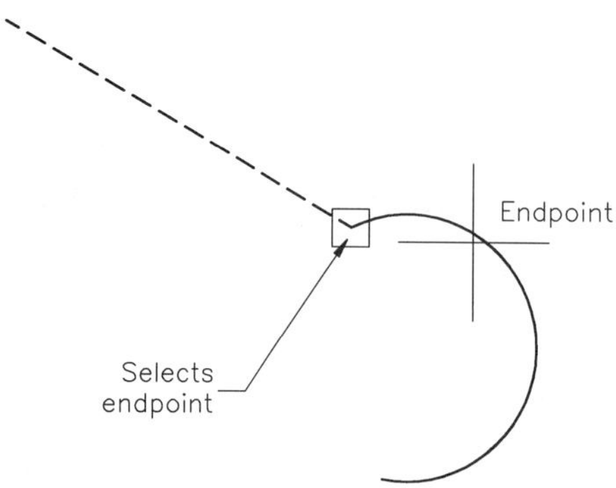

Figure 4-43 *Endpoint Object Snap mode*

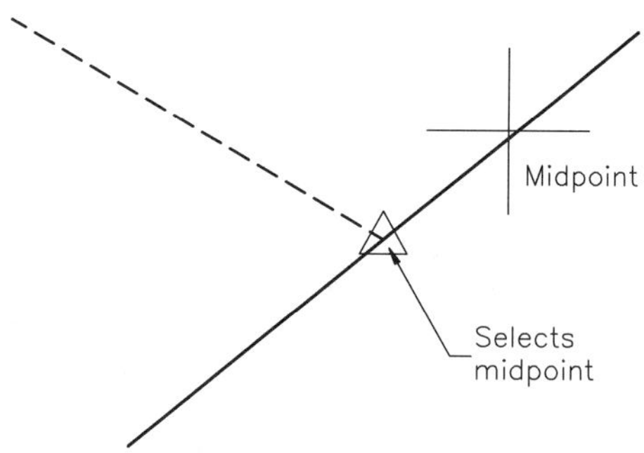

Figure 4-44 *Midpoint Object Snap mode*

Nearest

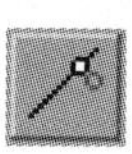

The **NEArest** Object Snap mode selects a point on an object (line, arc, circle, or ellipse) that is visually closest to the graphics cursor (crosshairs). To use this mode, enter the command, then choose the Nearest object snap. Move the crosshairs near the intended point on the object so as to display the marker at the desired point and then select the object. AutoCAD LT will grab a point on the line where the marker was displayed. For Figure 4-45, invoke the **LINE** command from the **Draw** toolbar. The following is the prompt sequence:

Specify first point: *Choose **Nearest** button from the **Object Snap** toolbar.*
_nea to *Select a point near an existing object.*
Specify next point or [Undo]: *Specify the endpoint of the line.*

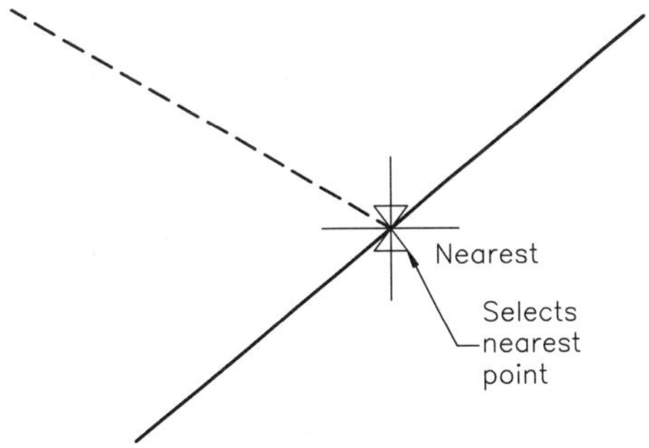

Figure 4-45 *NEArest Object Snap mode*

Center

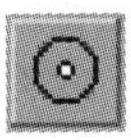

The CENter Object Snap mode allows you to snap to the center point of an ellipse, circle, or arc. After selecting this option, you must point to the visible part of the circumference of a circle or arc. For Figure 4-46, invoke the **LINE** command from the **Draw** toolbar. The following is the prompt sequence:

Specify first point: *Choose **Center** button from the **Object Snap** toolbar.*
_cen of *Move the cursor and select the circle.*
Specify next point or [Undo]: *Select the endpoint of the line.*

Tangent

The **TANgent** Object Snap allows you to draw a tangent to or from an existing ellipse, circle, or arc. To use this object snap, place the cursor on the circumference of the circle or arc to select it. For Figure 4-47, invoke the **LINE** command from the **Draw** toolbar. The following is the prompt sequence:

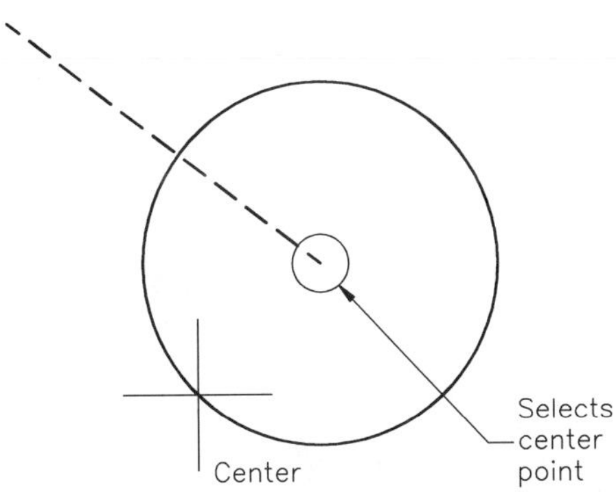

Figure 4-46 *Center Object Snap mode*

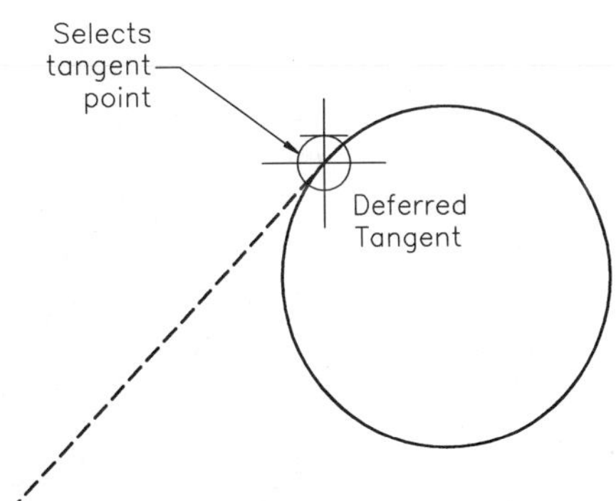

Figure 4-47 *Tangent Object Snap mode*

Specify first point: *Select the starting point of the line.*
Specify next point or [Undo]: *Choose the **Tangent** button from the **Object Snap** toolbar.*
_tan to *Move the cursor and select the circle.*
Specify next point or [Undo]: *Select the endpoint of the line (tangent of the circle).*

The Figure 4-48, shows the use of **Nearest**, **Endpoint**, **Midpoint**, and **Tangent** object snaps.

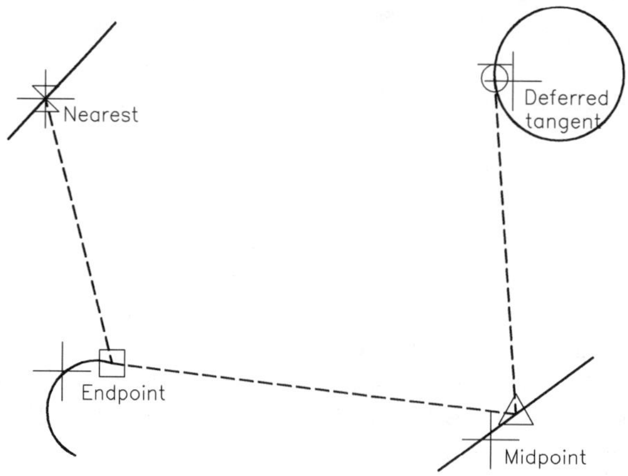

Figure 4-48 *Using Nearest, Endpoint, Midpoint, and Tangent Object Snap modes*

Quadrant

The **QUAdrant** Object Snap mode is used when you need to snap to a quadrant point of an ellipse, arc, or circle. A circle has four quadrants, and each quadrant subtends an angle of 90 degrees. The quadrant points are located at 0-, 90-, 180-, 270-degree positions. If the circle is inserted as a block (see Chapter 14, Blocks), and if it is rotated, the quadrant points are also rotated by the same amount. To use this object snap, position the cursor on the circle or arc closest to the desired quadrant. For Figure 4-49, invoke the LINE command. The prompt sequence for drawing a line from the third quadrant of a circle as shown in Figure 4-50 and Figure 4-51 is:

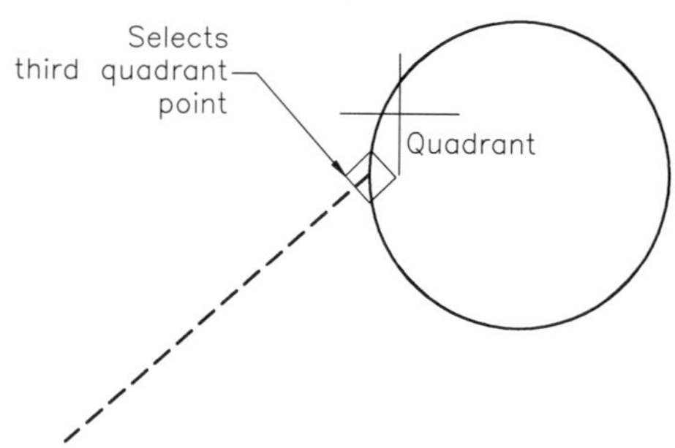

Figure 4-49 *Quadrant Object Snap mode*

Specify first point: *Choose **Quadrant** button from the **Object Snap** toolbar.*
_qua of *Move the cursor close to the third quadrant of the circle and select it.*
Specify next point or [Undo]: *Select the endpoint of the line.*

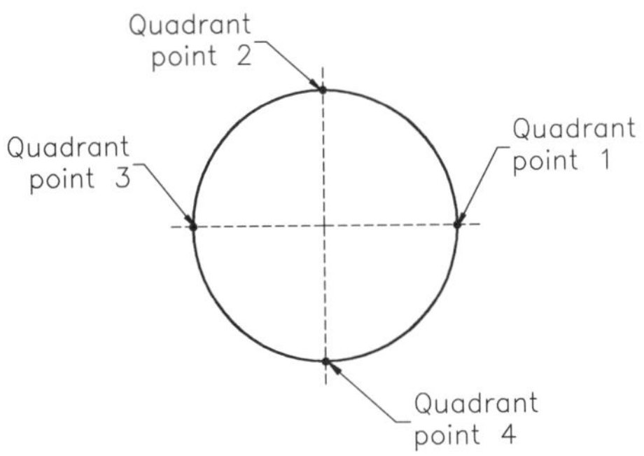

Figure 4-50 *Four quadrant of a circle*

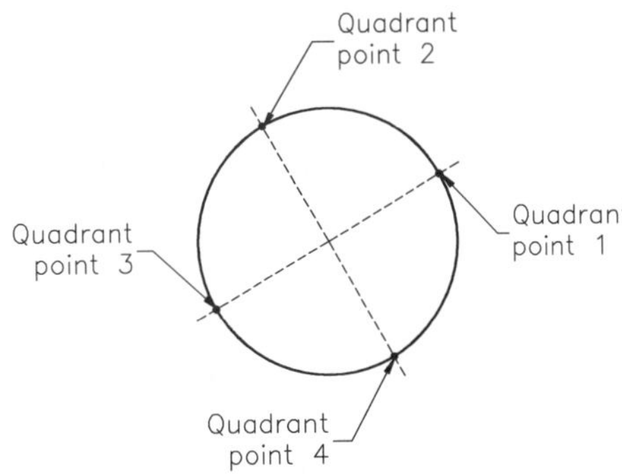

Figure 4-51 *Four quadrants of a circle*

Intersection

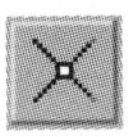

The **INTersection** Object Snap mode is used if you need to snap to a point where two or more lines, circles, ellipses or arcs intersect. To use this object snap, move the cursor close to the desired intersection so that the intersection is within the target box, and then specify that point. For Figure 4-52, invoke the LINE command. The prompt sequence is:

Specify first point: *Choose the **Intersection** button from the **Object Snap** toolbar.*
_int of *Position the cursor near the intersection and select it.*
Specify next point or [Undo]: *Select the endpoint of the line.*

After selecting the Intersection object snap, if your cursor is close to an object and not close to an actual intersection, the intersection marker displays ellipses [...] with it. This indicates an **extended intersection**. If you select this object now, AutoCAD LT prompts **and** for selection of another object. If your cursor is close to another object, AutoCAD LT marks the extended intersection point between these two objects. This mode selects projected or visual intersections of lines, arcs, circles, or ellipses (Figure 4-53). The projected intersections are those intersections that are not present on the screen, but are imaginary and formed if the line or arc is extended.

Apparent Intersection

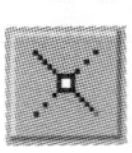

The **APParent Intersection** Object Snap mode selects projected or visual intersections of two objects in 3D space. Sometimes two objects appear to intersect one another in the current view, but in 3D space the two objects do not actually intersect. The Apparent Intersection snap mode selects such visual intersections. This mode works on wireframes in 3D space (see Chapter 21, Drawing and Viewing 3D Objects, for wireframe models).

Perpendicular

The **PERpendicular** Object Snap mode is used to draw a line perpendicular to or from another line, or normal to or from an arc or circle, or to an ellipse. When you use this mode and select an object, AutoCAD LT calculates the point on the selected object so that the previously selected point is perpendicular to the line. The object can be selected by positioning the cursor anywhere on the line. First invoke the **LINE** command; the

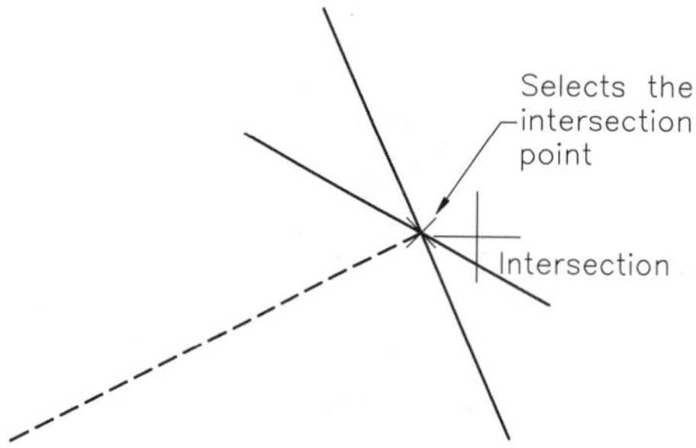

Figure 4-52 Intersection Object Snap mode

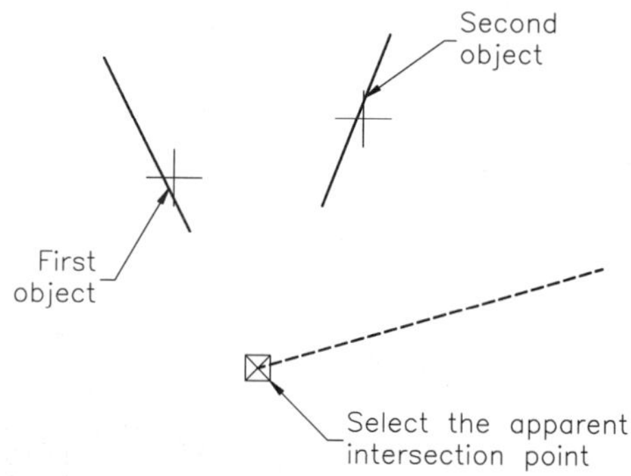

Figure 4-53 Extended Intersection Snap mode

prompt sequence is as follows to draw a line that is perpendicular to a given line (Figure 4-54):

Specify first point: *Select the starting point of the line.*
Specify next point or [Undo]: *Choose the **PERpendicular** button from the **Object Snap** toolbar.*
_per to *Select the line on which you want to draw perpendicular.*

When you select the line first, the rubber-band feature of the line is disabled. The line will appear only after the second point is selected. Invoke the **LINE** command. The prompt sequence for drawing a line perpendicular from a given line (Figure 4-55) is:

Specify first point: *Choose the **PERpendicular** button from the **Object Snap** toolbar.*
_per to *Select the line on which you want to draw perpendicular.*
Specify next point or [Undo]: *Select the endpoint of the line.*

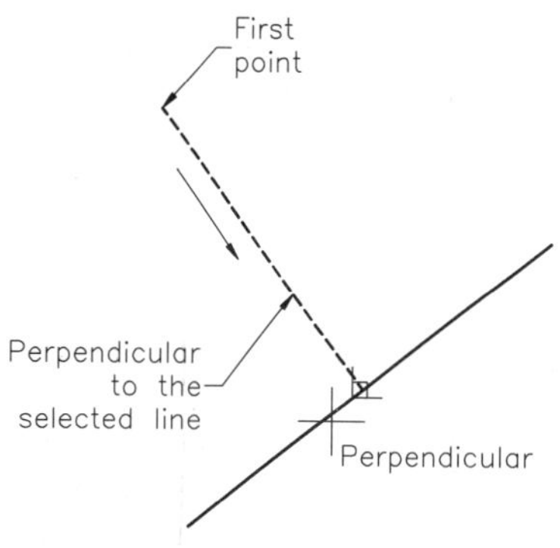

Figure 4-54 Selecting the start point, then the perpendicular snap

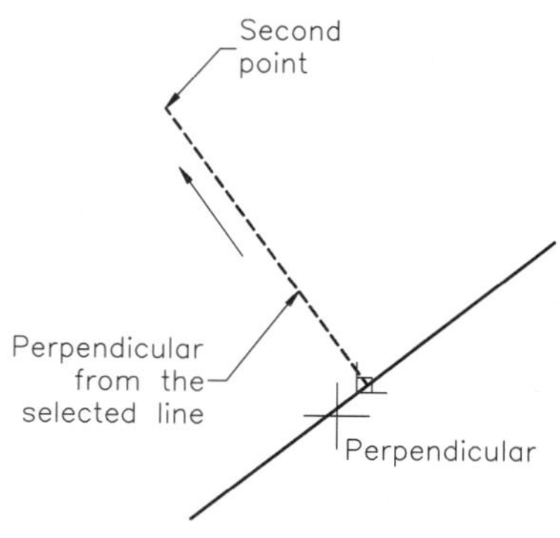

Figure 4-55 Selecting the perpendicular snap first

Figure 4-56, shows the use of the **CENter**, **INTersection**, and **PERpendicular** object snaps.

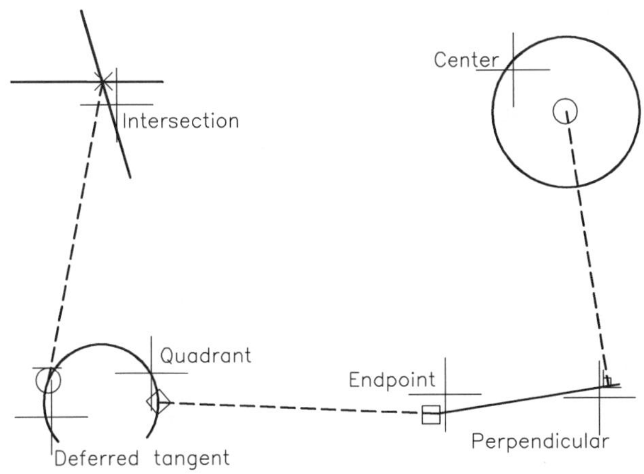

Figure 4-56 *CENter, Intersection, and PERpendicular Osnap modes*

Exercise 3 *General*

In Figure 4-57, P1 and P2 are the center points of the bottom and top arcs. The space between the dotted lines is 1 unit.

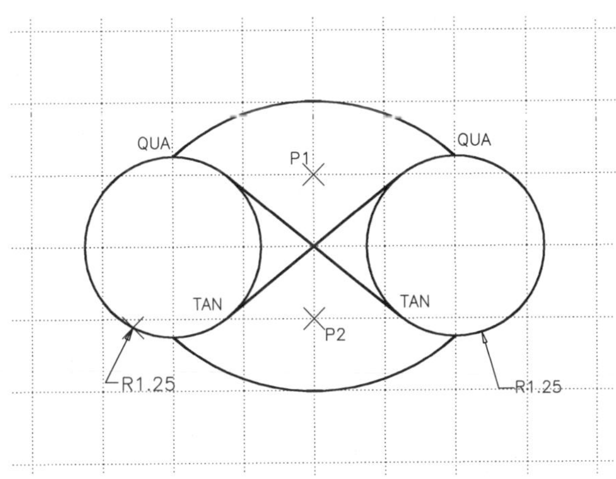

Figure 4-57 *Drawing for Exercise 3*

Node

You can use the **NODe** Object Snap to snap to a point object drawn using the **POINT** command, or placed using the **DIVIDE** or **MEASURE** commands. In Figure 4-58, three points have been drawn using the AutoCAD LT **POINT** command. You can snap to these points by using the NODe snap mode. Invoke the **LINE** command and the prompt sequence is as follows (Figure 4-59):

Specify first point: *Choose the **Node** button from the **Object Snap** toolbar.*
_nod of *Select point P1.*
Specify next point or [Undo]: *Choose the **Node** button from the **Object Snap** toolbar.*
_nod of *Select point P2.*
Specify next point or [Undo]: *Choose the **Node** button from the **Object Snap** toolbar.*
_nod of *Select point P3.*

Figure 4-58 *Point objects*

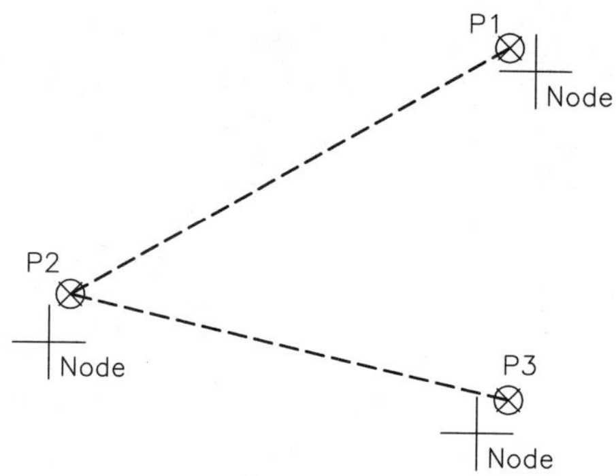

Figure 4-59 *Using the NODe object snap*

Insert

The **INSert** Object Snap mode is used to snap to the insertion point of a text, shape, block, attribute, or attribute definition. In Figure 4-60, the text **WELCOME** is left-justified and the text **AutoCAD LT** is center-justified. The point with respect to which the text is justified is the insertion point of that text string. If you want to snap to these insertion points or the insertion point of a block, you must use the Insert Object Snap mode. Invoke the **LINE** command and the following is the prompt sequence:

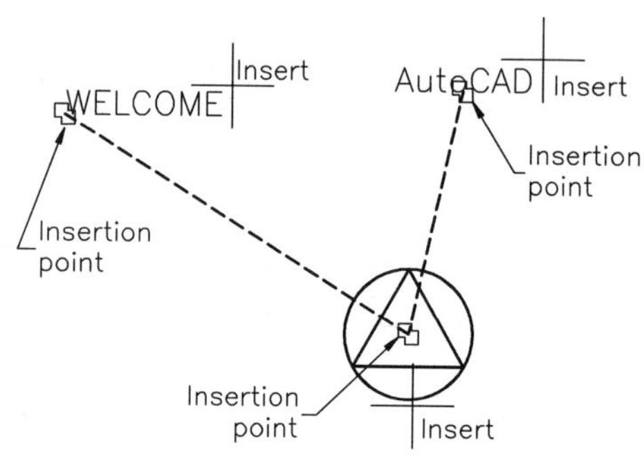

Figure 4-60 *Insert Object Snap mode*

Specify first point: *Choose the **Insert** button from the **Object Snap** toolbar.*
_ins of *Select WELCOME text.*
Specify next point or [Undo]: *Choose the Insert button from the **Object Snap** toolbar.*
_ins of *Select the block.*
Specify next point or [Undo]: *Choose the Insert button from the **Object Snap** toolbar.*
_ins of *Select AutoCAD LT text.*

None

The **NONe** Object Snap mode turns off any running object snap (see the section "Running Object Snap Mode" that follows.) for one point only. The following example illustrates the use of this Object Snap mode:

Invoke the **Drafting Settings** dialog box and select the **Object Snap** tab. Select the **Midpoint** and **Center** check boxes. This sets the object snaps to Mid and Cen. Now, suppose you want to draw a line whose starting point is a point closer to the endpoint of another line. After you invoke the **LINE** command and move the cursor to the desired position on the previous line, it automatically snaps to its midpoint. You can disable this using the **None** object snap by choosing it from the **Object Snap** toolbar. You can select the desired point on the line. Once you have selected a point it continues the command with the specified object snaps.

From

The **From** object snap can be used to locate a point relative to a given point (Figure 4-61). For example, if you want to locate a point that is 0.25 units up and 0.25 units right from the endpoint of a given line, you can use the **From** object snap as follows:

Command: **LINE**
Specify first point: *Choose the **From** button from the **Object Snap** toolbar.*
_from Base point: *Choose the **Endpoint** button from the **Object Snap** toolbar.*
_endp of *Specify the endpoint of the given line*
<Offset>: **@0.25,0.25**

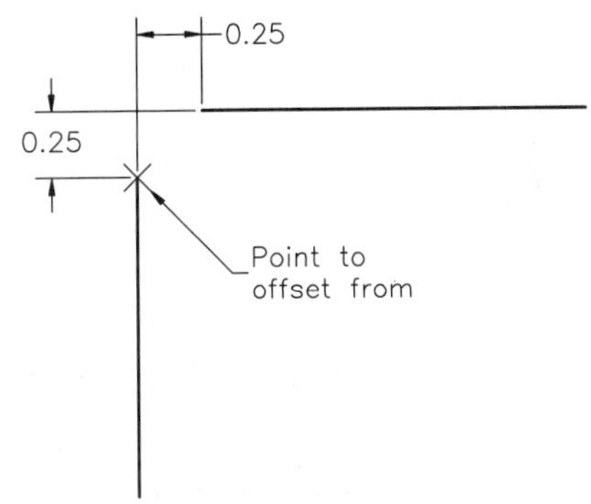

Figure 4-61 *Using the **From** object snap to locate a point*

Note
*The **From** object snap cannot be used as running object snap.*

Combining Object Snap Modes

You can also combine the snaps from the command line by separating the snap modes with a comma. AutoCAD LT will search for the specified modes and grab the point on the object that is closest to the point where the object is selected. The prompt sequence for using the Midpoint and Endpoint object snaps is:

Command: **LINE** Enter
Specify first point: **MID, END** Enter *(MIDpoint or ENDpoint object snap.)*
to *Select the object.*

Note
In the discussions of object snaps, "line" generally includes xlines, rays, and polyline segments, and "arc" generally includes polyarc segments.

Exercise 4 — *General*

Make the drawing in Figure 4-62. The space between the dotted lines is 1 unit.

Chapter 4

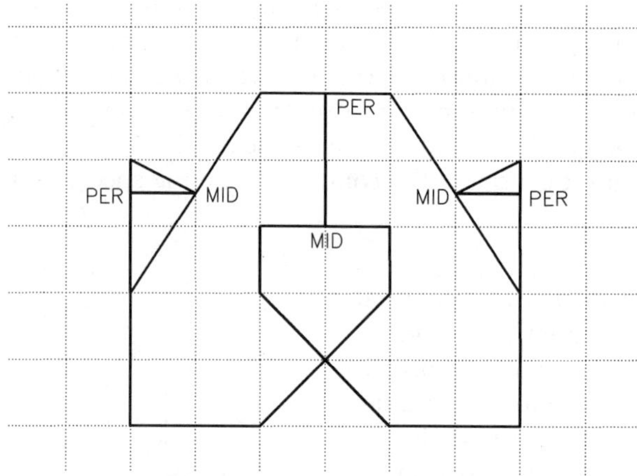

***Figure 4-62** Drawing for Exercise 4*

RUNNING OBJECT SNAP MODE

Toolbar:	Standard Toolbar > Tracking flyout, Object Snap > Object Snap Settings
Menu:	Tools > Drafting Settings
Command:	OSNAP

In the previous sections you have learned how to use the object snaps to snap to different points of an object. One of the drawbacks of these object snaps is that you have to select them every time you use them, even if it is the same snap mode. This can be solved by using **running object snaps**. The Running Osnap can be invoked from the **Object Snap** tab of the **Drafting Settings** dialog box (Figure 4-63). If you choose the **Object Snap Settings** button from the toolbar, or enter **OSNAP** at the Command line, or you choose **Settings** from the shortcut menu displayed when you right-click on the **Osnap** button in the status bar, the **Object Snap** tab is displayed in the **Drafting Settings** dialog box. In this tab you can set the running Object Snap modes by selecting the check boxes next to the snap modes. For example, if you want to set Endpoint as the running Object Snap mode, select the **Endpoint** check box and then the **OK** button in the dialog box.

Once you set the running Object Snap mode, you are automatically in that mode and the marker is displayed when you move the crosshairs over the snap points. If you had selected a combination of modes, AutoCAD LT selects the mode that is closest to the screen crosshairs. For example, you have selected the **Endpoint**, **Midpoint**, and **Center** check boxes in the dialog box. Now for selection of a point when you move the cursor over a previously drawn line and whenever your cursor is closer to the midpoint of the line the marker is displayed there. You can also move the cursor to the end of the line to invoke the Endpoint object snap. If you place the cursor over the circumference of a previously drawn circle, the Center osnap is invoked. The running Object Snap mode can be turned on or off, without losing the object snap settings, by choosing the **Osnap** button in the status bar. You can also accomplish this by pressing the function key F3 or CTRL+F keys.

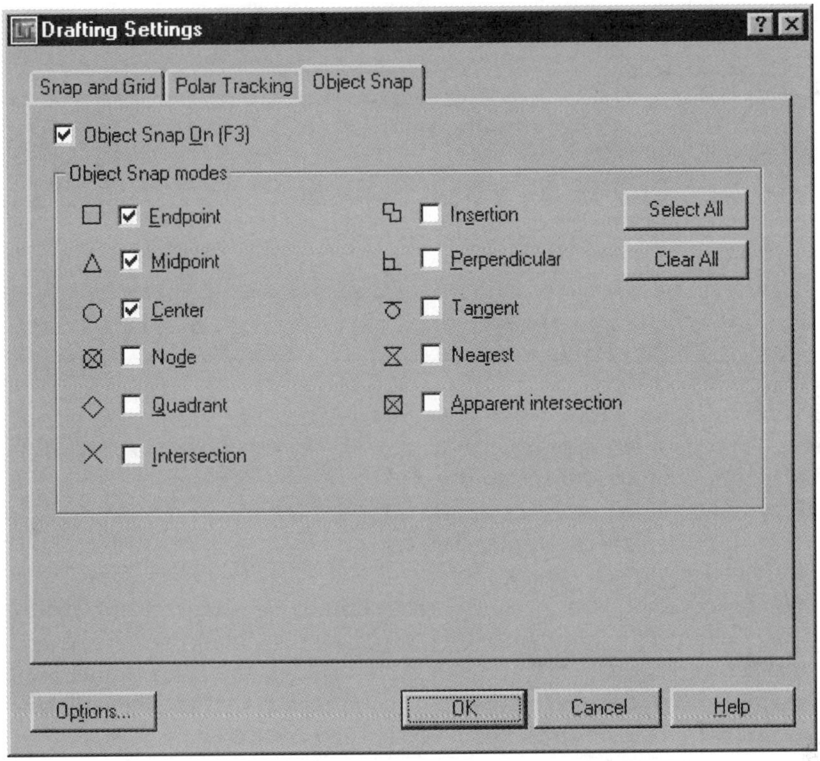

***Figure 4-63 Drafting Settings** dialog box (**Object Snap** tab)*

Note

When you use a temporary object snap and fail to specify a suitable point, AutoCAD LT issues an Invalid point prompt and reissues the pending command's prompt. However, with a running Object Snap mode, if you fail to specify a suitable point, AutoCAD LT accepts the point specified without snapping to an object.

*The running osnap works only when it is turned on by using the **Osnap** button in the status bar.*

Overriding the Running Snap

When you select the running object snaps, all other Object Snap modes are ignored unless you select another Object Snap mode. Once you select a different osnap mode, the running OSNAP mode is temporarily overruled. After the operation has been performed, the running OSNAP mode goes into effect again. If you want to discontinue the current running Object Snap modes totally, choose the **Clear all** button in the **Drafting Settings** dialog box. If you want to temporarily disable the running object snap, choose the **Osnap** button (off position) in he status bar.

If you are overriding the running OSNAP modes for a point selection and no point is found to satisfy the override Object Snap mode, AutoCAD LT displays a message to this effect. For example, if you specify an override Object Snap mode of Center and no circle, ellipse, or arc is found at that location, AutoCAD LT will display the message "No center found for specified point. Invalid point."

Cycling through Snaps

AutoCAD LT displays the geometric marker corresponding to the shapes shown in the **Obect snap settings** tab of the **Drafting Settings** dialog box. You can use the TAB key to cycle through the snaps. For example, if you have a circle with an intersecting rectangle as shown in Figure 4-64 and you want to snap to one of the geometric points on the circle, you can use the TAB key to cycle through the geometric points. The geometric points for a circle are the center point, quadrant points, and the intersecting points with the rectangle. To snap to one of these points, the first thing you need to do is to set the running object snaps (center, quadrant, and intersection object snaps) in the **Drafting Settings** dialog box. After entering a command, when you drag the cursor over the objects, AutoSnap displays a marker and a SnapTip. You can cycle through the snap points available for an object by pressing the TAB key. For example, if you press the TAB key while the aperture box is on the circle and the rectangle (near the lower left intersection point), AutoSnap will display the intersection, center, and quadrant points one by one. When a certain osnap is displayed, its pertaining object is highlighted by making it dashed. For example, when the center or quadrant snaps are displayed the circle becomes dashed, and when the intersection snap is displayed both the circle and the rectangle becomes dashed. By selecting a point you can snap to one of these points.

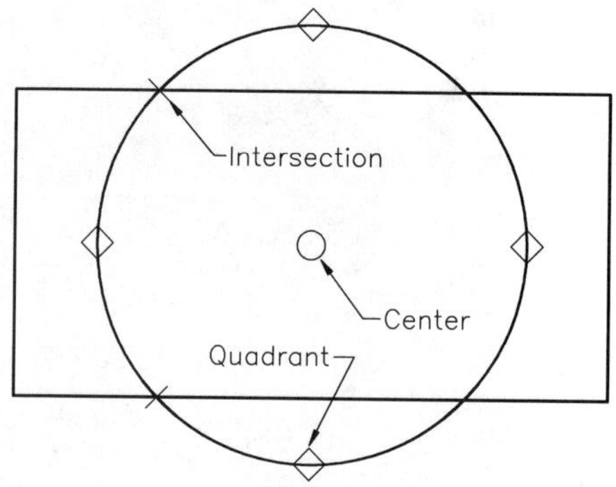

Figure 4-64 *Using the TAB key to cycle through the snaps*

Coordinate Entry Priority

Sometimes you may want the keyboard entry to take precedence over the running Object Snap modes. This is useful when you want to locate a point that is close to the running osnap. By default, when you specify the coordinates (by using the keyboard) of a point located near a running osnap, AutoCAD LT ignores the point and snaps to the running osnap. For example, you select the **Intersection** object snap in the **Object snap** tab of the **Drafting Settings** dialog box. Now if you enter the coordinates of the endpoint of a line very close to the intersection point (Figure 4-65a), the line will snap to the intersection point and ignore the keyboard entry point. You can set the priority between the keyboard entry and object snap through the **User Preferences** tab of the **Options** dialog box

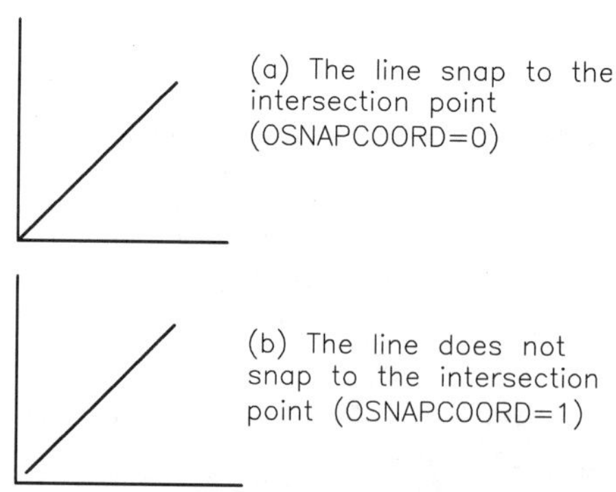

Figure 4-65 *Setting the coordinate entry priority*

(Figure 4-66). Select the **Keyboard Entry** radio button and then choose **OK**. Now if you enter the coordinates of the endpoint of a line very close to the intersection point (Figure 4-65b), the starting point will snap to the coordinates specified and the running osnap (intersection) is ignored. This setting is stored in the **OSNAPCOORD** system variable and you can also set the priority through this variable. A value **0** gives priority to running osnap, value **1** gives priority to keyboard entry, and value **2** gives priority to keyboard entry except scripts.

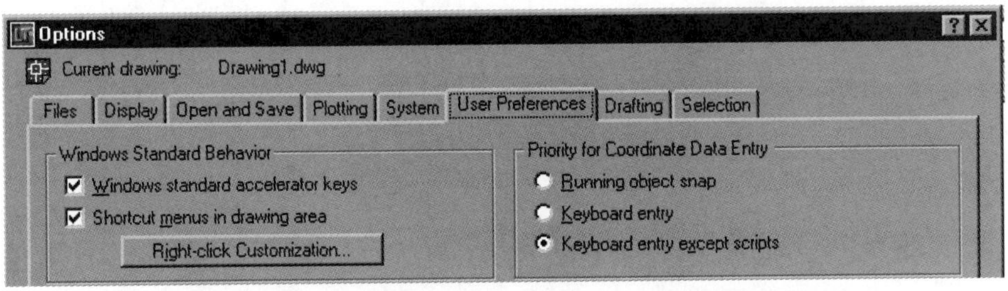

***Figure 4-66 User Preferences** tab of the **Options** dialog box*

Tip

*If you do not want the running osnap to take precedence over the keyboard entry, you can simply disable running osnap temporarily by selecting the **Osnap** button on the status bar to the off position. This lets you specify a point close to a running osnap. This way you do not have to change the settings for coordinate entry in the **Options** dialog box.*

USING AUTOTRACKING

When using AutoTracking, the cursor moves along temporary paths to locate key points in a drawing. It can be used to locate points with respect to other points or objects in the drawing. The Auto Tracking options in AutoCAD LT is the **Polar Tracking** option. Polar tracking is used to locate points on an angular alignment path. Polar tracking can be selected by choosing the **Polar** button on the status bar, by using the function key **F10** or by selecting the **Polar Tracking On** check box in the **Polar Tracking** tab of the **Drafting Settings** dialog box (Figure 4-67).

Polar Tracking constrains the movement of the cursor along a path that is based on the polar angle settings. For example, if the **Increment angle** list box value is set to **15** degrees in the **Polar Angle Settings** area, the cursor will move along alignment paths that are multiples of 15 degrees (0, 15, 30, 45, 60, and so on) and a tooltip will display a distance and angle. Selecting the **Additional angles** check box and choosing the **New** button allows you to add an additional angle value. The imaginary path will also be displayed at these new angles apart from the increments of the increment angle selected. For example, if the increment angle is set at **15** and you add an additional angle of **22**, the imaginary path will be displayed at 0, 15, 22, 30, 45, and the increments of 15. Polar tracking is on only when the Ortho mode is off.

In the **Drafting Settings** dialog box (**Polar Tracking** tab), you can set the polar tracking to absolute or relative to last segment. If you select the **Absolute** radio button, which is the default, the base angle is taken from 0. If you select **Relative to last segment** radio button, the base angle for the increments is set to the last segment drawn.

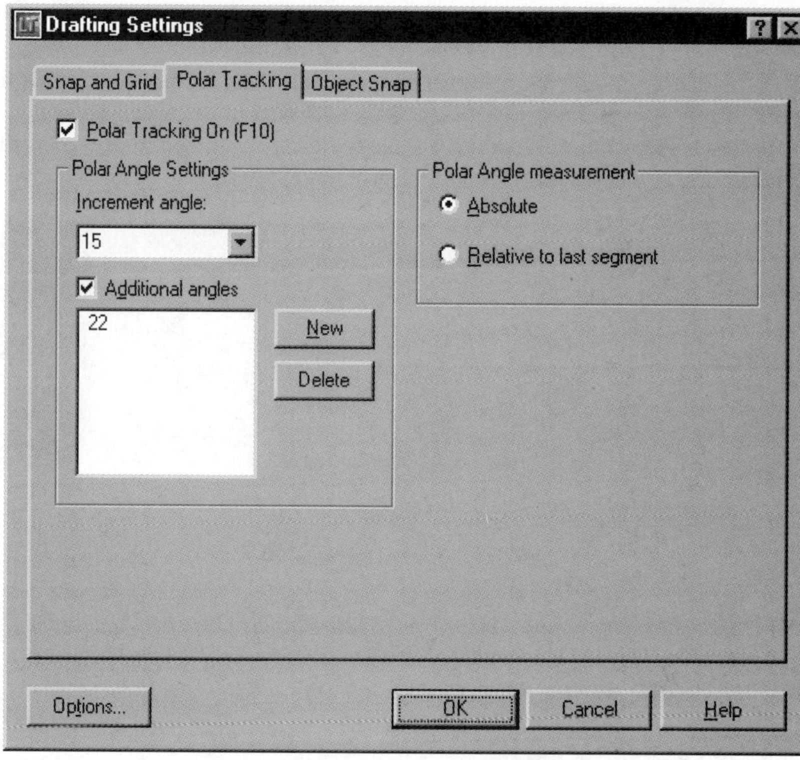

Figure 4-67 ***Polar Tracking*** *tab of the* ***Drafting Settings*** *dialog box*

The direction of the path is determined by the motion of the cursor or the point you select on the object. For example, if you want to draw a circle whose center is located in line with the center of two existing circles, you can use the Auto Tracking as follows:

Command: **CIRCLE** Enter
Specify center point for circle or [3P/2P/Ttr (tan tan radius)]: **TK** Enter
First tracking point: **CEN** Enter
of *(Select the first circle)*
Next point (Press ENTER to end tracking): **CEN** Enter
of *(Select the second circle)*
Next point (Press ENTER to end tracking): Enter
Specify radius of circle or [Diameter] <1.0799>: *Specify the radius of the circle.*

Similarly, you can use the Auto Tracking in combination with the midpoint object snap to locate the center of the rectangle and then draw a circle (Figure 4-68 and Figure 4-69).

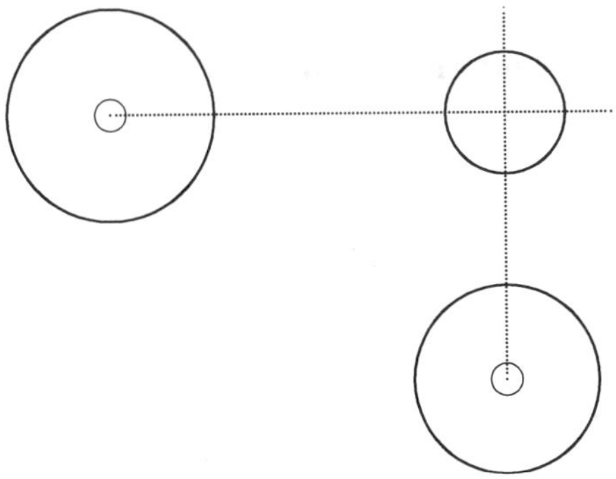

Figure 4-68 Using AutoTracking to locate a point (center of circle)

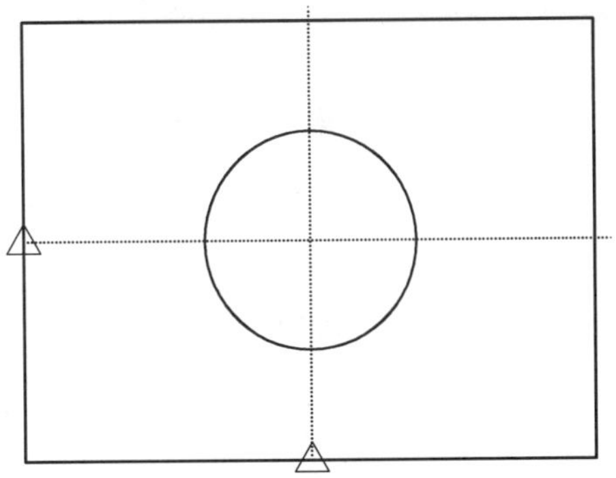

Figure 4-69 Using AutoTracking to locate a point (midpoint of rectangle)

AutoTrack Settings

You have different settings while working with Auto Tracking. These can be set in the **Drafting** tab of the **Options** dialog box (Figure 4-70). If you choose the **Options** button in the **Polar Tracking** tab of the **Drafting Settings** dialog box, the **Options** dialog box with the **Drafting** tab open is displayed. You can use the **Display polar tracking vector** check box to toggle the display of the angle alignment path for polar tracking. You can use the **Display Track tooltip** check box to toggle the display of tooltips with the paths. You can also use the **TRACKPATH** system variable to set the path display settings.

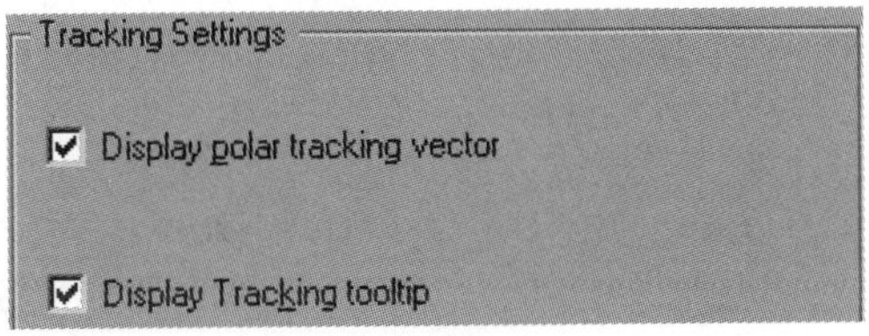

***Figure 4-70 AutoTrack Settings** in the **Options** dialog box (**Drafting** tab)*

Tip

*You need the **Options** dialog box quite frequently to change the different drafting settings. When you choose the **Options** button from the **Drafting Settings** dialog box, it directly opens the required tab. After making the changes in the dialog box, choose the **OK** button to get back to the **Drafting Settings** dialog box. The **Options** button is available in all the three tabs of the **Drafting Settings** dialog box.*

FUNCTION AND CONTROL KEYS

You can also use the function and control keys to change the status of coordinate display, Snap, Ortho, Osnap, tablet, screen, isometric planes, running Object Snap, Grid, and Polar tracking. The following is a list of function and control keys.

F1 Help
F2 Graphics Screen/Text Window
F3 Osnap On/Off **(CTRL+F)**
F4 Tablet mode On/Off **(CTRL+T)**
F5 Isoplane top/right/left **(CTRL+E)**
F6 Coordinate display On/Off **(CTRL+D)**
F7 Grid On/Off **(CTRL+G)**
F8 Ortho On/Off **(CTRL+L)**
F9 Snap On/Off **(CTRL+B)**
F10 Polar tracking On/Off

Self-Evaluation Test

Answer the following questions and then compare your answers to the correct answers given at the end of this chapter.

1. The layers that are turned off are displayed on the screen but cannot be plotted. (T/F)
2. The drawing in which you are working should be in named plot style mode (.stb) to make the plot style available in the **Layer Properties Manager** dialog box. (T/F)
3. The grid pattern appears within the drawing limits, which helps to define the working area. (T/F)
4. If the circle is inserted as a block, that is rotated, the quadrant points are not rotated by the same amount. (T/F)
5. You can also change the plot style mode from the command line by using the _________ system variable.
6. The _________ command enables you to set up an invisible grid that allows the cursor to move in fixed increments from one snap point to another.
7. The _________ snap works along with polar tracking only.
8. The _________ Object Snap mode selects projected or visual intersections of two objects in 3D space.
9. The _________ can be used to turn off any Object Snap mode for one point.
10. You can set the priority between the keyboard entry and object snap through the _________ tab of the **Options** dialog box.

Review Questions

Answer the following questions.

1. You cannot enter different values for horizontal and vertical grid spacing. (T/F)
2. You can lock a layer, which will prevent the user from accidentally editing the objects in that layer. (T/F)
3. When a layer is locked, you cannot use the objects in the locked layer for Osnaps. (T/F)
4. The thickness given to the objects in a layer using the Lineweight option is displayed on the screen and is plotted. (T/F)

5. When you select more than one layer using the SHIFT key, which of the following options is not displayed in the Layer shortcut menu in the **Layer Properties Manager** dialog box?

(a) **New Layer** (b) **Select All**
(c) **Make Current** (d) **Clear All**

6. Which of the following function keys acts as a toggle key for turning the grid on or off?

(a) F5 (b) F6
(c) F7 (d) F8

7. Which one of the following Object Snap modes turn off any running object snap for one point only?

(a) **Node** (b) **None**
(c) **From** (d) **Nearest**

8. Which one of the following object snaps cannot be used as running object snap?

(a) **Intersection** (b) **Quadrant**
(c) **From** (d) **Node**

9. Which of the following keys can be used to cycle through the different running object snaps?

(a) ENTER (b) SHIFT
(c) CTRL (d) TAB

10. While working on a drawing, at any point in time you can save all the layers with their present properties' settings under one name and then restore it anytime later using the __________ button in the **Layer Properties Manager** dialog box.

11. You can use the __________ window to locate drawing data with the help of search tools and then use it in your drawing.

12. The grid pattern appears within the drawing __________, which helps to define the working area.

13. The difference between the **Off** option and the **Freeze** option is that the frozen layers are not __________ by the computer while regenerating the drawing.

14. The size of the aperture is measured in __________, short for picture elements.

15. The __________ system variable controls the current color of the objects.

Exercises

Exercise 5

Mechanical

Set up layers with the following linetypes and colors. Then make the drawing as shown in Figure 4-71. The distance between the dotted lines is 1.0 unit.

Layer name	Color	Linetype
Object	Red	Continuous
Hidden	Yellow	Hidden
Center	Green	Center

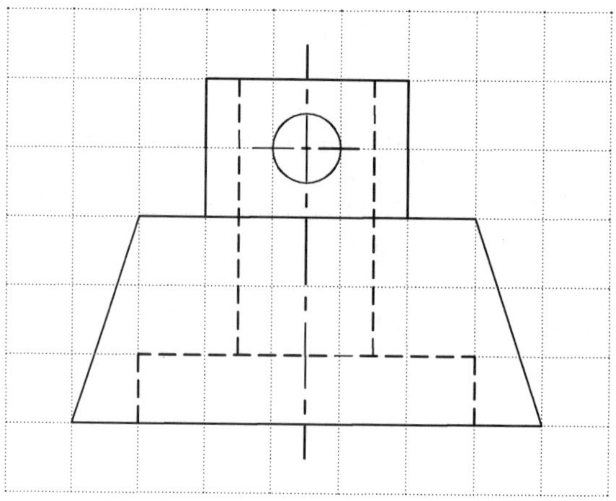

Figure 4-71 *Drawing for Exercise 5*

Exercise 6

Mechanical

Set up layers, linetypes, and colors as given in Exercise 5. Then make the drawing shown in Figure 4-72. The distance between the dotted lines is 1.0 unit.

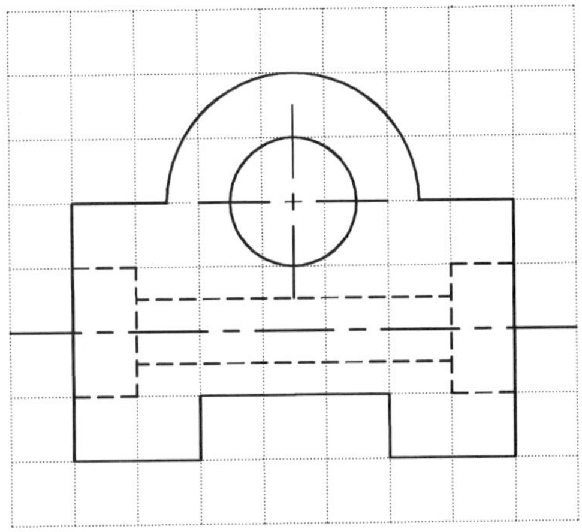

Figure 4-72 Drawing for Exercise 6

Exercise 7 *Mechanical*

Set up layers, linetypes, and colors and then make the drawing shown in Figure 4-73. The distance between the dotted lines is 1.0 unit.

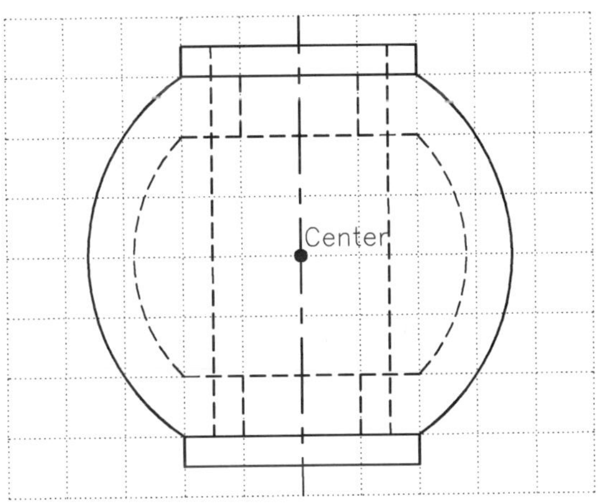

Figure 4-73 Drawing for Exercise 7

Exercise 8 *Mechanical*

Set up layers, linetypes, and colors and then make the drawing shown in Figure 4-74. Use the object snaps as indicated.

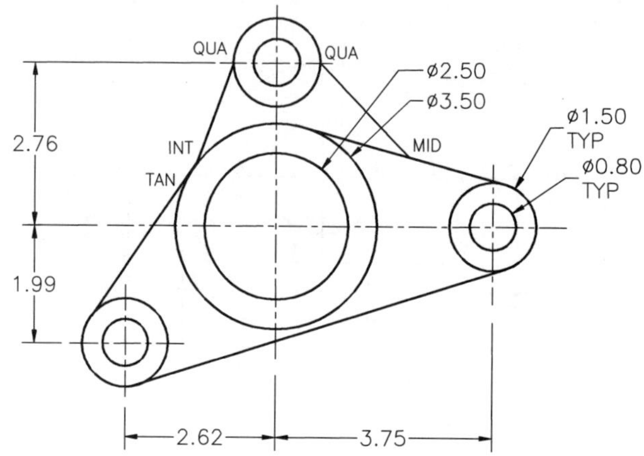

Figure 4-74 *Drawing for Exercise 8*

Problem Solving Exercise 1 — *Mechanical*

Make the drawing shown in Figure 4-75. First draw the lines and then draw the arcs using appropriate ARC command options.

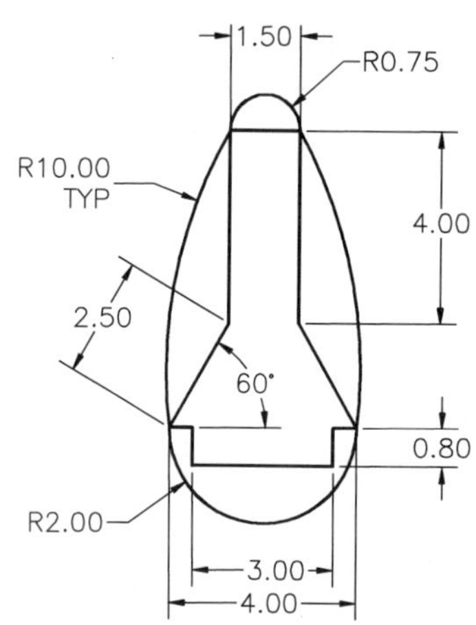

Figure 4-75 *Drawing for Problem Solving Exercise 1*

Problem Solving Exercise 2 | *Mechanical*

Make the drawing shown in Figure 4-76. First draw the front view (bottom left) and then the side and top views. Assume the missing dimensions.

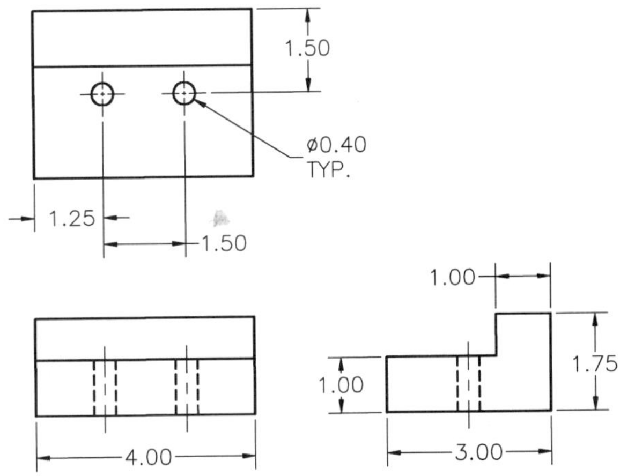

Figure 4-76 *Drawing for Problem Solving Exercise 2*

Answers to Self-Evaluation Test

1 - F, **2 -** T, **3 -** T, **4 -** F, **5 -** PSTYLEPOLICY, **6 -** SNAP, **7 -** Polar , **8 -** Apparent Intersection, **9 -** None, **10 -** User Preferences

Chapter 5

Editing Commands-I

Learning Objectives

After completing this chapter, you will be able to:

- *Create selection sets using various object selection options.*
- *Move the objects using the **MOVE** command.*
- *Make copies of existing objects using the **COPY** command.*
- *Copy objects with base point using the **COPYBASE** command.*
- *Use the **OFFSET** and **BREAK** commands.*
- *Fillet and chamfer objects using the **FILLET** and **CHAMFER** commands.*
- *Cut and extend objects using the **TRIM** and **EXTEND** commands.*
- *Stretch objects using the **STRETCH** command.*
- *Create polar and rectangular arrays using the **ARRAY** command.*
- *Use the **ROTATE**, **MIRROR**, and **SCALE** commands.*
- *Lengthen objects such as line, arc, and spline using the **LENGTHEN** command.*
- *Use the **MEASURE** and **DIVIDE** commands.*

CREATING A SELECTION SET

In Chapter 1, Getting Started-I, only two options of the selection set were discussed- Window, and Crossing. In this chapter you will learn additional selection set options that can be used to select objects. The following options are explained here.

Last	CPolygon	Add	Undo	Previous
Fence	BOX	SIngle	ALL	Group
AUto	WPolygon	Remove	Multiple	

Note

The quickest and easiest way to enter most selection options is by typing their required one- or more character abbreviations (shown in uppercase).

Last

This option is used to select the most recently drawn object that is partially or fully visible in the current display of the screen. This option is the most convenient option if you want to select the most recently drawn object that is visible on the screen. Keep in mind that if the last drawn object is not in the current display, the object in the current display that was drawn last will be selected. Although a selection set is being formed using the **Last** option, only one object is selected. However, you can use the Last option a number of times. You can use the Last selection option with any command that requires selection of objects (e.g., **COPY**, **MOVE**, and **ERASE**). After invoking the particular command, enter LAST or L at the **Select objects** prompt to use this object selection method.

Exercise 1 *General*

Using the **LINE** command, draw Figure 5-1(a). Then use the **ERASE** command with the **Last** option to erase the three most recently drawn lines to obtain Figure 5-1(d).

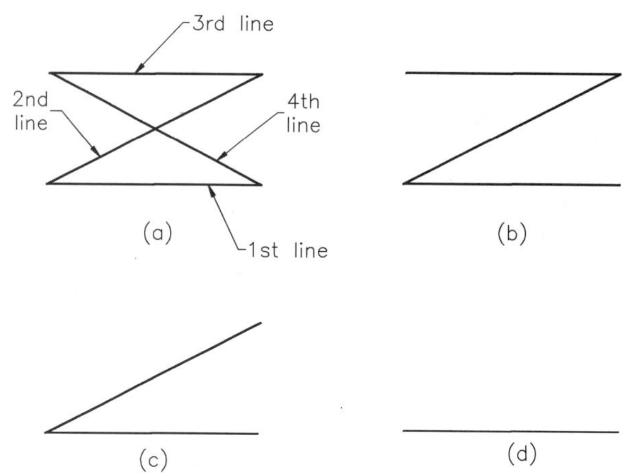

***Figure 5-1** Erasing objects using the **Last** selection option*

Previous

The Previous selection option automatically selects the objects in the most recently created selection set. To invoke this option you can enter **P** at the **Select objects** prompt. AutoCAD LT saves the previous selection set and lets you select it again by using this option. In other words, with the help of the Previous option you can edit the previous set without reselecting its objects individually. Another advantage of the Previous selection option is that you need not remember the objects if more than one editing operation has to be carried out on the same set of objects. For example, if you want to copy a number of objects and then move them, you can use the **Previous** option to select the same group of objects with the **MOVE** command. The prompt sequence will be as follows:

Command: Choose the **Copy** button.
Select objects: *Select the objects.*

Select objects: [Enter]
Specify base point or displacement, or [Multiple]: *Specify the base point.*
Specify second point of displacement or <use first point as displacement>: *Specify the point for displacement.*

Command: Choose the **Move** button.
Select objects: **P** [Enter]
found

The **Previous** option does not work with some editing commands, such as **STRETCH**. A previous selection set is cleared by the various deletion operations and the commands associated with them, such as **UNDO**. You cannot select the objects in model space and then use the same selection set in paper space, or vice versa. The reason for this is that AutoCAD LT keeps the record of the paper space or model space in which the individual selection set is created.

WPolygon

This option is similar to the **Window** option, except that in this option you can define a window that consists of an irregular polygon. You can specify the selection area by specifying points around the object you want to select (Figure 5-2). Similar to the window method, all the objects to be selected using this method should be completely enclosed within the polygon. The polygon is formed as you specify the points and can take any shape except the one that is self-intersecting. The last segment of the polygon is automatically drawn to close the polygon. The polygon can be created by specifying the coordinates of the points or by specifying the points with the help of a pointing device. With the **Undo** option, the most recently specified WPolygon point can be undone. To use the **WPolygon** option with object selection commands (like **ERASE**, **MOVE**, **COPY**), first invoke the particular command, and then enter WP at the **Select objects** prompt. The prompt sequence for selecting the objects using the **WPolygon** option is:

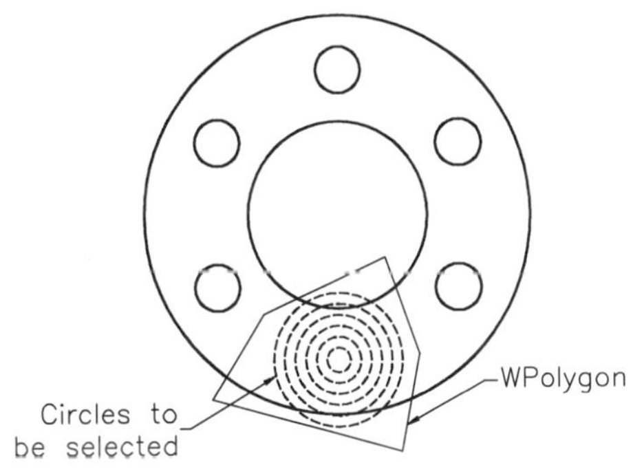

***Figure 5-2** Selecting objects using the **WPolygon** option*

Select objects: **WP** [Enter]
First polygon point: *Specify the first point.*
Specify endpoint of line or [Undo]: *Specify the second point.*
Specify endpoint of line or [Undo]: *Specify the third point.*
Specify endpoint of line or [Undo]: *Specify the fourth point.*
Specify endpoint of line or [Undo]: *Press ENTER after specifying the last point of polygon.*

Exercise 2 *General*

Draw a number of objects on the screen, and then erase some of them using the WPolygon to select the objects you want to erase.

CPolygon

This method of selection is similar to the WPolygon method except that **CPolygon** also selects those objects that are not completely enclosed within the polygon, but are touching the polygon boundaries. In other words, the object that is lying partially inside the polygon or is even touching it is also selected in addition to those objects that are completely enclosed within the polygon (Figure 5-3). CPolygon is formed as you specify the points. The points can be specified at the Command line or by specifying points with the pointing device. Just as in the **WPolygon** option, the crossing polygon can take any shape except the one in which it intersects itself. Also, the last segment of the polygon is drawn automatically, so the CPolygon is closed at all times. The prompt sequence for **CPolygon** option is:

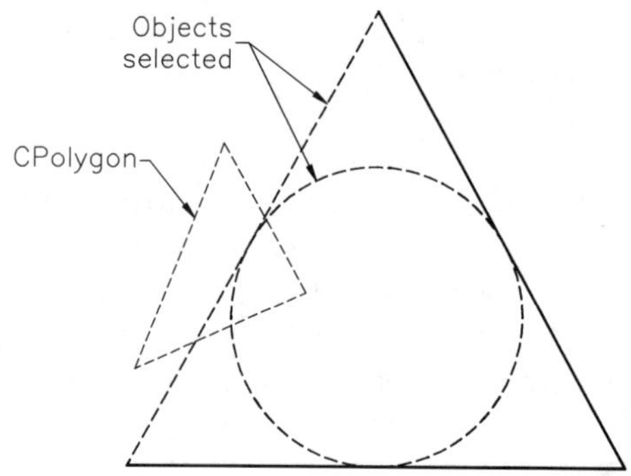

Figure 5-3 *Selecting objects using the **CPolygon** option*

Select objects: **CP** [Enter]
First polygon point: *Specify the first point.*
Specify endpoint of line or [Undo]: *Specify the second point.*
Specify endpoint of line or [Undo]: *Specify the third point.*
Specify endpoint of line or [Undo]: *Press ENTER after specifying the last point of the polygon.*

Remove

The **Remove** option is used to remove the objects from the selection set (but not from the drawing). After selecting many objects from a drawing by any selection method, there may be a need for removing some of the objects from the selection set. The following prompt sequence displays the use of the **Remove** option for removing the objects from the selection set shown in Figure 5-4.

Command: Choose the **Erase** button.
Select objects: *Select objects (a), (b), and (c).*
Select objects: **R**
Remove objects: *Select object (a).*
1 found, 1 removed, 2 total.
Select objects: [Enter]

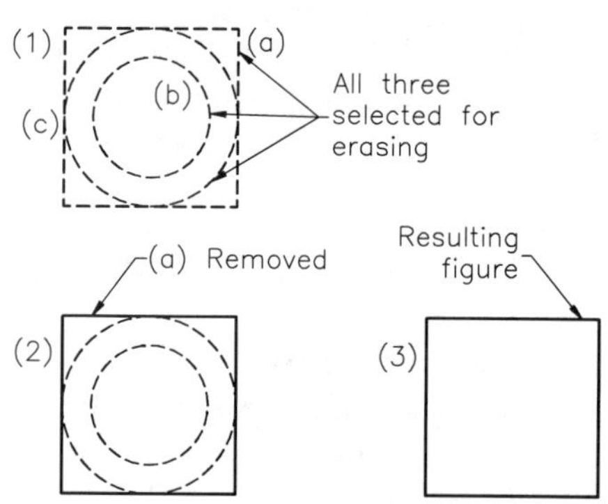

Figure 5-4 *Using the **Remove** option*

Tip

*The objects can also be removed from the selection set using the SHIFT key. For example, by pressing the SHIFT key and selecting object (a) with the pointing device will remove it from the selection set. AutoCAD LT will display the following message; **1 found, 1 removed, 2 total.***

Add

You can use the **Add** option to add objects to the selection set. When you begin creating a selection set, you are in **Add** mode. After you create a selection set by any selection method, you can add more objects by simply selecting them with the pointing device, when system variable **PICKADD** is set to 1 (default). When the system variable **PICKADD** is set to 0, to add objects to the selection set you will have to press the SHIFT key and then select the objects.

ALL

The **ALL** selection option is used to select all the objects in the current working environment of the current drawing. Note that the objects in the "OFF" layers are also selected with the **ALL** option. However, the objects that are in the frozen layers are not selected using this method. You can use this selection option with any command that requires object selection. After invoking the command, the ALL option can be entered at the **Select objects** prompt. Once you enter this option, all the objects drawn on the screen will be highlighted (dashed). For example, if there are four objects on the screen and you want to erase all of them, the prompt sequence is:

Command: Choose the **Erase** button.
Select objects: **ALL** [Enter]
4 found
Select objects: [Enter]

You can use this option in combination with other selection options. For example, consider there are five objects on the drawing screen and you want to erase three of them. After invoking the **ERASE** command, enter **ALL** at the **Select objects** prompt. Then press the SHIFT key and select the two objects you want to remove from the selection set; the remaining three objects are erased.

Fence

In the **Fence** option, a selection set is created by drawing an open polyline fence through the objects to be selected. Any object touched by the fence polyline is selected (Figure 5-5). The selection fence can be created by entering the coordinates at the Command line or by specifying the points with the pointing device. With this option, more flexibility for selection is provided because the fence can intersect itself. The Undo option can be used to undo the most recently selected fence point. Like the other selection options, this option is also used with the commands that need object selection. The prompt sequence for using the **Fence** option is:

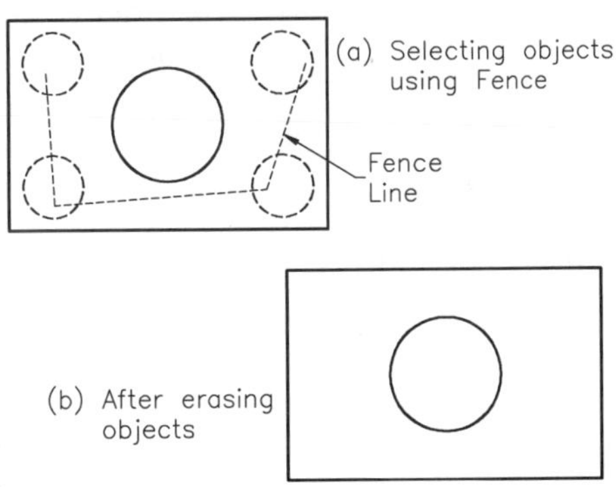

Figure 5-5 *Erasing objects using the* ***Fence*** *option*

Select objects: **F** [Enter]
First fence point: *Specify the first point.*
Specify endpoint of line or [Undo]: *Specify the second point.*
Specify endpoint of line or [Undo]: *Specify the third point.*
Specify endpoint of line or [Undo]: *Specify the fourth point.*
Specify endpoint of line or [Undo]: [Enter]

Group

The **Group** option enables you to select a group of objects by their group name. You can create a group and assign a name to it with the help of the **GROUP** command (see Chapter 18, Object Grouping and Editing Commands). Once a group has been created, you can select the group using the **Group** option for editing purposes. This makes the object selection process easier, since a set of objects is selected by entering just the group name. The prompt sequence is:

Command: **MOVE** [Enter]
Select objects: **G** [Enter]
Enter group name: *Enter the name of the predefined group you want to select.*
4 found
Select objects: [Enter]

Exercise 3 *General*

Draw six circles and select all of them to erase using the **ERASE** command with the **ALL** option. Now change the selection set contents by removing alternate circles from the selection set by using the fence so that alternate circles are erased.

BOX

When system variable **PICKAUTO** is set to 1 (default), the **BOX** selection option is used to select objects inside a rectangle. After you enter BOX at the **Select objects** prompt, you are required to specify the two corners of a rectangle at the **Specify first corner** and the **Specify opposite corner** prompts. If you define the Box from right to left, it is equivalent to the Crossing selection option. Therefore, it also selects those objects that are touching the rectangle boundaries in addition to those that are completely enclosed within the rectangle. If you define the Box from left to right, this option is equivalent to the Window option and selects only those objects that are completely enclosed within the rectangle. The prompt sequence is:

Select objects: **BOX** [Enter]
Specify first corner: *Specify a point.*
Specify opposite corner: *Specify opposite corner point of the box.*

AUto

The **AUto** option is used to establish automatic selection. You can select a single object, by selecting that object, as well as select a number of objects, by creating a window or a crossing. If you select a single object, it is selected; if you specify a point in the blank area you are automatically in the BOX selection option and the point you have specified becomes the first corner of the box. Auto and Add are default selections.

Multiple

When you enter **M** (Multiple) at the **Select objects** prompt, you can select multiple objects at a single **Select objects** prompt without the objects being highlighted. Once you give a null response to the **Select objects** prompt, all the selected objects are highlighted together.

Undo

This option removes the most recently selected object from the selection set.

SIngle

When you enter SI (SIngle) at the **Select objects** prompt, the selection takes place in the SIngle selection mode. Once you select an object or a number of objects using a Window or Crossing option, the **Select objects** prompt is not repeated. AutoCAD LT proceeds with the command for which the selection is made.

Command: *Choose the **Erase** button.*
Select objects: **SI** [Enter]
Select objects: *Select the object for erasing. The selected object is erased.*

You can also create a selection set using the **SELECT** command. The prompt sequence is:

Command: **SELECT** [Enter]
Select objects: *Use any selection method.*

EDITING COMMANDS

To use AutoCAD LT effectively, you need to know the editing commands and how to use them. In this section you will learn about the editing commands. These commands can be invoked from the toolbar, or menu, or can be entered at the Command prompt. Some of the editing commands such as **ERASE** and **OOPS** have been discussed in Chapter 1, Getting Started-I. The rest of the editing commands will be discussed in this section.

MOVE COMMAND

Toolbar:	Modify > Move
Menu:	Modify > Move
Command:	MOVE

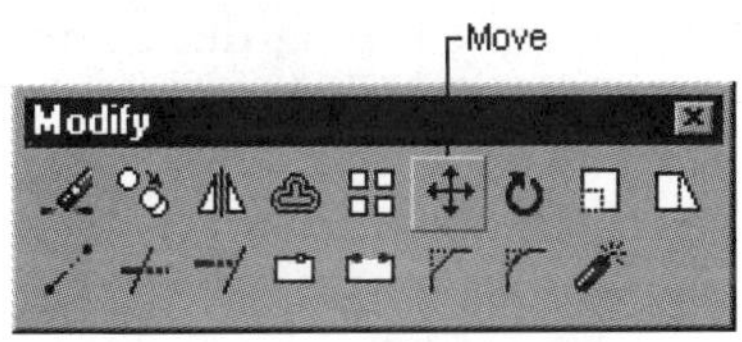

*Figure 5-6 Invoking the **MOVE** command from the **Modify** toolbar*

Sometimes, the objects are not located at the position where they actually should be. In these situations you can use the **MOVE** command (Figure 5-6). This command allows you to move one or more objects from their current location to a new location specified by you. This change in the location of the objects does not change their size or orientation. When you invoke this command, you will be prompted to select the objects to be moved. You can use any of the object selection techniques for selecting one or more objects. Once you have selected the objects, you will be prompted to specify the base point. This base point is the reference point with which the object will be picked and moved. It is advisable to select the base point on the object selected to move. Next, you will be prompted to specify the second point of displacement. This is the new location point where you want to move the object. When you specify this point, the selected objects will be moved to this point. The prompt sequence that will be followed when you choose the **Move** button from the **Modify** toolbar is:

Select objects: *Select the objects to be moved.*
Select objects: Enter
Specify base point or displacement: *Specify the base point for moving the selected object(s).*
Specify second point of displacement or <use first point as displacement>: *Specify second point or press ENTER to use the first point.*

If you press ENTER at the **Specify second point of displacement or <use first point as displacement>** prompt, AutoCAD LT interprets the first point as the relative value of the displacement in the X axis and Y axis direction. This value will be added in the X and Y axis coordinates and the object will be automatically moved to the resultant location. Figure 5-7 shows an object moved using the **MOVE** command.

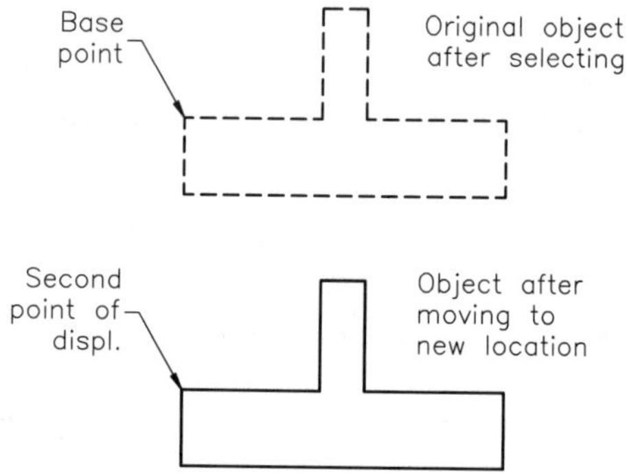

Figure 5-7 *Moving the objects to a new location*

COPY COMMAND

Toolbar:	Modify > Copy Object
Menu:	Modify > Copy
Command:	COPY

The **COPY** command is used to copy an existing object. This command is similar to the **MOVE** command in the sense that it makes copies of the selected objects and places them at specified locations, but the originals are left intact. Also, in this command you need to select the objects and then specify the base point. Then you are required to specify the second point, that is, where you want the copied object to be placed. Figure 5-8 shows the objects copied using this command. The prompt sequence that will be followed when you choose the **Copy Object** button from the **Modify** toolbar is:

Select objects: *Select the objects to copy.*
Select objects: Enter
Specify base point or displacement, or [Multiple]: *Specify the base point.*
Specify second point of displacement, or <use first point as displacement>: *Specify a new position on the screen using the pointing device or entering coordinates.*

Multiple Copies

This option of the **COPY** command is used to make multiple copies of the same object (Figure 5-9). To use this option, select the **Multiple** option or enter **M** at the **Specify base point or displacement, or [Multiple]** prompt. Next, AutoCAD LT prompts you to enter the second point. When you select a point, a copy is placed at this point, and the prompt is automatically repeated until you press ENTER to terminate the **COPY** command. The prompt sequence for the **COPY** command with the **Multiple** option is:

Specify base point or displacement, or [Multiple]: **M** [Enter]
Specify base point: *Specify the base point.*
Specify second point of displacement or <use first point as displacement>: *Specify a point for placement.*
Specify second point of displacement or <use first point as displacement>: *Specify another point for placement.*
Specify second point of displacement or <use first point as displacement>: *Specify another point for placement.*
Specify second point of displacement or <use first point as displacement>: [Enter]

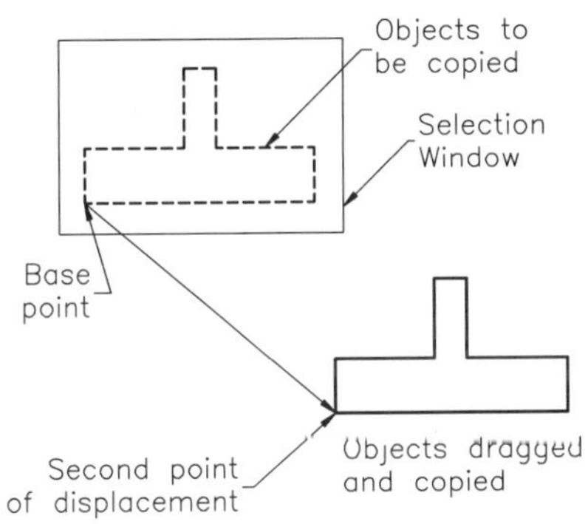

Figure 5-8 *Using the **COPY** command*

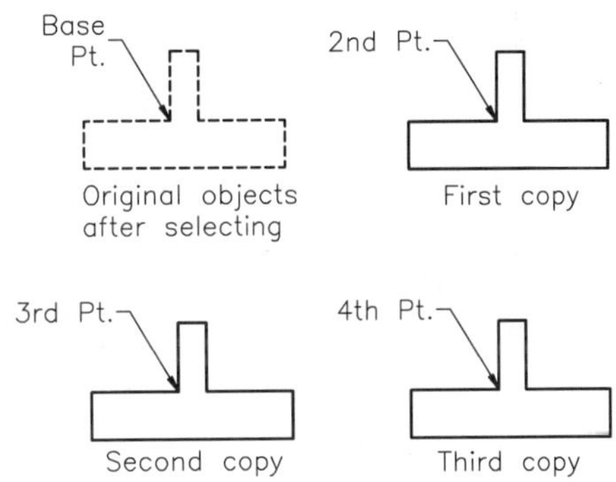

Figure 5-9 *Making multiple copies*

COPYBASE COMMAND

Menu:	Edit > Copy with Basepoint
Command:	COPYBASE

The command **COPYBASE** is used to specify the basepoint of objects to be copied. This command can also be invoked from the shortcut menu that is displayed upon selecting the object and right-clicking. Choose **Copy with Base Point** to invoke this command, see Figure 5-10. This command can be very useful while pasting an object very precisely in the same diagram or into another diagram. When you invoke this command from the shortcut menu, you will be prompted to specify the base point for copying the objects. Unlike the **COPY** command, where the objects are dragged and placed at the desired location, this command copies the selected object on to the Clipboard. Then from the Clipboard, the objects will be placed at the specified location. The Clipboard is defined as a medium for storing the data while transferring the data from one place to the other.

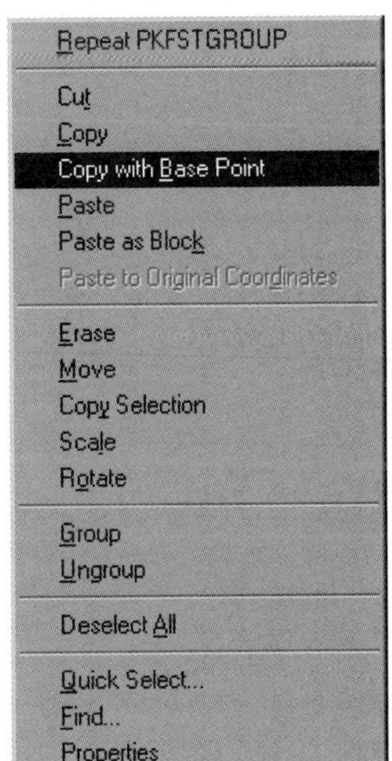

Figure 5-10 *Invoking the **COPYBASE** command from the shortcut menu*

Note that the contents of the Clipboard are not visible. The objects copied using this command can be copied from one drawing file to the other or from one working environment to the other. After copying objects with the **COPYBASE** command, you can paste the objects at the desired place with greater accuracy.

Note

You will learn more about the working environments and the Clipboard in the following chapters.

PASTEBLOCK COMMAND

Menu:	Edit > Paste as Block
Command:	PASTEBLOCK

The **PASTEBLOCK** command is used to paste the contents of the Clipboard into a new drawing or in the same drawing at a new location. You can also invoke the **PASTEBLOCK** command from the shortcut menu by right-clicking in the drawing area and choosing **Paste as Block**.

PASTEORIG COMMAND

Menu:	Edit > Paste to Original Coordinates
Command:	PASTEORIG

The **PASTEORIG** command is used to paste the contents of the Clipboard into a new drawing using the coordinates from the original drawing. You can also invoke the **PASTEORIG** command from the shortcut menu by right-clicking in the drawing area and choosing **Paste to Original Coordinates**. The **PASTEORIG** command is available only when the Clipboard contains AutoCAD LT data from a drawing other than the current drawing.

Exercise 4 *Mechanical*

In this exercise you will draw the object shown in Figure 5-11. Use the **COPY** command for creating the drawing.

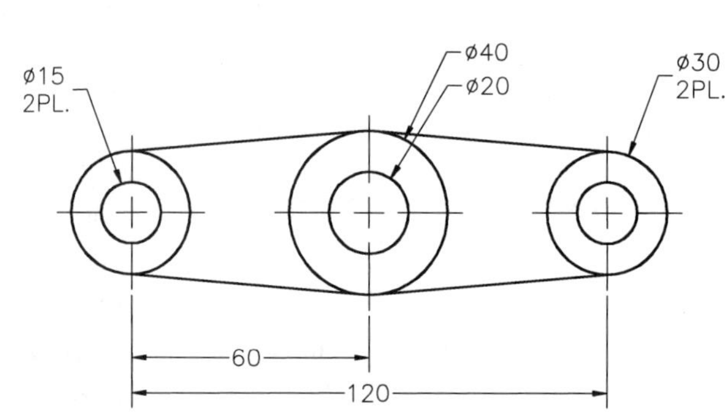

Figure 5-11 *Drawing for Exercise 4*

OFFSET COMMAND

Toolbar:	Modify > Offset
Menu:	Modify > Offset
Command:	OFFSET

If you want to draw parallel lines, polylines, concentric circles, arcs, curves, and so on, you can use the **OFFSET** command (Figure 5-12). This command creates another object that is similar to the selected one. Remember that you are allowed to select only one entity at a time to offset. When offsetting an object, you can specify the offset distance and the side to offset, or you can specify a distance through which you want to offset the selected object. Depending on the side to offset, you can create smaller or larger circles, ellipses, and arcs. If the offset side is toward the inner side of the perimeter, the arc, ellipse, or circle will be smaller than the original. The prompt sequence that will follow when you choose the **Offset** button from the **Modify** menu is:

Specify offset distance or [Through] <current>: *Specify the offset distance or press ENTER to use the **Through** option.*
Select object to offset or <exit>: *Select the object to offset.*

The offset distance can be specified by entering a value or by specifying two points with the pointing device. AutoCAD LT will measure the distance between these two points and use it as the offset distance. The **Through** option is generally used for creating the orthographic views. In this case, you do not have to specify a distance; you simply specify an offset point. See Figure 5-13.

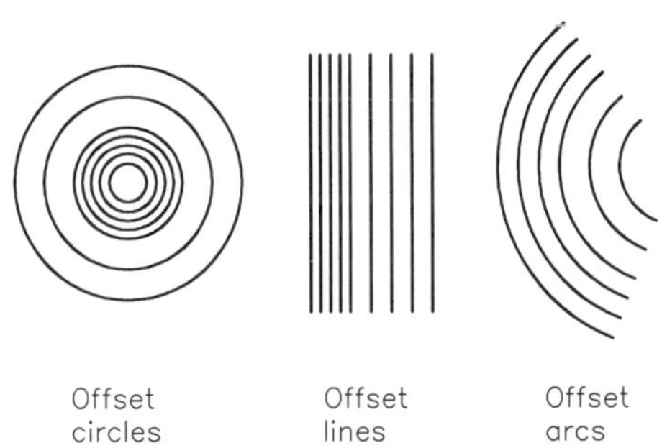

Figure 5-12 *Using the **OFFSET** command*

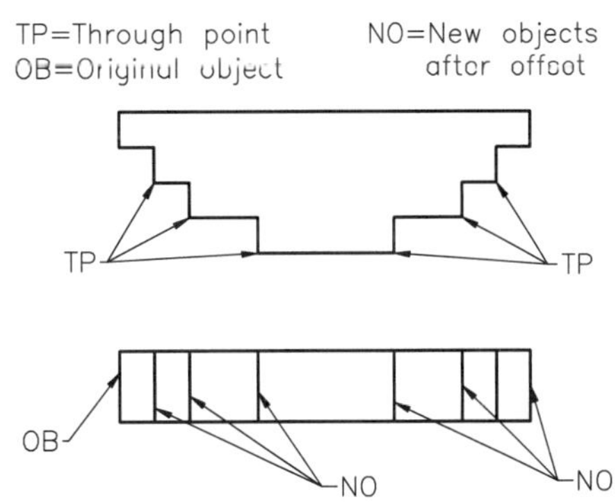

Figure 5-13 *Using the **Through** option*

The offset distance is stored in the **OFFSETDIST** system variable. A negative value indicates OFFSET is set to the Through option. You can offset lines, arcs, 2D polylines, xlines, circles, ellipses, elliptical arcs, rays, and planar splines. If you try to offset objects other than these, the message **Cannot offset that object** is displayed.

Exercise 5 — *General*

Use the **OFFSET** edit command to draw Figure 5-14 and Figure 5-15.

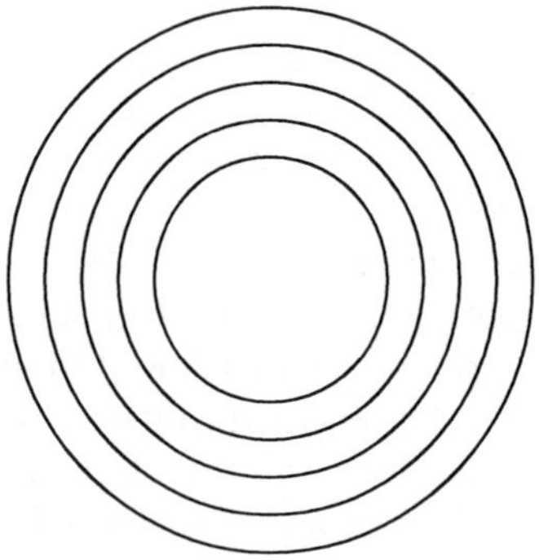

Figure 5-14 *Drawing for Exercise 5*

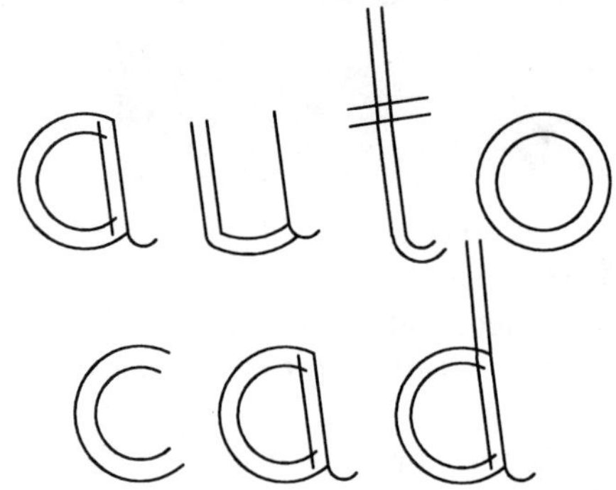

Figure 5-15 *Drawing for Exercise 5*

ROTATE COMMAND

Toolbar:	Modify > Rotate
Menu:	Modify > Rotate
Command:	ROTATE

While creating designs there are many occasions when you have to rotate an object or a group of objects. You can accomplish this by using the **ROTATE** command. When you invoke this command, AutoCAD LT will prompt you to select the objects and the base point about which the selected objects will be rotated. You should be careful when selecting the base point as it is easy to get confused if the base point is not located on a known object. After you specify the base point you are required to enter a rotation angle. By default, positive angles produce a counterclockwise rotation and negative angles produce a clockwise rotation (Figure 5-16). The **ROTATE** command can also be invoked from the shortcut menu by selecting the object and clicking the right button in the drawing area and choosing **Rotate**. The prompt sequence that will follow when you choose the **Rotate** button is:

Current positive angle in UCS: ANGDIR=*current* ANGBASE=*current*
Select objects: *Select the objects for rotation.*
Select objects: Enter
Specify base point: *Specify a base point about which the selected objects will be rotated.*
Specify rotation angle or [Reference]: *Enter a positive or negative rotation angle, or specify a point.*

If you need to rotate objects with respect to a known angle, you can do this in two different ways using the **Reference** option. The first way is to specify the known angle as the reference angle, followed by the proposed angle to which the objects will be rotated (Figure 5-17). Here the object is first rotated clockwise from the X axis, through the reference angle. Then the object is rotated through the new angle from this reference position in a counterclockwise direction. The prompt sequence is:

Current positive angle in UCS: ANGDIR=*current* ANGBASE=*current*
Select objects: *Select the objects for rotation.*

Select objects: Enter
Specify base point: *Specify the base point.*
Specify rotation angle or [Reference]: **R** Enter
Specify the reference angle <0>: **10** Enter
Specify the new angle: **90** Enter

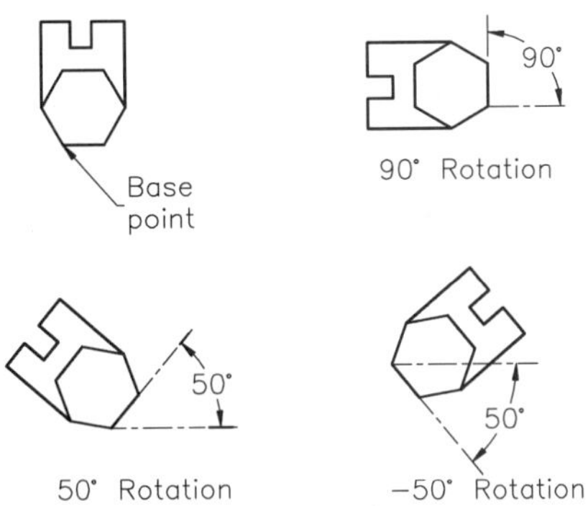

Figure 5-16 *Rotation of objects with different rotation angles*

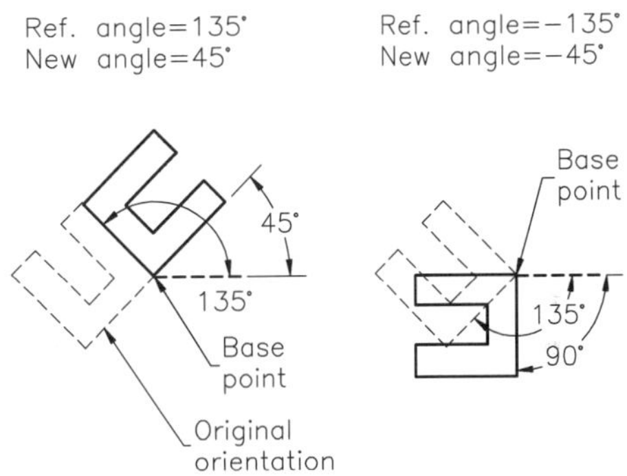

Figure 5-17 *Rotation using the* ***Reference*** *Angle option*

The other method is used when the reference angle and the new angle are not known. In this case you can use the edges of the original object and the reference object to specify the original object and reference angle, respectively. Figure 5-18 shows a model created at an unknown angle and a line also at an known angle. In this case this line will be used as a reference object for rotating the object. In such cases keep in mind that the base point should be taken on the reference object. The reason for this is that you can not define two points for specifying the new angle. You have to directly enter the angle value or specify only one point. Therefore, the base point will be taken as the first point and the second point can be defined for the new angle. Figure 5-19 shows the model after rotating with reference to the line such that the line and the model are inclined at similar angles. The prompt sequence to rotate the model in Figure 5-18 is:

Current positive angle in UCS: ANGDIR=*current* ANGBASE=*current*
Select objects: *Select the object for rotation.*
Select objects: Enter
Specify base point: *Specify the base point as the lower endpoint of the line. See Figure 5-18.*
Specify rotation angle or [Reference]: **R** Enter
Specify the reference angle <0>: *Specify the first point on the edge of the model. See Figure 5-18.*
Specify second point: *Specify the second point on the same edge of the model. See Figure 5-18.*
Specify new angle: *Select the other endpoint of the reference line. See Figure 5-18.*

SCALE COMMAND

Toolbar:	Modify > Scale
Menu:	Modify > Scale
Command:	SCALE

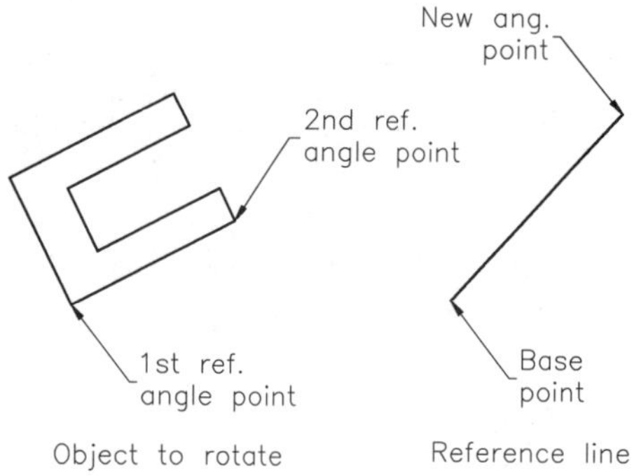

Figure 5-18 *Rotating the objects using a reference line*

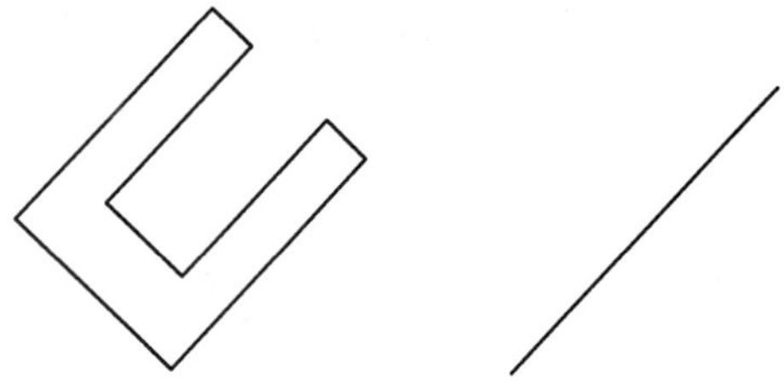

Figure 5-19 *Model after rotating with reference to the line*

Many times you will need to change the size of objects in a drawing. You can do this with the **SCALE** command. This command dynamically enlarges or shrinks the selected object about a base point, keeping the aspect ration of the object constant. This means that the size of the object will be increased or reduced equally in the X, Y, and Z direction. The dynamic scaling property allows you to view the object as it is being scaled. This can be viewed by moving the pointing device in the drawing area after selecting the base point. Application of the identical scale factor to the X, Y, and Z dimensions ensures that the shape of the objects being scaled do not change. This is a useful and time-saving editing command because instead of redrawing objects to the required size, you can scale the objects with a single **SCALE** command. Another advantage of this command is that if you have already put the dimensions on the drawing, they will also change according to the new scale. You can also invoke the **SCALE** command from the shortcut menu by right-clicking in the drawing area and choosing **Scale**. The prompt sequence that will follow when you choose the **Scale** button is:

Select objects: *Select objects to be scaled.*
Select objects: Enter
Specify base point: *Specify the base point, preferably a known point.*
Specify scale factor or [Reference]: *Specify the scale factor.*

The base point will not be moved from its position and the selected object(s) will be scaled around the base point as shown in Figure 5-20 and Figure 5-21. To reduce the size of an object, the scale factor should be less than 1 and to increase the size of an object, the scale factor should be greater than 1. You can enter a scale factor or select two points to specify a distance as a factor. When you select two points to specify a distance as a factor, the first point should be on the referenced object.

Sometimes it is time-consuming to calculate the relative scale factor. In such cases you can scale the object by specifying a desired size in relation to the existing size (a known dimension). In other words, you can use a **reference length**. This can be done by entering **R** at the **Specify scale factor or [Reference]** prompt. Then, you either specify two points to specify the length or enter a length. At the next prompt, enter the length relative to the reference length. For example, if a line is **2.5** units long and you want the length of the line to be **1.00** units, instead

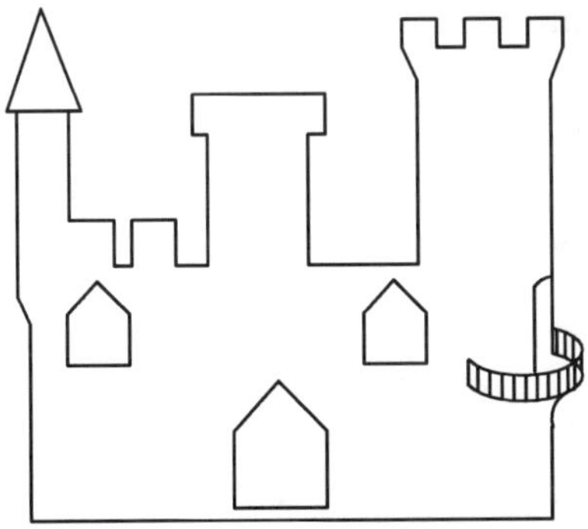

Figure 5-20 *Original object before scaling*

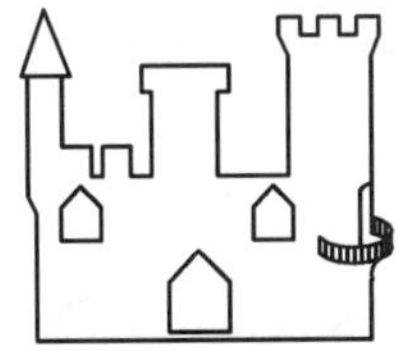

Figure 5-21 *After scaling to 0.5 of actual size*

of calculating the relative scale factor, you use the **Reference** option. The prompt sequence for using the **Reference** option is:

Select objects: *Select the object to scale.*
Select objects: Enter
Specify base point: *Specify the base point.*
Specify scale factor or [Reference]: **R** Enter
Specify reference length <1>: *Specify the reference length.*
Specify new length: *Specify the new length.*

Similar to the **ROTATE** command, here also you can scale one object using the reference of another object. Again, the base point has to be taken on the reference object as you can define only one point for the new length.

Tip

*If you have not used the required drawing units for a drawing, you can use the **Reference** option of the **SCALE** command to correct the error. Select the entire drawing with the help of the All selection option. Specify the **Reference option**, then select the endpoints of the object whose desired length you know. Specify the new length, and all objects in the drawing will be rescaled automatically to the desired size.*

FILLET COMMAND

Toolbar:	Modify > Fillet
Menu:	Modify > Fillet
Command:	FILLET

The **FILLET** command is used to create smooth round arcs to connect two objects. The edges in the design are generally filleted to reduce the area of stress concentration. The **FILLET** command helps you form round corners between any two entities by allowing you to define two entities that form a sharp vertex. A fillet can also be created between two intersecting parallel lines as well as nonintersecting and nonparallel lines, arcs, polylines, xlines, rays, splines, circles, and true ellipses. The fillet arc created will be tangent to both the selected entities. The radius of the arc to create the fillet has to be specified. The prompt sequence that will follow when you choose the **Line** button is:

Current Settings: Mode= TRIM, Radius= 0.5000
Select first object or [Polyline/Radius/Trim]:

Select First Object Option

This is the default method to fillet two objects. As the name implies, it prompts for the first object required for filleting. The prompt sequence to use this option is:

Current Settings: Mode= TRIM, Radius= 0.5000
Select first object or [Polyline/Radius/Trim]: *Specify first object.*
Select second object: *Select second object.*

The **FILLET** command can also be used to cap the ends of two parallel lines (Figure 5-22). The cap is a semicircle whose radius is equal to half the distance between the two parallel lines. The cap distance is calculated automatically when you select the two parallel lines for filleting. You can select lines with the Window, Crossing, or Last option, but to avoid unexpected results, select the objects by picking objects individually. Also, selection by picking objects is necessary in the case of arcs and circles that have the possibility of more than one fillet. They are filleted closest to the select points (Figure 5-23).

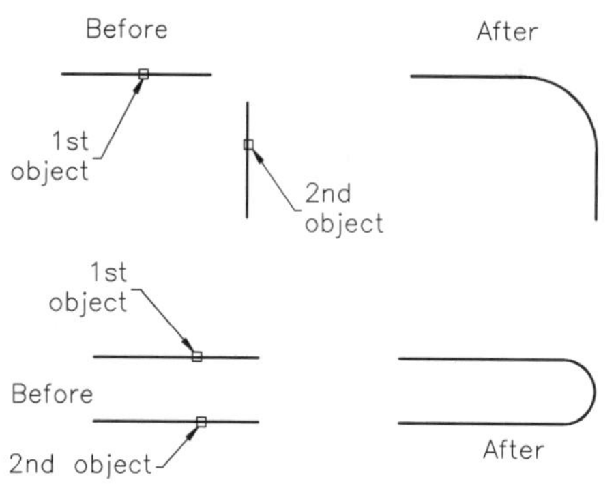

Figure 5-22 *Filleting the parallel and nonparallel lines*

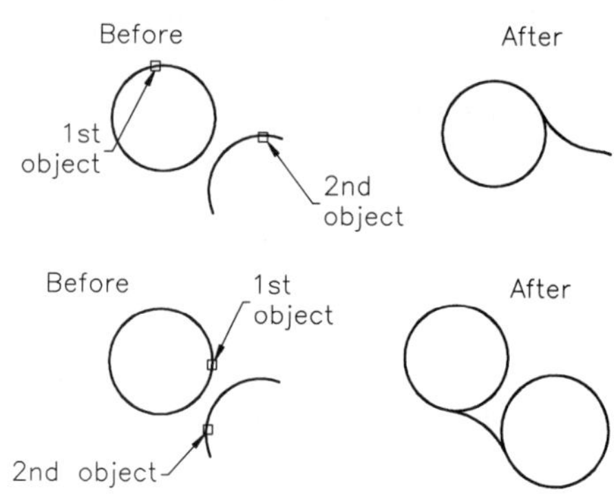

Figure 5-23 *Using the **FILLET** command on circles and arcs*

Radius Option

The fillet you create depends on the radius distance you specify. The default radius is 0.5000. You can enter a distance or two points. The new radius you enter becomes the default and remains in effect until changed. The prompt sequence is:

Select first object or [Polyline/Radius/Trim]: **R** Enter
Specify fillet radius <current>: *Enter a fillet radius or press ENTER to accept the current value.*

Note

*The **FILLETRAD** system variable controls and stores the current fillet radius.*

Tip

A fillet with a zero radius creates sharp corners and therefore can be used to clean up lines at corners if they overlap or have a gap.

Trim Option

When you create a fillet, an arc is created and the selected objects are either trimmed or extended at the fillet endpoint. This is because the **Trim** mode is set to trim. If it is set to **No Trim** they are left intact. Figure 5-24 shows a model filleted with the **Trim** mode set to **Trim** and to **No Trim**. The prompt sequence is:

Select first object or [Polyline/Radius/Trim]: **T** Enter
Specify trimmode option [Trim/No trim] <current>: *Enter* **T** *to trim the edges,* **N** *to leave them intact.*

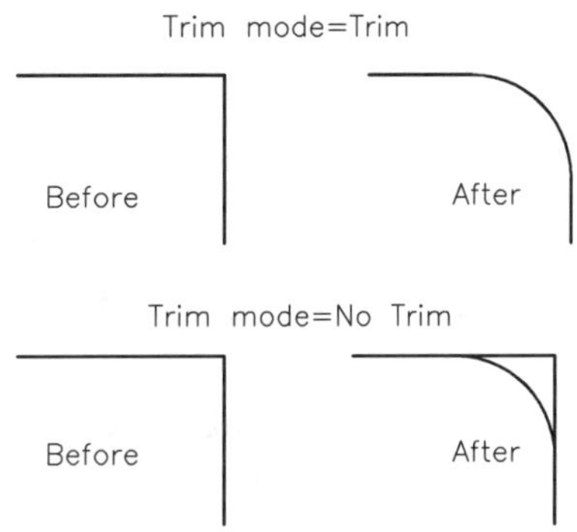

Figure 5-24** Filleting the lines with the **Trim** mode set to **Trim** and **No Trim

Polyline Option

Using this option you can fillet a number of entities that comprise a single polyline (Figure 5-25). The polylines can be created using the **POLYLINE** or the **RECTANG** command. If the object is created using the **POLYLINE** command, it must be closed using the **Close** option. Otherwise the last corner of the polyline will not be filleted. When you select this option, AutoCAD LT prompts you to select a polyline. All the vertices of the polyline are filleted when you select the polyline. The fillet radius for all the vertices will be the same. If the selected polyline is not closed, then the last corner is not filleted. The prompt sequence for using this option is:

Current Settings: Mode= *current*,
Radius= *current*
Select first object or [Polyline/Radius/Trim]: **P** Enter
Select 2D polyline: *Select the polyline.*

Filleting Objects with a Different UCS

The fillet command will also fillet objects that are not in the current UCS plane. To create a fillet for these objects, AutoCAD LT will automatically change the UCS transparently so that it can generate a fillet between the selected objects.

Note

Filleting the objects in different planes is discussed in Chapter 21, Drawing and Viewing 3D Objects.

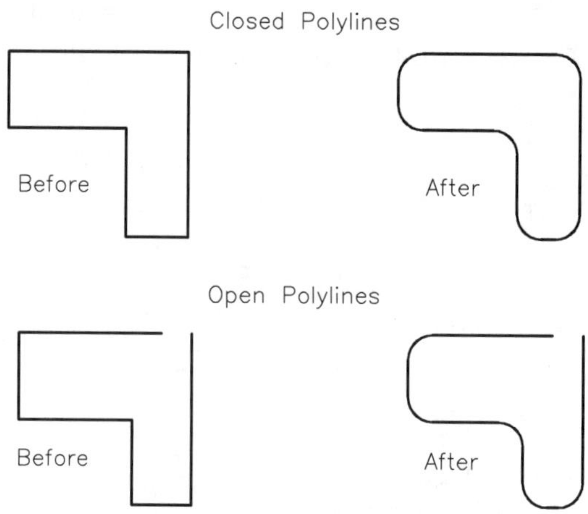

Figure 5-25 *Filleting closed and open polylines*

Setting the TRIMMODE System Variable

The **TRIMMODE** system variable is a variable that eliminates any size restriction on the **FILLET** command. By setting **TRIMMODE** to **0**, you can create a fillet of any size without actually cutting the existing geometry. Also, there is no restriction on the fillet radius. This means that the fillet radius can be larger than one or both objects that are being filleted. The default value of this variable is 1.

Note

TRIMMODE = 0 *Fillet or chamfer without cutting the existing geometry.*
TRIMMODE = 1 *Extend or trim the geometry.*

When you enter the ***FILLET*** *command, AutoCAD LT displays the current* **TRIMMODE** *and the current fillet radius.*

CHAMFER COMMAND

Toolbar:	Modify > Chamfer
Menu:	Modify > Chamfer
Command:	CHAMFER

Chamfering the sharp corners is another method of reducing the areas of stress concentration in the design. Chamfering is defined as the process in which the sharp edges or corners are beveled. In simple words, it is defined as the taper provided on a surface. A beveled line connects two separate objects to create a chamfer. The size of a chamfer depends on its distance from the corner. If a chamfer is equidistant from the corner in both directions, it is a 45-degree chamfer. A chamfer can be drawn between two lines that may or may not intersect. However, remember that the lines are not parallel because the parallel lines can not be chamfered. This command also works on a single polyline. In AutoCAD LT the chamfers can be created using two methods: by defining two distances, or by defining one distance and the chamfer angle. The prompt sequence that will follow when you choose the **Chamfer** button from the **Modify** toolbar is:

(TRIM mode) Current chamfer Dist1 = 0.5000, Dist2 = 0.5000
Select first line or [Polyline/Distances/Angle/Trim/Method]:

Select First Line Option

In this option you need to select two nonparallel objects so that they are joined with a beveled line. The size of the chamfer depends upon the values of the two distances. The default values of the distances are 0.5000 and 0.5000. The prompt sequence is:

(TRIM mode) Current chamfer Dist1 = 0.5000, Dist2 = 0.5000
Select first line or [Polyline/Distance/Angle/Trim/Method]: *Specify the first line.*
Select second line: *Select the second line.*

Distance Option

This option is used to enter the chamfer distance. This option can be invoked by entering **D** at the **Select first line or [Polyline/Distance/Angle/Trim/Method]** prompt. Next, enter the first and the second chamfer distances. The first distance is the distance of the corner calculated along the edge selected first. Similarly, the second distance is calculated along the edge that is selected last. The new chamfer distances remain in effect until you change them. Instead of entering the distance values, you can specify two points to indicate each distance (Figure 5-26). The prompt sequence is:

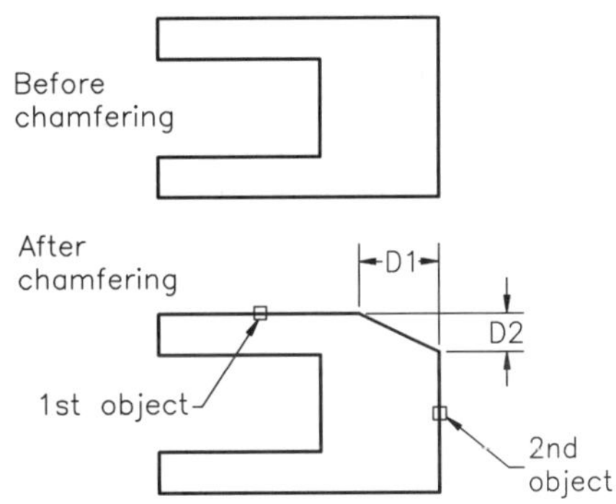

Figure 5-26 *Chamfering the model using the* ***Distance*** *option*

Select first line or [Polyline/Distance/Angle/Trim/Method]: **D** Enter
Specify first chamfer distance <current>: *Enter a distance value or specify two points.*
Specify second chamfer distance <current>: *Enter a distance value or specify two points.*

The first and second chamfer distances are stored in the **CHAMFERA** and **CHAMFERB** system variables.

Tip

If Dist 1 and Dist 2 are set to zero, the ***CHAMFER*** *command will extend or trim the selected lines so that they end at the same point.*

Polyline Option

You can use the **CHAMFER** command to chamfer all corners of a closed or open polyline (Figure 5-27). With a closed polyline, all the corners of the polyline are chamfered to the set distance values. Sometimes the polyline may appear closed. But if the **Close** option was not used to create it, it may not be. In this case, the last corner is not chamfered. The prompt sequence is:

(TRIM mode) Current chamfer Dist1 = 0.5000, Dist2 = 0.5000
Select first line or [Polyline/Distance/Angle/Trim/Method]: **P** Enter
Select 2D polyline: *Select the polyline.*

Angle Option

The second method of creating the chamfer is by specifying a distance and the chamfer angle (Figure 5-28). The chamfer angle can be defined using this option. The prompt sequence is:

Select first line or [Polyline/Distance/Angle/Trim/Method]: **A** Enter
Specify chamfer length on the first line <current>: *Specify a length.*
Specify chamfer angle from the first line <current>: *Specify an angle.*

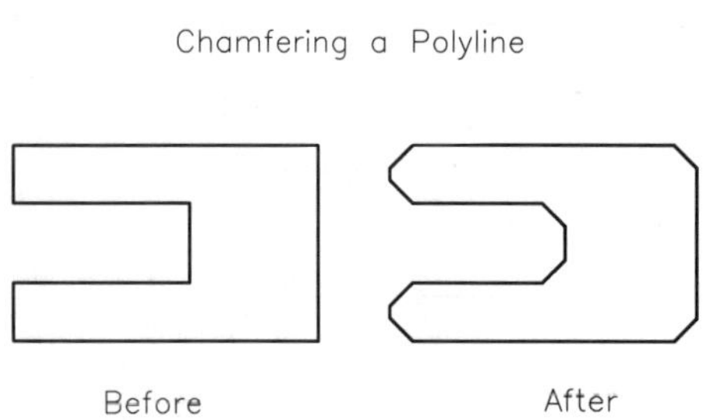

Figure 5-27 *Chamfering a polyline*

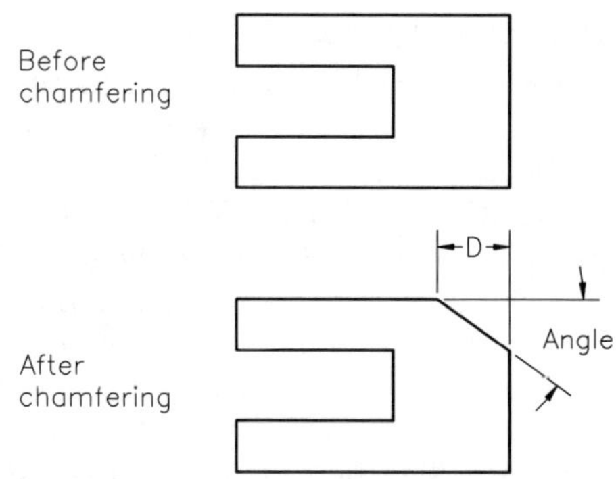

Figure 5-28 *Chamfering using the* ***Angle*** *option*

Trim Option

Depending on this option, the selected objects are either trimmed or extended to the endpoints of the chamfer line or left intact. The prompt sequence is:

Select first line or [Polyline/Distance/Angle/Trim/Method]: **T** Enter
Enter Trim mode option [Trim/No Trim] <current>:

Method Option

This option is used to toggle between the **Distance** method and the **Angle** method for creating the chamfer. The current settings of the selected method will be used for creating the chamfer. The prompt sequence is:

Select first line or [Polyline/Distance/Angle/Trim/Method]: **M** Enter
Enter trim method [Distance/Angle] <current>: *Enter* **D** *for Distance option,* **A** *for Angle option.*

Note

If you set the value of the ***TRIMMODE*** *system variable to* ***1*** *(default value), the objects will be trimmed or extended after they are chamfered and filleted. If* ***TRIMMODE*** *is set to zero, the objects are left untrimmed.*

Setting the Chamfering System Variables

The chamfer modes, distances, length, and angle can also be set by using the following variables:

CHAMMODE = 0	Distance/Distance (default)
CHAMMODE = 1	Length/Angle
CHAMFERA	Sets first chamfer distance on the first selected line (default = 0.5000)
CHAMFERB	Sets second chamfer distance on the second selected line (default = 0.5000)
CHAMFERC	Sets the chamfer length (default = 1.0000)
CHAMFERD	Sets the chamfer angle from the first line (default = 0)

TRIM COMMAND

Toolbar:	Modify > Trim
Menu:	Modify > Trim
Command:	TRIM

When creating a design, there are a number of places where you have to remove the unwanted and extending edges. Breaking individual objects takes time if you are working on a complex design with many objects. In such cases you can use the **TRIM** command. This command trims objects that extend beyond a required point of intersection. When you invoke this command, you will be prompted to select the cutting edges or boundaries. These edges can be lines, polylines, circles, arcs, ellipses, xlines, rays, splines, text, blocks or even viewports. There can be more than one cutting edge and you can use any selection method to select them. After the cutting edge or edges are selected, you must select each object to be trimmed. An object can be both a cutting edge and an object to trim. You can trim lines, circles, arcs, polylines, splines, ellipses, xlines, and rays. The prompt sequence is:

Current settings:Projection=UCS Edge = None
Select cutting edges...
Select objects: *Select the cutting edges.*
Select objects: Enter
Select object to trim or shift-select to extend or [Project/Edge/Undo]:

Select Object to trim Option

Here you have to specify the objects you want to trim and the side from which you select the object will be trimmed. This prompt is repeated until you press ENTER. This way you can trim several objects with a single **TRIM** command (Figure 5-29). The prompt sequence is:

Current settings: Projection= UCS Edge= None
Select cutting edges...
Select objects: *Select the first cutting edge.*
Select objects: *Select the second cutting edge.*
Select objects: Enter
Select object to trim or shift-select to extend or [Project/Edge/Undo]: *Select the first object.*
Select object to trim or shift-select to extend or [Project/Edge/Undo]: *Select the second object.*
Select object to trim or shift-select to extend or [Project/Edge/Undo]: Enter

Shift-select to extend Option

This option is used switch to the extend mode. It is used to extend the object instead of trimming it. In case the object to trim does not intersect with the cutting edge, you can press the SHIFT key and then select the object to trim. The selected edge will be extended taking the cutting edge as the boundary for extension.

Edge Option

This option is used whenever you want to trim those objects that do not intersect the cutting edges, but would intersect if the cutting edges were extended (Figure 5-30). The prompt sequence is:

Command: **TRIM** Enter
Current settings: Projection= UCS Edge= None
Select cutting edges...
Select objects: *Select the cutting edge.*
Select objects: Enter
Select object to trim or shift-select to extend or [Project/Edge/Undo]: **E** Enter
Enter an implied edge extension mode [Extend/No extend] <current>: **E** Enter
Select object to trim or shift-select to extend or [Project/Edge/Undo]: *Select object to trim.*

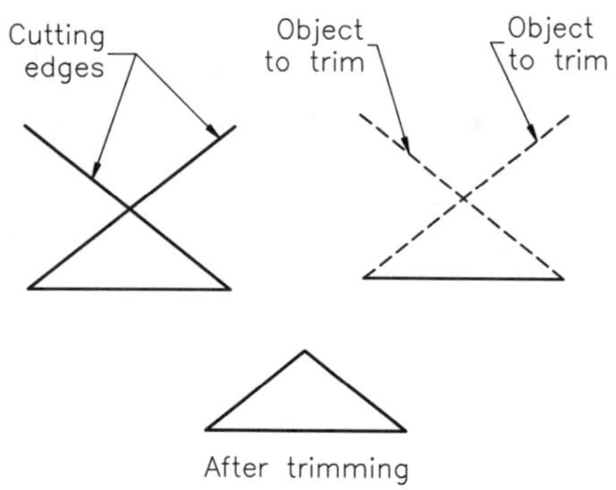

Figure 5-29 *Using the **TRIM** command*

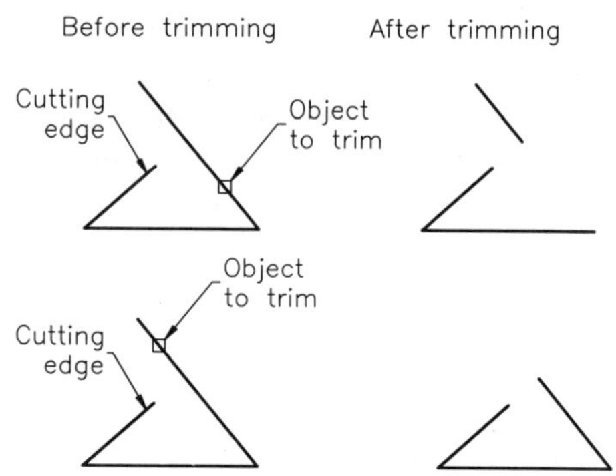

Figure 5-30 *Trimming an object using the **Edge** option (**Extend**)*

Project Option

In this option you can use Project mode while trimming objects. The prompt sequence is:

Select object to trim or shift-select to extend or [Project/Edge/Undo]: **P** Enter
Enter a projection option [None/Ucs/View] <current>:

The **None** option is used whenever the objects to trim intersect the cutting edges in 3D space. If you want to trim those objects that do not intersect the cutting edges in 3D space, but do visually appear to intersect in a particular UCS or the current view, use the **UCS** or **View** options. The UCS option projects the objects to the XY plane of the current UCS, while the **View** option projects the objects to the current view direction (trims to their apparent visual intersections).

Undo Option

If you want to remove the previous change created by the **TRIM** command, enter **U** at the **Select object to trim or [Project/Edge/Undo]:** prompt.

Exercise 6 *Mechanical*

Draw the top illustration in Figure 5-31. Then use the **FILLET**, **CHAMFER**, and **TRIM** commands to obtain the following figure. (Assume the missing dimensions.)

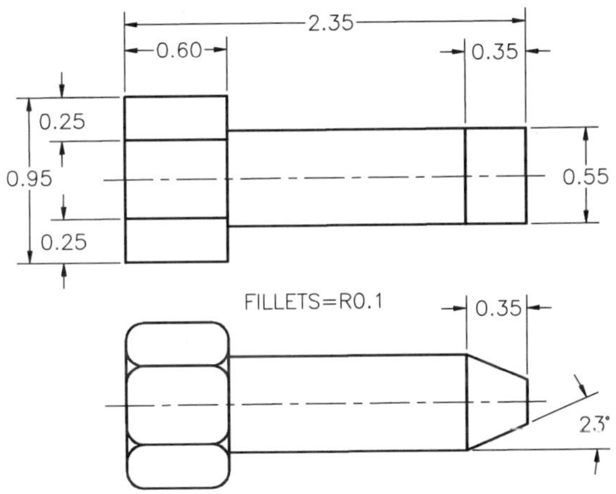

Figure 5-31 *Drawing for Exercise 6*

EXTEND COMMAND

Toolbar:	Modify > Extend
Menu:	Modify > Extend
Command:	EXTEND

This command may be considered the opposite of the **TRIM** command. In the **TRIM** command you trim objects; in the **EXTEND** command you can extend lines, polylines, rays, and arcs to meet other objects. This command does not extend closed loops. The command format is similar to that of the **TRIM** command. You are required to select the boundary edges first. The boundary edges are those objects that the selected lines or arcs extend to meet. These edges can be lines, polylines, circles, arcs, ellipses, xlines, rays, splines, text, blocks, or even viewports. The prompt sequence that will follow when you choose the **Extend** button is:

Current settings: Projection= UCS Edge= None
Select boundary edges ...
Select objects: *Select the boundary edges.*
Select objects: Enter
Select object to extend or shift-select to trim or [Project/Edge/Undo]:

Select object to extend Option

Here you have to specify the object you want to extend to the selected boundary (Figure 5-32). This prompt is repeated until you press ENTER. Now you can select a number of objects in a single **EXTEND** command.

Shift-select to trim Option

This option is used to switch to the trim mode in the **EXTEND** command. You can press the SHIFT key and then select the object to trim. In this case the boundary edges will be taken as the cutting edges.

Project Option

In this option you can use the projection mode while trimming objects. The prompt sequence that will follow when you choose the **Extend** button is:

Current settings: Projection= UCS Edge= None
Select boundary edges ...
Select objects: *Select the boundary edges.*
Select objects: Enter
Select object to extend or shift-select to trim or [Project/Edge/Undo]: **P** Enter
Enter a projection option [None/UCS/View] <current>:

The **None** option is used whenever the objects to be extended intersect with the boundary edge in 3D space. If you want to extend these objects that do not intersect the boundary edge in 3D space, use the **UCS** or **View** options. The UCS option projects the objects to the XY plane of the current UCS, while the View option projects the objects to the current view.

Edge Option

You can use this option whenever you want to extend objects that do not actually intersect the boundary edge, but would intersect its edge if the boundary edge were extended (Figure 5-33). If you enter **E** at the prompt, the selected object is extended to the implied boundary edge. If you enter **N** at the prompt, only those objects that would actually intersect the real boundary edge are extended (the default). The prompt sequence is:

Current settings: Projection= UCS Edge= None
Select boundary edges ...
Select objects: *Select the boundary edge.*
Select objects: Enter
Select object to extend or shift-select to trim or [Project/Edge/Undo]: **E** Enter
Enter an implied extension mode [Extend/No extend] <current>: **E** Enter
Select object to extend or shift-select to trim or [Project/Edge/Undo]: *Select the line to extend.*

Undo Option

If you want to remove the previous change created by the **EXTEND** command, enter **U** at the **Select object to extend or [Project/Edge/Undo]** prompt.

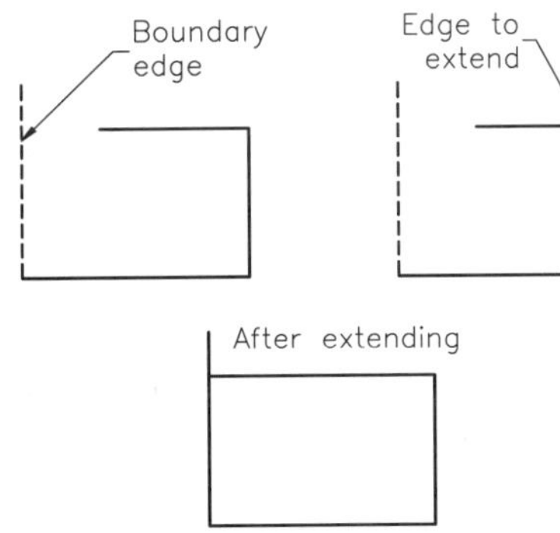

***Figure 5-32** Extending an edge*

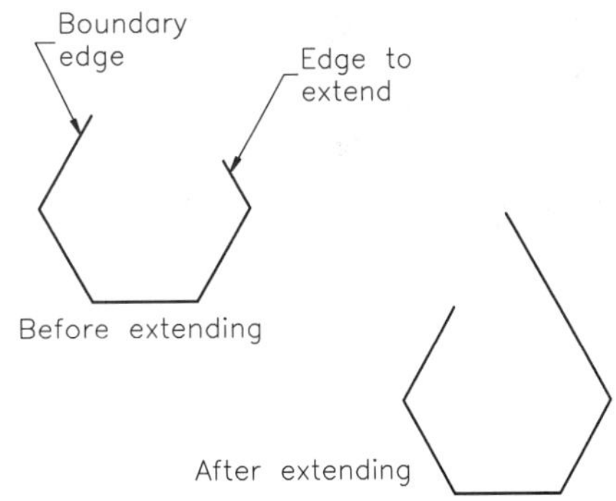

***Figure 5-33** Extending an edge using the **Edge** option (**Extend**)*

Trimming and Extending with Text, Region, or Spline

The **TRIM** and **EXTEND** commands can be used with text, regions, or splines as edges (Figure 5-34). This makes the **TRIM** and **EXTEND** commands two of the most useful editing commands in AutoCAD LT. The **TRIM** and **EXTEND** commands can also be used with arcs, elliptical arcs, splines, ellipses, 3D Pline, rays, and lines. See Figure 5-35.

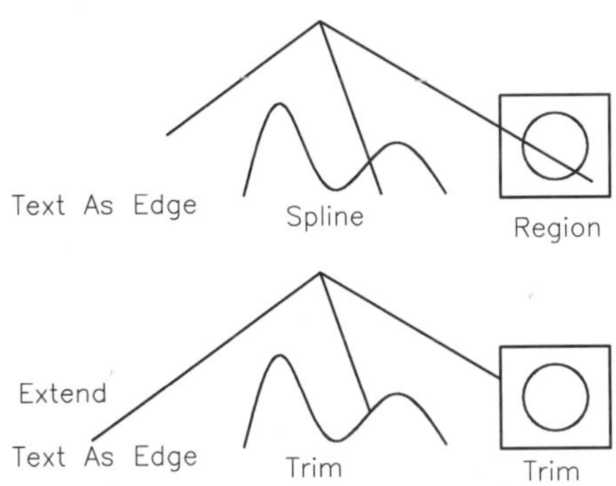

***Figure 5-34** Using the **TRIM** and **EXTEND** commands with text, spline, and region*

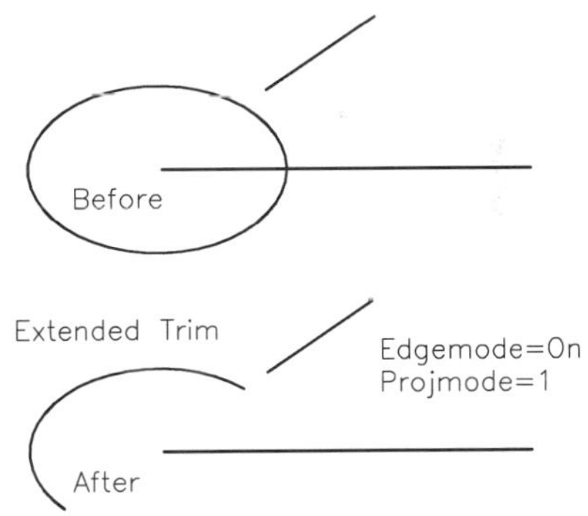

***Figure 5-35** Using the **Edge** option to do an implied trim*

The system variables **PROJMODE** and **EDGEMODE** determine how the **TRIM** and **EXTEND** commands are executed. The following are the values that can be assigned to these variables:

Value	PROJMODE	EDGEMODE
0	True 3D mode	Use regular edge without extension (default)
1	Project to current UCS XY plane (default)	Extend the edge to natural boundary
2	Project to current view plane	

STRETCH COMMAND

Toolbar:	Modify > Stretch
Menu:	Modify > Stretch
Command:	STRETCH

This command can be used to stretch objects, altering selected portions of the objects. With this command you can lengthen objects, shorten them, and alter their shapes (Figure 5-36). You must use a Crossing, or CPolygon selection to specify the objects to stretch. The prompt sequence that will follow when you choose the **Stretch** button is:

Select objects to stretch by crossing-window or crossing-polygon ...
Select objects: *Select the objects using the crossing window or polygon.*
Select objects: Enter

After selecting the objects, you have to specify the point of displacement. You should select only that portion of the object that needs stretching.

Specify base point or displacement: *Specify the base point.*
Specify second point of displacement or <use first point as displacement>: *Specify the displacement point or press ENTER to use the first point as the displacement.*

You normally use a Crossing or CPolygon selection with **STRETCH** because if you use a Window, those objects that cross the window are not selected, and the objects selected, because they are fully within the window, are moved, not stretched. The object selection and stretch specification process of **STRETCH** is a little unusual. You are really specifying two things: first, you are selecting objects. Second, you are specifying the portions of those selected objects to be stretched. You can use a Crossing or CPolygon selection to simultaneously specify both, or you can select objects by any method, and then use any window or crossing specification to specify what parts of those objects to stretch. Objects or portions of selected objects completely within the window or crossing specification are moved. If selected objects cross the window or crossing specification, their defining points within the window or crossing specification are moved, their defining points outside the window or crossing specification remain fixed, and the parts crossing the window or crossing specification are stretched. Only the last window or crossing specification made determines what is stretched or moved. Figure 5-36 illustrates using a crossing selection to simultaneously select the two angled lines and specify that their right ends will be stretched. Alternatively, you can select the lines by any method and then use a Crossing selection (which will not actually select anything) to specify that their right ends will be stretched.

Note

The regions can not be stretched. When you select them, they will be moved instead of stretched.

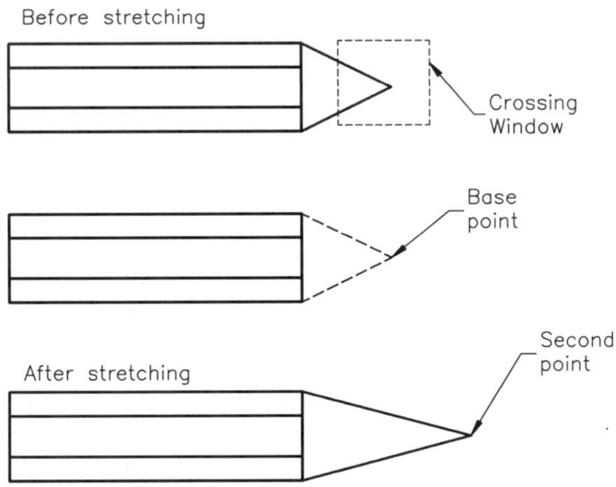

Figure 5-36 Stretching the entities

LENGTHEN COMMAND

Toolbar:	Modify > Lengthen
Menu:	Modify > Lengthen
Command:	LENGTHEN

Like the **TRIM** and **EXTEND** commands, the **LENGTHEN** command can be used to extend or shorten lines, polylines, elliptical arcs, and arcs. The **LENGTHEN** command has several options that allow you to change the length of objects by dynamically dragging the object endpoint, entering the delta valuc, cntering the percentage value, or entering the total length of the object. This command also allows the repeated selection of objects for editing. This command has no effect on closed objects such as circles. The selected entity will be increased or decreased from the endpoint closest to the selection point. The prompt sequence that will follow when you choose the **Lengthen** button is:

Select an object or [DElta/Percent/Total/DYnamic]:

Select an object Option

This is the default option that returns the current length or the included angle of the selected object. If the object is a line, AutoCAD LT returns only the length. However, if the selected object is an arc, AutoCAD LT returns the length and the angle. The same prompt sequence will be displayed after you select the object.

DElta Option

The **DElta** option is used to increase or decrease the length or angle of an object by defining the distance or angle by which the object will be extended. The delta value can be entered by entering a numerical value or by specifying two points. A positive value will increase (Extend) the length of the selected object and a negative value will decrease the length (Trim) (see Figure 5-37). The prompt sequence to use this option is:

Select an object or [DElta/Percent/Total/DYnamic]: **DE** Enter
Enter delta length or [Angle] <current>: *Specify the length or enter A for angle.*
Enter delta angle <current>: *Specify the delta angle.*
Select object to change or [Undo]: *Select object to be extended.*
Select object to change or [Undo]: Enter

Percent

The **Percent** option is used to extend or trim an object by defining the change as a percentage of the original length of the angle, as shown in Figure 5-38. For example, a positive number of 150 will increase the length by 50 percent and a positive number of 75 will decrease the length by 25 percent of the original value (negative values are not allowed).

Total

The **Total** option is used to extend or trim an object by defining the new total length or angle. For example, if you enter a total length of 1.25, AutoCAD LT will automatically increase or decrease the length of the object so that the new length of the object is 1.25. The value can be entered by entering a numerical value or by specifying two points. The object is shortened or lengthened with respect to the endpoint that is closest to the selection point. The selection point is determined by where the object was selected. The prompt sequence for using this option is:

Specify total length or [Angle] <current>: *Specify the length or enter* **A** *for angle.*

If you enter a numeric value in the previous prompt, the length of the selected objects is changed accordingly, (Figure 5-38). You can change the angle of an arc by entering **A** and giving the new value.

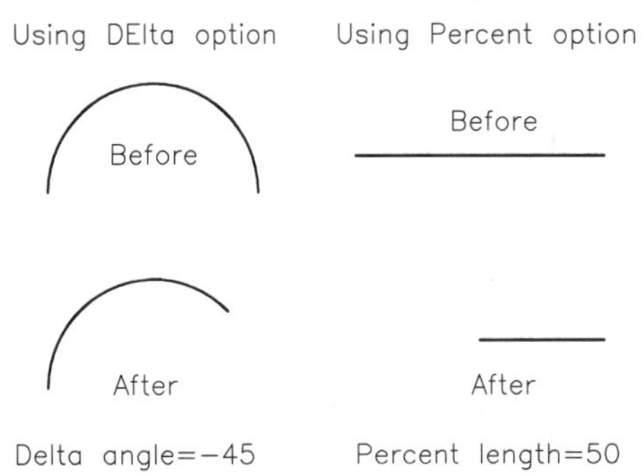

*Figure 5-37 Using the **DElta** and **Percent** option*

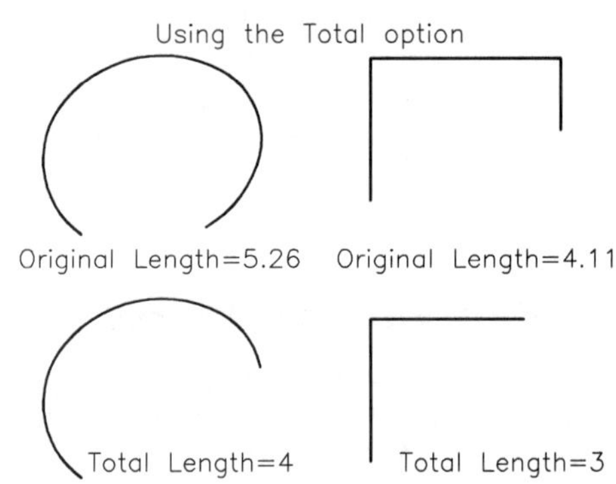

*Figure 5-38 Using the **Total** option*

DYnamic

The **DYnamic** option allows you to dynamically change the length or angle of an object by specifying one of the endpoints and dragging it to a new location. The other end of the object stays fixed and is not affected by dragging. The angle of lines, radius of arcs, and shape of elliptical arcs are unaffected.

ARRAY COMMAND

Toolbar:	Modify > Array
Menu:	Modify > Array
Command:	ARRAY

Array is defined as the method of creating multiple copies of the selected object and arranging them in the rectangular or circular fashion. In some drawings you may need to specify an object multiple times in a rectangular or circular arrangement. For example, suppose you have to draw six chairs around a table. This job can be accomplished by drawing each chair separately or by using the **COPY** command to make multiple copies of the chair. But it is a very tedious process and also the alignment of the chairs will have to be adjusted. This can be easily done using the **ARRAY** command. All you have to do is to create just one chair and the remaining five will be created and automatically arranged around the table by the **ARRAY** command. This method is more efficient and less time-consuming. This command allows you to make multiple copies of selected objects in a rectangular or polar fashion. Each resulting element of the array can be controlled separately. When you choose the **Array** button, the **Array** dialog box is displayed. You can use this dialog box for creating the rectangular array or for creating the polar array.

Rectangular Array

A rectangular array is formed by making copies of the selected object along the X and Y axes direction of an imaginary rectangle (along rows and columns). The rectangular array can be created by selecting the **Rectangular Array** radio button in the **Array** dialog box (Figure 5-39).

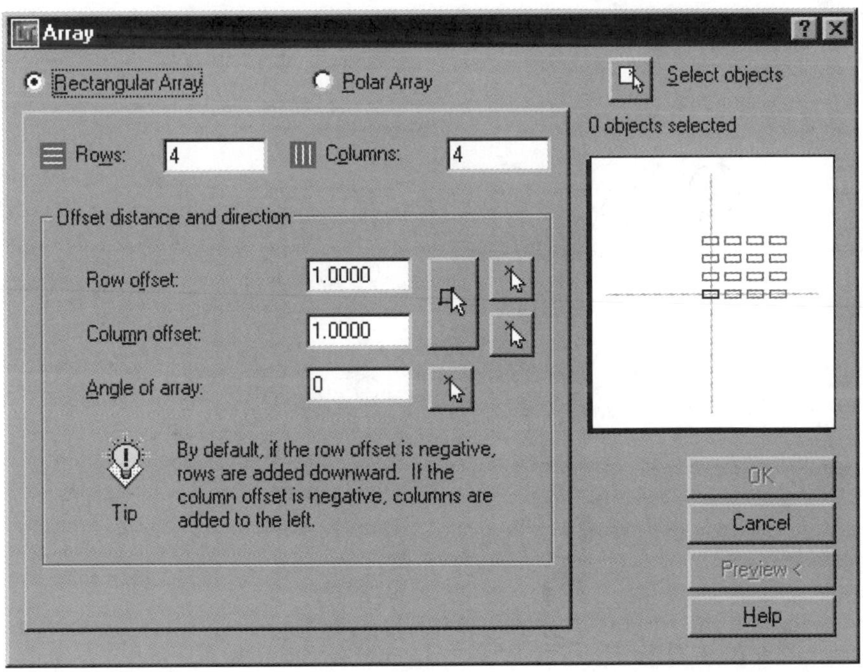

Figure 5-39 ***Rectangular Array*** *options in the* ***Array*** *dialog box*

Rows

This edit box is used to specify the number of rows in the rectangular array. The rows are arranged along the X axis of the current UCS.

Columns

This edit box is used to specify the number of columns in the rectangular array. The columns are arranged along the Y axis of the current UCS.

Offset distance and direction Area

The options under this area are used to define the distance between the rows and the columns and the angle of the array.

Row offset. This edit box is used to specify the distance between the rows (Figure 5-40). You can either enter the distance value in this edit box or choose the **Pick Row Offset** button to define the distance. This button is provided on the right of the **Row offset** edit box. When you choose this button, the **Array** dialog will be temporary closed and you will be prompted to specify two points on the screen to define the row offset distance. The positive value of row offset will create the rows along the positive X axis direction. The negative value of row offset will create the rows along the negative X axis direction (Figure 5-41).

Column offset. This edit box is used to specify the distance between the columns (Figure 5-40). You can either enter the distance value in this edit box or choose the **Pick Column Offset** button to define the distance. This button is provided on the right of the **Column offset** edit box. When you choose this button, the **Array** dialog will be temporary closed and you will be prompted to specify two points on the screen to define the column offset distance. The positive value of row offset will create the rows along the positive X axis direction. The negative value of row offset will create the rows along the negative X axis direction (Figure 5-41).

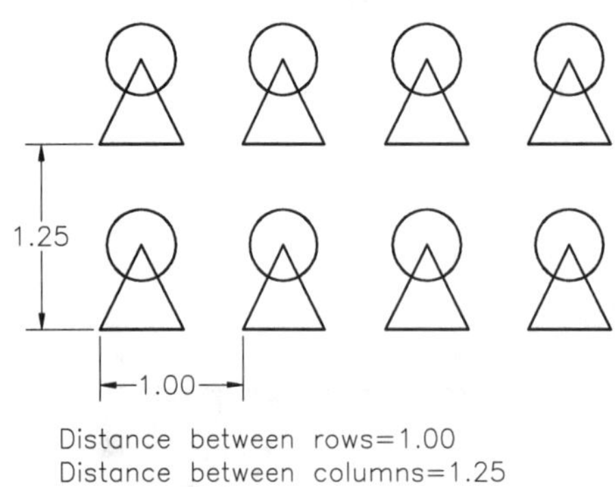

Figure 5-40 Rectangular array with row and column distance

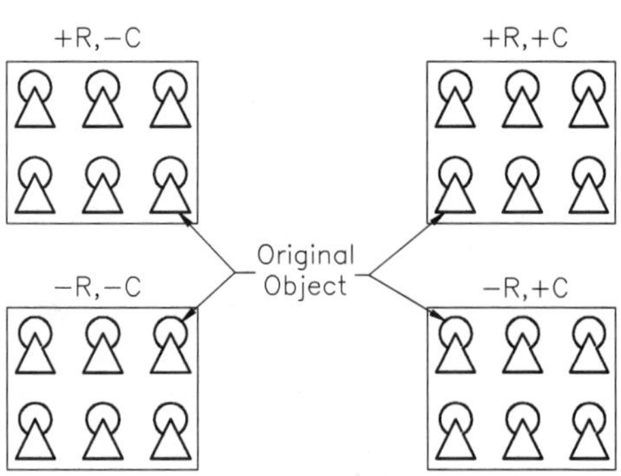

Figure 5-41 Specifying the direction of arrays

You can define both the row offset distance and the column offset distance simultaneously by choosing the **Pick Both Offset** button. When you choose this button, the **Array** dialog box will be temporarily closed and you will be prompted to define a unit cell by specifying two opposite corners. The distance along the X axis of the unit cell defines the distance between the rows and the distance along the Y axis of the unit cell defines the distance between the columns.

Angle of array. This edit box is used to define the angle of the array. This is the value by which the rows and the columns will be rotated. See Figure 5-42. The positive value will rotate them in the counterclockwise direction and the negative value will rotate them in the clockwise direction. You can also define the angle by choosing the **Pick Angle of Array** button and specifying two points on the screen.

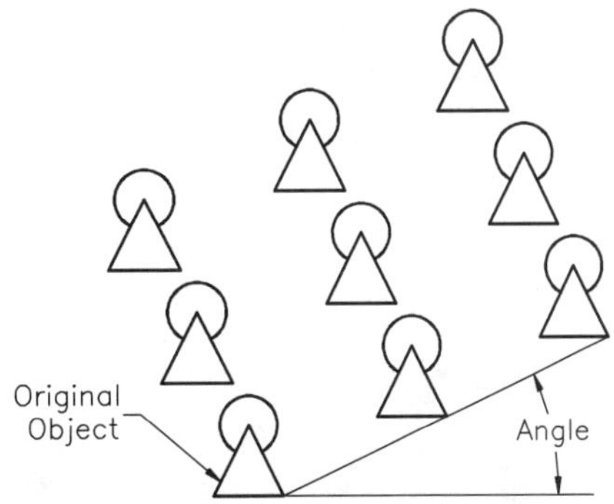

Figure 5-42 *Rotated rectangular array*

The object for array can be selected by choosing the **Select Objects** button. You can preview the creation of array by choosing the **Preview** button. When you choose this button, the **Array** dialog box will be displayed (Figure 5-43). If you are satisfied with the array, choose the **Accept** button, and the array will be created. If you are not satisfied with the array, choose the **Modify** button. The **Array** dialog box will be redisplayed on the screen and you can make the necessary changes.

Figure 5-43 Array *dialog box for accepting or modifying the array*

Polar Array

A polar array is an arrangement of objects around a point in a circular fashion. This kind of array is created by selecting the **Polar Array** radio button in the **Array** dialog box. See Figure 5-44.

Center point

The center point of the array is defined as the point around which the selected items will be arranged. It is considered as the center point of the imaginary circle on whose circumference the items will be placed. The coordinates of the center of the array can be specified in the **X** and **Y** edit boxes. You can either enter the values in these edit boxes or select the center point of the array from the screen. To select the center point of the array from the screen, choose the **Pick Center Point** button provided on the right side of the **Y** edit box. When you choose this button, the **Array** dialog box will be temporarily closed and you will be prompted to select the center of the array. The **Array** dialog box will be redisplayed when you select the center of the array. Figure 5-45 shows a polar array displaying the center of the imaginary circle about which the objects are rotated.

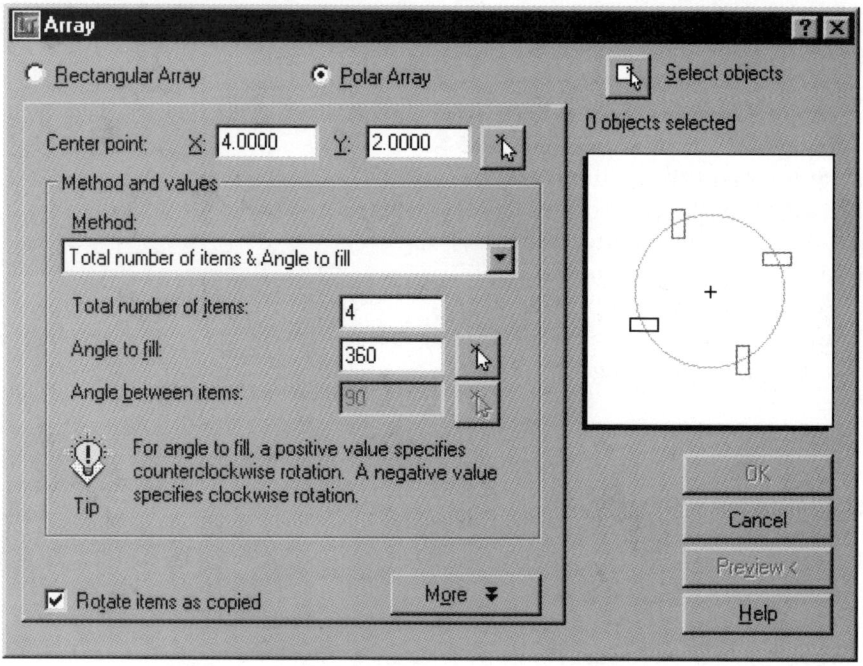

Figure 5-44 ***Polar Array*** *options in the* ***Array*** *dialog box*

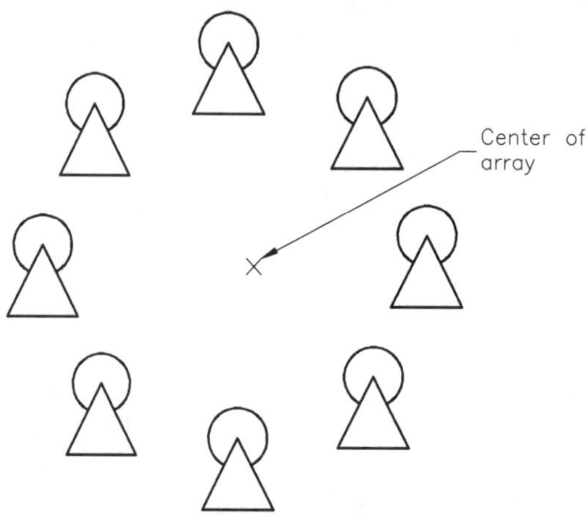

Figure 5-45 Center point of array

Methods and values Area

The options under this area are used to set the parameters related to the method that will be employed to create the polar array. The other options under this area will be available based on the method you select for creating the polar array.

Methods. This drop-down list provides you with three methods for creating the polar array. The three methods that are available are:

Total number of items & Angle to fill. This method is used to create a polar array by specifying the total number of items in the array and total included angle between the first and the last item of the array (Figure 5-46). The number of items and the angle to fill can be specified in the **Total number of items** and **Angle to fill** edit boxes, respectively. You can also specify the angle to fill on the screen. This is done by choosing the **Pick Angle to Fill** button provided on the right side of the **Angle to fill** edit box. The **Array** dialog box will be temporarily closed and you will be prompted to specify the angle to fill.

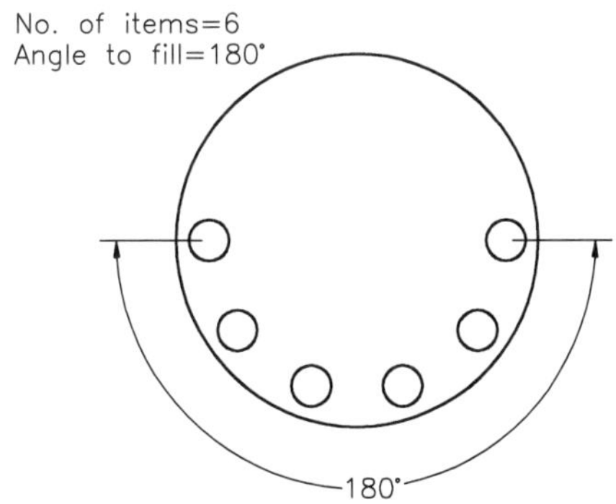

Figure 5-46 *Array created using number of items and angle to fill*

Total number of items & Angle between the items. This method is used when you want to create a polar array by specifying the total number of items in the array and the included angle between the two adjacent items of the array (Figure 5-47). The angle between the items is also called the incremental angle. The number of items and the angle between the items can be specified in the **Total number of items** and **Angle between items** edit boxes, respectively. You can also specify the angle between the items on the screen. This is done by choosing the **Pick Angle Between Items** button provided on the right side of the **Angle to fill** edit box. The **Array** dialog box will be temporarily closed and you will be prompted to specify the angle between the items.

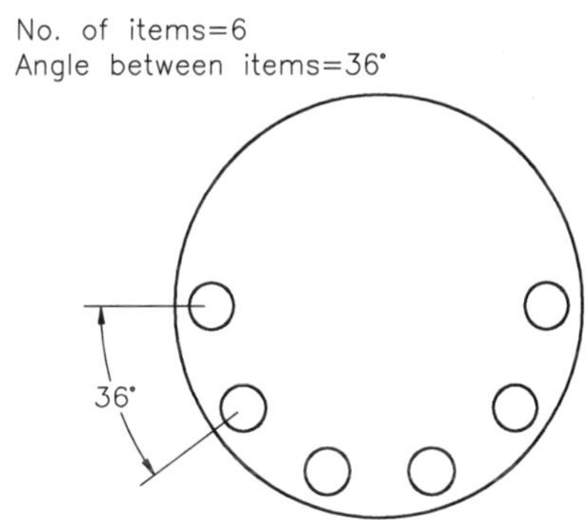

Figure 5-47 *Array created using number of items and angle between items*

Angle to fill & Angle between items. This method is used when you want to specify the angle between the items and total angle to fill (Figure 5-48). In this case the number of items are not specified. The number of items are automatically calculated using the total angle to fill in the array and the angle between the items. The angle to fill and the angle between the items can be entered in the respective edit boxes or can be specified on the screen using the respective buttons.

Rotate items as copied

This check box is selected to rotate the objects as they are copied around the center point. If this check box is cleared, the objects are not rotated as they are copied. This means that the replicated objects remain in the same orientation as the original object. You can have the items rotated around the center point as they are copied by selecting this check box. Here, the same

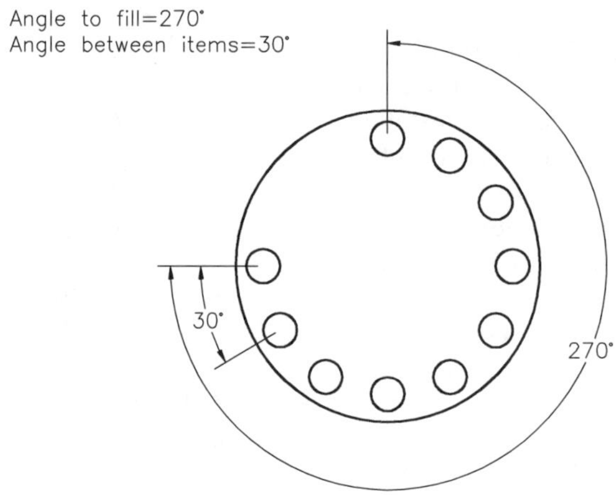

Figure 5-48 *Array created using angle to fill and angle between items*

face of each object points toward the pivot point. Figure 5-49 shows polar array with items rotated as copied and Figure 5-50 shows polar array with items not rotated as copied.

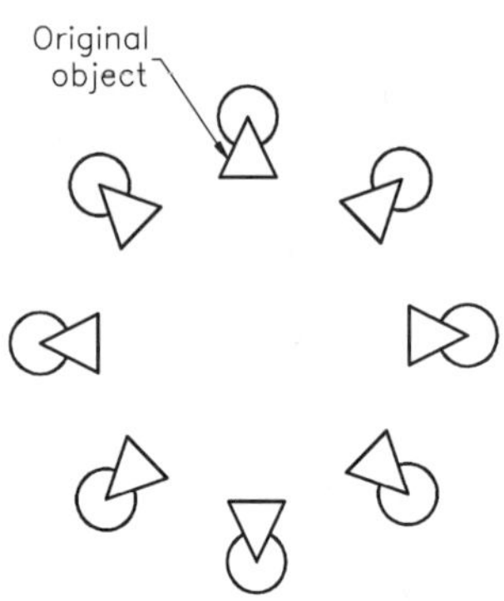

Figure 5-49 *Items rotated as copied*

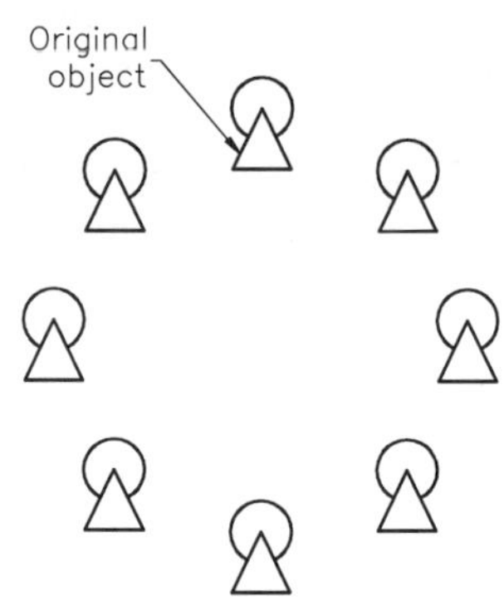

Figure 5-50 *Items not rotated as copied*

More

When this button is chosen, the **Array** dialog box expands providing the options of defining the base point of the objects (Figure 5-51). The base point will maintain a constant distance from the center of the array when the items are copied in polar fashion. You can use the object's default base points by selecting the **Set to object's default** check box. You can also specify the user-defined points by clearing this check box. Specify the X and Y coordinate values of the center of the array in the **X** and **Y** edit boxes in this area. You can also specify the base point on the screen by choosing the **Pick Base Point** button.

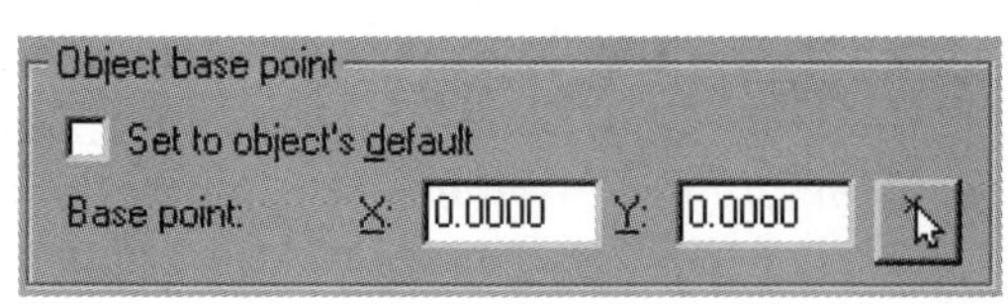

Figure 5-51 ***More*** *options of the* ***Array*** *dialog*

You can preview the array using the **Preview** button. The **Array** dialog box will be displayed and you can choose the **Accept** button if you are satisfied with the array or choose the **Modify** button if you want to make any modifications.

MIRROR COMMAND

Toolbar:	Modify > Mirror
Menu:	Modify > Mirror
Command:	MIRROR

The **MIRROR** command creates a mirror copy of the selected objects. The objects can be mirrored at any angle. This command is helpful in drawing symmetrical figures. When you invoke this command, AutoCAD LT will prompt you to select the objects and then the mirror line.

After you select the objects to be mirrored, AutoCAD LT prompts you to enter the first point of the mirror line and second point of the mirror line. A mirror line is an imaginary line about which objects are reflected. You can specify the endpoints of the mirror line by specifying the points on the screen or by entering their coordinates. The mirror line can be specified at any angle. After the first point on the mirror line has been selected, AutoCAD LT displays the selected objects as they would appear after mirroring. Next, you need to specify the second endpoint of the mirror line. Once this is accomplished, AutoCAD LT prompts you to specify whether you want to retain the original figure (Figure 5-52) or delete it and just keep the mirror image of the figure (Figure 5-53). The prompt sequence when you choose this button is:

Select objects: *Select the objects to be mirrored.*
Select objects: Enter
Specify first point of mirror line: *Specify the first endpoint.*
Specify second point of mirror line: *Specify the second endpoint.*
Delete source objects ? [Yes/No] <N>: *Enter* Y *for deletion,* N *for retaining the previous objects.*

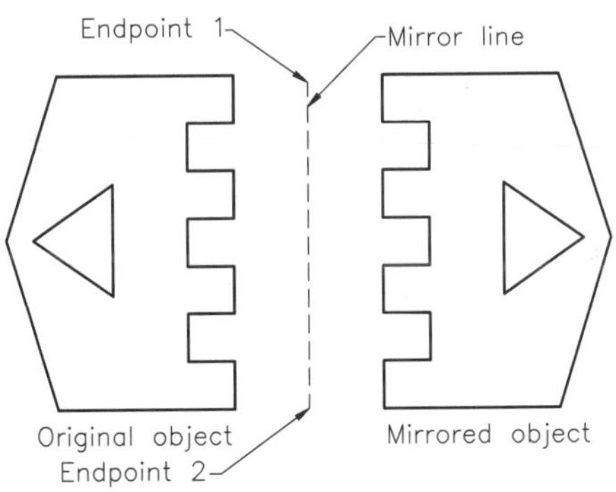

Figure 5-52 *Creating a mirror image of an object using the **MIRROR** command*

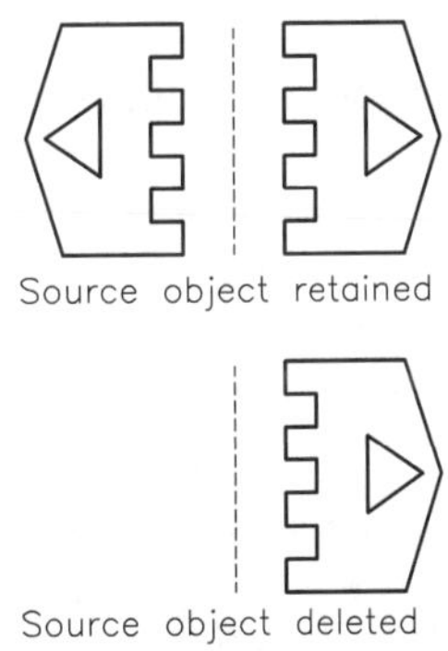

Figure 5-53 *Retaining and deleting old objects after mirroring*

In case you want to mirror the objects at some angle, define the mirror line accordingly. For example, if you want to mirror an object such that the mirrored object is placed at an angle of the 90-degrees from the original object, define the mirror line at an angle of 45-degrees

(Figure 5-54).

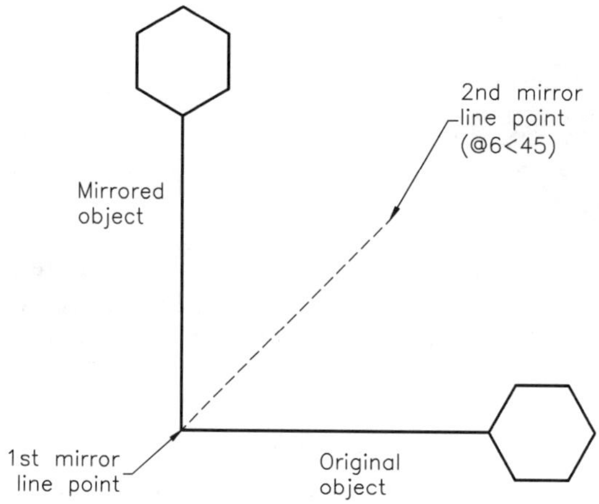

***Figure 5-54** Mirroring the object at an angle*

Text Mirroring

By default, the **MIRROR** command reverses all the objects including texts and dimensions. But, you may not want the text reversed (written backward). In such a situation, you should use the system variable **MIRRTEXT**. The **MIRRTEXT** variable has the following two values (Figure 5-55):

1 = Text is reversed in relation to the original object. This is the default value.
0 = Inhibits the text from being reversed with respect to the original object.

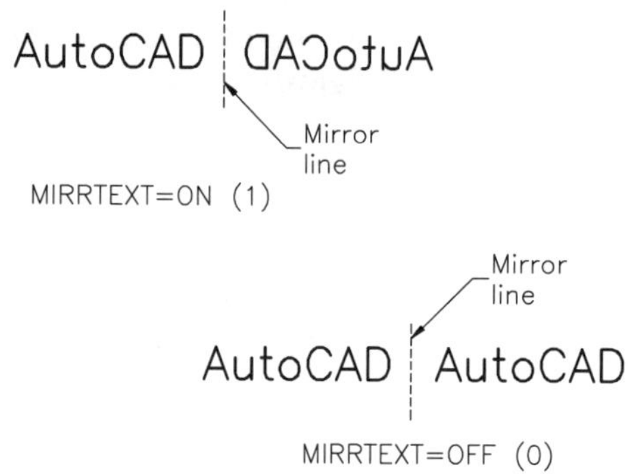

***Figure 5-55** Using the **MIRRTEXT** variable for mirroring the text*

Therefore, if you want the existing object to be mirrored, but at the same time you want the text to be readable and not reversed, set the value of the **MIRRTEXT** variable to **0**; then mirror the objects.

BREAK COMMAND

Toolbar:	Modify > Break at Point, Break
Menu:	Modify > Break
Command:	BREAK

The **BREAK** command breaks an existing object into two or erases portions of the objects. This command can be used to remove a part of the selected objects or to break objects such as lines, arcs, circles, ellipses, xlines, rays, splines, and polylines. You can break the objects using the following methods.

1 Point Option

This method of breaking the existing entities can be directly invoked by choosing the **Break at Point** button from the **Draw** toolbar. Using this method, you can break the object into two parts. When you choose this button, you will be prompted to select the object to be broken. Once you select the object to be broken, you will be prompted to specify the first break point. You will not be prompted to specify the second break point and the object will be automatically broken at the first point. The prompt sequence that will follow when you choose this button is:

Select object: *Select the object to be broken.*
Specify second break point or [First point]: f
Specify first break point: *Specify the point at which the object should be broken.*
Specify second break point: @

2 Points Option

This method allows you to break an object between two selected points. The portion of the object between the two selected point is removed. The point at which you select the object becomes the first break point and then you are prompted to enter the second break point. The prompt sequence that will follow when you choose this button is:

Select object: *Select the object to be broken.*
Specify second break point or [First point]: *Specify the second break point on the object.*

The object is broken between those two points and the in-between portion of the object is removed (Figure 5-56).

2 Points Select Option

This method is similar to the 2 Points option; the only difference is that instead of making the selection point as the first break point, you are allowed to specify a new first point (Figure 5-57). The prompt sequence that will follow when you choose the **Break** button is:

Select object: *Select the object to be broken.*
Enter second break point or [First point]: **F** Enter
Specify first break point: *Specify a new break point.*
Specify second break point: *Specify second break point on the object.*

If you need to work on arcs or circles, make sure that you work in a counterclockwise direction, or you may end up cutting the wrong part. In this case, the second point should be selected in a counterclockwise direction with respect to the first one (Figure 5-58). You can use the **2 Points** and **2 Points Select** method to break an object into two without removing a portion in between. This can be achieved by specifying the same point on the object as the first and the second break points. If you specify the first break point on the line and the second break point beyond the end of the line, one complete end starting from the first break point will be removed. The extrusion direction of the selected object need not be parallel to the Z axis of the UCS.

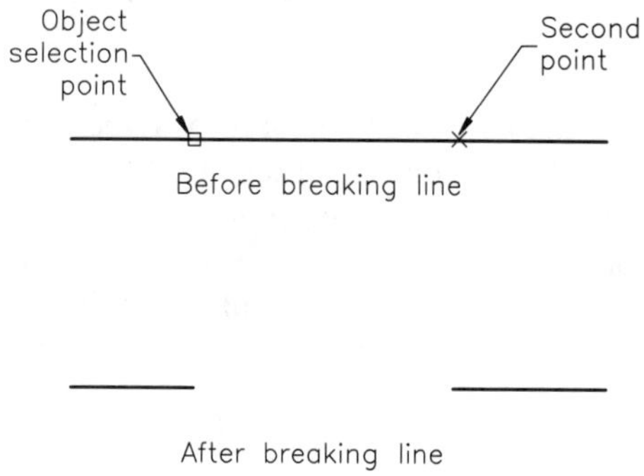

Figure 5-56 *Using the **2 Points** option for breaking the line*

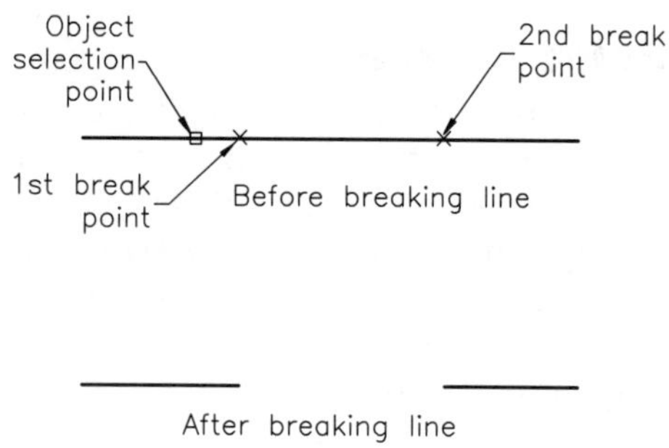

Figure 5-57 *Respecifying the first break point for breaking the lines*

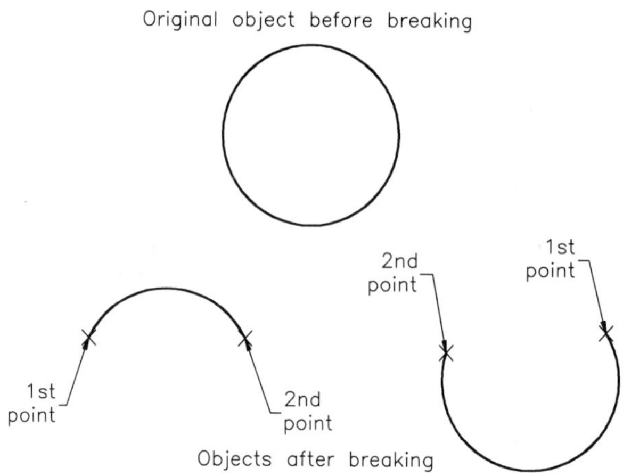

Figure 5-58 *Breaking a circle*

Exercise 7 *General*

Break a line at five different places and then erase the alternate segments. Next, draw a circle and break it into four equal parts.

MEASURE COMMAND

Menu:	Draw > Point > Measure
Command:	MEASURE

While drawing, you may need to segment an object at fixed distances without actually dividing it. You can use the **MEASURE** command to accomplish this. This command places points or blocks (nodes) on the given object at a specified distance. The shape of these points is determined by the **PDMODE** system variable and the size is determined by the **PDSIZE** variable. The **MEASURE** command starts measuring the object from the endpoint closest to where the object is selected. When a circle is to be measured, an angle from the center is formed that is equal to the Snap rotation angle. This angle becomes the starting point of measurement, and the markers are placed at equal intervals in a counterclockwise direction.

This command goes on placing markers at equal intervals of the specified distance without considering whether the last segment is the same distance or not. Instead of entering a value, you can also select two points that will be taken as the distance. In Figure 5-60 a line and a circle is measured. The Snap rotation angle is zero degrees. The **PDMODE** variable is set to 3 so that X marks are placed as markers. The prompt sequence that will follow when you invoke this option from the **Draw** menu is:

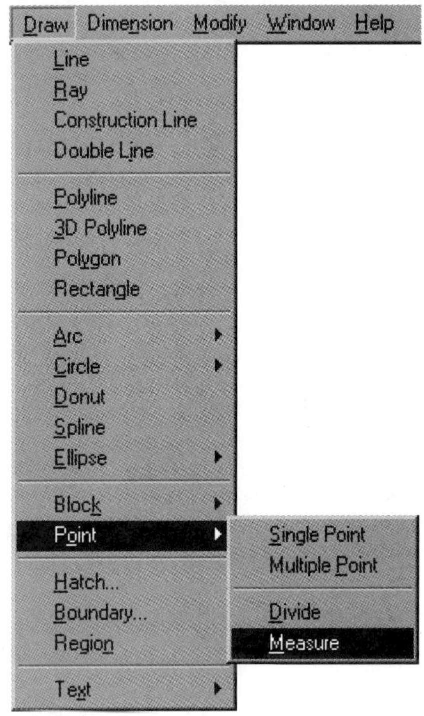

Figure 5-59 *Invoking the* ***MEASURE*** *command from the* ***Draw*** *menu*

> Select object to measure: *Select the object to be measured.*
> Specify length of segment or [Block]: *Specify the length for measuring the object.*

You can also place blocks as markers (Figure 5-61), but the block must already be defined within the drawing. You can align these blocks with the object to be measured. The prompt sequence is:

> Select object to measure: *Select the object to be measured.*
> Specify length of segment or [Block]: **B** Enter
> Enter name of block to insert: *Enter the name of the block.*
> Align block with object? [Yes/No] <Y>: *Enter* Y *to align,* N *to not align.*
> Specify length of segment: *Enter the measuring distance.*

Note
You will learn more about creating the blocks and inserting them in Chapter 14, Blocks.

DIVIDE COMMAND

Menu:	Draw > Point > Divide
Command:	DIVIDE

The **DIVIDE** command is used to divide an object into a specified number of equal length segments without actually breaking it. This command is similar to the **MEASURE** command except that here you do not have to specify the distance. The **DIVIDE** command calculates the full length of the object and places markers at equal intervals. This makes the last interval equal to the rest of the intervals. If you want a line to be divided, invoke this command and

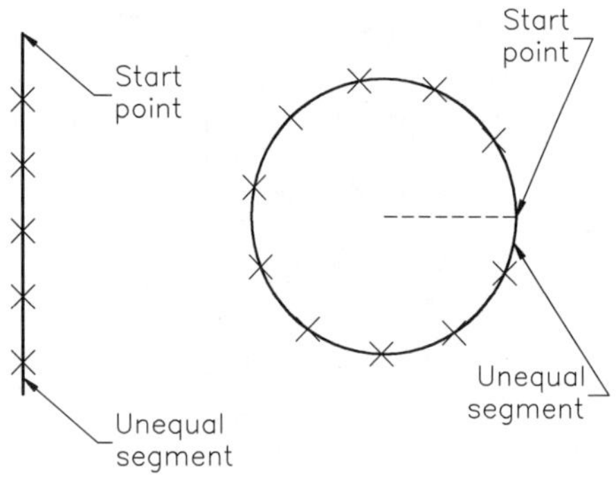

Figure 5-60 Using the ***MEASURE*** *command*

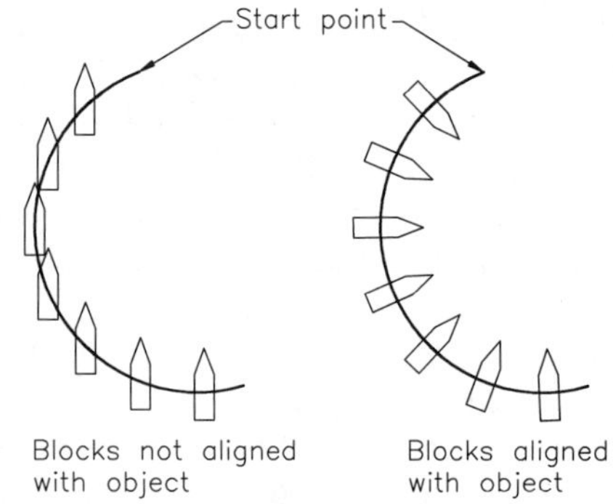

Figure 5-61 Using blocks with the ***MEASURE*** *command*

select the object to be divided. After this you enter the number of divisions or segments. The number of divisions entered can range from 2 to 32,767. The prompt sequence that will follow when you invoke this command is:

Select object to divide: *Select the object you want to divide.*
Enter number of segments or [Block]: *Specify the number of segments.*

You can also place blocks as markers, but the block must be defined within the drawing. You can align these blocks with the object to be measured. The prompt sequence is:

Select object to divide: *Select the object to be measured.*
Enter number of segments or [Block]: **B** Enter
Enter name of block to insert: *Enter the name of the block.*
Align block with object? [Yes/No] <Y>: *Enter Y to align, N to not align.*
Enter number of segments: *Enter the number of segments.*

Figure 5-62 shows a line divided using this command and Figure 5-63 shows the use of blocks for dividing the selected segment.

Note

The size and shape of points placed by the ***DIVIDE*** *and* ***MEASURE*** *commands are controlled by the* ***PDSIZE*** *and* ***PDMODE*** *system variables.*

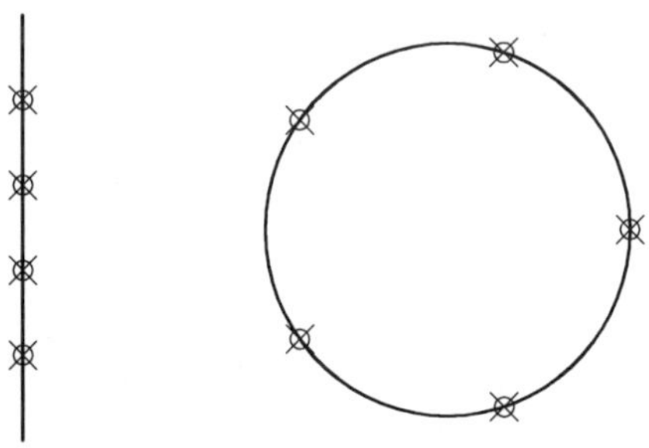

*Figure 5-62 Using the **DIVIDE** command*

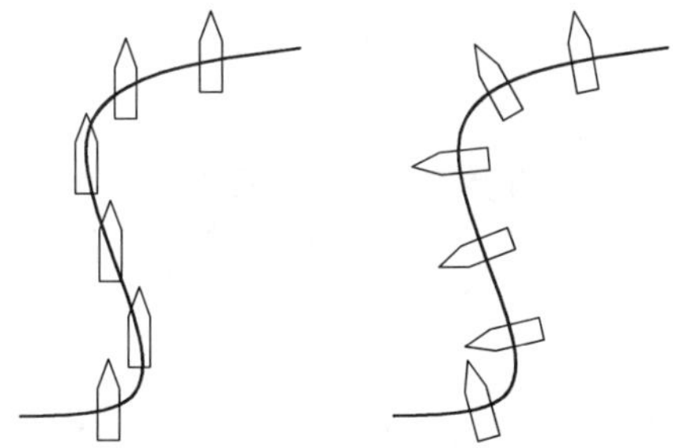

*Figure 5-63 Using blocks with the **DIVIDE** command*

Self-Evaluation Test

Answer the following questions and then compare your answers to the correct answers given at the end of this chapter.

1. When you shift a group of objects using the **MOVE** command, the size and orientation of these objects is changed. (T/F)

2. The **COPY** command makes copies of the selected object, leaving the original object intact. (T/F)

3. A fillet cannot be created between two parallel and nonintersecting lines. (T/F)

4. With the **BREAK** command, when you select an object, the selection point becomes the first break point. (T/F)

5. Depending on the side to offset, you can create smaller or larger circles, ellipses, and arcs with the _________ command.

6. The _________ command prunes objects that extend beyond a required point of intersection.

7. The offset distance is stored in the _________ system variable.

8. Instead of specifying the scale factor, you can use the _________ option to scale an object with reference of another object.

9. If the _________ system variable is set to a value of 1, the mirrored text is not reversed with respect to the original object.

Chapter 5

10. There are two types of arrays: _________ and _________.

Review Questions

Answer the following questions.

1. In the case of the **Through** option of the **OFFSET** command, you do not have to specify a distance; you simply have to specify an offset point. (T/F)

2. With the **BREAK** command, you cannot break an object in two without removing a portion in between. (T/F)

3. With the **FILLET** command, the extrusion direction of the selected object must be parallel to the Z axis of the UCS. (T/F)

4. A Rectangular array can be rotated. (T/F)

5. Which object selection allows you to create a polygon that selects all the objects that it touches or lies inside it?

   (a) Last (b) WPolygon
   (c) CPolygon (d) Fence

6. Which command is used to copy an existing object using a base point to another drawing?

   (a) **COPY** (b) **COPYBASE**
   (c) **MOVE** (d) **None**

7. Which command is used to change the size of an existing object with respect to an existing entity?

   (a) **ROTATE** (b) **SCALE**
   (c) **MOVE** (d) **None**

8. Which option of the **LENGTHEN** command modifies the length of the selected entity such that irrespective of the original length, the entity acquires the specified length?

   (a) **DElta** (b) **DYnamic**
   (c) **Percent** (d) **Total**

9. Which selection option is used to select the most recently drawn entity?

   (a) **Last** (b) **Previous**
   (c) **Add** (d) **None**

10. If an entire selected object is within a window or crossing specification, the _________ command works like the **MOVE** command.

11. The _________ option of the **EXTEND** command is used to extend objects to the implied boundary.

12. With the **FILLET** command, using the **Polyline** option, if the selected polyline is not closed, the _________ corner is not filleted.

13. When the chamfer distance is zero, the chamfer created is in the form of a _________.

14. AutoCAD LT saves the previous selection set and lets you select it again by using the _________ selection option.

15. The _________ selection option is used to select objects by touching the objects to be selected with the selection line.

Exercises

Exercise 8 *General*

Draw the object shown in Figure 5-64. Use the **DIVIDE** command to divide the circle and use the **NODE** object snap to select the points. Assume the dimensions of the drawing.

Figure 5-64 *Drawing for Exercise 8*

Exercise 9 *Mechanical*

Draw the object shown in Figure 5-65 and save the drawing. Assume the missing dimensions.

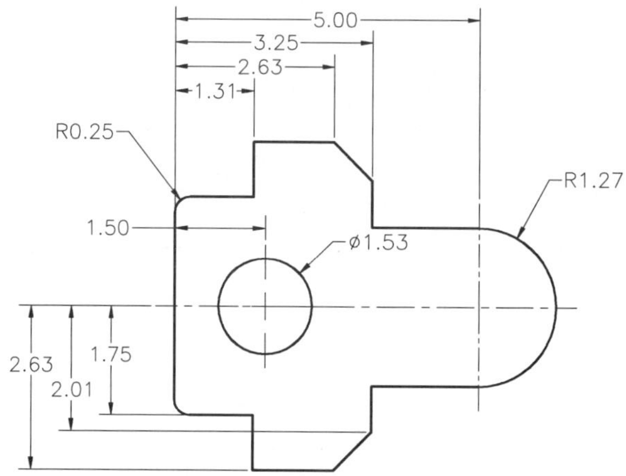

Figure 5-65 *Drawing for Exercise 9*

Exercise 10 *Mechanical*

Draw the drawing shown in Figure 5-66 and save the drawing.

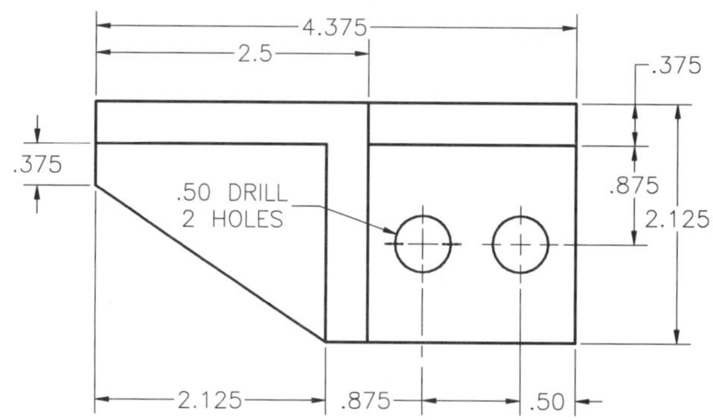

Figure 5-66 *Drawing for Exercise 10*

Exercise 11 *Mechanical*

Draw the object shown in Figure 5-67 and save the drawing.

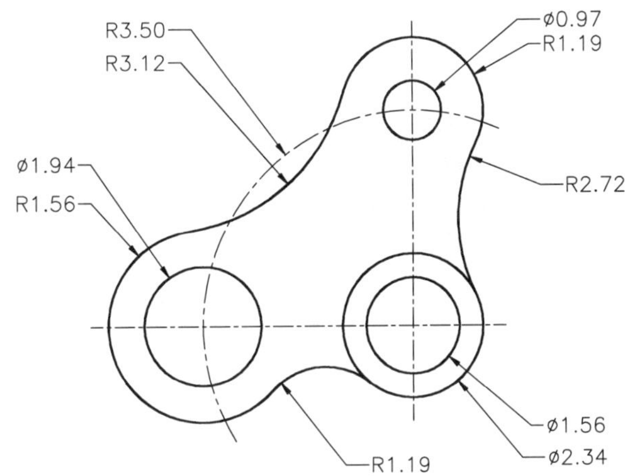

Figure 5-67 *Drawing for Exercise 11*

Exercise 12 *Mechanical*

Draw the object shown in Figure 5-68 and save the drawing. Assume the missing dimensions.

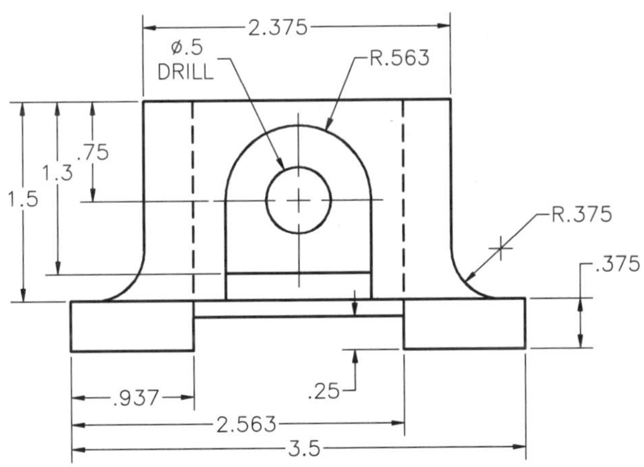

Figure 5-68 *Drawing for Exercise 12*

Problem Solving Exercise 1 — *Mechanical*

Draw the object shown in Figure 5-69 and save the drawing. Assume the missing dimensions.

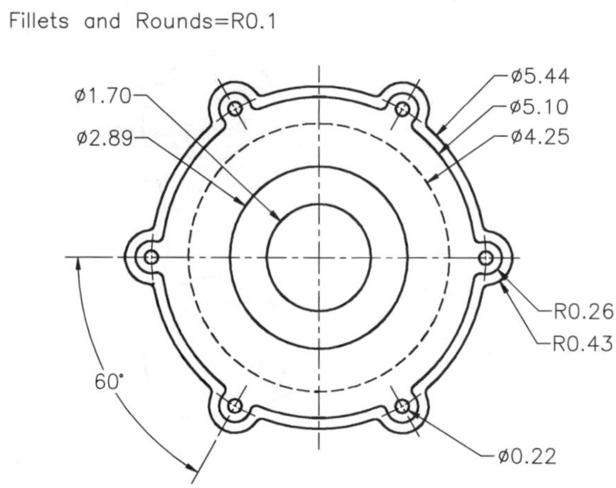

***Figure 5-69** Drawing for Problem Solving Exercise 1*

Problem Solving Exercise 2 — *Mechanical*

Draw the object shown in Figure 5-70 and save the drawing. Assume the missing dimensions.

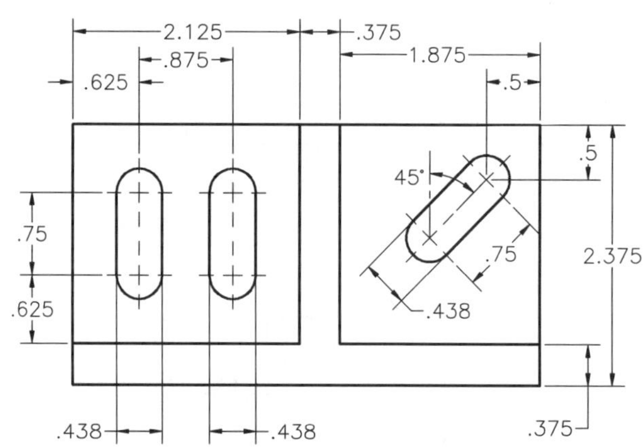

***Figure 5-70** Drawing for Problem Solving Exercise 2*

Answers to Self-Evaluation Test

1 - F, **2** - T, **3** - F, **4** - T, **5 - OFFSET**, **6 - TRIM**, **7 - OFFSETDIST**, **8 - Reference**, **9 - MIRRTEXT**, **10 - Rectangular, Polar**

Chapter 6

Editing Commands-II

Learning Objectives

After completing this chapter, you will be able to:
- *Understand the concept of grips and adjust grip settings.*
- *Stretch, move, rotate, scale, and mirror objects with grips.*
- *Use the* ***MATCHPROP*** *command to match the properties of the selected object.*
- *Use the* ***Properties*** *window for editing the objects.*
- *Use the* ***QSELECT*** *command to select the objects.*
- *Manage the contents using the* ***DesignCenter****.*
- *Use the various* ***Inquiry*** *commands.*

EDITING WITH GRIPS

Grips provide a convenient and quick means of editing objects. With grips you can stretch, move, rotate, scale, mirror objects, change properties, and load the Web browser. Grips are small squares that are displayed on an object at its definition points when the object is selected. The number of grips depends on the selected object. For example, a line has three grip points, a polyline segment has two, and an arc has three. Similarly, a circle has five grip points and a dimension (vertical) has five. When you select the **Enable grips** and the **Noun/Verb Selection** check boxes in the **Selection** tab of the **Options** dialog box, a small square (aperture box) at the intersection of the crosshair is displayed (Figure 6-1). The grip location of some of the objects is shown in Figure 6-2.

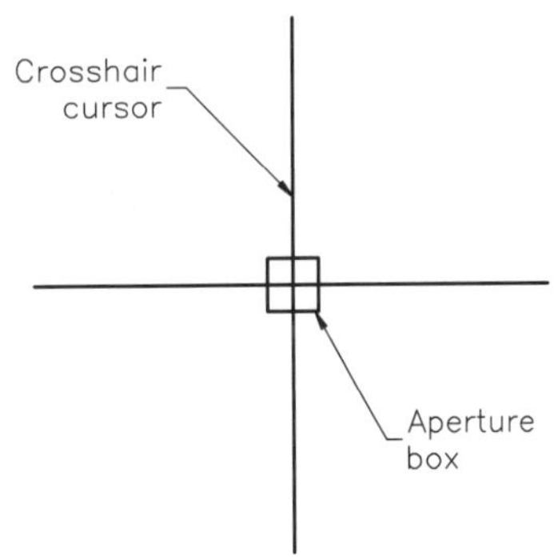

Figure 6-1 *Aperture box at the intersection of crosshair*

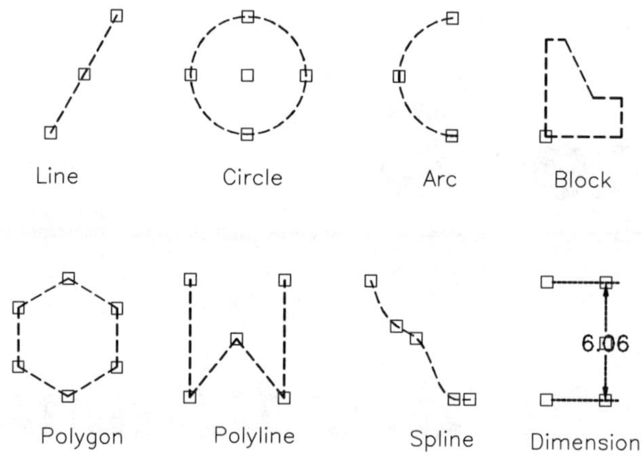

***Figure 6-2** Grip location of various objects*

Note
*AutoCAD LT also displays a small square (aperture box) at the intersection of crosshairs when the **PICKFIRST** (Noun/Verb Selection) system variable is set to 1 (On).*

GRIP TYPES

Grips can be classified into two types: unselected grips and selected grips. Selected grips are also called hot grips. When you select an object, the grips are displayed at the definition points of the object, and the object is highlighted by displaying it as a dashed line. These grips are called unselected grips (blue). Now, if you select a grip on this object, the grip becomes a hot grip (filled red square). Once the grip is hot, the object can be edited. To cancel the grip, press ESC. If you press ESC once, the hot grip changes to unselected grip, as shown in Figure 6-3. You can also snap to a cold grip.

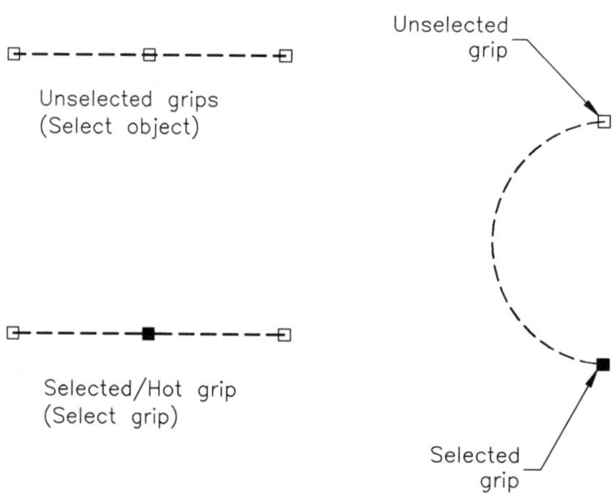

***Figure 6-3** Figure showing the selected and unselected grips*

ADJUSTING GRIP SETTINGS

Menu:	Tools > Options
Command:	OPTIONS

The grip settings can be adjusted using the options under the **Selection** tab of the **Options** dialog box. This dialog box can also be invoked by choosing **Options** from the shortcut menu (Figure 6-4). The shortcut menu is displayed upon right-clicking in the drawing area.

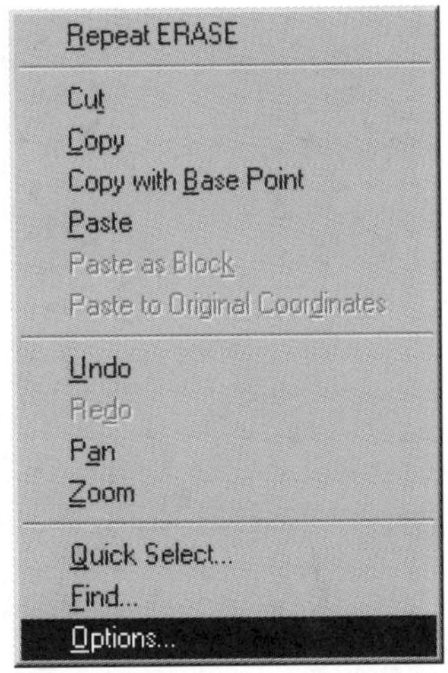

Figure 6-4 *Invoking the* ***Options*** *dialog box from the shortcut menu*

The options related to grips that are provided under the **Selection** tab of the **Options** dialog box (Figure 6-5) are:

Grips Area

The Grips area has two check boxes: **Enable grips** and **Enable grips within blocks**. The grips can be enabled by selecting the **Enable grips** check box. They can also be enabled by setting the **GRIPS** system variable to 1. The second check box, **Enable grips within blocks**, enables the grips within a block. If you select this box, AutoCAD LT will display grips for every object in the block. If you disable the display of grips within a block, the block will have only one grip at its insertion point. You can also

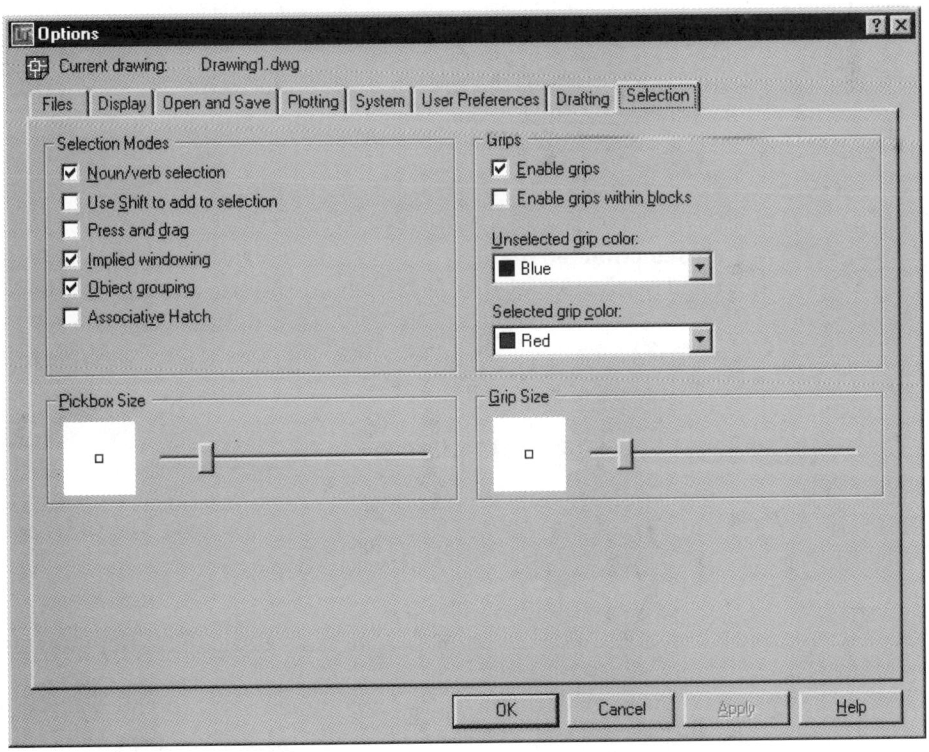

Figure 6-5 ***Selection*** *tab of the* ***Options*** *dialog box*

enable the grips within a block by setting the value of the **GRIPBLOCK** system variable to 1 (On). If **GRIPBLOCK** is set to 0 (Off), AutoCAD LT will display only one grip for a block at its insertion point (Figure 6-6).

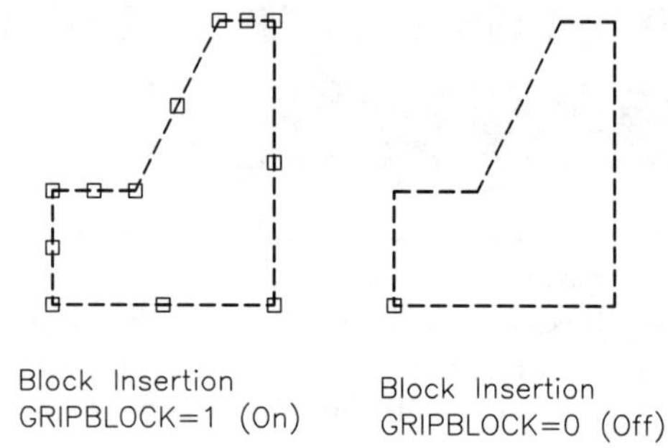

***Figure 6-6** Block insertion with **GRIPBLOCK** set to 1 and to 0*

Note

*If the block has a large number of objects, and if **GRIPBLOCK** is set to 1 (On), AutoCAD LT will display grips for every object in the block. Therefore, it is recommended that you set the system variable **GRIPBLOCK** to 0 or clear the **Enable grips within blocks** check box in the **Selection** tab of the **Options** dialog box.*

Unselected grip color

This drop-down list is used to set the color of the unselected grip. You can set the color by selecting it from this drop-down list or select the **More** option to display the **Select Color** dialog box. You can select the color for the unselected grip from this dialog box. This color can also be set using the **GRIPCOLOR** system variable.

Selected grip color

This drop-down list is used to set the color of the selected grip. You can set the color by selecting it from this drop-down list or select the **More** option to display the **Select Color** dialog box. You can select the color for the selected grip from this dialog box. This color can also be set using the **GRIPHOT** system variable.

Grip Size Area

The **Grip Size** area of the **Selection** tab of the **Options** dialog box consists of a slider bar and a rectangular box that displays the size of the grip. To adjust the size of the grip, move the slider box left or right. The size of the grip can also be adjusted by using the **GRIPSIZE** system variable. The **GRIPSIZE** variable is defined in pixels, and its value can range from 1 to 255 pixels.

EDITING THE OBJECTS WITH GRIPS

As mentioned earlier, you can perform different kinds of editing operations using the selected grip. The editing operations are discussed next.

Stretching Objects With Grips (Stretch Mode)

If you select an object, AutoCAD LT displays unselected grips at the definition points of the object. When you select a grip for editing, you are automatically in the **Stretch** mode. The Stretch mode has a function similar to the **STRETCH** command. When you select a grip, it acts as a base point and is called a base grip. You can also select several grips by holding the SHIFT key down and then select the grips. Now, release the SHIFT key and select one of the hot grips to stretch them simultaneously. The geometry between the selected base grips is not altered. You can also make copies of the selected objects or define a new base point. When selecting grips on text objects, blocks, midpoints of lines, centers of circles and ellipses, and point objects in the stretch mode, the selected objects are moved to a new location. The following example illustrates the use of the **Stretch** mode.

1. Use the **PLINE** command to draw a W-shaped figure as shown in Figure 6-7(a).

2. Select the object that you want to stretch [Figure 6-7(a)]. When you select the object, grips will be displayed at the endpoints of each object. A polyline has two grip points. If you use the **LINE** command to draw the object, AutoCAD LT will display three grips for each object.

3. Hold the SHIFT key down, and select the grips that you want to stretch [Figure 6-7(b)]. The selected grips will become hot grips, and the color of the grip will change from blue to red.

Note

You have to make sure that you hold the SHIFT key before selecting even the first grip. You can not hold the SHIFT key and select more grips if the first grip is selected without holding the SHIFT key.

4. Choose one of the selected (hot grip) grips, and specify a point to which you want to stretch the line [Figure 6-7(c)]. When you select a grip, the following prompt is displayed in the Command prompt area:

   STRETCH
   Specify stretch point or [Base point/Copy/Undo/eXit]:

   The Stretch mode has several options: **Base point**, **Copy**, **Undo**, and **eXit**. You can use the Base point option to define the base point and the Copy option to make copies.

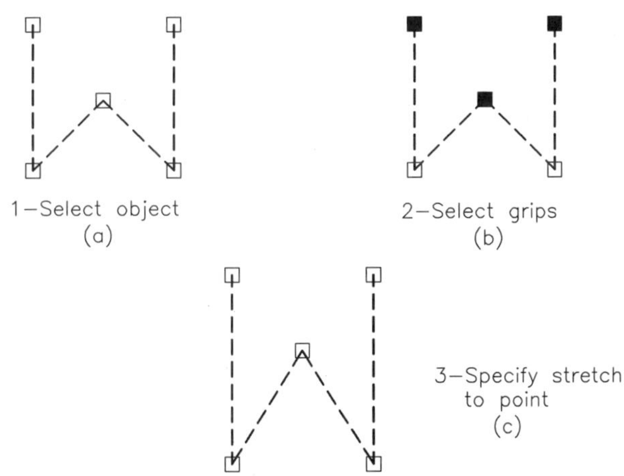

Figure 6-7 *Using the* ***Stretch*** *mode to stretch the lines*

5. Select the grip where the two lines intersect. Right-click to display the shortcut menu (Figure 6-8) and choose the **Copy** option. Select the points as shown in Figure 6-9(b). Each time you select a point, AutoCAD LT will make a copy.

   If you press the SHIFT key when specifying the point to which the object is to be stretched, without selecting the copy option, then also AutoCAD LT allows you to make multiple copies of the selected object. Also, if you press the SHIFT key again when specifying the next point, the cursor snaps to a point whose location is based on the distance between the first two points, that is, the distance between the selected object and the location of the copy of the selected object.

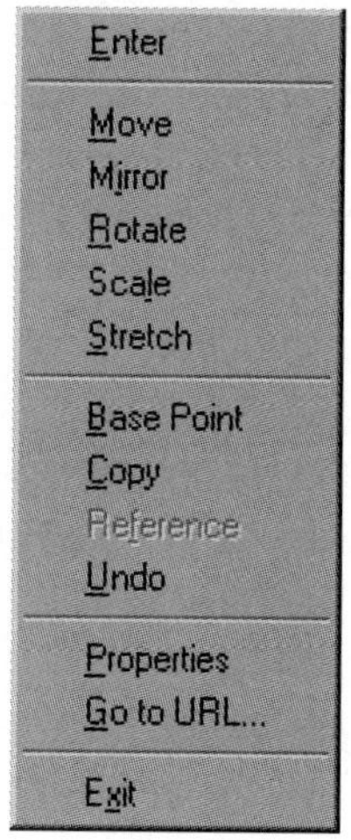

***Figure 6-8** Selecting different Grip options from the shortcut menu*

6. Make a copy of the drawing as shown in Figure 6-9(c). Select the object, and then select the grip where the two lines intersect. When AutoCAD LT displays the ****STRETCH**** prompt, select the **Base Point** option from the cursor menu or enter B at the Command: prompt. Select the bottom left grip as the base point, and then give the displacement point as shown in Figure 6-9(d).

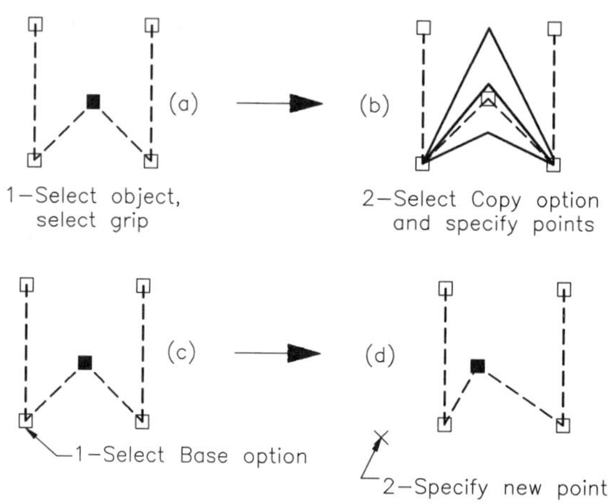

***Figure 6-9** Using the Stretch mode's **Copy** and **Base point** options*

7. To exit the grip mode, right-click to display the shortcut menu and then select **eXit**. You can also enter X at the Command: prompt or press ESC to exit.

Note

*You can select an option (**Copy** or **Base Point**) from the shortcut menu that can be invoked by right-clicking your pointing device after selecting a grip. The different modes can also be selected from the shortcut menu. You can also cycle through all the different modes by selecting a grip and pressing the ENTER key or the SPACEBAR.*

Moving Objects with Grips (Move Mode)

The Move mode lets you move the selected objects to a new location. When you move objects, the size of the objects and their angle do not change. You can also use this mode to make copies of the selected objects or to redefine the base point. The following example illustrates the use of the **Move** mode.

1. Use the **LINE** command to draw the shape as shown in Figure 6-10(a). When you select the objects, grips will be displayed at the definition points and the object will be highlighted.

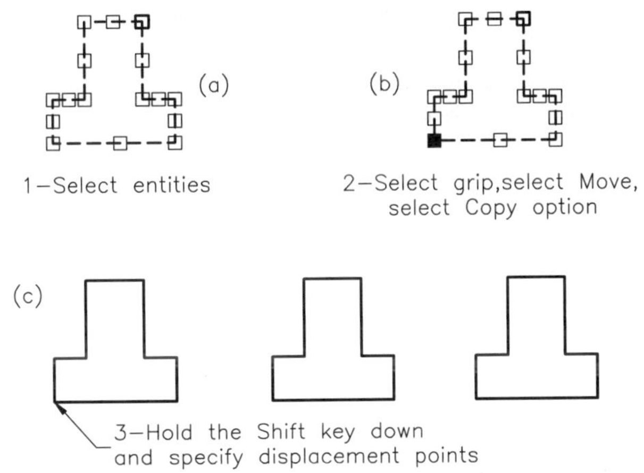

Figure 6-10 *Using the **Move** mode to move and make copies of the selected objects*

2. Select the grip located at the lower left corner, and then select **Move** from the shortcut menu. You can also invoke the Move mode by entering **MOVE** or **MO** at the keyboard or giving a null response by pressing the SPACEBAR or ENTER key. AutoCAD LT will display the following prompt in the Command: prompt area:

   MOVE
   Specify move point or [Base point/Copy/Undo/eXit]:

3. Hold down the SHIFT key, and then enter the first displacement point. The distance between the first and the second object defines the snap offset for subsequent copies. While holding down the SHIFT key, move the screen crosshair to the next snap point and select the point. AutoCAD LT will make a copy of the object at this location. If you release the SHIFT key, you can specify any point where you want to place a copy of the object. You can also enter coordinates to specify the displacement.

Rotating Objects with Grips (Rotate Mode)

The **Rotate** mode allows you to rotate objects around the base point without changing their size. The options of the Rotate mode can be used to redefine the base point, specify a reference angle, or make multiple copies that are rotated about the specified base point. You can access the Rotate mode by selecting the grip and then selecting **Rotate** from the shortcut menu, or by entering **ROTATE** or **RO** at the keyboard or giving a null response twice by pressing the SPACEBAR or the ENTER key. The following example illustrates the use of **Rotate** mode.

1. Use the **LINE** command to draw the shape as shown in Figure 6-11(a). When you select the objects, grips will be displayed at the definition points and the shape will be highlighted.

2. Select the grip located at the lower left corner and then invoke the Rotate mode. AutoCAD LT will display the following prompt:

   ROTATE
   Specify rotation angle or [Base point/Copy/Undo/Reference/eXit]:

3. At this prompt, enter the rotation angle. AutoCAD LT will rotate the selected objects by the specified angle [Figure 6-11(b)].

4. Make a copy of the original drawing as shown in Figure 6-11(c). Select the objects, and then select the grip located at the lower left corner of the object. Invoke the Rotate mode and then select the **Copy** option from the shortcut menu or enter **C** (Copy) at the Command prompt. Enter the rotation angle. AutoCAD LT will rotate a copy of the object through the specified angle [Figure 6-11(d)].

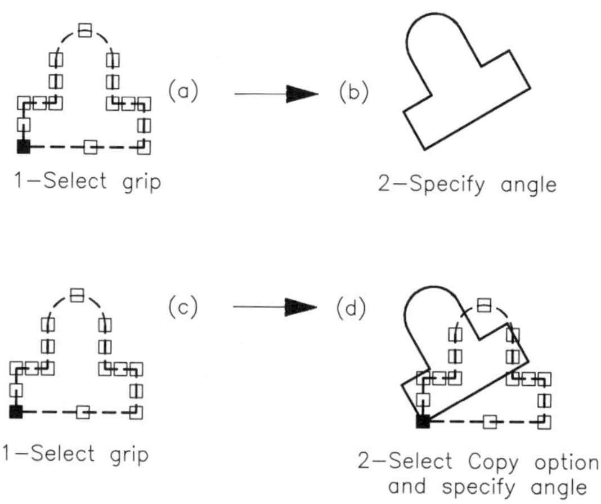

***Figure 6-11** Using the **ROTATE** mode to rotate and make copies of the selected objects*

5. Make another copy of the object as shown in Figure 6-12(a). Select the object, and then select the grip at point (P0). Access the Rotate mode and copy option as described earlier. Select **Reference** option from the shortcut menu or enter **R** (Reference) at the following

prompt:

**ROTATE (multiple) **
Specify rotation angle or [Base point/Copy/Undo/Reference/eXit]: **R**
Specify reference angle <0>: *Select the grip at (P1).*
Specify second point: *Select the grip at (P2).*
Specify new angle or [Base point/Copy/Undo/Reference/eXit]: **45**

In response to the **Specify reference angle <0>:** prompt, select the grips at points (P1) and (P2) to define the reference angle. When you enter the new angle, AutoCAD LT will rotate and insert a copy at the specified angle [Figure 6-12(c)]. For example, if the new angle is 45 degrees, the selected objects will be rotated about the base point (P0) so that the line P1P2 makes a 45-degree angle with respect to the positive X axis.

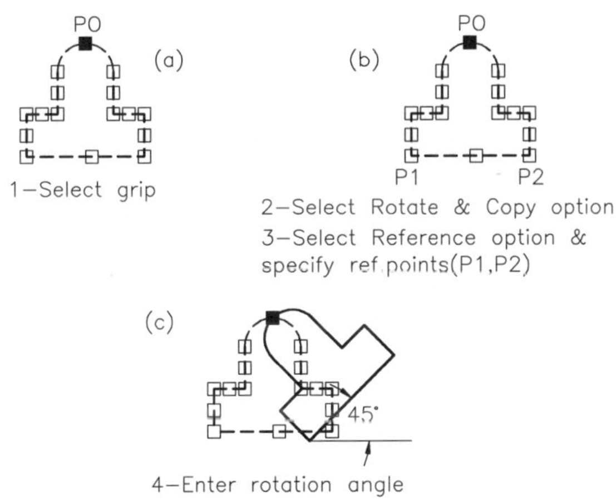

Figure 6-12 *Using the* ***ROTATE*** *mode to rotate by giving a reference angle*

Scaling Objects with Grips (Scale Mode)

The **Scale** mode allows you to scale objects with respect to the base point without changing their orientation. The options of Scale mode can be used to redefine the base point, specify a reference length, or make multiple copies that are scaled with respect to the specified base point. You can access the Scale mode by selecting the grip and then selecting Scale from the shortcut menu, or entering **SCALE** or **SC** on the keyboard, or giving a null response three times by pressing the SPACEBAR or the ENTER key. The following example illustrates the use of the **Scale** mode.

1. Use the **PLINE** command to draw the shape as shown in Figure 6-13(a). When you select the objects, grips will be displayed at the definition points, and the object will be highlighted.

2. Select the grip located at the lower left corner as the base grip, then invoke the **SCALE** mode. AutoCAD LT will display the following prompt in the Command prompt area:

SCALE
Specify scale factor or [Base point/Copy/Undo/Reference/eXit]:

3. At this prompt enter the scale factor or move the cursor and select a point to specify a new size. AutoCAD LT will scale the selected objects by the specified scale factor [Figure 6-13(b)]. If the scale factor is less than 1 (<1), the objects will be scaled down by the specified factor. If the scale factor is greater than 1 (>1), the objects will be scaled up.

4. Make a copy of the original drawing as shown in Figure 6-13(c). Select the objects, and then select the grip located at the lower left corner of the object. Invoke the Scale mode. At the following prompt, enter C (Copy), and then enter B for base point:

   **SCALE (multiple) **
   Specify scale factor or [Base point/Copy/Undo/Reference/eXit]: **B**

5. At the **Specify base point prompt:**, select the point (P0) as the new base point, and then enter **R** at the following prompt.

   **SCALE (multiple) **
   Specify scale factor or [Base point/Copy/Undo/Reference/eXit]: **R**
   Specify reference length <1.000>: *Select grips at (P1) and (P2).*

After specifying the reference length at the **Specify new length or [Base point/Copy/Reference/eXit]** prompt, enter the actual length of the line. AutoCAD LT will scale the objects so that the length of the bottom edge is equal to the specified value [Figure 6-13(c)].

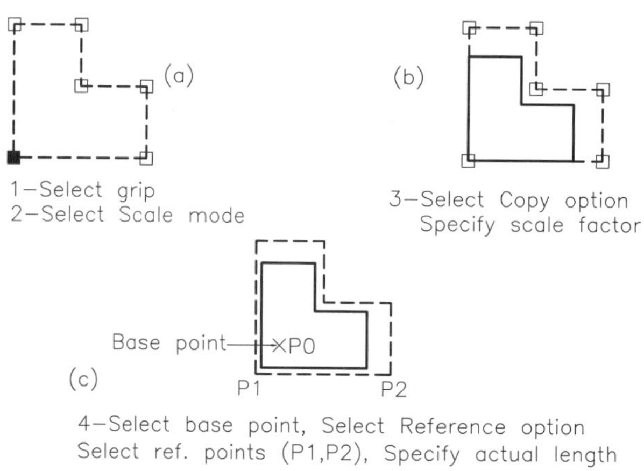

Figure 6-13 *Using the SCALE mode to scale and make copies of selected objects*

Mirroring Objects with Grips (Mirror Mode)

The **Mirror** mode allows you to mirror the objects across the mirror axis without changing the size of the objects. The mirror axis is defined by specifying two points. The first point is the base point, and the second point is the point that you select when AutoCAD LT prompts for the second point. The options of the Mirror mode can be used to redefine the base point and make a mirror copy of the objects. You can access the Mirror mode by selecting a grip and then choosing Mirror from the shortcut menu, or by entering **MIRROR** or **MI** at the keyboard, or giving a null response four times by pressing the SPACEBAR or the ENTER key four times. The following is the example for the **Mirror** mode.

1. Use the **PLINE** command to draw the shape as shown in Figure 6-14(a). When you select the object, grips will be displayed at the definition points and the object will be highlighted.

2. Select the grip located at the lower right corner (P1), and then invoke the Mirror mode. The following prompt is displayed:

   MIRROR
   Specify second point or [Base point/Copy/Undo/eXit]:

3. At this prompt, enter the second point (P2). AutoCAD LT will mirror the selected objects with line P1P2 as the mirror axis as shown in Figure 6-14(b).

4. Make a copy of the original figure as shown in Figure 6-14(c). Select the object, and then select the grip located at the lower right corner (P1) of the object. Invoke the Mirror mode and then select the Copy option to make a mirror image while retaining the original object. Alternatively, you can also hold down the SHIFT key and make several mirror copies by specifying the second point.

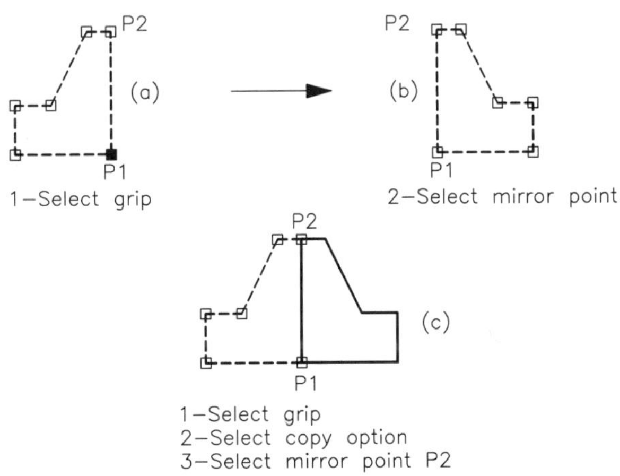

Figure 6-14 *Using the Mirror mode to create a mirror image of selected objects*

5. Select point (P2) in response to the prompt **Specify second point or [Base point/Copy/Undo/eXit]**. AutoCAD LT will create a mirror image, and the original object will be retained.

Note

*You can use some editing commands such as **ERASE, MOVE, ROTATE, SCALE, MIRROR,** and **COPY**, on an object with unselected grips. However, this is possible only if the **PICKFIRST** system variable is set to 1 (On).*

You cannot select an object when the grip is hot.

If you want to remove an object from the selection set displaying grips, press the SHIFT key and then select the particular object. This object, which is removed from the selection set, will now not be highlighted although the grips are still displayed (Cold grips).

LOADING HYPERLINKS

You can also use the grips to open a file associated with the hyperlink. For example, the hyperlink could start a word processor, or activate the Web browser and load a Web page that is embedded in the selected object. If you want to launch the Web browser that provides hyperlinks to other Web pages, select the URL embedded object and then right-click to display the shortcut menu. In the shortcut menu, select the **Hyperlink** option and AutoCAD LT will automatically load the Web browser. When you move the cursor over or near the object that contains a hyperlink, AutoCAD LT displays the hyperlink information with the cursor.

EDITING GRIPPED OBJECTS

You can also edit the properties of the gripped objects by using the **Object Properties** toolbar (Figure 6-15). The gripped objects are created when you select objects without invoking a command. The gripped objects are highlighted and will display grips (rectangular boxes) at their grip points. For example, if you want to change the color of the gripped objects, select the **Color** drop-down list in the **Object Properties** toolbar and then select a color. The color of the gripped objects will change to the selected color. Similarly, if you want to change the layer, lineweight, or linetype of the gripped objects, select the linetype, lineweight, or layer from the corresponding drop-down lists. If the gripped objects have been drawn in different layers or have different colors, linetypes, or lineweights, the Layer Control, Color Control, Linetype Control, and Lineweight Control boxes will appear blank. You can also change the Plot Style of the selected objects.

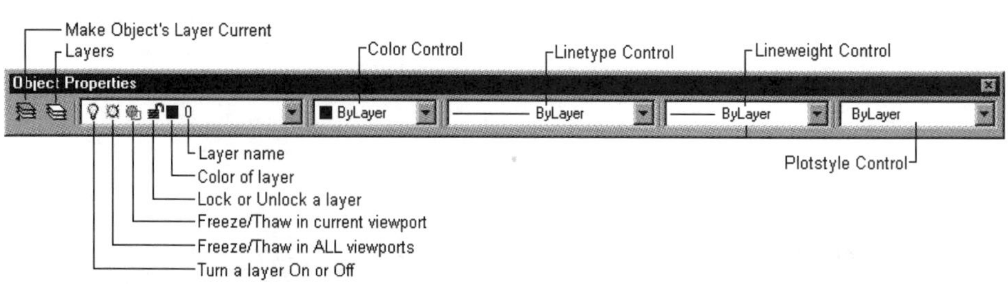

Figure 6-15 *Using the **Object Properties** toolbar to change properties of the gripped objects*

GRIP SYSTEM VARIABLES

System variable	Default	Setting	Function
GRIPS	1	1=On, 0=Off	Enables or disables Grip mode
GRIPBLOCK	0	1=On, 0=Off	Controls the display of grips in a block
GRIPCOLOR	5	1-255	Specifies the color of unselected grips
GRIPHOT	1	1-255	Specifies the color of selected grips
GRIPSIZE	3	1-255	Specifies the size of the grip box in pixels

PROPERTIES COMMAND

Toolbar:	Standard Toolbar > Properties
Menu:	Modify > Properties
Command:	PROPERTIES

You can also modify an object by using the **PROPERTIES** command. When you invoke this command, AutoCAD LT will display the **Properties** window on the screen (Figure 6-16). The **Properties** window can also be displayed when you double-click on the object to be edited. The contents of the **Properties** window changes according to the objects selected. For example, if you select text entity, the properties related to the text such as text height, text justification, style, rotation angle, obliquing factor, and so on, will be displayed.

The **Properties** window can also be invoked from the shortcut menu displayed when you right-click in the drawing area. Choose the **Properties** option to display the **Properties** window. If you select more than one object, the common properties of the selected objects will be displayed in the **Properties** window. To change properties of the selected objects, you can click in the cell next to the name of the property and change the values manually or you can choose from the available options in the drop-down list if one is available. You can cycle through the options by double-clicking in the property cell.

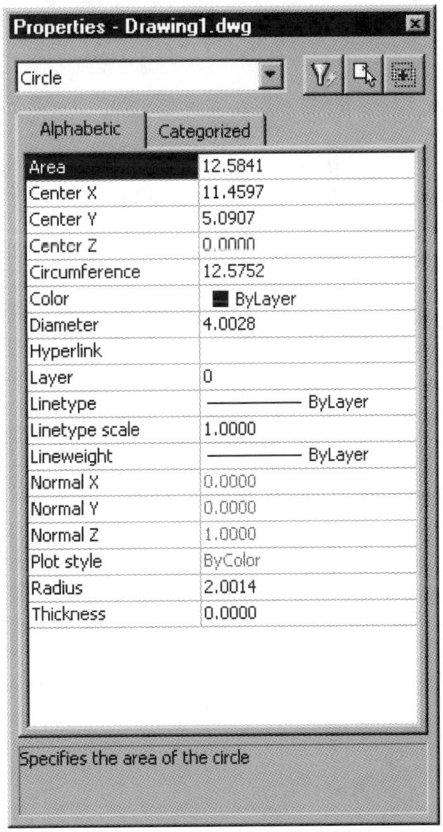

***Figure 6-16** Properties window for editing the properties of the objects*

Note

*Some of the options of the **Properties** window are explained in Chapter 4, Drawing Aids. Other options of the **Properties** window will explained in detail in Chapter 18, Object Grouping and Editing Commands.*

CHANGING PROPERTIES USING GRIPS

You can also use the grips to change the properties of a single or multiple object. To change the properties of an object, select the object to display the grips and then right-click to display the shortcut menu. In the shortcut menu, choose the **Properties** option to display the **Properties** window. If you select a circle, AutoCAD LT will display Circle in the Selection drop-down list in the **Properties** window. Similarly, if you select text, Text is displayed in the drop-down list. If you select several objects, AutoCAD LT will display all the objects in the selection drop-down list of the **Properties** window. You can use this window to change the properties (color, layer, linetype, linetypes scale, lineweight, plot style, thickness, and so on) of the gripped objects.

MATCHING PROPERTIES

Toolbar:	Standard Toolbar > Match Properties
Menu:	Modify > Match Properties
Command:	MATCHPROP

The **MATCHPROP** command can be used to change some properties like color, layer, linetype, and linetype scale of the selected objects. However, in this case you need a source object whose properties will be forced on the destination objects. When you invoke this command, AutoCAD LT will prompt you to select the source object and then the destination objects. The properties of the destination objects will be changed to that of the source object. This command is a transparent command and can be used inside another command. The prompt sequence that will follow when you choose the **Match Properties** button from the **Standard Toolbar** is:

Select Source Object: *Select the source object.*
Current active settings: Color Layer Ltype Ltscale Lineweight Thickness Plotstyle Text Dim Hatch
Select destination object(s) or [Settings]:

If you select the destination object in the **Select destination object(s) or [Settings]** prompt, the properties of the source object will be forced on it. If you select the Setting option, AutoCAD LT displays the **Properties settings** dialog box (Figure 6-17). The properties displayed in this dialog box are those of the source object. You can use this dialog box to edit the properties that are copied from source to destination objects.

The table shown next lists the properties associated with different types of objects.

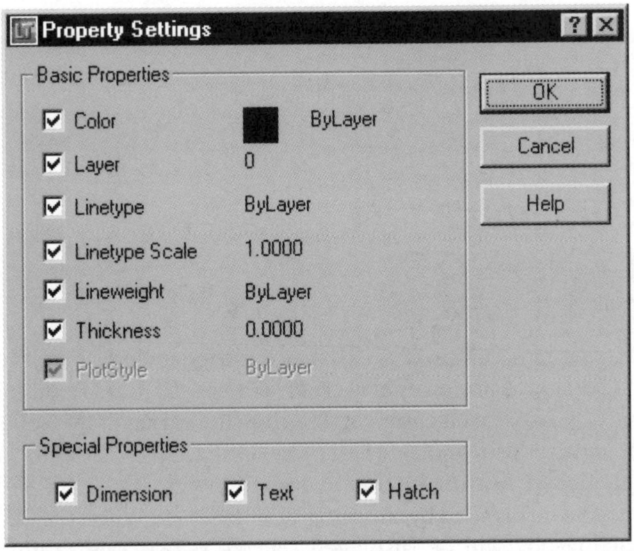

***Figure 6-17 Property Settings** dialog box*

Object Plot	Color & layer	Linetype & Linetype Scale	Line-weight	Thick-ness	Text	Dime-nsion	Hatch style
Arc	x	x	x	x			x
AttDef	x			x	x		x
Body	x	x					x
Circle	x	x	x	x			x
Dimension	x	x			x		x
Ellipse	x	x	x				x
Hatch	x		x			+	x
Image	x	x					x
Insert	x	x	x				x
Leader	x	x	x		x		x
Line	x	x	x	x			x
Mtext	x				x		x
OLE object							
Point	x		x	x			x
2D polyline	x	x	x		x	x	
3D polyline	x	x	x			x	
3D mesh	x	x	x			x	
Pface mesh	x	x	x			x	
Ray	x	x	x			x	
Region	x	x	x	x		x	
2D solid	x	x	x			x	
Spline	x	x	x			x	
Text	x	x	x	x			x

Tolerance	x	x	x	x		x
Viewport	x		x			x
Xline	x	x	x		x	
Xref	x	x	x		x	
Zombie	x	x	x		x	

QSELECT COMMAND

Menu:	Tools > Quick Select
Command:	QSELECT

This command creates a new selection set that will either include or exclude all objects that match the specified object type and property criteria. The **QSELECT** command can be applied to the entire drawing or existing selection set. If a drawing is partially opened, **QSELECT** does not consider the objects that are not loaded. The **QSELECT** command can be invoked by choosing the **Quick Select** button in the **Properties** window. In the shortcut menu, the **QSELECT** command can be invoked by choosing **Quick Select**. When you invoke this command, the **Quick Select** dialog box will be displayed (Figure 6-18). The **Quick Select** dialog box specifies the object filtering criteria and creates a selection set from that criteria.

Apply to

The **Apply to** drop-down list specifies whether to apply the filtering criteria to the entire drawing or to the current selection set. If there is an existing selection set, the **Current selection** is the default value. Otherwise the entire drawing is the default value. You can select the objects to create a selection set by choosing the **Select objects** button provided on the right side of this drop-down list. The **Quick Select** dialog box is temporarily closed when you choose this button and you will be prompted to select the objects. The dialog box will be redisplayed once a selection set is made.

Object type

This drop-down list specifies the type of object to be filtered. It lists all the available object types and if some objects are selected, it lists all the selected object types. **Multiple** is the default setting.

Properties

This list box displays the properties to be filtered. All the properties related to the object type will be displayed in this list box. The property selected from this list box will define the options that will be available in the **Operator** and **Value** drop-down list.

Operator

This drop-down list specifies the range of the filter for the chosen property. The filters that are available are:

- Equals =
- Not Equal <>
- Greater than >

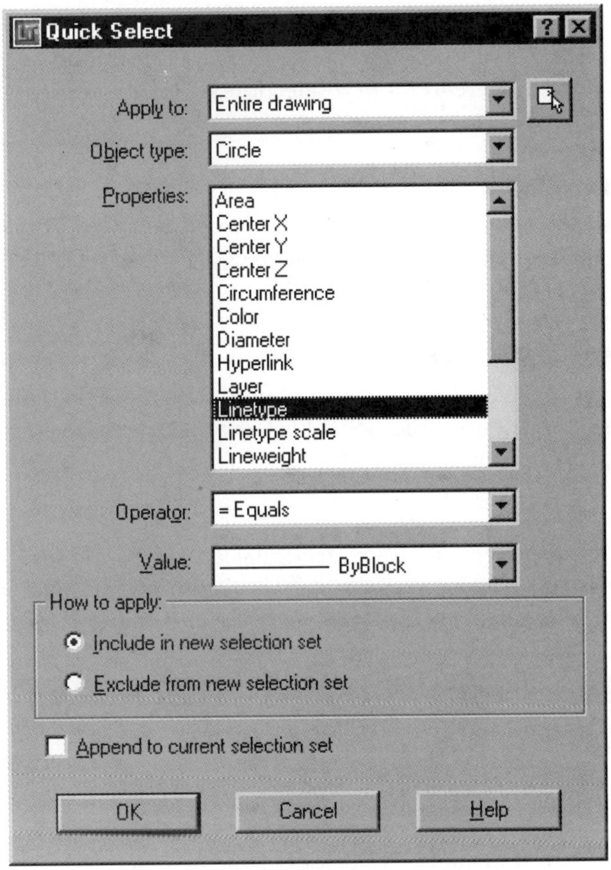

Figure 6-18 ***Quick Select*** *dialog box*

- Less than <
- Select All

Note

The ***Value*** *drop-down list will not be available when you select* ***Select All*** *from the* ***Operator*** *drop-down list.*

Value

This drop-down list specifies the property value of the filter. If the values are known, it becomes a list of the available values from which you can select a value. Otherwise you can enter a value.

How to apply Area

The options under this area are used to specify whether the filtered entities will be included or excluded from the new selection set. This area provides the following two radio buttons:

Include in new selection set

If this radio button is selected, the filtered entities will be included in the new selection set. If selected, this radio button creates a new selection set only composed of those objects that conform to the filtering criteria.

Exclude from new selection set

If this radio button is selected, the filtered entities will be excluded from the new selection set. This radio button creates a new selection set of objects that does not conform to the filtering criteria.

Append to current selection set

This creates a cumulative selection set by using multiple uses of the Quick Select. It specifies whether the objects selected using the **QSELECT** command replace the current selection set or append the current selection set.

Tip

*Quick Select supports custom objects (objects that are created by some other applications) and their properties. If custom objects have other properties than AutoCAD LT, then the source application of the object should be running for the properties to be available by the **QSELECT**.*

MANAGING CONTENTS USING THE AUTOCAD DESIGNCENTER

Toolbar:	Standard Toolbar > AutoCAD DesignCenter
Menu:	Tools > AutoCAD DesignCenter
Command:	ADCENTER

When you choose the **AutoCAD DesignCenter** button from the **Standard Toolbar**, the **DesignCenter** window (Figure 6-19) is displayed. In its default position it is docked to the left of the screen. It can be moved to any other location by picking and dragging it with the grab bars located on the top of the window to any other position on the screen. You can also resize it by clicking the borders and dragging them to the right or left. Right-clicking on the title bar of the window displays a shortcut menu that gives options to **Allow docking**, **Hide**, **Move**, and **Resize** the **DesignCenter** window. The CTRL+2 key acts as a toggle between hiding the window and displaying it. Also, double-clicking on the title bar of the window docks the **DesignCenter** window if it is undocked and vice versa. To use this option, make sure that the **Allow Docking** option is selected.

The **DesignCenter** window is used for locating and organizing drawing data, and for inserting blocks, layers, external references, and other customized drawing content. These contents can be selected from either your own files, local drives, a network, or the Internet, and for viewing contents of a file. You can even access and use the contents between files or from the Internet. You can use the DesignCenter to conveniently drag and drop any information that has been previously created into a current drawing. This is a powerful tool that reduces repetitive tasks of again creating information that already exists.

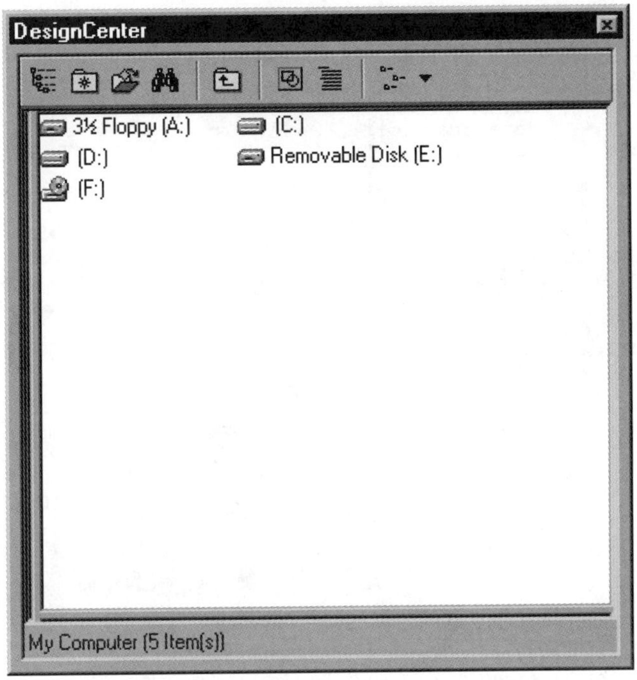

***Figure 6-19 DesignCenter** window at the default position*

Figure 6-20 shows the **DesignCenter** toolbar buttons. When you choose the **Tree View Toggle** button on the **DesignCenter** toolbar, it displays the **Tree View** (Left Pane) with a tree view of the contents of the drives. If the tree view is not displayed, you can also right-click in the window and choose **Tree** from the shortcut menu that is displayed. Now, the window is divided into two parts, the **Tree View** (left pane) and the **Palette** (right pane). The Palette displays folders, files, objects in a drawing, images, Web-based content, and custom content. You can also resize both the Tree View and the Palette by clicking and dragging the bar between them to the right or the left.

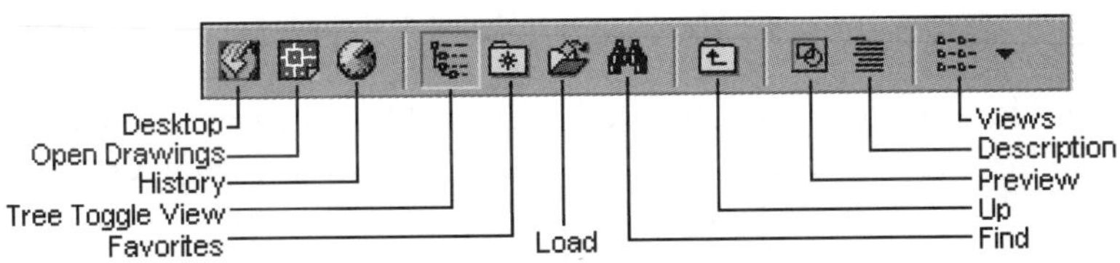

***Figure 6-20 DesignCenter** toolbar buttons*

In the Tree View you can browse the contents of any folder by clicking on the plus sign (+) adjacent to it, to expand the view. Further expanding the contents of a file displays the categories such as **Blocks**, **Dimstyles**, **Layers**, **Linetypes**, **Textstyles**, **Layout**, and **External References**. Clicking on any one of these categories in the Tree View displays the listing under the selected category in the Palette (Figure 6-21). Alternately, right-clicking a particular folder, file, or category of the file contents displays a shortcut menu. The **Explore** option in

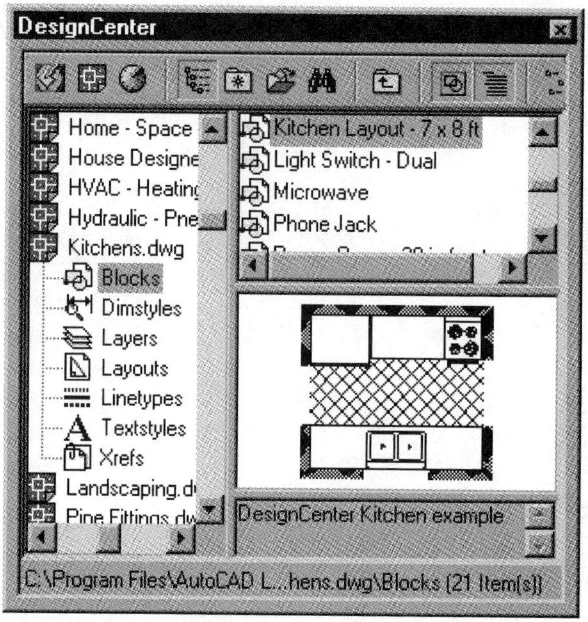

***Figure 6-21 DesignCenter** window displaying Tree pane, Palette, Preview pane, and the Description box*

this shortcut menu also further expands the selected folder, file, or category of contents to display the listing of contents respectively. Choosing the **Preview** button from the toolbar displays an image of the selected object or file in a **Preview pane** below the Palette. Choosing the **Description** button displays a brief text description of the selected item if it has one in the **Description** box. When you click on a specific block name in the palette, its preview image and description that was defined earlier when creating the block are displayed in the Preview pane and the Description box respectively.

You can drag and drop any of the contents into the current drawing, or add them by double-clicking on the contents. These are then reused as part of the current drawing. When you double-click on specific Xrefs and blocks, AutoCAD LT displays the **External Reference** dialog box and the **Insert** dialog box respectively to help in attaching the external reference and inserting the block, respectively. Right-clicking a block displays the options of **Insert Block** or **Copy** and right-clicking an Xref displays the options of **Attach Xref** or **Copy** in the shortcut menus. Similarly, when you double-click a layer, text style, dimstyle, layout, or linetstyle, they also get added to the current drawing. If any of these named objects already exist in the current drawing, duplicate definition is ignored and it is not added again. When you right-click on a specific linetype, layer, textstyle, layout or dimstyle, in the palette, a shortcut menu is displayed that gives you an option to **Add** or **Copy**. Using the **Add** option, directly adds the selected named object to the current drawing. Using the **Copy** option, copies the specific named object to the clipboard from where you can then paste it into a particular drawing.

Note

You will learn more about inserting the blocks in Chapter 14, Blocks.

The other options available in the shortcut menu are **Add to Favorites**, **Organize Favorites**, and **Open in Window**. Choosing **Add to Favorites** adds the selected file or folder to the **Favorites** folder, which contains the most often accessed files and folders. **Organize Favorites** allows you to reorganize the contents of the **Favorites** folder. The **Open in window** option opens the selected drawing file as a current drawing. This option is available when you right-click on a file in the palette.

Choosing the **Desktop** button in the **AutoCAD DesignCenter** toolbar, lists all the local and network drives. The **Open Drawings** button lists the drawings that are open, including the current drawing which is being worked on. The **History** button lists the most recent locations accessed through the **AutoCAD DesignCenter**. The **Tree View Toggle** button displays or hides the tree pane with the tree view of the contents in a hierarchial form. This option is disabled when you choose the **History** button. Choosing the **Favorites** button displays short-cuts to files and folders that are accessed frequently by you and are stored in the **Favorites** folder. This reduces the time you take to access these files or folders from their normal location. Choosing the **Load** button displays the **Load DesignCenter Palette** dialog box (Figure 6-22) whose options are similar to the standard **Select file** dialog box. When you select a file here and choose the **Open** button, AutoCAD LT displays the selected file and its contents in the **DesignCenter** window.

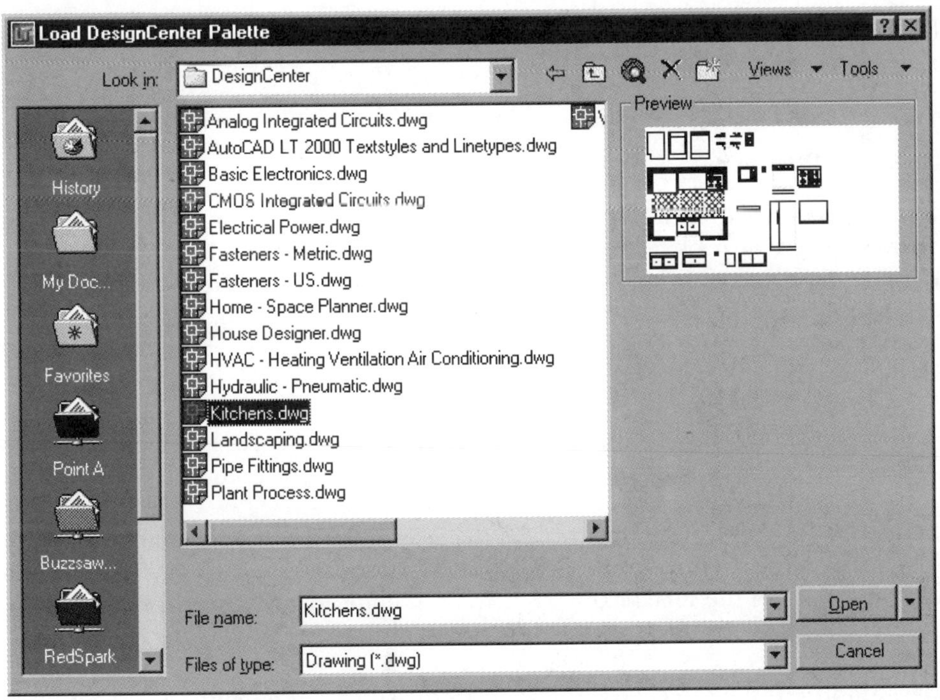

***Figure 6-22 Load DesignCenter Palette** dialog box*

Choosing the **Find** button displays the **Find** dialog box (Figure 6-23). By using this dialog box, you can search for drawings or named objects such as blocks, textstyles, dimstyles, layers, layouts, external references, or linetypes. You can select the type of object you want to search

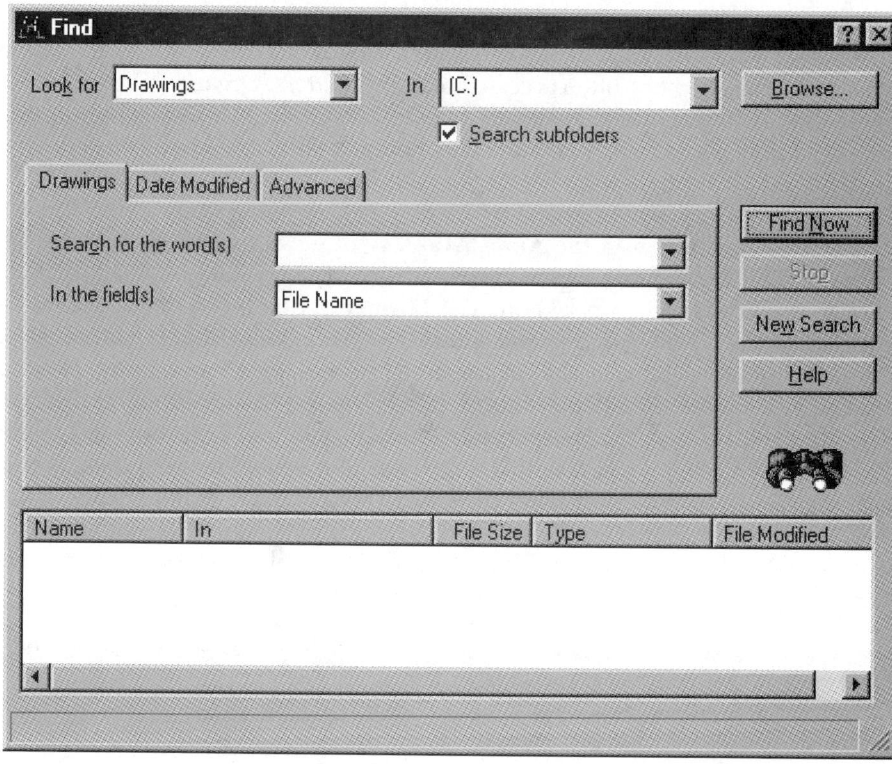

Figure 6-23 ***Find*** *dialog box*

for from the **Look for** drop-down list. You can look for a drawing or a named object in your hard drives. If you are looking for a named object, you can first select the drive where you think it might be located from the **In** drop-down list or use the **Browse** button to locate a folder. The **Search subfolders** check box is selected by default. Then, you can enter the name of the named object you are looking for in the **Search for the name** edit box and then choose the **Find Now** button to start the search operation.

If you are looking for drawing files, three tabs are available in the Find dialog box that provide additional search criteria. You can conduct a search based on the following additional criteria:

1. In the **Drawings** tab, enter the words in the **Search for the word(s)** edit box. The field in which the particular words are to be located can be selected from the **In the field(s)** drop-down list of the **Find** dialog box. You can enter text that might be part of the File name, title, subject, author, or description text of a particular drawing.

2. **Date Modified**. You can choose the **Date Modified** tab in the **Find** dialog box and select the appropriate radio buttons there.

3. In the **Advanced** tab, you can select the category to look for from the **Containing** drop-down list of the **Find** dialog box. You can choose either of the following categories on which to base your search: **Block name**, **Block and drawing description**, **Attribute tag**, or **Attribute**

value. If the named object or drawing you are searching for does not contain any blocks or attributes, and does not have any drawing description, you need not select any of the previous from the **Containing** drop-down list. In such a case, the Containing text edit box is not available. The text blocks, attributes, or drawing description can be entered in the **Containing text** edit box. You can also specify the **At least** or **At most** sizes of the file you are searching for in the **Size is** edit boxes to further narrow down the search.

After you have entered the appropriate criteria for searching the specific file, choose the **Find Now** button. The name, location, file size, type, and date last modified on is displayed at the bottom of the dialog box under the **Name**, **In**, **File Size Type**, and **File Modified** columns of the list box respectively.

Now, when you double-click on the drawing name, it gets highlighted in the tree view of the **AutoCAD DesignCenter**, and its contents are displayed in the palette, or it gets loaded in the palette. You can also right-click on the drawing name in the list box of the **Find** dialog box to display a shortcut menu. This shortcut menu provides you with the options of loading a drawing in the palette, inserting it as a block, attaching it as an external reference, opening it as a current drawing, or copying it to the pasteboard.

In the **AutoCAD DesignCenter**, the **Up** button moves one level up in the tree structure from the current location. The **Views** button gives four display format options for the contents of the palette: **Large icons**, **Small icons**, **List**, and **Details**. The **List** option lists the contents in the palette while the **Details** option gives a detailed list of the contents in the palette with the name, file size, and type.

Right-clicking in the palette displays a shortcut menu with all the options provided in the **AutoCAD DesignCenter** in addition to the **Add to Favorites**, **Organize favorites**, and **Refresh** options. The **Refresh** option refreshes the palette display if you have made any changes to it. The following example will illustrate how to use the AutoCAD DesignCenter to locate a drawing and then use its contents into a current drawing.

Example 1 *Architectural*

Use the **AutoCAD DesignCenter** window to locate and view contents of the drawing KITCHENS. Also, use the **DesignCenter** to insert a block from this drawing and also import the layer **100** and textstyle **6** from the **Cottage Plan and Elevation.dwg** located in the **Sample** folder. Use these to make a drawing of a Kitchen plan and then add text to it as shown in Figure 6-24. Save this drawing with the name **MyKitchen.dwg** in the default AutoCAD LT directory.

1. Open a new drawing using the **Start from Scratch** option. Make sure to select the **English (feet and inches)** option in the **AutoCAD LT Today** window.

2. Change the units to **Architectural**. Increase the limits to 10',10'. Invoke the **ALL** option of the **ZOOM** command so that you are able to view the entire drawing.

3. Choose the **AutoCAD DesignCenter** button from the **Standard Toolbar**. The **DesignCenter**

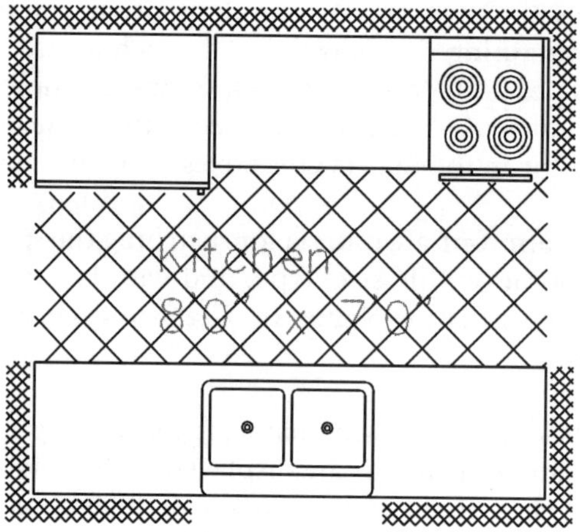

Figure 6-24 *Drawing for Example 1*

window is displayed at its default location.

4. In the **DesignCenter** toolbar, choose the **Tree View Toggle** button to display the Tree View and the Palette (if not already displayed). Also, choose the **Preview** button. You can resize the window, if need be, to view both the Tree View and Palette, conveniently.

5. Choose the **Find** button in the **DesignCenter**. The **Find** dialog box will be displayed. In this dialog box, select **Drawings** from the **Look For** drop-down list and **C:** from the **In** drop-down list. Select the **Search subfolders** check box. In the **Drawings** tab, type KITCHENS in the **Search for the word(s)** edit box and select **File Name** from the **In the field(s)** drop-down list. Now, choose the **Find Now** button to commence the search. After the drawing has been located, its details and path are displayed in a list box at the bottom of the dialog box.

6. Now, right-click on **Kitchens.dwg** in the list box of the **Find** dialog box and choose **Load into Palette** from the shortcut menu. You will notice that the drawing and its contents are displayed in the Tree view.

7. Close the **Find** dialog box.

8. Double-click on **Kitchens.dwg** in the Tree View to display its contents in case they are not displayed. You can also expand the contents by clicking on the plus sign adjacent to the File name in the Tree View.

9. Select **Blocks** in the Tree View to display the list of blocks in the drawing in the Palette. Using the left mouse button, drag and drop the block **Kitchen Layout - 7x8 ft** in the current drawing.

10. Now, double-click on **Cottage Plan and Elevation.dwg** located in the **Sample** folder in the

same directory to display its contents in the Palette.

11. Select **Layers** in the Tree View to display the layers in the drawing. Drag and drop or double-click the layer **100** from the Palette to the current drawing. Now, you can use this layer for placing the text in the current drawing after making it the current layer.

12. Select **Textstyles** to display the list of Textstyles in the Palette. Select **6** in the Palette and drag and drop it in the current drawing. You can use this textstyle for adding text to the current drawing.

13. Use the imported data to add text to the current drawing and complete it as shown in the Figure 6-24.

14. Save the current drawing as **MyKitchen.dwg**.

MAKING INQUIRIES ABOUT A DRAWING (INQUIRY COMMANDS)

When you create a drawing or examine an existing one, you often need some information about the drawing. In manual drafting, you inquire about the drawing by performing measurements and calculations manually. Similarly, when drawing in an AutoCAD LT environment, you will need to make inquiries about data pertaining to your drawing. The inquiries can be about the distance from one location on the drawing to another, the area of an object like a polygon or circle, coordinates of a location on the drawing, and so on. AutoCAD LT keeps track of all the details pertaining to a drawing. Since inquiry commands are used to obtain information about the selected objects, these commands do not affect the drawings in any way. The following is the list of Inquiry commands:

AREA	**DIST**	**ID**	**LIST**	**TIME**
DWGPROPS	**MASSPROP**			

Note

*The **MASSPROP** command will be discussed in Chapter 21, Drawing and Viewing 3D Objects.*

For most of the INQUIRY commands, you are prompted to select objects. Once the selection is complete, AutoCAD LT switches from graphics mode to text mode, and all the relevant information about the selected objects is displayed. For some commands information is displayed in the AutoCAD LT Text Window. The display of the text screen can be tailored to your requirements with the help of a pointing device. Therefore, by moving the text screen to one side, you can view the drawing screen and the text screen simultaneously. If you select the **minimize** button or select the close button, you will return to the graphics screen. You can also return to the graphics screen by entering **GRAPHSCR** command at the Command prompt. Similarly, you can return to the AutoCAD LT Text Window by entering **TEXTSCR** at the Command prompt.

Calculating the Area of the Objects (AREA Command)

Toolbar:	Inquiry > Area
Menu:	Tools > Inquiry > Area
Command:	AREA

Finding the area of a shape or an object manually is time-consuming. In AutoCAD LT, the **AREA** command is used to automatically calculate the area of an object in square units. This command saves time when calculating the area of shapes, especially when the shapes are complicated.

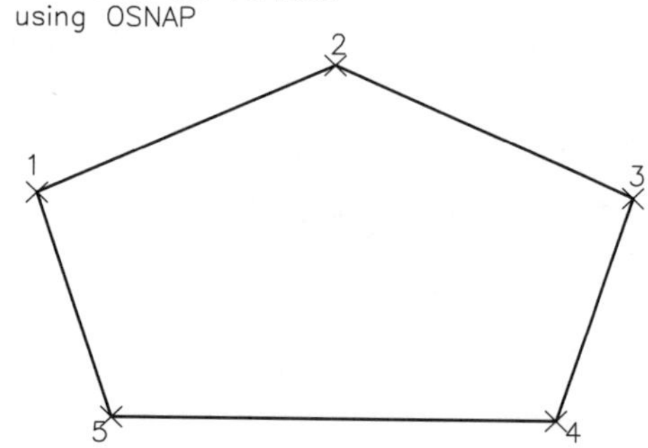

Figure 6-25 *Using the* ***AREA*** *command*

You can use the default option of the **AREA** command to calculate the area and perimeter or circumference of the space enclosed by the sequence of specified points. For example, to find the area of an object (one which is not formed of a single object) you have created with the help of the **LINE** command (Figure 6-25), you need to select all the vertices of that object. By selecting the points, you define the shape of the object whose area is to be found. This is the default method for determining the area of an object. The only restriction is that all the points you specify should be in a plane parallel to the XY plane of the current UCS. You can make the best possible use of object snaps such as ENDpoint, INTersect, and TANgent, or even use running Osnaps, to help you select the vertices quickly and accurately. For AutoCAD LT to find the area of a shape, the shapes need not have been drawn with polylines; nor do the lines need to be closed. However, curves must be approximated with short straight segments. In such cases, AutoCAD LT computes the area by assuming that the first point and the last point are joined. The prompt sequence that will follow when you choose the **Area** button from the **Inquiry** toolbar is:

Specify first corner point or [Object/Add/Subtract]: *Specify first point.*
Specify next corner point or press ENTER for total: *Specify the second point.*
Specify next corner point or press ENTER for total: *Continue selecting until all the points enclosing the area have been selected.*
Specify next corner point or press ENTER for total: Enter
Area = X, Perimeter = Y

Here, X represents the numerical value of the area and Y represents the circumference/perimeter. It is not possible to accurately determine the area of a curved object, such as an arc, with the default (Point) option. However, the approximate area under an arc can be calculated by specifying several points on the given arc. If the object whose area you want to find is not closed (formed of independent segments) and has curved lines, you should use the following steps to determine the accurate area of such an object:

1. Convert all the segments in that object into polylines using the **PEDIT** command.

2. Join all the individual polylines into a single polyline. Once you have performed these operations, the object becomes closed and you can then use the **Object** option of the **AREA** command to determine the area.

If you specify two points on the screen, the **AREA** command will display the value of the area as 0.00; the perimeter value is the distance between the two points.

Object Option

You can use the **Object** option of the **AREA** command to find the area of objects such as polygons, circles, polylines, regions, solids, and splines. If the selected object is a polyline or polygon, AutoCAD LT displays the area and perimeter of the polyline. In case of open polylines, the area is calculated assuming that the last point is joined to the first point but the length of this segment is not added to the polyline length unlike the default option. If the selected object is a circle, ellipse, or planar closed spline curve, AutoCAD LT will provide information about its area and circumference. For a solid, the surface area is displayed. For a 3D polyline, all vertices must lie in a plane parallel to the XY plane of the current UCS. The extrusion direction of a 2D polyline whose area you want to determine should be parallel to the Z axis of the current UCS. In case of polylines which have a width, the area and length of the polyline is calculated using the centerline. If any of these conditions is violated, an error message is displayed on the screen. The prompt sequence that will follow when you choose the **Area** button is:

Specify first corner point or [Object/Add/Subtract]: O Enter
Select objects : Select an object Enter
Area = (X), Circumference = (Y)

The X represents the numerical value of the area, and Y represents the circumference/perimeter.

Tip
*In many cases, the easiest and most accurate way to find the area of an area enclosed by multiple objects is to use the **BOUNDARY** command to create a polyline, then use the **AREA Object** option.*

Add Option

Sometimes you want to add areas of different objects to determine a total area. For example, in the plan of a house, you need to add the areas of all the rooms to get the total floor area. In such cases, you can use the **Add** option. Once you invoke this option, AutoCAD LT activates the **Add** mode. By using the **First corner point** option at the **Specify first corner point or [Object/Subtract]:** prompt, you can calculate the area and perimeter by selecting points on the screen. Pressing ENTER after you have selected points defining the area which is to be added, calculates the total area, since the **Add** mode is on. The command prompt is as follows:

Specify next corner point or press ENTER for total (ADD mode):

If the polygon whose area is to be added, is not closed, the area and perimeter are calculated assuming a line is added which connects the first point to the last point to close the polygon. The length of this area is added in the perimeter. The **Object** option adds the areas and perimeters of selected objects. While using this option, if you select an open polyline, the area is calculated considering the last point is joined to the first point but the perimeter does not consider the length of this assumed segment, unlike the **First corner point** option. When you select an object, the area of the selected object is displayed on the screen. At this time the total area is equal to the area of the selected object. When you select another object, AutoCAD LT displays the area of the selected object as well as the combined area (total area) of the previous object and the currently selected object. In this manner you can add areas of different objects. Until the **Add** mode is active, the string ADD mode is displayed along with all subsequent object selection prompts to remind you that the **Add** mode is active. When the **AREA** command is invoked, the total area is initialized to zero.

Subtract Option

The action of the **Subtract** option is the reverse of that of the **Add** option. Once you invoke this option, AutoCAD LT activates the **Subtract** mode. The **First corner point** and **Object** options work similar to the way they work in the ADD mode. When you select an object, the area of the selected object is displayed on the screen. At this time, the total area is equal to the area of the selected object. When you select another object, AutoCAD LT displays the area of the selected object as well as the area obtained by subtracting the area of the currently selected object from the area of the previous object. In this manner, you can subtract areas of objects from the total area. Until the **Subtract** mode is active, the string SUBTRACT mode is displayed along with all subsequent object selection prompts, to remind you that the SUBTRACT mode is active. To exit the **AREA** command, press ENTER (null response) at the **Specify first corner point or [Object/Add/Subtract]** prompt. The prompt sequence for these two modes for Figure 6-26 is:

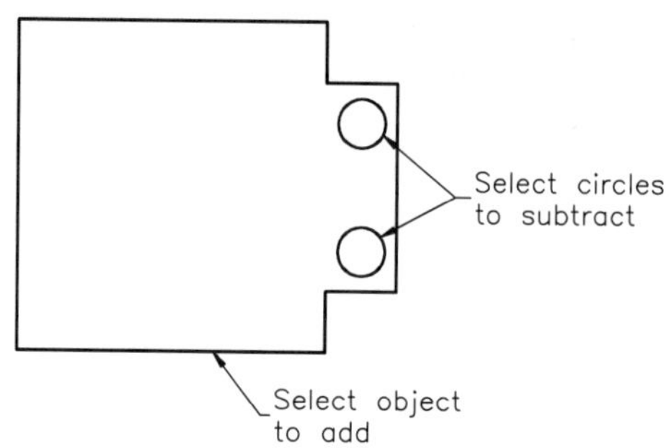

Figure 6-26 *Using the **Add** and **Subtract** options*

Specify first corner point or [Object/Add/Subtract]: A Enter
Specify first corner point or [Object/Subtract]: O Enter
(ADD mode) Select objects: *Select the polyline.*
Area = 2.4438, Perimeter = 6.4999
Total area = 2.4438
(ADD mode) Select objects: Enter
Specify first corner point or [Object/Subtract]: S Enter
Specify first corner point or [Object/Add]: O Enter
(SUBTRACT mode) Select object: *Select the circle.*
Area = 0.0495, Circumference = 0.7890
Total area = 2.3943
(SUBTRACT mode) Select objects: *Select the second circle.*

Area = 0.0495, Circumference = 0.7890
Total area = 2.3448
(SUBTRACT mode) Select object: Enter
Specify first corner point or [Object/Add]: Enter

The **AREA** and **PERIMETER** system variables hold the area and perimeter (or circumference in the case of circles) of the previously selected polyline (or circle). Whenever you use the **AREA** command, the **AREA** variable is reset to zero.

Calculating Distance between Two Points (DIST Command)

Toolbar:	Inquiry > Distance
Menu:	Tools > Inquiry > Distance
Command:	DIST

The **DIST** command is used to measure the distance between two selected points (Figure 6-27). The angles that the selected points make with the X axis and the XY plane are also displayed. The measurements are displayed in current units. Delta X (horizontal displacement), delta Y (vertical displacement), and delta Z are also displayed. The distance computed by the **DIST** command is saved in the **DISTANCE** variable. The prompt sequence that will follow when you choose the **Distance** button is:

Specify first point: *Specify a point.*
Specify second point: *Specify a point.*

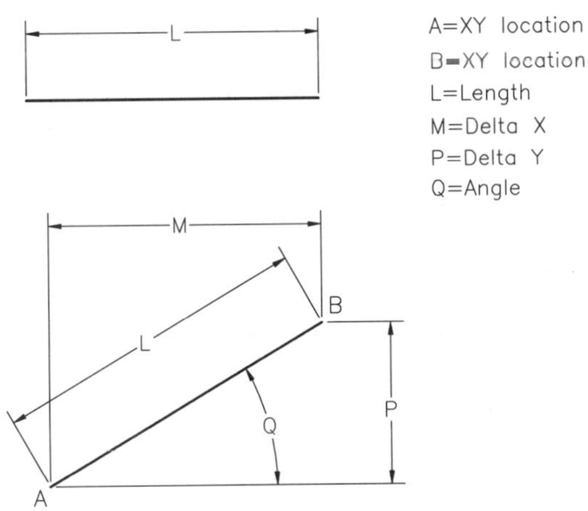

Figure 6-27 *Using the* ***DIST*** *command*

AutoCAD LT returns the following information:

Distance = *Calculated distance between the two points.*
Angle in XY plane = *Angle between the two points in the XY plane.*
Angle from XY plane = *Angle the specified points make with the XY plane.*
Delta X = *Change in X,* Delta Y = *Change in Y,* Delta Z = *Change in Z.*

If you enter a single number or fraction at the **Specify first point** prompt, AutoCAD LT will convert it into the current unit of measurement and display it in the command line.

Command: **DIST**
Specify first point: 3-3/4 *(Enter a number or a fraction.)*
Distance = 3.7500

Note
The Z coordinate is used in 3D distances. If you do not specify the Z coordinates of the two points between which you want to know the distance, AutoCAD LT takes the current elevation as the Z coordinate value.

Identifying a Position on the Screen (ID Command)

Toolbar:	Inquiry > Locate Point
Menu:	Tools > Inquiry > ID Point
Command:	ID

The **ID** command is used to identify the position of a point you specify by displaying the X, Y, and Z coordinates of the point. The prompt sequence that will follow when you choose the **Locate Point** button from the **Inquiry** toolbar is

Specify point: *Specify the point to be identified.*
X = X coordinate Y = Y coordinate Z = Z coordinate

AutoCAD LT takes the current elevation as the Z coordinate value. If an **Osnap** mode is used to snap to a 3D object in response to the **Specify point** prompt, the Z coordinate displayed will be that of the selected feature of the 3D object. You can also use the **ID** command to identify the location on the screen. This can be realized by entering the coordinate values you want to locate on the screen. AutoCAD LT identifies the point by drawing a blip mark at that location if the **BLIPMODE** system variable is on. For example, the prompt sequence to find the position on the screen where X = 2.345, Y = 3.674, and Z = 1.0000 is located is:

Specify point: 2.345,3.674,1.00 Enter
X = 2.345 Y = 3.674 Z = 1.0000

The coordinates of the point specified in the **ID** command are saved in the **LASTPOINT** system variable. You can locate a point with respect to the **ID** point by using the relative or polar coordinate system. You can also snap to this point by typing @ when AutoCAD LT prompts for a point.

Listing Information about Objects (LIST Command)

Toolbar:	Inquiry > List
Menu:	Tools > Inquiry > List
Command:	LIST

The **LIST** command displays all the information pertaining to the selected objects. The information is displayed in the AutoCAD LT Text Window. The prompt sequence that follows when you choose the **List** button from the **Inquiry** toolbar is:

Select objects: *Select objects whose data you want to list.*
Select objects: Enter

Once you select the objects to be listed, AutoCAD LT shifts you from the graphics screen to the AutoCAD LT Text Window. The information displayed (listed) varies from object to object. The information on an object's type, its coordinate position with respect to the current UCS (user coordinate system), the name of the layer on which it is drawn, and whether the object is in model space or paper space is listed for all types of objects. If the color, lineweight, and the linetype are not BYLAYER, they are also listed. Also, if the thickness of the object is greater than 0, that is also displayed. The elevation value is displayed in the form of a Z coordinate (in the case of 3D objects). If an object has an extrusion direction different from the Z axis of the current UCS, the object's extrusion direction is also provided.

More information based on the objects in the drawing is also provided. For example, for a line the following information is be displayed:

1. The coordinates of the endpoints of the line.
2. Its length (in 3D).
3. The angle made by the line with respect to the X axis of the current UCS.
4. The angle made by the line with respect to the XY plane of the current UCS.
5. Delta X, delta Y, delta Z: this is the change in each of the three coordinates from the start point to the endpoint.
6. The name of the layer in which the line was created.
7. Whether the line is drawn in Paper space or Model space.

The center point, radius, true area, and circumference of circles is displayed. For polylines, this command displays the coordinates. In addition, for a closed polyline, its true area and perimeter are also given. If the polyline is open, AutoCAD LT lists its length and also calculates the area by assuming a segment connecting the start point and endpoint of the polyline. In the case of wide polylines, all computation is done based on the centerlines of the wide segments. For a selected viewport, the **LIST** command displays whether the viewport is on and active, on and inactive, or off. Information is also displayed about the status of Hideplot and the scale relative to paper space.

Checking the Time-related Information (TIME Command)

Menu:	Tools > Inquiry > Time
Command:	TIME

The time and date maintained by your system are used by AutoCAD LT to provide information about several time factors related to the drawings. Hence, you should be careful about setting the current date and time in your computer. The **TIME** command can be used to display information pertaining to time related to a drawing and the drawing session. The display obtained by invoking the **TIME** command is similar to the following:

Command: **TIME**

Current time: Friday, August 10, 2001 at 6:40:42:480 PM

Times for this drawing:
Created: Friday, August 10, 2001 at 6:12:48:460 PM
Last updated: Friday, August 10, 2001 at 6:12:48:460 PM
Total editing time: 0 days 00:27:54.120
Elapsed timer (on): 0 days 00:27:54.120
Next automatic save in: 0 days 01:36:27.734

Enter option [Display/ON/OFF/Reset]:

The foregoing display gives you information on the following.

Current Time

Provides today's date and the current time.

Created

Provides the date and time that the current drawing was created. The creation time for a drawing is set to the system time when the **NEW**, **WBLOCK**, or **SAVE** command is used to create that drawing file.

Last updated

Provides the most recent date and time you saved the current drawing. In the beginning, it is set to the drawing creation time, and it is modified every time you use the **QUIT** or **SAVE** command to save the drawing.

Total editing time

This specifies the total time spent on editing the current drawing since it was created. If you terminate the editing session without saving the drawing, the time you have spent on that editing session is not added to the total time spent on editing the drawing. Also, the last update time is not revised.

Elapsed timer

This timer operates while you are in AutoCAD LT. You can stop this timer by entering OFF at the **Enter an option [Display/ON/OFF/Reset]** prompt. To activate the timer, enter ON. If you want to know how much time you have spent on the current drawing or part of the drawing in the current editing session, use the **Reset** option as soon as you start working on the drawing or part of the drawing. This resets the user-elapsed timer to zero. By default, this timer is ON. If you turn this timer OFF, the time accumulated in this timer up to the time you turned it OFF will be displayed.

Next automatic save in

This specifies when the next automatic save will be performed. The automatic save time interval can be set in the **Options** dialog box (**Open and Save** tab) or with the **SAVETIME** system variable. If the time interval has been set to zero, the **TIME** command displays the following message:

Next automatic time save in: <disabled>

If the time interval is not set to zero, and no editing has taken place since the previous save, the **TIME** command displays the following message:

Next automatic time save in: <no modification yet>

If the time interval is not set to zero, and editing has taken place since the previous save, the **TIME** command displays the following message:

Next automatic time save in: 0 days hh:mm:ss:msec

hh stands for hours
mm stands for minutes
ss stands for seconds
msec stands for milliseconds

At the end of the display of the **TIME** command, AutoCAD LT prompts:

Enter an option [Display/ON/OFF/Reset]:

The information displayed by the **TIME** command is static. This means that the information is not updated dynamically on the screen. If you respond to the last prompt with DISPLAY (or D), the display obtained by invoking the **TIME** command is repeated. This display contains updated time values.

With the ON response, the user-elapsed timer is started, if it was off. As mentioned earlier, when you enter the drawing editor, by default the timer is on. The OFF response is just the opposite of the ON response and stops the user-elapsed time, if it is on. With the **Reset** option, you can set the user-elapsed time to zero.

Displaying Drawing Properties (DWGPROPS Command)

Menu:	File > Drawing Properties
Command:	DWGPROPS

On choosing **Drawing Properties** from the **File** Menu, the **Drawing Properties** dialog box is displayed (Figure 6-28). This dialog box has four tabs under which information about the drawing is displayed. The information displayed in this dialog box helps you look for the drawing more easily. The tabs are as follows:

General

This tab displays general properties about the drawing like the **Type**, **Size**, and **Location**.

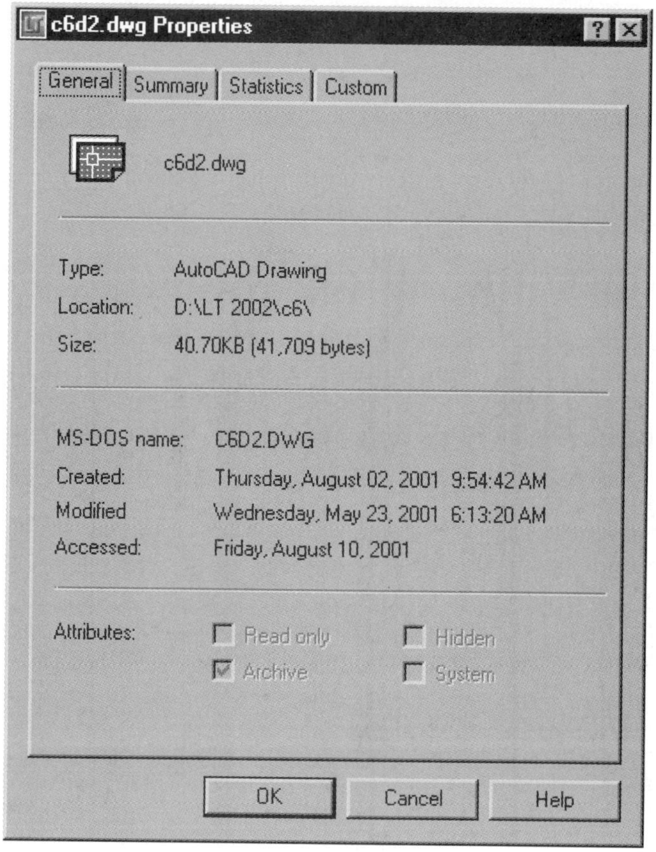

***Figure 6-28 Drawing Properties** dialog box*

Summary

The **Summary** tab displays predefined properties such as the Author, title, and subject.

Statistics

This tab stores and displays data such as the file size and data such as the dates when the drawing was last saved on or modified on.

Custom

This tab displays custom file properties including values assigned by you.

Self-Evaluation Test

Answer the following questions, and then compare your answers to the correct answers given at the end of this chapter.

1. The number of grips depends on the selected object. (T/F)
2. You can use the **Options** dialog box to modify the grips parameters. (T/F)
3. You need at least one source object while using the **MATCHPROP** command. (T/F)
4. You can not drag and drop the entities from the AutoCAD DesignCenter. (T/F)
5. A grip is a small square that is displayed on an object at its _________ points.
6. A line has ________ grip points and a polyline has _________ .
7. You can enable grips within a block by setting the system variable _________ to 1 (On).
8. The color of the unselected grips can also be changed by using the _________system variable.
9. You can access the Mirror mode by selecting a grip and then entering _________ or _________ from the keyboard or giving a null response four times by pressing the SPACEBAR four times.
10. The _________ drop-down list will not be available if you select **Select All** from the **Operator** drop-down list in the **Quick Select** dialog box.

Review Questions

Answer the following questions.

1. If you select a grip of an object, the grip becomes a hot grip. (T/F)
2. To cancel the grip, press the ESC key once. (T/F)
3. The Rotate mode allows you to rotate objects around the base point without changing their size. (T/F)
4. You can drag and drop the contents of any drawing using the AutoCAD DesignCenter. (T/F)

5. Which system variable is used to modify the color of the selected grip?

   (a) **GRIPCOLOR** (b) **GRIPHOT**
   (c) **GRIPCOLD** (d) **GRIPBLOCK**

6. Which system variable is used to enable the display of the grips inside the blocks?

   (a) **GRIPCOLOR** (b) **GRIPHOT**
   (c) **GRIPCOLD** (d) **GRIPBLOCK**

7. Which system variable is used to modify the size of the grips?

   (a) **GRIPCOLOR** (b) **GRIPSIZE**
   (c) **GRIPCOLD** (d) **GRIPBLOCK**

8. By holding down which key you can select and make more than one grips hot?

   (a) SHIFT (b) CTRL
   (c) ESC (d) ALT

9. Which system variable is used to enable the grip mode?

   (a) **GRIPCOLOR** (b) **GRIPHOT**
   (c) **GRIPS** (d) **GRIPBLOCK**

10. When you double-click on a circle, the ________ window is displayed.

11. The **GRIPSIZE** is defined in pixels, and its value can range from ________ to ________ pixels.

12. When you select a grip for editing, you are automatically in the ________ mode.

13. The ________ mode lets you move the selected objects to a new location.

14. The ________ mode allows you to scale the objects with respect to the base point without changing their orientation.

15. The Mirror mode allows you to mirror the objects across the ________ without changing the size of the objects.

Exercises

Exercise 1 *General*

1. Use the **LINE** command to draw the shape as shown in Figure 6-29(a).
2. Use grips (Stretch mode) to get the shape as shown in Figure 6-29(b).
3. Use the Rotate and Stretch modes to get the copies as shown in Figure 6-29(c)

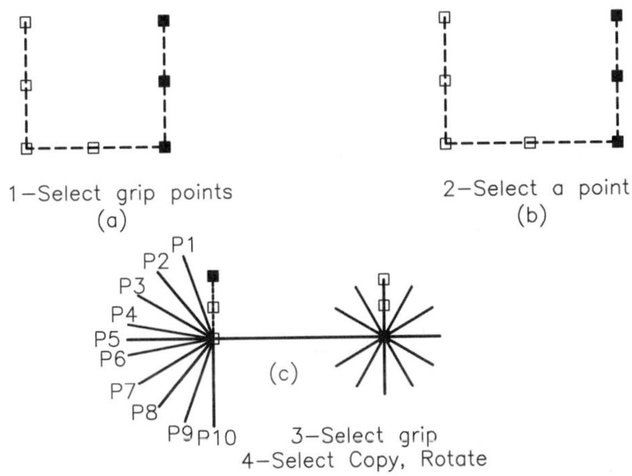

***Figure 6-29** Drawing for Exercise 1*

Exercise 2 *Mechanical*

Use the drawing and editing commands to make the drawing shown in Figure 6-30.

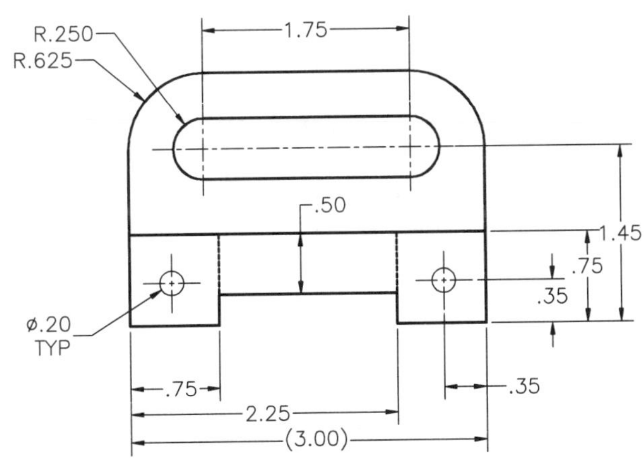

***Figure 6-30** Drawing for Exercise 2*

Exercise 3 *Mechanical*

Use the drawing and editing commands to make the drawing shown in Figure 6-31.

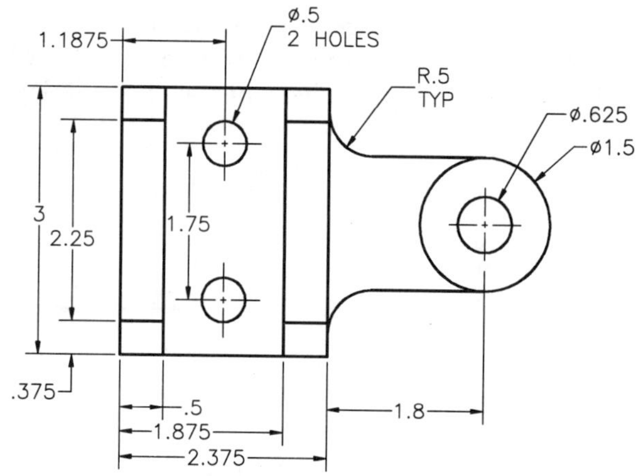

Figure 6-31 *Drawing for Exercise 3*

Problem Solving Exercise 1 *Mechanical*

Draw Figure 6-32 using the drawing and editing commands. Use the **MIRROR** command to mirror the shape 9 units across the Y axis so that the distance between two center points is 9 units. Mirror the shape across the X axis and then reduce the mirrored shape by 75 percent. Join the two ends to complete the shape of the open end spanner. Save the file. Assume the missing dimensions. Note that this is not a standard size spanner.

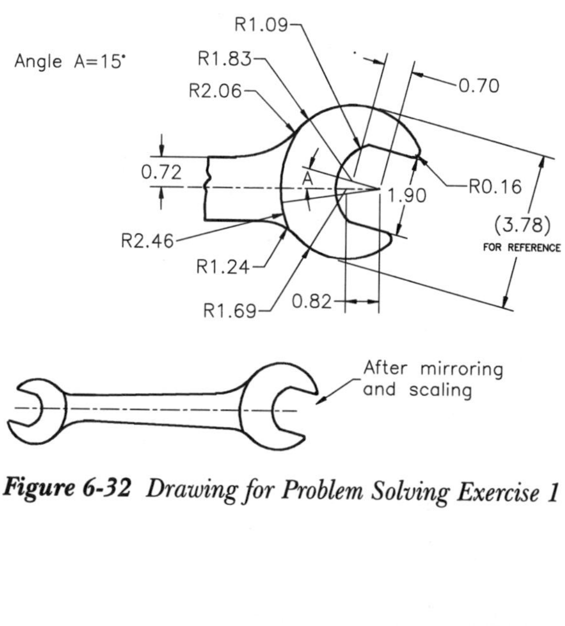

Figure 6-32 *Drawing for Problem Solving Exercise 1*

Problem Solving Exercise 2 *Mechanical*

Use the drawing and editing commands to make the drawing shown in Figure 6-33.

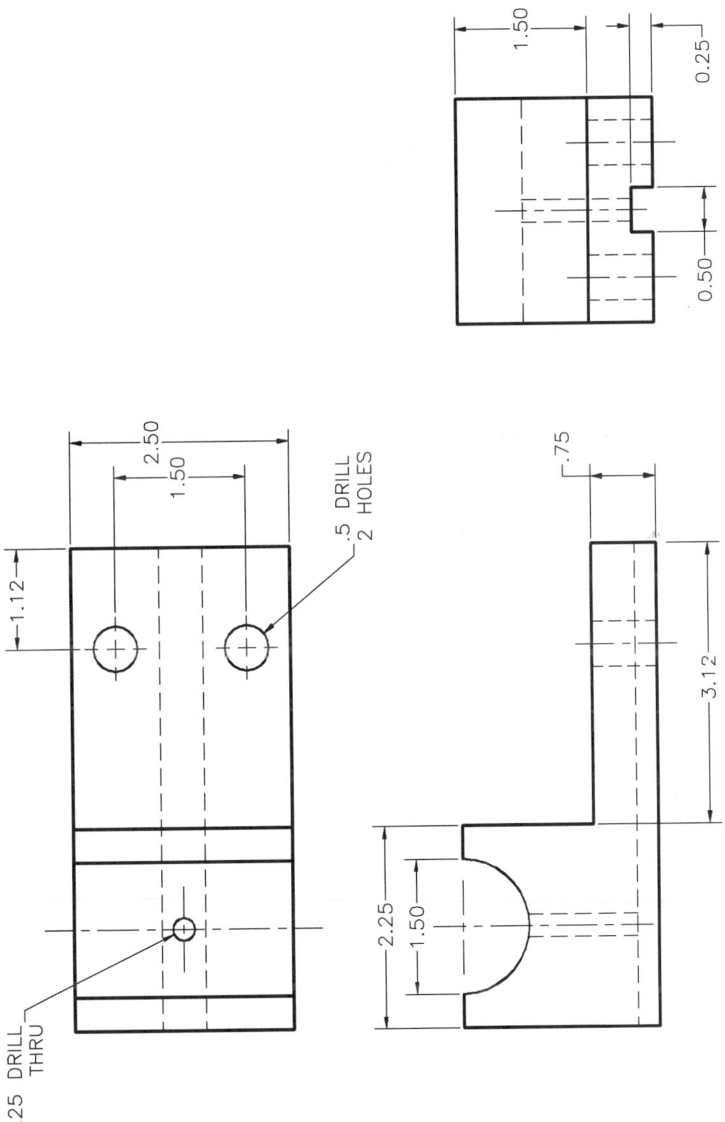

Figure 6-33 *Drawing for Problem Solving Exercise 2*

Problem Solving Exercise 3

Mechanical

Use the drawing and editing commands to make the drawing shown in Figure 6-34. Assume the missing dimensions.

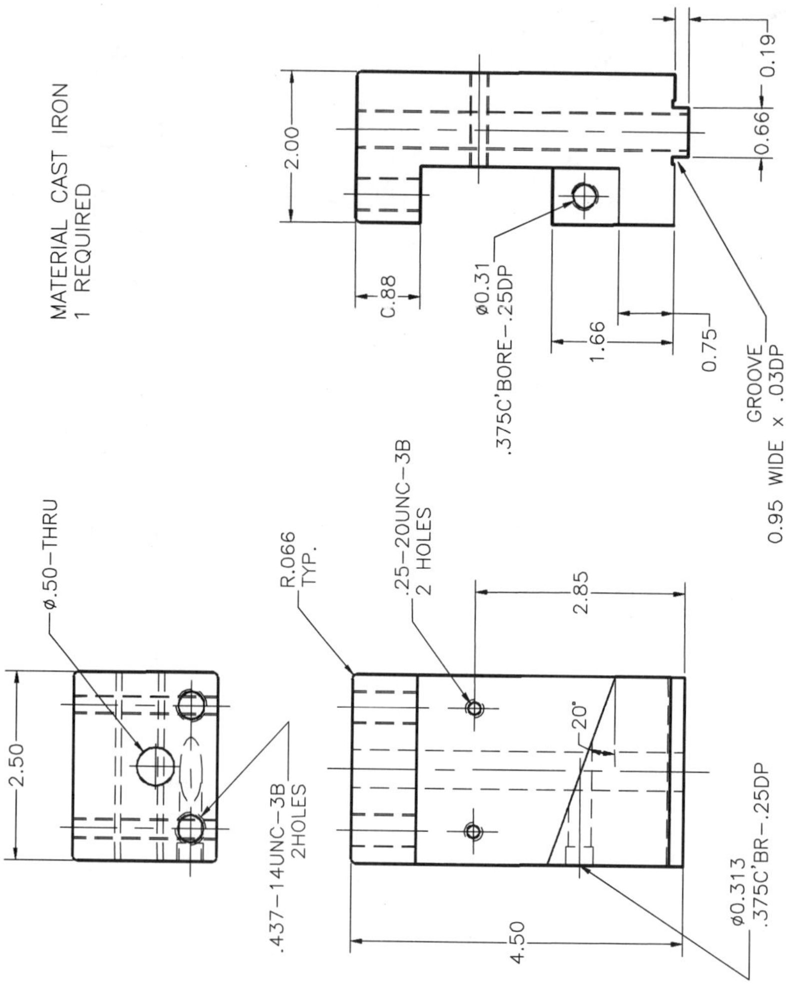

***Figure 6-34** Drawing for Problem Solving Exercise 3*

Answers to Self-Evaluation Test

1 - T, **2** - T, **3** - T, **4** - F, **5** - definition, **6** - three, two, **7** - **GRIPBLOCK**, **8** - **GRIPCOLOR**, **9** - **MIRROR, MI**, **10** - **Value**

Chapter 7

Controlling the Drawing Display and Creating Text

Learning Objectives

After completing this chapter, you will be able to:

- *Use the **REDRAW** and **REGEN** commands.*
- *Use the **ZOOM** command and its options.*
- *Understand the **PAN** and **VIEW** commands.*
- *Understand the use of the **Aerial View** window.*
- *Draw text using the **TEXT** command.*
- *Create paragraph text using the **MTEXT** command.*
- *Edit text using the **DDEDIT** command.*
- *Use the **Properties** window to change the properties of the text.*
- *Substitute fonts and specify alternate default fonts.*
- *Create text styles using the **STYLE** command.*
- *Determine text height.*
- *Check spellings and find and replace text.*

BASIC DISPLAY OPTIONS

Drawing in AutoCAD LT is much simpler than manual drafting in many ways. Sometimes while drawing, it is very difficult to see and alter minute details. In AutoCAD LT, you can overcome this problem by viewing only a specific portion of the drawing. For example, if you want to display a part of the drawing on a larger area, you can use the **ZOOM** command, which lets you enlarge or reduce the size of the drawing displayed on the screen. Similarly, you can use the **REGEN** command to regenerate the drawing and **REDRAW** to refresh the screen. In this chapter you will learn some of the drawing display commands such as **REDRAW**, **REGEN**, **PAN**, **ZOOM**, and **VIEW**. These commands can also be used in the transparent mode. Transparent commands are commands that can be used while another command is in progress. Once you have completed the process involved with a transparent command, AutoCAD LT automatically returns you to the command with which you were working before you invoked the transparent command.

REDRAW COMMAND

Menu:	View > Redraw
Command:	REDRAW

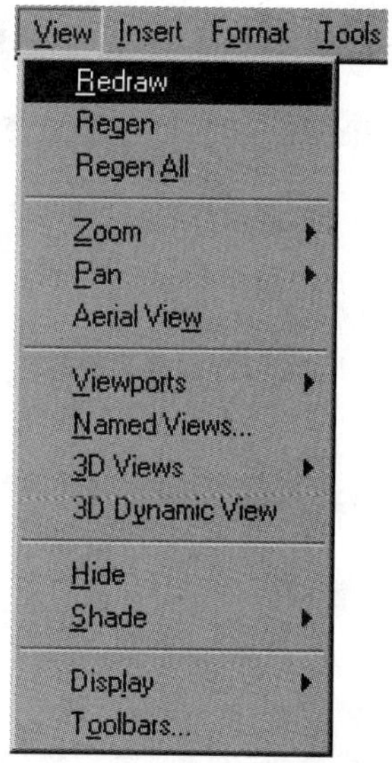

***Figure 7-1** Invoking the **REDRAW** command from the **View** menu*

The **REDRAW** command (Figure 7-1) redraws the screen and is used to remove the small cross marks (blips) that appear when a point is specified on the screen when BLIPMODE is set to on. The blip mark is not treated as an element of the drawing. It also redraws the objects that do not display on the screen as a result of editing some other object. In AutoCAD LT, several commands redraw the screen automatically (for example, when a grid is turned off), but it is sometimes useful to redraw the screen explicitly. In AutoCAD LT, the **REDRAW** command can also be used in the transparent mode. This can be done by typing an apostrophe in front of the command. The apostrophe appended to a command indicates that the command is to be used as a transparent command (Command: **'REDRAW**). Use of the **REDRAW** command does not involve a prompt sequence; instead, the redrawing process takes place without any prompting for information.

The **REDRAW** command affects only the current viewport. If you have more than one viewport you can use the **REDRAWALL** command to redraw all the viewports. Remember that the **Redraw** in the **View** menu is the **REDRAWALL** command.

Tip

*While working on complex drawings, it may be better to set the **BLIPMODE** variable as **Off** (default) instead of using the **REDRAW** command to clear blips.*

REGEN COMMAND

Menu:	View > Regen
Command:	REGEN

The **REGEN** command makes AutoCAD LT regenerate the entire drawing to update it. The need for regeneration usually occurs when you change certain aspects of the drawing. All the objects in the drawing are recalculated and redrawn in the current viewport. One of the advantages of this command is that the drawing is refined by smoothing out circles and arcs. To use this command, enter **REGEN** at the Command prompt. AutoCAD LT displays the message **Regenerating model** while it regenerates the drawing. The **REGEN** command affects only the current viewport. If you have more than one viewport, you can use the **REGENALL** command to regenerate all the viewports. The **REGEN** command can be aborted by pressing the ESC key. This saves time if you are going to use another command that causes automatic regeneration.

Tip

*Under certain conditions, the **ZOOM** and **PAN** commands automatically regenerate the drawing. Some other commands also perform regenerations under certain conditions.*

ZOOM COMMAND

Toolbar:	Zoom toolbar or Standard > Zoom flyout
Menu:	View > Zoom
Command:	ZOOM

Creating drawings on the screen would not be of much use if you could not magnify the drawing view to work on minute details. Getting close to or away from the drawing is the function of the **ZOOM** command. In other words, this command enlarges or reduces the view of the drawing on the screen, but it does not affect the actual size of the objects. In this way the **ZOOM** command functions like the zoom lens on a camera. When you magnify the apparent size of a section of the drawing, you see that area in greater detail. On the other hand, if you reduce the apparent size of the drawing, you see a larger area.

The ability to zoom in, or magnify, has been helpful in creating the minuscule circuits used in the electronics and computer industries. The **ZOOM** command is one of the most frequently used commands (Figure 7-2). Also, this command can be used transparently. This means that it can be used while working in other commands. The **ZOOM** command can be invoked from the shortcut menu by right-clicking in the drawing area and choosing **Zoom** from the menu (Figure 7-3). This command has several options and can be used in a number of ways.

Command: **ZOOM** Enter
Specify corner of window, enter a scale factor (nX or nXP) , or
[All/Center/Dynamic/Extents/Previous/Scale/Window] <real time>:

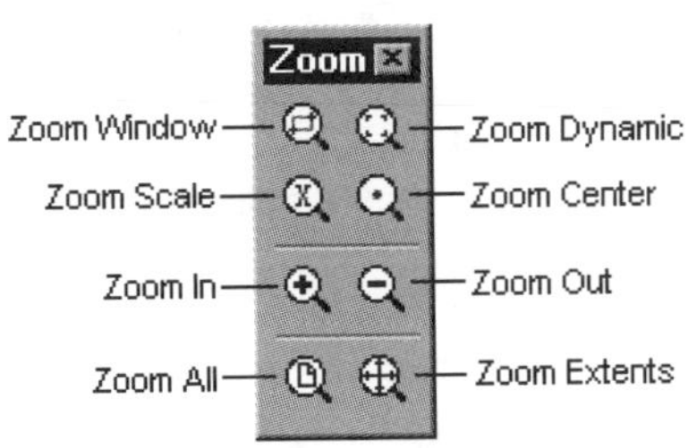

***Figure 7-2** Selecting **Zoom** options from the **Zoom** toolbar*

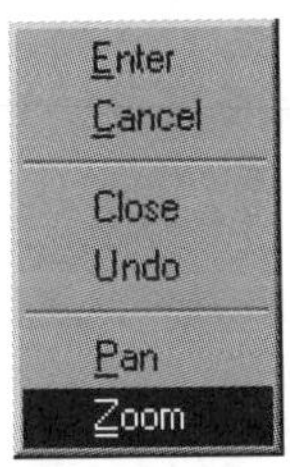

***Figure 7-3** Invoking the **ZOOM** command from the **shortcut** menu when in **LINE** command*

Realtime Zooming

You can use the **Realtime Zoom** to zoom in and zoom out interactively. To zoom in, invoke the command, then hold the pick button down and move the cursor up. If you want to zoom in further, release the pick button and bring the cursor down. Specify a point and move the cursor up again. Similarly, to zoom out, hold the pick button down and move the cursor down. If you move the cursor vertically up from the midpoint of the screen to the top of the window, the drawing is magnified by 100% (zoom in 2x magnification). Similarly, if you move the cursor vertically down from the midpoint of the screen to the bottom of the window, the drawing display is reduced 100% (zoom out 0.5x magnification). Realtime zoom is the default setting for the **ZOOM** command. Pressing ENTER after entering the **ZOOM** command automatically invokes the realtime zoom.

When you use the realtime zoom, the cursor becomes a magnifying glass and displays a plus sign (+) and a minus sign (–). When you reach the zoom out limit, AutoCAD LT does not display the minus sign (–) with the cursor when you hold the pick button down. Similarly, when you reach the zoom in limit, AutoCAD LT does not display the plus sign (+) with the cursor. To exit the realtime zoom, press ENTER, ESC, or select **Exit** from the shortcut menu.

All Option

This option of the **ZOOM** command adjusts the display area based on the drawing limits (Figure 7-4) or extents of the object, whichever is greater. Even if the objects are not within the limits, they are still included in the display. Hence, with the help of the **All** option, you can view the entire drawing in the current viewport (Figure 7-5).

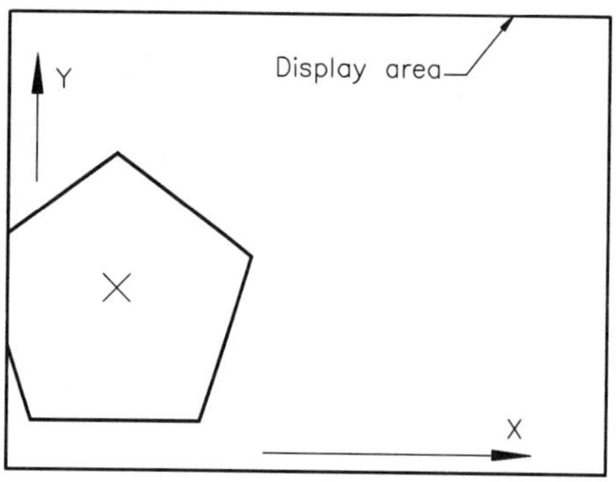

***Figure 7-4** Drawing showing limits*

***Figure 7-5** The **Zoom All** option*

Center Option

This option lets you define a new display window by specifying its center point (Figure 7-6 and Figure 7-7) and the magnification or height. Here, you are required to enter the **center** and the **height** of the subsequent screen display. If you press ENTER instead of entering a new center point, the center of the view will remain unchanged. Instead of entering a height, you can enter the **magnification factor** by typing a number. If you press ENTER at the height prompt, or if the height you enter is the same as the current height, magnification does not take place. For example, if the current height is 2.7645 and you press ENTER at the **Enter magnification or height <2.7645>** prompt, magnification will not take place. The smaller the value, the greater the enlargement of the image. You can also enter a number followed by **X**. This indicates the change in magnification, not as an absolute value, but as a value relative to the current screen. The following is the prompt sequence:

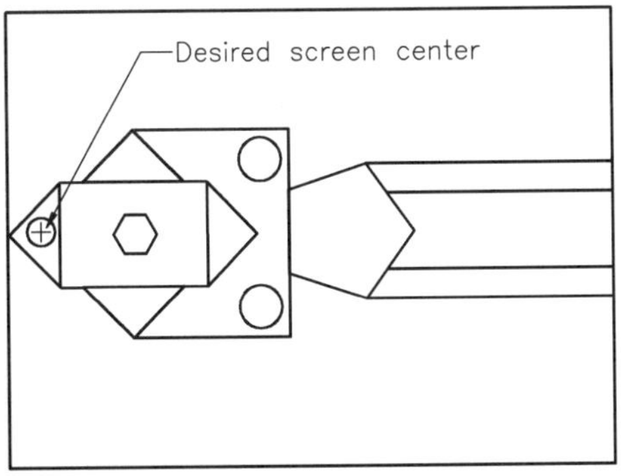

*Figure 7-6 Drawing before using the **ZOOM Center** option*

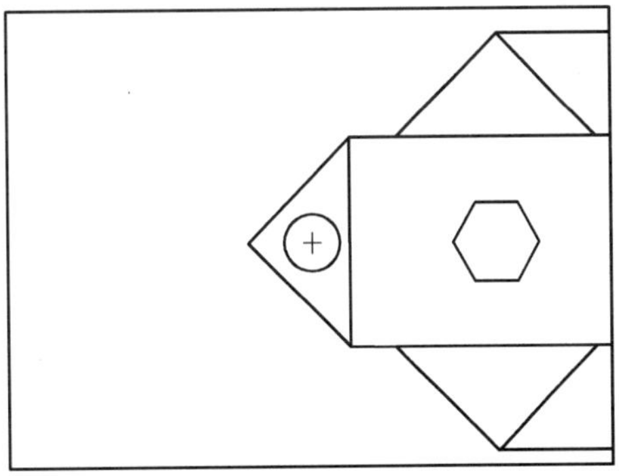

*Figure 7-7 Drawing after using the **ZOOM Center** option*

Command : **ZOOM** Enter
Specify corner of window, enter a scale factor (nX or nXP), or
[All/Center/Dynamic/Extents/Previous/Scale/Window] <real time>: **C** Enter
Specify center point : *Specify a center point.*
Enter magnification or height <current>: 5X Enter

In Figure 7-7, the current magnification height is 5X, which magnified the display five times. If you enter a value of 2, the size (height and width) of the zoom area changes to 2 x 2 around the specified center. In Figure 7-8, if you enter .12 as the height after specifying the center point as the circle's center, the circle will zoom to fit in the display area since its diameter is 0.12. The prompt sequence is:

Command: **ZOOM** Enter
Specify corner of window, enter a scale factor (nX or nXP), or
[All/Center/Dynamic/Extents/Previous/Scale/Window]<real time>:**C** Enter
Center point: *Select the center of the circle.*
Magnification or Height <5.0>:**.12** Enter

Extents Option

As the name indicates, this option lets you zoom to the extents of the biggest object in the drawing. The extents of the drawing comprise the area that has the drawings in it. The rest of the empty area is neglected. With this option, all the objects in the drawing are magnified to the largest possible display (Figure 7-9).

Dynamic Option

This option displays the portion of the drawing that you have already specified. The prompt sequence for using this option of the **ZOOM** command is:

Command: **ZOOM** Enter
Specify corner of window, enter a scale factor (nX or nXP), or
[All/Center/Dynamic/Extents/Previous/Scale/Window]<real time>: **D** Enter

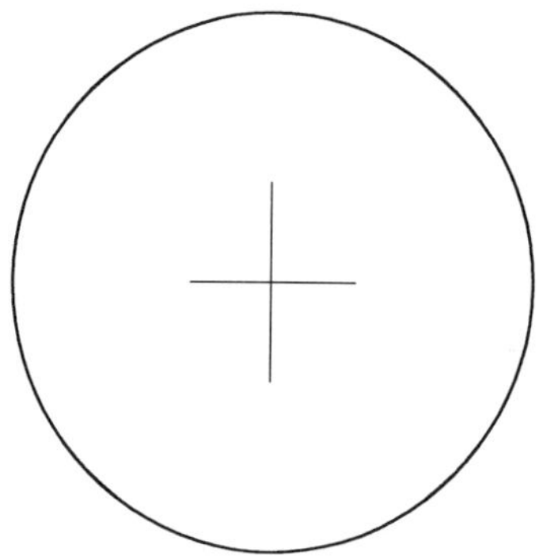

Figure 7-8 *Drawing after using the* ***ZOOM Center*** *option*

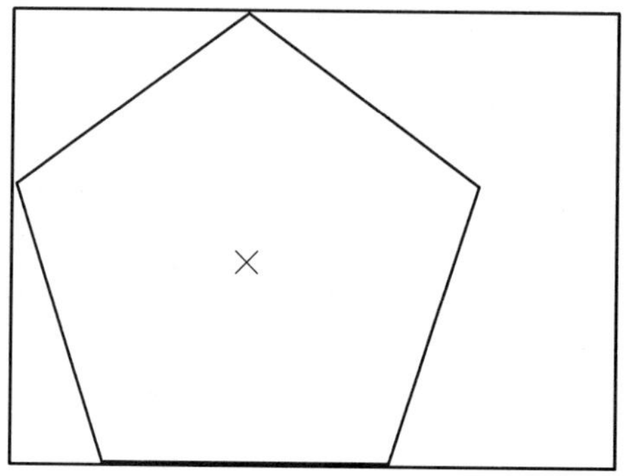

Figure 7-9 *The* ***ZOOM Extents*** *option*

You can then specify the area you want to be displayed by manipulating a view box representing your viewport. This option lets you enlarge or shrink the view box and move it around. When you have the view box in the proper position and size, the current viewport is cleared by AutoCAD LT and a special view selection screen is displayed. This special screen comprises information regarding the current view as well as available views. In a color display, the different viewing windows are very easy to distinguish because of their different colors, but in a monochrome monitor, they can be distinguished by their shape.

Blue dashed box representing drawing extents. Drawing extents are represented by a dashed blue box (Figure 7-10), which constitutes the larger of the drawing limits or the actual area occupied by the drawing.

Green dashed box representing the current view. A green dashed box is formed to represent the area that the current viewport comprises when the **Dynamic** option of the **ZOOM** command is invoked (Figure 7-11).

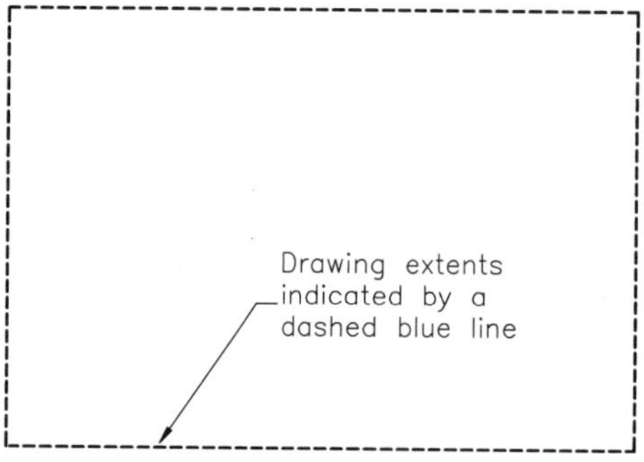

Figure 7-10 *Box representing drawing extents*

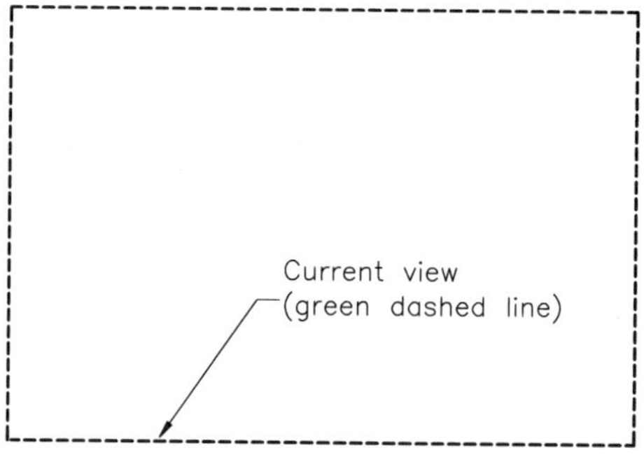

Figure 7-11 *Representation of the current view*

Panning view box (X in the center). A view box initially of the same size as the current view box is displayed with an X in the center (Figure 7-12). You can move this box with the help of your pointing device. This box, known as the **panning view box**, helps you to find the center point of the zoomed display you want. When you have found the center, you press the pick button to make the zooming view box appear.

Zooming View Box (arrow on the right side). After you press the pick button in the center of the panning view box, the X in the center of the view box is replaced by an arrow pointing to the right edge of the box. This **zooming view box** (Figure 7-13) indicates the ZOOM mode. You can now increase or decrease the area of this box according to the area you want to zoom into. To shrink the box, move the pointer to the left; to increase it, move the pointer to the right. The top, right, and bottom sides of the zooming view box move as you move the pointer, but the left side remains fixed, with the zoom base point at the midpoint of the left side. You can slide it up or down along the left side. When you have the zooming view box in the desired size for your zoom display, press ENTER to complete the command, and zoom into the desired area of the drawing. Before pressing ENTER, if you want to change the position of the zooming view box, click the pick button of your pointing device to make the panning view box reappear. After repositioning, press ENTER.

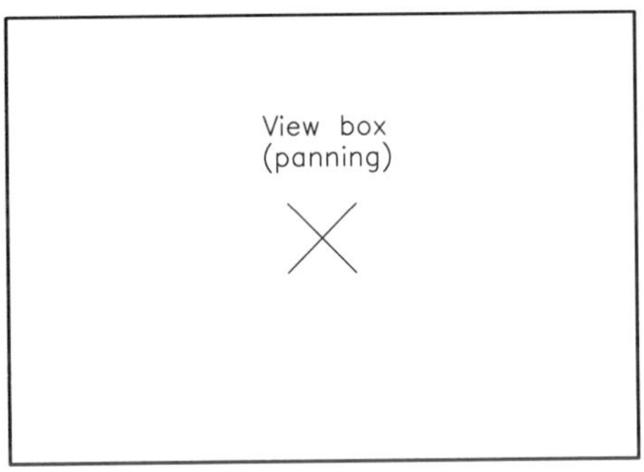

Figure 7-12 The panning view box

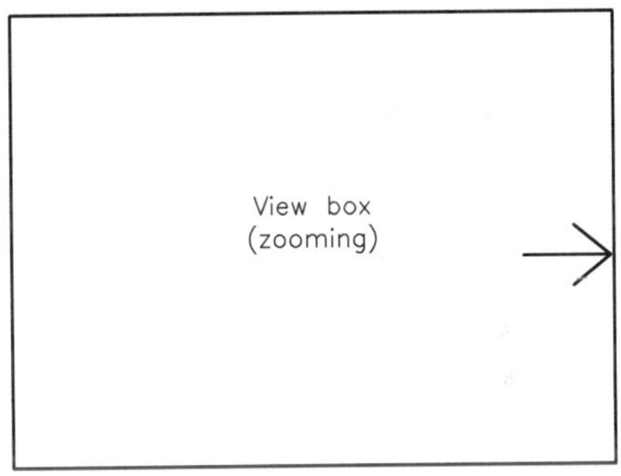

Figure 7-13 The zooming view box

Previous Option

While working on a complex drawing, you may need to zoom in on a portion of the drawing to edit some minute details. Once the editing is over you may want to return to the previous view. This can be done using the **Previous** option of the **ZOOM** command. Without this option it would be very tedious to zoom back to previous views. AutoCAD LT saves the view specification of the current viewport whenever it is being altered by any of the ZOOM options or by the **PAN**, **VIEW Restore**, **DVIEW**, or **PLAN** commands (which are discussed later). Up to ten views are saved for each viewport. The prompt sequence for this option is:

Command: **ZOOM** [Enter]
Specify corner of window, enter a scale factor (nX or nXP), or
[All/Center/Dynamic/Extents/Previous/Scale/Window]<real time>: **P** [Enter]

Successive **ZOOM,** P commands can restore up to ten previous views. **VIEW** here refers to the area of the drawing defined by its display extents. If you erase some objects and then issue a **ZOOM** Previous command, the previous view is restored, but the erased objects are not.

Window Option

This is the most commonly used option of the **ZOOM** command. It lets you specify the area you want to zoom in on, by letting you specify two opposite corners of a rectangular window. The center of the specified window becomes the center of the new display screen. The area inside the window is magnified or reduced in size to fill the display as completely as possible. The points can be specified either by selecting them with the help of the pointing device or by entering their coordinates. The prompt sequence is:

Command: **ZOOM** Enter
Specify corner of window, enter a scale factor (nX or nXP), or
[All/Center/Dynamic/Extents/Previous/Scale/Window]<real time>: *Specify a point.*
Specify opposite corner: *Specify another point.*

Whenever the **ZOOM** command is invoked, the window method is one of two default options. This is illustrated by the previous prompt sequence where you can specify the two corner points of the window without invoking any option of the **ZOOM** command. The **Window** option can also be used by entering **W**. In this case the prompt sequence is:

Command: **ZOOM** Enter
Specify corner of window, enter a scale factor (nX or nXP), or
[All/Center/Dynamic/Extents/Previous/Scale/Window]<real time>: **W** Enter
Specify first corner: *Specify a point.*
Specify opposite corner: *Specify another point.*

Scale Option

The **Scale** option of the **ZOOM** command is a very versatile option. It can be used in the following ways.

Scale: Relative to full view. This option of the **ZOOM** command lets you magnify or reduce the size of a drawing according to a scale factor (Figure 7-14). A scale factor equal to 1 displays an area equal in size to the area defined by the established limits. This may not display the entire drawing if the previous view was not centered on the limits or if you have drawn outside the limits. To get a magnification relative to the full view, you can enter any other number. For example, you can type 4 if you want the displayed image to be enlarged four times. If you want to decrease the magnification relative to the full view, you need to enter a number that is less than 1. In Figure 7-15, the image size decreased because the scale factor is less than 1. In other words, the image size is half of the full view because the scale factor is 0.5. The prompt sequence is:

Command: **ZOOM** Enter
Specify corner of window, enter a scale factor (nX or nXP), or
[All/Center/Dynamic/Extents/Previous/Scale/Window]<real time>: **S** Enter
Enter a scale factor (nX or nXP) : **0.5** Enter

Scale: Relative to current view. The second way to scale is with respect to the current view (Figure 7-16). In this case, instead of entering only a number, enter a number followed by an

Figure 7-14 *Drawing before* ***ZOOM Scale*** *option*

Figure 7-15 *Drawing after* ***ZOOM Scale*** *option*

X. The scale is calculated with reference to the current view. For example, if you enter **0.25X**, each object in the drawing will be displayed at one-fourth (¼) of its current size. The following example increases the display magnification by a factor of 2 relative to its current value (Figure 7-17):

Command: **ZOOM** Enter
Specify corner of window, enter a scale factor (nX or nXP), or
[All/Center/Dynamic/Extents/Previous/Scale/Window]<real time>: **2X** Enter

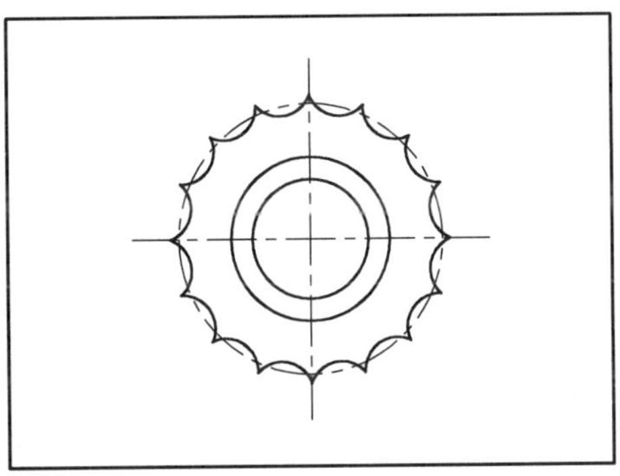

Figure 7-16 *Drawing before* ***ZOOM Scale*** *(X) option*

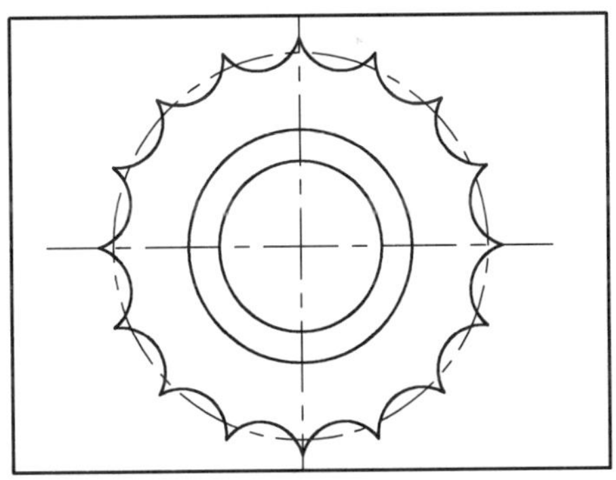

Figure 7-17 *Drawing after* ***ZOOM Scale*** *(X) option*

Scale: Relative to paper space units. The third method of scaling is with respect to paper space. You can use paper space in a variety of ways and for various reasons. For example, you can array and plot various views of your model in paper space. To scale each view relative to paper space units, you can use the **ZOOM XP** option. Each view can have an individual scale. The drawing view can be at any scale of your choice in a model space viewport. For example, to display a model space at one-fourth (¼) the size of the paper space units, the prompt sequence is:

Command: **ZOOM** Enter
Specify corner of window, enter a scale factor (nX or nXP), or
[All/Center/Dynamic/Extents/Previous/Scale/Window]<real time>: **1/4XP** Enter

Note

For details about the paper space, refer to Chapter 11, Model Space Viewports, Paper Space Viewports, and Layouts.

Zoom In and Out

You can also zoom into the drawing using the **In** option, which doubles the image size.

Similarly, you can use the **Out** option to decrease the size of the image by half. To invoke these options from the command line, enter **ZOOM 2X** for the **In** option or **ZOOM .5X** for the **Out** option at the Command prompt.

Exercise 1 — *Mechanical*

Draw Figure 7-18 according to the given dimensions. Use the **ZOOM** command to get a bigger view of the drawing. Do not dimension the drawing.

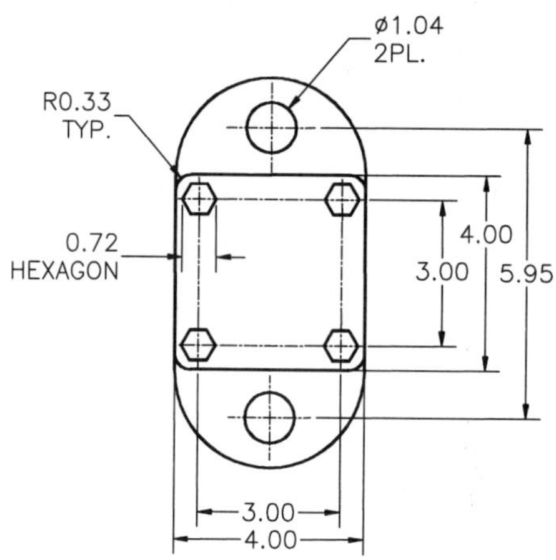

Figure 7-18 *Drawing for Exercise 1*

PAN COMMAND

Toolbar:	Standard Toolbar > Pan
Menu:	View > Pan
Command:	PAN

You may want to view or draw on a particular area outside the current viewport. You can do this using the **PAN** command. If done manually, this would be like holding one corner of the drawing and dragging it across the screen. The **PAN** command allows you to bring into view portions of the drawing that are outside the display area of the current viewport. This is done without changing the magnification of the drawing. The effect of this command can be illustrated by imagining that you are looking at a big drawing through a window (**display window**) that allows you to slide the drawing right, left, up, and down to bring the part you want to view inside this window. You can invoke the **PAN** command from the shortcut menu also.

Panning in Realtime

You can use the **Realtime PAN** to pan the drawing interactively. To pan a drawing, invoke the command and then hold the pick button down and move the cursor in any direction. When you select the realtime pan, AutoCAD LT displays an image of a hand indicating that you are in PAN mode. **Realtime pan** is the default setting for the **PAN** command. Choosing the **Pan Realtime** button in the toolbar and entering **PAN** at the Command prompt automatically invokes the realtime pan. To exit the realtime pan, press ENTER, ESC, or choose **Exit** from the shortcut menu.

There are different options for the **PAN** command, that can be used to pan the drawing in a particular direction. These items can be invoked only from the menu (Figure 7-19).

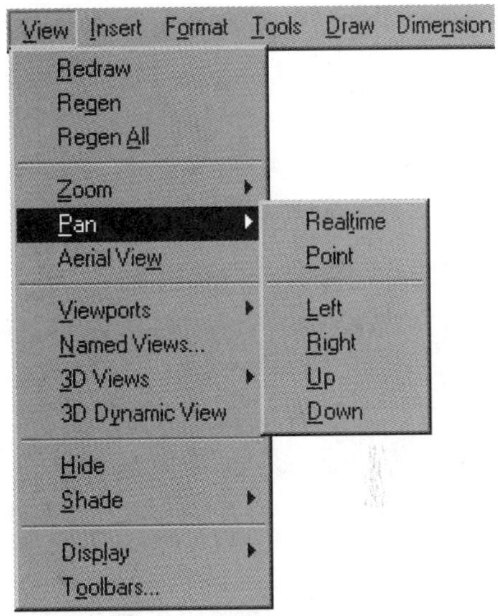

***Figure 7-19** Invoking the **PAN** command options from the **View** menu*

Point. In this option you are required to specify the displacement. To do this you need to specify in what direction to move the drawing and by what distance. You can give the displacement either by entering the coordinates of the points or by specifying the coordinates by using a pointing device. The coordinates can be entered in two ways. One way is to specify a single coordinate pair. In this case, AutoCAD LT takes it as a relative displacement of the drawing with respect to the screen. For example, in the following case the **PAN** command would shift the displayed portion of the drawing 2 units to the right and 2 units up.

Command: *Select the **Point** option from the **View** menu.*
Specify base point or displacement: **2,2** [Enter]
Specify second point: [Enter]

In the second case, you can specify two coordinate pairs. AutoCAD LT computes the displacement from the first point to the second. Here, displacement is calculated between point (3,3) and point (5,5).

Command: *Specify the **Point** item from the **View** menu.*
Specify base point or displacement: **3,3** [Enter] *(Or specify a point.)*
Specify second point: **5,5** [Enter] *(Or specify a point.)*

Left. Moves the drawing left so that some of the right portion of the drawing is brought into view.

Right. Moves the drawing right so that some of the left portion of the drawing is brought into view.

Up. Moves the drawing up so that some of the bottom portion of the drawing is brought into view.

Down. Moves the drawing down so that some of the top portion of the drawing is brought into view.

Tip

*You can use the scroll bars to pan the drawing vertically or horizontally. The scroll bars are located at the right side and the bottom of the drawing area. You can control the display of the scroll bars in the **Display** tab of the **Options** dialog box.*

CREATING VIEWS

Toolbar:	View> Named Views Standard > Named Views
Menu:	View > Named Views
Command:	View

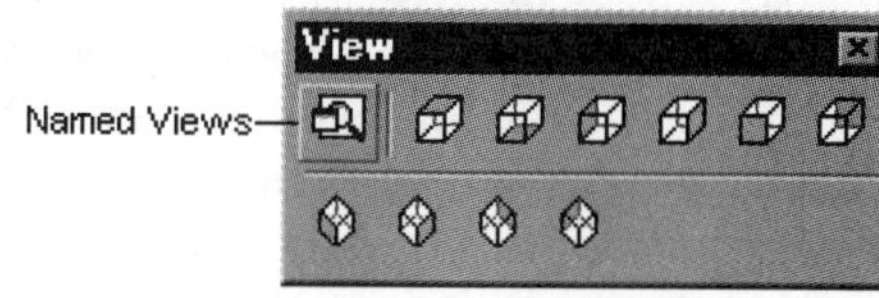

***Figure 7-20** Invoking Named Views from the **View** toolbar*

While working on a drawing, you may frequently be working with the **ZOOM** and **PAN** commands, and you may need to work on a particular drawing view (some portion of the drawing) more often than others. Instead of wasting time by recalling your zooms and pans and selecting the same area from the screen over and over again, you can store the view under a name and restore the view using the name you have given it (Figure 7-20). The **View** dialog box (Figure 7-21) is used to save the current view under a name so that you can restore (display) it later. It does not save any drawing object data, only the view parameters needed to redisplay that portion of the drawing.

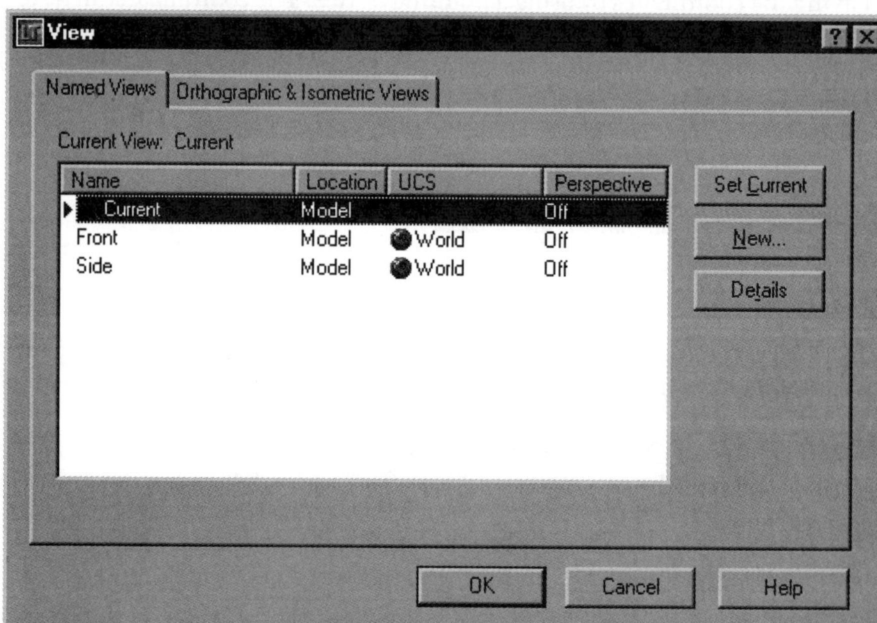

***Figure 7-21** View dialog box*

View Dialog Box

You can save and restore the views from the **View** dialog box (Figure 7-21). This dialog box is very useful when you are saving and restoring many view names. With this dialog box you can name the current view or restore some other view. The **Named Views** tab lists all the created named views . The **Orthographic & Isometric Views** tab lists all the preset views of the drawing, and allows you to set current any of those views. The following are the different options in the **Named Views** tab of the **View** dialog box.

Current View

The View list box displays a list of the named views in the drawing. The list appears with the names of all saved views and the space in which each was defined.

New

The **New** button allows you to create a new view and save it by giving it a name. When you choose the **New** button, the **New View** dialog box (Figure 7-22) is displayed. Enter a name for the view in the **View name** edit box. If you want to save the current view, then activate the **Current display** radio button. If you want to save a rectangular portion of the current drawing as a view (without first zooming in on that area), then choose the **Define window** radio button to activate it. You can specify two points on the screen to describe the window by choosing the **Define View Window** button or you can enter the X and Y coordinates in the **Specify first corner** and **Specify opposite corner** in the command lines. After specifying the view, choose the **OK** button to save the named view.

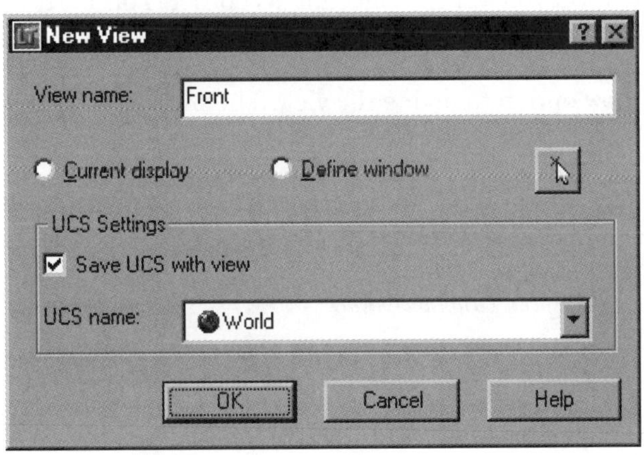

Figure 7-22 New View dialog box

Set Current

The **Set Current** button allows you to replace the current viewport by the view you specify. AutoCAD LT uses the center point and magnification of each saved view and executes a **ZOOM Center** with this information when a view is restored.

Note

*If **TILEMODE** is on, you cannot restore a Paper space view. For the details about Paper space, refer to Chapter 11, Model Space Viewports, Paper Space Viewports, and Layouts.*

Details

You can also see the description of the general parameters of a view by selecting the particular view and then choosing the **Details** button.

Tip

You can use the shortcut menu to rename and delete any named view in the dialog box. Highlight the named view and then right-click to display the shortcut menu.

Using the Command Prompt

You can also use the **-VIEW** command to work with views at the Command prompt.

Command: **-VIEW** [Enter]
Enter an option [?/Orthographic/Delete/Restore/Save/UCS/Window]:

You can use the different options to save, restore, delete, or list the views. If you try to restore a Model space while working in Paper space, AutoCAD LT automatically switches to floating model space. In this case AutoCAD LT will prompt you further as follows:

Enter view name to restore: *Select viewport for restoring.*

You can select the viewport you want by selecting its border. This particular viewport must be on and active. The restored viewport also becomes the current one.

You can use the Window option to define the view. The prompt sequence is:

Command: **-VIEW** [Enter]
Enter an option [?/Orthographic/Delete/Restore/Save/UCS/Window]: **W** [Enter]
Enter view name to save: *Enter the name.*
Specify first corner: *Specify a point.*
Specify other corner: *Specify another point.*

Tip

*The different options of the **VIEW** command can be used transparently by entering **'-VIEW** at the Command prompt. The **VIEW** command that invokes the dialog box cannot be used transparently.*

AERIAL VIEW

Menu: View > Aerial View
Command: DSVIEWER

As a navigational tool, AutoCAD LT provides you with the option of opening another drawing display window along with the graphics screen window you are working on. This window, called the **Aerial View** window, can be used to view the entire drawing and select those portions you want to quickly zoom or pan. The AutoCAD LT graphics screen window resets itself to display the portion of the drawing you have selected for Zoom or Pan in the Aerial View window. You can keep the Aerial View window open as you work on the graphics screen, or minimize it so that it stays on the screen as a button which can be restored when required (Figure 7-23). As with all windows, you can resize or move it. The Aerial View window can also be invoked when you are in the midst of any command other than the **DVIEW** command.

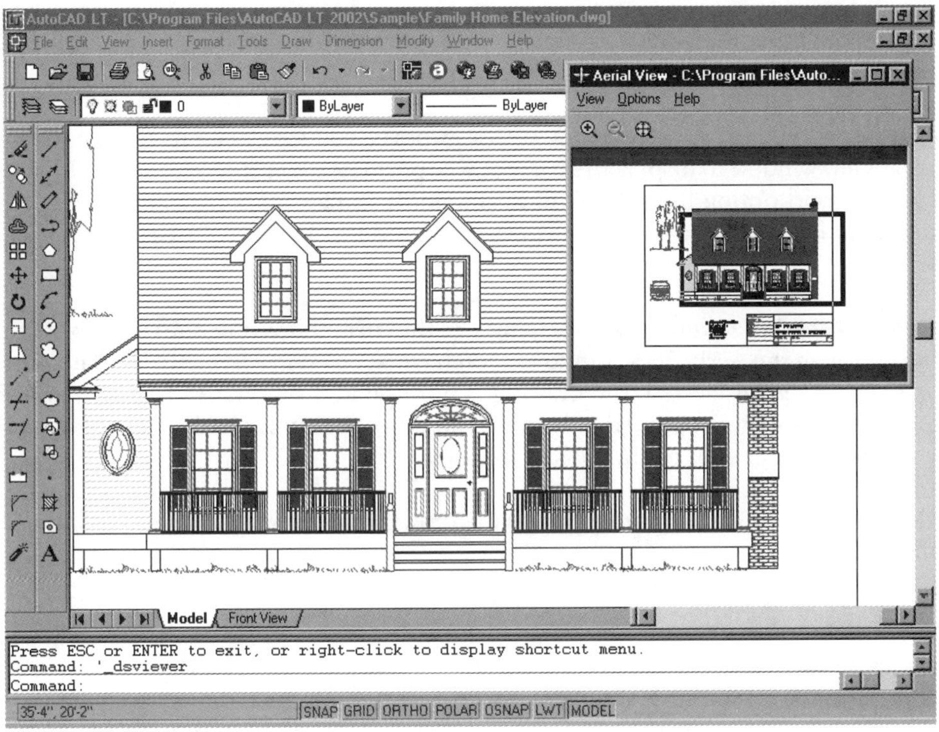

Figure 7-23 *AutoCAD LT graphics screen with the Aerial View window*

The Aerial View window has two menus, **View** and **Options**. It also has a toolbar containing zoom in, zoom out, and zoom global options. The following is a description of the available options in the Aerial View.

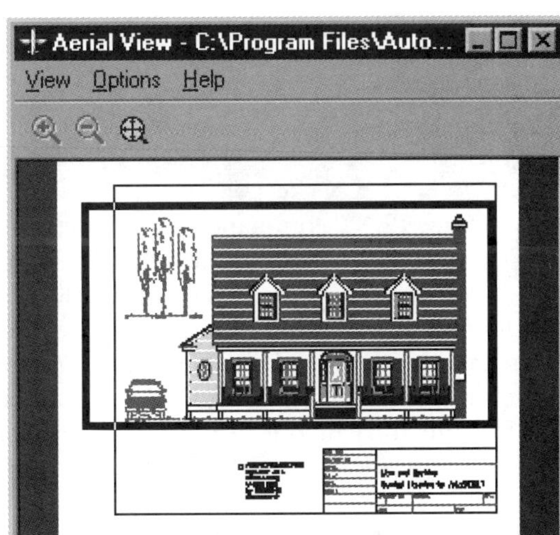

Figure 7-24 *Aerial View window*

Toolbar Buttons

The Aerial View window has three buttons: **Zoom In**, **Zoom Out**, and **Zoom Global** (Figure 7-24).

Zoom In. This option leads to the magnification by a factor of **2** which is centered on the current view box.

Zoom Out. This option leads to the reduction by **half** centered on the current view box.

Zoom Global. The Global option displays the complete drawing in the Aerial Viewport.

Menus

The menus available in the menu title bar are **View** and **Options**. The **View** menu contains **Zoom In**, **Zoom Out**, and **Global** options that are the same as their respective buttons in the toolbar. The **Options** menu has the following options:

Auto Viewport. When you are working with viewports, you may need to change the view in the Aerial View window to display the current viewport view. You can achieve this by selecting the Auto Viewport option from the **Options** menu. If you are working in multiple viewports and you zoom or pan in the Aerial Viewport window, only the view in the current viewport is affected.

Dynamic Update. When you make any changes in the current drawing, the view box is updated simultaneously in the Aerial View window if you have selected the Dynamic Update option. If Dynamic Update is unselected, the drawing is not updated simultaneously. The drawing in the Aerial View window is updated only when you move the cursor in the Aerial View window or invoke any of the menus or the toolbar buttons.

Realtime Zoom. If on, this option updates the drawing area when you are zooming in Aerial view in realtime.

CREATING TEXT

In manual drafting, lettering is accomplished by hand using a lettering device, pen, or pencil. This is a very time-consuming and tedious job. Computer-aided drafting has made this process extremely simple. Engineering drawings invoke certain standards to be followed in connection with the placement of a text in a drawing. In this section, you will learn how text can be added in a drawing by using **TEXT** and **MTEXT** commands.

CREATING SINGLE LINE TEXT

Toolbar:	Text > Single Line Text
Menu:	Draw > Text > Single Line Text
Command:	TEXT

The **TEXT** command lets you write text on a drawing and allows you to delete what has been typed by using the BACKSPACE key. You can enter multiple lines in one command. The **TEXT** command displays a line in the drawing area where you specified the start point after entering the height and the rotation angle. This line identifies the start point and the size of the text height entered. The characters appear on the screen as you enter them. When you press ENTER after typing a line, the cursor automatically places itself at the start of the next line and repeats the prompt **Enter text**. You can end the command by pressing the ENTER key or ESC key at the **Enter text** prompt.

The screen crosshairs can be moved irrespective of the cursor line for the text. If you specify a point, this command will complete the current line of the text and move the cursor line to the point you selected. This cursor line can be moved and placed anywhere on the screen; therefore, multiple lines of text can be entered at any desired locations in the drawing area with a single **TEXT** command. By pressing BACKSPACE you can delete one character to the left of the current position of the cursor box. Even if you have entered several lines of the text, you can use BACKSPACE and go on deleting until you reach the start point of the first line entered. Upon deletion of an entire line, **TEXT** displays a ***Deleted*** message in the Command prompt area.

This command can be used with most of the text alignment modes, although it is most useful in the case of left-justified texts. In the case of aligned texts, this command assigns a height appropriate for the width of the first line to every line of the text. Irrespective of the **Justify** option chosen, the text is first left-aligned at the selected point. After the **TEXT** command ends, the text is momentarily erased from the screen and regenerated with the requested alignment. For example, if you use the **Middle** option, the text will first appear on the screen as left-justified, starting at the point you designated as the middle point. After you end the command and press ENTER, it is regenerated with the proper alignment. The prompt sequence that follows when you choose this command is:

> Specify start point of text or [Justify/Style]: *Specify the start point.*
> Specify height <0.25>: **0.15** Enter
> Specify rotation angle of text <0>: Enter
> Enter text: *Enter first line of the text.*
> Enter text: *Enter second line of the text.*
> Enter text: Enter

Start Point Option

This is the default and the most commonly used option in the **TEXT** command. By specifying a start point, the text is left-justified along its baseline starting from the location of the starting point. Before AutoCAD LT draws the text, it needs the text height, the rotation angle for the baseline, and the text string to be drawn. It prompts you for all this information. The prompt sequence is:

> Specify height <default>:
> Specify rotation angle of text <default>
> Enter text:

The **Specify height** prompt determines the distance by which the text extends above the baseline, measured by the capital letters. This distance is specified in drawing units. You can specify the text height by specifying two points or entering a value. In the case of a null response, the default height, that is, the height used for the previous text drawn in the same style, will be used.

The **Specify rotation angle of text** prompt determines the angle at which the text line will be drawn. The default value of the rotation angle is **0 degrees** (3 o'clock, or east), and in this case the text is drawn horizontally from the specified start point. The rotation angle is measured in a counterclockwise direction. The last angle specified becomes the current rotation angle, and if you give a null response the last angle specified will be used as default. You can also specify the rotation angle by specifying two points. The text is drawn upside down if a point is specified at a location to the left of the start point.

As a response to the **Enter text** prompt, enter the text string. Spaces are allowed between words. After entering the text, press ENTER.

Justify Option

AutoCAD LT offers different options to align the text. **Alignment** refers to the layout of the text. The main text alignment modes are **left, center,** and **right**. You can align a text using a combination of modes; for example, top/middle/bottom and left/center/right (Figure 7-25). **Top** refers to the line along which lie the top points of the capital letters; **Baseline** refers to the line along which lie their bases. Letters with descenders (such as p, g, y) dip below the baseline to the bottom.

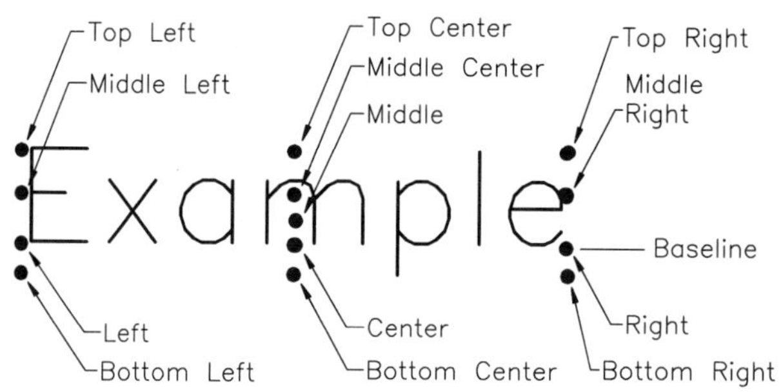

Figure 7-25 *Text alignment positions*

When the **Justify** option is invoked, the user can place text in one of the fourteen various alignment types by selecting the desired alignment option. The orientation of the text style determines the command interaction for Text Justify. (Text styles and fonts are discussed later in this chapter). For now, assume that the text style orientation is horizontal. The prompt sequence that will follow when you choose the **Single Line Text** button to use this option is:

Specify start point of text or [Justify/style]: **J** Enter
Enter an option [Align/Fit/Center/Middle/Right/TL/TC/TR/ML/MC/MR/BL/BC/ BR]: *Select any of these options.*

If the text style is vertically oriented (refer to the "**STYLE** Command" section later in this chapter), only four alignment options are available. The prompt sequence is as follows:

Specify start point of text or [Justify/Style]: **J** Enter
Enter an option [Align/Center/Middle/Right]:

If you know what justification you want, you can enter it directly at the **Specify start point of text or [Justify/Style]** prompt instead of first entering J to display the justification prompt. If you need to specify a style as well as a justification, you must specify the style first. The different alignment options are as follows.

Align option. In this option the text string is written between two points (Figure 7-26). You must specify the two points that act as the endpoints of the baseline. The two points may be specified horizontally or at an angle. AutoCAD LT adjusts the text width (compresses or expands) so that it fits between the two points. The text height is also changed, depending on the distance between points and number of letters.

Specify start point of text or [Justify/Style]: **J** Enter
Enter an option [Align/Fit/Center/Middle/Right/TL/TC/TR/ML/MC/MR/BL/BC/BR]: **A** Enter
Specify first endpoint of text baseline: *Specify a point.*
Specify second endpoint of text baseline: *Specify a point.*
Enter text: *Enter the text string.*

Fit option. This option is very similar to the previous one. The only difference is that in this case you select the text height, and it does not vary according to the distance between the two points. AutoCAD LT adjusts the letter width to fit the text between the two given points, but the height remains constant (Figure 7-26). The Fit option is not accessible for vertically oriented text styles. If you try the Fit option on the vertical text style, you will notice that the text string does not appear in the prompt. The prompt sequence is:

Enter an option [Align/Fit/Center/Middle/Right/TL/TC/TR/ML/MC/MR/BL/BC/BR]: **F**
Specify first endpoint of text baseline: *Specify a point.*
Specify second endpoint of text baseline: *Specify a point.*
Specify height<current>: *Enter the height.*
Enter text: *Enter the text.*

Note

*You do not need to select the **Justify** option (J) for selecting the text justification. You can enter the text justification by directly entering justification when AutoCAD LT prompts "Specify start point of text or [Justify/Style]:"*

Center option. You can use this option to select the midpoint of the baseline for the text. This option can be invoked by entering **Justify** and then **Center** or **C**. After you select or specify the center point, you must enter the letter height and the rotation angle (Figure 7-26).

Specify start point of text or [Justify/Style]: C Enter
Specify center point of text: *Specify a point.*
Specify height<current>: **0.15** Enter
Specify rotation angle of text<0>: Enter
Enter text: **CENTER JUSTIFIED TEXT** Enter

Middle option. Using this option you can center text not only horizontally, as with the previous option, but also vertically. In other words, you can specify the middle point of the text string (Figure 7-26). You can alter the text height and the angle of rotation to your requirement. The prompt sequence that will follow when you choose the **Single Line Text** button from the **Text** toolbar is:

Specify start point of text or[Justify/Style]: **M** Enter
Specify middle point of text : *Specify a point.*
Specify height<current>: **0.15** Enter
Specify rotation angle of text<0>: Enter
Enter text: **MIDDLE JUSTIFIED TEXT** Enter

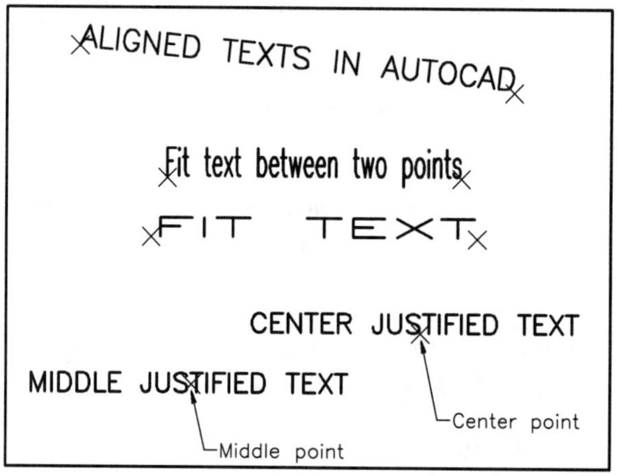

*Figure 7-26 Writing the text using **Align**, **Fit**, **Center**, and **Middle** options*

Right Option. This option is similar to the default left-justified Start point option. The only difference is that the text string is aligned with the lower right corner (the endpoint you specify); that is, the text is **right-justified** (Figure 7-27). The prompt sequence that will follow when you choose this button is:

Specify start point of text or [Justify/Style]: R Enter
Specify right endpoint of text baseline : *Specify a point.*
Specify height<current>: **0.15** Enter
Specify rotation angle of text<0>: Enter
Enter text: **RIGHT JUSTIFIED TEXT** Enter

TL option. In this option the text string is justified from the **top left** (Figure 7-27). The prompt sequence is:

Specify start point of text or[Justify/Style]: **TL** Enter
Specify top-left point of text : *Specify a point.*
Specify height<current>: **0.15** Enter
Specify rotation angle of text <0>: Enter

Enter text: **TOP/LEFT JUSTIFIED TEXT**

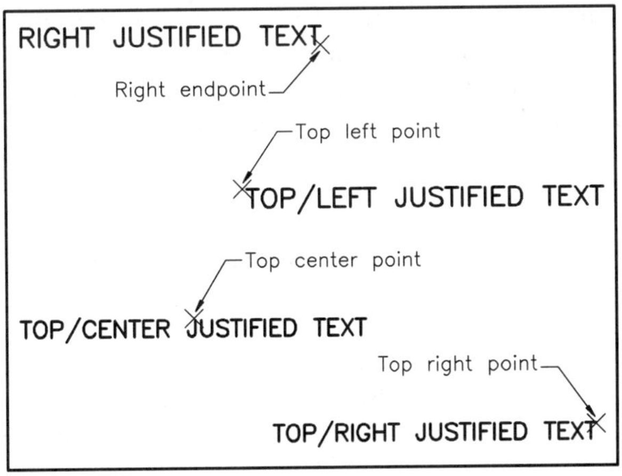

Figure 7-27 *Writing text using the Right, Top-Left, Top-Center, and Top-Right options*

Note

The rest of the text alignment options are similar to those just discussed, and you can try them on your own. The prompt sequence is almost the same as those given for the previous examples.

Style Option

With this option you can specify another existing text style. Different text styles can have different text fonts, heights, obliquing angles, and other features. This option can be invoked by entering **TEXT** and then S at the next prompt. The prompt sequence that will follow when you choose the **Single Line Text** button for using this option is:

Specify start point of text or [Justify/Style]: **S**
Enter style name or [?] <current>:

If you want to work in the previous text style, just press ENTER at the last prompt. If you want to activate another text style, enter the name of the style at the last prompt. You can also choose from a list of available text styles, which can be displayed by entering ?. After you enter ?, the next prompt is:

Enter text style(s) to list <*>:

Press ENTER to display the available text style names and the details of the current styles and commands in the **AutoCAD LT Text Window**.

Note

*With the help of the **Style** option of the **TEXT** command, you can select a text style from an existing list. If you want to create a new style, use the **STYLE** command, which is explained later in this chapter.*

DRAWING SPECIAL CHARACTERS

In almost all drafting applications, you need to draw **special characters** (symbols) in the normal text and in the dimension text. For example, you may want to draw the degree symbol (°) or the diameter symbol (ø), or you may want to underscore or overscore some text. This can be achieved with the appropriate sequence of control characters (control code). For each symbol, the control sequence starts with a percent sign written twice (%%). The character immediately following the double percent sign depicts the symbol. The control sequences for some of the symbols are:

Control sequence	Special character
%%c	Diameter symbol (ø)
%%d	Degree symbol (°)
%%p	Plus/minus tolerance symbol (±)
%%o	Toggle for overscore mode on/off
%%u	Toggle for underscore mode on/off
%%%	Single percent sign (%)

For example, if you want to draw **25° Celsius**, you need to enter 25%%dCelsius. If you enter **43.0%%c**, you get **43.0ø** on the drawing screen. To underscore (underline) text, use the **%%u** control sequence followed by the text to be underscored. For example, to underscore the text: **UNDERSCORED TEXT IN AUTOCAD LT**, enter **%%uUNDERSCORED TEXT IN AUTOCAD LT** at the prompt asking for text to be entered. To underscore and overscore a text string, include **%%u%%o** at the text string.

Note

*The special characters %%O and %%U act as toggles. For example, if you enter "**This %%Utoggles%%U the underscore**", the word **toggles** will be underscored (toggles).*

None of these codes will be translated in the **TEXT** command until this command is complete. For example, to draw the degree symbol, you can enter **%%d**. As you are entering these symbols, they will appear as %%d on the screen. After you have completed the command and pressed ENTER, the code %%d will be replaced by the degree symbol (°).

You may be wondering why a percent sign should have a control sequence when a percent sign can easily be entered at the keyboard by pressing the percent (%) key. The reason is that sometimes a percent symbol is immediately followed (without a space) by a control sequence. In this case, the **%%%** control sequence is needed to draw a single percent symbol. To make the concept clear, assume you want to draw **67%±3.5.** Try drawing this text string by entering **67%%%p3.5.** The result will be **67%p3.5,** which is wrong. Now enter **67%%%%%p3.5** and notice the result on the screen. Here you obtain the correct text string on the screen, that is, **67%±3.5**. If there were a space between the 67% and ±3.5, you could enter 67% %%p3.5 and the result would be **67% ±3.5**.

In addition to the control sequences shown earlier, you can use the %%nnn control sequence to draw special characters. The nnn can take a value in the range of 1 to 126. For example, to draw the & symbol, enter the text string **%%038**.

CREATING PARAGRAPH TEXT

Toolbar:	Text > Multiline Text
Menu:	Draw > Text > Multiline Text
Command:	MTEXT

Figure 7-28 *Invoking* ***Multiline Text*** *from the* ***Text*** *toolbar*

You can use the **MTEXT** command to write a paragraph text whose width can be specified by defining two corners of the text boundary or by entering a width, using coordinate entry (Figure 7-28). The text created by the **MTEXT** command is a single object regardless of the number of lines it contains. After specifying the width, you have to enter the text in the **Multiline Text Editor** dialog box (Figure 7-29). The following is the prompt sequence:

Command: **MTEXT** [Enter]
Current text style: STANDARD. Text height: 0.2000
Specify first corner: *Select a point to specify first corner.*
Specify opposite corner or [Height/Justify/Line spacing/Rotation/Style/Width]: *Select an option or select a point to specify other corner.*

After selecting the first corner you can drag the pointing device so that a box that shows the location and size of the paragraph text is formed. An arrow is displayed within the boundary, which indicates the direction of the text flow. When you define the text boundary, it does not mean that the text paragraph will fit within the defined boundary. AutoCAD LT only uses the width of the defined boundary as the width of the text paragraph. The height of the text boundary has no effect on the text paragraph. Once you have defined the boundary of the paragraph text, AutoCAD LT will display the **Multiline Text Editor** dialog box (Figure 7-29).

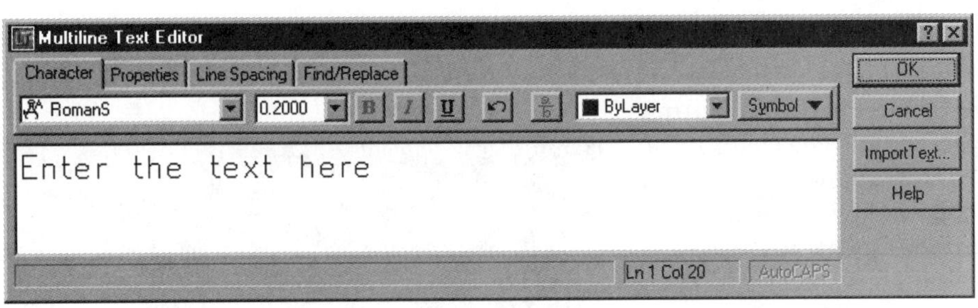

Figure 7-29 ***Multiline Text Editor*** *dialog box (Character tab)*

Note

Although the box boundary that you specify controls the width of the paragraph, a single word is not broken to adjust inside the boundary limits. This means that if you write a single word whose width is more than the box boundary specified, AutoCAD LT will write the word irrespective of the box width, and therefore, will exceed the boundary limits.

In addition to a Text Box, the **Multiline Text Editor** dialog box has four tabs: **Character**, **Properties**, **Line Spacing**, and **Find/Replace**. The following is a description of these tabs and the options available in each tab.

Text Box

The Text Box area displays the text you enter. The width of the active text area is determined by the specified width of paragraph text. The initial height of the text box is approximately equal to three lines of text. You can increase the size of the dialog box by dragging the edges of the box. You can also use the scroll bar to move up or down to display the text.

Character Tab

The options provided under this tab are as follows:

Font. If you choose the font drop-down arrow, AutoCAD LT will display a list of different fonts available (Figure 7-30). You can select any font to apply to the text.

Font height. You can enter the new value of the text height in the **Text Height** edit box to change the text height of the entire paragraph text. Once you specify the height, AutoCAD LT retains that value unless you change it. The **MTEXT** height does not affect the size specified for the **TEXT** command.

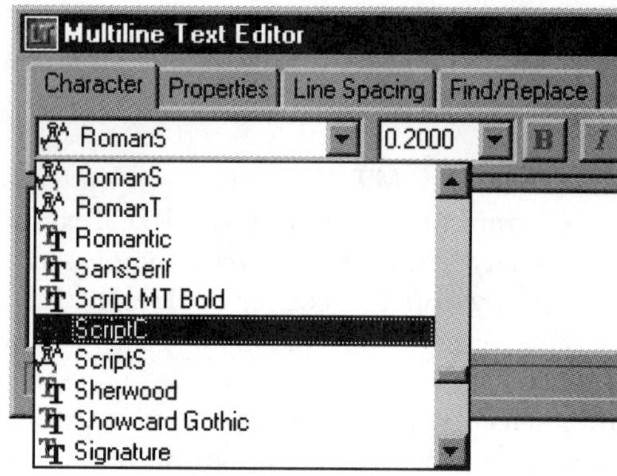

***Figure 7-30 Font** drop-down list*

Bold, italic, underline. You can use the appropriate tool buttons located at the top of the text box to make the selected text boldface, italics, or create underlined text. Boldface and italics are not supported by SHX fonts and hence will not be available for the particular fonts. These three buttons toggle between on and off.

Stacking text. To create a fraction text you must use the stack button with special characters /, ^, and #. The character/stacks the text vertically with a line, and the character ^ stacks the text vertically without a line (tolerance stack). The character # stacks the text with a diagonal line. After you enter the text with the required special character between them, select the text, and then select the **Stack/Unstack** button. The stacked text is displayed equal to 70 percent of the actual height. If you enter two numbers separated by / or ^ and then press the Enter key, AutoCAD LT displays the **AutoStack Properties** dialog box (Figure 7-31). You can use this dialog box to control the stacking properties.

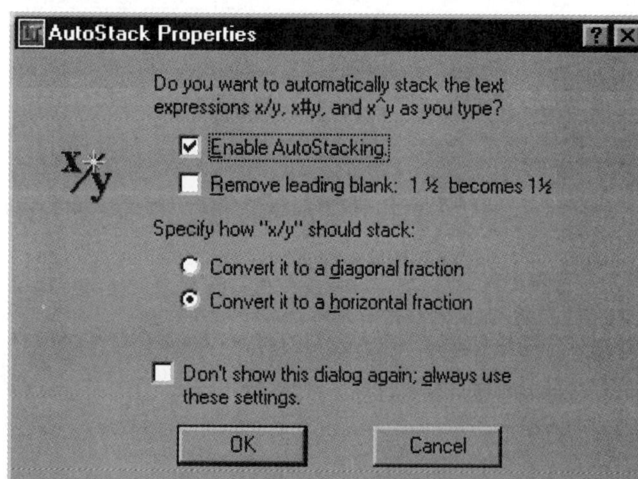

***Figure 7-31 AutoStack Properties** dialog box*

Text color. You can choose the color for text from the Text color list box. You can also choose the color from the **Select Color** dialog box that is displayed by selecting **Other** in the list box.

Symbol. You can select the special characters from the **Symbol** menu that you can insert in the text. You can select **Others** to display **Character Map** dialog box for other special characters. To insert the characters from the dialog box, select the character you want to copy and then choose the **Select** button. Once you are done selecting the special characters, choose the **Copy** button and then close the dialog box. In the Text box of the Multiline Text Editor, position the cursor where you want to insert the special characters and right-click to display the shortcut menu. Select **Paste** to insert the selected special character in the MultiLine Text Editor.

Shortcut Menu

In the Multiline Text Editor, right-click to display the Shortcut menu (Figure 7-32). You have different options to edit the text while you are in the text editor. Select the text that you want to edit and then right-click such that the different editing options in the menu become available. The **Cut** and **Copy** options can be used to move or copy the text from the text editor to any other windows application. Similarly using the **Paste** option, you can paste text from any windows text-based application to the Multiline Text Editor. You can use the **Remove Formatting** option to remove any additional formatting from the selected text. You have options to change the case of the text and join the paragraphs together. When you have a stacked text selected two additional options, **Unstack** and **Properties**, are displayed in the shortcut menu. The Properties option displays the **Stack Properties** dialog box where you can change the text values, size, and the fraction style of the stacked text.

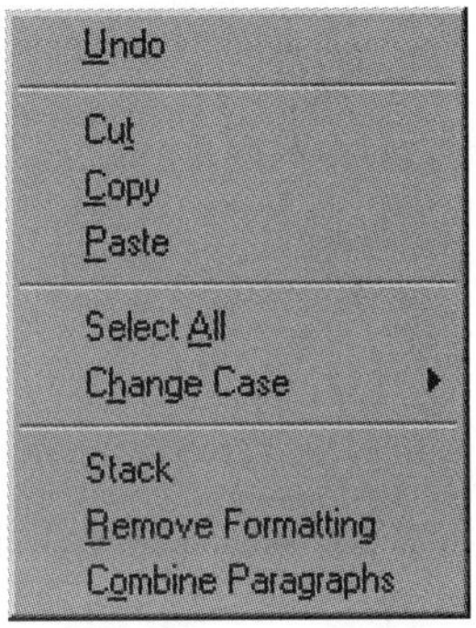

Figure 7-32 Shortcut menu

Properties Tab

From the **Properties** tab (Figure 7-33) you can use the following options that can be applied to the text.

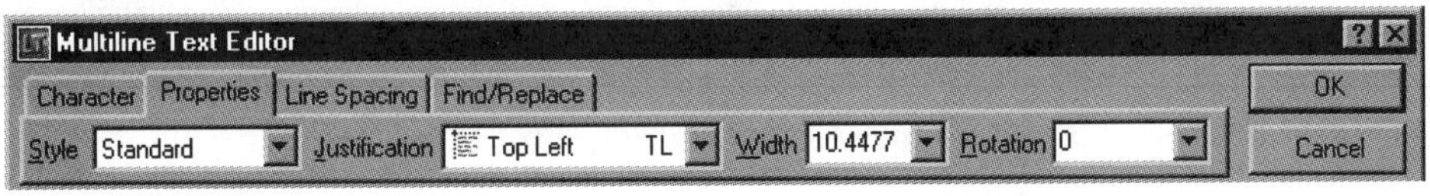

*Figure 7-33 **Multiline Text Editor** dialog box (Properties tab)*

Style. You can select the drop-down arrow to display the predefined styles and select a new text style. The text style that you want to use must be predefined.

Justification. You can choose the drop-down arrow to display the predefined text justifications and then choose a new justification from the list box. The justification is used to control the justification of the text paragraph. For example, if the text justification is bottom-right (BR), the text paragraph will spill to the left and above the insertion point, regardless of how you define the width of the paragraph. The different justifications are TL, ML, BL, TC, MC, BC, TR, MR, BR. Figure 7-34 shows different text justifications for **MTEXT**.

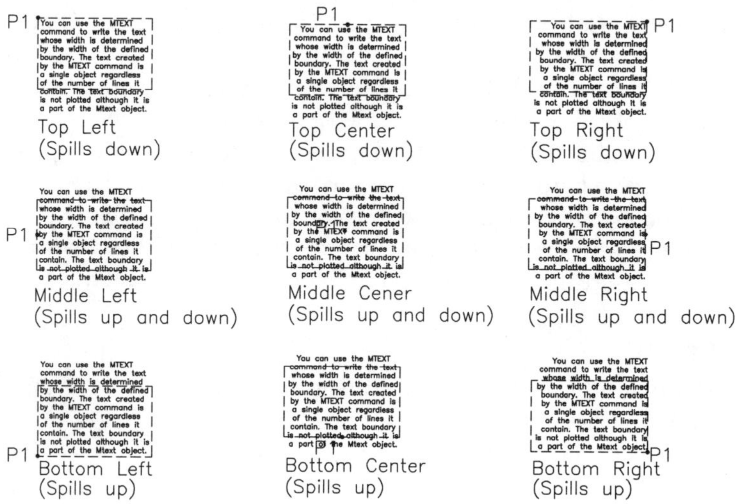

Figure 7-34 Text justifications for ***MTEXT****. P1 is the text insertion point.*

Width. You can enter a new width for the paragraph text or select the width from the previously defined values. The (no wrap) option is the same as entering a width of 0.

Rotation. You can choose a rotation for the text from the drop-down list box. The angles are in 15-degree increments. The rotation angle specifies the rotation of the text. For example, if the rotation angle is 15 degrees, the text will rotate 15 degrees counterclockwise.

Line Spacing Tab

From the **Line Spacing** tab (Figure 7-35) you can use the following options that you can apply to change the text.

Figure 7-35 Multiline Text Editor *dialog box (****Line Spacing*** *tab)*

Line Spacing. This drop-down list has two options. **Atleast** is the default setting in AutoCAD LT where spacing is based upon the height of the largest character in the line, while the **Exactly** option forces the line spacing for all the lines to be uniform. This can be used effectively for making the tables.

Single[1.0x]. This option determines the vertical distance between the two lines in multiline text (Figure 7-36). Single gap is the default setting. Single line spacing is equal to 1.66 times the height of the text. The available options include Single (1x), 1.5 lines (1.5x), and Double (2x). You can also enter an absolute value for line spacing in the range of 0.0833 (0.25x) to 1.333 (4x) by entering a value in the edit box.

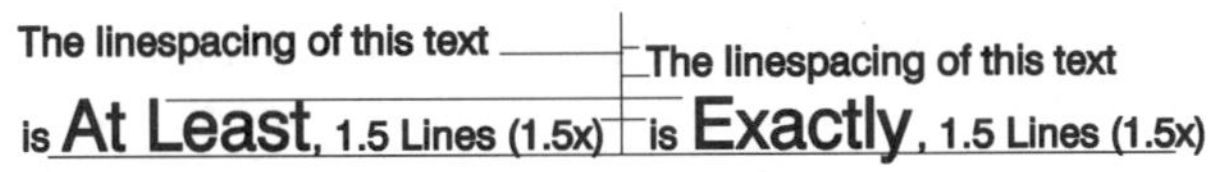

Figure 7-36 Specifying linespacing in multiline text

Find/Replace Tab

From the **Find/Replace** tab (Figure 7-37) you can use the following options that you can apply to the text.

***Figure 7-37 Multiline Text Editor** dialog box (**Find/Replace** tab)*

Find. In this edit box, you can define the text string to be searched. You can locate a part of a word or the complete word.

Replace with. If you want some text to be replaced, enter new text for replacement in this edit box.

Replace button. Select this button to replace the highlighted text with the text in the **Replace with** edit box.

Match Case. Activate this check box if you want to find the text, only if all characters in text object are identical to the text in the Find box.

Whole Word. Activate this check box if you want to match the text in the Find box, only if it is a single complete word.

Import Text

When you choose this option, AutoCAD LT displays the **Select File** dialog box. In this dialog box you can select any text file you want to import in the **Multiline Text Editor** dialog box. The imported text is displayed in the text area. Note that only the ASCII or RTF files are interpreted properly.

Example 1 *General*

In this example you will use the **Multiline Text Editor** dialog box to write the following text on the screen.

For long, complex entries, create multiline text using the MTEXT option. The angle is 10 degrees. Diameter = 1/2" and Length = 32 1/2".

The font of the text is **Swis721 BT,** text height is 0.20, color Red, and written at an angle of 10 degrees with Middle-Left justification. Make the word "multiline" bold, underline the text "multiline text", and make the word "angle" italic. The line spacing type and line spacing between the lines are AtLeast and 1.5x respectively. Use the symbol for degrees and replace the word "option" with "command".

1. The first step is to choose the **Multiline Text** button from the **Draw** toolbar. You can also invoke this command from the **Draw** menu or enter the command at the Command prompt. After invoking the command, specify two corners on the screen to define the paragraph text boundary. The prompt sequence is :

   Current text style: STANDARD. Text height: 0.1000
   Specify first corner: *Select a point to specify first corner.*
   Specify opposite corner or [Height/Justify/Line spacing/Rotation/ Style/ Width]: *Select a point to specify other corner.*
   The **Multiline Text Editor** dialog box is displayed with the **Character** tab activated (default).

2. Choose the arrow in the **Font** list box to display the drop-down list. Select **Swis721 BT** true type font from the list.

3. Enter **0.20** in the **Font height** edit box.

4. Choose the **Text Color** swatch button to display the list box and select Red from the list.

5. Now enter the text in the **Multiline Text Editor** (Figure 7-38). To put the degrees symbol, choose the **Symbol** button to display the menu and select **Degrees.** When you type 1/2 and then press the SPACEBAR, AutoCAD LT displays the **AutoStack Properties** dialog box. Select **Convert it to a diagonal fraction**, if it is not already selected, and then choose **OK**. Make sure the **Enable AutoStacking** check box is checked. You can also use the # character instead of the / character. Highlight the numbers and the # character and then choose the **Stack/Unstack** button. This stacks the text with a diagonal line without using the **AutoStack Properties** dialog box.

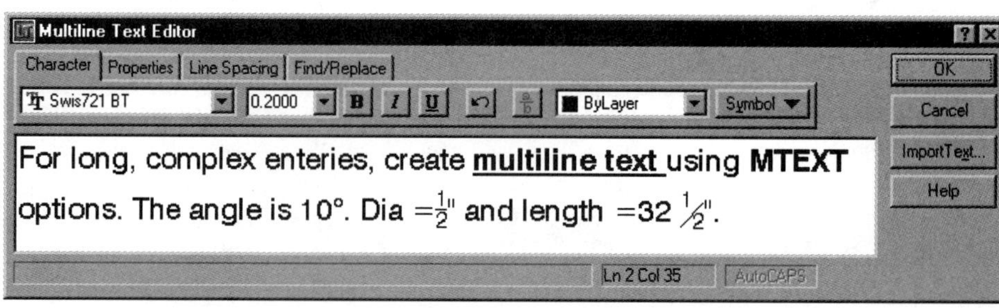

***Figure 7-38 Multiline Text Editor** dialog box (**Character** tab)*

6. Double-click on the word "multiline" to select it and then choose the **Bold** button to make it boldface. As it is already selected select the **Underline** button to underline it.

7. Similarly, highlight the word "text" and then select the **Underline** button to underline it.

8. Highlight on the word "angle" by double-clicking on it (or pick and drag to select the text) and then choose the **Italic** button (Figure 7-38).

9. Select the **Properties** tab to display its options.

10. From the **Justification** drop-down list choose **ML**, and from the **Rotation** drop-down list choose **10**. You can also enter 10 in the edit box if it is not listed.

11. Select the **Line Spacing** tab to change the line spacing.

12. In the **Line spacing type** drop-down list choose the **At Least** option if not already selected and in the **Spacing** drop-down list choose the **1.5 Lines (1.5x)** option.

13. Now select the **Find/Replace** tab (Figure 7-39) to display its different options.

14. In the **Find** edit box, enter **option** and in the **Replace** edit box, enter **command**.

15. Choose the **Find** button. AutoCAD LT finds the word "option" (Figure 7-39). Choose the **Replace** button to replace **option** by **command**.

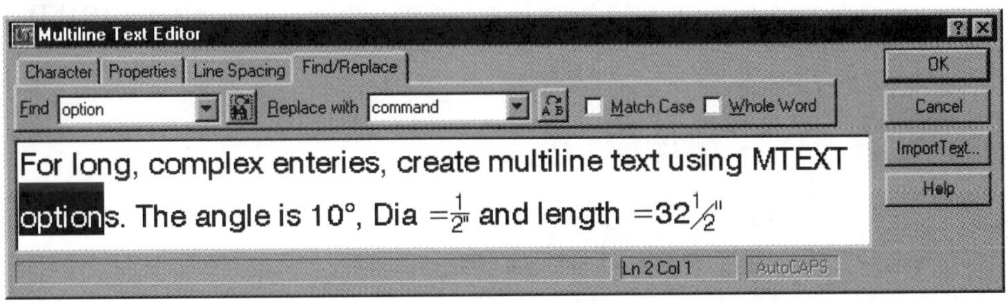

***Figure 7-39 Multiline Text Editor** dialog box (**Find/Replace** tab)*

16. Choose the **OK** button to exit the dialog box and the text is displayed in the selected width on the screen (Figure 7-40).

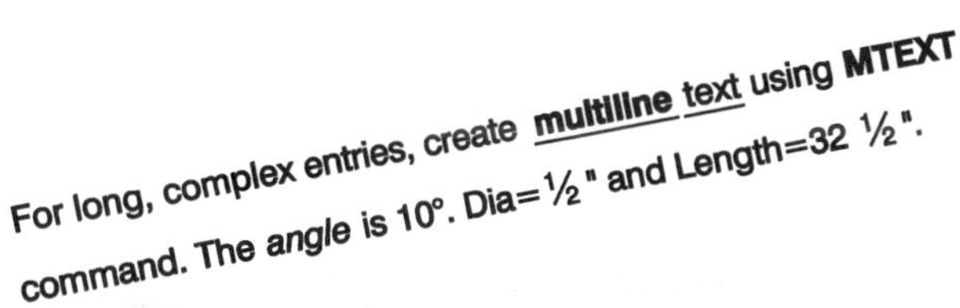

***Figure 7-40** Multiline text for Example 1*

Tip

*The text in the **Multiline Text Editor** can be selected by double-clicking on the word to select a word, by holding down the pick button of the pointing device and then dragging the cursor, or by triple-clicking on the text to select the entire line or paragraph.*

Exercise 2 *General*

Draw the text on the screen as shown in Figure 7-41 and Figure 7-42. Use the special characters and text justification options shown in the drawing using the **TEXT** and **MTEXT** commands. The text height is 0.1 and 0.15 for Figure 7-41 and Figure 7-42 respectively.

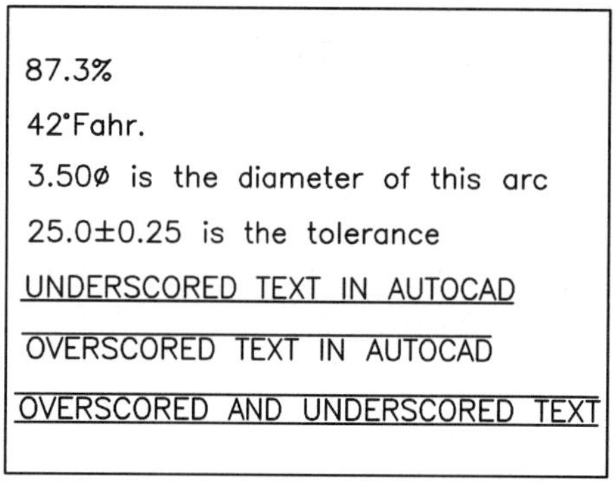

Figure 7-41 *Drawing with special characters*

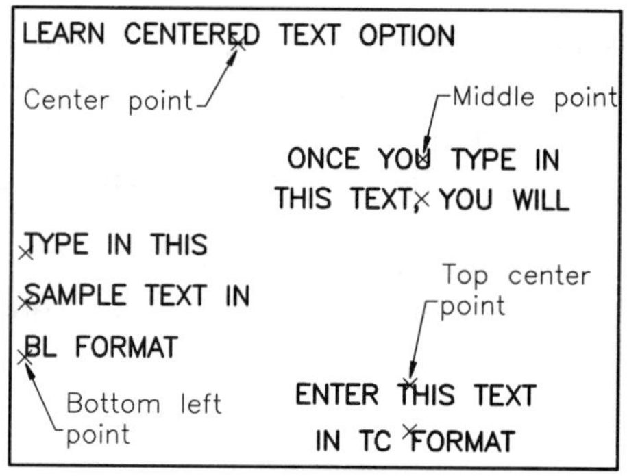

Figure 7-42 *Drawing for Exercise 2*

EDITING TEXT

The contents of **MTEXT**, and **TEXT**, object can be edited by using the **DDEDIT** and **PROPERTIES** commands. You can also use the AutoCAD LT editing commands, such as **MOVE**, **ERASE**, **ROTATE**, **COPY**, **MIRROR**, and **GRIPS** with any text object.

Editing Text Using DDEDIT Command

Toolbar:	Text > Edit Text
Menu:	Modify > Object > Text > Edit
Command:	DDEDIT

You can use the **DDEDIT** command to edit the text. If you select **TEXT** object, AutoCAD LT displays the **Edit Text** dialog box (Figure 7-43), where the selected text is displayed in the **Text:** edit box. You can make the changes in the text string only and then choose the **OK** button. For the TEXT object you cannot modify any of its properties in the **Edit Text** dialog box. If you select the paragraph text created by the **MTEXT** command, AutoCAD LT displays the text in the **Multiline Text Editor** dialog box. You can make the changes using the different options in the dialog box. Apart from changing the text string you are also allowed to change the properties of the paragraph text.

Figure 7-43 *Using the **Edit Text** dialog box to edit text*

Command: **DDEDIT**
Select an annotation object or [Undo]: *Select a text object*

You can also invoke the **Edit Text** dialog box and the **Multiline Text Editor** from the shortcut menu. Select the text for editing and then right-click in the drawing area. A shortcut menu is displayed and depending upon the text object you have selected, the options Text Edit or Mtext Edit are present in the menu. Selecting the options displays the respective dialog boxes.

Tip
*You can simply double-click on a TEXT object or a MTEXT object to display the **Edit Text** dialog box or the **Multiline Text Editor** respectively without selecting or entering the **DDEDIT** command.*

Editing Text Using the PROPERTIES Window

Using the **DDEDIT** command with the TEXT object, you can only change the text string and not its properties of the height, angle, and so on. You can use the **PROPERTIES** command for changing the properties. Select the text and invoke the **PROPERTIES** command (**Properties** button in the **Standard** toolbar). AutoCAD LT displays the **Properties** window (Figure 7-44) with all the properties of the selected text. Here you can change any value and also the text string. If you are editing a single line text you can change the text string in the window, but for a multiline text you must choose the **Full editor** button in the **Contents** edit box. AutoCAD LT automatically switches to the **Multiline Text Editor** dialog box, where you can make changes to the paragraph text.

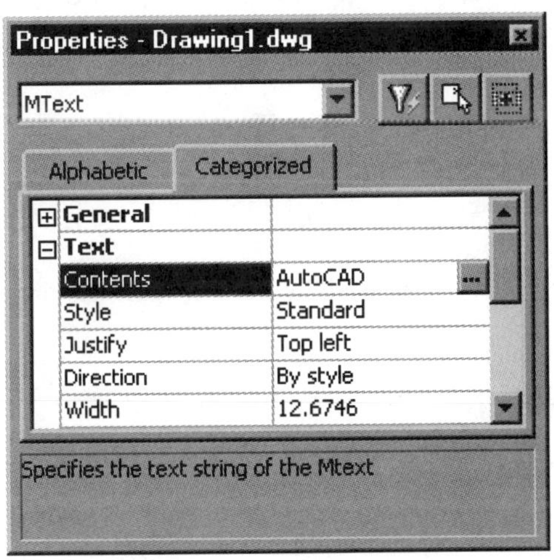

Figure 7-44 ***Properties*** *window*

SUBSTITUTING FONTS

AutoCAD LT provides the facility to designate the fonts that you want to substitute for other fonts used in the drawing. The information about font mapping is specified in the font mapping file (ACLT.FMP). The font mapping has the following advantages:

1. You can specify a font to be used when AutoCAD LT cannot find a font used in the drawing.

2. You can enforce the use of a particular font in your drawings. If you load a drawing that uses different fonts, you can use the font mapping to substitute the desired font for the fonts used in the drawing.

3. You can use .shx fonts when creating or editing a drawing. When you are done and ready to plot the drawing, you can substitute other fonts for .shx fonts.

The font mapping file is an ASCII file with FMP extension containing one font mapping per line. The format on the line is:

Base name of the font file;Name of the substitute font with extension (ttf, shx, etc.)

For example, if you want to substitute the ROMANC font for SWISS.TTF; the entry is:

SWISS;ROMANC.SHX

You can enter this line in the ACLT.FMP file or create a new file. If you create a new font mapping file, you need to specify this new file. You can use the **Options** dialog box to specify the new font map file. To specify a font mapping table in the **Options** dialog box, choose **Options** from the **Tools** menu to display the **Options** dialog box. Choose the **Files** tab and click on the **plus** sign next to **Text Editor, Dictionary, and Font File Names**. Now, click on the plus sign next to **Font Mapping File** to display the path and the name of the font mapping file. Double-click on the file to display the **Select a file** dialog box. Select the new font mapping file and exit the **Options** dialog box. At the Command prompt enter **REGEN** to convert the existing text font to the font as specified in the new font mapping file. You can also use the **FONTMAP** system variable to specify the new font map file.

Command: **FONTMAP** Enter
Enter new value for FONTMAP, or . for none <"Aclt.fmp">: *Enter the name of the font mapping file.*

The following file is a partial listing of ACLT.FMP file with the new font mapping line added (swiss;romanc.shx).

swiss;romanc.shx
cibt;CITYB___.TTF
cobt;COUNB___.TTF
eur;EURR____.TTF
euro;EURRO___.TTF
par;PANROMAN.TTF
rom;ROMANTIC.TTF
romb;ROMAB___.TTF
romi;ROMAI___.TTF
sas;SANSS___.TTF
sasb;SANSSB__.TTF
sasbo;SANSSBO_.TTF
saso;SANSSO__.TTF

Note

The text styles that were created using the PostScript fonts are substituted with an equivalent TrueType font and plotted using the substituted font.

Specifying an Alternate Default Font

When you open a drawing file that specifies a font file that is not on your system or is not specified in the font mapping file, AutoCAD LT, by default, substitutes the simplex.shx font file. You can specify a different font file in the **Options** dialog box or by changing the **FONTALT** system variable.

Command: **FONTALT** [Enter]
New value for FONTALT, or . for none <"simplex.shx">: *Enter the font file name.*

STYLE COMMAND

Toolbar:	Text > Text Style
Menu:	Format > Text Style
Command:	STYLE

You can change the current text styles in the **MTEXT** or **TEXT** commands using the **Style** option, but to create new text styles and modify the existing style you need the **STYLE** command. Choose this button to display the **Text Style** dialog box (Figure 7-45).

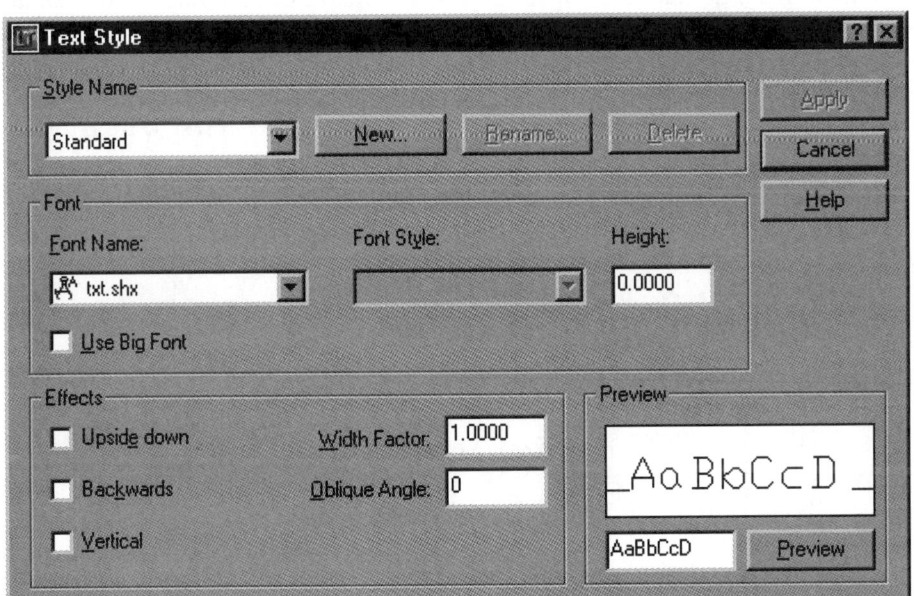

***Figure 7-45** **Text Style** dialog box*

In the **Style Name** edit box the default style (STANDARD) will be displayed. For creating a new style, choose the **New** button to display the **New Text Style** dialog box (Figure 7-46) and enter the name of the style you want to create.

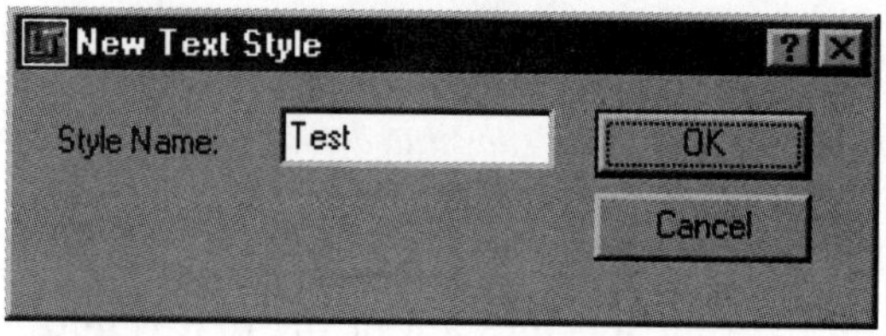

***Figure 7-46** New Text Style dialog box*

A new style having the entered name and the properties present in the **Text Style** dialog box will be created. To modify this style, select the style name from the list box and then change the different settings by entering new values in the appropriate boxes. You can change the font by selecting a new font from the **Font** drop-down list. Similarly, you can change the text height, width, and oblique angle. With the **MTEXT** command you can modify all the predefined properties of that style for the present MTEXT object using the **Multiline Text Editor**. With the **TEXT** command, if you have already specified the height of the text in the Text Style dialog box, AutoCAD LT will not prompt you to enter the text height. The text will be created using the height specified in the text style. If you want AutoCAD LT to prompt you for the text height, specify 0 text height in the dialog box.

For **Width factor,** 1 is the default. If you want the letters expanded, give a width factor greater than 1; for compressed letters, give a width factor less than 1. Similarly, for the **Oblique angle,** 0 is the default. If you want the slant of the letters toward the right, the value should be greater than 0; to slant the letters toward the left, the value should be less than 0. You can also force the text to be written upside down (Figure 7-47), backwards and vertically by checking their respective check boxes. As you make the changes, you can see their effect in the **Preview** box. After making the desired changes, choose the **Apply** button and the **Close** button to exit the dialog box. You can also use the **-STYLE** command to define a new test style from the command line.

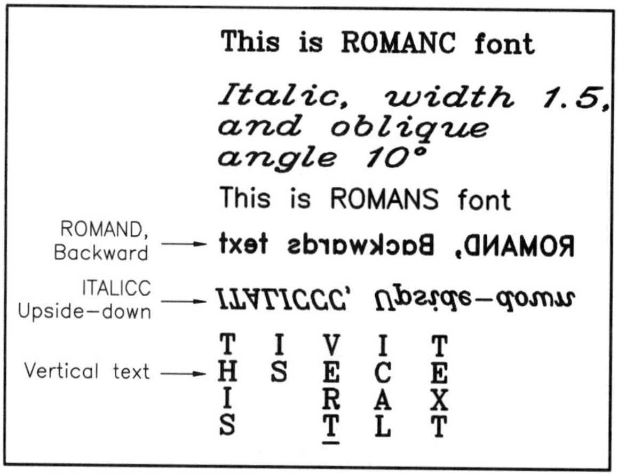

***Figure 7-47** Specifying different features to text style files*

DETERMINING TEXT HEIGHT

The actual text height is equal to the product of the **scale factor** and the **plotted text height**. Therefore, scale factors are important numbers for plotting the text at the correct height. This factor is a reciprocal of the drawing plot scale. For example, if you plot a drawing at a scale of ¼ = 1, you calculate the scale factor for text height as follows:

¼" = 1" (i.e., the scale factor is 4)

The scale factor for an architectural drawing that is to be plotted at a scale of ¼" = 1'0" is calculated as:

¼" = 1'0", or ¼" = 12", or 1 = 48

Therefore, in this case, the scale factor is 48.

For a civil engineering drawing with a scale 1"= 50', the scale factor is as follows:

1" = 50', or 1" = 50X12", or 1 = 600

Therefore, the scale factor is 600.

Next, calculate the height of the AutoCAD LT text. If it is a full-scale drawing (1=1) and the text is to be plotted at 1/8" (0.125), it should be drawn at that height. However, in a civil engineering drawing a text drawn 1/8" high will look like a dot. This is because the scale for a civil engineering drawing is 1"= 50', which means that the drawing you are working on is 600 times larger. To draw a normal text height, multiply the text height by 600. Now the height will be:

0.125" x 600 = 75

Similarly, in an architectural drawing, which has a scale factor of 48, a text that is to be 1/8" high on paper must be drawn 6 units high, as shown in the following calculation:

0.125 x 48 = 6.0

It is very important to evaluate scale factors and text heights before you begin a drawing. It would be even better to include the text height in your prototype drawing by assigning the value to the **TEXTSIZE** system variable.

CHECKING SPELLING

Menu:	Tools > Spelling
Command:	SPELL

You can check the spelling of text (text generated by the **TEXT** or **MTEXT** commands) by using the **SPELL** command. The prompt sequence is:

Command: **SPELL**
Select object: *Select the text for spell check or enter ALL to select all text objects.*

If no misspelled words are found in the selected text, then AutoCAD LT displays a message. If the spelling is not correct for any word in the selected text, AutoCAD LT displays the **Check Spelling** dialog box (Figure 7-48). The misspelled word is displayed under **Current word**, and correctly spelled alternate words are listed in the **Suggestions** box. The dialog box also displays the misspelled word with the surrounding text in the **Context** box. You may select a word from the list, ignore the correction, and continue with the spell check, or accept the change. To add the listed word in the custom dictionary, choose the **Add** button. The dictionary can be changed from the **Check Spelling** dialog box or by specifying the name in the **DCTMAIN** or **DCTCUST** system variables. Dictionary must be specified for spell check.

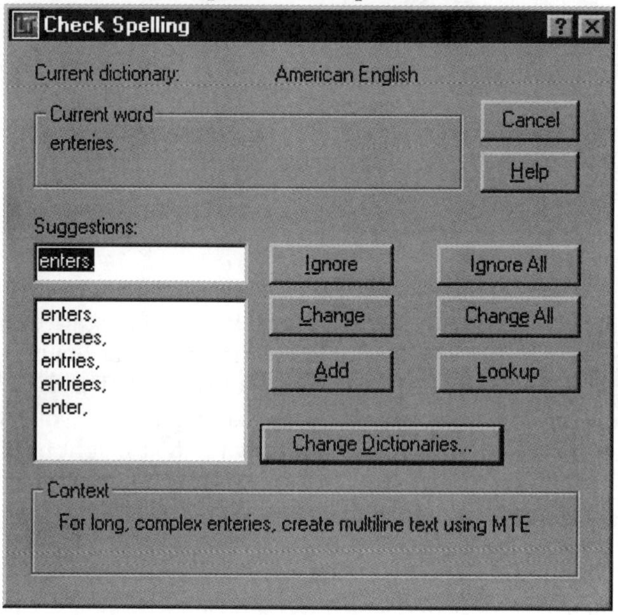

***Figure 7-48 Check Spelling** dialog box*

Note

*You can also rename and use the Word 97 dictionary or any other dictionary. Choose the **Change Dictionaries** button to display the **Change Dictionaries** dialog box and enter the new dictionary name with the .cus extension.*

TEXT QUALITY AND TEXT FILL

AutoCAD LT supports **TrueType fonts**. You can use your own TrueType fonts by adding them to the Fonts directory. You can also keep your fonts in a separate directory, in which case you must specify the location of your fonts directory in the AutoCAD LT search path.

The resolution and text fill of the TrueType font text is controlled by the **TEXTFILL** and **TEXTQLTY** system variables. If **TEXTFILL** is set to 1, the text will be filled. If the value is set to 0, the text will not be filled. On the screen the text will appear filled, but when it is plotted, the text will not be filled. The **TEXTQLTY** variable controls the quality of the TrueType font text. The value of this variable can range from 0 to 100. The default value is 50, which gives a resolution of 300 dpi (dots per inch). If the value is set to 100, the text will be drawn at 600 dpi. The higher the resolution, the more time it takes to regenerate or plot the drawing.

FINDING AND REPLACING TEXT

Toolbar:	Standard Toolbar > Find and Replace
Menu:	Edit > Find
Command:	FIND

You can use the **FIND** command to find and replace the text. The text could be a line text created by **TEXT** command, paragraph text created by **MTEXT** command, dimension annotation text, block attribute value, hyperlinks, or hyperlink description. When you invoke this command, AutoCAD LT displays the **Find and Replace** dialog box, Figure 7-49. You can use this dialog box to perform the following functions.

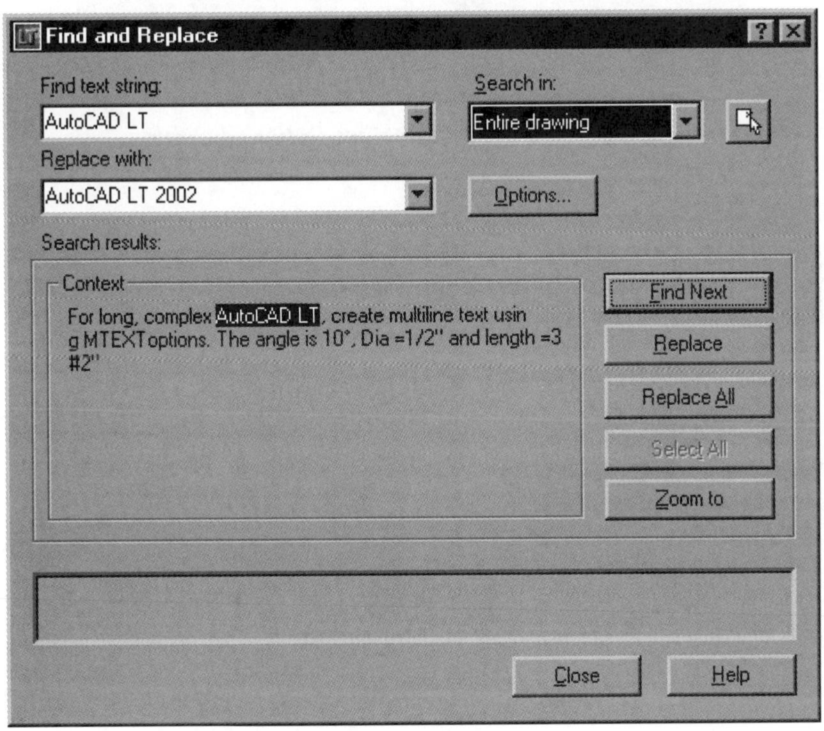

Figure 7-49 ***Find and Replace*** *dialog box*

Find Text

To find the text, enter the text you want to find in the **Find text string** edit box. You can do the search in the entire drawing or confine the search to selected text. To select text, choose the **Select Objects** button. AutoCAD LT temporarily switches to drawing window. Once you are done selecting text, the **Find and Replace** dialog box reappears. In the **Search in** drop-down list you can specify if you want to search the entire drawing or current selection. If you choose the Options button, AutoCAD LT displays the **Find and Replace Options** dialog box,

Figure 7-50. In this dialog box you can specify whether to find the whole word and whether to match the case of the specified text. To find the text, choose the **Find** button. AutoCAD LT displays the found text with the surrounding text in the Context area. To find the next occurrence of the text, choose the **Find Next** button.

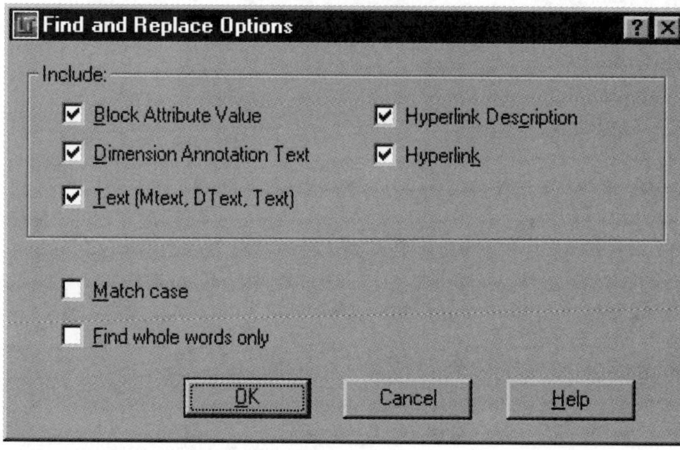

***Figure 7-50 Find and Replace Options** dialog box*

Replacing Text

If you want to replace the specified text with new text, enter the new text in the **Replace with** edit box. Now, if you choose the **Replace** button, only the found text will be replaced. If you choose the **Replace All** button, all occurrences of the specified text will be replaced with the new text.

Self-Evaluation Test

Answer the following questions and then compare your answers to the answers given at the end of this chapter.

1. Transparent commands cannot be used while another command is in progress. (T/F)

2. The **ALL** option of the **ZOOM** command displays the drawing limits or extents, whichever is greater. (T/F)

3. While using the ZOOM scale option with respect to the current view you have to enter a number followed by an **X**. (T/F)

4. You can use the scroll bars to pan the drawing in any direction. (T/F)

5. The **VIEW** command does not save any drawing object data, only the __________ parameters needed to redisplay that portion of the drawing.

6. Multiple lines of text can be entered at any desired locations in the drawing area with a single __________ command.

7. With the __________ justification option of the **TEXT** command, AutoCAD LT adjusts the letter width to fit the text between the two given points, but the height remains constant.

8. In the **MTEXT** command, the height specified in the Multiline Text Editor does not affect the __________ system variable.

9. With the __________ command, if you are editing a single line text you can change the text string in the window, but for a multiline text you must choose the **Full editor** button in the **Content** edit box.

10. You can use the __________ system variable to specify the new font map file.

Review Questions

Answer the following questions.

1. After completion of a transparent command, AutoCAD LT returns you to the Command prompt. (T/F)

2. If the **BLIPMODE** variable is set to On, blip marks do not appear on the screen. (T/F)

3. With the **ZOOM** command, the actual size of the object changes. (T/F)

4. The **REDRAW** command can be used as a transparent command. (T/F)

5. The **TEXT** command does not allow you to see the text on the screen as you type it. (T/F)

6. Which command recalculates all the objects in a drawing and redraws the current viewport only?

   (a) **REDRAW** (b) **REDRAWALL**
   (c) **REGEN** (d) **REGENALL**

7. Which of the following commands cannot be used transparently?

   (a) **ZOOM** (b) **PAN**
   (c) **VIEW** (d) **REDRAW**

8. How many views are saved with the **Previous** option of the **ZOOM** command?

   (a) 6 (b) 8
   (c) 10 (d) 12

9. In the **Multiline Text Editor**, which of the following characters is used to stack the text with a diagonal line without using the **Autostack Properties** dialog box?

   (a) ^ (b) /
   (c) # (d) @

10. Which command can be used to create a new text style and modify the existing ones?

    (a) **TEXT** (b) **MTEXT**
    (c) **STYLE** (d) **SPELL**

11. The four main text alignment modes are __________, __________, __________, and __________.

12. You can use the __________ command to write a paragraph text whose width can be specified by defining the __________ of the text boundary.

13. When the **Justify** option is invoked, the user can place text in one of the __________ various alignment types by selecting the desired alignment option.

14. The text created by the __________ command is a single object regardless of the number of lines it contains.

15. Using the **MTEXT** command, the character __________ stacks the text vertically without a line (tolerance stack).

16. If you want to edit text, select it and then right-click such that the different editing options __________ in the menu become available.

17. You can view the entire drawing (even if it is beyond limits) with the help of the __________ option.

18. In the **ZOOM Window** option, the area inside the window is __________ to completely __________ the current viewport.

Exercises

Exercise 3 — General

Draw the text on the screen as shown in Figure 7-51. Use the text justification that will produce the text as shown in the drawing. Assume a value for text height. Use the **PROPERTIES** window to change the text as shown in Figure 7-52.

Figure 7-51 *Drawing for Exercise 3*

Figure 7-52 *Drawing for Exercise 3 (after changing the text)*

Exercise 4 — General

Draw the text on the screen shown in Figure 7-53. You must first define text style files using the **STYLE** command with the attributes as shown in the drawing. The text height is 0.25 units.

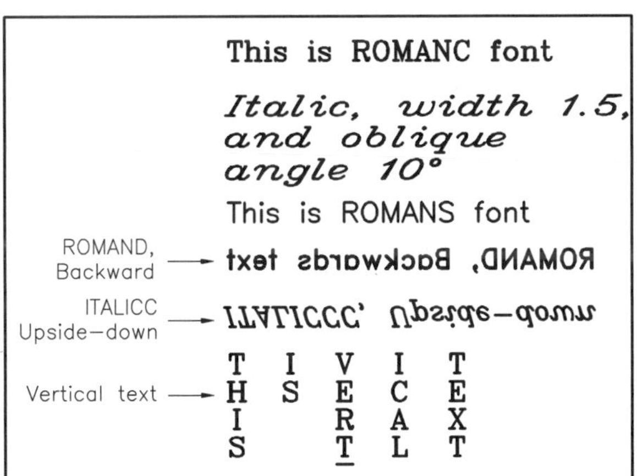

Figure 7-53 *Drawing for Exercise 4*

Exercise 5 *Mechanical*

Draw Figure 7-54 using of the Draw, Edit, and Display commands. Do not dimension the drawing.

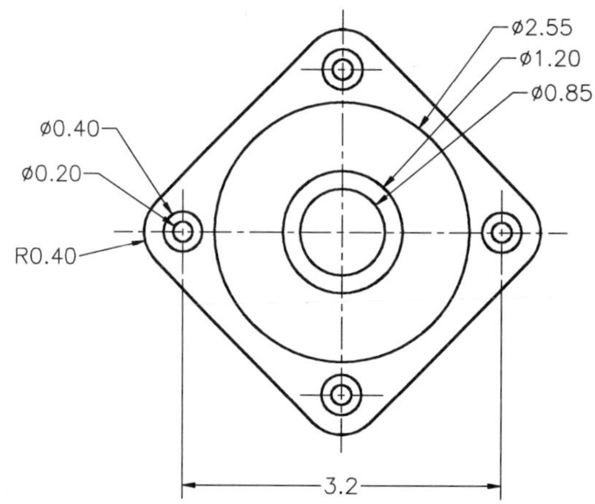

Figure 7-54 *Drawing for Exercise 5*

Exercise 6 *Mechanical*

Draw Figure 7-55 using the **MIRROR** command to duplicate the features that are identical. Use the display commands to facilitate the process. Do not dimension the drawing.

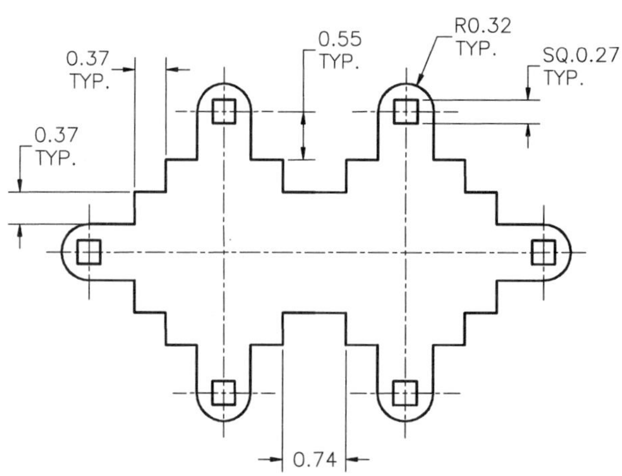

Figure 7-55 *Drawing for Exercise 6*

Exercise 7 *Mechanical*

Draw Figure 7-56. Use the Display commands to facilitate the process. Do not dimension the drawing.

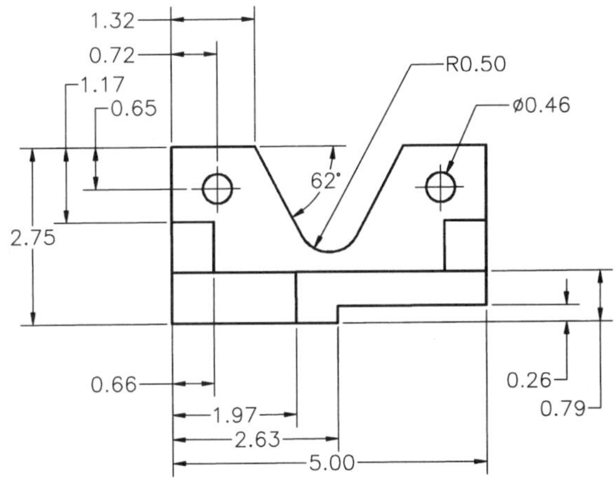

Figure 7-56 Drawing for Exercise 7

Exercise 8 *Mechanical*

Draw Figure 7-57 using the **MIRROR** command to duplicate the features that are identical. Use the Display commands to facilitate the process. Do not dimension the drawing.

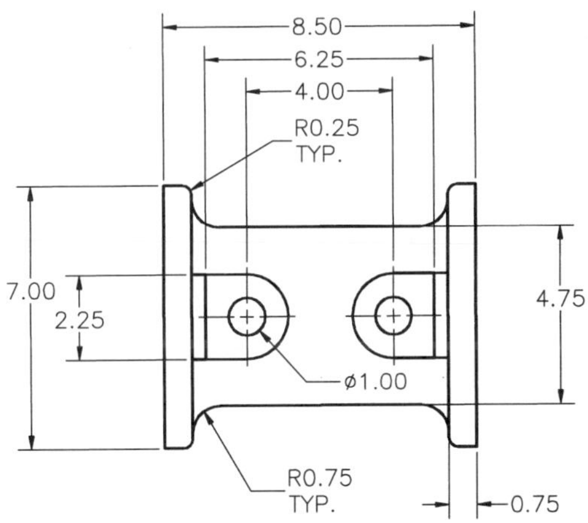

Figure 7-57 Drawing for Exercise 8

Exercise 9 *Architectural*

Draw Figure 7-58 using AutoCAD LT's Draw, Edit, and Display commands. Assume the missing dimensions. Do not dimension the drawing.

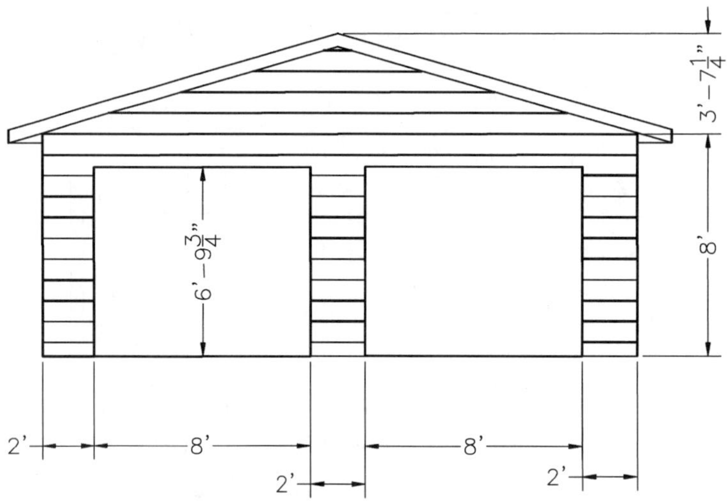

***Figure 7-58** Drawing for Exercise 9*

Answers to Self-Evaluation Test

1 - F, **2** - T , **3** - T, **4** - T , **5** - view , **6** - **TEXT**, **7** - Fit, **8** - TEXTSIZE, **9** - PROPERTIES, **10** - FONTMAP

Chapter 8

Basic Dimensioning, Geometric Dimensioning, and Tolerancing

Learning Objectives

After completing this chapter, you will be able to:

- *Understand the need for dimensioning in drawings.*
- *Understand the fundamental dimensioning terms.*
- *Understand associative dimensioning.*
- *Create different types of dimensions in the drawing.*
- *Create center marks and centerlines.*
- *Attach leaders to the objects.*
- *Use geometric tolerancing, feature control frames, and characteristics symbols.*
- *Combine geometric characteristics and create composite position tolerancing.*
- *Use feature control frames with leaders.*
- *Use the projected tolerance zone.*

NEED FOR DIMENSIONING

To make designs more informative and practical, the drawing must convey more than just the graphic picture of the product. To manufacture an object, the drawing must contain size descriptions such as the length, width, height, angle, radius, diameter, and location of features. All this information is added to the drawing with the help of **dimensioning**. Some drawings also require information about tolerances with the size of features. This information conveyed through dimensioning is vital and often just as important as the drawing itself. With the advances in computer-aided design/drafting and computer-aided manufacturing, it has become mandatory to draw the part to actual size so that the dimensions reflect the actual size of the features. At times it may not be necessary to draw the object of the same size as the actual object would be when manufactured, but it is absolutely essential that the dimensions be accurate. Incorrect dimensions will lead to manufacturing errors.

By dimensioning, you are not only giving the size of a part, you are also giving a series of instructions to a machinist, an engineer, or an architect. The way the part is positioned in a machine, the sequence of machining operations, and the location of different features of the part depend on how you dimension the part. For example, the number of decimal places in a dimension (2.000) determines the type of machine that will be used to do that machining operation. The machining cost of such an operation is significantly higher than for a dimension that has only one digit after the decimal (2.0). Similarly, whether a part is to be forged or cast, the radii of the edges and the tolerance you provide to these dimensions determine the cost of the product, the number of defective parts, and the number of parts you get from a single die.

DIMENSIONING IN AUTOCAD LT

The objects that can be dimensioned in AutoCAD LT range from straight lines to arcs. The dimensioning commands provided by AutoCAD LT can be classified into four categories:

Dimension Drawing Commands
Dimension Style Commands
Dimension Editing Commands
Dimension Utility Commands

While dimensioning an object, AutoCAD LT automatically calculates the length of the object or the distance between two specified points. Also, settings such as a gap between the dimension text and the dimension line, the space between two consecutive dimension lines, arrow size, and text size are maintained and used when the dimensions are being generated for a particular drawing. The generation of arrows, lines (dimension lines, extension lines), and other objects that form a dimension are automatically performed by AutoCAD LT to save the user's time. This also results in uniform drawings. However, you can override the default measurements computed by AutoCAD LT and change the settings of various standard values. The modification of dimensioning standards can be achieved through the dimension variables.

The dimensioning functions offered by AutoCAD LT provide you with extreme flexibility in dimensioning by letting you dimension various objects in a variety of ways. This is of great help because different industries, such as architectural, mechanical, civil, or electrical, have different standards for the placement of dimensions.

FUNDAMENTAL DIMENSIONING TERMS

Before studying AutoCAD LT's dimensioning commands, it is important to know and understand various dimensioning terms that are common to linear, angular, radius, diameter, and ordinate dimensioning. Figure 8-1 and Figure 8-2 shows various dimensioning parameters.

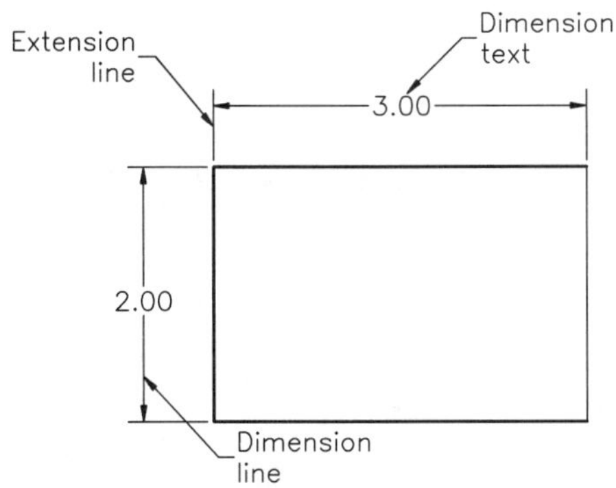

Figure 8-1 *Figure showing various dimension parameters*

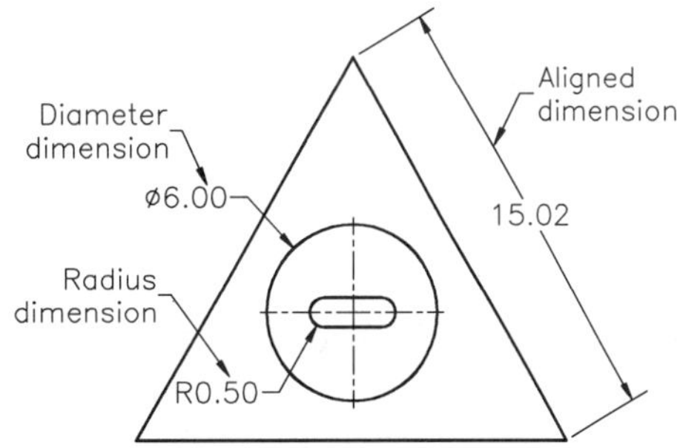

Figure 8-2 *Figure showing various dimension parameters*

Dimension Line

The **dimension line** indicates which distance or angle is being measured. Usually this line has arrows at both ends, and the dimension text is placed along the dimension line. By default the dimension line is drawn between the extension lines (Figure 8-1). If the dimension line does not fit inside, two short lines with arrows pointing inward are drawn outside the extension lines. The dimension line for angular dimensions (which are used to dimension angles) is an arc. You can control the positioning and various other features of the dimension lines by setting the parameters in the dimension styles. (The dimension styles are discussed in Chapter 10, Dimension Styles and Dimensioning System Variables.)

Dimension Text

Dimension text is a text string that reflects the actual measurement (dimension value) between the selected points as calculated by AutoCAD LT. You can accept the value that AutoCAD LT returns or enter your own value. In case you use the default text, AutoCAD LT can be supplied with instructions to append the tolerances to it. Also, you can attach prefixes or suffixes of your choice to the dimension text.

Arrowheads

An **arrowhead** is a symbol used at the end of a dimension line (where dimension lines meet the extension lines). Arrowheads are also called **terminators** because they signify the end of the dimension line. Since the drafting standards differ from company to company, AutoCAD LT allows you to draw arrows, tick marks, closed arrows, open arrows, dots, right angle arrows, or user-defined blocks (Figure 8-3). The user-defined blocks at the two ends of the dimension line can be customized to your requirements. The size

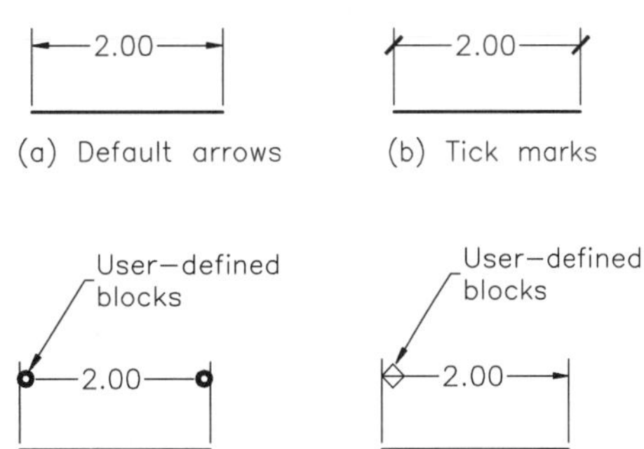

Figure 8-3 *Using arrows, tick marks, and user-defined blocks*

of the arrows, tick marks, user blocks, and so on, can be regulated by using the dimension variables.

Extension Lines

Extension lines are drawn from the object measured to the dimension line (Figure 8-4). These lines are also called **witness lines.** Extension lines are used in linear and angular dimensioning. Generally, extension lines are drawn perpendicular to the dimension line. However, you can make extension lines incline at an angle by using the **DIMEDIT** command (Oblique option) or by selecting **Dimension Edit** from the **Dimension** toolbar. AutoCAD LT also allows you to suppress either one or both extension lines in a dimension (Figure 8-5). Other aspects of the extension line can be controlled by using the dimension variables (these variables are also discussed in Chapter 10, Dimension Styles and Dimensioning System Variables).

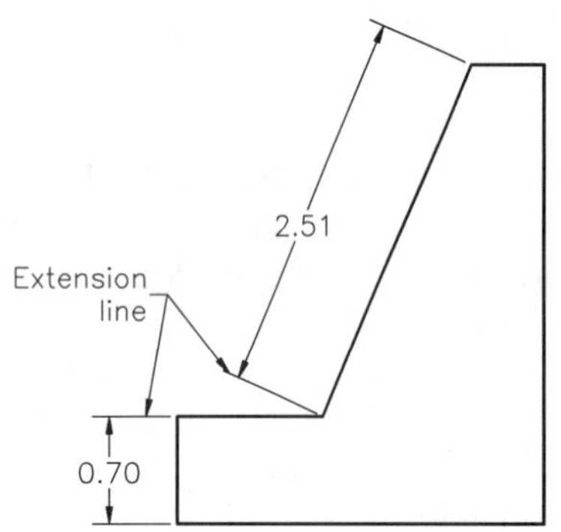

Figure 8-4 *Extension lines*

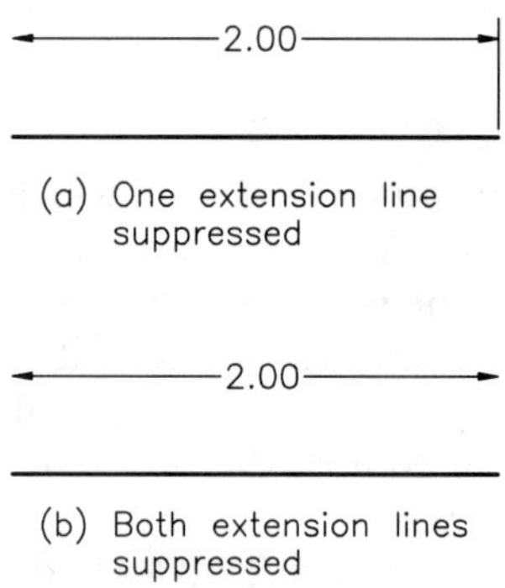

Figure 8-5 *Extension line suppression*

Leader

A **leader** is a line that stretches from the dimension text to the object being dimensioned. Sometimes the text for dimensioning and other annotations does not adjust properly near the object. In such cases, you can use a leader and place the text at the end of the leader line. For example, the circle shown in Figure 8-6 has a keyway slot that is too small to be dimensioned. In this situation, a leader can be drawn from the text to the keyway feature. Also, a leader can be used to attach annotations such as part numbers, notes, and instructions to an object.

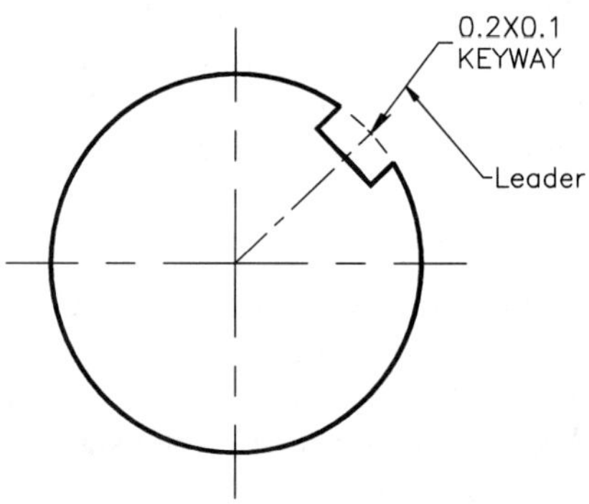

Figure 8-6 *Leader used to attach annotation*

Center Mark and Centerlines

The **center mark** is a cross mark that identifies the center point of a circle or an arc. Centerlines are mutually perpendicular lines passing through the center of the circle/arc and intersecting the circumference of the circle/arc. A center mark or the centerlines are automatically drawn when you dimension a circle or arc (see Figure 8-7). The length of the center mark and the extension of the centerline beyond the circumference of the circle is determined by the value assigned to the **DIMCEN** dimension variable or the value assigned to Center Mark for Circle in the **Dimension Style Manager** dialog box.

Alternate Units

With the help of **alternate units** you can generate dimensions for two systems of measurement at the same time (Figure 8-8). For example, if the dimensions are in inches, you can use the alternate units dimensioning facility to append metric dimensions to the dimensions (controlling the alternate units through the dimension variables is discussed in Chapter 10).

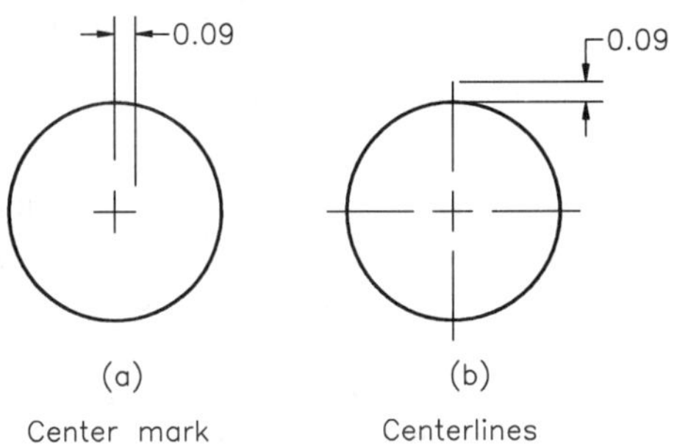

Figure 8-7 Center mark and centerlines

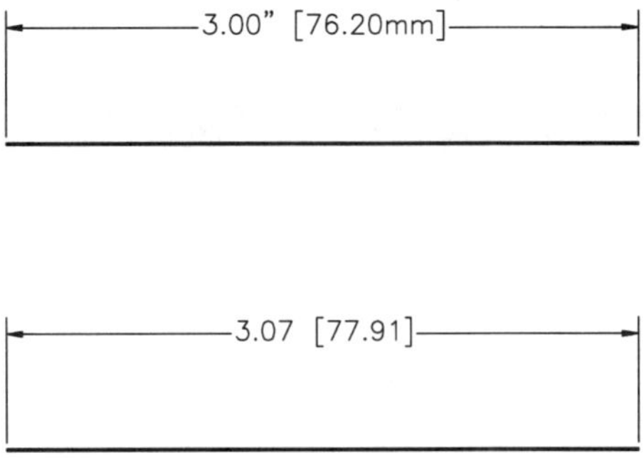

Figure 8-8 Using alternate units dimensioning

Tolerances

Tolerance is the amount by which the actual dimension can vary (Figure 8-9). AutoCAD LT can attach the plus/minus tolerances to the dimension text (actual measurement computed by AutoCAD LT). This is also known as **deviation tolerance**. The plus and minus tolerance that you specify can be the same or different. You can use the dimension variables to control the tolerance feature (these variables are discussed in Chapter 10).

Limits

Instead of appending the tolerances to the dimension text, you can apply the tolerances to the measurement itself (Figure 8-10). Once you define the tolerances, AutoCAD LT will automatically calculate the upper and lower **limits** of the dimension. These values are then displayed as a dimension text.

For example, if the actual dimension as computed by AutoCAD LT is 2.610 units and the tolerance values are +0.025 and -0.015, the upper and lower limits are 2.635 and 2.595. After calculating the limits, AutoCAD LT will display them as dimension text, as shown in Figure 8-10. The dimension variables that control the limits are discussed in Chapter 10.

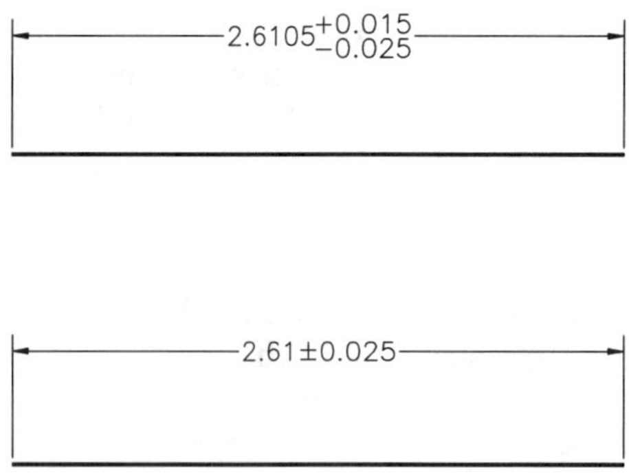

Figure 8-9 Using tolerances with dimensions

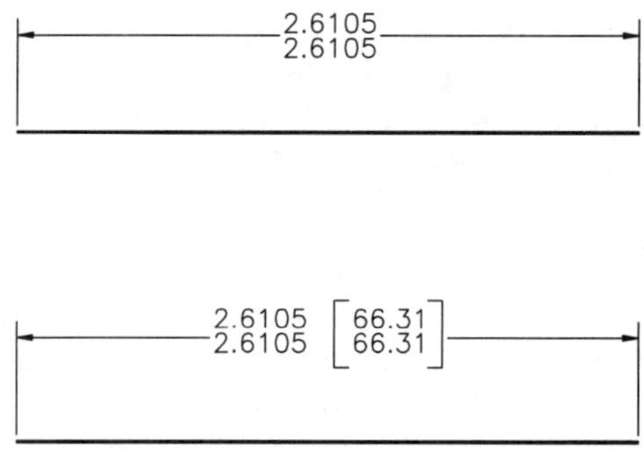

Figure 8-10 Using limits dimensioning

ASSOCIATIVE DIMENSIONS

Associative dimensioning is a method of dimensioning in which the dimension is associated with the object that is dimensioned. In other words, the dimension is influenced by the changes in the size of the object. In the previous releases of AutoCAD LT, the dimensions were not truly associative, but were related to the objects being dimensioned by definition points on the DEFPOINTS layer. To cause the dimension to be associatively modified, these definition points had to be adjusted along with the object being changed. If, for example, you use the **SCALE** command to change an object's size and select the object, the dimensions will not be modified. If you select the object and its defpoints (using the Crossing selection method), then the dimension will be modified. However, in AutoCAD LT 2002, a new concept called the **true associative dimensioning** is introduced. This concept ensures that if the dimensions are associated to the object, then if the object changes its size, the dimensions will also change automatically. With the introduction of the true associative dimensions, there is no need to select the definition points along with the object. This eliminates the use of definition points for updating the dimensions.

The associative dimensions automatically update their values and location if the value or location to the object is modified. For example, if you edit an object using simple editing operations such as breaking using the **BREAK** command, then the true associative dimension will be modified, automatically. The dimensions can be converted into the true associative dimensions using the **DIMREASSOCIATE** command. The association of the dimensions with the objects can be removed using the **DIMDISASSOCIATE** command. Both of these commands will be discussed later in this chapter.

The dimensioning variable **DIMASSOC** controls associativity of dimensions. The default value of this variable is **1**, which means turned on. When **DIMASSOC** is turned off, the dimension will be placed in the exploded format. This means that the dimensions will now be placed as a combination of individual arrowheads, dimension lines, extension lines, and text. Also, note that the exploded dimensions can not be associated to any object.

DEFINITION POINTS

Definition points are the points drawn at the positions used to generate a dimension object. The definition points are used by the dimensions to control their updating and rescaling. AutoCAD LT draws these points on a special layer called **DEFPOINTS.** These points are not plotted by the plotter because AutoCAD LT does not plot any object on the **DEFPOINTS** layer. If you explode a dimension (which is as good as turning **DIMASSOC** off), the definition points are converted to point objects on the **DEFPOINTS** layer. In Figure 8-11, the small circles indicate the definition points for different objects.

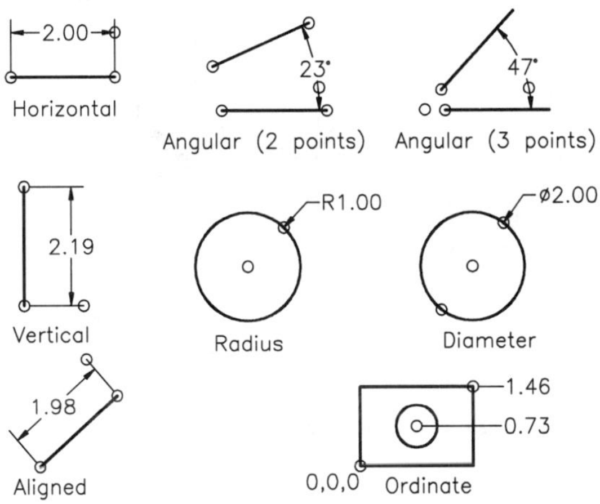

Figure 8-11 *Definition points of linear, angular, and ordinate dimensions*

The definition points for linear dimensions are the points used to specify the extension lines and the point of intersection of the first extension line and the dimension line. The definition points for the angular dimension are the endpoints of the lines used to specify the dimension and the point used to specify the dimension line arc. For example, for three-point angular dimension, the definition points are the extension line endpoints, the angle vertex, and the point used to specify the dimension line arc.

The definition points for the radius dimension are the center point of the circle or arc, and the point where the arrow touches the object. The definition points for the diameter dimension are the points where the arrows touch the circle. The definition points for the ordinate dimension are the UCS origin, the feature location, and the leader endpoint.

Note

In addition to the definition points just mentioned, the middle point of the dimension text serves as a definition point for all types of dimensions.

SELECTING DIMENSIONING COMMANDS

Using the Toolbar and the Dimension Menu

You can select the dimension commands from the **Dimension** toolbar by choosing the desired dimension button (Figure 8-12), or from the **Dimension** menu (Figure 8-13). The **Dimension** toolbar can also be displayed by right-clicking on any toolbar and choosing **Dimensions** from the shortcut menu (Figure 8-13).

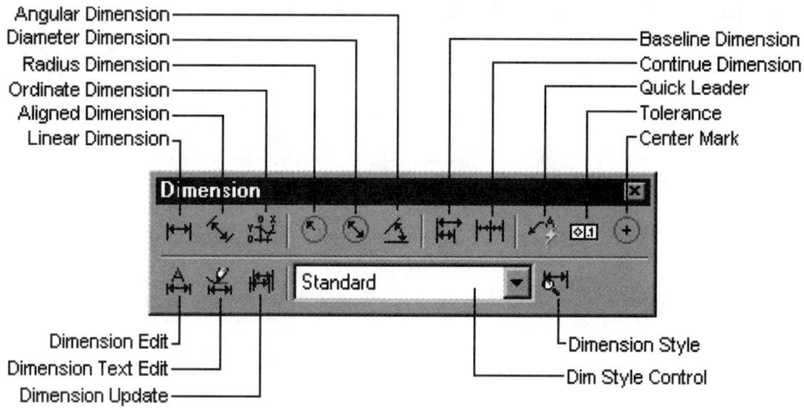

***Figure 8-12* *Dimension* toolbar**

Using the Command Line

You can directly enter a dimensioning command in the Command line or use the **DIM** or the **DIM1** commands to invoke the dimensioning commands. Both of these methods for using the command line are discussed next.

Using Dimensioning Commands

The first mcthod for using the Command line is to directly enter the dimensioning command in the Command line. For example, if you want to draw the linear dimension, the **DIMLINEAR** command can be entered directly at the Command prompt.

Command: **DIMLINEAR**
Specify first extension line origin or <select object>: *Select a point or press ENTER.*
Specify second extension line origin: *Select second point.*
Specify dimension line location or [Mtext/Text/Angle/Horizontal/Vertical/Rotated]: *Select a point to locate the position of the dimension.*
Command: *(After you have finished dimensioning, AutoCAD LT returns to the Command prompt.)*

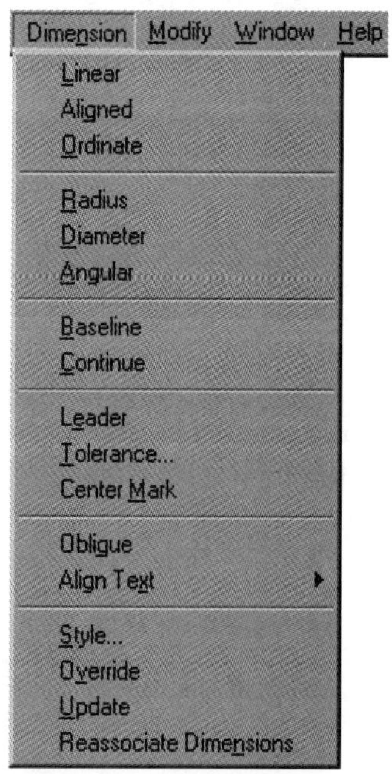

Figure 8-13 *Selecting dimensions from the **Dimension** menu*

DIM and DIM1 Commands

Since dimensioning has several options, it also has its own command mode. The **DIM** command keeps you in the dimension mode, and the **Dim:** prompt is repeated after each dimensioning command until you exit the dimension mode to return to the normal AutoCAD LT Command prompt. To exit the dimension mode, enter EXIT (or just E) at the **Dim:** prompt. You can also exit by pressing ESC. The previous command will be repeated if you press the SPACEBAR or ENTER at the **Dim** prompt. In the dimension mode, it is not possible to execute the normal set of AutoCAD LT commands, except function keys, object snap overrides, control key combinations, transparent commands, dialog boxes, and menus.

Command: **DIM**
Dim: **Hor**
Specify first extension line origin or <select object>: *Select a point or press ENTER.*
Specify second extension line origin: *Select the second point.*
Specify dimension line location or [MText/Text/Angle]: *Select a point to locate the position of the dimension.*
Enter dimension text <default>: *Press ENTER to accept the default dimension.*
Dim: *(After you have finished dimensioning, AutoCAD LT returns to the* ***Dim:*** *prompt.)*

The **DIM1** command is similar to the **DIM** command. The only difference is that **DIM1** lets you execute a single dimension command and then automatically takes you back to the normal Command prompt.

Command: **DIM1**
Dim: **Hor**
Specify first extension line origin or <select object>: *Select a point or press ENTER.*
Specify second extension line origin: *Select the second point.*
Specify dimension line location or [MText/Text/Angle]: *Select a point to locate the position of the dimension.*
Enter dimension text <default>: *Press ENTER to accept the default dimension.*
Command: *(After you are done dimensioning, AutoCAD LT returns to the Command prompt.)*

AutoCAD LT has provided the following fundamental dimensioning types.

Linear dimensioning	**Diameter dimensioning**	**Radius dimensioning**
Angular dimensioning	**Ordinate dimensioning**	

Figure 8-14 and Figure 8-15 shows all these fundamental dimension types.

Note

The ***DIMDEC*** *variable sets the number of decimal places for the value of primary dimension and the* ***DIMADEC*** *variable for angular dimensions. For example, if* ***DIMDEC*** *is set to 3, AutoCAD LT will display the decimal dimension up to three decimal places (2.037).*

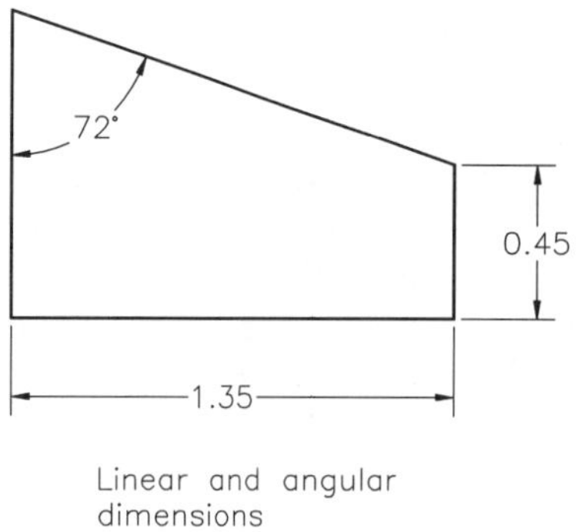

Figure 8-14 Linear and angular dimensions

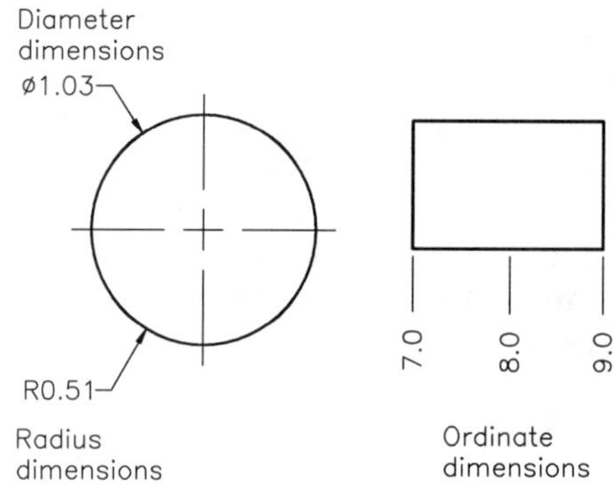

Figure 8-15 Radius, diameter, and ordinate dimensions

LINEAR DIMENSIONING

Toolbar:	Dimension > Linear Dimension
Menu:	Dimension > Linear.
Command:	DIMLINEAR

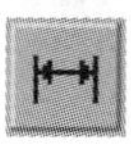

Linear dimensioning applies to those dimensioning commands that measure the shortest distance between two points. You can directly select the object to dimension or select two points. The points can be any two points in the space, the endpoints of an arc or line, or any set of points that can be identified. To achieve accuracy, points must be selected with the help of object snaps or by selecting an object to dimension. If the object selected is aligned, the linear dimensions will add **Horizontal** or **Vertical** dimensions to the object. The prompt sequence that will follow when you choose this button is:

Specify first extension line origin or <select object>: Enter
Select object to dimension: *Select the object.*
Specify dimension line location or
[Mtext/Text/Angle/Horizontal/Vertical/
Rotated]: *Select a point to locate the position of the dimension.*

Instead of selecting the object, you can also select the two endpoints of the line that you want to dimension (Figure 8-16). Usually the points on the object are selected by using the **object snaps** (endpoints, intersection, center, etc.). The prompt sequence is as follows:

Specify first extension line origin or <select object>: *Select a point.*
Specify second extension line origin: *Select second point.*

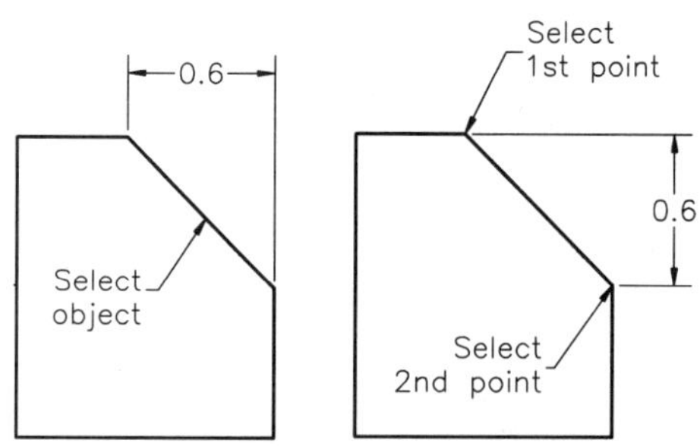

Figure 8-16 Drawing linear dimension

Specify dimension line location or
[Mtext/Text/Angle/Horizontal/Vertical/Rotated]: *Select a point to locate the position of the dimension.*

When using the **DIMLINEAR** command, you can obtain the horizontal or vertical dimension by simply defining the appropriate dimension location point. If you select a point above or below the dimension, AutoCAD LT creates a horizontal dimension. If you select a point that is on the left or right of the dimension, AutoCAD LT creates a vertical dimension through that point.

DIMLINEAR Command Options

The options provided under this command are discussed next.

Mtext Option

The **Mtext** option allows you to override the default dimension text and also change the font, height, and so on, using the **Multiline Text Editor** dialog box. When you enter **M** at the **Specify dimension line location or [Mtext/Text/Angle/Horizontal/Vertical/Rotated]** prompt, the **Multiline Text Editor** is displayed. By default, it includes the <> code to represent the measured dimension text. You can change the text by entering a new text and deleting <> code. You can also use the different options of the dialog box (explained in Chapter 7, Controlling the Drawing Display and Creating Text) and then choose **OK**. However, if you override the default dimensions, the dimensional associativity of the **dimension text** is lost. This means that if you modify the object using the definition points, AutoCAD LT will not recalculate the dimension text. Even if the dimension is a true associative dimension, the text will not be recalculated when the object is modified. The prompt sequence to invoke this option is:

Specify first extension line origin or <select object>: *Specify a point.*
Specify second extension line origin: *Specify second point.*
Specify dimension line location or
[Mtext/Text/Angle/Horizontal/Vertical/Rotated]: **M** *(Enter dimension text in the **Mutiline Text Editor** dialog box and then choose **OK**.)*
Specify dimension line location or
[Mtext/Text/Angle/Horizontal/Vertical/Rotated]: *Specify the dimension location.*

Text Option

This option also allows you to override the default dimension. However, this option will prompt you to specify the new text value in the Command prompt itself (Figure 8-17). The prompt sequence to invoke this option is:

Specify first extension line origin or <select object>: *Select a point.*
Specify second extension line origin: *Select second point.*
Specify dimension line location or
[Mtext/Text/Angle/Horizontal/Vertical/Rotated]: **T**
Enter dimension text <Current>: *Enter new text.*
Specify dimension line location or
[Mtext/Text/Angle/Horizontal/Vertical/Rotated]: *Specify the dimension location.*

Angle Option

This option lets you change the angle of the dimension text (Figure 8-17).

Horizontal Option

This option lets you create a horizontal dimension regardless of where you specify the dimension location (Figure 8-17).

Vertical Option

This option lets you create a vertical dimension regardless of where you specify the dimension location (Figure 8-17).

Rotated Option

This option lets you create a dimension that is rotated at a specified angle. See Figure 8-17.

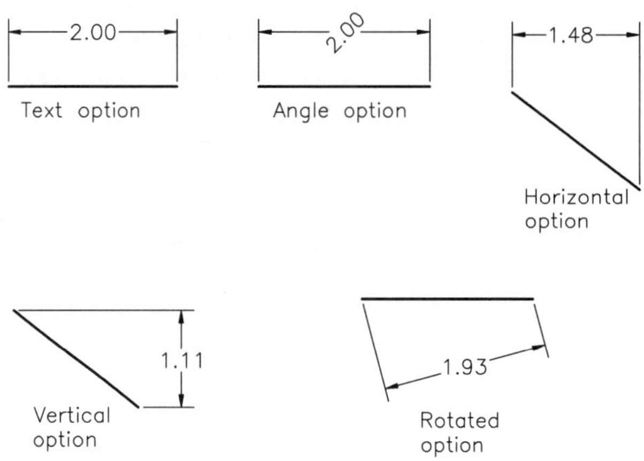

Figure 8-17 *Text, Angle, Horizontal, Vertical, and Rotated options*

Note

If you override the default dimensions, the dimensional associativity of the dimension text is lost and AutoCAD LT will not recalculate the dimension when the object is scaled.

Example 1 — *General*

In this example, you will use the linear dimensioning to dimension a horizontal line of 4 units length. The dimensioning will be done by selecting the object and by specifying the first and second extension line origins. Using the **Multiline Text Editor** modify the default text such that the dimension is underlined.

Selecting the Object

1. Choose the **Linear Dimension** button from the **Dimension** toolbar. The prompt sequence is as follows:

Specify first extension line origin or <select object>: Enter
Select object to dimension: *Select the line.*
Specify dimension line location or
[Mtext/Text/Angle/Horizontal/Vertical/Rotated]: **M**

*The **Multiline Text Editor** will be displayed as shown in Figure 8-18. The default dimension value will be displayed in the **<>** code. Select this code and then choose the **Underline** button to underline the text.*

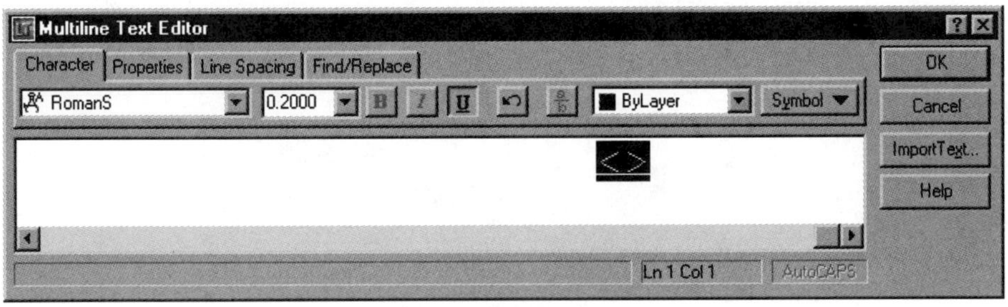

Figure 8-18 Multiline Text Editor

Specify dimension line location or
[Mtext/Text/Angle/Horizontal/Vertical/Rotated]: *Place the dimension.*
Dimension text = 4.00

Specifying the Extension Line Origins

2. Choose the **Linear Dimension** button from the **Dimension** toolbar. The prompt sequence is as follows:

Specify first extension line origin or <select object>: *Select the first endpoint of the line using the Endpoint object snap. See Figure 8-19.*
Specify second extension line origin: *Select the second endpoint of the line using the Endpoint object snap. See Figure 8-19.*
Specify dimension line location or
[Mtext/Text/Angle/Horizontal/Vertical/Rotated]: **M**

*Select the code and then choose the **Underline** button to underline the text in the **Multiline Text Editor**.*

Specify dimension line location or
[Mtext/Text/Angle/Horizontal/Vertical/Rotated]: *Place the dimension.*
Dimension text = 4.00

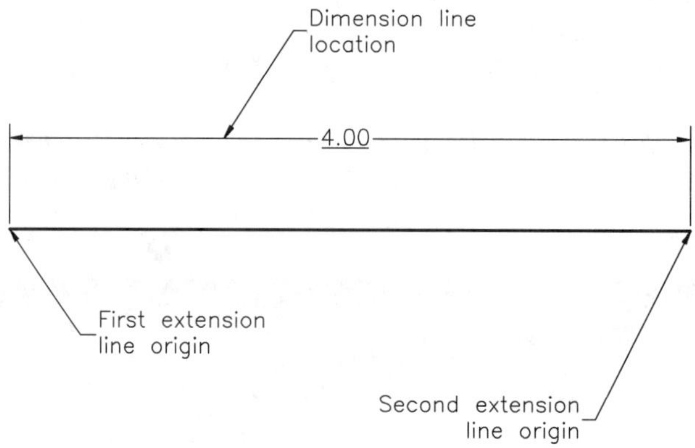

***Figure 8-19** Line for Example 1 after dimensioning*

ALIGNED DIMENSIONING

Toolbar:	Dimension > Aligned Dimension
Menu:	Dimension > Aligned
Command:	DIMALIGNED

Generally, the drawing consists of various objects that are neither parallel to the X axis nor to the Y axis. Dimensioning of such objects can be done using the **aligned dimensioning**. In horizontal or vertical dimensioning, you can only measure the shortest distance from the first extension line origin to the second extension line origin along the horizontal or vertical axis, respectively. Whereas, with the help of aligned dimensioning, you can measure the true aligned distance between the two points. The working of the **ALIGNED** dimension command is similar to that of the other linear dimensioning commands. The dimension created with the **ALIGNED** command is **parallel to the object being dimensioned**. The prompt sequence that will follow when you choose this button is:

Specify first extension line origin or <select object>: *Specify the first point or press ENTER to select the object.*
Specify second extension line origin: *Specify second point.*
Specify dimension line location or
[Mtext/Text/Angle]: *Specify the location for the dimension line.*
Dimension text = Current.

The options provided under this command are similar to those under the **DIMLINEAR** command. Figure 8-20 illustrates aligned dimensioning.

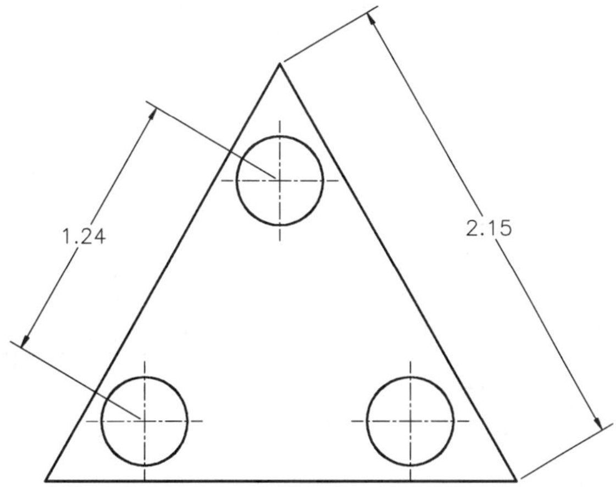

Figure 8-20 *Aligned dimensioning*

Exercise 1 *Mechanical*

Draw Figure 8-21, and then use the linear and aligned dimensioning to dimension the part. The distance between the dotted lines is 0.5 units. The dimensions should be up to two decimal places. To get dimensions up to two decimal places, enter DIMDEC at the Command prompt and then enter 2. (There will be more information about dimension variables in Chapter 10, Dimension Styles and Dimensioning System Variables.)

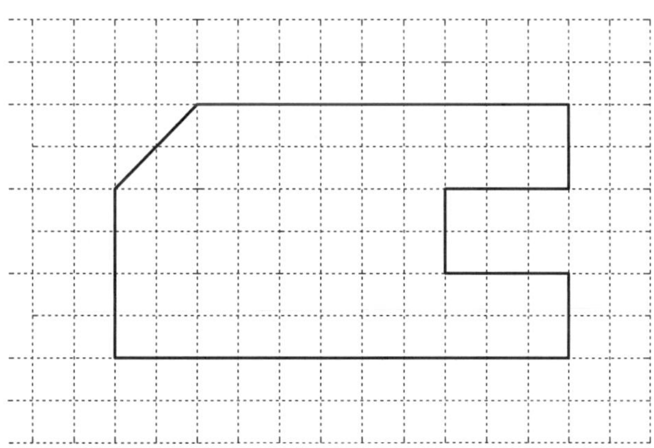

Figure 8-21 *Drawing for Exercise 1*

ROTATED DIMENSIONING

Rotated dimensioning is used when you want to place the dimension line at an angle (if you do not want to align the dimension line with the extension line origins selected) (Figure 8-22). The **ROTATED** dimension option will prompt you to specify the dimension line angle. You can invoke this command by using **DIMLINEAR** (**Rotate** option) or by entering **ROTATED**

at the **Dim:** Command prompt. The prompt sequence is:

Dim: **ROTATED**
Specify angle of dimension line <0>: **110**
Specify first extension line origin or <select object>: *Select the lower right corner of the triangle.*
Specify second extension line origin: *Select the top corner.*
Specify dimension line location or [MText/Text/Angle]: *Select the location for the dimension line.*
Enter dimension text <2.0597>: Enter

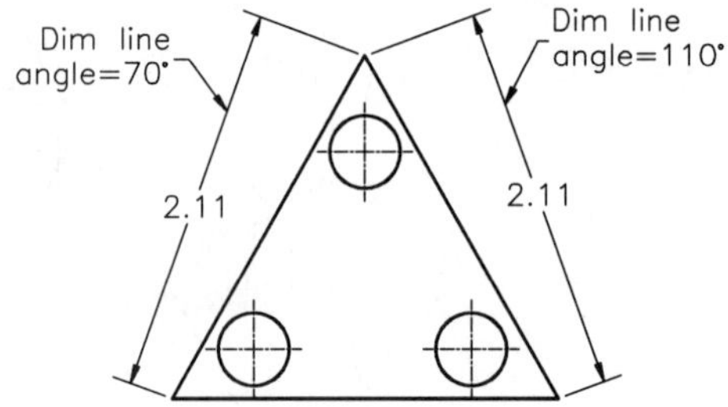

Figure 8-22 *Rotated dimensioning*

Tip
You can draw horizontal and vertical dimensioning by specifying the rotation angle of 0 degrees for horizontal dimensioning and 90 degrees for vertical dimensioning.

BASELINE DIMENSIONING

Toolbar:	Dimension > Baseline Dimension
Menu:	Dimension > Baseline
Command:	DIMBASELINE

Sometimes in manufacturing, you may want to locate different points and features of a part with reference to a fixed point (base point or reference point). This can be accomplished by using **BASELINE dimensioning** (Figure 8-23). With this command you can continue a linear dimension from the first extension line origin of the first dimension. The new dimension line is automatically offset by a fixed amount to avoid overlapping of the dimension lines. Remember that **there must already exist a linear, ordinate, or angular associative dimension to use the Baseline dimensions.** When you choose the **Baseline Dimension** button, the last linear, ordinate, or angular dimension that was created will be selected and used as baseline. The prompt sequence that will follow when you choose this button is:

Specify a second extension line origin or [Undo/Select] <Select>: *Select the origin of the second extension line.*
Dimension text = current
Specify a second extension line origin or [Undo/Select] <Select>: *Select the origin of the second extension line.*
Dimension text = current
Specify a second extension line origin or [Undo/Select] <Select>: *Select the origin of the second extension line or press ENTER.*
Select base dimension: Enter

When you use the **DIMBASELINE** command, you cannot change the default dimension text. However, the **DIM** mode commands allows you to override the default dimension text.

Command: **DIM**
Dim: **HOR**
Specify first extension line origin or <select object>: *Select left corner (P1, Figure 8-23). (Use Endpoint object snap.)*
Specify second extension line origin: *Select the origin of the second extension line (P2).*
Specify dimension line location or [MText/Text/Angle]: **T**
Enter dimension text <1.0000>: **1.0**
Specify dimension line location or [MText/Text/Angle]: *Select the dimension line location.*

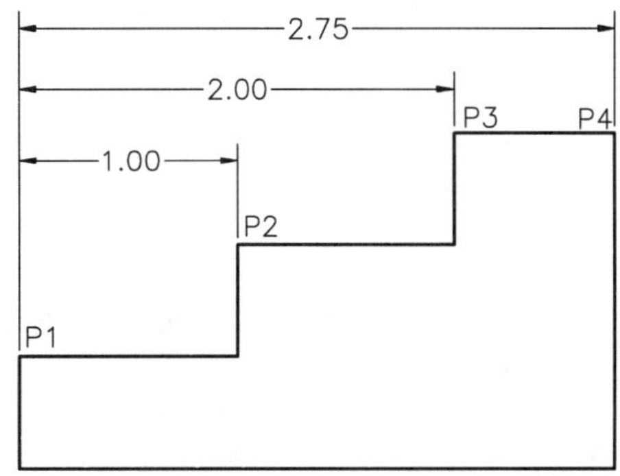

Figure 8-23 *Baseline dimensioning*

Dim: **BASELINE (or BAS)**
Specify a second extension line origin or [Select]<Select>: *Select the origin of the next extension line (P3).*
Enter dimension text <2.0000>: **2.0**
Dim: **BAS**
Specify a second extension line origin or [Select]<Select>: *Select the origin of the next extension line (P4).*
Enter dimension text <3.000>: **2.75**

The next dimension line is automatically spaced and drawn by AutoCAD LT.

CONTINUE DIMENSIONING

Toolbar:	Dimension > Continue Dimension
Menu:	Dimension > Continue
Command:	DIMCONTINUE

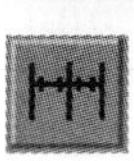

With this command you can continue a linear dimension from the second extension line of the previous dimension. This is also called as **chained** or **incremental dimensioning**. **There must exist linear, ordinate, or angular associative dimension to use the Continue dimensions**. The prompt sequence that will follow when you choose this button is:

Specify a second extension line origin or [Undo/Select] <Select>: *Select the origin of the second extension line or press ENTER to select the existing dimension.*
Select continued dimension: *Select the dimension.*
Specify a second extension line origin or [Undo/Select] <Select>: *Specify the point on the origin of the second extension line.*
Dimension text = current
Specify a second extension line origin or [Undo/Select] <Select>: *Specify the point on the origin of the second extension line.*

Dimension text = current
Specify a second extension line origin or [Undo/Select] <Select>: Enter
Select continued dimension: Enter

In this case also the **DIM** command should be used if you want to change the default dimension text.

Command: **DIM**
Dim: **HOR**
Specify first extension line origin or <select object>: *Select left corner (P1, see Figure 8-24). (Use the Endpoint object snap.)*
Specify second extension line origin: *Select the origin of the second extension line (P2, see Figure 8-24).*
Specify dimension line location or [MText/Text/Angle]: **T**
Enter dimension text <current>: **0.75**
Specify dimension line location or [MText/Text/Angle]: *Select the dimension line location.*

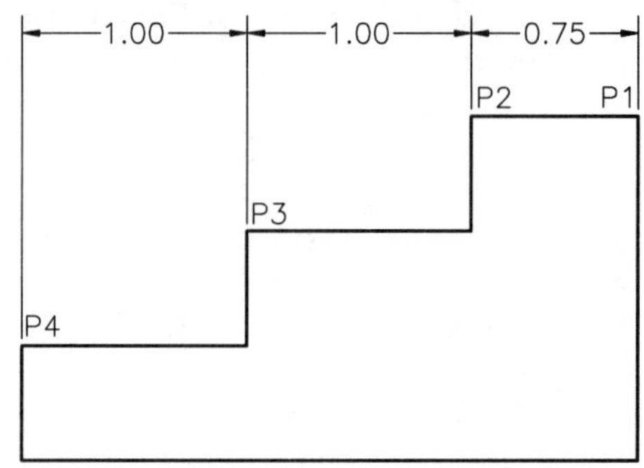

Figure 8-24 *Continue dimensioning*

Dim: **CONTINUE**
Specify a second extension line origin or [Select]<Select>: *Select the origin of the next extension line (P3, see Figure 8-24).*
Enter dimension text <current>: Enter
Dim: **CONTINUE**
Specify a second extension line origin or [Select]<Select>: *Select the origin of the next extension line (P4, see Figure 8-24).*
Enter dimension text <current>: Enter

The default base (first extension line) for the dimensions created with the **CONTINUE** command is the previous dimension's second extension line. You can override the default by pressing ENTER at the **Specify a second extension line origin or [Select]<Select>** prompt, and then specifying the other dimension. The extension line origin nearest to the selection point is used as the origin for the first extension line.

Tip

*You can use the **Select** option of the **DIMBASELINE** or the **DIMCONTINUE** command to select any other existing dimension to be used as baseline or continue dimensions.*

Exercise 2 — *Mechanical*

Draw the object shown in Figure 8-25. Then use the **baseline** dimensioning to dimension the top half and **continue dimensioning** to dimension the bottom half. The distance between the dotted lines is 0.5 units.

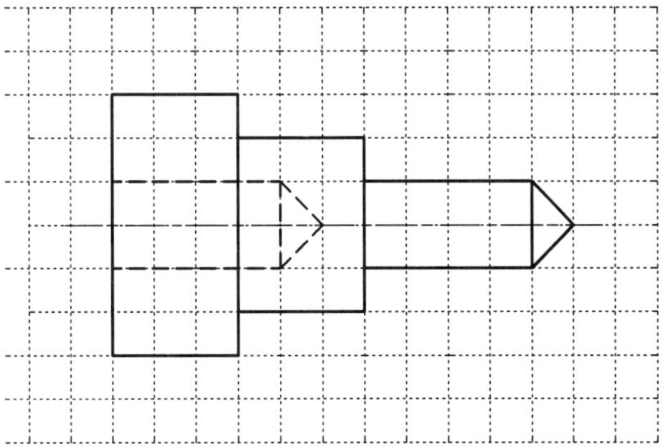

Figure 8-25 *Drawing for Exercise 2*

ANGULAR DIMENSIONING

Toolbar:	Dimension > Angular Dimension
Menu:	Dimension > Angular
Command:	DIMANGULAR

Angular dimensioning is used when you want to dimension an angle. This command generates a dimension arc (dimension line in the shape of an arc with arrowheads at both ends) to indicate the angle between two nonparallel lines. This command can also be used to dimension the vertex and two other points, a circle with another point, or the angle of an arc. For every set of points there exists one acute angle and one obtuse angle (inner and outer angles). If you specify the dimension arc location between the two points, you will get the acute angle; if you specify it outside the two points, you will get the obtuse angle. Figure 8-26 shows the four ways to dimension two nonparallel lines. The prompt sequence that will follow when you choose this button is:

> Select arc, circle, line, or <specify vertex>: *Select the object or press ENTER to select a vertex point where two segments meet.*
> Select second line: *Select the second object.*
> Specify dimension arc line location or [Mtext/Text/Angle]: *Place the dimension or select an option.*
> Dimension text = current

The methods of dimensioning different entities using this command are discussed next.

Dimensioning the Angle Between Two Nonparallel Lines

The angle between two nonparallel lines or two straight line segments of a polyline can be dimensioned with the **DIMANGULAR** dimensioning command. The vertex of the angle is taken as the point of intersection of the two lines. The location of the extension lines and dimension arc is determined by how you specify the dimension arc location.

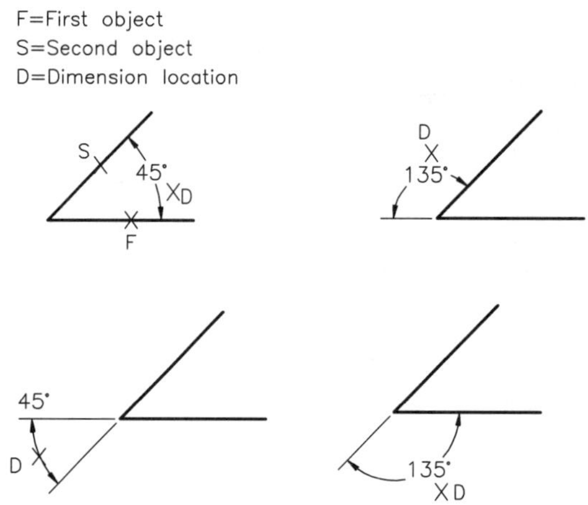

Figure 8-26 *Angular dimensioning between two nonparallel lines*

The following example illustrates the dimensioning of two nonparallel lines using the **DIMANGULAR** command invoked using the **Angular Dimension** button:

> Select arc, circle, line, or <specify vertex>: *Select the first line.*
> Select second line: *Select the second line.*
> Specify dimension arc line location or [Mtext/Text/Angle]: **M** *(Enter the new value in the **Multiline Text Editor** dialog box.)*
> Specify dimension arc line location or [Mtext/Text/Angle]: *Specify the dimension arc location.*

Dimensioning the Angle of an Arc

Angular dimensioning can also be used to dimension the angle of an arc. In this case, the center point of the arc is taken as the vertex and the two endpoints of the arc are used as the extension line origin points for the extension lines (Figure 8-27). The following example illustrates the dimensioning of an arc using the **DIMANGULAR** command.

> Select arc, circle, line, or <specify vertex>: *Select the arc.*
> Specify dimension arc line location or [Mtext/Text/Angle]: *Specify a location for the arc line or select an option.*

Angular Dimensioning of Circles

The angular feature associated with the circle can be dimensioned by selecting a circle object at the **Select arc, circle, line, or <specify vertex>** prompt. The center of the selected circle is used as the vertex of the angle. The first point selected (when the circle is selected for angular dimensioning) is used as the origin of the first extension line. In a similar manner, the second point selected is taken as the origin of the second extension line (Figure 8-28). The following is the prompt sequence for dimensioning a circle:

Select arc, circle, line, or <specify vertex>: *Select the circle at the point where you want the first extension line.*
Specify second angle endpoint: *Select the second point on or away from the circle.*
Specify dimension arc line location or [Mtext/Text/Angle]: *Select the location for the dimension line.*

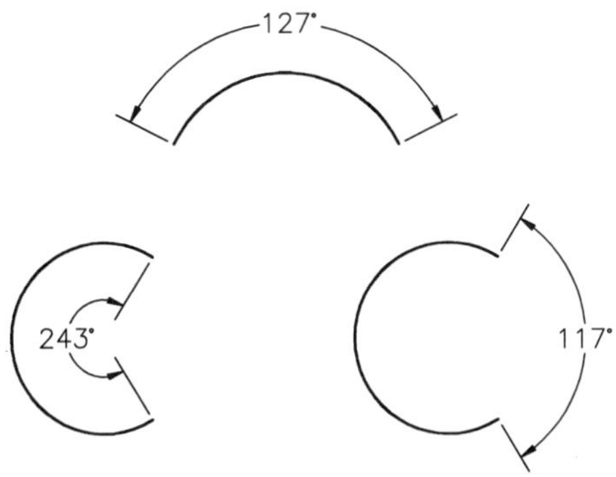

Figure 8-27 *Angular dimensioning of arcs*

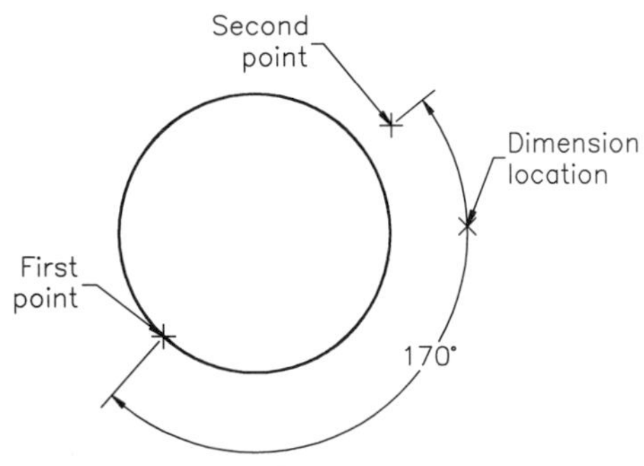

Figure 8-28 *Angular dimensioning of circles*

Angular Dimensioning Based on Three Points

If you press ENTER at the **Select arc, circle, line, or <specify vertex>** prompt, AutoCAD LT allows you to select three points to create an angular dimension. The first point is the vertex point, and the other two points are the first and second angle endpoints of the angle (Figure 8-29). The coordinate specifications of the first and the second angle endpoints must not be identical. However, the angle vertex coordinates and one of the angle endpoint coordinates can be identical. The following example illustrates angular dimensioning by defining three points:

Select arc, circle, line, or <specify vertex>: Enter
Specify angle vertex: *Specify the first point, vertex. This is the point where the two segments meet. If the two segments do not meet actually, use the **Apparent Intersection** object snap.*
Specify first angle endpoint: *Specify the second point. This point will be the origin of the first extension line.*
Specify second angle endpoint: *Specify the third point. This point will be the origin of the first extension line.*
Specify dimension arc line location or [Mtext/Text/Angle]: *Select the location for the dimension line.*

Note

*If you use the **DIMANGULAR** command, you cannot specify the text location. With the **DIMANGULAR** command, AutoCAD LT positions the dimensioning text automatically. If you want to manually define the position of dimension text, use the **ANGULAR** option of the **DIM** command.*

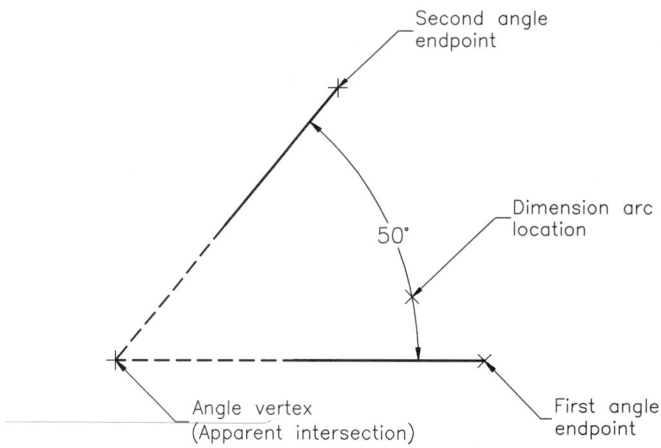

Figure 8-29 *Angular dimensioning for three points*

Exercise 3 *Mechanical*

Make the drawing in Figure 8-30. Then use the **angular dimensioning** to dimension all angles of the part. The distance between the dotted lines is 0.5 units.

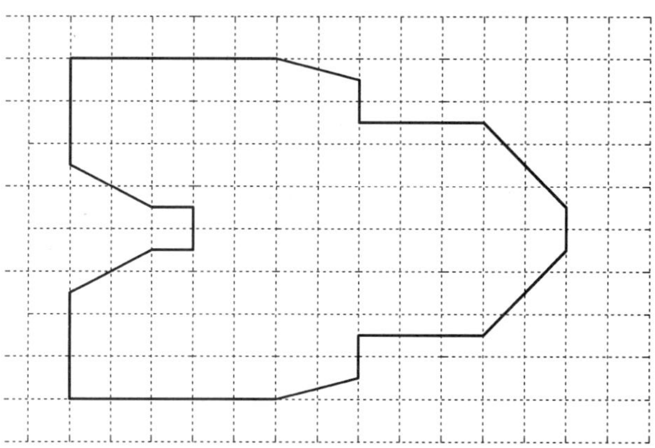

Figure 8-30 *Drawing for Exercise 3*

DIAMETER DIMENSIONING

Toolbar:	Dimension > Diameter Dimension
Pull-down:	Dimension > Diameter
Command:	DIMDIAMETER

Diameter dimensioning is used to dimension a circle or an arc. Here, the measurement is done between two diametrically opposite points on the circumference of the circle or arc (Figure 8-31). The dimension text generated by AutoCAD LT commences with the Ø symbol, to indicate a diameter dimension. The prompt sequence that will follow when you choose this button is:

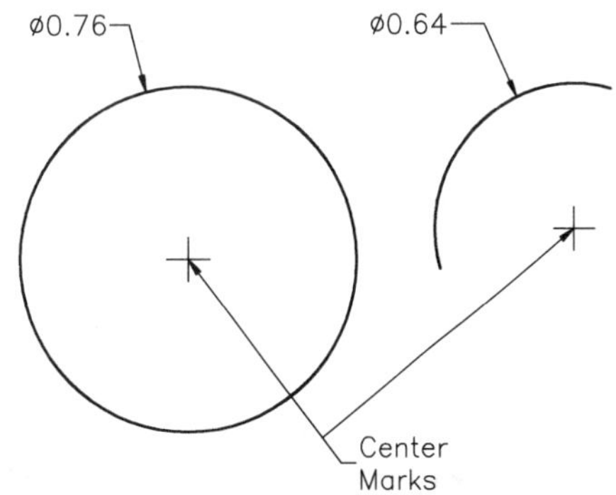

Figure 8-31 *Diameter dimensioning*

Select arc or circle: *Select an arc or circle by selecting a point anywhere on its circumference.*
Dimension text= Current
Specify dimension line location or [Mtext/Text/Angle]: *Specify a point to position the dimension.*

If you want to override the default value of the dimension text, use the Mtext or Text option. The control sequence %%C is used to obtain the diameter symbol ø. It is followed by the dimension text that you want to appear in the diameter dimension. For example, if you want to write a text that displays a value ø20, then enter %%c20 in the text prompt.

Tip

The control sequence %%d can be used to generate the degree symbol "°". Therefore, if you type 45%%d, the result that will be displayed on the screen will be 45°.

Chapter 8

RADIUS DIMENSIONING

Toolbar:	Dimension > Radius Dimension
Menu:	Dimension > Radius
Command:	DIMRADIUS

Radius dimensioning is used to dimension a circle or an arc (Figure 8-32). Radius and diameter dimensioning are similar; the only difference is that instead of the diameter line, a radius line is drawn (half of the diameter line) which is measured from the center to any point on the circumference. The dimension text generated by AutoCAD LT is preceded by the letter **R** to indicate a radius dimension. If you want to use the default dimension text (dimension text generated automatically by AutoCAD LT), simply specify a point to position the dimension at the **Specify dimension line location or [Mtext/Text/Angle]** prompt. You can also enter a new value or specify a prefix or suffix, or suppress the entire text by entering a blank space following the **Enter**

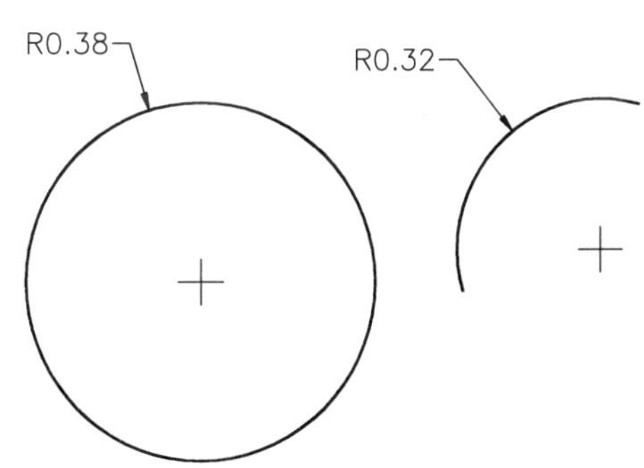

Figure 8-32 *Radius dimensioning*

dimension text <current> prompt. A center mark for the circle/arc is drawn automatically, provided the center mark value controlled by the **DIMCEN** variable is not 0. The prompt sequence that will follow when you choose this button is:

Select arc or circle: *Select the object you want to dimension.*
Dimension text= Current
Specify dimension line location or [Mtext/Text/Angle]: *Specify the dimension location.*

If you want to override the default value of the dimension text, use the **Text** or the **Mtext** option.

GENERATING CENTER MARKS AND CENTERLINES

Toolbar:	Dimension > Center Mark
Pull-down:	Dimension > Center Mark
Command:	DIMCENTER

When circles or arcs are dimensioned with the **DIMRADIUS or DIMDIAMETER** command, a small mark known as a center mark, or line known as centerline, may be drawn at the center of the circle/arc. Sometimes you may want to mark the center of a circle or an arc without using these dimensioning commands. This can be achieved with the help of the **DIMCENTER** command. You can invoke this command by choosing the **Center Mark** button from the **Dimension** toolbar or by entering **CENTER** (or **CEN**) at the **Dim** prompt. When you invoke this command, you will be prompted to select the arc or the circle. The result of this command will depend upon the value of the **DIMCEN** variable. If the value of this variable is positive, center marks are drawn (Figure 8-33) and if the value is negative, centerlines are drawn (Figure 8-34).

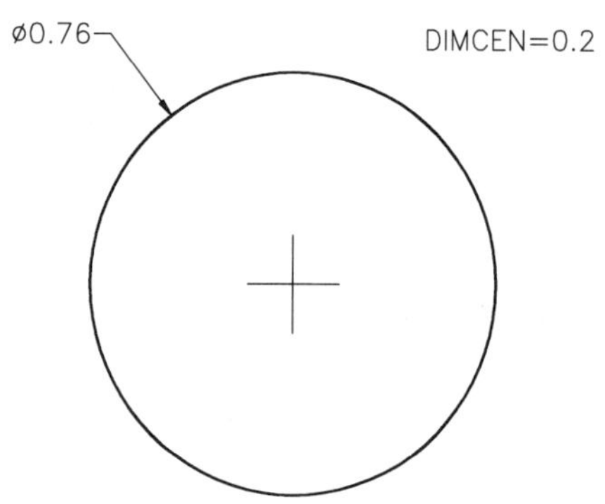

Figure 8-33 *Using a positive value for* ***DIMCEN***

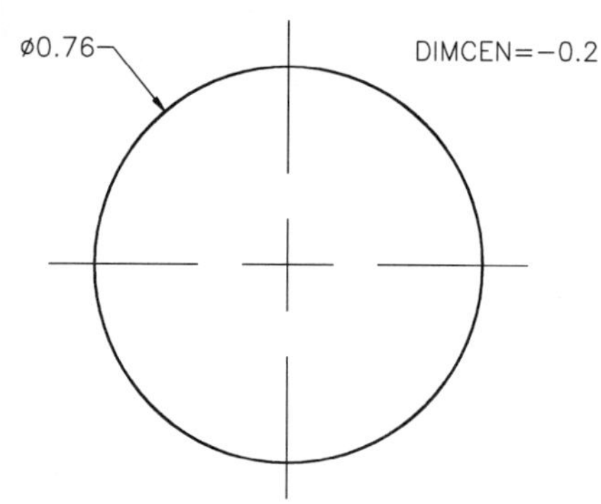

Figure 8-34 *Using a negative value for* ***DIMCEN***

Note

The center marks created by ***DIMCENTER*** *or* ***DIM, CENTER*** *are lines, not associative dimensioning objects, and they have an explicit linetype.*

Exercise 4 — *Mechanical*

Draw the model in Figure 8-35. Then use the radius and diameter dimensioning commands to dimension the part. Use the **DIMCENTER** command to draw the centerlines through the circles. The distance between the dotted lines is 0.5 units.

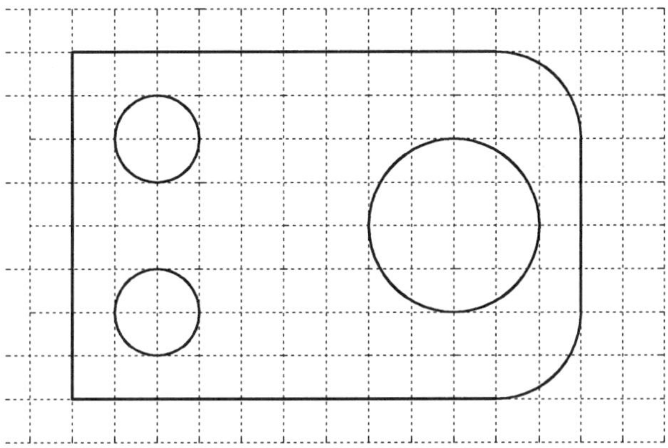

Figure 8-35 *Figure for Exercise 4*

ORDINATE DIMENSIONING

Toolbar:	Dimension > Ordinate Dimension
Menu:	Dimension > Ordinate
Command:	DIMORDINATE

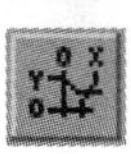

The **Ordinate dimensioning** is used to dimension the X and the Y coordinates of the selected point. This type of dimensioning is also known as **arrowless** dimensioning because no arrowheads are drawn in this type of dimensioning. The ordinate dimensioning is also called datum dimensioning because all dimensions are related to a common base point. The current UCS (user coordinate system) origin becomes the reference or the base point for ordinate dimensioning. With ordinate dimensioning you can determine the X or Y displacement of a selected point from the current UCS origin.

In ordinate dimensioning, AutoCAD LT automatically places the dimension text (X or Y coordinate value) and the leader line along the X or Y axis (Figure 8-36). Since the ordinate dimensioning pertains to either the X coordinate or the Y coordinate, you should keep ORTHO on. When ORTHO is off, the leader line is automatically given a bend when you select the second leader line point that is offset from the first point. This allows you to generate offsets and avoid overlapping text on closely spaced dimensions. In ordinate dimensioning, only one extension line (leader line) is drawn.

The leader line for an X coordinate value will be drawn perpendicular to the X axis, and the leader line for a Y coordinate value will be drawn perpendicular to the Y axis. Since you cannot override this, the leader line drawn perpendicular to the X axis will have the dimension text aligned with the leader line. The dimension text is the X datum of the selected point. The leader line drawn perpendicular to the Y axis will have the dimension text, which is the Y datum of the selected point, aligned with the leader line. Any other alignment specification for the dimension text is nullified. Therefore, changes in Text Alignment in the **Dimension Style Manager** dialog box (**DIMTIH** and **DIMTOH** variables) have no effect on the alignment of dimension text. You can specify the coordinate value you want to dimension at the **Specify leader endpoint or [Xdatum/Ydatum/MText/Text/Angle]** prompt.

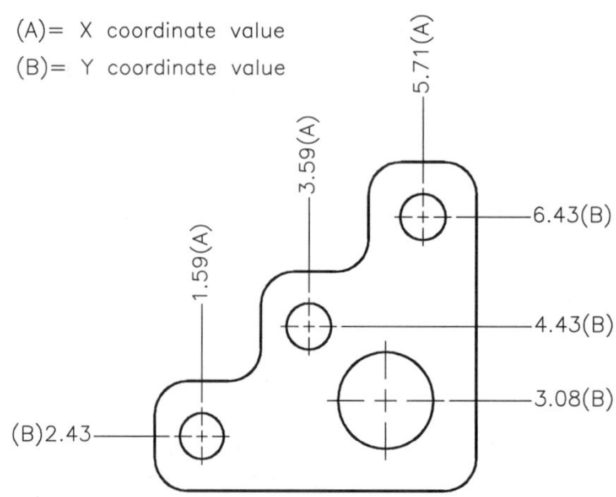

***Figure 8-36** Ordinate dimensioning*

If you select or enter a point, AutoCAD LT checks the difference between the feature location and the leader endpoint. If the difference between the X coordinates is greater, the dimension measures the Y coordinate; otherwise, the X coordinate is measured. In this manner AutoCAD LT determines whether it is an X or Y type of ordinate dimension. However, if you enter Y instead of specifying a point, AutoCAD LT will dimension the Y coordinate of the selected feature. Similarly, if you enter X, AutoCAD LT will dimension the X coordinate of the selected point. The prompt sequence that follows when you choose this button is:

Specify feature location: *Select a point on an object.*
Specify leader endpoint or [Xdatum/Ydatum/Mtext/Text/Angle]: *Enter the endpoint of the leader.*

You can override the default text with the help of the **Mtext** or the **Text** option. If you use the **Mtext** option, the **Multiline Text Editor** will be displayed. If you use the **Text** option, you will be prompted to specify the new text in the Command line itself.

Exercise 5 *General*

Draw the model shown in Figure 8-37. Then use the ordinate dimensioning to dimension the part. The distance between the dotted lines is 0.5 units.

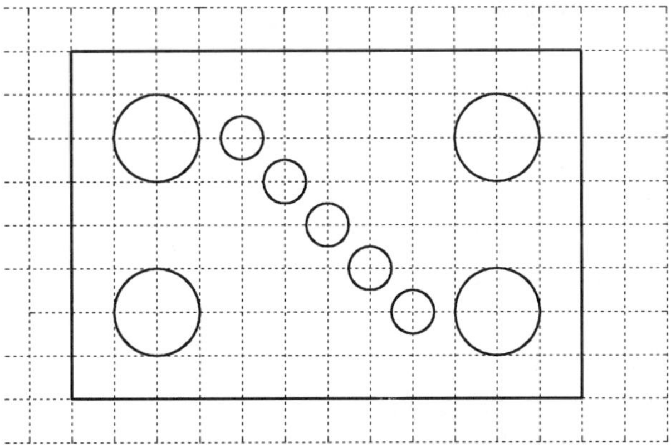

*Figure 8-37 Using the **DIMORDINATE** command to dimension the part*

CREATING TRUE ASSOCIATIVE DIMENSIONS*

The true associative dimensions are the dimensions that are automatically modified when the objects to which they are associated are modified. As mentioned earlier, this concept is introduced in AutoCAD LT 2002. If the dimension attached to the object is a true associative dimension, then the dimension will be modified automatically when the object is modified. In this case you do not have to select the definition points of the dimensions. Any dimension can be converted into a true associative dimension with the help of the **DIMREASSOCIATE** command. This command is discussed next.

DIMREASSOCIATE Command*

Menu:	Dimension > Reassociate Dimensions
Command:	DIMREASSOCIATE

This command is used to create a true associative dimension by associating the selected dimension to the specified object. When you invoke this command, you will be prompted to select the objects. These objects are the dimensions to be associated. Once you select the dimensions to be associated, a cross is displayed and you are prompted to select the feature location. This cross implies that the dimension is not associated. You can define a new association point for the dimensions by selecting the objects or by using the object snaps. If you select a dimension that has already been associated to an object, the cross will be displayed inside a box. The prompt sequence that is displayed varies depending upon the type of dimension selected. In case of **linear**, **aligned**, **radius**, and **diameter** dimensions, you can directly select the object to associate the dimension. If the arcs or circles are assigned the angular dimensions using three points, then you can also select these arcs or circles directly for associating the dimensions. For rest of the dimension types, you can use the object snaps to specify the point to associate the dimensions.

Chapter 8

REMOVING THE DIMENSION ASSOCIATIVITY (DIMDISASSOCIATE COMMAND)*

Command: DIMDISASSOCIATE

This command is used to remove the associativity of the dimensions from the object to which they are associated using the **DIMREASSOCIATE** command. When you invoke this command, you will be prompted to select the dimensions to disassociate. The true association of the selected dimensions is automatically removed once you exit this command. The number of dimensions disassociated is displayed in the Command prompt.

DRAWING LEADERS

Toolbar: Dimension > Quick Leader
Menu: Dimension > Leader
Command: QLEADER

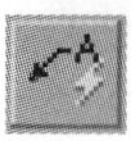

The leader line is used to attach annotations to an object or when the user wants to show a dimension without using another dimensioning command. Sometimes leaders of the dimensions of circles or arcs are so complicated that you need to construct a leader of your own. The leaders can be created using the **QLEADER** command. The leaders drawn by using this command create the arrow and the leader lines as a single object. The text is created as a separate object. This command can create multiline annotations and offer several options such as copying existing annotations, and so on. You can customize the leader and annotation by selecting the **Settings** option at the **Specify first leader point, or [Settings]<Settings>** prompt. The prompt sequence that will follow when you choose this button is:

Specify first leader point, or [Settings]<Settings>: *Specify the start point of the leader.*
Specify next point: *Specify endpoint of the leader.*
Specify next point: *Specify next point.*
Specify text width <current>: *Enter text width of multiline text.*
Enter first line of annotation text <MText>: *Press ENTER; AutoCAD LT displays the* ***Multiline Text Editor*** *dialog box. Enter text in the dialog box and then choose* ***OK*** *to exit.*

If you press ENTER at the **Specify first leader point or [Settings]<Settings>** prompt then the **Leader Settings** dialog box is displayed. The **Leader Settings** dialog box gives you a number of options for the leader line and the text attached to it. It has the following tabs:

Annotation Tab

This tab provides you with various options to control annotation features. See Figure 8-38.

Annotation Type Area

Mtext. When selected, AutoCAD LT uses the **Multiline Text Editor** to create annotation. Three Mtext options are available.

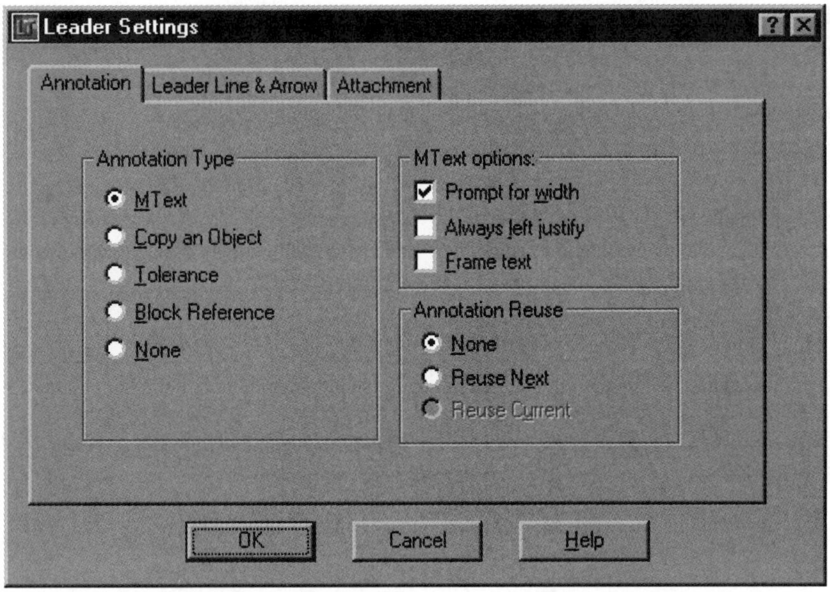

***Figure 8-38 Leader Settings** dialog box, **Annotation** tab*

Copy an Object. This option allows you to copy an existing annotation object (like mtext, single line text, tolerance, or block) and attach it at the end of the leader. For example, if you have a text string in the drawing that you want to place at the end of the leader, you can use the **Copy an Object** option to place it at the end of the leader.

Tolerance. When you select the **Tolerance** option, AutoCAD LT displays the **Geometric Tolerance** dialog box on the screen. Specify the tolerance in the dialog box and choose **OK** to exit. AutoCAD LT will place the specified Geometric Tolerance with feature control frame at the end of the leader (Figure 8-39).

Block Reference. The block option allows you to insert a predefined block at the end of the leader. When you select this option, AutoCAD LT will prompt you to enter the block name, and insertion point.

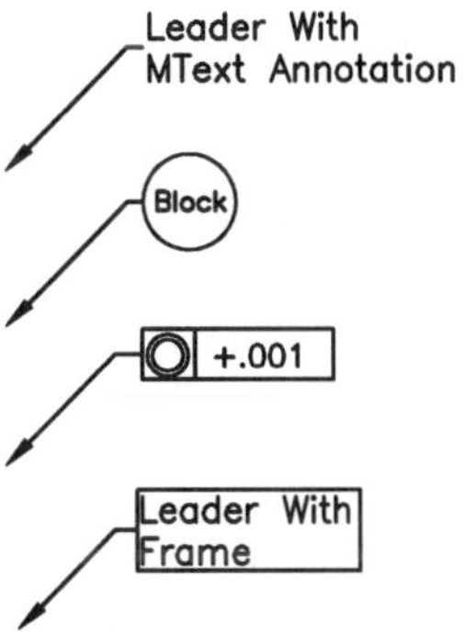

***Figure 8-39** Leader with MText, Block Reference, Tolerance, and MText with Frame annotations*

None. Creates a leader without placing any annotation at the end of the leader.

MText options Area

The options under this area will be available only if the **MText** radio button is selected from the **Annotation Type** area. This area provides you with the following options:

Prompt for width. Selecting this check box allows you to specify the width of the mtext annotation.

Always left justify. Left justifies the mtext annotation in all situations. Selecting this check box makes the Prompt for width option unavailable.

Frame text. Selecting this check box draws a box around the mtext annotation.

Annotation Reuse Area

These option allows you to reuse the annotation.

None. When selected, AutoCAD LT does not reuse the leader annotation.

Reuse Next. This option allows you to reuse the annotation that you are going to create next for all subsequent leaders.

Reuse Current. This option allows you to reuse the current annotation for all subsequent leaders.

Leader Line & Arrow Tab

The options under this area are related to the leader parameters. See Figure 8-40.

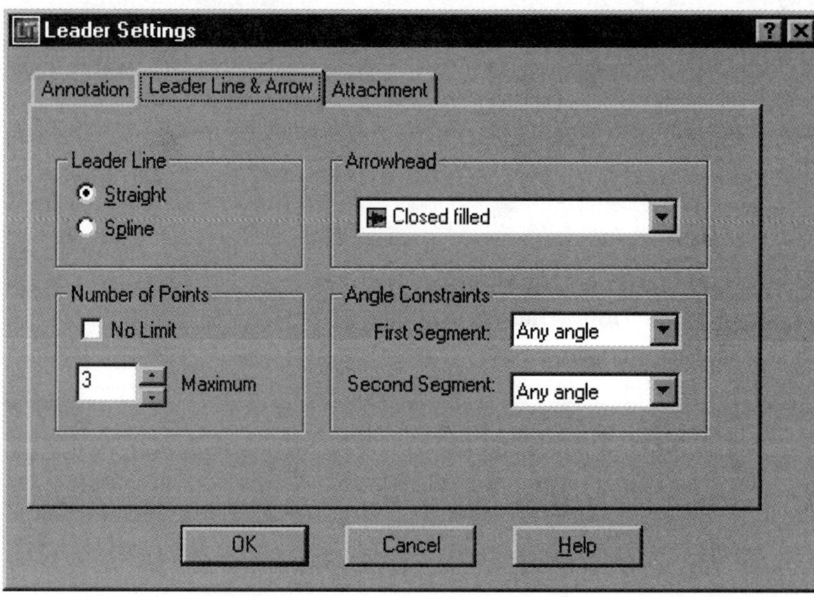

Figure 8-40 ***Leader Settings*** *dialog box,* ***Leader Line & Arrow*** *tab*

Leader Line Area

Gives the options for the **leader line type** such as straight or spline. The **Spline** option draws a spline through the specified leader points and the **Straight** option draws straight lines. Figure 8-42 shows straight and spline leader lines.

Number of Points Area

No Limit. If this check box is selected, you can define as many number of points in the leader line. AutoCAD LT will keep prompting you for the next point until you press ENTER at this prompt.

Maximum. This spinner is used to specify the maximum number of points on the leader line. The default value is 3 which means that by default, there will be only three points in the leader line. You can specify the maximum number of points using this spinner. Keep in mind that the start point of the leader is the first leader point. This spinner will be available only if the **No Limit** check box is cleared. Figure 8-41 shows a leader line with five number of points.

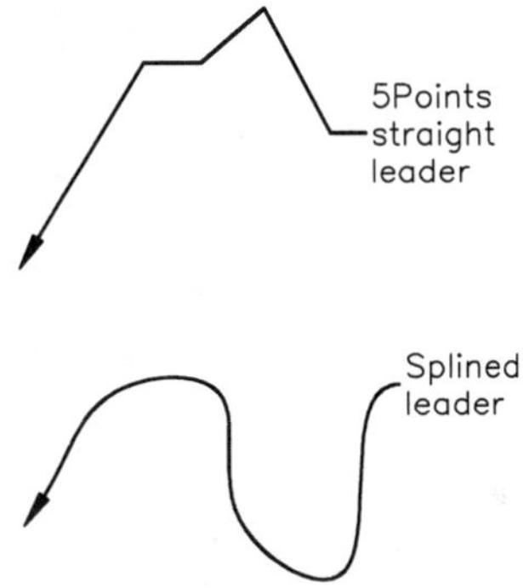

***Figure 8-41** Figure showing the splined and straight leaders*

Arrowheads Area

The drop-down list provided under this area allows you to define a leader arrowhead. The arrowhead available here is the same as the one available for dimensioning. You can also use the user-defined arrows by selecting **User Arrow** from the drop-down list.

Angle Constraint Area

The options provided under this area are used to define the angle for the segments of the leader lines.

First Segment. This drop-down list is used to specify the angle at which the first leader line segment will be drawn. You can select the predefined values from this drop-down list.

Second Segment. This drop-down list is used to specify the angle at which the second leader line segment will be drawn.

Attachment Tab

The **Attachment** tab (Figure 8-42) will be available only if you have selected **MText** from the **Annotation Type** area of the **Annotation** tab. The options provided under this tab are used for attaching the Multiline Text to the leader. It has two columns: **Text on left side** and **Text on right side**. Both of these columns have five radio buttons below them. Each radio button corresponds to the option of attaching the multiline text. If you draw a leader from right to left, AutoCAD LT uses the settings under **Text on left side**. Similarly, if you draw a leader from left to right, AutoCAD LT uses the settings as specified under **Text on right side**. This area also provides you with the **Underline bottom line** check box. If this check box is selected, the last line of the multiline text will be underlined.

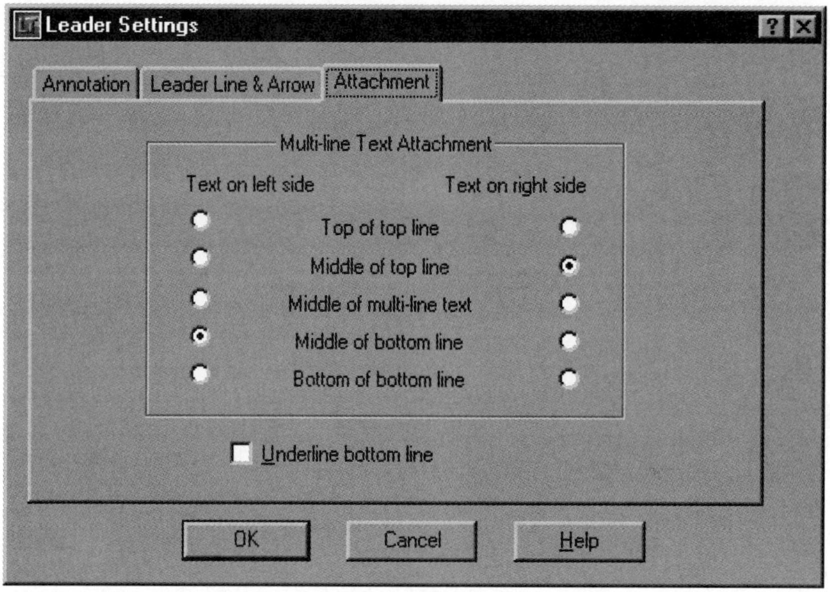

***Figure 8-42 Leader Settings** dialog box, **Attachment** tab*

Exercise 6 *Mechanical*

Make the drawing in Figure 8-43. Then use the **QLEADER** command to dimension the part as shown. The distance between the dotted lines is 0.5 units.

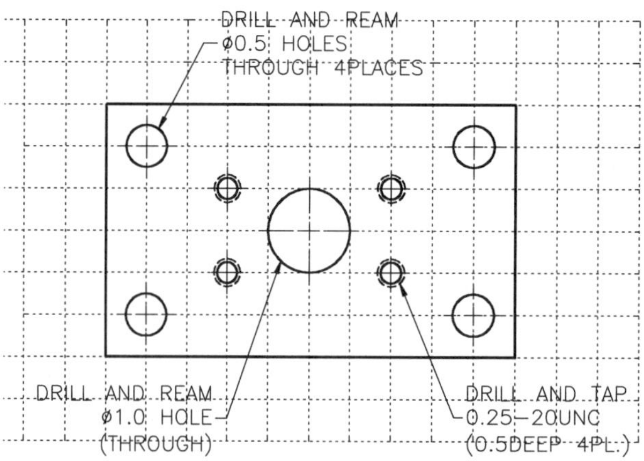

***Figure 8-43** Drawing for Exercise 6*

Note

*You can use the Command line to create the leaders with the help of the **LEADER** command. The options provided under this command are similar to those under the **QLEADER** command. The only difference is that the **LEADER** command uses the Command line.*

USING LEADER WITH THE DIM COMMAND

You can also draw a leader by using the **Leader** option of the **DIM** command. The **DIM, LEADER** command creates the arrow, leader lines, and the text as separate objects. This command has the feature of defaulting to the most recently measured dimension. Once you invoke the **DIM, LEADER** command and specify the first point, the prompt sequence is similar to that of the **LINE** command. The start point of the leader should be specified at the point closest to the object being dimensioned. After drawing the leader, enter a new dimension text or keep the default one.

Command: **DIM**
Leader (or **L**)
Leader start: *Specify the starting point of the leader.*
To point: *Specify the endpoint of the leader.*
To point: *Specify the next point.*
To point: Enter
Dimension text <current>: *Enter dimension text.*
Dim: **EXIT**

The value between the angle brackets is the current value that is the measurement of the most recently dimensioned object. If you want to retain the default text, press ENTER at the **Dimension text <current>** prompt. You can enter text of your choice, specify a prefix/suffix, or suppress the text. The text can be suppressed by pressing the SPACEBAR, and then pressing ENTER at the **Dimension text <current>** prompt. An arrow is drawn at the start point of the leader segment if the length of the segment is greater than two arrow lengths. If the length of the line segment is less than or equal to two arrow lengths, only a line is drawn.

GEOMETRIC DIMENSIONING AND TOLERANCING

One of the most important parts of the design process is giving the dimensions and tolerances, since every part is manufactured from the dimensions given in the drawing. Therefore, every designer must understand and have a thorough knowledge of the standard practices used in industry to make sure that the information given on the drawing is correct and can be understood by other people. Tolerancing is equally important, especially in the assembled parts. Tolerances and fits determine how the parts will fit. Incorrect tolerances could result in a product that is not usable. In addition to dimensioning and tolerancing, the function and the relationship that exists between the mating parts is important if the part is to perform the way it was designed. This aspect of the design process is addressed by **geometric dimensioning and tolerancing**, generally known as **GDT**.

Geometric dimensioning and tolerancing is a means to design and manufacture parts with respect to actual function and the relationship that exists between different features of the same part or the features of the mating parts. Therefore, a good design is not achieved by just giving dimensions and tolerances. The designer has to go beyond dimensioning and think of the intended function of the part and how the features of the part are going to affect its function. For example, Figure 8-44 shows a part with the required dimensions and tolerances. In this drawing there is no mention of the relationship between the pin and the plate. Is the pin perpendicular to the plate? If it is, to what degree should it be perpendicular? Also, it does not mention on which surface the perpendicularity of the pin is to be measured. A design like this is open to individual interpretation based on intuition and experience. This is where geometric dimensioning and tolerancing play an important part in the product design process.

Figure 8-45 has been dimensioned using geometric dimensioning and tolerancing. The feature symbols define the datum (reference plane) and the permissible deviation in the perpendicularity of the pin with respect to the bottom surface. From a drawing like this, the chances of making a mistake are minimized.

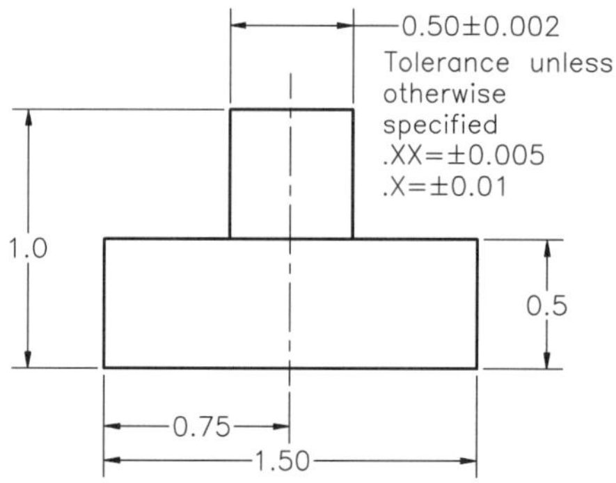

Figure 8-44 Using the traditional dimensioning and tolerancing technique

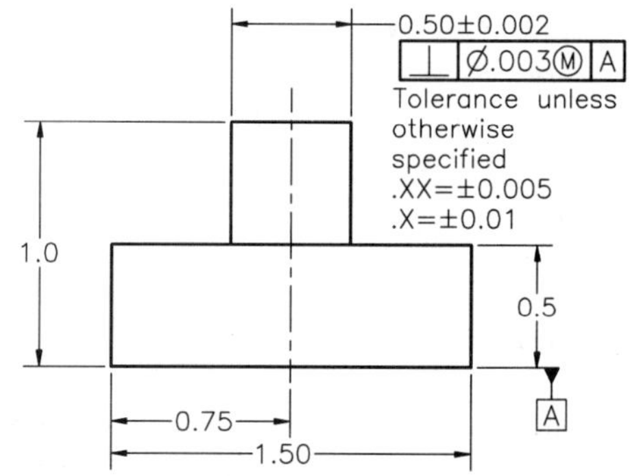

Figure 8-45 Using geometric dimensioning and tolerancing

GEOMETRIC CHARACTERISTICS AND SYMBOLS

Before discussing the application of AutoCAD LT commands in geometric dimensioning and tolerancing, you need to understand the following feature symbols and tolerancing components. Figure 8-46 shows the geometric characteristics and symbols used in geometric dimensioning and tolerancing. These symbols are the building blocks of geometric dimensioning and tolerancing.

KIND OF FEATURE	TYPE OF FEATURE	CHARACTERISTICS	
RELATED	LOCATION	Position	⌖
		Concentricity or Coaxiality	◎
		Symmetry	⌯
	ORIENTATION	Parallelism	//
		Perpendicularity	⊥
		Angularity	∠
INDIVIDUAL	FORM	Cylindricity	⌭
		Flatness	⏥
		Circularity or Roundness	○
		Straightness	—
INDIVIDUAL or RELATED	PROFILE	Surface Profile	⌓
		Line Profile	⌒
RELATED	RUNOUT	Circular Runout	↗
		Total Runout	⌰

Figure 8-46 Characteristics and symbols used in GTOL

ADDING GEOMETRIC TOLERANCE (TOLERANCE COMMAND)

Toolbar: Dimension > Tolerance
Menu: Dimension > Tolerance
Command: TOLERANCE

Geometric Tolerance displays the deviations of profile, orientation, form, location and runout of a feature. In AutoCAD LT, geometrical tolerancing is displayed by feature control frames. The frames contain all the information about tolerances for a single dimension. To display feature control frames with the various tolerancing parameters, the specifications are entered in the **Geometric Tolerance** dialog box (Figure 8-47).

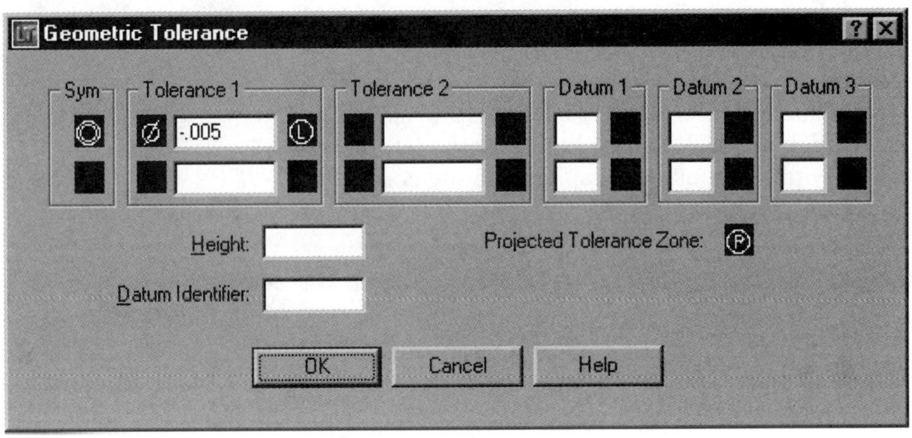

***Figure 8-47** Geometric Tolerance dialog box*

The various components that constitute the GTOL are shown in Figure 8-48 and Figure 8-49.

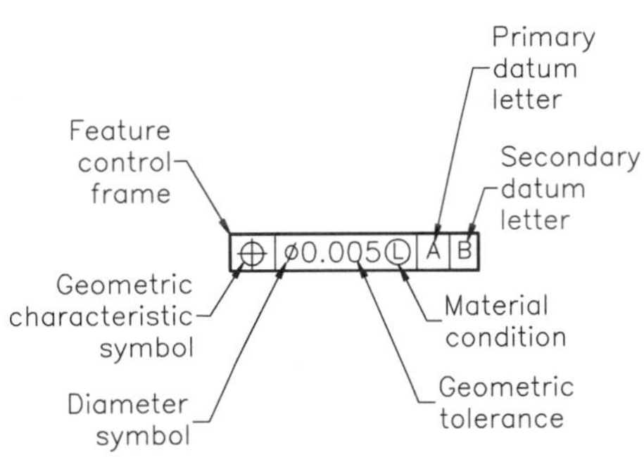

***Figure 8-48** Components of GTOL*

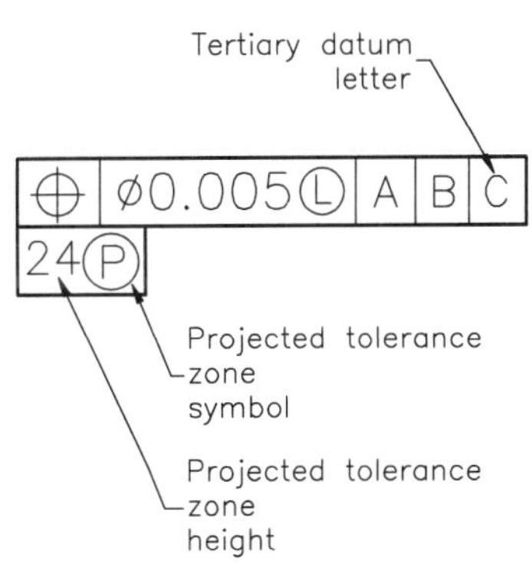

***Figure 8-49** Components of GTOL*

Feature Control Frame

The **feature control frame** is a rectangular box that contains the geometric characteristics symbols and tolerance definition. The box is automatically drawn to standard specifications; you do not need to specify its size. You can copy, move, erase, rotate, and scale the feature control frame. You can also snap to them using various Object snap modes. You can edit feature control frames using the **DDEDIT** command or you can also edit them using **GRIPS**. System variable **DIMCLRD** controls the color of the feature control frame. System variable **DIMGAP** controls the gap between the feature control frame and the text.

Geometric Characteristics Symbol

The geometric characteristics symbols indicate the characteristics of a feature like straightness, flatness, perpendicularity and so on. You can select the required symbol from the **Symbol** dialog box (Figure 8-50). This dialog box is displayed by selecting the box provided in the **Sym** area of the **Geometric Tolerance** dialog box. To select the required symbol, just pick the symbol using the left mouse button. The symbol will now be displayed in the box under the **Sym** area.

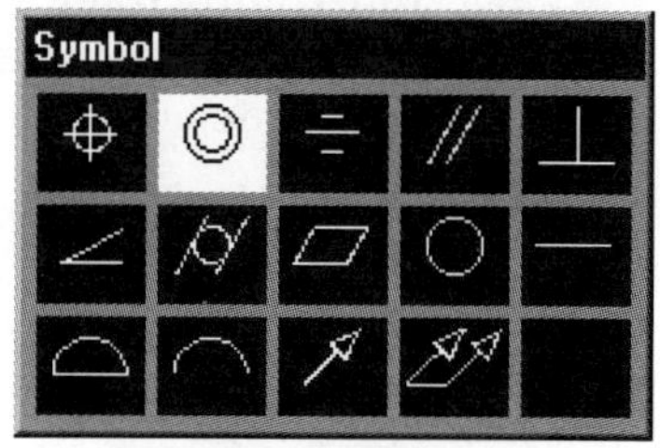

***Figure 8-50* *Symbol* dialog box**

Tolerance Value and Tolerance Zone Descriptor

The tolerance value specifies the tolerance on the feature as indicated by the tolerance zone descriptor. For example, a value of .003 indicates that the feature must be within a 0.003 tolerance zone. Similarly, ϕ.003 indicates that this feature must be located at true position within 0.003 diameter. The tolerance value can be entered in the edit box provided under the **Tolerance 1** or the **Tolerance 2** area of the **Geometric Tolerance** dialog box. The tolerance zone descriptor can be invoked by selecting the box, located to the left of the edit box. The system variable **DIMCLRT** controls the color of the tolerance text, variable **DIMTXT** controls the tolerance text size, and variable **DIMTXSTY** controls the style of the tolerance text. Using the **Projected Tolerance Zone**, inserts a projected tolerance zone symbol, which is an encircled P, after the projected tolerance zone value.

Material Condition Modifier

The **material condition modifier** specifies the material condition when the tolerance value takes effect. For example, ϕ.003(M) indicates that this feature must be located at true position within a 0.003 diameter at maximum material condition (MMC). The material condition modifier symbol can be selected from the **Material Condition** dialog box (Figure 8-51). This dialog box can be invoked by selecting the boxes located on the right side of the edit boxes under the **Tolerance 1**, **Tolerance 2**, **Datum 1**, **Datum 2**, and **Datum 3** area of the **Geometric Tolerance** dialog box.

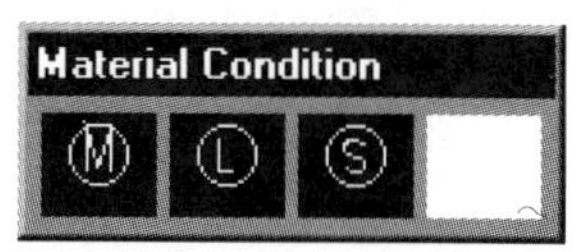

***Figure 8-51 Material Condition* dialog box**

Datum

The datum is the origin, surface, or feature from which the measurements are made. The datum is also used to establish the geometric characteristics of a feature. The datum feature symbol consists of a reference character enclosed in a feature control frame. You can create the datum feature symbol by entering characters (like -A-) in the **Datum Identifier** edit box in the **Geometric Tolerance** dialog box and then selecting a point where you want to establish this datum.

You can also combine datum references with geometric characteristics. AutoCAD LT automatically positions the datum references on the right end of the feature control frame.

Example 2 *Mechanical*

In the following example, you will create a feature control frame to define perpendicularity specification (see Figure 8-52).

1. Chose the **Tolerance** button from the **Dimension** toolbar to display the **Geometric Tolerance** dialog box. Choose the upper box from the **Sym** area to display the **Symbol** dialog box. Select the **perpendicularity** symbol. It will now be displayed in the **Sym** area.

2. Select the box available on the left of the upper edit box under the **Tolerance 1** area. A diameter symbol will appear to denote a cylindrical tolerance zone.

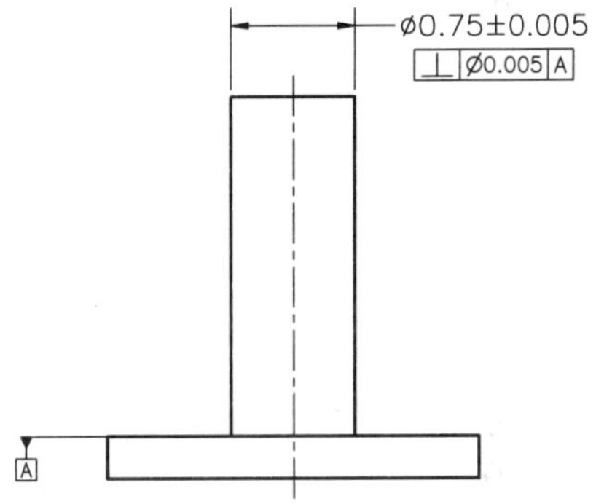

Figure 8-52 *Drawing for Example 2*

3. Enter **0.005** in the upper edit box under the **Tolerance 1** area.

4. Enter **A** in the edit box under the **Datum 1** area. Choose the **OK** button to accept the changes made in the **Geometric Tolerance** dialog box.

5. The **Enter tolerance location** prompt is displayed in the Command line area and the **Feature Control Frame** is attached to the cursor at its middle left point. Select a point to insert the frame.

6. To place the datum symbol, use the **TOLERANCE** command to display the **Geometric Tolerance** dialog box. In the **Datum Identifier** edit box, enter **A.**

7. Choose the **OK** button to accept the changes to the **Geometric Tolerance** dialog box, and then select a point to insert the frame.

COMPLEX FEATURE CONTROL FRAMES

Combining Geometric Characteristics

Sometimes it is not possible to specify all geometric characteristics in one frame. For example, Figure 8-53 shows the drawing of a plate with a hole in the center.

In this part, it is determined that surface C must be perpendicular to surfaces A and B within 0.002 and 0.004, respectively. Therefore, we need two frames to specify the geometric characteristics of surface C. The first frame specifies the allowable deviation in perpendicularity of surface C with respect to surface A. The second frame specifies the allowable deviation in perpendicularity of surface C with respect to surface B. In addition to these two frames, we need a third frame that identifies datum surface C.

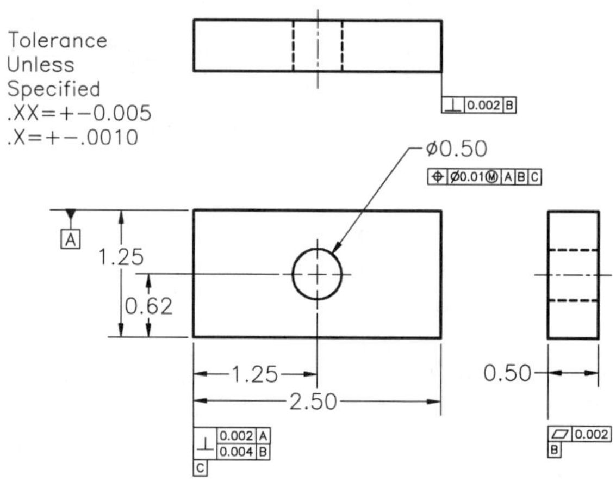

Figure 8-53 *Combining feature control frames*

All the three feature control frames can be defined in one instance of the **TOLERANCE** command.

1. Choose the **Tolerance** button to invoke the **Geometric Tolerance** dialog box. Select box under the **Sym** area to display the **Symbol** dialog box. Select the **perpendicular** symbol. AutoCAD LT will display the selected symbol in the first row of the **Sym** area.

2. Enter **0.002** in the edit box of the first row under the **Tolerance 1** area and enter **A** in the first row edit box under the **Datum 1** area.

3. Select the second row box under the **area** to display the **Symbol** dialog box. Select the **perpendicular** symbol. AutoCAD LT will display the selected symbol in the second row box of the **Sym** area.

4. Enter **.004** in the second row edit box under the **Tolerance 1** area and enter **B** in the second row edit box under the **Datum 1** area.

5. In the **Datum Identifier** edit box enter **C**, and then choose the **OK** button to exit the dialog box.

6. In the graphics screen, select the position to place the frame.

7. Similarly, create the remaining feature control frames.

COMPOSITE POSITION TOLERANCING

Sometimes the accuracy required within a pattern is more important than the location of the pattern with respect to the datum surfaces. To specify such a condition, composite position tolerancing may be used. For example, Figure 8-54 shows four holes (pattern) of diameter 0.15. The design allows a maximum tolerance of 0.025 with respect to datums A, B, and C at the maximum material condition (holes are smallest). The designer wants to maintain a closer positional tolerance (0.010 at MMC) between the holes within the pattern. To specify this requirement, the designer must insert the second frame. This is generally known as composite position tolerancing. AutoCAD LT provides the facility to create the two composite position tolerance frames by means of the **Geometric Tolerance** dialog box. The composite tolerance frames can be created as follows:

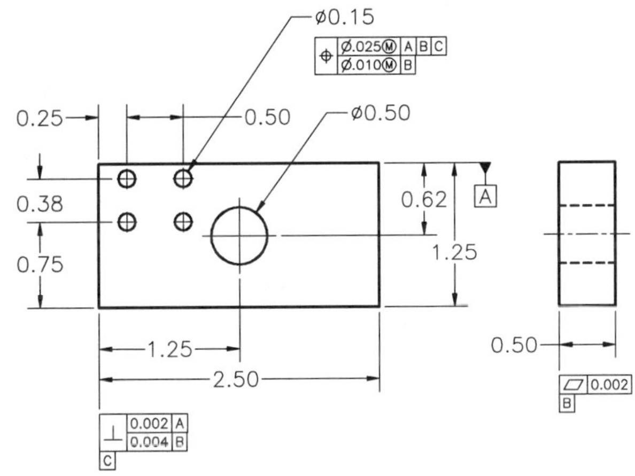

Figure 8-54 *Using composite position tolerancing*

1. Invoke the **TOLERANCE** command to display the **Geometric Tolerance** dialog box. Select box under the **Sym** area to display the **Symbol** dialog box. Select the **position** symbol. AutoCAD LT will display the selected symbol in the first row of the **Sym** area.

2. In the first row of the **Geometric Tolerance** dialog box, enter the geometric characteristics and the datum references required for the first position tolerance frame.

3. In the second row of the **Geometric Tolerance** dialog box, enter the geometric characteristics and the datum references required for the second position tolerance frame.

4. When you have finished entering the values, choose the **OK** button in the **Geometric Tolerance** dialog box, and then select the point where you want to insert the frames. AutoCAD LT will create the two frames and automatically align them with the common position symbol, as shown in Figure 8-54.

Chapter 8

USING FEATURE CONTROL FRAMES WITH LEADERS

The **Leader Settings** dialog box invoked using the **QLEADER** command has the Tolerance option, that allows you to create the feature control frame and attach it to the end of the leader extension line, (Figure 8-55). The following is the prompt sequence for using the **QLEADER** command with the Tolerance option:

Specify first leader point, or [Settings] <Settings>: *Press ENTER to display the* ***Leader Settings*** *dialog box. Choose the* ***Annotation*** *tab. Now, select the* ***Tolerance*** *radio button from the* ***Annotation Type*** *area. Choose* ***OK****.*
Specify first leader point, or [Settings] <Settings>: *Specify the start point of the leader line.*
Specify next point: *Specify the second point.*
Specify next point: *Specify the third point to display the* ***Geometric Tolerance*** *dialog box.*

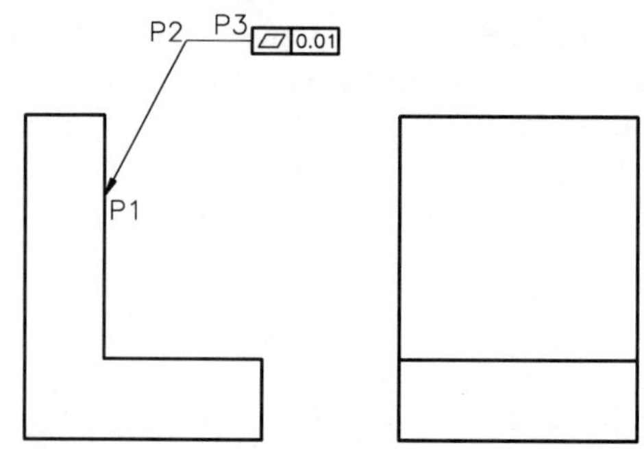

Figure 8-55 *Using the feature control frame with leaders*

PROJECTED TOLERANCE ZONE

Figure 8-56 shows two parts joined with a bolt. The lower part is threaded, and the top part has a drilled hole. When these two parts are joined, the bolt that is threaded in the lower part will have the orientation error that exists in the threaded hole. In other words, the error in the threaded hole will extend beyond the part thickness, which might cause interference, and the parts may not assemble. To avoid this problem, projected tolerance is used. The projected tolerance establishes a tolerance zone that extends above the surface. In Figure 8-56, the position tolerance for the threaded hole is 0.010, which extends 0.1 above the surface (datum A). By using the projected tolerance, you can ensure that the bolt is within the tolerance zone up to the specified distance.

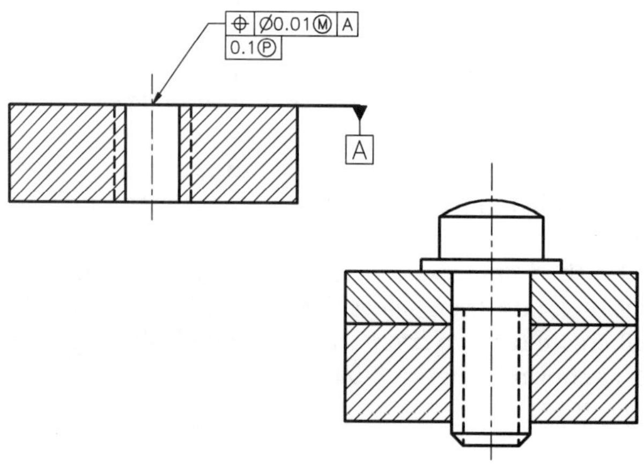

Figure 8-56 *Projected tolerance zone*

You can use the AutoCAD LT GDT feature to create feature control frames for the projected tolerance zone as follows:

1. Invoke the **QLEADER** command and then press ENTER at the **Specify first leader point, or [Settings] <Settings>** prompt to display the **Leader Settings** dialog box.

2. Choose the **Annotation** tab and then select the **Tolerance** radio button from the **Annotation Type** area. Choose **OK** to return to the Command line. Specify the first, second, and the third leader point as shown in Figure 8-56. As soon as you specify the third point, the **Geometric Tolerance** dialog box will be displayed. Choose the position symbol from the **Symbol** dialog box.

3. In the first row of the **Geometric Tolerance** dialog box, enter the geometric characteristics and the datum references required for the first position tolerance frame.

4. In the **Height** edit box, enter the height of the tolerance zone (0.1 for the given drawing) and select the box on the right of Projected Tolerance Zone. The projected tolerance zone symbol will be displayed in the box.

5. Choose the **OK** button in the **Geometric Tolerance** dialog box. AutoCAD LT will create the two frames and automatically align them, as shown in Figure 8-56.

6. Similarly, attach the datum symbol also.

Example 3 — *Mechanical*

In the following example, you will create a leader with a combination feature control frame to control runout and cylindricity (Figure 8-57).

1. Choose the **Quick Leader** button from the **Dimension** toolbar to invoke the **QLEADER** command. The prompt sequence is as follows:

   Specify first leader point, or [Settings] <Settings>: *Press ENTER to display the **Leader Settings** dialog box. Choose the **Annotation** tab. Select the **Tolerance** radio button from the **Annotation Type** area. Choose **OK**.*
   Specify first leader point, or [Settings] <Settings>: *Specify the leader start point as shown in Figure 8-57.*
   Specify next point: *Specify the second point of the leader.*
   Specify next point: *Specify the third point of the leader line to display the **Tolerance** dialog box.*

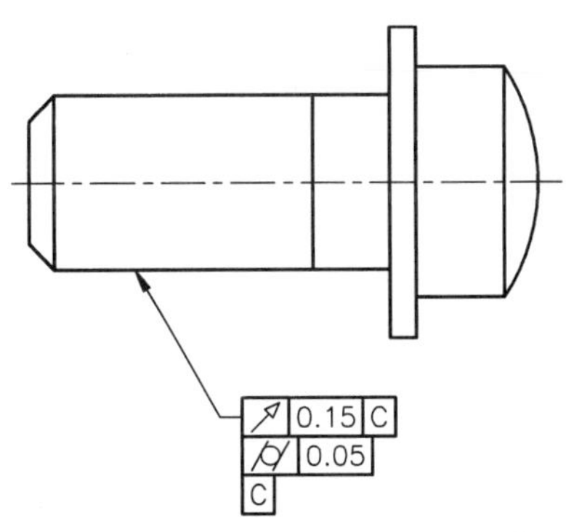

Figure 8-57 *Drawing for Example 3*

2. Choose the runout symbol from the **Symbol** dialog box. The **runout** symbol will be displayed on the first row of the **Sym** area. Enter **0.15** in the first row edit box under the Tolerance 1 area.

3. Enter **C** in the edit box under the **Datum 1** area.

4. Select the edit box on the second row of the **Sym** area and select the **cylindricity** symbol. The **cylindricity** symbol will be displayed in the second row of the **Sym** area.

5. Enter **0.05** in the second row edit box of the Tolerance 1 area.

6. Enter **C** in the **Datum Identifier** edit box.

7. Choose the **OK** button to accept the changes to the **Geometric Tolerance** dialog box. The control frames will be automatically attached at the end of the leader.

Self-Evaluation Test

Answer the following questions, and then compare your answers to the answers given at the end of this chapter.

1. You can specify dimension text of your own or accept the measured value computed by AutoCAD LT. (T/F)

2. The rotated dimension can be specified to get the effect of horizontal or vertical dimension. (T/F)

3. The center point of arc/circle is taken as the vertex angle in the angular dimensioning of the arc/circle. (T/F)

4. You can not combine GTOL with the leaders. (T/F)

5. The __________ command is used to dimension only the X coordinates of the selected object.

6. __________ symbols are the building blocks of geometric dimensioning and tolerancing.

7. In the __________ dimensions, the dimension text is, by default, aligned with the object being dimensioned.

8. The dimensions that are automatically updated when the object to which they are assigned change, are called __________.

9. The __________ point is taken as the vertex point of the angular dimensions while dimensioning the arc or circle.

10. The feature control frame is __________ in shape.

Review Questions

Answer the following questions.

1. Only inner angles (acute angles) can be dimensioned with angular dimensioning. (T/F)

2. In addition to the most recently drawn dimension (the default base dimension), you can use any other linear dimension as the base dimension. (T/F)

3. In continued dimensions, the base for successive continued dimensions is the base dimension's first extension line. (T/F)

4. In rotated dimensioning, the dimension line is always aligned with the object being dimensioned. (T/F)

5. The dimensions can be converted into true associative dimensions using which command?

   (a) **DIMREASSOCIATE** (b) **DIMASSOCIATE**
   (c) **DIMDISASSOCIATE** (d) **None**

6. How many GTOL symbols are available in AutoCAD LT?

   (a) 10 (b) 11
   (c) 12 (d) 14

7. You can dimension different points and features of a part with reference to a fixed point using which command?

   (a) **DIMBASELINE** (b) **DIMANGULAR**
   (c) **DIMRADIUS** (d) **DIMALIGNED**

8. You can add geometric dimensions and tolerance to the current drawing using which command?

   (a) **GTOL** (b) **TOLERANCE**
   (c) **TEXT** (d) **None**

9. Which option of the **QLEADER** command allows you to make the settings for attaching tolerances to the leader?

   (a) **Settings** (b) **Set**
   (c) **Undo** (d) **None**

10. Geometric dimensioning and tolerancing is generally known as __________.

11. Give three examples of geometric characteristics that indicate the characteristics of a feature. ____.

12. The two ways to return to the Command prompt from the **Dim** prompt (dimensioning mode) are __________, and __________.

13. The fundamental dimensioning types provided by AutoCAD LT are ____________________ ____.

14. Horizontal dimensions measure displacement along the __________.

15. Vertical dimensions measure displacement along the __________.

Exercises

Exercise 7 *Mechanical*

Draw the object shown in Figure 8-58, and then dimension it. Save the drawing as a DIMEXR7 drawing file.

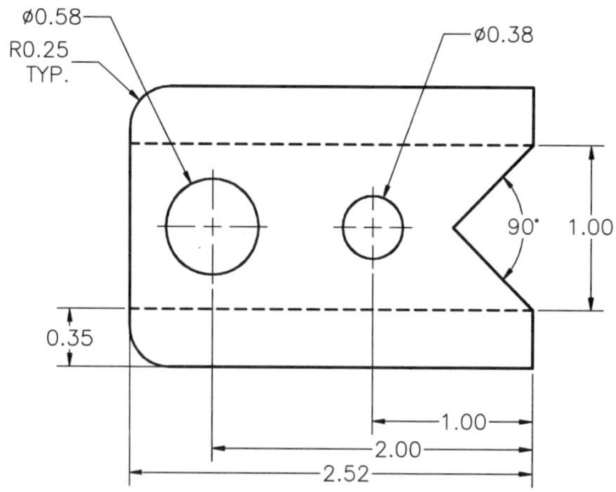

Figure 8-58 *Drawing for Exercise 7*

Exercise 8 *Mechanical*

Draw and dimension the object shown in Figure 8-59. Save the drawing as DIMEXR8.

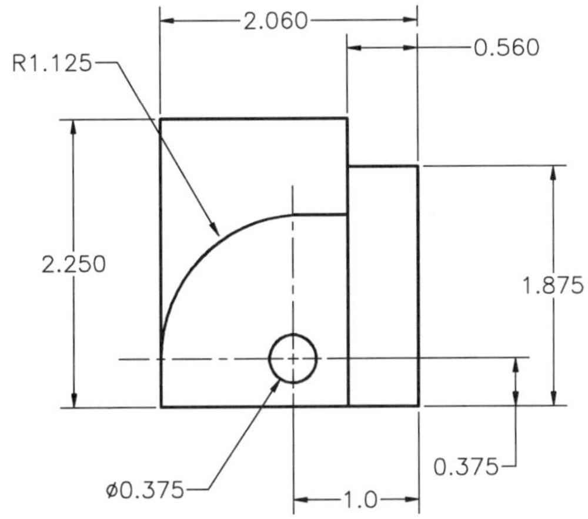

Figure 8-59 *Drawing for Exercise 8*

Exercise 9 *Mechanical*

Draw the object shown in Figure 8-60, then dimension it. Save the drawing as DIMEXR9.

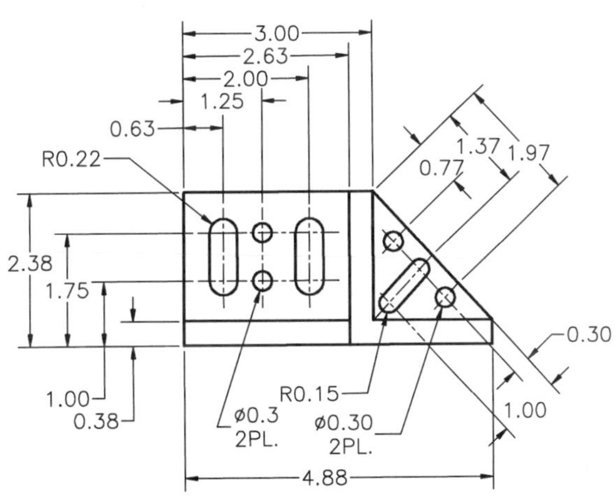

Figure 8-60 *Drawing for Exercise 9*

Exercise 10 *Mechanical*

Draw the object shown in Figure 8-61, then dimension it. Save the drawing as DIMEXR10.

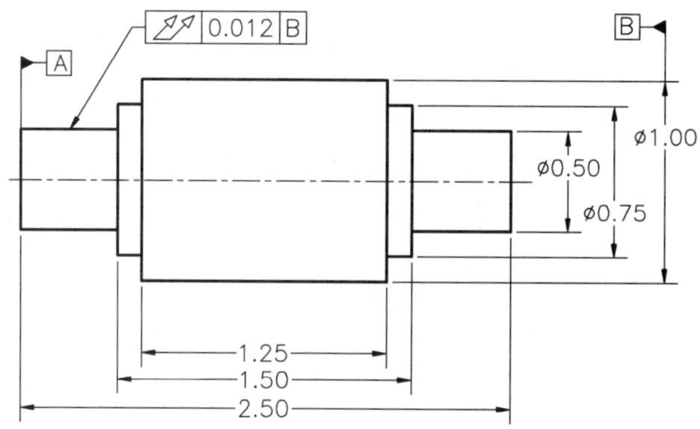

Figure 8-61 *Drawing for Exercise 10*

Problem Solving Exercise 1 *Mechanical*

Draw Figure 8-62 and dimension it as shown in the drawing. The **FILLET** command should be used where needed. Save the drawing as DIMPSE1.

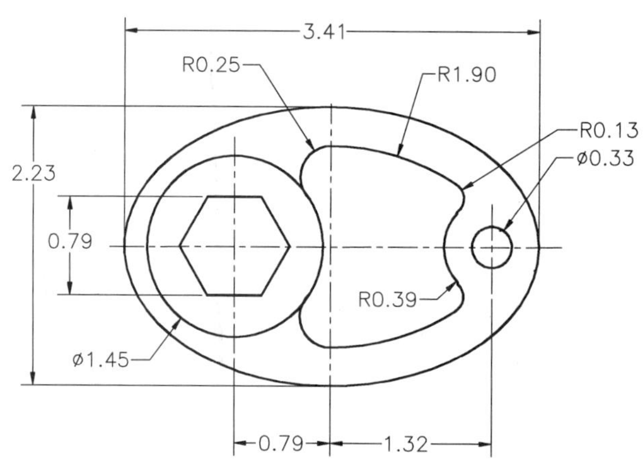

Figure 8-62 *Drawing for Problem Solving Exercise 1*

Problem Solving Exercise 2

Mechanical

Draw Figure 8-63 and then dimension it as shown in the drawing. Save the drawing as DIMPSE2.

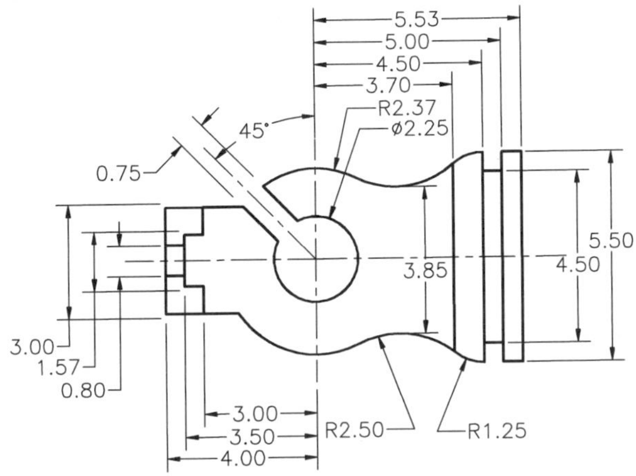

Figure 8-63 *Drawing for Problem Solving Exercise 2*

Problem Solving Exercise 3

Mechanical

Draw Figure 8-64 and then dimension it as shown in the drawing. Save the drawing as DIMPSE3.

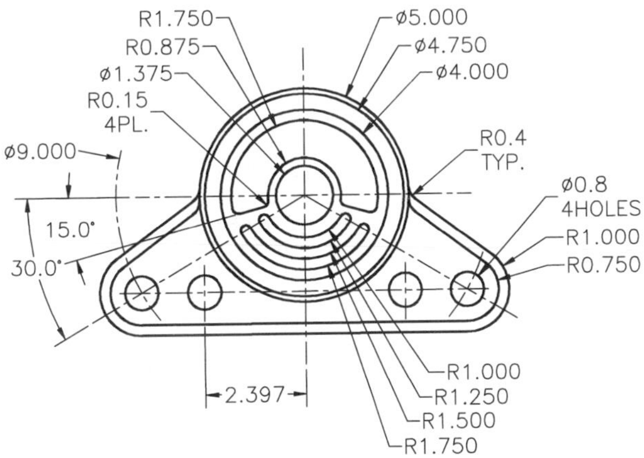

Figure 8-64 *Drawing for Problem Solving Exercise 3*

Problem Solving Exercise 4

Mechanical

Draw the three orthographic views of an object as shown in Figure 8-65 and then dimension it as shown in the drawing. Save the drawing as DIMPSE4.

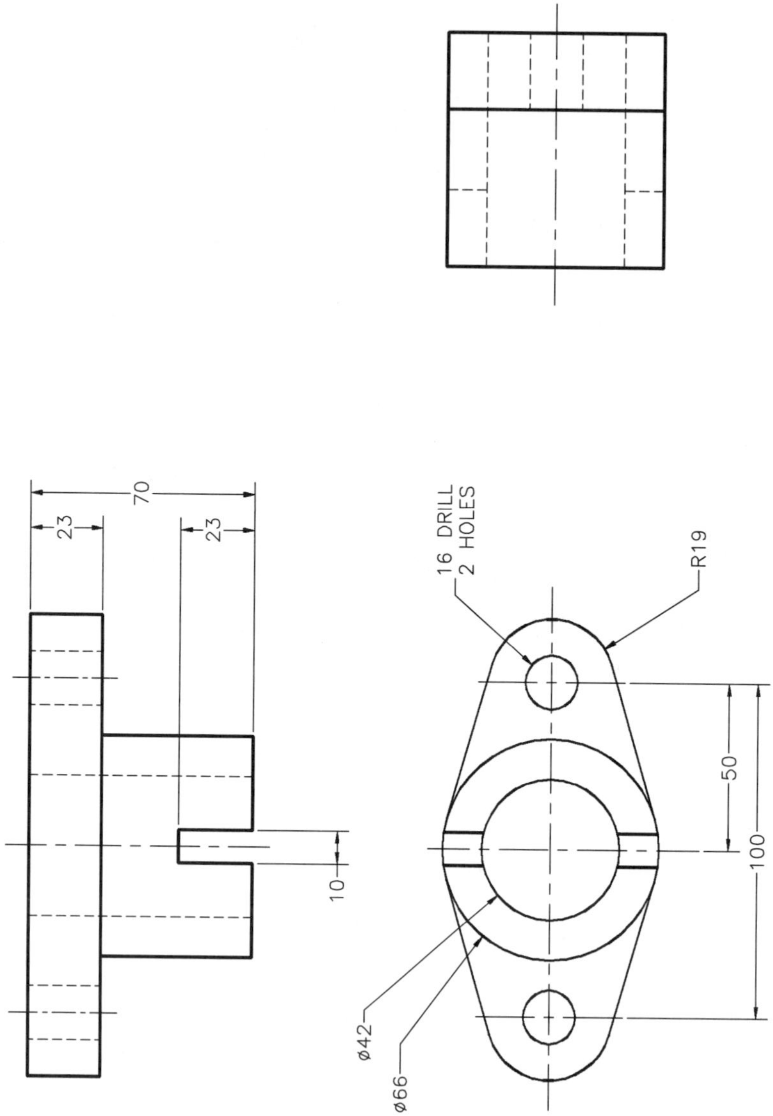

Figure 8-65 *Drawing for Problem Solving Exercise 4*

Problem Solving Exercise 5 — *Architectural*

Draw the floor plan as shown in Figure 8-66 and then give the dimension as shown in the drawing. Save the drawing as DIMPSE5.

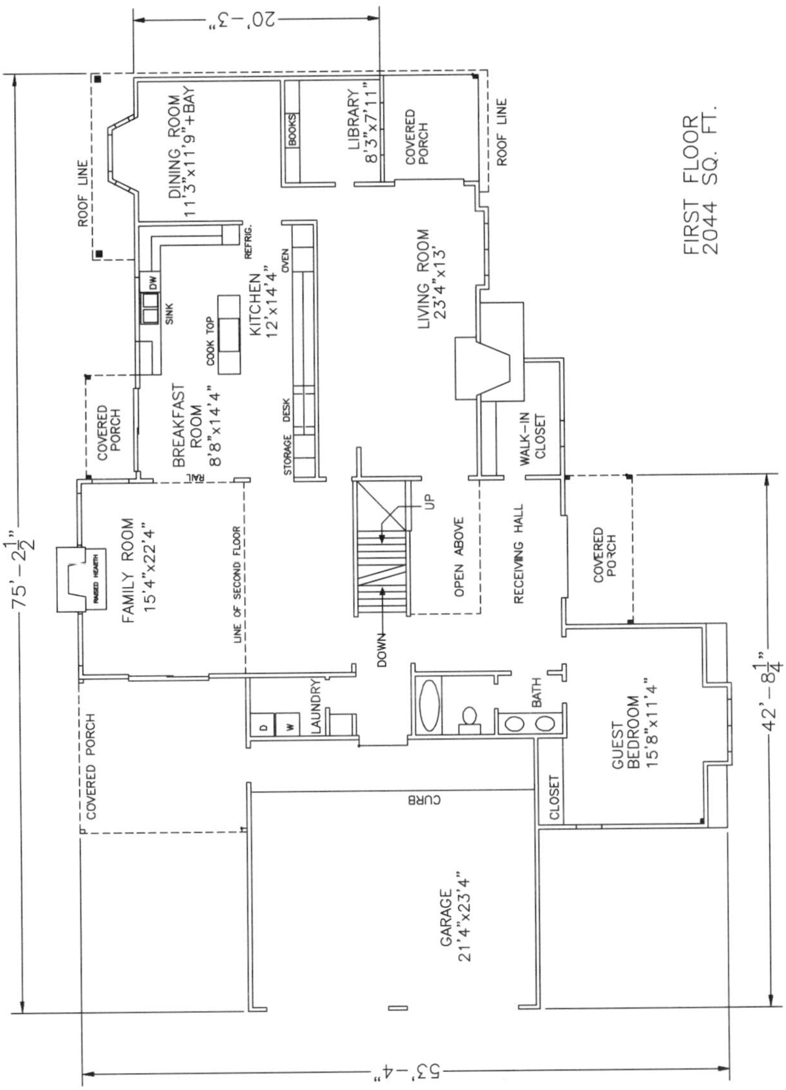

Figure 8-66 *Drawing for Problem Solving Exercise 5*

Answers to Self-Evaluation Test

1 - T, **2** - T, **3** - T, **4** - F, **5** -**DIMORDINATE**, **6** - geometric characteristics, **7** - aligned, **8** - true associative dimensions, **9** - center, **10** - rectangular

Chapter 9

Editing Dimensions

Learning Objectives

After completing this chapter, you will be able to:

- *Edit dimensions.*
- *Stretch, extend, and trim dimensions.*
- *Use the **DIMEDIT** and **DIMTEDIT** command options to edit dimensions.*
- *Update dimensions using the **DIM** command, update **DIMSTYLE**, and apply commands.*
- *Use the **PROPERTIES** command to edit dimensions.*
- *Dimension in model space and paper space.*

EDITING DIMENSIONS

For editing dimensions, AutoCAD LT has provided some special editing commands that work with dimensions. These editing commands can be used to define new dimension text, return to home text, create oblique dimensions, and rotate and update the dimension text. You can also use the **TRIM**, **STRETCH**, and **EXTEND** commands to edit the dimensions. In case the dimension assigned to the object is a true associative dimension, it will be automatically updated if the object is modified. However, if the dimension is not a true associative dimension, you will have to include the dimensions along with the objects in the edit selection set. The properties of the dimensioned objects can also be changed using the **Properties** window or **Dimension Style Manager**.

Editing the Dimensions by Stretching

You can edit the dimension object by stretching it. However, to stretch a dimension object, appropriate definition points must be included in the selection crossing or window. As the middle point of the dimension text is a definition point for all types of dimensions, you can easily stretch and move the dimension text to any location you want. When you stretch the dimension text, the gap in the dimension line gets filled automatically. When editing, the definition points of the dimension being edited must be included in the selection crossing box. The dimension is automatically calculated when you stretch the dimensions.

Note

The dimension type remains the same after stretching. For example, the vertical dimension maintains itself as a vertical dimension and measures only the vertical distance even after the line it dimensions is modified and converted into an inclined line. The following example illustrates the stretching of object lines and dimensions.

Example 1 *Mechanical*

In this example you will stretch the objects and the dimensions shown in Figure 9-1 to a new location using grips. The new location of the lines and dimension is at a distance of 0.5 in the positive Y axis direction. See Figure 9-2.

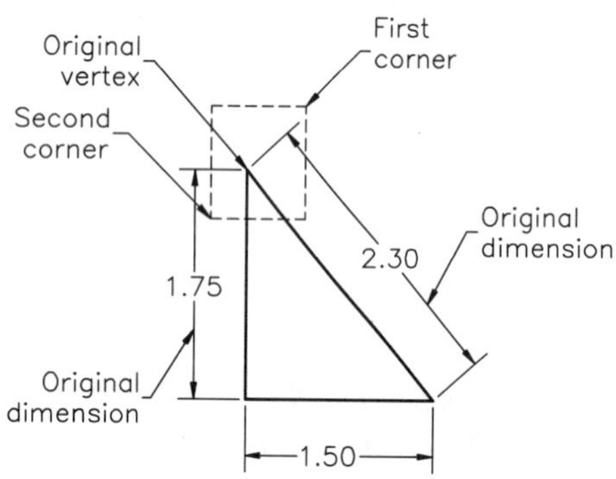

Figure 9-1 *Original location of lines and dimensions*

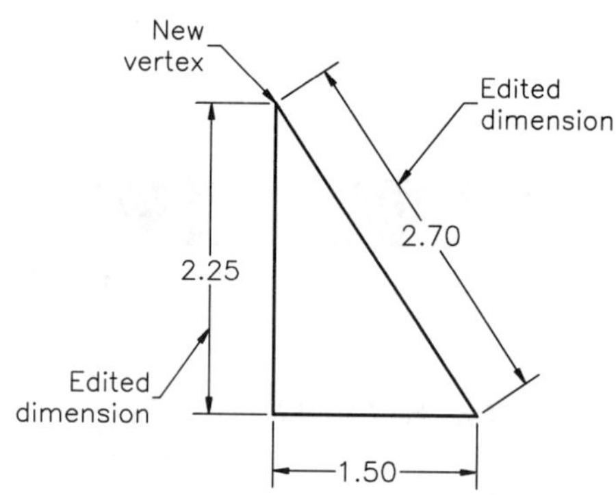

Figure 9-2 *New location of lines and dimensions*

1. Chose the **Stretch** button from the **Modify** toolbar. The prompt sequence is as follows:

   Select objects to stretch by crossing-window or crossing-polygon
   Select objects: Specify opposite corner: *Define a crossing window using the first and second corner as shown in Figure 9-1.*
   Select objects: Enter
   Specify base point or displacement: *Select original vertex using the osnaps as the base point.*
   Specify second point of displacement or <use first point as displacement>: **@0.5<90**

2. The selected entities will be stretched to the new location. The dimension that was initially 1.75, will become 2.25 and the dimension that was initially 2.30, will become 2.70 (Figure 9-2). Press ESC to remove the grip points from the objects.

Exercise 1 *General*

The two dimensions in Figure 9-3(a) are too close. Fix the drawing by stretching the dimension as shown in Figure 9-3(b).

1. Stretch the outer dimension to the right so that there is some distance between the two dimensions.
2. Stretch the dimension text of the outer dimension so that the dimension text is staggered (lower than the first dimension).

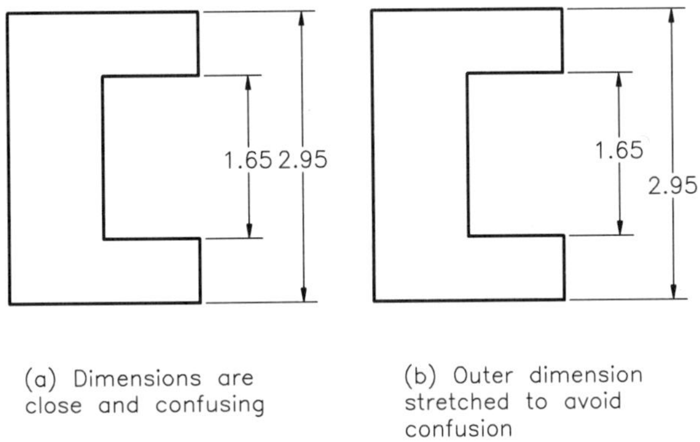

Figure 9-3 *Drawing for Exercise 1, stretching dimensions*

Editing Dimensions by Trimming and Extending

Trimming and extending operations can be carried out with all types of linear dimensions (horizontal, vertical, aligned, rotated) and the ordinate dimension. Even if the dimensions are true associative, you can trim and extend them. AutoCAD LT trims or extends a linear dimension between the extension line definition points and the object used as a boundary or trimming edge. To extend or trim an ordinate dimension, AutoCAD LT moves the feature location (location of the dimensioned coordinate) to the boundary edge. To retain the original ordinate value, the boundary edge to which the feature location point is moved should be orthogonal to the measured ordinate. In both cases, the imaginary line drawn between the two extension line definition points is trimmed or extended by AutoCAD LT, and the dimension is adjusted automatically. Figure 9-4 and Figure 9-5 shows the dimensions edited by trimming and extending.

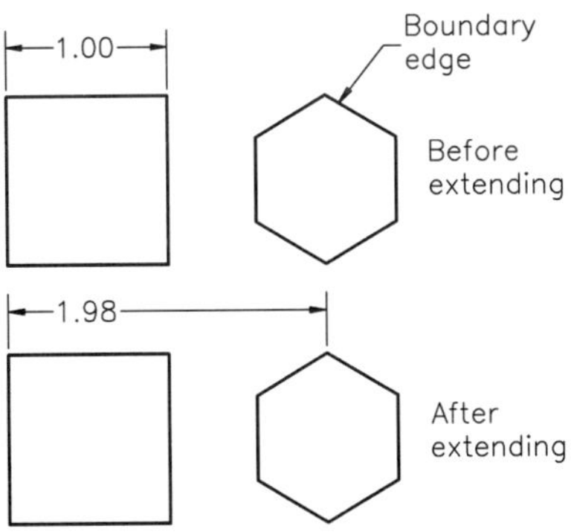

Figure 9-4 *Dimensions edited by extending*

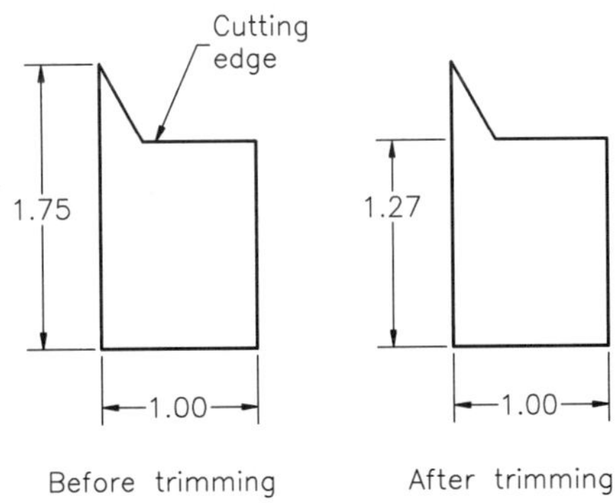

Figure 9-5 *Dimensions edited by trimming*

Chapter 9

Exercise 2 — Mechanical

Use the **Edgemode > Extend** option of the **TRIM** command to trim the dimension in Figure 9-6(a) so it looks like Figure 9-6(b).

1. Make the drawing and dimension it as shown in Figure 9-6(a). Assume the dimensions where necessary.
2. Trim the dimensions by setting the **Edgemode** option of the **TRIM** command to **Extend**.

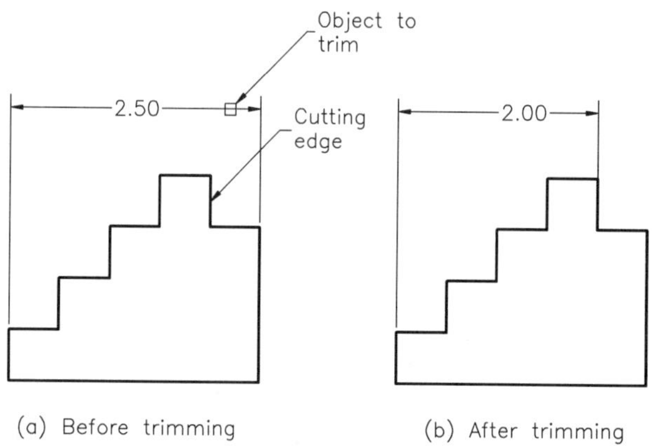

Figure 9-6 *Drawing for Exercise 2*

EDITING THE DIMENSIONS

Toolbar:	Dimension > Dimension Edit
Command:	DIMEDIT

The dimension text can be edited by using the **DIMEDIT** command (Figure 9-7). This command has four options: New, Rotate, Home, and Oblique. The prompt sequence that will follow when you choose this button is:

Enter type of dimension editing (Home/New/Rotate/Oblique) <Home>: *Enter an option.*

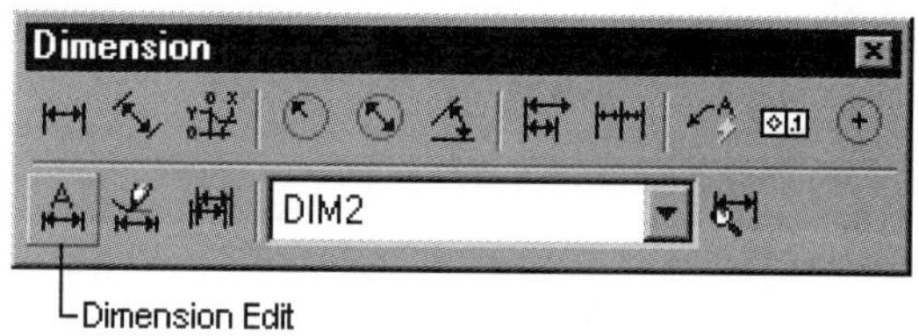

Figure 9-7 *Invoking the **DIMEDIT** command from the toolbar*

New

The **New** option is used to replace the existing dimension text with a new text string. When you invoke this option, the **Multiline Text Editor** will be displayed (see Figure 9-8). The default text will be displayed as angular brackets (<>). You can change the default text or add

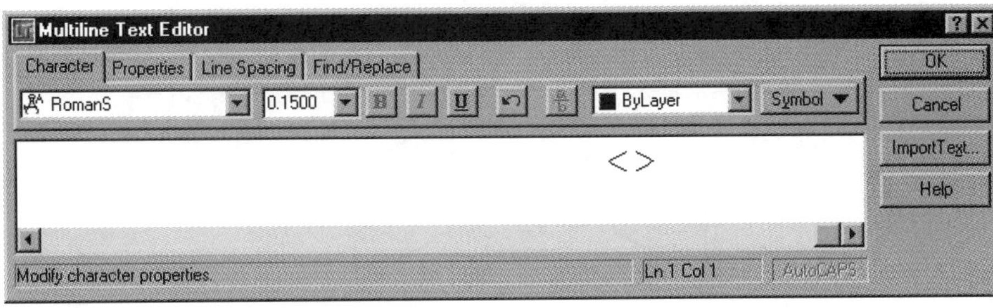

Figure 9-8 Multiline Text Editor *for editing the default text*

prefix or suffix to the text using this **Multiline Text Editor**. Once you have edited the dimension in the editor and chosen **OK**, you will be prompted to select the dimension to be changed. The <> prefix or suffix facility can also be used when you enter the new text string. In this case, the prefix/suffix is appended before or after the dimension measurement that is placed instead of the <> characters. For example, if you enter Ref-Dim before <> and mm after <> (Ref-Dim<>mm), AutoCAD LT will attach Ref-Dim as prefix and mm as suffix (Ref-Dim1.1257mm).

If information about the dimension style is available on the selected dimension, it is used to redraw the dimension, or the prevailing variable settings are used for the redrawing process.

Rotate

The **Rotate** option is used to position the dimension text at a specified angle. With this option, you can change the orientation (angle) of the dimension text of any number of associative dimensions. The angle can be specified by entering its value at the **Specify angle for dimension text** prompt or by specifying two points at the required angle. Once you have specified the angle, you will be prompted to select the dimension text to be rotated. You will notice that the text rotates around its middle point. See Figure 9-9.

Home

The Home option restores the text of a dimension to its original (home/default) location if the position of the text has been changed by stretching or editing. See Figure 9-9.

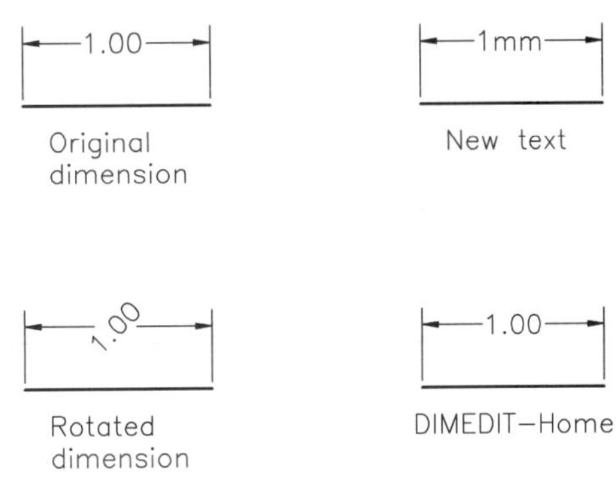

Figure 9-9 *Using the* ***DIMEDIT*** *command to edit dimensions*

Oblique

In the linear dimensions, extension lines are drawn perpendicular to the dimension line. The **Oblique** option bends the linear dimensions. It draws extension lines at an oblique angle (Figure 9-10). This option is particularly important to create isometric dimensions and can be used to resolve conflicting situations due to the overlapping of extension lines with other

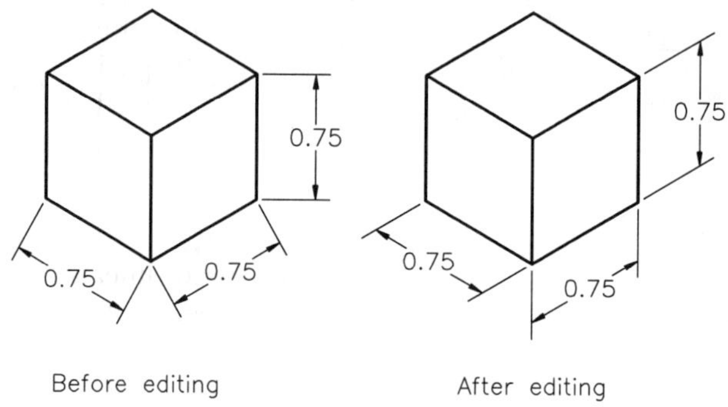

Figure 9-10 *Using the* ***Oblique*** *option to edit dimensions*

objects. Making an existing dimension oblique by specifying an angle oblique to it does not affect the generation of new linear dimensions. The oblique angle is maintained even after performing most editing operations. When you invoke this option, you will be prompted to select the dimension to be edited. After selecting the dimensions you will be prompted to specify the obliquing angle. The extensions lines will be bent at the angle specified. You can also invoke this option by choosing **Oblique** from the **Dimension** menu.

EDITING DIMENSION TEXT

Toolbar:	Dimension > Dimension Text Edit
Menu:	Dimension > Align Text
Command:	DIMTEDIT

The dimension text can also be edited by using the **DIMTEDIT** command. This command is used to edit the placement and orientation of a single existing dimension. You can apply this command, for example, in cases where dimension text of two or more dimensions are too close together, creating confusion. In such cases, the **DIMTEDIT** command is invoked to move the dimension text to some other location so that there is no confusion. The prompt sequence that will follow when you choose this button is:

Select dimension: *Select the dimension to modify.*
Specify new location for dimension text or [Left/Right/Center/Home/Angle]:

Left

With this option, you can left-justify the dimension text along the dimension line. The vertical placement setting determines the position of the dimension text. The horizontally aligned text is moved to the left and the vertically aligned text is moved down as shown in Figure 9-11. This option can be used only with the linear, diameter, and radius dimensions.

Right

With this option, you can right-justify the dimension text along the dimension line. Similar to the Left option, the vertical placement setting determines the position of the dimension text. The horizontally aligned text is moved to the right, and the vertically aligned text is moved up (Figure 9-11). This option can be used only with linear, diameter, and radius dimensions.

Center

With this option you can center-justify the dimension text for linear, and aligned dimensions (Figure 9-11). The vertical setting controls the vertical position of the dimension text.

Home

The **Home** option is used to restore (move) the dimension text of a dimension to its original (home/default) location if the position of the text has been changed (Figure 9-11).

Angle

With the Angle option, you can position the dimension text at the angle you specify (Figure 9-11). The angle can be specified by entering its value at the **Specify angle for dimension text** prompt or by specifying two points at the required angle. You will notice that the text rotates around its middle point. If the dimension text alignment is set to Orient Text Horizontally, the dimension text is aligned with the dimension line. If information about the dimension style is available on the selected dimension, AutoCAD LT uses it to redraw the dimension, or the prevailing dimension variable settings are used for the redrawing process. Entering 0 degree angle changes the text to its default orientation.

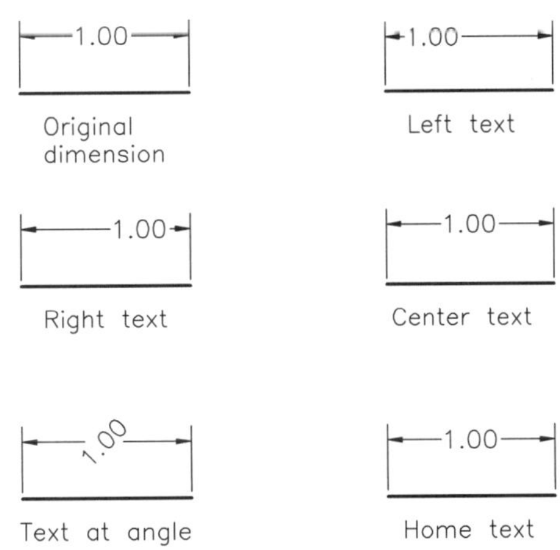

Figure 9-11 *Using the* ***DIMTEDIT*** *command to edit dimensions*

UPDATING DIMENSIONS

Toolbar:	Dimension > Dimension Update
Menu:	Dimension > Update

Choosing the **Dimension Update** button from the **Dimension** toolbar regenerates (updates) prevailing dimension objects (such as arrows and text height) using current settings for the dimension variables, dimension style, text style, and units. When you choose this button, you will be prompted to select the dimensions to be updated. You can select all the dimensions or specify the dimensions that should be updated.

EDITING DIMENSIONS WITH GRIPS

You can also edit dimensions by using GRIP editing modes. GRIP editing is the easiest and quickest way to edit dimensions. You can perform the following operations with GRIPS:

1. Position the text anywhere along the dimension line. You cannot move the text and position it above or below the dimension line.
2. Stretch a dimension to change the spacing between the dimension line and the object line.
3. Stretch the dimension along the length. When you stretch a dimension, the dimension text automatically changes.
4. Move, rotate, copy, or mirror the dimensions.
5. Relocate a dimension origin.
6. Change properties such as color, layer, linetype, and linetype scale.
7. Load Web browser (if any *Universal Resource Locator* is associated with the object).

EDITING DIMENSIONS USING THE PROPERTIES WINDOW

Toolbar:	Standard Toolbar > Properties
Menu:	Modify > Properties
Command:	PROPERTIES

You can also modify a dimension or leader by using the **Properties** window. The **Properties** window is displayed when you choose the **Properties** button from the **Standard Toolbar**. The **Properties** window can also be invoked by double-clicking on the dimension to be edited. All the properties of the selected object are displayed in the **Properties** window (see Figure 9-12).

Properties Window (Dimension)

You can use the **Properties** window (Figure 9-12) to change the properties of a dimension, change the dimension text style, or change the geometry, format, and annotation-related features of the selected dimension. The changes takes place dynamically in the drawing. The **Object Properties Window** is classified **Alphabetic** as well as **Categorized** under the respective tabs. The different categories for modification of dimensions are as follows.

General

In the general category, the various parameters displayed are **Color**, **Layer**, **Linetype**, **Linetype scale**, **Plot style**, **Lineweight**, **Associativity,** and **Hyperlink** with their current values. For example, if you want to change the color of the selected object, select Color property and then select the required color from the drop-down list. Similarly, Layer, Plot style, Linetype and

Lineweight can be changed from the respective drop-down lists. The Linetype scale can be changed manually at the corresponding cell.

Misc

This category displays the dimension style by name (for **the DIMSTYLE** system variable, use **SETVAR**). You can change the dimension style from the drop-down list for the selected dimension object.

Lines & Arrows

The various parameters of the lines and arrows in the dimension objects such as arrowhead size, type, arrow lineweight, and so on can be changed in this category.

Text

The different parameters that control the text in the dimension object such as text color, text height, vertical position text offset, and so on can be changed in this category.

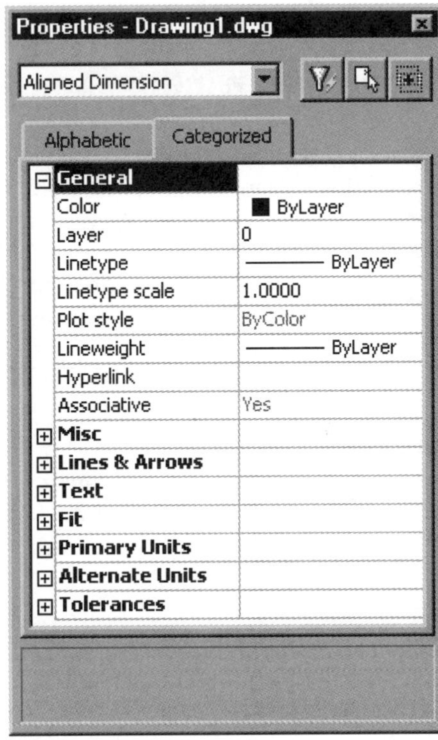

*Figure 9-12 The **Properties** window for dimensions*

Fit

In the fit category, the various parameters are **Dim line forced**, **Dim line inside**, **Dim scale overall**, **Fit**, **Text inside**, and **Text movement**. All the parameters can be changed by the drop-down list except Dim scale overall (which can be changed manually).

Primary Units

In the primary units category, the various parameters displayed are **Decimal separator**, **Dim prefix**, **Dim suffix**, **Dim roundoff**, **Dim scale linear**, **Dim units**, **Suppress leading zeroes**, **Suppress trailing zeroes**, **Suppress zero feet**, **Suppress zero inches** and **Precision**. Among these properties, Dim units, Suppress leading zeroes, Suppress trailing zeroes, Suppress zero feet, Suppress zero inches, and Precision properties can be changed with the corresponding drop-down lists and the other parameters can be changed manually.

Alternate Units

In the alternate unit category, there are various parameters for the alternate units. They can be changed only if the **Alt enabled** parameter is **on**. The parameters such as **Alt format**, **Alt precision**, **Alt suppress leading zeroes**, and **Alt suppress trailing zeroes**, **Alt suppress zero feet**, **Alt suppress zero inches** can be changed from the respective drop-down lists and others can be changed manually.

Tolerances

The parameters of this category can be changed only if the Tolerances display parameter has some mode of tolerance selected. The various parameters are available and correspond to the mode of tolerance selected.

Properties Window (Leader)

The **Properties** window for **Leader** can be invoked by selecting a Leader and then choosing the **Properties** button from the **Standard Toolbar**. You can also invoke the **Properties** window (Figure 9-13) from the shortcut menu by right-clicking in the drawing area and choosing **Properties**. This window can also be invoked by double-clicking in the leader to be edited. The various properties under the **Properties** window (Leader) are described as follows.

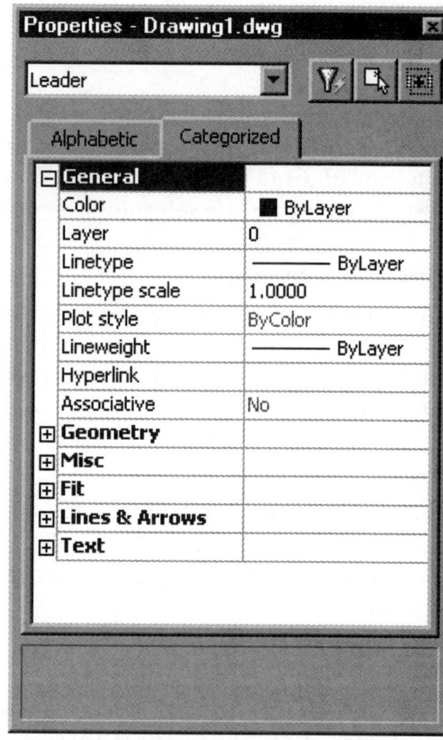

Figure 9-13 ***Properties*** *window for leaders*

General

The parameters in the general category are the same as those discussed in the previous section (**Object Properties Window** for dimensions).

Geometry

This category displays the coordinates of the Leader. The various parameters under this category are **Vertex**, **Vertex X**, **Vertex Y**, and **Vertex Z**. You can choose any vertex of the leader and change its coordinates.

Misc

This category displays the **Dim Style** and **Type** of the Leader. You can change the style name and the type of the Leader by using these properties.

Fit

This category displays the **Dim scale overall** property that specifies the overall scale factor applied to size, distances, or the offsets of the Leader.

Lines & Arrows

Lines and Arrows displays the different specifications of the arrowheads and linetypes for the Leader.

Text

Text category displays the **Text offset** to the dimension line and the vertical position (**Text pos vert**) of the dimension text and can be changed accordingly.

Example 2 *Mechanical*

In this example you will modify the dimensions in Figure 9-14 so that they match Figure 9-15.

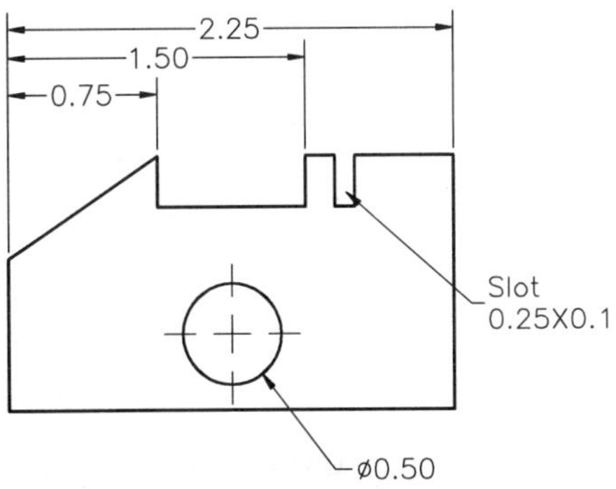

Figure 9-14 Drawing for Example 2

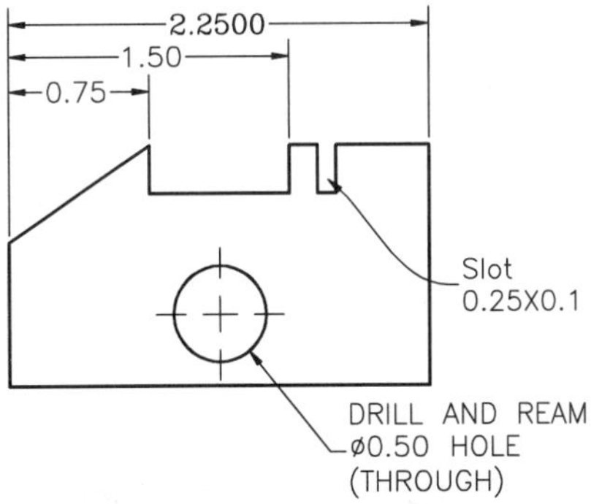

Figure 9-15 Drawing after editing the dimensions

1. Choose **Text Style** from the **Format** menu and create a style with the name **ROMANC**. Select **romanc.shx** as the font for the style.

2. Double-click on the dimension 2.25 to display the **Properties** window.

3. In the **Text** category, select the **Text style** drop-down list and then select **ROMANC** style from this drop-down list. The changes will take place dynamically.

4. Select **0.0000** from the **Precision** drop-down list in the **Primary Units** area.

5. Once all the required changes are made in the linear dimension, choose the **Select Object** button in the **Properties** window. You will be prompted to select the object. Select the leader line and then press ENTER.

6. The **Properties** window will not display the leader options. Select **Spline with arrow** from the **Type** drop-down list in the **Misc** category. The straight line will be converted into a spline with an arrow dynamically.

7. Close the **Properties** window.

8. Choose the **Dimension Edit** button from the **Dimension** toolbar. Enter **N** in the prompt sequence to display the **Multiline Text Editor**.

9. Enter **DRILL AND REAM %%C0.25 HOLE (THROUGH)** in the text editor and then choose **OK**.

10. You will be prompted to select the object to change. Select the diameter dimension and then press ENTER. The diameter dimension will be modified to the new value.

MODEL SPACE AND PAPER SPACE DIMENSIONING

Dimensioning objects can be drawn in model space or paper space. If the drawings are in model space, associative dimensions should also be created in model space. If the drawings are

Chapter 9

in model space and the associative dimensions are in paper space, the dimensions will not change when you perform such editing operations as stretching, trimming, and extending, or such display operations as zoom and pan in the model space viewport. The definition points of a dimension are located in the space where the drawing is drawn. You can select the **Scale dimensions to layout (paper space)** check box under the **Scale for Dimension Features** area in the **Fit** tab of the **Modify**, **New**, or **Override** dialog boxes in the **Dimension Style Manager** dialog box, depends on whether you want to modify the present style or you want to create a new style (see Figure 9-16). Choose **OK/OK** to exit from both the dialog boxes. AutoCAD LT calculates a scale factor that is compatible with the model space and the paper space viewports. Choose **Update** from the **Dimension** menu and select the dimension objects for updating.

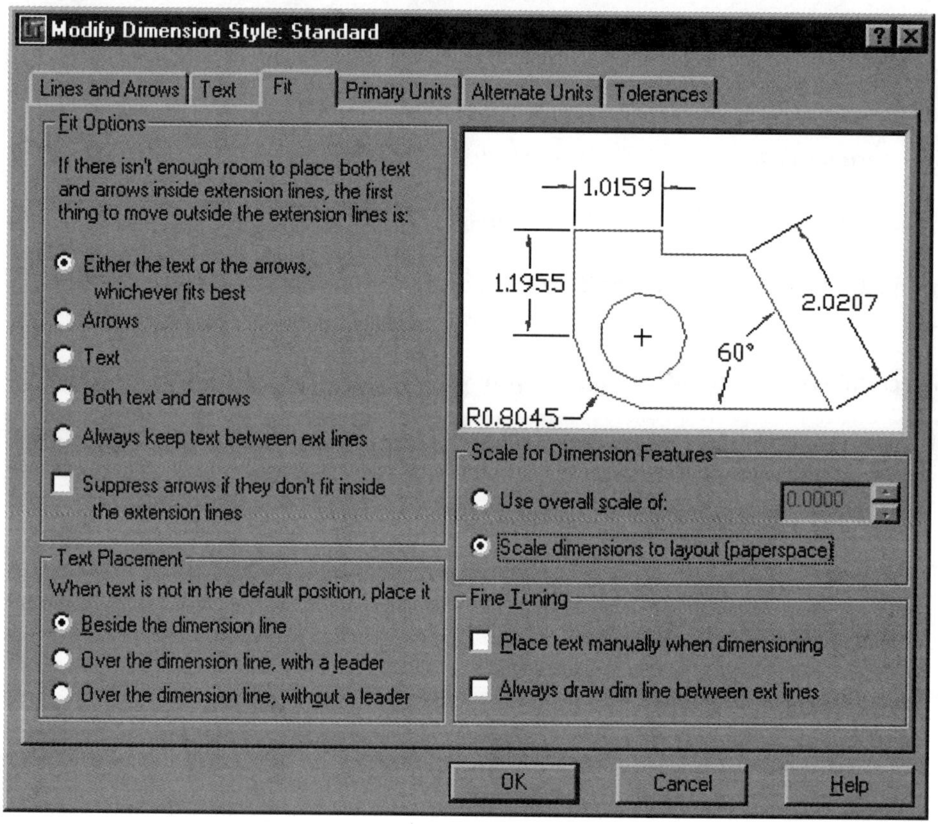

***Figure 9-16** Selecting paper space scaling in the **Modify Dimension Style** dialog box*

The drawing shown (Figure 9-17) uses paper space scaling. The main drawing and detail drawings are located in different floating viewports (paper space). The zoom scale factors for these viewports are different: 0.3XP, 1.0XP, and 0.5XP, respectively. When you use paper scaling, AutoCAD LT automatically calculates the scale factor for dimensioning so that the dimensions are uniform in all the floating viewports (model space viewports).

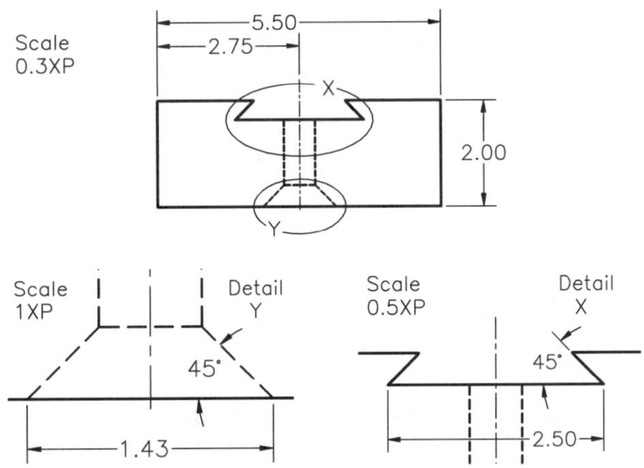

***Figure 9-17** Dimensioning in paper model space viewports using paper space scaling or setting **DIMSCALE** to 0*

Self-Evaluation Test

Answer the following questions, and then compare your answers to the correct answers given at the end of this chapter.

1. In associative dimensioning, the items constituting a dimension (such as dimension lines, arrows, leaders, extension lines, and dimension text) are drawn as a **single object.** (T/F)

2. If the value of the variable **DIMDASSOC** is set to zero, the dimension lines, arrows, leaders, extension lines, and dimension text are drawn as **independent** objects. (T/F)

3. You can not edit dimensions using the grips. (T/F)

4. The true associative dimensions can not be trimmed or extended. (T/F)

5. You can use the __________ command to break the dimensions into individual entities.

6. The __________ option of the **DIMTEDIT** command is used to justify the dimension text towards the left side.

7. The __________ option of the **DIMTEDIT** command is used to justify the dimension text to the center of the dimension.

8. The _________ option of the **DIMEDIT** command is used to create a new text string.

9. The _________ option of the **DIMEDIT** command is used to bend the extension lines through the specified angle.

10. The _________ button from the **Dimension** toolbar is used to update the dimensions.

Review Questions

Answer the following questions.

1. The horizontal, vertical, aligned and rotated dimensions can not be edited using the grips.

2. Trimming and extending operations can be carried out with all types of linear (horizontal, vertical, aligned, and rotated) dimensions and with the ordinate dimension. (T/F)

3. To extend or trim an ordinate dimension, AutoCAD LT moves the feature location (location of the dimensioned coordinate) to the boundary edge. (T/F)

4. Once moved from the original location, the dimension text can not be restored at its original position. (T/F)

5. With the _________ or _________ commands you can edit the dimension text.

6. The _________ command is particularly important for creating isometric dimensions and is applicable in resolving conflicting situations due to overlapping of extension lines with other objects.

7. The _________ command is used to edit the placement and orientation of a single existing dimension.

8. The _________ command regenerates (updates) prevailing associative dimension objects (like arrows, and text height) using the current settings for the dimension variables, dimension style, text style, and units.

9. Explain when to use the **EXTEND** command and how it works with dimensions. _________
____.

10. Explain the use and working of the **PROPERTIES** command for editing dimensions.
____.

Exercises

Exercise 3 *General*

1. Make the drawing as shown in Figure 9-18. Assume the dimensions where necessary.
2. Dimension the drawing as shown in Figure 9-18.
3. Edit the dimensions so that they match the dimensions shown in Figure 9-19.

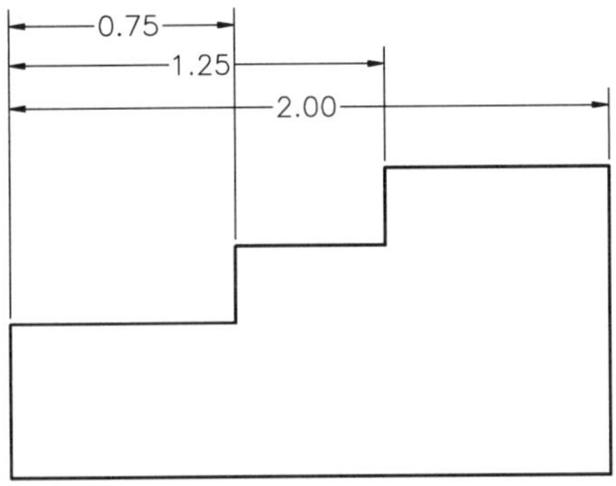

Figure 9-18 *Drawing for Exercise 3, before editing dimensions*

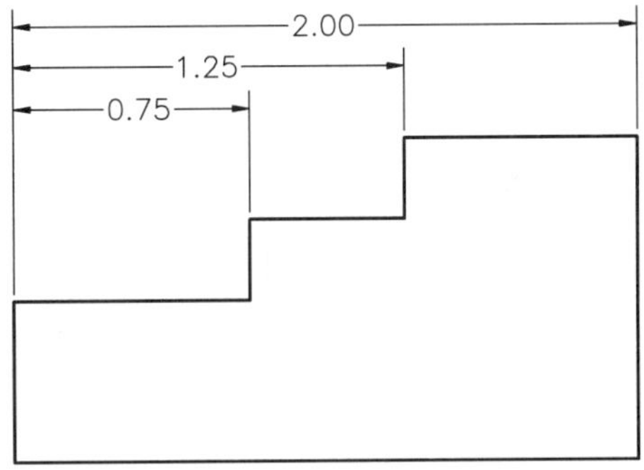

Figure 9-19 *Drawing for Exercise 3, after editing dimensions*

Exercise 4 *Mechanical*

1. Make the drawing as shown in Figure 9-20 (a). Assume the dimensions where necessary.
2. Dimension the drawing and edit them as shown in Figure 9-20 (b).

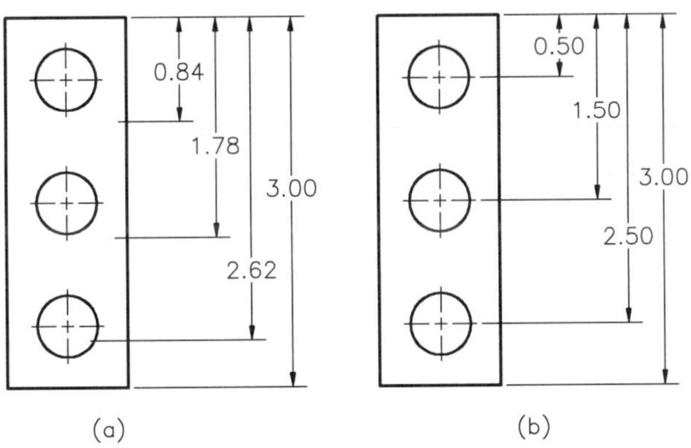

Figure 9-20 *Drawings for Exercise 4*

Exercise 5 *Architectural*

Make the drawing shown in Figure 9-21 and then dimension it. Assume the dimensions wherever necessary. After dimensioning the drawing, edit the dimensions so that they match the dimensions shown in Figure 9-21

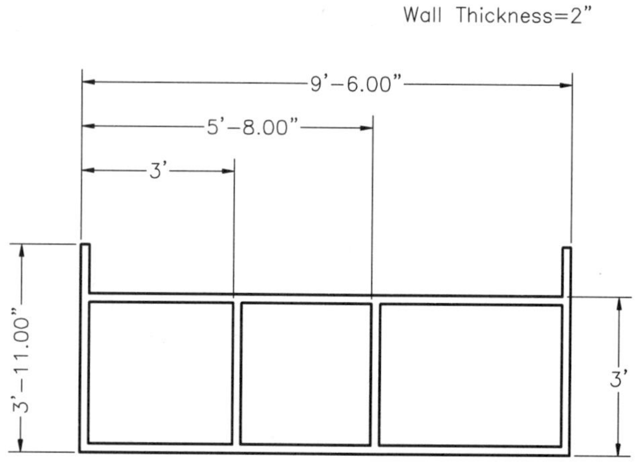

Figure 9-21 *Drawing for Exercise 5*

Problem Solving Exercise 1 *Mechanical*

Make the drawing in Figure 9-22 and then dimension it as shown. Edit the dimensions so that the dimensions are positioned as shown in the drawing. You may change the dimension text height and arrow size to 0.08 units.

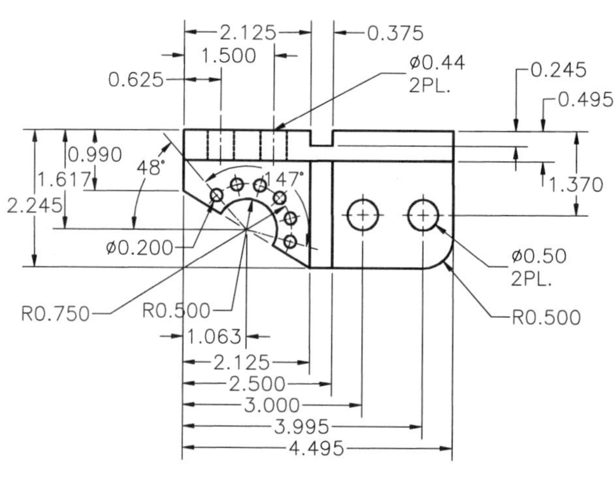

Figure 9-22 *Drawing for Problem Solving Exercise 1*

Problem Solving Exercise 2 *Mechanical*

Make the front and side view of an object shown in Figure 9-23 and then dimension the two views. Edit the dimensions so that the dimensions are positioned as shown in the drawing. You may change the dimension text height, and arrow size to 0.08 units.

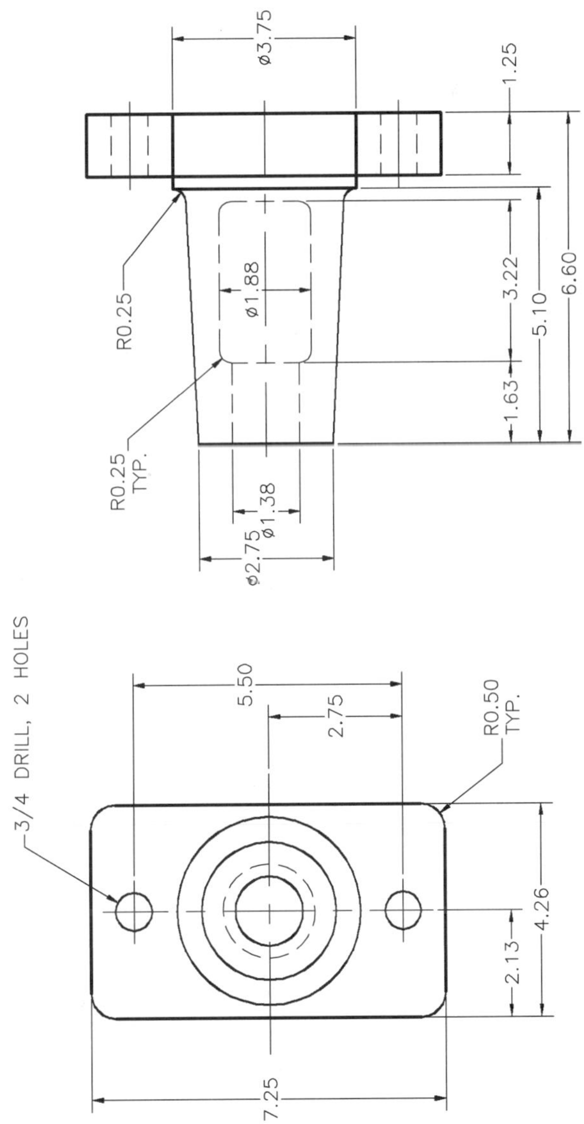

Figure 9-23 *Drawing for Problem Solving Exercise 2*

Problem Solving Exercise 3 *Architectural*

Make the drawing of the floor plan shown in Figure 9-24 and then give the dimensions as shown. Edit the dimensions, if needed, so that they are positioned as shown in the drawing.

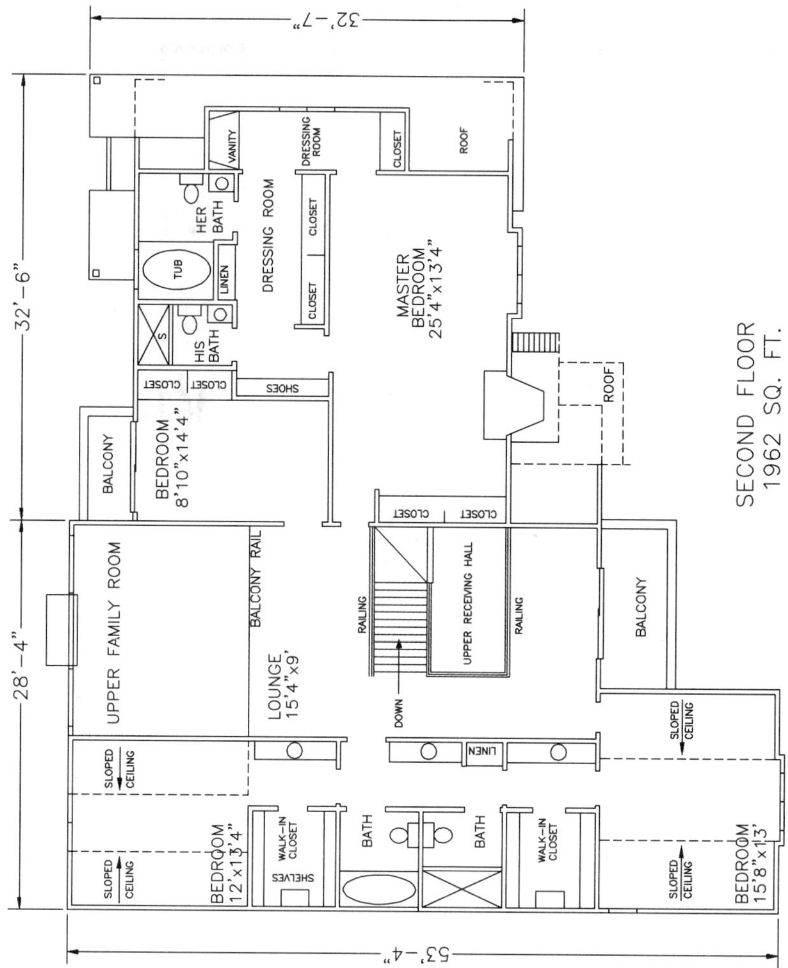

Figure 9-24 Drawing for Problem Solving Exercise 3

Answers to Self-Evaluation Test

1 - T, 2 - T, 3 - F, 4 - F, 5 - EXPLODE, 6 - Left, 7 - Center, 8 - New, 9 - Oblique, 10 - Dimension Update

Chapter 10

Dimension Styles and Dimensioning System Variables

Learning Objectives

After completing this chapter, you will be able to:

- *Use styles and variables to control dimensions.*
- *Create dimensioning styles.*
- *Set dimension variables using the different tabs of the* ***New, Modify,*** *and* ***Override Dimension Style*** *dialog boxes.*
- *Set other dimension variables that are not in dialog boxes.*
- *Understand dimension style families and how to apply them in dimensioning.*
- *Use dimension style overrides.*
- *Compare and list dimension styles.*
- *Import externally referenced dimension styles.*

USING STYLES AND VARIABLES TO CONTROL DIMENSIONS

In AutoCAD LT, the appearance of dimensions on the drawing screen and the manner in which they are saved in the drawing database are controlled by a set of dimension variables. The dimensioning commands use these variables as arguments. The variables that control the appearance of the dimensions can be managed with dimension styles. You can use the **Dimension Style Manager** dialog box to control dimension styles and dimension variables through a set of dialog boxes.

CREATING AND RESTORING DIMENSION STYLES

Toolbar:	Dimension > Dimension Style
Menu:	Dimension > Style
	or Format > Dimension style
Command:	DIMSTYLE

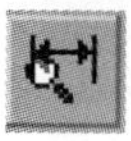

The dimension styles (Figure 10-1) control the appearance and positioning of dimensions and leaders in the drawing. If the default dimensioning style (STANDARD) does not meet your requirements, you can select another existing dimensioning style or create a new one that does. The default dimension style file name is **STANDARD**. Dimension styles can be created by using the **Dimension Style Manager** dialog box (Figure 10-2). This dialog box is displayed upon choosing the **Dimension Style** button from the **Dimension** toolbar. You can also invoke the **Dimension Style Manager** dialog box by choosing **Style** from the **Dimension** menu or choosing **Dimension Style** from the **Format** menu.

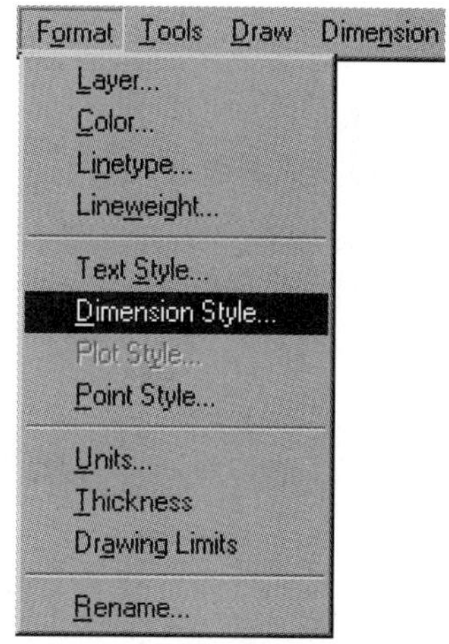

***Figure 10-1** Choosing **Dimension Style** from the **Format** menu*

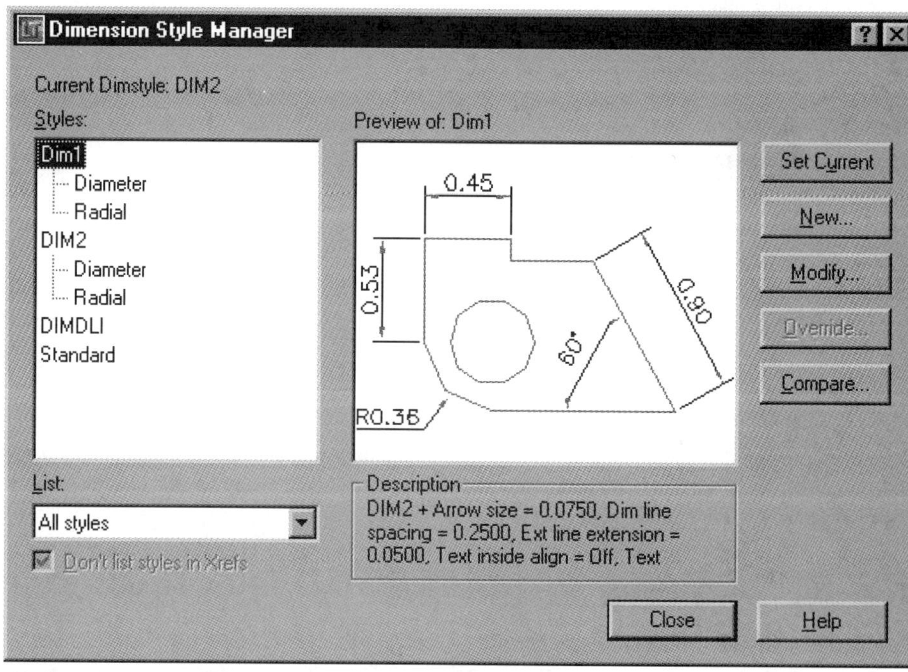

***Figure 10-2 Dimension Style Manager** dialog box*

In the **Dimension Style Manager** dialog box, choose the **New** button to display the **Create New Dimension Style** dialog box (Figure 10-3). Enter the dimension style name in the **New Style Name** text box and then select a style on which you want to base your style from the **Start With** drop-down list. The **Use for** drop-down list allows you to select the dimension type to which you want to apply the new dimension style. For example if you wish to use the new style for only the diameter dimension, you can select **Diameter dimension** from the **Use for** drop-down list. Choose the **Continue** button to display the **New Dimension Style** dialog box where you can define the new style.

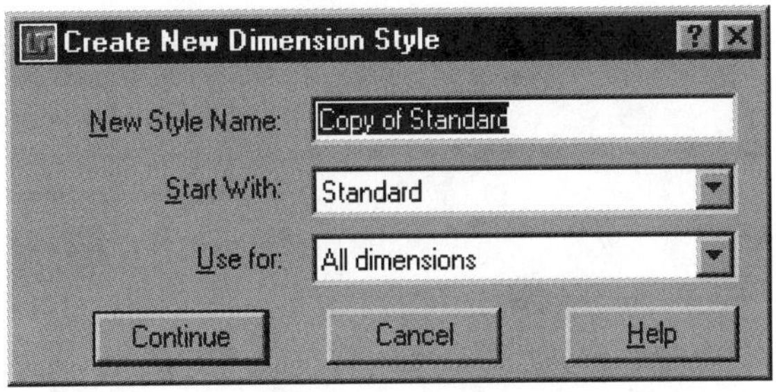

***Figure 10-3 Create New Dimension Style** dialog box*

In the **Dimension Style Manager** dialog box, the current dimension style name is shown in front of **Current dimstyle** and is also shown highlighted in the **Styles** list box. A brief description of the current style (its differences from the default settings) is also displayed in the **Description** area. The **Dimension Style Manager** dialog box also has a **Preview of** window which displays a preview of the current dimension style. A style can be made current (restored) by selecting the name of the dimension style you want to make current from the list of defined dimension styles and choosing the **Set Current** button. You can also make a style current by double-clicking on the style name in the **Styles** list box. The drop-down list in the **Dimension** toolbar also displays the dimension styles. Selecting a Dimension style from this list also sets it current. The list of dimension styles displayed in the **Styles** list box is dependent on the option selected from the **List** drop down list. If you select the **Styles in use** option, only the dimension styles in use will be listed in the **Style** list box. If you right click on a style in the **Style** list box, a shortcut menu is displayed that provides you with the options to **Set current**, **Rename**, or **Delete** a dimension style. Selecting the **Don't list styles in Xrefs** check box does not list the names of Xref styles in the **Styles** list box. Choosing the **Modify** button displays the **Modify Dimension Style** dialog box where you can modify an existing style. Choosing the **Override** button, displays the **Override Current Style** dialog box where you can define overrides to an existing style (discussed later in this chapter). Both these dialog boxes along with the **New Dimension Style** dialog box have identical properties. Choosing the **Compare** button displays the **Compare Dimension Styles** dialog box (also discussed later in this chapter) that allows you to compare two existing styles.

NEW DIMENSION STYLE DIALOG BOX

The **New Dimension Style** dialog box can be used to specify the dimensioning attributes (variables) that affect the various properties of the dimensions. The various tabs provided under the **New Dimension Style** dialog box are discussed next.

Lines and Arrows Tab

The options under the **Lines and Arrows** tab (Figure 10-4) of the **New Dimension Style** dialog box can be used to specify the dimensioning attributes (variables) that affect the format of the dimension lines and arrows. For example, the appearance and behavior of the dimension lines, extension lines, arrowheads, and center marks can be changed with this tab. If the settings of the dimension variables have not been altered in the current editing session, the settings displayed in the dialog box are the default settings.

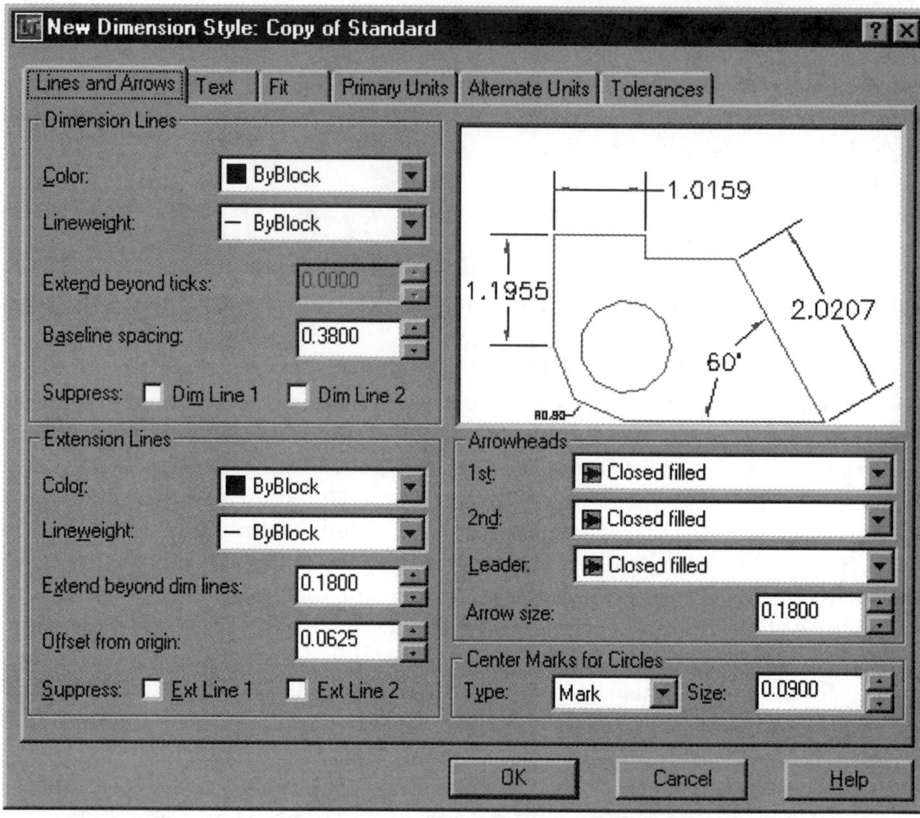

***Figure 10-4 Lines and Arrows** tab of the **New Dimension Style** dialog box*

Dimension Lines Area

This area provides you with the options of controlling the display of the dimension lines and leader lines. These options are discussed next.

Color. This drop-down list is used to set the colors for the dimension lines and arrowheads. The dimension arrowheads have the same color as the dimension line because arrows constitute a part of the dimension line. The color you set here will also be assigned to the leader lines and arrows. The default color for the dimension lines and arrows is BYBLOCK. You can specify the color of the dimension line by selecting it from the **Color** drop-down list. You can also select **Other** from the **Color** drop-down list to display the **Select Color** dialog box where you can choose a specific color. The color number or the special color label is stored in the **DIMCLRD** variable.

Lineweight. This drop-down list is used to specify the Lineweight for the dimension line. You can select the required lineweight by selecting it from this drop-down list. This value is also stored in the **DIMLWD** variable. The default value is BYBLOCK. Keep in mind that you can not assign the lineweight to the arrowheads using this drop-down list.

Extend beyond ticks. The **Extend beyond ticks** (Oblique tick extension) spinner will be available only when you select the oblique, Architectural tick, or any such arrowhead type in the **1st** and **2nd** drop-down lists in the **Arrowheads** area. This spinner is used to specify the distance by which the dimension line will extend beyond the extension line. The extension value entered in the **Extend beyond ticks** edit box gets stored in the **DIMDLE** variable. By default, this edit box is disabled because the oblique arrowhead type is not selected.

Baseline spacing. The **Baseline Spacing** (Baseline Increment) spinner is used to control the spacing between successive dimension lines drawn using the baseline dimensioning (Figure 10-5). You can specify the dimension line increment to your requirement by specifying the desired value using the **Baseline Spacing** spinner. The default value displayed in the **Baseline Spacing** spinner is 0.38 units. This spacing value is also stored in the **DIMDLI** variable.

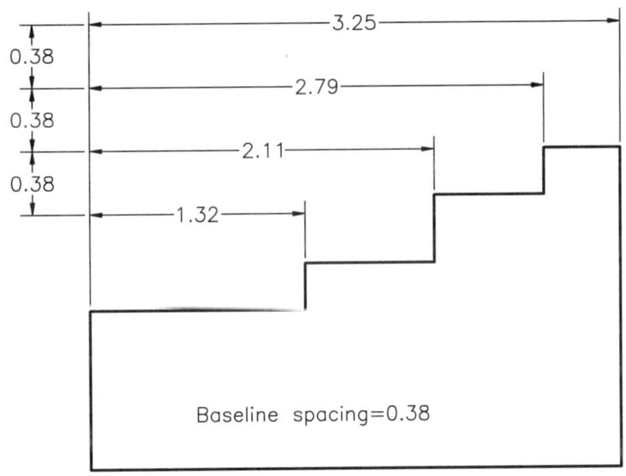

Figure 10-5 *Figure showing baseline increment*

Suppress. The **Suppress** check boxes control the display of the first and second dimension lines. By default, both dimension lines will be drawn. You can suppress one or both the dimension lines by selecting their corresponding check boxes. The values of these check boxes are stored in the **DIMSD1** and **DIMSD2** variables.

Note

The first and second dimension lines are determined by how you select the extension line origins. If the first extension line origin is on the right, the first dimension line is also on the right.

Extension Lines Area

Color. This drop-down list is used to control the color of the extension lines. The default extension line color is BYBLOCK. You can assign a new color to the extension lines by selecting it from this drop-down list. The color number or the color label is saved in the **DIMCLRE** variable.

Lineweight. This drop-down list is used to modify the lineweight of the extension lines. The default value is BYBLOCK. You can change the lineweight value by selecting a new value from this drop-down list. The value for lineweight is stored in the **DIMLWE** variable.

Extend beyond dim lines. This is the distance by which the extension lines extends past the dimension lines (Figure 10-6). You can change the extension line offset using the **Extend beyond dim lines** spinner. This value is also stored in the **DIMEXE** variable. The default value for extension distance is 0.1800 units.

Offset from origin. This is the distance by which the extension line is offsetted from the point you specify as the origin of the extension line, as shown in Figure 10-7. You may need to override this setting for specific dimensions when dimensioning curves and angled lines. You can specify an offset distance of your choice using this spinner. AutoCAD LT stores this value in the **DIMEXO** variable. The default value for this distance is 0.0625.

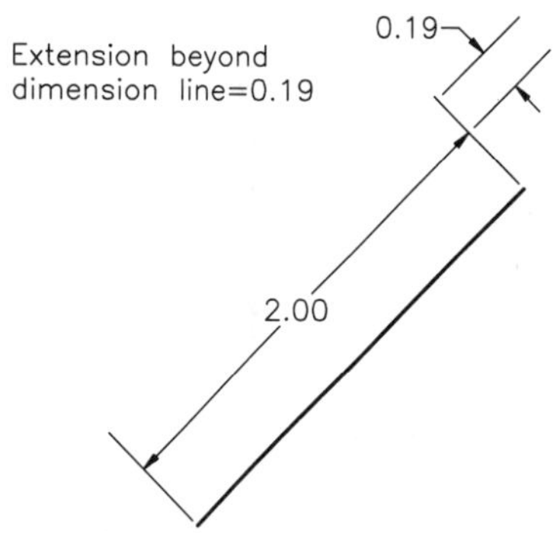

Figure 10-6 *Extension beyond dimension lines*

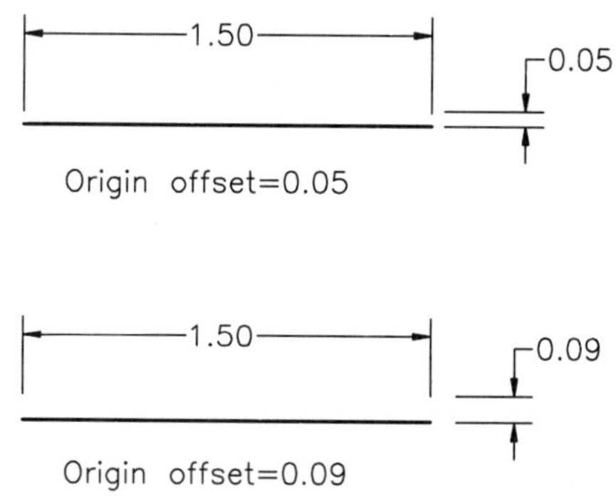

Figure 10-7 *Figure showing the origin offset*

Suppress. The **Suppress** check boxes area used to control the display of the extension lines. By default, both extension lines will be drawn. You can suppress one or both extension lines by selecting the corresponding check boxes (Figure 10-8). The values of these check boxes are stored in the **DIMSE1** and **DIMSE2** variables.

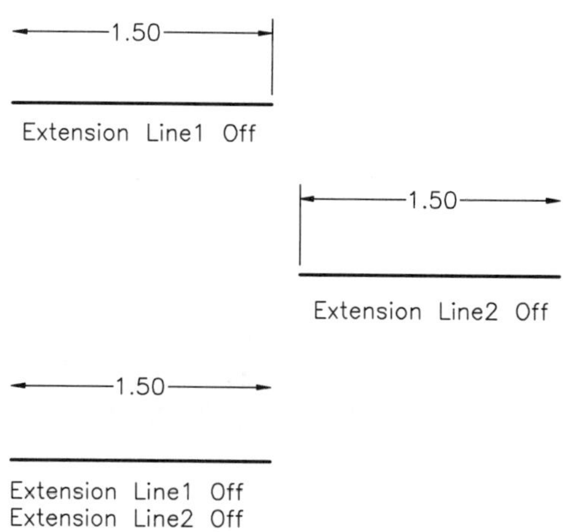

Figure 10-8 *Suppressing the extension lines*

Note

The first and second extension lines are determined by how you select the extension line origins. If the first extension line origin is on the right, the first extension line is also on the right.

Arrowheads Area

1st/2nd. When you create a dimension, AutoCAD LT draws the terminator symbols at the two ends of the dimension line. These terminator symbols, generally referred to as **arrowheads**, represent the beginning and end of a dimension. AutoCAD LT has provided nineteen standard termination symbols you can apply at each end of the dimension line. In addition to these, you can create your own arrows or terminator symbols. By default, the same arrowhead type is applied at both ends of the dimension line. If you select the first arrowhead, it is automatically applied to the second by default. However, if you want to specify a different arrowhead at the second dimension line endpoint, you must select the desired arrowhead type from the **2nd** drop-down list. The first endpoint of the dimension line is the intersection point of the first extension line and the dimension line. The first extension line is determined by the first extension line origin. However, in angular dimensioning the second endpoint is located in a counterclockwise direction from the first point, regardless of how the points were selected when creating the angular dimension. The specified arrowhead types are selected from the **1st** and **2nd** drop-down lists. The first arrowhead type is saved in the **DIMBLK1** system variable, and the second arrowhead type is saved in the **DIMBLK2** system variable.

AutoCAD LT provides you with an option of specifying a user-defined arrowhead. To define a user-defined arrow, you must create one as a block (See Chapter 14, Blocks, for information regarding blocks). Now, from the **1st** or the **2nd** drop-down list, select **User Arrow**. The **Select Custom Arrow Block** dialog box will be displayed. See Figure 10-9.

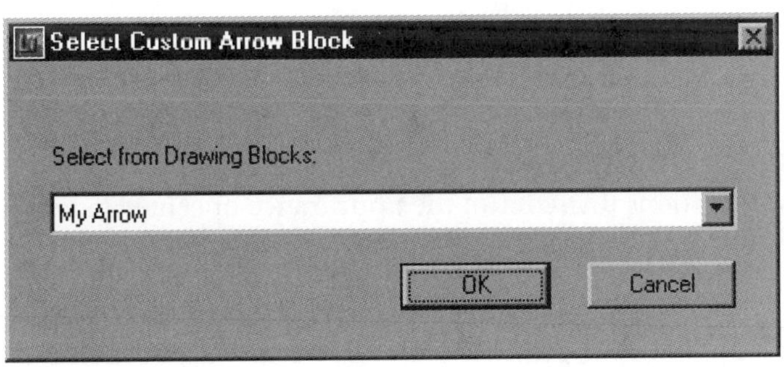

***Figure 10-9 Select Custom Arrow Block** dialog box*

All the blocks available in the current drawing will be available in the **Select from Drawing Blocks** drop-down list. You can select the desired block from this drop-down list and it will become the current arrowhead.

Creating an Arrowhead Block

1. To create a block for an arrowhead, you have to use a 1 x 1 box, as shown in Figure 10-10. AutoCAD LT automatically scales the block's X and Y scale factors to arrowhead size multiplied by the overall scale. The **DIMASZ** variable controls the overall length of the arrowhead. For example, if **DIMASZ** is set to 0.25, the length of the arrow will be

0.25 units. Also, if the length of the arrow is not 1 unit, it will leave a gap between the dimension line and the arrowhead block.

2. The arrowhead must be drawn as it would appear on the right side of the dimension line. Choose the **Make Block** button from the **Draw** toolbar to convert it into a block.

3. The insertion point of the arrowhead block must be the point that will coincide with the extension line. See Figure 10-10.

Leader. The **Leader** drop-down list displays arrowhead types for the Leader arrow. Here also you can either select the standard arrowheads from the drop-down list or select **User Arrow** that allows you to define and use a user-defined arrowhead type.

Arrow size. This spinner is used to define the size of the arrowhead, as shown in Figure 10-11. The default value is 0.18 units. This value is stored in the **DIMASZ** system variable.

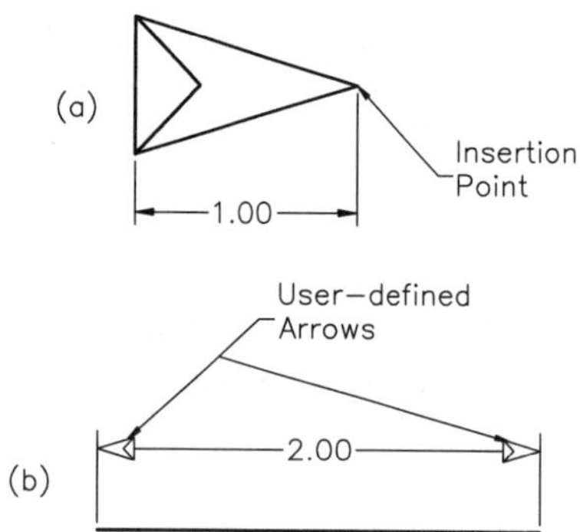

Figure 10-10 Creating user-defined arrows

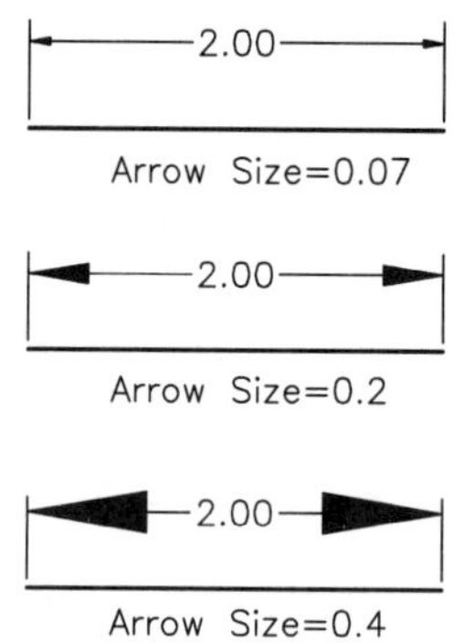

Figure 10-11 Defining arrow sizes

Center Marks for Circles Area

This area deals with options that control the appearance of center marks and centerlines in the radius and diameter dimensioning. However, keep in mind that the center marks or the centerlines will be drawn only when dimensions are placed outside the circle.

Type. The **Type** drop-down list provides you with an option of drawing a center mark or a centerline when you create radius or diameter dimensions. If you select **Mark**, a mark will be drawn at the center of the circle and if you select **Line**, centerlines will be drawn at the center of the circles. If you select **None**, no mark or line will be drawn at the center of the circle. If you use the **DIMCEN** command, a positive value will create a center mark, and a negative value will create a centerline. If the value is 0, AutoCAD LT does not create center marks or centerlines.

Size. This spinner is used to set the size of the center marks or centerlines. You can specify the required size of the center marks or centerlines using the **Size** spinner. This value is stored in the **DIMCEN** variable. The default value of the **DIMCEN** variable is 0.09.

Note

*Unlike specifying a negative value for the **DIMCEN** variable, you cannot enter a negative value in the **Size** spinner. Selecting **Line** from the **Type** drop-down list automatically treats the value in the **Size** spinner as the size for the centerlines and sets **DIMCEN** to the negative of the value shown.*

Exercise 1 *Mechanical*

Draw Figure 10-12. Then set the values in the **Lines and Arrows** tab of the **New Dimension Style** dialog box to dimension the drawing as shown in this figure. (Baseline Spacing = 0.25, Extension beyond dimension line = 0.10, Offset from Origin = 0.05, Arrowhead size= 0.09.) Assume the missing dimensions.

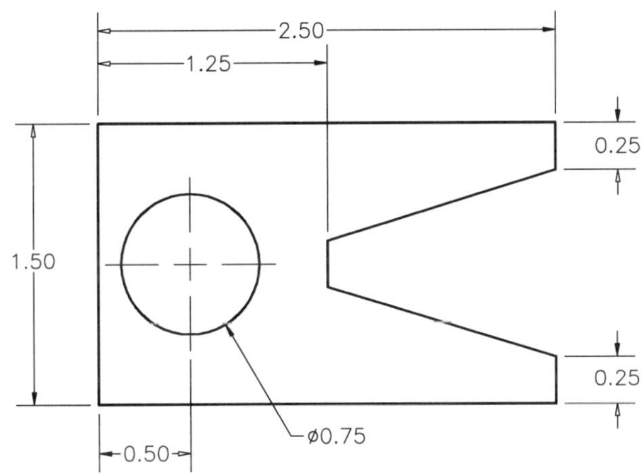

***Figure 10-12** Drawing for Exercise 1*

CONTROLLING DIMENSION TEXT FORMAT

Text Tab

You can control the dimension text format through the **Text** tab of the **New Dimension Style** dialog box (Figure 10-13). In the **Text** tab, you can control the parameters such as the placement, appearance, horizontal and vertical alignment of dimension text, and so on. For example, you can force AutoCAD LT to align the dimension text along the dimension line. You can also force the dimension text to be displayed at the top of the dimension line. You can save the settings in a dimension style file for future use. The **New Dimension Style** dialog box has a Preview window that updates dynamically to display the text placement as the settings are changed. Individual items of the this tab and related dimension variables are described next.

Text Appearance Area

Text style. This drop-down list displays the names of the predefined text styles. From this list you can select the style name that you want to use for dimensioning. You must define the text style before you can use it in dimensioning (see **STYLE** Command in Chapter 7, Controlling the Drawing Display and Creating Text). Choosing the [...] button, displays the **Text Style**

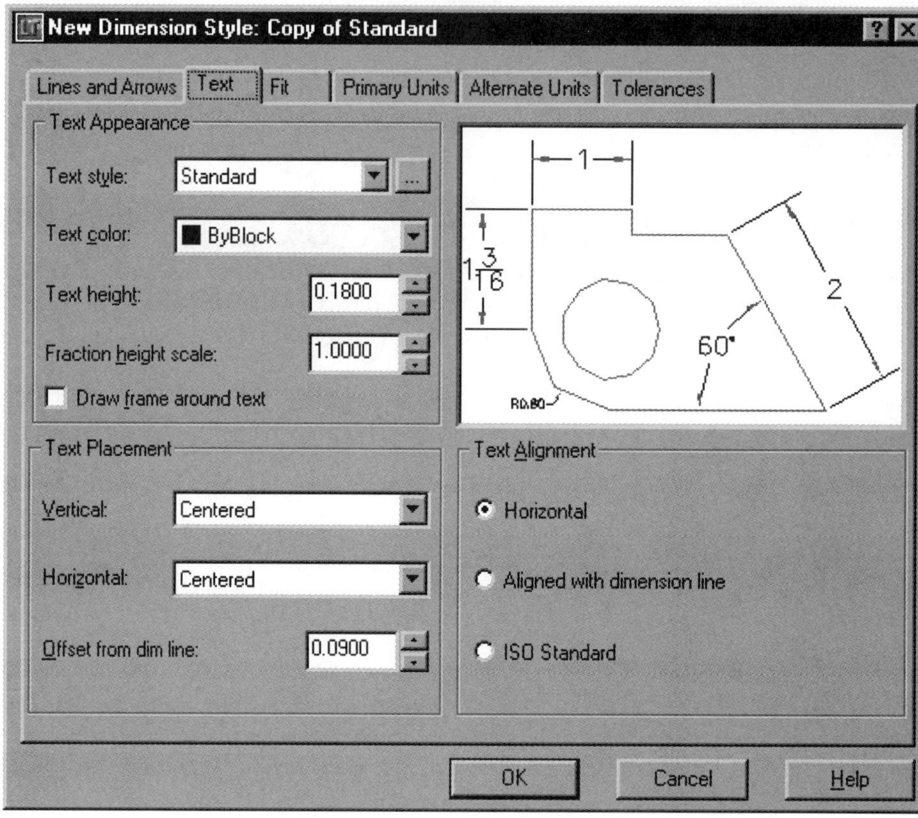

Figure 10-13 ***Text*** *tab of the* ***New Dimension Style*** *dialog box*

dialog box that allows you to create a new or modify an existing text style. The value of this setting is stored in the **DIMTXSTY** system variable. The change in dimension text style does not affect the text style you are using to draw other text in the drawing.

Text color. This drop-down list is used to modify the color of the dimension text. The default color is BYBLOCK. If you choose the **Other** option from the **Text Color** drop-down list, the **Select Color** dialog box is displayed where you can choose a specific color. This color or color number is stored in the **DIMCLRT** variable.

Text height. This spinner is used to modify the height of the dimension text (Figure 10-14). You can change the dimension text height only when the current text style does not have a fixed height. In other words, the text height specified in the **STYLE** command should be zero. The reason for this is that a predefined text height (specified in the **STYLE** command) overrides any other setting for the dimension text height. This value is stored in the **DIMTXT** variable. The default text height is 0.1800 units.

Fraction height scale. This spinner is used to set the scale of the fractional units in relation to the Dimension text height. This spinner will be available only when you select a format for the primary units in which you can define the values in fractions, such as architectural or fractional. This value is stored in the **DIMTFAC** variable.

Draw frame around text. When selected this check box draws a frame around the dimension text, see Figure 10-15. This value is stored as a negative value in the **DIMGAP** system variable.

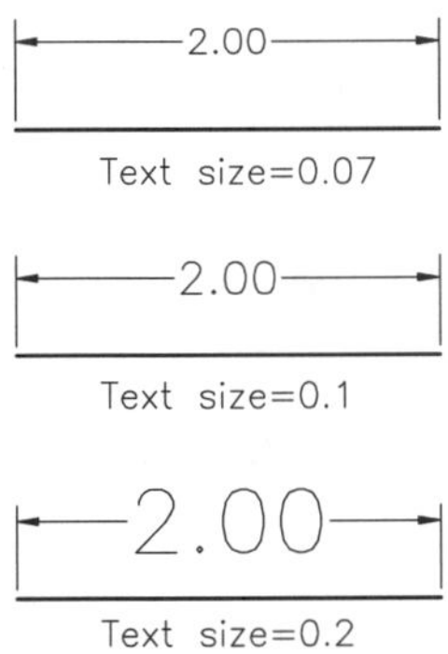

Figure 10-14 Changing the dimension height

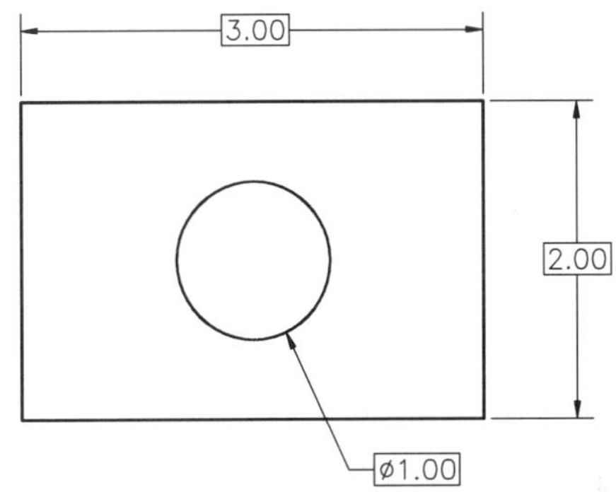

Figure 10-15 Dimension text inside frame

Text Placement Area

Vertical. The **Vertical** drop-down list displays the options that control the vertical placement of the dimension text. The current setting is highlighted. Controlling the vertical placement of the dimension text is possible only when the dimension text is drawn in its normal (default) location. This setting is stored in the **DIMTAD** system variable. The options provided under this drop-down list are:

Centered. If this option is selected, the dimension text gets positioned on the dimension line in such a way that the dimension line is split to allow for placement of the text (Figure 10-16). If the **1st** or **2nd Extension Line** option is selected in the **Horizontal** drop-down list, this centered setting will position the text on the extension line, not on the dimension line.

Above. If this option is selected, the dimension text is placed above the dimension line, except when the dimension line is not horizontal and the dimension text inside the extension lines is horizontal. The distance of the dimension text from the dimension line is controlled by the **DIMGAP** variable. This results in an unbroken solid dimension line being drawn under the dimension text. See Figure 10-16.

Outside. This option places the dimension text on the side of the dimension line,.

JIS. This option lets you place the dimension text to conform to **JIS** (Japanese Industrial Standards) representation.

Note

The Horizontal and Vertical Placement options selected are reflected in the dimensions shown in the Preview window.

Horizontal. This drop-down list is used to control the horizontal placement of the dimension text. You can select the required horizontal placement from this list. However, remember that these options will be useful only when the **Place text manually when dimensioning** check box in the **Fine Tuning** area of the **Fit** tab is cleared. The options provided under this drop-down list are:

Centered. This option places the dimension text between the extension lines. This is the default option.

At Ext Line 1. This option is selected to place the text near the first extension line along the dimension line. See Figure 10-17.

At Ext Line 2. This option is selected to place the text near the second extension line along the dimension line. See Figure 10-17.

Over Ext Line 1. This option is selected to place the text over the first extension line and also along the first extension line. See Figure 10-17.

Over Ext Line 2. This option is selected to place the text over the second extension line and also along the second extension line. See Figure 10-17.

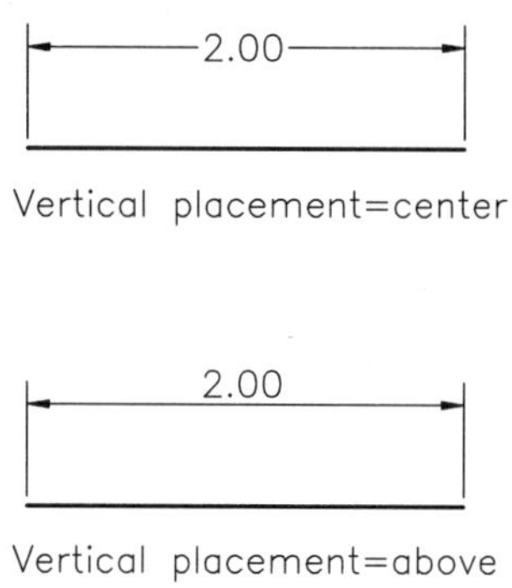

Figure 10-16 *Vertical text placement*

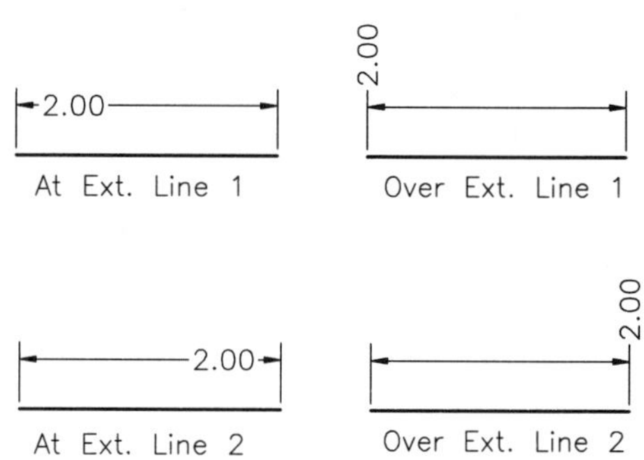

Figure 10-17 *Horizontal text placement*

Offset from dim line. This spinner is used to specify the distance between the dimension line and the dimension text (Figure 10-18). You can set the text gap you need using this spinner. The Text Gap value is also used as the measure of minimum length for the segments of the dimension line and in basic tolerance. The default value specified in this box is 0.09 units. The value of this setting is stored in the **DIMGAP** system variable.

Text Alignment Area

Horizontal. This is the default option and if selected, the dimension text is drawn horizontally with respect to the current UCS (user coordinate system). The alignment of the dimension line does not affect text alignment. Selecting this option turns both the **DIMTIH** and **DIMTOH** system variables **on**. The text is drawn horizontally even if the dimension line is at an angle.

Aligned with dimension line. Selecting this radio button aligns the text with the dimension line and both the system variables **DIMTIH** and **DIMTOH** are off.

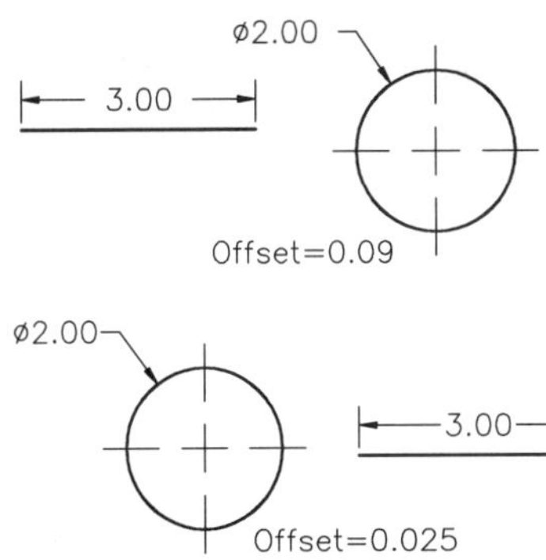

Figure 10-18 Offset from dimension line

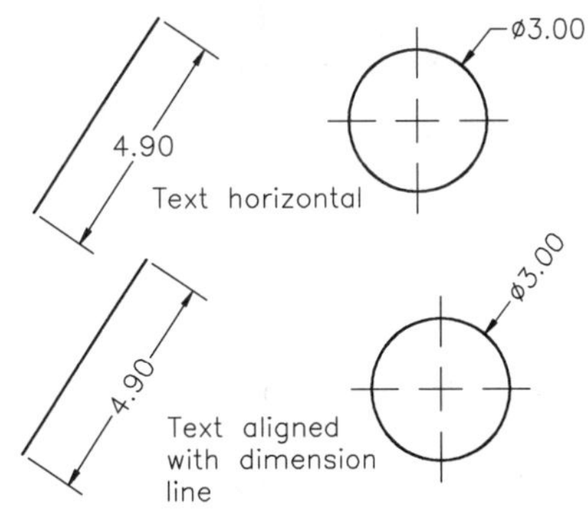

Figure 10-19 Specifying test alignment

ISO Standard. If you select the **ISO Standard** radio button, the dimension text is aligned with the dimension line only when the dimension text is inside the extension lines. Selecting this option turns the system variable **DIMTOH** on; that is, the dimension text outside the extension line is horizontal regardless of the angle of the dimension line.

Exercise 2 *Mechanical*

Draw Figure 10-20. Then set the values in the **Lines and Arrows** and **Text** tab of the **New Dimension Style** dialog box to dimension the drawing as shown in the figure. (Baseline Spacing = 0.25, Extension beyond dimension lines = 0.10, Offset from Origin = 0.05, Arrow size = 0.09, Text height = 0.08).

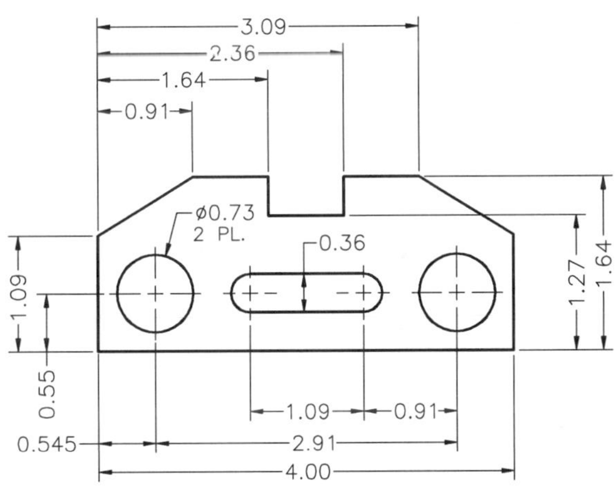

Figure 10-20 Drawing for Exercise 2

FITTING DIMENSION TEXT AND ARROWHEADS

Fit Tab

The **Fit** tab provides you with the options that control the placement of dimension lines, arrowheads, leader lines, text, and the overall dimension scale (Figure 10-21).

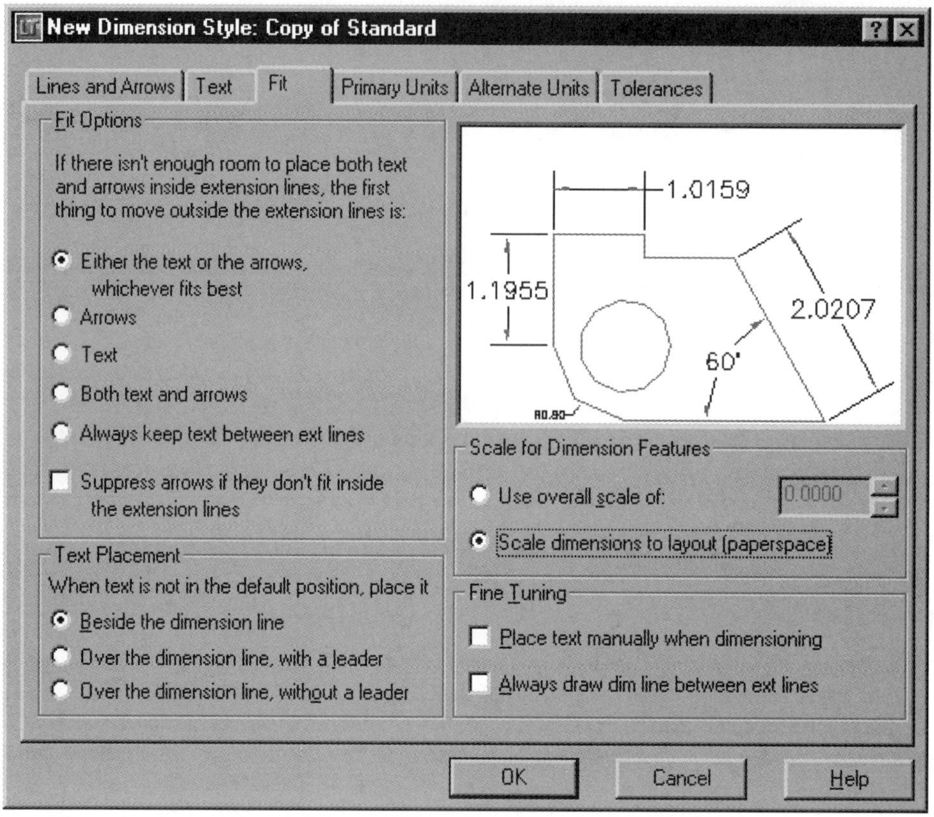

Figure 10-21 ***Fit*** *tab of the* ***New Dimension Style*** *dialog box*

Fit Options Area

These options are used to set the priorities for moving the text and arrowheads outside the extension lines, if the space between the extension lines is not enough to fit both of them.

Either the text or the arrows, whichever fits best. This is the default option. In this option, AutoCAD LT places the dimension where it fits best between the extension lines.

Arrows. When you select this option, AutoCAD LT places the text and arrowheads inside the extension lines if there is enough space to fit both. If space is not available for both the arrows and text, the arrows are moved outside the extension lines. If there is not enough space for text, both text and arrowheads are placed outside the extension lines.

Text. When you select this option, AutoCAD LT places the text and arrowheads inside the extension lines if there is enough space to fit both. If there is enough space to fit the arrows, the arrows will be placed inside the extension lines and the dimension text moves outside the extension lines. If there is not enough space for either text or arrowheads, both are placed outside the extension lines.

Both text and arrows. If you select this option, AutoCAD LT will place the arrows and dimension text between the extension lines if there is enough space available to fit both. Otherwise, both text and arrowheads are placed outside the extension lines.

Always keep text between ext lines. This option always keeps the text between extension lines even in cases where AutoCAD LT would not do so. Selecting this radio button does not affect radius and diameter dimensions. The value is stored in the **DIMTIX** variable and the default value is **off**.

Suppress arrows if they don't fit inside the extension lines. If you select this check box, the arrowheads are suppressed if the space between the extension lines is not enough to adjust them. The value is stored in the **DIMSOXD** variable and the default value is **off**.

Text Placement Area

This area provides you with the options to position the dimension text when it is moved from the default position. The value is stored in the **DIMTMOVE** variable. The options under this area are as follows.

Beside the dimension line. This option places the dimension text beside the dimension line.

Over the dimension line, with a leader. Selecting this option places the dimension text away from the dimension line and a leader line is created, which connects the text to the dimension line. But, if the dimension line is too close to the text, a leader is not drawn. The Horizontal placement decides whether the text is placed to the right or left of the leader.

Over the dimension line, without a leader. In this option, AutoCAD LT does not create a leader line if there is insufficient space to fit the dimension text between the extension lines. The dimension text can be moved freely, independent of the dimension line.

Scale for Dimension Features Area

Sets the value for overall Dimension scale or scaling to paper space.

Use overall scale of. The current general scaling factor that pertains to all of the size-related dimension variables such as text size, center mark size, and arrowhead size is displayed in the **Use overall scale of** spinner. You can alter the scaling factor to your requirement by entering the scaling factor of your choice in this spinner. Altering the contents of this box alters the value of the **DIMSCALE** variable, since the current scaling factor is stored in it. The overall scale (**DIMSCALE**) is not applied to the measured lengths, coordinates, angles, or tolerance. The default value for this variable is 1.0. In this condition, the dimensioning variables assume their preset values and the drawing is plotted at full scale. The scale factor is the reciprocal of the drawing size so the drawing is to be plotted at half size. The overall scale factor (**DIMSCALE**) will be the reciprocal of ½, which is 2.

Note

*If you are in the middle of the dimensioning process and you change the **DIMSCALE** value and save the changed setting in a dimension style file, the dimensions with that style will be updated.*

Tip

When you increase the limits of the drawing, increase the overall scale of the drawing using the ***Use overall scale of*** *spinner before dimensioning. This will save the time required in changing the individual scale factors of all the dimension parameters.*

Scale dimensions to layout (paperspace). If you select the **Scale dimensions to layout (paper space)** radio button, the scale factor between the current model space viewport and floating viewport (paper space) is computed automatically. Also, by selecting this radio button, you disable the **Use overall scale of** spinner (it is disabled in the dialog box) and the overall scale factor is set to 0. When the overall scale factor is assigned a value of 0, AutoCAD LT calculates an acceptable default value based on the scaling between the current model space viewport and paper space. If you are in paper space (**TILEMODE=0**), or are not using the **Scale dimensions to layout (paper space)** feature, AutoCAD LT sets the overall scale factor to 1; otherwise, AutoCAD LT calculates a scale factor that makes it possible to plot text sizes, arrow sizes, and other scaled distances at the values in which they have been previously set. (For further details regarding the model space and layouts, refer to Chapter 11, Model Space Viewoprts, Paper Space Viewports and Layouts.)

Fine Tuning Area

The **Fine Tuning** area provides additional options governing placement of dimension text. The options are as follows.

Place text manually when dimensioning. When you dimension, AutoCAD LT places the dimension text in the middle of the dimension line (if there is enough space). If you select the **Place text manually when dimensioning** check box, you can position the dimension text anywhere along the dimension line. You will also notice that when you select this check box, the **Horizontal Justification** is ignored. This setting is saved in the **DIMUPT** system variable. The default value of this variable is **off**. Selecting this check box enables you to position the dimension text anywhere along the dimension line.

Always draw dim line between ext lines. This check box is selected when you want the dimension line to appear between the extension lines, even if the text and dimension lines are placed outside the extension lines. When you select this option in the radius and diameter dimensions (when the default text placement is horizontal), the dimension line and arrows are drawn inside the circle or arc, and the text and leader are drawn outside. When you select the **Always draw dimension line between extension lines** check box, the **DIMTOFL** variable is set to on by AutoCAD LT. The default setting is off.

FORMATTING PRIMARY DIMENSIONS UNITS

Primary Units Tab

You can use the **Primary Units** tab of the **New Dimension Style** dialog box to control the dimension text format and precision values (Figure 10-22). You can use the options under this tab to control Units, Dimension Precision, and Zero Suppression for dimension measurements. AutoCAD LT lets you attach a user-defined prefix or suffix to the dimension text. For example, you can define the diameter symbol as a prefix by entering %%C in the **Prefix** edit box; AutoCAD LT will automatically attach the diameter symbol in front of the dimension text.

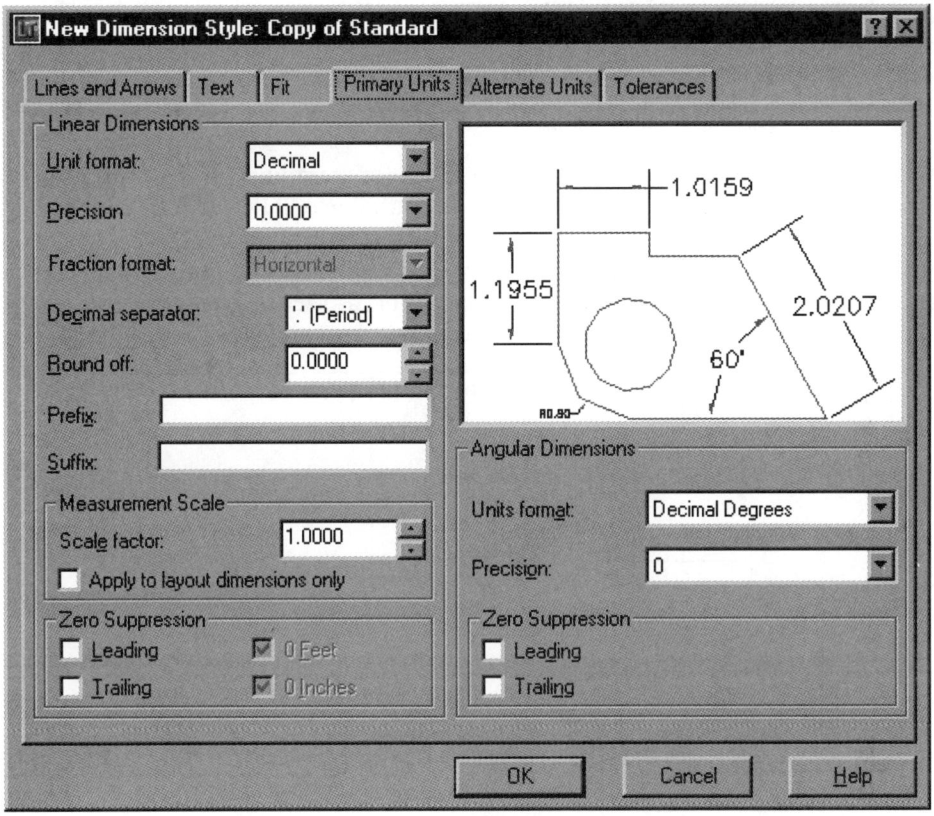

Figure 10-22 ***Primary Units*** *tab of the* ***New Dimension Style*** *dialog box*

Similarly, you can define a unit type such as **mm**, as a suffix; AutoCAD LT will then attach **mm** at the end of every dimension text. This tab also enables you to define zero suppression, precision, and dimension text format.

Linear Dimensions Area

Unit format. This drop-down list provides you with the options of specifying the units for the primary dimension units. The formats that are available include Decimal, Scientific, Architectural, Engineering, Fraction, and Windows Desktop. Keep in mind that by selecting a dimension unit format, the drawing units (which you might have selected by using the **UNITS** command) are not affected. The unit setting for linear dimensions is stored in the **DIMLUNIT** system variable.

Precision. This drop-down list is used to control the number of decimal places for the primary units. The setting for precision (number of decimal places) is saved in the **DIMDEC** variable.

Fraction format. This drop-down list is used to set the fraction format. The options are Diagonal, Horizontal, and not stacked. This drop-down list will be available only when you select **Architectural** or **Fractional** from the **Unit format** drop-down list. The value is stored in the **DIMFRAC** variable.

Decimal separator. This drop-down list is used to select an options that will be used as the decimal separators. For example, Period [.], Comma [,] or Space [ ]. If you have selected Windows desktop units in the **Unit Format:** drop-down list, AutoCAD LT uses the Decimal symbol settings. The value is stored in the **DIMDSEP** variable.

Round off. The **Round off** spinner sets the value for rounding off the dimension values. For example, if the **Round off** spinner is set to 0.05, a value of 0.06 will round off to 0.10. The number of decimal places of the round off value you enter in the edit box should be less than or equal to the value in the **Precision:** edit box. The value is stored in the **DIMRND** variable and the default value in the **Round Off:** edit box is 0. Also see Figure 10-23.

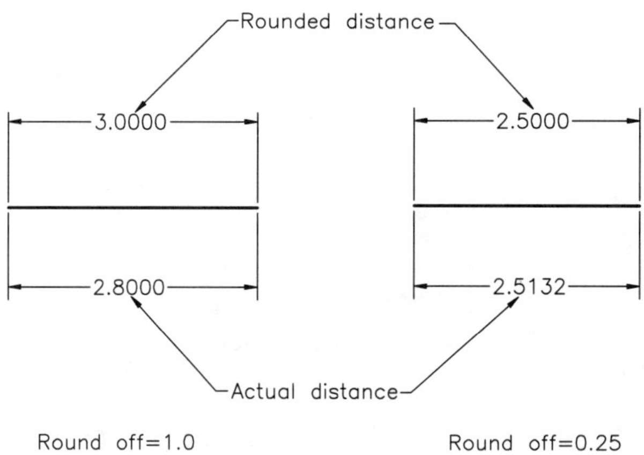

Figure 10-23 *Rounding off the dimension measurements*

Prefix. You can append a prefix to the dimension measurement by entering the prefix in this edit box. The dimension text is converted into **Prefix<dimension measurement>** format. For example, if you enter the text "Ht" in the Prefix edit box, "Ht" will be placed in front of the dimension text (Figure 10-24). The prefix string is saved in the **DIMPOST** system variable.

Note

*Once you specify a prefix, default prefixes such as **R** in radius dimensioning and ø in diameter dimensioning are canceled.*

Suffix. Just like appending a prefix, you can append a suffix to the dimension measurement by entering the desired suffix in this edit box. For example, if you enter the text cm in the **Suffix** edit box, the dimension text will have <dimension measurement>cm format (Figure 10-24). AutoCAD LT stores the suffix string in the **DIMPOST** variable.

Tip

*The **DIMPOST** variable is used to define both prefix and suffix to the dimension text. This variable takes a string value as its argument. For example, if you want to have a suffix for millimeters, set **DIMPOST** to mm. To establish a prefix to a dimension text, type the prefix text string and then "<>".*

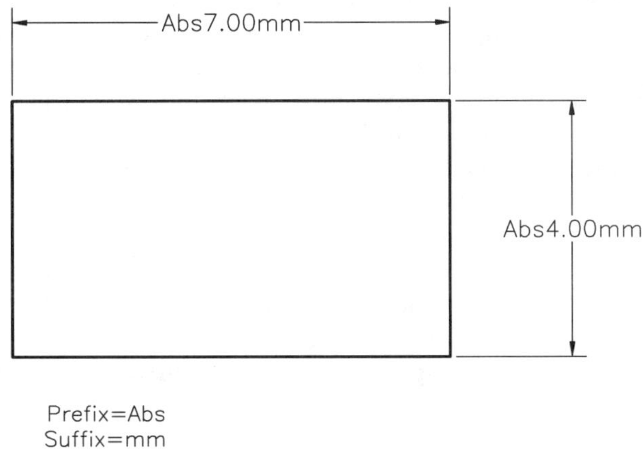

Figure 10-24 *Adding prefix and suffix to the dimensions*

Measurement Scale Area

Scale factor. You can specify a global scale factor for linear dimension measurements by setting the desired scale factor in the **Scale factor** spinner. All the linear distances measured by dimensions, which includes radii, diameters, and coordinates, are multiplied by the existing value in this spinner. For example, if the value of the **Scale factor** spinner is set to 2, two unit segments will be dimensioned as 4 units (2 x 2). However, the angular dimensions are not affected. In this manner, the value of the linear scaling factor affects the contents of the default (original) dimension text (Figure 10-25). Default value for linear scaling is 1. With the default value, the dimension text generated is the actual measurement of the object being dimensioned. The linear scaling value is saved in the **DIMLFAC** variable.

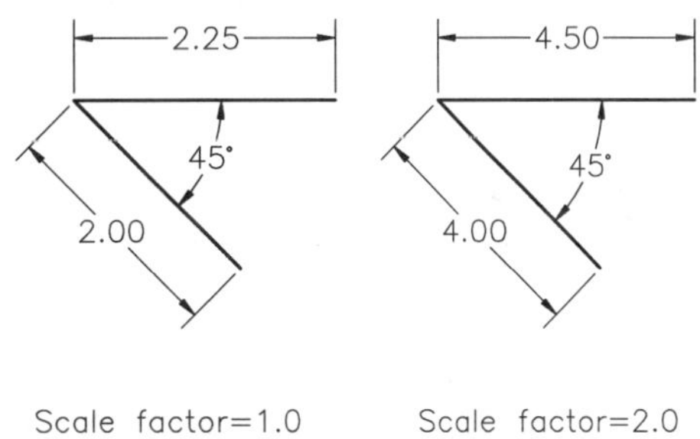

Figure 10-25 *Identical figures dimensioned using different scale factors*

Note

The linear scaling value is not exercised on rounding a value or on plus or minus tolerance values. Therefore, changing the linear scaling factor will not affect the tolerance values.

Apply to layout dimensions only. When you select the **Apply to layout dimensions only** check box, the scale factor value is applied only to the dimensions in the layout. The value is stored as a negative value in the **DIMLFAC** variable. If you change the **DIMLFAC** from the **Dim:** prompt, AutoCAD LT displays the viewport option to calculate the **DIMLFAC**. First, the **TILEMODE** should be set to 0 (paper space), and then invoke the **MVIEW** command to get the **Viewport** option.

Zero Suppression Area. The options under this area are used to suppress the leading or trailing zeros in the dimensioning. This area provides you with four check boxes. These check boxes can be selected to suppress the leading or trailing zeros or zeros in the feet and inches. The **0 Feet** and the **0 Inches** check boxes will be available only when you select **Engineering** or **Architectural** from the **Unit format** drop-down list. When Architectural units are being used, the **Leading** and **Trailing** check boxes are disabled. For example, if you select the **0 Feet** check box, the dimension text 0'-8 ¾" becomes 8 ¾". By default, the 0 Feet and 0 Inches value is suppressed. If you want to suppress the inches part of a feet-and-inches dimension when the distance in the feet portion is an integer value and the inches portion is zero, select the **0 Inches** check box. For example, if you select the **0 Inches** check box, the dimension text 3'-0" becomes 3'. Similarly, if you select **Leading** check box, the dimension that was initially 0.53 will become .53. If you select **Trailing** check box, the dimension that was initially 2.0 will become 2.

Angular Dimensions Area

This area provides you with the options to control the units format, precision, and zero suppression for Angular units.

Units format. The **Units format** drop-down list displays a list of unit formats for the angular dimensions. The default value in which the angular dimensions are displayed is **Decimal Degrees**. The value governing the unit setting for angular dimensions is stored in the **DIMAUNIT** variable.

Precision. You can select the number of decimal places for the angular dimensions from this drop-down list. This value is stored in the **DIMADEC** variable.

Zero Suppression. Similar to linear dimensions, you can suppress the **leading**, **trailing**, and neither or both zeros in the angular dimensions by selecting the respective check boxes in this area. The value is stored in the **DIMAZIN** variable.

FORMATTING ALTERNATE DIMENSION UNITS

Alternate Units Tab

By default, the **Alternate Units** tab of the **New Dimension Style** dialog box is disabled and the value of the **DIMALT** variable is turned off. If you want to perform alternate units dimensioning, select the **Display Alternate Units** check box. By doing so, AutoCAD LT activates various options in this area (Figure 10-26). This tab sets the format, precision, angles, placement, scale, and so on, for the alternate units in use. In this tab you can specify the values that will be applied to alternate dimensions.

Alternate Units Area

The options under this area are identical to those under the **Linear Dimensions** area of the **Primary Units** tab. This area provides you with the options to set the format for all dimension types except Angular.

Unit format. You can select a unit format to apply to the alternate dimensions from this drop-down list. The options under this drop-down list include Scientific, Decimal, Engineering,

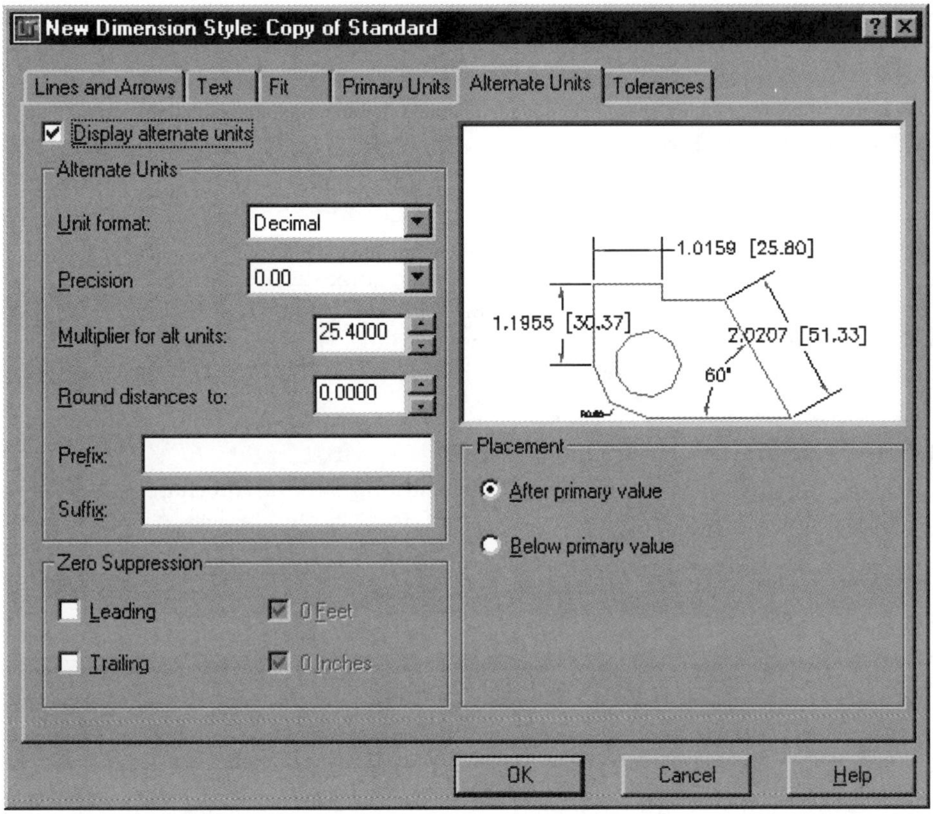

Figure 10-26 ***Alternate Units*** *tab of the* ***New Dimension Style*** *dialog box*

Architectural stacked, Fractional stacked, Architectural, Fractional, and Windows Desktop. The value is stored in the **DIMALTU** variable. The relative size of fractions is governed by the **DIMTFAC** variable.

Precision. You can select the number of decimal places for the alternate units from the **Precision** drop-down list. The value is stored in the **DIMALTD** variable.

Multiplier for alt units. To generate a value in the alternate system of measurement, you need a factor with which all the linear dimensions will be multiplied. The value for this factor can be set using the **Multiplier for alt units** spinner. The default value of 25.4 is for dimensioning in inches with alternate units in millimeters. This scaling value (contents of the **Multiplier for alt units** spinner) is stored in the **DIMALTF** variable.

Round distances to. This spinner is used to set a value to which you want all your measurements (made in alternate units) rounded off to. This value is stored in the **DIMALTRND** system variable. For example, if you set the value of the **Round distances to** spinner to 0.25, all the alternate dimensions get rounded off to the nearest .25 unit.

Prefix/Suffix. The **Prefix** and **Suffix** edit boxes are similar to the edit boxes in the **Linear**

Dimensions area of the **Primary Units** tab. You can enter the text or symbols that you want to precede or follow the alternate dimension text. The value is stored in the **DIMAPOST** variable. You can also use control codes and special characters to display special symbols.

Zero Suppression Area

This area allows you to suppress the leading or trailing zeros in decimal unit dimensions by selecting either, both, or none of the **Trailing** and **Leading** check boxes. Similarly, selecting the **0 Feet** check box, suppresses the zeros in the feet area of the dimension, when the dimension value is less than a foot. Selecting the **0 inches** check box suppresses the zeros in the inches area of the dimension. For example, 1'-0" becomes 1'. The **DIMALTZ** variable controls the suppression of zeros for alternate unit dimension values. The values that are between 0 to 3 affect feet-and-inch dimensions only.

Placement Area

This area provides the options that control the positioning of the Alternate units. The value is stored in the **DIMAPOST** variable.

After primary value. Selecting the **After primary value** radio button places the alternate units dimension text after the primary units. This is the default option (Figure 10-27).

Below primary value. Selecting the **Below primary value** radio button places the alternate units dimension text below the primary units (Figure 10-27).

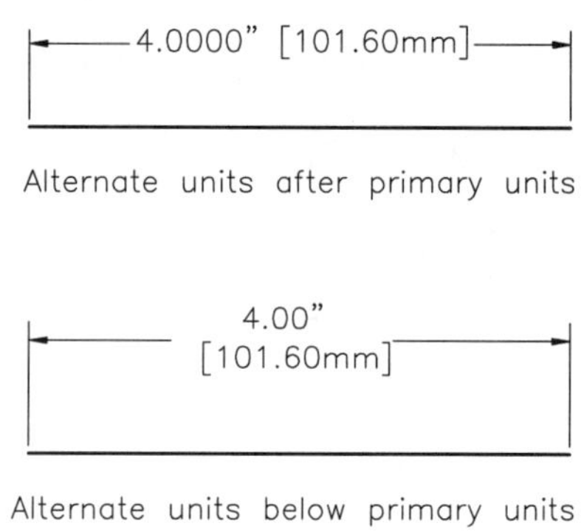

Figure 10-27 Placements of alternate units

Figure 10-27 illustrates the result of entering information in the **Alternate Units** tab. The decimal places get saved in the **DIMALTD** variable, the scaling value (contents of the **Multiplier for alt units** spinner) in the **DIMALTF** variable, and the suffix string (contents of the **Suffix** edit box) in the **DIMAPOST** variable. Similarly, the units format for alternate units are in **DIMALTU**, and suppression of zeros for alternate units decimal values are in **DIMALTZ**.

FORMATTING THE TOLERANCES

Tolerances Tab

The **Tolerances** tab (Figure 10-28) allows you to set the parameters for options that control the format and display of tolerance dimension text. These include the alternate unit tolerance dimension text.

Tolerance Format Area

The **Tolerance format** area of the **Tolerances** tab (Figure 10-28) lets you specify the tolerance method, tolerance value, position of tolerance text, precision and the height of the tolerance

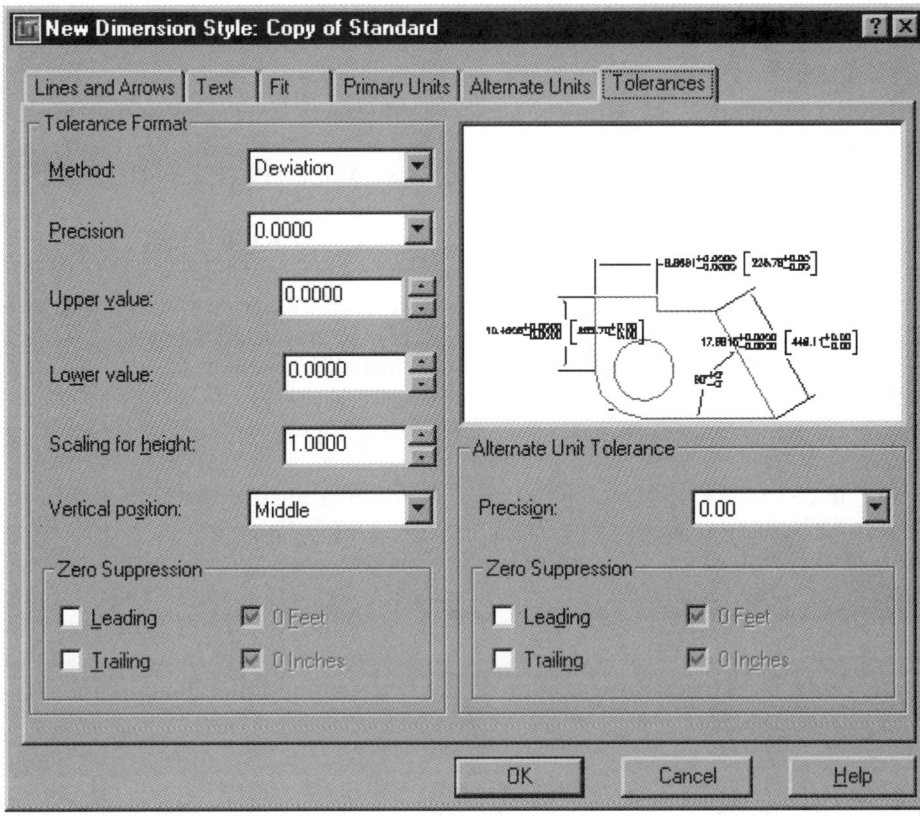

Figure 10-28 ***Tolerances*** *tab of the* ***New Dimension Style*** *dialog box*

text. For example, if you do not want a dimension to deviate more than plus 0.01 and minus 0.02, you can specify this by selecting **Deviation** from the **Method** drop-down list and then specifying the plus and minus deviation in the **Upper Value** and the **Lower Value** edit boxes. When you dimension, AutoCAD LT will automatically append the tolerance to the dimension. The **DIMTP** variable sets the maximum (or upper) tolerance limit for the dimension text and **DIMTM** variable sets the minimum (or lower) tolerance limit for dimension text. Different settings and their effects on relevant dimension variables are explained in following sections.

Method. The **Method** drop-down list lets you select the tolerance method. The tolerance methods supported by AutoCAD LT are **Symmetrical**, **Deviation**, **Limits**, and **Basic**. These tolerance methods are described next.

None. Selecting the **None** option sets the **DIMTOL** variable to 0 and does not add tolerance values to the dimension text, that is, the **Tolerances** tab is disabled.

Symmetrical. This option is used to specify the symmetrical tolerances. When you select this option, the **Lower Value** spinner is disabled and the value specified in the **Upper Value** spinner is applied to both plus and minus tolerance. For example, if the value specified in the **Upper Value** spinner is 0.05, the tolerance appended to the dimension text is ±0.05 (Figure 10-29). The value of **DIMTOL** is set to 1 and the value of **DIMLIM** is set to 0.

Deviation. If you select the **Deviation** tolerance method, the values in the **Upper Value** and **Lower Value** spinners will be displayed as plus and minus dimension tolerances. If you enter values for the plus and minus tolerances, AutoCAD LT appends a plus sign (+) to the positive values of the tolerance and a negative sign (–) to the negative values of the tolerance. For example, if the upper value of the tolerance is 0.005 and the lower value of the tolerance is 0.002, the resulting dimension text generated will have a positive tolerance of 0.005 and a negative tolerance of 0.002 (Figure 10-29). Even if one of the tolerance values is 0, a sign is appended to it. On specifying the deviation tolerance, AutoCAD LT sets the **DIMTOL** variable value to 1 and the **DIMLIM** variable value to 0. The values in the **Upper Value** and **Lower Value** edit boxes are saved in the **DIMTP** and **DIMTM** system variables, respectively.

Limits. If you select the **Limits** tolerance method from the **Method** drop-down list, AutoCAD LT adds the upper value (contents of the **Upper Value** spinner) to the dimension text (actual measurement) and subtracts the lower value (contents of the **Lower Value** spinner) from the dimension text. The resulting values are displayed as the dimension text (Figure 10-29). Selecting the **Limits** tolerance method results in setting the **DIMLIM** variable value to 1 and the **DIMTOL** variable value to 0. The numeral values in the **Upper Value** and **Lower Value** edit boxes are saved in the **DIMTP** and **DIMTM** system variables, respectively.

Basic. A basic dimension text is dimension text with a box drawn around it (Figure 10-29). The basic dimension is also called a reference dimension. Reference dimensions are used primarily in geometric dimensioning and tolerances. The basic dimension can be realized by selecting the basic tolerance method. The distance provided around the dimension text (distance between dimension text and the rectangular box) is stored as a negative value in the **DIMGAP** variable. The negative value signifies basic dimension. The default setting is off, resulting in the generation of dimensions without the box around the dimension text.

Precision. The **Precision** drop-down list is used to select the number of decimal places for the tolerance dimension text. The value is stored in **DIMTDEC** variable.

Upper value/Lower value. In the **Upper value** spinner the positive upper or maximum value is specified. If the method of tolerances is symmetrical, the same value is used as the lower value also. The value is stored in the **DIMTP** variable. In the **Lower** spinner the lower or minimum value is specified. The value is stored in the **DIMTM** variable.

Scaling for height. The **Scaling for height** spinner is used to specify the height of the dimension tolerance text relative to the dimension text height. The default value is 1 which means the height of the tolerance text is the same as the dimension text height. If you want the tolerance text to be 75 percent of the dimension height text, enter 0.75 in the **Scaling for height** edit box. The ratio of the tolerance height to the dimension text height is calculated by AutoCAD LT and then stored in the **DIMTFAC** variable. **DIMTFAC = Tolerance Height/Text Height.**

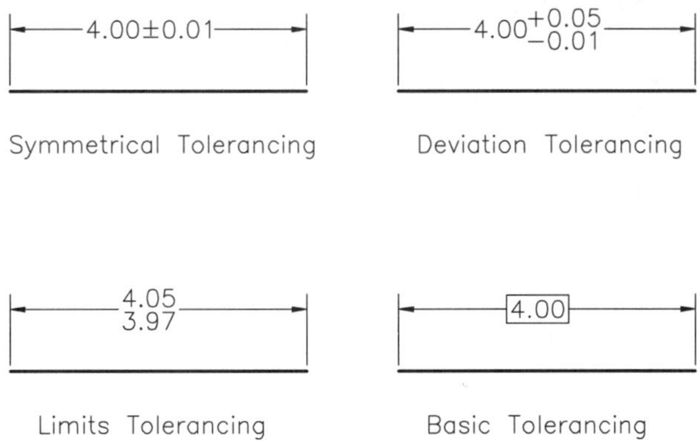

***Figure 10-29** Figure showing different tolerancing methods*

Vertical position. This drop-down list allows you to specify the location of the tolerance text for deviation and symmetrical methods only. The three alignments that are possible are with the **Bottom**, **Middle**, or **Top** of the main dimension text. The settings are saved in the **DIMTOLJ** system variable (Bottom=0, Middle=1, and Top=2).

Zero Suppression Area

This area controls the zero suppression in the tolerance text depending on which of the check boxes are selected. Selecting the **Leading** check box suppresses the leading zeros in all decimal tolerance text. For example, 0.2000 becomes .2000. Selecting the **Trailing** check box suppresses the trailing zeros in all decimal tolerance text. For example, 0.5000 becomes 0.5. Similarly, selecting both the boxes suppresses both the trailing and leading zeros and selecting none, suppresses none. If you select **0 Feet** check box, the zeros in the feet portion of the tolerance dimension text are suppressed if the dimension value is less than one foot. Similarly, selecting the **0 Inches** check box, suppresses the zeros in the inches portion of the dimension text. The value is stored in the **DIMTZIN** variable.

Alternate Unit Tolerance Area

Defines the precision and zero suppression settings for Alternate unit tolerance values. The options under this area will be available only when you are displaying the alternate units along with the primary units.

Precision. This drop-down list is used to set the number of decimal places to be displayed in the tolerance text of the alternate dimensions. This value is stored in the **DIMALTTD** variable.

Zero Suppression Area. Selecting the respective check boxes controls the suppression of the **Leading** and **Trailing** zeros in decimal values and the suppression of zeros in the Feet and Inches portions for dimensions in the feet and inches format. The value is stored in the **DIMALTTZ** variable.

Exercise 3 *Mechanical*

Draw Figure 10-30. Then set the values in the various tabs of the **New Dimension Style** dialog box to dimension it as shown. (Baseline Spacing = 0.25, Extension beyond dim lines = 0.10, Offset from origin = 0.05, Arrowhead size =0.07, Text height = 0.08). Assume the missing dimensions.

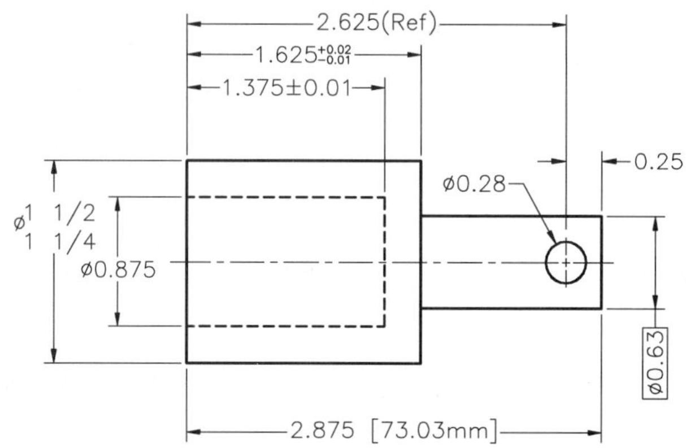

***Figure 10-30** Drawing for Exercise 3*

OTHER DIMENSIONING VARIABLES

Positioning Dimension Text (DIMTVP)

You can position the dimension text with respect to the dimension line by using the **DIMTVP** system variable (Figure 10-31). In certain cases, **DIMTVP** is used with **DIMTAD** to control the vertical position of the dimension text. The **DIMTVP** value applies only when the **DIMTAD** is off. To select the vertical position of the dimension text to meet your requirement (over or under the dimension line), you must first calculate the numerical value by which you want to offset the text from the dimension line. The vertical placing of the text is done by offsetting the dimension text. The magnitude of the offset of dimension text is a product of text height and **DIMTVP** value. If the value of **DIMTVP** is 1, **DIMTVP** acts as **DIMTAD**. For example, if you want to position the text 0.25 units from the dimension line, the value of **DIMTVP** is calculated as follows:

DIMTVP=Relative Position value/Text Height value
DIMTVP=0.25/ 0.09=2.7778

The value 2.7778 is stored in the dimension variable **DIMTVP**. If the absolute value is less than 0.70, the dimension line is broken to accommodate the dimension text. Relative positioning is not effective on angular dimensions.

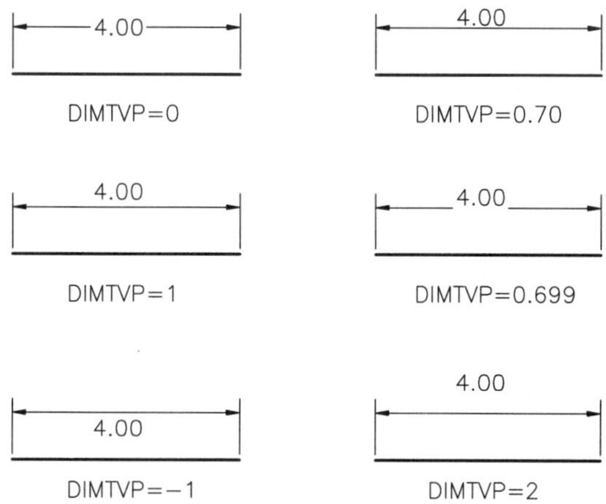

Figure 10-31 *Vertical positioning of dimension text*

DIMENSION STYLE FAMILIES

The dimension style feature of AutoCAD LT lets the user define a dimension style with values that are common to all dimensions. For example, the arrow size, dimension text height, or color of the dimension line are generally the same in all types of dimensioning such as linear, radial, diameter, and angular. These dimensioning types belong to the same family because they have some characteristics in common. In AutoCAD LT, this is called a **dimension style family,** and the values assigned to the family are called **dimension style family values.**

After you have defined the dimension style family values, you can specify variations on it for other types of dimensions such as radial and diameter. For example, if you want to limit the number of decimal places to two in radial dimensioning, you can specify that value for radial dimensioning. The other values will stay the same as the family values to which this dimension type belongs. When you use the radial dimension, AutoCAD LT automatically uses the style that was defined for radial dimensioning; otherwise, it creates a radial dimension with the values as defined for the family. After you have created a dimension style family, any changes in the parent style are applied to family members if the particular property is the same. Special suffix codes are appended to the dimension style family name that correspond to different dimension types. For example, if the dimension style family name is MYSTYLE and you define a diameter type of dimension, AutoCAD LT will append $4 at the end of the dimension style family name. The name of the diameter type of dimension will be MYSTYLE$4. The following are the suffix codes for different types of dimensioning:

Suffix Code	Dimension Type	Suffix Code	Dimension Type
0	Linear	2	Angular
3	Radius	4	Diameter
6	Ordinate	7	Leader

Chapter 10

Example 1 *Mechanical*

The following example illustrates the concepts of dimension style families, Figure 10-32.

1. Specify the values for the dimension style family.
2. Specify the values for linear dimension.
3. Specify the values for diameter and radius dimension.
4. After creating the dimension style, use it to dimension the given drawing.

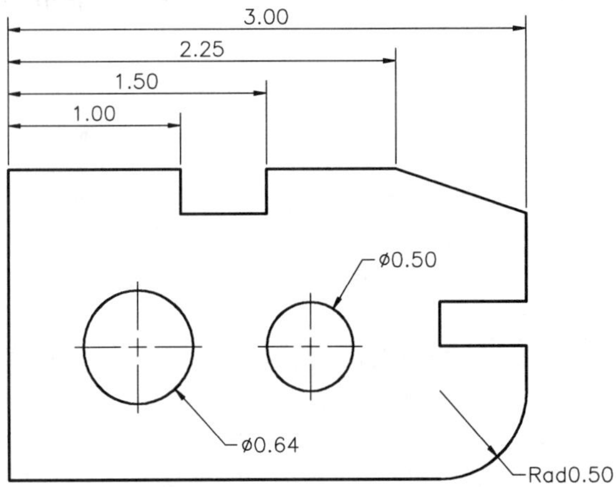

Figure 10-32 Drawing for Example 1

1. Open a new file and then draw the object shown in Figure 10-32.

2. Invoke the **Dimension Style Manager** dialog box by choosing the **Dimension Style** button from the **Dimension** toolbar. AutoCAD LT will display **Standard** in the **Styles** list box. Select the **Standard** style from the **Styles** list box.

3. Choose the **New** button to display the **Create New Dimension Style** dialog box. In this dialog box, enter **MyStyle** in the **New Style Name** edit box. Select **Standard** from the **Start With** drop-down list. Also select **All dimensions** from the **Use for** drop-down list. Now, choose the **Continue** button to display the **New Dimension Style: MyStyle** dialog box. In this dialog box, choose the **Lines and Arrows** tab and enter the following values:

   Baseline Spacing: 0.15 | **Extension beyond dim line: 0.07**
   Offset from origin: 0.03 | **Arrow size: 0.09**
   Center Mark for circle, Size: 0.05 | **Center Mark for Circles Type: Line**

4. Choose the **Text** tab and change the following values:

   Text Height: 0.09 | **Offset from dimension line: 0.03**

5. Choose the **Fit** tab and set the value of the **Use overall scale of** spinner in the **Scale for Dimension Features** area to **1**.

6. After entering the values, choose the **OK** button to return to the **Dimension Style Manager** dialog box. This dimension style contains the values that are common to all dimension types.

7. Now, choose the **New** button again in the **Dimension Style Manager** dialog box to display the **Create New Dimension Style** dialog box. AutoCAD LT displays **Copy of MyStyle** in the **New Style name** edit box. Select **MyStyle** from the **Start with** drop-down list if it is not already selected. From the **Use for** drop-down list, select **Linear dimensions**. Choose the **Continue** button to display the **New Dimension Style: MyStyle: Linear** dialog box and set the following values in the **Text** tab:

   a. Select the **Aligned with dimension line** radio button in the **Text Alignmen**t area.
   b. In the **Text Placement** area, from the **Vertical** drop-down list, select **Above**.

8. In the **Primary Units** tab, set the precision to two decimal places. Choose the **OK** button to return to the **Dimension Style Manager** dialog box.

9. Choose the **New** button again to display the **New Dimension Style** dialog box. Select **MyStyle** from the **Start with** drop-down list. Also select **Diameter dimension** type from the **Use for** drop-down list. Choose the **Continue** button to display the **New dimension Style: MyStyle: Diameter** dialog box.

10. Choose the **Primary units** tab and set the **Precision** to two decimal places. In the **Lines and Arrows** tab, select **Line** from the **Type** drop-down list in the **Center mark for circle** area. Choose the **OK** button to return to the **Dimension Style Manager** dialog box.

11. In this dialog box, choose the **New** button to display the **Create New Dimension Style** dialog box. Select **MyStyle** from the **Start with** drop-down list and **Radius dimensions** from the **Use for** drop-down list. Choose the **Continue** button to display the **New Dimension Style: MyStyle: Radial** dialog box.

12. Choose the **Primary Units** tab and set the precision to two decimal places. Enter **Rad** in the **Prefix** edit box.

13. In the **Fit** tab, select the **Text** radio button in the **Fit Options** area. Choose the **OK** button to return to the **Dimension Style Manager** dialog box.

14. Select **MyStyle** from the **Styles** list box and choose the **Set current** button. Choose the **Close** button to exit the dialog box.

15. Use the linear and baseline dimensioning to draw the linear dimensions as shown in Figure 10-32. You will notice that when you enter any linear dimensioning, AutoCAD LT automatically uses the values that were defined for the linear type of dimensioning.

16. Use the diameter dimensioning to dimension the circles as shown in Figure 10-32. Again, notice that the dimensions are drawn according to the values specified for the diameter type of dimensioning.

17. Now, use the radius dimensioning to dimension the fillet as shown in Figure 10-32.

USING DIMENSION STYLE OVERRIDES

Most of the dimension characteristics are common in a production drawing. The values that are common to different dimensioning types can be defined in the dimension style family. However, at times you might have different dimensions. For example, you may need two types of linear dimensioning: one with tolerance and one without. One way to draw these dimensions is to create two dimensioning styles. You can also use the dimension variable overrides to override the existing values. For example, you can define a dimension style (**MyStyle**) that draws dimensions without tolerance. Now, to draw a dimension with tolerance or update an existing dimension, you can override the previously defined value. You can override the values through the **Dimension Style Manager** dialog box or by setting the variable values at the Command prompt. The following example illustrates how to use the dimension style overrides.

Example 2 *Mechanical*

In this example, you will update the overall dimension (3.00) so that the tolerance is displayed with the dimension. You will also add two linear dimensions, as shown in Figure 10-33.

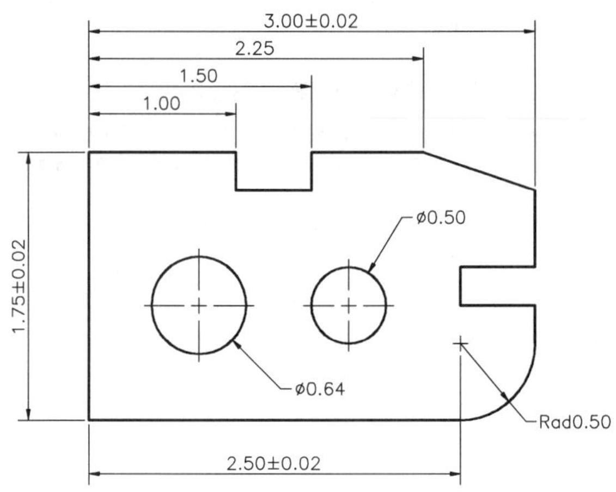

Figure 10-33 *Drawing for Example 2*

This problem can be solved by dimension style overrides as well as using the **Properties** window. However, here only the dimension style overrides method will be discussed.

1. Invoke the **Dimension Style Manager** dialog box. Select **MyStyle** from the **Styles** list box and choose the **Override** button to display the **Override Current Style: My Style** dialog box. The options in this dialog box are identical to the **New Dimension Style** dialog box discussed earlier in the chapter.

2. Choose the **Tolerance** tab and select **Symmetrical** from the **Method** drop-down list.

3. Set the value of the **Precision** spinner to two decimal places. Set the value of the **Upper value** spinner to **0.02**. Choose the **OK** button to exit the dialog box (this does not save the style). You will notice that the **<style overrides>** is displayed under **MyStyle** in the **Style** list box, indicating that the style overrides the **MyStyle** dimension style.

4. This **<style overrides>** is displayed until you save it under a new name or under the Style it is displayed under, or until you delete it. Select the **<style overrides>** and right- click to display the shortcut menu. Choose the **Save to current Style** option from the shortcut menu to save the overrides to the current style. Choosing the **Rename** option allows you to rename the style override and save it as a new style.

5. Choose **Update** from the **Dimension** menu and select the dimension that measures **3.00**. It will now display the symmetrical tolerance.

6. Draw the remaining two linear dimensions. They will automatically appear with the tolerances (Figure 10-33).

Tip

*You can also use the **DIMOVERRIDE** command to apply the change to the existing dimensions. Apply the changes to the **DIMTOL**, **DIMTP**, and **DIMTM** variables.*

COMPARING AND LISTING DIMENSION STYLES

Choosing the **Compare** button in the **Dimension Style Manager** dialog box displays the **Compare Dimension Styles** dialog box where you can compare the settings of two dimensions styles or list all the settings of one of them (Figure 10-34).

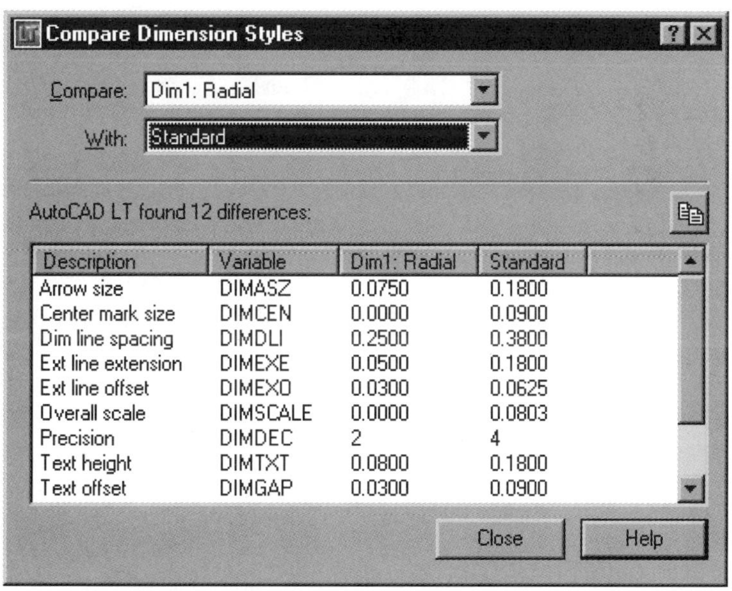

***Figure 10-34** **Compare Dimension Styles** dialog box*

The **Compare** and the **With** drop-down lists display the dimension styles in the current drawing and selecting dimension styles from the respective lists compare the two styles. In the **With** drop-down list, if you select **None** or the same style as selected from the **Compare** drop-down list, all the properties of the selected style are displayed. The comparison results are displayed under four headings: **Description** of the Dimension Style property, the **System Variable** controlling a particular setting, and the **values of the variable for both the dimension styles** which differ in the two styles in comparison. The number of differences between the selected dimension styles are displayed below the **With** drop-down list. The button provided in this dialog box prints the comparison results to the Windows clipboard from where they can be pasted to other Windows applications.

USING EXTERNALLY REFERENCED DIMENSION STYLES

The externally referenced dimensions cannot be used directly in the current drawing. When you Xref a drawing, the drawing name is appended to the style name and the two are separated by the vertical bar (|) symbol. It uses the same syntax as other externally dependent symbols. For example, if the drawing (FLOOR) has a dimension style called DECIMAL and you Xref this drawing in the current drawing, AutoCAD LT will rename the dimension style to FLOOR|DECIMAL. You cannot make this dimension style current, nor can you modify or override it. However, you can use it as a template to create a new style. To accomplish this, invoke the **Dimension Style Manager** dialog box. If the **Don't list styles in Xrefs** check box is selected, the styles in the Xref are not displayed. Clear this check box to display the Xref dimension styles and choose the **New** button. In the **New Style Name** edit box of the **New Dimension Style** dialog box, enter the name of the dimension style. AutoCAD LT will create a new dimension style with the same values as those of the externally referenced dimension style (FLOOR|DECIMAL).

Self-Evaluation Test

Answer the following questions, and then compare your answers to the correct answers given at the end of this chapter.

1. You can invoke the **Dimension Style Manager** dialog box using both **Format** menu and **Dimension** menu. (T/F)

2. The size of the arrow block is determined by the value stored in the **Arrow size** edit box. (T/F)

3. The default dimension style file name is **Drawing**. (T/F)

4. The size of the tolerance text with respect to the dimensions can be defined. (T/F)

5. The **DIMTVP** variable is used to control the __________ position of the dimension text.

6. When you select the **Arrows** option, AutoCAD LT places the text and arrowheads __________.

7. A basic dimension text is dimension text with a __________drawn around it.

8. The **Suppress** check boxes in the **Dimension Lines** area control the display of __________ and __________.

9. You can specify the tolerancing using __________ number of methods.

10. The __________ button in the **Dimension Style Manager** dialog box is used to override the current dimension style.

Review Questions

Answer the following questions.

1. You cannot replace the default arrowheads at the end of the dimension lines. (T/F)

2. When the **DIMTVP** variable has a negative value, the dimension text is placed below the dimension line. (T/F)

3. Dimension style name cannot be changed. (T/F)

4. The named dimension style associated with the dimension being updated by overriding is not updated. (T/F)

5. To use a dimension style for dimensioning you will have to first make it active using which button?

   (a) **Set Current** (b) **New**
   (c) **Override** (d) **Modify**

6. To add a suffix **mm** to the dimensions, which tab of the **Dimension Style Manager** dialog box will you use?

   (a) **Fit** (b) **Text**
   (c) **Primary Units** (d) **Alternate Units**

7. If you want to place the dimension text manually every time you create a dimension, which tab of the **Dimension Style Manager** dialog box will you use?

   (a) **Fit** (b) **Text**
   (c) **Primary Units** (d) **Alternate Units**

8. The size of the _________ is determined by the value stored in the **Arrow size** edit box.

9. When **DIMSCALE** is assigned a value of _________, AutoCAD LT calculates an acceptable default value based on the scaling between the current model space viewport and paper space.

10. If you use the **DIMCEN** command, a positive value will create a center mark, and a negative value will create a _________.

11. If you select the _________ check box, you can position the dimension text anywhere along the dimension line.

12. You can append a prefix to the dimension measurement by entering the desired prefix in the **Prefix** edit box of the _________ dialog box.

13. If you select the **Limits** tolerance method from the **Method** drop-down list, AutoCAD LT _________ the upper value to the dimension and _________ the lower value from the dimension text.

14. You can also use the _________ command to override a dimension value.

15. What is the dimension style family, and how does it help in dimensioning? ___________ ____.

Exercises

Exercises 4 through 9 — *Mechanical*

Make the drawings as shown in Figure 10-35 through Figure 10-40. You must create dimension style files and specify values for different dimension types such as linear, radial, diameter, and ordinate. Assume the missing dimensions.

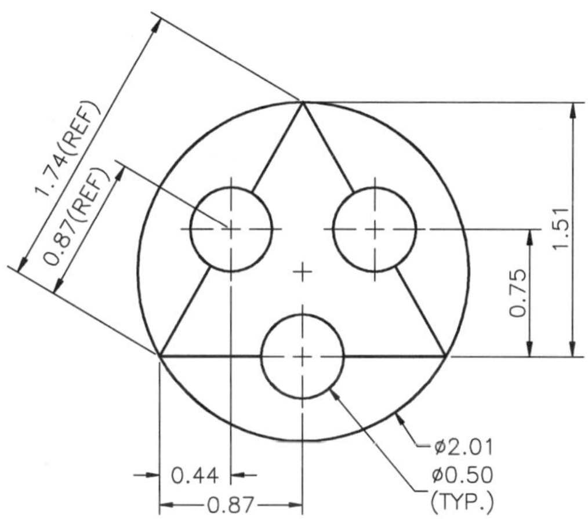

***Figure 10-35** Drawing for Exercise 4*

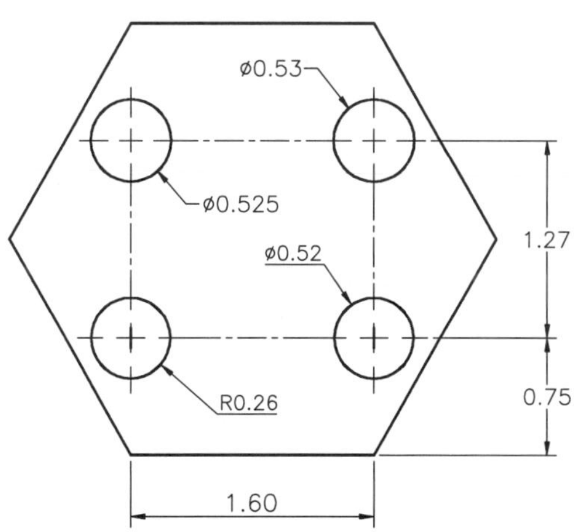

***Figure 10-36** Drawing for Exercise 5*

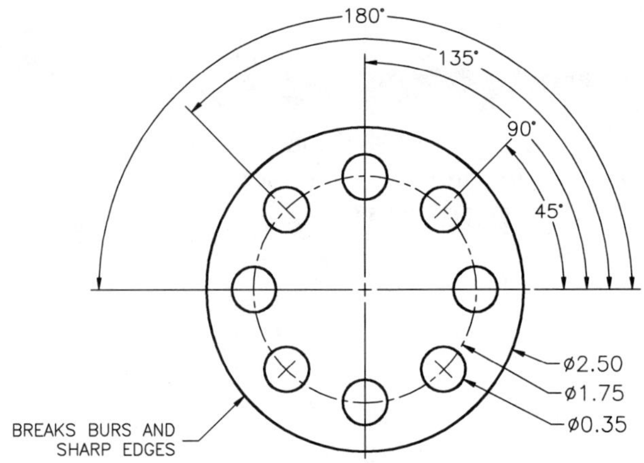

Figure 10-37 *Drawing for Exercise 6*

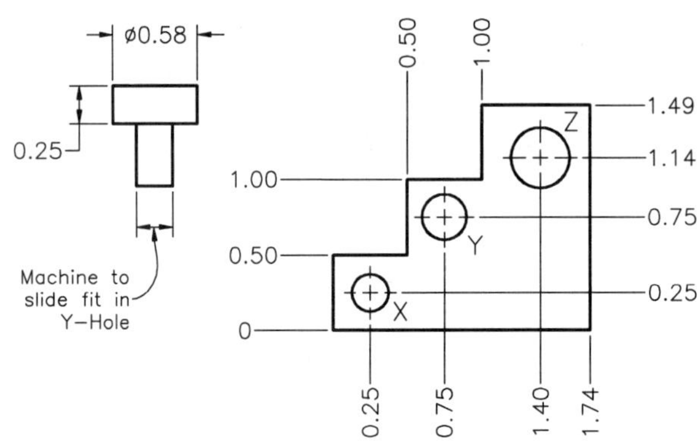

HOLE	X	Y	Z
RAD.	0.125	0.15	0.20

Figure 10-38 *Drawing for Exercise 7*

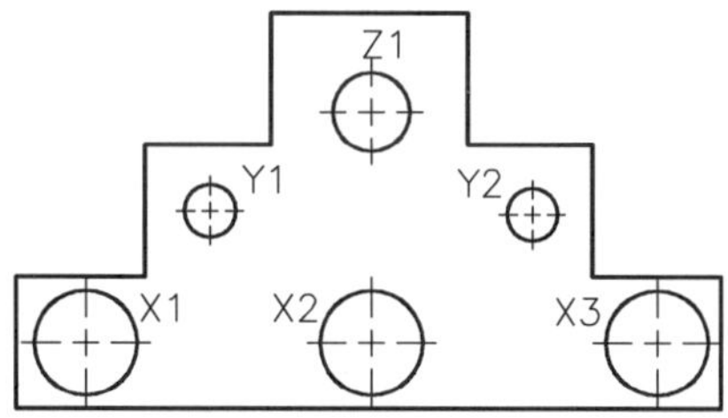

HOLE	X1	X2	X3	Y1	Y2	Z1
DIM.	R0.2	R0.2	R0.2	R0.1	R0.1	R0.15
QTY.	1	1	1	1	1	1
X	0.25	1.375	2.50	0.75	2.0	1.375
Y	0.25	0.25	0.25	0.75	0.75	1.125
Z	THRU	THRU	THRU	1.0	1.0	THRU

***Figure 10-39** Drawing for Exercise 8*

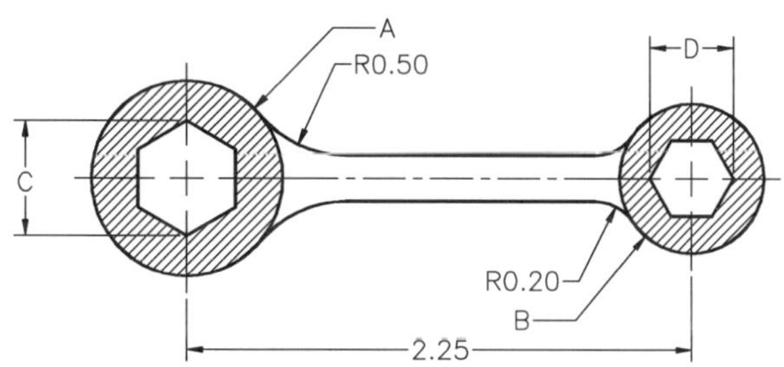

SPANNER NO.	A	B	C	D
S1	0.85	0.65	0.50	0.38
S2	1	0.75	0.59	0.44
S3	1.15	0.88	0.67	0.52
S4	1.25	0.95	0.74	0.56

***Figure 10-40** Drawing for Exercise 9*

Exercise 10 *Mechanical*

Draw the objects shown in Figure 10-41. You must create the dimension style and specify different dimensioning parameters. Also, suppress the leading and trailing zeros in the dimension style. Assume the missing dimensions.

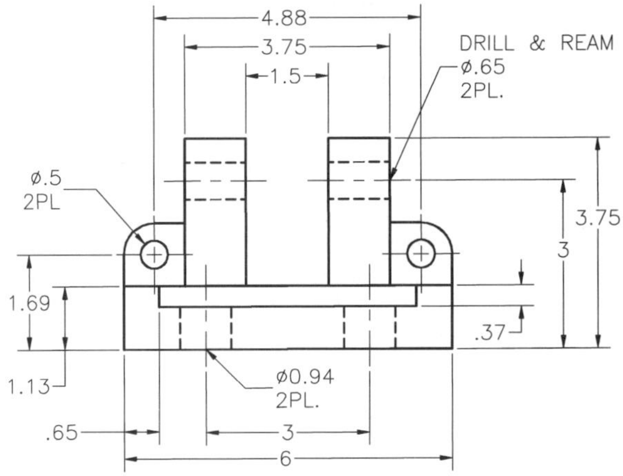

Figure 10-41 Drawing for Exercise 10

Exercise 11 — *Mechanical*

Draw the objects shown in Figure 10-42. You must create the dimension style and specify different dimensioning parameters in the dimension style. Assume the missing dimensions.

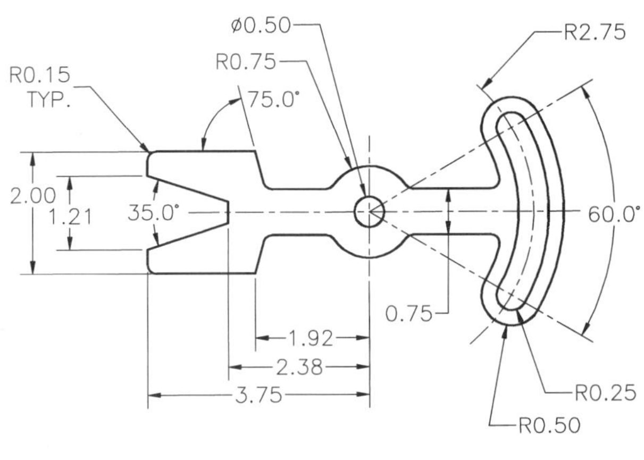

Figure 10-42 Drawing for Exercise 11

Exercise 12 — *Mechanical*

Draw the objects shown in Figure 10-43. You must create the dimension style and specify different dimensioning parameters in the dimension style. Assume the missing dimensions.

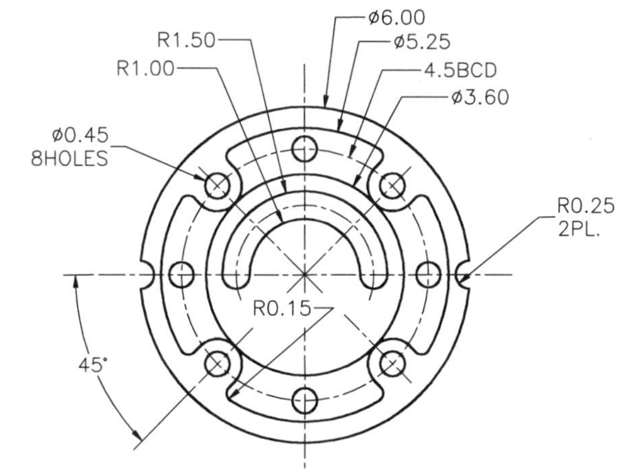

***Figure 10-43** Drawing for Exercise 12*

Problem Solving Exercise 1

Mechanical

Draw the objects shown in Figure 10-44. You must create the dimension style and specify different dimensioning parameters in the dimension style. Also, suppress the leading and trailing zeros in the dimension style. Assume the missing dimensions.

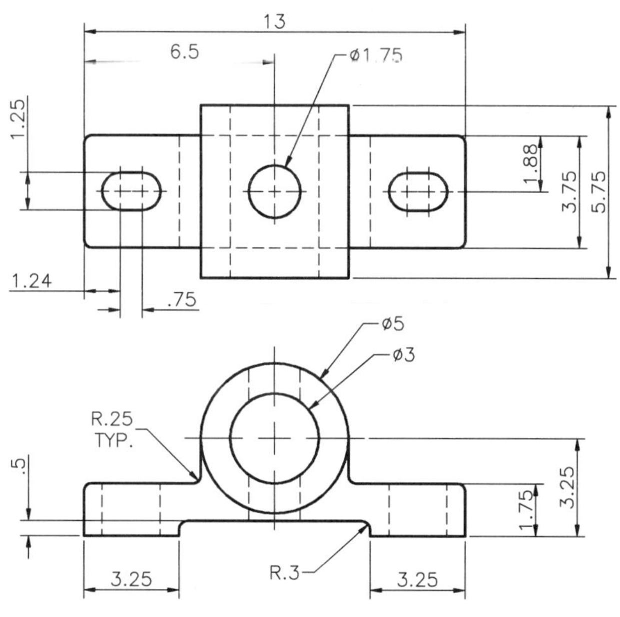

***Figure 10-44** Drawing for Problem Solving Exercise 1*

Problem Solving Exercise 2 — *Mechanical*

Draw the shaft shown in Figure 10-45. You must create the dimension style and specify the dimensioning parameters based on the given drawing. Assume the missing dimensions.

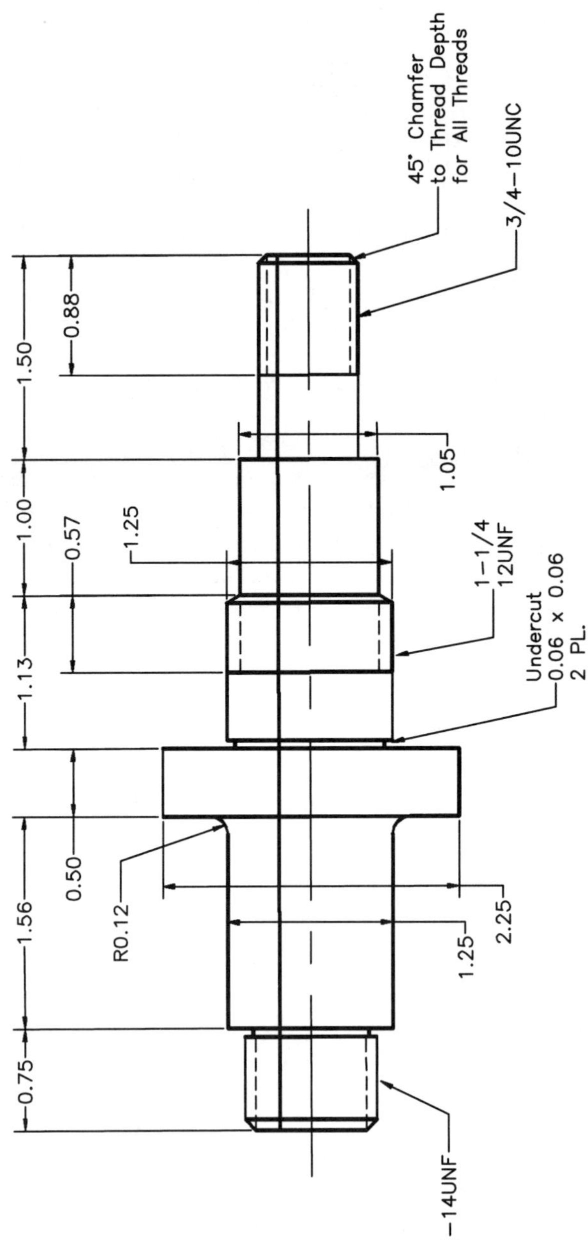

Figure 10-45 *Drawing for Problem Solving Exercise 2*

Problem Solving Exercise 3 — *Mechanical*

Draw the connecting rod shown in Figure 10-46. You must create the dimension style and specify the dimensioning parameters based on the given drawing. Assume the missing dimensions.

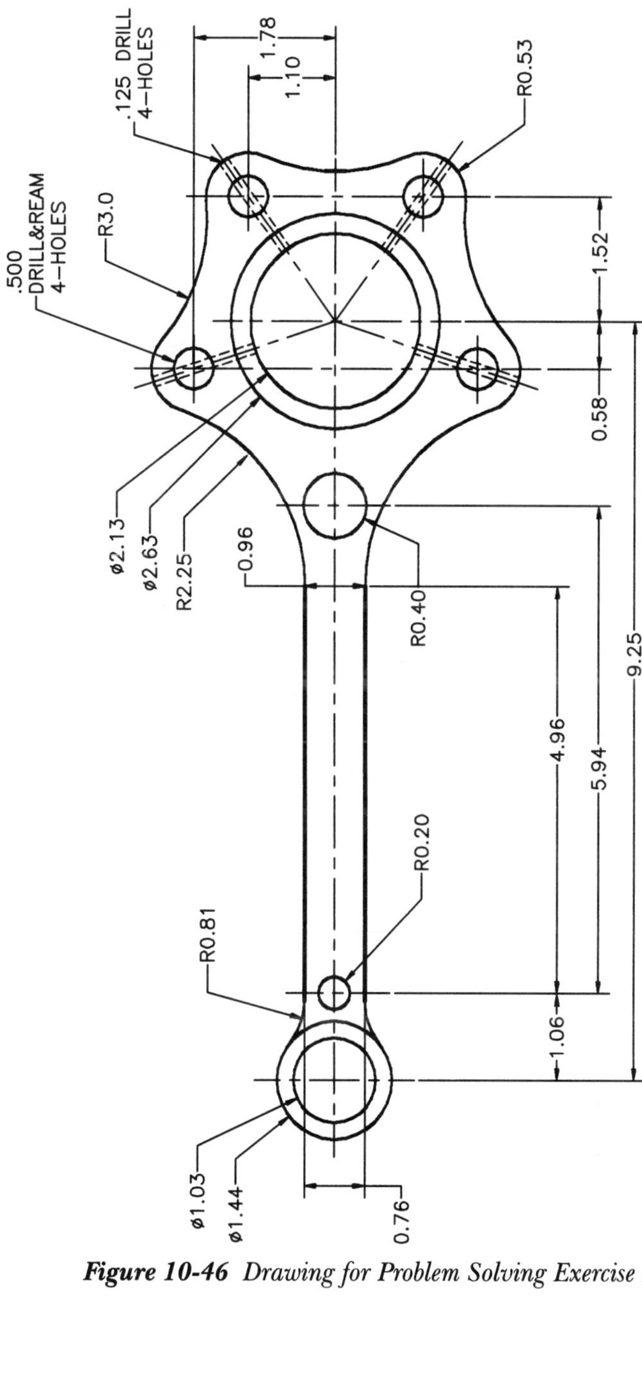

Figure 10-46 *Drawing for Problem Solving Exercise 3*

Problem Solving Exercise 4 *Architectural*

Draw the elevation of the house shown in Figure 10-47. Assume the missing dimensions.

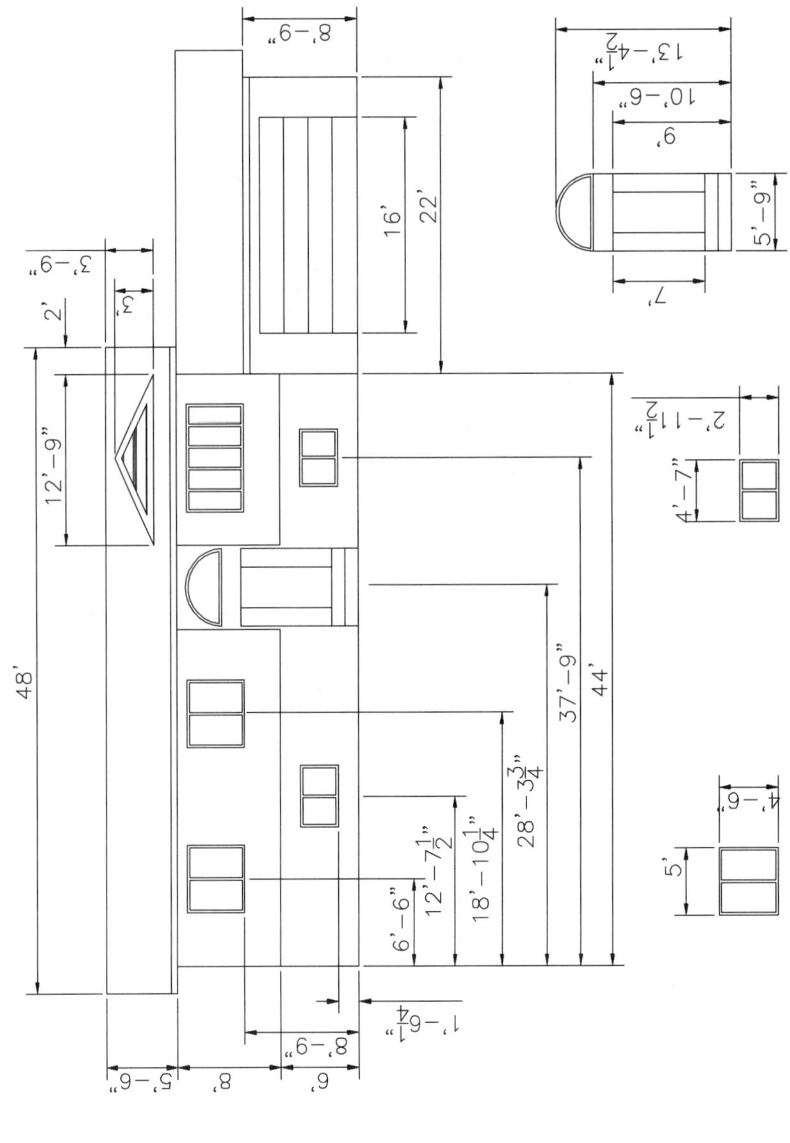

Figure 10-47 *Drawing for Problem Solving Exercise 4*

Answers to Self-Evaluation Test

1 - T **2** - T, **3** - **F**, **4** - T, **5** - vertical, **6** - inside, **7** - frame, **8** - first, second dimension lines, **9** - four, **10** - **Override**

Chapter 11

Model Space Viewports, Paper Space Viewports, and Layouts

Learning Objectives

After completing this chapter, you will be able to:
- *Understand the concepts of model space and paper space.*
- *Create tiled viewports in the model space using the various command commands.*
- *Create floating viewports in layouts using various commands.*
- *Shift from paper space to model space using the* ***MSPACE*** *command.*
- *Shift from model space to paper space using the* ***PSPACE*** *command.*
- *Control the visibility of viewport layers with the* ***VPLAYER*** *command.*
- *Set linetype scaling in paper space using the* ***PSLTSCALE*** *command.*

MODEL SPACE AND PAPER SPACE/LAYOUTS

For ease in designing, AutoCAD LT provides two different types of environments. These environments are the model space and the paper space. The paper space is also called layout. The model space is basically used for the designing or the drafting work. This is the default environment that is active when you start using AutoCAD LT. Almost the entire design is created in the model space. The other environment is the paper space. This environment is used for plotting the drawings. By default, AutoCAD LT provides you with two layouts. A layout can be considered a sheet of paper on which you can place the design created in the model space and then print it. You can also assign different plotting parameters to these layouts for plotting. Almost all the commands of the model space also work in the layouts. Note that under certain conditions you can not select the drawing objects created in model space when they are displayed in the viewports in layouts. However, you can snap on to the different points of the drawing objects such as the endpoints, midpoints, center points, and so on, using the **OSNAP** options.

You can shift from one environment to the other environment by choosing the **Model** tab or the **Layout1/Layout2** tabs provided at the bottom left corner of the drawing window. You can also shift from one environment to the other using the **TILEMODE** system variable. The default value of this variable is **1**. If the value of this system variable is set to **0**, you will be shifted to the layouts and if its value is set to **0**, you will be shifted to model space. The viewports created in the model space are called **tiled viewports** and the viewports created in the layouts are called **floating viewports**.

MODEL SPACE VIEWPORTS (TILED VIEWPORTS)

A viewport in the model space is defined as a rectangular area of the drawing window in which you can create the design. When you start AutoCAD LT, only one viewport is displayed in the model space. You can create multiple nonoverlpping viewports in the model space that can display different views of the same object (Figure 11-1). Each viewport will become an individual drawing area. This is generally used while creating the complex designs. You can view the same design from different positions by creating the tiled viewports and define the distinct coordinate system configuration for each viewport. You can also use the **PAN** or **ZOOM** command to display different portions or different levels of detail of the drawing in each viewport. The tiled viewports can be created using the **VPORTS** command.

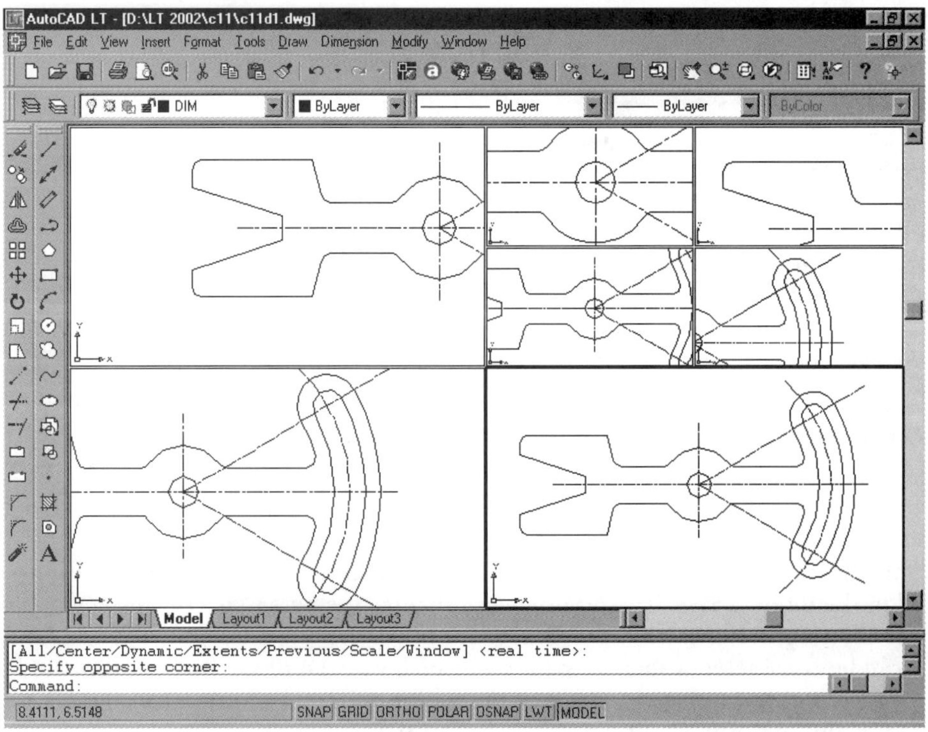

Figure 11-1 Screen display with multiple tiled viewports

Creating Tiled Viewports

Toolbar:	Viewports > Display Viewports Dialog
Menu:	View > Viewports > New Viewports
Command:	VPORTS

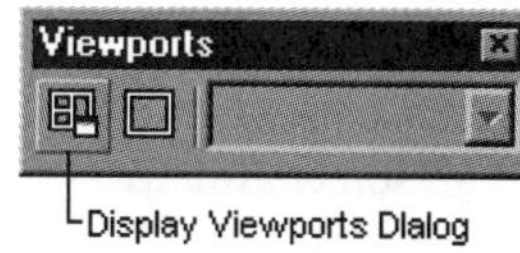

***Figure 11-2** Displaying the **Viewports** dialog box using the **Viewports** toolbar*

As mentioned earlier, the display screen in the model space can be divided into multiple nonoverlapping tiled viewports. All these viewports are created only in the rectangular shape. The number of these viewports depends on the equipment and the operating system on which AutoCAD LT is running. Each tiled viewport contains a view of the drawing. The tiled viewports touch each other at the edges without overlapping. While using tiled viewports, you are not allowed to edit, rearrange, or turn individual viewports on or off. These viewports are created using the **VPORTS** command when the system variable **TILEMODE** is set to 1 or the **Model** tab is active. The **Viewports** dialog box is displayed when you choose the **Display Viewports Dialog** button (Figure 11-2). You can use this dialog box to create new viewport configurations and save them. The options under both the tabs of the **Viewports** dialog box (Figure 11-3) are discussed next.

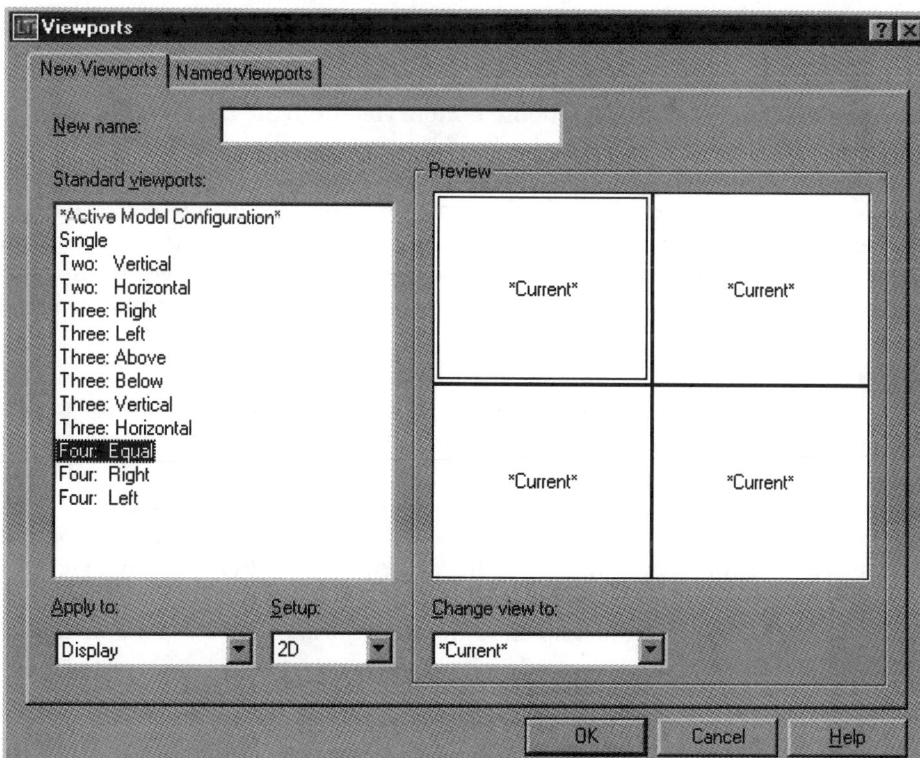

***Figure 11-3** The **New Viewports** tab of the **Viewports** dialog box*

New Viewports Tab

This tab of the **Viewports** dialog box provides the options related to the standard viewport configurations that can be used. You can also save a user-defined configuration using this tab. The name for the viewport configuration you wish to create can be specified in the **New name** edit box. If you do not enter a name in this edit box, the viewport configuration you create is not saved and, therefore, cannot be used later. A list of standard viewport configurations is listed in the **Standard viewports** list box. This list also contains the ***Active Model Configuration***, which is the current viewport configuration. From the **Standard viewports** list, you can select and apply any one of the listed standard viewport configurations. A preview image of the selected configuration is displayed in the **Preview** window. The **Apply to** drop-down list has the **Display** and **Current viewport** options. Selecting the **Display** option applies the selected viewport configuration to the entire display and selecting the **Current viewport** option applies the selected viewport configuration to only the current viewport. With this option you can create more viewports inside the existing viewports. The changes will be applied to the current viewport and the new viewports will be created inside the current viewport.

From the **Setup** drop-down list, you can select **2D** or **3D**. When you selecting the **2D** option it creates the new viewport configuration with the current view of the drawing in all the viewports initially. Using the **3D** option applies the standard orthogonal and isometric views to the viewports. For example, if your configuration has four viewports, they are assigned the **Top**, **Front**, **Left**, and **South East Isometric** views respectively. You can also modify these standard orthogonal and isometric views by selecting from the **Change view to** drop-down list and replacing the existing view in the selected viewport. For example, you can select the viewport that is assigned the **Top view** and then choose **Bottom view** from the **Change view to** drop-down list to replace it. The Preview image in the **Preview** window reflects the changes you make. If you are using the **2D** option, you can select a named viewport configuration to replace the selected one. Choose **OK** to exit the dialog box and apply the created or selected configuration to the current display in the drawing. When you name and save a viewport configuration, it saves information about the number and position of viewports, the viewing direction and zoom factor, and the grid, snap, coordinate system and UCS icon settings of the viewports.

Named Viewports Tab

The **Named Viewports** tab of the **Viewports** dialog box (Figure 11-4) displays the name of the current viewport next to **Current name**. The names of all the saved viewport configurations in a drawing are displayed in the **Named viewports** list box. You can select any one of the named viewport configurations and apply it to the current display. A preview image of the selected configuration is displayed in the **Preview** window. Choose **OK** to exit the dialog box and apply the selected viewport configuration to the current display. In the **Named viewports** list box, you can select a name and right-click to display a shortcut menu. Choosing **Delete** deletes the selected viewport configuration and choosing **Rename** allows you to rename the selected viewport configuration.

MAKING A VIEWPORT CURRENT

The viewport you are working in is called the current viewport. You can display several model space viewports on the screen, but you can work in only one of them at a time. You can switch from one viewport to another even when you are in the middle of a command. For example,

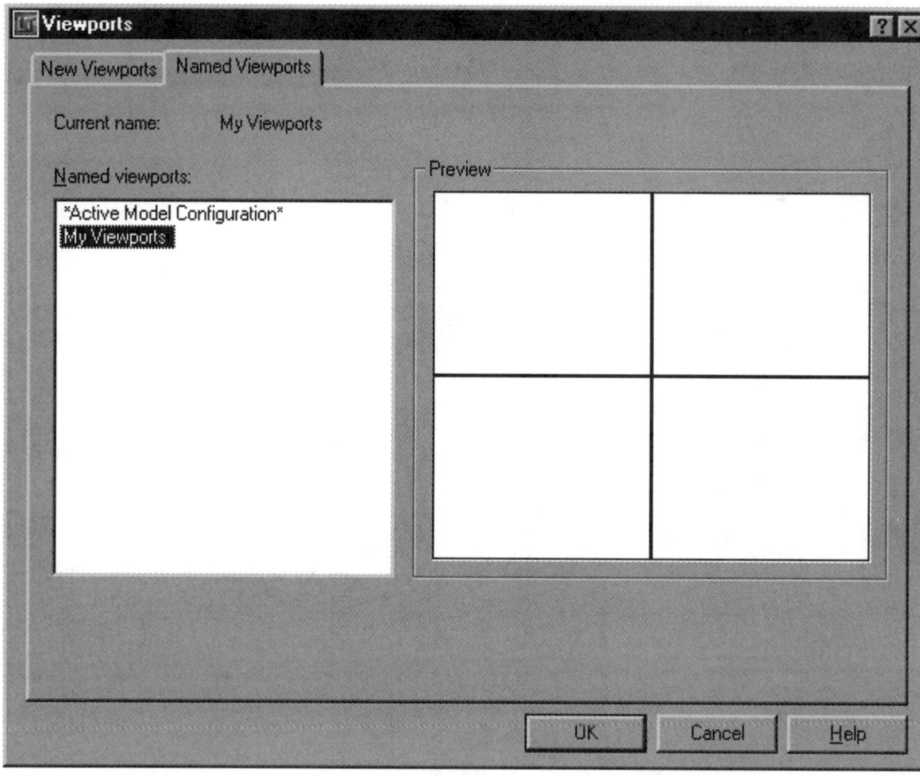

***Figure 11-4** The **Named Viewports** tab of the **Viewports** dialog box*

you can specify the start point of the line in one viewport and the endpoint of the line in the other viewport. The current viewport is indicated by a border that is heavy compared to the borders of the other viewports. Also, the graphics cursor appears as a drawing cursor (screen crosshairs) only when it is within the current viewport. Outside the current viewport this cursor appears as an arrow cursor. You can enter points and select objects only from the current viewport. To make a viewport current, you can select it with the pointing device. Another method of making a viewport current is by assigning its identification number to the **CVPORT** system variable. The identification numbers of the named viewport configurations are not listed in the display.

JOINING TWO ADJACENT VIEWPORTS

AutoCAD LT provides you with an option of joining two adjacent tiled viewports. However, remember that the viewports you wish to join should result in a rectangular-shaped viewport only. As mentioned earlier, the viewports in model space can only be in rectangular shape. Therefore, you will not be able to join two viewports in case they do not result in a rectangular shape. The viewports can be joined by choosing **Viewports > Join** from the **View** menu. Upon choosing this, you will be prompted to select the dominant viewport. A dominant viewport is one whose display will be retained after joining. After selecting the dominant viewport, you will be prompted to select the viewport to join. Figure 11-5 shows a viewport configuration before joining and Figure 11-6 shows a viewport configuration after joining.

Figure 11-5 Selecting the dominant viewport and the viewport to join

Figure 11-6 Viewports after joining

Tip

You can also use the Command line to create, save, restore, delete, or join the viewport configurations. This is done with the help of the ***-VPORTS*** *command.*

Note

By dividing, subdividing, and joining viewports, you can create nearly any viewport configuration you want, so long as all viewports are rectangular.

OBTAINING PLAN VIEWS AUTOMATICALLY (UCSFOLLOW SYSTEM VARIABLE)

You can obtain the plan view of an object by using the **PLAN** command. If you have different viewports, you can obtain the plan view relative to the current UCS by activating a viewport, setting the UCS to your requirement, and using the default option of the **PLAN** command. This operation can be automated by using the **UCSFOLLOW** system variable. When **UCSFOLLOW** is assigned a value of 1, a plan view is automatically generated if you change the UCS. By default, this variable is assigned the value 0. All of the viewports have the **UCSFOLLOW** facility; therefore, you need to specify the **UCSFOLLOW** setting separately for each viewport. The prompt sequence is:

Command: **UCSFOLLOW** Enter
Enter new value for UCSFOLLOW <0>: **1** Enter

PAPER SPACE VIEWPORTS (FLOATING VIEWPORTS)

As mentioned earlier, the viewports created in the layouts are called floating viewports. The viewports in the layouts are called floating viewports because unlike model space, the viewports in the layouts can be overlapping. In layouts there is no restriction of the location of the viewports. You can even place the viewports one over the other. This property of the floating viewports is very useful when you need to display a detailed view of the design. You can create one main viewport and then create multiple floating viewports over this main viewport where you can show the details of the design. Figure 11-7 shows one such arrangement with a number of floating viewports. In this figure you will see that the viewports in the layouts can be created one over the other, thus the name floating viewports.

Similar to the tiled viewports, the floating viewports can also be created using the **Viewports** dialog box. This dialog box can be invoked using the **VPORTS** command. The method of creating the floating viewports is discussed next.

Creating Floating Viewports (VPORTS Command)

Toolbar:	Viewports > Display Viewports Dialog
Menu:	View > Viewports > New Viewports
Command:	VPORTS

This command is used for creating the floating viewports in layouts. However, when you invoke this command in the layouts, the dialog box displayed is slightly modified. Instead of the **Apply to** drop-down list in the **New Viewports** tab, the **Viewport Spacing**

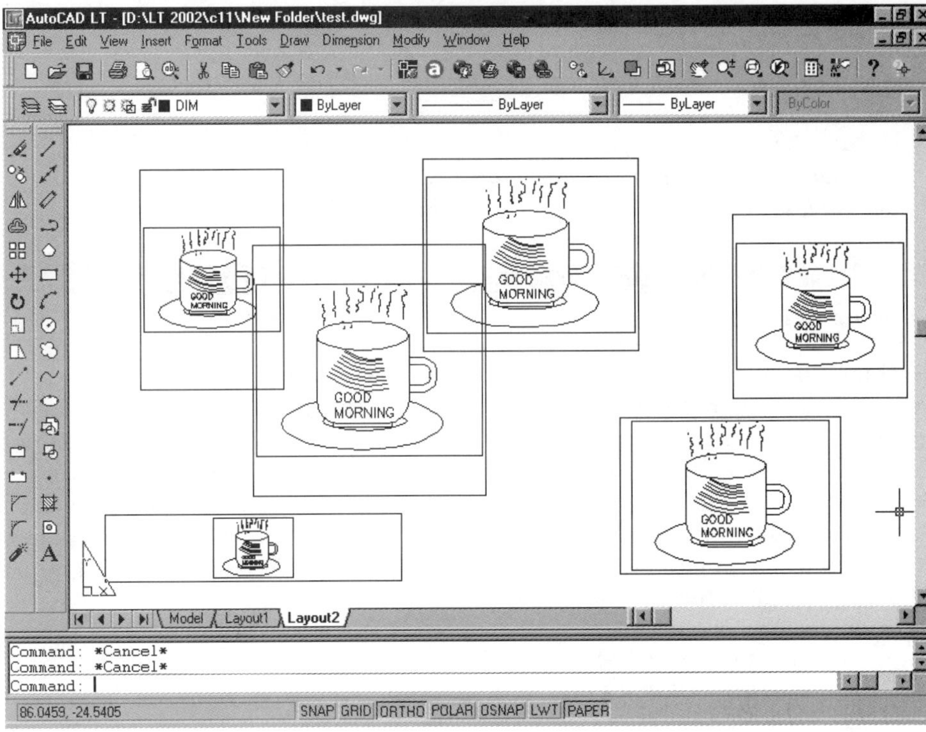

Figure 11-7 Screen display with multiple floating viewports

spinner is displayed (Figure 11-8). This spinner is used to set the spacing between the adjacent viewports. The rest of the options in both **New Viewports** and the **Named Viewports** tabs of the **Viewports** dialog box are the same as those discussed under the tiled viewports. When you select a viewport configuration and choose **OK**, you will be prompted to specify the first and the second corner of a box that will act as a reference for placing the viewports. You will also be given an option of **Fit**. This option fits the configuration of viewports such that they fit exactly in the current display.

Tip

*You can also use the **+VPORTS** command to display the **Viewports** dialog box. When you invoke this command, you will be prompted to specify the **Tab Index**. Enter **0** to display the **New Viewports** tab and enter **1** to display the **Named Viewports** tab directly.*

Note

*You can also use the Command line for creating the floating viewports. This can be done using the **-VPORTS** command or the **MVIEW** command. The working of both these commands in the layouts is same.*

*When you shift to the layouts, the **Page Setup** dialog box is displayed for printing and a rectangular viewport is created that fits the drawing area. If you want, you can retain this viewport or you can also delete this viewport using the **ERASE** command.*

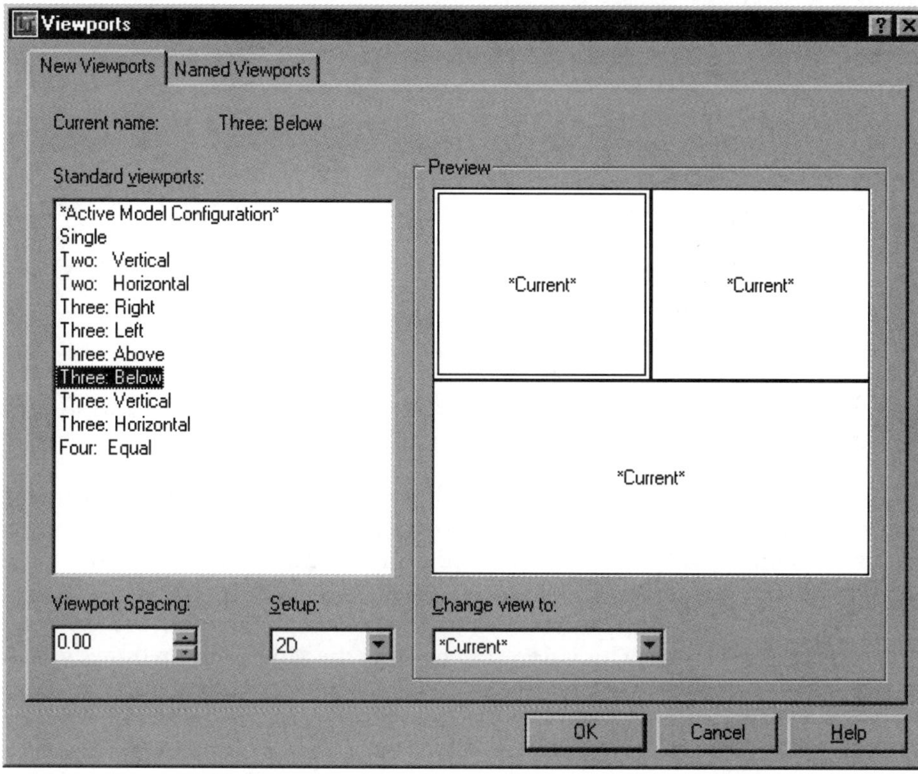

Figure 11-8 *The* ***New Viewports*** *tab of the* ***Viewports*** *dialog box displayed in layouts*

TEMPORARY MODEL SPACE

Sometimes, when you create a floating viewport in the layout, the drawing display in the viewport is not what you desire. For example, you want to create a viewport that should display the detailed view of a design. But when you create the viewport, the entire design is displayed in it as shown in Figure 11-9. In such cases you need to zoom or pan the drawings to display the desired portion of the design. But when you invoke any of the **ZOOM** or the **PAN** commands in the layouts, the display of entire layout is modified instead of the display inside the viewport. Now, to change the display of the viewports, you will have to switch to the **temporary model space**. The temporary model space is defined as a state when the model space is activated in the layouts. The temporary model space is exactly similar to the actual model space and you can make any kind of modifications in the drawing from temporary model space. Therefore, the main reason for invoking the

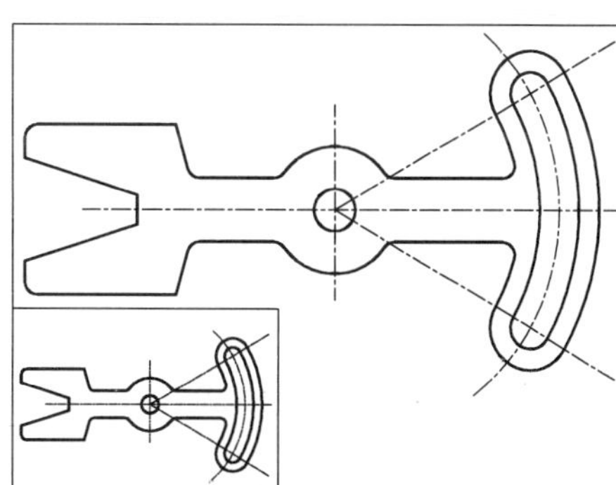

Figure 11-9 *Both the viewports displaying the complete design*

temporary model space is that you can modify the display of the drawing in the viewports. You can modify the display of the objects using any of the drawing display commands. Figure 11-10 shows two viewports with different design display. The design display in these viewports is modified by switching to the temporary model space. The temporary model space can be invoked by choosing the **Paper** button from the status bar. You can also switch to the temporary model space by double-clicking inside the viewports. You will see that the model space UCS icon is automatically displayed when you switch to the temporary model space. When you switch to the temporary model space, the extents of the viewport become the extents of the drawing. You can use the **ZOOM** and **PAN** commands to fit the model inside the viewport. The temporary model space can also be invoked using the **MSPACE** command.

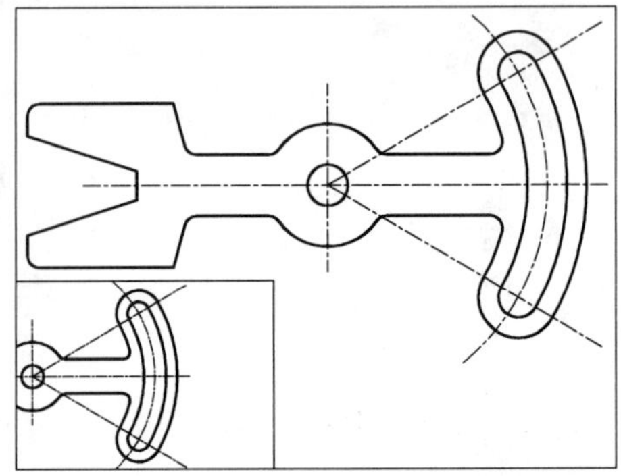

Figure 11-10 *Both the viewports displaying the different portion of design*

Once you have modified the display of the drawing in the temporary model space, you have to switch back to the paper space. This is done by choosing the **Model** button from the status bar. You can also switch back to the paper space by double-clicking anywhere in the layout outside the viewport. This can also be done using the **PSPACE** command. You can easily distinguish between the temporary model space and paper space by looking at the UCS icon. The temporary model space displays the actual model space UCS icon inside the viewports and the paper space displays the paper space icon on the lower left corner of the drawing window, see Figure 11-11 and Figure 11-12.

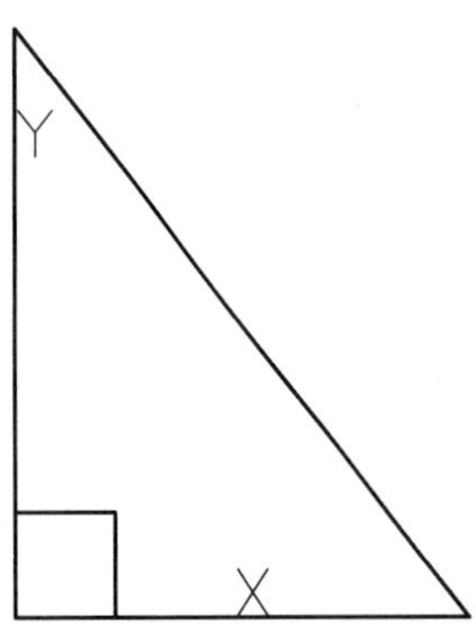

Figure 11-11 *Paper space UCS icon*

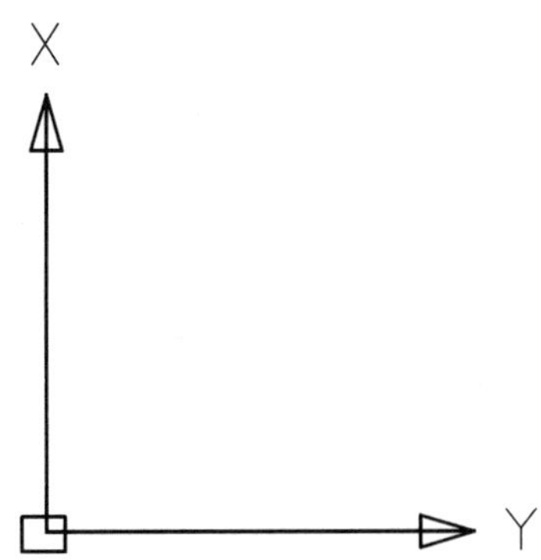

Figure 11-12 *Model space UCS icon*

EDITING THE VIEWPORTS

You can perform different editing operations on the viewports. For example, you can control the visibility of the objects in the viewports, lock the display of the viewports, hide the display of the hidden lines in the printouts, and so on. All these editing operations are discussed next.

Controlling the Display of the Objects in the Viewports

The display of the objects in the viewports can be turned on or off. If the display is turned off, the objects will not be displayed in the viewport. However, the object in the model space is not affected by this editing operation. To control the display of the objects, select the viewport entity in the layout and right-click to display the shortcut menu. In this menu choose **Display Viewport Objects > No** to turn off the display. Similarly, you can again turn on the display of the objects by choosing **Display Viewport Objects > On** from the shortcut menu. This shortcut menu is displayed upon selecting the viewport and right-clicking. This editing operation can also be done using the **OFF** option of the **MVIEW** or the **-VPORTS** command.

Locking the Display in the Viewports

To avoid the accidental modification of the display of objects in the viewports, you can lock their display. If the display of a viewport is locked, the commands such as **ZOOM** and **PAN** does not work in it. For example, if the display of a viewport is locked, you can not zoom or pan the display in that viewport even if you switch to the temporary model space. To lock the display of the viewports, select the viewport and right-click on it to display the shortcut menu. In this menu choose **Display Locked > Yes**. Now, the display of this viewport will not be modified. However, you can draw objects or delete objects in this viewport by switching to the temporary model space. Similarly, you can unlock the display of the objects in the viewports by choosing **Display Locked > No** from the shortcut menu. You can also lock or unlock the display of the viewports using the **Lock** option of the **MVIEW** command or **-VPORTS** command.

Controlling the Display of the Hidden Lines in the Viewports

While working with three-dimensional surface or solid models, there are number of occasions where you have to plot the models such that the hidden lines are not displayed. If you are plotting the models in the model space (Tilemode=1), it can be easily done by selecting the **Hide objects** check box in the **Plot options** area of the **Plot Settings** tab of the **Plot** dialog box. However, if you are plotting the drawings in the layouts, then selecting this check box does not hide the hidden lines. In this case you will have to control the display of the hidden lines in the viewports. To control the display of hidden lines select the viewport and right-click to display the shortcut menu. In this menu choose **Hide Plot > Yes**. Although the hidden lines will be displayed in the viewports, now they will not be plotted. The display of the hidden lines can also be controlled using the **Hideplot** option of the **MVIEW** command or the **-VPORTS** command.

Note

Plotting is explained in detail in Chapter 12, Plotting Drawings.

Tip

Apart from the previously mentioned editing operations, you can also move, copy, rotate, stretch, scale, or mirror the viewports using the respective commands. You can also use the grips to edit the viewports using the grips.

MANIPULATING THE VISIBILITY OF VIEWPORT LAYERS

Command: VPLAYER

You can control the visibility of layers inside a floating viewport with the **VPLAYER** or **LAYER** command. The **On/Off** or **Freeze/Thaw** option of the **LAYER** command controls the visibility of layers globally, including the viewports. However, with the **VPLAYER** command you can control the visibility of layers in individual floating viewports. For example, you can use the **VPLAYER** command to freeze a layer in the selected viewport. The contents of this layer will not be displayed in the selected viewports, although in the other viewports, the contents are displayed. This command can be used from either temporary model space or paper space. The only restriction is that **TILEMODE** be set to 0 (Off); that is, you can use this command only in the **Layout** tab.

Command: **VPLAYER** Enter
Enter an option [?/Freeze/Thaw/Reset/Newfrz/Vpvisdflt]:

? Option

You can use this option to obtain a listing of the frozen layers in the selected viewport. When you enter ?, you will be prompted to select the viewport. The AutoCAD LT text window will be displayed showing all the layers that are frozen in the current layer. If you invoke this option when you are in temporary model space, AutoCAD LT will temporarily shift you to paper space to let you select the viewport.

Freeze Option

The **Freeze** option is used to freeze a layer (or layers) in one or more viewports. When you select this option, you will be prompted to specify the name(s) of the layer(s) you want to freeze. If you want to specify more than one layer, the layer names must be separated by commas. You can also select an object whose layer you want to freeze. Once you have specified the name of the layer(s), AutoCAD LT prompts you to select the viewport(s) in which you want to freeze the specified layer(s). You can select one or all the viewports. The layers will be frozen after you exit this command.

Thaw Option

With this option, you can thaw the layers that have been frozen in viewports using the **VPLAYER** command or the **LAYER** command. Layers that have been frozen, thawed, turned on, or turned off globally are not affected by **VPLAYER Thaw**. For example, if a layer has been frozen, the objects on the frozen layer are not regenerated on any viewport even if **VPLAYER Thaw** is used to thaw that layer in any viewport. If you want to thaw more then one layer, they must be separated by commas. You can thaw the specified layers in the current, selected, or all the viewports.

Reset Option

With the **Reset** option, you can set the visibility of layer(s) in the specified viewports to their current default setting. The visibility defaults of a layer can be set by using the **Vpvisdflt** option of the **VPLAYER** command. When you invoke this option, you will be prompted to specify the names of the layers to be reset. You can reset the layers in the current viewport, in selected viewports or in all the viewports.

Newfrz (New Freeze) Option

With this option, you can create new layers that are frozen in all the viewports. This option is used mainly where you need a layer that is visible only in one viewport. This can be accomplished by creating the layer with the **Newfrz** option and then thawing that particular layer in the viewport where you want to make the layer visible. When you invoke this option, you will be prompted to specify the name(s) of the new layer(s) that will be frozen in all the viewports. If you want to specify more than one layer, separate the layer names with commas. After you specify the name(s) of the layer(s), AutoCAD LT creates frozen layers in all viewports. Also, the default visibility setting of the new layer(s) is set to Frozen; therefore, if you create any new viewports, the layers created with **VPLAYER Newfrz** are also frozen in the new viewports.

Vpvisdflt (Viewport Visibility Default) Option

With this option, you can set a default for the visibility of layer(s) in the subsequent new viewports. When a new viewport is created, the frozen/thawed status of any layer depends on the **Vpvisdflt** setting for that particular layer. When you invoke this option, you will be prompted to specify the names of the layer(s) whose visibility is to be changed. After specifying the name(s), you will be prompted to specify whether the layers should be frozen or thawed in the new viewports.

CONTROLLING THE LAYERS IN THE VIEWPORTS USING THE LAYER PROPERTIES MANAGER DIALOG BOX

You can use the **Layer Properties Manager** dialog box (Figure 11-13) to perform certain functions of the **VPLAYER** command such as freezing/thawing layers in viewports.

Current VP Freeze

When the **TILEMODE** is turned off, you can freeze or thaw the selected layers in the current floating viewport by selecting the **Current viewport Freeze** option. The frozen layers will still be visible in other viewports.

New VP Freeze

If you want to freeze some layers in the new floating viewports, then select the **New VP Freeze** toggle icon. AutoCAD LT will freeze the layers in subsequently created new viewports without affecting the viewports that already exist. If you start drawing on the frozen layer, objects drawn in the new viewport will not be displayed in the new viewport. However, in other viewports, the objects drawn in the new viewport will be displayed.

Note

*For more information about the **Layer Properties Manager** dialog box, see Chapter 4, Drawing Aids.*

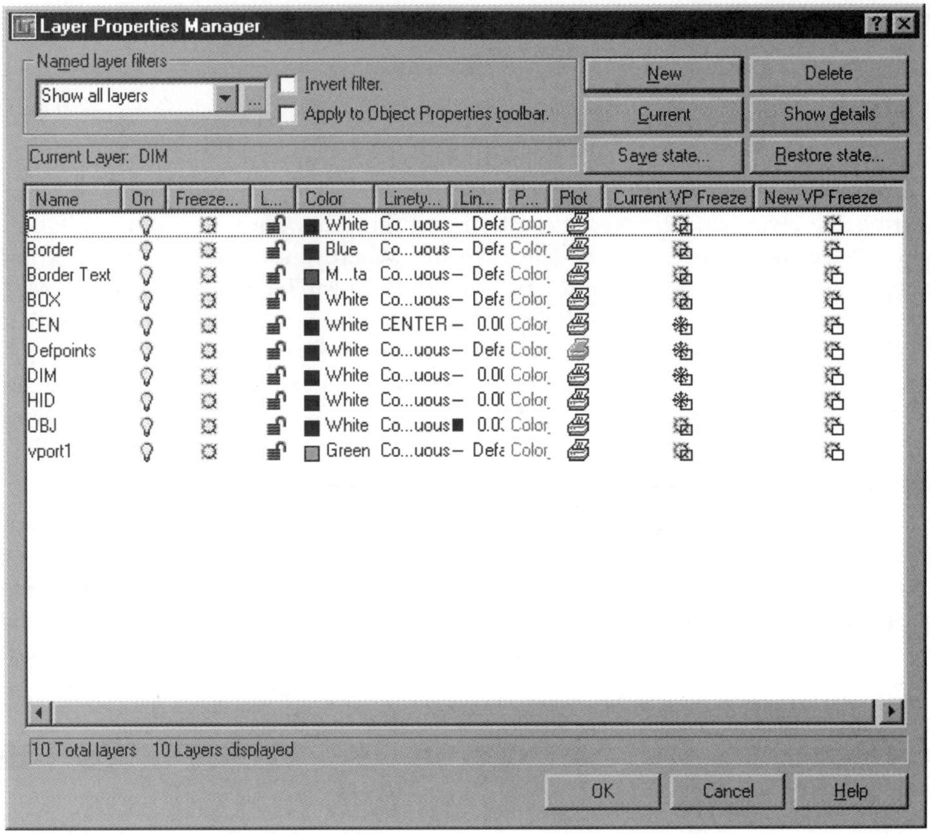

***Figure 11-13** Controlling the visibility of layers in viewports using the **Layer Properties Manager** dialog box*

PAPER SPACE LINETYPE SCALING (PSLTSCALE SYSTEM VARIABLE)

By default, the linetype scaling is controlled by the **LTSCALE** system variable. Therefore, the display size of the dashes depends on the **LTSCALE** factor, on the drawing limits, or on the drawing units. If you have different viewports with different zoom (XP) factors, the size of the dashes will be different for these viewports. Figure 11-14 shows three viewports with different sizes and different zoom (XP) factors. You will notice that the dash length is different in each of these three viewports.

Generally, it is desirable to have identical line spacing in all viewports. This can be achieved with the **PSLTSCALE** system variable. By default, **PSLTSCALE** is set to 0. In this case, the size of the dashes depends on the **LTSCALE** system variable and on the zoom (XP) factor of the viewport where the objects have been drawn. If you set **PSLTSCALE** to 1 and **TILEMODE** to 0, the size of the dashes for objects in the model space are scaled to match the **LTSCALE** of objects in the paper space viewport, regardless of the zoom scale of the viewports. In other words, if **PSLTSCALE** is set to 1, even if the viewports are zoomed to different sizes, the

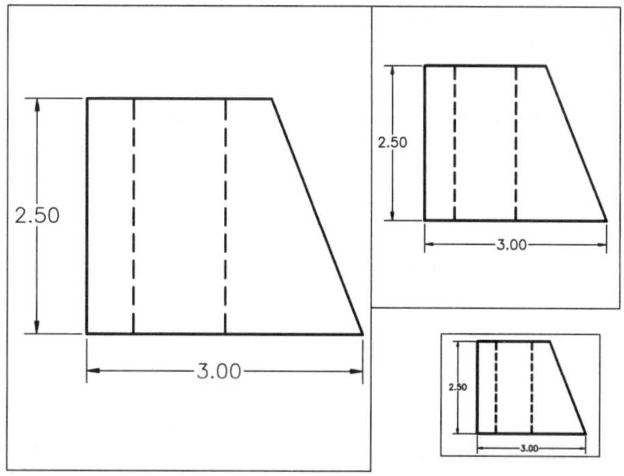

Figure 11-14 Varying sizes of dashed lines with ***PSLTSCALE*** *= 0*

length of the dashes will be identical in all viewports. Figure 11-15 shows three viewports with different sizes. Notice that the dash length is identical in all three viewports.

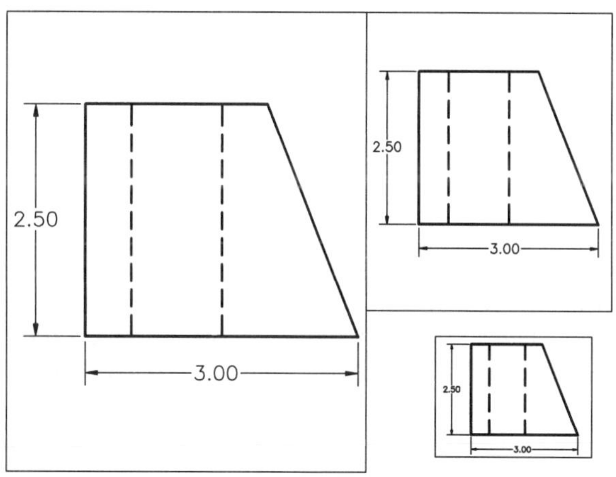

Figure 11-15 Similar sizes of dashed lines with ***PSLTSCALE*** *= 1*

Tip

You can control the scale factor for displaying the objects in the viewports using the ***Viewport Scale Control*** *drop-down list. This drop-down list is available in the* ***Viewports*** *toolbar.*

CREATING AND WORKING WITH THE LAYOUTS

Toolbar:	Layouts > New Layout
Menu:	Insert > Layout > New Layout
Command:	LAYOUT

The **LAYOUT** command is used to create a new layout. It also allows you to rename, copy, save, and delete existing layouts. A drawing designed in the **Model** tab can be composed for plotting in the **Layout** tab. The prompt sequence is as follows:

Enter layout option [Copy/Delete/New/Template/Rename/SAveas/Set/?]<Set>:

New Option

This option is used to create a new layout. When you invoke this command, you will be prompted to specify the name of the new layout. A new tab with the new layout name appears in the drawing.

Copy Option

This option copies a layout. When you invoke this option, you will be prompted to specify the layout that has to be copied. Upon specifying the layout to be copied, you will be prompted to specify the name of the new layout. If you do not enter a name for the new copied layout, the name of the copied layout is assumed with an incremental number in brackets next to it. For example, Layout 1 is copied as Layout1 (2). The name of the new layout appears as a new tab next to the copied **layout** tab.

Delete Option

This option deletes an exiting layout. When you invoke this option, you will be prompted to specify the name of the layout to be deleted. The current layout is the default layout for deleting. Remember that the **Model** tab cannot be deleted.

Template Option

This option creates a new template based on an existing layout template in either **.DWG** or **.DWT** files. You can directly invoke this option by choosing the **Layout from Template** button from the **Layouts** toolbar. You can also choose **Layout > Layout from Template** from the **Insert** menu to directly invoke this option. This option invokes the **Select Template From File** dialog box (Figure 11-16). The layout and geometry from the specified template or drawing file is inserted into the current drawing. After the DWT or DWG file is selected, the **Insert Layout(s)** dialog box is displayed, as shown in Figure 11-17.

Rename Option

This option allows you to rename a layout. When you invoke this option, you will be prompted to specify the name of the layout to be renamed. Upon specifying the name of the layout to be renamed, you will be prompted to specify the new name of the layout. The layout names have to be unique and can contain up to 255 characters, out of which only 32 are displayed in the tab. The characters in the name are not case sensitive.

SAveas Option

This option is used to save a layout in the drawing template file. When you invoke this option,

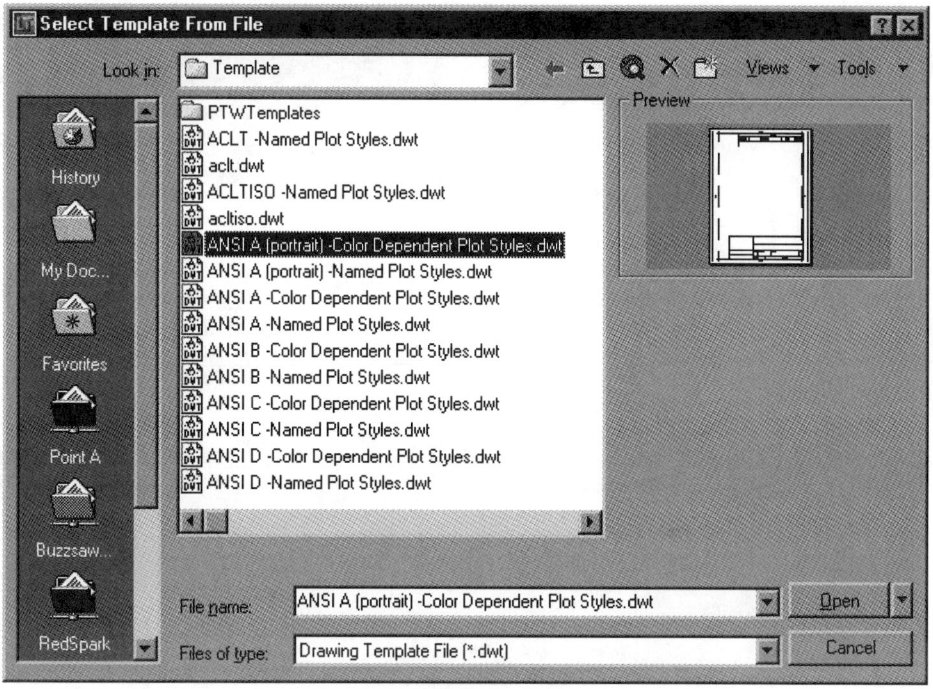

***Figure 11-16** Selecting the template file for creating the layout*

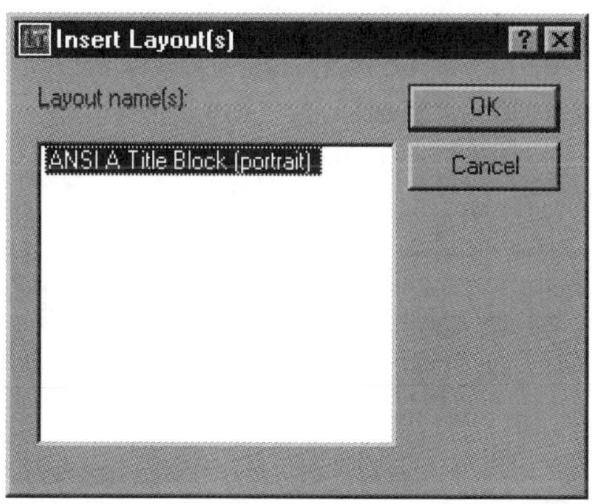

***Figure 11-17** **Insert Layout(s)** dialog box*

you will be prompted to specify the layout that has to be saved. The **Create Drawing File** dialog box appears (Figure 11-18) when you specify the name of the layout that has to be saved. In this dialog box you can enter the name of the template in the **File name** edit box. The layout templates can be saved in the **.dwg**, **.dwt**, or the **.dxf** format.

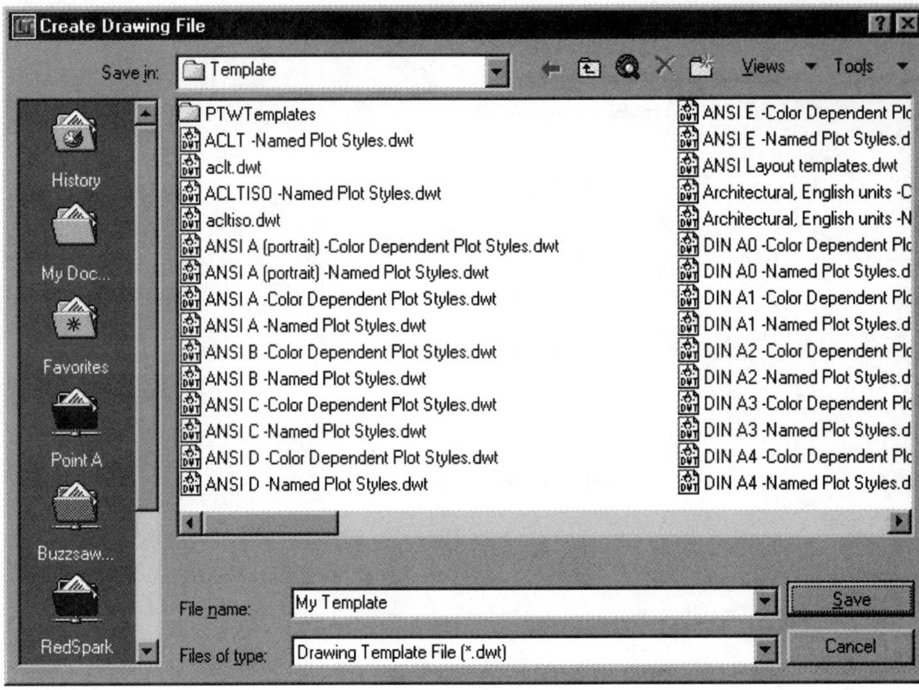

***Figure 11-18 Create Drawing File** dialog box for saving the layouts in the drawing templates*

Set Option

Sets a layout as the current layout. When you invoke this option, you will be prompted to specify the name of the layout that has to be made current.

? Option

This option lists all the layouts that are available in the current drawing. The list will be displayed in the Command line. You can open the AutoCAD LT Text Window to view the list by pressing the F2 key.

Tip

You can right-click on the layout tabs to display the shortcut menu. This shortcut menu can be used for creating a layout, creating a layout using template, deleting, renaming, plotting, moving, or copying the layouts.

PAGESETUP COMMAND

Toolbar:	Layouts > Page Setup
Menu:	File > Page Setup
Command:	PAGESETUP

You can also right-click the **Model** or any of the **Layout** tabs and choose **Page Setup** from the shortcut menu which is displayed. This command allows you to specify the Layout and Plot device settings for each new layout. When you invoke this command, the **Page Setup** dialog box is displayed, the details of which will be discussed in Chapter 12, Plotting Drawings.

Every time you select a new **Layout** tab, the **Page Setup** dialog box is displayed automatically. These default settings are controlled in the **Layout elements** area of the **Display** tab of the **Options** dialog box.

LAYOUTWIZARD COMMAND

Menu:	Insert > Layouts > Layout Wizard
	Tools > Wizards > Create Layout
Command:	LAYOUTWIZARD

This command displays the **Layout Wizard** which guides you step-by-step through the process of creating a new layout. Different pages of this wizard will allow you to specify the name of the layout, select the printer that will be used to print in the new layout, specify the paper size and the orientation of the paper in the layout, insert a title block in the new layout, and define the viewports in the layout.

Tip

*You can use the **Define Viewports** page of the **Create Layout** wizard for creating an array of viewports in the layouts.*

Example 1 — *Mechanical*

In this example, you will create a drawing in the model space and then learn how to use the paper space to plot the drawing. The drawing to be used is shown in Figure 11-19.

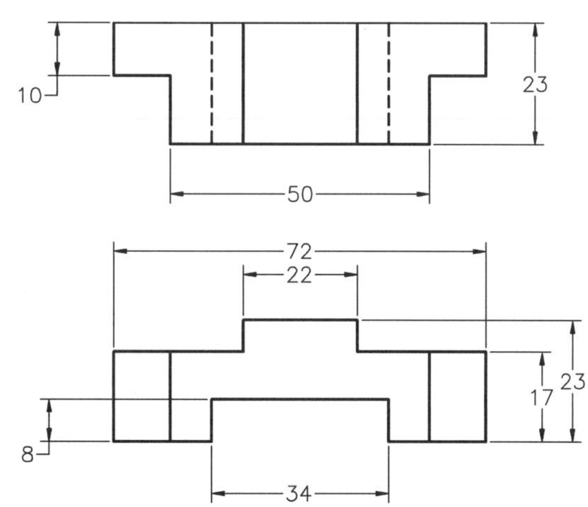

Figure 11-19 *Drawing for Example 1*

1. Based on the drawing size, calculate the limits. Since the drawing is 72 x 23, the limits should be about 75 x 75.

2. Set the limits and layers, then draw the plan and front view as shown in Figure 11-19.

3. Choose the **Layout1** tab; AutoCAD LT displays the **Page Setup - Layout1** dialog box. Select the **Plot Device** tab and then the select the printer or plotter that you want to use from the **Name** drop-down list. In this example, **HP LaserJet 4000** is used. Choose the **Layout Settings** tab and then select the paper size that is supported by your plotting device from the **Paper size** drop-down list. In this example, the paper size is **A4 (210 x 297mm)**. Choose the **OK** button to accept the settings and exit the dialog box. AutoCAD LT displays **Layout1** on the screen with the default viewport. Delete this viewport.

4. Choose the **Single Viewport** button from the **Viewports** toolbar. The prompt sequence is as follows:

   Specify corner of viewport or
   [ON/OFF/Fit/Hideplot/Lock/Object/Polygonal/Restore/2/3/4] <Fit>: *Specify the first corner of the viewport near the bottom left corner of the drawing window.*
   Specify opposite corner: **@297,210**
   Regenerating model.

5. Use the **ZOOM** command in the paper space to zoom to the extents of the viewport.

6. Double-click in the viewport to switch to the temporary model space and use the **ZOOM** command to zoom the drawing to 2XP. In this example, it is assumed that the scale factor is 2:1, therefore, the zoom factor is 2XP.

7. Create the dimension style with the text height of 0.08 and arrowhead height of 0.075. Define all the other parameters and then select the **Scale dimensions to layout (paperspace)** radio button from the **Scale for Dimension Features** area of the **Fit** tab of the **New Dimension Style** dialog box.

8. Switch to the temporary model space and then dimension the drawing. Make sure that you do not change the scale factor. You can use the **PAN** command to adjust the display.

9. Double-click in the paper space to switch back to the paper space. Choose the **Plot (Ctrl+P)** button from the **Standard Toolbar** to display the **Plot** dialog box.

10. Choose the **Window** button from the **Plot area** of the **Plot Settings** tab. The **Plot** dialog box will be temporarily removed from the screen and you will be prompted to specify the first and the second corner of the window. Define a window such that the viewport is not included in the window.

11. As soon as you define both the corners of the window, the **Plot** dialog box will be redisplayed on the screen. Select **1:1** from the **Scale** drop-down list of the **Plot scale** area.

12. Select the **Center the plot** check box from the **Plot offset** area.
13. Choose the **Full Preview** button to display the plot preview. You can make any adjustments if required by redefining the window.
14. After you are satisfied with the preview, choose the **OK** button. The drawing will be printed with the scale factor of 2:1. This means that two plotted units will be equal to one actual unit. Save this drawing with the name **Example1.dwg**.

Self-Evaluation Test

Answer the following questions, and then compare your answers to the correct answers given at the end of this chapter.

1. The viewports in the model space can be of any shape. (T/F)
2. The viewports in the model space can overlap each other. (T/F)
3. Two different tiled viewports can be joined. (T/F)
4. You cannot insert any additional layout in the current drawing. (T/F)
5. The _________ and _________ commands can be used to create the tiled viewports.
6. The viewports in the layouts are called _________ viewports.
7. The two working environments provided by AutoCAD LT are _________ and _________.
8. The _________ command can be used to create a layout using a wizard.
9. When you join two adjacent viewports, the resultant viewport is _________ in shape.
10. The default viewport that is created in the new layout is _________ in shape.

Review Questions

Answer the following questions.

1. The viewports created in the layouts can be polygonal in shape. (T/F)
2. You can not lock the display of a floating viewport. (T/F)
3. An existing closed loop can be converted into a viewport in the model space. (T/F)

4. You can create an array of viewports using the **LAYOUTWIZARD** command in the model space. (T/F)

5. Which command can be used to control the display of the objects in the viewports?

   (a) **MVIEW** (b) **DVIEW**
   (c) **LAYOUT** (d) **MSPACE**

6. Which command can be used to switch to the temporary model space?

   (a) **MVIEW** (b) **DVIEW**
   (c) **LAYOUT** (d) **MSPACE**

7. Which option of the **MVIEW** command can be used to hide the hidden lines of the solid models in the printing?

   (a) **Hide** (b) **Hideplot**
   (c) **Create** (d) **None**

8. Which command can be used for clipping an existing floating viewport?

   (a) **MVIEW** (b) **DVIEW**
   (c) **VPCLIP** (d) **Cannot clip a floating viewport**

9. Which option of the **VPLAYER** command is used to create a layer that will be frozen in all the viewports?

   (a) **Freeze** (b) **Thaw**
   (c) **Newfrz** (d) **Reset**

10. The _________ page of the **LAYOUTWIZARD** command can be used to insert title blocks.

11. You can work only in the _________ tiled viewport.

12. The _________ command can be used to set similar linetype scale for all the viewports.

13. The _________ command is used to switch back to paper space from the temporary model space.

14. The _________ dialog box is used to save a viewport configuration in the model space.

15. Layers that have been frozen, thawed, switched on, or switched off globally are not affected by the _________ command.

Exercise

Exercise 1 *Mechanical*

In this exercise you will perform the following operations:

a. In the model space, make the drawing of the shaft shown in Figure 11-20.

b. Create three tiled viewports in the model space and then display the drawing in all three tiled viewports.

c. Create a new layout with the name **Title Block** and insert a title block of ANSI A size in this layout.

d. Create two viewports, one for the drawing and one for the detail "A". See Figure 11-20 for approximate size and location. The dimensions in detail "A" viewport must not show up in the other viewport. Also adjust the LTSCALE factor for hidden and centerlines.

e. Plot the drawing.

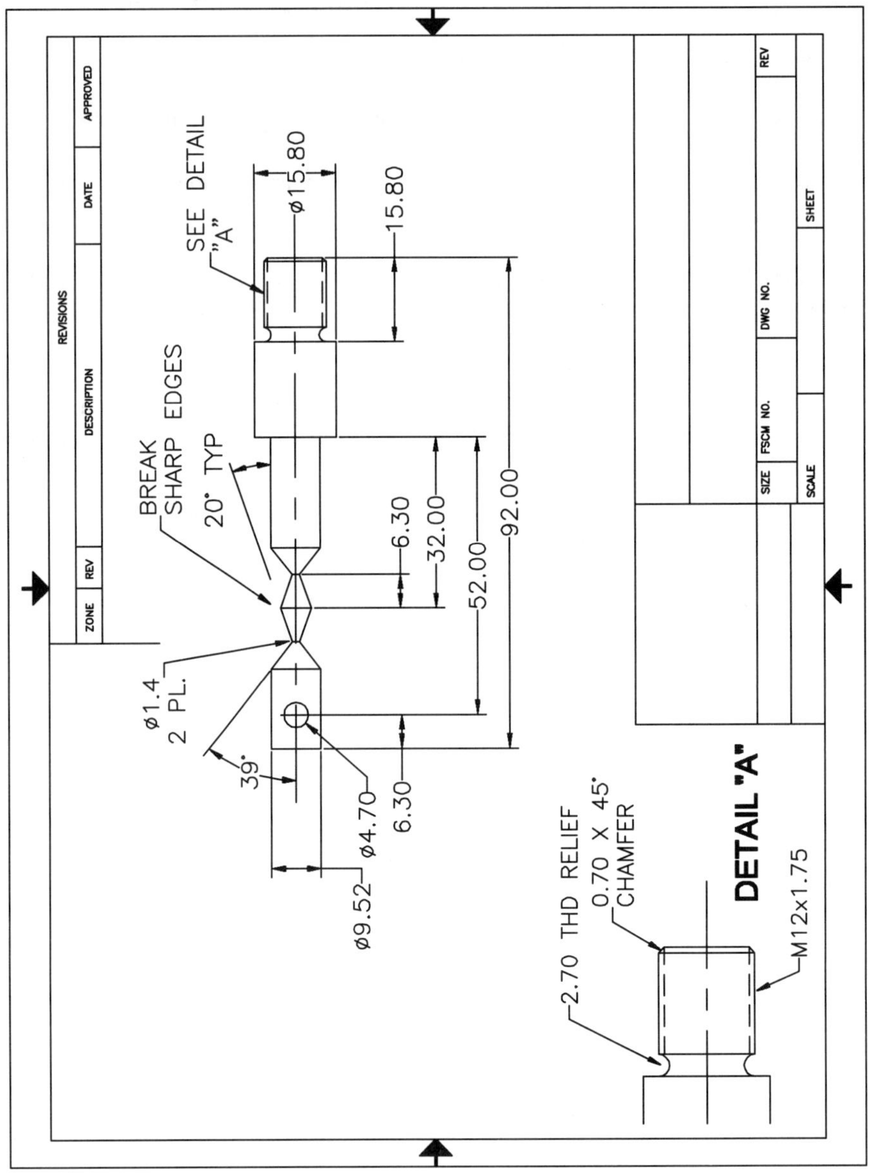

Figure 11-26 Drawing for Exercise 1

Answers to Self-Evaluation Test

1 - F, **2** - F, **3** - T, **4** - F, **5** - **VPORTS**, **-VPORTS**, **6** - floating, **7** - model space and paper space/ layouts, **8** - **LAYOUTWIZARD**, **9** - rectangular, **10** - rectangular

Chapter 12

Plotting Drawings

Learning Objectives

After completing this chapter, you will be able to:
- *Set plotter specifications.*
- *Plot the drawings using the PLOT command.*
- *Configure plotters and then edit the plotter configuration files.*
- *Create, use, and modify plot styles and plot style tables.*

PLOTTING DRAWINGS IN AUTOCAD LT

When you are done with a drawing, you can store it on the computer storage device such as the hard drive or diskettes. However, to get a hard copy of the drawing, you should plot the drawing on a sheet of paper using a plotter or printer. A hard copy is a handy reference for professionals working on site. With the help of pen plotters, you can obtain a high-resolution drawing. Basic plotting has already been discussed in Chapter 2, Getting Started-II. You can plot drawings in the **Model** tab or any of the layout tabs. A drawing has a **Model** and two layout tabs (**Layout 1**, **Layout 2**) by default. Each of these tabs has its own settings and can be used to create different plots. You can also create new layout tabs using the **LAYOUT** command. This has been discussed later in Chapter 11, Model Space Viewports, Paper Space Viewports, and Layouts.

PLOT COMMAND

Toolbar:	Standard > Plot
Menu:	File > Plot
Command:	PLOT

The **PLOT** command is used to plot a drawing. The **Plot** dialog box (Figure 12-1) is displayed when you invoke the **PLOT** command. You can also right-click on the **Model**

tab or any of the layouts tabs to display the shortcut menu and choose **Plot** to invoke the **Plot** dialog box.

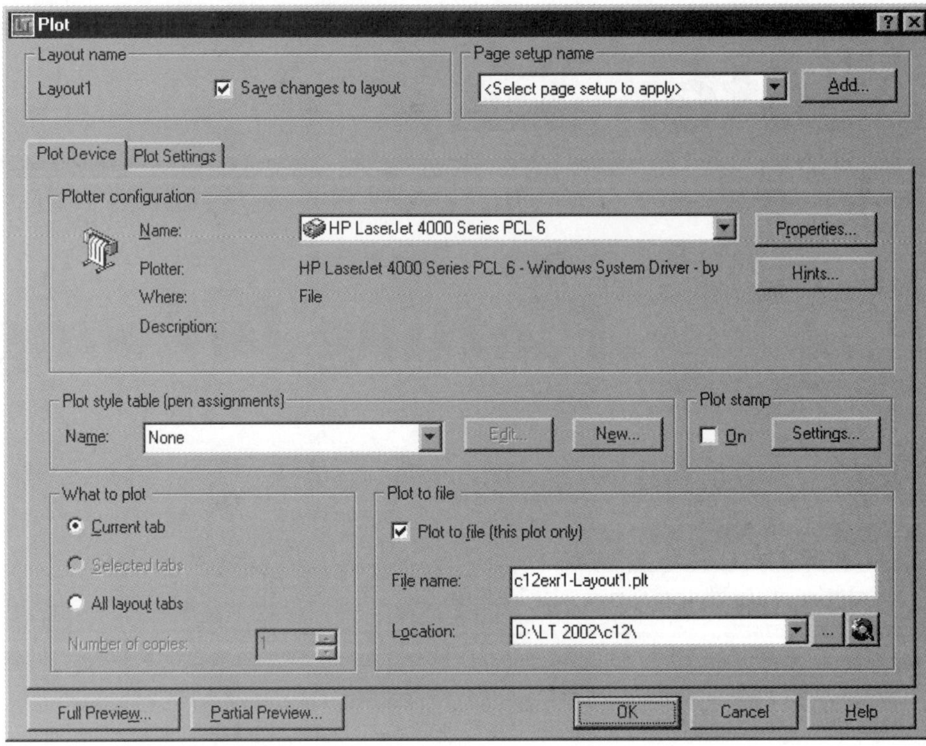

Figure 12-1 ***Plot Device*** *tab of the* ***Plot*** *dialog box*

Some values in this dialog box were set when AutoCAD LT was first configured. You can examine these values and if they conform to your requirements, you can start plotting directly. If you want to alter the plot specifications, you can do so through the options provided in the **Plot** dialog box. The available plot options are described next.

Layout name Area

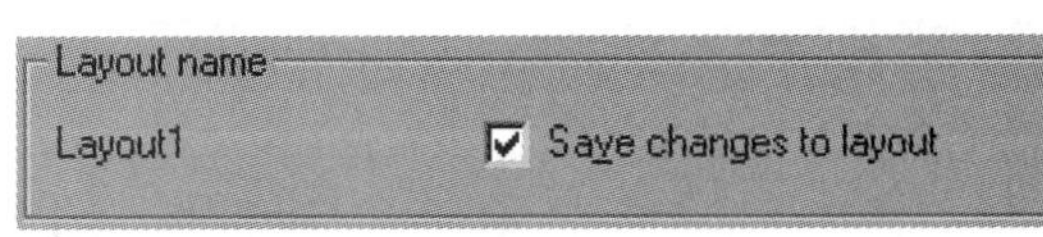

Figure 12-2 ***Layout name*** *area of the* ***Plot*** *dialog box*

This area (Figure 12-2) displays the current layout name or displays **Selected layouts** if multiple tabs have been selected. For example, if the **Model** tab is currently selected and you invoke the **PLOT** command, the **Layout name** area displays **Model**. A layout can be defined as the way in which the drawing has been arranged for plotting. A drawing can have many layouts, represented by different tabs at the bottom of the drawing area and each of these layouts can have different plotting settings. When you choose a layout tab for the first time, a **Page Setup** dialog box is displayed. This dialog box is the same as the **Plot** dialog box, except that it does not allow you to preview the plotted drawing. This dialog box is used to create layout settings, which can be saved as page setups and can be used later for plotting drawings. Page setups will be discussed later in this chapter. At this stage you can choose **OK** or **Cancel** to exit the dialog box. This area also provides the **Save changes to layout** check box. This check box is selected by default. If this check box is selected, any changes made to the layout such as zooming and so on will be reflected even after the **Plot** command is exited. If you select more than one layout for plotting, this option is not available.

Page setup name Area

The **Page setup name** drop-down list displays all the saved and named page setups. A page setup is the way a drawing is laid out on a sheet of paper to create a layout. It consists of all settings related to the plotting of a drawing and include the plot device being used, the scale, the pen settings, and so on. These settings can be saved as a named page setup, which can be later selected from this drop-down list and then be used for plotting a drawing. If you select **Previous Plot** from the drop-down list, the settings used for the last drawing plotted are applied to the current drawing. You can choose the base for the current page setup on a named page setup, or you can add a new named page setup by choosing the **Add** button that is located next to the drop-down list. When you choose this button, AutoCAD LT displays the **User Defined Page Setups** dialog box (Figure 12-3). In this dialog box, you can create, delete, rename or import page setups. All the settings applied to the current layout are saved in the user-defined page setup file.

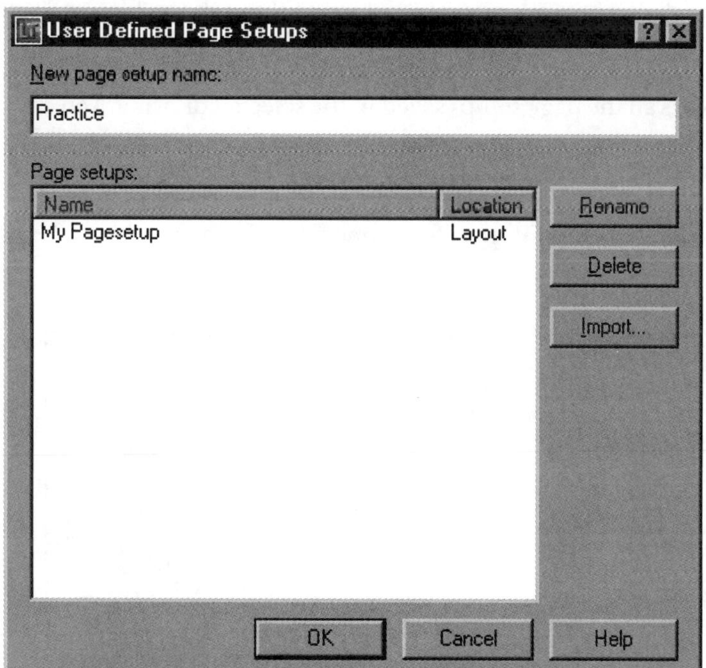

***Figure 12-3 User Defined Page Setups** dialog box*

The options under **User Defined Page Setups** dialog box are discussed next.

New page setup name

You can enter a name for the new user-defined page setup in this box. The current page setup can then be based on this named page setup.

Page Setups

This list box lists all the named and saved user-defined page setups. You can select a named page setup from the list box and use it for the current layout.

Rename

You can select any user-defined page setup from the list box and choose the **Rename** button to rename the page setup.

Delete

You can select a user-defined page setup from the list box and choose the **Delete** button to delete it.

Import

You can import a user-defined page setup from another drawing and apply it to a layout in the current drawing. When you choose the **Import** button, AutoCAD LT displays the **Select Page Setup From File** dialog box, where you can browse through the files and select a file for the page setup selection. After you have selected a file, choose the **Open** button. The **Import user defined page setup(s)** dialog box is displayed (Figure 12-4). The **Page setups** list box in this dialog box displays all the page setups saved in the selected drawing. You can select any one of the page setups listed there and choose **OK** to return to the **User Defined Page Setups** dialog box. You will notice that the page setup you imported is listed in the current **Page setups** list box. You can select it here and choose **OK** to get back to the **Plot** dialog box. The imported page setup is now listed in the **Page setup name** drop-down list and can be selected and used for the current plot.

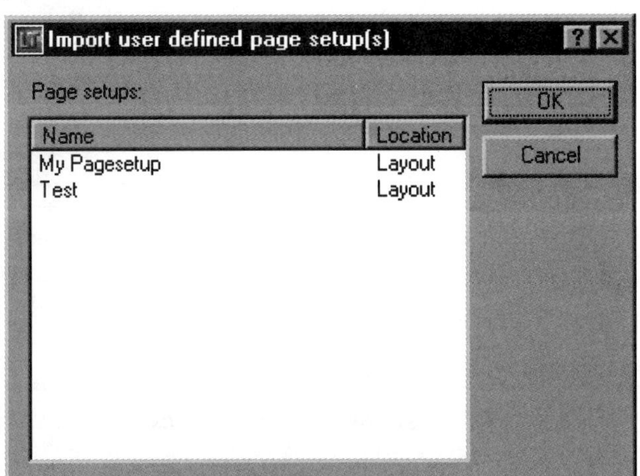

Figure 12-4 Import user defined page setup(s) *dialog box*

Tip

*Select an existing page setup from the **Page setup name** drop-down list and make modifications in it. Then choose the **Add** button to create a new page setup based on an existing one.*

The **Plot** dialog box has two tabs. They are **Plot Device** and **Plot Settings**. Both these tabs and the options available in them are discussed next.

Plot Device Tab (Plot Dialog Box)

In this tab (see Figure 12-1), information about the current configured plotter, plot style table, the layout or layouts to be plotted, and information about plotting to a file is displayed. All the options and areas in this tab are discussed next.

Plotter configuration Area

This area displays all the information about the configured printers and plotters currently selected from the **Name** drop-down list. It displays the plotter driver being used and the printer port being used. It also displays some description text about the selected plotter or printer. All the plotters that are currently configured are displayed in the **Name** drop-down list.

Note

*To add plotters and printers to the **Name** drop-down list, choose **Plotter Manager** from the **File** menu to display the **Plotters** window. Double-click on the **Add-A-Plotter Wizard** icon in this window to display the **Add Plotter** wizard. You can use this wizard to add a plotter to the list of configured plotters and a plotter configuration file (PC3) for the plotter is created. This file consists of all the settings needed by the specific plotter to plot. The **Plotters** window will be discussed later in this chapter in the section "**PLOTTERMANAGER** Command".*

Properties. If you want to check information about a configured printer or plotter, choose the **Properties** button. When you choose this button, the **Plotter Configuration Editor** is displayed. This dialog box lists all the details of the selected plotter under three tabs; **General**, **Ports**, and **Device and Document Settings**. The **Plotter Configuration Editor** will be discussed later in the "**Editing Plotter Configuration**" section of this chapter.

Hints. When you choose the **Hints** button, AutoCAD LT displays a help topic that helps you decide which plotter driver to use for the selected plotter. It also displays the criteria to select a plotter driver, and to select a system or nonsystem printer driver.

Plot style table (pen assignments) Area

This area in the **Plot** dialog box allows you to view and select a plot style table, edit the current plot style table, or create a new plot style table. A plot style table is a collection of plot styles. A plot style is a group of pen settings that are assigned to an object or layer and that determine the color, thickness, line ending, and the fill style of drawing objects when they are plotted. It is a named file that allows you to control the pen settings for a plotted drawing. You can select **None** from the **Name** drop-down list in this area of the dialog box if you want to plot a drawing without using any plot styles. You can assign different plot style tables to a drawing and plot the same drawing differently each time. Using plot styles will be discussed later in this chapter.

Name. The **Name** drop-down list lists all the available plot style tables. From this drop-down list, you can select a plot style table that is assigned to the current **Model** tab or layout tab. Also, if you have selected more than one layout tab, each with different plot style tables assigned to them, the **Name** edit box displays **Varies**.

Edit. You can edit a plot style table that you have selected from the **Name** drop-down list by choosing the **Edit** button. This button is not available when you have selected **None** from the **Name** drop-down list. When you choose the **Edit** button, AutoCAD LT displays the **Plot Style Table Editor**, where you can edit the selected plot style table. This dialog box has three tabs. They are: **General**, **Table View**, and **Form View**. The **Plot Style Table Editor** will be discussed later in the "**Using Plot Styles**" section of this chapter.

New. When you choose the **New** button, AutoCAD LT displays the **Add Named Plot Style Table** wizard, which is used to create a new plot style table. Using this wizard will be discussed later in this chapter.

Plot stamp Area

The options under this area are used to set the plot stamp. Plot stamp is user-defined information that will be displayed on the sheet after plotting. You can set the plot stamping when you use the **On** check box. You can set the parameters for the plot stamp by using the **Plot Stamp** dialog box that is displayed when you choose the **Settings** button.

What to plot Area

This area in the **Plot** dialog box defines what you want to plot. You can decide to plot the **Model** tab, a layout tab, or multiple layout tabs. In this area of the dialog box you can also determine the number of copies of the drawing to be plotted. The options available are as follows.

Current tab. If this radio button is selected, AutoCAD LT plots the current **Model** or **Layout** tab. If multiple tabs are selected in the drawing, the tab that displays the viewing area gets plotted. You can select multiple tabs by holding down the CTRL key and then choosing the tabs. The selected tabs are highlighted.

Selected tabs. This radio button is not available if only the **Model** or only one of the layout tabs is selected. If multiple tabs have been selected, hold down the CTRL key and this option becomes available. You can then select it if you want to plot the multiple tabs.

All layout tabs. This option in the **Plot** dialog box plots all layout tabs, irrespective of the tabs that have been selected.

Number of copies. You can enter the number of copies that you want plotted in this edit box. You can also use the spinners to change the value in this edit box and plot the desired number of plots. If multiple layouts and copies are selected and some of the layouts are set for plotting to a file or AutoSpool, they will produce a single plot. Autospool allows you to send a file for plotting while you are working on another program.

Plot to file area

The options in this area are used for plotting a drawing to a file instead of the plotter. By default, the options in this area are not available unless you select the **Plot to file** check box.

Plot to file. If you select this check box, AutoCAD LT plots the output to a file rather than to the plotter. The file is plotted in the **.PLT** format. The remaining options in the **Plot to file** area will be available only when you select this check box.

File Name. You can specify the file name of the plot in this field. The plot file name is the drawing name with the tab name, separated by a hyphen, and it has a .plt file extension.

Location. This drop-down list displays the location of the directory where the plot file will be stored. The default location of the plot file is the directory where the drawing file resides. The last six locations that have been used to store the plot files previously are saved in this drop-down list. This way, whenever you need to store another plot file in a folder where you have stored an earlier plot file, you can select the path from the drop-down list, instead of typing the path or locating the folder again.

[...]. If you choose the [...] button, AutoCAD LT displays a standard **Browse for Folder** dialog box. You can choose a location of the directory where you want to store the plot file using this dialog box.

Browse the Web. When you choose this button, the **Browse the Web** dialog box is displayed. Here, you can browse the Web and find a Web location where you can save the plot file.

Plot Settings Tab (Plot Dialog Box)

The **Plot Settings** tab in the **Plot** dialog box (Figure 12-5) controls the paper size, orientation, scale of plotting, plot offset, and other plotting options. The description of the options available in this tab is as follows.

Paper size and paper units Area

The **Paper size and paper units** area provides options that help you determine the paper size on which plotting takes place. The options available in this area are discussed next.

Plot device. The device that was selected for plotting in the **Plot Device** tab is displayed here.

Paper size. The **Paper size** drop-down list displays all the available standard paper sizes for the selected plotting device. You can select any size from the list to make it current. If **None** has been selected currently from the **Name** drop-down list in the **Plotter configuration** area of the **Plot Device** tab, AutoCAD LT displays the entire standard paper size list to select from.

Printable area. This is the actual dimension of the area on the paper that is used for plotting. It is based on the current paper size selected and is indicated by the width (in the X-axis direction) and the height (in Y-axis direction), when the default drawing orientation **Landscape** is selected in the **Drawing orientation** area.

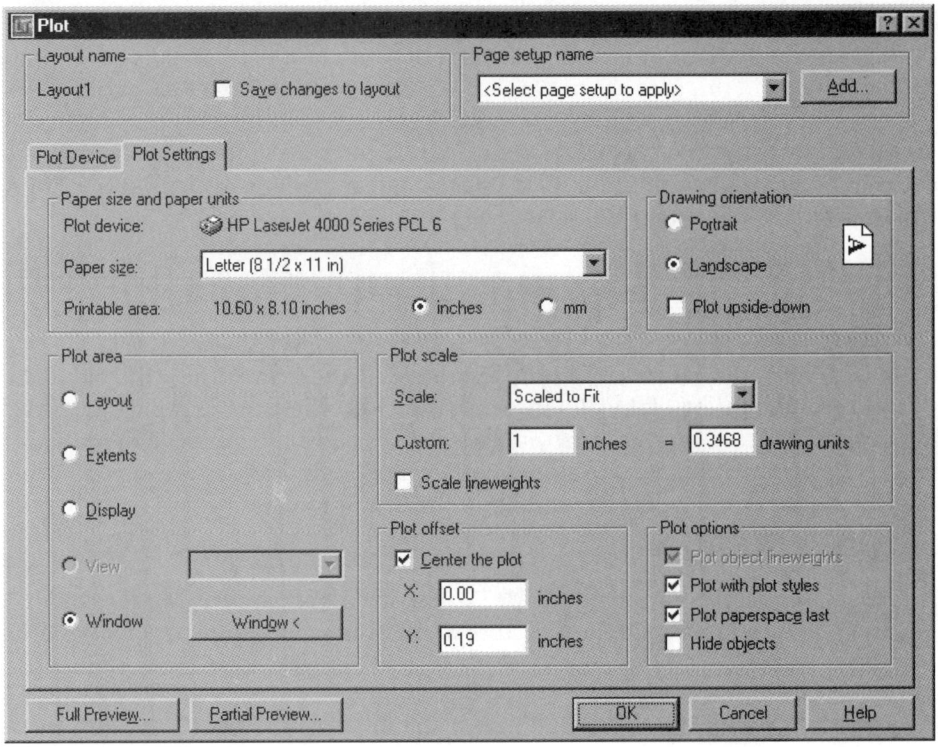

***Figure 12-5 Plot Settings** tab of the **Plot** dialog box*

inches/mm. These radio buttons specify the units to be used for plotting.

Note

*The paper size selected here is saved with the layout and overrides the PC3 file settings created by the **Add Plotter** wizard (discussed later in the "**PLOTTERMANAGER** Command" section of this chapter).*

Drawing orientation Area

This area provide options that help you specify the orientation of the drawing on the paper for the plotters that support landscape or portrait orientation. You can change the drawing orientation by selecting the **Portrait** or **Landscape** radio buttons, with or without selecting the **Plot Upside-Down** check box. The paper icon displayed on the right side of this area indicates the media orientation of the selected paper and the letter icon (A) on it indicates the orientation of the drawing on the page. The **Landscape** radio button is selected by default for AutoCAD LT drawings and orients the length of the paper along the X axis, that is horizontally. If we assume this orientation to be at a rotation angle of 0 degrees, when selecting the **Portrait** radio button, the plot is oriented with the width along the X axis, that is equivalent to the plot being rotated through a rotation angle of 90 degrees. Similarly, if you select both the **Landscape** radio button and the **Plot upside-down** check box at the same time, the plot gets rotated through a rotation angle of 180 degrees and if you select both the **Portrait** radio button and the **Plot upside-down** check box at the same time, the plot gets rotated through a rotation angle of 270 degrees. The AutoCAD LT screen conforms to the landscape orientation by default.

Plot Area

In this area you can specify the portion of the drawing to be plotted. You can also control the way the plotting will be carried out. The five options available in this area are described next.

Layout/Limits. If you are plotting a layout tab, the first radio button available in this area is **Layout** and if you are plotting in the **Model** tab, the first radio button available is **Limits**. When you select the **Layout** radio button, AutoCAD LT plots everything within the margins of the specified paper size. In the case of layouts, the origin is calculated at 0,0. If you select the **Limits** radio button, the whole area defined within the drawing limits is plotted.

Extents. If you select this option, the section of drawing that currently contains the objects is plotted. In this respect, this option resembles the **Extents** option of the **ZOOM** command. If you add objects to the drawing, they are also included in the plot and the extents of the drawing are recalculated. If you reduce the drawing extents by erasing, moving, or scaling the objects, the extents of the drawing are again recalculated. You can use the **Extents** option of the **ZOOM** command to determine what objects shall be plotted. If you use the **Extents** option when the perspective view is on and the position of the camera is not outside the drawing extents, the following message is displayed: **Plot of perspective view has been scaled to fit available area**. When no objects are drawn in the drawing area, the **Extents** object is not available.

Display. If you select this radio button the portion of the drawing that is currently being displayed on the screen is plotted.

Note

*To be able to clearly view the differences between the three previously listed plotting options, it may be a good idea to make sure that the default scale options have been selected. If not, select **Scaled to Fit** from the **Scale** drop-down list in the **Plot scale** area of the dialog box if you are in the **Model** tab, and select 1:1 if you are working in any one of the **layouts** tabs.*

View. Selecting the **View** radio button in the **Plot area** of the dialog box enables you to plot a view that was created with the **VIEW** command. The view must be defined in the current drawing. If no view has been created, the **View** radio button and the drop-down list are not available. The **View** radio button and the drop-down list can be activated by creating a view in the current drawing. Once views have been created in the current drawing, you can select a view for plotting from the **View** drop-down list (Figure 12-6) and then choose **OK** in the **Plot** dialog box. When using the **View** option, the specifications of the plot depend on the specifications of the named view.

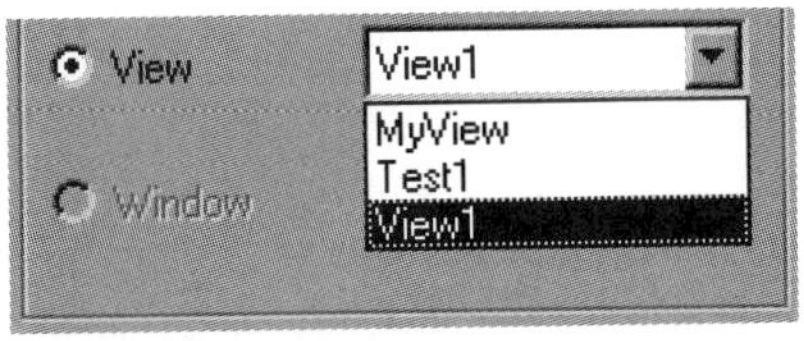

Figure 12-6 *Selecting a named view for plotting*

Window. With this option, you can specify the section of the drawing to be plotted by defining a window. The section of the drawing contained within the window defined by selecting a lower left corner and an upper right corner is plotted. To define a window, choose the **Window**

button that is adjacent to the **Window** radio button. The **Window** radio button is not available until you have defined a window by choosing the **Window** button. The **Plot** dialog box is temporarily closed and you are allowed to specify two points on the screen that define a window, the area within which shall be plotted. Once you have defined the window, the **Plot** dialog box reappears on the screen. You will notice that the **Window** radio button is displayed as selected now. You can choose the **OK** button in the dialog box if you want to plot the drawing.

Note

*Sometimes, when using the **Window** option, the area you have selected may appear clipped off. This may happen because the objects are too close to the window you have defined on the screen. You need to redefine the window in this situation. Such errors can be avoided by using the preview options discussed later.*

Plot scale Area

This area controls the drawing scale of the plot area. The **Scale** drop-down list has thirty-one architectural and decimal scales apart from **Custom** and **Scaled to Fit** options. The default scale setting is **1:1** when you are plotting a layout. But if you are plotting in a **Model** tab, the default setting is **Scaled to Fit**. The **Scaled to Fit** option allows you to automatically fit the entire drawing on the paper. It is useful when you have to print a large drawing using a printer that uses a smaller size of paper or when you want to plot the drawing on a small sheet.

Whenever you select a standard scale from the drop-down list, the scale is displayed in the **Custom** edit boxes as a ratio of the plotted units to the drawing units. You can also change the scale factor manually in the **Custom** edit boxes. When you do so, the **Scale** edit box displays **Custom**. For example, for an architectural drawing, that is to be plotted at the scale 1/4"=1'-0", you can either enter 1/4"=1'-0" or 1=48 in the **Custom** edit boxes.

Note

*The **PSLTSCALE** system variable controls the paper space linetype scaling and has a default value of 1. This implies that irrespective of the zoom scale of the viewports, the linetype scale of the objects in the viewports remains the same. If you want linetype scale of objects in different viewports with different magnification factors to appear different, you should set the value of the **PSLTSCALE** variable to 0. This has been discussed in detail in Chapter 11, Model Space Viewports, Paper Space Viewports, and Layouts.*

The **Scale lineweights** check box is available only if you are plotting in a layout tab. This option is not available in the **Model** tab. If you select the **Scale lineweights** check box, you can scale lineweights in proportion to the plot scale. Lineweights generally specify the linewidth of the printable objects and are plotted with the original linewidth size regardless of the plot scale.

Plot offset Area

This area of the **Plot Settings** tab allows you to specify an offset of the plotting area from the lower left corner of the paper. The lower left corner of a specified plot area is positioned at the lower left margin of the paper by default. If you select the **Center the plot** check box, AutoCAD LT automatically centers the plot on the paper by calculating the X and Y offset values. This check box is not available when you are working in one of the layout tabs and the **Layout** radio button is selected in the **Plot area** of the dialog box. You can specify an offset from the origin by entering positive or negative values in the **X** and **Y** edit boxes. For example, if you want the drawing to be plotted 4 units to the right and 4 units above the origin point, enter 4 in both the **X** and **Y** edit boxes. Depending on the units you had specified in the **Paper size and paper units** area of the dialog box, the offset values are either in inches or millimeters.

Plot options Area

This area displays four additional plotting options that can be selected as per the plot requirements. They are described next.

Plot object lineweights. This check box is not available if the **Plot with plot styles** check box is selected. To activate this option, clear the **Plot with plot styles** check box. This check box is selected by default and AutoCAD LT plots the drawing with the specified lineweights. To plot the drawing without the specified line widths, clear this check box.

Plot with plot styles. When you select the **Plot with plot styles** check box, AutoCAD LT plots using the plot styles applied to objects in the drawing and defined in the plot style table. The different property characteristics associated with the different style definitions are stored in the plot style tables and can be easily attached to geometry. This setting replaces pen mapping in earlier versions of AutoCAD LT.

Plot paperspace last. This check box is not available when you are in the **Model** tab because no paper space objects are present in the **Model** tab. When you are working in a layout tab, this option is available. By selecting the **Plot paper space last** check box, you get an option of plotting model space geometry before paper space objects. Usually paper space geometry is plotted before model space geometry. This option is also useful when there are multiple tabs selected for plotting and you want to plot the model space geometry before the layout tabs.

Hide objects. When you have created a 3D object in the **Model** tab, you can select the **Hide objects** check box to plot the drawing without the hidden geometry. Hidden geometry consists of objects that lie behind the facing geometry and displays the object as it would be seen in reality. If this check box is not selected, the drawing is plotted with the hidden objects in it.

Similarly, when you are in a layout tab and in paper space, you can use this option to remove hidden geometry from the plot. Suppose you draw a 3D object in the layout paper space (although it will be displayed in plan view) such that it lies partially over another 3D object or a viewport. You can now select the **Hide objects** check box if you want to remove the geometry of the hidden portion of the 3D object that lies below the other 3D object. This option is not applicable to objects in the layout model space. To plot such objects without the hidden geometry, you should turn on the hideplot option of the viewports. This has been discussed in Chapter 11, Model Space Viewports, Paper Space Viewports, and Layouts.

Plot Preview

You can view the plot on the specified paper size before actually plotting it by selecting the **Full Preview** or **Partial Preview** buttons in the **Plot** dialog box. Previews help you find errors and correct them prior to plotting, thus saving stationery and time. AutoCAD LT provides two types of plot previews. They are partial and full.

Full Preview

When you choose the **Full Preview** button AutoCAD LT displays the drawing on the screen just as it would be plotted on the paper. Full preview takes more time than the partial preview because regeneration of the drawing takes place. A smaller file size results in a faster regeneration. Once regeneration is performed, the dialog boxes on the screen are removed temporarily, and an outline of the paper size is shown. In the plot preview (Figure 12-7), the cursor is replaced with the Zoom Realtime icon. This Realtime icon can be used to zoom in and out interactively by holding the pick button down and then moving the pointing device. You can right-click to display a shortcut menu and then choose **Exit** to exit the preview or press the ENTER or ESC keys to return to the dialog box. You can also choose **Plot** to plot the drawing right away or choose the other zooming options available.

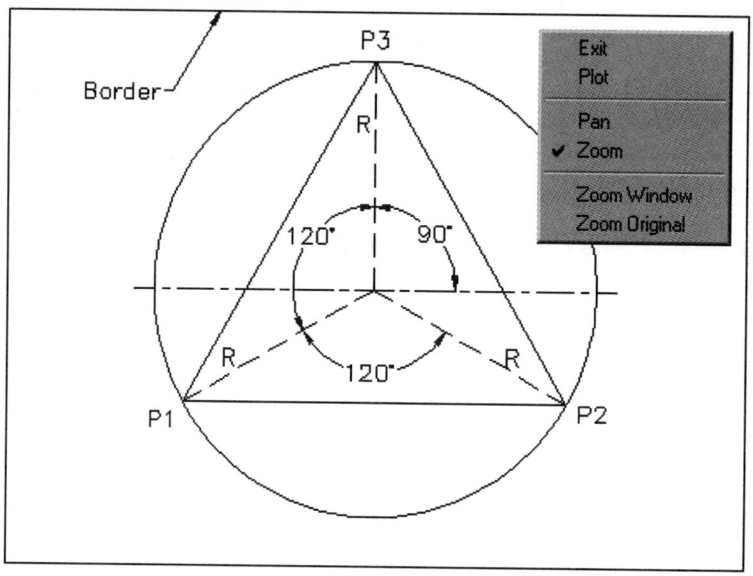

***Figure 12-7** Full plot preview with the shortcut menu*

Partial Preview

To generate a partial preview of a plot, choose the **Partial Preview** button. The **Partial Plot Preview** dialog box (Figure 12-8) is displayed. The paper size is graphically represented by the white paper icon. The dashed rectangle shows the printable area. The blue hatched rectangle is the section in the paper that is used by the image. This area is also known as the **effective area**. The dimensions of the paper, printable area, and effective area are also displayed numerically. If you define the origin of the plot so that the effective area is not accommodated in the graphic area of the dialog box, AutoCAD LT displays a warning. Hence, with the help of a partial preview, you can accurately see how the plot will fit on the paper. If there is something wrong with the specifications of the plot, AutoCAD LT provides warning messages so that corrections can be made before the actual plotting takes place.

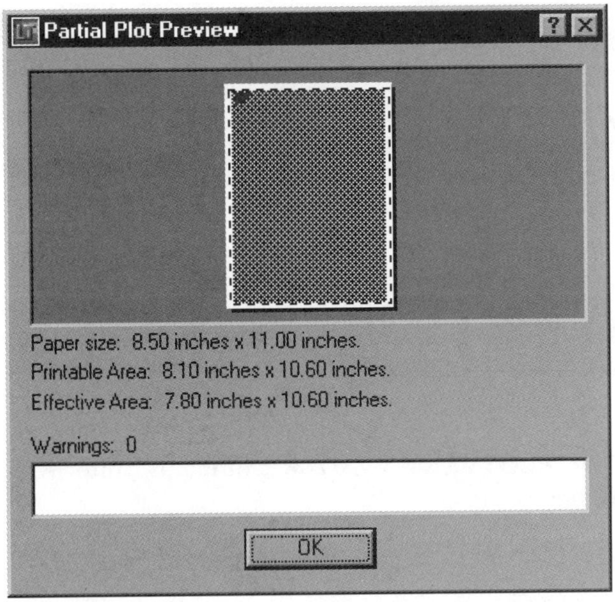

***Figure 12-8 Partial Plot Preview** dialog box*

The red triangle icon located at one of the corners of the effective area represents the orientation of the plot. If it is displayed in the upper left corner of the effective area, it means the plot has the default landscape orientation. In case of Portrait orientation, the red triangle icon is located at the lower left corner of the effective area. Similarly, the icon is displayed at the lower right corner in case of upside-down landscape orientation and is displayed at the upper right corner of the effective area in case of an upside-down portrait orientation.

Tip

*You can also choose **Plot Preview** from the **File** menu to see a full preview of the plot, and at the same time, bypass the plot dialog box.*

After you have finished with all the settings and other parameters, if you choose the **OK** button in the **Plot** dialog box, AutoCAD LT starts plotting the drawing in the file or plotters as specified. AutoCAD LT displays the **Plot Progress** dialog box (Figure 12-9), where you can view the actual progress in plotting.

ADDING PLOTTERS

In AutoCAD LT, the plotters can be added using the **PLOTTERMANAGER** command. This command is discussed next.

PLOTTERMANAGER Command

Menu:	File > Plotter Manager
Command:	PLOTTERMANAGER

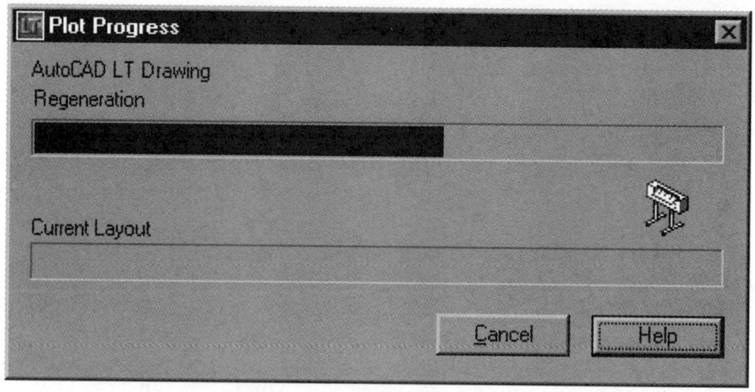

Figure 12-9 ***Plot Progress*** *dialog box*

When you invoke the **PLOTTERMANAGER** command, AutoCAD LT will display the **Plotters** window (Figure 12-10).

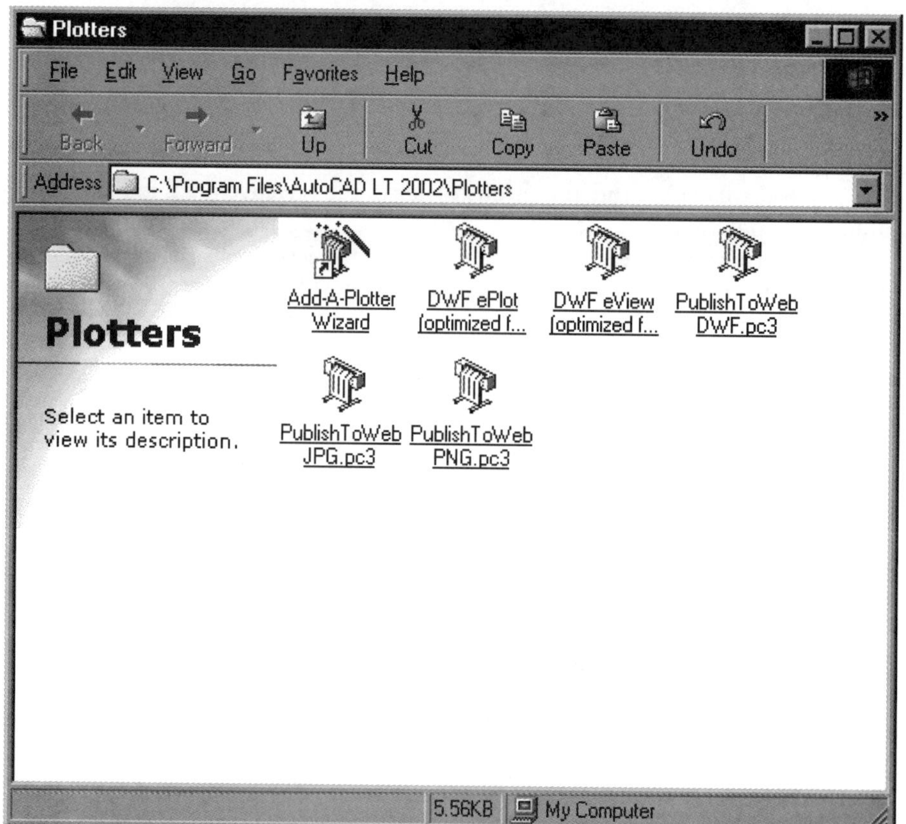

Figure 12-10 ***Plotters*** *window*

The **Plotters** window is basically a Windows Explorer window. It displays all the configured plotters and the **Add-A-Plotter-Wizard** icon. You can right-click on any one of the icons belonging to plotters that already have been configured to display a shortcut menu. You can choose **Delete** from the shortcut menu to remove a plotter from the list of available plotters in the **Name** drop-down list in the **Plot/Page Setup** dialog box. You can also choose **Rename** from the shortcut menu to rename the name of the plotter configuration file or choose **Properties** to view the properties of the configured device.

Add-A-Plotter-Wizard

If you double-click on the **Add-A-Plotter-Wizard** icon in the **Plotters** window, AutoCAD LT guides you to configure a nonsystem plotter for plotting your drawing files. AutoCAD LT stores all the information of a configured plotter in configured plot (PC3) files. The **PC3** files are stored in **AutoCAD LT 2002\Plotters** folder by default. The steps for configuring a new plotter using the **Add-A-Plotter-Wizard** is as follows:

1. Open the **Plotters** window by choosing **Plotter Manager** from the **File** menu.
2. In the **Plotters** window, double-click on the **Add-A-Plotter-Wizard** icon.
3. In the **Add-A-Plotter-Wizard**, carefully read the **Introduction Page**, and then choose the **Next** button to advance to the **Add Plotter - Begin** page.
4. On the **Add Plotter - Begin** page, choose the **My Computer** radio button (selected by default). Choose the **Next** button. The **Add Plotter - Plotter Model** page is displayed.
5. On this page, select a manufacturer and model of your nonsystem plotter from the Manufacturers and Models list box respectively. Now, choose the **Next** button. The **Add Plotter -Import Pcp and Pc2** page is displayed. If your plotter is not present in the list of available plotters, and you have a driver disk for your plotter, choose the **Have Disk** button to locate the **HIF** file from the driver disk, and install the driver supplied with your plotter.
6. In the **Add Plotter - Import Pcp and Pc2** page, if you want to import configuring information from a **PCP** or a **PC2** file created with a previous version of AutoCAD LT, you can choose the **Import File** button and select the file. Otherwise, simply choose the **Next** button to advance to the next page.
7. On the **Add Plotter - Ports** page, select the port from the list to use when plotting and choose **Next**.
8. On the **Add Plotter - Plotter Name** page, you can specify the name of the currently configured plotter or the default name will be entered automatically. Choose **Next**.
9. When you reach the **Add Plotter - Finish** page, you can choose the **Finish** button to exit the **Add-A-Plotter-Wizard**.
   You can also choose the **Edit Plotter Configuration** button to display the **Plotter Configuration Editor** where you can edit the current plotter's configuration. Also, in this page you can choose the **Calibrate Plotter** button to display the **Calibrate Plotter** wizard. This wizard allows you to calibrate your plotter by setting up a test measurement. After the test plotting it compares the plot measurements with the actual measurements and computes a correction factor.

Once you have chosen **Finish** to exit the wizard, a PC3 file for the newly configured plotter will be displayed in the **Plotters** window. This PC3 file contains all the settings needed by the plotter to plot. Also the newly configured plotter name is added to the **Name** drop-down list in the **Plotter configuration** area of the **Plot Device** tab in the **Plot** dialog box. You can now use the plotter for plotting.

EDITING PLOTTER CONFIGURATION

You can modify the properties of a selected plot device by using the **Plotter Configuration Editor**. This dialog can be invoked in several ways. As discussed earlier, when using the **Plot/Page Setup** dialog boxes, you can choose the **Properties** button in the **Plotter configuration** area to display the **Plotter Configuration Editor** (Figure 12-11). You can modify the default settings for a plotter while configuring it, by choosing the **Edit Plotter Configuration** button on the **Add Plotter - Finish** page of the **Add Plotters** wizard. You can also select the PC3 file for editing in the **Plotters** window using the Windows Explorer (by default, PC3 files are stored in the **AutoCAD LT 2002\Plotters** folder) and double-click the file or right-click the file and choose **Open** from the shortcut menu.

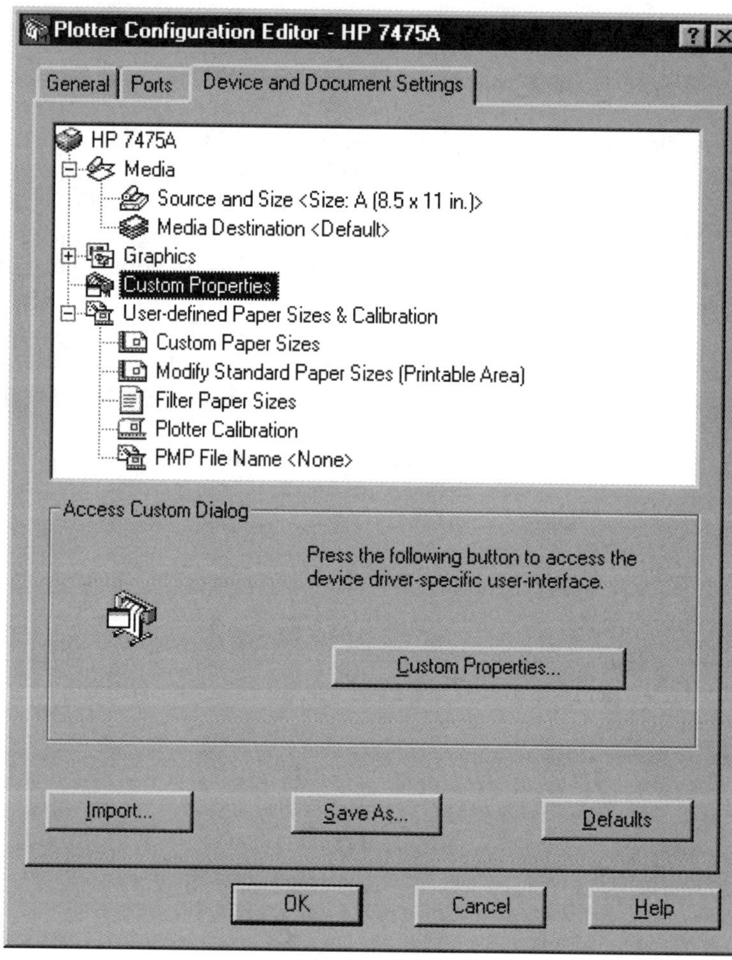

Figure 12-11 Plotter Configuration Editor

The three tabs available in the **Plotter Configuration Editor** are discussed next.

General Tab

This tab contains basic information about the configured plotter or the PC3 file. You can make changes only in the **Description** area. The rest of the information in the tab is read only. This tab contains information of the configured plotter file name, plotter driver type, HDI driver file version number, name of the system printer (if any), and the location and name of the PMP file (if any calibration file is attached to the PC3 file).

Ports Tab

This tab contains information about the communication between the plotting device and your computer. You can choose between a serial (local), parallel (local), or network port. The default settings for parallel ports is **LPT1** and **COM1** for serial ports. You can also change the port name if your device is connected to a different port. You can also select the **Plot to File** radio button, if you want to save the plot as a file. You can select **Autospool**, if you want that plotting to occur automatically while you continue work on another application.

Device and Document Settings Tab

This tab contains plotting options specific to the selected plotter, displayed as a tree view in a window. For example, if you configure a nonsystem plotter you have the option to modify the pen characteristics. You can select any plotter properties from the tree view displayed in the window to change the values as required. Whenever you select an icon from the tree view in the window, corresponding information is displayed in an area below. For example, if you select **PMP File Name <None>** in the tree view in the window, a **PMP file** area is displayed below. This area contains the current settings and options to modify it. Information that is displayed within brackets (<>) can be modified. By default, **Custom properties** is displayed as highlighted in the window, since it contains properties that are modified commonly and the **Access Custom Dialog** area is displayed below the window. Choosing the **Custom Properties** button in this area displays a properties dialog box specific to the selected plotter. This properties dialog box has several properties of the selected plotter grouped and displayed under various tabs and can be modified here.

Once you have made the desired changes, choose **OK** to exit the properties dialog box and then choose **Save As** in the **Plotter Configuration Editor** if you want to save the changes you just made to the PC3 file. You can also import old plot configuration files from previous releases of AutoCAD LT using the **Import** button and save some of the information from these settings as a new PC3 file. When you choose the **Import** button, the **Plotting Components** dialog box is displayed. This dialog box displays what to use to import AutoCAD Release 14 information into an AutoCAD LT 2002 drawing. For example, it tells you that to import PCP or PC2 file settings into a drawing in AutoCAD LT 2002, you should use the **Import PCP or PC2 Plot Settings Wizard**. Choose **OK** to exit this dialog box and the **Import** dialog box is displayed. Here you can select the file to import and then choose the **Import** button.

Tip

It is better to create a new PC3 file for a plotter and keep the original file as it is so that you encounter no error while using the specific printer later. The PC3 files determine the proper function of a plotter and any modifications may lead to errors.

IMPORTING PCP/PC2 CONFIGURATION FILES

If you want to import a **PCP** or **PC2** configuration file or plot settings created by previous releases of AutoCAD LT into the **Model** tab or current layout for the drawing, you can also use **PCINWIZARD** command to display the **Import PCP or PC2 Plot Settings Wizard**. All the information from a **PCP** or **PC2** file regarding plot area, rotation, plot offset, plot optimization, plot to file, paper size, plot scale, and pen mapping can be imported. Read the **Introduction** page of the wizard that is displayed carefully and then choose the **Next** button. The Browse File Name page is displayed. Here you can either enter the name of the PCP or PC2 directly in the **PC2 or PCP file name** edit box or then choose the **Browse** button to display the **Import** dialog box, where you can select the file to import. After you specify the file for importing, choose **Import** to return to the wizard. Choose the **Next** button to display the **Finish** page. After importing the files you can modify the rest of the plot settings for the current layout.

SETTING THE PLOT PARAMETERS

Before starting with the drawing, you can set various plotting parameters in the **Model** tab or in the **Layouts**. The plot parameters that can be set include the plotter to be used, plot style table, the paper size, units so on. All these parameters can be set using the **PAGESETUP** command discussed next.

PAGESETUP Command

Toolbar:	Layouts > Page Setup
Menu:	File > Page Setup
Command:	PAGESETUP

As discussed earlier, page setup contains settings required to plot a drawing. Each layout as well as the **Model** tab can have a unique page setup attached to it. You can use the **PAGESETUP** command to create named page setups that can be used later. A page setup consists of specifications for the layout page, plotting device, paper size, and settings for the layouts to be plotted. The **PAGESETUP** command can also be invoked from the shortcut menu by right-clicking in the current **Model** or any of the **layout** tabs and choosing **Page Setup**. AutoCAD LT displays the **Page Setup** dialog box (Figure 12-12).

The **Page Setup** dialog box is similar to the **Plot** dialog box that has been described earlier except that it does not have the preview buttons and the **What to plot** and **Plot to file** areas. The first time you choose a layout tab in a drawing session, the **Page Setup** dialog box is automatically displayed and specifies the page layout and current plotting device settings. The layout settings are stored with the layout. The current Layout name is displayed in the **Layout Name** box, and in the **Page Setup Name** area, a list of all the named and saved page setups are available. You can choose to base the current page setup on a named page setup or you can add a new named page setup by choosing the **Add** button, AutoCAD LT displays the **User Defined Page Setups** dialog box (discussed earlier in this chapter in "**PLOT** Command"). The **Page Setup** dialog box has two tabs that are described next.

Plot Device Tab

This tab specifies the current plotting device for plotting the layout. This tab is similar to

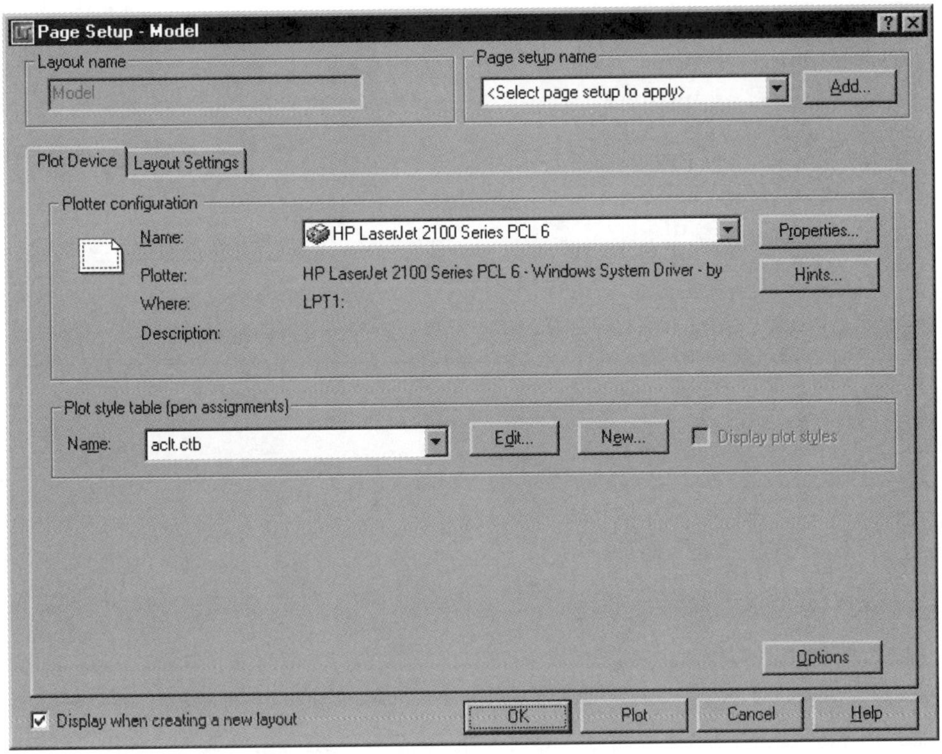

*Figure 12-12 **Page Setup** dialog box when displayed in the **Model** tab*

the **Plot Device** tab of the **Plot** dialog box, excluding for the **What to Plot** and **Plot to file** areas. The additional feature in this tab is the **Options** button. If you choose the **Options** button, AutoCAD LT displays the **Plotting** tab of the **Options** dialog box (Figure 12-13). This tab has three areas that are discussed here.

Default plot settings for new drawings area. This area provides options to set the default plot settings. You can either decide to use the settings that were used successfully for the last plot created by selecting the **Use last successful plot settings** radio button or you can select a device from the **Use as default output device** drop-down list and use it as the default device every time you create a new plot. You can also choose the **Add or Configure Plotters** button to display the **Plotters** window and use the **Add-A-Plotter-Wizard** to configure a new plotter and then use it as the default plotter.

General plot options area. This area provides options that control the general plotting environment. The **Keep the layout paper size if possible** radio button is selected by default and allows you to keep plotting on the paper size selected from the **Paper size** drop-down list in the **Layout Settings** tab of the **Page Setup** dialog box irrespective of the plotting device selected as long as possible. When no longer possible, AutoCAD LT will display a warning saying that the paper size is not supported by the current device and will use the paper size that is specified in the PC3 file of the plotter or the default printer settings of the printer. This is also controlled by the system variable **PAPERUPDATE**. When this option is selected the

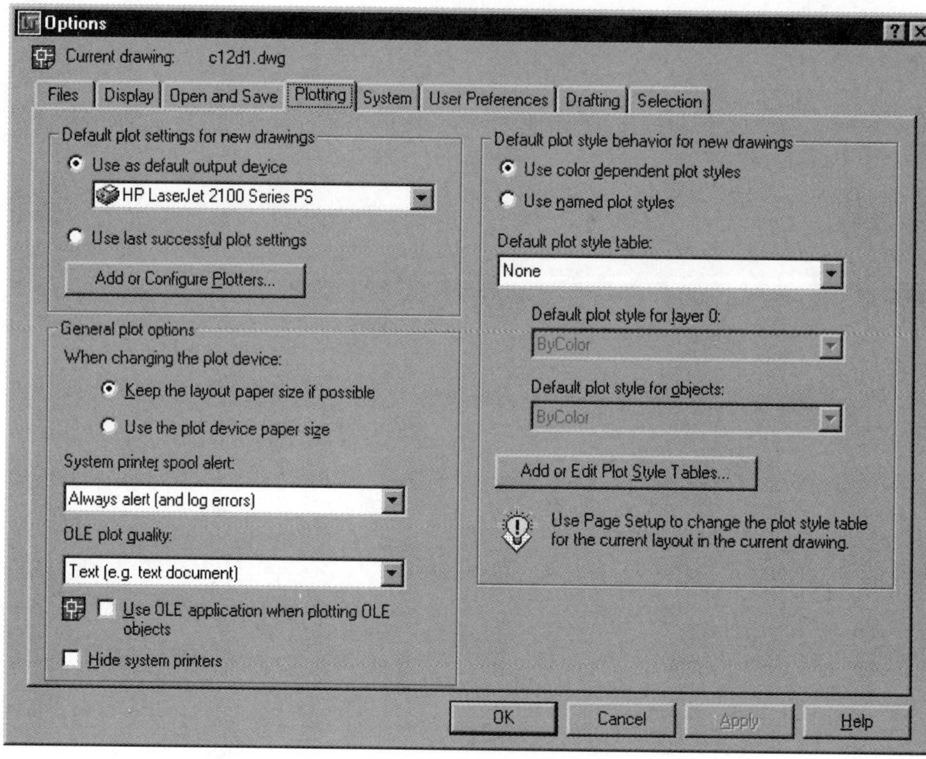

***Figure 12-13 Plotting** tab of the **Options** dialog box*

value of this variable is set to 0. If you select the **Use the plot device paper size** radio button, only the paper size specified in the plotter configuration file (PC3) of the selected device is used and the value of the **PAPERUPDATE** variable is set to 1.

The **System printer spool alert** drop-down list displays options that determine whether AutoCAD LT will display an alert message if there is an error (due to conflict between the output or input port) and the drawing is spooled through the system printer and when it will do so. The options also determine if AutoCAD LT will log the error.

The **OLE plot quality** drop-down list provides options that control the quality of OLE objects, when they are plotted. The **OLEQUALITY** system variable also controls this option. OLE stands for Object Linking and Embedding and will be discussed later in Chapter 19, Data Exchange, and Object Linking and Embedding. Selecting the **Use OLE application when plotting OLE objects** check box loads applications used to create OLE objects and is useful when you want to improve the quality of OLE objects. This option is also controlled by the **OLESTARTUP** system variable.

Default plot style behavior for new drawings area. This area provides options that allow you to set the default plot style mode and table for new drawings. These options shall be discussed later in this chapter in the section "Using Plot Styles".

Layout Settings Tab

This tab specifies the layout settings of the drawing such as paper size, drawing orientation, plot area, plot scale, plot offset, and other plotting options. This tab is similar to the **Plot Settings** tab of the **Plot** dialog box (discussed earlier in the chapter).

The settings specified in the **Page Setup** dialog box are stored with the layout. You can plot the current layout from the **Page Setup** dialog box by choosing the **Plot** button after specifying all the necessary parameters in the dialog box. You can also control whether the plotting should include lineweights and whether the lineweights are in proportion to the plot scale.

PSETUPIN Command

Command: PSETUPIN

As discussed earlier, it is possible for you to import a user-defined page setup from an existing drawing and use it in the current drawing or base the current page setup for the drawing on it. This option is available through the **Plot/Page Setup** dialog box. It is also possible to bypass the **Plot/Page Setup** dialog box and directly import a page setup from an existing drawing into a new drawing layout by using the **PSETUPIN** command. This command facilitates importing a saved and named page setup from a drawing into a new drawing. The settings of the named page setup can be applied to layouts in the new drawing. When you invoke this command, a standard file selection dialog box is displayed, where you can locate a drawing (.dwg) file whose pagesetups have to be imported. After you select a drawing file, AutoCAD LT displays the **Import User Defined Page Setups** dialog box (Figure 12-14). You can also enter **-PSETUPIN** at the Command prompt to display prompts at the command line.

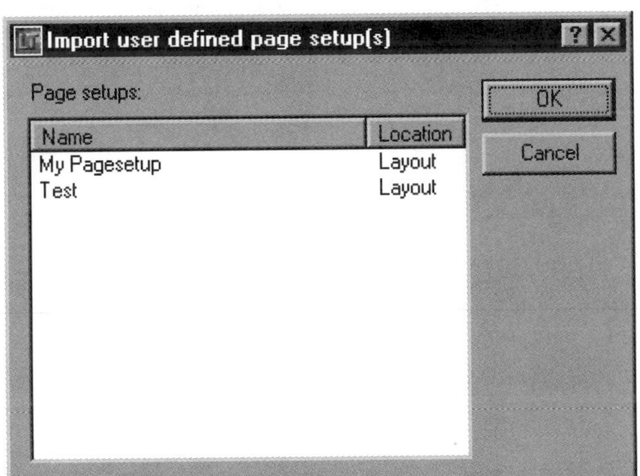

*Figure 12-14 **Import user defined page setup(s)** dialog box*

USING PLOT STYLES

The plot styles can change the complete look of a plotted drawing. You can use this feature to override a drawing object's color, linetype, and lineweight. For example, if an object is drawn on a layer that is assigned the color red and no plot style is assigned to it, it will be plotted as red. But, if you have assigned a plot style to the object with the color blue, the object will be plotted as blue irrespective of the layer color it was drawn on. Similarly, you can change the end, join and fill styles of the drawing, and also change the output effects such as dithering, gray scales, pen assignments, and screening. Basically, you can use **Plot Styles** effectively to plot the same drawing in various ways.

Every object and layer in the drawing has a plot style property. The plot style characteristics are defined in the plot style tables attached to the **Model** tab, layouts, and viewports within the layouts. You can attach and detach different plot style tables to get different looks for your plots. Generally, there are two plot style modes. They are **Color-Dependent** and **Named**. The **Color-dependent** plot styles are based on object color and there are **255** color-dependent plot styles. It is possible to assign each color in the plot style a value for the different plotting properties and these settings are then saved in a color-dependent plot style table file that has a .CTB extension. Similarly, **Named** plot styles are independent of object color and you can assign any plot style to any object regardless of that object's color. These settings are saved in a named plot style table that has a .STB extension. Every drawing in AutoCAD LT 2002 is in either one of the plot style modes.

STYLESMANAGER Command

Menu:	File > Plot Style Manager
Command:	STYLESMANAGER

All plot styles are saved in the **AutoCAD LT 2002\Plot Styles** folder. If you enter **STYLESMANAGER** at the Command prompt, AutoCAD LT displays the **Plot Styles** window (Figure 12-15). This window display icons for all the available plot styles in addition to the **Add-A-Plot Style Table Wizard** icon. You can double-click on any of the plot style icons to display the **Plot Style Table Editor** dialog box and edit the selected plot style. When you double-click on the **Add-A-Plot Style Table Wizard** icon, the **Add Plot Style Table** wizard is displayed and you can use it to create a new plot style.

Add-A-Plot Style Table Wizard

If you want to add a new plot style table to your drawing, double-click on the **Add-A-Plot Style Table Wizard** in the **Plot Styles Manager** window to display the **Add Plot Style Table** wizard. You can also invoke the wizard by choosing **Add Plot Style Table** from the **Tools > Wizards** menu. The following are the steps for creating a new plot style table using the wizard:

1. Read the **Introduction** page carefully and choose the **Next** button.

2. In the **Begin** page, select the **Start from scratch** radio button and choose **Next**. Selecting this option creates an absolutely new plot style table and does not use an existing file. Therefore, the **Browse File** page is not available.

   In addition to the **Start from scratch** option, this page has three more options. They are: **Use an existing plot style table**, **Use My R14 Plotter Configuration (CFG)**, and **Use a PCP or PC2 file**. When you use the **Use an existing plot style table** option, an existing plot style table is used as a base for the new plot style table you are creating. In such a

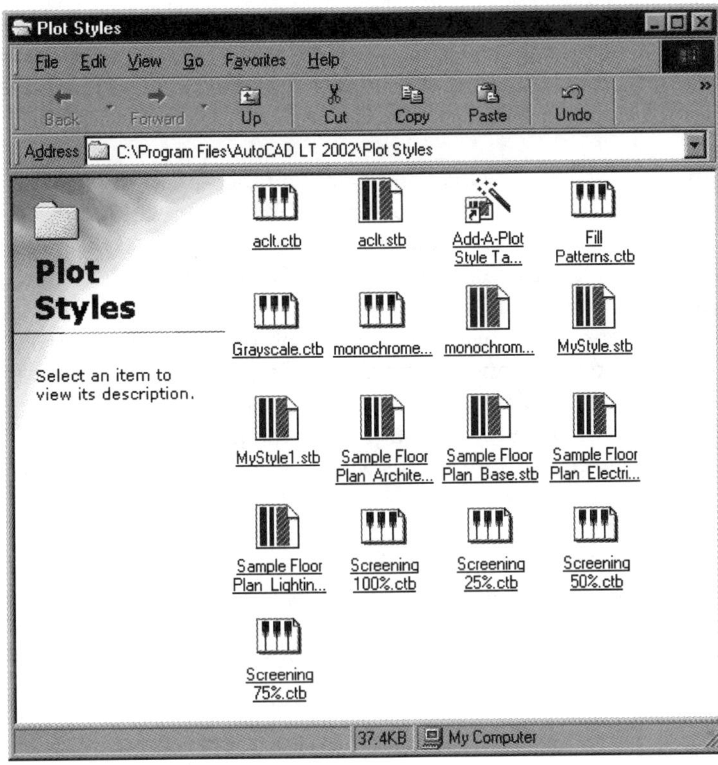

Figure 12-15 ***Plot Styles*** *window*

situation, the **Table Type** page of the wizard is not available and is not displayed because the table type will be based on the existing plot style table you are using to a create a new one. With the **Use My R14 Plotter Configuration (CFG)** option, the pen assignments from the **aclt2002.cfg** file is used as a base for the new table you are creating. If you are using the **Use a PCP or PC2** option, pen assignments saved earlier in a Release 14 PCP or PC2 file are used to create the new plot style.

3. In the **Pick Plot Style Table** page, select the **Named Plot Style Table** or the **Color-Dependent Plot Style Table** according to your requirement, and then choose the **Next** button. The **Named Plot Style Table** radio button is selected by default.

4. Since, you have selected the **Start from scratch** radio button in the **Begin** page, the **Browse File** page is not available and the **File name** page is displayed. However, if you had selected any of the other three options available on the **Begin** page, the **Browse File** page is displayed. You can select an existing file from the drop-down list available in this page or choose the **Browse** button to display the **Select File** dialog box. You can then browse and select a file from a specific folder and choose **Select** to return to the wizard. You can also enter the name of the existing plot style table you want to base the new plot style table on, directly in the edit box. After you have specified the file name, choose **Next** to display the **File name** page of the wizard. In the **File name** page, enter a file name for the new plot style table and choose **Next**.

Note

*If you are using the pen assignments from the Release 14 aclt.cfg file to define the new plot style table, you have to also specify the printer or plotter to use from the drop-down list available in the **Browse File** page of the wizard.*

5. The **Finish** page is displayed. This page gives you the option of choosing the **Plot Style Table Editor** button to display the **Plot Style Table Editor** and then edit the plot style table you have created. If you select the **Use this plot style table for new and pre-AutoCAD LT 2002 drawings** check box in this page of the wizard, the plot style table that you have created will become the default plot style table for all the drawings you create. This check box is available only if the plot style mode you have selected in the wizard is the same as the one you have specified as the default plot style mode in the **Default plot style behavior for new drawings** area in the **Plotting** tab of the **Options** dialog box. Choose **Finish** in the **Finish** page of the wizard to exit the wizard. A new plot style table gets added to the **Plot Styles** window and can be used for plotting.

Note

*You can also choose **Add Named Plot Style Table/ Add Color-Dependent Plot Style Table** from the **Tools > Wizards** menu to display wizards that are similar to the **Add Plot Style Table** wizard except that in the **Begin** page, the **Use an existing plot style table** option is not available. Also the **Table Type** page is not there and the **Finish** page has an additional option to use the new plot style table for the current drawing.*

Plot Style Table Editor

When you double-click on any of the plot style table icons available in the Plotters window, the **Plot Style Table Editor** is displayed where you can edit the particular plot style table. You can also choose the **Plot Style Table Editor** button in the **Finish** page to display the **Plot Style Table Editor** and choose the **Edit** button adjacent to the **Name** drop-down list in the **Plot style table (pen assignments)** area in the **Plot Device** tab of the **Plot** or **Page Setup** dialog box to display the **Plot Style Table Editor**.

The **Plot Style Table Editor** has three tabs: **General**, **Table View**, and **Form View**. You can edit all the properties of an existing plot style table using the different tabs. The description of the tabs are as follows.

General Tab

This tab provides the information about the file name, location of the file, version, and scale factor. All the information except the description are read only. You can enter a description about the plot style table in the **Description** text box here. If you select the **Apply global scale factor to non-ISO linetypes** check box, all the non-ISO linetypes in a drawing are scaled by the scale factor specified in the **Scale factor** edit box placed below the check box. If this check box is cleared (by default), the **Scale factor** edit box is not available.

Table View Tab

This tab displays all the plot styles, available with their properties in tabular form, that can be edited individually here (Figure 12-16). In the case of a named plot style table, you can edit

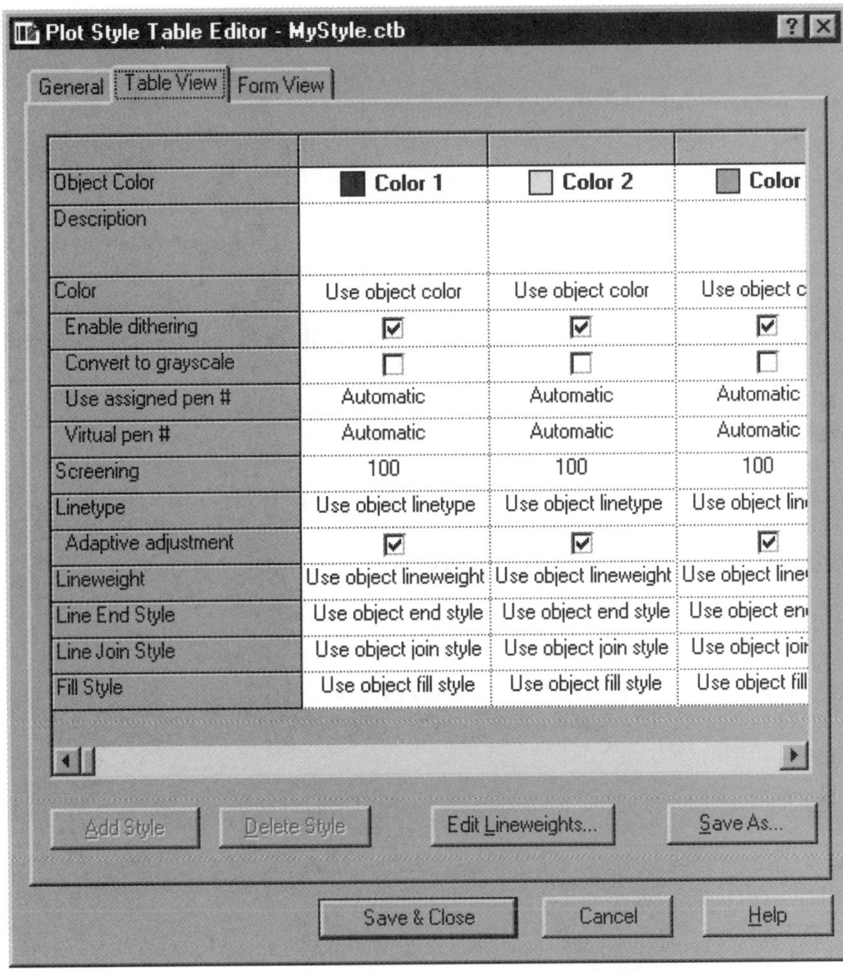

***Figure 12-16 Plot Style Table Editor (Table View** tab)*

the existing styles or add new styles by choosing the **Add Styles** button. A new column with default style name **Style1** will be added in the table. You can change the style name if you want to. You can edit the various properties in the table by selecting a particular value you want to modify and a corresponding drop-down list is displayed. You can select a value from this drop-down list. This manner of editing is similar to the one we use when editing properties of objects using the **Properties** window. The **Normal** plot style table is not available and therefore, cannot be edited. This plot style is assigned to layers by default.

You can select a particular plot style by clicking on the gray bar placed above the column and the entire column gets highlighted. You can select several plot styles by pressing the SHIFT key and selecting more plot styles. All the plot styles that are selected are highlighted. If you choose the **Delete Style** button now, the selected plot styles are removed from the table. Choosing the **Edit Lineweights** button displays the **Edit Lineweights** dialog box. You can select the units for specifying the lineweights in the Units for listing area of the dialog box, and can use either **Millimeters** or **Inches**. You can also edit the value of a lineweight by selecting it in the **Lineweights** list box and choosing the **Edit Lineweight** button. After you have edited lineweights, choose the **Sort Lineweights** button to rearrange the lineweight values in the list box. Choose **OK** to exit the dialog box and return to the **Plot Style Table Editor**.

With a color dependent plot style table, the **Table View** tab displays all 255 plot styles, one for each color, and the properties can be edited in the table in the same way as discussed for named plot styles. The only difference is that you cannot add a new plot style or delete an existing one and therefore, the **Delete Style** and **Add Style** buttons are not available. The properties that can be defined in a plot style are discussed next.

Color. The color you assign to plot style overrides the color of the object in the drawing. The default value is **Use object color**. The other properties related to color are as follows:

Enable dithering. Dithering is described as a mixing of various colored dots to produce a new color. You can enable or disable this property by selecting or clearing the check box available in this field. Dithering is enabled by default and is independent of the color selected.

Convert to grayscale. If you are using plotter that supports grayscaling, selecting this option applies grayscale to the objects color. By default, this check box is not selected and the object's color is used.

Use assigned pen #. This property is applied only to pen plotters. The pens range from 1 to 32. The default value is **0** or **Automatic**, which implies that the pen used will be based on the plotter configuration.

Virtual pen #. Nonpen plotters can behave like pen plotters using virtual pens. The value of this property lies between 1 and 255. The default value is **0** or **Automatic**. This implies that AutoCAD LT will assign a virtual pen automatically from the AutoCAD LT Color Index (ACI).

Screening. This property of a plot style indicates the amount of ink used while plotting. The value ranges between 0 and 100. The default value is **100**, which creates the plot in its full intensity. Similarly, a value of **0** produces white color. This may be useful when plotting on a colored background.

Linetype. Like the property of color, the linetype assigned to a plotstyle overrides the object linetype on plotting. The default value is **Use object linetype**.

Adaptive adjustment. This property is applied by default and implies that the linetype scale of a linetype is applied such that on plotting, the linetype pattern will be completed.

Lineweight. The lineweight value assigned here overrides the value of the object lineweight on plotting. The default value is **Use object lineweight**.

Line End Style. This determines the manner in which a plotted line ends. The effect of this property is more noticeable when the thickness of the line is substantial. The line can end in a **Butt**, **Square**, **Round**, or **Diamond** shape. The default value is **Use object end style**.

Line Join Style. You can select the manner in which two lines join in a plotted drawing. The available options are: **Miter**, **Bevel**, **Round**, and **Diamond**. The default value is **Use object join style**.

Fill Style. The fill style assigned to a plot style overrides the objects fill style, when plotted. The available options are: **Solid**, **Checkerboard**, **Crosshatch**, **Diamonds**, **Horizontal Bars**, **Slant Left**, **Slant Right**, **Square Dots**, and **Vertical Bars**.

Form View Tab

This tab displays all the properties in a form, as shown in Figure 12-17. All the available plot styles are displayed in the **Plot Styles** list box. You can select any style in the list box and then edit its properties in the **Properties** area.

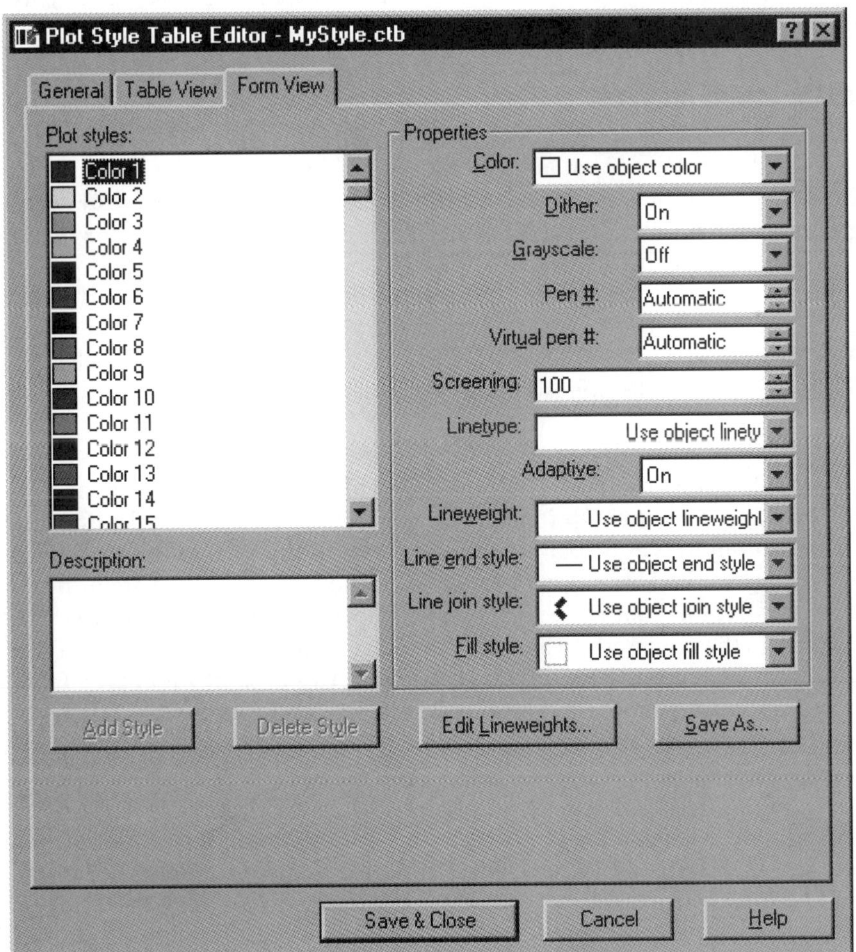

Figure 12-17 Plot Style Table Editor (Form View tab)

With a named plot styles table, if you want to add a new plot style, choose the **Add Style** button and AutoCAD LT will display the **Add Plot Style** dialog box (Figure 12-18) with default plot style name **Style 1**. You can change the name of the plot style in the Plot Style edit box. Now, when you choose the **OK** button, the new style will be added in the **Plot Styles** list box and you can select it for editing. If you want to delete a style, select a style and then choose the **Delete Style** button. With color-dependent plot styles the **Form View** tab also does not provide options that allow you to add or delete plot styles.

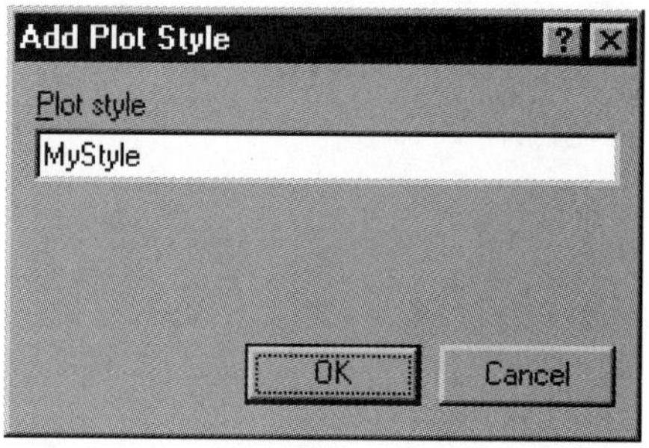

***Figure 12-18 Add Plot Style** dialog box*

After editing, choose the **Save & Close** button to save and return to the **Plot Styles** window. You can also choose the **Save As** button to display the **Save As** dialog box and save a plot style table with another name. If you want to remove the changes you made to a plot style table, choose the **Cancel** button.

Applying Plot Styles

The **Model**, or any of the layout tabs can be assigned a plot style table. The plot style table can be either named or color-dependent as discussed earlier. These plot style modes for a new drawing can be determined in the **Default plot style behavior for new drawings** area of the **Plotting** tab of the **Options** dialog box (see Figure 12-13). The **Use color dependent plot styles** option is selected by default and the drawings are assigned a color-dependent plot style. If you select the **Use named plot styles** radio button, the new drawings (not the current drawing) will be assigned a named plot style. The **PSTYLEPOLICY** system variable also controls the default plot style modes of the new drawings. A value of 0 implies a named plot style mode and a value of 1 implies a color-dependent plot style mode.

You can select a plot style table that you want to use as a default for the drawings from the **Default plot style table** drop-down list in the **Default plot style behavior for new drawings** area of the **Plotting** tab of the **Options** dialog box. If you select **None**, the drawing is plotted with the object properties as displayed on the screen. In this area of the **Options** dialog box, only when you select the **Use named plot styles** radio button, are the **Default plot style for layer 0** and the **Default plot style for objects** drop-down lists available. You can select the default plot styles that you want to assign to the Layer 0 and to the objects in a drawing from these drop-down lists respectively. If you choose the **Add or Edit Plot Style Table** button in this area of the **Plotting** tab of the **Options** dialog box, the **Plot Styles** window is displayed. Here, you can double-click on any plot style table icon available and edit it using the **Plot Style Table Editor** that is displayed.

To change the plot style table for a current layout, you have to invoke the **Page Setup** or **Plot** dialog box and then select a plot style table from the **Name** drop-down list in the **Plot style table (pen assignments)** area of the **Plot Device** tab of the dialog box. A color-dependent plot style table can be selected and applied to a tab only if the default plot style mode already has been set to color dependent. Similarly, if you want to apply a named plot style table to a tab, the **Use named plot styles** radio button should have been selected in the **Plotting** tab of the **Options** dialog box.

A color-dependent plot style cannot be applied to objects or layers and therefore the **Plot**

Style Control drop-down list in the **Object Properties** toolbar is not available when a drawing has a color-dependent plot style mode. The plot styles also appear grayed out in the **Layer Properties Manager** dialog box and cannot be selected and changed. But, named plot styles can be applied to objects and layers. A plot style applied to an object overrides the plot style applied to the layer on which the object is drawn. To apply a plot style to a layer, invoke the **Layer Properties Manager** dialog box where all the layers in the selected tab are displayed. Select a layer to which you want to apply a plot style and select the default plot style (Normal) currently applied to the layer. The **Select Plot Style** dialog box (Figure 12-19) is displayed. The **Plot styles** list box in this dialog box displays all the plot styles present in the plot style table attached to the current tab. You can select another plot style table to attach to the current tab from the **Active plot style table** drop-down list. You will notice that all the plot styles in the selected plot style table are displayed in the list box now. You can also choose the **Editor** button to display the **Plot Style Table Editor** to edit plot style tables, as discussed earlier. The **Select Plot Style** dialog box also displays the name of the original plot style assigned to the object adjacent to **Original**. Also, the new plot style to be assigned to the selected object is displayed next to **New**.

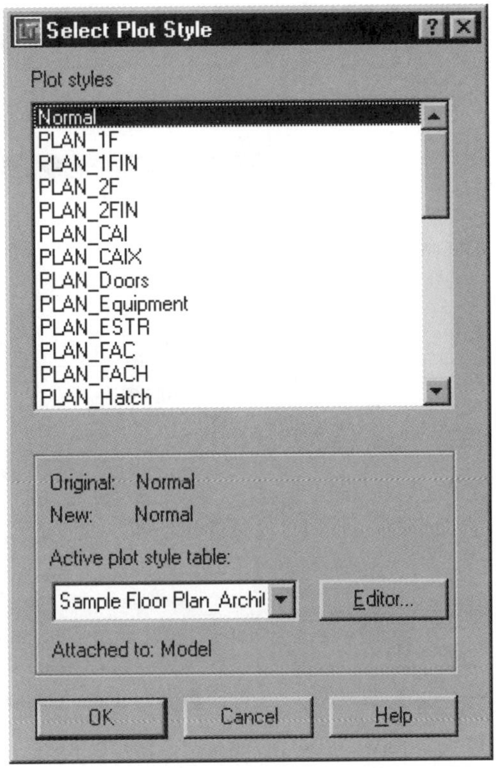

***Figure 12-19 Select Plot Style** dialog box*

You can apply a named plot style to an object using the **Plot Style Control** drop-down list in the **Object Properties** toolbar or the **Properties** window. The process is the same as that for layers, colors, linetypes, and lineweights are applied. This named plot style is applied to an object irrespective of the tab on which it is drawn. If the plot style assigned to an object is present in the plot style table of the tab in which it is present, the object is plotted with the specified plot style. But, if the plot style assigned to the object is not present in the plot style table assigned to the tab on which it is drawn, the object will be plotted with the properties that are displayed on the screen. The default plot style assigned to an object is **Normal** and the default plot style assigned to a layer is **ByLayer**.

PLOTSTYLE Command

You can use the **PLOTSTYLE** command to set the current plot style for new objects or of selected objects. When you enter **PLOTSTYLE** at the Command prompt, and if no object has been selected in the drawing, AutoCAD LT displays the **Current Plot Style** dialog box (Figure 12-20). But if any object selection is there in the drawing, then AutoCAD LT displays the **Select Plot Style** dialog box (Figure 12-19), that has been discussed earlier. You can select a plot style from the list box and choose **OK** to assign it to the selected objects in the drawing.

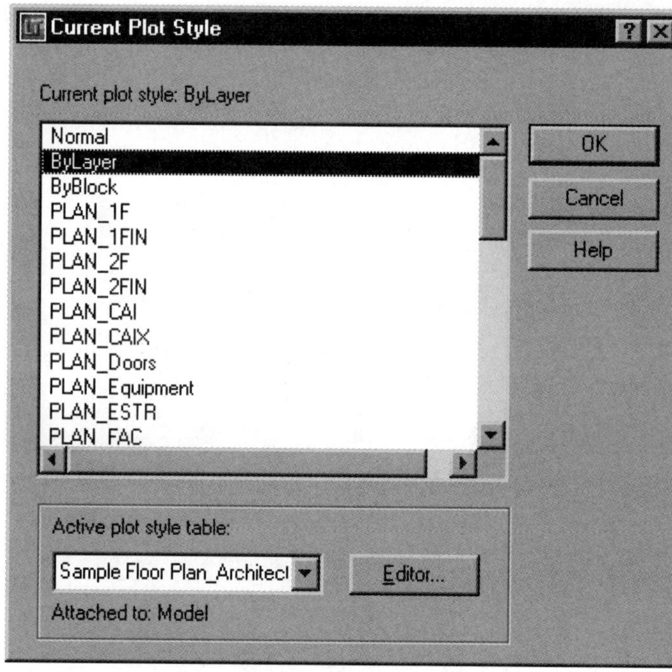

***Figure 12-20 Current Plot Style** dialog box*

All the plot styles present in the current plot style table that are assigned to the current tab are displayed in the list box in the **Current Plot Style** dialog box. You can select any one of these plot styles and choose **OK**. Now, when you create new objects, they will have the plot style that you had set current in the **Current Plot Style** dialog box. The parameters of this dialog box are described next.

Current plot style

The name of the current plot style is displayed adjacent to this label.

Plot style list box

This list box lists all the available plot styles that can be assigned to an object, including the default plot style, **Normal**.

Active plot style table

This drop-down list displays the names of all the available plot style tables. The current plot style table attached to the current layout or viewport is displayed in the edit box.

Editor

If you choose the **Editor** button, adjacent to the drop-down list, AutoCAD LT displays the **Plot Style Table Editor** to edit the selected plot style table.

Attached to

The tab to which the selected plot style table is attached, **Model** or any one of the layout tabs, is displayed next to this label.

Note

You can also assign a current plot style using the command line. This can be done with the help of the ***-PLOTSTYLE*** *command.*

Exercise 1 *Mechanical*

Make the drawing shown in Figure 12-21. Create a named plot style table My Named Table.stb with three plot styles: Style1, Style2, and Style3, in addition to the Normal plot style. The Normal plot style is used for plotting the object lines. These three styles have the following specifications.

Style 1. This style has a value of Screening = 50. The dimensions, dimension lines, and the text in the drawing must be plotted with this style.

Style 2. This style has a value of Lineweight = 0.800. The border and title block must be plotted with this style.

Style 3. This style has a linetype of Medium Dash. The centerlines must be plotted with this plot style.

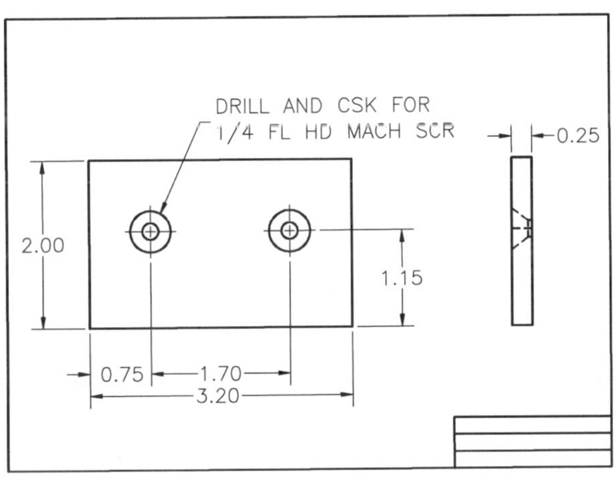

Figure 12-21 *Drawing for Exercise 1*

Self-Evaluation Test

Answer the following questions, and then compare your answers to the correct answers given at the end of this chapter.

1. All the settings about a plotter are saved in the .PC3 file. (T/F)
2. Different objects in the same drawing can be plotted in different colors, with different linetypes and line widths. (T/F)
3. Full preview takes less time than partial preview. (T/F)
4. The **PSLTSCALE** system variable controls the paper space linetype scaling and has a default value of 1. (T/F)
5. The size of a plot can be specified by selecting any paper size from the __________ drop-down list in the **Plot** dialog box.
6. If you want to store the plot in a file and not have it printed directly on a plotter, select the __________ check box in the **Plot Device** tab of the **Plot** dialog box.
7. The scale for the plot can also be specified in the __________ edit boxes in the **Plot** dialog box.
8. If you select the __________ radio button in the **Plot** dialog box, the portion of the drawing that is in the current display is plotted.
9. In the Partial Plot Preview dialog box, the __________ is graphically represented by a white rectangle. The blue hatched rectangle is the section in the paper that is used by the __________ .
10. The __________ command displays the **Plot Styles** window that can also be used to edit the existing plot style tables.

Review Questions

Answer the following questions.

1. When you choose a layout tab for the first time in a drawing, the **Page Setup** dialog box is displayed. (T/F)

2. By selecting the **View** radio button in the **Plot area** of the **Page Setup/Plot** dialog box, you can plot a view that was created with the **VIEW** command in the current drawing. (T/F)

3. If you do not want the hidden lines of a 3D object created in the **Model** tab, select the **Hide Objects** check box in the **Plot options** area of the **Plot Settings** tab of the **Plot** dialog box. (T/F)

4. The orientation of the drawing can be changed using the **Plot** dialog box. (T/F)

5. Which check box in the **Plot options area** of the Plot/ Page Setup dialog box is available only if you are plotting in a layout tab?

   (a) **Plot paper space last** (b) **Hide objects**
   (c) **Plot with plot styles** (d) **None**

6. Which command when invoked displays the **Plotters** window?

   (a) **PLOTTER** (b) **PLOTTERMANAGER**
   (c) **PLOTSTYLE** (d) **None**

7. With which command is it possible to bypass the **Plot/Page Setup** dialog box and directly import a page setup from an existing drawing into a new drawing layout?

   (a) **PSETUPIN** (b) **PLOTTERMANAGER**
   (c) **PLOTSTYLE** (d) **None**

8. Which command is used to create a new plot style?

   (a) **STYLE** (b) **STYLESMANAGER**
   (c) **PLOTSTYLE** (d) **None**

9. Which command can be used to import a **PCP** file or **PC2** files?

   (a) **PCINWIZARD** (b) **PCIN**
   (c) **PLOTSTYLE** (d) **None**

10. You can view the plot on the specified paper size before actually plotting it by selecting the __________ button in the **Plot** dialog box.

11. In the **Partial Plot Preview** dialog box, the red triangle icon located at one of the corners of the effective area (hatched blue rectangle) represents the __________ of the plot.

12. You can modify the properties of a selected plotting device by using the __________ .

13. The plot style modes available are __________ and __________ .

14. The **Plot Style Table Editor** has three tabs. They are: __________ , __________ , and __________ .

15. The __________ check box in the **Plot Settings** tab is selected to replace the pen mapping in the earlier versions of AutoCAD LT.

Exercises

Exercise 2 *Mechanical*

Make the drawing shown in Figure 12-22 and plot it according to the following specifications. Create and use a plot style table with the specified plot styles.

1. The drawing is to be plotted on 10 x 8-inch paper.

2. The object lines must be plotted with a plot style Style 1. Style 1 must have a value of lineweight = 0.800 mm.

3. The dimension lines must be plotted with plot style Style 2. Style 2 must have a value of screening = 50.

4. The centerlines must be plot with plot style Style 3. Style 3 must have a linetype of Medium Dash and screening = 50.

5. The border and title block must be plotted with plot style Style 4. The value of lineweight =0.25 mm.

Exercise 3 *Mechanical*

Make the drawing shown in Figure 12-23 and plot the drawing according to your specifications.

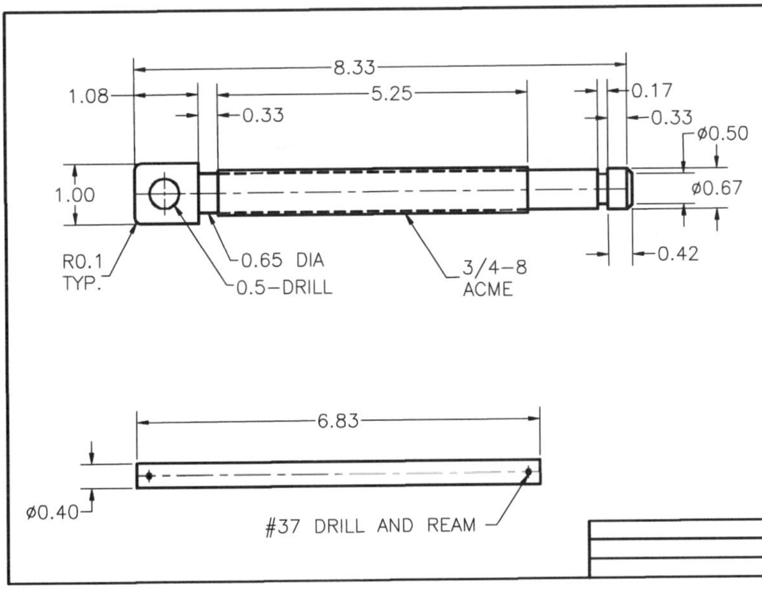

Figure 12-22 Drawing for Exercise 2

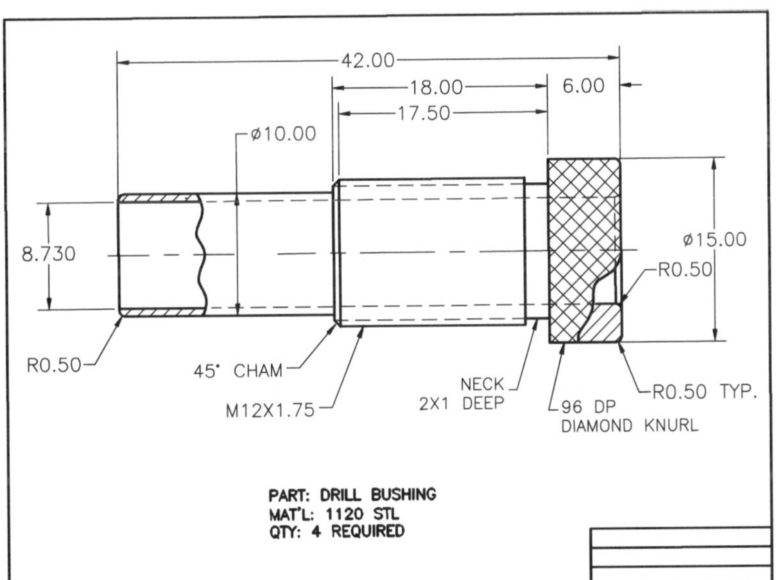

Figure 12-23 Drawing for Exercise 3

Problem Solving Exercise 1 — *Mechanical*

Make the drawing shown in Figure 12-24 and plot it according to your specifications.

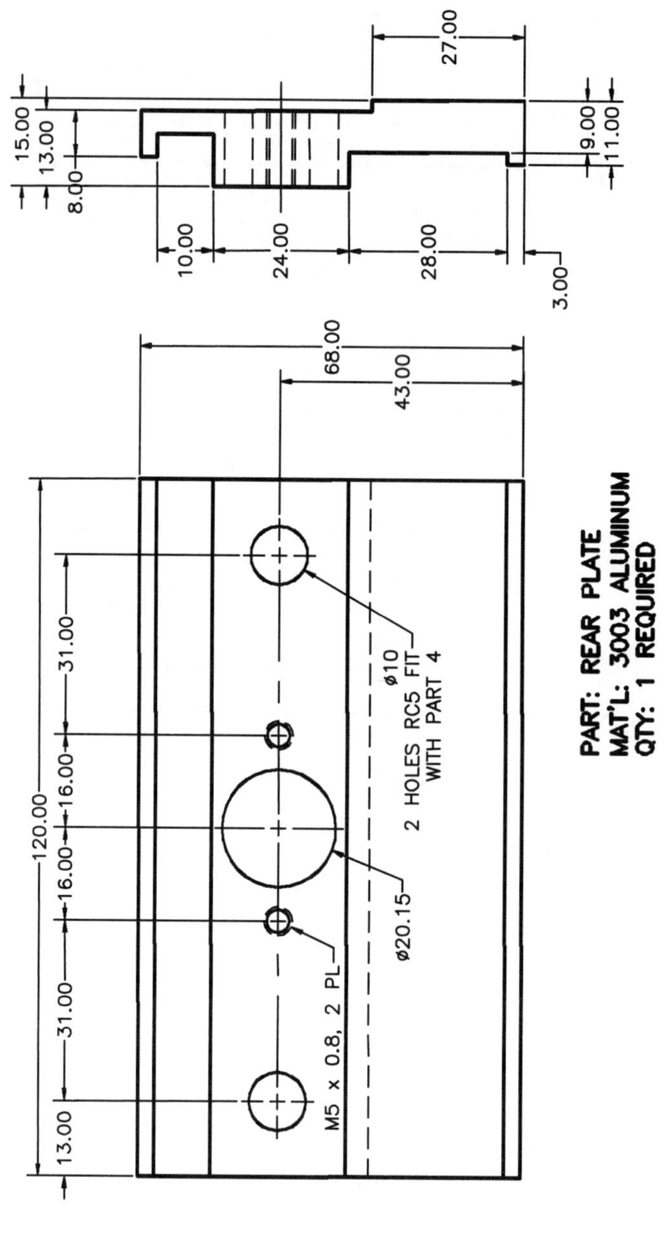

Figure 12-24 *Drawing for Problem Solving Exercise 1*

Answers to Self-Evaluation Test

1 - T, **2** - T, **3** - F, **4** - T, **5** - **Paper size**, **6** - **Plot to file**, **7** - **Custom**, **8** - **Display**, **9** - paper size, image, **10** - **STYLESMANAGER**

Chapter 13

Hatching

Learning Objectives

After completing this chapter, you will be able to:
- *Use the **BHATCH** command to hatch an area using various patterns.*
- *Use boundary hatch with predefined, user-defined, and custom hatch patterns as options.*
- *Specify pattern properties.*
- *Preview and apply hatching.*
- *Use advanced hatching options and ray casting.*
- *Edit associative hatch and hatch boundary using **HATCHEDIT** command, Properties window, and Grips.*
- *Hatch inserted blocks.*
- *Align hatch lines in adjacent hatch areas.*
- *Hatch by using the **HATCH** command at the Command prompt.*

HATCHING

In many drawings such as sections of solids, the sectioned area needs to be filled with some pattern. Different filling patterns make it possible to distinguish between different parts or components of an object. Also, the material the object is made of can be indicated by the filling pattern. You can also use these filling patterns in graphics for rendering architectural elevations of buildings, or indicating the different levels in terrain and contour maps. Filling objects with a pattern is known as hatching (Figure 13-1). This hatching process can be accomplished by using the **BHATCH** or **HATCH** commands.

Before using the **BHATCH** and **HATCH** commands, you need to understand some terms that are used in hatching. The following subsection describes some of the terms.

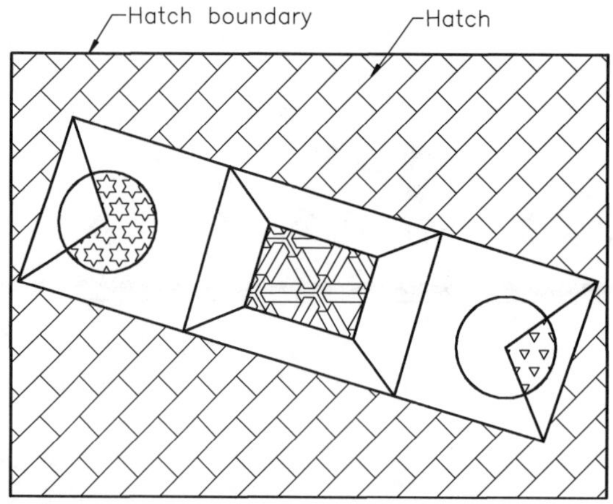

Figure 13-1 Illustration of hatching

Hatch Patterns

AutoCAD LT supports a variety of hatch patterns (Figure 13-2). Every hatch pattern is comprised of one or more hatch lines or a solid fill. The lines are placed at specified angles and spacing. You can change the angle and the spacing between the hatch lines. These lines may be broken into dots and dashes, or may be continuous, as required. The hatch pattern is trimmed or repeated, as required, to fill exactly the specified area. The lines comprising the hatch are drawn in the current drawing plane. The basic mechanism behind hatching is that the line objects of the pattern you have specified are generated and incorporated in the desired area in the drawing. Although a hatch can contain many lines, AutoCAD LT normally groups them together into an internally generated object and treats them as such for all practical purposes. For example, if you want to perform an editing operation such as erasing the hatch, all you need to do is select any point on the hatch and when you press ENTER, the entire pattern is deleted. If you want to break a pattern into individual lines to edit individual lines, you can use AutoCAD LT's **EXPLODE** command.

Figure 13-2 Some hatch patterns

Hatch Boundary

Hatching can be used on parts of a drawing enclosed by a boundary. This boundary may be lines, circles, arcs, polylines, or other objects, and at least part of each bounding object must be displayed within the active viewport. The **BHATCH** command automatically defines the boundary, whereas in the case of the **HATCH** command you have to define the boundary by selecting the objects that form the boundary of the hatch area.

BHATCH COMMAND

Toolbar:	Draw > Hatch
Menu:	Draw > Hatch
Command:	BHATCH

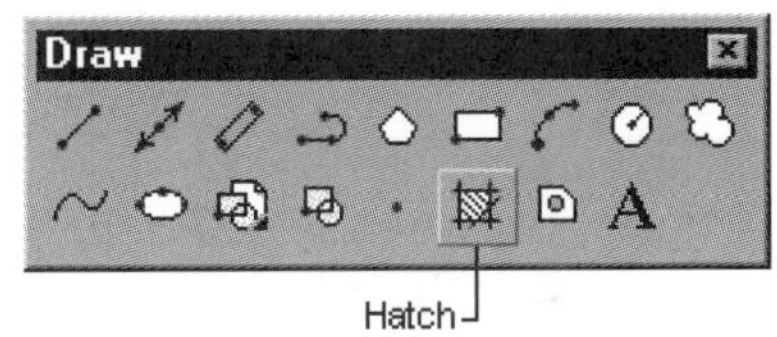

***Figure 13-3** Invoking the **BHATCH** command from the **Draw** toolbar*

The **BHATCH** (boundary hatch) command (Figure 13-3) allows you to hatch a region enclosed within a boundary (closed area) by selecting a point inside the boundary or by selecting the objects to be hatched. This command automatically designates a boundary and ignores any other object (whole or partial) that may not be a part of this boundary. When you invoke the **BHATCH** command, the **Boundary Hatch** dialog box is displayed through which you can perform the hatching operation (Figure 13-4).

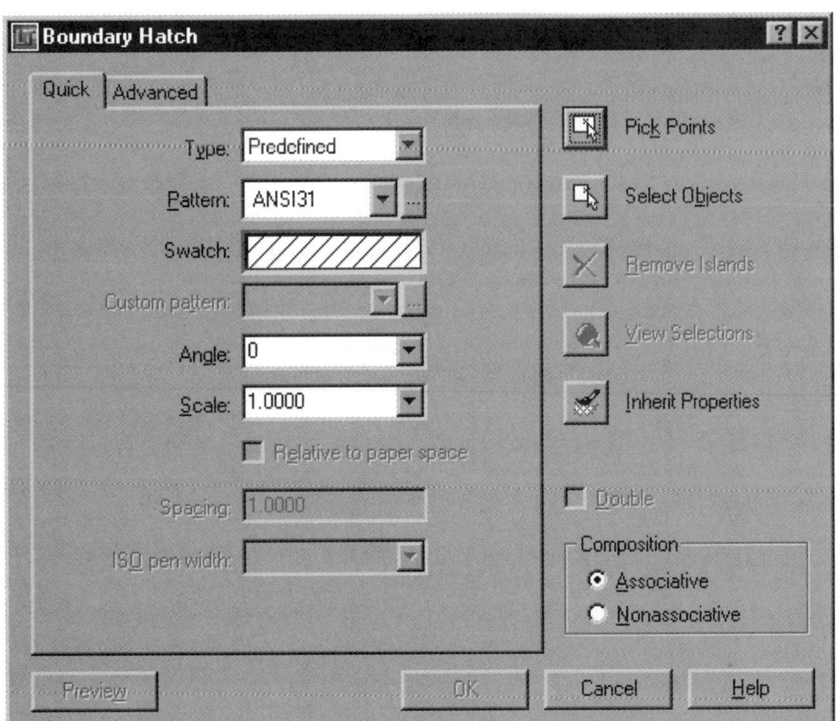

***Figure 13-4** **Boundary Hatch** dialog box*

One of the advantages of this command is that you do not have to select each object comprising the boundary of the area you want to hatch as in the case of the **HATCH** command (discussed later). This is because this command defines a boundary by creating a polyline from all the objects comprising the boundary. By default, this polyline gets deleted; however, if you want to retain it, you can specify that. This command also allows you to preview the hatch before actually applying it.

Example 1 *General*

In this example, you will hatch a circle using the default hatch settings. Later in the chapter you will learn how to change the settings to get a desired hatch pattern.

1. Invoke the **Boundary Hatch** dialog box by choosing the **Hatch** button on the **Draw** toolbar.

2. Choose the **Pick Points** button in the **Boundary Hatch** dialog box. The dialog box temporarily closes.

3. Select a point inside the circle (P1) (Figure 13-5) and press ENTER. The **Boundary Hatch** dialog box is displayed again. You can now preview the hatching by choosing the **Preview** button. After specifying an internal point, you can also right-click to display a shortcut menu and select the **Preview** option.

4. After you are satisfied with the preview, press ENTER or ESC to complete the point specification, and the dialog box reappears. You can also right-click to return to the dialog box.

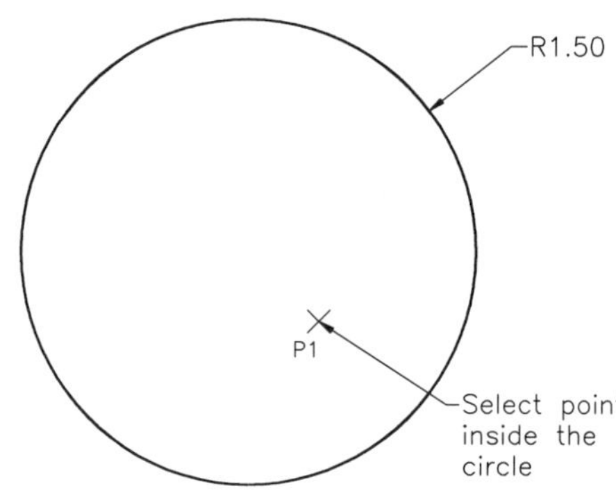

***Figure 13-5** Using the **BHATCH** command to hatch an object*

5. Choose **OK** to apply the hatch to the selected object.

Tip

*The **FILLMODE** system variable is set to On by default and therefore, the hatch patterns are displayed. If you have large hatching areas, you can set the **FILLMODE** to Off, so that the hatch pattern is not displayed and regeneration time is saved.*

BOUNDARY HATCH OPTIONS

The **Boundary Hatch** dialog box has several options that let you control various aspects of hatching, such as pattern type, scale, angle, boundary parameters, associativity, and so on. The following is a description of these options. The options have been grouped by the tab name in which they are found in the **Boundary Hatch** dialog box. The **Boundary Hatch** dialog box has two tabs. They are **Quick** and **Advanced**. The options under both these tabs are discussed next.

Quick Tab

The **Quick** tab (**Boundary Hatch** dialog box) provides options that define the appearance of the hatching patterns that can be applied (Figure 13-4). For quick hatching of an object this tab of the dialog box can be used easily.

Type

The **Type** drop-down list (Figure 13-6) displays the types of the patterns that can be used for hatching drawing objects. This list lets you choose the type of hatch pattern you want. The three types of hatch patterns available are: **Predefined**, **User defined**, and **Custom**. The Predefined type of patterns come with AutoCAD LT and are stored in the aclt.pat and acltiso.pat files. The predefined type of hatch pattern is the default type. When you select **User defined** from the **Type** drop-down list a pattern of lines which is based on the current linetype is created. You can set the angle of the lines in the pattern and also the spacing between them. When you select **Custom** from the **Type** drop-down list, you can select a pattern that has been defined in a custom PAT file.

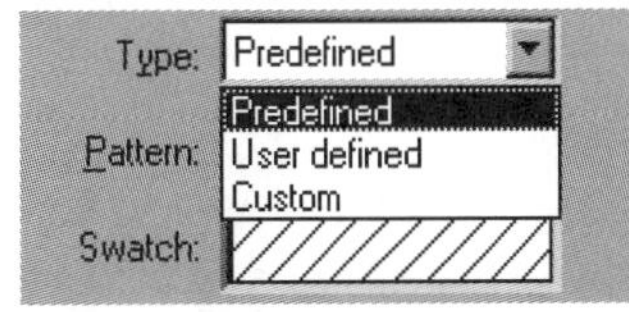

Figure 13-6 The ***Type*** *drop-down list displaying hatch pattern types*

Note

When using the patterns defined in custom PAT files, make sure that their search path has been added in the ***Files*** *tab of the* ***Options*** *dialog box.*

Pattern

The **Pattern** drop-down list (Figure 13-7) displays the names of all the predefined patterns available in AutoCAD LT. You can select any pattern from this list and the selected pattern is displayed in the **Swatch** box. The selected pattern is stored in the **HPNAME** system variable. ANSI31 is the default pattern in the **HPNAME** system variable. The **Pattern** drop-down list is available only if you have selected **Predefined** from the **Type** drop-down list. The six most recently used hatch patterns are displayed at the top of the **Pattern** drop-down list.

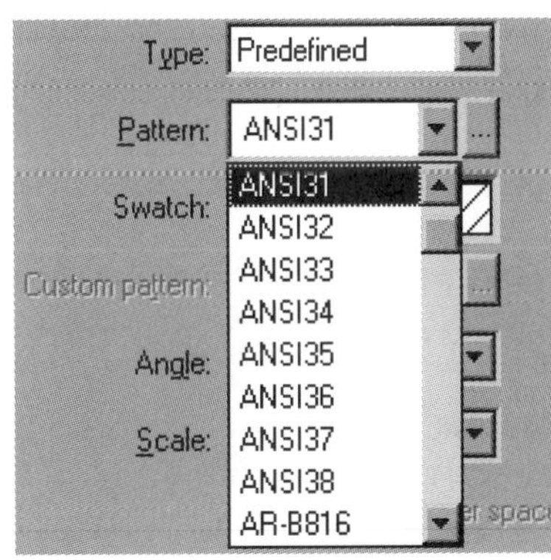

Figure 13-7 Pattern *drop-down list*

To select a particular hatch pattern, you can also choose the [...] button that displays the **Hatch Pattern Palette** dialog box (Figure 13-8). This dialog box shows the images and names of the available hatch patterns. This dialog box is also displayed by clicking on the pattern displayed in the **Swatch** box. The **Hatch Pattern Palette** dialog box has four tabs. They are described as follows.

ANSI. This tab displays all the ANSI (American National Standards Institute) defined hatch patterns that come with AutoCAD LT. ANSI31 is the default pattern.

ISO. This tab displays all the ISO (International Standard Organization) defined hatch patterns

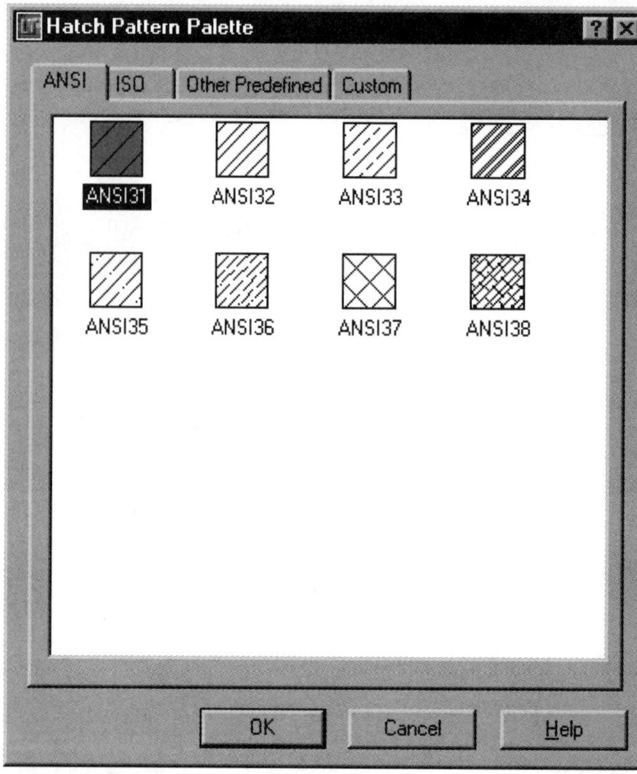

Figure 13-8 ***Hatch Pattern Palette*** *dialog box (**ANSI** tab)*

that come with AutoCAD LT.

Other Predefined. This displays all the predefined hatch patterns other than ANSI or ISO defined patterns.

Custom. This displays all custom patterns described in any custom PAT file that is added in the search path of AutoCAD LT 2002.

Tip

*Using the **Hatch Pattern Palette** dialog box for selecting a hatch pattern is more convenient since you can view all the pattern names and the corresponding patterns simultaneously.*

Swatch

This box displays the selected pattern. When you click in the **Swatch** box, the **Hatch Pattern Palette** dialog box is displayed.

Custom Pattern

This drop-down list is available only if you have selected **Custom** from the **Type** drop-down list. It displays all the available custom hatch patterns defined in the custom PAT files that have been added to the AutoCAD LT search path. The recently used custom hatch patterns in a current drawing session appear in the list. The selected pattern is stored in **HPNAME** system variable. You can also choose the **[...]** button, the **Hatch Pattern Palette** dialog box (**Custom** tab) appears. This tab displays the custom pattern paths.

Angle

The **Angle** drop-down list (Figure 13-9) displays angles (at an increment of 15 degrees) by which you can rotate the hatch pattern with respect to the X axis of the current UCS. You can select any angle from this drop-down list or you can also enter an angle of rotation of your choice in the Angle edit box. The angle value is stored in the **HPANG** system variable. The angle of hatch lines of a particular hatch pattern is governed by the values specified in the hatch definition. For example, in the **ANSI31** hatch pattern definition, the specified angle of hatch lines is 45 degrees. If you select an angle of 0, the angle of hatch lines will be 45 degrees. If you enter an angle of 45 degrees, the angle of the hatch lines will be 90 degrees.

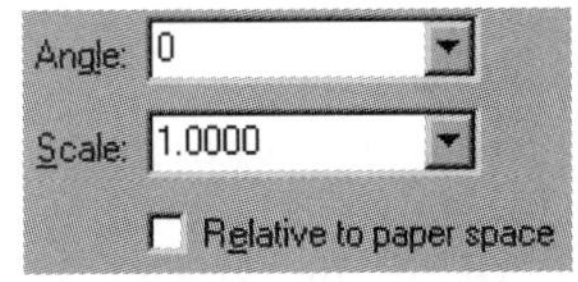

***Figure 13-9** The **Angle** and **Scale** drop-down lists*

Scale

The **Scale** drop-down list (Figure 13-9) displays scale factors by which you can expand or contract the selected hatch pattern. The scale factors that are displayed range between 0.25 to 2 at increments of 0.25. You can also enter the scale factor of your choice in the **Scale** edit box. The **Scale** drop-down list is not available when you are using user-defined hatch patterns. The scale value is stored in the **HPSCALE** system variable. A value of 1 does not mean that the distance between the hatch lines is 1 unit. The distance between the hatch lines and other parameters of a hatch pattern is governed by the values specified in the hatch definition. For example, in the ANSI31 hatch pattern definition, the specified distance between the hatch lines is 0.125. If you select a scale factor of 1, the distance between the lines will be 0.125. If you enter 0.5 scale factor, the distance between the hatch lines will be 0.5 x 0.125 = 0.0625.

Relative to Paper Space

If this check box (Figure 13-9) is selected then AutoCAD LT 2002 will automatically scale the hatched pattern relative to the paper space units. This option can be used to display the hatch pattern at a scale that is appropriate for your layout. This option is available only in a layout.

Spacing

The **Spacing** edit box is available only when you have selected **User defined** from the **Type** drop-down list. The value you enter in the **Spacing** edit box sets the spacing between the lines in a user-defined hatch pattern. This spacing is stored in the **HPSPACE** system variable. The default spacing value is 1.0000.

ISO pen width

The **ISO pen width** drop-down list is available only for ISO hatch patterns. You can select the desired pen width value from the **ISO pen width** drop-down list (Figure 13-10). The value selected specifies the ISO-related pattern scaling.

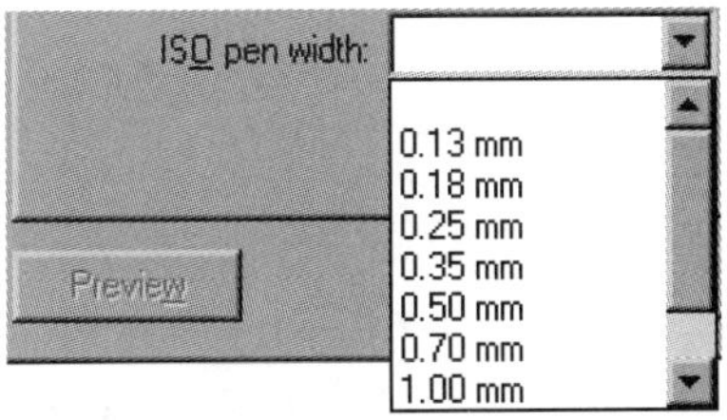

***Figure 13-10** The **ISO pen width** drop-down list*

User defined Hatch Patterns

To define a simple pattern, you can select **User defined** from the **Type** drop-down list. You will notice that the **Angle** and **Spacing** edit boxes, and the **Double** check box are available in the dialog box. The **Scale** drop-down list is not available.

Angle

Just as with predefined patterns, you can specify an angle with respect to the X axis of the current UCS through which the hatch pattern is rotated in the **Angle** edit box. This value is stored in the **HPANG** system variable.

Spacing

This edit box is available only for user-defined patterns and allows you to specify the space between the hatch lines in the pattern. The entered spacing value is stored in the **HPSPACE** system variable.

Double

This check box is available only for user-defined patterns. When you select this check box, AutoCAD LT doubles the original pattern by drawing a second set of lines at right angles to the original lines in the hatch pattern. For example, if you have a parallel set of lines as a user-defined pattern and if you select the **Double** check box, the resulting pattern has two sets of lines intersecting at 90 degrees. You can notice the effect of selecting the **Double** check box in Figure 13-11. If the **Double** box is selected, the **HPDOUBLE** system variable is set to 1.

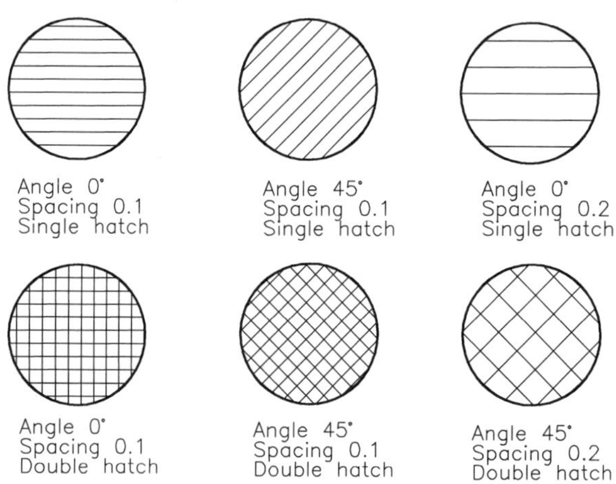

Figure 13-11 *Specifying angle and spacing for user-defined hatch patterns*

Custom Hatch Patterns

When you select **Custom** from the **Type** drop-down list in the Boundary hatch dialog box, you will notice that the **Custom pattern** drop-down list is available. The **Angle** and **Scale** drop-down lists are also available. See Figure 13-12.

Custom Pattern

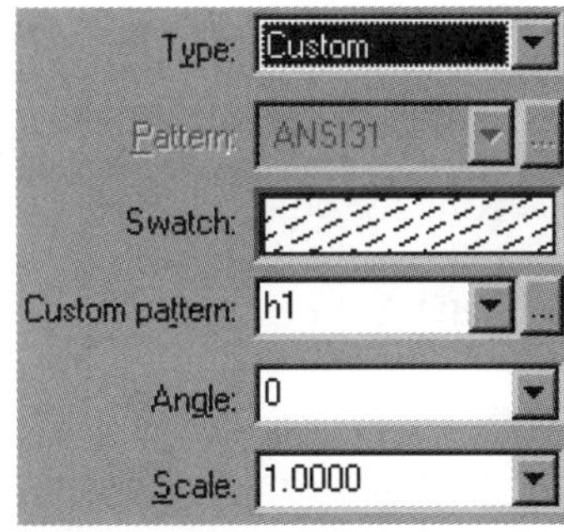

Figure 13-12 Selecting a Custom pattern

In AutoCAD LT, hatch patterns are normally stored in the file named **ACLT.PAT**. You can select a hatch pattern from an individual file by selecting **Custom** from the **Type** drop-down list (Figure 13-12). The **Custom pattern** drop-down list is available now. You can choose the [...] button to display the **Custom** tab of the **Hatch Pattern Palette** dialog box. Here, the list box on the left side of the dialog box displays the paths of the custom hatch patterns. When you select a path, the corresponding pattern is displayed in a preview window on the left. Choose **OK** to return to the **Boundary Hatch** dialog box. The corresponding pattern is displayed in the **Swatch** window. You can also select a pattern from the **Custom pattern** drop-down list, or enter the name of the custom pattern in the field if you know the name of the hatch pattern. The name is held in the **HPNAME** system variable. If AutoCAD LT does not locate the entered pattern in the **ACLT.PAT** file, it searches for it in a file with the same name as the pattern. You can specify an angle and scale for the custom pattern using the **Angle** and **Scale** drop-down lists as in the case of predefined patterns.

Tip

When you are hatching large areas, it is better to use a larger scale factor for the hatch pattern since you save time regenerating and plotting. Also, when using a larger scale factor for the hatch pattern, the drawing appears to be neater and not too cluttered.

Exercise 1 — *Mechanical*

In this exercise, you will hatch the given drawing using the hatch pattern named STEEL. Set the hatch scale and the angle to match the drawing shown in Figure 13-13.

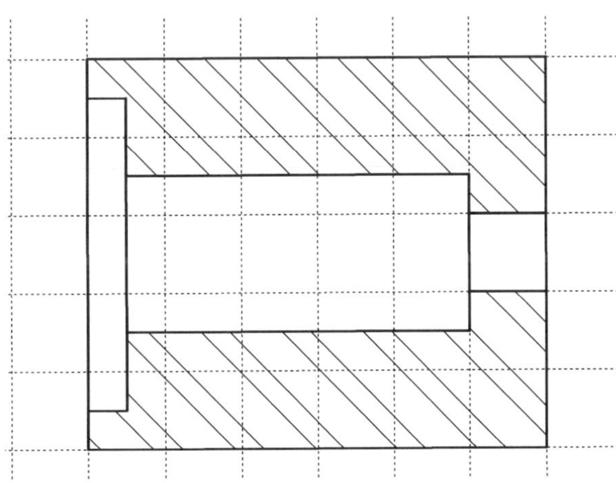

Figure 13-13 Drawing for Exercise 1

Chapter 13

Selecting Hatch Boundary

The boundary selection options available in the **Boundary Hatch** dialog box allow you to define the hatch boundary. This is done by selecting a point inside the area to be hatched or by selecting the objects. There are other options also available here for removing islands, viewing the selection, using properties from an existing hatched object, and controlling associativity of the hatch pattern.

Pick Points

When you choose this button, AutoCAD LT automatically constructs a boundary from the objects that form an enclosed area. This is the easiest method to specify a region for hatching. When you choose the **Pick Points** button, the dialog box disappears temporarily to allow you to select a point within the region to be hatched. Select a point inside the object and a boundary is defined around the selected point. The following prompts appear when you choose the **Pick Points** button in the **Boundary Hatch** dialog box.

Select internal point: *Select a point inside the object to hatch.*
Selecting everything visible...
Analyzing the selected data...
Analyzing internal islands...
Select internal point: *Select another internal point or press ENTER to end selection.*

Once the internal points have been selected, press ENTER and the dialog box reappears. Choose **OK** to apply the hatch pattern to the region. If you want to hatch an object and leave another object contained in it alone, select a point inside the object you want to hatch. By default, when using the **BHATCH** command, selecting a point creates multiple boundaries, a boundary of the internal object is also created, and the region between these two boundaries is selected for hatching. In Figure 13-14, the **Pick Points** button has been used to hatch the square but not the triangle simply by selecting the point inside the square, but outside the triangle (using the default settings of the dialog box).

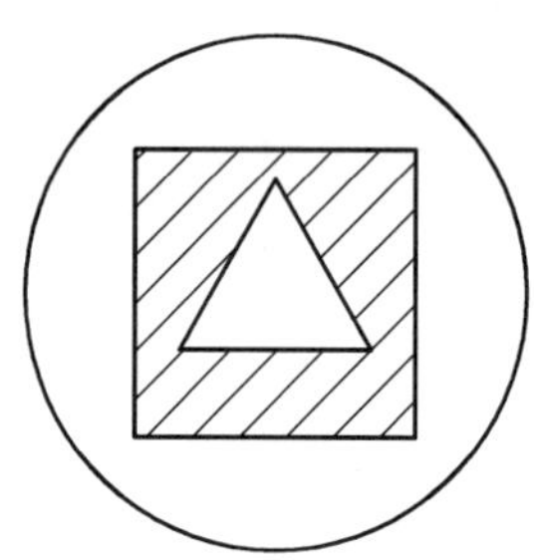

***Figure 13-14** Defining multiple hatch boundaries by selecting a point*

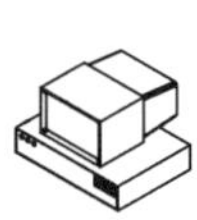

Tip

*When you have selected a point within an area to hatch and you realize you have selected the wrong area, you can enter **U** or **UNDO** at the Command prompt to undo the last selection. You can also undo a hatch pattern you have already applied to an area by entering **UNDO** at the Command prompt.*

Boundary Definition Error. This dialog box (Figure 13-15) is displayed if AutoCAD LT finds that the selected point is not inside the boundary or that the boundary is not closed. Choose **OK** and select another point inside the boundary or ensure that the boundary is closed.

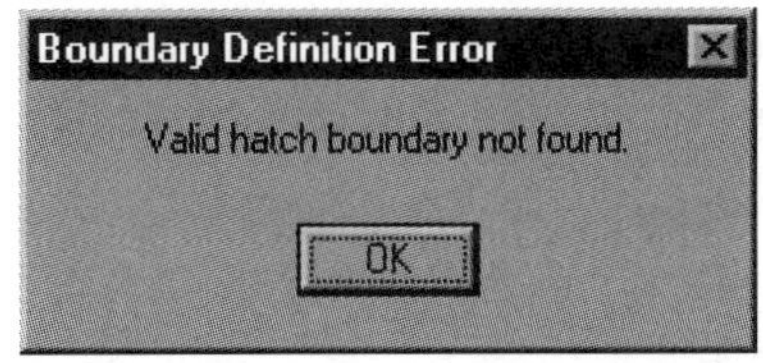

***Figure 13-15 Boundary Definition Error** dialog box*

Note

*When selecting internal points for hatching, you should select only one point for one region. If you select another internal point in the same hatch region, a **Boundary Definition Error** dialog box is displayed with the message: **Boundary duplicates an existing boundary**. Choose **OK** to close the dialog box.*

Select Objects

This option in the **Boundary Hatch** dialog box lets you select objects that form the boundary for hatching. It is useful when you have to hatch an object and disregard objects that lie inside it or intersect with it. When you select this option, AutoCAD LT will prompt you to select objects. You can select the objects individually or use other object selection methods. In Figure 13-16, the **Select Objects** button is used to select the triangle. This uses the triangle as the hatch boundary, and everything inside it. The text inside the triangle, in this case, also gets hatched. To avoid hatching of such internal objects, select them at the next **Select Objects** prompt. In the Figure 13-17, the text is selected to exclude it from hatching. While, when using the **Pick Points** option, the text automatically gets selected to be excluded from the hatching.

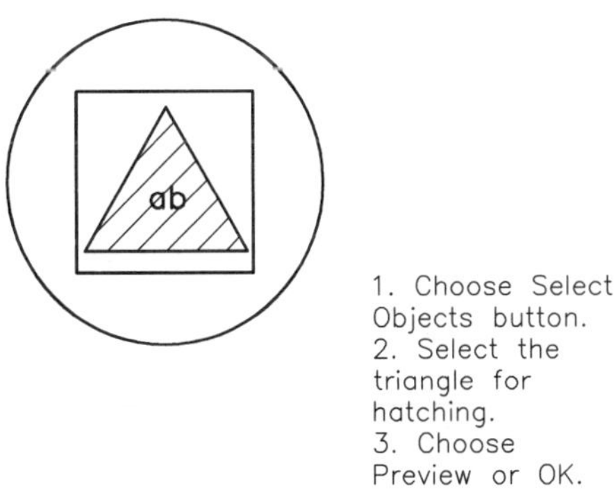

***Figure 13-16** Using the **Select Objects** button to specify the hatching boundary*

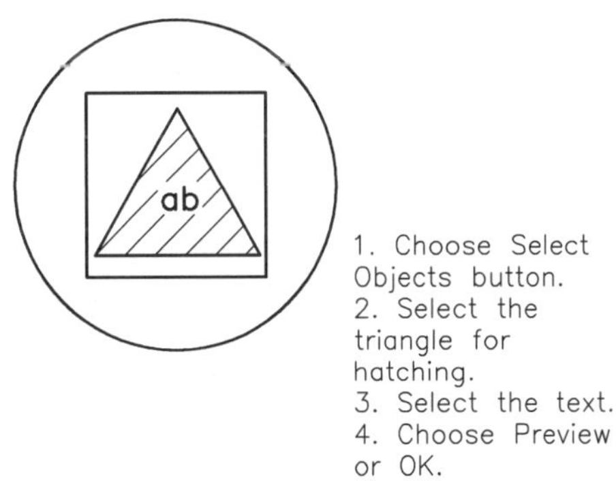

***Figure 13-17** Using the **Select Objects** button to exclude the text from hatching*

Tip

If you have an area to be hatched that has many objects intersecting with it, it is easier to select the entire object to be hatched rather than choosing internal points within each of the smaller regions created by the intersection.

You can also turn off layers that contain text or lines that make it difficult for you to select hatch boundaries.

Chapter 13

Remove Islands

This option is used to remove any islands from the hatching area. Boundaries inside another boundary are known as islands. If you choose the **Pick Points** button to hatch an area, islands are not hatched by default. But, if you want to hatch the islands, you can choose the **Remove Islands** button in the **Boundary Hatch** dialog box. This applies, for example, if you have a rectangle with circles within, as shown in Figure 13-18. You can use the **Pick Points** option to select a point inside the rectangle; AutoCAD LT will select both the circles and the rectangle. To remove the circles (islands), you can use the **Remove Islands** option. When you select this option, AutoCAD LT will prompt you to select the islands to remove. Select the islands to remove and press ENTER to return to the dialog box. Choose **OK** to apply the hatch.

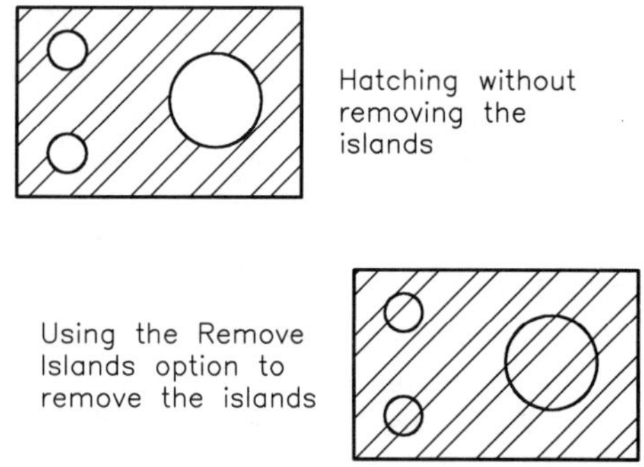

***Figure 13-18** Using the **Remove Islands** option to remove islands from the hatch area*

Tip

*It may be a good idea to use the **Select Objects** option to select the object containing islands if you want to remove the islands from the hatching area.*

View Selections

This option lets you view the selected boundaries before you apply the hatch pattern. This option is not available when you have not selected any points or boundaries. When you choose the **View selections** button, the dialog box closes temporarily and the selected boundaries in the drawing are displayed as highlighted.

Inherit Properties

This option hatches the specified boundaries using the hatch properties of an existing object. If you select a hatch object whose properties you want the current hatch to inherit, you can right-click in the drawing area to use the shortcut menu to toggle between the **Select Objects** and **Pick Internal Point** options to specify boundaries. When you choose the **Inherit Properties** button, the following command sequence appears:

Select associative hatch object: *Select the object from which you have to inherit the hatch pattern.*
Inherited Properties: Name <current>, Scale <current>, Angle <current>
Select Internal point: *Select a point inside the object to be hatched.*

When you press ENTER, the **Boundary Hatch** dialog box reappears with the name of the inherited hatch pattern in the **Pattern** edit box. The inherited hatch pattern is also displayed in the **Swatch** box. The selected pattern now becomes the current hatch pattern. If you then want to adjust the hatch properties, such as the angle or scale of the pattern, you can do so.

Composition

The **Composition** area of the dialog box provides options that control whether the hatch is associative or nonassociative with the hatch boundary. By default, the **Associative** radio button is selected, which implies that when you modify the boundary of a hatch object, the hatch patterns will be automatically updated to fill up the new area. If you select the **Nonassociative** radio button, the hatch pattern becomes independent of the boundary, which means that when you edit the boundary of a hatched object, the hatch pattern does not change with it. One of the major advantages with the associative hatch feature is that you can edit the hatch pattern or edit the geometry that is hatched without having to modify the associated pattern or boundary separately. After editing, AutoCAD LT automatically regenerates the hatch and the hatch geometry to reflect the changes. The hatch pattern can be edited by using the **HATCHEDIT** command, and the hatch geometry can be edited by using grips or AutoCAD LT editing commands.

Preview

You can use this option after you have selected the area to be hatched to see the hatching before it is actually applied. When you choose the **Preview** button, the dialog box is cleared and the object selected for hatching is temporarily filled with the specified hatch pattern. Previewing can also be done by right-clicking in the drawing area to display the shortcut menu, from where you can choose **Preview**. Pressing ENTER or right-clicking redisplays the **Boundary Hatch** dialog box so that the hatching operation can be completed and modifications made, if required. This option is available only after you have selected the area to be hatched.

OK

By choosing the **OK** button, you can apply hatching in the area specified by the hatch boundary. This option works whether or not you have previewed your hatch. This button is disabled if the hatch boundary is not defined.

Exercise 2 *Mechanical*

In this exercise, you will hatch the front view section of the drawing in Figure 13-19 using the hatch pattern for brass. Two views, top and front, are shown. In the top view, the cutting plane line indicates how the section is cut and the front view shows the full section view of the object. The section lines must be drawn only where the material is actually cut.

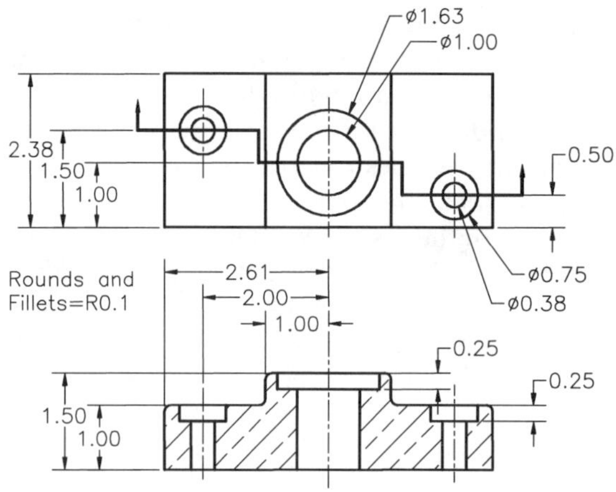

***Figure 13-19** Drawing for Exercise 2*

Advanced Tab (Boundary Hatch Dialog Box)

You can choose the **Advanced** tab of the **Boundary Hatch** dialog box (Figure 13-20) to specify options that improve the hatching speed. Defining a boundary by specifying an internal point is quite simple in the case of small and less complicated drawings. It may take more time in the case of large, complicated drawings because AutoCAD LT examines everything that is visible

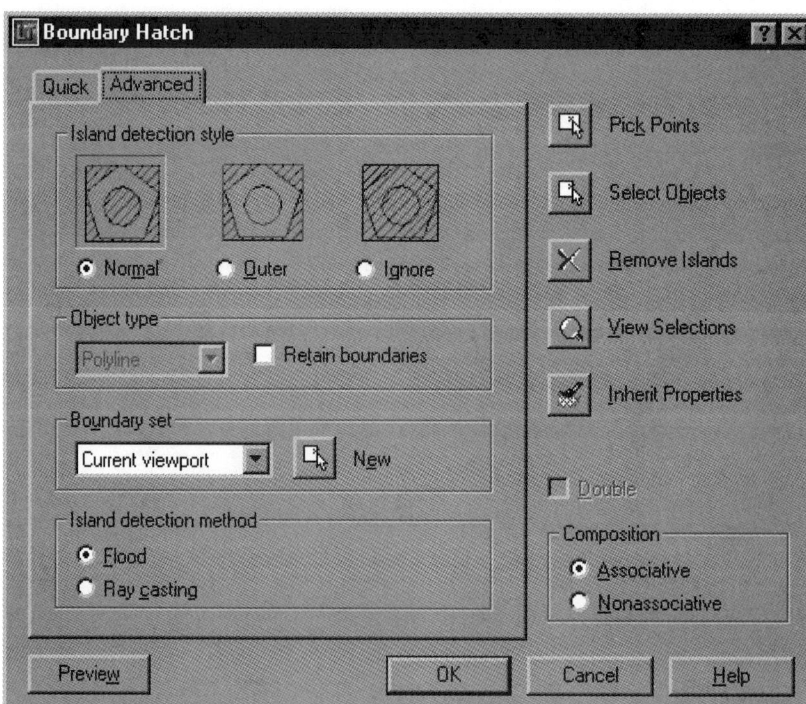

***Figure 13-20 Advanced** tab of the **Boundary Hatch** dialog box*

in the current viewport. Thus, the larger and more complicated the drawing, the more time it takes to locate the boundary edges. In such cases, setting parameters in the **Advanced** tab of the **Boundary Hatch** dialog box improves the speed of hatching.

Island detection style

The **Island detection style** area provides options to select the style of island detection during hatching (Figure 13-21). There are three styles available: **Normal**, **Outer**, and **Ignore**. To select a particular style, you can select the corresponding radio button. The effect of using the particular style is displayed in the form of an illustration in the image tile placed above the particular radio button. You can also choose the image tiles to select a particular island detection style. The results of using each of these styles is discussed next (Figure 13-22).

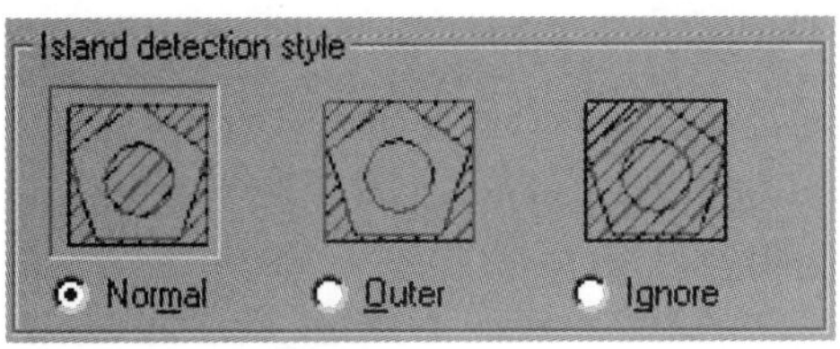

Figure 13-21 Island detection styles

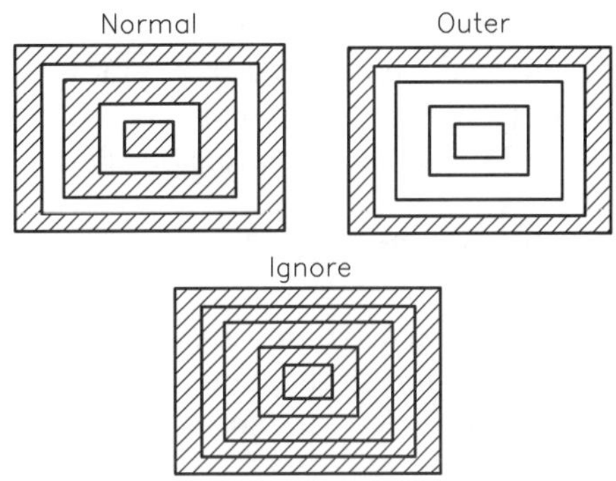

Figure 13-22 Using hatching styles

Note

The selection of an island detection style carries meaning only if the objects to be hatched are nested (that is, one or more selected boundaries is within another boundary).

Normal. The Normal style is selected by default. This style hatches inward starting at the outermost boundary. If it encounters an internal boundary, it turns off the hatching. An internal boundary causes the hatching to turn off until another boundary is encountered. In this manner, alternate areas of the selected object are hatched, starting with the outermost area. Thus, areas separated from the outside of the hatched area by an odd number of boundaries are hatched, while those separated by an even number of boundaries are not.

Outer. This particular option also lets you hatch inward from an outermost boundary, but the hatching is turned off if an internal boundary is encountered. Unlike the previous case, it does not turn the hatching on again. The hatching process in this case starts from both ends of each hatch line; only the outermost level of the structure is hatched, therefore, the name **Outer**.

Ignore. In this option, all areas bounded by the outermost boundary are hatched. The option ignores any hatch boundaries that are within the outer boundary. All islands are completely ignored and everything within the selected boundary is hatched.

Tip

*It is also possible to set the pattern and the island detection style at the same time by using the **HPNAME** system variable. AutoCAD LT stores Normal style code by adding **N** to the pattern name. Similarly, **O** is added for Outer style, and **I** for the Ignore style to the value of the*

*HPNAME system variable. For example, if you want to apply the BOX pattern using the outer style of island detection, you should enter the **HPNAME** value to be **BOX, O**. Now, when you apply the hatch pattern, the BOX pattern is applied using the outer style of hatching.*

Note

*The **Normal**, **Outer**, and **Ignore** options are also available in the shortcut menu when you select an internal point in the drawing object and right-click in the drawing area. The shortcut menu shall be discussed later.*

Object type Area

When you select an internal point in a region to be hatched, a boundary is created around it. By default, these boundaries are removed as soon as the hatch pattern is applied. The **Object type** area provides options that allow you to specify whether these boundaries are to be retained as objects or not. You can also specify the type of object it can be saved as.

Retain boundaries. The **Retain boundaries** check box is cleared by default, which implies that the hatch boundaries are not saved. If you select this check box, you can retain the defined boundary. When a hatching is successful and you want to keep the boundary as a polyline or region so that you can use it again, you can select the **Retain boundaries** check box. When you select this check box, the **Object Type** drop-down list is available.

Object type drop-down list. The **Object type** drop-down list is available when you select the **Retain boundaries** check box. From this list you can select the type of object AutoCAD LT will create when you define a boundary for hatching. It has two options: **Polyline** and **Region**. **Polyline** is selected by default and the boundary created around the hatch area is a polyline. Similarly, when you select **Region**, the boundary of the hatch area is a region. Regions are two-dimensional areas that can be created from closed shapes or loops.

Boundary Set

When hatching takes place normally, AutoCAD LT evaluates the entire drawing visible on the screen to determine the boundary. For larger drawings this may take a lot of time and therefore, slow down the process of hatching. The **Boundary set** area provides options that allow you to specify what is considered when hatching. You can either select an option from the drop-down list or you can create a new boundary set. The boundary set comprises the objects that the **BHATCH** command uses when constructing the boundary. The default boundary set is **Current viewport**, which comprises everything that is visible in the current viewport. Boundary hatching is made faster by specifying a boundary set because, in this case, AutoCAD LT does not have to examine everything on screen.

New. This option allows you to define a boundary area so that only a specific portion of the drawing is considered for hatching. You use this option to create a new boundary set. When you choose the **New** button, the dialog box is cleared from the screen temporarily to allow selection of objects to be included in the new boundary set. While constructing the boundary set, AutoCAD LT uses only those objects that you select and that are hatchable. If a boundary set already exists, it is replaced by the new one. If you do not select any hatchable objects, no new boundary set is created. AutoCAD LT retains the current set if there is one. Once you have selected the objects to form the boundary set, press ENTER and the dialog box reappears on the screen. You will notice that **Existing Set** gets added to the drop-down list. When you invoke the **BHATCH** command and you have not formed a boundary set, there is only one option, **Current viewport**, available in the drop-down list. The benefit of creating a selection set is that when you select a point or select the objects to define the hatch boundary, AutoCAD LT will search only for the objects that are in the selection set. By confining the search to the objects in the selection set, the hatching process is faster. If you select an object that is not a part of the selection set, AutoCAD LT ignores them. When a boundary set is formed, it becomes the default for hatching until you exit the **BHATCH** command or select the **Current viewport** option from the drop-down list.

Tip

To improve hatching speed in large drawings, you should zoom into the area to be hatched, so that defining the boundary to be hatched is easier. Also, since AutoCAD LT does not have to search the entire drawing to find hatch boundaries, the hatch process gets faster.

Island Detection Method

The **Island detection method** area of the **Boundary Hatch** dialog box allows you to select a method of island detection. The two methods available are: **Flood** and **Ray Casting**. These methods determine whether to include the objects that lie within the outermost boundary, referred to as islands, as boundary objects or not.

Flood. This method of defining the hatch boundary includes the islands as boundary objects. This is the default method for island detection.

Ray casting. This method casts rays in all directions or a particular direction from the point specified to the nearest object and then traces the boundary in a counterclockwise direction, thus excluding islands as boundary objects. Ray casting is discussed next.

RAY CASTING

You can use the **ray casting** technique to define a hatch boundary. In ray casting, AutoCAD LT casts an imaginary ray in all directions or a particular specified direction and locates the object that is nearest to the point you have selected. After locating the object, it takes a left turn to trace the boundary and the boundary gets highlighted. If the point you have specified lies within the highlighted boundary, the islands within are ignored and the entire object is hatched. But, if the point lies outside the highlighted boundary, an error message is displayed and you are asked to select another point. Ray casting has five options. They are: **Nearest**, **+X**, **-X**, **+Y**, and **-Y**. All these options are available when you use the **-BHATCH** command. But when you are using the **BHATCH** command, only the **Nearest** option can be used.

Using the BHATCH Command

When you invoke the **BHATCH** command, the **Boundary Hatch** dialog box is displayed. Choose the **Advanced** tab in the dialog box and select the **Ray casting** radio button. Choose the **Pick points** button. The dialog box closes temporarily to allow you to specify a point for ray casting. In Figure 13-23, depending upon the location of the specified point, a boundary object is selected. If you select a point close to the outer circle, the circle gets highlighted as the hatching boundary. This is because imaginary rays are cast from the specified point and the circle is the nearest closed boundary available. When you press ENTER and the dialog box returns, choose the **Preview** button to see the preview of the hatching, you will notice that the entire circle is hatched and the inner objects (rectangle and triangle) have been ignored (Figure 13-24).

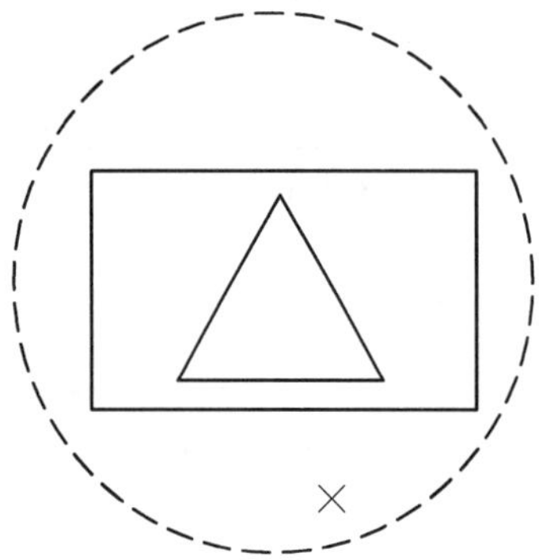

Figure 13-23 *Specifying a point closer to the outer circle during hatching using ray casting*

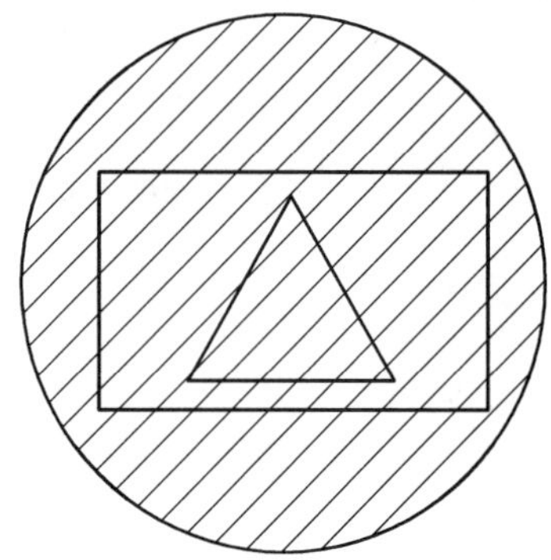

Figure 13-24 *The result of using the* ***Nearest*** *ray casting option*

Now, for the next time, select a point inside the circle but closer to the inner rectangle. The inner rectangle gets highlighted as it is nearest to the specified point and a **Boundary Definition Error** dialog box (Figure 13-25) is displayed since the specified point is not within the highlighted boundary. The **Boundary Definition Error** dialog box has two buttons. They are: **OK** and **Look at it**. If you choose **OK**, the dialog box is closed and you can select another point. If you choose **Look at it**, AutoCAD LT displays the highlighted boundary and the position of the specified point is shown marked with a (X) marker. An **AutoCAD LT Message** box is displayed with the message: **Press ENTER to continue**. You can press ENTER or choose **OK** to continue and specify another point.

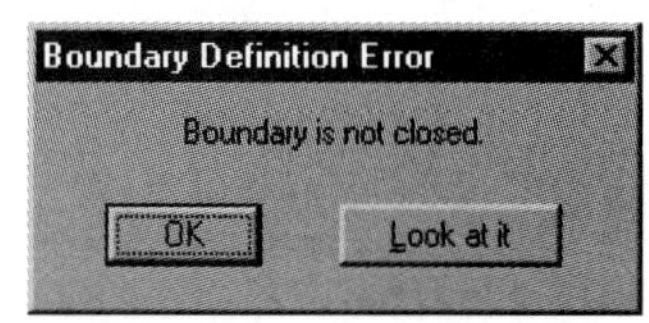

Figure 13-25 *The* ***Boundary Definition Error*** *dialog box*

Using the -BHATCH Command

You can also create hatch patterns from the command line using the **-BHATCH** command. The **-BHATCH** command is especially useful when you want to avoid boundary definition errors, by forcing the ray casting in a particular direction, or want to use other ray casting options that are not available through the dialog box. These options control the direction in which AutoCAD LT casts the ray to form a hatch boundary. When using the **-BHATCH** command, the dialog box is bypassed and the available options are **Nearest**, **+X**, **-X**, **+Y**, and **-Y**. The prompt sequence is:

Command: **-BHATCH**
Current hatch Pattern: ANSI31
Specify internal point or [Properties/Select/Remove islands/Advanced]: **A**
Enter an option [Boundary set/Retain boundary/Island detection/Style/Associativity]: **I**
Do you want island detection? [Yes/No] <Y>: **N**
Enter type of ray casting [Nearest/+X/-X/+Y/-Y/Angle] <Nearest>: *Specify an option.*

Enter an option [Boundary set/Retain boundary/Island detection/Style/Associativity]: Enter
Current hatch pattern: ANSI31
Specify internal point or [Properties/Select/Remove islands/Advanced]: *Select a point.*

Nearest. The **Nearest** option is the default option. When this option is used, AutoCAD LT casts an imaginary ray from the specified point to the nearest hatchable object, then takes a turn to the left and continues the tracing process in an effort to form a boundary around the internal point. To make this process work properly, the point you select inside the boundary should be closer to the boundary than any other object that is visible on the screen.

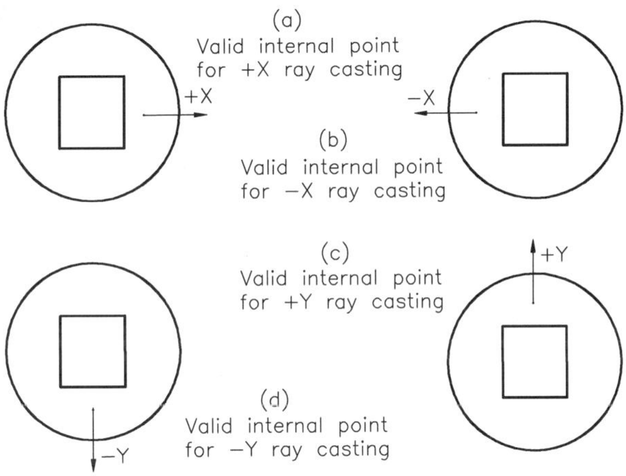

Figure 13-26 *Valid points for casting a ray in different directions to select the circle as boundary*

+X, -X, +Y, and -Y options. These options can be explained better by studying Figure 13-26. Figure 13-26(a) shows that the points that can be selected are those that are nearest to the right edge of the circumference of the circle because the ray casting takes place in the direction of the positive X. If you select a point on the left of the rectangle that is placed inside the circle, no valid boundary is detected. This is because when you have selected this point, rays are cast horizontally from it in the positive X direction and the rectangle is encountered which is invalid, since this point is not within the highlighted boundary.

Figure 13-26(b) shows that the points that can be used for selection are those that are nearest to the left edge of the circle because the ray casting takes place in the negative X direction. Any point selected on the right of the rectangle casts rays towards the negative X direction and encounters the rectangle, which is again an invalid boundary for the selected point.

In Figure 13-26(c), the ray casting takes place in the positive Y direction, so the points that can be used for selection are the ones closest to the upper edge of the circle. If a point is selected below the rectangle which is placed within the circle, rays cast in the positive direction and a rectangle is encountered, which is an invalid boundary for the selected point.

In Figure 13-26(d), the ray casting takes place in the negative Y direction, so the internal points that can be selected are the ones closest to the lower edge of the circle. When a point is selected above the rectangle, rays are cast in the negative Y direction and the rectangle is encountered, which is an invalid boundary for the selected point.

The valid areas for selecting points for the **+X**, **-X**, **+Y**, and **-Y** options of ray casting are shown in Figure 13-27 and Figure 13-28 as shaded. Points selected within the unshaded portions result in boundary errors.

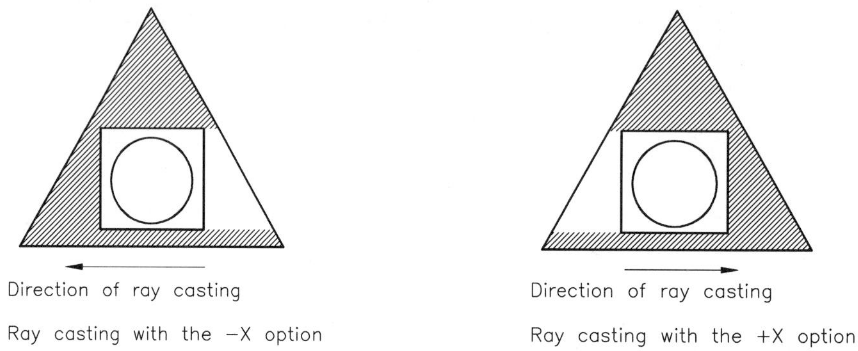

***Figure 13-27** Valid areas (shaded) for selecting points for the **-X** and **+X** options of ray casting*

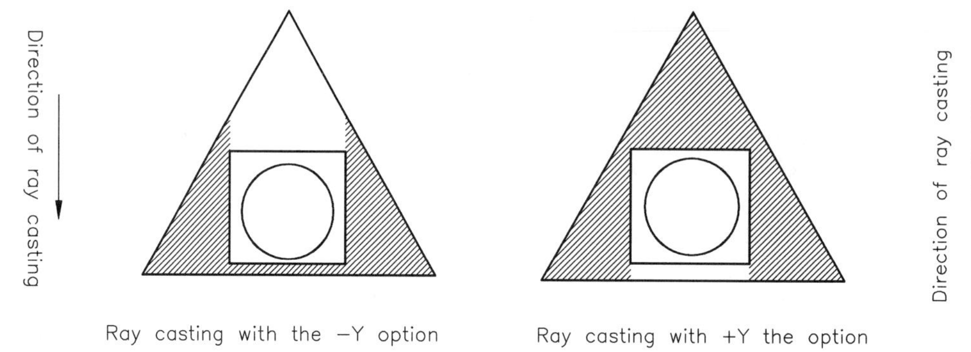

***Figure 13-28** Valid areas (shaded) for selecting points for the **-Y** and **+Y** options of ray casting*

The BHATCH Shortcut Menu

After invoking the **BHATCH** command and when you have selected a point that lies in the interior of the object to be hatched, or the object to be hatched itself, you can right-click to display the shortcut menu (Figure 13-29). This shortcut menu displays options that are also available through the **Boundary Hatch** dialog box. If you choose **Preview**, you can preview the hatching before applying it. If you choose **Enter**, the dialog box reappears. Choosing **Clear All** removes all the previous selections made. You can also select another island detection mode and also change from one method of hatching to another.

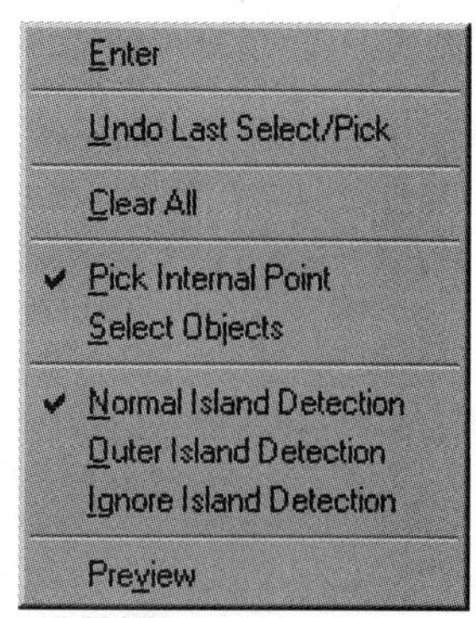

***Figure 13-29** The **BHATCH** shortcut menu*

Tip

Using the shortcut menu is more convenient and saves time.

HATCHING AROUND TEXT, ATTRIBUTES, SHAPES, AND SOLIDS

When you select a point within a boundary to be hatched and if it contains text, by default, the hatch lines do not pass through the text, attributes, or shapes present in the object being hatched. AutoCAD LT places an imaginary box around these objects that does not allow the hatch lines to pass through it. If you are using the select objects option to select objects to hatch, you must select the text/attribute/shape along with the object in which it is placed when defining the hatch boundary. If multiple line text is to be written, the **MTEXT** command is used. You can also select both the boundary and the text when using the window selection method. Figure 13-30 shows you how hatching takes place around text and a polyline with width.

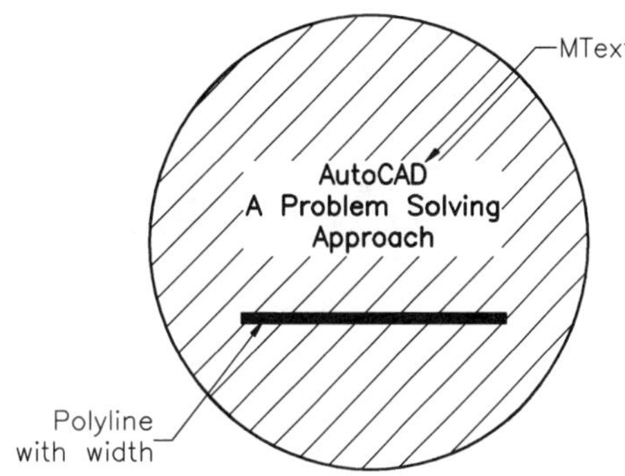

Figure 13-30 *Hatching around text and polyline*

Note

In case of polylines with width, the hatching stops exactly at the outline of the polyline unlike in the case of text where the box formed around the text allows a small gap between the text and the hatch pattern. This gap makes the text easily readable.

EDITING HATCH PATTERNS

Using the HATCHEDIT Command

Toolbar:	Modify II > Edit Hatch
Menu:	Modify > Object > Hatch
Command:	HATCHEDIT

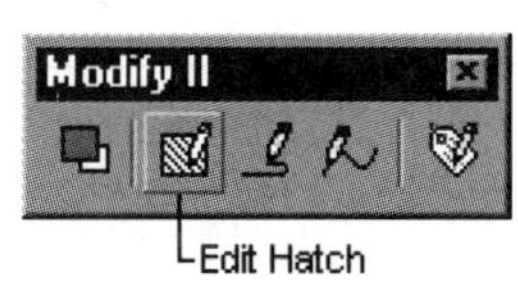

Figure 13-31 *Choosing* ***Edit Hatch*** *from the* ***Modify II*** *toolbar*

The **HATCHEDIT** command (Figure 13-31) can be used to edit the hatch pattern. When you invoke this command and select the hatch for editing, the **Hatch Edit** dialog box (Figure 13-32) is displayed on the screen. The **Hatch Edit** dialog box can also be invoked from the shortcut menu or by double-clicking on the hatch pattern. After selecting a hatched object, right-click in the drawing area, and choose **Hatch Edit** from the shortcut menu. The command sequence is as follows:

Command: **HATCHEDIT**
Select associative hatch objects: *Select a hatch object to edit.*

AutoCAD LT will display the **Hatch Edit** dialog box, where the hatch pattern can be changed or modified accordingly (Figure 13-32). This dialog box is the same as the **Boundary Hatch** dialog box except that only the options that control the hatch pattern are available. These available options work in the same way as they do in the **Boundary Hatch** dialog box.

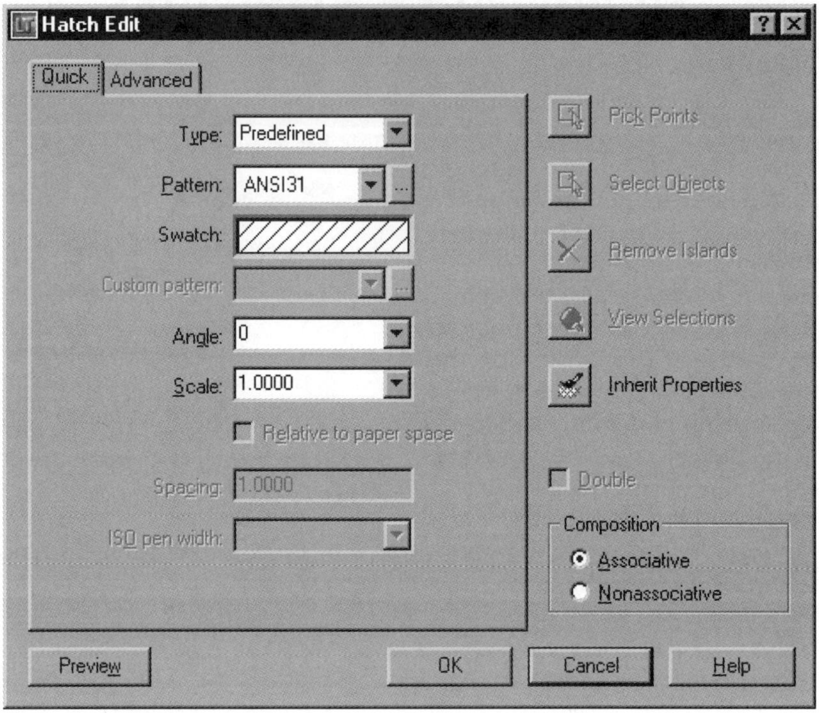

***Figure 13-32** The **Quick** tab of the **Hatch Edit** dialog box*

The **Hatch Edit** dialog box has two tabs just like the **Boundary Hatch** dialog box. They are **Quick** and **Advanced**. In the **Quick** tab, you can redefine the type of hatch pattern by selecting another type from the **Type** drop-down list. If you are using **Predefined** pattern, you can select a new hatch pattern name from the **Pattern** drop-down list. You can also change the scale or angle by entering new values in the **Scale** or **Angle** edit boxes. When using **User defined** pattern, you can redefine the spacing between the lines in the hatch pattern by entering a new value in the **Spacing** edit box. If you are using **Custom** pattern type, you can select another pattern from the **Custom pattern** drop-down list. You can also redefine the island detection style by selecting either of the **Normal**, **Outer**, or **Ignore** styles in the **Advanced** tab of the dialog box. If you want to copy the properties from an existing hatch pattern, choose the **Inherit Properties** button, and then select the hatch whose properties you want to inherit. If the **Associative** radio button in the **Composition** area of the dialog box is selected, the pattern is associative. This implies that whenever you modify the hatch boundary, the hatch pattern is automatically updated.

Note

If a hatch pattern is associative, the hatch boundary can be edited using grips and editing commands and the associated pattern is modified accordingly. This is discussed later.

In Figure 13-33, the object is hatched using the ANSI31 hatch pattern. With the **HATCHEDIT** command, you can edit the hatch using the **Hatch Edit** dialog box to change the hatch to the one shown in Figure 13-34. You can also edit an existing hatch through the command line by entering **-HATCHEDIT** at the Command prompt.

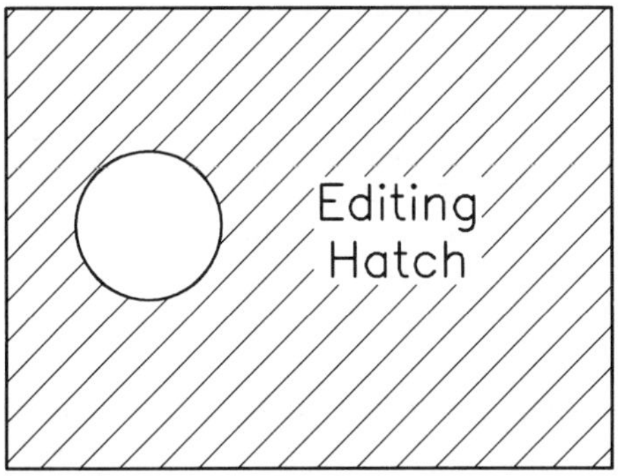

***Figure 13-33 ANSI31** hatch pattern*

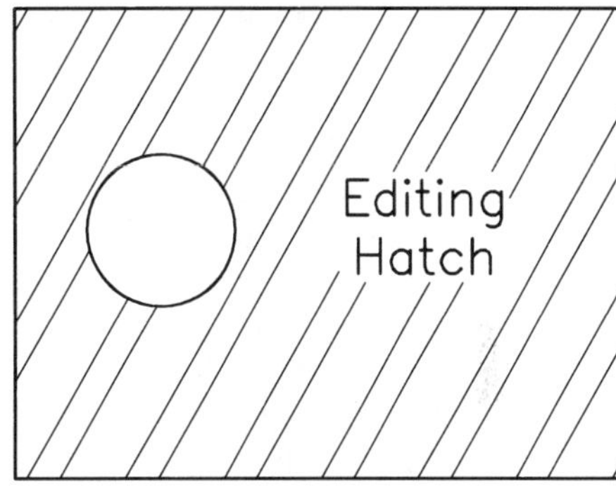

***Figure 13-34** Using the **HATCHEDIT** command to edit the hatch pattern*

Using the PROPERTIES Command

Toolbar:	Standard Toolbar > Properties
Menu:	Modify > Tools > Properties
Command:	PROPERTIES

You can also use the **PROPERTIES** command to edit the hatch pattern. When you select a hatch pattern for editing, and invoke the **PROPERTIES** command, AutoCAD LT displays the **Properties** window (Hatch) (Figure 13-35). You can also invoke the **Properties** window by selecting the hatch pattern and right-clicking to display a shortcut menu. Choose **Properties** here, the **Properties** window is displayed. This window displays the properties of the selected pattern under the following categories:

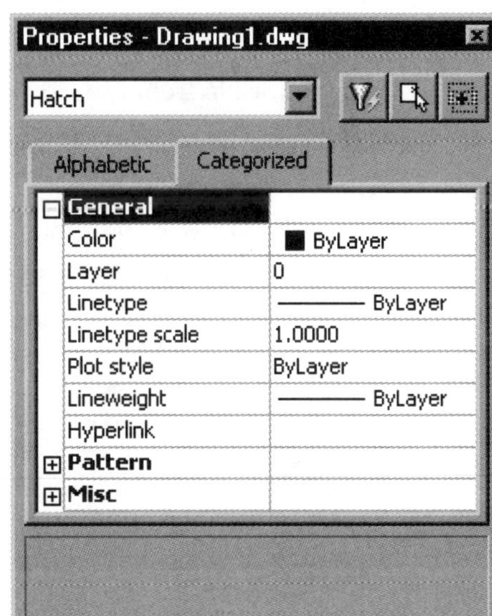

***Figure 13-35 Properties** window (Hatch)*

General

In this category, you can change the general pattern properties like Color, Layer, Linetype, Linetype scale, Lineweight, and so on. When you click on the field corresponding to a particular property, a drop-down list is displayed from where you can select an option or value. Whenever a drop-down list does not get displayed, you can enter a value in the field itself.

Pattern

In this category you can change the Pattern Type, Pattern name, Angle, and Scale of the pattern. In the case of ISO patterns, you can redefine the ISO pen width of the pattern and you can also modify the spacing of **User defined** patterns. You can also determine whether you want to have a **User defined** pattern as double or not. When you choose the [...] button in the **Type** property field, AutoCAD LT displays the **Hatch Pattern Type** dialog box (Figure 13-36). Here, you can select the type of hatch pattern from the **Pattern Type** drop-down list. If you have selected a **Predefined** pattern type, you can select a pattern name from the **Pattern** drop-down list or choose the **Pattern** button to display the **Hatch Pattern Palette** dialog box. You can then select a pattern here in one of the tabs and then choose **OK** to exit the dialog box. Now, choose **OK** in the **Hatch Pattern Type** dialog box to return to the **Properties** window. You will notice that the pattern name you have selected is displayed in the **Pattern** name field. Similarly, if you select the **Custom** pattern type, you can enter the name of the custom pattern in the **Custom pattern** edit box. If you select a **User defined** pattern type, both the **Pattern** drop-down list and the **Custom** edit box are not available. When you click in the **Pattern name** property field, the [...] button is displayed. When you choose the [...] button, the **Hatch Pattern Palette** dialog box is displayed with the default pattern ANSI31 selected. The values of the angle, scale, and spacing properties can be entered in the corresponding fields. A value for the ISO pen width can be selected from the **ISO pen width** drop-down list. You can select **Yes** or **No** from the **Double** drop-down list to determine if you want the hatch pattern to double at right angles to the original pattern or not.

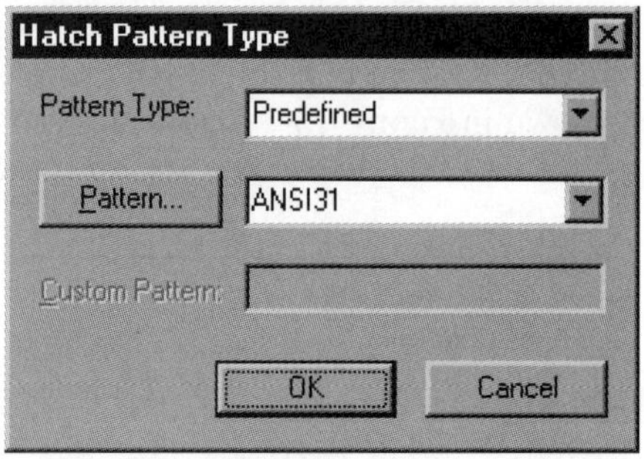

***Figure 13-36 Hatch Pattern Type** dialog box*

Tip

*It is more convenient to choose the [...] button in the **Pattern Type** edit box to display the **Hatch Pattern Type** dialog box and enter both the type and name of the pattern at the same time.*

Misc

In this category, the elevation, associativity, and island detection style properties of the hatch pattern can be changed. A new value of elevation for the hatch pattern can be entered in the **Elevation** field. You can select **Yes** or **No** from the **Associative** drop-down list to determine if a pattern is associative or not. Similarly, you can select an island detection style from the **Island detection style** drop-down list.

EDITING HATCH BOUNDARY

Using Grips

One of the ways you can edit the hatch boundary is by using grips. You can select the hatch pattern or the hatch boundaries. If you select the hatch pattern, the hatch highlights and a grip is displayed at the centroid of the hatch. A centroid for a region that is coplanar with the XY plane is the center of that particular area. However, if you select an object that defines the hatch boundary, the object grips are displayed at the defining points of the selected object. Once you change the boundary definition, and if the hatch pattern is associative, AutoCAD LT will reevaluate the hatch boundary and then hatch the new area. When you edit the hatch boundary, make sure that there are no open spaces in the hatch boundary. AutoCAD LT will not create a hatch if the outer boundary is not closed. Figure 13-37 shows the result of moving the circle and text, and shortening the bottom edge of the hatch boundary of the object in Figure 13-34. The objects were edited by using grips and since the pattern is associative, when modifications are made to the boundary object, the pattern automatically fills up the new area. If you select the hatch pattern, AutoCAD LT will display the hatch object grip at the centroid of the hatch. You can select the grip to make it hot and then edit the hatch object. You can stretch, move, scale, mirror, or rotate the hatch pattern. Once an editing operation takes place on the hatch pattern, it may be moved, rotated, or scaled; then, the associativity is lost and AutoCAD LT displays the message: "**Hatch boundary associativity removed**."

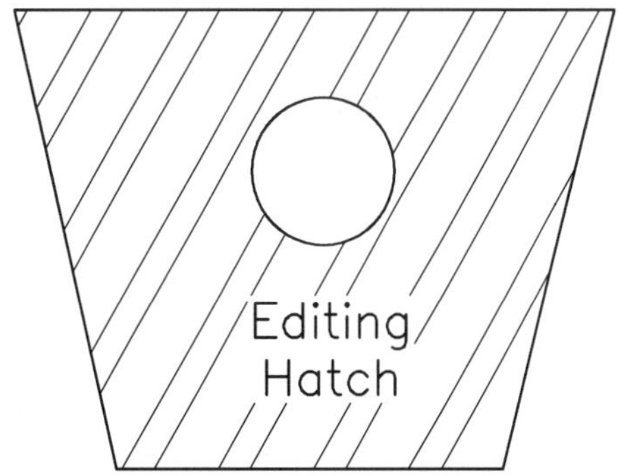

Figure 13-37 *The result of using grips to edit the hatch boundary of the object shown in Figure 13-34*

The **Retain Boundaries** check box in the **Boundary Hatch** dialog box is cleared when a hatch is created by default. Therefore, all objects selected or found by the selection or point selection processes during the hatch creation are associated with the hatch as boundary objects. Editing any of them may affect the hatch. If, however, the **Retain Boundaries** check box was selected when the hatch was created, only the retained polyline boundary created by **BHATCH** is associated with the hatch as its boundary object. Editing it will affect the hatch, and editing the objects selected or found by selection or point selection processes during the hatch created will have no effect on the hatch. You can, of course, select and simultaneously edit the objects selected or found by the selection or point selection processes, as well as the retained polyline boundary created by the **BHATCH** command.

Using AutoCAD LT Editing Commands

When you use the editing commands such as **MOVE**, **ROTATE**, **SCALE**, and **STRETCH**, associativity is maintained, provided all objects that define the boundary are selected for editing. If any of the boundary-defining objects is missing, the associativity will be lost and AutoCAD LT will display the message **Hatch boundary associativity removed**. When you rotate or scale an associative hatch, the new rotation angle and the scale factor are saved with the hatch object data. This data is then used to update the hatch. When using the **ARRAY**, **COPY**, and **MIRROR** commands, you can array, copy, or mirror just the hatch boundary, without the hatch pattern. Similarly, you can erase just the hatch boundary, using the **ERASE** command and associativity is removed. If you explode an associative hatch pattern, the associativity between the hatch pattern and the defining boundary is removed. Also, when the hatch object is exploded, each line in the hatch pattern becomes a separate object.

When the original boundary objects are being edited, the associated hatch gets updated, but new boundary objects do not have any effect. For example, when you have a square island within a circular boundary and then you erase the square, the hatch pattern is updated to fill up the entire circle. But, once the island is removed, another island cannot be added, since it was not calculated as a part of the original set of boundary objects.

HATCHING BLOCKS AND XREF DRAWINGS

When you apply hatching to inserted blocks and xref drawings, their internal structure is treated as if the block or xref drawing were composed of independent objects. This means that if you have inserted a block that consists of a circle within a rectangle and you want the internal circle to be hatched, all you have to do is invoke the **Boundary Hatch** dialog box and then choose the **Pick Points** button to select a point within the circle to generate the desired hatch (Figure 13-38). If you want to hatch a block using the **Select Objects** option, the objects inside the blocks should be parallel to the current UCS. If the block is comprised of objects such as arcs, circles, or polyline arc segments, they need not be uniformly scaled for hatching (Figure 13-38).

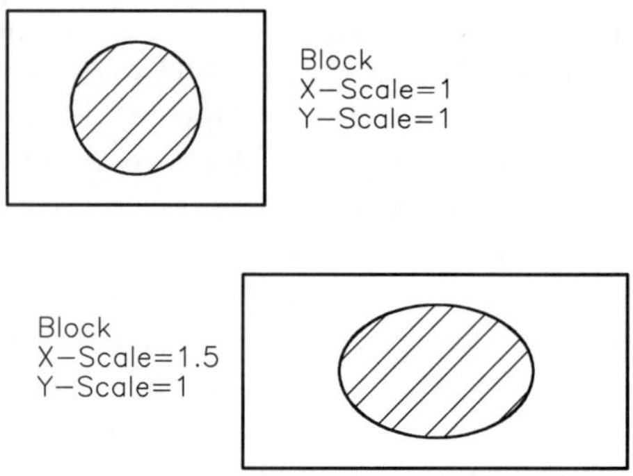

***Figure 13-38** Hatching inserted blocks*

When you xref a drawing, you can hatch any part of the drawing that is visible. If you use the **XCLIP** command to clip the xref drawing, the hatch pattern is not clipped, although the hatch boundary associativity is removed. Similarly, when you detach the xref drawing, the hatch pattern and its boundaries are not detached, although the hatch boundary associativity is removed.

PATTERN ALIGNMENT DURING HATCHING (SNAPBASE Variable)

Pattern alignment is an important feature of hatching, since on many occasions you need to hatch adjacent areas with similar or sometimes identical hatch patterns while keeping the adjacent hatch patterns properly aligned. Proper alignment of hatch patterns is taken care of automatically by generating all lines of every hatch pattern from the same reference point. The reference point is normally at the origin point (0,0). Figure 13-39 shows two adjacent hatch areas. The area on the right is hatched using the pattern ANSI32 at an angle of 0 degrees and the area on the left is hatched using the same pattern at an angle of 90 degrees. When you hatch these areas, the hatch lines may not be aligned, as shown in Figure 13-39 (a). To align

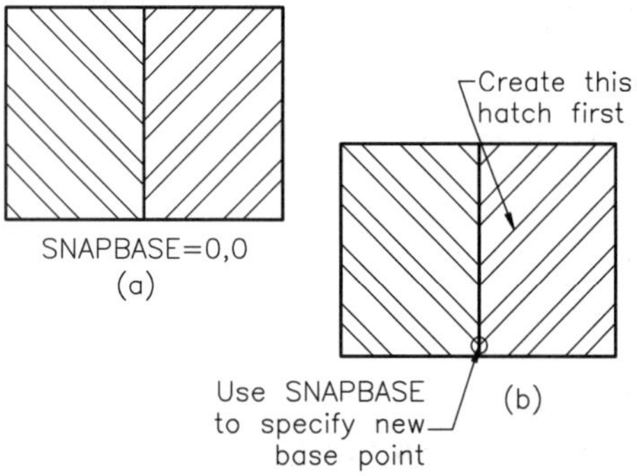

***Figure 13-39** Aligning hatch patterns using the **SNAPBASE** variable*

them, you can redefine the value of the **SNAPBASE** variable before you apply the hatch to the area on the left, so that the hatch lines are aligned, as shown in Figure 13-39 (b). The reference point for hatching can be changed using the system variable **SNAPBASE**. The command prompt sequence is:

Command: **SNAPBASE**
Enter new value for SNAPBASE <0.0000, 0.0000>: *Enter the new reference point.*

In this case, the endpoint of the hatch in the area on the right is selected as the new reference point for hatching the area on the left. See Figure 13-39 (b).

Note
*The value of the **SNAPBASE** variable can also be set using the **X base** and **Y base** edit boxes in the **Snap** area of the **Drafting Settings** dialog box (**Snap and Grid** tab).*

Tip
You should set the base point back to (0,0) when you are done hatching the object; otherwise, the most recent value gets stored and is used the next time you apply a hatch pattern.

Exercise 3 — *Mechanical*

In this exercise, you will hatch the given drawing using the hatch pattern ANSI31. Use the **SNAPBASE** variable to align the hatch lines shown in the drawing (Figure 13-40).

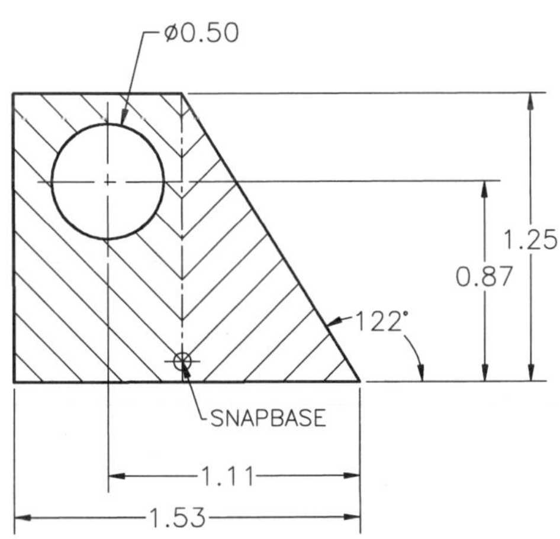

Figure 13-40 *Drawing for Exercise 3*

BOUNDARY COMMAND

Menu:	Draw > Boundary
Command:	BOUNDARY

The **BOUNDARY** command is used for creating a polyline or region around a selected point within a closed area, in a manner similar to the one used for defining a hatch boundary. This command works exactly like the **BHATCH** command, when the **Retain boundaries** check box is selected, except that the boundary object created is not hatched. When this command is entered, AutoCAD LT displays the **Boundary Creation** dialog box, shown in Figure 13-41.

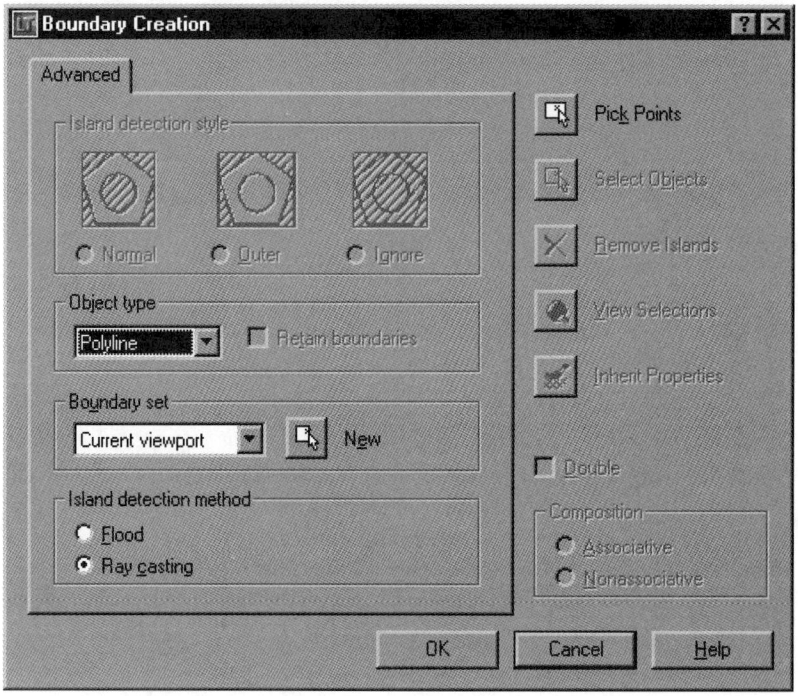

*Figure 13-41 The **Boundary Creation** dialog box*

The options in the **Boundary Creation** dialog box are similar to the options provided in the **Advanced** tab of the **Boundary Hatch** dialog box discussed earlier. Although, only the options related to boundary selection and not hatch patterns are available. The **Pick Points** button in the **Boundary Creation** dialog box is used to create a boundary by selecting a point inside the object whose boundary you want to create.

When you select the **Pick Points** option, the dialog box is cleared from the screen temporarily and the following prompt is displayed:

Select internal point: *Select a point that lies within the boundary of the object.*

Once you select an internal point, a polyline or a region is formed around the boundary (Figure 13-42). To end this process, press ENTER at the **Select internal point** prompt. Whether the boundary created is a polyline or region is determined by the option you have selected from the drop-down list in the **Object type** area of the **Boundary Creation** dialog box. **Polyline** is the default option. You can edit the boundary that has been created with the editing commands. The boundary is selected by using the last object selection option.

Just like when you used the **BHATCH** command, you can also define a boundary set here by choosing the **New** button in the **Boundary set** area of the dialog box. The dialog box will temporarily close to allow you to select objects to be used to create a boundary. The default boundary set is **current viewport**, which means everything visible in the current viewport consists of the boundary set. As discussed earlier, the boundary set option is especially useful when you are working with large and complex drawings, where examining everything on the screen becomes a time-consuming process.

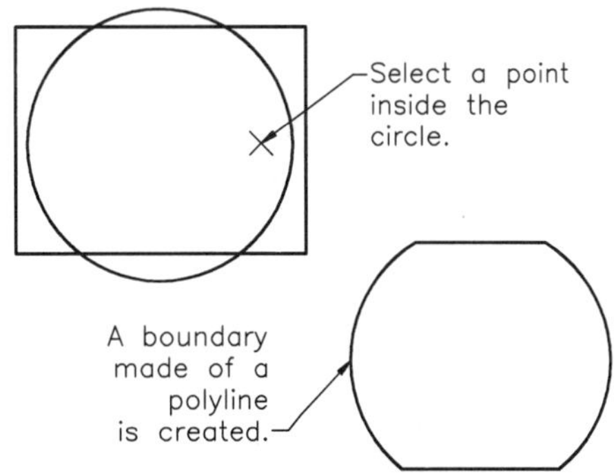

***Figure 13-42** Using the **BOUNDARY** command to create a polyline boundary*

The **Boundary Creation** dialog box also provides you the option of determining the island detection method. While **Flood** is the default option and islands are included as boundary objects, you can also select **Ray casting**, where rays shall be cast from the selected internal point in all directions (by default, using the dialog box). The nearest closed object it encounters is used for boundary creation. If the selected point does not lie within the encountered boundary, a **Boundary Definition Error** dialog box is displayed and you have to select another internal point.

The **-BOUNDARY** command can be used if you want to bypass the dialog box and use the command line. When using the command line, all the options (**Nearest**, **+X**, **-X**, **+Y**, and **-Y**) of ray casting are available. When using the dialog box, only the **Nearest** option is available. The different options of ray casting have been described earlier in the chapter.

Note

*The **HPBOUND** system variable controls the type of boundary object created using the **BOUNDARY** command. Its default value of 1 creates a polyline and if the value is 0, the object created is a region.*

Exercise 4 — *Mechanical*

First see Figure 13-43. Then make the drawing shown in Figure 13-44 using the **BOUNDARY** command to create a hatch boundary. Copy the boundary from the drawing shown in Figure 13-43, and then hatch.

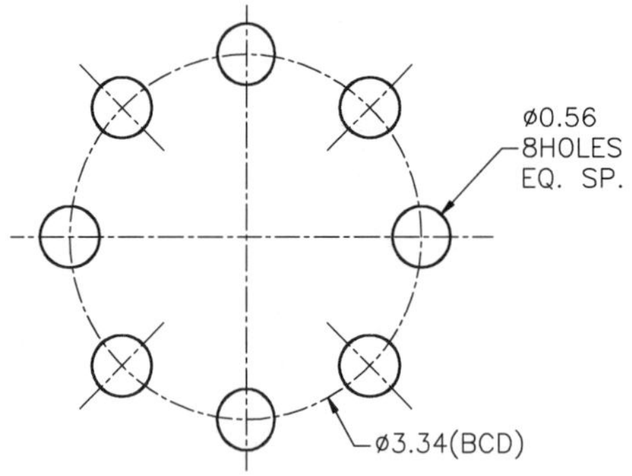

Figure 13-43 *Drawing for Exercise 4*

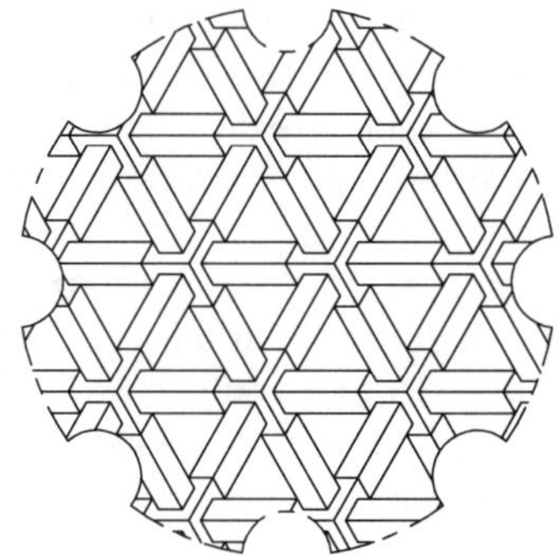

Figure 13-44 *Final drawing for Exercise 4*

OTHER FEATURES OF HATCHING

1. In AutoCAD LT, the hatch patterns are separate objects. The objects store the information about the hatch pattern boundary with reference to the geometry that defines the pattern for each hatch object.

2. When you save a drawing, the hatch patterns are stored with the drawing. Therefore, the hatch patterns can be updated even if the hatch pattern file that contains the definitions of hatch patterns is not available.

3. If the system variable **FILLMODE** is 0 (Off), the hatch patterns are not displayed. To see the effect of a changed **FILLMODE** value, you must use the **REGEN** command to regenerate the drawing after the value has been changed.

4. You can edit the boundary of a hatch pattern even when the hatch pattern is not visible (**FILLMODE**=0).

5. The hatch patterns created with earlier releases are automatically converted to AutoCAD LT 2002 hatch objects when the hatch pattern is edited.

6. When you save an AutoCAD LT 2002 drawing in LT 97 format, the hatch objects are automatically converted to LT 97 hatch blocks.

7. You can use the **CONVERT** command to change the pre-LT 97 hatch patterns to AutoCAD LT 2002 objects. You can also use this command to change the pre-LT 97 polylines to the AutoCAD LT 2002 optimized format to save memory and disk space.

HATCHING BY USING THE HATCH COMMAND

Hatches can also be created with the **HATCH** command. However, such hatches are not associated with the boundary. Also, the **HATCH** command does not create its own boundary and therefore, you have to select objects that will form the boundary for hatching. You can invoke this command only by entering **HATCH** at the AutoCAD LT Command prompt.

Note

*Although the **Draw** menu displays **Hatch** option and the **Draw** toolbar displays the **Hatch** button, choosing them invokes the **BHATCH** command and not the **HATCH** command.*

When you enter the **HATCH** command, it does not display a dialog box as in the case of the **BHATCH** command. The prompt sequence is:

Command: **HATCH**
Enter a pattern name or [?/Solid/User defined] <ANSI31>: *Enter a pattern name or ?*

If you know the name of the pattern, enter it or you can use the ? option. You can also use the following options in response to this prompt:

? Option

If you enter **?** as the response to the **Enter a pattern name or [?/Solid/User defined] <ANSI31>:** prompt, the following prompt is displayed:

Enter pattern(s) to list <*>: Enter

When you press ENTER at the last prompt, the names and descriptions of all the defined hatch patterns are displayed in the AutoCAD LT Text window. If you are not sure about the name of the pattern you want to use, you can select this option before entering a name. You can also list a particular pattern by specifying it at the prompt.

Solid Option

This option is used to fill the selected boundary with solid. If you enter **S** as the response to the **Enter a pattern name or [?/Solid/User defined] <ANSI31>:** prompt, the following prompts are displayed:

Select objects to define hatch boundary or <direct hatch>,
Select objects: *Select a boundary for filling or press ENTER for direct hatch.*

If you select a boundary object for filling, it gets highlighted and when you press ENTER, the selected area is solid filled. But, if you press ENTER at the **Select objects:** prompt, you can use the direct hatch option and you will be allowed to create a boundary for filling. The prompt sequence for the direct hatch option is:

Select objects to define hatch boundary or <direct hatch>,
Select objects: Enter
Retain polyline boundary [Yes/No] <N>: *Enter **Y** to retain the polyline, **N** to lose it.*

When using the direct hatch option, a polyline boundary is created around the area to be hatched. You can retain this polyline if you enter Y at the previous prompt and if you do not want to retain the polyline boundary, enter N at the previous prompt. The prompt sequence that follows is similar to that of the **PLINE** command:

Specify start point: *Select the first point of the boundary.*
Specify next point or [Arc/ Length/ Undo]: *Select the next point of the boundary.*
Specify next point or [Arc/ Close/ Length/ Undo]: *Select the next point of the boundary.*
Specify next point or [Arc/ Close/ Length/ Undo]: **C** (*Enter* **C** *to close the boundary.)*
Specify start point for new boundary or <apply hatch>: Enter

The created boundary will be solid-filled. You can also specify another boundary by selecting another set of points and not pressing ENTER at the last Command prompt.

Note

*A solid-filled or hatched area using the direct hatch option need not be drawn within an existing area. You can create new solid-filled/hatched areas, wherever you want. This is not possible using the **BHATCH** command.*

Tip

*The **HATCH** command allows you to create a polyline hatch boundary, even if the objects selected do not form a closed area. This is useful when you need to hatch areas that do not form a closed area.*

User defined Option

This option is used for creating a user-defined array or double-hatched grid of straight lines. To invoke this option, enter **U** at the **Enter pattern name or [?/Solid/User defined] <ANSI31>:** prompt. After you enter **U**, the following prompts are displayed:

Specify angle for crosshatch lines <0>: *Specify the angle for hatching.*
Specify spacing between the lines <1.0000>: *Specify the spacing between hatch lines.*
Double hatch area? [Yes/No] <N> *Press ENTER if you do not want double hatching or enter Y.*

At the **Specify angle for crosshatch lines <0>:** prompt, you need to specify the angle at which you want the hatch to be drawn. This angle is measured with respect to the X axis of the current UCS. You can also specify the angle by selecting two points on the screen. This value is stored in the **HPANG** system variable.

At the second prompt, **Specify spacing between the lines <1.0000>**, you can specify the space between the hatch lines. The default value for spacing is 1.0000. The specified spacing value is stored in the **HPSPACE** system variable.

At the third prompt, **Double hatch area? [Yes/No] <N>**, you can specify if you want double hatching. The default is no (**N**), and in this case, double hatching is disabled. You can enter Yes (**Y**) if you want double hatching. In that case, AutoCAD LT draws a second set of lines at right angles to the original lines. If the double hatch is on, AutoCAD LT sets the **HPDOUBLE** system variable to 1. The prompts that follow are the same as mentioned earlier with solid-filled areas.

Hatching by Specifying the Pattern Name

When you invoke the **HATCH** command, you can also specify the hatch pattern name at the

Enter a pattern name or [?/Solid/User defined] <ANSI31>: prompt. With the hatch name you can also specify the hatch style by adding style option to the pattern name. For example, for the pattern ANSI32 which needs to be applied with Outer style, the pattern shall be entered as ANSI32,O. The format is as follows:

Command: **HATCH**
Enter a pattern name or [?/Solid/User defined] <ANSI31>: **Stars, I** *(Enter a pattern name proceeded by a comma and the style.)*
Specify a scale for the pattern <1.0000>: *Specify a scale.*
Specify an angle for the pattern <0>: *Specify the angle for hatching.*

By specifying **Stars,I** you are specifying the hatch pattern as Stars and the hatch style as Ignore. There are three hatch styles: **Normal** (N), **Outer** (O), and **Ignore** (I). They have been discussed earlier in this chapter under the section "Advanced Tab (Boundary Hatch Dialog Box)".

All hatch patterns drawn using hatch names are drawn as blocks, which means that the entire hatch pattern acts as a single object and you cannot erase the lines in the pattern individually. If you want to edit the lines in the pattern individually, you can type an asterisk (*) before the hatch pattern name. For example, you can enter *ANSI31 at the **Enter a pattern name or [?/Solid/User defined] <ANSI31>:** prompt to edit the lines in the pattern individually.

All the stored hatch patterns are assigned an initial scale (size) and a rotation of 0 degrees with respect to the positive X axis. You can expand or contract according to your needs. You can scale the pattern by specifying a scale factor. You can also rotate the pattern with respect to the positive direction of the current UCS by entering a desired angle in response to the **Specify an angle for the pattern <0>:** prompt. You can even specify the angle manually by selecting or entering two points. For example, Figure 13-45 illustrates the hatch pattern ANSI31 with different scales and angles. The scale and angle are shown with the figures. Once you are finished specifying the pattern, AutoCAD LT prompts you to select objects for hatching or to use the direct hatch option (discussed earlier).

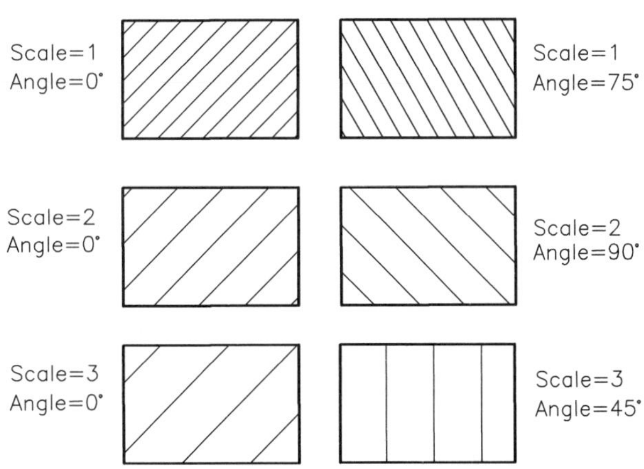

Figure 13-45 *Using different values for scale and angle*

Select objects to define hatch boundary or <direct hatch>,
Select objects: *Select a boundary for hatching or press ENTER for direct hatch.*

Note

You can also reference the scale for the pattern to paper space. This is useful when plotting drawings, since the scale factor is then based on the plotted scale of the drawing. Enter XP after the scale in the ***Specify a scale for the pattern <1.000>:*** *prompt. For example, if you enter 1XP as the scale factor, AutoCAD LT will calculate the actual scale in Model space such that it matches a value of 1 in paper space and this model space equivalent value is stored as the default value for the next time you use the* ***HATCH*** *command.*

Defining the boundaries in the **HATCH** command is a bit tedious compared with defining them with the **BHATCH** command. In the **BHATCH** command, the boundaries are automatically defined, whereas with the **HATCH** command you need to define the boundaries by selecting the objects. You can use any object selection method to select the objects that define the boundary of the area to be hatched and the objects that you want to exclude from hatching. If the boundary objects you select do not form a closed boundary, depending upon the angle and type of hatch pattern, the resulting hatch varies. For example, in Figure 13-46 (a), hatching does not take place. Here, only two lines lie completely within the window and the boundary is not closed. The pattern you have selected starts getting applied from one end of one of the lines to another line, if it is able to find it using its current angle and direction. When AutoCAD LT is unable to find another object to end the pattern, it displays the following message: **Unable to hatch the boundary**. When you change the angle of the pattern (ANSI31) from **0** degrees (default) to **135** degrees, see Figure 13-46 (b), the hatch pattern starts getting applied from one end and continues until another end is being found. Some patterns, like TRIANG, need more than two edges to be completely drawn and therefore, are drawn incomplete in the absence of the required boundary edges.

In the Figure 13-47 (a), the boundary is selected by selecting objects. In this illustration, the lines LQ and MR extend beyond the hatch boundary, which makes the hatching begin outside the intended hatch area. You must break the top and bottom line at points Q and R so that lines LQ, QR, RM, and ML define the hatch boundary. Now, if you select the objects, AutoCAD LT will hatch the area bound by the four selected objects, as shown Figure 13-47 (b).

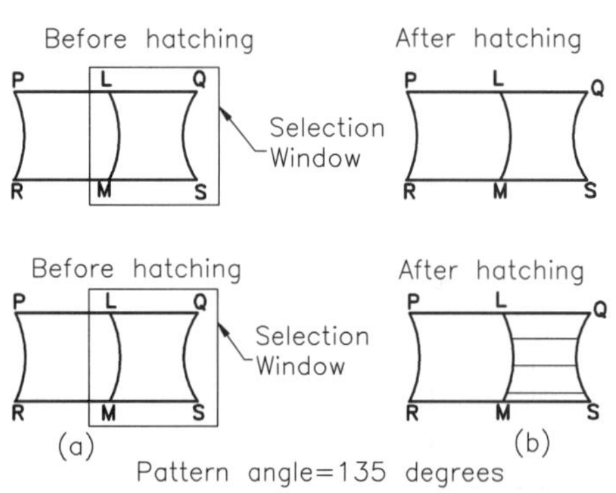

Figure 13-46 Selecting objects using the ***HATCH*** *command*

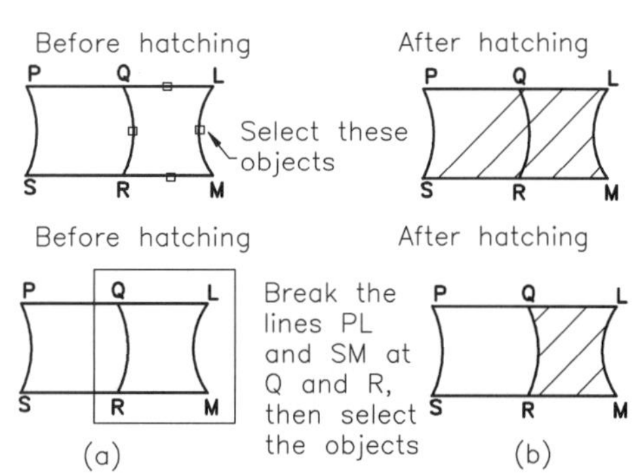

Figure 13-47 Selecting objects using the ***HATCH*** *command*

Note

Patterns drawn using the ***HATCH*** *command are nonassociative. This implies that whenever the hatch boundary is modified, the hatch pattern is not updated automatically.*

Self-Evaluation Test

Answer the following questions, and then compare your answers to the correct answers given at the end of this chapter.

1. Different filling patterns make it possible to distinguish between different parts or components of an object. (T/F)

2. If AutoCAD LT does not locate the entered pattern in the ACLT.PAT file, it searches for it in a file with the same name as the pattern. (T/F)

3. When a boundary set is formed, it does not become the default for hatching. (T/F)

4. You can edit the hatch boundary and the associated hatch pattern by using grips. (T/F)

5. The __________ command automatically defines the hatch boundary, whereas in the case of the __________ command you have to define the hatch boundary by selecting the objects that form the boundary of the hatch area.

6. One of the ways to specify a hatch pattern from the group of stored hatch patterns is by selecting one from the __________ drop-down list.

7. The value you enter in the __________ edit box lets you rotate the hatch pattern with respect to the X axis of the current UCS.

8. The __________ option in the **Boundary Hatch** dialog box lets you select objects that form the boundary for hatching.

9. You can use the __________ option after you have selected the area to be hatched to see the hatching before it is actually applied.

10. The selection of a hatch style carries meaning only if the objects to be hatched are __________.

Review Questions

Answer the following questions.

1. The **BHATCH** command does not allow you to hatch a region enclosed within a boundary (closed area) by selecting a point inside the boundary. (T/F)

2. A **Boundary Definition Error** dialog box is displayed if AutoCAD LT finds that the selected point is not inside the boundary or that the boundary is not closed. (T/F)

3. You can improve the speed of hatching by setting parameters in the **Advanced** tab of the **Boundary Hatch** dialog box. (T/F)

4. When you use editing commands such as **MOVE**, **SCALE**, **STRETCH**, and **ROTATE**, associativity is lost, even if all the boundary objects are selected. (T/F)

5. The hatching procedure in AutoCAD LT does not work on inserted blocks. (T/F)

6. Patterns drawn using the **HATCH** command are associative. (T/F)

7. If the **Double** hatch box is selected, which system variable is set to 1?

   (a) **HPSPACE** (b) **HPDOUBLE**
   (c) **HPANG** (d) **HPSCALE**

8. The **Iso Pen Width** drop-down list is available only for which hatch patterns?

   (a) **ANSI** (b) **Predefined**
   (c) **Custom** (d) **ISO**

9. If you want to have the same hatching pattern, style, and properties as that of an existing hatch on the screen, which button should you choose?

   (a) **Inherit Properties** (b) **Select Objects**
   (b) **Pick Points** (d) **Remove Islands**

10. Which variable can you use to align the hatches in adjacent hatch areas?

    (a) **SNAPBASE** (b) **FILLMODE**
    (c) **SNAPANG** (d) **HPANG**

11. The specified hatch spacing value is stored in which system variable?

    (a) **HPDOUBLE** (b) **HPANG**
    (c) **HPSCALE** (d) **HPSPACE**

12. One of the advantages of using the __________ command is that you don't have to select

each object comprising the boundary of the area you want to hatch, as in the case of the __________ command.

13. To select a custom hatch pattern, first select **Custom** from the **Type** drop-down list and then select the name of a previously stored hatch pattern from the __________ drop-down list.

14. There are three hatching styles from which you can choose: __________, __________, and __________.

15. If you select the __________ style, all areas bounded by the outermost boundary are hatched, ignoring any hatch boundaries that lie within the outer boundary.

Exercises

Exercise 5 *Mechanical*

Hatch the drawings in Figure 13-48 and Figure 13-49 using the hatch pattern to match.

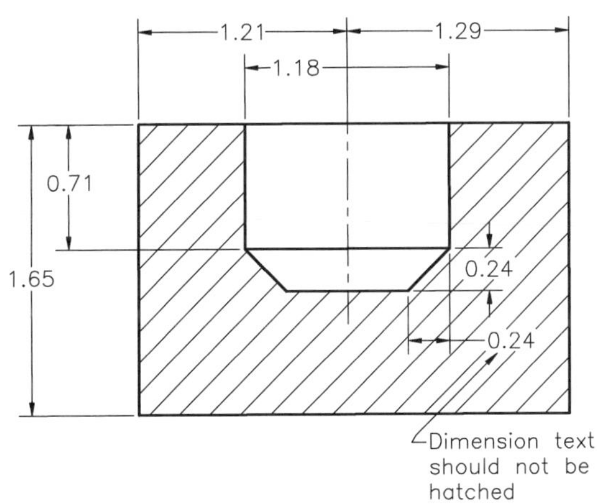

Figure 13-48 *Drawing for Exercise 5*

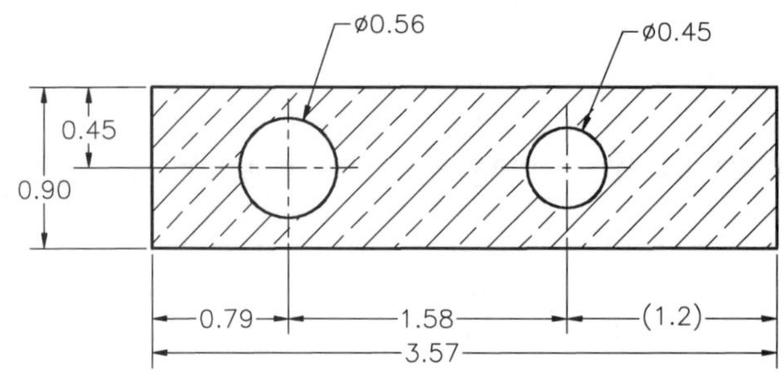

Figure 13-49 Drawing for Exercise 5

Exercise 6 — *General*

Hatch the drawing in Figure 13-50 using the hatch pattern ANSI31. Use the **SNAPBASE** variable to align the hatch lines as shown in the drawing.

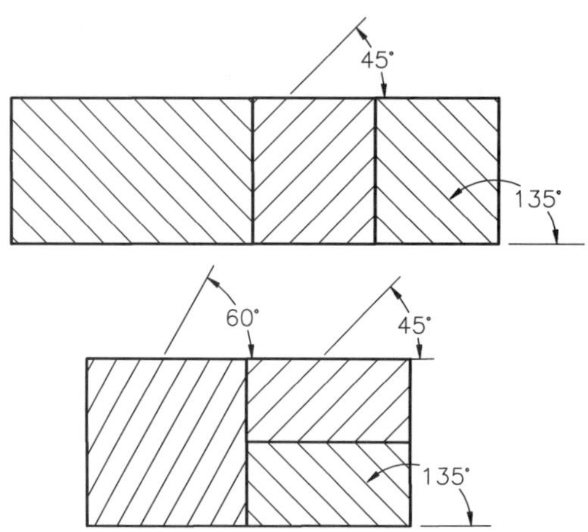

Figure 13-50 Drawing for Exercise 6

Exercise 7 — *Mechanical*

Figure 13-51 shows the top and front views of an object. It also shows the cutting plane line. Based on the cutting plane line, hatch the front views in section. Use the hatch pattern of your choice.

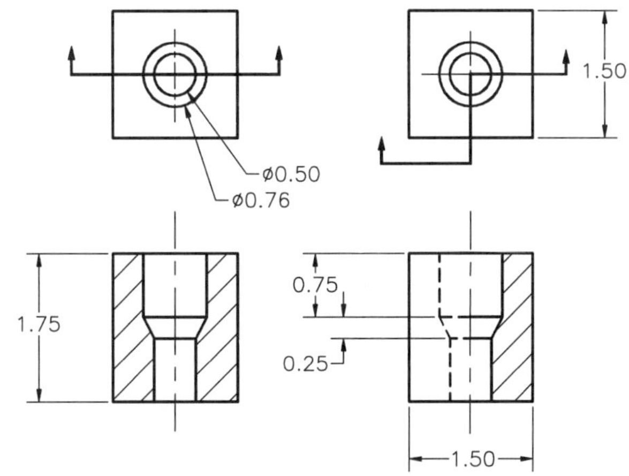

Figure 13-51 *Drawing for Exercise 7*

Exercise 8 *Mechanical*

Figure 13-52 shows the top and front views of an object. Hatch the front view in full section. Use the hatch pattern of your choice.

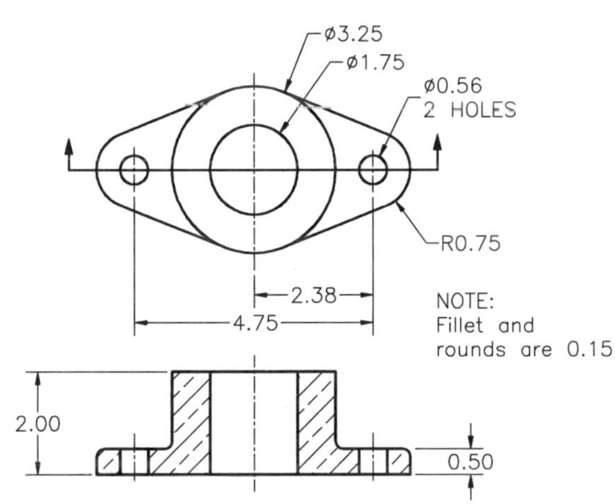

Figure 13-52 *Drawing for Exercise 8*

Exercise 9 *Mechanical*

Figure 13-53 shows the front and side views of an object. Hatch the side view in halfsection. Use the hatch pattern of your choice.

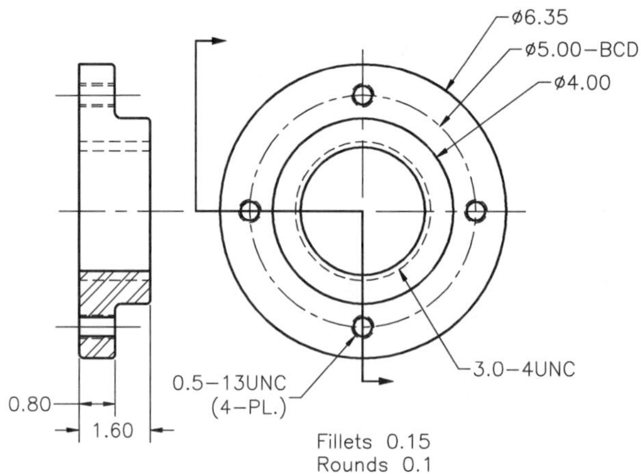

Figure 13-53 Drawing for Exercise 9

Exercise 10 *Mechanical*

Figure 13-54 shows the front view with broken section and top views of an object. Hatch the front view as shown. Use the hatch pattern of your choice.

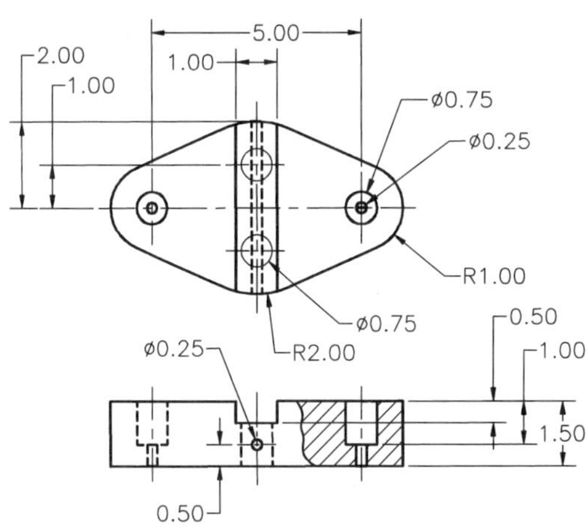

Figure 13-54 Drawing for Exercise 10

Exercise 11 *Mechanical*

Figure 13-55 shows the front, top, and side views of an object. Hatch the side view in section using the hatch pattern of your choice.

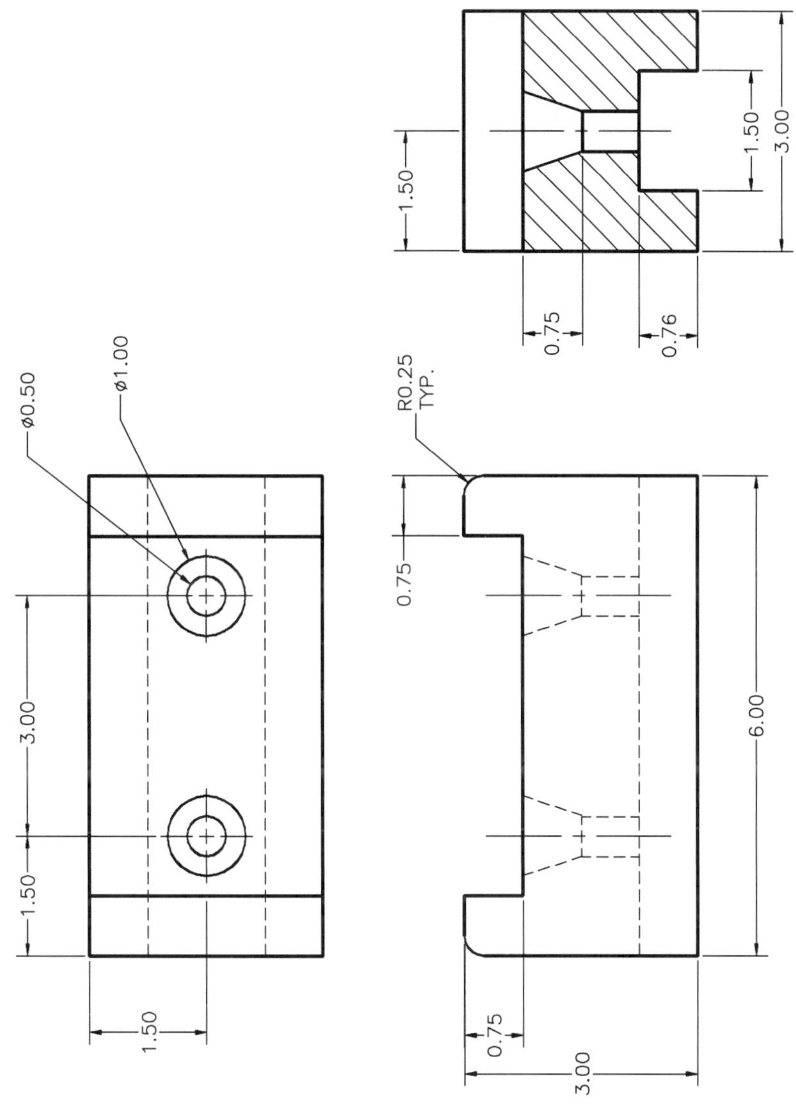

Figure 13-55 *Drawing for Exercise 11*

Problem Solving Exercise 1 — *Mechanical*

Figure 13-56 shows an object with front and side view with aligned section. Hatch the side view using the hatch pattern of your choice

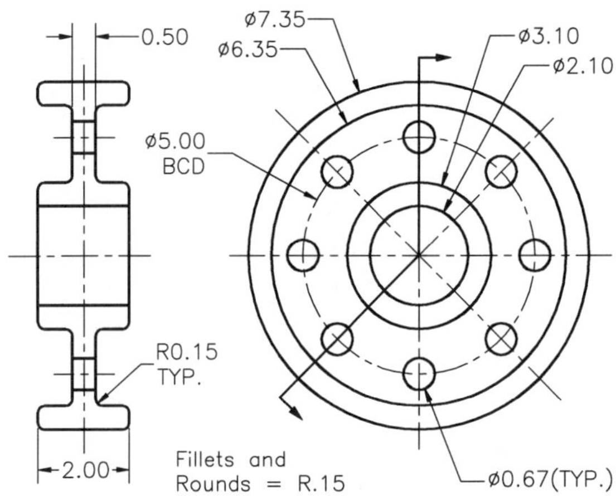

Figure 13-56 *Drawing for Problem Solving Exercise 1*

Problem Solving Exercise 2 *Mechanical*

Figure 13-57 shows the front view, side view, and the detail "A" of an object. Hatch the side view and draw the detail drawing as shown.

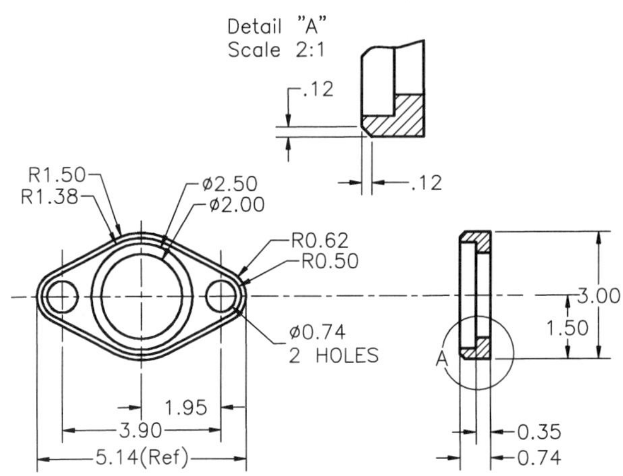

Figure 13-57 *Drawing for Problem Solving Exercise 2*

Answers to Self-Evaluation Test

1 - T, **2** - T, **3** - F, **4** - T, **5** - **BHATCH, HATCH**, **6** - **Pattern**, **7** - **Angle**, **8** - **Select Objects**, **9** - **Preview**, **10** - nested

Chapter 14

Blocks

Learning Objectives

After completing this chapter, you will be able to:
- *Create blocks with the **BLOCK** command.*
- *Insert blocks with the **INSERT** command.*
- *Create drawing files using the **WBLOCK** command.*
- *Use the **DesignCenter** to locate, preview, copy, or insert blocks and existing drawings.*
- *Perform editing operations on the blocks.*
- *Split a block into individual objects using the **EXPLODE** and **XPLODE** commands.*
- *Rename blocks and delete unused blocks.*

THE CONCEPT OF BLOCKS

The ability to store parts of a drawing, or the entire drawing, so that they need not be redrawn when required in the same drawing or another drawing is a great benefit to the user. These parts of a drawing, entire drawings, or symbols (also known as blocks) can be placed (inserted) in a drawing at the location of your choice, in the desired orientation, with the desired scale factor. The block is given a name (block name) and the block is referenced (inserted) by its name. All the objects within a block are treated as a single object. You can **MOVE**, **ERASE**, or **LIST** the block as a single object, that is, you can select the entire block simply by selecting anywhere on it. As for the edit and inquiry commands, the internal structure of a block is immaterial, since a block is treated as a primitive object, like a polygon. If a block definition is changed, all references to the block in the drawing are updated to incorporate the changes made to the block.

A block can be created using the **BLOCK** command. You can also save objects in a drawing, or an entire drawing as a drawing file using the **WBLOCK** command. The main difference between the two is that a wblock can be inserted in any other drawing, but a block can be inserted only in the drawing file in which it was created.

Another feature of AutoCAD LT is that instead of inserting a symbol as a block (which results in adding the content of the referenced drawing to the drawing in which it is inserted), you can reference drawings (Xref). This means that the contents of the referenced drawing are not added to the current drawing file, although they become part of that drawing on the screen. This is explained in detail in Chapter 16, External References.

Advantages of Using Blocks

Blocks offer many advantages. Here are some of them.

1. Drawings often have some repetitive features. Instead of drawing the same feature again and again, you can create a block of this feature and insert it wherever required. This helps you to reduce drawing time and better organize your work.
2. Another advantage of using blocks is that they can be drawn and stored for future use. You can thus create a custom library of objects required for different applications. For example, if your drawings are concerned with gears, you could create blocks of gears and then integrate these blocks with custom menus (see Chapter 25, for writing menus). In this manner, you could create an application environment of your own in AutoCAD LT.
3. The size of a drawing file increases as you add objects to it. AutoCAD LT keeps track of information about the size and position of each object in the drawing; for example, the points, scale factors, radii; and so on. If you combine several objects into a single object by forming a block with the **BLOCK** command, there will be a single scale factor, rotation angle, position, and so on, for all objects in the block, thereby saving storage space. Each object repeated in multiple block insertions needs to be defined only once in the block definition. Ten insertions of a gear made of forty-three lines and forty-one arcs require only ninety-four objects (10+43+41), while ten individually drawn gears require 840 objects [10 x (43+41)].
4. If the specification for an object changes, the drawing needs to be modified. This is a very tedious task if you need to detect each location where the change is to be made and edit it individually. But if this object has been defined as a block, you can redefine it, and everywhere the object appears, it is revised automatically. This has been discussed in Chapter 15, Defining Block Attributes.
5. Attributes (textual information) can be included in blocks. Different attribute values can be specified in each insertion of a block.
6. You can create symbols and then store them as blocks with the **BLOCK** command. Later on, with the **INSERT** command, you can insert the blocks in the drawing in which they were defined. There is no limit to the number of times you can insert a block in a drawing.
7. You can store symbols as drawing files using the **WBLOCK** command and later insert these symbols into any other drawing using the **INSERT** command.
8. Blocks can have varying X, Y, and Z scale factors and rotation angles from one insertion to another (Figure 14-1).

FORMATION OF BLOCKS

Drawing Objects for Blocks

The first step in creating blocks is to draw the object(s) to be converted into a block. You can consider any symbol, shape, or view that may be used more than once for conversion into a block (Figure 14-2). Even a drawing that is to be used more than once can be inserted as a block.

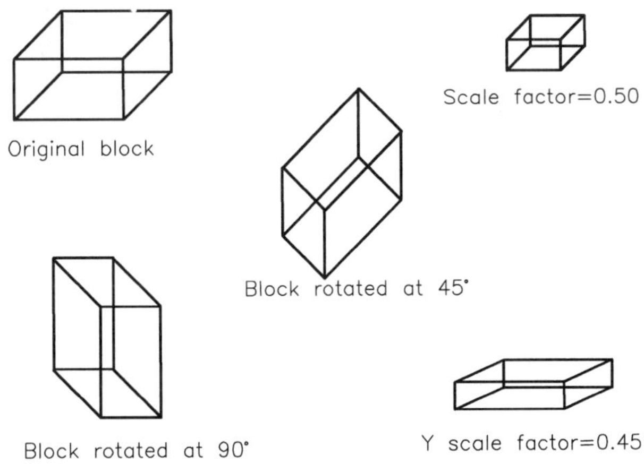

Figure 14-1 *Block with different specifications*

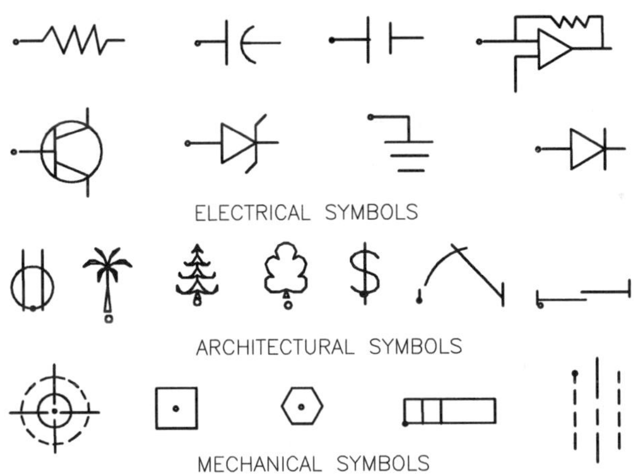

Figure 14-2 *Some common drafting symbols created to store in a library*

Tip

Examine the sketch of the drawing you are about to begin and examine it carefully to look for any shapes, symbols, and components that are to be used more than once. They can then be drawn once and be stored as blocks.

Checking Layers

You should be particularly careful about the layers on which the objects to be converted into a block are drawn. The objects must be drawn on layer 0 if you want the block to inherit the linetype and color of the layer on which it is inserted. If you want the block to retain the linetype and color of the layer on which it was drawn, draw the objects defining the block on that layer. For example, if you want the block to have the linetype and color of layer OBJ, draw the objects comprising the block on layer OBJ. Now, even if you insert the block on any other layer, the linetype and color of the block will be those of layer OBJ. If the objects in a block are drawn with the color, linetype, and lineweight as ByBlock, this block when inserted assumes the properties of the current layer. To set the color to ByBlock, select **ByBlock** from the **Color Control** drop-down list of the **Object Properties** toolbar. You can also choose **Color** from the **Format** menu to display the **Select Color** dialog box and then choose the **ByBlock** button in the **Logical Colors** area of the dialog box. Choose **OK** to close the dialog box. To set the linetype to ByBlock, select **ByBlock** from the **Linetype Control** drop-down list of the **Object Properties** toolbar. You can also select **ByBlock** in the **Linetype Manager** dialog box that is displayed when you choose **Linetype** from the **Format** menu and then choose the **Current** button. The current linetype is now set to **ByBlock**. Similarly, you can also specify the lineweight as **ByBlock**.

Note

*As discussed previously, before you make a block you should check the layers on which the objects are made because these layers will determine the color and linetype of the block when inserted. To change the layer associated with an object, use the **Properties** window, the **Object Properties** toolbar, **Properties** option of the **CHANGE** command, or the **CHPROP** command.*

CONVERTING OBJECTS INTO A BLOCK

BLOCK Command

Toolbar:	Draw > Make Block
Menu:	Draw > Block > Make
Command:	BLOCK

When you invoke the **BLOCK** command (Figure 14-3), the **Block Definition** dialog box (Figure 14-4) is displayed. You can use the **Block Definition** dialog box to save any object or objects as a block.

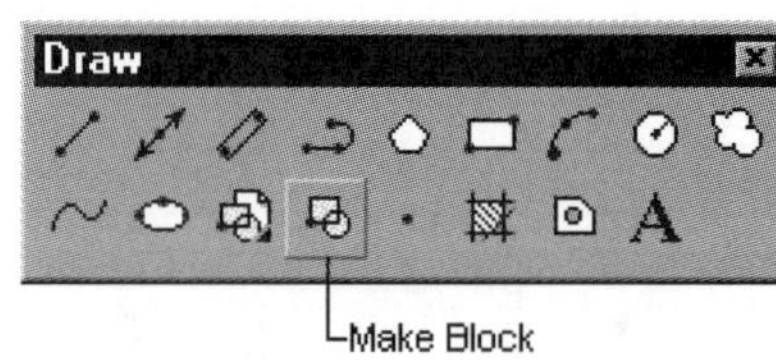

*Figure 14-3 **Make Block** button in the **Draw** toolbar*

In the **Block Definition** dialog box, you can enter the name of the block you want to create in the **Name** edit box. All the block names present in a current drawing are displayed in the **Name** drop-down list in an alphabetical and numerical order. If a drawing contains more than six blocks, a scroll bar appears on the right side of the **Name** drop-down list which can be moved up and down to view all the blocks in the drawing. This way you can verify whether a block you have defined has been saved. By default, the block name can have up to 255 characters. The name can contain letters, digits, blank spaces as well as any special characters including the $ (dollar sign), - (hyphen), and _ (underscore), provided they are not being used for any other purpose by AutoCAD LT or Windows.

Note

*The block name is controlled by the **EXTNAMES** system variable and the default value is 1. If the **EXTNAMES** is set to 0, the block name will be only thirty-one characters long and will not include spaces or any other special characters apart from a $ (dollar sign), - (hyphen), or _ (underscore).*

If a block already exists with the block name that you have specified in the **Name** edit box, an AutoCAD LT dialog box, warning that the block name is already defined, is displayed when you choose the **OK** button in the **Block Definition** dialog box. In this dialog box, you can

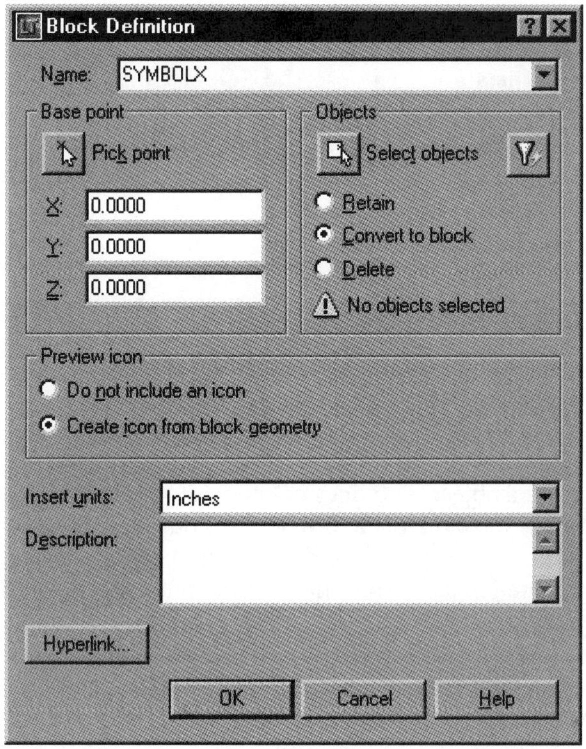

***Figure 14-4 Block Definition** dialog box*

either redefine the existing block by choosing the **Yes** button or you can exit it by choosing the **No** button. You can then use another name for the block in the dialog box for the block.

After you have specified a block name, you are required to specify the insertion base point. This point is used as a reference point to insert the block. Usually, either the center of the block or the lower left corner of the block is defined as the insertion base point. Later on, when you insert the block, you will notice that the block appears at the insertion point, and you can insert the block with reference to this point. The point you specify as the insertion base point is taken as the origin point of the block's coordinate system. You can specify the insertion point by choosing the **Pick point** button in the **Base point** area of the dialog box. The dialog box is temporarily removed, and you can select a base point on the screen. You can also enter the coordinates in the **X:**, **Y:**, and **Z:** edit boxes instead of selecting a point on screen. Once the insertion base point is specified, the dialog box reappears and the X, Y, and Z coordinates of the insertion base point are displayed in the **X**, **Y**, and **Z** edit boxes respectively. If no insertion base point is selected, AutoCAD LT assumes the insertion point coordinates to be 0,0,0, which are the default coordinates.

Tip

In drawings where the insertion point of a block is important, for example, in the case of inserting the block of a door symbol in architectural plans, it may be a good idea to first specify the requisite scale factors and rotation angles and then the insertion point. This way, you are able to preview the block symbol as it is to be inserted.

After specifying the insertion base point, you are required to select the objects that will constitute the block. Until the objects are selected, AutoCAD LT displays a warning: **No objects selected**, at the bottom of the **Objects** area. Choose the **Select Objects** button. The dialog box is temporarily removed from the screen. You can select the objects on the screen using any selection method. After completing the selection process, right-click or press the ENTER key to return to the dialog box. The number of objects selected are displayed at the bottom of the **Objects** area of the dialog box.

The **Objects** area of the **Block Definition** dialog box also has a **Quick Select** button. If you choose this button, the **Quick Select** dialog box is displayed that allows you to define a selection set based on properties of objects. **Quick Select** is used in cases where the drawings are very large and complex. The **Quick Select** dialog box has been discussed earlier in Chapter 6, Editing Commands-II.

In the **Objects** area, if you select the **Retain** radio button it retains the selected objects that form the block as individual objects and does not convert them into a block after the block has been defined. If you select the **Convert to block** radio button (default option) it converts the selected objects into a block after you have defined the block. If you select the **Delete** radio button it deletes the selected objects from the drawing after the block has been defined.

Tip

*When you select the **Delete** radio button in the **Objects** area when defining a block, the selected objects are removed from the drawing. If you want them back, you can use the **OOPS** command. Entering **U** at the Command prompt, or choosing the **Undo** button from the **Standard Toolbar** will undo the block definition from the drawing.*

The **Preview icon** area gives you the following options. Selecting the **Do not include an icon** radio button, does not create a preview icon and whenever a block name is selected from the name drop-down list, the **Preview Image box** is not displayed at all in this area of the dialog box. If you select the **Create icons from block geometry** radio button creates a **Preview icon** that is saved with the block definition in the drawing. So, whenever you select a block name from the **Name** drop-down list of the **Block Definition** dialog box or the **DesignCenter**, this preview image is displayed. The **DesignCenter** has been discussed earlier in Chapter 6, Editing Commands-II. Similarly, in the **Description** text box, if you have entered a textual description of the block, it is stored with the block definition and displayed on selecting the block name from the **Name** drop-down list of the **Block Definition** dialog box or the **AutoCAD DesignCenter**.

Tip

Saving a preview image and the description text with the block definition makes it easier to identify a block and is an easy reference.

The **Insert units** drop-down list displays the units the **AutoCAD DesignCenter** will use when inserting the particular block. The default value is Inches. Choose the **OK** button to complete defining the block.

Tip

It is a good idea to create a unit block, especially for drawings where the same block is to be inserted with various scale factors. A unit block can be defined as a block that is created within a unit square area (1 unit x 1 unit). This way, everytime you specify a scale factor, the block simply gets enlarged or reduced by the said amount directly, since the multiplication factor is 1.

-BLOCK Command

You can use the **-BLOCK** command to create blocks from the command line. The following prompt sequence saves a symbol as a block named SYMBOLX.

Command: **-BLOCK**
Enter block name or [?]: **SYMBOLX**
Specify insertion base point: *Select a point.*
Select objects: *Select all the objects comprising the block.*
Select objects: *Press ENTER when selection is complete.*

AutoCAD LT acknowledges the creation of the block by removing the objects defining the block from the screen. However, you can get the objects back at the same position by entering **OOPS**. You can use the **?** option to list all the blocks created in the drawing. The prompt sequence is:

Command: **-BLOCK**
Enter block name or [?]: ?
Enter block(s) to list <*>: Enter

Information similar to the following is displayed in the AutoCAD LT Text Window:

Defined blocks
"BLK1"
"BLK2"
"SYMBOLX"

User Blocks	*External References*	*Dependent Blocks*	*Unnamed Blocks*
n	*n*	*n*	*n*

Here **n** denotes a numeric value.

Defined blocks. The names of all the blocks defined in the drawing.

User blocks. The number of blocks created by the user.

External references. The number of external references. These are the drawings that are referenced with the **XREF** command and are discussed in Chapter 16, External Reference.

Dependent blocks. The number of externally dependent blocks. These are the blocks that are present in a referenced drawing.

Unnamed blocks. The number of unnamed blocks in the drawing. These are objects such as associative dimensions and hatch patterns.

Exercise 1 *General*

Draw a unit radius circle, and use the **BLOCK** command to create a block. Name the block CIRCLE. Figure 14-5 shows the steps involved in creating a block. Select the center as the insertion point. Verify if the block CIRCLE was saved.

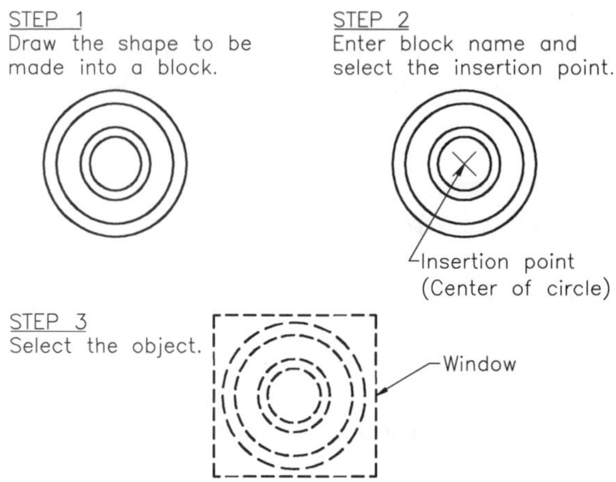

***Figure 14-5** Using the **BLOCK** command*

INSERTING BLOCKS (INSERT COMMAND)

Toolbar:	Insert > Insert Block
	Draw > Insert
Menu:	Insert > Block
Command:	INSERT

Insertion of a predefined block is possible with the **INSERT** command (Figure 14-6). An inserted block is called a block reference. You should determine the layer on which you want to insert the block, the location where you want to insert it, its size, and the angle by which you want the block to be rotated, prior to insertion. If the layer on which you want to insert the block is not current, you can use the **Layer Control** drop-down list in the **Object Properties** toolbar or choose the **Current** button in the **Layer Properties Manager** dialog box to make it current. When you invoke the **INSERT** command, the **Insert** dialog box (Figure 14-7) is displayed. You can specify the different parameters of the block or external file to be inserted in the **Insert** dialog box.

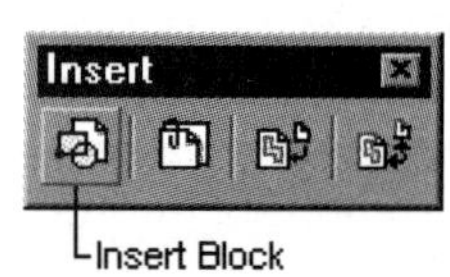

***Figure 14-6** Invoking the **INSERT** command from the **Insert** toolbar*

Name

This drop-down is used to specify the name of the block to be inserted. Select the name of the block to be inserted from this drop-down list. You can also enter the name of the block to be

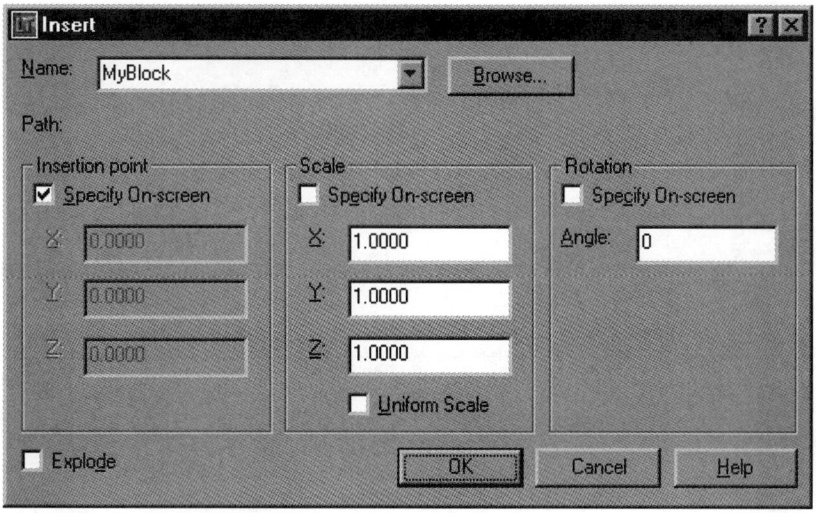

***Figure 14-7 Insert** dialog box*

inserted. All the blocks that are created in the current drawing are available in the **Name** drop-down list.

Note

*The last block name selected from the **Name** drop-down list and inserted becomes the default name for the next insertion and is displayed in this drop-down list.*

The **Browse** button is used to insert external files. When you choose this button, the **Select Drawing File** dialog box (Figure 14-8) is displayed. This dialog box works in a manner similar to the standard **Select File** dialog box, which has been discussed earlier in Chapter 2, Getting Started-II. You can select a drawing file from the listed drawing files existing in a current directory, and can change the directory by selecting the desired directory in the **Look in** drop-down list. Once you select the drawing file in this dialog box and choose the **Open** button, the drawing file name is displayed in the **Name** drop-down list of the **Insert** dialog box. **Path** displays the path of the external file selected to be inserted. Now, if you want to change the block name, just change the name in the **Name** drop-down list. In this manner, the drawing can be inserted with a different name. Note that changing the original drawing does not affect the inserted drawing.

Note

*If the name you have specified in the **Name** edit box does not exist as a block in the current drawing, AutoCAD LT will search the drives and directories on the path (specified in the **Options** dialog box) for a drawing of the same name and if it finds it, it will insert it.*

*Also, suppose you have inserted a block in a drawing and then you want to insert a drawing with the same name as the block, AutoCAD LT will display a message saying that the specified block already exists and asks if you want to redefine it. If you choose **Yes**, the block in the drawing gets replaced by the drawing with the same name, that is, the block gets redefined.*

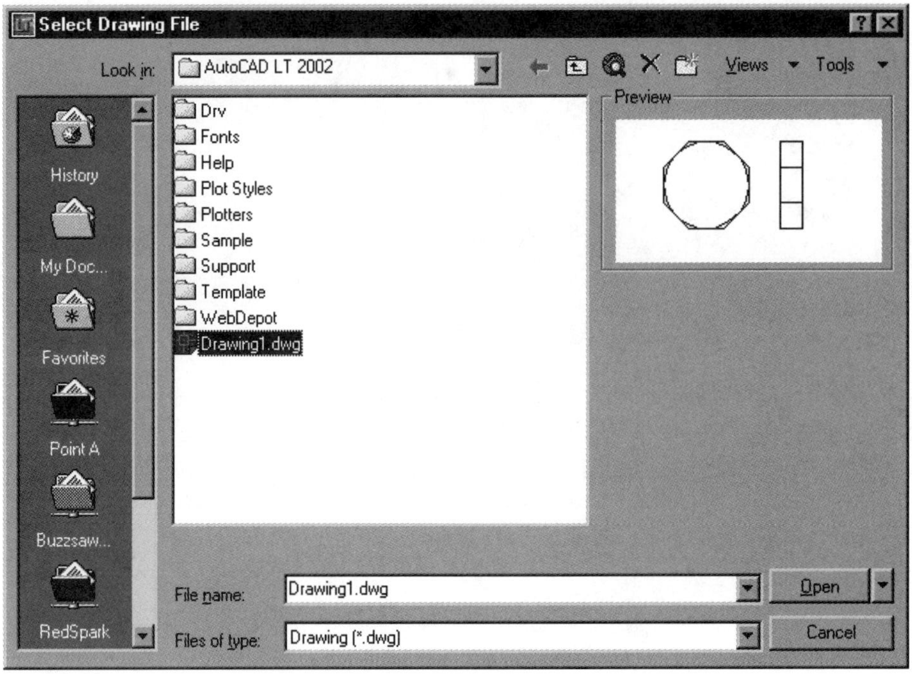

***Figure 14-8 Select Drawing File** dialog box*

Tip

*You can create a block in a current drawing from an existing drawing file. This saves time by avoiding redrawing the object as a block. Locate and select an existing file using the **Browse** button and rename it with a name that you want to assign to the block in the **Name** edit box. When you choose **OK**, you can insert the block in the drawing or press the ESC key. The new block gets added into the current drawing.*

Insertion Point

When a block is inserted, the coordinate system of the block is aligned parallel to the current UCS. In the **Insertion point** area, you can specify the X, Y, and Z coordinate locations of the block insertion point in the **X**, **Y**, and **Z** edit boxes respectively. If you select the **Specify On-screen** check box, the **X**, **Y**, and **Z** edit boxes are not available and you can specify the insertion point on the screen. By default, the **Specify On-screen** check box is selected and therefore, you can specify the insertion point on screen.

Scale

In this area, you can specify the X, Y, and Z scale factors of the block to be inserted in the **X**, **Y**, and **Z** edit boxes respectively. By selecting the **Specify On-screen** check box, you can specify the scale of the block at which it has to be inserted on the screen. The **Specify On-screen** check box is cleared by default and the block is inserted with the scale factors of 1 along the three axes. Also, if the **Uniform Scale** check box is selected, the X scale factor value is assumed for the Y and Z scale factors also. This means that if this check box is selected, you need to specify only the X scale factor. All the dimensions in the block are multiplied by the X, Y, and Z scale factors you specify. These scale factors allow you to stretch or compress a block along the X and Y axes, respectively, according to your needs. You can also insert 3D objects into a drawing by specifying the third scale factor (since 3D objects have three dimensions), the Z scale factor. Figure 14-9 shows variations of the X and Y scale factors for the block, CIRCLE, during insertion.

Tip

By specifying negative scale factors, you can insert a mirror image of a block along the particular axis. A negative scale factor for both the X and Y axes is the same as rotating the block reference through 180 degrees, since it mirrors the block reference in the opposite quadrant of the coordinate system. In Figure 14-10, the effect of negative scale factor on block (DOOR) can be marked by a change in position of insertion point marked with an (X).

Also, specifying a scale factor of less than 1 inserts the block reference smaller than the original size and a scale factor greater than 1 inserts the block reference larger than its original size.

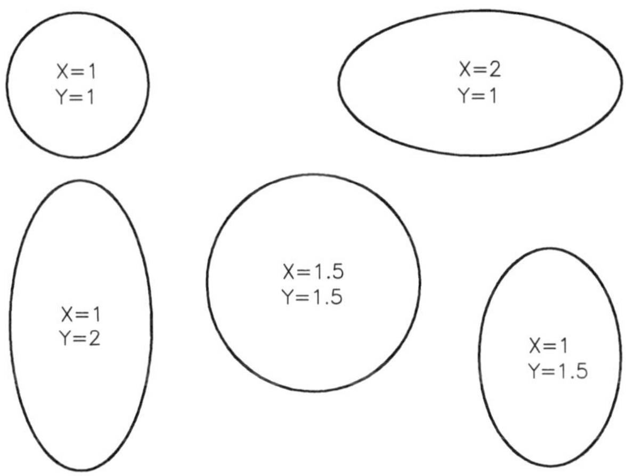

Figure 14-9 *Block inserted with different scale factors*

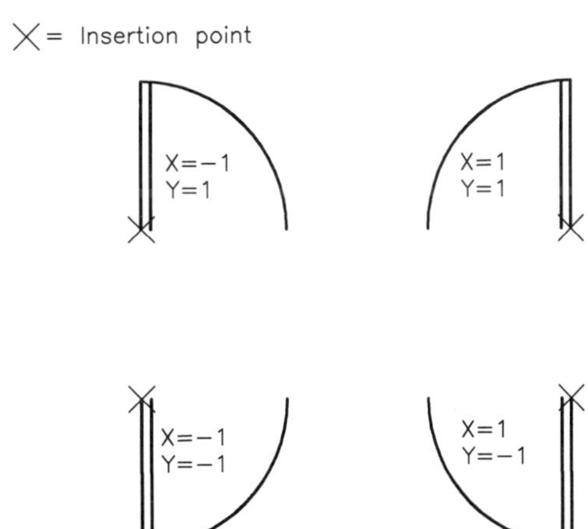

Figure 14-10 *Block inserted using negative scale factors*

Rotation

You can enter the angle of rotation for the block to be inserted in the **Angle** edit box. The insertion point is taken as the location about which rotation takes place. Selecting the **Specify On-screen** check box allows you to specify the angle of rotation on the screen. This check box is cleared by default and the block is inserted at an angle of zero degrees.

Explode Check Box

By selecting this check box, the inserted block is inserted as a collection of individual objects. The function of the **Explode** check box is identical to that of the **EXPLODE** command. Once a block is exploded, the X, Y, and Z scale factors are identical. Hence, you are provided access to only one scale factor edit box (**X** edit box), the **Y** and **Z** edit boxes are not available, and the **Uniform Scale** check box also gets selected. The X scale factor is assigned to the Y and Z scale factors too. You must enter a positive scale factor value.

Note

If you select the ***Explode*** *check box before insertion, the block reference is exploded and the objects are inserted with their original properties such as layer, color, and linetype.*

Once you have entered the relevant information in the dialog box, choose the **OK** button. Once you do so, the **Insert** dialog box is removed from the screen and if you have selected the **Specify On-Screen** check boxes, you can specify the insertion point, scale, and angle of rotation with a pointing device. Whenever the insertion point is to be specified on screen (by default), you can specify scale factors and rotation angle values that will override the values specified in the dialog box, using the command line. Before you specify the insertion point on screen, you can right-click to display the shortcut menu, which has all the options available through the command line. However, if you have already specified an insertion point in the dialog box and have selected the **Specify on-screen** check boxes in the **Scale** or **Rotation** areas of the **Insert** dialog box, you are allowed to specify only the scale factors or the rotation angle through the command line. You can also use the command line where the following prompts appear:

Specify insertion point or [Scale/X/Y/Z/Rotate/PScale/PX/PY/PZ/PRotate]:
Enter X scale factor, specify opposite corner, or [Corner/XYZ] <1>:
Enter Y scale factor <use X scale factor>:
Specify rotation angle <0>:

Tip

Using the command line allows you to override scale factors and rotation angles already specified in the dialog box.

The command line options available are:

Scale	**X**	**Y**	**Z**	**Rotate**
PScale	**PX**	**PY**	**PZ**	**PRotate**

Scale

If you enter S at the **Specify Insertion point** prompt, you are asked to enter the scale factor.

After entering the scale factor, the block assumes the specified scale factor, and AutoCAD LT lets you drag the block until you locate the insertion point on the screen. The X, Y, and Z axes are uniformly scaled by the specified scale factor. Using this option the prompt sequence is:

Specify insertion point or [Scale/X/Y/Z/Rotate/PScale/PX/PY/PZ/PRotate]: **S**
Specify scale factor for XYZ axes: *Enter a value to preset general scale factor.*
Specify insertion point: *Select the insertion point.*
Specify rotation angle <0>: *Specify angle of rotation.*

X, Y, Z

With X, Y, or Z as the response to the **Specify insertion point** prompt, you can specify the X, Y, or Z scale factors before specifying the insertion point. The prompt sequence when you use the **X** option is:

Specify insertion point or [Scale/X/Y/Z/Rotate/PScale/PX/PY/PZ/PRotate]: **X**
Specify X scale factor: *Enter a value for the X scale factor.*
Specify insertion point: *Select the insertion point.*
Specify rotation angle <0>: *Specify angle of rotation.*

Rotate

If you enter R at the **Specify insertion point** prompt, you are asked to specify the rotation angle. You can enter the angle of rotation or specify the angle by specifying two points on the screen. Dragging is resumed only when you specify the angle, and the block assumes the specified rotation angle. The prompt sequence for this option is:

Specify insertion point or [Scale/X/Y/Z/Rotate/PScale/PX/PY/PZ/PRotate]: **R**
Specify rotation angle: *Enter a value for the rotation angle.*
Specify insertion point: *Select the insertion point.*
Enter X scale factor, specify opposite corner, or [Corner/XYZ] <1>: *Specify the X scale factor.*
Enter Y scale factor <Use X scale factor>: *Specify the Y scale factor.*

PScale (Preview Scale)

The **PScale** option is similar to the **Scale** option. The difference is that in the case of the **PScale** option, the scale factor specified reflects only in the display of the block while it is dragged to a desired position. This implies that you can preview the effect of this scale factor on the block before actually applying it. After you have specified the insertion point for the block, you are again prompted for a scale factor. If you do not enter the scale factor, the block is drawn exactly as it was when created. The prompt sequence for this option is:

Specify insertion point or [Scale/X/Y/Z/Rotate/PScale/PX/PY/PZ/PRotate]: **PS**
Specify preview scale factor for XYZ axes: *Enter a value to preview the general scale factor.*
Specify insertion point: *Specify the insertion point for the block.*
Enter X scale factor, specify opposite corner, or [Corner/XYZ] <1>: *Specify the X scale factor.*
Enter Y scale factor <Use X scale factor>: *Specify the Y scale factor.*
Specify rotation angle <0>: *Specify angle of rotation.*

PX, PY, PZ (Preview X, Y, Z)

These options are similar to the **PScale** option. The only difference is that the X, Y, or Z scale factors specified are reflected only in the display of the block while it is dragged to a desired position. After you have specified the insertion point for the block, you are again prompted for the X scale factor and the Y scale factor. If you do not enter the scale factor, the block is drawn exactly as it was when created.

Specify insertion point or [Scale/X/Y/Z/Rotate/PScale/PX/PY/PZ/PRotate]: **PX**
Specify preview X scale factor: *Enter a value to preview the X scale factor.*
Specify insertion point: *Specify the insertion point for the block.*
Enter X scale factor, specify opposite corner, or [Corner/XYZ] <1>: *Specify the X scale factor.*
Enter Y scale factor <Use X scale factor>: *Specify the Y scale factor.*
Specify rotation angle <0>: *Specify angle of rotation.*

Note

If you have already specified scale factors in the ***X, Y,*** *and* ***Z*** *edit boxes of the* ***Scale*** *area of the* ***Insert*** *dialog box, the* ***PScale, PX, PY,*** *and* ***PZ*** *options work equivalent to the* ***Scale, X, Y,*** *and* ***Z*** *options respectively and allow you to insert the block with the PScale, PX, PY, or PZ values you specified.*

PRotate (Preview Rotate)

The **PRotate** option is similar to the **Rotate** option. The difference is that the rotation angle specified reflects only in the display of the block while it is dragged to a desired position. After you have specified the insertion point for the block, you are again prompted for an angle of rotation. If you do not enter the rotation angle, the block is drawn exactly as it was when created (with the same angle of rotation as specified on the creation of the block). The prompt sequence is:

Specify insertion point or [Scale/X/Y/Z/Rotate/PScale/PX/PY/PZ/PRotate]: **PR**
Specify preview rotation angle: *Specify the preview value for rotation.*
Specify insertion point: Specify the insertion point for the block.
Enter X scale factor, specify opposite corner, or [Corner/XYZ]: *Specify the X scale factor.*
Enter Y scale factor < Use X scale factor>: *Specify the Y scale factor.*
Specify rotation angle <0>: *Specify angle of rotation.*

Note

If you have already specified rotation angle in the ***Angle*** *edit box of the* ***Rotation*** *area of the* ***Insert*** *dialog box, the* ***PRotate*** *option works equivalent to the* ***Rotate*** *option and allows you to insert the block with the PRotate value you specified.*

Tip

The preview options (PScale, PX, PY, PZ and PRotate) that are preceded with a P allow you to view the effect of the scale factor or rotation angle specified before actually applying it to the block reference.

The XYZ and Corner Options

When you are specifying the insertion point and scale factors on screen, the following prompt appears after you have selected an insertion point:

Enter X scale factor, specify opposite corner or [Corner/XYZ]: *Enter X scale factor, specify opposite corner, or select an option.*

Here, the **XYZ** option can be used to enter a 3D block reference and the successive prompts allow you to specify the X, Y, and Z scale factors individually.

With the **Corner** option, you can specify both X and Y scale factors at the same time. When you invoke this option, and **DRAGMODE** is turned off, you are prompted to specify the other corner (the first corner of the box is the insertion point). You can also enter a coordinate value instead of moving the cursor. The length and breadth of the box are taken as the X and Y scale factors for the block. For example, if the X and Y dimensions (length and width) of the box are the same as that of the block, the block will be drawn without any change. Points selected should be above and to the right of the insertion block. Mirror images will be produced if points selected are below or to the left of the insertion point. The prompt sequence is:

Enter X scale factor, specify opposite corner or [Corner/XYZ]: **C**
Specify opposite corner: *Select a point as the corner.*

The **Corner** option also allows you to use a dynamic scaling technique when **DRAGMODE** is turned to Auto (default). You can also move the cursor at the **Enter X scale factor...** prompt to change the block size dynamically and, when the size meets your requirements, select a point.

Note

*You should try to avoid using the **Corner** option if you want to have identical X and Y scale factors because it is difficult to select a point whose X distance equals its Y distance, unless the **SNAP** mode is on. If you use the **Corner** option, it is better to specify the X and Y scale factors explicitly or select the corner by entering coordinates.*

-INSERT Command

Predefined blocks can also be inserted in a drawing with the **-INSERT** command, using the command line. When using the **-INSERT** command, the dialog box is not displayed. To insert a drawing file as a block, enter the file name at the **Enter block name or** [**?**] prompt. You can enter ? if you do not know the name of the blocks in the drawing. To create a block with a different name from the drawing file, enter **blockname= file name** at the **Enter block name or[?]** prompt. If you enter **(~)** tilde character at the **Enter block name or [?]** prompt, the **Select Drawing File** dialog box is displayed on the screen. If you enter an **(*)** asterisk before entering the block name, the block is automatically exploded upon insertion. All the command line options discussed earlier are available through the **-INSERT** command. Also, if you make changes to a block and want to update the block in the existing drawing, enter **blockname=** at the **Specify insertion point** prompt. AutoCAD LT will display the prompt:

Block "current" already exists. Redefine it? [Yes/No] <No>: *Enter y to redefine it.*

Note

If you create a template file that has blocks created and stored in it and when you use this template file to create a new drawing, the blocks are available in the new drawing also.

Also when you insert a drawing into a current drawing, all the blocks belonging to the inserted drawing are also brought into the current drawing.

Example 1 — *General*

Using the dialog box, insert a block SQUARE into a drawing. The block insertion point is (1,2). The X scale factor is 2 units, the Y scale factor is 2 units, and the angle of rotation is 35 degrees. It is assumed that the block SQUARE is already defined in the current drawing.

Choose the **Insert Block** button from the **Draw** toolbar. The **Insert** dialog box is displayed on the screen. In the **Name** edit box, enter SQUARE or select the Block SQUARE from the **Name** drop-down list. Clear the **Specify On-screen** check box. Now, in the **Insertion point** area, enter 1 in the **X** edit box and 2 in the **Y** edit box. In the **Scale** area, enter 2 in the **X** edit box and 2 in the **Y** edit box. In the **Rotation** area, enter 35 in the **Angle** edit box. Choose the **OK** button to insert the block on the screen.

Exercise 2 — *General*

a. Insert the block CIRCLE created in Exercise 1. Use different X and Y scale factors to get different shapes after inserting this block.
b. Insert the block CIRCLE created in Exercise 1. Use the **Corner** option to specify the scale factor.

Exercise 3 — *General*

a. Construct a triangle and form a block of it. Name the block TRIANGLE. Now, set the Y scale factor of the inserted block as 2.
b. Insert the block TRIANGLE with a rotation angle of 45 degrees. After defining the insertion point, enter an X and a Y scale factor of 2.

USING AUTOCAD DESIGNCENTER TO INSERT BLOCKS

You can use the **AutoCAD DesignCenter** to locate, preview, copy or insert blocks or existing drawings into a current drawing. On the **Standard** toolbar, choose the **AutoCAD DesignCenter** button to display the **Design Center** window (Figure 14-11). In this window, choose the **Tree View Toggle** button to display the **tree pane** on the left side, if not already displayed. By default, the tree pane is displayed. Expand **My Computer** to display the C:/Program Files/AutoCAD LT 2002/Sample/DesignCenter folder by clicking on the plus (+) signs on the left of the respective folders. Click on the plus sign adjacent to the folder to display its contents. Select a drawing file you wish to use to insert blocks from and then click on the plus sign adjacent to the drawing again. All the icons depicting the different types of content in the selected drawing are displayed, that is, Blocks, Dimstyles, Layers, Linetypes, Textstyles, and so on. Select **Blocks** by clicking on it and a list of blocks in the drawing are displayed in the palette. Select the block you wish to insert, and drag and drop the block into the current drawing. Later, you can move it in the drawing to a desired location. You can also right-click on the block name to display a shortcut menu. Select **Insert Block** from the shortcut menu. The **Insert** dialog box is displayed. Here, you can specify the **Insertion point**, **scale**, **rotation angle** in the respective edit boxes. If you select the **Specify On-Screen** check boxes in the **Insert** dialog box, you are allowed to specify these parameters on the screen. The selected block is inserted into the current drawing based on the type of units specified when the block was created. For example, if you had created a block of 1" x 1", and specified feet as units at that time, the block when inserted using the AutoCAD DesignCenter, will be 12" x 12". You can also choose **Copy** from the shortcut menu; the selected block is copied to the clipboard. Now, you can right-click in the drawing area to display a shortcut menu and choose **Paste**. The selected block is displayed attached to the cursor and as you move the cursor, the block also moves with it. You can now select a point where you want to paste the block. The advantage of this option over drag and drop is that you can select an insertion point at the time of pasting. In the case of a drag and drop operation, you need to move the inserted block to a specific point later and it is an additional step.

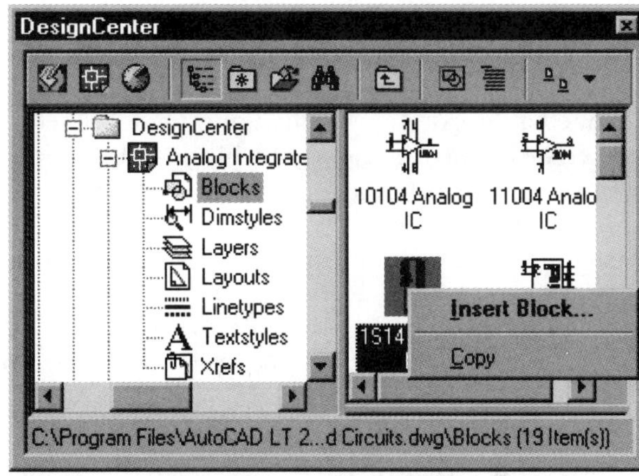

Figure 14-11 *Using the DesignCenter to insert blocks*

If you want to insert an entire drawing, select the folder in which the drawing resides by double-clicking on the folder. All the contents in the folder are displayed in the palette. Drag and drop the desired drawing into the current drawing. You can also select the drawing and right-click to display a shortcut menu. Choose **Insert as Block** from the shortcut menu. The **Insert** dialog box is displayed, where you can specify the Insertion point, scale factors, and rotation angle, if required. You can also select the **Specify On-screen** check boxes to select everything on the screen.

Note

For a detailed explanation of the DesignCenter, see Chapter 6, Editing Commands-II.

LAYERS, COLORS, LINETYPES, AND LINEWEIGHTS FOR BLOCKS

A block possesses the properties of the layer on which it is drawn. The block may be comprised of objects drawn on several different layers, with different colors, linetypes, and lineweights. All this information is preserved in the block. At the time of insertion, each object in the block is drawn on its original layer with the original linetype, lineweight, and color, irrespective of the current drawing layer, object color, object linetype, and object lineweights. You may want all instances of a block to have identical layers, linetype properties, lineweight, and color. This can be achieved by allocating all the properties explicitly to the objects forming the block. On the other hand, if you want the linetype and color of each instance of a block to be set according to the linetype and color of the layer on which it is inserted, draw all the objects forming the block on layer 0 and set the color, lineweight, and linetype to BYLAYER. Objects with a BYLAYER color, linetype, and lineweight can have their colors, linetypes, and lineweights changed after insertion by changing the layer settings. If you want the linetype, lineweight, and color of each instance of a block to be set according to the current explicit linetype, lineweight, and color at the time of insertion, set the color, lineweight, and linetype of its objects to BYBLOCK. You can use the **PROPERTIES**, **CHPROP,** or **CHANGE** command to change some of the characteristics associated with a block (such as layer).

Note

The block is inserted on the layer that is current, but the objects comprising the block are drawn on the layers on which they were drawn when the block was being defined.

For example, assume block B1 includes a square and a triangle that were originally drawn on layer X and layer Y, respectively. Let the color assigned to layer X be red and to layer Y be green. Also, let the linetype assigned to layer X be continuous and for layer Y be hidden. Now, if you insert B1 on layer L1 with color yellow and linetype dot, block B1 will be on layer L1, but the square will be drawn on layer X with color red and linetype continuous, and the triangle will be drawn on layer Y with color green and linetype hidden.

The **BYLAYER** option instructs AutoCAD LT to assign objects within the block the color and linetype of the layers on which they were created. There are three exceptions:

1. If objects are drawn on a special layer (layer 0), they are inserted on the current layer. These objects assume the characteristics of the current layer (the layer on which the block is inserted) at the time of insertion, and can be modified after insertion by changing that layer's settings.

2. Objects created with the special color BYBLOCK are generated with the color that is current at the time of insertion of the block. This color may be explicit or BYLAYER. You are thus allowed to construct blocks that assume the current object color.

3. Objects created with the special linetype BYBLOCK are generated with the linetype that is prevalent at the time the block is inserted. Blocks are thus constructed with the current object linetype, which may be BYLAYER or explicit.

Note

If a block is created on a layer that is frozen at the time of insertion, the block is not shown on the screen.

Tip

If you are providing drawing files to others for their use, using only BYLAYER settings provide the greatest compatibility with varying office standards for layer/color/linetype/lineweight, because they can be changed more easily after insertion.

NESTING OF BLOCKS

The concept of having one block within another block is known as the **nesting of blocks**. For example, you can insert several blocks by selecting them and then with the **BLOCK** command, create another block. Similarly, if you use the **INSERT** command to insert a drawing that contains several blocks into the current drawing, it creates a block containing nested blocks in the current drawing. There is no limit to the degree of nesting. The only limitation in nesting of blocks is that blocks that reference themselves cannot be inserted. The nested blocks must have different block names. Nesting of blocks affects layers, colors, and linetypes. The general rule is as follows.

If an inner block has objects on layer 0, or objects with linetype or color BYBLOCK, these objects may be said to behave like fluids. They "float up" through the nested block structure until they find an outer block with fixed color, layer, or linetype. These objects then assume the characteristics of the fixed layer. If a fixed layer is not found in the outer blocks, then the objects with color or linetype BYBLOCK are formed; that is, they assume the color white and the linetype CONTINUOUS.

Example 2 *General*

To make the concept of nested blocks clear, consider the following example:

1. Draw a rectangle on layer 0, and form a block of it named X.
2. Change the current layer to OBJ, and set its color to red and linetype to hidden.
3. Draw a circle on OBJ layer.
4. Insert the block X in the OBJ layer.
5. Combine the circle with the block X (rectangle) to form a block Y.
6. Now, insert block Y in any layer (say, layer CEN) with color green and linetype continuous.

You will notice that block Y is generated in color red and linetype hidden. Normally, block X, which is nested in block Y and created on layer 0, should have been generated in the color (green) and linetype (continuous) of the layer CEN. The reason for this is that the object (rectangle) on layer 0 floated up through the nested block structure and assumed the color and linetype of the first outer block (Y) with a fixed color (red), layer (OBJ), and linetype (hidden). If both the blocks (X and Y) were on layer 0, the objects in block Y would assume the color and linetype of the layer on which the block was inserted.

Example 3 *General*

1. Change the color of layer 0 to red.
2. Draw a circle and form a block of it, B1, with color BYBLOCK. It appears white because the color is set to BYBLOCK (Figure 14-12).
3. Set the color to BYLAYER and draw a square. The color of the rectangle is red.
4. Insert block B1. Notice that the block B1 (circle) assumes red color.
5. Create another block B2 consisting of Block B1 (circle) and rectangle.
6. Create a layer L1 with green color and hidden linetype. Insert block B2 in layer L1.
7. Explode block B2. Notice the change.
8. Explode block B1, circle. You will notice that the circle changes to white because it was drawn with color set to BYBLOCK.

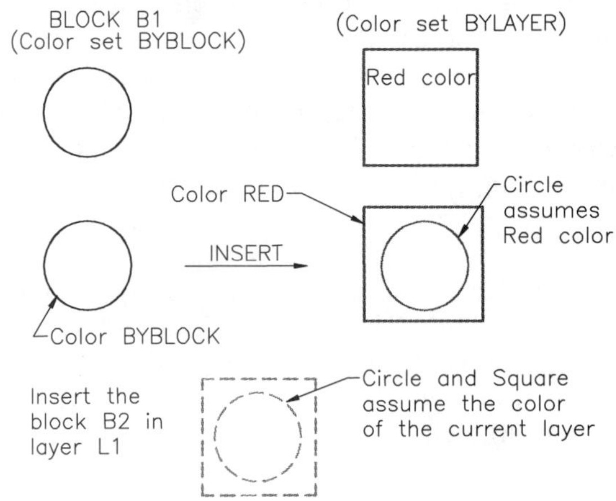

***Figure 14-12** Blocks versus layers and colors*

Exercise 4 *General*

Part A

1. Draw a unit square on layer 0 and make it a block named B1.
2. Make a circle with radius 0.5 and change it into block B2.
3. Insert block B1 into a drawing with an X scale factor of 3 and a Y scale factor of 4.
4. Now, insert the block B2 in the drawing and position it at the top of B1.
5. Make a block of the entire drawing and name it PLATE.
6. Insert the block Plate in the current layer.
7. Create a new layer with different colors and linetypes and insert blocks B1, B2, and Plate.

Keep in mind the layers on which the individual blocks and the inserted block were made.

Part B

Try nesting the blocks drawn on different layers and with different linetypes.

Part C

Change the layers and colors of the different blocks you have drawn so far.

CREATING DRAWING FILES (WBLOCK COMMAND)

Command: WBLOCK

The blocks or symbols created by the **BLOCK** command can be used only in the drawing in which they were created. This is a shortcoming because you may need to use a particular block in different drawings. The **WBLOCK** command is used to export symbols by writing them to new drawing files that can then be inserted in any drawing. With the **WBLOCK** command, you can create a drawing file (.DWG extension) of specified blocks, selected objects in the current drawing, or the entire drawing. All the used named objects (linetypes, layers, styles, and system variables) of the current drawing are inherited by the new drawing created with the

WBLOCK command. This block can then be inserted in any drawing.

When the **WBLOCK** command is invoked, the **Write Block** dialog box is displayed (Figure 14-13). This dialog box converts blocks into drawing files and also saves objects as drawing files. You can also save the entire current drawing as a new drawing file.

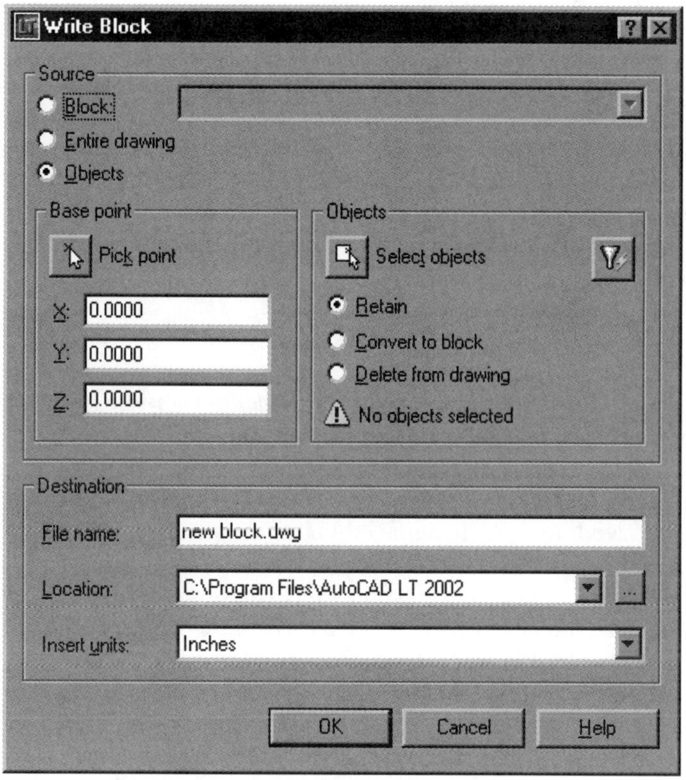

***Figure 14-13 Write Block** dialog box*

The **Write Block** dialog box has two main areas: **Source** and **Destination**. The **Source** area allows you to select objects and blocks, specify insertion base points and convert them into drawing files. In this area of the dialog box, depending on the selection you make, different default settings are displayed. By default, the **Objects** radio button is shown as selected and in the **Destination** area, the **File name** edit box displays **new block.dwg** as the new file name. The **Block** drop-down list is also not available. Now, you can select objects in a drawing and save them as a wblock, and can enter a name and a path for the file. If the current drawing consists of blocks and if you want to save a block as a wblock, you can select the **Block** radio button which is available. When the **Block** radio button is selected, the **Block** drop-down list is available. The **Block** drop-down list displays all the block names in the current drawing and you can select a name of the block that you want to convert into a wblock. The **Base point** and **Objects** areas are not available, since the insertion points and objects have already been saved with the block definition. Also, you will notice that in the **Destination** area, the **File Name** edit box displays the name of the selected block by default. This means that you can keep the name of the wblock the same as the selected block or you can change the name if you want to. Selecting the **Entire drawing** radio button selects the current drawing as a block and saves it as a new file. When you use this option, the **Base point** and **Objects** areas are not available.

The **Base point** area allows you to specify the base point of a wblock which it uses as an insertion point. You can either enter values in the **X**, **Y**, and **Z** edit boxes or choose the **Pick Insertion Base Point** button to select it on the screen. The default value is 0,0,0. The **Objects** area allows you to select objects to save as a file. You can use the **Select objects** button to select objects or use the **Quick Select** button to set parameters in the **Quick Select** dialog box to select objects in the current drawing. The number of objects selected is displayed at the bottom of the **Objects** area. If the **Retain** radio button is selected in the **Objects** area, the selected objects in the current drawing are kept as such after they have been saved as a new file. If the **Convert to block** radio button is selected, the selected objects in the current drawing are converted into a block with the same name as the new file after being saved as a new file. Selecting the **Delete** radio button deletes the selected objects from the current drawing after they have been saved as a file.

Note

*Both the **Base point** and **Objects** areas are available in the **Write Block** dialog box only when the **Objects** radio button is selected in the **Source** area of the dialog box.*

The **Destination** area sets the File name, location, and units of the new file in which the selected objects are saved. In the **File name** edit box you can specify the file name of the block or the selected objects. From the **Location** drop-down list you can select a path for the new file. You can also choose the **[...]** button to display the **Browse for folder** window, where you can specify a path for where the new file will be saved. From the **Insert Units** drop-down list, you can select the units the new file will use when inserted as a block. The settings for units are stored in the **INSUNITS** system variable and the default option Inches has a value of 1. On specifying the required information in the dialog box, choose **OK**. The objects or the block is saved as a new file in the path specified by you. A **WBLOCK Preview** window with the new file contents is displayed. This preview image is stored and displayed in the AutoCAD DesignCenter, when using it to insert drawings and blocks.

Note

Whenever a drawing is inserted into a current drawing, it acts as a single object. It cannot be edited unless exploded.

Tip

*Using the **Entire drawing** radio button in the **Write Block** dialog box to save the current drawing as a new drawing is a good way to reduce the drawing file size, since all unused blocks, layers, linetypes, text styles, dimension styles, multiline styles, shapes, and so on are removed from the drawing. The new drawing does not contain any information which is no longer needed. This **Entire drawing** option is faster than the **PURGE** command (which also removes unused named objects from a drawing file) and can be used whenever you have completed a drawing and want to save it. A part of the **PURGE** command is discussed in this chapter. Note that the **PURGE** command is discussed in detail in Chapter 18, Object Grouping and Editing Commands.*

Using the -WBLOCK Command

The **-WBLOCK** command can be used when you want to use the command line. When this command is invoked, AutoCAD LT displays the **Create Drawing File** dialog box, except when the system variable **FILEDIA** is set to 0. This dialog box displays a list of all the drawing files in the current directory. Enter the name of the output drawing file in the **File name** edit box and then choose the **Save** button. If the file name you specify is identical to some other file in the same directory path, AutoCAD LT displays a message saying that a file with the same name already exists and asks if you want to replace it. After choosing the **Save** button, the dialog box is cleared from the screen, and AutoCAD LT issues the following prompt asking you for the name of an existing block that you want to convert into a permanent symbol:

Enter name of existing block or
[=(block=output file)/*(whole drawing)] <define new drawing>: *Enter the name of a predefined block.*

The drawing file with the name specified in the dialog box is created from the predefined block.

If you want to assign the same name to the output drawing file as that of the existing block, enter an (=) equal sign as a response to the **Enter name of existing block** prompt and enter the same file name in the **File name** edit box of the **Create Drawing File** dialog box. If a block by the name you have already assigned to the output file does not exist in the current drawing, AutoCAD LT gives an appropriate message that no block by that name exists in the current drawing, and the **Enter name of existing block or** prompt is repeated. This option is the same as the **Block** option of the **WBLOCK** command.

If you have not yet created a block, but want to create an output drawing file of the objects in the current drawing, press ENTER at the **Enter name of existing block** prompt. You are then required to select an insertion point and select the objects to be incorporated in the drawing file. This option is the same as the **Objects** option of the **WBLOCK** command.

You can also store an entire drawing as a wblock. In other words, the current file is copied into a new one specified in the **File name** edit box. To do so, respond to the prompt **Enter name of existing block** with an asterisk (*). Here, the entire drawing is saved in the file in the directory you have specified. This option is the same as the **Entire drawing** option of the **WBLOCK** command.

Export Data Dialog Box

You can also use the **Export Data** dialog box (Figure 14-15) to create drawing files. This dialog box can be invoked by choosing **Export** from the **File** menu, or by entering **EXPORT** at the Command prompt. In this dialog box, you can select **Block (*.dwg)** from the **Save as type** drop-down list, enter the name of the drawing file in the **File name** edit box, and then choose the **Save** button. The dialog box is cleared from the screen and AutoCAD LT issues the same prompts as the **-WBLOCK** command discussed earlier.

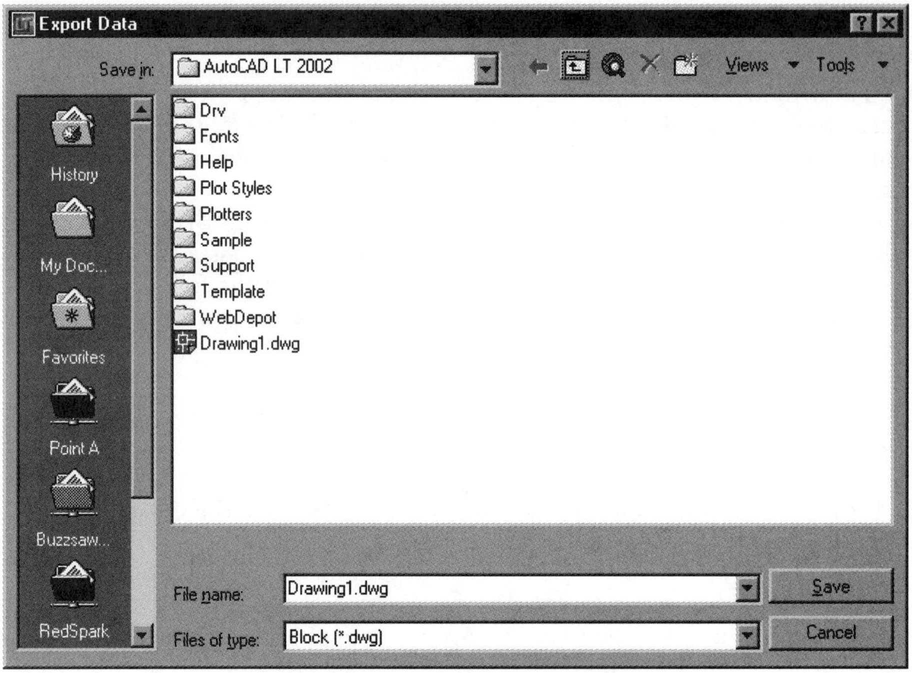

***Figure 14-14 Export Data** dialog box*

DEFINING THE INSERTION BASE POINT (BASE COMMAND)

Menu:	Draw > Block > Base
Command:	BASE

The **BASE** command lets you set the insertion base point for a drawing just as you set the base insertion point using the **BLOCK** command. This base point is defined so that when you insert the drawing into some other drawing, the specified base point is placed on the insertion point. By default, the base point is at the origin (0,0,0). When a drawing is inserted on a current layer, it does not inherit the color, linetype, or thickness properties of the current layer. When you invoke the **BASE** command, the prompt sequence is:

Enter base point <0.0000,0.0000,0.0000>: *Specify a base point or press ENTER to accept the default.*

Tip

*You can insert a drawing that you want to refer to for checking dimensions or certain features into the current drawing for reference. Then later you can use the **UNDO** command to delete the inserted drawing. This way you can save stationery and time required for printing.*

Note

Drawings inserted into a current drawing become part of the current drawing and increase the file size. Also, when the drawing is modified it does not get updated in the drawing; it has been inserted automatically. Therefore, it is better to xref a drawing into the current drawing instead. External references are discussed in Chapter 16, External References.

Exercise 5 — *General*

a. Create a drawing file named CHAIR using the **WBLOCK** command. Get a listing of your .DWG files and make sure that CHAIR.DWG is listed. Quit the drawing editor.
b. Begin a new drawing and insert the drawing file into the drawing. Save the drawing.

BREAKING UP A BLOCK

A block may be comprised of different basic objects such as lines, arcs, polylines, and circles. All these objects are grouped together in the block and are treated as a single object. To edit any particular object of a block, the block needs to be "exploded," or split into independent parts. This is especially useful when an entire view or drawing has been inserted and a small detail needs to be corrected. This can be done in the following ways.

Selecting the Explode Check Box in the Insert Dialog Box

As discussed earlier, individual objects within a block cannot be edited unless the block is broken up (exploded). You can explode a block at the time of insertion by simply selecting the **Explode** check box before you choose **OK** in the **Insert** dialog box. Now, when the block is inserted, it would have been already exploded into its component objects.

Entering * as the Prefix of Block Name

To insert a block as a collection of individual objects, you can also enter an asterisk before its name, when using the command line. The prompt sequence for exploding the block BLOCKPLAN while inserting is:

Command: **-INSERT**
Enter block name or [?] <current>: ***BLOCKPLAN**
Specify insertion point or [Scale/X/Y/Z/Rotate/PScale/PX/PY/PZ/PRotate]: *Select the insertion point.*
Scale factor <1>: [Enter]
Specify rotation angle <0>: [Enter]

The inserted object will no longer be treated as a block. The figure consists of various objects that can be edited separately.

Note

Blocks with different X, Y, and Z values or nonuniformly scaled blocks can also be exploded. This was not possible in releases prior to the Release 13 of AutoCAD LT.

Using the EXPLODE Command

Toolbar:	Modify > Explode
Menu:	Modify > Explode
Command:	EXPLODE

As mentioned before, the other method to break a block into individual objects is by using the **EXPLODE** command (Figure 14-15). When you invoke the **EXPLODE** command you are prompted to select the block you want to explode.

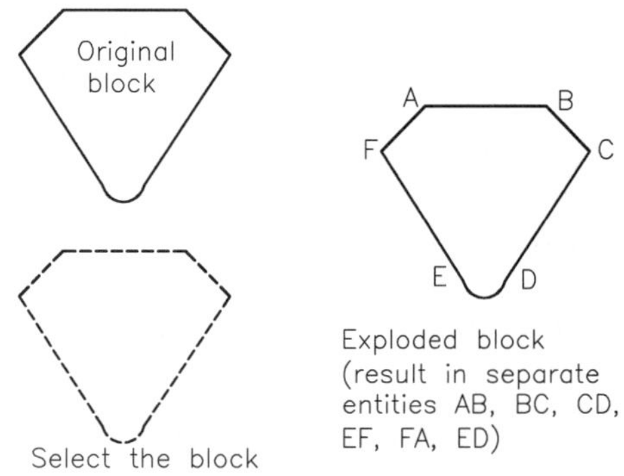

Figure 14-15 *Using the **EXPLODE** command*

Once you have selected the block you want to explode, press ENTER. The **EXPLODE** command explodes a block into its component objects regardless of the scale factors. It also does not have control over the properties such as, layer, linetype, lineweight, and color of the component objects. When a block is exploded, there may be no visible change in the drawing. The drawing is identical except that the color and linetype may have changed because of floating layers, colors, or linetypes. The exploded block is now a group of objects that can be edited individually. To check whether the breaking of the block has taken place, select any object that was formerly a part of the block; only that particular object should be highlighted. After a block is exploded, the block definition continues to be in the block symbol table. After exploding a block you can modify it and then redefine it by using the **BLOCK** command. The **Block Definition** dialog box is displayed where you can select the name of the block you wish to redefine from the **Name** drop-down list. Then, use the dialog box options and redefine it. You can also use the **-BLOCK** command to redefine a block. Once you redefine a block, it automatically gets updated in the current drawing and for future use.

Note

Blocks inserted with External references cannot be exploded.

While redefining a block, if you are using one of its previous insertions, you should explode the block into its basic components. If you do not explode the block, AutoCAD LT displays a warning message that selected block references itself and the block will not be redefined.

Using the XPLODE Command

Command:	XPLODE

With the **XPLODE** command, you can explode a block or blocks into component objects and simultaneously control their properties such as layer, linetype, color, and lineweight. The prompts sequence that will follow when you invoke this command is as follows:

Command: **XPLODE**
Select objects: *Use any object selection method and select objects and then press ENTER.*

On pressing ENTER, AutoCAD LT reports the total number of objects selected and also the number of objects that cannot be exploded. It also further prompts for an option that decides whether changes in the properties of the component objects will be made individually or globally. The prompt is:

> Enter an option [Individually/Globally] <Globally>: *Enter i, g, or press ENTER to accept the default option.*

If you enter **i** at the above prompt, AutoCAD LT will modify each object individually, one at a time. The next prompt is:

> Enter an option [All/Color/LAyer/LType/Inherit from parent block/Explode] <Explode>: *Select an option.*

All the options available are discussed next.

All

Sets all the properties such as color, layer, linetype, and lineweight of selected objects after exploding them. AutoCAD LT prompts you to enter new color, linetype, lineweight, and layer name for the exploded component objects.

Color

Sets the color of the exploded objects. The prompt is:

> Enter new color for exploded objects [Red/Yellow/Cyan/Blue/Magenta/White/BYLayer/BYBlock]<BYLAYER>: *Enter a color, option, or press ENTER.*

When you enter BYLAYER, the component objects take on the color of the exploded object's layer and when you enter BYBLOCK, they take on the color of the exploded object.

Layer

Sets the layer of the exploded objects. The default option is inheriting the current layer. The command prompt is as follows:

> Enter new layer name for exploded objects <current>: *Enter an existing layer name or press ENTER.*

LType

Sets the linetype of the components of the exploded object. The command prompt is:

> Enter new linetype name for exploded objects <BYLAYER>: *Enter a linetype name or press ENTER to accept the default.*

Inherit from parent block

Sets the properties of the component objects to that of the exploded parent object, provided the component objects are drawn on layer 0 and the color, lineweight, and linetype are BYBLOCK.

Explode

Explodes the selected object exactly as in the **EXPLODE** command.

Selecting the **Globally** option, applies changes to all the selected objects at the same time and the options are similar to the ones discussed in the **Individually** option.

Tip

Using the ***XPLODE*** *command is better than the* ***EXPLODE*** *command since it gives you an added advantage of determining the properties of the exploded components of the block.*

RENAMING BLOCKS

Menu:	Format > Rename
Command:	RENAME

Blocks can be renamed with the **RENAME** command. AutoCAD LT displays the **Rename** dialog box (Figure 14-16) which allows you to modify the name of an existing block.

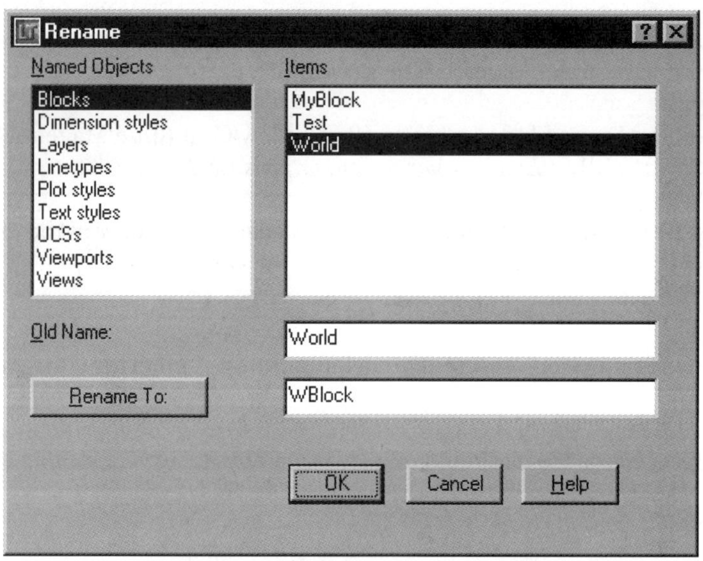

***Figure 14-16 Rename** dialog box*

In the **Rename** dialog box, the **Named Objects** list box displays the categories of object types that can be renamed, such as blocks, layers, dimension styles, linetypes, plot and text styles, UCSs, views, and viewports. You can rename all of these except layer 0 and continuous linetype. When you select **Blocks** from the **Named Objects** list, the **Items** list box displays all the block names in the current drawing. When you select a block name you want to rename from the **Items** list box, it is displayed in the **Old Name** edit box. Enter the new name to be assigned to the block in the **Rename To** edit box. Choosing the **Rename To** button applies the change in name to the old name. Choose **OK** to exit the dialog box. For example, if you want to rename a block named Drawing1 to Drawing2, select Drawing1 from the Items list box. It is displayed in the **Old Name** edit box. Enter Drawing2 in the **Rename To** edit box and choose the **Rename To** button. Drawing2 appears in the **Items** list box. Now choose **OK** to exit the dialog box.

You can also rename a block using the command line, if you enter **-RENAME** at the Command prompt.

Note
*The layer 0 and Continuous linetype cannot be renamed and, therefore do not appear in the **Items** list box, when Layers and Linetypes is selected in the named **Objects** list box.*

DELETING UNUSED BLOCKS (PURGE Command)

Menu:	File > Drawing Utilities > Purge
Command:	PURGE

Sometimes after completing a drawing, you may notice that the drawing contains several named objects, such as dimstyles, textstyles, layers, blocks and so on which are not being used. Since these unused named objects unnecessarily occupy disk space, you may want to remove them. Unused blocks can be deleted with the **PURGE** command. For example, if you want to delete an unused block named WBlock, invoke the **PURGE** command. When you invoke this command, the **Purge** dialog box will be displayed as shown in Figure 14-17.

Click on the + sign in front of the **Blocks** to display all the blocks in the current drawing. Select the block with the name **WBlock** from this list and then choose the **Purge** button to remove this block from the current drawing.

Tip
*You can also use the command line to remove the unused blocks from the drawing. This can be done using the **-PURGE** command.*

Note
*If you precede the name of the wblock with an (*) asterisk when entering the name of the wblock while using the **-WBLOCK** command, or select the **Entire drawing** radio button in the **Write Drawing** dialog box when creating a wblock using the **WBLOCK** command, it will have the same effect as using the **PURGE** command. But, the **WBLOCK** command is faster and deletes the unused named objects automatically, while the **PURGE** command allows you to select the type of named objects you want to delete, and it also gives you an option to verify the objects before deletion occurs.*

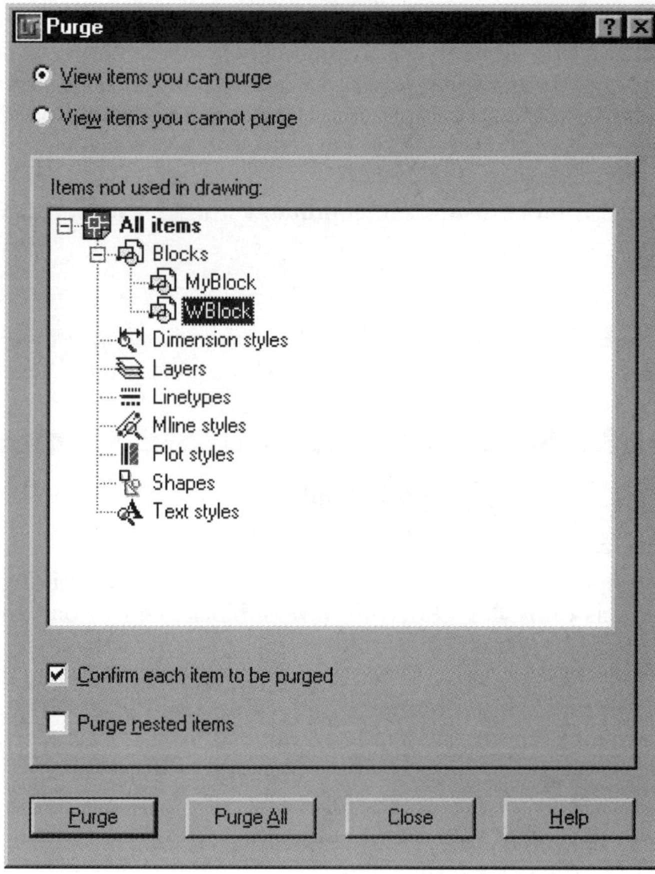

***Figure 14-17 Purge** dialog box to remove the unused blocks*

Self-Evaluation Test

Answer the following questions, and then compare your answers with the correct answers given at the end of this chapter.

1. Individual objects in a block can be erased. (T/F)
2. A block can be mirrored by providing a scale factor of -1 for X. (T/F)
3. Blocks created by the **BLOCK** command can be used in any drawing. (T/F)
4. An existing block cannot be redefined. (T/F)
5. The ____________ command lets you create a drawing file (.DWG extension) of a block defined in the current drawing. (T/F)

6. The ____________ command can be used to change the name of a drawing file. (T/F)

7. The _________ command is used to place a previously created block in a drawing.

8. You can delete unreferenced blocks with the _________ command.

9. You can use the _________ to locate, preview, copy, and insert blocks or existing drawings into the current drawing.

10. The _________ command is used for creating the blocks using the command line.

Review Questions

Answer the following questions.

1. An entire drawing can be converted into a block. (T/F)

2. The objects in a block possess the properties of the layer on which they are drawn, such as color and linetype. (T/F)

3. If the objects forming a block were drawn on layer 0 with color and linetype BYLAYER, then at the time of the insertion, each object that makes up a block is drawn on the current layer with the current linetype and color. (T/F)

4. Objects created with the special color BYBLOCK are generated with the color that is current at the time the block was inserted. (T/F)

5. The concept of having one block inside the other block is called nesting of blocks. (T/F)

6. The block name is controlled by which system variable?

   (a) **EXTNAMES** (b) **EXTMIN**
   (c) **EXTMAX** (d) **SAVENAME**

7. Which command should you use if you want to get back the objects that consist of the block and that have been removed from the drawing?

   (a) **OOPS** (b) **BLIPS**
   (c) **BLOCK** (d) **UNDO**

8. By what amount is a block rotated if the values of both the X and Y scale factors is -1?

   (a) 90 (b) 180
   (c) 270 (d) 360

9. When you insert a drawing into a current drawing, how many of the blocks belonging to the inserted drawing are brought into the current drawing?

   (a) One (b) None
   (c) All (d) Two

10. Which option of the **INSERT** command allows you to preview the rotation angle applied to the object without actually inserting it?

    (a) **Rotate** (b) **PRotate**
    (c) **PY** (d) **PX**

11. **The WBLOCK**-asterisk method or the **Entire drawing** option of the **WBLOCK** command have the same effect as the **PURGE** command. The only difference is that with the **PURGE** command, ______________________________.

12. Using the __________ command has the added advantage of being able to control the properties of the exploded components of the block, besides being able to break up the block into its component objects.

13. Applying the __________ scale factor inserts the mirror image of the block.

14. The blocks created inside one unit square are called __________.

15. The Layer 0 and the Continuous linetype __________ be renamed using the **RENAME** command.

Exercises

Exercise 6 *Mechanical*

Draw part (a) of Figure 14-18 and define it as a block named A. Then, using the block insert command, insert the block in the plate as shown.

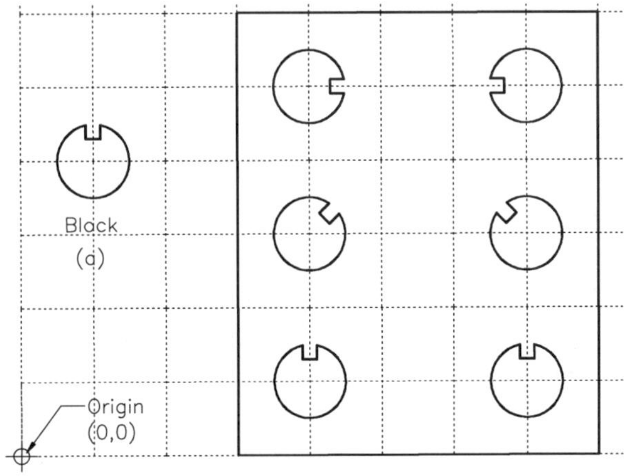

Figure 14-18 Drawing for Exercise 6

Exercise 7 *Piping*

Draw the diagrams in Figure 14-19 using blocks.

a. Create a block for the valve shown in Figure 14-19(a).

b. Use a thick polyline for the flow lines.

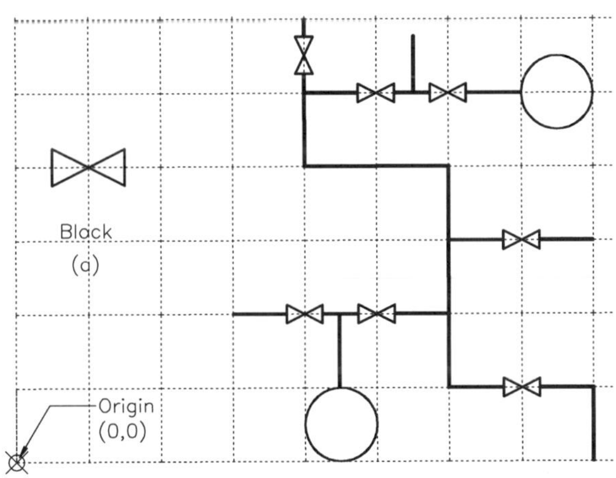

Figure 14-19 Drawing for Exercise 7

Exercise 8 *General*

Draw part (a) of Figure 14-20 and define it as a block named B. Then, using the relevant insertion method, generate the pattern as shown. Note that the pattern is rotated at 30 degrees.

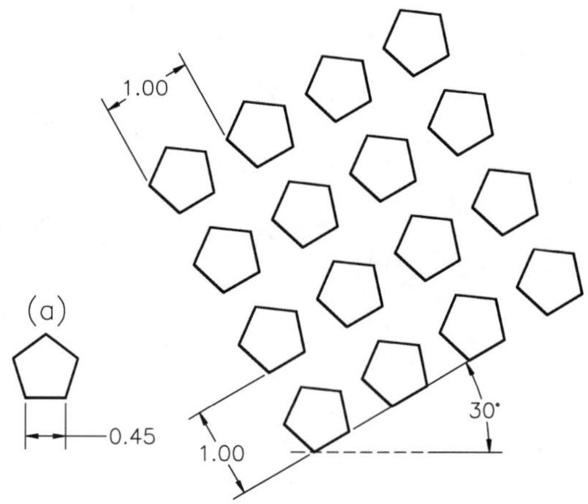

Figure 14-20 Drawing for Exercise 8

Hints:
Create the object as shown in the Figure 14-20(a) and then convert it into a block. Now, insert the block using the **INSERT** command and then create a rectangular array of the blocks using the **Array** dialog box.

Answers to Self-Evaluation Test
1 - F, **2** - T, **3** - F, **4** - F, **5** - **WBLOCK**, **6** - **RENAME**, **7** - **INSERT**, **8** - **PURGE**, **9** - **DesignCenter**, **10** - **-BLOCK**

AutoCAD LT Part II

Author's Web Sites

For Faculty: Please contact the author at **stickoo@calumet.purdue.edu** or **tickoo@cadcim.com** to access the Web site that contains the following:

1. PowerPoint presentations, programs, and drawings used in this textbook.
2. Syllabus, chapter objectives and hints, and questions with answers for every chapter.

For Students: You can download drawing-exercises, tutorials, programs, and special topics by accessing the author's Web site at **www.cadcim.com** or **www.calumet.purdue.edu/public/mets/tickoo/index.htm**.

Chapter 15

Defining Block Attributes

Learning Objectives

After completing this chapter, you will be able to:
- *Understand what attributes are and how to define attributes with a block.*
- *Edit attribute tag names.*
- *Insert blocks with attributes and assign values to attributes.*
- *Extract attribute values from the inserted blocks.*
- *Control attribute visibility.*
- *Perform global and individual editing of attributes.*
- *Insert a text file in a drawing to create bill of material.*

ATTRIBUTES

AutoCAD LT has provided a facility that allows the user to attach information to blocks. This information can then be retrieved and processed by other programs for various purposes. For example, you can use this information to create a bill of material for a project, find the total number of computers in a building, or determine the location of each block in a drawing. Attributes can also be used to create blocks (such as title blocks) with prompted or preformatted text to control text placement. The information associated with a block is known as **attribute value** or simply **attribute**. AutoCAD LT references the attributes with a block through tag names.

Before you can assign attributes to a block, you must create an attribute definition by using the **ATTDEF** command. The attribute definition describes the characteristics of the attribute. You can define several attribute definitions (tags) and include them in the block definition. Each time you insert the block, AutoCAD LT will prompt you to enter the value of the attribute. The attribute value automatically replaces the attribute tag name. The information (attribute values) assigned to a block can be extracted and written to a file by using AutoCAD LT's **ATTEXT** command. This file can then be inserted in the drawing as a table or processed by other programs to analyze the data. The attribute values can be edited by using the **ATTEDIT** command. The display of attributes can be controlled with the **ATTDISP** command.

DEFINING ATTRIBUTES

Menu:	Draw > Block > Define Attributes
Command:	ATTDEF

When you invoke the **ATTDEF** command, the **Attribute Definition** dialog box (Figure 15-1) is displayed. The block attributes can be defined through this dialog box. When you create an attribute definition, you must define the mode, attributes, insertion point, and text information for each attribute. All this information can be entered in the dialog box. The following is a description of each area of the **Attribute Definition** dialog box.

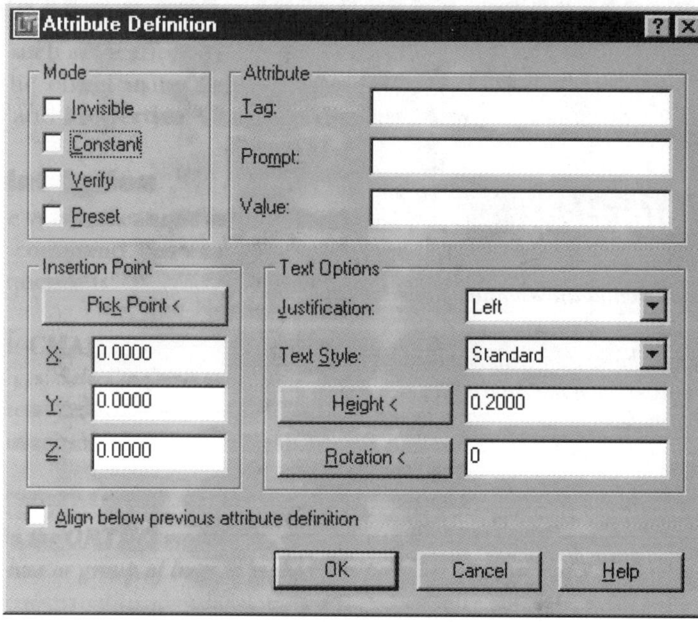

***Figure 15-1 Attribute Definition** dialog box*

Mode Area

The **Mode** area of the **Attribute Definition** dialog box (Figure 15-2) has four check boxes: **Invisible**, **Constant**, **Verify**, and **Preset**. These options determine the display and edit features of the block attributes. For example, if you select the **Invisible** check box, the attribute becomes invisible; that is, it is not displayed on the screen. Similarly, if the **Constant** check box is selected, the attribute becomes constant. This means that its value is predefined and cannot be changed. These options are described next.

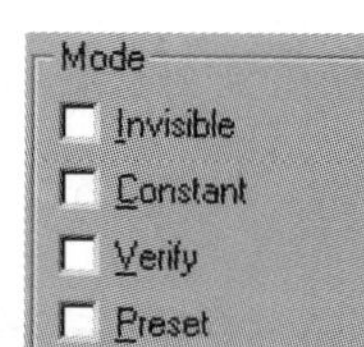

***Figure 15-2 Mode** area of the **Attribute Definition** dialog box*

Invisible

This option lets you create an attribute that is not visible on the screen, by default. Clear this check box if you want the attribute to be visible.

Tip

The Invisible mode is especially useful when you do not want the attribute values to be displayed on the screen to avoid cluttering the drawing. Also, if the attributes are invisible, it takes less time to regenerate the drawing.

You can make the invisible attribute visible by using the **ATTDISP** command discussed later in this chapter in the section, "Controlling Attribute Visibility" (ATTDISP Command).

Constant

This option lets you create an attribute that has a fixed value and cannot be changed after block insertion. When you select this mode, the **Prompt** edit box and the **Verify** and **Preset** check boxes are disabled. Since the value is constant, there is no need to be prompted for new values. This check box is cleared by default and you can use different attribute values for the blocks.

Verify

This option allows you to verify the attribute value you have entered when inserting a block by asking you twice for the data. If the value is incorrect, you can correct it by entering the new value. If this check box is cleared, you are not prompted for verification of the attribute values.

Preset

This option allows you to create an attribute that is automatically set to default value. The attribute values are not requested when you insert a block and default values are used. But unlike a constant attribute, the preset attribute value can be edited later.

Note

*Not selecting any of the check boxes in the **Mode** area displays all the prompts at the command line and the values will be visible on the screen. This is also referred to as the Normal mode.*

Attribute Area

The **Attribute** area (Figure 15-3) of the **Attribute Definition** dialog box has three edit boxes: **Tag**, **Prompt**, and **Value**, where you can enter values. You can enter up to 256 characters in these edit boxes. If the first character to be entered in any one of these edit boxes is a space, you should start with a backslash (\). But if the first character is a backslash (\), you should start the value to be entered with two backslashes (\\). The three edit boxes have been described next.

***Figure 15-3 Attribute** area of the **Attribute Definition** dialog box*

Tag

This is like a label that is used to identify an attribute. For example, the tag name COMPUTER can be used to identify an item. Here you can enter the tag names as uppercase, lowercase, or both, but all lowercase letters are automatically converted into uppercase when displayed. The tag name cannot be null. Also, the tag name must not contain any blank spaces.

Tip

It is advisable to specify a tag name that reflects the contents of the item being tagged. For example, the tag name COMP or COMPUTER is an appropriate name for labeling computers.

Prompt

The text that you enter in the **Prompt** edit box is used as a prompt when you insert a block that contains the defined attribute. For example, if COMPUTER is the tag, you can enter **Enter the Memory** or **Enter Memory** in the **Prompt** edit box. AutoCAD LT will then prompt you with this same statement when you insert the block with which the attribute is defined. If you have selected the **Constant** check box in the **Mode** area, the **Prompt** edit box is not available because no prompt is required if the attribute is constant. If you do not enter anything in the **Prompt** edit box, the entry made in the **Tag** edit box is used as the prompt.

Value

The entry in the **Value** edit box defines the default value of the specified attribute. If you do not enter a value, it is used as the value for the attribute. The entry of a value is optional.

Insertion Point Area

The **Insertion Point** area of the **Attribute Definition** dialog box (Figure 15-4) lets you define the insertion point of block attribute text. You can define the insertion point by entering the values in the **X**, **Y**, and **Z** edit boxes or by choosing the **Pick Point** button. When you choose this button, the dialog box is temporarily closed and you can select an insertion point on the screen or enter the X, Y, and Z values of the insertion point at the command line. Once you have specified the insertion point, the **Attribute Definition** dialog box reappears.

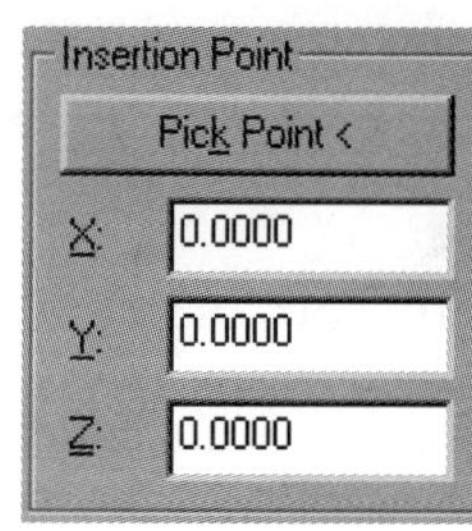

Figure 15-4 ***Insertion Point*** *area of the* ***Attribute Definition*** *dialog box*

Just below the **Insertion Point** area of the dialog box is a check box labeled **Align below previous attribute definition**. This check box is not available when you use the **Attribute Definition** dialog box for the first time. After you have defined an attribute and when you press ENTER to display the **Attribute Definition** dialog box again, this check box is available. You can select this check box to place the subsequent attribute text just below the previously defined attribute automatically. When you select this check box, the **Insertion Point** area and the **Text Options** areas of the dialog box are not available and AutoCAD LT assumes previously defined values for text such as text height, text style, text justification, and text rotation. The text is automatically placed on the following line.

Text Options Area

The **Text Options** area of the **Attribute Definition** dialog box (Figure 15-5) lets you define the justification, text style, height, and rotation of the attribute text. To set the text justification, select a justification type from the **Justification** drop-down list. The default option is **Left**. Similarly, you can use the **Text Style** drop-down list to select a text style. All the text styles defined in the current drawing are displayed in the **Text Style** drop-down list. The default text style is **Standard**. You can specify the text height and text rotation in the **Height** and **Rotation** edit boxes. You can also define the text height by choosing the **Height** button. When you choose this button, AutoCAD LT temporarily exits the dialog box and lets you enter the height value by selecting points on the screen or from the command line. Once you have defined the height on the screen, the dialog box reappears and the defined text height is displayed in the edit box. Similarly, you can define the text rotation by choosing the **Rotation** button and then selecting points on the screen or by entering the rotation angle at the command line.

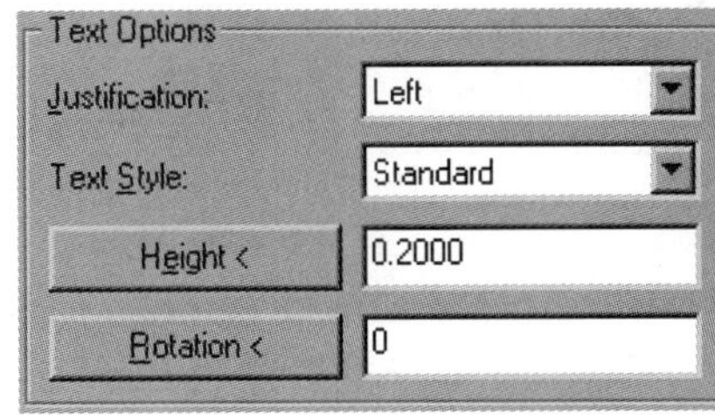

Figure 15-5 Text Options *area of the* ***Attribute Definition*** *dialog box*

Note

The text style must be defined before it can be used to specify the text style.

If you select a style that has the height predefined, AutoCAD LT automatically disables the ***Height*** *edit box.*

If you have selected the ***Align*** *option from the* ***Justification*** *drop-down list, the* ***Height*** *and* ***Rotation*** *edit boxes are disabled.*

If you have selected the ***Fit*** *option from the* ***Justification*** *drop-down list, the* ***Rotation*** *edit box is disabled.*

After you complete the settings in the **Attribute Definition** dialog box and choose **OK**, the attribute tag text is inserted in the drawing at the specified insertion point. Now, you can use the **BLOCK** or **WBLOCK** commands to select all the objects and attributes to define a block.

Note

You can use the ***-ATTDEF*** *command to display prompts on the Command line. All the options available in the* ***Attribute Definition*** *dialog box are available through the Command line too.*

Example 1 *General*

In this example, you will define the following attributes for a computer and then create a block using the **BLOCK** command. The name of the block is COMP.

Mode	Tag name	Prompt	Default value
Constant	ITEM		Computer
Preset, Verify	MAKE	Enter make:	CAD-CIM
Verify	PROCESSOR	Enter processor type:	Unknown
Verify	HD	Enter Hard-Drive size:	100MB
Invisible,Verify	RAM	Enter RAM:	4MB

1. Draw the computer as shown in Figure 15-6. Assume the dimensions, or measure the dimensions of the computer you are using for AutoCAD LT.

2. Invoke the **ATTDEF** command. The **Attribute Definition** dialog box is displayed.

3. Define the first attribute as shown in the preceding table. Select **Constant** check box in the **Mode** area because the mode of the first attribute is constant. In the **Tag** edit box, enter the tag name, ITEM. Similarly, enter COMPUTER in the **Value** edit box. Note that the **Prompt** edit box is not available because the mode is constant.

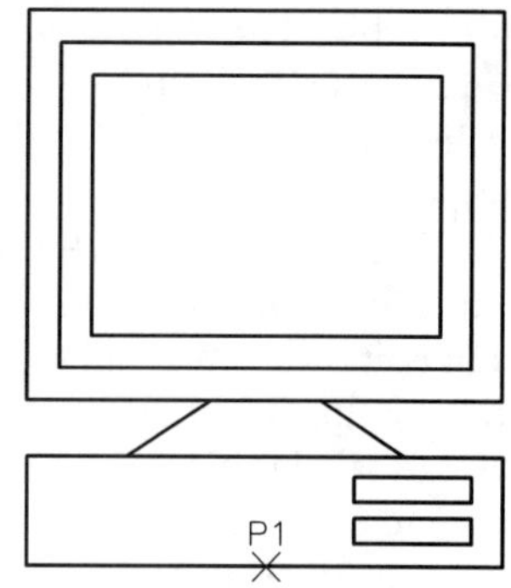

Figure 15-6 *Drawing for Example 1*

4. In the **Insertion Point** area, choose the **Pick Point** button to define the text insertion point. Select a point below the insertion base point (P1) of the computer to place the text.

5. In the **Text Options** area, specify the justification, style, height, and rotation of the text.

6. Choose the **OK** button once you have entered information in the **Attribute Definition** dialog box.

7. Press ENTER to invoke the **Attribute Definition** dialog box again. Enter the mode and attribute information for the second attribute as shown in the table at the beginning of Example 1. You need not define the insertion point and text options again. Select the **Align below previous attribute definition** check box that is located just below the **Insertion Point** area. You will notice that when you select this check box, the **Insertion Point** and **Text Options** areas are not available. Now, choose the **OK** button. AutoCAD LT places the attribute text just below the previous attribute text.

8. Similarly, define the remaining attributes also (Figure 15-7).

9. Now, use the **BLOCK** command to create a block. The name of the block is COMP, and the insertion point of the block is P1, midpoint of the base. When you select the objects for the block, make sure you also select the attributes.

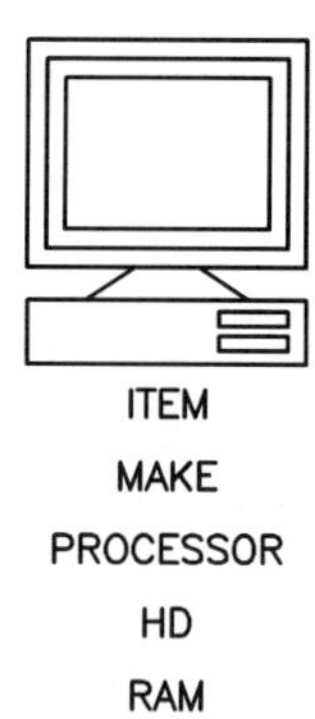

Figure 15-7 *Define attributes below the computer drawing*

Note

The order of prompts is the same as the order of attributes selection.

EDITING ATTRIBUTE DEFINITION

Toolbar:	Text > Edit Text
Menu:	Modify > Object > Text > Edit
Command:	DDEDIT

The **DDEDIT** command lets you edit text and attribute definitions, before you define the block. After invoking this command, AutoCAD LT will prompt you to **select an annotation object or [Undo]**. If you select an attribute created using the **Attribute Definition** dialog box, the **Edit Attribute Definition** dialog box is displayed and lists the tag name, prompt, and default value of the attribute (Figure 15-8).

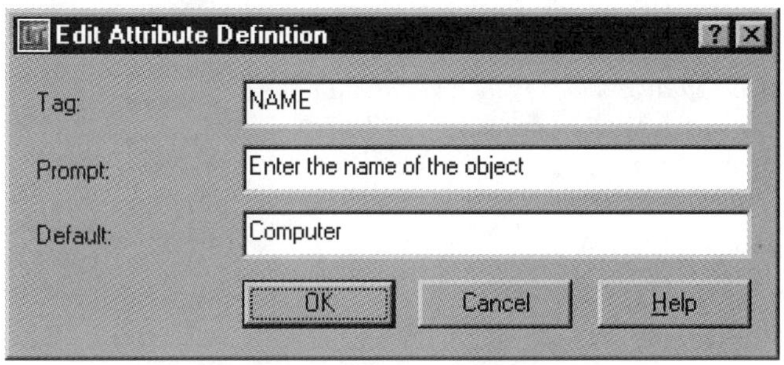

***Figure 15-8 Edit Attribute Definition** dialog box*

You can enter the new values in the respective edit boxes. Once you have entered the changed values, choose the **OK** button in the dialog box. After you exit the dialog box, AutoCAD LT will continue to prompt you to select another text or attribute object (Attribute tag). If you have finished editing and do not want to select another attribute object to edit, press ENTER to return to the Command prompt.

Using the Properties Window

The **PROPERTIES** command has been already discussed in Chapter 4, Drawing Aids. It can also be used to edit defined attributes. Select the attribute to modify and right-click to display a shortcut menu. Choose **Properties** here and the **Properties** window is displayed (Figure 15-9). You will notice that **Attribute** is displayed in the text box located at the top of the window. Also, under the **Categorized** tab you will find that all the properties of the selected attribute are displayed under four headings. They are **General**, **Text**, **Geometry**, and **Misc**. You can change these values in their corresponding fields. For example, you can modify the color, layer, linetype, thickness, linetype scale, and so on, of the selected attribute under the **General** head. Similarly, you can modify the tag name, prompt, and value of the selected attribute in the **Tag**, **Prompt**, and **Value** fields under the **Text** heading. Under the **Text** head, you can also modify the text style, justification, text height, rotation angle, width factor, and obliquing angle values of the selected attribute. Under the **Geometry** heading you can redefine the insertion point of the selected attribute by choosing the button with an arrow icon which is displayed when you select either of the **Position X**, **Position Y**, or **Position Z** fields. When you choose the button with the arrow icon, you are allowed to reposition an existing insertion point by selecting on the screen. You can also determine if you want the attribute text to appear upside down or backwards or not under the **Misc** heading. Here, you can also modify the attribute modes which have been already defined. When you select a particular mode, for example **Invisible**, a drop-down list is available in the corresponding field. This list displays two options, **Yes** and **No**. If you select **Yes**, the attribute is made invisible and if you select **No**, the attribute is not made invisible. To select a particular mode, you should choose **Yes** from their corresponding drop-down lists.

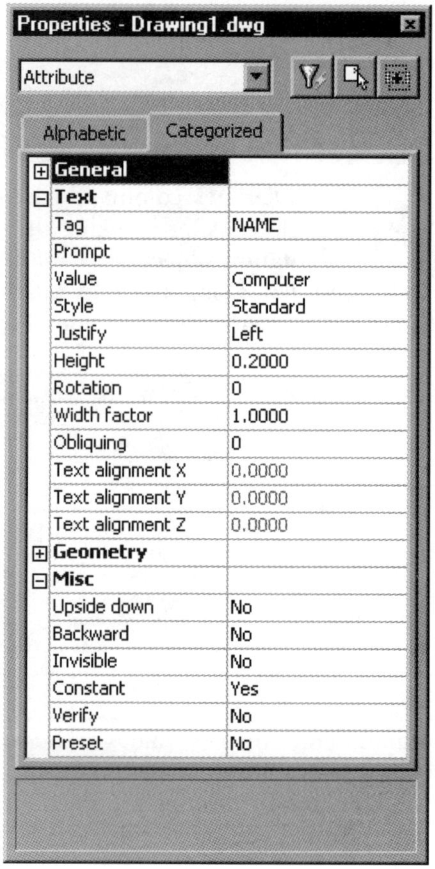

Figure 15-9 *Using the* ***Properties*** *window to modify attribute definitions*

Note

Remember that if you change the justification of the first attribute below which you have aligned the remaining attributes, the remaining attributes are not updated automatically. You will have to change the justifications of the remaining attributes individually to align them below the first attribute.

You can also use the ***CHANGE*** *command to edit the attribute objects using the command line.*

INSERTING BLOCKS WITH ATTRIBUTES

Using the Dialog Box

The value of the attributes can be specified during block insertion, either at the command line or in the **Enter Attributes** dialog box. When you use the **INSERT** command or the **-INSERT** command to insert a block in a drawing (discussed earlier in Chapter 14, Blocks), and after you have specified the insertion point, scale factors, and rotation angle, the **Enter Attributes** dialog box (Figure 15-10) is displayed if the system variable **ATTDIA** is set to **1**. The default value for **ATTDIA** is **0**, which disables the dialog box. The prompts and their default values, which you had specified with the attribute definition, are then displayed on the command line after you have specified the insertion point, scale, and rotation angle for the block to be inserted.

In the **Enter Attributes** dialog box, the prompts that were entered at the time of attribute definition in the dialog box are displayed with their default values in corresponding fields. If an attribute has been defined with the **Constant** mode, it is not displayed in the dialog box because a constant attribute value cannot be edited. You can enter the attribute values in the fields located next to the attribute prompt. If no new values are specified, the default values are displayed. Eight attribute values are displayed at a time in the dialog box. If there are more attributes, they can be accessed by using the **Next** or **Previous** buttons. The block name is displayed at the top of the dialog box. After entering the new attribute values, choose the **OK** button. AutoCAD LT will place these attribute values at the specified location.

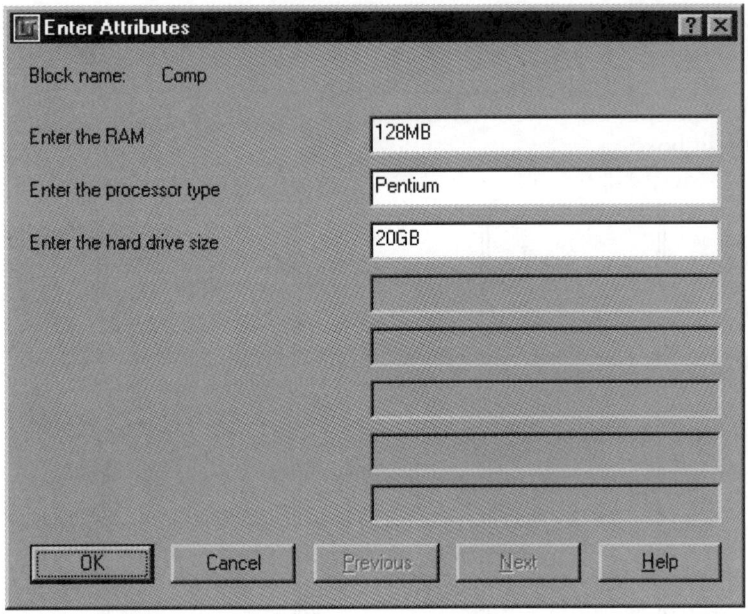

***Figure 15-10** Enter attribute values in the **Enter Attributes** dialog box*

Tip

*It is more convenient to use the **Edit Attributes** dialog box, since you can view all the attribute values at a glance and can correct them here itself, before placing. Therefore, it is a good idea to set the **ATTDIA** value to 1, before you insert a block with attributes.*

*Attributes can also be defined from the command line by setting system variable **ATTDIA** to 0 (default value). Now, when you use the **INSERT** command, AutoCAD LT does not display the **Enter Attributes** dialog box. Instead, AutoCAD LT will prompt you to enter the attribute values for various attributes that have been defined in the block at the command line.*

Example 2 — *General*

In this example, you will insert the block (COMP) that was defined in Example 1. The following is the list of the attribute values for computers.

ITEM	MAKE	PROCESSOR	HD	RAM
Computer	Gateway	486-60	150MB	16MB
Computer	Zenith	486-30	100MB	32MB
Computer	IBM	386-30	80MB	8MB
Computer	Del	586-60	450MB	64MB
Computer	CAD-CIM	Pentium-90	100 Min	32MB
Computer	CAD-CIM	Unknown	600MB	Standard

1. Make the floor plan drawing as shown in Figure 15-11 (assume the dimensions).

2. Set the system variable **ATTDIA** to 1. Use the **INSERT** command to insert the blocks.

When you invoke the **INSERT** command, the **Insert** dialog box is displayed. Enter COMP in the **Name** edit box and choose **OK** to exit the dialog box. Select an insertion point on screen to insert the block. After you specify the insertion point, the **Enter Attributes** dialog box is displayed, where you can change the attribute values, if you need to, in the different edit boxes.

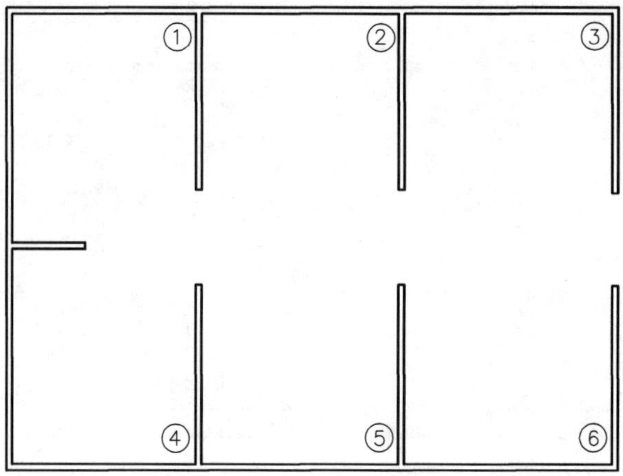

Figure 15-11 *Floor plan drawing for Example 2*

3. Repeat the **INSERT** command to insert other blocks, and define their attribute values as shown in Figure 15-12.

Figure 15-12 *The floor plan after inserting blocks and defining their attributes*

4. Save the drawing for further use.

EXTRACTING THE ATTRIBUTES (ATTEXT COMMAND)

Menu:	Tools > Attribute Extraction
Command:	ATTEXT

The **ATTEXT** command allows you to use the **Attribute Extraction** dialog box (Figure 15-13) for extracting the attributes. The information about the **File format**, **Template file**, and **Output** file must be entered in the dialog box to extract the defined attribute. Also, you must select the blocks whose attribute values you want to extract. If you do not specify a particular block, all the blocks in the drawing are used.

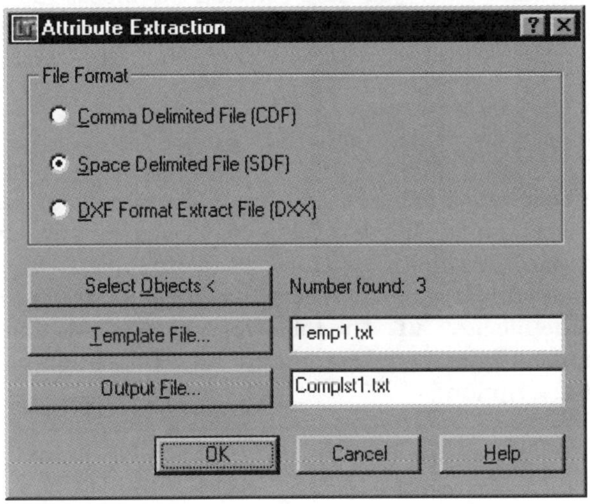

***Figure 15-13 Attribute Extraction** dialog box*

File Format Area

This area of the dialog box lets you select the file format. You can select either of the three radio buttons available in this area. They are **Comma Delimited File (CDF)**, **Space Delimited File (SDF)**, or **DXF Format Extract File (DXX)**. The format selection is determined by the application that you plan to use to process the data. Both CDF and SDF formats can be used with database software. All of these formats are printable.

Comma Delimited File (CDF)

When you select this radio button, the extracted attribute information is displayed in a CDF format. Here, each character field is enclosed in single quotes, and the records are separated by a delimiter (comma by default). A CDF file is a text file with the extension **.TXT**. Of the three formats available, this is the most cumbersome.

Space Delimited File (SDF)

In SDF format, the records are of fixed width as specified in the template file. The records are separated by spaces and the character fields are not enclosed in single quotes. The SDF file is a text file with the extension **.TXT**. This file format is the most convenient and easy to use.

DXF Format Extract File (DXX)

If you select this file format, you will notice that the **Template File** button and edit box in the **Attribute Extraction** dialog box are not available. This is because extraction in this file format does not require any template. The file created by this option contains only block references, attribute values, and end-of-sequence objects. This is the most complex of the three file formats available and is related to programming. The extension of these files is **.DXX**.

Select Objects

Select the blocks with attributes whose attribute information you want to extract. You can use any object selection method. Once you have selected the blocks that you want to use for attribute extraction, right-click or press ENTER. The **Attribute Extraction** dialog box is redisplayed and the number of objects you have selected is displayed adjacent to **Number found**.

Template File

When you choose the **Template File** button, the **Template File** dialog box is displayed where you are allowed to select a template file that has been defined already. After you have selected a template file, choose the **Open** button to return to the **Attribute Extraction** dialog box. The name of the selected file is displayed in the **Template File** edit box. The template file is saved with the extension of the file as **.TXT**. The following are the fields that you can specify in a template file (the comments given on the right are for explanation only; they must not be entered with the field description):

BL:LEVEL	Nwww000	(Block nesting level)
BL:NAME	Cwww000	(Block name)
BL:X	Nwwwddd	(X coordinate of block insertion point)
BL:Y	Nwwwddd	(Y coordinate of block insertion point)
BL:Z	Nwwwddd	(Z coordinate of block insertion point)
BL:NUMBER	Nwww000	(Block counter)
BL:HANDLE	Cwww000	(Block's handle)
BL:LAYER	Cwww000	(Block insertion layer name)
BL:ORIENT	Nwwwddd	(Block rotation angle)
BL:XSCALE	Nwwwddd	(X scale factor of block)
BL:YSCALE	Nwwwddd	(Y scale factor of block)
BL:ZSCALE	Nwwwddd	(Z scale factor of block)
BL:XEXTRUDE	Nwwwddd	(X component of block's extrusion direction)
BL:YEXTRUDE	Nwwwddd	(Y component of block's extrusion direction)
BL:ZEXTRUDE	Nwwwddd	(Z component of block's extrusion direction)
Attribute tag		(The tag name of the block attribute)

The extract file may contain several fields. For example, the first field might be the item name and the second field might be the price of the item. Each line in the template file specifies one field in the extract file. Any line in a template file consists of the name of the field, the width of the field in characters, and its numerical precision (if applicable). For example:

ITEM N015<u>002</u>

BL:NAME C015000

Where **BL:NAME** ------	Field name
Blankspaces ---	Blank spaces (must not include the tab character)
C ------------------	Designates a character field
N ------------------	Designates a numerical field
015 ---------------	Width of field in characters
002 ---------------	Numerical precision

BL:NAME or **ITEM** Indicates the field names; can be of any length.

C Designates a character field; that is, the field contains characters or it starts with characters. If the file contains numbers or starts with numbers, then C will be replaced by N. For example, **N015002**.

015 Designates a field that is fifteen characters long.

002 Designates the numerical precision. In this example, the numerical precision is 2, or two places following the decimal. The decimal point and the two digits following the decimal are **included in the field width**. In the next example, (000), the numerical precision, is not applicable because the field does not have any numerical value (the field contains letters only).

After creating a template file, when you choose the **Template File** button, the **Template File** dialog box (Figure 15-14) is displayed, where you can browse and select a template file.

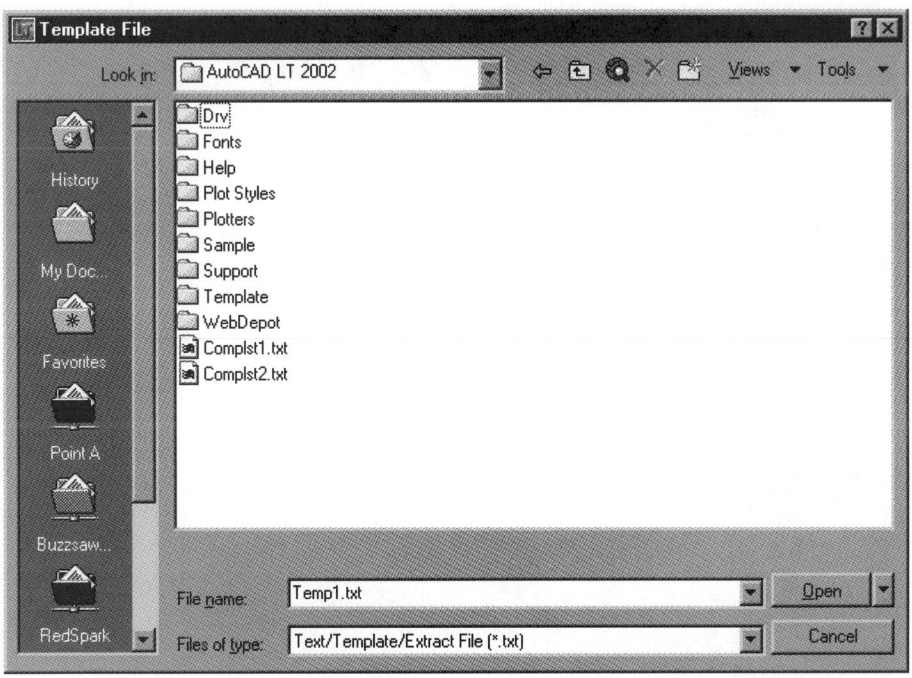

***Figure 15-14 Template File** dialog box to select a template file*

Note

You can put any number of spaces between the field name and the character C or N (ITEM N015002). However, you must not use the tab characters. Any alignment in the fields must be done by inserting spaces after the field name.

In the template file, a field name must not appear more than once. The template file name and the output file name must be different.

The template file must contain at least one field with an attribute tag name because the tag names determine which attribute values are to be extracted and from which blocks. If several blocks have different block names but the same attribute tag, AutoCAD LT will extract attribute values from all selected blocks. For example, if there are two blocks in the drawing with the attribute tag PRICE, then when you extract the attribute values, AutoCAD LT will extract the value from both blocks (if both blocks were selected). To extract the value of an attribute, the tag name must match the field name specified in the template file. AutoCAD LT automatically converts the tag names and the field names to uppercase letters before making a comparison.

Output File

When you choose the **Output File** button, the **Output File** dialog box (Figure 15-15) is displayed. You can select an existing file here, if you want the extracted or output file to be saved as an existing file. You can enter a name in the **File Name** edit box in the **Output File** dialog box and then choose the **Save** button, if you want to save the output file as a new file. By default, the output file has the same name as the drawing name. For example a drawing named Drawing1.dwg will have an output file by the name Drawing1.txt by default. Once a name for the output file is

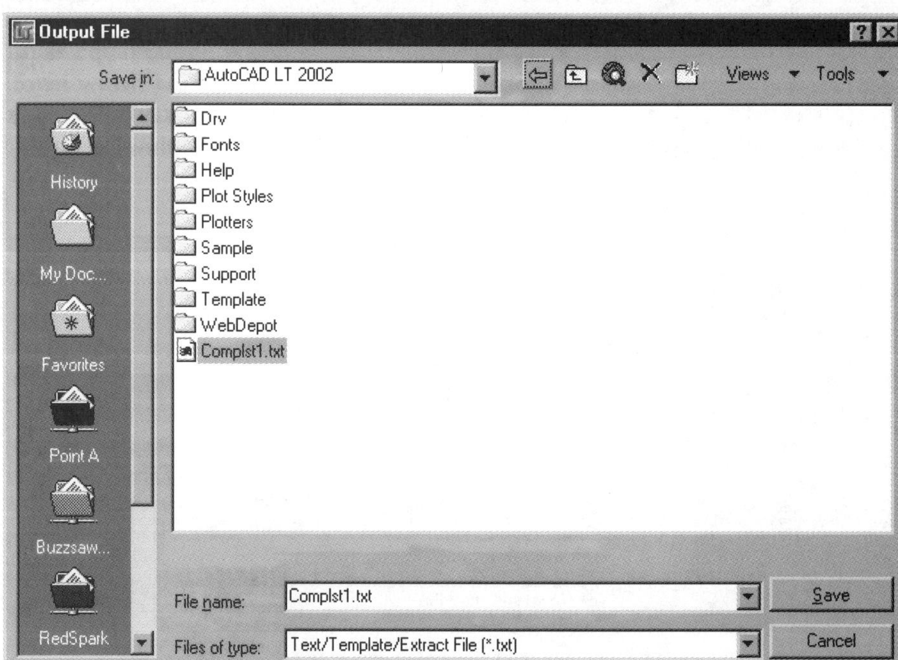

Figure 15-15 ***Output File*** *dialog box to save an output file*

specified, it is displayed in the **Output File** edit box in the **Attribute Extraction** dialog box. You can also enter the file name in this edit box. As discussed earlier, AutoCAD LT appends **.TXT** file extension for CDF or SDF files and **.DXX** file extension for DXF files.

Note

*You can also use the **-ATTEXT** command to extract attributes using the command line. Here, you are prompted to specify a file format for the extract information and you can also select specific blocks to extract their attribute information. You can specify a template file using the **Select Template File** dialog box and an extract file using the **Create extract file** dialog box.*

Example 3 *General*

In this example, you will write a template file for extracting the attribute values as defined in Example 2. These attribute values must be written to a file **Complst1.txt** and the values arranged as shown in the following table:

Field width in characters

< 10 >	< 12 >	< 10 >	< 12 >	< 10 >	< 10 >
COMP	Computer	Gateway	486-60	150MB	**16MB**
COMP	Computer	Zenith	486-30	100MB	**32MB**
COMP	Computer	IBM	386-30	80MB	**8MB**
COMP	Computer	Del	586-60	450MB	**64MB**
COMP	Computer	**CAD-CIM**	Pentium-90	100 Min	**32MB**
COMP	Computer	**CAD-CIM**	**Unknown**	600MB	Standard

1. Load the drawing you saved in Example 2.

2. Use the Windows Notepad to write the following template file. You can use any text editor or word processor to write the file. After writing the file, save it as an ASCII file under the file name **Temp1.txt** in the default AutoCAD LT 2002 directory. Exit the Notepad and access AutoCAD.

BL:NAME	C010000	(Block name, 10 spaces)
Item	C012000	(Item, 12 spaces)
Make	C010000	(Computer make, 10 spaces)
Processor	C012000	(Processor type, 12 spaces)
HD	C010000	(Hard drive size, 10 spaces)
RAM	C010000	(RAM size, 10 spaces)

3. Use the **ATTEXT** command to invoke the **Attribute Extraction** dialog box, and select the Space Delimited File (SDF) radio button.

4. Choose the **Select Objects** button to select the objects (blocks) present on the screen. You can select the objects by using the Window or Crossing option. After selection is complete, right-click your pointing device to display the dialog box again.

5. Choose the **Template File** button to display the **Template File** dialog box. Select the

template file **Temp1.txt**.

6. Choose the **Output File** button to display the **Output File** dialog box (Figure 15-16). Enter the name of the output file as **Complst1.txt** in the **File name** edit box. Choose the **Save** button to save this output file. The **Output File** dialog box will be closed and the **Attribute Extraction** dialog box will be redisplayed on the screen.

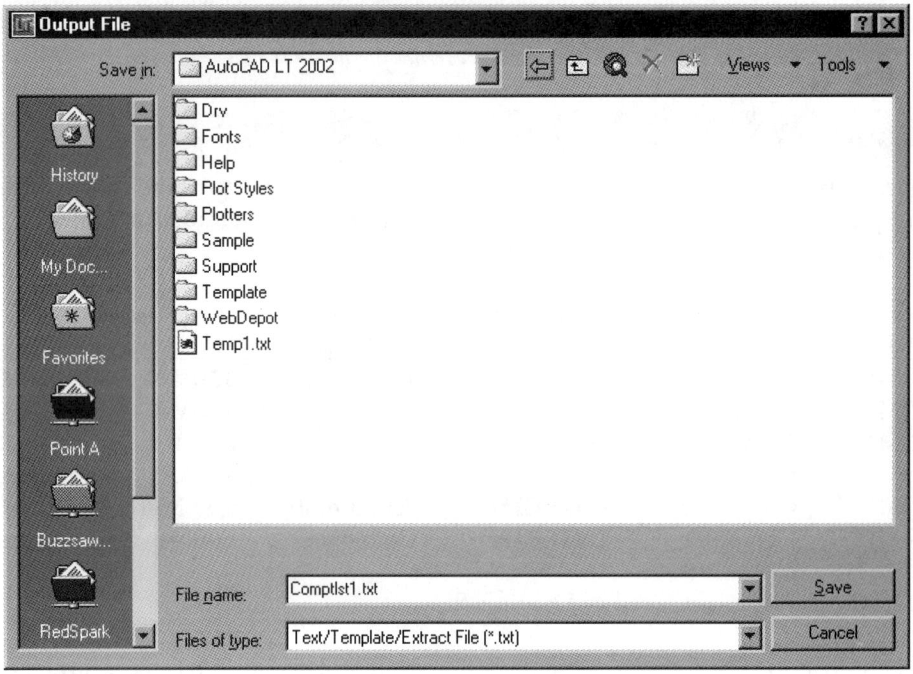

Figure 15-16 Creating the output file for Example 3

7. Choose the **OK** button in the **Attribute Extraction** dialog box. The Space delimited file will be created in the default AutoCAD LT 2002 directory with the name **Complst1.txt.**

8. You can view the resultant output file in the Notepad. Open the Notepad using the **Start** menu and then specify the location of the **Complst1.txt** file. As you decided to create the space delimited file, therefore, the fields in the output file will be separated by spaces.

Note

You can also extract the attributes using the command line. This is done using the ***-ATTEXT*** *command. Note that even while using the command line, the* ***Select Template File*** *and the* ***Create extract file*** *dialog boxes will be displayed for selecting the template file and saving the output file if the value of the* ***FILEDIA*** *variable is set to* ***1*** *(default value). The options under both these dialog boxes are similar to the standard* ***Select File*** *and* ***Save Drawing As*** *dialog boxes.*

CONTROLLING ATTRIBUTE VISIBILITY

Menu:	View > Display > Attribute Display
Command:	ATTDISP

The **ATTDISP** command allows you to change the visibility of all attribute values. Normally, the attributes are visible unless they are defined invisible by using the **Invisible** mode. The invisible attributes are not displayed, but they are a part of the block definition. You can select any one of the options of the **ATTDISP** command to turn the display of the attributes completely on or off. You can also select the **Normal** option where the attributes that were created using the **Invisible** mode continue to be invisible. The options can be selected from the **View** > **Display** > **Attribute Display** cascading menu or can be entered on the command line. The prompt sequence for the command is:

Command: **ATTDISP**
Enter attribute visibility setting [Normal/ON/OFF] <Normal>: *Specify an option and press ENTER.*

When you select **ON**, all attribute values will be displayed, including the attributes that were defined in the **Invisible** mode. If you select **OFF**, all attribute values will become invisible. Similarly, if you select N (**Normal**), AutoCAD LT will display the attribute values the way they were defined, that is, the attributes that were defined invisible will stay invisible and the attributes that were not defined in the **Invisible** mode are visible.

In Example 2, the RAM attribute was defined with the Invisible mode. Therefore, the RAM values are not displayed with the block. If you want to make the RAM attribute values visible (Figure 15-17), choose **On** from the **View > Display > Attribute Display** menu.

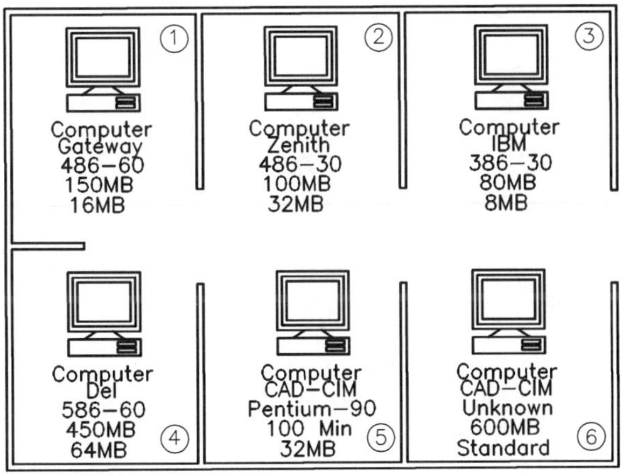

Figure 15-17 *Using the **ATTDISP** command to make the RAM attribute values visible*

Tip

*After you have defined the attribute values and saved them with the block definition, it may be a good idea to use the **Off** option of the **ATTDISP** command. By doing this, the drawing is simplified and also regeneration time is reduced.*

EDITING BLOCK ATTRIBUTE VALUES

You can edit single block at a time using the dialog box or can edit more then one blocks at a time. Both the methods of editing the attribute values are discussed next.

Editing Attributes Using the ATTEDIT Command

Toolbar:	Modify II > Edit Attributes
Menu:	Modify > Object > Attribute > Single
Command:	ATTEDIT

The **ATTEDIT** command allows you to edit the block attribute values through the **Edit Attributes** dialog box. When you invoke this command, AutoCAD LT prompts you to select the block whose values you want to edit. After selecting the block, the **Edit Attribute** dialog box (Figure 15-18) is displayed. The dialog box is similar to the **Enter Attributes** dialog box and shows the prompts and the attribute values of the selected block. If an attribute was defined with the **Constant** mode, it is not displayed in the dialog box because a constant attribute value cannot be edited. To make any changes, select the existing value and enter a new value in the corresponding edit box. After you have made the modifications, choose the **OK** button. The attribute values will be updated in the selected block.

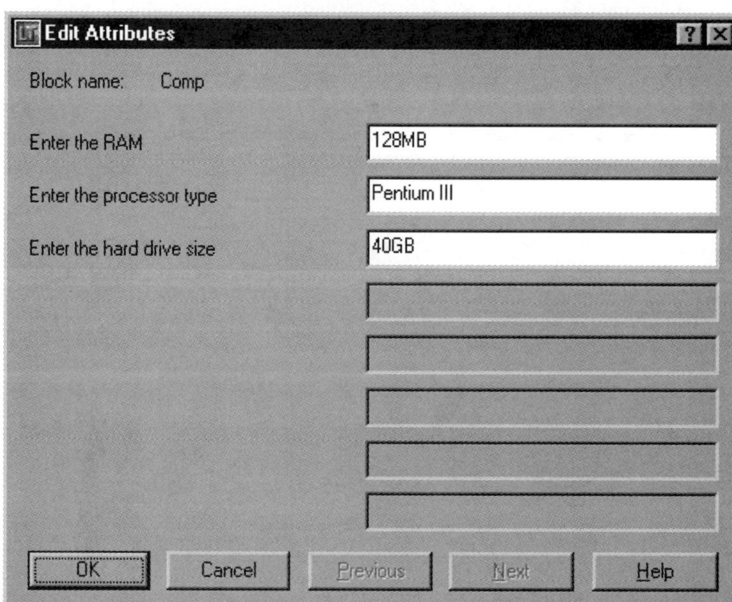

***Figure 15-18** Editing attribute values using the **Edit Attributes** dialog box*

If a selected block has no attributes, AutoCAD LT will display the alert message **"That block has no editable attributes"**. Similarly, if the selected object is not a block, AutoCAD LT again displays the alert message **"That object is not a block"**.

Note

*You cannot use the **ATTEDIT** command to do global editing of attribute values, or to modify position, height, or style of the attribute value.*

Example 4 — *General*

In this example you will use the **ATTEDIT** command to change the attribute of the first computer (150 MB to 2.1 GB), which is located in Room-1.

1. Open the drawing that was created in Example 2. The drawing has six blocks with attributes. The name of the block is COMP, and it has five defined attributes, one of them invisible. Zoom in so that the first computer is displayed on the screen (Figure 15-19).

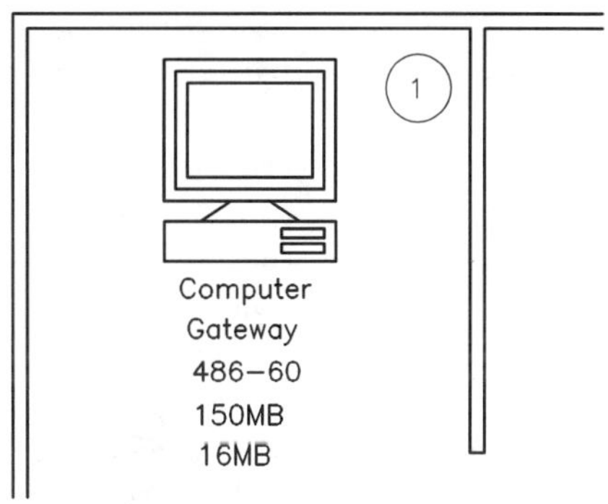

Figure 15-19 *Zoomed view of the first computer*

2. Invoke the **ATTEDIT** command. AutoCAD LT will prompt you to select a block. Select the computer located in Room-1. AutoCAD LT will display the **Edit Attribute** dialog box. All the editable attribute values will be displayed in this dialog box. Change the value **150 MB** to **2.1 GB**.

3. Choose the **OK** button in the dialog box. When you exit the dialog box, the attribute values are updated.

Tip

*You can also edit attributes with the **FIND** command. This command can be invoked by choosing the **Find and Replace** button in the **Standard Toolbar** or by entering **FIND** at the Command prompt. When you invoke this command, the **Find and Replace** dialog box is displayed.*

*When you know that a set of attributes in a drawing may have to be changed in the future, you can make a group out of them by using the **GROUP** command. Then later you can choose the **Select objects** button in the **Find and Replace** dialog box and enter G at the **Select objects** prompt and enter the name of the group. All the objects in the group will get selected.*

Global Editing of Attributes

Menu:	Modify > Object > Attribute > Global
Command:	-ATTEDIT

The **-ATTEDIT** command allows you to edit the attribute values independently of the blocks that contain the attribute reference. For example, if there are two blocks, COMPUTER and TABLE, with the attribute value PRICE, you can globally edit this value (PRICE) independently of the block that references these values. You can also edit the attribute values one at a time. For example, you can edit the attribute value (PRICE) of the block TABLE without affecting the value of the other block, COMPUTER. When you enter the **-ATTEDIT** command, AutoCAD LT displays the following prompt:

Command: **-ATTEDIT**
Edit attributes one at a time? [Yes/No]: N
Performing global editing of attribute values

If you enter **N** at this prompt, it means that you want to do the global editing of the attributes. However, you can restrict the editing of attributes by block names, tag names, attribute values, and visibility of attributes on the screen.

Editing Visible Attributes Only

After you select global editing, AutoCAD LT will display the following prompt:

Edit only attributes visible on screen? [Yes/No] <Y>: **Y**

If you enter Y at this prompt, AutoCAD LT will edit only those attributes that are visible and displayed on the screen. The attributes might not have been defined with the Invisible mode, but if they are not displayed on the screen they are not visible for editing. For example, if you zoom in, some of the attributes may not be displayed on the screen. Since the attributes are not displayed on the screen, they are invisible and cannot be selected for editing.

Editing All Attribute

If you enter N at the previously mentioned prompt, AutoCAD LT flips from graphics to text screen and displays the following message on the screen:

Drawing must be regenerated afterwards.

Now, AutoCAD LT will edit all attributes even if they are not visible or displayed on the screen. Also, changes that you make in the attribute values are not reflected immediately. Instead, the attribute values are updated and the drawing is regenerated after you are done with the command.

Editing Specific Blocks

Although you have selected global editing, you can confine the editing of attributes to specific blocks by entering the block name at the prompt. For example:

Enter block name specification <*>: **COMP**

When you enter the name of the block, AutoCAD LT will edit the attributes that have the given block (COMP) reference. You can also use the wild-card characters to specify the block names. If you want to edit attributes in all blocks that have attributes defined, press ENTER.

Editing Attributes with Specific Attribute Tag Names

Like blocks, you can confine attribute editing to those attribute values that have the specified tag name. For example, if you want to edit the attribute values that have the tag name MAKE, enter the tag name at the following AutoCAD LT prompt:

Enter attribute tag specification <*>: **MAKE**

When you specify the tag name, AutoCAD LT will not edit attributes that have a different tag name, even if the values being edited are the same. You can also use the wild-card characters to specify the tag names. If you want to edit attributes with any tag name, press ENTER.

Editing Attributes with a Specific Attribute Value

Like blocks and attribute tag names, you can confine attribute editing to a specified attribute value. For example, if you want to edit the attribute values that have the value 100 MB, enter the value at the following AutoCAD LT prompt:

Enter attribute value specification <*>: **100MB**

When you specify the attribute value, AutoCAD LT will not edit attributes that have a different value, even if the tag name and block specification are the same. You can also use the wild-card characters to specify the attribute value. If you want to edit attributes with any value, press ENTER.

Sometimes the value of an attribute is null, and these values are not visible. If you want to select the null values for editing, make sure you have not restricted the global editing to visible attributes. To edit the null attributes, enter \ at the following prompt:

Enter attribute value specification <*>: \

After you enter this information, AutoCAD LT will prompt you to select the attributes. You can select the attributes by selecting individual attributes or by using one of the object selection options (Window, Crossing, or individually).

Select Attributes: *Select the attribute values parallel to the current UCS only.*

After you select the attributes, AutoCAD LT will prompt you to enter the string you want to change and the new string. A string is a sequence of consecutive characters. It could also be a portion of the text. AutoCAD LT will retrieve the attribute information, edit it, and then update the attribute values.

Enter string to change: *Enter the value that has to be modified.*
Enter new string: *Enter the new value.*

The following is the complete Command prompt sequence of the **-ATTEDIT** command. It is assumed that the editing is global and for visible attributes only.

Command: **-ATTEDIT**
Edit attributes one at a time? [Yes/No] <Y>: N
Performing global editing of attribute values.
Edit only attributes visible on screen? [Yes/No] <Y>: N
Drawing must be regenerated afterwards.
Enter block name specification <*>: Enter
Enter attribute tag specification <*>: Enter
Enter attribute value specification <*>: Enter
Enter string to change: *Enter the value to be modified.*
Enter new string: *Enter the new value.*

Note
AutoCAD LT regenerates the drawing at the end of the command automatically unless system variable ***REGENAUTO*** *is off, which controls automatic regeneration of the drawing. Also, if you select an attribute defined with the* ***Constant*** *mode, using the* ***-ATTEDIT*** *command, prompt displays 0 found, since attributes with constant mode is uneditable.*

Example 5 *General*

In this example, you will use the drawing from Example 2 to edit the attribute values that are **highlighted** in the following table. The tag names are given at the top of the table (ITEM, MAKE, PROCESSOR, HD, RAM). The RAM values are invisible in the drawing.

	ITEM	MAKE	PROCESSOR	HD	RAM
COMP	Computer	Gateway	486-60	150MB	**16MB**
COMP	Computer	Zenith	486-30	100MB	**32MB**
COMP	Computer	IBM	386-30	80MB	**8MB**
COMP	Computer	Del	586-60	450MB	**64MB**
COMP	Computer	**CAD-CIM**	Pentium-90	100 Min	**32MB**
COMP	Computer	**CAD-CIM**	**Unknown**	600MB	Standard

Make the following changes in the **highlighted** attribute values (Figure 15-20).

1. Change Unknown to Pentium.

2. Change CAD-CIM to Compaq.

3. Change MB to Meg for all attribute values that have the tag name RAM. (No changes should be made to the values that have the tag name HD.)

The following are the steps required to change the attribute value from **Unknown** to **Pentium**.

1. Enter the **-ATTEDIT** command at the Command prompt. The prompt sequence is as follows:

   Command: **-ATTEDIT**

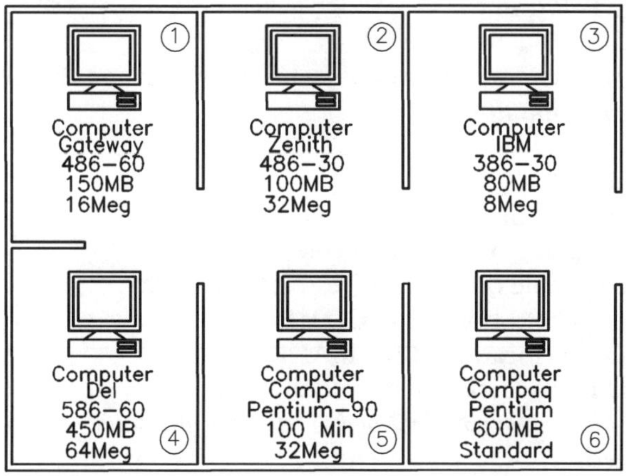

Figure 15-20 *Using **-ATTEDIT** to change the attribute values*

Edit attributes one at a time? [Yes/No] <Y>: **N**
Performing global editing of attribute values.

2. Since only the attributes that are visible on the screen are to be modified, therefore, press ENTER at the following prompt:

   Edit only attributes visible on screen? [Yes/No] <Y>: Enter

3. As shown in the table, the attributes belong to a single block, COMP. In a drawing, there could be more blocks. To confine the attribute editing to the COMP block only, enter the name of the block (COMP) at the next prompt.

   Enter block name specification <*>: **COMP**

4. At the next two prompts, enter the attribute tag name and the attribute value specification. When you enter these two values, only those attributes that have the specified tag name and attribute value will be edited.

   Enter attribute tag specification<*>: **Processor**
   Enter attribute value specification<*>: **Unknown**

5. Next, AutoCAD LT will prompt you to select attributes. Use the Crossing option to select all blocks. AutoCAD LT will search for the attributes that satisfy the given criteria (attributes belong to the block COMP, the attributes have the tag name Processor, and the attribute value is Unknown). Once AutoCAD LT locates such attributes, they will be highlighted.

6. At the next two prompts, enter the string you want to change, and then enter the new string.

Enter string to change: **Unknown**
Enter new string: **Pentium**

7. The following is the Command prompt sequence to change the make of the computers from **CAD-CIM** to **Compaq**.

   Command: **-ATTEDIT**
   Edit attributes one at a time? [Yes/No] <Y>: **N**
   Performing global editing of attribute values.
   Edit only attributes visible on screen? [Yes/No] <Y>: Enter
   Enter block name specification <*>: **COMP**
   Enter attribute tag specification <*>: **MAKE**
   Enter attribute value specification <*>: Enter
   Select Attributes: *Use any selection method to select the attributes.*
   n attributes selected.
   Select Attributes: Enter
   Enter string to change: **CAD-CIM**
   Enter new string: **Compaq**

8. The following is the Command prompt sequence to change **MB** to **Meg**.

   Command: **-ATTEDIT**
   Edit attributes one at a time? [Yes/No] <Y>: **N**
   Performing global editing of attribute values.

   At the next prompt, you must enter **N** because the attributes you want to edit (tag name, RAM) are not visible on the screen.

   Edit only attributes visible on screen? [Yes/No] <Y>: **N**
   Drawing must be regenerated afterwards.
   Enter block name specification <*>: **COMP**

   At the next prompt, about the tag specification, you must specify the tag name because the text string MB also appears in the hard drive size (tag name, HD). If you do not enter the tag name, AutoCAD LT will change all MB attribute values to Meg.

   Enter attribute tag specification <*>: **RAM**
   Enter attribute value specification <*>:
   n Attributes selected
   Enter string to change: **MB**
   Enter new string: **Meg**

9. Choose **On** from the **View > Display > Attribute Display** menu to display the invisible attributes on the screen.

Individual Editing of Attributes

Menu:	Modify > Object > Attribute > Global
Command:	-ATTEDIT

The **-ATTEDIT** command can also be used to edit the attribute values individually. When you enter this command, AutoCAD LT will prompt **Edit attributes one at a time? [Yes/No] <Y>:**. At this prompt, press ENTER to accept the default or enter **Y.** The next three prompts are about block specification, attribute tag specification, and attribute value specification, which were discussed in the previous section in this chapter. These options let you limit the attributes for editing. For example, if you specify a block name, AutoCAD LT will limit the editing to those attributes that belong to the specified block. Similarly, if you also specify the tag name, AutoCAD LT will limit the editing to the attributes in the specified block and with the specified tag name.

Command: **-ATTEDIT**
Edit attributes one at a time? [Yes/No] <Y>: Enter
Enter block name specification<*>: Enter
Enter attribute tag specification<*>: Enter
Enter attribute value specification<*>: Enter
Select Attributes:

At the **Select Attributes:** prompt, select the objects by choosing the objects or by using an object selection option such as Window, Crossing, WPolygon, CPolygon, or Box. By using these options you can further limit the attribute values selected for editing. After you select the objects, AutoCAD LT will mark the first attribute it can find with an **X**. The next prompt is:

Enter an option [Value/Position/Height/Angle/Style/Layer/Color/Next] <N>:

Value

The **Value** option lets you change the value of an attribute. To change the value, enter **V** at this prompt. AutoCAD LT will display the following prompt:

Enter type of value modification [Change/Replace] <R>:

The **Change** option allows you to change a few characters in the attribute value. To select the **Change** option, enter Change or **C** at the prompt. AutoCAD LT will display the next prompt:

Enter string to change:
Enter new string:

At the **Enter string to change**: prompt, enter the characters you want to change and press ENTER. At the next prompt, **Enter new string:**, enter the new string.

Note

*You can use ? and * in the string value. When these characters are used in string values, AutoCAD LT does not interpret them as wild-card characters.*

To use the **Replace** option, enter **R** or press ENTER at the **Enter type of value modification [Change/Replace] <R>** prompt. AutoCAD LT will display the following prompt:

Enter new Attribute value:

At this prompt, enter the new attribute value. AutoCAD LT will replace the string bearing the **X** mark with the new string. If the new attribute is null, the attribute will be assigned a null value.

Position, Height, Angle

You can change the position, height, or angle of an attribute value by entering, respectively, **P**, **H**, or **A** at the following prompt:

Enter an option [Value/Position/Height/Angle/Style/Layer/Color/Next] <N>:

The **Position** option lets you define the new position of the attribute value. AutoCAD LT will prompt you to enter the new starting point, center point, or endpoint of the string. If the string is aligned, AutoCAD LT will prompt for two points. You can also define the new height or angle of the text string by entering, respectively, **H** or **A** at the prompt.

Style

This option allows you to modify the text style of the attributes. You can specify the style to be used at the **Enter new text style <Standard>** prompt.

Layer and Color

The **Layer** and **Color** options allow you to change the layer and color of the attribute. For a color change, you can enter the new color by entering a color number (1 through 255), a color name (red, green, and so on), BYLAYER, or BYBLOCK.

Example 6 *General*

In this example, you will use the drawing in Example 2 to edit the attributes individually. Make the following changes in the attribute values.

a. Change the attribute value 100 Min to 100 MB.
b. Change the height of all attributes with the tag name RAM to 0.075 units.

1. Load the drawing that you had saved in Example 2.

2. At the AutoCAD LT Command prompt, enter the **-ATTEDIT** command. The following is the command prompt sequence to change the value of 100 Min to 100 MB.

   Command: **-ATTEDIT**
   Edit attributes one at a time? [Yes/No] <Y>: Enter
   Enter block name specification <*>: **COMP**
   Enter attribute tag specification <*>: Enter
   Enter attribute value specification <*>: Enter

Select Attributes: *Select the attribute.*
Enter an option [Value/Position/Height/Angle/Style/Layer/Color/Next] <N>: **V**
Enter type of value modification [Change/Replace] <R>: **C**
Enter string to change: \ **Min**
Enter new string: **100MB**

When AutoCAD LT prompts **Enter string to change:**, enter the characters you want to change. In this example, the characters **Min** are preceded by a space. If you enter a space, AutoCAD LT displays the next prompt, **Enter new string**. If you need a leading blank space, the character string must start with a backslash (\), followed by the desired number of blank spaces.

3. To change the height of the attribute text, enter the **-ATTEDIT** command as just shown. When AutoCAD LT displays the following prompt, enter **H** for height.

Enter an option [Value/Position/Height/Angle/Style/Layer/Color/Next] <N>: **H**
Specify new height <current>: **0.075**

After you enter the new height and press ENTER, AutoCAD LT will change the height of the text string that has the **X** mark. AutoCAD LT will then repeat the last prompt. Use the **Next** option to move the **X** mark to the next attribute. To change the height of other attribute values, repeat these steps. The drawing after editing the attributes is shown in Figure 15-21.

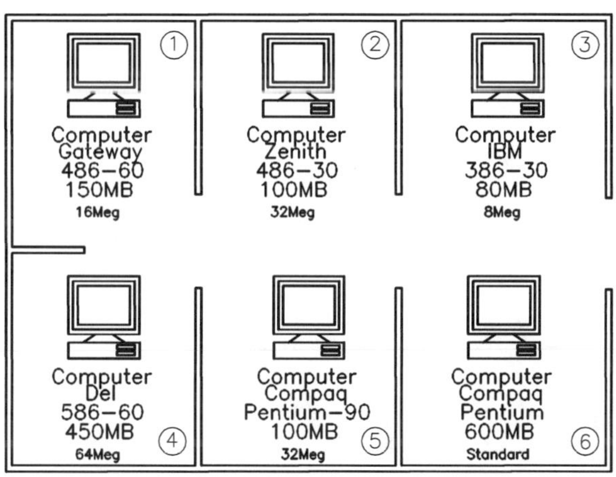

***Figure 15-21** Using **-ATTEDIT** to change the attribute values individually*

Tip

*When you are defining attributes and you have certain attributes whose values are not known at that time, you should enter AAAA or something similar. Later on you can replace such text with the values that you have obtained, using the **ATTEDIT** or **-ATTEDIT** commands. This is easier than adding an attribute later after the block has already been defined.*

INSERTING TEXT FILES IN THE DRAWING

Toolbar:	Text > Multiline Text
Menu:	Draw > Text > Multiline Text
Command:	MTEXT

After you have extracted attribute information into a file, you may want to insert this text into a drawing. You can insert this text file in a drawing by using the **Import Text** button available in the **Multiline Text Editor**. This text editor is displayed when using the **MTEXT** command.

After you have invoked the **MTEXT** command, AutoCAD LT prompts you to enter the insertion point and other corner of the paragraph text box, within which the text file will be placed. After you specify these points, the **Multiline Text Editor** dialog box appears on screen. To insert the text file **Complst1.txt** (created in Example 3), choose the **Import Text** button in the **Multiline Text Editor**. AutoCAD LT displays the **Select File** dialog box (Figure 15-22).

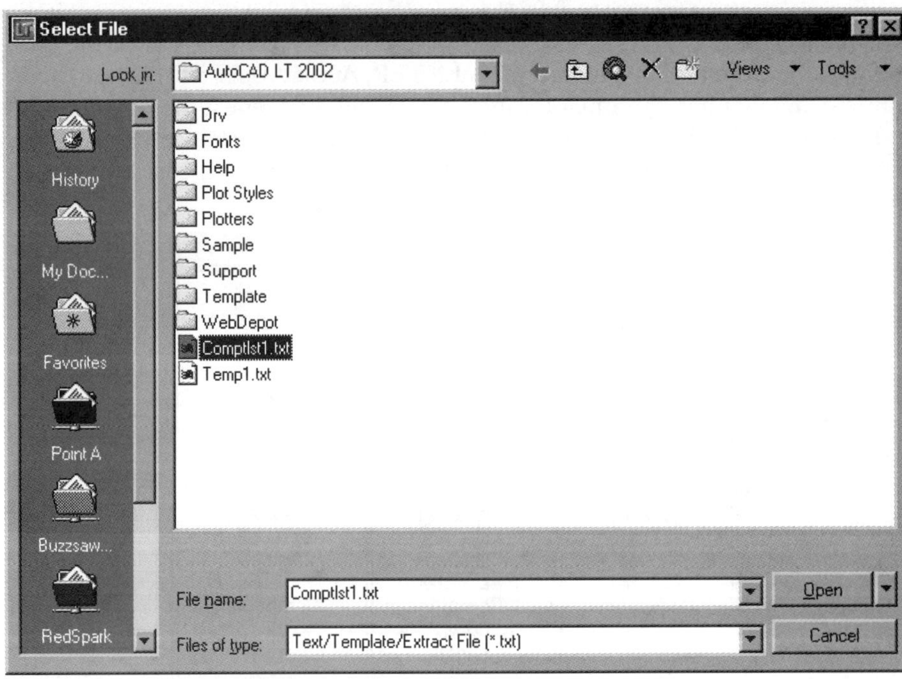

Figure 15-22 *Selecting the file for inserting the text*

In the **Select File** dialog box, you can select the text file **COMPLST1** from the list box and then choose the **Open** button. The imported text is displayed in the text area of the **Multiline Text Editor** dialog box (Figure 15-23). Note that only ASCII files are properly interpreted. Now choose the **OK** button to get the imported text in the selected area on the screen (Figure 15-24).

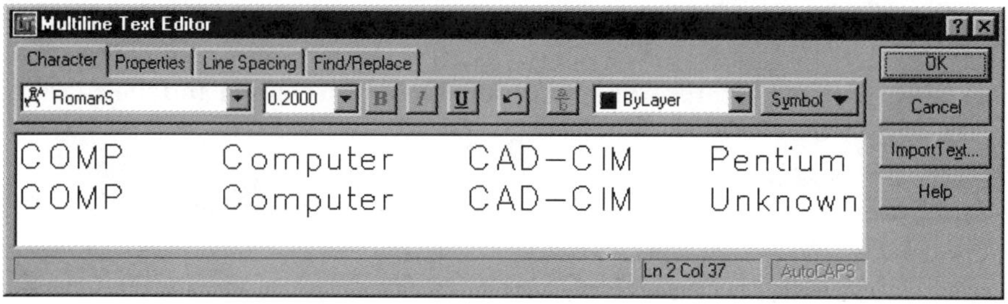

Figure 15-23 ***Multiline Text Editor*** *dialog box displaying the imported text*

COMP	COMPUTER	GATEWAY	486-60	150MB	16MB
COMP	COMPUTER	ZENITH	486-30	100MB	32MB
COMP	COMPUTER	IBM	386-30	80MB	8MB
COMP	COMPUTER	cad-cim	UNKNOWN	600MB	STANDARD
COMP	COMPUTER	cad-cim	PENTIUM-90	100MB	32MB
COMP	COMPUTER	DEL	586-60	450MB	64MB

Figure 15-24 *Imported text file on the screen*

You can also use the **Multiline Text Editor** to change the text style, height, direction, width, rotation, line spacing, and attachment.

Self-Evaluation Test

Answer the following questions, and then compare your answers to the correct answers given at the end of this chapter.

1. Like a constant attribute, the **Preset** attribute cannot be edited. (T/F)

2. For tag names, any lowercase letters are automatically converted to uppercase. (T/F)

3. You can use the **ATTEDIT** command to modify the justification, height, or style of the attribute value. (T/F)

4. If you select the **On** option of the **ATTDISP** command, even the attributes defined with the **Invisible** mode are displayed. (T/F)

5. The entry in the **Value** edit box of the **Attribute Definition** dialog box defines the __________ of the specified attribute.

6. If you have selected the **Align** option from the **Justification** drop-down list in the **Text Options** area of the **Attribute Definition** dialog box, the **Height** and **Rotation** edit boxes are __________.

7. You can use the __________ command or the **Properties** window to edit text or attribute definitions.

8. The default value of the **ATTDIA** variable is __________, which disables the dialog box.

9. In the __________ File, the records are not separated by a comma and the character fields are not enclosed in single quotes.

10. The __________ command allows you to edit the attributes independently of the blocks that contain the attribute reference.

Review Questions

Answer the following questions.

1. If you do not enter anything in the **Prompt** edit box, the entry made in the **Tag** edit box is used as the prompt. (T/F)

2. You can also use the **Find and Replace** dialog box to modify the attribute values. (T/F)

3. The **Constant** mode lets you define an attribute that has a constant value and cannot be edited later. (T/F)

4. You can use ? and * in the string value. When these characters are used in string values, AutoCAD LT does not interpret them as wild-card characters. (T/F)

5. The template file name and the output file name can be the same. (T/F)

6. Not selecting any of the check boxes in the **Mode** area of the **Attribute Definition** dialog box displays all the prompts at the command line and the values will be visible on the screen. This is also referred to as which mode?

   (a) Formal (b) Normal
   (c) Abnormal (d) None of the above

7. Which of the following system variables when set to 0 will suppress the display of prompts for new values?

   (a) **ATTDIA** (b) **ATTREQ**
   (c) **ATTMODE** (d) **ATTDEF**

8. Selecting the **Off** option of which command turns off the visibility of all attribute values?

   (a) **ATTDISP** (b) **ATTEXT**
   (c) **ATTEDIT** (d) **ATTDEF**

9. AutoCAD LT regenerates the drawing at the end of the **-ATTEDIT** command, unless which of the following is turned off?

   (a) **AUTOSNAP** (b) **REGENAUTO**
   (c) **ATTDIA** (d) **ATTMODE**

10. If you need a leading blank space in the string to change, the character string must start with which one of the following characters?

    (a) space ( ) (b) backlash (\)
    (c) asterisk (*) (d) colon (:)

11. You can insert the text file in the drawing by choosing the _________ button in the **Multiline Text Editor** dialog box displayed when using the **MTEXT** command.

12. The function of the **Preset** option is _______________________________________.

13. If you select the **Constant** check box in the **Mode** area of the **Attribute Definition** dialog box, the **Prompt** edit box is _________.

14. You should select the _________ check box in the **Attribute Definition** dialog box to automatically place the subsequent attribute text just below the previously defined attribute.

15. The template file required to extract the attributes is saved as a _________ file.

Exercises

Exercise 1 *Electronics*

In this exercise, you will define the following attributes for a resistor and then create a block using the **BLOCK** command. The name of the block is RESIS.

Mode	Tag name	Prompt	Default value
Verify	RNAME	Enter name	RX
Verify	RVALUE	Enter resistance	XX
Verify, Invisible	RPRICE	Enter price	00

1. Draw the resistor as shown in Figure 15-25.

2. Enter **ATTDEF** at the AutoCAD LT Command prompt to invoke the **Attribute Definition** dialog box.

3. Define the attributes as shown in the preceding table, and position the attribute text as shown in Figure 15-25.

4. Use the **BLOCK** command to create a block. The name of the block is RESIS, and the

insertion point of the block is at the left end of the resistor. When you select the objects for the block, make sure you also select the attributes.

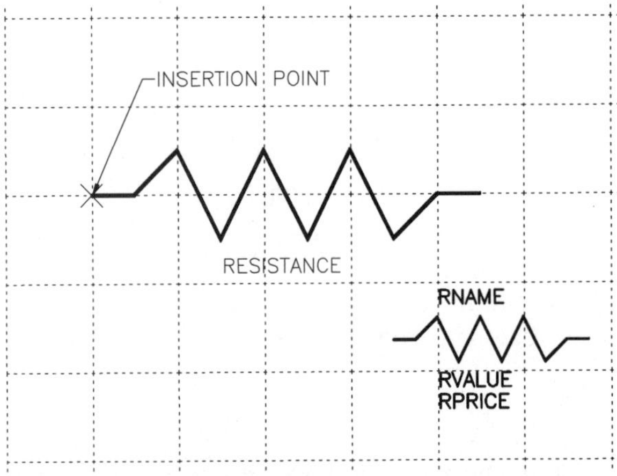

***Figure 15-25** Drawing of a resistor for Exercise 1*

Exercise 2 — *Electronics*

In this exercise, you will use the **INSERT** command to insert the block that was defined in Exercise 1 (RESIS). The following is the list of the attribute values for the resistances in the electric circuit.

RNAME	RVALUE	RPRICE
R1	35	0.32
R2	27	0.25
R3	52	0.40
R4	8	0.21
RX	10	0.21

1. Draw the electric circuit diagram as shown in Figure 15-26 (assume the dimensions).

2. Set the system variable **ATTDIA** to 1. Use the **INSERT** command to insert the blocks, and define the attribute values in the **Enter Attributes** dialog box.

3. Repeat the **INSERT** command to insert other blocks, and define their attribute values as given in the table. Save the drawing as **ATTEXR2.DWG** (Figure 15-27).

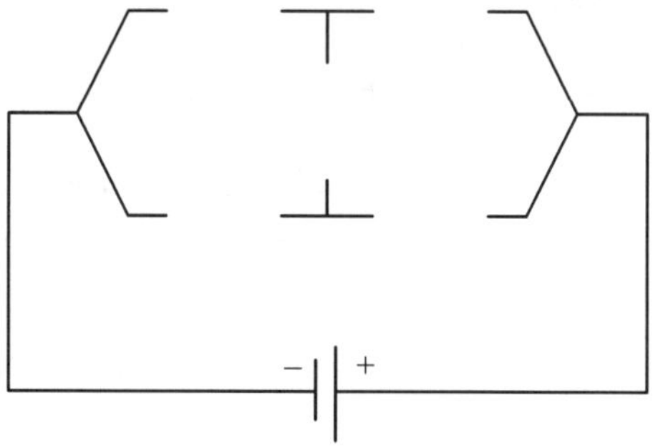

Figure 15-26 Electric circuit diagram without resistors for Exercise 2

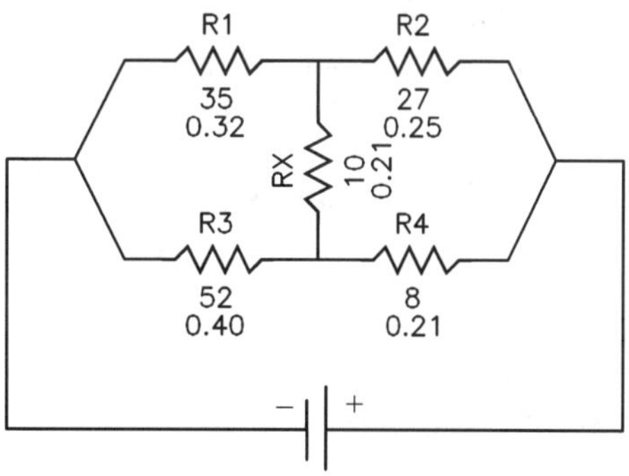

Figure 15-27 Electric circuit diagram with resistors for Exercise 2

Exercise 3 *Electronics*

In this exercise, you will extract the attribute values that were defined in Exercise 2. Extract the values of RNAME, RVALUE, and RPRICE. These attribute values must be saved in a Space Delimited File format named **RESISLST** and arranged as shown in the following table:

RESIS	1	R1	35	0.32
RESIS	1	R2	27	0.25
RESIS	1	R3	52	0.40
RESIS	1	R4	8	0.21
RESIS	1	RX	10	0.21

1. Load the drawing **ATTEXR2** that you saved in Exercise 2.
2. Create the template file in Windows Notepad. After creating the template file, save it under the name **Temp2.txt**.
3. Invoke the **Attribute Extraction** dialog box and select the **Space Delimited File (SDF)** radio button.
4. Select the objects by choosing the **Select Objects** button.
5. Choose the **Template File** button and select **Temp2.txt** as the template file for extracting the attributes.
6. Choose the **Output File** button and specify the name of the output file in the **Output File** dialog box.
7. Choose the **OK** button in the **Attribute Extraction** dialog box to create the attribute extraction file with the required values.

Exercise 4 *Electronics*

In this exercise, you will change the attributes of the resistances that are highlighted in the following table. You will also extract the attribute values and insert the text file in the drawing.

1. Load the drawing **ATTEXR2** that was created in Exercise 2. The drawing has five resistances with attributes. The name of the block is RESIS, and it has three defined attributes, one of them invisible.

2. Use the **ATTEDIT** command or the **-ATTEDIT** command to edit the values that are **highlighted** in the following table.

RESIS	R1	**40**	0.32
RESIS	R2	**29**	0.25
RESIS	R3	52	**0.45**
RESIS	R4	8	**0.25**
RESIS	**R5**	10	0.21

3. Extract the attribute values, and write the values to a text file.

4. Use the **MTEXT** command to insert the text file in the drawing.

Exercise 5 *Electronics*

Use the information given in Exercise 3 to extract the attribute values, and write the data to the output file. The data in the output file should be Comma Delimited CDF. Use the **ATTEXT** and **-ATTEXT** commands to extract the attribute values.

Exercise 6 *Electronics*

In this exercise you will create the blocks with the required attributes. Next, draw the circuit diagram shown in Figure 15-28 and then extract the attributes to create a bill of materials.

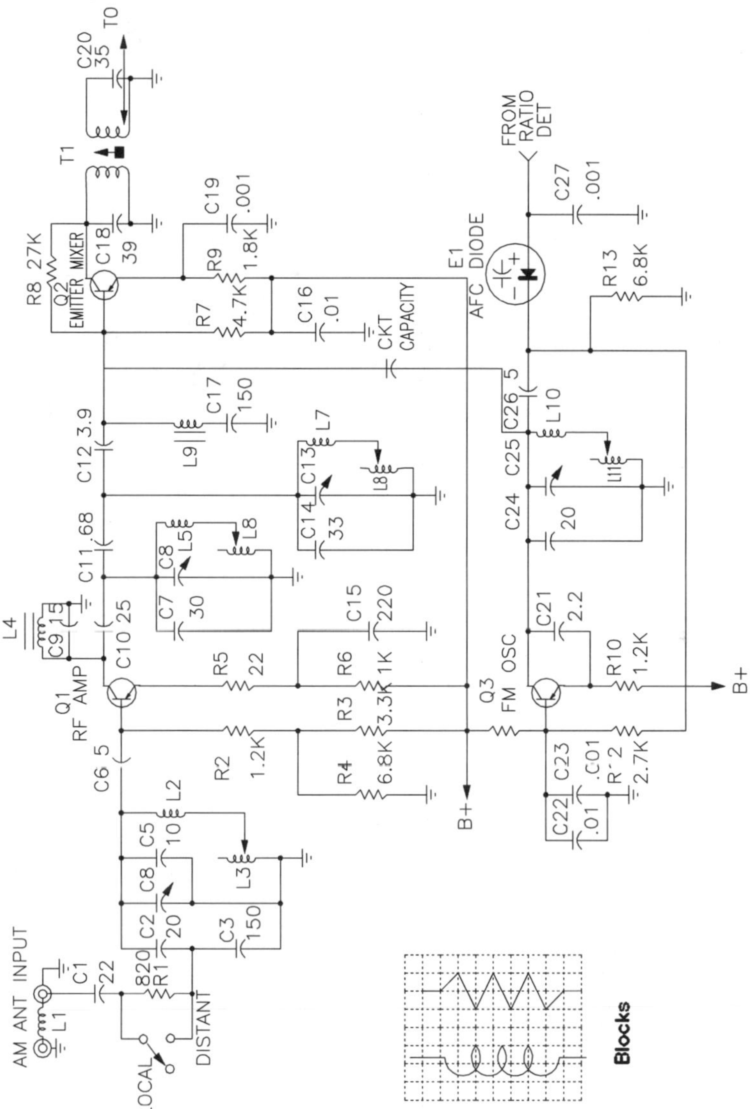

Figure 15-28 Drawing of the circuit diagram for Exercise 6

Exercise 7 *Electronics*

In this exercise you will create the blocks with the required attributes. Next, draw the circuit diagram shown in Figure 15-29 and then extract the attributes to create a bill of materials.

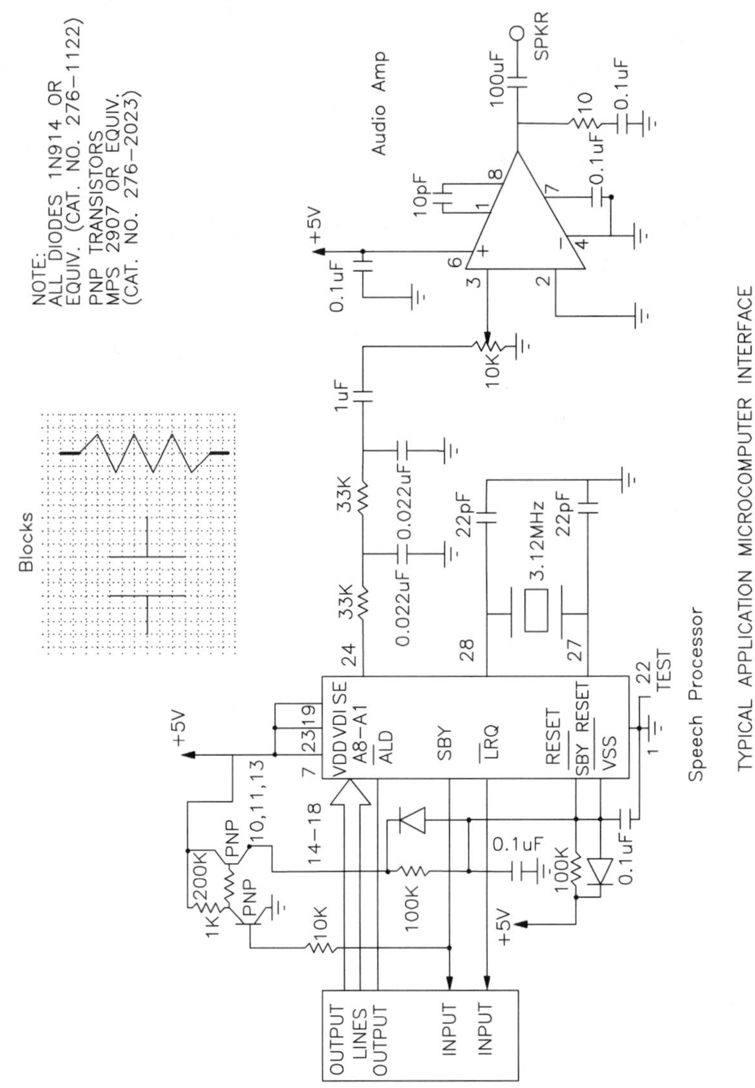

Figure 15-29 Drawing for Exercise 7

Answers to Self-Evaluation Test

1 - F, **2** - T, **3** - F, **4** - T, **5** - default value, **6** - disabled, **7** - **DDEDIT**, **8** - 0, **9** - Space Delimited, **10** - **-ATTEDIT**

Chapter 16

External References

Learning Objectives

After completing this chapter, you will be able to:

- *Understand external references and their applications.*
- *Understand dependent symbols.*
- *Use the **XREF** command and its options.*
- *Use the **Attach**, **Unload**, **Reload**, **Detach**, and **Bind** options.*
- *Edit the path of an xref.*
- *Understand the difference between the **Overlay** and **Attachment** options.*
- *Use the **XATTACH** command.*
- *Use the **AutoCAD DesignCenter** to attach a drawing as an xref.*
- *Use the **XBIND** command to add dependent symbols.*
- *Understand demand loading.*

EXTERNAL REFERENCES

The external reference feature allows you to reference an external drawing without making that drawing a permanent part of the existing drawing. For example, assume that we have an assembly drawing ASSEM1 that consists of two parts, SHAFT and BEARING. The SHAFT and BEARING are separate drawings created by two CAD operators or provided by two different vendors. We want to create an assembly drawing from these two parts. One way to create an assembly drawing is to insert these two drawings as blocks by using the **INSERT** command. Now assume that the design of BEARING has changed due to customer or product requirements. To update the assembly drawing we have to make sure that we insert the BEARING drawing after the changes have been made. If we forget to update the assembly drawing, then the assembly drawing will not reflect the changes made in the piece part drawing. In a production environment, this could have serious consequences.

You can solve this problem by using the **external reference** facility, which lets you link the piece part drawings with the assembly drawing. If the xref drawings (piece part) get updated, the changes are automatically reflected in the assembly drawing. This way, the assembly drawing stays updated no matter when the changes were made in the piece part drawings. There is no limit to the number of drawings that you can reference. You can also have **nested references**. For example, the piece part drawing BEARING could be referenced in the SHAFT drawing, then the SHAFT drawing could be referenced in the assembly drawing ASSEM1. When you open or plot the assembly drawing, AutoCAD LT automatically loads the referenced drawing SHAFT and the nested drawing BEARING. When using external references, several people working on the same project can reference the same drawing and all the changes made are displayed everywhere the particular drawing is being used.

If you use the **INSERT** command to insert the piece parts, the piece parts become a permanent part of the drawing and, therefore, the drawing has a certain size. However, if you use the external reference feature to link the drawings, the piece part drawings are not saved with the assembly drawing. AutoCAD LT only saves the reference information with the assembly drawing; therefore, the size of the drawing is minimized. Like blocks, the xref drawings can be scaled, rotated, or positioned at any desired location, but they cannot be exploded. You can also use only a part of the xref by making clipped boundary of xrefs.

Tip

External referenced drawings are useful for creating parts or subassemblies and then putting them together in one drawing to create the main assembly. You can also use it for laying out the contents of a drawing with multiple views before plotting.

DEPENDENT SYMBOLS

If you use the **INSERT** command to insert a drawing, the information about the named objects is lost if the names are duplicated. If they are unique, it is imported. The **named objects** are entries such as blocks, layers, text styles, and layers. For example, if the assembly drawing has a layer HIDDEN with blue color and HIDDEN linetype, and the piece part BEARING has a layer HIDDEN with green color and HIDDEN2 linetype, then when you insert the BEARING drawing in the assembly drawing, the values set in the assembly drawing will override the values of the inserted drawing (Figure 16-1). Therefore, in the assembly drawing, the layer HIDDEN will retain blue color and HIDDEN linetype, ignoring the layer

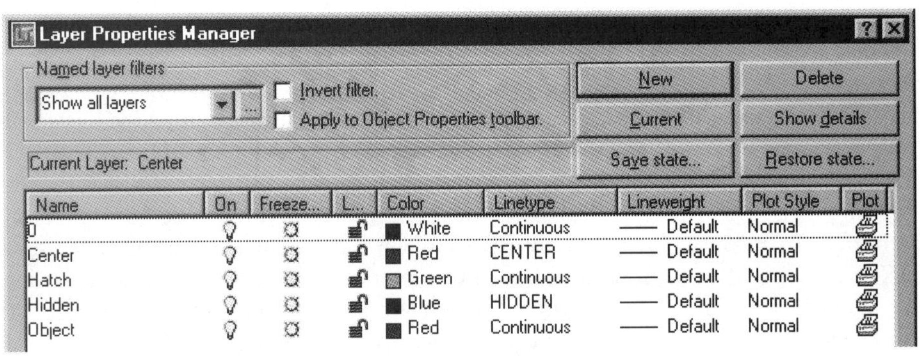

Figure 16-1 *Layer settings of the current drawing override the layers of the inserted drawing*

settings of the inserted drawing. Only those layers that have the same names are affected. Remaining layers that have different layer names are added to the current drawing.

In the xref drawings, the information about named objects is not lost because AutoCAD LT will create additional named objects such as the specified layer settings as shown in Figure 16-2. For xref drawings, these named objects become dependent symbols (features such as layers, linetypes, object color, text style, and so on).

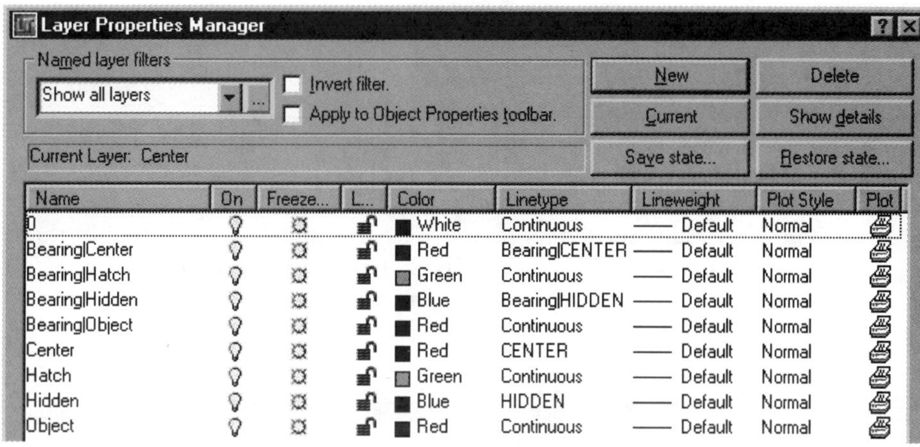

Figure 16-2 *Xref creates additional layers*

The layer HIDDEN of the xref drawing (BEARING) is appended with the name of the xref drawing BEARING, and the two are separated by the vertical bar symbol (|). Also, the name of these layers appears faded in the **Layer Control** drop-down list of the **Object Properties** toolbar and cannot be selected. The layer name HIDDEN changes to BEARING|HIDDEN. Similarly, CENTER is renamed BEARING|CENTER and OBJECT is renamed BEARING|OBJECT (Figure 16-2). The information added to the current drawing is not permanent. It is added only when the xref drawing is loaded. If you detach the xref drawing, the dependent symbols are automatically erased from the current drawing.

When you xref a drawing, AutoCAD LT does not let you reference the symbols directly. For example, you cannot make the dependent layer, BEARING|HIDDEN, current. Therefore, you cannot add any objects to that layer. However, you can change the color, linetype, lineweight, plotstyle, or visibility (on/off, freeze/thaw) of the layer in the current drawing. If the **Retain changes to Xref layers** check box in the **External References (Xrefs)** area of the **Open and Save** tab of the **Options** dialog box is cleared, which also implies that the system variable **VISRETAIN** is set to 0, the settings are retained only for the current drawing session. This means that when you save and exit the drawing, the changes are discarded and the layer settings return to their default status. If this check box is selected (default) which also implies that the **VISRETAIN** variable is set to 1, layer settings such as color, linetype, on/off, and freeze/thaw are retained and the settings are saved with the drawing and are used when you open the drawing the next time. Whenever you open a drawing or plot it, AutoCAD LT reloads each xref in the drawing and as a result, the latest updated version of the drawing is loaded automatically.

Note

You cannot make the xref dependent layers current in a drawing. Only when the xref drawing is binded to the current drawing using the ***XBIND*** *command can you make the xref dependent layers a permanent part of the current drawing and use them. The* ***XBIND*** *command will be discussed later in this chapter.*

MANAGING EXTERNAL REFERENCES IN A DRAWING

Toolbar:	Insert > External Reference Reference > External Reference
Menu:	Insert > Xref Manager
Command:	XREF

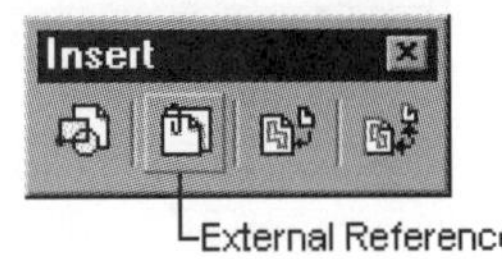

Figure 16-3 *The* ***Insert*** *toolbar*

When you invoke the **XREF** command (Figure 16-3), AutoCAD LT displays the **Xref Manager** dialog box (Figure 16-4). The **Xref Manager** dialog box displays the status of each Xref in the current drawing and the relation between the various Xrefs. It allows you to attach a new xref, detach, unload, load an existing one, change an attachment to an overlay, or an overlay to an attachment. It also allows you to edit an xref's path and bind the xref definition to the drawing.

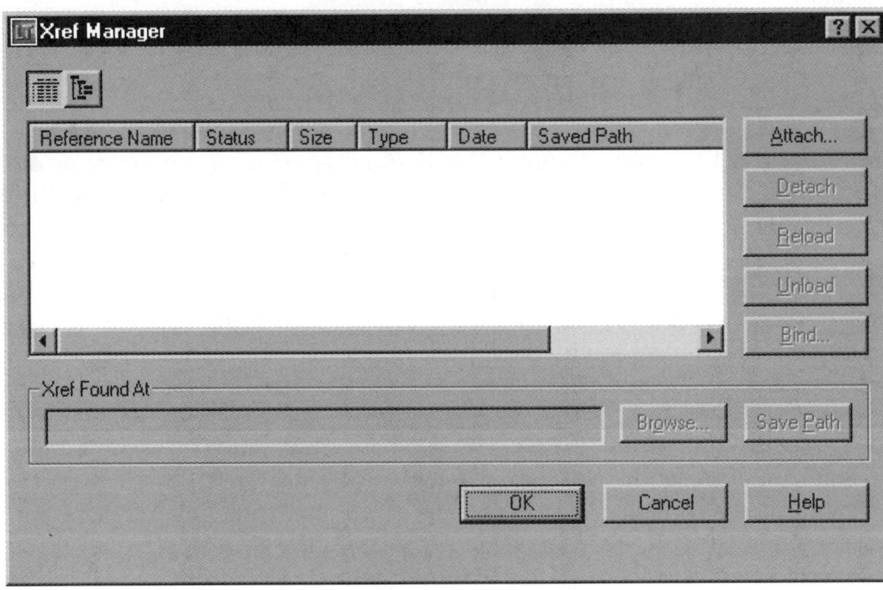

Figure 16-4 Xref Manager *dialog box*

Apart from the methods mentioned in the command box, you can also invoke the **Xref Manager** dialog box by selecting an xref in the current drawing and then right-clicking in the drawing area to display a shortcut menu. Now, choose **Xref Manager** from the shortcut menu; the **Xref Manager** dialog box is displayed.

The upper left corner of the dialog box has two buttons: **List View** and **Tree View**.

List View

Choosing the **List View** button displays the xrefs present in the drawing in alphabetical order. This is the default view. The list view displays information about xrefs in the current drawing under the following headings.

Reference Name

This column lists the name of all existing references in the current drawing.

Status

This column lists the current status of each xref in the drawing. It lists whether a xref is loaded, unloaded, unreferenced, not found, orphaned, unresolved, or marked to be reloaded. An loaded xref implies that the xref is attached to the current drawing. You can then unload it and then reload it using the options in this dialog box (this will be discussed later). An xref selected to be unloaded or reloaded displays **Unload** and **Reload** respectively under the **Status** column. If the xref has nested references that cannot be found, the status is **Unreferenced** and if the parent of the nested reference gets unloaded, or cannot be found, its status is described as **Orphaned**. An unreferenced xref will not be displayed. If the xref is not found in the search paths defined, its status is **Not Found**. A missing xref or one that cannot be found is **Unresolved**.

Size

The file size of each xref is listed here.

Type

This column lists whether the xref is an attachment or overlay.

Date

This column lists the date on which the xref drawing was last saved.

Saved Path

This column lists the path of the xref, that is, the route taken to locate the particular referenced drawing.

Choosing any of these headings sorts and lists the xrefs in the current drawing according to that particular title. For example, choosing **Reference Name**, sorts and lists the xrefs as per name. The column widths can be increased or decreased as per requirements. When you place your cursor at the edge of a column title button, the cursor changes to a horizontal resizing cursor. Now, press the pick button of your mouse and drag the column edge to increase or decrease its width. After you increase the column widths, it is possible that the width of the columns extend beyond the list box width. In such a case, a horizontal scroll bar appears at the bottom of the list box. You can use the scroll bar to view the columns that extend beyond the width of the list box.

Choosing the **Tree View** button, displays the xrefs in the drawing in a hierarchical tree view. It displays information on nested xrefs and their relationship with one another. Xrefs are indicated by an icon of a paper with a paper clip. This icon appears faded when the xref has been unloaded and if there is a missing xref, a question mark appears. Similarly, an upward arrow indicates that the xref was reloaded and an arrow pointing downwards indicates that the xref is unloaded. You can also choose **List View** and **Tree View** by pressing the F3 and F4 keys respectively.

Attaching an Xref Drawing (Attach Option)

The **Attach** button of the **Xref Manager** dialog box is used to attach an xref drawing to the current drawing. This option can be invoked by choosing the **Attach** button in the **Xref Manager** dialog box. The following examples illustrate the process of attaching an xref to the current drawing. In this example, it is assumed that there are two drawings, SHAFT and BEARING. SHAFT is the current drawing that is loaded on the screen (Figure 16-5) and the BEARING drawing is saved on the disk. We want to xref the BEARING drawing in the SHAFT drawing.

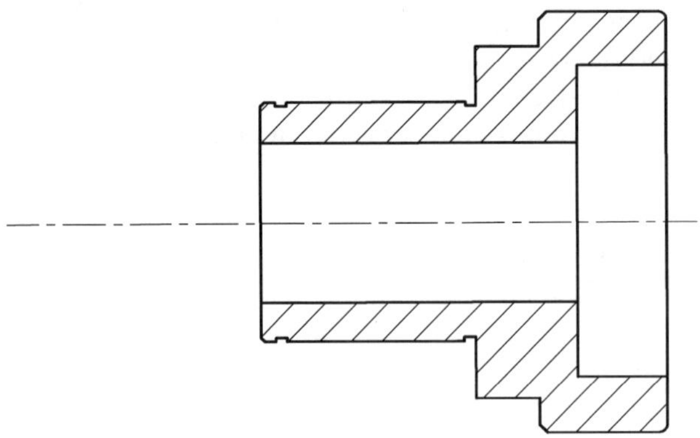

Figure 16-5 *Current drawing, SHAFT*

1. The first step is to make sure that the SHAFT drawing is on the screen (draw the shaft drawing with assumed dimensions).

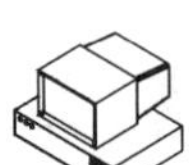

Tip

One of the drawings need not be on the screen. You could attach both drawings, BEARING and SHAFT, to an existing drawing, even if it is a blank drawing.

2. Invoke the **XREF** command to display the **Xref Manager** dialog box. In this dialog box, choose the **Attach** button. You can also choose **External Reference** from the **Insert** menu. The **Select Reference File** dialog box (Figure 16-6) is displayed. Select the drawing that you want to attach (BEARING), and then choose the **Open** button. The **External reference** dialog box is displayed on the screen (Figure 16-7).

   In the **External Reference** dialog box (Figure 16-7), the name of the file you have selected to be attached to the current drawing as an xref is displayed in the **Name** edit box. You can also select a name of the file to attach from the **Name** drop-down list. The path of the file is also displayed adjacent to **Path**, located below the **Name** edit box. The **Retain Path**

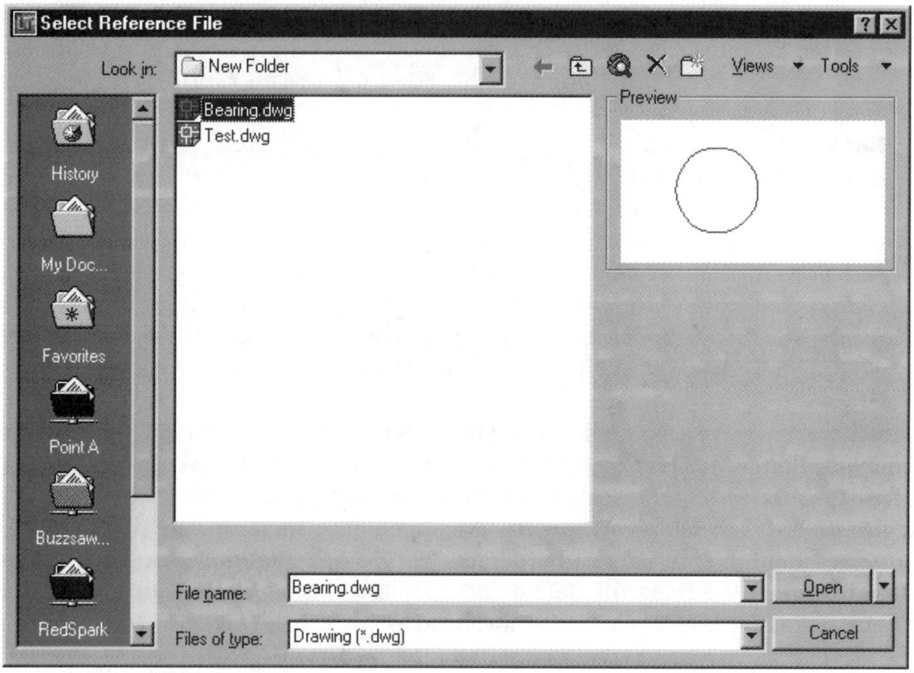

Figure 16-6 ***Select Reference File*** *dialog box*

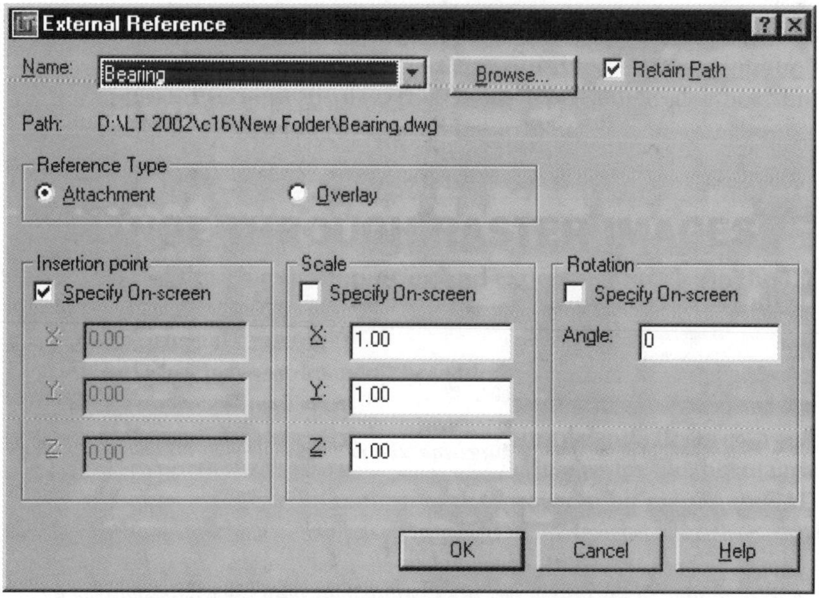

Figure 16-7 ***External Reference*** *dialog box for inserting the xrefs*

check box is selected by default and saves the location path of the xref. If this check box is not selected, AutoCAD LT searches for the referenced file in the paths defined in the **Support File Search Path** folder in the **Files** tab of the **Options** dialog box only.

Note

AutoCAD LT also searches for the xref file in the paths defined in the ***Project Files Search Path*** *folder in the* ***Files*** *Tab of the* ***Options*** *dialog box. This folder does not have any paths defined in it and displays* ***Empty*** *when the tree view is expanded. To define a search path, highlight* ***Empty*** *and choose the* ***Add*** *button in the dialog box. You can enter a project name here, if you want to. Now, expand the tree view again and select* ***Empty*** *again and choose the* ***Browse*** *button. The* ***Browse for Folder*** *window is displayed. Select the folder that is to be searched for the file and choose OK. Then choose Apply and OK in the Options dialog box to exit it.*

In the **Reference Type** area, select the **Attachment** radio button if it is not already selected (default option). The **Overlay** option will be discussed later in this chapter. You can either specify the insertion point, scale factors, and rotation angle in the respective **X**, **Y**, **Z** and **Angle** edit boxes or select the **Specify On-screen** check boxes to use the pointing device to specify them on the screen. By default, the X, Y and Z scale factors are 1 and the rotation angle is 0. Accept the default values and choose the **OK** button in the **External Reference** dialog box. Specify the insertion point on the screen. After attaching the BEARING drawing as an xref, save the current drawing with the file name SHAFT (Figure 16-8).

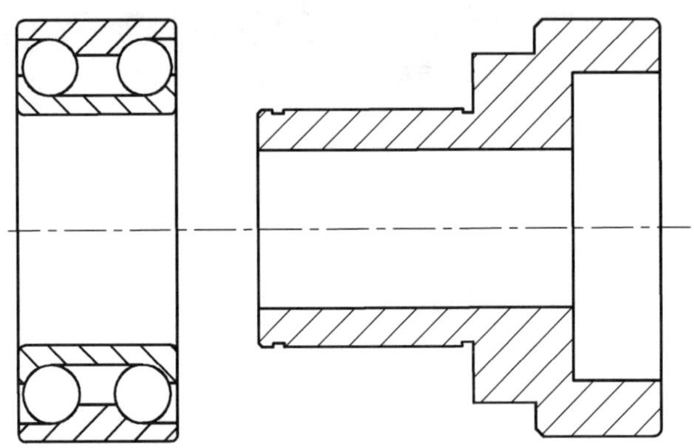

Figure 16-8 *Attaching xref drawing BEARING*

Note

You can also use the ***-XREF*** *command to attach a drawing from the command line. All the options available in the dialog box are available through the command line too.*

3. Load the drawing BEARING and make the changes shown in Figure 16-9 (draw polylines on the sides). Now, save the drawing with the file name BEARING.

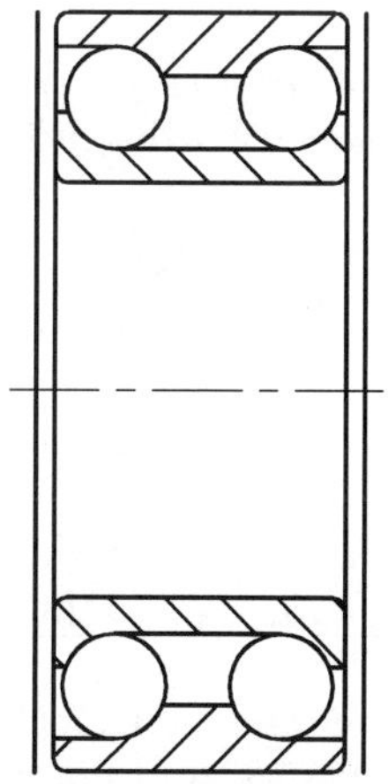

Figure 16-9 *Modifying the xref drawing BEARING*

4. Load the drawing SHAFT (Figure 16-10) on the screen. You will notice that the xref drawing BEARING is automatically updated. This is the most useful feature of the **XREF** command. You could also have inserted the BEARING drawing as a block, but if you had updated the BEARING drawing, the drawing in which it was inserted would not have been updated automatically.

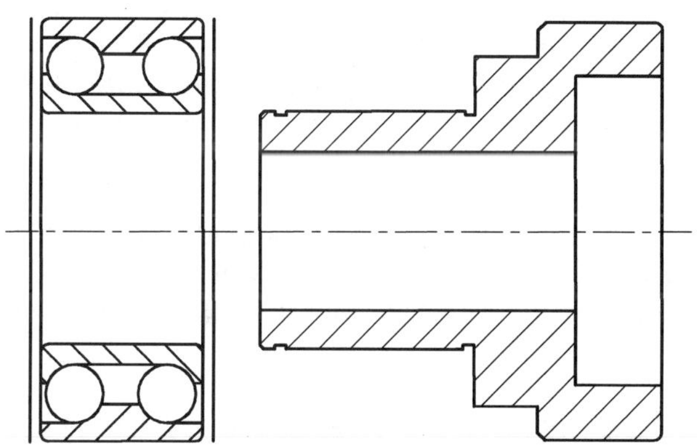

Figure 16-10 *After loading the drawing SHAFT, BEARING is automatically updated*

When you attach an xref drawing, AutoCAD LT remembers the name of the attached drawing. If you xref the drawing again, AutoCAD LT displays a message to that effect as shown:

Xref "BEARING" has already been defined.
Using existing definition.
Specify insertion point or [Scale/X/Y/Z/Rotate/PScale/PX/PY/PZ/PRotate]: *Specify where to place another copy of the xref.*

Note

If the xref drawing you want to attach is currently being edited, AutoCAD LT will attach the drawing that was last saved through the ***SAVE, WBLOCK,*** *or* ***QUIT*** *command.*

Points to Remember about Xref

1. When you enter the name of the xref drawing, AutoCAD LT checks for block names and xref names. If a block exists with the same name as the name of the xref drawing in the current drawing, the **XREF** command is terminated and an error message is displayed.

2. When you xref a drawing, the objects that are in the model space are attached. Any objects that are in the paper space are not attached to the current drawing.

3. The layer 0, DEFPOINTS, and the linetype CONTINUOUS are treated differently. The current drawing layers 0, DEFPOINTS, and linetype CONTINUOUS will override the layers and linetypes of the xref drawing. For example, if the layer 0 of the current drawing is white and the layer 0 of the xref drawing is red, the white color will override the red.

4. The xref drawings can be nested. For example, if the BEARING drawing contains the reference INRACE and you xref the BEARING drawing to the current drawing, the INRACE drawing is automatically attached to the current drawing. If you detach the BEARING drawing, the INRACE drawing gets detached automatically.

5. You can rename an xref under the **Reference** column name in the list box of the **Xref Manager** dialog box by highlighting the xref and then clicking on it again. You can now enter a new name. An AutoCAD LT warning is displayed: **Caution! "XXXX" is an externally referenced block. Renaming it will also rename its dependent symbols**.

6. When you xref a drawing, AutoCAD LT stores the name and path of the drawing by default. If the name of the xref drawing or the path where the drawing was originally stored has changed or you cannot find it in the path specified in the **Options** dialog box, AutoCAD LT cannot load the drawing, plot it, or use the **Reload** option of the **XREF** command.

Detaching an Xref Drawing (Detach Option)

The **Detach** option can be used to detach or remove the xref drawings. If there are any nested xref drawings defined with the xref drawings, they are also detached. Once a drawing is detached, it is erased from the screen. To detach an xref drawing, select the file name in the **Xref Manager** dialog box list box to highlight it and then choose the **Detach** button. When you choose **OK** in the dialog box, the xref is completely removed from the current drawing. If you feel that you do not want to remove the specified xref from the current drawing, and have already selected the xref in the list box and chosen the **Detach** button, you can simply choose the **Cancel** button to cancel the detach operation.

You can also use the **-XREF** command to detach the xref drawings. When AutoCAD LT prompts for an xref file name to detach, you can enter the name of one xref drawing or the name of several drawings separated by commas. You can also enter * (asterisk), in which case all referenced drawings, including the nested drawings, will be detached.

Updating an Xref Drawing (Reload Option)

When you load a drawing, AutoCAD LT automatically loads the referenced drawings. The **Reload** option of the **XREF** command lets you update the xref drawings and nested xref drawings at any time. You do not need to exit the drawing editor and then reload the drawing. To reload the xref drawings, invoke the **Xref Manager**, select the drawings in the list box, and then choose the **Reload** button. AutoCAD LT will scan for the referenced drawings and the nested xref drawings and load the most recently saved version of the drawing.

The **Reload** option is generally used when the xref drawings are currently being edited and you want to load the updated drawings. The xref drawings are updated based on what is saved on the disk. Therefore, before reloading an xref drawing, you should make sure that the xref drawings that are being edited have been saved. If AutoCAD LT encounters an error while loading the referenced drawings, the **XREF** command is terminated, and the entire reload operation is canceled.

You can also reload the xref drawings by using the **-XREF** command. When you enter the **-XREF** command, AutoCAD LT will prompt you to enter the name of the xref drawing. You could enter the name of one xref drawing or the name of several drawings separated by commas. If you enter * (asterisk), AutoCAD LT will reload all xref and nested xref drawings.

Unloading an Xref Drawing (Unload Option)

The **Unload** option allows you to temporarily remove the definition of an xref drawing from a current drawing. However, AutoCAD LT retains the pointer to the xref drawings. When you unload the xref drawings, the drawings are not displayed on the screen. You can reload the xref drawings by using the **Reload** option.

Tip

It is recommended that you unload the referenced drawings if they are not being used. After unloading the xref drawings, the drawings load much faster and need less memory.

Adding an Xref Drawing (Bind Option)

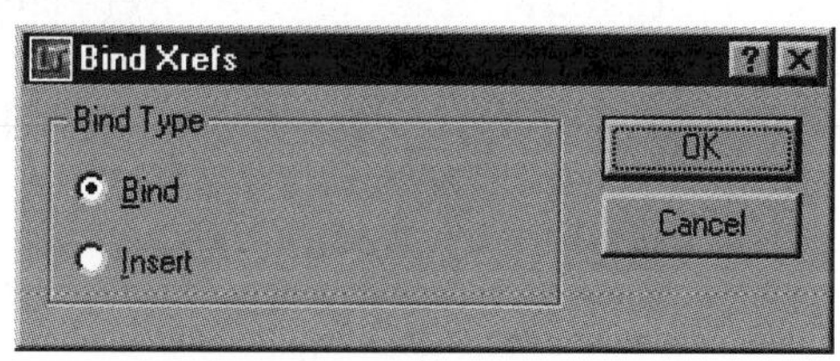

***Figure 16-11 Bind Xrefs** dialog box*

The **Bind** option lets you convert the xref drawings to blocks in the current drawing. The bound drawings, including the nested xref drawings (that are no longer xrefs), become a permanent part of the current drawing. The bound drawing cannot be detached or reloaded. You can use this option when you want to send a copy of your drawing to a customer for review. Since all the xref drawings are a part of the existing drawing, you do not need to include the xref drawings or the path information. You can also use this option to safeguard the master drawing from accidental editing of the piece parts. To bind the xref drawings, select the file names in the **Xref Manager** dialog box list box and then choose the **Bind** button. The **Bind Xrefs** dialog box (Figure 16-11) is displayed. AutoCAD LT provides two methods to bind the xref drawing in the **Bind Type** area of the dialog box. These methods are discussed next.

Bind

When you use the **Bind** option, AutoCAD LT binds the selected xref definition to the current drawing. All the xrefs are converted to blocks and the named objects are renamed. For example, if you xref the drawing BEARING with a layer named OBJECT, a new layer BEARING|OBJECT is created in the current drawing. When you bind this drawing, the xref dependent layer BEARING|OBJECT will become a locally defined layer BEARING0OBJECT (Figure16-12). If the BEARING0OBJECT layer already exits, AutoCAD LT will automatically increment the number, and the layer name becomes BEARING1OBJECT.

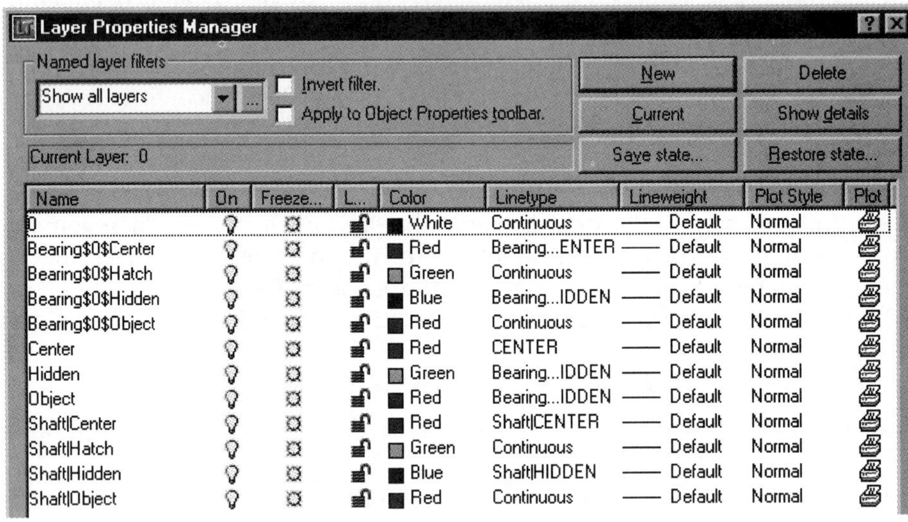

*Figure 16-12 **Layer Properties Manager** dialog box*

Insert

When you use the **Insert** option, AutoCAD LT inserts the xref drawing. The xrefs get converted into blocks. For example, if you xref the drawing BEARING with a layer named OBJECT, a new layer BEARING|OBJECT is created in the current drawing. If you use the **Insert** option to bind the xref drawing, the layer name BEARING|OBJECT is renamed as OBJECT (Figure 16-12). If the object layer already exists, then the values set in the current drawing override the values of the inserted drawing.

Note

It is possible to bind only an individual or several xref dependent named objects instead of the entire drawing into a current drawing. This will be discussed later in this chapter.

*You can also use the **-XREF** command to bind the xref drawings. Prompts are displayed on the command line.*

Editing an Xref's Path

By default, AutoCAD LT saves the path of the referenced drawing and displays it in the **Saved Path** column in the **Xref Manager** dialog box. As mentioned earlier, when AutoCAD LT loads the drawing containing a referenced file, and if it is not able to find the file at the location specified in the **Saved Path** column of the **Xref Manager** dialog box, it searches for the file in the current directory, and in the **Support File Search Path** locations specified in the **Files** tab of the **Options** dialog box. If a file with the same name is found here, it is loaded. Now when you invoke the **Xref Manager** dialog box, you will notice that when you select an xref name in the list box to highlight it, the path displayed in the **Saved Path** column for the xref file is different from the one displayed in the **Xref Found At** edit box. To update the path of the xref file, choose the **Save Path** button. The new path is saved and displayed in the **Saved Path** column.

If AutoCAD LT cannot locate the specified file even in the directories specified in the **Files** tab of the **Options** dialog box, it will display an error message saying that it cannot find the specified file. The path of the file is displayed as a marker text in the current drawing. Now, when you invoke the **Xref Manager** dialog box, the status of the drawing is shown as **Not Found**. To specify a new path for the xref file, select the xref file name in the list box of the dialog box and choose the **Browse** button. The **Select new path** dialog box is displayed where you can locate the drawing to be used as xref. Once you have found the file, choose the **Open** button to return to the **Xref Manager** dialog box. The new path is displayed in the **Saved Path** column and the **Xref Found At** edit box. The specified xref file is reloaded and replaces the marker text in the drawing when you choose the **OK** button in the **Xref Manager** dialog box. If you remember the new location of the xref file, you can also enter it in the **Xref Found At** edit box. For example, if the drawing was originally in the C:\CAD\Proj1 subdirectory and the drawing has been moved to A:\Parts directory, the path must be edited so that AutoCAD LT is able to load the xref drawing.

You can also use the **-XREF** command to change the path using the prompts on the command line. When AutoCAD LT prompts you to enter the name of the xref whose path you want to edit, you can enter the name of one xref drawing or the names of several drawings separated by commas. You can also enter * (asterisk), in which case AutoCAD LT will prompt you for the path name of each xref drawing. The path name stays unchanged if you press ENTER when AutoCAD LT prompts for a new path name.

OVERLAY OPTION

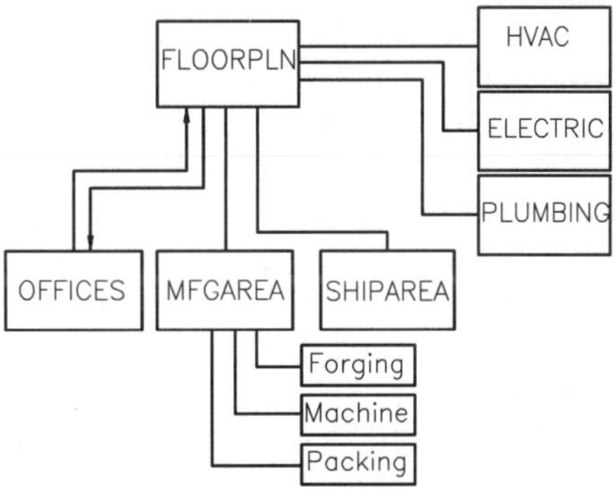

Figure 16-13 *Drawing files hierarchy*

As discussed earlier, when you are attaching an xref to a drawing, the **External Reference** dialog box is displayed. The **Reference Type** area of this dialog box has two radio buttons. They are **Attachment** and **Overlay**. **Attachment** is the default option. One of the problems with the **Attachment** option is that you cannot have circular reference. For example, assume you are designing the plant layout of a manufacturing unit. One person is working on the floor plan (see Figure 16-13), and the second person is working on the furniture layout in the offices (Figure 16-14). The names of the drawings are FLOORPLN and OFFICES, respectively. The person working on the office layout uses the **Attachment** option to insert the FLOORPLN drawing so that he or she has the latest floor plan drawing. The person who is working on the

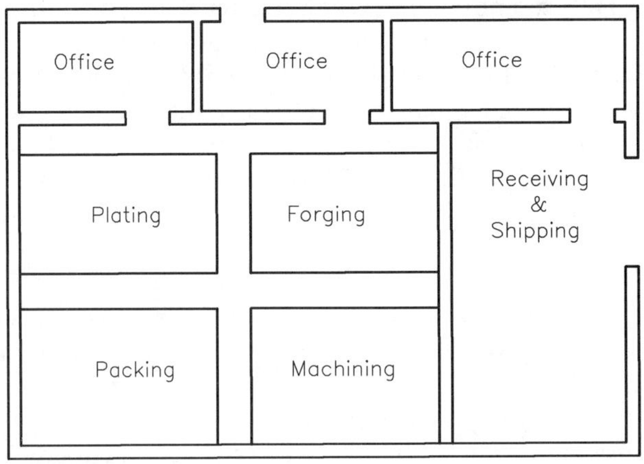

Figure 16-14 *Sample plant layout drawing*

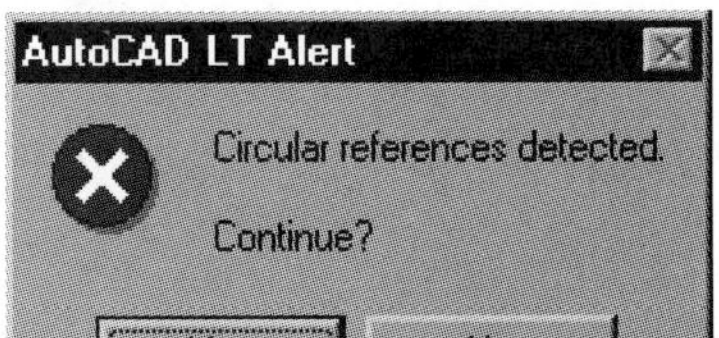

Figure 16-15 AutoCAD LT Alert *message*

floor plan wants to reference the OFFICES drawing. Now, if the **Attachment** option is used to reference the drawings, AutoCAD LT displays an error message because by attaching the OFFICES drawing a circular reference is created (Figure 16-15). The AutoCAD LT message displayed is: "**Circular references detected. Continue?**" If you choose the **No** button, the **XREF** command is canceled and no drawing is referenced. But, if you choose the **Yes** button, the following message is displayed: **Breaking circular reference from "offices" to "current drawing"** and the particular file you wanted to reference is referenced.

However, to overcome this problem of circular reference, you can use the **Overlay** option to overlay the OFFICES drawing. This is a very useful option because the **Overlay** option lets different operators avoid circular reference and share the drawing data without affecting the drawing. Overlaying allows you to view a referenced drawing without having to attach it to the current drawing. This option can be invoked by selecting the **Overlay** radio button in the **External Reference** dialog box which is displayed after you have selected a drawing to reference. Also, when a drawing that has a nested overlay is overlaid, the nested overlay is not visible in the current drawing. This is another difference between attaching an xref and overlaying an xref to a drawing. This feature is especially useful when you want to reference a drawing that another user who is referencing your drawing does not need. While the attachment will reference the nested reference too, the **overlay** option ignores nested references.

You can also use the **-XREF** command to display prompts on the command line and enter the **Overlay** option. Selecting the **Overlay** option, displays the **Enter Name of file to Overlay** dialog box, where you can select the file you want to overlay.

Example 1 *Architectural*

In this example you will use the **Attachment** and **Overlay** options to attach and reference the drawings. Two drawings, PLAN and PLANFORG, are given. The PLAN drawing (Figure 16-16) consists of the floor plan layout, and the PLANFORG drawing (Figure 16-17) has the details of the forging section only. The CAD operator who is working on the PLANFORG drawing wants to xref the PLAN drawing for reference. Also, the CAD operator working on the PLAN drawing should be able to xref the PLANFORG drawing to complete the project. The following steps illustrate how to accomplish the defined task without creating a circular reference.

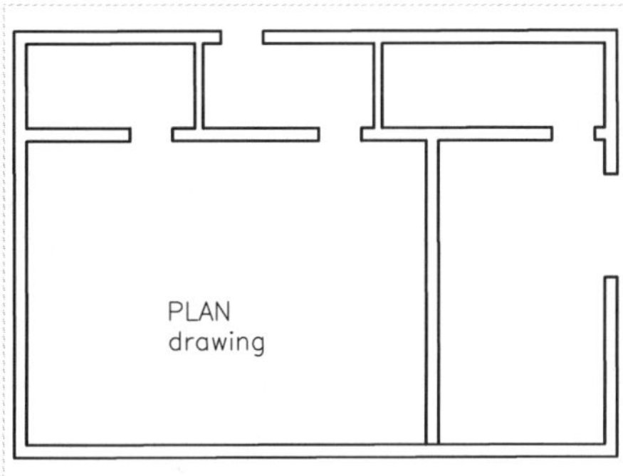

Figure 16-16 *PLAN drawing*

Figure 16-17 *PLANFORG drawing*

How circular reference is caused:

1. Load the drawing PLANFORG, invoke the **XREF** command, and choose the **Attach** button in the **Xref Manager** dialog box. Select the PLAN drawing in the list box of the **Select Reference File** dialog box and choose the **Open** button. The **External Reference** dialog box is displayed. The name of the PLAN drawing is displayed in the **Name** edit box and the **Attachment** radio button is selected by default in the **Reference Type** area of the dialog box. Choose **OK** to exit the dialog box and specify an insertion point on screen. Now the drawing consists of PLANFORG and PLAN. Save the drawing.

2. Open the drawing file PLAN, and invoke the **XREF** command and attach the PLAN drawing using the same steps described in Step 1. AutoCAD LT will display the message that circular reference has been detected and ask you if you want to continue. If you choose to continue by choosing **Yes** in the AutoCAD LT message box, the circular reference is broken and you are allowed to reference the specific drawing.

   Another possible solution is for the operator working on the PLANFORG drawing to detach the PLAN drawing. This way, the PLANFORG drawing does not contain any reference to the PLAN drawing and would not cause any circular reference. The other solution is to use the **Overlay** option, as follows.

How to prevent circular reference:

3. Open the drawing PLANFORG (Figure 16-18) and select the **Overlay** radio button in the **External reference** dialog box which is displayed after you have selected the PLAN drawing to reference. The PLAN drawing is overlaid on the PLANFORG drawing (Figure 16-19).

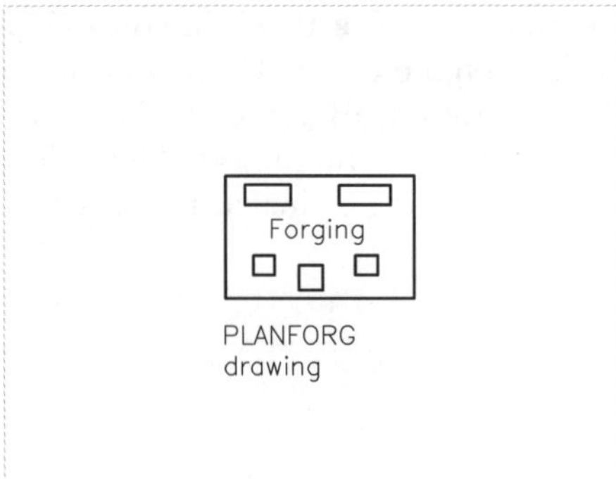

Figure 16-18 PLANFORG drawing

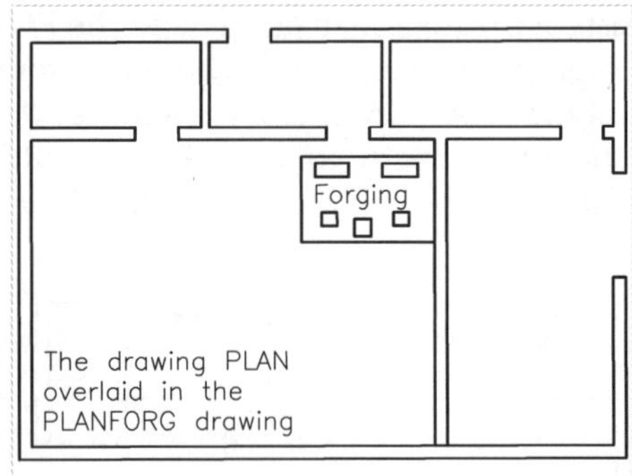

Figure 16-19 PLANFORG drawing after overlaying the PLAN drawing

4. Open the drawing file PLAN (Figure 16-20), and choose the **Attachment** radio button in the **External References** dialog box, which is displayed when you have selected the PLANFORG drawing in the **Select Reference** dialog box to attach as an xref to the PLAN drawing. You will notice that only the PLANFORG drawing is attached (Figure 16-21). The drawing that was overlaid in the PLANFORG drawing (PLAN) does not appear in the current drawing. This way, the CAD operator working on the PLANFORG drawing can overlay the PLAN drawing, and the CAD operator working on the PLAN drawing can attach the PLANFORG drawing, without causing a circular reference.

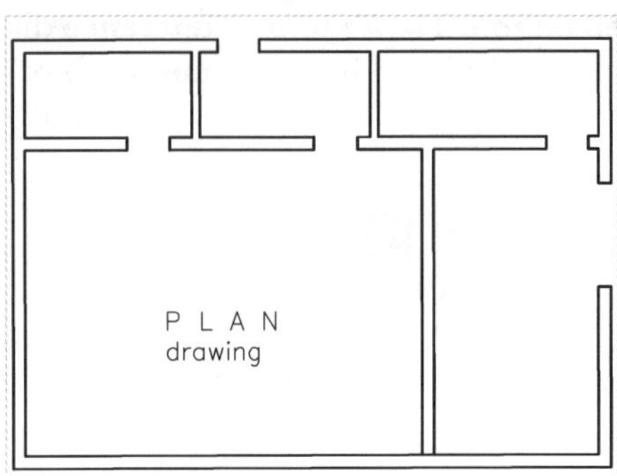

Figure 16-20 PLAN drawing

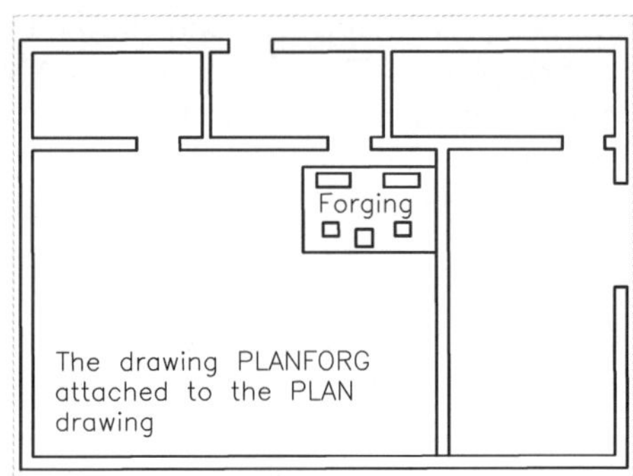

Figure 16-21 PLAN drawing after attaching the PLANFORG drawing

XATTACH COMMAND

Toolbar:	Reference > External Reference Attach
Menu:	Insert > External Reference
Command:	XATTACH

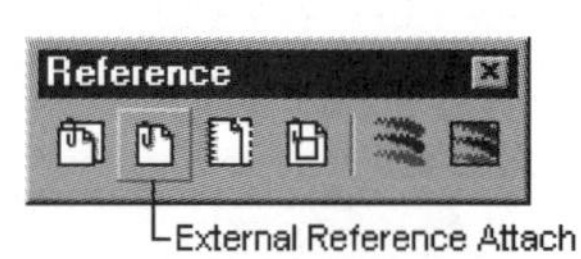

Figure 16-22 ***Reference*** *toolbar*

If you want to attach a drawing without invoking the **Xref Manager** dialog box, you can use the **XATTACH** command (Figure 16-22). When you invoke this command, AutoCAD LT displays the **Select Reference file** dialog box. This command makes it easier to attach a drawing, since most of the xref operations involve simply attaching a drawing file. After you have selected the drawing file to attach, choose the Open button. The **External Reference** dialog box is displayed. Select the **Attachment** radio button under the **Reference Type** area. You can specify the insertion point, scale, and rotation angle on screen or in the respective edit boxes.

Tip

*When you attach or reference a drawing that has a drawing order created by using the **DRAWORDER** command, the drawing order is not maintained in the xref. To correct the drawing order, first open the xref drawing and specify the drawing order in it. Now, use the **WBLOCK** command to convert it into a drawing file and the **XATTACH** command to attach the newly created drawing file to the current drawing. This way the drawing order will be maintained.*

Note

*AutoCAD LT maintains a log file (.XLG) for xref drawings if the **XREFCTL** system variable is set to 1. This file lists information about the date and time of loading and other xref operations to be completed. This .XLG file is saved with the current drawing with the same name as the current drawing and is updated each time the drawing is loaded or any xref operations are carried out.*

Using AutoCAD DesignCenter to Attach a Drawing as Xref

The AutoCAD DesignCenter can also be used to attach an xref to a drawing. Choose the **AutoCAD DesignCenter** button in the **Standard Toolbar** to display the **DesignCenter** window (Figure 16-23). In the **DesignCenter** toolbar, choose the **Tree View Toggle** button, if it not chosen already, to display the **tree pane**. Expand the Tree view and double-click the folder whose contents you want to view. The contents of the selected folder are displayed in the palette on the right side. From the list of drawings displayed in the palette, right-click the drawing you wish to attach as an xref. A shortcut menu is displayed. Choose **Attach as Xref**, the **External Reference** dialog box is displayed. You can also use the right button of your mouse to drag and drop the drawing into the current drawing. A shortcut menu is displayed again. Choose **Attach as Xref** and the **External Reference** dialog box is displayed.

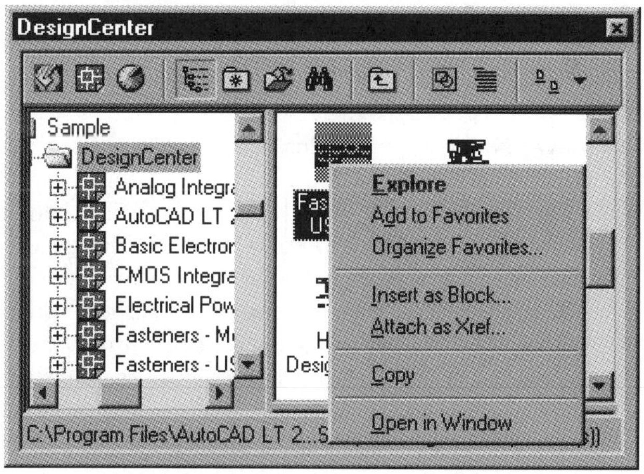

***Figure 16-23** The **DesignCenter** window*

The **Name** edit box displays the name of the selected file to be inserted as an xref and **Path** displays the path of the file. Select the **Attachment** radio button in the **Reference type** area, if not already selected. Specify the **Insertion point**, **Scale**, and **Rotation** in the respective edit boxes or select the **Specify On-screen** check boxes to specify this information on the screen. Choose **OK** to exit the dialog box.

ADDING DEPENDENT SYMBOLS TO A DRAWING

Toolbar:	Reference > External Reference Bind
Menu:	Modify > Object > External Reference > Bind
Command:	XBIND

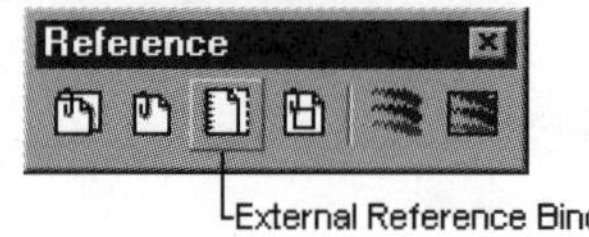

Figure 16-24 Reference toolbar

You can use the **XBIND** command to add the selected dependent symbols of the xref drawing to the current drawing. The following example describes how to use the **XBIND** command (Figure 16-24).

1. Load the drawing BEARING that was created earlier when the **Attach** option of the **XREF** command was discussed. Make sure the drawing has the following layer setup. Otherwise, create the following layers, using the **Layer Properties Manager** dialog box.

Layer Name	**Color**	**Linetype**
0	White	Continuous
Object	Red	Continuous
Hidden	Blue	Hidden2
Center	White	Center2
Hatch	Green	Continuous

2. Draw a circle and use the **BLOCK** command to create a block of it. The name of the block is SIDE. Save the drawing as BEARING.

3. Start a new drawing with the following layer setup.

Layer Name	**Color**	**Linetype**
0	White	Continuous
Object	Red	Continuous
Hidden	Green	Hidden

4. Use the **XATTACH** command or the **Attach** option of the **XREF** command and attach the BEARING drawing to the current drawing. When you xref the drawing, the layers will be added to the current drawing, as discussed earlier in this chapter.

5. Now, invoke the **XBIND** command. The **Xbind** dialog box (Figure 16-25) is displayed on the screen. This dialog box has two areas with list boxes. They are **Xrefs** and **Definitions to Bind**. If you want to bind the blocks defined in the xref drawing BEARING, first click on the plus sign adjacent to the xref BEARING. Icons for named objects in the drawing are displayed in a tree view. Click on the plus sign next to the Block icon. AutoCAD LT lists the blocks defined in the xref drawing (BEARING). Select the block BEARING|SIDE and then choose the **Add** button. The block name will be added to the **Definitions to Bind** area list box. Choose **OK** to exit the dialog box. AutoCAD LT will bind the block with the current drawing and a message at the command line is displayed: **1 Block(s) bound**. The name of the block will change to BEARING0SIDE. You can invoke the **Block Definition** dialog box using the **BLOCK** command and check the **Name** drop-down list to see if the block with the name BEARING0SIDE has been added to the drawing.

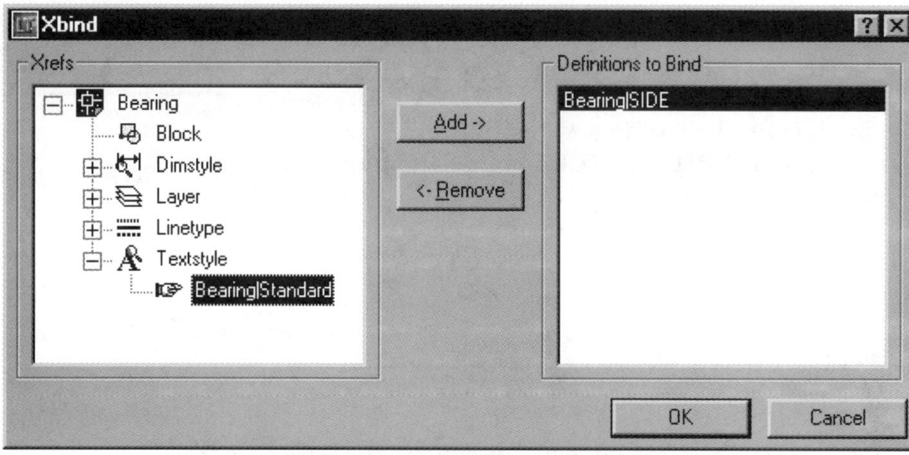

***Figure 16-25** The **Xbind** dialog box*

If you want to insert the block, you must enter the new block name (BEARING0SIDE). You can also rename the block to another name that is easier to use.

If the block contains a reference to another xref drawing, AutoCAD LT binds that xref drawing and all its dependent symbols to the current drawing also. Once you **XBIND** the dependent symbols, AutoCAD LT does not delete them, not even when the xref drawing is detached. For example, the block BEARING0SIDE will not be deleted even when you detach the xref drawing or end the drawing session.

You can also use the **-XBIND** command to bind the selected dependent symbols of the xref drawing, using the command line.

6. Similarly, you can bind the dependent symbols, BEARING|STANDARD (textstyle), BEARING|HIDDEN, and BEARING|OBJECT layers of the xref drawing. Click on the plus signs adjacent to the respective icons to display the contents and then select the layer or textstyle you want to bind and choose the **Add** button. The selected named objects are displayed in the **Definitions to Bind** area list box. If you have selected a named object that you do not want to bind, select it in the **Definitions to Bind** area list box and choose the **Remove** button. Once you have finished selecting the named objects that you want to bind to the current drawing, choose **OK**. A message indicating the number of named objects that are bound to the current drawing is displayed at the command line.

   Once bound, the layer names will change to BEARING0HIDDEN and BEARING0OBJECT. If the layer name BEARING0HIDDEN was already there, the layer will be named BEARING1HIDDEN. These two layers become a permanent part of the current drawing. Even if the xref drawing is detached or the current drawing session is closed, the layers are not discarded.

DISPLAYING CLIPPING FRAME

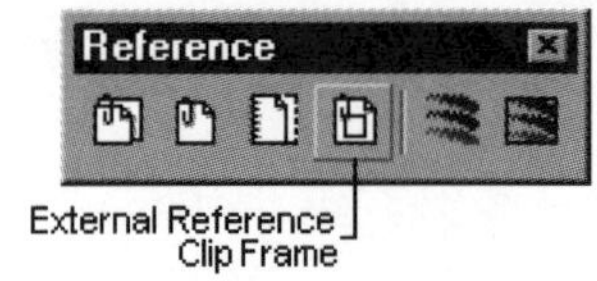

*Figure 16-26 **Reference** toolbar*

You can use the **XCLIPFRAME** system variable to turn the clipping boundary on or off. The clipping boundary is invisible by default. This system variable can be set by choosing the **External Reference Clip Frame** button from the **Reference** toolbar (Figure 16-26), or by entering **XCLIPFRAME** at the Command prompt. When the value is 0 (default), the clipping boundary is not displayed. When it is 1, the clipping boundary is displayed.

Note

*The polyline generated when you use the **generate Polyline** option of the **XCLIP** command should not be confused with the clipping boundary. Although they both coincide with each other, you can select and edit only the polyline. When you select the clipping boundary, the xref object also gets selected.*

DEMAND LOADING

The demand loading feature loads only that part of the referenced drawing that is required in the existing drawing. For example, demand loading provides a mechanism by which objects on frozen layers are not loaded. Also, only the clipped portion of the referenced drawing can be loaded. This makes the xref operation more efficient since less disk space is used, especially when the drawing is reopened.

Demand loading is enabled by default. You can modify its settings in the **External references (XRefs)** area of the **Open and Save** tab of the **Options** dialog box (Figure 16-27). You can select any of the three settings available in the **Demand load Xrefs** drop-down list in this dialog box. They are **Disabled**, **Enabled**, and **Enabled with copy**. These options correspond to a value of **0**, **1**, and **2** of the **XLOADCTL** system variable, respectively, and are discussed next.

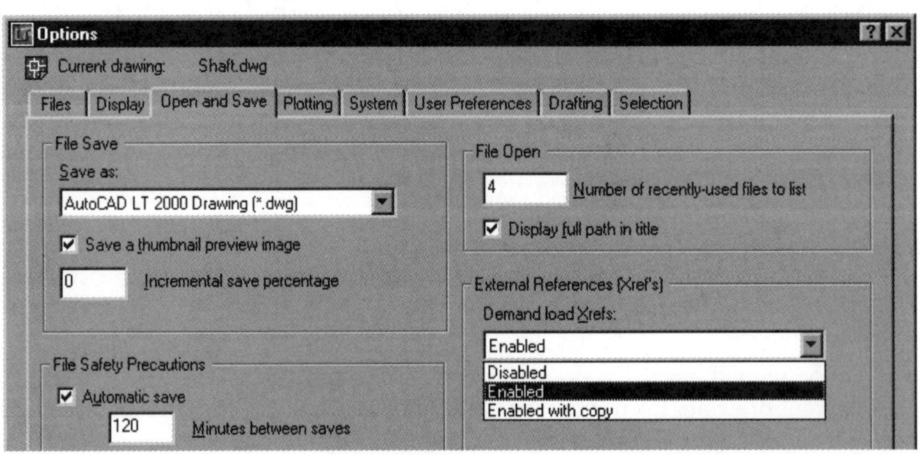

***Figure 16-27** Setting **XLOADCTL** using the **Options** dialog box*

Setting	Value of XLOADCTL	Features
Disabled	0	1. Turns off demand loading. 2. Loads entire xref drawing file. 3. The File is available on the server and other users can edit the xref drawing.
Enabled	1	1. Turns on demand loading. 2. The referenced file is kept open. 3. Makes the referenced file read-only for other users.
Enabled with copy	2	1. Turns on demand loading with the copy option. 2. A copy of a referenced drawing is opened. 3. Other users can access and edit the original referenced drawing file.

You can also set the value of **XLOADCTL** on the command line. When you are using the **Enabled with copy** option of demand loading, the temporary copies of the xref are saved in the AutoCAD LT temporary files directory (defined in the **Temporary File Location** folder in the **Files** tab of the **Options** dialog box) or in a user-specified directory. The **XLOADPATH** system variable creates a temporary path to store demand loaded xrefs.

SPATIAL AND LAYER INDEXES

As mentioned previously, the demand loading improves performance when the drawing contains referenced files. To make it work effectively and to take full advantage of demand loading, you must store a drawing with Layer and Spatial indexes. The layer index maintains a list of objects in different layers and the spatial index contains lists of objects based on their location in 3D space.

The **INDEXCTL** variable controls the creation of layer index and spatial index and its value can be set using the Command line. The following are the settings of the **INDEXCTL** system variable:

Setting	Features
0	No index created.
1	Layer index created.
2	Spatial index created.
3	Layer and spatial index created.

Self-Evaluation Test

Answer the following questions, and then compare your answers to the correct answers given at the end of this chapter.

1. If the assembly drawing has been created using **Xref > Insert**, the drawing will be updated automatically if a change is made in the drawing that was inserted. (T/F)

2. The external reference facility helps you keep the drawing updated no matter when the changes were made in the piece part drawings. (T/F)

3. Objects can be added to a dependent layer. (T/F)

4. While using the **Attachment** option during referencing a drawing, the drawing will reference the nested references too, while the **Overlay** option ignores nested references. (T/F)

5. The __________ are entries such as blocks, layers, and text styles.

6. If you use the **INSERT** command to insert a drawing, the information about the named objects is lost if the names are ____________ and if the names are ___________, the drawing is imported.

7. The __________ button of the **Xref Manager** dialog box is used to attach an xref drawing to the current drawing.

8. The ________________ feature loads only that part of the referenced drawing that is required in the existing drawing.

9. The _________ option can be used to overcome the problem of circular reference.

10. If the **Retain changes to Xref layers** is __________, in the **External References (Xrefs)** area of the **Open and Save** tab of the **Options** dialog box, the layer settings such as color, linetype, on/off, and freeze/thaw are retained. The settings are saved with the drawing and are used when you xref the drawing the next time.

Review Questions

Answer the following questions.

1. If the xref drawings get updated, the changes are not automatically reflected in the assembly drawing when you open the assembly drawing. (T/F)

2. There is a limit to the number of drawings you can reference. (T/F)

3. It is not possible to have nested references. (T/F)

4. Like blocks, the xref drawings can be scaled, rotated, or positioned at any desired location. (T/F)

5. You can change the color, linetype, or visibility (on/off, freeze/thaw) of the dependent layer. (T/F)

6. Which of the following features lets you reference an external drawing without making this drawing a permanent part of the existing drawing?

   (a) demand loading (b) external reference
   (c) external clipping (d) insert drawing

7. If the xref has nested references that cannot be found, which of the following will be displayed under the status heading of the **List View** button in the **Xref Manager** dialog box?

   (a) Orphaned (b) Not found
   (c) Unreferenced (d) Unresolved

8. Which of the following commands can be used from the command line to attach a drawing?

   (a) **-XBIND** (b) **-XREF**
   (c) **XCLIP** (d) None of the above

9. Which of the following system variables, when set to 1, will allow AutoCAD LT to maintain a log file (.XLG) for xref drawings?

   (a) **XLOADCTL** (b) **XLOADPATH**
   (c) **XREFCTL** (d) **INDEXCTL**

10. Which of the following system variables when set to 0 will not allow the clipping boundary to be displayed?

    (a) **XCLIPFRAME** (b) **XLOADCTL**
    (c) **INDEXCTL** (d) **XREFCTL**

11. In the __________ drawings the information regarding dependent symbols is not lost.

12. What is the function of the **XCLIPFRAME** system variable? Explain. ______________
    ____.

13. AutoCAD LT maintains a log file for xref drawings if the __________ variable is set to 1.

14. The value of the **INDEXCTL** if set to __________ creates the layer and spatial index.

15. You can use the __________ command to add selected dependent symbols from the xref drawing to the current drawing.

Exercises

Exercise 1 *Mechanical*

In this exercise, you will start a new drawing and xref the drawings Part-1 and Part-2. You will also edit one of the piece parts to correct the size and use the **XBIND** command to bind some of the dependent symbols to the current drawing. The following are detailed instructions for completing this exercise.

1. Start a new drawing, Part-1, and set up the following layers.

Layer Name	**Color**	**Linetype**
0	White	Continuous
Object	Red	Continuous
Hidden	Blue	Hidden2
Center	White	Center2
Dim-Part1	Green	Continuous

2. Draw Part-1 with dimensions as shown in Figure 16-28. Save the drawing as Part-1.

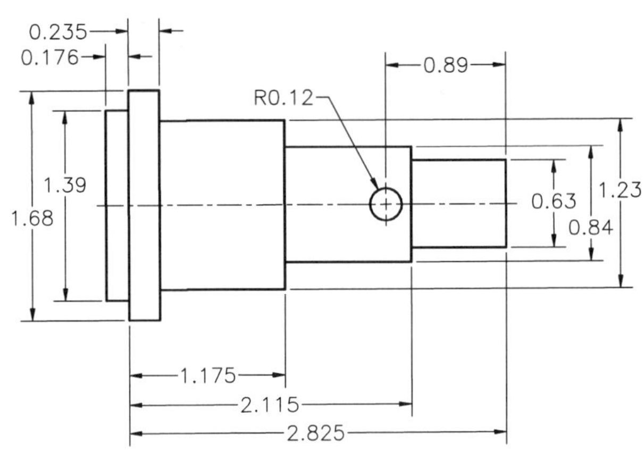

Figure 16-28 *Part-1, Drawing for Exercise 1*

3. Start a new drawing, Part-2, and set up the following layers.

Layer Name	**Color**	**Linetype**
0	White	Continuous
Object	Red	Continuous
Hidden	Blue	Hidden
Center	White	Center
Dim-Part2	Green	Continuous
Hatch	Magenta	Continuous

4. Draw Part-2 with dimensions as shown in the Figure 16-29. Save the drawing as Part-2.

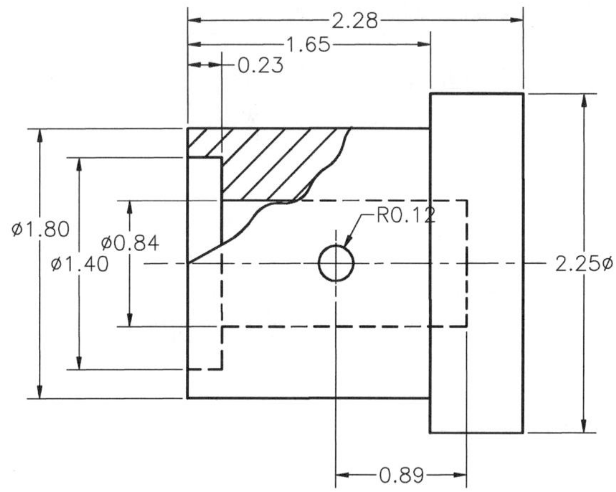

Figure 16-29 Part-2, Drawing for Exercise 1

5. Start a new drawing, ASSEM1, and set up the following layers.

Layer Name	Color	Linetype
0	White	Continuous
Object	Blue	Continuous
Hidden	Yellow	Hidden

6. Xref the two drawings Part-1 and Part-2 so that the centers of the two drilled holes coincide. Notice the overlap as shown in Figure 16-30. Save the assembly drawing as ASSEM1.

7. Open the drawing Part-1 and correct the mistake so that there is no overlap. You can do it by editing the line (1.175 dimension) so that the dimension is 1.160.

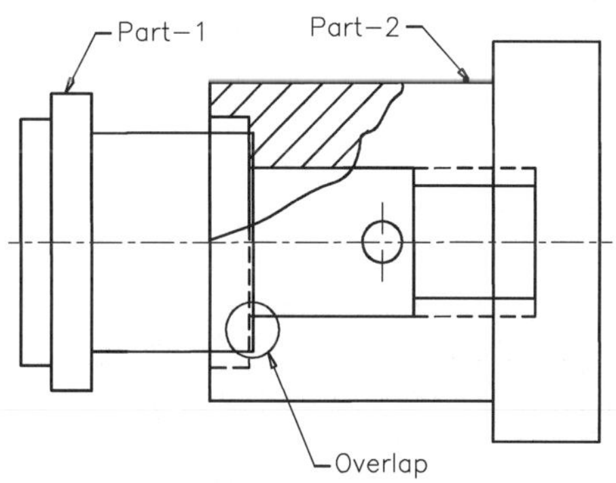

Figure 16-30 Assembly drawing after attaching Part-1 and Part-2

8. Open the assembly drawing ASSEM1 and notice the change in the overlap. The assembly drawing gets updated automatically.

9. Study the layers and notice how AutoCAD LT renames the layers and linetypes assigned to each layer. Check to see if you can make the layers belonging to Part-1 or Part-2 current.

10. Use the **XBIND** command to bind the OBJECT and HIDDEN layers that belong to drawing Part-1. Check again to see if you can make one of these layers current.

11. Use the **Detach** option to detach the xref drawing Part-1. Study the layer again, and notice that the layers that were edited with the **XBIND** command have not been erased. Other layers belonging to Part-1 are erased.

12. Use the **Bind** option of the **XREF** command to bind the xref drawing Part-2 with the assembly drawing ASSEM1. Open the xref drawing Part-1 and add a border or make any changes in the drawing. Now, open the assembly drawing ASSEM1 and check to see if the drawing is updated.

Answers to Self-Evaluation Test

1 - F, **2** - T, **3** - F, **4** - T, **5** - named objects, **6** - duplicated, unique, **7** - **Attach**, **8** - demand loading, **9** - **Overlay**, **10** - selected

Chapter 17

More Draw Commands

Learning Objectives

After completing this chapter, you will be able to:

- *Create double lines using the* ***DLINE*** *command.*
- *Create revision clouds using the* ***REVCLOUD*** *command.*
- *Draw* ***NURBS*** *using the* ***SPLINE*** *command.*
- *Edit* ***NURBS*** *using the* ***SPLINEDIT*** *command.*
- *Digitize drawings.*

CREATING DOUBLE LINES (USING THE DLINE COMMAND)

Toolbar:	Draw > Double Line
Menu:	Draw > Double Line
Command:	DLINE

The AutoCAD LT Double Line feature allows you to create double lines that consist of two parallel lines or arcs. Double lines are drawn like lines. You can also draw double connected lines. AutoCAD LT treats each line segment forming a double line as an individual object. The prompt sequence that will follow when you choose the **Double Line** button (Figure 17-1) to create a simple double line is:

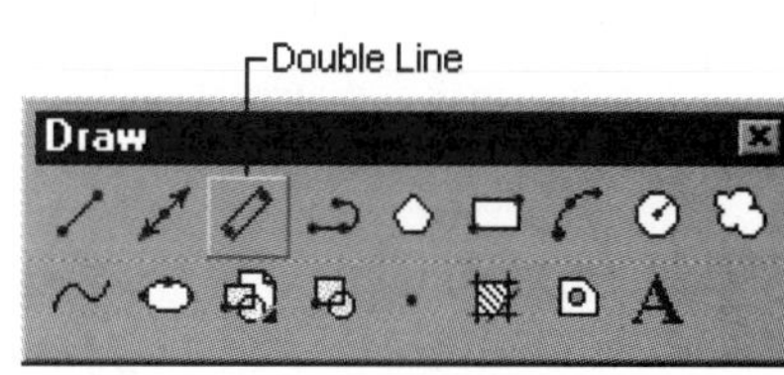

Figure 17-1 *Selecting the* ***DLINE*** *command from the toolbar*

Break/Caps/Dragline/Offset/Snap/Undo/Width/<start point>: *Specify the start point of the double line.*
Arc/Break/CAps/CLose/Dragline/Snap/Undo/Width/<next point>: *Specify the next point or select an option.*

The different options of the **DLINE** command are described in the following.

Start Point

This is the default option and requires you to specify a point either by entering its coordinates or by picking the point on the screen with the pointing device (example mouse).

Break

If you require that a line should be broken when it intersects a double line, as shown in Figure 17-2, set **Break** to ON. The default option is ON. Set **Break** OFF if you do not want to have breaks at double line intersections. The prompt sequence to set **Break** to ON is :

Command : **DLINE** Enter
Break/Caps/Dragline/Offset/Snap/Undo/Width/<start point>: **B**
Break Dline's at start and end points? OFf/<ON>: Enter

Caps

With this option you can close the ends of a double line and obtain clean corners as shown in the Figure 17-3.

Command: **DLINE** Enter
Break/Caps/Dragline/Offset/Snap/Undo/Width/<start point>: **C** Enter
Draw which endcaps? Both/End/None/Start/<Auto>: Enter

At this prompt you can enter any of these options. If you press ENTER, that is if you select **Auto**, AutoCAD LT caps all the ends of the double line except those that are snapped to an object. With the rest of the options, AutoCAD LT caps the ends specified even if they are snapped to an object. Select **Both** if you want both ends of a double line to be capped. To specify the end of the double line which you want capped, select **End** or **Start**. **Start** caps the double line at its beginning while **End** caps the end you pick last. If you do not want any ends to be capped, select **None**.

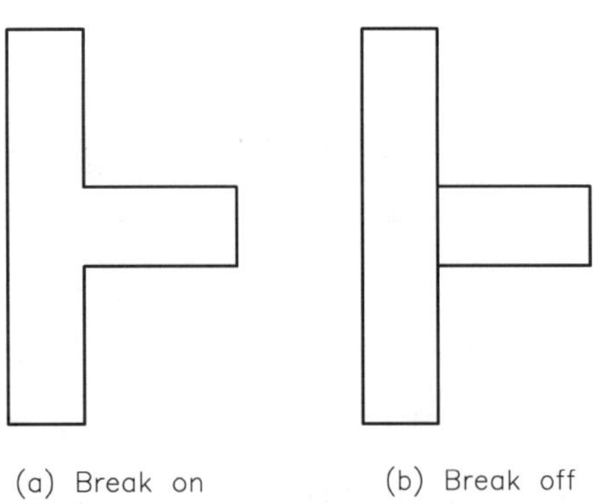

Figure 17-2 *Using the Break option*

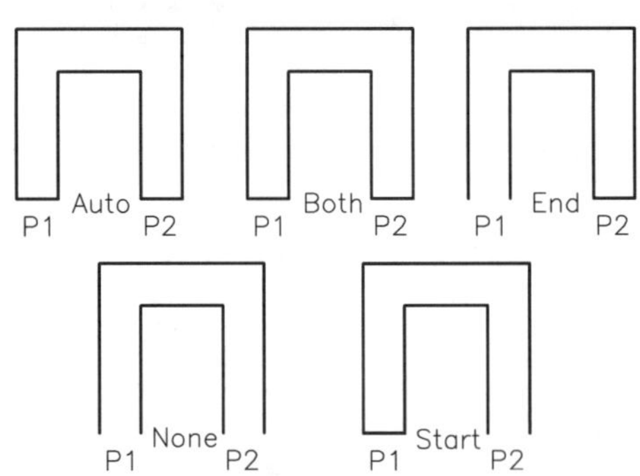

Figure 17-3 *Using the Caps option*

Dragline

This option is used to set the location of the pick point with respect to the double line. You can set the pick point to be located at the center, or on the left or right leg of the double line as shown in Figure 17-4. To ascertain the left and right sides of the double line, imagine looking towards the end of the double line from its start point. The default option is set to center of the double line. The prompt sequence to set the pick point to be on the right leg of the double line is:

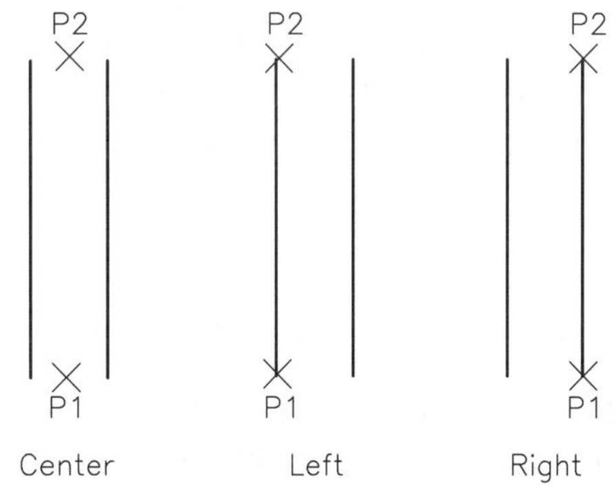

***Figure 17-4** Using the **Dragline** option*

Command: **DLINE** [Enter]
Break/Caps/Dragline/Offset/Snap/Undo/Width/<start point>: **D** [Enter]
Set dragline position to Left/Center/Right/<Offset from center = 0.0000>: **R** [Enter]

Another way of setting the location of the pick point is by setting the offset distance from the center. You can enter any negative value to set the pick point to the left of the center of the double line. Any positive value sets the pick point to the right of the center of the double line.

Offset

With this option you can start a double line at a specified distance and direction from a base point. Enter O at the prompt you receive after entering the **DLINE** command. The subsequent prompts are as follows:

Offset from: *Specify the point from which you want to offset.*
Offset toward: *Specify point.*
Enter the offset distance <Default>: [Enter]

At this prompt you can also enter a distance or pick point.

Snap

If you want a double line to start or end at an existing entity, set the **snap** option of the **DLINE** command. AutoCAD LT prompts you as follows:

Set snap size or snap On/Off. Size/OFF/<ON>:

You can set the snap on or off or set a snap size. To set a snap size enter S at the above prompt.

New snap size (1 - 10) <3>: *Select a number between 1 and 10.*

This size is in pixels and refers to the search area set up by AutoCAD LT to find an object to snap to. The larger the number you can enter, the farther the cursor can be made from the object, that is, AutoCAD LT sets up a larger area to search for the object. AutoCAD LT breaks up the object if the **Break** option is on.

Undo

If you draw one or more double lines by mistake, you can undo them in the **DLINE** command itself. With the **Undo** option, you can erase the last drawn double line segment. Thus, you can undo a series of double line segments while you are still in the **DLINE** command.

Width

You can assign any width to the double line using the **Width** option of the **DLINE** command. The prompt sequence to set the width of 0.5 for the double line is:

Command: **DLINE** Enter
Break/Caps/Dragline/Offset/Snap/Undo/Width/<start point>: **W** Enter
New DLINE width <default>: **1** Enter

Arc

You can draw double line arcs with the **DLINE** command. The prompt sequence is :

Command: **DLINE** Enter
Break/Caps/Dragline/Offset/Snap/Undo/Width/<start point>: *Specify the start point of the double line.*
Arc/Break/CAps/CLose/Dragline/Snap/Undo/Width/<next point>: <Ortho off> **A** Enter
Break/CAps/CEnter/CLose/Dragline/Endpoint/Line/Snap/Undo/Width/<second point>:

Here, you pick the next point of the double line arc. The prompt sequence specifying its third point is:

Endpoint: *Select endpoint of the arc.*

The prompt sequence to draw an arc by specifying its center point is:

Break/CAps/CEnter/CLose/Dragline/Endpoint/Line/Snap/Undo/Width/<second point>: **CE** Enter
Center point: *Specify the center point of the arc.*
Angle/<Endpoint>: Enter
Endpoint: *Specify the endpoint of the arc.*

You can also specify the included angle by entering **A** at the **Angle/<Endpoint>**: prompt.

Angle/<Endpoint>: **A** Enter
Included angle: *Specify the included angle of the arc.*

Another method of drawing a double line arc is by specifying the Endpoint first. The prompt sequence to draw an arc by this method is:

Command: **DLINE** Enter
Break/Caps/Dragline/Offset/Snap/Undo/Width/<start point>: *Specify the start point of the double line.*

Arc/Break/CAps/CLose/Dragline/Snap/Undo/Width/<next point>: **A** Enter
Break/CAps/CEnter/CLose/Dragline/Endpoint/Line/Snap/Undo/Width/<second point>: **E** Enter
Endpoint: *Specify the endpoint of the arc.*
Included angle: *Specify the included angle of the arc.*

Line

The **Line** option allows you to change from double line arcs to drawing double lines. AutoCAD LT continues to make double line arcs once you select the **Arc** option. To resume drawing double lines enter **L** at the **Break/CAps/CEnter/CLose/Dragline/Endpoint/Line/Snap/Undo/Width/<second point>** prompt.

Close

You can close a double line with this option of the **DLINE** command. When you select this option, AutoCAD LT closes the double line with an arc or a line depending on the mode which is active. The **Close** option works only if the double line consists of two or more line segments or one or more arcs. After closing the double line, AutoCAD LT exits the **DLINE** command. The prompt sequence for closing an arc segment is:

Command: **DLINE** Enter
Break/Caps/Dragline/Offset/Snap/Undo/Width/<start point>: *Specify the start point.*
Arc/Break/CAps/CLose/Dragline/Snap/Undo/Width/<next point>: **A** Enter
Break/CAps/CEnter/CLose/Dragline/Endpoint/Line/Snap/Undo/Width/<second point>: *Specify the second point of the arc.*
Endpoint: *Specify the endpoint of the arc.*
Break/CAps/CEnter/CLose/Dragline/Endpoint/Line/Snap/Undo/Width/<second point>: **CL** Enter

CREATING REVISION CLOUDS (REVCLOUD COMMAND)

Toolbar:	Draw > Revcloud
Menu:	Tools > Revision Cloud
Command:	REVCLOUD

This command creates a polyline of sequential arcs to form a cloud shape. The prompt sequence that will follow when you choose the **Revcloud** button from the **Draw** toolbar is as follows:

Current arc length: 0.5000
Specify start point or [Arc length]: *Specify the start point of the revision cloud.*
Guide crosshairs along cloud path...

As you drag and draw, different arcs of the cloud with varied radius are drawn. When the start point and endpoints meet, the revision cloud is completed and you get a message.

Revision Cloud Finished.

If you choose the **Arc Length** option, you can define the length of the arcs to be drawn and all the arcs drawn are of constant length.

The resulting object is a polyline. Figure 17-5 shows the use of the **REVCLOUD** command.

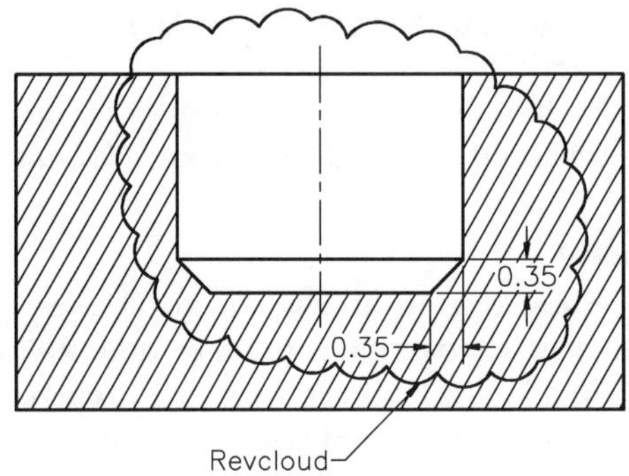

***Figure 17-5** Using **REVCLOUD** command*

Note

***REVCLOUD** stores the last used arc length in the system registry. The value is multiplied by **DIMSCALE** to provide consistency when the program is used with drawings that have different scale factors.*

CREATING NURBS

Toolbar:	Draw > Spline
Menu:	Draw > Spline
Command:	SPLINE

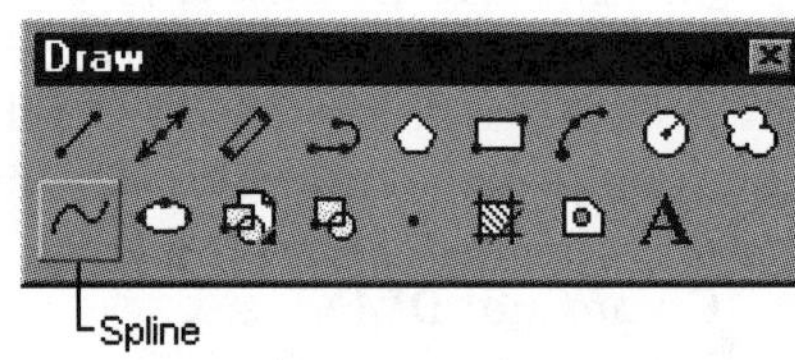

***Figure 17-6** Invoking the **SPLINE** command*

The NURBS is an acronym for **NonUniform Rational Bezier-Spline**. These splines are considered true splines. In AutoCAD LT, you can create NURBS using the **SPLINE** command (Figure 17-6). The spline created with the **SPLINE** command is different from the spline created using the **PLINE** command. The nonuniform aspect of the spline enables the spline to have sharp corners because the spacing between the spline elements that constitute a spline can be irregular. Rational means that irregular geometry such as arcs, circles, and ellipses can be combined with free-form curves. The Bezier-spline (B-spline) is the core that enables accurate fitting of curves to input data with Bezier's curve-fitting interface. Not only are spline curves more accurate compared to smooth polyline curves, but they also use less disk space. The following is the prompt sequence for creating the spline shown in Figure 17-7:

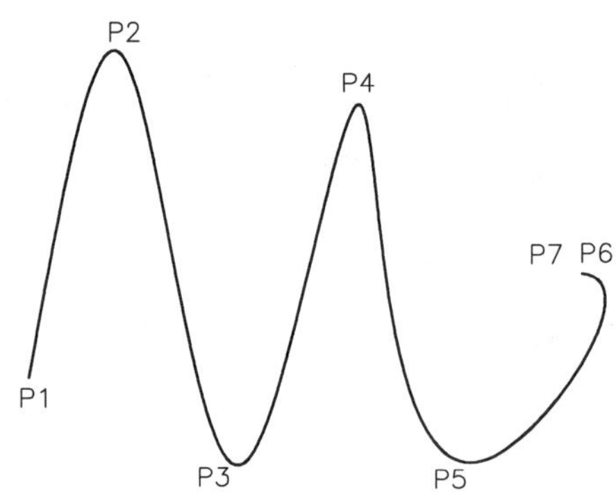

***Figure 17-7** Creating the spline*

Specify first point or [Object]: *Select point (P1).*
Specify next point: *Select the second point (P2).*
Specify next point or [Close/Fit tolerance] <start tangent>: *Select point (P3).*
Specify next point or [Close/Fit tolerance] <start tangent>: *Select point (P4).*
Specify next point or [Close/Fit tolerance] <start tangent>: *Select point (P5).*
Specify next point or [Close/Fit tolerance] <start tangent>: *Select point (P6).*
Specify next point or [Close/Fit tolerance] <start tangent>: *Press ENTER to end the process of point specification.*

Specify start tangent: *Press ENTER to accept the default start tangent.*
Specify end tangent: *Select point (P7).*

You can specify the start and end tangents to change the direction in which the spline curve starts and ends at the **Specify start tangent:** and **Specify end tangent:** prompts. The **SPLINE** command options are discussed next.

Options for Creating the Splines

The options provided under this command for creating the splines are as follows.

Object

This option allows you to change a 2D or 3D splined polyline into a NURBS. The original splined polyline is deleted if the system variable **DELOBJ** is set to 1, which is the default value of the variable. You can change a polyline into a splined polyline using the **Spline** option of the **PEDIT** command.

Close

This option allows you to create closed NURBS. When you use this option, AutoCAD LT will automatically join the endpoint of the spline with the start point, and you will be prompted to define the start tangent only.

Fit Tolerance

This option allows you to control the fit of the spline between specified points. By default, this value is zero and so the spline passes through the points through which it is created. Using this option you can specify some tolerance value that will govern the spline creation, (see Figure 17-8). The splines will be offsetted from the specified point by a distance equal to the tolerance value. The smaller the value of the tolerance, the closer the spline will be to the specified points.

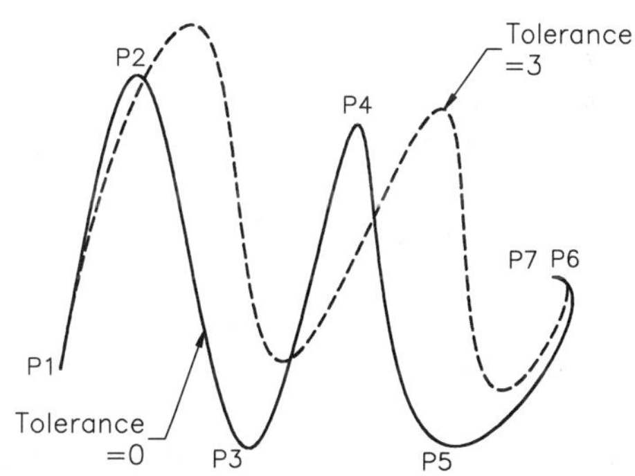

***Figure 17-8** Creating a spline with a **Fit Tolerance** of 0 and 3*

Start and End Tangents

This allows you to control the tangency of the spline at the start point and endpoint of the spline. If you press ENTER at these prompts, AutoCAD LT will use the default value. By default, the tangency is determined by the slope of the spline at the specified point.

EDITING SPLINES

Toolbar:	Modify II > Edit Spline
Menu:	Modify > Object > Spline
Command:	SPLINEDIT

The NURBS can be edited using the **SPLINEDIT** command (Figure 17-9). With this command,

Chapter 17

you can fit data in the selected spline, close or open the spline, move vertex points, and refine or reverse a spline. Apart from the ways mentioned in the preceding command box, you can also invoke the **SPLINEDIT** command by choosing **Spline Edit** from the shortcut menu that is displayed when you select a spline and right-click. The prompt sequence that will follow when you choose the **Edit Spline** button is:

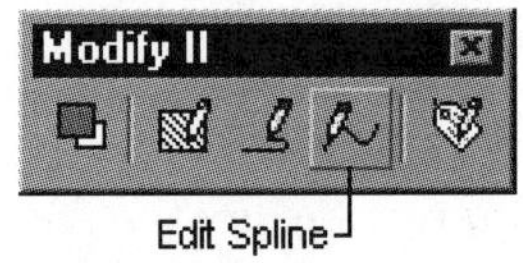

***Figure 17-9** Choosing the **Edit Spline** button from the **Modify II** toolbar*

Select spline: *Select the spline that is to be edited if not selected already using the previously mentioned shortcut menu.*
Enter an option [Fit data/Close/Move vertex/Refine/rEverse/Undo]: *Select any one of the options.*

Options for Editing the Splines

The options provided under this command for editing the splines are described next.

Fit Data

When you draw a spline, the spline is fit to the specified points (data points). The **Fit Data** option allows you to edit these points. You can add, delete, or move the data points. These data points or control points are also referred to as fit points. For example, if you want to redefine the start and end tangents of a spline, select the **Fit Data** option, then select the **Tangents** option. The prompt sequence that will follow when you invoke this option is:

Select spline: *Select the spline that is to be edited.*
Enter an option [Fit data/Close/Move vertex/Refine/rEverse/Undo]: **F**
Enter a fit data option
[Add/Close/Delete/Move/Purge/Tangents/toLerance/eXit] <eXit>:

The start and end tangent points can be selected or its coordinates can be entered. The options available within the **Fit data** option are as follows.

Add. You can use this option to add new fit points to the spline. When you invoke this option, you will be prompted to specify the control point. This control point should be one of the existing control points on the spline. After selecting the existing control point, you will be prompted to specify the location of the new control points. The fit point you select and the next fit point appear as selected grips. You can now add a fit point between these two selected fit points as shown in Figure 17-10. If you select the start point or endpoints of the spline, only those points are highlighted. When you select the start point of the spline, you are prompted to specify whether you want to add the new fit point before or after the start point of the spline. AutoCAD LT will continue prompting for the location of new control points until you press ENTER at the **Specify new point <exit>** prompt.

Close/Open. This option allows you to close an open spline or open a closed spline (Figure 17-11). If the spline is open, the **Close** option is available and if the spline is closed, the **Open** option is displayed.

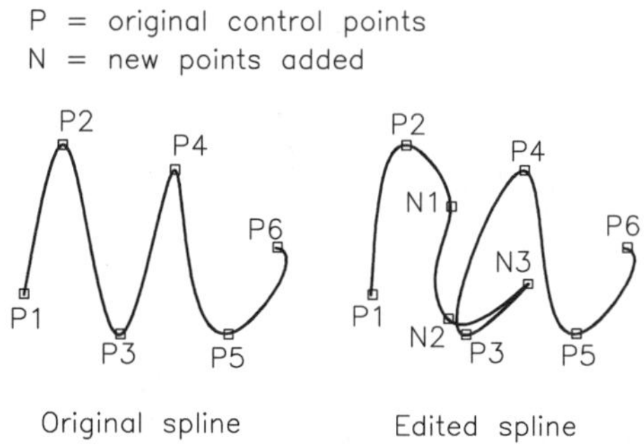

Figure 17-10 Original spline and edited spline after adding the data points

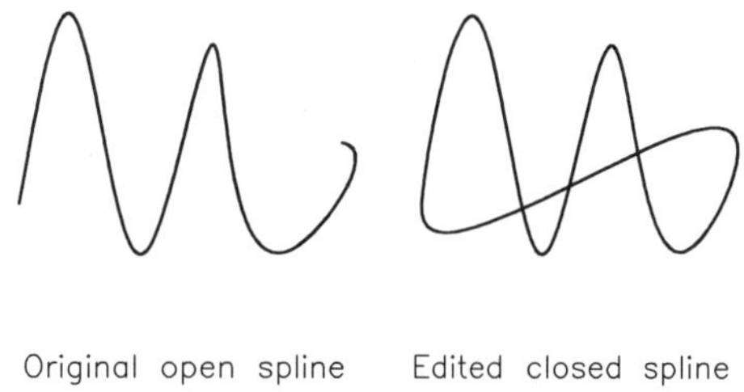

Figure 17-11 An open and a closed spline

Delete. This option allows you to delete a selected fit point from the spline. You can continue deleting fit points from a spline until only two fit points are left in the spline.

Move. You can move fit points by using this option. When you invoke this option, the start point of the spline is highlighted and the prompt sequence is as follows:

Specify new location or [Next/Previous/Select point/eXit] <N>: *Select a new location for the start point using the mouse pick button or enter any one of these options.*

You can enter **N** if you want to select the next point, **P** if you want to select a previous point, or **S** if you want to select any other point. If you enter **X** at the preceding prompt, you can exit the command. Figure 17-12 shows moving of data points in a spline. Remember that this option is used to move only the data points on the spline and not the control points on the Bezier control frame.

Purge. This option allows you to remove fit point data from a spline. This reduces the file size which is useful when a drawing, for example a landscape contour map, contains too many splines. Purging simplifies the spline definition and the drawing file size. But, once fit point data has been removed from a spline, editing a spline gets difficult. Also, the **Fit Data** option is not displayed at all when you again use the **SPLINEDIT** command on the edited spline.

Tangents. This option allows you to modify the tangents of the start and endpoints of the selected spline (see Figure 17-13). When you invoke this option, you will be prompted to specify the start and end tangents for the spline. You can specify the start and end tangents or use the systems default tangents.

toLerance. This option allows you to modify the fit tolerance values of a selected spline. As discussed while creating the splines, different tolerance values produces different splines. A smaller tolerance value creates a spline that passes very closely through the definition points of a spline. When you invoke this option, you will be prompted to specify the tolerance value for the spline.

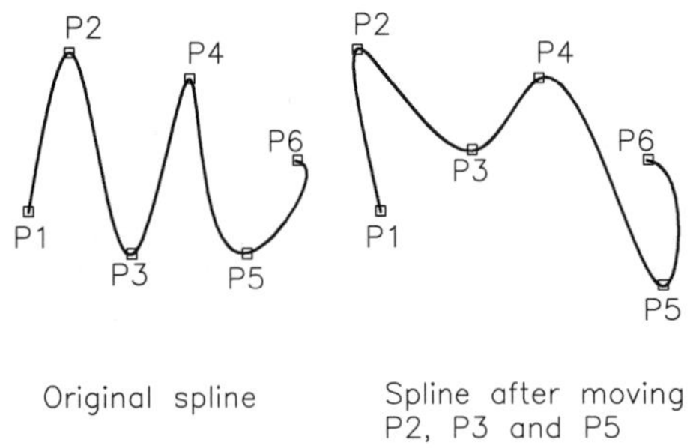

Figure 17-12 Original spline and edited spline after moving the data points

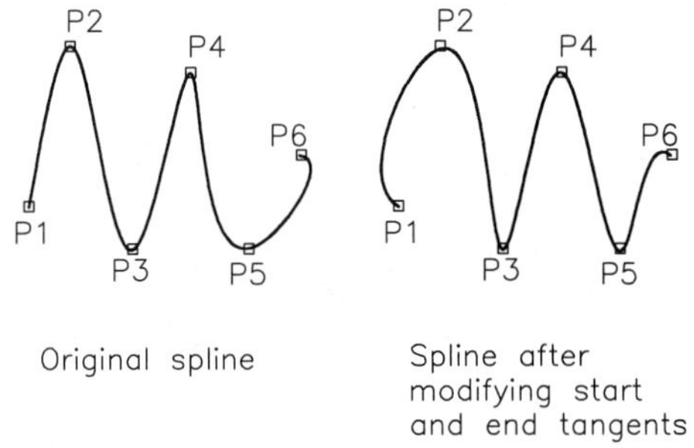

Figure 17-13 Original spline and the spline after modifying the start and end tangents

eXit. This option allows you to exit from the **Fit data** option of editing the splines.

Close/Open

This option allows you to close an open spline or open a closed spline. When you select the **Close** option, AutoCAD LT lets you open, move the vertex, or refine or reverse the spline.

Move vertex

When you draw a spline, it is associated with the Bezier control frame. The **Move vertex** option allows you to move the vertices of the control frame. To display this frame with the spline, set the value of the **SPLFRAME** system variable to 1. The default value of this system variable is 0 and the frames are not displayed with the spline. The **Move vertex** option is similar to the **Move** option of the **Fit data** option of editing the splines. The prompt sequence that follows when you invoke this option is:

Enter an option [Fit data/Close/Move vertex/Refine/rEverse/Undo]: **M**
Specify new location or [Next/Previous/Select point/eXit] <N>:

Refine

This option allows you to refine a spline by adding more control points in the spline, elevating the order, or adding weight to vertex points. The prompt sequence that will follow when you invoke this option is:

Enter an option [Fit data/Close/Move vertex/Refine/rEverse/Undo]: **R**
Enter a refine option [Add control point/Elevate order/Weight/eXit] <eXit>:

Add control point. This option allows you to add more control points on the spline. When you invoke this option, you will be prompted to specify the location of the new point on the spline. You can directly specify the location of the new point on the spline using the left mouse button. A new control point will be added at the location you specified.

Elevate order. This option allows you to increase the order of a spline curve. An order of the curve can be defined as the highest power of the algebraic expression that defines the spline plus 1. For example, the order of a cubic spline will be 3 + 1 = 4. Using the **Elevate order** option you can increase this order for a selected spline. This results in more control points on the curve and a greater possibility of controlling the spline. The value of the spline order varies from 4 to 26. You can only increase the order and not decrease it. For example, if a spline order is 18, you can elevate its order to any value greater than 18, but not less than 18.

Weight. You can also add weight to any of the vertices of the spline by using this option. When weight is added to a particular vertex, the spline gets pulled more towards it. Similarly, a lower value of weight of a particular spline vertex will result in the spline getting pulled less towards that particular vertex. In other words, adding weight to a particular point will force the selected point to maintain its tangency with the point. The more weight added to the point, the more the distance through which the spline will remain tangent to the point. By default, the spline gets pulled equally towards the vertices of the spline. The default value of weight provided to each control point is 1.0 and can have only positive values. Once you have added the weight to a point, you can proceed to the next point. You can also directly select the point to which the weight has to be added. The prompt sequence for using this option is:

Enter a refine option [Add control point/Elevate order/Weight/eXit] <eXit>: **W**
Enter new weight (current = 1.0000) or [Next/Previous/Select point/eXit] <N>:

rEverse

This option allows you to reverse the spline direction. This implies that when you apply this option to a spline, its start point becomes its endpoint and vice versa.

Undo

This option will undo the previous editing operation applied to a spline within the current session of the **SPLINEDIT** command. You can continue to use this option until you reach the spline as it was when you started to edit it.

DIGITIZING DRAWINGS

In digitizing, the information in the drawing is transferred to a CAD system by locating the points on the drawing that is placed on the top surface of the digitizing tablet. Sometimes it is easier to digitize an existing drawing than to create the drawing again on a CAD system. The digitizing of drawings is an important application in industry. The advantages of a drawing based on a CAD system have prompted companies to digitize drawings.

If you want to digitize a drawing, you should have a digitizer as large as the size of the largest drawing to be digitized so that you do not have to realign the drawing sheet. But situations often arise in which the digitizer is not as large as the drawing to be digitized. In such cases, after digitizing a part of the drawing (as large as the digitizer can accommodate), digitize the other parts until the entire drawing is digitized. Digitizing can be carried out in the Tablet mode. In the Tablet mode, AutoCAD LT uses the digitizing tablet as a digitizer, not as a screen pointing device. In this mode, the coordinate system of the paper drawings can be mapped directly into AutoCAD LT. The Tablet mode can be turned on or off with the **TABLET** command or by pressing function key F4. The Command prompt sequence is as follows:

Command: **TABLET**
Enter an option [ON/OFF/CAL/CFG]: **ON** (or **OFF**)

TABLET Command

Menu:	Tools > Tablet
Command:	TABLET

The first step in digitizing a drawing is to configure the tablet so that the maximum possible area on the tablet can be used. You can do this with the **TABLET** command.

At the next prompt, enter CFG (Configuration option) to configure the tablet.

Enter an option [ON/OFF/CAL/CFG]: **CFG**

After this, you are prompted to specify the number of tablet menus you want. When digitizing drawings, the whole tablet area is used for digitizing the drawing; therefore, there are no tablet menus.

Enter the number of tablet menus desired (0-4) <current>: **0**

At the next prompt, enter **Y** so that reconfiguration of the tablet is possible:

Do you want to respecify the Fixed Screen pointing area? [Yes/No] <N>: **Y**
Digitize lower left corner of Fixed Screen pointing area: *Specify the lower left corner of the pointing area.*
Digitize upper right corner of Fixed Screen pointing area: *Specify the upper right corner of the pointing area.*
Do you want to specify the Floating Screen pointing area? [Yes/No] <N>: **N**

The next step in digitizing a drawing is to calibrate the tablet. For this, position the drawing to be digitized to the tablet with the help of adhesive tape. Calibration is possible in model space and paper space. But, before you proceed with digitization of an existing drawing, there are some points you need to determine. They are:

1. Set the limits according to the drawing to be digitized.
2. Set the drawing units such that they relate to the drawing to be digitized.
3. Set the grid and snap values as per convenience.
4. Set the drawing origin.

Once you have established the above settings, invoke the **TABLET** command as follows:

Command: **TABLET**
Enter an option [ON/OFF/CAL/CFG]: **CAL**

After you select or enter **CAL**, the Tablet mode is turned on and the screen crosshairs cursor disappears. The process of calibration involves digitizing two or more points on the drawing and then entering their coordinate values. This shall help in alignment of the drawing with the tablet. Selecting more points for digitizing during calibration leads to greater accuracy of the information that is transferred from an existing drawing. The points to be selected can be anywhere on the drawing. Once you have specified the position of the points and the coordinate values of these positions, AutoCAD LT automatically calibrates the digitizing area. The prompt sequence is as follows:

Digitize point #1: *Select the first point.*
Enter coordinates for point #1: *Enter the coordinates of the first point selected.*
Digitize point #2: *Select the second point.*
Enter coordinates for point #2: *Enter the coordinates of the second point selected.*
Digitize point #3 (or RETURN to end): *Select the third point.*
Enter coordinates for point #3: *Enter the coordinates of the second point selected.*
Digitize point #4 (or RETURN to end): *Press ENTER to end.*

AutoCAD LT will display the Text Window displaying the transformation information.

Enter transformation type [Orthogonal/Affine/Projective/Repeat table] <Repeat>: **A**

You can enter any number of points. The more points you enter, the more accurate is the digitizing. If you have entered only two points, AutoCAD LT will automatically compute orthogonal transformation. This is the most basic type of transformation type. If you have entered three or more points, AutoCAD LT will compute orthogonal, affine, or projective transformations and determine which best fits the configuration of the selected points. After configuring and calibrating the tablet, the crosshairs cursor is redisplayed on the screen. Now you can use AutoCAD LT commands to digitize the drawing.

Floating Screen Pointing Area

You can configure the tablet for two pointing areas: the fixed pointing area and the floating pointing area. Before you configure the tablet, the entire tablet surface is the fixed screen pointing area. At this point the digitizer tablet performs like a pointing device. You can use the **CFG** option of the **TABLET** command to specify the new fixed screen pointing area. When you enter this command, AutoCAD LT will prompt you to specify the two points: lower left corner and upper right corner of the new screen pointing area. Now, if you move the digitizer puck within the fixed pointing area, the cursor moves on the entire screen because there is one-to-one correspondence between the fixed pointing area on the digitizer and the screen. You can also define a floating screen pointing area that is the same as the screen pointing area or you can specify a different area. The following is the prompt sequence to specify a floating screen pointing area:

Command: **TABLET**
Enter an option [ON/OFF/CAL/CFG]: **CFG**
Enter the number of tablet menus desired (0-4)<>: 0
Do you want to respecify the Fixed Screen pointing area? [Yes/No] <N>: **N**
Do you want to specify the Floating Screen pointing area? [Yes/No]<N>: **Y**
Do you want the Floating Screen area to be the same size as the Fixed Screen pointing area? [Yes/No] <Y>: **N**
Digitize lower left corner of the Floating Screen pointing area: *Specify the lower left corner*

of the pointing area.
Digitize upper right corner of the Floating Screen pointing area: *Specify the upper right corner of the pointing area.*

Next, AutoCAD LT will prompt you to specify the toggle key. By default, you can use the function key **F12** to toggle between the floating and fixed screen pointing area or you can define a button of the digitizer puck as the toggle key. You can still use the function key **F4** to turn the Tablet mode on or off to digitize a drawing.

Self-Evaluation Test

Answer the following questions, and then compare your answers to the correct answers given at the end of this chapter.

1. You can specify whether or not the double line will be open or closed at the point where it intersects another double line. (T/F)

2. The **Dragline** option of the **DLINE** command can be used to specify the location of the pick point with respect to the double line. (T/F)

3. All the double line segments drawn using single **DLINE** command becomes a single entity. (T/F)

4. You can also draw an arc from within the **DLINE** command. (T/F)

5. The __________ command can be used to draw a revision cloud.

6. The __________ option of the **DLINE** command is used to specify whether or not the double line will be open or closed at the start and endpoints.

7. The revision cloud drawn using the **REVCLOUD** command is a __________ entity.

8. The __________ option of the **Fit data** option of the **SPLINEDIT** command reduces the file size by removing the fit data of the selected splines.

9. NURBS is an acronym for _________.

10. The _________ option of the **SPLINE** command is used to convert a splined polyline into an actual spline.

Review Questions

Answer the following questions.

1. The revision cloud is closed automatically when you take the cursor close to the start point of the revision cloud. (T/F)

2. Once the **Purge** option of the **Fit Data** option of the **SPLINEDIT** command is used on a spline, editing it gets difficult and the **Fit Data** option of the **SPLINEDIT** command is no longer available for the particular spline. (T/F)

3. You cannot specify the included angle for the arc created using the **DLINE** command. (T/F)

4. If the **Break** option is set to **ON**, the double line will be opened at the point where it intersects another double line. (T/F)

5. By default, what is the value of the **Dragline** option of the **DLINE** command?

   (a) Center (b) Left
   (c) Right (d) Top

6. Which option of the **DLINE** command is used to modify the distance between the two lines that comprise the double line?

   (a) Caps (b) Offset
   (c) Width (d) Snap

7. Which sub-option of the **Caps** option of the **DLINE** command caps all the ends of the double line except the ends that are snapped to an object?

   (a) Both (b) Left
   (c) Right (d) Auto

8. Which command is used to configure a tablet on your system?

(a) **TAB** (b) **TABLET**
(c) **CONFIGURE** (d) **None**

9. Which option can be used to reverse the direction of spline creation while editing the splines?

(a) **Fit data** (b) **Refine**
(c) **Reverse** (d) **Weight**

10. The __________ option is used to edit the data points of the spline.

11. The value of the __________ variable is set to 1 to display the frames of the splines.

12. If the Fit tolerance value of a spline is __________ , the spline passes exactly through the fit points of the spline.

13. The __________ option is used to pull the spline more towards the vertex.

14. The __________ sub-option of the **Refine** option of the **SPLINEDIT** command is used to increase the order of the spline curve.

15. The __________ sub-option of the **Fit Data** option is used to remove the fit points from the spline.

Exercises

Exercise 1 *General*

Draw the illustration shown at the top of Figure 17-14. Then use the required commands to obtain the illustration shown at the bottom. You can also use grips to obtain this illustration. Assume the missing dimensions.

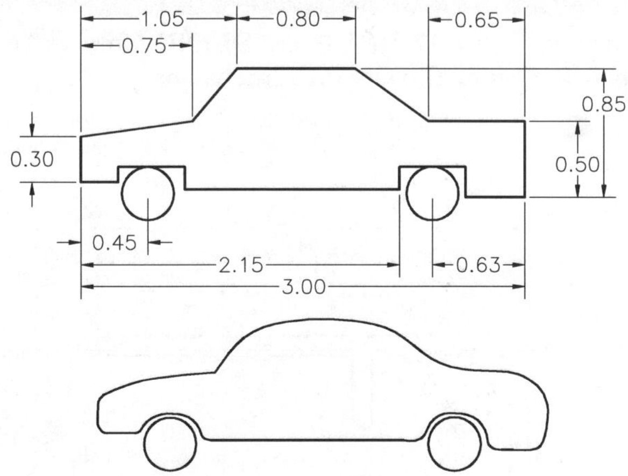

Figure 17-14 *Drawing for Exercise 1*

Exercise 2 *Mechanical*

Make the drawing shown in Figure 17-15. Assume the missing dimensions.

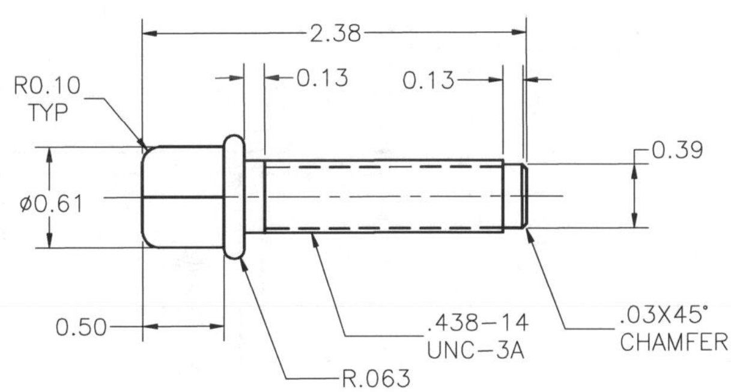

Figure 17-15 *Drawing for Exercise 2*

Exercise 3 *Architecture*

Make the drawing shown in Figure 17-16. Use the **DLINE** command to draw the walls. The wall thickness is 12 inches. Assume the missing dimensions.

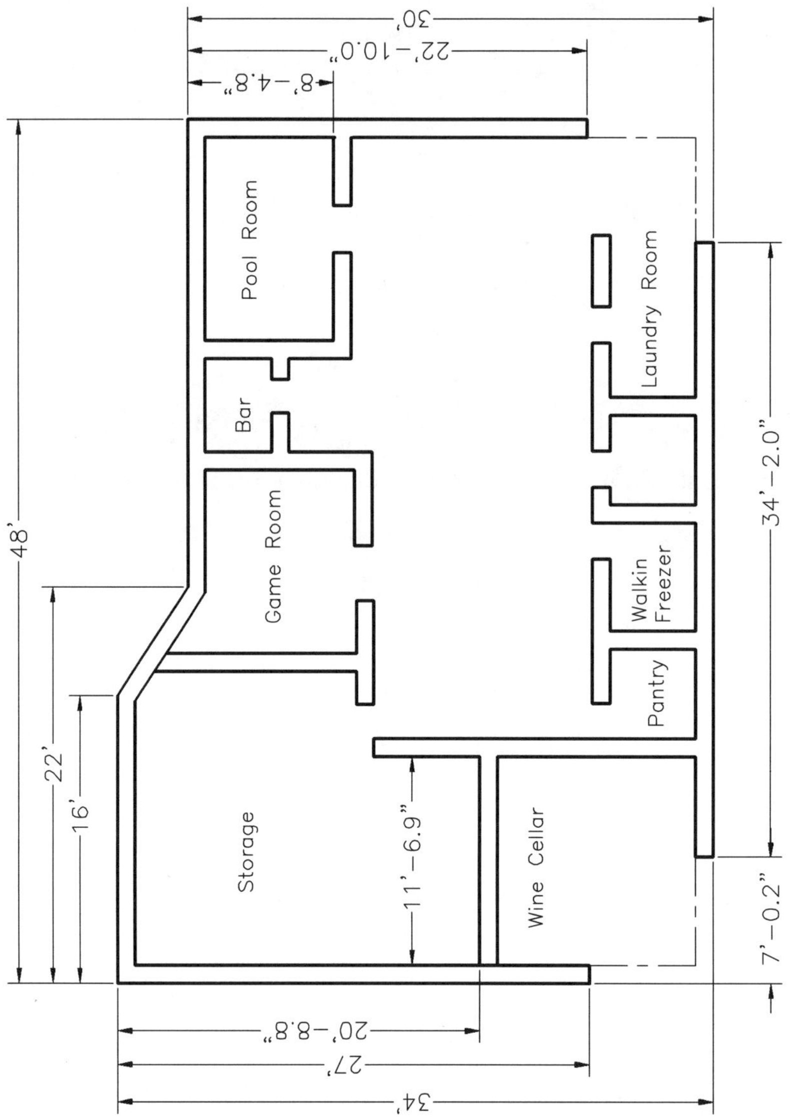

Figure 17-16 *Drawing for Exercise 3*

Problem Solving Exercise 1 — *General*

Make the drawing shown in Figure 17-17. Some of the reference dimensions are given in the drawing. Assume the missing dimensions so that the drawing looks similar to the given drawing.

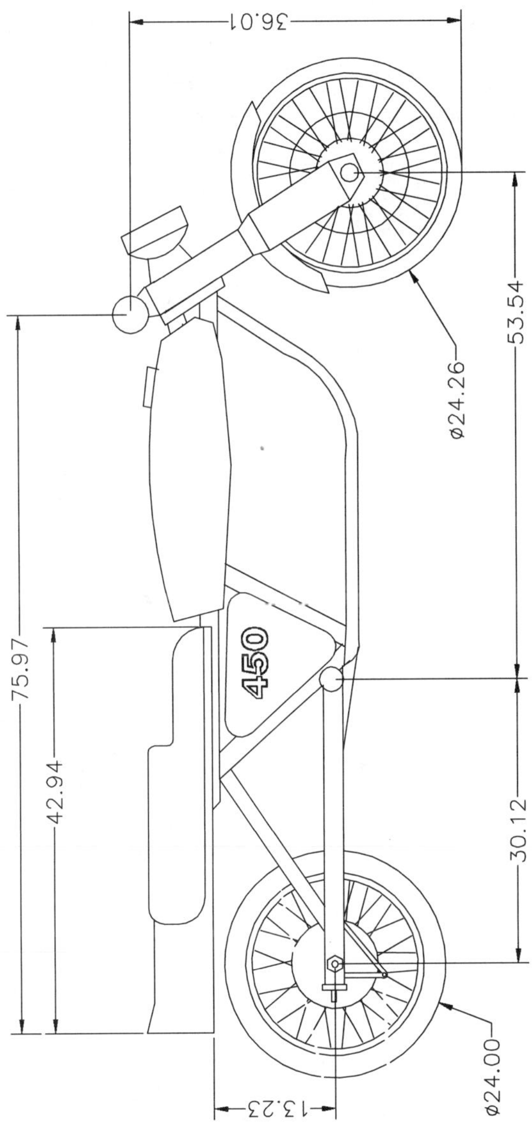

Figure 17-17 *Drawing for Problem Solving Exercise 1*

Problem Solving Exercise 2

General

Make the drawing shown in Figure 17-18. Some of the reference dimensions are given in the drawing. Assume the missing dimensions so that the drawing looks similar to the given drawing.

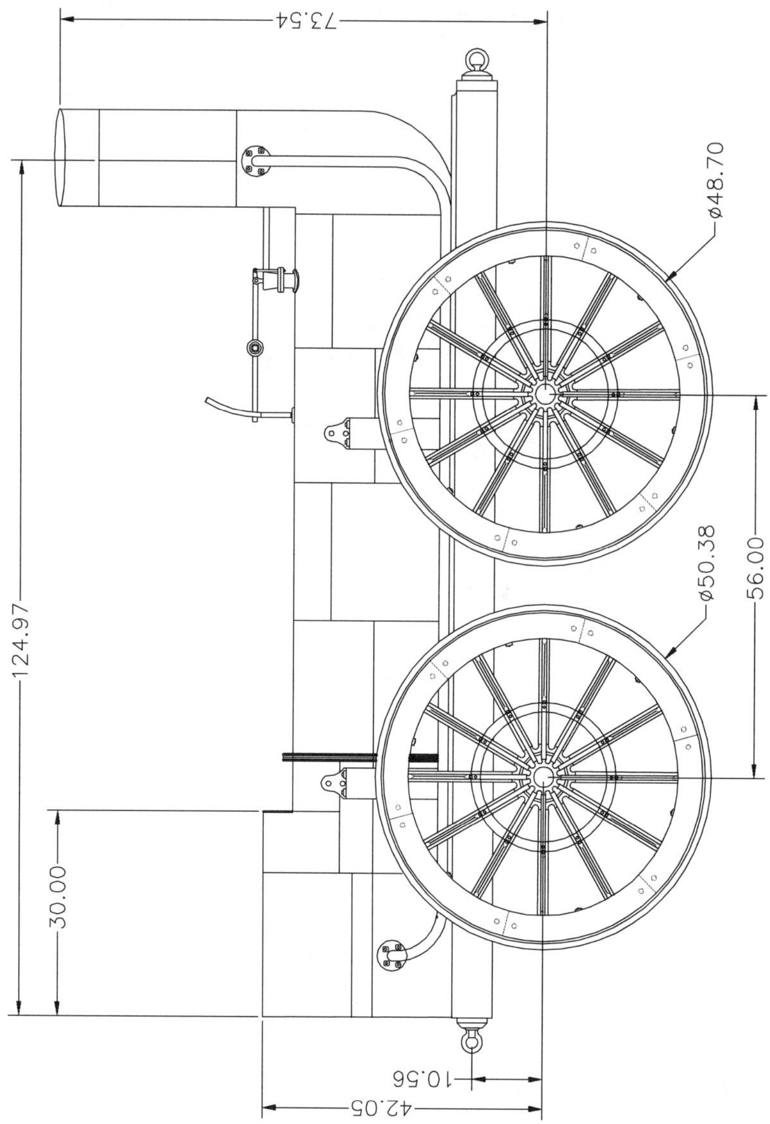

Figure 17-18 *Drawing for Problem Solving Exercise 2*

Problem Solving Exercise 3 *General*

Make the drawing shown in Figure 17-19. Some of the reference dimensions are given in the drawing. Assume the missing dimensions so that the drawing looks similar to the given drawing.

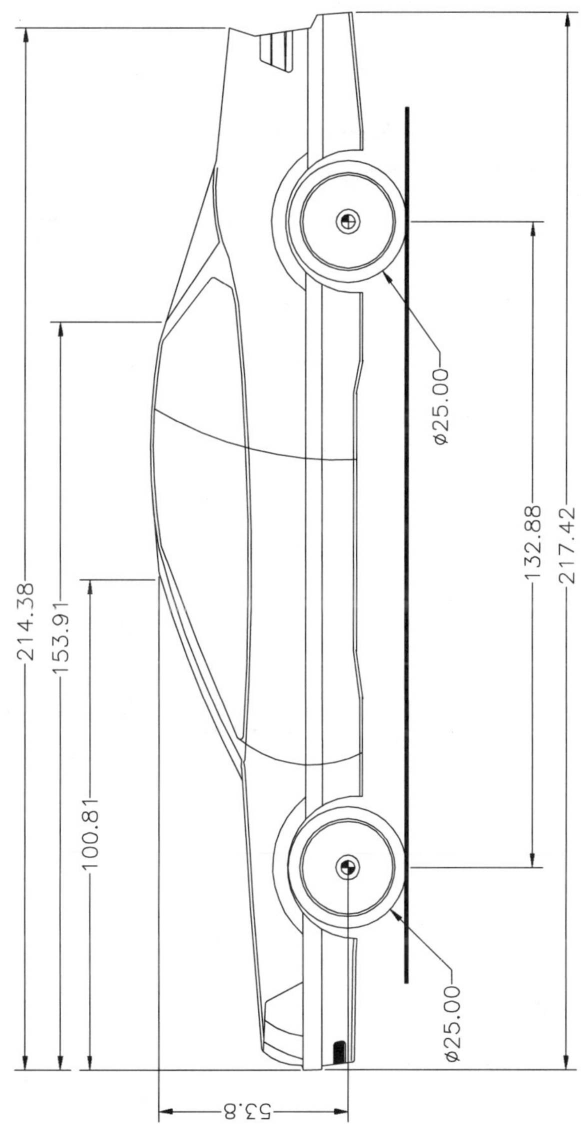

Figure 17-19 *Drawing for Problem Solving Exercise 3*

Problem Solving Exercise 4

Architecture

Make the drawing shown in Figure 17-20. Some of the reference dimensions are given in the drawing. Assume the missing dimensions so that the drawing looks similar to the given drawing.

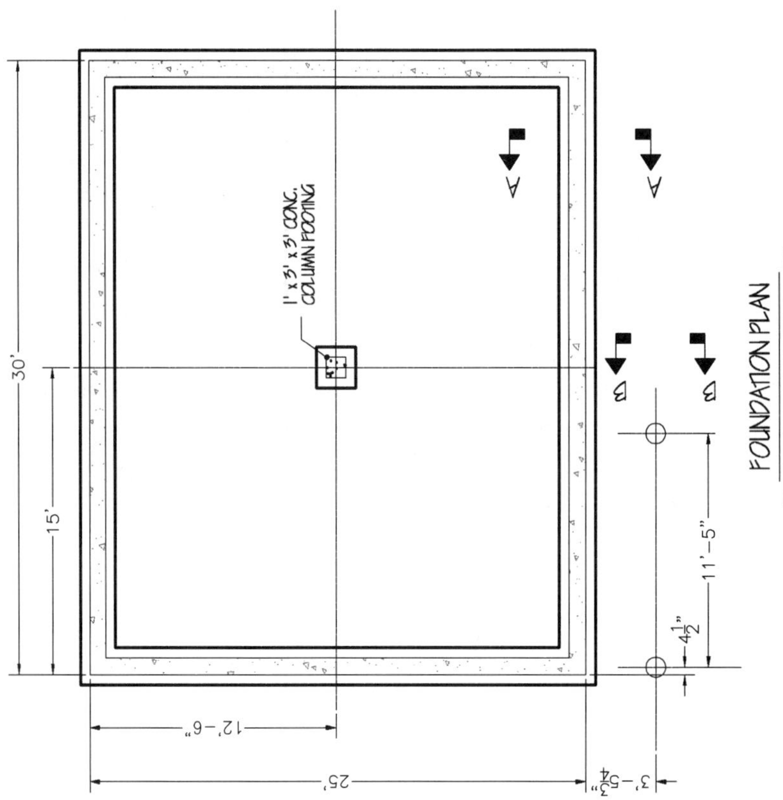

Figure 17-20 *Drawing for Problem Solving Exercise 4*

Problem Solving Exercise 5

Architecture

Make the drawing shown in Figure 17-21. Some of the reference dimensions are given in the drawing. Assume the missing dimensions so that the drawing looks similar to the given drawing.

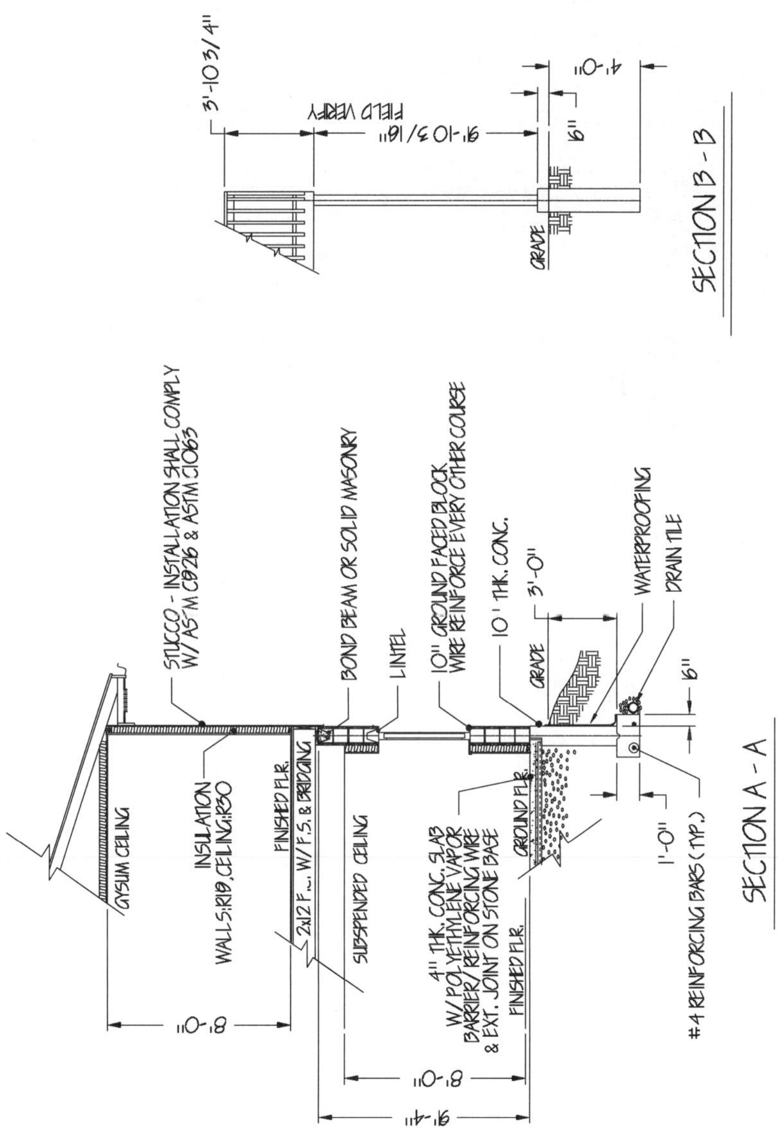

Figure 17-21 Drawing for Problem Solving Exercise 5

Problem Solving Exercise 6 *Architecture*

Make the drawing shown in Figure 17-22. Some of the reference dimensions are given in the drawing. Assume the missing dimensions so that the drawing looks similar to the given drawing.

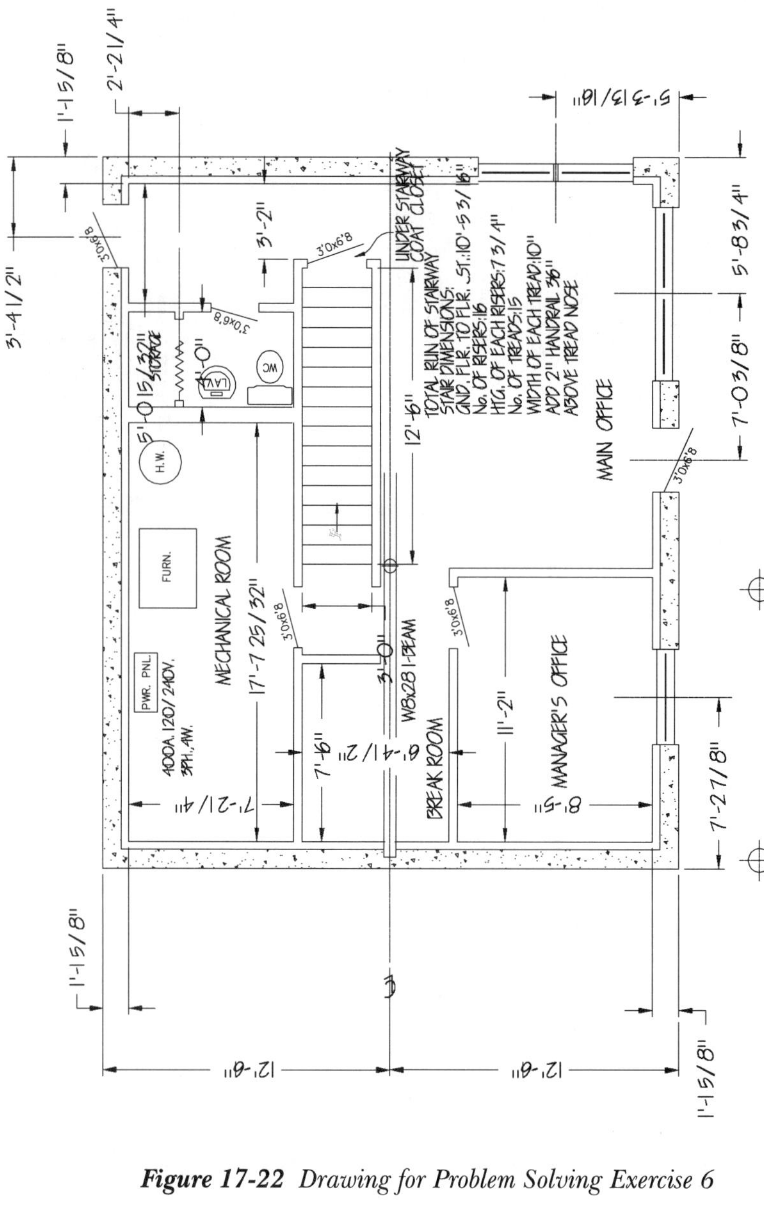

Figure 17-22 *Drawing for Problem Solving Exercise 6*

Problem Solving Exercise 7

Architecture

Make the drawing shown in Figure 17-23. Some of the reference dimensions are given in the drawing. Assume the missing dimensions so that the drawing looks similar to the given drawing.

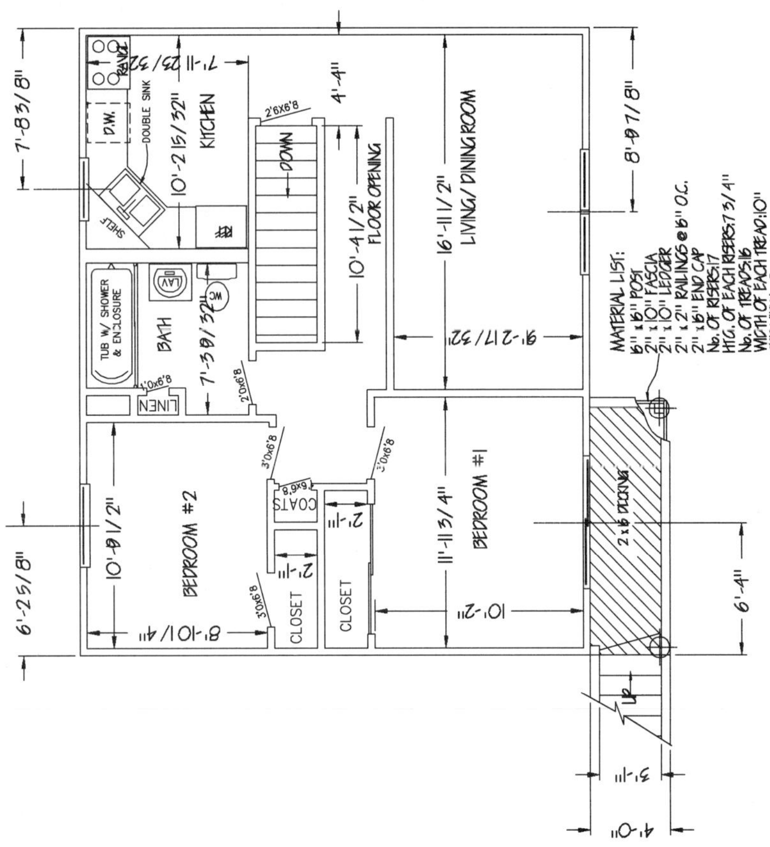

Figure 17-23 Drawing for Problem Solving Exercise 7

Problem Solving Exercise 8 — *Architecture*

Make the drawing shown in Figure 17-24. Some of the reference dimensions are given in the drawing. Assume the missing dimensions so that the drawing looks similar to the given drawing.

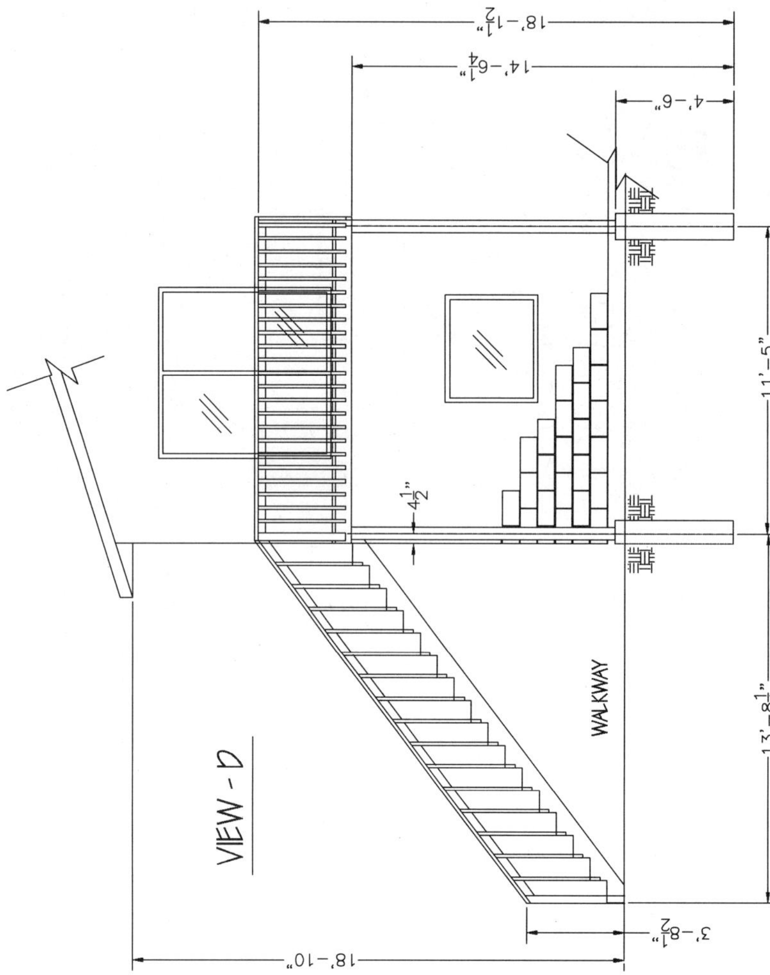

***Figure 17-24** Drawing for Problem Solving Exercise 8*

Answers to Self-Evaluation Test

1 - T, **2** - T, **3** - F, **4** - T, **5** - **REVCLOUD**, **6** - **Caps**, **7** - single, **8** - **Purge**, **9** - NonUniformal Rational Bezier Spline, **10** - **Object**

Chapter 18

Object Grouping and Editing Commands

Learning Objectives

After completing this chapter, you will be able to:

- *Use the **GROUP** command to group objects.*
- *Select and cycle through defined groups.*
- *Change properties and points of objects using the **CHANGE** and **PROPERTIES** commands.*
- *Perform editing operations on polylines using the **PEDIT** command.*
- *Explode compound objects using the **EXPLODE** command.*
- *Undo previous commands using the **UNDO** command.*
- *Rename named objects using the **RENAME** command.*
- *Remove unused named objects using the **PURGE** command.*

GROUP MANAGER (GROUP COMMAND)

Toolbar:	Group > Group Manager
Menu:	Tools > Group Manager
Command:	GROUP

Groups are simply a mechanism that enables you to form groups and edit objects by groups. Creating groups (Figure 18-1) makes the object selection process easier and faster. Objects can be members of several groups. You can use the **Group Manager** dialog box (Figure 18-2) to group AutoCAD LT objects and assign a name to the group. Once you have created groups, you can select the objects by group name. The individual characteristics of an object are not affected by forming groups.

***Figure 18-1** Invoking **Group Manager** from the **Group** toolbar*

GROUP MANAGER DIALOG BOX OPTIONS

The **Group Manager** dialog box (Figure 18-2) provides the following options.

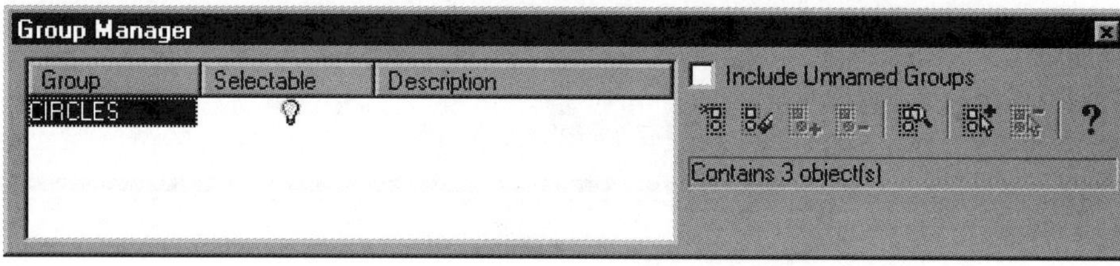

Figure 18-2 ***Group Manager*** *dialog box*

Group

The **Group** list in the **Group Manager** dialog box displays the names of the existing groups. The list box also displays whether a group is **selectable** or not. A **selectable** group is one, in which all members are selected on selecting a single member. It also displays description regarding the group.

Note

If the value of the ***PICKSTYLE*** *System Variable is set to 1 or 3, entire group is selected upon selecting any one object from the group. If the value is set to 0, entire group is not selected upon selecting any one object from the group.*

Create Group

Choosing this button allows you to create a new group consisting of the objects you have selected. Names assigned to the groups are displayed under the **Group** heading. Choose this button and then enter the name of the group. Now, select the objects to be included in this group and choose the **Add to Group** button. You can also select the objects first and then choose this button. This way you do not have to choose the **Add to Group** button to add the objects in the group.

Ungroup

This button of the **Group Manager** dialog box deletes the selected group from the **Group** list and also removes the association of the objects in the group. This button is chosen when you want to delete a group.

Add to Group

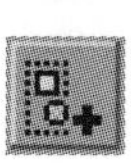

Choosing this button allows you to add the selected objects to the group already created and selected under **Group** list. This option is used when you want to add some more objects to the group already created.

Remove from Group

Choosing this button removes the selected objects from the group which is selected in the **Group** list. You can remove a few selected objects from the group of number of objects by choosing this button.

Note

Objects can be added or removed from a selection set by the ***PICKADD*** *system variable. If the value of this system variable is set to 1, you can add or remove the objects from the group. If the value is set to 0, the current selection set replaces the previous selection set.*

Details

Choosing this button displays the **Group Manager - Details** dialog box (Figure 18-3). This dialog box display all the details about the selected group.

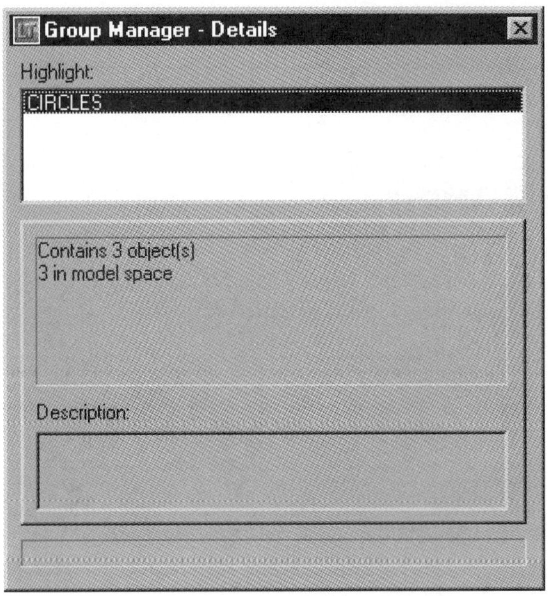

Figure 18-3 Group Manager - Details *dialog*

Select Group

As you choose this button, all of the objects in the specified group are highlighted. You have to first select the group name from the **Group** list and then choose the **Select Group** button to highlight the objects in the selected group.

Deselect Group

As you choose this button, the specified group is removed from the selection set. This option is used when more than one groups are selected at the same time. In this case you can remove a particular group from the selection set by selecting the group from the **Group** list and then choosing this button.

Include Unnamed

The **Include Unnamed** check box is used to display the names of the unnamed groups in the **Object Grouping** dialog box. The unnamed groups are created when you copy the objects that belong to a group. AutoCAD LT automatically groups and assigns a name to the copied objects. The format of the name is A*n* (for example, *A1, *A2, *A3). If you select the **Include Unnamed** check box, the unnamed group names (*A1, *A2, *A3, ...) will also be displayed in the **Group Name** list box of the dialog box.

Tip

The unnamed groups can also be created by not assigning a name to the group. This is done by pressing the ENTER key when you have to specify the name of the group. AutoCAD LT automatically assigns a name to the new group. If the ***Include Unnamed Groups*** *check box is selected, the unnamed group will be displayed under the* ***Group*** *list.*

Note

You can also use the Command line for creating the groups. This is done using the ***-GROUP*** *command. All the options of the* ***Group Manager*** *dialog box are also available when you use the command line.*

SELECTING GROUPS

You can select a group by name by entering **G** at the **Select objects** prompt. For example, if you have to move a particular group, choose the **Move** button from the **Object Properties** toolbar and the prompt sequence will be as follows:

Select objects: **G**
Enter group name: *Enter the groups name.*
n found
Select objects: Enter

Instead of entering **G** at the **Select Objects** prompt, if you select any member of a selectable group, all the group members get selected. Make sure that the **PICKSTYLE** system variable is set to 1 or 3. Also, the group selection can be turned on or off by pressing SHIFT+CTRL+A. If the group selection is off and you want to erase a group from a drawing, invoke the **ERASE** command from the **Modify** toolbar. The prompt sequence is as follows:

Select objects: SHIFT+CTRL+A
<Groups on> *Select an object that belongs to a group. (If the group has been defined as selectable, all objects belonging to that group will be selected.)*

If the group has not been defined as selectable, you cannot select all objects in the group, even if you turn the group selection on by pressing the SHIFT+CTRL+A keys. This setting can also be changed in the **Selection** tab of the **Options** dialog box. You can select or clear the **Object grouping** check box in the **Selection Modes** area of the dialog box to turn group selectability on or off.

Tip

The combination of SHIFT, CTRL, and A keys is used as a toggle to turn the group selection on or off.

CYCLING THROUGH GROUPS

When you use the **GROUP** command to form object groups, AutoCAD LT lets you sequentially highlight the groups of which the selected object is a member. For example, assume that an object belongs to two different groups, and you want to highlight the objects in those groups. To accomplish this, press CTRL at the **Select objects** prompt of any command and select the object that belongs to different groups. AutoCAD LT will highlight the objects in one of the groups. Press the pick button of your pointing device to cycle through the groups (Figure 18-4). Choose the **ERASE** command, the prompt sequence is:

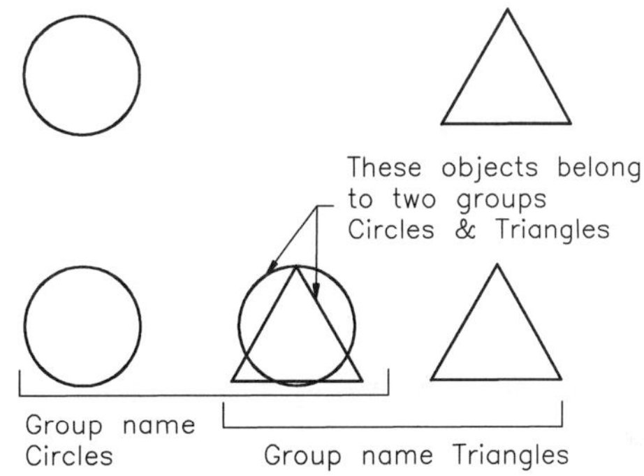

Figure 18-4 *Object selection cycling*

Select objects: *Hold down CTRL.*
<Cycle on> *Select the object that belongs to different groups. (Press the pick button on your pointing device to cycle through the groups.)*
Select objects: Enter

Also, if objects are very close or directly on top of one another, such that they lie within the same selection pickbox, you can press CTRL at the **Select objects:** prompt and repeatedly click your pointing device until the object you want is highlighted. Press ENTER to select it.

CHANGING PROPERTIES OF AN OBJECT

Using the Properties Window

Toolbar:	Standard > Properties
Menu:	Tool > Properties
Command:	PROPERTIES

As discussed earlier in Chapter 4, Drawing Aids, and Chapter 6, Editing Commands-II, invoking the **PROPERTIES** command displays the **Properties** window (Figure 18-5) that can be used for changing properties of objects. The window has two tabs. They are **Alphabetic** and **Categorized**. While the **Alphabetic** tab lists the various properties of a selected object alphabetically, the **Categorized** tab displays the properties of the selected object under categories. Depending on the type of object selected, different categories are displayed. The **General** category displays the general properties of objects such as color, layer, linetype, linetype scale, plot style, lineweight, hyperlink, and thickness. Changing the general properties of objects using the **Properties** window has already been discussed earlier, in Chapter 4, Drawing Aids. Here, the **Geometry** category will be discussed that contains properties that control the geometry of an object. Depending on the type of object selected, the **Geometry** category will contain a set of different properties. If you have selected many types of objects in a drawing, the edit box at the top of the **Properties** window displays **All** and only the **General** category is available in the window. You can select a type of object from the drop-down list and the corresponding categories of object properties are displayed in the window.

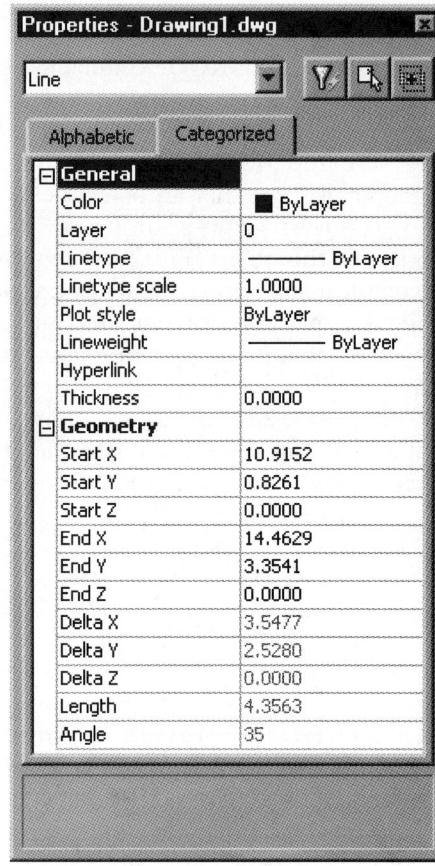

Figure 18-5 *The **Properties** window*

Selecting Line

If you have selected a line in the drawing, the **Properties** window appears similar to what is shown in Figure 18-5, where you can change the properties of the selected object.

Start X/ Start Y/ Start Z. The **Start X/ Start Y/ Start Z** fields in the **Geometry** category, display the start point coordinates of the selected line. You can enter new values in these fields or use the **Pick Point** button that appears in the field you click in. When you choose the **Pick Point** button, a rubber-band line gets attached to the cursor on its start point and the endpoint of the line is fixed. You can now move the cursor and specify a new start point for the line.

End X/ End Y/ End Z. Similar to the start point you can also modify the location of the endpoint of a line by entering new coordinate values in the **End X**, **End Y**, **End Z** fields. You can also choose the **Pick Point** button in the respective fields and select a new location for the endpoint of the selected line. This **Pick Point** button appears in the fields when you click in them.

Note

*The **Delta X**, **Delta Y**, **Delta Z**, **Length**, and **Angle** fields cannot be changed but the values in these fields get updated as the start and endpoints of the line are changed.*

Selecting Circle

When you select a circle, the **Geometry** category displays the categories as displayed in Figure 18-6. You can modify the values in these fields and the effects are discussed next.

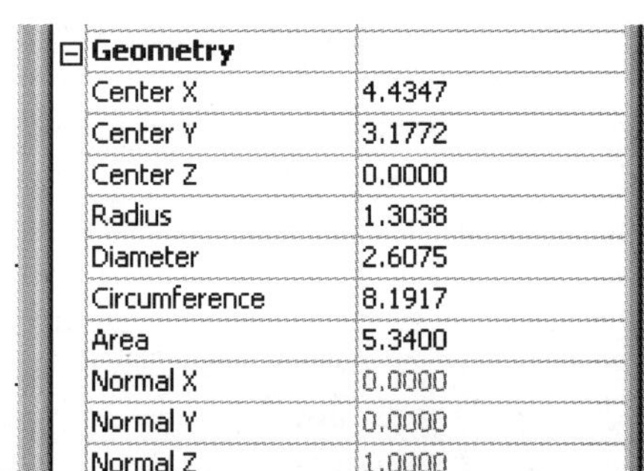

Figure 18-6 *The **Geometry** category of the **Properties** window for a circle*

Center X/ Center Y/ Center Z. Here under **Center X/ Center Y/ Center Z fields of Geometry category,** you can specify a new location of the center point of the circle by entering new coordinate values in the **Center X**, **Center Y**, and **Center Z** fields. When you click in any of these fields, the **Pick Point** button is displayed. You can choose this button to locate a new location of the center point of the circle. The circle will be drawn at the new location with the same radius as before. It is as if the circle has moved from the old location to a new one.

Radius/ Diameter. You can enter a new value for the radius or diameter of the circle in the **Radius** or **Diameter** fields respectively. The circle's radius or diameter is modified as per the new values you have entered. As you modify the values of the radius or the diameter, the values in the **Circumference** and **Area** fields get modified accordingly.

Note

You can also enter new values of the circumference or area in the ***Circumference*** *and* ***Area*** *fields respectively if you want to create a circle with a given value of circumference or area. You will notice that the radius and diameter values are modified accordingly. The* ***Normal X, Normal Y,*** *and* ***Normal Z*** *fields are not available for modifications.*

Selecting Text

If you select text written in a drawing and invoke the **Properties** window, apart from the **General** category, two more categories of properties are displayed. They are **Text** and **Geometry**. The properties under these two categories (Figure 18-7) and how they can be modified are discussed next.

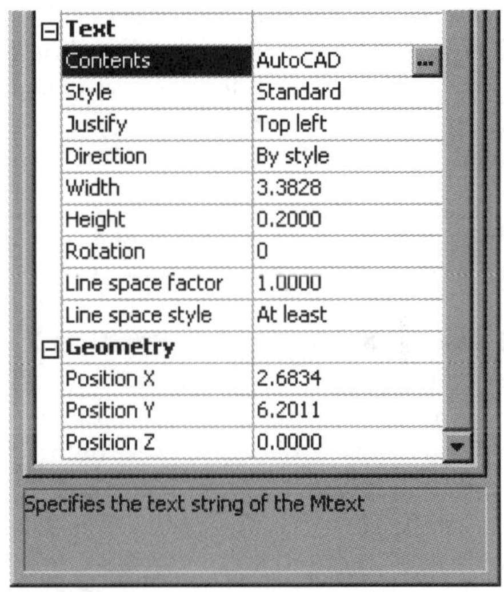

Figure 18-7 *The* ***Geometry*** *and* ***Text*** *categories of the* ***Properties*** *window for text*

Contents. The current contents of the text are displayed in the **Contents** field under the **Text** category of the **Properties** window. To modify the contents click in the field and the **[...]** button is displayed at the right corner of the field. You can then choose the **[...]** button to display the **Multiline Text Editor** where you can make the necessary modifications using the various options available in the dialog box. Once you have made the changes, choose **OK** to exit the **Multiline Text Editor** and return to the Properties window. The changes you made are reflected in the value of the **Contents** field of the **Properties** window and in the drawing.

Style. When you click in the **Style** field in the **Properties** window, a drop-down list of the defined text styles is displayed. By default only the Standard style is displayed. If you had defined more text styles in the drawing, they can be selected from the drop-down list and be applied to the current text in the drawing.

Justify. All the text justification options are available in the **Justify** drop-down list. This drop-down list is available when you click in the **Justify** field. By default, the Top left justification is applied to the text. You can select a text justification from the drop-down list to be applied to the selected text in the drawing.

Width. In the **Width** field you can enter a new value for the paragraph text width. The changes are evident when you have multiline text in the drawing. You can increase or decrease the value of the paragraph text.

Height. You can enter a new value for the selected text in the drawing. The default text height is 0.2. But, you can make the text smaller or larger in the drawing by simply entering a new value in the **Height** field of the **Properties** window.

Rotation. This field displays the current angle of the selected text and is by default, **0** degrees. You can change the angle of rotation of the text by entering an angle by which you want to rotate the selected text in the **Rotation** field; then press ENTER. The effect is immediately visible in the drawing.

Direction. This drop-down list displays the possible directions for the text. This option is generally available for the multiline text objects.

Line space factor. The effect of changing the value of this field is more evident when you have selected paragraph or multiline text. This value controls the spacing between lines in a paragraph or multiline text. A scale factor of 1 is the default value. You can increase or decrease the spacing between the lines of the selected paragraph text by entering a new scale factor of your choice and pressing ENTER. The result is immediately visible in the drawing.

Line space style. This field when selected displays the two line spacing options in the corresponding drop-down list. They are **Exactly** and **Atleast**. **Atleast** is the default option.

Position X/ Position Y/ Position Z. You can change the location of the selected text by entering new X/Y/Z coordinates for the text in the **Position X/ Position Y/ Position Z** fields under the **Geometry** category in the **Properties** window. You can also use the **Pick Points** button that is displayed in these fields when you click in them. When you choose the **Pick Points** button, a rubber-band line is attached between the cursor and the selected text and you can move the cursor and specify a new location for the text on the screen. The values of the three fields get updated automatically.

Selecting Block Reference

When you select a block in the drawing and invoke the **Properties** window, apart from the **General** category, two more categories containing the properties of the block are available. They are **Geometry** and **Misc**. The properties under these categories are discussed next.

Position X/ Position Y/ Position Z. Here under the **Position X/ Position Y/ Position Z** field of **Geometry** category you can enter new X, Y, and Z coordinates for the selected block. The block reference shall move to the new location as you specify. You can also click in any one of the fields to display the **Pick Points** button and choose it to specify a new location for the block on the screen.

Scale X/ Scale Y/ Scale Z. You can specify new X, Y, and Z scale factors for the selected block in the **Scale X**, **Scale Y**, and **Scale Z** fields respectively. The current scale factors for the block are displayed in the **Scale X**, **Scale Y**, and **Scale Z** fields and as you change them, the modifications are reflected in the drawing.

Name. This field under the **Misc** category displays the name of the selected block reference and is not available for modifications.

Rotation. This field specifies the current angle of rotation of the selected block. You can specify a new angle of rotation for the selected block in this field. After specifying an angle, when you press ENTER, the selected block in the drawing is rotated through the specified angle.

Selecting Attribute Definition

When you select an attribute definition in the drawing, apart from the **General** category, the other three categories under which the properties of an attribute are displayed are **Text**, **Geometry**, and **Misc**. Using the **Properties** window to edit an attribute has already been discussed in Chapter 15, Defining Block Attributes.

Tag. You can modify the text of the tag by entering a new value in the **Tag** field under the **Text** category. After you have modified the tag text and pressed ENTER, the old attribute text in the drawing is replaced by the new text.

Prompt. You can also change the text of the prompt and enter a new prompt value in the **Prompt** field.

Value. This field displays the default value of the selected attribute. You can change this default value by entering a new value in the **Value** field.

Style. When you click on this field, the **Style** drop-down list is available. You can select a style for the attribute text from this drop-down list. Only text styles that already have been defined are available in this drop-down list and can be selected.

Justify. You can select an attribute text justification from the **Justification** drop-down list.

Height. You can change the height of the selected attribute text by entering a new value in this field.

Rotation. You can also rotate the selected attribute text through an angle that you specify in the **Rotation** field.

Width Factor. You can modify the width factor of the attribute text by entering a new value in the **Width Factor** field of the **Properties** window.

Obliquing. The attribute text can also be made to slant by entering an obliquing angle in the **Obliquing** field of the **Properties** window.

Note

*The **Text alignment** fields of the **Properties** window are not available for modifications.*

Position X/ Position Y/ Position Z. As discussed earlier in Chapter 15, Defining Block Attributes you can enter new X, Y, and Z coordinates in the **Position X**, **Position Y**, and **Position Z** fields of the **Geometry** category to relocate the attribute text. You can also choose the **Pick Points** button that is displayed in these fields when you click in them to relocate the selected attribute text. After you have selected a point in the drawing to place the selected attribute, the coordinate values in the **Position X**, **Position Y**, and **Position Z** fields are updated accordingly.

Misc. As discussed earlier, under the **Misc** category, you can redefine the attribute modes that have been defined at the time of the attribute definition. You can also make the attribute text appear upside-down or backwards by selecting **Yes** from the **Upside-down** drop-down list and the **Backward** drop-down list respectively.

CHANGE Command

Command: CHANGE

With the help of the **CHANGE** command, you can change some of the characteristics associated with an object such as location, color, layer, lineweight, and linetype. You can also change the geometry of the object using this command. The **CHANGE** command has two options, **Change point** and **Properties**. These are discussed next.

Change Point Option

You can change various features and the location of an object with the **Change point** option of the **CHANGE** command. For example, to change the endpoint of a line or a group of lines, the prompt sequence is:

Command: **CHANGE**
Select objects: *Select the line(s).*
Select objects: Enter
Select Change point or [Properties]: *Specify a point to be used as a new endpoint.*

Tip

*When the **ORTHO** mode is on, you can use the **CHANGE** command to change the endpoint of a line or group of lines to make them parallel to either the X axis or the Y axis.*

To change various features associated with the text, the prompt sequence is:

Command: **CHANGE**
Select objects: *Select the text.*
Select objects: Enter
Specify change point or [Properties]: Enter
Specify new text insertion point <no change>: *Specify the new text insertion point (location of text) or press ENTER for no change.*
Enter new text style <current>: *Enter the name of the new text style or press ENTER to accept the current style.*
Specify new height <current>: *Specify the new text height or press ENTER to accept the current value.*
Specify new rotation angle <current>: *Specify the new rotation angle or press ENTER to accept the current value.*

Enter new text <current>: *Enter the new text or press ENTER to accept the current text.*

The properties of an attribute definition text can also be changed just as you change the properties of text. The prompt sequence for changing the properties of an attribute definition text is as follows:

Command: **CHANGE**
Select objects: *Select the attribute definition text.*
Select objects: Enter
Specify new change point or [Properties]: Enter
Specify new text insertion point <no change>: *Specify the new attribute definition text insertion point (location) or press ENTER if you do not want to modify the location of the attribute text.*
Enter new text style <current>: *Enter the name of the new text style or press ENTER to accept the current style.*
Specify new height <current>: *Specify the new attribute definition text height or press ENTER to accept the current value.*
Specify new rotation angle <current>: *Specify the new rotation angle or press ENTER to accept the current value.*
Enter new tag <current>: *Enter a new tag or press ENTER to accept the current value of tag.*
Enter new prompt <current>: *Enter a new prompt or press ENTER to accept the current value.*
Enter new default value <current>: *Enter the new default value or press ENTER to accept the current value.*

You can also change the position of an existing block and specify a new rotation angle to it using the **CHANGE** command. The prompt sequence is as follows:

Command: **CHANGE**
Select objects: *Select the block.*
Select objects: Enter
Specify change point or [Properties]: *Specify the new block insertion point or press ENTER.*
Specify new block insertion point <no change>: *Specify the new block insertion point or press ENTER to accept the current location of the block.*
Specify new block rotation angle <current>: *Specify the new rotation angle for the block or press ENTER to accept the current rotation angle.*

The radius of a circle can also be changed with the **Change point** option of the **CHANGE** command by specifying the new radius in the case of the circle. To change the radius of a circle, the prompt sequence is:

Command: **CHANGE**
Select objects: *Select the circle.*
Select objects: Enter
Specify change point or [Properties]: *Select a point to specify the radius of the circle.*

If more than one circle is selected, after the first circle is modified, the same prompts are repeated for the next circle.

Properties Option

The **Properties** option can be used to change the characteristics associated with an object.

Changing the layer of an object. If you want to change the layer on which an object exists and other characteristics associated with layers, you can use the **LAyer** option of the **Properties** option. The prompt sequence is:

Command: **CHANGE**
Select objects: *Select the object whose layer you want to change.*
Select objects: *If you have finished selection, press ENTER.*
Specify change point or [Properties]: P
Enter property to change [Color/Elev/LAyer/LType/ltScale/LWeight/Thickness/PLotstyle]: LA
Enter new layer name <0>: *Enter a new layer name.*
Enter property to change [Color/Elev/LAyer/LType/ltScale/LWeight/Thickness]: Enter

The selected object is now placed on the desired layer.

Changing the color of an object. Similarly, you can change the color of the selected object with the **Color** option of the **CHANGE** command. If you want to change the color and linetype to match the layer, you can use the **Properties** option of the **CHANGE** command and then, change the color or linetype by entering BYLAYER at the **Enter new color:** prompt. The Command prompt sequence for this option is similar to the one displayed previously.

Note
Blocks originally created on layer 0 assume the color of the new layer. Otherwise (if it was not created on layer 0), it will retain the color of the layer on which it was created.

Changing the thickness of an object. You can also change the thickness of the selected object in a similar manner. After you have invoked the **CHANGE** command and selected the object to change, use the **Properties** option and enter T at the **Enter property to change** prompt. You are then prompted to enter the new thickness you want to assign to the selected object.

Changing the lineweight of an object. You can similarly also change the lineweight of the selected object by using the **LWeight** option of the **CHANGE** command. Since the lineweight values are predefined, if you enter a value that is not predefined, a predefined value of lineweight, which comes closest to the specified value, is assigned to the selected object.

Tip
*You can see the effect of changing the thickness of an object in a 3D view. Choose **SE Isometric** from the **View > 3D Views** menu to view the effect of the changed thickness value of an object. To get back to the original view, simply choose the **Undo** button in the **Standard Toolbar**.*

Changing the linetype/ Linetype scale of an object. Similarly, you can change the linetype, and linetype scale of the selected objects by entering the appropriate option at the **Enter**

property to change prompt.

Changing the elevation of an object. You can also change the Z-axis elevation of a selected object by entering E at the **Enter property to change** prompt, provided all the points in the particular object have the same Z value. After changing the elevation value of an object, you can view the changes in a 3D view.

CHPROP Command

Command:	CHPROP

You can also use the **CHPROP** command to change the properties of an object using the command line. The prompt sequence is as follows:

> Command: **CHPROP**
> Select objects: *Select the object whose properties you want to change.*
> Select objects: Enter
> Enter property to change [Color/LAyer/LType/ltScale/LWeight/Thickness]: *Specify the property to be modified.*

Note

*Apart from the commands discussed, you can also use the **Object Properties** toolbar to change some of the properties of the selected objects such as, layer, color, linetype, lineweight, and plot style. These options are discussed in Chapter 6, Editing Commands-II.*

Exercise 1 — *General*

Draw a hexagon on layer OBJ in red color. Let the linetype be hidden. Now, use the **PROPERTIES** command to change the layer to some other existing layer, the color to yellow, and the linetype to continuous. Also, in the **Properties** window, under the **Geometry** category, specify a vertex in the **Vertex** field or select one using the arrow (Next or Previous) buttons and then relocate it. The values of the **Vertex X** and **Vertex Y** fields change as the coordinate values of the vertices change. Also, notice the change in the value in the **Area** field. You can also assign a start and end lineweight to each of the hexagon segments between the specified vertices. Use the **LIST** command to verify that the changes have taken place.

EXPLODE COMMAND

Toolbar:	Modify > Explode
Menu:	Modify > Explode
Command:	EXPLODE

This command has been discussed earlier in Chapter 14, Blocks, regarding editing blocks. The **EXPLODE** command is used to split compound objects such as blocks, polylines, regions, polyface meshes, polygon meshes, multilines, bodies, or dimensions into the objects that make them up (Figure 18-8). For example, if you explode a polyline or a 3D polyline, the result will be ordinary lines or arcs (tangent specification and width are not considered). When a region is exploded, it turns into lines, ellipses, splines, or arcs.

Two-dimensional polylines lose their width and tangent specifications and 3D polylines explode into lines. When a body is exploded, it changes into single-surface bodies, curves, or regions. When a leader is exploded, the components are lines, splines, solids, block inserts, text or Mtext, tolerance objects and so on. Mtext explodes into single line text. This command is especially useful when you have inserted an entire drawing and you need to alter a small detail. After you invoke the **EXPLODE** command, you are prompted to select the objects you want to explode. After selecting the objects, press ENTER or right-click to explode the selected objects and then end the command. Note that the basic AutoCAD entities line the lines, circles, arcs, ellipses, and splines cannot be exploded using the **EXPLODE** command.

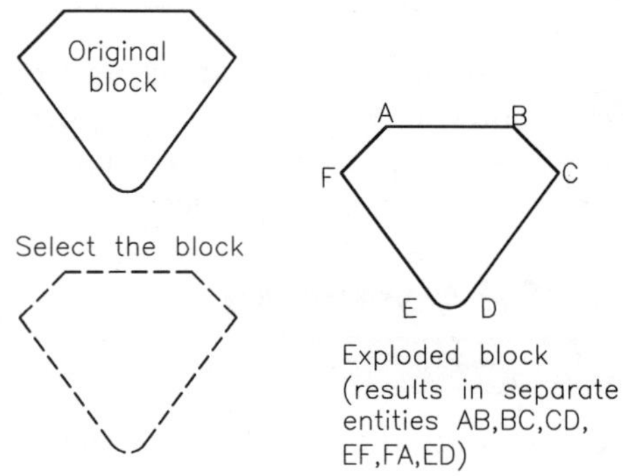

Figure 18-8 *Using the* ***EXPLODE*** *command*

When a block or dimension is exploded, there is no visible change in the drawing. The drawing remains the same except that the color and linetype may have changed because of floating layers, colors, or linetypes. The exploded block is turned into a group of objects that can be modified separately. To check whether the explosion of the block has taken place, select any object that was a part of the block. If the block has been exploded, only that particular object will be highlighted. With the **EXPLODE** command, only one nesting level is exploded at a time. Therefore, if there is a nested block or a polyline in a block and you explode it, the inner block or the polyline will not be exploded. Attribute values are deleted when a block is exploded, and the attribute definitions are redisplayed.

Note

The drawings inserted using the ***XREF*** *command cannot be exploded.*

Tip

If you want to insert a block in the form of its separate components, while using the ***INSERT*** *command select the* ***Explode*** *check box in the* ***Insert*** *dialog box. Also, if you are using the* ***-INSERT*** *command, type * in front of the block name at the* ***Enter block name or [?]:*** *prompt. The block will be inserted in the drawing as separate component objects and not as an entire block.*

POLYLINE EDITING

A polyline can assume various characteristics such as width, linetype, joined polyline, and closed polyline. You can edit polylines, polygons, or rectangles to attain the desired characteristics using the **PEDIT** command. In this section, we will be discussing how to edit simple 2D polylines. The following are the editing operations that can be performed on an existing polyline using the **PEDIT** command. All these editing operations are discussed in detail later in this chapter.

1. A polyline of varying widths can be converted to a polyline of uniform width.
2. An open polyline can be closed and a closed one can be opened.
3. All bends and curved segments between two vertices can be removed to make a straight polyline.
4. A polyline can be split up into two.
5. Individual polylines or polyarcs connected to one another can be joined into a single polyline.
6. Multiple polylines can be edited.
7. The appearance of a polyline can be changed by moving and adding vertices.
8. Curves of arcs and B-spline curves can be fit to all vertices in a polyline, with the specification of the tangent of each vertex being optional.
9. The linetype generation at the vertices of a polyline can be controlled.
10. Multiple polylines can be joined together to form a single polyline.

PEDIT Command

Editing Single Polyline

Toolbar:	Modify II > Edit Polyline
Menu:	Modify > Object > Polyline
Command:	PEDIT

Apart from the methods displayed in the **PEDIT** command box as shown, you can also invoke the **PEDIT** command by choosing **Polyline Edit** from the shortcut menu that is displayed when you select a polyline and right-click. You can use the **PEDIT** command to edit any type of polyline. When you invoke this command you are prompted to select a polyline. If the selected line is an arc or a line (not a polyline), AutoCAD LT issues the following prompt:

Select polyline or [Multiple]:

If the selected entity is not a polyline then the following message is displayed:

Object selected is not a polyline.
Do you want to turn it into one? <Y>:

If you want to turn the object into a polyline, respond by entering a Y and pressing ENTER or by simply pressing ENTER. To let the object be as it is, enter N. AutoCAD LT will then prompt you to select another polyline or object to edit. The subsequent prompts and editing options depend on the type of polyline that has been selected. AutoCAD LT provides you the option of either selecting a single polyline or multiple polylines. In the case a single 2D polyline is selected, the next prompt displayed is:

> Enter an option [Close/Join/Width/Edit vertex/Fit/Spline/Decurve/Ltype gen/Undo]: *Enter an option or press ENTER to end command.*

The options available in **Single** polyline selection method are discussed next:

C (Close) option. This option is used to close an open polyline. **Close** creates the segment that connects the last segment of the polyline to the first. You will get this option only if the polyline is not closed. Figure 18-9 illustrates this option.

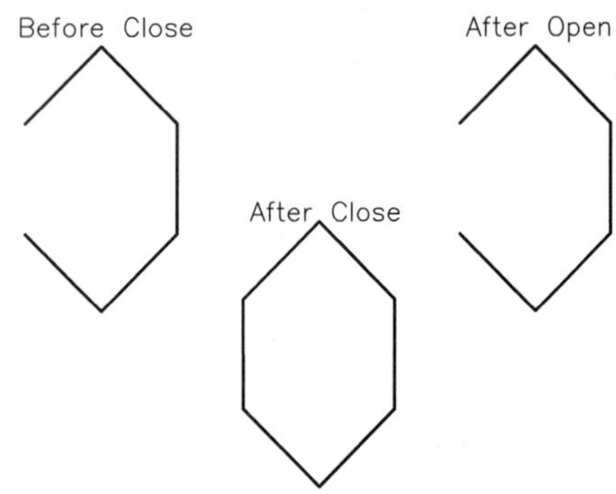

Figure 18-9 The ***Close*** *and the* ***Open*** *options*

O (Open) option. If the selected polyline is closed, the **Close** option is replaced by the **Open** option. Entering **O**, for open, removes the closing segment (Figure 18-10).

J (Join) option. This option appends lines, polylines, or arcs whose endpoints meet a selected polyline at any of its endpoints and adds (joins) them to it (Figure 18-10). This option can be used only if a polyline is open. After this option has been selected, AutoCAD LT prompts you to select objects. Once you have chosen the objects to be joined to the original polyline, AutoCAD LT examines them to determine whether any of them has an endpoint in common with the current polyline, and joins such an object with the original polyline. The search is then repeated using new endpoints. They will not join if the endpoint of the object does not exactly meet the polyline. The line touching a polyline at its endpoint to form a T will not be joined. If two lines meet a polyline in a Y shape, only one of them will be selected, and this selection is unpredictable. To verify which lines have been added to the polyline, use the **LIST** command or select a part of the object. All the segments that are joined to the polyline will be highlighted.

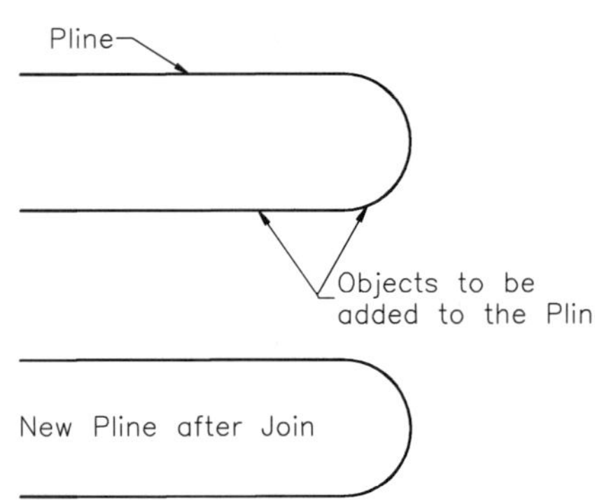

Figure 18-10 *Using the* ***Join*** *option*

W (Width) option. The **Width** option allows you to define a new, unvarying width for all segments of a polyline (Figure 18-11). It changes the width of a polyline with a constant or a varying width. The desired new width can be specified either by entering the width at the keyboard or by specifying the width as the distance between two specified points. Once the width has been specified, the polyline assumes it. Here, you will change the width of the given polyline in the figure from 0.02 to 0.05 (Figure 18-12). After you invoke the **PEDIT** command, the next prompts are:

Select polyline or [Multiple]: *Select the polyline.*
Enter an option [Close/Join/Width/Edit vertex/Fit/Spline/Decurve/Ltype gen/Undo]: **W**
Specify new width for all segments: **0.05**

Note
*Circles drawn using the **CIRCLE** command cannot be changed to polylines. However, polycircles can be drawn using the **Pline Arc** option (by drawing two semicircular polyarcs) or by using the **DONUT** command (you can modify the thickness using the **PEDIT Width** option).*

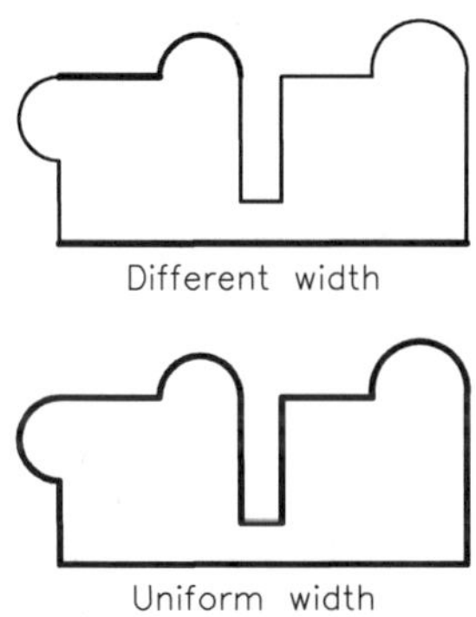

Figure 18-11 *Making the width of a polyline uniform*

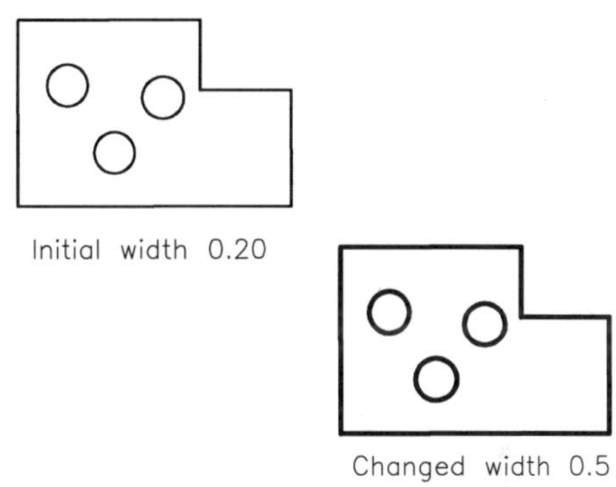

Figure 18-12 *Entering a new width for all segments*

E (Edit Vertex) option. The **Edit vertex** option lets you select a vertex of a polyline and perform different editing operations on the vertex and the segments following it. A polyline segment has two vertices. The first one is at the start point of the polyline segment; the other one is at the endpoint of the segment. When you invoke this option, an X marker appears on the screen at the first vertex of the selected polyline. If a tangent direction has been specified for this particular vertex, an arrow is generated in that direction. After this option has been selected, the next prompt appears with a list of options for this prompt. The prompt sequence after you invoke the **PEDIT** command is:

Select polyline or [Multiple]: *Select the polyline to be edited.*
Enter an option [Close/Join/Width/Edit vertex/Fit/Spline/Decurve/Ltype gen/Undo]: **E**
Enter a vertex editing option
[Next/Previous/Break/Insert/Move/Regen/Straighten/Tangent/Width/eXit]<N>: *Enter an editing option or press ENTER to accept default.*

All the options available for the **Edit vertex** option are discussed next.

Next and Previous options. These options move the X marker to the next or the previous vertex of a polyline. The default value in the **Edit vertex** is one of these two options. The option that is selected as default is the one you chose last. In this manner, the **Next** and **Previous** options help you to move the X marker to any vertex of the polyline by selecting one of these two options, and then pressing ENTER to reach the desired vertex. These options cycle back and forth between first and last vertices, but cannot move past the first or last vertices, even if the polyline is closed (Figure 18-13).

The prompt sequence for using this option after you invoke the **PEDIT** command is:

Select polyline or [Multiple]: *Select the polyline to be edited.*
Enter an option [Close/Join/Width/Edit vertex/Fit/Spline/Decurve/Ltype gen/Undo]: **E**
Enter a vertex editing option
[Next/Previous/Break/Insert/Move/Regen/Straighten/Tangent/Width/eXit] <N>: *Enter N or P to move to the next or previous vertices respectively.*

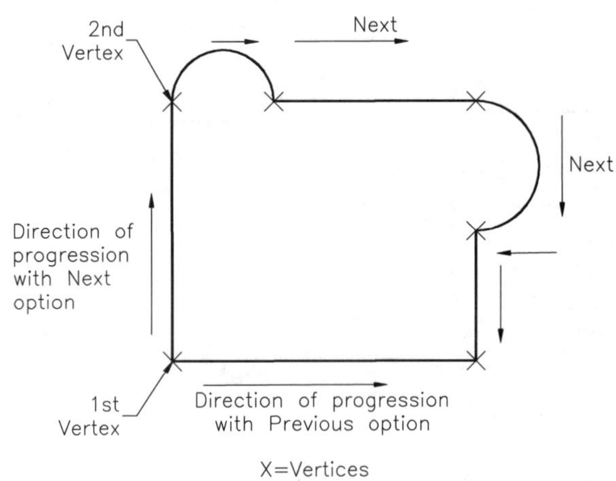

Figure 18-13 *The **Next** and **Previous** options*

Break option. With the **Break** option, you can remove a portion of the polyline as shown in Figure 18-14 or break it at a single point. The breaking of the polyline can be specified between any two vertices. By specifying two different vertices, all the polyline segments and vertices between the specified vertices are erased. If one of the selected vertices is at the endpoint of the polyline, the **Break** option will erase all the segments between the first vertex and the endpoint of the polyline. The exception to this is that AutoCAD LT does not erase the entire polyline if you specify the first vertex at the start point (first vertex) of the polyline and the second vertex at the endpoint (last vertex) of the polyline. If both vertices are at the endpoint of the polyline, or only one vertex is specified and its location is at the endpoint of the polyline, no change is made to the polyline. The last two selections of vertices are treated as invalid by AutoCAD LT, which acknowledges this by displaying the message ***Invalid***.

To use the **Break** option, first you need to move the marker to the first vertex where you want the split to start. Placement of the marker can be achieved with the help of the **Next** and **Previous** options. Once you have selected the first vertex to be used in the **Break** operation, invoke the **Break** option by entering **B** at the Command prompt. AutoCAD LT takes the vertex where the marker (X) is placed as the first point of breakup. The next prompt asks you to specify the position of the next vertex for breakup. You can enter **GO** at this prompt if you want to split the polyline at one vertex only. Otherwise, use the **Next** or **Previous** option to specify the position of next vertex and then enter **GO**. The polyline segments between the two selected vertices is erased. The prompt sequence after you have invoked the **PEDIT** command is:

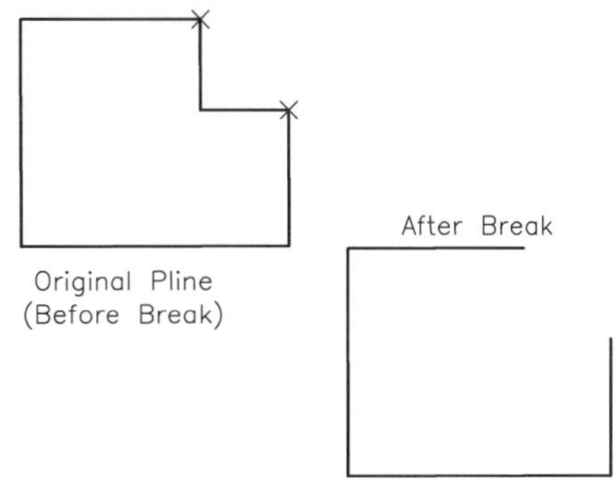

Figure 18-14 *Using the **Break** option*

Select polyline or [Multiple]: *Select the polyline to be edited.*
Enter an option [Close/Join/Width/Edit vertex/Fit/Spline/Decurve/Ltype gen/Undo]: **E**
Enter a vertex editing option
[Next/Previous/Break/Insert/Move/Regen/Straighten/Tangent/Width/eXit]<N>: *Enter N or P to locate the first vertex for the **Break** option.*
Enter a vertex editing option
[Next/Previous/Break/Insert/Move/Regen/Straighten/Tangent/Width/eXit]<N>: **B**

Once you invoke the **Break** option, AutoCAD LT treats the vertex where the marker (X) is displayed as the first point for splitting the polyline. The next prompt issued is:

Enter an option [Next/Previous/Go/eXit] <N>: *Enter G if you want to split the polyline at one vertex only or move the X marker using the **Next** or **Previous** options to specify the position of next vertex for breakup.*

After you have specified the next position of the X marker using the **Next** and **Previous** options, entering **Go** deletes the polyline segment between the two markers specified. Now, exit the **Enter a vertex editing option** prompt using the **eXit** option.

Insert option. The **Insert** option allows you to define a new vertex and add it to the polyline (Figure 18-15). You can invoke this option by entering I for Insert. You should invoke this option only after you have moved the marker (X) to the vertex after which the new vertex is to be added. The new vertex is inserted immediately after the vertex with the X mark. After you have invoked the **PEDIT** command, the prompt sequence for using the **Insert** option is as follows:

Select polyline or [Multiple]: *Select the polyline to be edited.*
Enter an option [Close/Join/Width/Edit vertex/Fit/Spline/Decurve/Ltype gen/Undo]: **E**
Enter a vertex editing option
[Next/Previous/Break/Insert/Move/Regen/Straighten/Tangent/Width/eXit]<N>: *Move the marker to the vertex after which the new vertex is to be inserted.*

Enter a vertex editing option
[Next/Previous/Break/Insert/Move/Regen/Straighten/Tangent/Width/eXit]<N>: **I**
Specify location for new vertex: *Move the cursor and select to specify the location of the new vertex.*

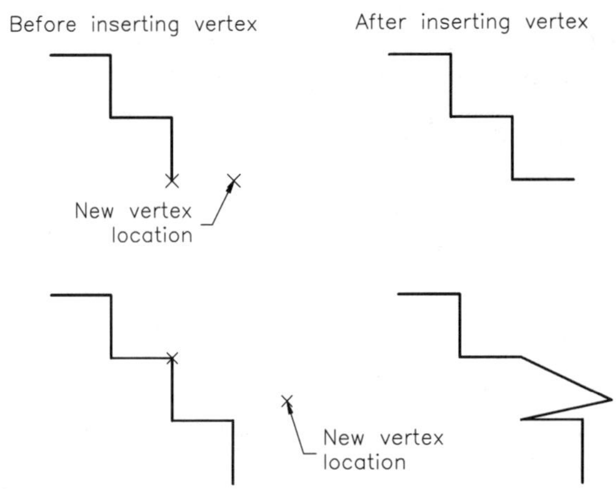

Figure 18-15 *Using the **Insert** option to define new vertex points*

Move option. This option lets you move the X-marked vertex to a new position (Figure 18-16). Before invoking the **Move** option, you must move the X marker to the vertex you want to relocate by selecting the **Next** or **Previous** option. The prompt sequence for relocating a vertex after you invoke the **PEDIT** command is:

Select polyline or [Multiple]: *Select the polyline to be edited.*
Enter an option [Close/Join/Width/Edit vertex/Fit/Spline/Decurve/Ltype gen/Undo]: **E**
Enter a vertex editing option
[Next/Previous/Break/Insert/Move/Regen/Straighten/Tangent/Width/eXit]<N>: *Enter N or P to move the X marker to the vertex you want to relocate.*
Enter a vertex editing option
[Next/Previous/Break/Insert/Move/Regen/Straighten/Tangent/Width/eXit]<N>: **M**
Specify new location for marked vertex: *Specify the new location for the selected vertex by selecting a new location using the pick button of your pointing device or by entering its coordinate values.*
Enter a vertex editing option
[Next/Previous/Break/Insert/Move/Regen/Straighten/Tangent/Width/eXit]<N>: *Enter an option or enter X to exit.*

Regen option. The **Regen** option regenerates the polyline to display the effects of edits you have made, without having to exit the vertex mode editing. It is used most often with the **Width** option.

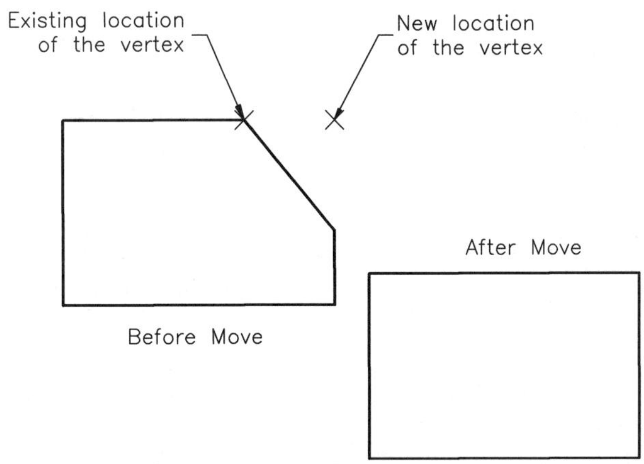

***Figure 18-16** The **Move** option*

Straighten option. The **Straighten** option can be used to straighten polyline segments or arcs between specified vertices (Figure 18-17). It deletes the arcs, line segments, or vertices between the two specified vertices and substitutes them with one polyline segment.

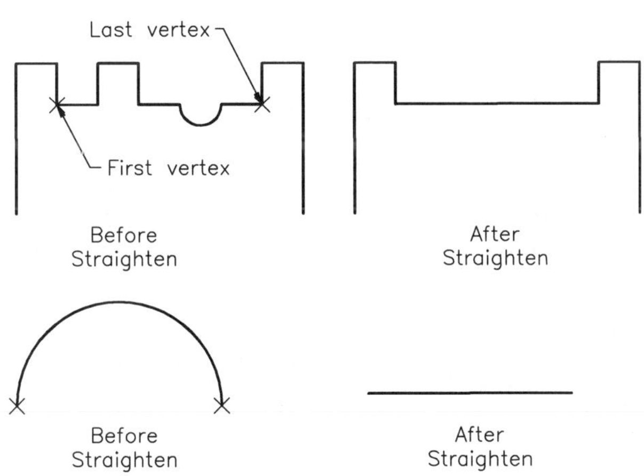

***Figure 18-17** Using the **Straighten** option to straighten polylines*

The prompt sequence for using this option after you have invoked the **PEDIT** command is as follows:

Select polyline or [Multiple]: *Select the polyline to be edited.*
Enter an option [Close/Join/Width/Edit vertex/Fit/Spline/Decurve/Ltype gen/Undo]: **E**
Enter a vertex editing option
[Next/Previous/Break/Insert/Move/Regen/Straighten/Tangent/Width/eXit]<N>: *Move the marker to the desired vertex from where you want to start applying the **Straighten** option with the*

Next *or* ***Previous*** *option.*
Enter a vertex editing option
[Next/Previous/Break/Insert/Move/Regen/Straighten/Tangent/Width/eXit]<N>: **S**
Enter a vertex editing option
[Next/Previous/Break/Insert/Move/Regen/Straighten/Tangent/Width/eXit]<N>: *Move the marker to the next desired vertex, until you reach the vertex you want to straighten.*
Enter an option [Next/Previous/Go/eXit] <N>: **G**

The polyline segments between two marker locations are replaced by a single straight line segment. If you specify a single vertex, the segment following the specified vertex is straightened, if it is curved.

Tangent option. The **Tangent** option is used to associate a tangent direction to the current vertex (marked by X). The tangent direction is used in curve fitting or the **Fit** option of the **PEDIT** command. This option is discussed in detail in the subsequent section on curve fitting. The prompt issued on using the **Tangent** option is:

Specify direction of vertex tangent: *Specify a point or enter an angle.*

You can specify the direction of the vertex tangent by entering an angle at the previous prompt or by selecting a point to express the direction with respect to the current vertex. You can then move the marker to another vertex using the **Next** or **Previous** options and change its direction of tangent or enter X to exit the **Enter a vertex editing option** prompt.

Width option. The **Width** option lets you change the starting and the ending widths of a polyline segment that follows the current vertex (Figure 18-18). By default, the ending width is equal to the starting width and therefore, you can get a polyline segment of uniform width by pressing ENTER at the **Specify ending width for next segment <starting width>** prompt. You can also specify different starting and ending widths to get a varying-width polyline. The prompt sequence is:

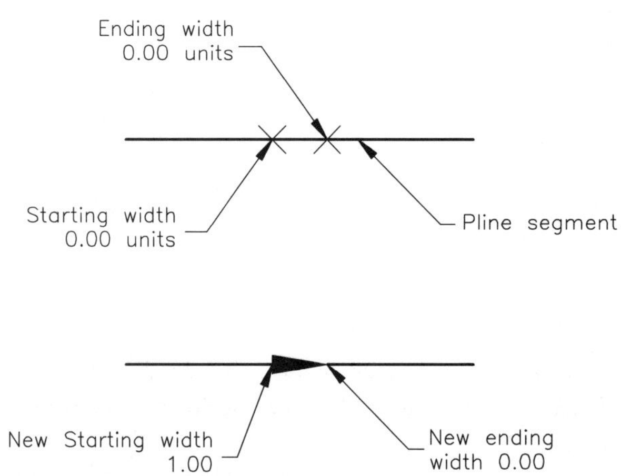

Figure 18-18 *Using the* ***Width*** *option to change the width of a polyline*

Enter an option [Close/Join/Width/Edit vertex/Fit/Spline/Decurve/Ltype gen/Undo]: E
Enter a vertex editing option
[Next/Previous/Break/Insert/Move/Regen/Straighten/Tangent/Width/eXit]<N>: *Move the marker to the starting vertex of the segment whose width is to be altered, using the* ***Next*** *or* ***Previous*** *options.*
Enter a vertex editing option
[Next/Previous/Break/Insert/Move/Regen/Straighten/Tangent/Width/eXit]<N>: W
Specify starting width for next segment <current>: *Enter the revised starting width.*
Specify ending width for next segment <starting width>: *Enter the revised ending width or press ENTER to accept the default option of keeping the ending width of the segment equal to the starting width.*

If no difference is noticed in the appearance of the polyline, you may need to use the **Regen** option. The segment with the revised widths is redrawn after invoking the **Regen** option or when you exit the vertex mode editing.

eXit option. This option lets you exit the Vertex mode editing and return to the main **PEDIT** prompt.

Fit option. The **Fit** or **Fit curve** option generates a curve that passes through all the corners (vertices) of the polyline, using the tangent directions of the vertices (Figure 18-19). The curve is composed of a series of arcs passing through the corners (vertices) of the polyline. This option is used when you draw a polyline with sharp corners and need to convert it into a series of smooth curves. An example of this is a graph. In a graph, we need to show a curve by joining a series of plotted points. The process involved is called curve fitting; therefore, the name of this option. The vertices of the polyline are also known as the control points. The closer together these control points are, the smoother the curve. Therefore, if the **Fit** option does not give optimum results, insert more vertices into the polyline or edit the tangent directions of vertices and then, use the **Fit** option on the polyline. Before using this option you may give each vertex a tangent direction. The curve is then constructed, according to the

Chapter 18

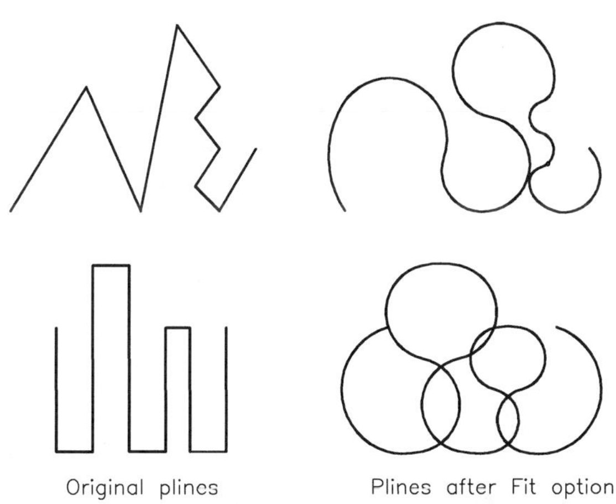

Figure 18-19 *The* ***Fit*** *option*

tangent directions you have specified. The following prompt sequence illustrates the **Fit** option after you invoke the **PEDIT** command:

Select polyline: *Select the polyline to be edited.*
Enter an option [Close/Join/Width/Edit vertex/Fit/Spline/Decurve/Ltype gen/Undo]: **F**

If the tangent directions need to be edited, use the **Edit vertex** option of the **PEDIT** command. Move the X marker to each of the vertices that need to be changed. Now, you can invoke the **Tangent** option, and either enter the tangent direction in degrees or select points. The chosen direction is expressed by an arrow placed at the vertex. The prompt sequence after you invoke the **PEDIT** command is:

Select polyline or [Multiple]: *Select the polyline to be edited.*
Enter an option [Close/Join/Width/Edit vertex/Fit/Spline/Decurve/Ltype gen/Undo]: **E**
Enter a vertex editing option
[Next/Previous/Break/Insert/Move/Regen/Straighten/Tangent/Width/eXit]<N>: **T**
Specify direction of vertex tangent: *Specify a direction in + or - degrees, or select a point in the desired direction. Press ENTER.*

Once you specify the tangent direction are specified, use the **eXit** option to return to the previous prompt and use its **Fit** option.

Spline option. The **Spline** option (Figure 18-20) also smooths the corners of a straight segment polyline, as does the **Fit** option, but the curve passes through only the first and the last control points (vertices), except in the case of a closed polyline. The spline curve is stretched toward the other control points (vertices) but does not pass through them, as in the case of the **Fit** option. The greater the number of control points, the greater the force with which the curve is stretched toward them. The prompt sequence after you invoke the **PEDIT** command is as follows:

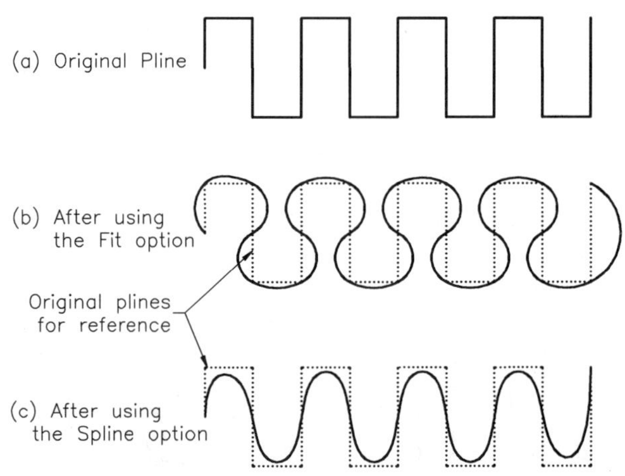

Figure 18-20 *The **Spline** option*

Select polyline or [Multiple]: *Select the polyline.*
Enter an option [Close/Join/Width/Edit vertex/Fit/Spline/Decurve/Ltype gen/Undo]: **S**

The generated curve is a B-spline curve. The **frame** is the original polyline without any curves in it. If the original polyline has arc segments, these segments are straightened when the spline's frame is formed. A frame that has width produces a spline curve that tapers smoothly from the width of the first vertex to that of the last. Any other width specification between the first width specification and the last is neglected. When a spline is formed from a polyline, the frame is displayed as a polyline with zero width and continuous linetype. Also, AutoCAD LT saves its frame so that it may be restored to its original form. Tangent specifications on control point vertices do not affect spline construction.

By default, the spline frames are not shown on screen, but you may want them displayed for reference. In this case, the system variable **SPLFRAME** needs to be manipulated. The default value for this variable is zero. If you want to see the spline frame as well, set it to 1.

Now, whenever the **Spline** option is used on a polyline, the frame will also be displayed. Most editing commands such as **MOVE**, **ERASE**, **COPY**, **MIRROR**, **ROTATE**, and **SCALE** work similarly for both the Fit curves and Spline curves. They work on both the curve and its frame, whether the frame is visible or not. The **EXTEND** command changes the frame by adding a vertex at the point the last segment intersects with the boundary. If you use any of the previously mentioned commands and then use the **Decurve** option to decurve the Spline curve and later use the **Spline** option again, the same Spline curve is generated. The **BREAK**, **TRIM**, and **EXPLODE** commands delete the frame. The **DIVIDE**, **MEASURE**, **FILLET**, **CHAMFER**, **AREA**, and **HATCH** commands recognize only the Spline curve and do not consider the frame. The **STRETCH** command first stretches the frame and then fits the Spline curve to it.

When you use the **Join** option of the **PEDIT** command, the Spline curve is decurved and the original spline information is lost. The **Next** and **Previous** options of the **Edit vertex** option of the **PEDIT** command moves the marker only to points on the frame whether visible or not. The **Break** option discards the spline curve. **Insert**, **Move**, **Straighten** and **Width** options refit the Spline curve. Object Snaps consider the Spline curve and not the frame; therefore, if you wish to snap to the frame control points, restore the original frame.

There are two types of spline curves:

Quadratic B-spline
Cubic B-spline

Both of them pass through the first and the last control points, which is characteristic of the spline curve. Cubic curves are very smooth. A cubic curve passes through the first and last control points, and the curve is closer to the other control points. Quadratic curves are not as smooth as the cubic ones, but they are smoother than the curves produced by the **Fit curve** option. A quadratic curve passes through the first and last control points, and the rest of the curve is tangent to the polyline segments between the remaining control points (Figure 18-21).

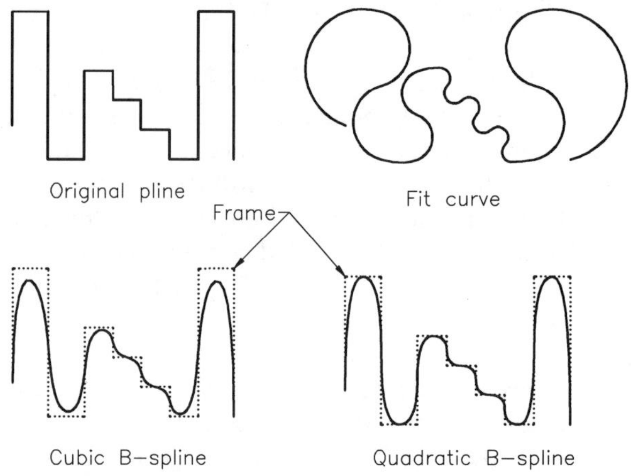

Figure 18-21 *Comparison of fit curve, quad B-spline, and cubic B-spline*

Generation of different types of spline curves. If you want to edit a polyline into a B-spline curve, you are first required to enter a relevant value in the **SPLINETYPE** system variable. A value of 5 produces the quadratic curve, whereas 6 produces a cubic curve. The default value is 6, which implies that when we use the **Spline** option of the **PEDIT** command, a cubic curve is produced by default. You can change the value of the **SPLINETYPE** variable using the command line.

SPLINESEGS. The system variable **SPLINESEGS** governs the number of line segments used to construct the spline curves, so you can use this variable to control the smoothness of the curve. The default value for this variable is 8. With this value, a reasonably smooth curve that does not need much regeneration time is generated. The greater the value of this variable, the smoother the curve, but greater regeneration time, and more space occupied by the drawing file.

Figure 18-22 shows cubic curves with different values for the **SPLINESEGS** parameter.

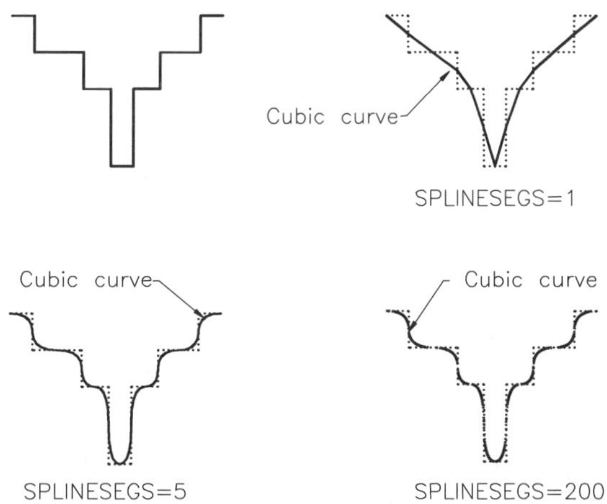

Figure 18-22 *Using the* ***SPLINESEGS*** *variable*

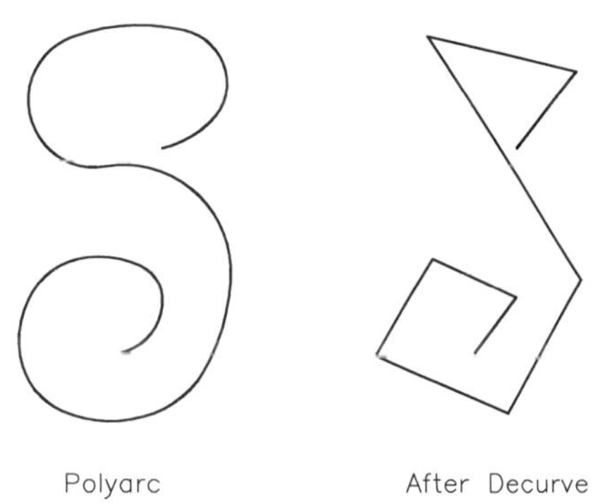

Figure 18-23 *The* ***Decurve*** *option*

Decurve option. The **Decurve** option straightens the curves generated after using the **Fit** or **Spline** option on a polyline. They return to their original shape (Figure 18-23). The polyline segments are straightened using the **Decurve** option. The vertices inserted after using the **Fit** or **Spline** option are also removed. Information entered for tangent reference is retained for use in future fit curve operations. You can also use this command to straighten out any curve drawn with the help of the **Arc** option of the **PLINE** command. Enter **D** at the **Enter an option [Close/Join/Width/Edit vertex/Fit/Spline/Decurve/Ltype gen/Undo]:** prompt to invoke the **Decurve** option of the **PEDIT** command.

Note

If you use edit commands such as the ***BREAK*** *or* ***TRIM*** *commands on spline curves, the* ***Decurve*** *option cannot be used.*

Ltype gen option. You can use this option to control the linetype pattern generation for linetypes other than Continuous with respect to the vertices of the polyline. This option has two modes ON and OFF. If turned off, the break in the noncontinuous linetypes will be avoided at the vertices of the polyline and a continuous segment will be displayed at the vertices (Figure 18-24). If turned on, this option generates the linetype in a continuous pattern such that the gaps may be displayed at the vertices. This option is not applicable to polylines with tapered segments.

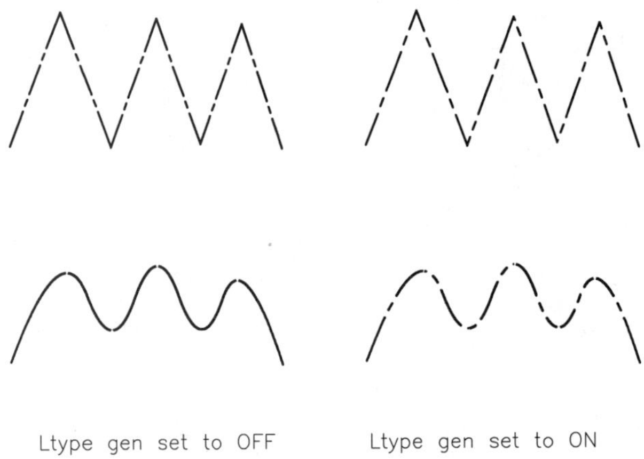

***Figure 18-24** Using the **Ltype gen** option*

The command prompt displayed when you select a polyline and invoke the **Ltype gen** option of the **PEDIT** command is as follows:

> Enter polyline linetype generation option [ON/OFF] <Off>: *Enter ON or press ENTER to accept the default value.*

The linetype generation for new polylines can also be controlled with the help of the **PLINEGEN** system variable, which acts as a toggle. The default value is 0 (off). If you enter a value of 1, this turns on the **Ltype gen** option.

U (Undo) option. The **Undo** option negates the effect of the most recent **PEDIT** operation. You can go back as far as you need to in the current **PEDIT** session by using the **Undo** option repeatedly until you get the desired screen. If you started editing by converting an object into a polyline, and you want to change the polyline back to the object from which it was created, the **Undo** option of the **PEDIT** command will not work. In this case, you will have to exit to the Command prompt and use the **UNDO** command to undo the operation.

Editing Multiple Polyline

Selecting the **Multiple** option of the **Pedit** command allows you to select more then one polyline for editing. You can select the polylines using any of the objects selection techniques. After the objects to be edited are selected, press ENTER or right-click to proceed with the command. The prompt sequence that will follow is:

> Enter an option [Close/Open/Join/Width/Fit/Spline/Decurve/Ltype gen/Undo]:

Join. This option is used to join more than one polylines that may or may not be in contact with each other. Even the polylines that are not coincident can be joined using this option. The polylines are joined using **Fuzz distance**. After you select the **Join** option, the sequence of prompts are as follows:

Join Type = Add segment
Enter fuzz distance or [Jointype] <current>: *Enter a distance or* ***J*** *for changing the* ***Jointype***.

The prompt sequence to follow when you enter **J** at the above prompt is:

Enter join type [Extend/Add/Both] <Add>:

Extend. This option extends or trims the selected polylines at the endpoints that are nearest to each other, see (Figure 18-25). Keep in mind that the segments of selected polylines that are nearest to each other should not be parallel. This means that the segments that are nearest to each other should intersect at some point when extended.

Add. Joins the selected polylines by drawing a straight line between the nearest endpoints, see (Figure 18-26). The fuzz distance should be greater then the actual distance between the two endpoints to be joined.

Both. This option joins the endpoints by extending or trimming if possible; otherwise, it adds a line segment between the two endpoints.

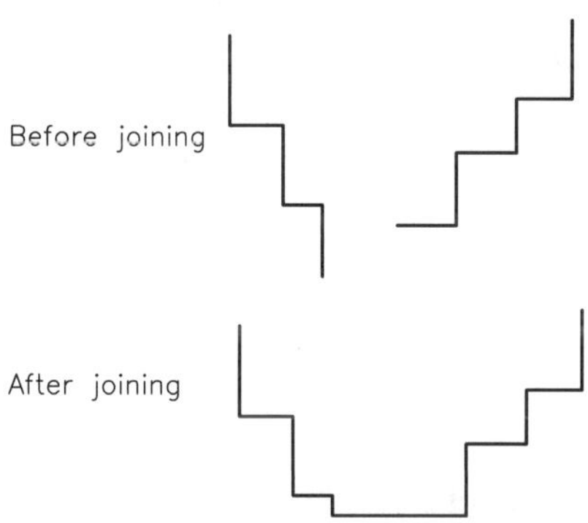

Figure 18-25 *Joining two polylines using the* ***Extend*** *option*

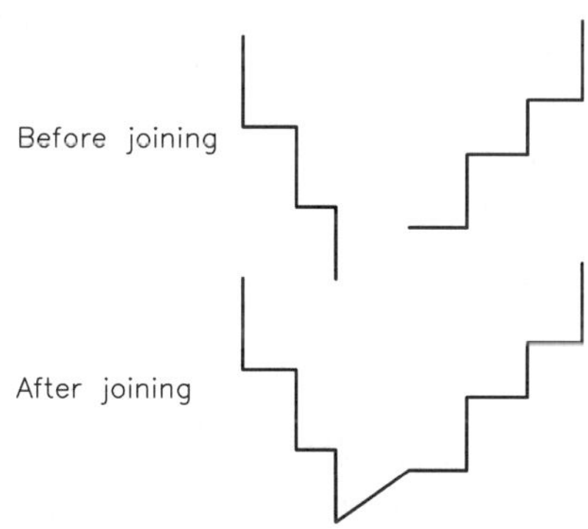

Figure 18-26 *Joining two polylines using the* ***Add*** *option*

Exercise 2 *General*

a. Draw a line from point (0,0) to point (6,6). Convert the line into a polyline with starting width 0.30 and ending width 0.00.

b. Draw different polylines of varying width segments and then use the **Join** option to join all the segments into one polyline. After joining the different segments, make the width of the segments uniform, with the **Width** option.

Exercise 3 *General*

a. Draw a staircase-shaped polyline, then use different options of the **PEDIT** command to generate a fit curve, a quadratic B-spline, and a cubic B-spline. Then, convert the curves back into the original polyline.

b. Draw square-wave-shaped polylines, and use different options of the **Edit vertex** option to navigate around the polyline, split the polyline into two at the third vertex, insert more vertices at different locations in the original polyline, and convert the square-wave-shaped polyline into a straight-line polyline.

UNDOING COMMANDS

Command: UNDO

The **LINE** and **PLINE** commands have the **Undo** option, which can be used to undo (nullify) the changes made within these commands. The **UNDO** command is used to undo a previous command or to undo more than one command at one time. This command can be invoked by entering **UNDO** at the Command prompt. **Undo** is also available from the **Edit** menu and the **Standard** toolbar, but this can only undo the previous command and only one command at a time. If you right-click in the drawing area a shortcut menu is displayed. Choose **Undo**. It is equivalent to **UNDO 1** or to enter **U** at the Command prompt. The **U** command can undo only one operation at a time. You can use **U** as many times as you want until AutoCAD LT displays the message: **Nothing to undo**. When an operation can not be undone, the message name is displayed but no action takes place. External commands such as **PLOT** can not be undone. The prompt sequence for the **UNDO** command is:

Command: **UNDO**
Enter the number of operations to undo or [Auto/Control/BEgin/End/Mark/Back] <1>: *Enter a positive number, an option or press ENTER to undo a single operation.*

The various options of this command are discussed next.

Number (N) Option

This is the default option. This number represents the number of previous command sequences to be deleted. For example, if the number entered is 1, the previous command is undone; if the number entered is 4, the previous four commands are undone, and so on. This is identical to invoking the **U** command four times, except that only one regeneration takes place. AutoCAD LT lets you know which commands were undone by displaying a message after you press ENTER. The prompt sequence is:

Command: **UNDO**
Enter the number of operations to undo or [Auto/Control/BEgin/End/Mark/Back] <1>: 3
PLINE LINE CIRCLE

Control (C) Option

This option lets you determine how many of the options are active in the **UNDO** command. You can disable the options you do not need. With this option you can even disable the **UNDO** command. To access this option, type C. You will get the following prompt:

Enter an UNDO control option [ALL/None/One] <All>: *Enter an option or press ENTER to accept the default.*

ALL

The **All** option activates all the features (options) of the **UNDO** command.

None (N)

This option turns off **UNDO** and the **U** command. If you have used the **BEgin** and **End** options or **Mark** and **Back** options to create **UNDO** information, all of that information is lost. The prompt sequence for invoking the **Control** option is as follows:

Command: **UNDO**
Enter the number of operations to undo [Auto/Control/BEgin/End/Mark/Back] <1>: **C**
Enter an UNDO control option [ALL/None/One] <All>: **N**

If you try to use the **U** command now, while the **UNDO** command has been disabled, AutoCAD LT gives you the following message:

Command: **U**
U command disabled. Use **UNDO** command to turn it on.

The prompt sequence for the **UNDO** command after issuing the **None** option is:

Command: **UNDO**
Enter an UNDO control option [All/None/One] <All>: *Enter O or press ENTER to accept the default.*

To enable the **UNDO** options again, you must enter the **All** or **One** (one mode) option.

One (O)

This option restrains the **UNDO** command to a single operation. All **UNDO** information saved earlier during editing is scrapped. The prompt sequence is as follows:

Enter the number of operations to undo or [Auto/Control/BEgin/End/Mark/Back] <1>: **C**
Enter an UNDO control option [All/None/One] <All>: **O**

If you then enter the **UNDO** command, you will get the following prompt:

Command: **UNDO**
Enter an option [Control] <1>: *Press ENTER to undo the last operation or enter C to select another control option.*

In response to this last prompt, you can now either press ENTER to undo only the previous command, or go into the **Control** options by entering C. AutoCAD LT acknowledges undoing the previous command with messages such as:

Command: **UNDO**

Enter an option [Control] <1>: Enter
CIRCLE
Everything has been undone

Tip

AutoCAD LT stores all information about all the ***UNDO*** *operations taking place in a drawing session. When you use the* ***None*** *control option of the* ***UNDO*** *command, this information is removed and valuable disk space is made available. In case of limited disk space availability, it may be a good idea to use the* ***One*** *suboption of the* ***UNDO Control*** *command. This allows you to use both the* ***U*** *and* ***UNDO*** *commands and at the same time makes the disk space available.*

Auto (A) Option

Enter A to invoke this option. The following prompt is displayed:

Enter UNDO Auto mode [ON/OFF] <current>: *Select ON or OFF or press ENTER to accept the default.*

This option is On by default and every menu item is a single operation. When you use the **Begin** and **End** options to group a series of commands as a single operation, entering **U** at the Command prompt removes the objects individually and not as a group, if the Auto option is On. But if the **Auto** option is Off, entering **U** at the Command prompt will remove the entire group of objects, you had grouped using the **Begin** and **End** options in one single operation.

Also, if you have put a marker in between the begin and end operations and if **Auto** is On, the group of commands till the marker is undone altogether on using the **Back** option are considered to be a single group and are undone as a single command. Although, entering **U** at the Command prompt removes the objects individually until the marker is encountered. But, if the **Auto** option is off, entering **U** at the Command prompt or using the **Back** option of the **UNDO** command, undoes the entire group of operations irrespective of the marker being there.

BEgin (BE) and End (E) Options

A group of commands is treated as one command for the **U** and **UNDO** commands (provided **Auto** option is Off) by embedding the commands between the **BEgin** and **End** options of the **UNDO** command. If you anticipate the removal of a group of successive commands later in a drawing, you can use this option, since all of the commands after the **BEgin** option and before the **End** option are treated as a single command by the **U** command (if the **Auto** option is Off). For example, the following sequence illustrates the possibility of removal of two commands:

Command: **UNDO**
Enter the number of operations to undo or [Auto/Control/BEgin/End/Mark/Back] <1>: **BE**
Command: **CIRCLE**
Specify center point for Circle or [3P/2P/Ttr (tan tan radius)]: *Specify the center.*
Specify radius of circle or [Diameter] <current>: *Specify the radius of the circle.*

Command: **PLINE**
Specify start point or [Arc/Close/Halfwidth/Length/Undo/Width]: *Select first point.*
Specify next point: *Select the next point.*
Specify start point or [Arc/Close/Halfwidth/Length/Undo/Width]: Enter
Command: **UNDO**
Enter the number of operations to undo or [Auto/Control/BEgin/End/Mark/Back]: **E**
Command: **U**

To start the next group once you are finished specifying the current group, use the **End** option to end this group. Another method is to enter the **BEgin** option to start the next group while the current group is active. This is equivalent to issuing the **End** option followed by the **BEgin** option. The group is complete only when the **End** option is invoked to match a **BEgin** option. If **U** or the **UNDO** command is issued after the **BEgin** option has been invoked and before the **End** option has been issued, only one command is undone at a time until it reaches the juncture where the **BEgin** option has been entered. If you want to undo the commands issued before the **BEgin** option was invoked, you must enter the **End** option so that the group is complete. This is demonstrated in the following Example.

Example 2 *General*

Enter the following commands in the same sequence as given, and notice the changes that take place on screen.

CIRCLE
POLYGON
UNDO BEgin (Make sure that the Auto is OFF)
PLINE
TRACE
U
DONUT
UNDO End
TEXT
U
U
U
U

The first **U** command will undo the **TRACE** command. If you repeat the **U** command, the **PLINE** command will be undone. Any further invoking of the **U** command will not undo any previously drawn object (**POLYGON** and **CIRCLE**, in this case), because after the **PLINE** is undone you have an **UNDO BEgin**. Only after you enter **UNDO End** can you undo the **POLYGON** and the **CIRCLE**. In the example, the second **U** command will undo the **TEXT** command, the third **U** command will undo the **DONUT** and **PLINE** commands (these are enclosed in the group), the fourth **U** command will undo the **POLYGON** command, and the fifth **U** command will undo the **CIRCLE** command. Whenever the commands in a group are undone, the name of each command or operation is not displayed as it is undone. Only the name, **GROUP**, is displayed.

Tip

*You can use the **BEgin** and **End** options only when the **UNDO Control** is set to **All** and the **Auto** option is Off.*

Mark (M) and Back Options

The **Mark** option installs a marker in the Undo file. The **Back** option lets you undo all the operations until the mark. In other words, the **Back** option returns the drawing to the point where the previous mark was inserted. For example, if you have completed a portion of your drawing and do not want anything up to this point to be deleted, you insert a marker and then proceed. Then, even if you use the **UNDO Back** option, it will work only until the marker. You can insert multiple markers, and with the help of the **Back** option you can return to the successive mark points. The following prompt sequence is used to introduce a mark point:

Command: **UNDO**
Enter the number of operations to undo or [Auto/Control/BEgin/End/Mark/Back]: **M**

The prompt sequence for using the **Back** option is:

Command: **UNDO**
Enter the number of operations to UNDO or [Auto/Control/BEgin/End/Mark/Back]: **B**

Once all the marks have been exhausted with the successive **Back** options, any further invoking of the **Back** option displays the message: **This will undo everything. OK? <Y>**. If you enter Y (Yes) at this prompt, all the operations carried out since you entered the current drawing session will be undone. If you enter N (No) at this prompt, the **Back** option will be disregarded.

You cannot undo certain commands and system variables, for example, **DIST**, **LIST**, **DELAY**, **NEW**, **OPEN**, **QUIT**, **AREA**, **HELP**, **PLOT**, **QSAVE** and **SAVE**, among many more. Actually, these commands have no effect that can be undone. Commands that change operating modes (**GRID**, **UNITS**, **SNAP**, **ORTHO**) can be undone, though the effect may not be apparent at first. This is the reason why AutoCAD LT displays the command names as they are undone.

REDO COMMAND

Toolbar:	Standard > Redo
Menu:	Edit > Redo
Command:	REDO

If you right-click in the drawing area, a shortcut menu is displayed. Choose **Redo** to invoke the **REDO** command. The **REDO** command brings back the process you removed previously using the **U** and **UNDO** commands. This command undoes the last **UNDO** command, but it must be entered immediately after the **UNDO** command and the objects previously undone reappear on the screen.

RENAMING NAMED OBJECTS

Menu: Format > Rename

You can edit the names of named objects such as blocks, dimension styles, layers, linetypes, styles, UCS, views, and viewports using the **Rename** dialog box (Figure 18-27). You can select the type of named object from the list provided in the **Named Objects** area of the dialog box. Corresponding names of all the objects of the specified type that can be renamed are displayed in the **Items** area. For example, if you want to rename the layer named LOCKED to HID, the process will be as follows:

1. Select **Layers** in the **Named Objects** list box. All the layer names in the current drawing that can be renamed are displayed in the **Items** list box.

2. Select LOCKED in the **Items** list box to highlight it. LOCKED gets displayed in the **Old Name** edit box.

3. Enter HID in the **Rename To** edit box, and choose the **OK** button.

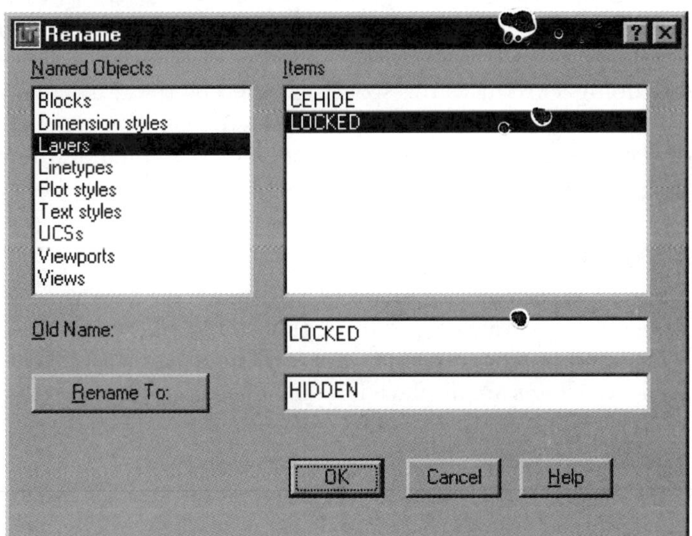

***Figure 18-27 Rename** dialog box*

Now the layer named LOCKED is renamed to HID. You can invoke the **Rename** dialog box again, view the **Layer Control** drop-down list in the **Object Properties** toolbar, or invoke the **Layer properties Manager** dialog box to notice this change. You can rename blocks, dimension styles, linetypes, styles, UCS, views, and viewports in the same way.

Note

*The **RENAME** command cannot be used to rename drawing files created using the **WBLOCK** command. You can change the name of a drawing file by using the **Rename** option of your Operating System.*

*The Layer **0** and the **Continuous** linetype cannot be renamed and therefore do not appear in the **Items** list box of the **Rename** dialog box.*

REMOVING UNUSED NAMED OBJECTS

Menu:	File > Drawing Utilities > Purge

This is another editing operation used for deletion and it was discussed earlier in relation to blocks. You can delete unused named objects such as blocks, layers, dimension styles, linetypes, text styles, and shapes with the help of the **PURGE** command. When you create a new drawing or open an existing one, AutoCAD LT records the named objects in that drawing and notes other drawings that reference the named objects. Usually only a few of the named objects in the drawing (such as layers, linetypes, and blocks) are used. For example, when you create a new drawing, the prototype drawing settings may contain various text styles, blocks, and layers which you do not want to use. Also, you may want to delete particular unused named objects such as unused blocks, in an existing drawing. Deleting inactive named objects is important and useful because doing so reduces the space occupied by the drawing. With the **PURGE** command, you can select the named objects you want to delete. You can use this command at any time in the drawing session. When you invoke the **Purge** command, the **Purge** dialog box is displayed (Figure 18-28).

View items you can purge

When this radio button is selected, AutoCAD LT displays the tree view of all the named objects that are in the current drawing and those that can be purged. When this radio button is chosen, **Items not used in drawing** area, and two check boxes **Confirm each item to be purged**, the **Purge nested items** are displayed in the dialog box. These are discussed as follows.

Items not used in drawing

This area lists all the named objects that can be purged. You can list the items of any object type by choosing the plus (+) sign or by double-clicking on the named object in the tree view. Select them to purge. You can select more then one item by holding down the shift key and selecting the items.

Confirm each item to be purged

This check box is selected to confirm before purging the selected item. The **Confirm Purge** dialog box is displayed when items are selected to purge and after the **Purge** or **Purge All** buttons are chosen. You can confirm or cancel the items to be purged in this dialog box.

Purge nested items

This check box, when selected, removes all the nested items not in use. This removes the nested items only when you select the **Purge All** button or select blocks.

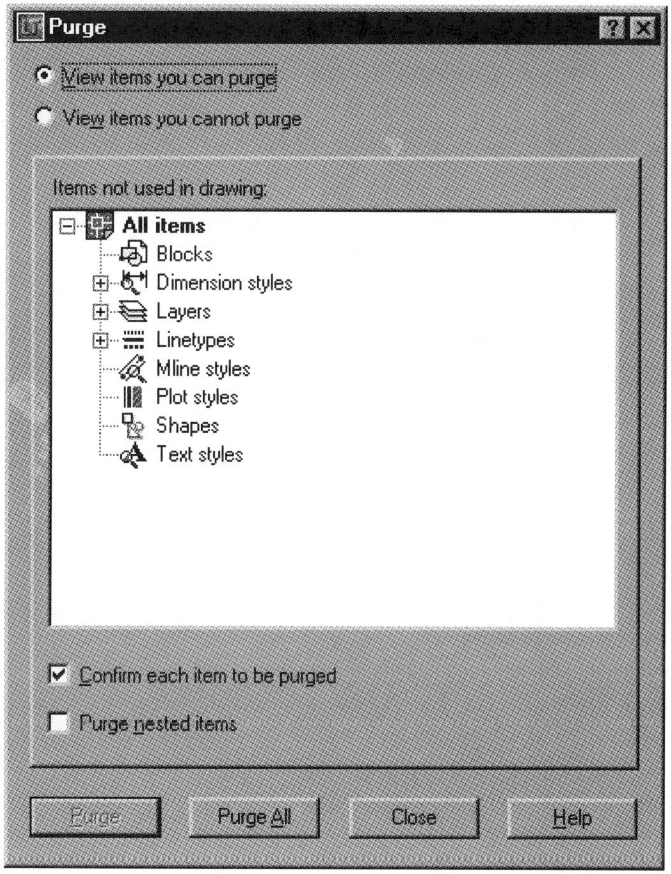

***Figure 18-28 Purge** dialog box with items that can be purged*

View items you cannot purge

This radio button is selected to display the tree view of the items you cannot purge. These items are those that are used in the current drawing or are default items that cannot be removed (Figure 18-29). Choosing this radio button displays the tree view in the **Items currently used in drawing** area of the dialog box. When you select an object in this tree view, information about why you cannot purge this object is displayed below the tree view.

Note

*The **Entire Drawing** option of the **Write Block** dialog box that is invoked on using the **WBLOCK** command or the **-WBLOCK asterisk** option that have been discussed earlier have the same effect as the **PURGE All** command. The only difference is that the unused named objects are removed automatically without any verifications, though these methods are much faster.*

*Standard objects created by AutoCAD LT (such as layer 0, STANDARD text style, and linetype CONTINUOUS) cannot be removed by the **PURGE** command, even if these objects are not used.*

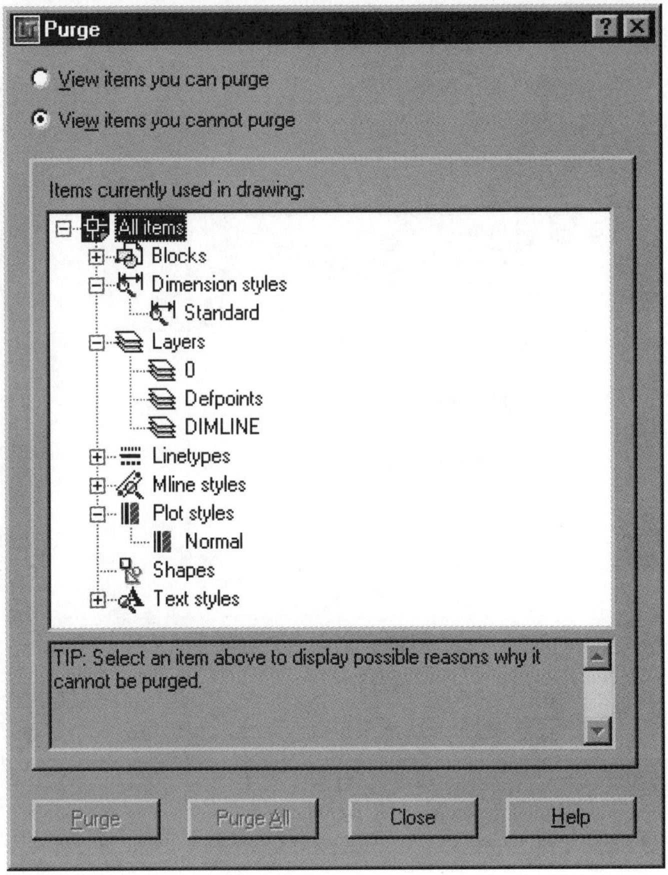

***Figure 18-29 Purge** dialog box with items that cannot be removed*

OBJECT SELECTION MODES

Menu:	Tools > Options
Command:	OPTIONS

When you select a number of objects, the selected objects form a **selection set**. Selection of the objects is controlled in the **Options** dialog box (Figure 18-30) that is invoked by the **OPTIONS** command. Six selection modes are provided in the **Selection** tab of this dialog box. You can select any one of these modes or a combination of various modes.

Noun/Verb Selection

By selecting the **Noun/Verb Selection** check box, you can select the objects (noun) first and then specify the operation (verb) (command) to be performed on the selection set. This mode is active by default. For example, if you want to move some objects when the mode **Noun/Verb Selection** is enabled, first select the objects to be moved, and then invoke the **MOVE** command. The objects selected are highlighted automatically when the **MOVE** command is invoked, and AutoCAD LT does not issue any **Select objects** prompt. The following commands can be used

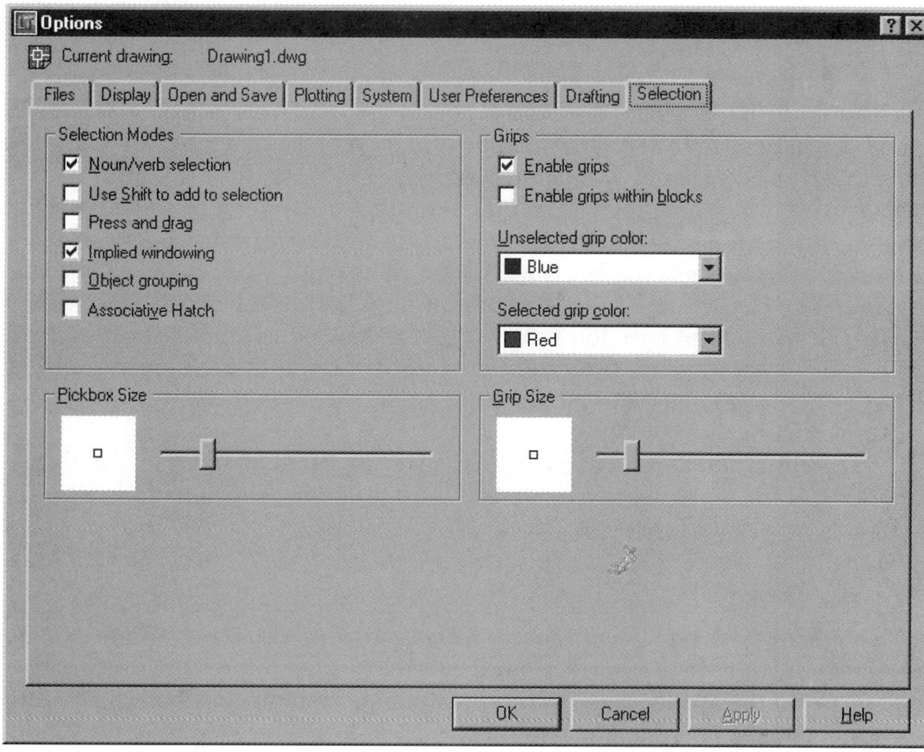

***Figure 18-30** The **Selection** tab of the **Options** dialog box*

on the selected objects when the **Noun/Verb Selection** mode is active:

ARRAY **BLOCK** **CHANGE** **CHPROP** **COPY**
DVIEW **EXPLODE** **ERASE** **LIST** **MIRROR**
MOVE **PROPERTIES** **ROTATE** **SCALE** **WBLOCK**

The following are some of the commands that are not affected by the **Noun/Verb Selection** mode. You are required to specify the objects (noun) on which an operation (command/verb) is to be performed, after specifying the command (verb):

BREAK **CHAMFER** **DIVIDE** **EXTEND**
FILLET **MEASURE** **OFFSET** **TRIM**

When the **Noun/Verb Selection** mode is active, the **PICKFIRST** system variable is set to 1 (On). In other words, you can also activate the **Noun/Verb Selection** mode by setting the **PICKFIRST** system variable to 1 (On).

Use Shift to Add to Selection

The next option in the **Selection Modes** area of the **Selection** tab of the **Options** dialog box is **Use Shift to add to selection**. Selecting this option establishes the additive selection mode, which is the normal method of most Windows programs. In this mode, you have to hold down the SHIFT key when you want to add objects to the selection set. For example, suppose X, Y, and Z are three objects on the screen and you have selected the **Use Shift to add to selection** check box in the **Selection Modes** area of the **Options** dialog box. Select object X. It is highlighted and put in the selection set. After selecting X, and while selecting object Y, if you do not hold down the SHIFT key, object Y only is highlighted and it replaces object X in the selection set. On the other hand, if you hold down the SHIFT key while selecting Y, it is added to the selection set (which contains X), and the resulting selection set contains both X and Y. Also, both X and Y are highlighted. To summarize the concept, objects are added to the selection set only when the SHIFT key is held down while objects are selected. Objects can be discarded from the selection set by reselecting these objects while the SHIFT key is held down. If you want to clear an entire selection set quickly, draw a blank selection window anywhere in a blank drawing area. You can also right-click to display a shortcut menu and choose **Deselect All**. All selected objects in the selection set are discarded from it.

When the **Use Shift to add to selection** mode is active, the **PICKADD** system variable is set to 0 (Off). In other words, you can activate the **Use Shift to add to selection** mode by setting the **PICKADD** system variable to Off.

Press and Drag

This selection mode is used to govern the way you can define a selection window or crossing window. When you select this option, you can create the window by pressing the pick button to select one corner of the window and continuing to hold down the pick button and dragging the cursor to define the other diagonal point of the window. When you have the window you want, release the pick button. If the **Press and Drag** mode is not active, you have to select twice to specify the two diagonal corners of the window to be defined. This mode is not active by default. This implies that to define a selection window or crossing window, you have to select twice to define their two opposite corner points.

When the **Press and Drag** mode is active, the **PICKDRAG** system variable is set to 1 (On). In other words, you can activate the **Press and Drag** mode by setting **PICKDRAG** to On.

Implied Windowing

By selecting this option, you can automatically create a Window or Crossing selection when the **Select objects** prompt is issued. The selection window or crossing window in this case is created in the following manner: At the **Select objects** prompt, select a point in empty space on the screen. This becomes the first corner point of the selection window. After this, AutoCAD LT asks you to specify the other corner point of the selection window. If the first corner point is to the right of the second corner point, a Crossing selection is defined; if the first corner point is to the left of the second corner point, a Window selection is defined. The **Implied Windowing** check box is selected by default.

If this option is not active, or if you need to select the first corner in a crowded area where selecting would select an object, you need to specify Window or Crossing at the **Select objects** prompt, depending on your requirement.

When the **Implied Windowing** mode is active, the **PICKAUTO** system variable is set to 1 (On). In other words, you can activate **Implied Windowing** mode by setting the value of the **PICKAUTO** system variable to 1 (On).

Object Grouping

This turns the automatic group selection on and off. When this option is on and you select a member of a group, the whole group is selected. You can also activate this option by setting the value of the **PICKSTYLE** system variable to 1. (Groups were discussed earlier in this chapter.)

Tip

*You can clear the **Object Grouping** check box in the **Selection Modes** area of the **Selection** tab of the **Options** dialog box to be able to select the objects of a group individually for editing without having to explode the group.*

Associative Hatch

If the **Associative Hatch** check box is selected in the **Selection modes** area of the **Selection** tab of the **Options** dialog box, the boundary object is also selected when an associative hatch is selected. You can also select this option by setting the value of the **PICKSTYLE** system variable to 2 or 3. Hatching and boundaries have been discussed in Chapter 13, Hatching. It is recommended that you select the **Associative Hatch** check box for most drawings.

Pickbox Size

The **Pickbox** slider bar controls the size of the pickbox. The size ranges from 0 to 20. The default size is 3. You can also use the **PICKBOX** system variable.

Self-Evaluation Test

Answer the following questions, and then compare your answers to the answers given at the end of this chapter.

1. Even if a group is defined as selectable, if the **PICKSTYLE** system variable is set to 0, you will not be able to select the entire group by selecting one of its members. (T/F)

2. Only the **CHANGE** command can be used to change the properties associated with an object. (T/F)

3. The **Properties** window can be used to modify the geometry of objects apart from their general properties. (T/F)

4. The **RENAME** command can be used to change the name of a drawing file created by the **WBLOCK** command. (T/F)

5. A group of commands is treated as one command for the **U** and **UNDO** commands by embedding them between the ________ and ________ options of the **UNDO** command.

6. If the **Press and Drag** mode is not active, you have to select ________ to specify the two diagonal points of a selection window.

7. The ________ option of the **PEDIT** command's main prompt can be used to change the starting and ending widths of a polyline separately to a desired value.

8. The ________ option of the **PEDIT** command allows you to select more then one polyline.

9. You can move past the first and last vertices in a closed polyline by using either the ______ option or the ________ option.

10. While using the ______ command, if you select a line or arc, AutoCAD LT provides you with the option of converting them into a polyline first.

Review Questions

Answer the following questions

1. When you use the **GROUP** command to form object groups, AutoCAD LT lets you sequentially highlight the groups of which the selected object is a member. (T/F)

2. The color, linetype, lineweight, and layer on which a block is drawn can be changed with the help of the **CHANGE** command. (T/F)

3. After exploding an object, the object remains identical except that the color and linetype may change because of floating layers, colors, or linetypes. (T/F)

4. The **PURGE** command has the same effect as the **WBLOCK Entire drawing** or the **-WBLOCK asterisk** method. The only difference is that, with the **PURGE** command, deletion takes place automatically. (T/F)

5. If you have made a copy of a group without naming the copy then its name is displayed in which of the following notations in the **Group Name** list box?

   (a) $A1 (b) @A1
   (c) %A1 (d) *A1

6. Which of the following options in the **Group Manager** dialog box deletes the selected group from the drawing?

   (a) **Remove** (b) **Explode**
   (c) **Ungroup** (d) **Rename**

7. Which of the following system variable controls the smoothness of the curve?

   (a) **SPLINETYPE** (b) **SPLINESEGS**
   (c) **PLINEGEN** (d) **PICKFIRST**

8. Which option of the **PEDIT** command can straighten a curve drawn with the help of the **PLINE** command?

   (a) Join (b) Close
   (c) Decurve (d) None of the above

9. Which of the following properties cannot be changed using the **CHANGE** command?

   (a) Lineweight (b) Color
   (c) Plotstyle (d) Thickness

10. The __________ command can be used to change the name of a block.

11. If the __________ option of the **UNDO** command is off, any group of commands grouped together using the **Begin** and **End** options of the **UNDO** command are undone together.

12. Circles drawn using the **CIRCLE** command can be changed to __________.

13. You can also set the **Implied Windowing** mode by setting the value of the __________ system variable to **1**.

14. A group can be selected by entering __________ at the AutoCAD LT **Select objects**: prompt.

15. The __________ option of the **UNDO** command disables the **UNDO** and **U** commands entirely.

Exercises

Exercise 4 *Graphics*

Draw part (a) in Figure 18-31 and then, using the **PROPERTIES** and relevant **PEDIT** command options, convert it into parts (b), (c), and (d). The linetype used in part (d) is HIDDEN.

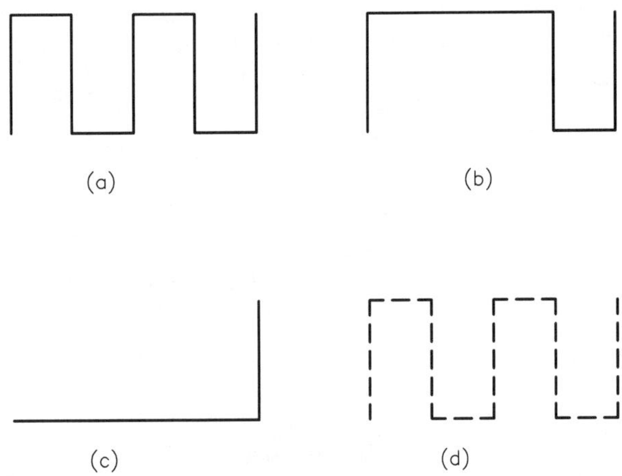

Figure 18-31 Drawing for Exercise 4

Exercise 5 *Mechanical*

Draw the object in Figure 18-32 using the **LINE** command. Change the object to a polyline with a width of 0.05.

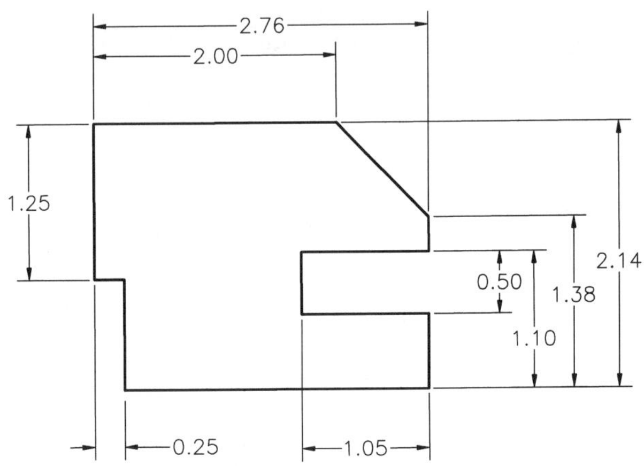

Figure 18-32 Drawing for Exercise 5

Exercise 6 *Graphics*

Draw part (a) in Figure 18-33; then, using the relevant **PEDIT** command options, convert it into drawing (b).

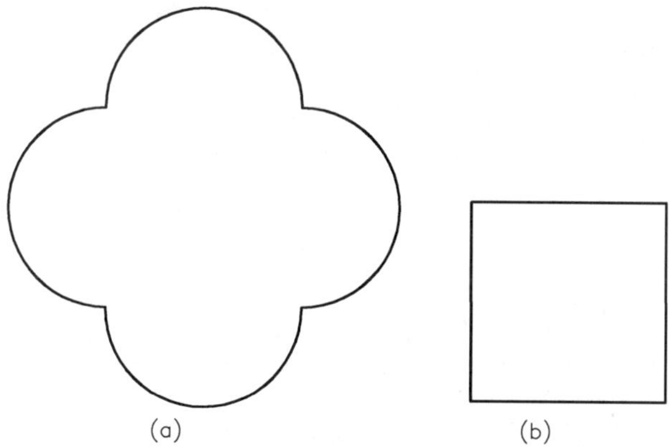

Figure 18-33 *Drawing for Exercise 6*

Exercise 7

Graphics

Draw part (a) in Figure 18-34; then, using the relevant **PEDIT** options, convert it into drawings (b), (c), and (d). Identify the types of curves.

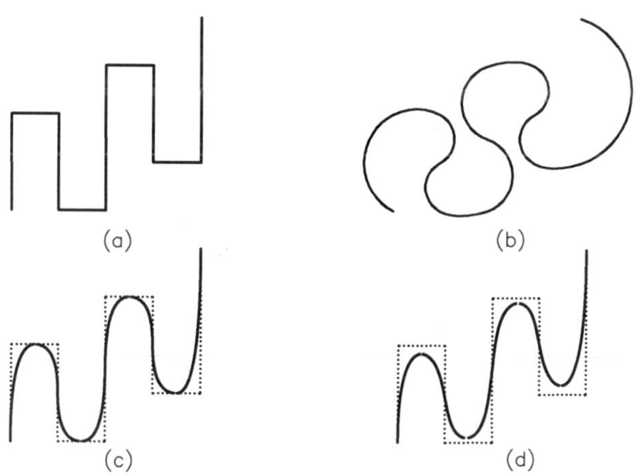

Figure 18-34 *Drawing for Exercise 7*

Exercise 8 *Mechanical*

Make the drawing shown in Figure 18-35. Use the splined polylines to draw the break lines. Dimension the drawing as shown.

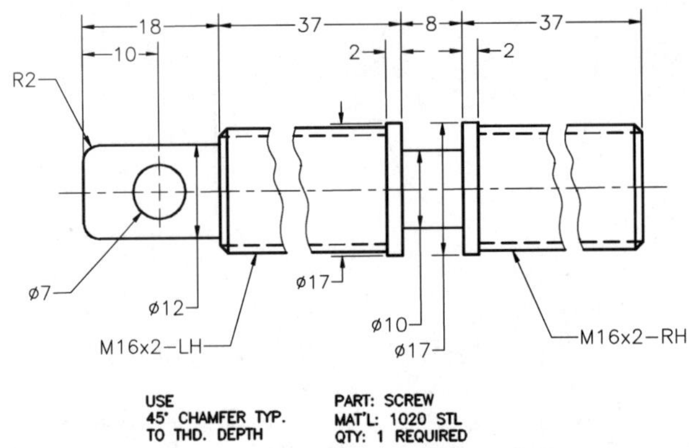

Figure 18-35 *Drawing for Exercise 8*

Answers to Self-Evaluation Test

1 - T, **2** - **F**, **3** - T, **4** - F, **5** - **Begin**, **End**, **6** - Twice, **7** - Width, **8** - **Multiple**, **9** - **Next**, **Previous**, **10** - **PEDIT**

Chapter 19

Data Exchange, and Object Linking and Embedding

Learning Objectives

After completing this chapter, you will be able to:

- *Understand the concept of Data Exchange in AutoCAD LT.*
- *Import and export .dxf files using the* ***SAVEAS*** *and* ***OPEN*** *commands.*
- *Create and use Windows Metafiles.*
- *Understand the concept of Raster Images.*
- *Manage the raster images.*
- *Edit the raster images.*
- *Understand the embedding and the linking functions of the* ***OLE*** *feature of Windows.*
- *Embed objects in AutoCAD LT.*

DATA EXCHANGE IN AUTOCAD LT

Different companies have developed different software for applications such as CAD, desktop publishing, and rendering. This nonstandardization of software has led to the development of various data exchange formats that enable transfer (translation) of data from one data processing software to another. This chapter will cover various data exchange formats provided in AutoCAD LT. AutoCAD LT uses the .DWG format to store drawing files. This format is not recognized by most other CAD software such as Intergraph, CADKEY, and MicroStation. To eliminate this problem so that files created in AutoCAD LT can be transferred to other CAD software for further use, AutoCAD LT provides various data exchange formats such as DXF (data interchange file) and DXB (binary drawing interchange).

DXF FILE FORMAT (DATA INTERCHANGE FILE)

The DXF file format generates a text file in ASCII code from the original drawing. This allows any computer system to manipulate (read/write) data in a DXF file. Usually, DXF format is used for CAD packages based on microcomputers. For example, packages like SmartCAM use DXF files. Some desktop publishing packages, such as Pagemaker and Ventura Publisher, also use DXF files.

Creating a Data Interchange File

The **SAVE** or the **SAVEAS** command is used to create an ASCII file with a .DXF extension from an AutoCAD LT drawing file. Once you invoke any of these commands, the **Save Drawing As** dialog box (Figure 19-1) is displayed. By default, the DXF file to be created assumes the name of the drawing file from which it will be created. However, you can specify a file name of your choice for the DXF file by typing the desired file name in the **File name** edit box. Select the extension as **DXF [*dxf]**. This can be observed in the **Files of type** drop-down list where you can select the output file format. You can also use the **-WBLOCK** command. The **Create Drawing** dialog box is displayed, where you can enter the name of the file in the **File name** edit box and select **DXF [*dxf]** from the **File of type** drop-down list.

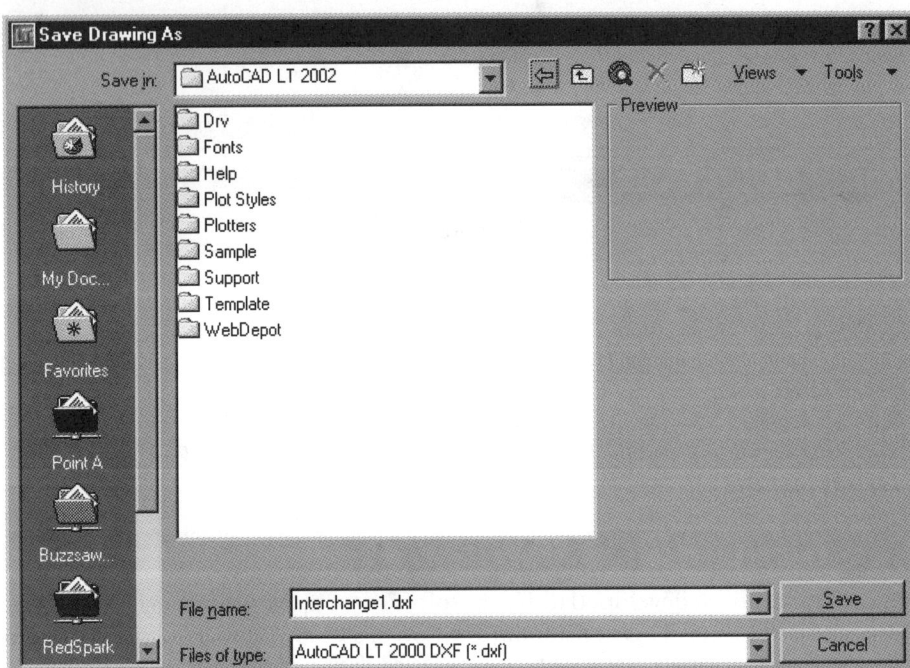

***Figure 19-1 Save Drawing As** dialog box*

Choose the **Tools** button and then choose **Options** to display the **Saveas Options** dialog box (Figure 19-2). In this dialog box, choose the **DXF Options** tab and enter the degree of accuracy for the numeric values. The default value for the degree of accuracy is sixteen decimal places. You can enter a value between 0 and 16 decimal places.

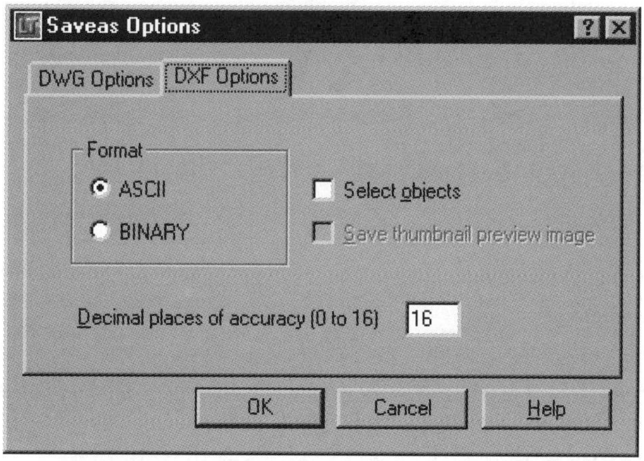

***Figure 19-2 Saveas Options** dialog box*

In this dialog box, you can also select the **Select Objects** check box, which allows you to specify objects you want to include in the DXF file. In this case, the definitions of named objects such as block definitions, text styles, and so on, are not exported. Selecting the **Save thumbnail preview image** check box, saves a preview image with the file that can be previewed in the **Preview** window of the **Select File** dialog box.

Select the **ASCII** radio button. Choose **OK** to return to the **Save Drawing As** dialog box. Choose the **Save** button here. Now an ASCII file with a .DXF extension has been created, and this file can be accessed by other CAD systems. This file contains data on the objects specified. By default, DXF files are created in ASCII format. However, you can also create binary format files by selecting the **Binary** radio button in the **Saveas Options** dialog box. Binary DXF files are more efficient and occupy only 75 percent of the ASCII DXF file. You can access a file in binary format more quickly than the same file in ASCII format.

Information in a DXF File

The DXF file contains data on the objects specified using the **Select objects** option in the **Save Drawing As** dialog box. You can change the data in this file to your requirement. To examine the data in this file, load the ASCII file in word processing software. A DXF file is composed of the following parts.

Header

In this part of the drawing database, all the variables in the drawing and their values are displayed.

Classes

This section deals with the internal database.

Tables

All the named objects such as linetypes, layers, blocks, text styles, dimension styles, and views, are listed in this part.

Blocks

The objects that define blocks and their respective values are displayed in this part.

Entities

All the entities in the drawing such as Circle, and so forth, are listed in this part.

Objects

Objects in the drawing are listed in this part.

Converting DXF Files into Drawing Files

You can import a DXF file into a new AutoCAD LT drawing file with the **OPEN** command. After you invoke the **OPEN** command, the **Select File** dialog box (Figure 19-3) is displayed. From the **Files of type** drop-down list, select **DXF [*.dxf]**. In the **File name** edit box enter the name of the file you want to import into AutoCAD LT or select the file from the list. Choose the **Open** button. Once this is done, save the drawing in the .DWG format. The specified DXF file is converted into a standard DWG file, regeneration is carried out, and the file is inserted into the new drawing. Now you can perform different operations on this file just as with other drawing files.

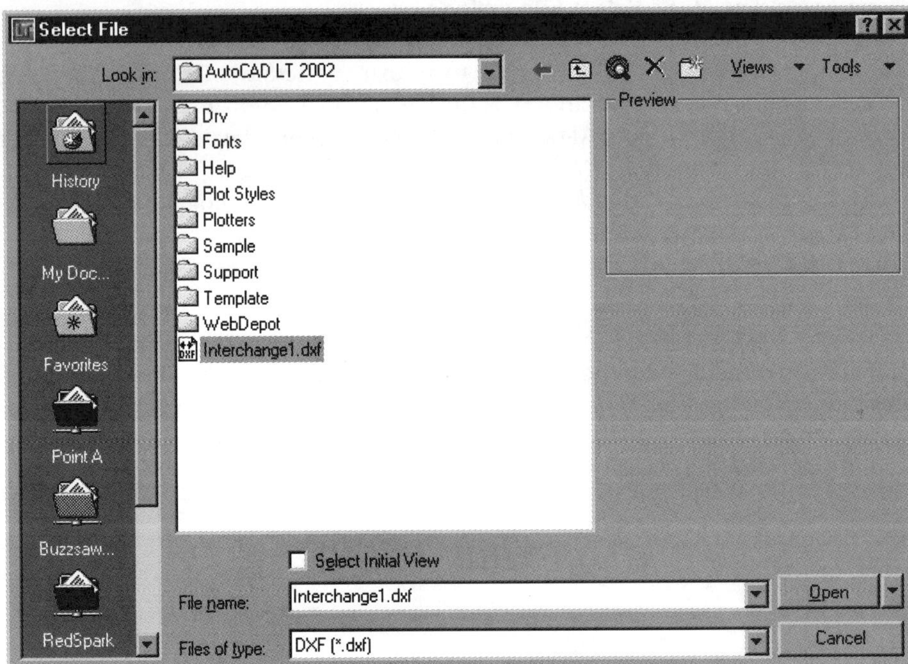

***Figure 19-3 Select File** dialog box*

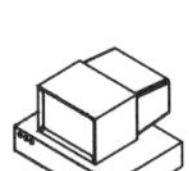

Tip

*A data interchange file (DXF) can also be inserted in the current drawing using the **INSERT** command.*

OTHER DATA EXCHANGE FORMATS

The other formats that can be used to exchange the data from one data processing format to the other are discussed next.

Creating and Using a Windows Metafile

Menu:	File > Export
Command:	EXPORT, WMFOUT

The Windows Metafile format (WMF) file contains screen vector and raster graphics format. In the **Export Data** dialog box or the **Create WMF file** dialog box, enter the file name. Select the objects you want to save in this file format. Select **Metafile [*.wmf]** from the **Files of type** drop-down list. The extension .wmf is appended to the file name.

The **WMFIN** command displays the **Import WMF** dialog box. Window metafiles are imported as blocks in AutoCAD LT. Select the wmf file you want to import and choose the **Open** button. Specify an insertion point, rotation angle, and scale factor. Specify scaling by entering a **scale factor** using the **corner** option to specify an imaginary box whose dimensions correspond to the scale factor, or by entering **xyz** to specify 3D scale factors. You can also invoke the **WMFIN** command by choosing **Windows Metafile** from the **Insert** menu.

Creating a BMP File

Menu:	File > Export
Command:	BMPOUT

This is used to create bitmap images of the objects in your drawing. Entering the **BMPOUT** command at the Command prompt displays the **Create BMP File** dialog box. Enter the file name and choose **Save**. Select the objects to save as bitmap. The file extension .bmp is appended to the file name.

DATA EXCHANGE THROUGH RASTER IMAGES

In the previous section of this chapter importing and exporting the files in the DXF formats was discussed. To mention the accuracy of the drawings, the DXF files include almost all the information about the original drawing file. The accuracy is maintained at the expense of the size of the DXF file and also the complexity of these files. However, there are many applications where the accuracy is not very important, such as desktop publishing. In such applications only the image presentation is of prime concern. A very simple and effective way of storing an image for import and export is in the form of **raster files**. In a raster file the image is stored in the dot patterns on the screen. This dot pattern is also known as **bit map**. AutoCAD LT allows you to add raster images to the vector base AutoCAD LT drawing, and view and plot the resulting file.

RASTER IMAGES

A raster image consists of small square-shaped dots known as pixels. In a colored image, the color is determined by the color of pixels. The raster images can be moved, copied, or clipped, and used as a cutting edge with the **TRIM** command. They can also be modified by using grips. You can also control the image contrast, transparency, and quality of the image. AutoCAD LT stores images in a special temporary image swap file whose default location is the Windows **Temp** directory. You can change the location of this file modifying it under **Temporary File Location** in the **Files** tab of the **Options** dialog box.

The images can be 8-bit gray, 8-bit color, 24-bit color, or bitonal. When image transparency is set to On, the image file formats with transparent pixels is recognized by AutoCAD LT and transparency is allowed. The transparent images can be in color or grayscale. AutoCAD LT supports the following file formats.

Image Type	File Extension	Description
BMP	.bmp, .dib, .rle	Windows and OS/2 Bitmap Format
CALS-I	.gp4, .mil, .rst	Mil-R-Raster I
FLIC	.flc, .fli	Flic Autodesk Animator Animation
GEOSPOT	.bil	GeoSPOT (BIL files must be accompanied with HDR and PAL files with connection data in the same directory.)
IG4	.ig4	Image Systems group 4
IGS	.igs	Image Systems Grayscale
JFIF or JPEG	.jpg, jpeg	Joint Photographics Expert group
PCX	.pcx	Picture PC Paintbrush Picture
PICT	.pct	Picture Macintosh Picture
PNG	.png	Portable Network Graphic
RLC	.rlc	Run-length Compressed
TARGA	.tga	True Vision Raster based Data format
TIFF/LZW .tif	Taffed image file format	

When you store images as Tiled Images, that is, in the Tagged Image File Format [TIFF], you can edit or modify any portion of the image; only the modified portion is regenerated thus saving time. Tiled images load much faster compared to nontiled images.

Managing Raster Images (Image Manager Dialog Box)

Toolbar:	Reference > Image
Command:	IMAGE

When you choose this button, AutoCAD LT displays the **Image Manager** dialog box (Figure 19-4). You can also invoke this command by right-clicking on the image and choosing **Image > Image Manager** from the shortcut menu. You can view image information either as a list or as a tree view by choosing the respective buttons located at the upper left corner of the **Image Manager** dialog box. The **List View** displays the names of all the images in the drawing, its loading status, size, date last modified on, and its search path. The **Tree View** displays the images in a hierarchy which shows its nesting levels within blocks and xrefs. The **Tree View** does not display the status, size, or any other information about the image file. You can rename an image file in this dialog box. The different options in the **Image Manager** dialog box are as follows.

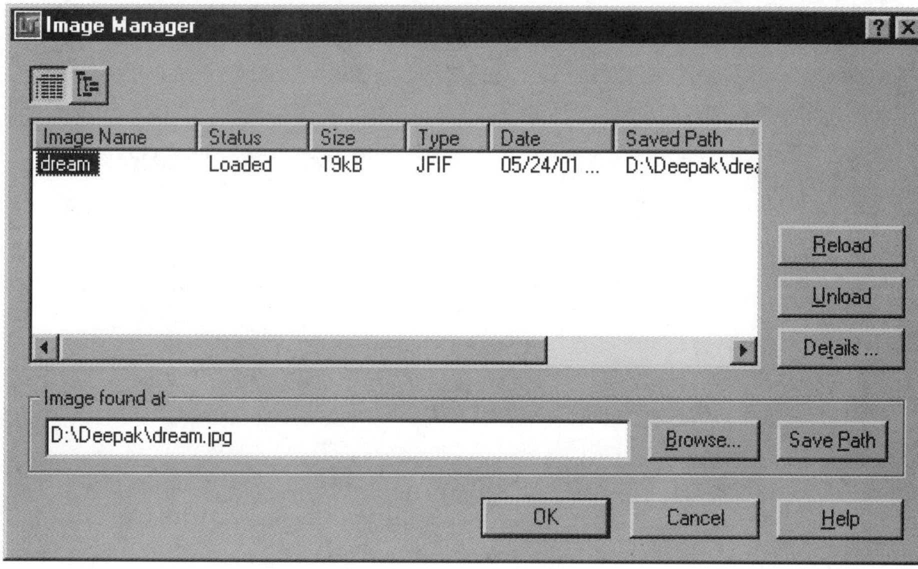

***Figure 19-4 Image Manager** dialog box*

Reload

This button simply reloads an image. The changes made to the image since the last insert will be loaded on the screen. You can change the status of the image file in the **Image Manager** dialog box by double-clicking on the current status. It changes from **Unload** to **Reload** and vice versa.

Unload

The Unload option unloads an image. An image is unloaded when it is not needed in the current drawing session. When you unload an image, AutoCAD LT retains information about the location and size of the image in the database of the current file. Unloading a file does not unlink the file from the drawing and the image frame is displayed. If you reload the image, the image will appear at the same point and in the same size as the image was before unloading. Unloading the raster images enhances AutoCAD LT performance. Also, the unloaded images are not plotted. If multiple images are to be loaded and the memory is insufficient, AutoCAD LT automatically unloads them.

Details

Details displays the **Image File Details** dialog box as shown in Figure 19-5. This dialog box lists information about the image such as file name, saved path, file creation date, file size, file type, color and color path, pixel width and height, resolution, and default size. It also displays the image in the preview box.

Image found at Area

The edit box in this area displays the path of the selected image. You can edit this path and choose the **Save path** button to save the new path. If you have changed the path of an image, choose the **Browse** button to display the **Select Image File** dialog box. In this dialog box,

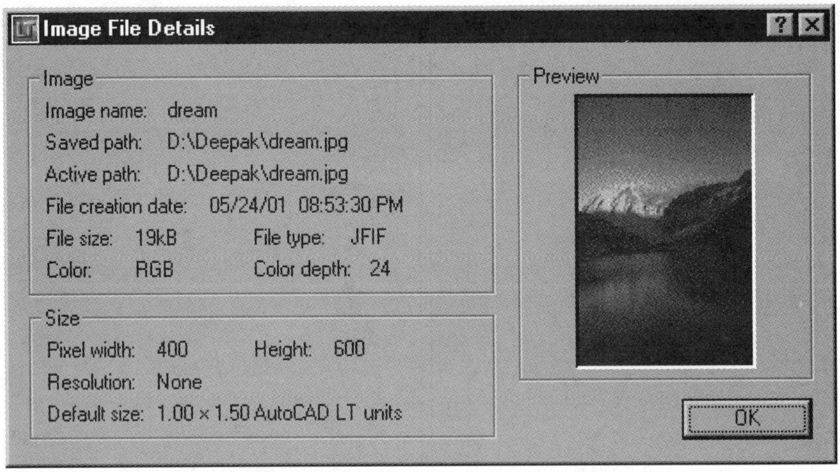

***Figure 19-5 Image File Details** dialog box*

locate the image file and then choose the **Open** button. The new path is displayed in this edit box. Choose the **Save Path** button to save this new path. This path is also displayed in the **Saved path** column of the dialog box.

Browse
Displays the **Select Image File** dialog box to select the image file.

Save Path
Saves the current path of the image.

Tip
*You can also use the command line for managing the image files using the **-IMAGE** command. All the options of the **Image Manager** dialog box will be available in the command line.*

EDITING RASTER IMAGE FILES
Controlling the Display of Image Frames

Toolbar:	Reference > Image Frame
Menu:	Modify > Object > Image Frame
Command:	IMAGEFRAME

The **IMAGEFRAME** command is used to turn the image boundary on or off. If the image boundary is off, the image cannot be selected with the pointing device and therefore, cannot be accidentally moved or modified.

Changing the Display Order

Toolbar:	Modify II > Draworder
Menu:	Tools > Display Order
Command:	DRAWORDER

This command is used to change the display order of images and other objects. When you invoke this command, you will be prompted to select the object. The prompt sequence that will follow when you select the object is as follows.

Enter object ordering option [Above object/Under object/Front/Back] <Back>:

If you specify **Above object**, the selected object will be moved above the referenced object. If you select the **Under object**, the selected object will be moved under the referenced object. If you select **Front**, the selected object will be moved to the top of the draworder. If you select **Back**, the selected object will be moved to the bottom of the draworder. Figure 19-6 shows two images with original order of display. Now, if you want to bring the image that is placed below, change the display order using the **DRAWORDER** command. See Figure 19-7.

Figure 19-6 *Original display order of the images*

Figure 19-7 *Display order modified using the* ***DRAWORDER*** *command*

Other Editing Commands

You can use other editing commands such as **COPY**, **MOVE**, **ROTATE**, and **STRETCH** to edit the raster image. You can also use the image as the trimming edge for trimming objects. However, you cannot trim an image. You can insert the raster image several times or make multiple copies of it. You can also edit the image using grips. You can use the **Properties** window to change the image layer, and color.

Scaling Raster Images

The scale of the inserted image is determined by the actual size of the image and the unit of measurement (inches, feet, and so on). For example, if the image is 1" x 1.26" and you insert this image with a scale factor of 1, the size of the image on the screen will be 1 AutoCAD LT unit by 1.26 AutoCAD LT units. If the scale factor is 5, the image will be five times larger. The image that you want to insert must contain the resolution information (DPI). If the image does not contain this information, AutoCAD LT treats the width of the image as one unit.

POSTSCRIPT FILES

PostScript is a page description language developed by Adobe Systems. It is used mostly in

DTP (desktop publishing) applications. AutoCAD LT allows you to work with PostScript files. You can create and export PostScript files from AutoCAD LT so that these files can be used for DTP applications. PostScript images have higher resolution than raster images. The extension for these files is .EPS (Encapsulated PostScript).

Creating the PostScript Files

Command: PSOUT

As just mentioned, any AutoCAD LT drawing file can be converted into a PostScript file. This can be accomplished with the **PSOUT** command. When you invoke this command, the **Create PostScript File** dialog box is displayed (Figure 19-8). You can also use the **EXPORT** command to display the **Export Data** dialog box. The options are similar in both the dialog boxes.

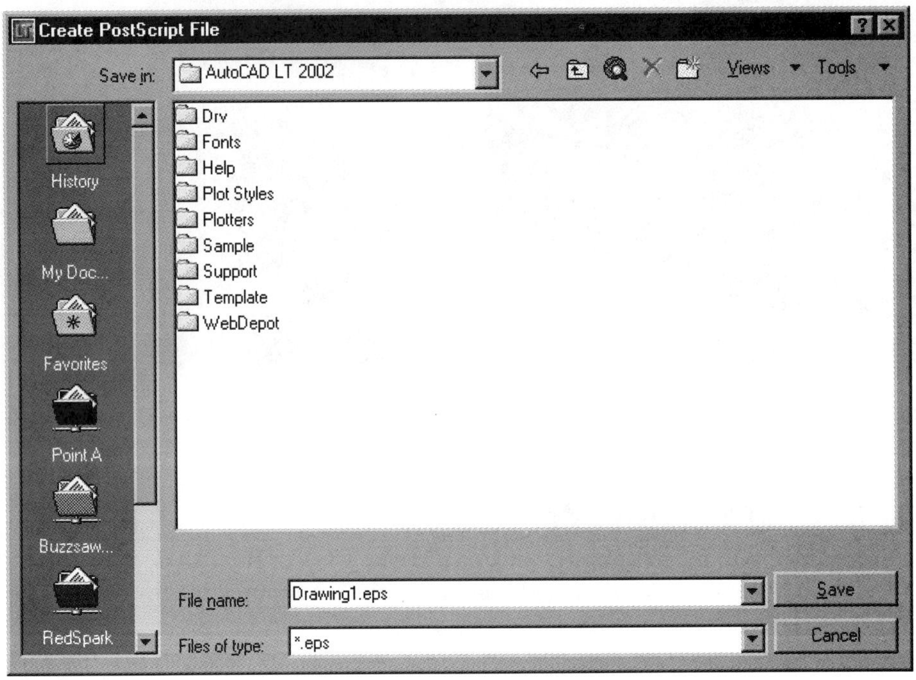

***Figure 19-8 Create PostScript File** dialog box*

In the **File name** edit box, enter the name of the PostScript (EPS) file you want to create. Choose the **Save** button to accept the default setting and create the PostScript file. You can also choose the **Tools > Options** in this dialog box to change the settings through the **PostScript Out Options** dialog box (Figure 19-9) and then save the file. The **PostScript Out Options** dialog box provides the following options.

Prolog Section Name

In this edit box, you can assign a name for a prolog section to be read from the aclt.psf file.

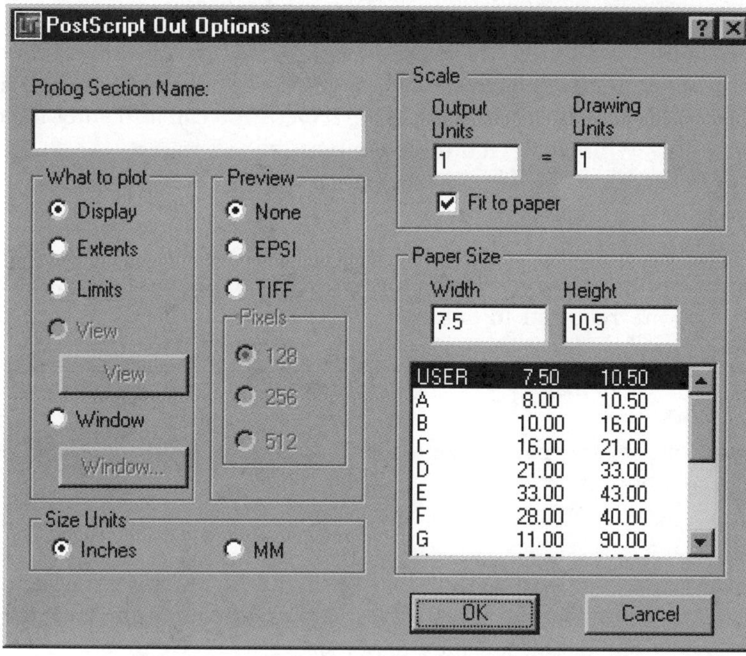

***Figure 19-9 PostScript Out Options** dialog box*

What to plot Area

The **What to plot** area of the dialog box has the following options.

Display. If you specify this option when you are in model space, the display in the current viewport is saved in the specified EPS file. Similarly, if you are in layouts, the current view is saved in the specified EPS file.

Extents. If you use this option, the PostScript file created will contain the section of the AutoCAD LT drawing that currently holds objects. In this way, this option resembles the **ZOOM Extents** option. If you add objects to the drawing, they are also included in the PostScript file to be created because the extents of the drawing are also altered. If you reduce the drawing extents by erasing, moving, or scaling objects, then you must use the **ZOOM Extents** or **ZOOM All** option. Only then does the **Extents** option of the **PSOUT** command understand the extents of the drawing to be exported. If you are in model space, the PostScript file is created in relation to the model space extents; if you are in paper space, the PostScript file is created in relation to the paper space extents.

Limits. With this option, you can export the whole area specified by the drawing limits. If the current view is not the plan view [viewpoint (0,0,1)], the **Limits** option exports the area just as the **Extents** option would.

View. Any view created with the **VIEW** command can be exported with this option. This option will be available only when you have created a view using the **VIEW** command. Choose the **View** button to display the **View Name** dialog box from where you can select the view.

Window. In this option, you need to specify the area to be exported with the help of a window. When this radio button is selected, the **Window** button is also available. Choose the **Window** button to display the **Window Selection** dialog box where you can select the **Pick** button and then specify the two corners of the window on the screen. You can also enter the coordinates of the two corners in the **Window Selection** dialog box.

Preview Area

The **Preview** area of the dialog box has two types of formats for preview images: **EPSI** and **TIFF**. If you want a preview image with no format, select the **None** radio button. If you select **TIFF** or **EPSI**, you are required to enter the pixel resolution of the screen preview in the **Pixels** area. You can select a preview image size of 128 x 128, 256 x 256, or 512 x 512.

Size Units Area

In this area, you can set the paper size units to **Inches** or **Millimeters** by selecting their corresponding radio buttons.

Scale Area

In this area, you can set an explicit scale by specifying how many drawing units are to be output per unit. You can select the **Fit to paper** check box so that the view to be exported is made as large as possible for the specified paper size.

Paper Size Area

You can select a size from the list or enter a new size in the **Width** and **Height** edit boxes to specify a paper size for the exported PostScript image.

OBJECT LINKING AND EMBEDDING

With Windows, it is possible to work with different Windows-based applications by transferring information between them. You can edit and modify the information in the original Windows application, and then update this information in other applications. This is made possible by creating links between the different applications and then updating those links, which in turn updates or modifies the information in the corresponding applications. This linking is a function of the OLE feature of Microsoft Windows. The OLE feature can also join together separate pieces of information from different applications into a single document. AutoCAD LT and other Windows-based applications such as Microsoft Word, Notepad, and Windows WordPad support the Windows OLE feature.

For the OLE feature, you should have a source document where the actual object is created in the form of a drawing or a document. This document is created in an application called a **server** application. AutoCAD LT for Windows and Paintbrush can be used as server applications. Now this source document is to be linked to (or embedded in) the **compound** (destination) document, which is created in a different application, known as the **container** application. AutoCAD LT for Windows, Microsoft Word, and Windows WordPad can be used as container applications.

Clipboard

The transfer of a drawing from one Windows application to another is performed by copying the drawing or the document from the server application to the Clipboard. The drawing or document is then pasted in the container application from the Clipboard. Therefore, a Clipboard is used as a medium for storing the documents while transferring them from one Windows application to another. The drawing or the document on the Clipboard stays there until you copy a new drawing, which overwrites the previous one, or until you exit Windows. You can save the information present on the Clipboard with the .CLP extension.

Object Embedding

You can use the embedding function of the OLE feature when you want to ensure that there is no effect on the source document even if the destination document has been changed through the server application. Once a document is embedded, it has no connection with the source. Although editing is always done in the server application, the source document remains unchanged. Embedding can be accomplished by means of the following steps. In this example, AutoCAD LT for Windows is the server application and MS Word is the container application.

1. Create a drawing in the server application (AutoCAD LT).

2. Open MS Word (container application) by double-clicking on its shortcut icon at the desktop of the computer.

3. It is preferable to arrange both the container and the server windows so that both are visible as shown in Figure 19-10.

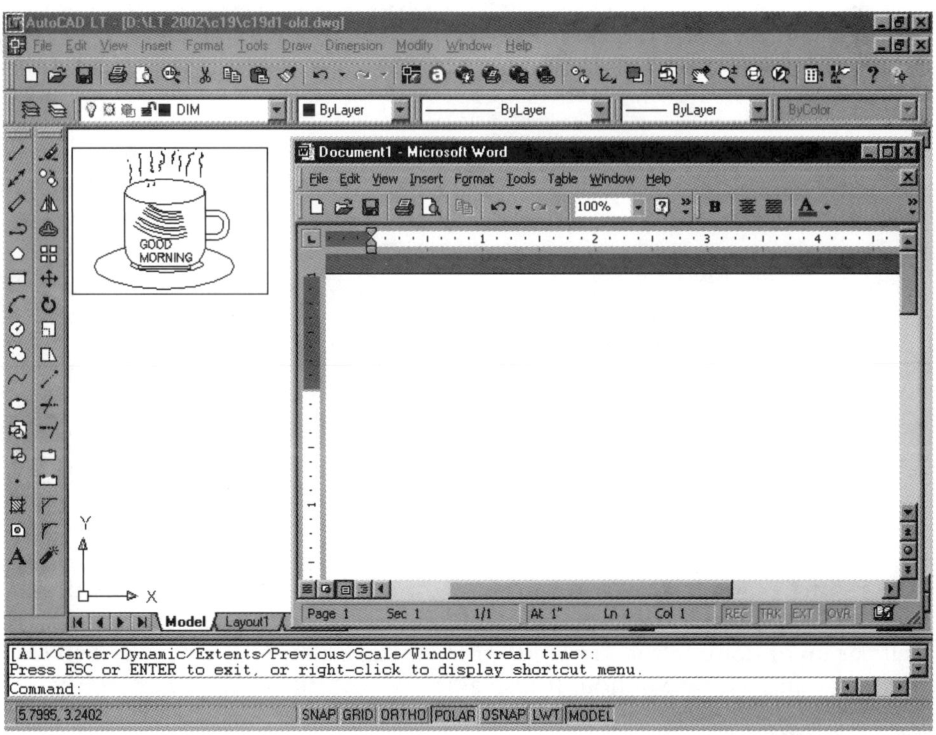

Figure 19-10 *AutoCAD LT graphics screen with the MS Word window*

4. In the AutoCAD LT graphics screen, choose the **Copy to Clipboard (Ctrl+C)** button from the **Standard Toolbar** to invoke the **COPYCLIP** command. This command can be used in AutoCAD LT for embedding the drawings. This command can also be invoked from the **Edit** menu (Choose **Copy**), or by entering **COPYCLIP** at the command line. The next prompt, **Select objects**, allows you to select the entities you want to transfer. You can either select the full drawing by entering **ALL** or select some of the entities by selecting them. You can use any of the object selection methods for selecting the objects. With this command the selected objects are automatically copied to the Windows Clipboard.

5. After the objects are copied to the Clipboard, make the MS Word window active. To get the drawing from the Clipboard to the MS Word application (client), select the **Paste** in the Word application. Choose **Paste** from the **Edit** menu in MS Word (Figure 19-11). You can also use **Paste Special** from the **Edit** menu, which will display the **Paste Special** dialog box (Figure 19-12). In this dialog box, select the **Paste** radio button (default) for embedding, and then choose **OK**. The drawing is now embedded in the MS Word window.

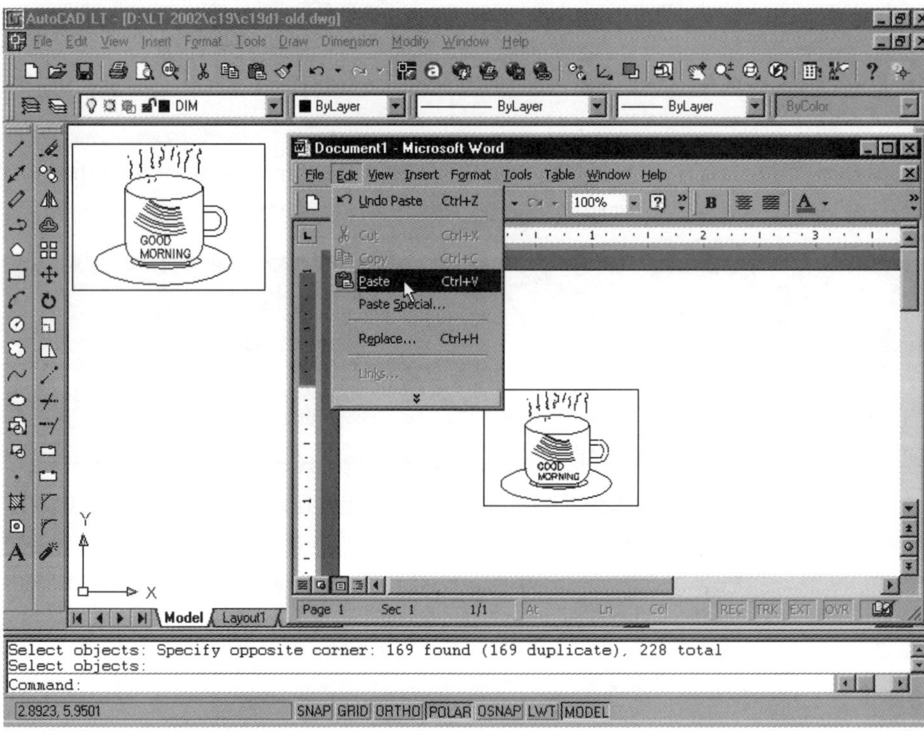

Figure 19-11 *Pasting a drawing to the MS Word application*

6. Your drawing is now displayed in the MS Word window, but it may not be displayed at the proper position. You can get the drawing in the current viewport by moving the scroll bar up or down in the MS Word window. You can also save your embedded drawing by choosing the **Save** button. It displays a **Save As** dialog box where you can enter a file name. You can now exit AutoCAD LT.

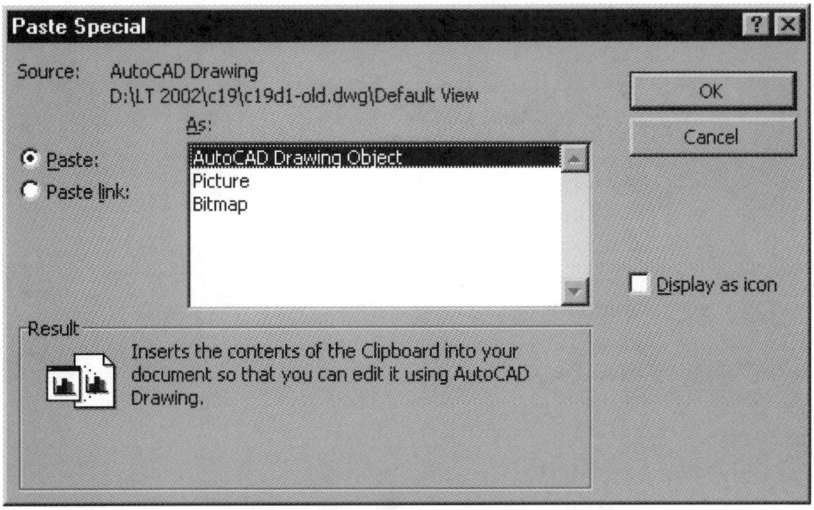

***Figure 19-12 Paste Special** dialog box*

7. You can now edit your embedded drawing. Editing is performed in the server application, which in this case is AutoCAD LT for Windows. You can get the embedded drawing into the server application (AutoCAD LT) directly from the container application (MS Word) by double-clicking on the drawing in MS Word. The other method is by choosing **Edit > Drawing Object** in the **Edit** menu after selecting the drawing. (This menu item has replaced **Object**, which was present before pasting the drawing.)

8. Now you are in AutoCAD LT, with your embedded drawing displayed on the screen, but as a temporary file with a file name such as [Drawing in Document]. Here you can edit the drawing by changing the color and linetype or by adding and deleting text, entities, and so on. In Figure 19-13, the cup and plate have been hatched.

9. After you have finished modifying your drawing, choose **Update Microsoft Word** from the **File** menu in the server (AutoCAD LT). This menu item has replaced the previous **Save** menu item. When you choose **Update Microsoft Word**, AutoCAD LT automatically updates the drawing in MS Word (container application). Now you can exit this temporary file in AutoCAD LT.

Note

Do not zoom or pan the drawing in the temporary file. If you do so, this will be included in the updating and the new file will display only that portion of the drawing that lies inside the original area.

10. This completes the embedding function so you can exit the container application. While exiting, a dialog box that asks whether or not to save changes in MS Word is displayed.

Linking Objects

The linking function of OLE is similar to the embedding function. The only difference is that here a link is created between the source document and the destination document. If you

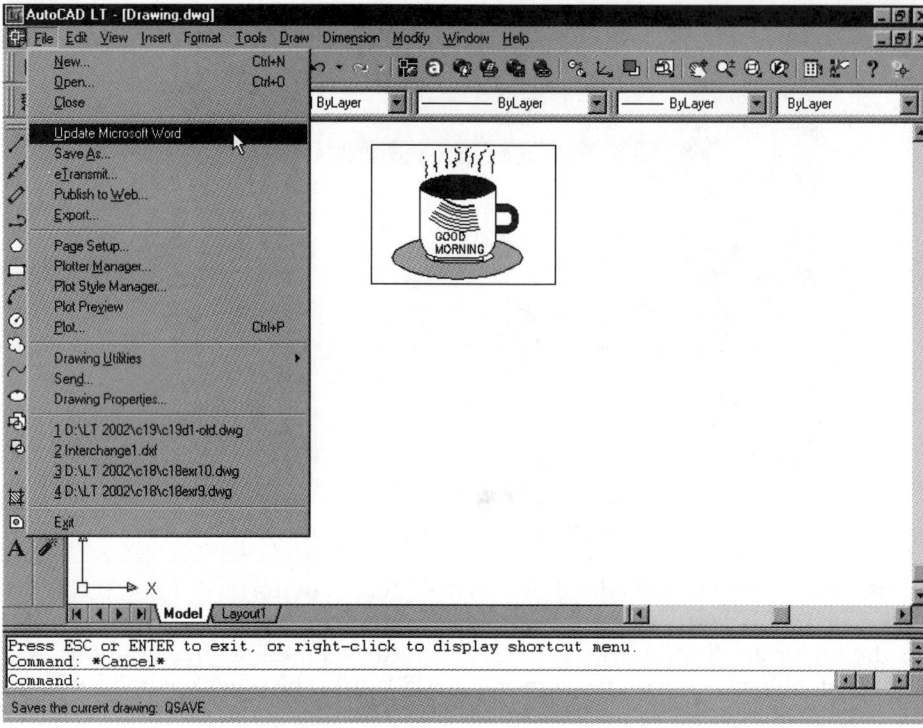

***Figure 19-13** Updating the drawing in AutoCAD LT*

edit the source, you can simply update the link, which automatically updates the client. This allows you to place the same document in a number of applications, and if you make a change in the source document, the clients will also change by simply updating the corresponding links. Consider AutoCAD LT for Windows to be the server application and MS Word to be the container application. Linking can be performed by means of the following.

1. Open a drawing in the server application (AutoCAD LT). If you have created a new drawing, then you must save the drawing before you can link it with the container application.

2. Open MS Word (container application) by double-clicking on its shortcut icon at the desktop of the computer.

3. It is preferable to arrange both the container and the server windows so that both are visible.

4. In the AutoCAD LT graphics screen, choose **Copy Link** from the **Edit** menu to invoke the **COPYLINK** command. This command can be used in AutoCAD LT for linking the drawing. This command copies all the objects that are displayed in the current viewport directly to the Clipboard. Here, you cannot select the objects for linking. If you want only a portion of the drawing to be linked, you can zoom into that view so that it is displayed in the current viewport prior to invoking the **COPYLINK** command. This command also creates a new view of the drawing having a name OLE1.

5. Make the MS Word window active. To get the drawing from the Clipboard to the Write (container) application, choose **Paste Special** from the **Edit** menu. This will display the **Paste Special** dialog box. In this dialog box, select the **Paste Link** radio button for linking. Choose **OK**. Note that the **Paste Link** radio button in the **Paste Special** dialog box will not be available if you have not saved the drawing.

6. The drawing is now displayed in the MS Word window and is linked to the original drawing. You can also save your linked drawing by choosing **Save** from the **File** menu. It displays a **Save As** dialog box where you can enter a file name.

7. You can now edit your original drawing. Editing can be performed in the server application, which in this case is AutoCAD LT for Windows. You can edit the drawing by changing the color and linetype or by adding and deleting text, entities, and so on. Then save your drawing in AutoCAD LT by using the **QSAVE** command. You can now exit AutoCAD LT.

8. You will notice that the drawing is automatically updated, and the changes made in the source drawing are present in the destination drawing also. This automatic updating is dependent on the selection of the **Automatic** radio button (default) in the **Links** dialog box (Figure 19-14). The **Links** dialog box can be invoked by choosing **Links** from the **Edit** menu. For updating manually, you can select the **Manual** radio button in the dialog box. In the manual case, after making changes in the source document and saving it, you need to invoke the **Links** dialog box and then choose the **Update Now** button. This will update the drawing in the container application and display the updated drawing in the MS Word document.

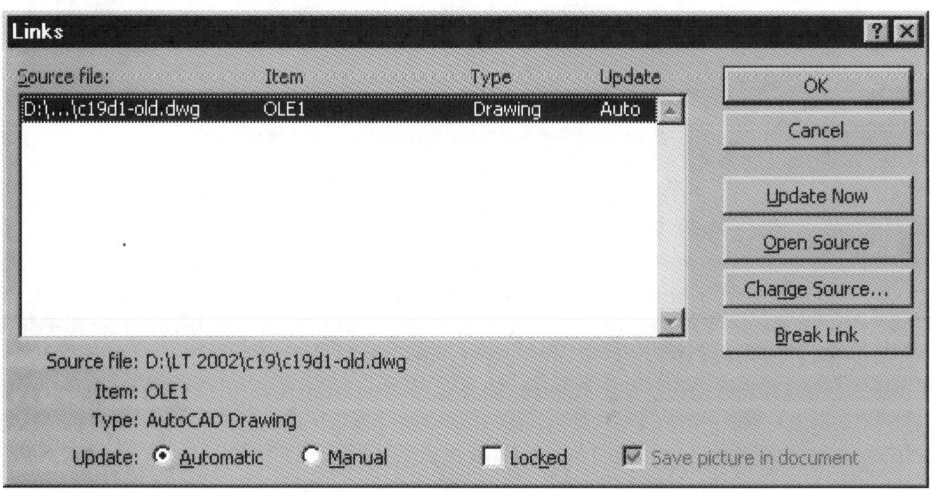

***Figure 19-14 Links** dialog box*

9. Exit the container application after saving the updated file.

Linking Information into AutoCAD LT

Similarly, you can also embed and link information from a server application into an

AutoCAD LT drawing. You can also drag selected OLE objects from another application into AutoCAD LT, provided this application supports Microsoft Activex and the application is running and visible on the screen. Dragging and dropping is like cutting and pasting. If you press the CTRL key while you drag the object, it is copied to AutoCAD LT. Dragging and dropping an OLE object into AutoCAD LT embeds it into AutoCAD LT.

Linking Objects into AutoCAD LT

Start any server application like the MS Word and open a document in it. Select the information you wish to use in AutoCAD LT with your pointing device and choose the **Copy** button to copy this data to the Clipboard. Open the AutoCAD LT drawing you wish to link this data to. Choose **Paste Special** from the **Edit** menu or use the **PASTESPEC** command. The **Paste Special** dialog box is displayed (Figure 19-15).

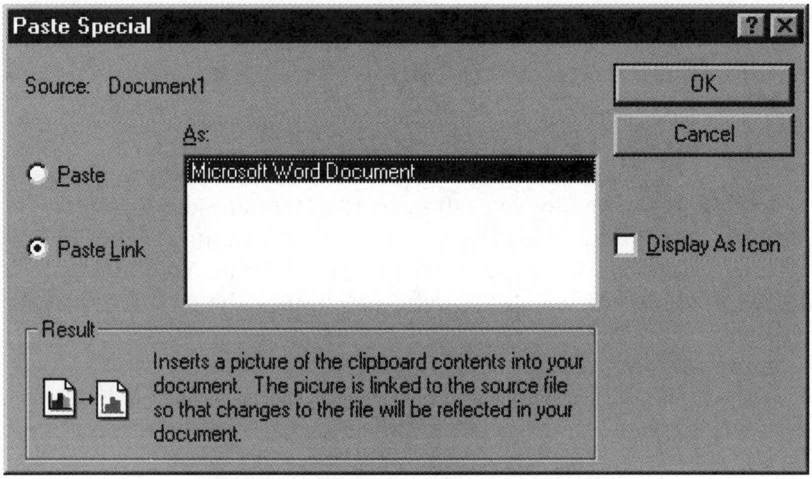

***Figure 19-15 Paste Special** dialog box*

In the **As** list box, select the data format you wish to use. For example, for a MS Word document, select **Microsoft Word Document**. Picture format uses a Metafile format. Select the **Paste Link** radio button to paste the contents of the Clipboard to the current drawing. If you select the **Paste** radio button, the data is embedded and not linked. Choose **OK** to exit the dialog box. The data is displayed in the drawing and can be positioned as needed. You can also use the **INSERTOBJ** command by entering **INSERTOBJ** at the Command prompt, by choosing **Ole Object** from the **Insert** menu, or by choosing the **OLE Object** button in the **Insert** toolbar. This command links an entire file to a drawing from within AutoCAD LT. Using this command displays the **Insert Object** dialog box (Figure 19-16).

Select the **Create from File** radio button. Also select the **Link** check box. Choosing the **Browse** button, displays the **Browse** dialog box. Select a file you want to link from the list box or enter a name in the **File name** edit box and choose the **Insert** button. The path of the file is displayed in the **File** edit box of the **Insert Object** dialog box. If you select the **Display as icon** check box, an icon is also displayed in the dialog box. Choose **OK** to exit the dialog box, the selected file is linked to the AutoCAD LT drawing.

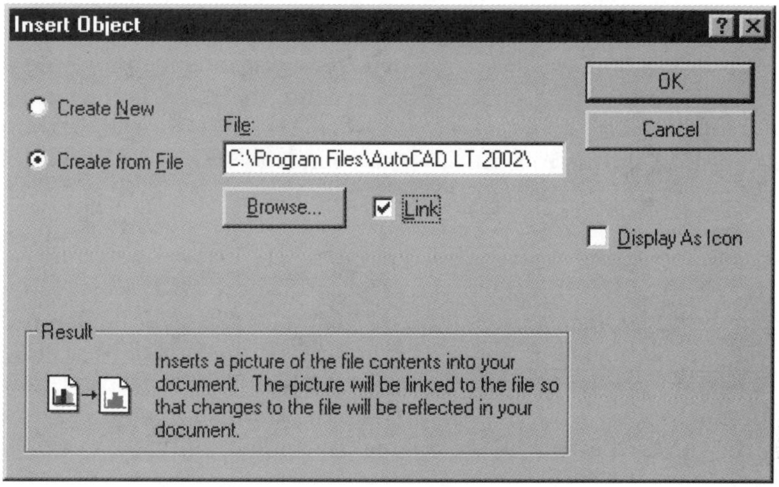

***Figure 19-16 Insert Object** dialog box*

AutoCAD LT updates the links automatically by default, whenever the server document changes, but you can use the **OLELINKS** command to display the **Links** dialog box (Figure 19-17) where you can change these settings. This dialog box can also be displayed by choosing **OLE Links** from the **Edit** menu. In the **Links** dialog box, select the link you want to update and then choose the **Update Now** button. Then choose the **Close** button. If the server file location changes or if it is renamed, you can choose the **Change Source** button in the **Links** dialog box to display the **Change Source** dialog box. In this dialog box, locate the server file name and location and choose the **Open** button. You can also choose the **Break Link** button in the **Links** dialog box to disconnect the inserted information from the server application. This is done when the linking between the two is no longer required.

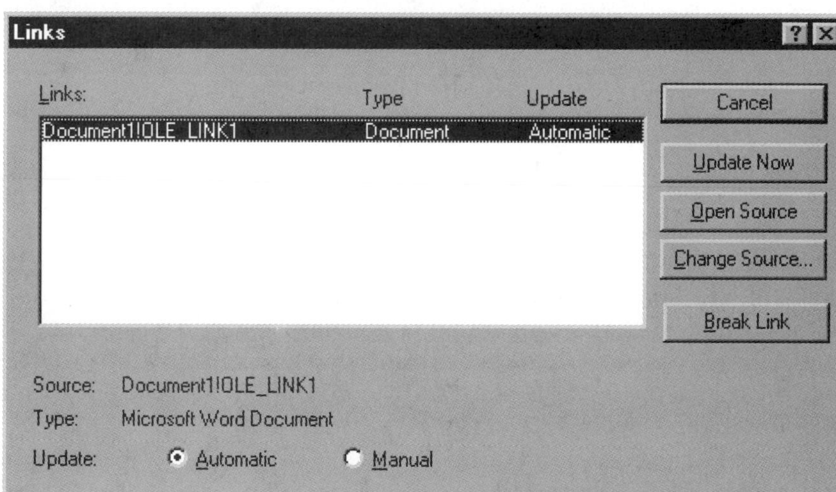

***Figure 19-17 Links** dialog box*

Embedding Objects into AutoCAD LT

Open the server application and select the data you want to embed into the AutoCAD LT drawing. Copy this data to the Clipboard by choosing the **Copy** button in the toolbar or choosing **Copy** from the **Edit** menu. Open the AutoCAD LT drawing and choose **Paste** from the AutoCAD LT **Edit** menu. You can use the command **PASTECLIP** also. The selected information is embedded into the AutoCAD LT drawing.

You can also create and embed an object into an AutoCAD LT drawing starting from AutoCAD LT itself using the **INSERTOBJ** command. You can choose **OLE Object** from the **Insert** menu or choose the **OLE Object** button from the **Insert** toolbar, and the **Insert Object** dialog box is displayed (Figure 19-18). In this dialog box, select the **Create New** radio button and select the application you wish to use from the **Object Type** list box. Choose **OK**. The selected application opens. Now you can create the information you wish to insert in the AutoCAD LT drawing and save it before closing the application.

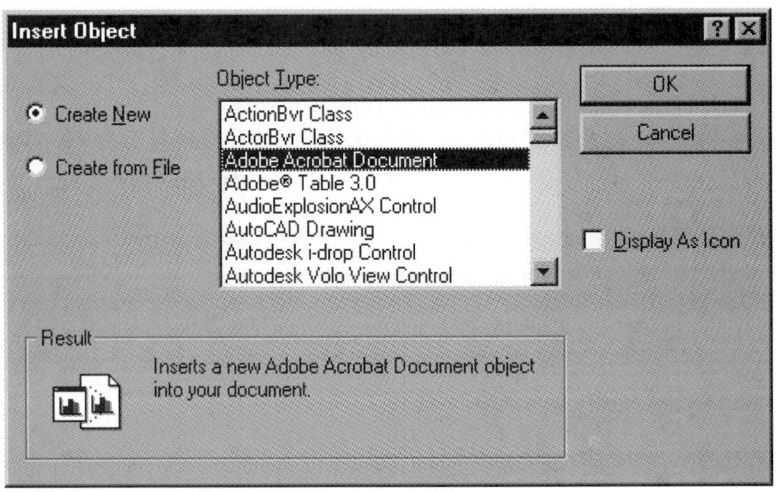

Figure 19-18 ***Insert Object*** *dialog box*

The **OLE Properties** dialog box is displayed (Figure 19-19) where you can resize and rescale the inserted objects. This dialog is displayed by default. If you do not want to display this dialog box, clear the **Display dialog when pasting OLE objects** check box in the **OLE Properties** dialog box. You can also clear the **Display OLE properties dialog** check box in the **Systems** tab of the **Options** dialog box. You can edit information embedded in the AutoCAD LT drawing by opening the server application by double-clicking on the inserted OLE object. You can also select the object and right-click to display a shortcut menu. Choose **Object** > **Edit**. After editing the data in the server application, choose **Update** from the **File** menu to reflect the modifications in the AutoCAD LT drawing.

Working with OLE Objects

Select an OLE Object and right-click to display a shortcut menu; choose **Properties**. The **OLE Properties** dialog box is displayed. You can also select the OLE Object and use the

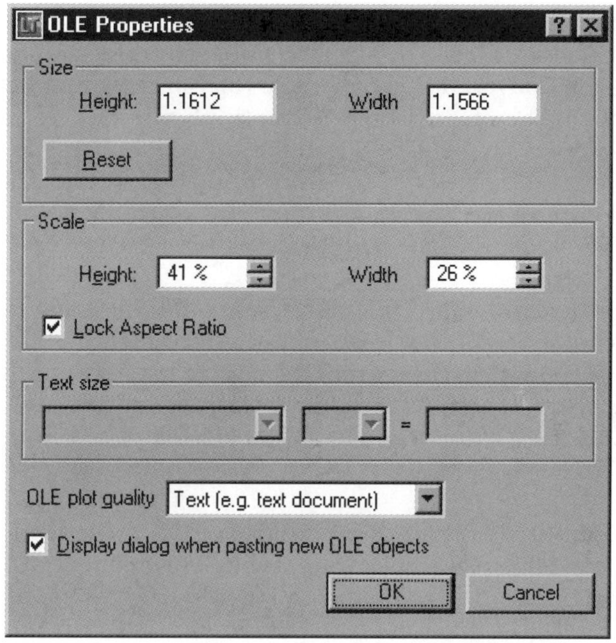

Figure 19-19 ***OLE Properties*** *dialog box*

OLESCALE* command. Specify a new height and a new width in the **Height** and **Width** edit boxes in the **Size** area or Under **Scale**, enter a value in percentage of the current values in the **Height** and **Width** edit boxes. Here, if you select the **Lock aspect Ratio** check box, whenever you change either the height or the width under **Scale** or **Size**, the respective width or height changes automatically to maintain the aspect ratio. If you want to change only the height or only the width, clear this check box. Choose **OK** to apply changes.

Choosing the **Reset** button restores the selected OLE objects to their original size, that is, the size they were when inserted. If the AutoCAD LT drawing contains an OLE object with text with different fonts and you wish to select and modify specific text, you can select a particular font and point size from the drop-down lists under the **Text size** area and enter the value in drawing units in the box after the equal (=) sign. For example if you wish to select text in the Times Roman font, of point size 10 and modify it to size 0.5 drawing units, select Times Roman and 10 point size from the drop-down lists and in the text box after the equal (=) sign enter .5. The text that was in Times Roman and was of point size 10 will change to 0.5 drawing units in height.

The pointing device can also be used to modify and scale an OLE Object in the AutoCAD LT drawing. Selecting the object displays the object frame and the move cursor. The move cursor allows you to select and drag the object to a new location. The middle handles allow you to select the frame and stretch it. They do not scale objects proportionately. The corner handles scale the object proportionately.

Select an OLE object and right-click to display the shortcut menu. Choosing **Cut** removes the object from the drawing and pastes it on the Clipboard; **Copy** places a copy of the selected object on the clipboard; and **Clear** removes the object from the drawing and does not place it on the clipboard. Choosing **Object** displays the **Convert**, **Open**, and **Edit** options. Choosing **Convert** displays the **Convert** dialog box where you can convert objects from one type to another and **Edit** opens the object in the Server application where you can edit it and update it in the current drawing. **Undo** cancels the last action. **Bring to Front** and **Send to Back** options place the OLE objects in the front of or back of the AutoCAD LT objects. The **Selectable** option turns the selection of the OLE object on or off. If the **Selectable** option is on, the object frame is visible and the object is selected.

If you want to change the layer of an OLE object, select the object and right-click to display the shortcut menu; choose **Cut**. The selected object is placed on the clipboard. Change the current layer to the one you want to change the OLE object's layer to using the **Layer Properties Manager** dialog box. Now, choose **Paste** from the **Edit** menu to paste the contents of the clipboard in the AutoCAD LT drawing. The OLE object is pasted in the New layer, in its original size.

The **OLEHIDE** system variable controls the display of OLE objects in AutoCAD LT. The default value is 0, that is, all the OLE objects are visible. The different values and their effects are as follows.

0 All OLE objects are visible
1 OLE objects are visible in paper space only
2 OLE objects are visible in model space only
3 No OLE objects are visible

The **OLEHIDE** system variable affects both screen display and printing.

Self-Evaluation Test

Answer the following questions, and then compare your answers to the answers given at the end of this chapter.

1. The DXF file format generates a text file in ASCII code. (T/F)

2. You can directly open a DXF file in AutoCAD LT using the **OPEN** command. (T/F)

3. An image attached to AutoCAD LT can not be detached. (T/F)

4. The file in the container application is automatically modified only when a link is maintained between the server application and container application. (T/F)

5. The __________ command can be used in AutoCAD LT for embedding the drawing.

6. You can edit the embedded drawing in the _________ application.

7. You can modify the size and text font of the linked objects using the _________ dialog box.

8. The _________ is used as a medium for storing the documents while transferring them from one Windows application to another.

9. You can get an embedded drawing into the server application directly from the container application by _________ on the drawing.

10. The **COPYLINK** command copies a drawing in the _________ to the Clipboard.

Review Questions

Answer the following questions.

1. The **Image Manager** dialog box can also be used to attach the image in AutoCAD LT. (T/F)

2. An AutoCAD LT drawing can be saved in DXF format using the **Save Drawing As** dialog box. (T/F)

3. You can export an AutoCAD LT drawing as a 3D Studio MAX. (T/F)

4. You can modify the brightness of the images using the **Image Manager** dialog box. (T/F)

5. Which command can be used to open a DXF format file?

   (a) **OPEN** (b) **NEW**
   (c) **START** (d) **None**

6. Which command is used to unload an attached image file?

   (a) **IMAGE** (b) **DETACH**
   (c) **IMAGEDETACH** (d) **None**

7. Which dialog box can be used to specify whether the DXF file will be in ACSII format or the BINARY format?

   (a) **Options** (b) **Save**
   (c) **Saveas Options** (d) **None**

8. Which command is used to modify the order of display of the images?

(a) **IMAGE** (b) **IMAGEATTACH**
(c) **IMAGECLIP** (d) **DRAWORDER**

9. Which command is used to control the display of frames of the images?

(a) **DRAWORDER** (b) **FRAMES**
(c) **IMAGEFRAME** (d) **None**

10. The __________ command is used to create an ASCII format file with the .DXF extension from AutoCAD LT drawing files.

11. Binary DXF files are __________ efficient and occupy only 75 percent of the ASCII DXF file.

12. File access for files in binary format is __________ than for the same file in ASCII format.

13. In a __________ file, information is stored in the form of a dot pattern on the screen. This bit pattern is also known as __________ .

14. The __________ dialog box can be used to modify the method of updating the linked file from automatic to manual.

15. You can choose __________ from the **Create PostScript File** dialog box to display the **PostScript Out Options** dialog box.

Exercise

Exercise 1 *General*

In this exercise you will create a cup and a plate as shown in Figure 19-10. Below this cup and plate, write the text in MS Word and then using OLE, paste it in the current drawing. The text to be written is:

These objects are drawn in AutoCAD LT 2002.

Answers to Self-Evaluation Test

1 - T, **2** - T, **3** - F, **4** - T, **5** - COPYLINK, **6** - server, **7** - **OLE Properties**, **8** - Clipboard, **9** - double-clicking, **10** - current display

Chapter 20

The User Coordinate System

Learning Objectives

After completing this chapter, you will be able to:
- *Understand the conventions in AutoCAD LT.*
- *Understand the concept of the world coordinate system (WCS).*
- *Understand the concept of the user coordinate system (UCS).*
- *Control the display of the UCS icon using the* ***UCSICON*** *command.*
- *Change the current UCS icon type using the* ***UCSICON*** *command.*
- *Use the* ***UCS*** *command.*
- *Understand different options of changing the UCS using the* ***UCS*** *command.*
- *Manage the UCS through a dialog box using the* ***UCSMAN*** *command.*
- *Understand the different system variables related to the UCS and the UCS icon.*

CONVENTIONS IN AUTOCAD LT

Before starting with the user coordinate, it is important for you to know the conventions in AutoCAD LT. AutoCAD LT follows these three conventions:

1. Whenever you open a new drawing in AutoCAD LT, it is in the Plan view and whatever you draw, it will be in the world XY plane using the world coordinate system.

2. The right-hand rule is followed in AutoCAD LT to identify the X, Y, and Z axes directions. The right-hand rule states that if you keep the thumb, index finger, and the middle finger of the right hand mutually perpendicular to each other as shown in Figure 20-1, then the thumb of the right hand displays the direction of the positive X axis, the index finger

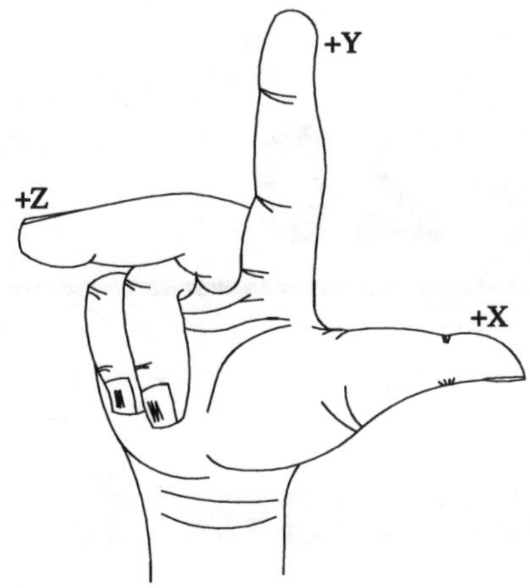

Figure 20-1 *Figure showing the orientation of the fingers*

displays the direction of the positive Y axis and the middle finger of the right hand displays the direction of the positive Z axis (Figure 20-2).

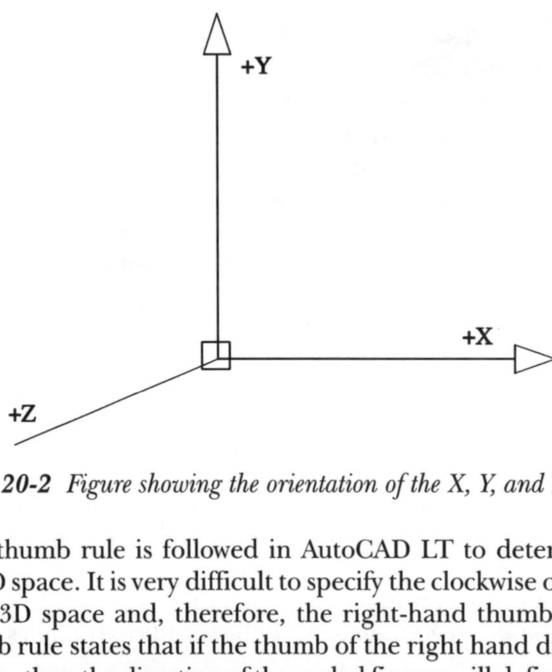

Figure 20-2 *Figure showing the orientation of the X, Y, and Z axes*

3. The right-hand thumb rule is followed in AutoCAD LT to determine the direction of rotation in the 3D space. It is very difficult to specify the clockwise or the counterclockwise direction in the 3D space and, therefore, the right-hand thumb rule is followed. The right-hand thumb rule states that if the thumb of the right hand displays the direction of the axis of rotation, then the direction of the curled fingers will define the positive direction of rotation. See Figure 20-3.

This rule will be used during the rotation of UCS.

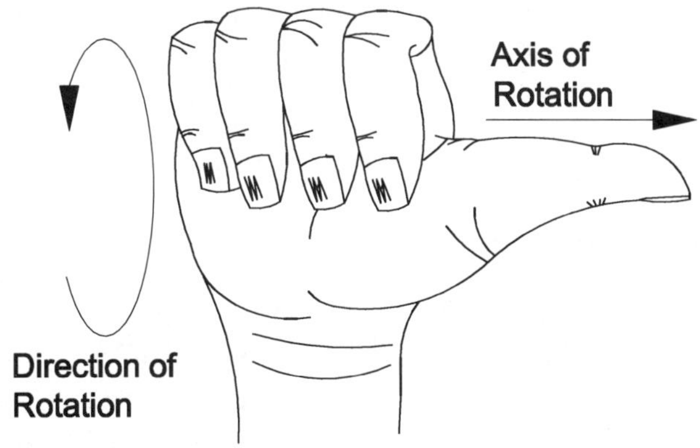

Figure 20-3 *Figure showing the axis and direction of rotation*

WORLD COORDINATE SYSTEM (WCS)

As discussed earlier, when you start a new AutoCAD LT drawing, by default, the world coordinate system (WCS) is established. The objects you have drawn until now use the WCS. In the WCS, the X, Y, and Z coordinates of any point are measured with reference to the fixed origin (0,0,0). This origin is located at the lower left corner of the screen by default. This coordinate system is fixed and cannot be moved. The WCS is generally used in 2D drawings, wireframe models, and surface models. However, it is not possible to create a solid model keeping the origin and the orientation of the X, Y, and Z axes at the same place. The reason for this is that in case you want to create a feature on the top face of an existing model, you can do it easily by shifting the working plane using the **Elevation** option of the **ELEV** command (Figure 20-4). But in case you want to create a feature on the faces other than the top and the bottom faces of an existing model, it is not possible using the **ELEV** command (Figure 20-5).

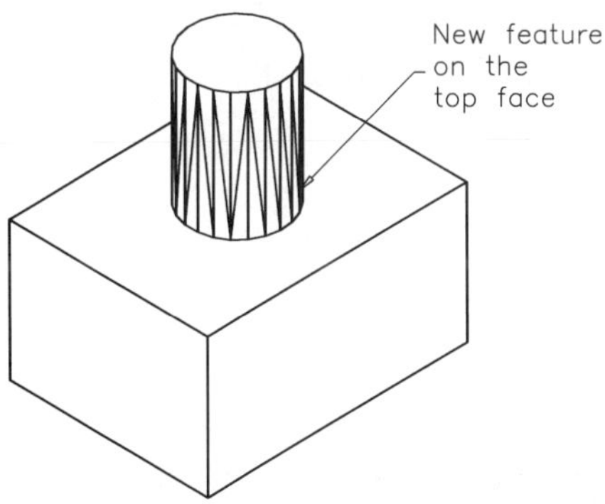

Figure 20-4 *Creating a feature on the top face*

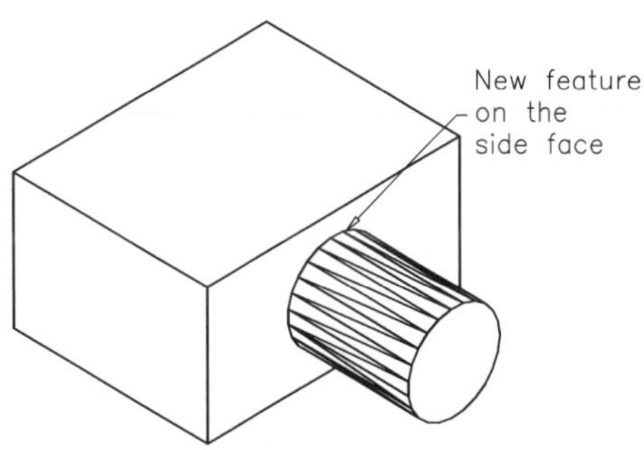

Figure 20-5 *Creating a feature on the side face*

Note

*The **ELEV** command is discussed in Chapter 21, Drawing and Viewing 3D Objects.*

This problem can be solved using a concept called the user coordinate system (UCS). The UCS allows you to relocate and reorient the origin and X, Y, and Z axes and establish your own coordinate system depending upon your requirement. The UCS is mostly used in 3D drawings, where you may need to specify points that vary from each other along the X, Y, and Z axes. It is also useful for relocating the origin or rotating the XY axes in 2D work such as ordinate dimensioning, drawing auxiliary views, or controlling hatch alignment. The UCS can be modified using the UCSICON and the **UCS** commands.

UCSICON COMMAND*

Menu:	View > Display > UCS Icon
Command:	UCSICON

This command is used to control the visibility and the location of the UCS icon. The UCS icon is a geometric representation of the current X, Y and Z axes direction. AutoCAD LT displays different UCS icons in model space and paper space as shown in Figure 20-6 and Figure 20-7. By default the UCS icon is displayed near the bottom left corner of the drawing area. You can change the location and visibility of this icon using the **UCSICON** command.

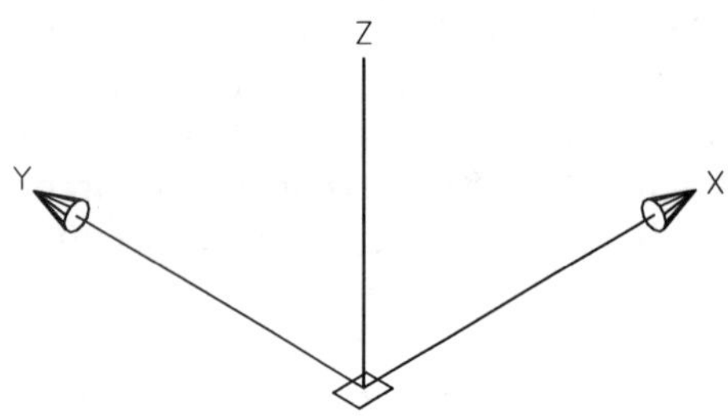

Figure 20-6 *Model space UCS icon*

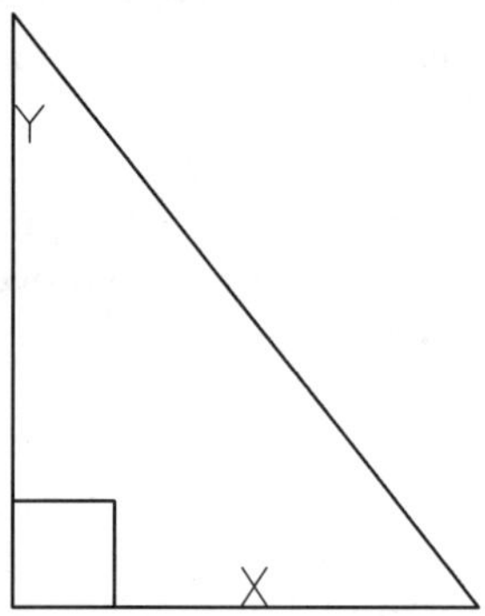

Figure 20-7 *Paper space UCS icon*

The following prompt sequence is issued when you invoke this command:

Enter an option [ON/OFF/All/Noorigin/ORigin/Properties] <ON>:

ON

This option is used to display the UCS icon on the screen.

OFF

This option is used to make the UCS icon invisible from the screen. The UCS icon will no longer be displayed on the screen. You can again turn on the display of the UCS icon using the **On** option of the **UCSICON** command.

All

This option is used to apply the changes to UCS icons in all the active viewports. If this option is not used, the changes will be applied only to the current viewport.

Noorigin

This option is used so that the UCS icon is displayed at the lower left corner of the viewport irrespective of the actual location of the origin of the current UCS.

ORigin

This option is used to place the UCS icon at the origin of the current UCS.

Properties

When you invoke this option, the **UCS Icon** dialog box will be displayed as shown in Figure 20-8. The options provided in this dialog box are discussed next.

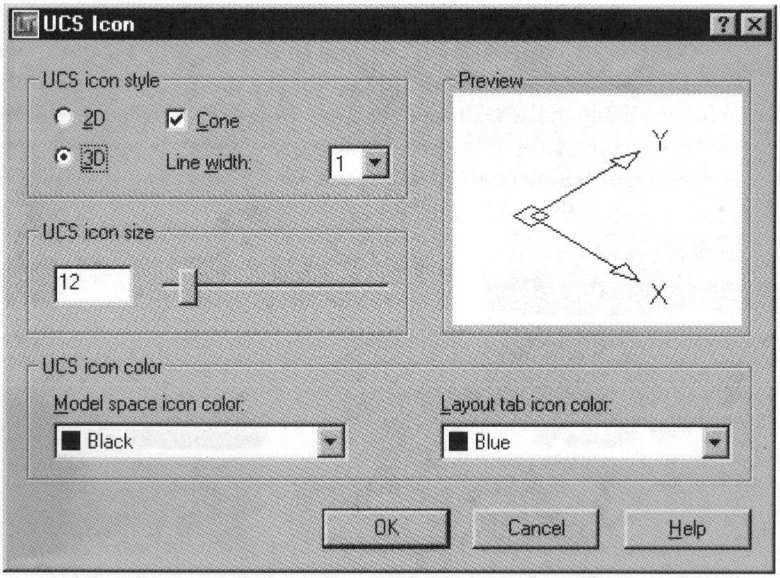

***Figure 20-8** The **UCS Icon** dialog box*

UCS icon style area

2D. If this radio button is selected, the 2D UCS icon will be displayed on the screen instead of the 3D UCS icon. (Figure 20-9).

3D. If this radio button is selected, the 3D UCS icon will be displayed on the screen. This is the default option and AutoCAD LT displays the 3D UCS icon by default.

Cone. If this check box is cleared, the cones at the end of the X and Y axes of the 3D UCS icon will not be displayed. Instead, the arrows will be displayed.

Line width. This drop-down list provides the width values that can be assigned to the 3D UCS icon. The default value for the line width is **1**. This drop-down list will not be available if the **2D** radio button is selected.

UCS icon size area

The slider bar provided in this area is used to increase the size of the UCS icon. You can also enter the value of the size of the UCS icon directly in the edit box, but the value of the UCS icon size can vary between **5** and **95**. The default value of the size of the UCS icon is **12**.

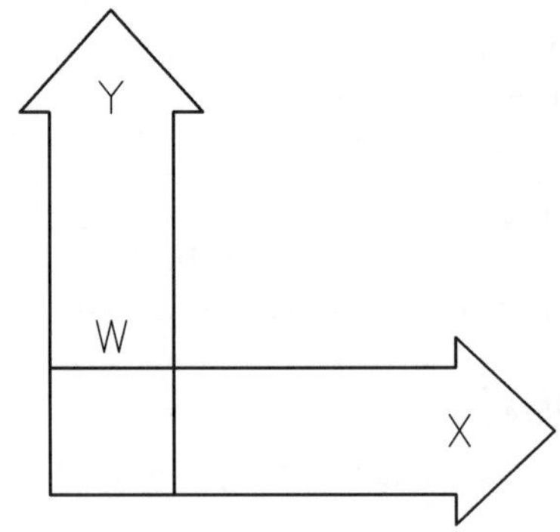

Figure 20-9 *2D UCS icon at the World position*

UCS icon color area

This area provides the drop-down lists that are used to change the color of the UCS icon in the **Model** space as well in the **Layout** tab. By default the color in the **Model** space is black and in the **Layout** is blue. You can assign any desired colors to the UCS icon. By default there are seven colors that are available in these drop-down lists. However, you can also select the desired color from the **Select Color** dialog box. This dialog box will be displayed if you select **More** from the **Model space icon color** or the **Layout tab icon color** drop-down list.

UCS COMMAND

Toolbar:	UCS
Menu:	Tools > New UCS
Command:	UCS

The **UCS** command (Figure 20-10) is used to set a new coordinate system by shifting the working plane (XY plane) to the desired location. For certain views of the drawing it is better to have the origin of measurements at some other point on or relative to your drawing objects. This makes locating the features and dimensioning the objects easier. The change in the UCS can be viewed by the change in the position and orientation of the UCS icon, that is by default placed at the lower left corner of the drawing window. The origin and orientation of a coordinate system can be redefined using the **UCS** command. The prompt sequence is:

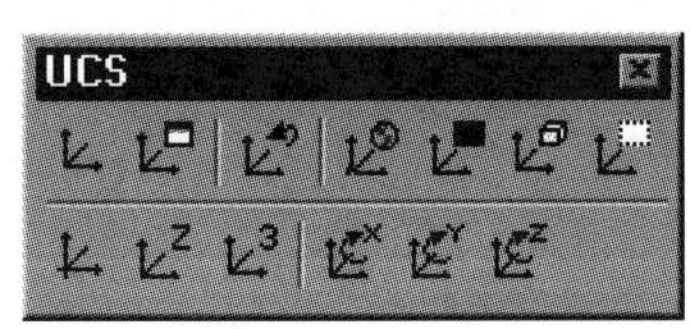

Figure 20-10 *Figure showing the* ***UCS*** *toolbar*

Command: **UCS**
Current ucs name: *WORLD*
Enter an option [New/Move/orthoGraphic/Prev/Restore/Save/Del/?/World]<World>: *Select an option.*

If the **UCSFOLLOW** system variable is set to 0, any change in the UCS does not affect the drawing view.

W (World) Option

With this option, you can set the current UCS back to the world coordinate system, which is the default position. When the UCS is placed at the world position, a small rectangle is displayed at the point where all three axes meet in the UCS icon (see Figure 20-11). If the UCS is moved from its default position, this rectangle is no longer displayed indicating that the UCS is not at the world position as shown in Figure 20-12.

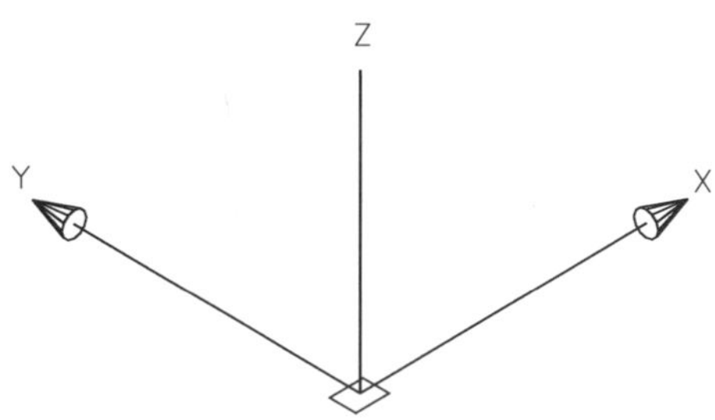

Figure 20-11 *UCS at the world position*

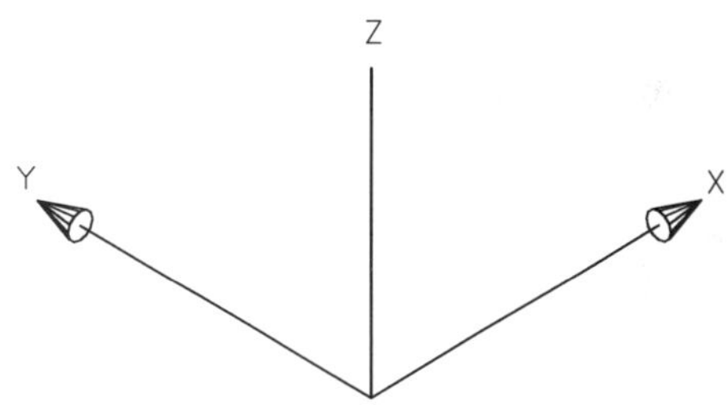

Figure 20-12 *UCS not at the world position*

Tip

In case you have selected the 2D UCS instead of the 3D UCS icon, the W will not be displayed if the UCS is not at the world position.

N (New) Option

The **New** option is used to define a new UCS by using various suboptions provided under this option. The prompt sequence is:

Command: **UCS**
Enter an option [New/Move/orthoGraphic/Prev/Restore/Save/Del/?/World]<World>: **N**
Specify origin of new UCS or [ZAxis/3point/OBject/Face/View/X/Y/Z] <0,0,0>:

O (Origin) Option

This option is used to define a new UCS by changing the origin of the current UCS. The directions of the X, Y, and Z axes remain unaltered. The new point defined will now be the origin point (0,0,0) for all the coordinate entries from this point on, until the origin is changed again. You can specify the coordinates of the new origin or simply pick it on the screen using the pointing device. The prompt sequence that will follow when you choose this button is:

Specify new origin point <0,0,0>: *Specify the origin point as shown in Figure 20-13.*

Figure 20-14 shows the UCS after defining a new UCS origin.

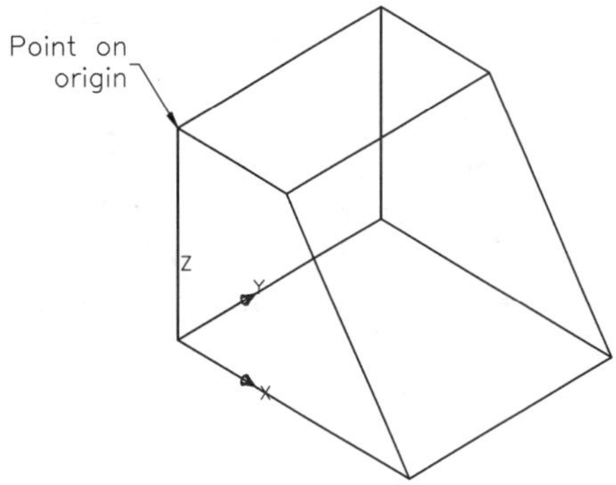

Figure 20-13 *Defining a new origin for the UCS*

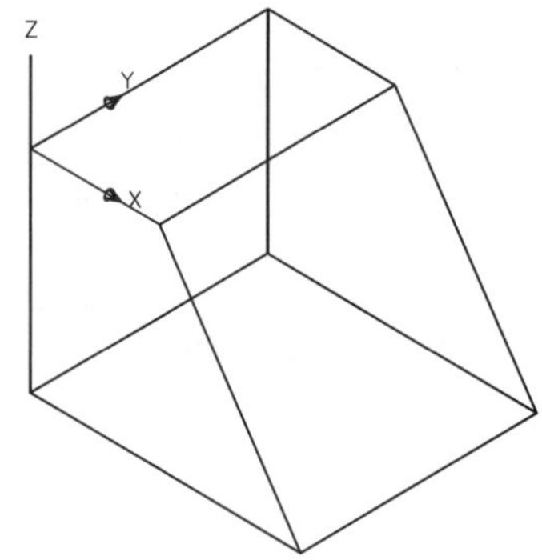

Figure 20-14 *Relocating UCS to the new origin*

Tip

If you do not provide a Z coordinate for the origin, the Z coordinate is assigned the current elevation value.

Note

The ***Origin, ZAxis, 3points, OBject, Face, View, X, Y*** *and* ***Z*** *options can also be invoked by entering* ***N*** *at the* ***Enter an option [New/Move/orthoGraphic/Prev/Restore/Save/Del/?/World]<World>:*** *prompt.*

ZA (ZAxis) Option

This option is used to change the coordinate system by selecting the origin point of the XY plane and a point on the positive Z axis. After you specify a point on the Z axis, AutoCAD LT determines the X and Y axes of the new coordinate system accordingly. The prompt sequence that will follow when you choose this button is:

Specify new origin point <0,0,0>: *Specify the origin point as shown in Figure 20-15.*
Specify point on positive portion of Z-axis <default>: **@0,-1,0**

The front face of the model will now become the new work plane (Figure 20-16) and all the new objects will be oriented accordingly. If you give a null response for the **Specify point on the positive portion of Z axis <current>** prompt, the Z axis of the new coordinate system will be parallel to (in the same direction as) the Z axis of the previous coordinate system. Null responses to the origin point and the point on the positive Z axis establish a new coordinate system in which the direction of the Z axis is identical to that of the previous coordinate system; however, the X and Y axes may be rotated around the Z axis. The positive Z axis direction is also known as the extrusion direction.

3 (3point) Option

With this option you can establish a new coordinate system by specifying a new origin point, a point on the positive side of the new X axis, and a point on the positive side

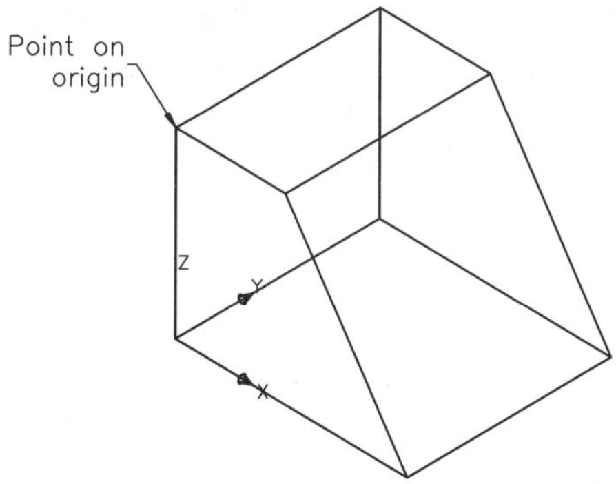

Figure 20-15 *Specifying point on origin*

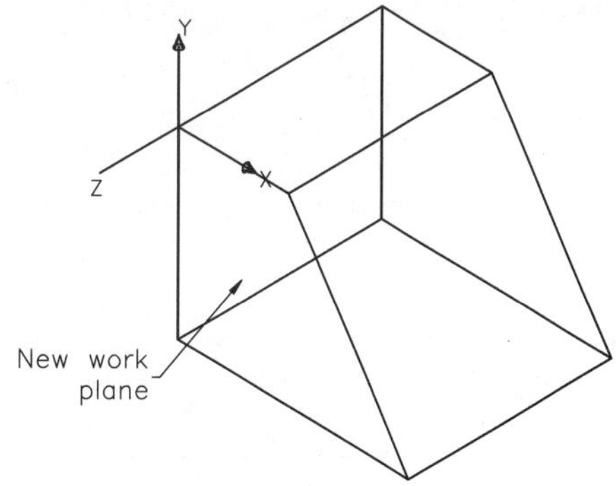

Figure 20-16 *Relocating the UCS using the* ***ZA*** *option*

of the new Y axis (Figure 20-17). The direction of the Z axis is determined by applying the right-hand rule. The **3point** option of the **UCS** command changes the orientation of the UCS to any angled surface. The prompt sequence that will follow when you choose this button is:

> Specify new origin point <0,0,0>: *Specify the origin point of the new UCS.*
> Specify point on positive portion of X-axis <1.0000,0.0000,0.0000>: *Specify a point on the positive portion of the X axis.*
> Specify point on positive-Y portion of the UCS XY plane <0.0000,1.0000,0.0000>: *Specify a point on the positive portion of the Y axis.*

A null response to the **Specify new origin point <0,0,0>** prompt will lead to a coordinate system in which the origin of the new UCS is identical to that of the previous UCS. Similarly, null responses to the Point on the X or Y axis prompt will lead to a coordinate system in which the X or Y axis of the new UCS is parallel to that of the previous UCS. In Figure 20-18, the UCS has been relocated by specifying three points (the origin point, a point on the positive portion of the X axis, and a point on the positive portion of the Y axis).

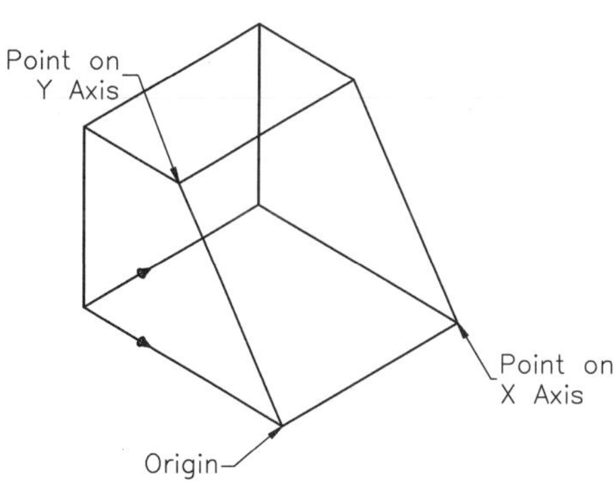

Figure 20-17 *Relocating the UCS using three points*

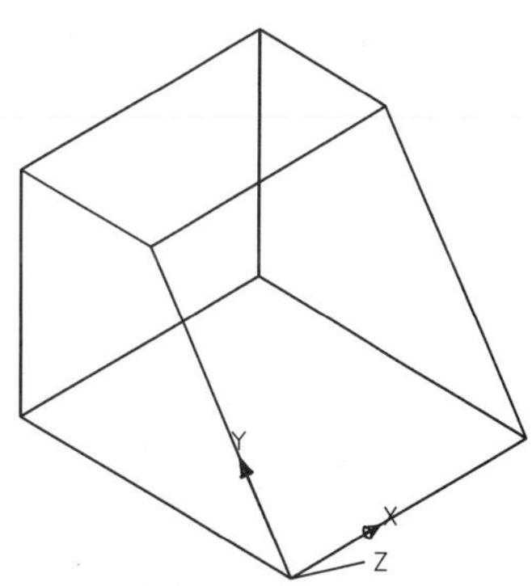

Figure 20-18 *UCS at new position*

Example 1

In this example you will draw a tapered rectangular block. After drawing the tapered block you will align the UCS on the inclined face of the block using the **3point** option of the **UCS** command. Then you will draw a circle at the inclined face. The dimensions for the block and the circle are given in Figure 20-19.

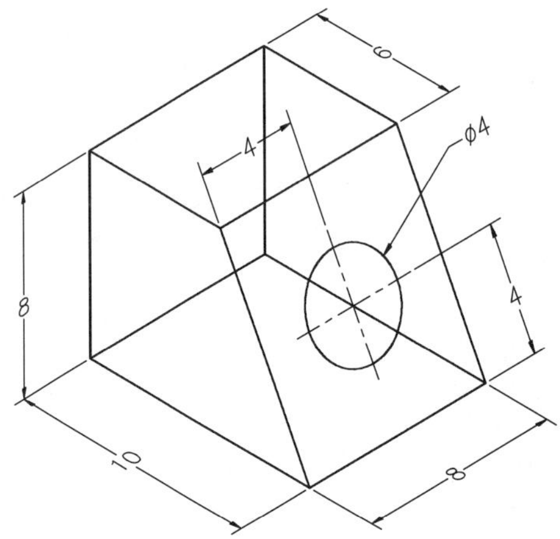

Figure 20-19 *Model for Example 1*

This problem can be broken into the following steps:

- Draw the edges on the bottom face of the tapered block.
- Draw the edges on the top face of the tapered block.
- Draw the remaining edges of the block.
- Align the UCS on the inclined face.
- Draw the circle on the inclined face.

Creating the model using these steps is discussed next.

1. Open a new drawing and choose the **Line** button from the **Draw** toolbar. The prompt sequence is as follows:

   Specify first point: **1,1**
   Specify next point or [Undo]: **@10,0**
   Specify next point or [Undo]: **@0,8**
   Specify next point or [Close/Undo]: **@-10,0**
   Specify next point or [Close/Undo]: **C**

2. The top face of the model is at a distance of 8 units from the bottom face, in the positive Z direction. Therefore, you will have to define a UCS at a distance of 8 units to create the top face. This can be done by choosing the **Origin UCS** button from the **UCS** toolbar. The prompt sequence is as follows:

Specify new origin point <0,0,0>: **0,0,8**

3. Choose the **Line** button from the **Draw** toolbar. The prompt sequence is as follows:

   Specify first point: **1,1**
   Specify next point or [Undo]: **@6,0**
   Specify next point or [Undo]: **@0,8**
   Specify next point or [Close/Undo]: **@-6,0**
   Specify next point or [Close/Undo]: **C**

4. By default, when you open a new drawing, you view the model from the top view. Therefore, when viewing from the top, the three edges of the top side overlap with the corresponding bottom edges. Therefore, you will not be able to see them. However, you can change the viewpoint to clearly view the model in 3D. Choose the **SE Isometric View** button from the **View** toolbar to proceed to the SE isometric view. You can now see the 3D shape of the objects. See Figure 20-20.

5. Join the remaining edges of the model using the **Line** button. The model after joining all the edges should look similar to the one shown in Figure 20-21.

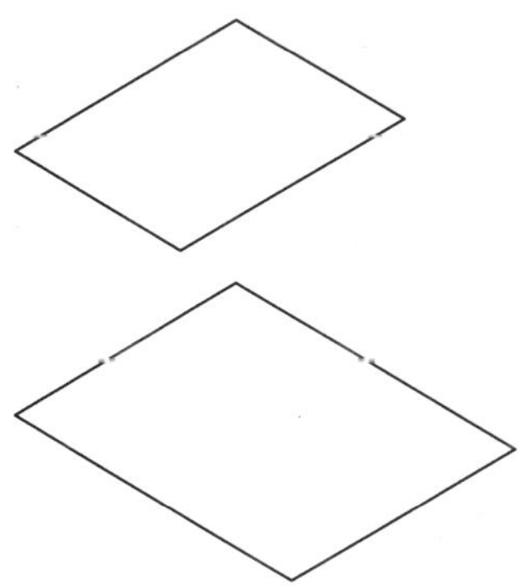

Figure 20-20 *3D view of the objects*

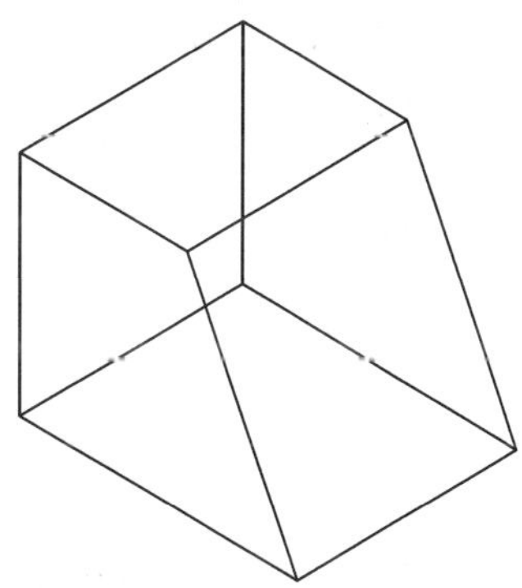

Figure 20-21 *Model after joining all the edges*

6. Now, you have to draw the circle at the inclined face. But to draw the circle at the inclined face, you will have to make the inclined face as the current working plane. Therefore, choose the **3 Point UCS** button from the **UCS** toolbar. The prompt sequence is as follows:

   Specify new origin point <0,0,0>: *Select point P1 as shown in Figure 20-22.*
   Specify point on positive portion of X-axis <default>: *Select point P2 as shown in Figure 20-22.*
   Specify point on positive-Y portion of the UCS XY plane <default>: *Select point P3 as shown in Figure 20-22.*

7. Choose the **Circle** button from the **Draw** toolbar and draw the circle at the inclined face.

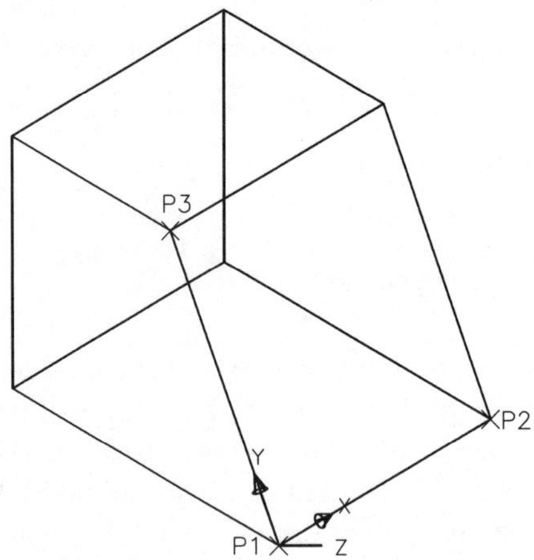

Figure 20-22 *Model after aligning the UCS on the inclined face*

8. The final model should look similar to the one shown in Figure 20-23.

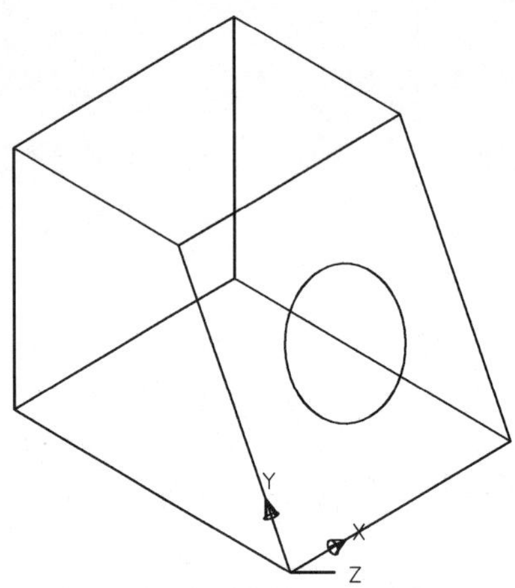

Figure 20-23 *Final model for Example 1*

OB (OBject) Option

With the **OB** (**OBject**) option of the **UCS** command, you can establish a new coordinate system by pointing to any object in an AutoCAD LT drawing. However, the objects that can not be used as an object for defining a UCS are a 3D polyline, 3D mesh, viewport object, or xline. The positive Z axis of the new UCS is in the same direction as the positive Z axis of the object selected. If the X and Z axes are given, the new Y axis is determined by the right-hand rule. The prompt sequence that will follow when you choose this button is:

Select object to align UCS: *Select the object to align the UCS.*

In Figure 20-24, the UCS is relocated using the **OBject** option and is aligned to the circle. The origin and the X axis of the new UCS are determined by the following rules.

Arc. When you select an arc, the center of the arc becomes the origin for the new UCS. The X axis passes through the endpoint of the arc that is closest to the point selected on the object.

Circle/Ellipse. The center of the circle becomes the origin for the new UCS, and the X axis passes through the point selected on the object (Figure 20-25).

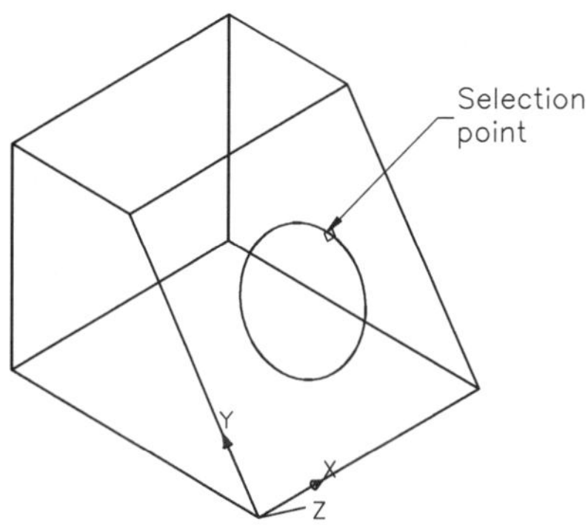

Figure 20-24 *Relocating the UCS using a circle*

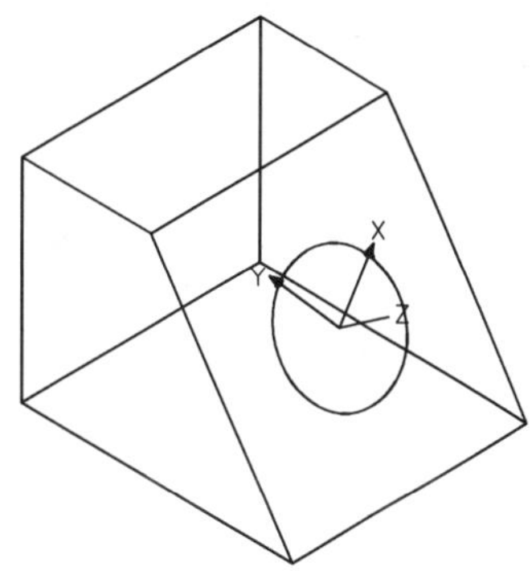

Figure 20-25 *UCS at new location*

Line/Dline/Ray/Leader. The new UCS origin is the endpoint of the line nearest the point selected on the line. The X axis is defined so that the line lies in the XY plane of the new UCS. Therefore, in the new UCS the Y coordinate of the second endpoint of the line is 0.

Tip

The linear edges of the regions are considered as individual lines when selected to align a UCS.

Spline. The origin of the new UCS is the endpoint of the spline that is nearest to the point selected on the spline. An imaginary line will be drawn between the two endpoints of the spline and the X axis will be aligned along this imaginary line.

Dimension. The middle point of the dimension text becomes the new origin. The X axis direction is identical to the direction of the X axis of the UCS that existed when the dimension was drawn.

Point. The position of the point is the origin of the new UCS. The direction of the X, Y, and Z axes will be same as those of the previous UCS.

Solid. The origin of the new UCS is the first point of the solid. The X axis of the new UCS lies along the line between the first and second points of the solid.

2D Polyline. The start point of the polyline or polyarc is treated as the new UCS origin. The X axis extends from the start point to the next vertex.

Shape/ Text/ Insert/ Attribute/ Attribute definition. The insertion point of the object becomes the new UCS origin. The new X axis is defined by the rotation of the object around its positive Z axis. Therefore, the object you select will have a rotation angle of zero in the new UCS.

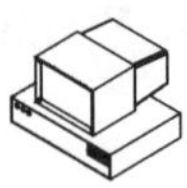

Tip

The XY plane of the new UCS will be parallel to the XY plane existing when the object was drawn; however, X and Y axes may be rotated.

F (Face) Option

This option aligns the new UCS with the selected face of the solid object. The prompt sequence that will follow when you choose this button is:

Select face of solid object: *Select the face to align the UCS.*
Enter an option [Next/Xflip/Yflip] <accept>:

The **Next** option, locates the new UCS on the next, adjacent face or the back face of the selected edge. **Xflip** rotates the new UCS by 180 degrees about the X axis and **Yflip** rotates it about the Y axis. Pressing ENTER at the **Enter an option [Next/Xflip/Yflip] <accept>** accepts the location of the new UCS as specified.

V (View) Option

The **V (View)** option of the **UCS** command lets you define a new UCS whose XY plane is parallel to the current viewing plane. The current viewing plane in this case is the screen of the monitor. Therefore, a new UCS is defined that is parallel to the screen of the monitor. The origin of the UCS defined in this option remains unaltered. This option is used mostly to view a drawing from an oblique viewing direction or for writing the text for the objects on the screen. As soon as you choose this button, a new UCS is defined parallel to the screen of the monitor.

X/Y/Z Options

With these options, you can rotate the current UCS around a desired axis. You can specify the angle by entering the angle value at the required prompt or by selecting two points on the screen with the help of a pointing device. You can specify a positive or a negative angle. The new angle is taken relative to the X axis of the existing UCS. The **UCSAXISANG** system variable stores the default angle by which the UCS is rotated around the specified axis, by using the **X/ Y/ Z** options of the **New** option of the **UCS** command. The right-hand thumb rule is used to determine the positive direction of rotation of selected axis.

X Option

In Figure 20-26, the UCS is relocated using the X option by specifying an angle about the X axis. The first model shows the UCS setting before the UCS was relocated and the second model shows the relocated UCS. The prompt sequence that will follow when you choose this button is:

Specify rotation angle about X axis <90>: *Specify the rotation angle.*

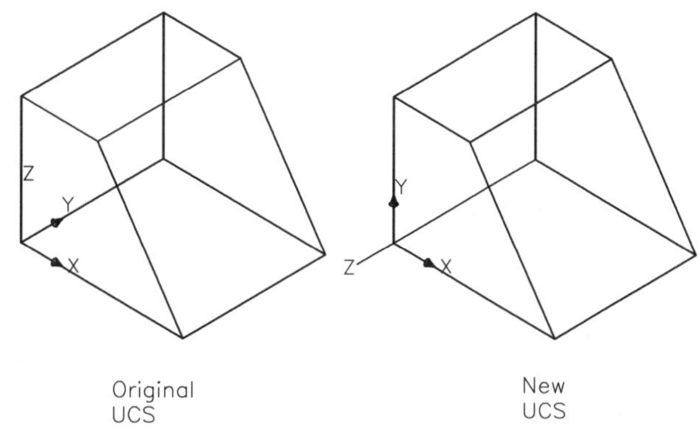

Figure 20-26 *Figure showing UCS rotated about the X axis*

Y Option

In Figure 20-27, the UCS is relocated using the **Y** option by specifying an angle about the Y axis. The first model shows the UCS setting before the UCS was relocated and the second model shows the relocated UCS. The prompt sequence that will follow when you choose this button is:

Specify rotation angle about Y axis <90>: *Specify the angle.*

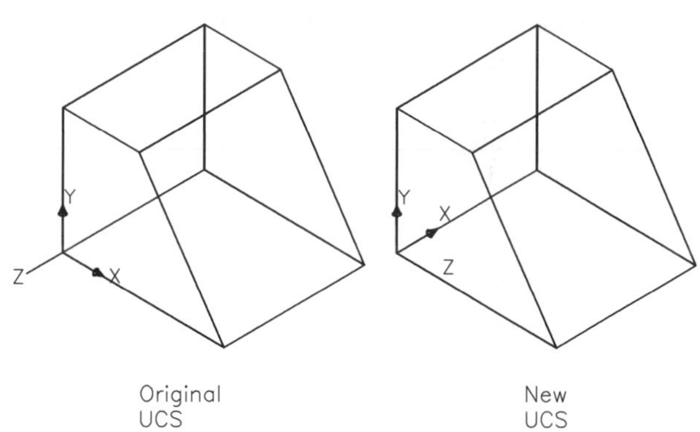

Figure 20-27 *Figure showing UCS rotated about the Y axis*

Z Option

In Figure 20-28, the UCS is relocated using the **Z** option by specifying an angle about the Z axis. The first model shows the UCS setting before the UCS was relocated and the second model shows the relocated UCS. The prompt sequence that will follow when you choose this button is:

Specify rotation angle about Y axis <90>: *Specify the angle.*

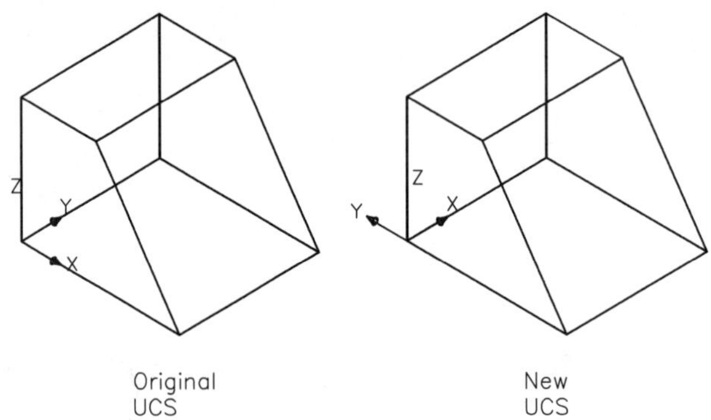

Figure 20-28 *Figure showing UCS rotated about the Z axis*

Tip

Rotation about the Z axis can be used as an alternative to using the ***SNAP*** *command to rotate the snap and grid.*

M (Move) Option

This option redefines the UCS by moving the UCS Origin in the positive or negative direction along the Z axis with respect to the current UCS Origin. The orientation of the XY plane remains the same. Bear in mind that this button is not available in the **UCS I** toolbar. Instead, this button is available in the **UCS II** toolbar. The prompt sequence that will follow when you choose this button is:

Specify new origin point or [Zdepth]<0,0,0>: *Specify the point or the depth.*

If you enter a point at the previously mentioned prompt, then the origin of the current UCS is moved to the specified point. The prompt sequence that will follow when you enter **Z** at this prompt is:

Specify Zdepth<0>: *Specify the depth in the Z axis direction.*

The **Zdepth** option specifies the distance by which the UCS origin moves along the Z axis. When you use the **Move UCS** option, in case there are multiple viewports in the drawing, the change in the UCS is applied only to the current viewport. Note that the **Move** option is not added to the **Previous** list.

G (Orthographic) Option

This option allows you to set current any one of the six Orthographic UCSs provided with AutoCAD LT. They are **Top**, **Bottom**, **Front**, **Back**, **Left**, and **Right**. You can specify any one of these UCSs by choosing from the **UCS II** toolbar drop-down list of predefined UCSs (Figure 20-29). The Orthographic UCSs are used normally when one is viewing and editing models.

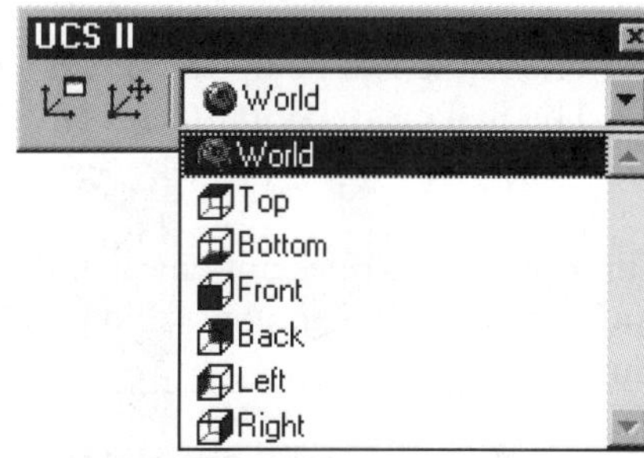

Figure 20-29 *Selecting orthographic UCS from the **UCS II** toolbar*

The Orthographic UCSs are set with respect to the WCS. The **UCSBASE** system variable controls the UCS upon which the orthographic settings are based, and the initial value is the world coordinate system.

P (Previous) Option

The **P** (**Previous**) option restores the current UCS settings to the previous UCS settings. The last ten user coordinate system settings are saved by AutoCAD LT. You can go back to the previous ten UCS settings in the current space using the **Previous** option. If **TILEMODE** is off, the last ten coordinate systems in paper space and in model space are saved. When you choose this button, the previous UCS settings are automatically restored.

R (Restore) Option

With this option of the UCS command, you can restore a previously saved named UCS. Once a saved UCS is restored, it becomes the current UCS. The viewing direction of the saved UCS is not restored. You can also restore a named UCS by selecting it from the **UCS II** toolbar drop-down list. Because this option does not have a button for it, you have to invoke this option using the **UCS** button in the **UCS** toolbar. The prompt sequence that will follow when you choose the **UCS** button is:

Current ucs name: *WORLD*
Enter an option [New/Move/orthoGraphic/Prev/Restore/Save/Del/?/World] <World>: **R**
Enter name of UCS to restore or [?]:

You can specify the name of the UCS to be restored or list the UCSs that can be restored by pressing ENTER at the previously mentioned prompt. The prompt sequence that will be followed is:

Enter UCS name(s) to list <*>: *Specify the name of the UCS to list or give a null response to list all the available UCSs.*

If you give a null response at the previously mentioned prompt, then the AutoCAD LT text window will be opened listing all the available UCSs.

S (Save) Option

With this option, you can name and save the current UCS settings. When you are naming the UCS, remember the following:

1. The name can be up to 255 characters long.
2. The name can contain letters, digits, blank spaces, and the special characters $ (dollar), - (hyphen), and _ (underscore).

This option also does not have a button. Therefore, this option will be invoked using the **UCS** button in the **UCS** toolbar. The prompt sequence that will follow when you choose the **UCS** button is:

Current ucs name: *WORLD*
Enter an option [New/Move/orthoGraphic/Prev/Restore/Save/Del/?/World] <World>: **S**
Enter name to save current UCS or [?]:

Enter a valid name for the UCS at this prompt and AutoCAD LT saves it as a UCS. You can also list the previously saved UCSs by pressing ENTER at this prompt. The next prompt sequence is:

Enter UCS name(s) to list <*>:

Enter the name of the UCS to list or give a null response to list all the available UCSs.

D (Delete) Option

The **D** (**Delete**) option is used to delete the selected UCS from the list of saved coordinate systems. This option also does not have a button. Therefore, this option will be invoked using the **UCS** button in the **UCS** toolbar. The prompt sequence that will follow when you choose the **UCS** button is:

Current ucs name: *WORLD*
Enter an option [New/Move/orthoGraphic/Prev/Restore/Save/Del/?/World]<World>: **D**
Enter UCS name(s) to delete <none>: *Specify the name of the UCS to delete.*

The UCS name you enter at this prompt is deleted. You can delete more than one UCS by separating the UCS names with commas or by using wild cards. If you delete a UCS that is current, it is renamed to **UCS Unnamed**.

? Option

By invoking this option, you can list the name of the specified UCS. This option gives you the name, origin, and X, Y, and Z axes of all of the coordinate systems relative to the existing UCS. If the current UCS has no name, it is listed as WORLD or UNNAMED. The choice between these two names depends on whether the current UCS is the same as the WCS. Choose the **UCS** button to invoke this option. The prompt sequence that will follow is:

Current ucs name: *WORLD*
Enter an option [New/Move/orthoGraphic/Prev/Restore/Save/Del/?/World] <World>: **?**
Enter UCS name(s) to list <*>:

MANAGING UCS THROUGH DIALOG BOX (UCSMAN COMMAND)

Toolbar:	UCS > Display UCS Dialog UCS II > Display UCS Dialog
Menu:	Tools > Named UCS
Command:	UCSMAN

The **UCSMAN** command displays the **UCS** dialog box (Figure 20-30). This dialog box can be used to restore the saved and orthographic UCSs, specify UCS icon settings, and rename UCSs. This dialog box has three tabs: the **Named UCSs** tab, the **Orthographic UCSs** tab, and the **Settings** tab.

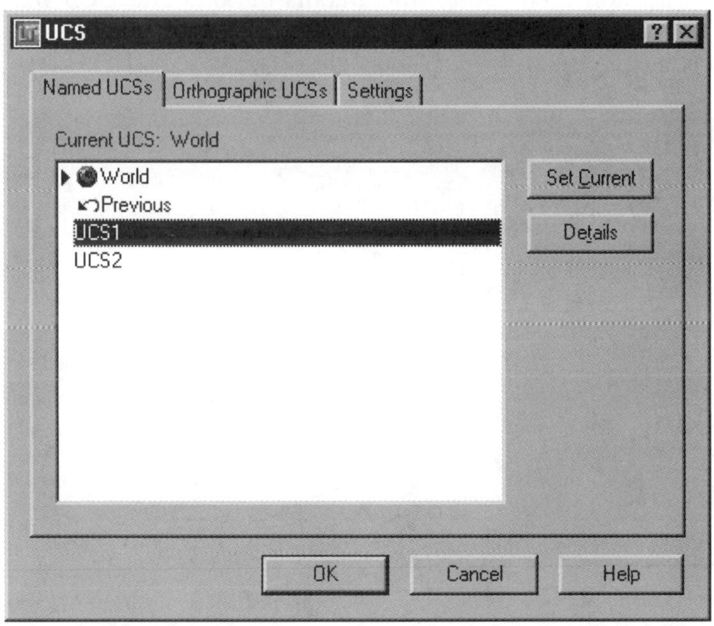

***Figure 20-30 Named UCSs** tab of the **UCS** dialog box*

Named UCSs Tab

The list of all the coordinate systems defined (saved) on your system is displayed in the list box of this tab. The first entry in this list is always **World**, which means the world coordinate system. The next entry is **Previous**. If you have defined any other coordinate systems in the current editing session then it will also be displayed in the list box. Selecting the **Previous** entry and then choosing the **OK** button repeatedly enables you to go backward through the coordinate systems defined in the current editing session. **Unnamed** is the next entry in the list if you have not named the current coordinate system. If there are a number of viewports and unnamed settings then only the current viewport UCS name is displayed in the list. The current coordinate system is indicated by a small pointer icon to the left of the coordinate system name. The current UCS name is also displayed next to **Current UCS**. If you want to make some other coordinate system current, select that coordinate system name in the UCS Names list, and then choose the **Set Current** button. To delete a coordinate system, select that coordinate system name, and then right-click to display a shortcut menu. The options are **Set Current**, **Rename**, **Delete**, and **Details**. Choose the **Delete** button to delete the selected UCS name. To rename a coordinate system, select that coordinate system name, and then right-click to display the shortcut menu and choose **Rename**. Now, enter the desired new name. You can also double-click the name to be modified and then change the name.

Tip
*All the changes and updating of the UCS information in the drawing are carried out only after you choose the **OK** button.*

If you want to check the current coordinate system's origin and X, Y, and Z axis values, select a UCS from the list and then choose the **Details** button. The **UCS Details** dialog box (Figure 20-31) containing that information is then displayed. You can also choose **Details** from the shortcut menu displayed when you right-click a specific UCS name in the list box.

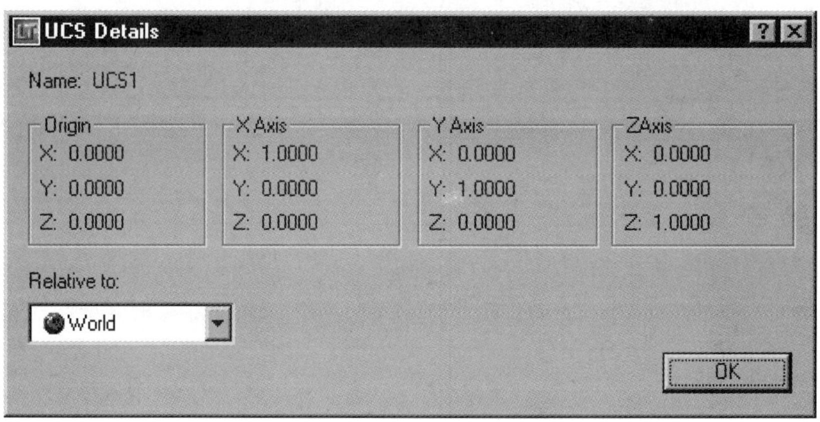

***Figure 20-31** **UCS Details** dialog box*

Setting UCS to Preset Orthographic UCSs Using the Orthographic UCSs Tab

The **UCS** dialog box with the **Orthographic UCSs** tab is displayed also on choosing **Orthographic UCS** > **Preset** from the **Tools** menu. You can select any one of the preset orthographic UCSs from the list in the **Orthographic UCSs** tab of the **UCS** dialog box (Figure 20-32). Selecting **Top** and choosing the **Set Current** button results in creation of the UCS icon in the top view, also known as the plan view. You can double-click on the UCS name you want to set as current to make it current. You can also right-click the specific orthographic UCS name in the list and choose **Set Current** from the shortcut menu displayed. The **Current UCS** displays the name of the current UCS name. If the current settings have not been saved and named, the **Current UCS** name displayed is **Unnamed**.

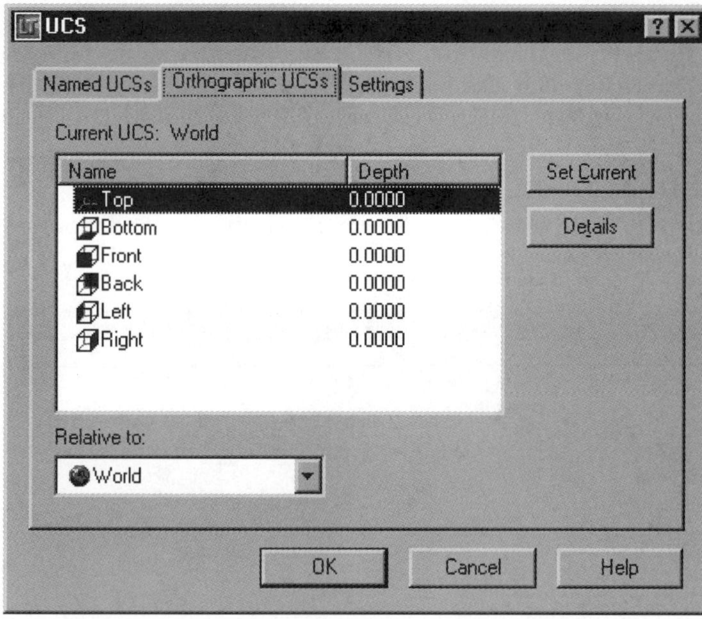

Figure 20-32 ***Orthographic UCSs*** *tab of the* ***UCS*** *dialog box*

The **Depth** field in the list box of this dialog box displays thc distance between the XY plane of the selected orthographic UCS setting and a parallel plane passing through the origin of the UCS base setting. The system variable **UCSBASE** stores the name of the UCS which is considered the base setting; that is, it defines the origin and orientation. You can enter or modify values of **Depth** by double-clicking the Depth value of the selected UCS in the list box to display the **Orthographic UCS depth** dialog box (Figure 20-33), where you can enter new depth values. You can also right-click a specific Orthographic UCS in the list box to display a shortcut menu. Choose **Depth** to display the **Orthographic UCS depth** dialog box. Enter a value in the **Top Depth** edit box or choose the **Select new origin** button to specify a new origin or depth on the screen.

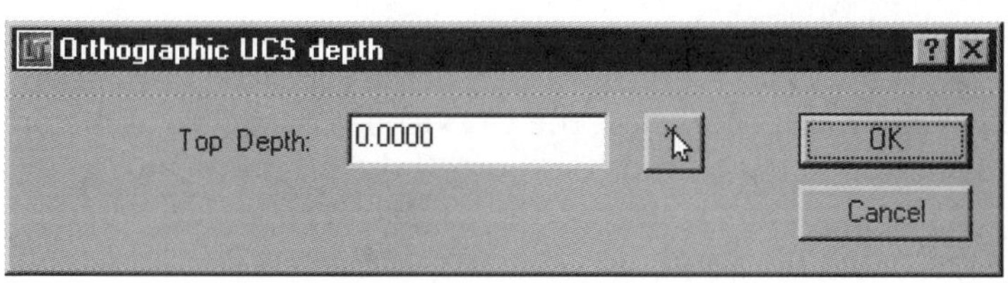

Figure 20-33 ***Orthographic UCS depth*** *dialog box*

In the **Orthographic UCSs** tab of the **UCS** dialog box, you are given the choice of establishing a new UCS relative to a named UCS or to the WCS. This choice is offered from the **Relative to** drop-down list. This list lists all the named UCSs in the current drawing. By default **World** is the base coordinate system. If you have set current any one of the preset orthographic UCSs and the you select a UCS other than World in the **Relative to:** drop-down list, the Current UCS changes to **Unnamed**. Choosing the **Details** button, displays the **UCS**

Details dialog box with the origin and the X, Y, Z coordinate values of the selected UCS. You can also choose **Details** from the shortcut menu that is displayed on right-clicking a UCS in the list box. The shortcut menu also has a **Reset** option. Choosing **Reset** from the shortcut menu restores the origin of the selected orthographic UCS to its default location (0,0,0).

Settings Tab

The **Settings** tab of the **UCS** dialog box (Figure 20-34) displays and modifies UCS and UCS icon settings of a specified viewport.

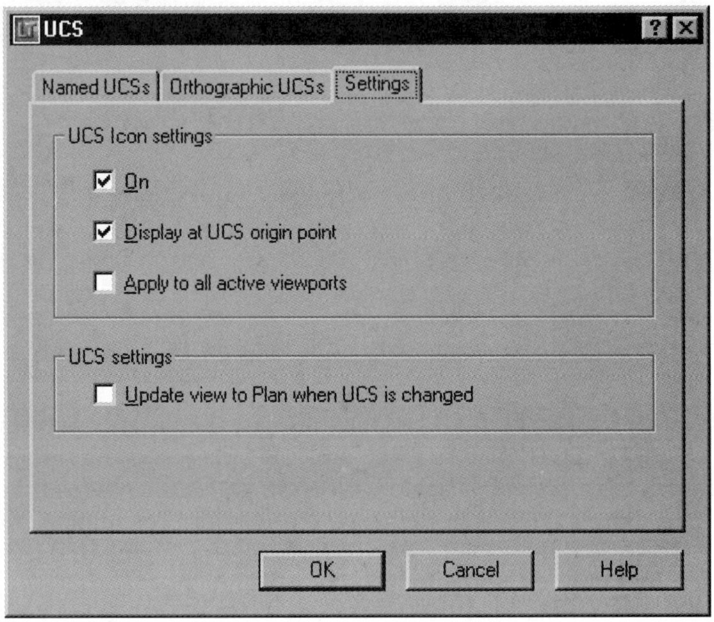

Figure 20-34 ***Settings** tab of the **UCS** dialog box*

UCS Icon settings Area

Selecting the **On** check box displays the UCS icon in the current viewport. It is similar to using the **UCSICON** command to set the display of the UCS icon to on or off. If you select the **Display at UCS origin point** check box, the UCS icon is displayed at the origin point of the coordinate system in use in the current viewport. If the origin point is not visible in the current viewport or if the check box is cleared, the UCS icon is displayed in the lower left corner of the viewport. Selecting the **Apply to all active viewports** check box applies the current UCS icon settings to all active viewports in the current drawing.

UCS settings Area

This area of the **UCS** dialog box **Settings** tab, specifies UCS settings for the current viewport. Selecting the **Save UCS with viewport** check box saves the UCS with the viewport. The value is stored in the **UCSVP** system variable. If you clear this check box, the **UCSVP** variable is set to 0 and the UCS of the viewport reflects the UCS of the current viewport. When you select the **Update view to plan** check box, the plan view is restored when the UCS in the viewport is changed. Also, when the selected UCS is restored, the plan view is restored. The value is stored in the **UCSFOLLOW** system variable.

SYSTEM VARIABLES

The coordinate value of the origin of the current UCS is held in the **UCSORG** system variable. The X and Y axis directions of the current UCS are held in the **UCSXDIR** and **UCSYDIR** system variables, respectively. The name of the current UCS is held in the **UCSNAME** variable. All these variables are read-only. If the current UCS is identical to the WCS, the **WORLDUCS** system variable is set to 1; otherwise, it holds the value 0. The current UCS icon setting can be examined and manipulated with the help of the **UCSICON** system variable. This variable holds the UCS icon setting of the current viewport. If more than one viewport is active, each one can have a different value for the **UCSICON** variable. If you are in paper space, the **UCSICON** variable will contain the setting for the UCS icon of the paper space. The **UCSFOLLOW** system variable controls the automatic display of a plan view when you switch from one UCS to another. If **UCSFOLLOW** is set to 1, a plan view is automatically displayed when you switch from one UCS to another. The **UCSAXISANG** variable stores the default angle value for the X, Y, or Z axis around which the UCS is rotated using the **X**, **Y**, **Z** options of the **New** option of the UCS command. The **UCSBASE** variable stores the name of the UCS that acts as the base; that is, it defines the origin and orientation of the orthographic UCS setting. The **UCSVP** variable decides whether the UCS settings are stored with the viewport or not.

Self-Evaluation Test

Answer the following questions and then compare your answers with the correct answers given at the end of this chapter.

1. The world coordinate system can be moved from its position. (T/F)

2. The **View** option of the **New** option of the **UCS** command allows you to define a new UCS whose XY plane is parallel to the current viewing plane. (T/F)

3. The ellipse can not be used as an object for defining a new UCS. (T/F)

4. If you do not specify the Z coordinates of the new point while moving the UCS, the previous Z coordinates will be taken for the new value. (T/F)

5. By default the _________ is established as the coordinate system in the AutoCAD LT environment.

6. The _________ coordinate system can be moved and rotated to any desired location.

7. Once a saved UCS is restored, it becomes the __________ UCS.

8. The __________ command controls the display of the UCS icon.

9. You can change the 3D UCS icon to a 2D UCS icon using the __________ option of the **UCSICON** command.

10. The __________ point of the 3D face defines the origin of the new UCS.

Review Questions

Answer the following questions.

1. The insertion point of an attribute becomes the origin of the new UCS. (T/F)

2. The line width of the UCS icon can be changed. (T/F)

3. The size of the UCS icon can vary between 5 to 100. (T/F)

4. The name of the current UCS is stored in the **UCSNAME** variable. (T/F)

5. Which option is used to restore the previous UCS settings?

   (a) ZAxis (b) Restore
   (c) Previous (d) Save

6. Which option is used to list the names of all the saved UCSs?

   (a) ZAxis (b) Restore
   (c) ? (d) Save

7. Which variable is used to control the automatic display of the plan view when you switch to the new UCS?

   (a) **UCSORG** (b) **UCSFOLLOW**
   (c) **UCS** (d) **UCSBASE**

8. Which variable stores the coordinates of the origin of the current UCS?

   (a) **UCSICON** (b) **UCSORG**
   (c) **UCSVP** (d) **UCSFOLLOW**

9. Which command is used to manage the UCS using a dialog box?

   (a) **UCS** (b) **UCSMAN**
   (c) **UCSICON** (d) **UCSBASE**

10. The _________ option of the **UCS** command is used to change the origin of the current UCS.

11. The _________ direction is known as the extrusion direction.

12. The lineweight of the UCS icon can be changed using the _________ option of the **UCSICON** command.

13. The _________ variable is used to control the X axis direction of the current UCS.

14. The _________ variable is used to control the Y axis direction of the current UCS.

15. The _________ variable is used to store the default angle value of the by which the UCS will be rotated using the **X**, **Y** and **Z** option of the **New** option of the **UCS** command.

Exercises

Exercise 1 *Mechanical*

In this exercise you will draw the model shown in Figure 20-35. Assume the missing dimensions for the model.

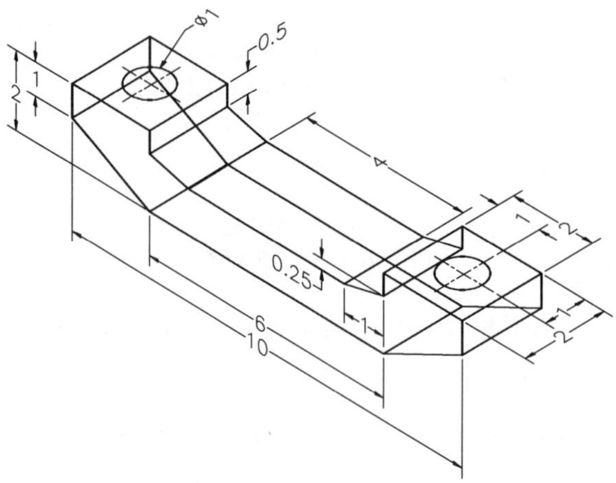

Figure 20-35 *Model for Exercise 1*

Exercises 2 and 3 *Mechanical*

Make the drawings shown in Figure 20-36 and Figure 20-37. Assume the missing dimensions for the model. Use the **UCS** command to align the ucsicon and then draw the objects.

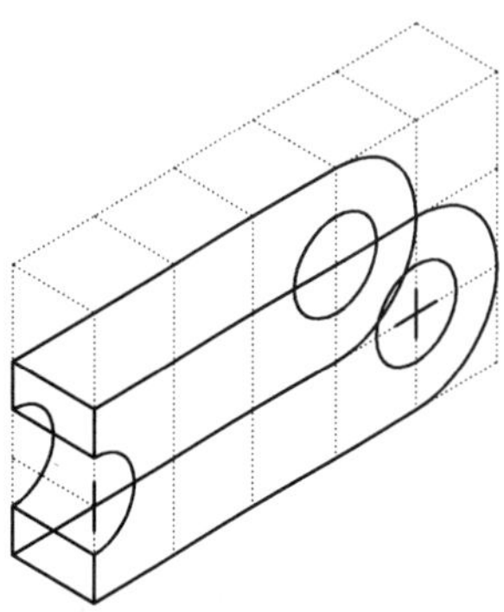

Figure 20-36 *Drawing for Exercise 2*

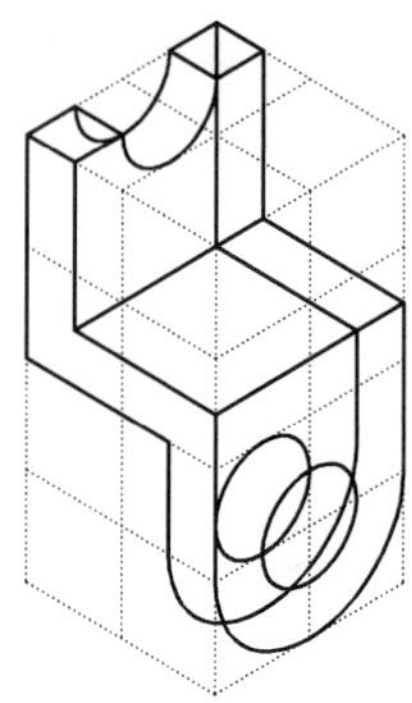

Figure 20-37 *Drawing for Exercise 3*

Exercises 4 and 5 *Mechanical*

Make the drawings shown in Figure 20-38 and Figure 20-39. Assume the missing dimensions for the model. Use the **UCS** command to align the ucsicon and then draw the objects.

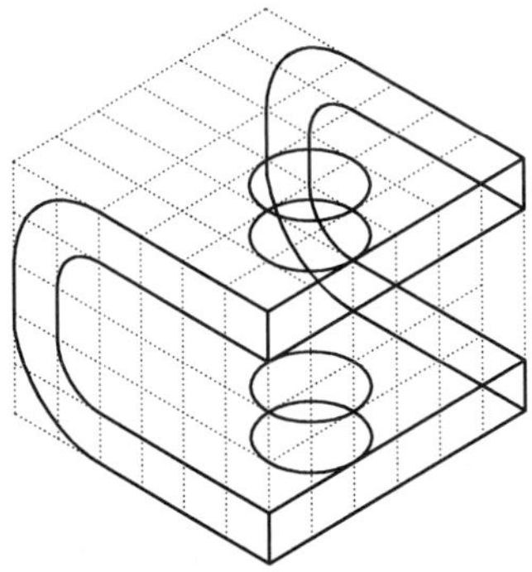

Figure 20-38 Drawing for Exercise 4

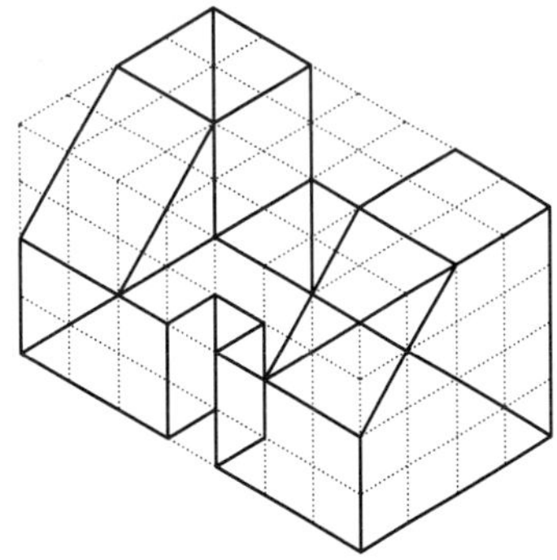

Figure 20-39 Drawing for Exercise 5

Exercise 6 *General*

Make the drawing shown in Figure 20-40. Assume the missing dimensions for the model. Use the **UCS** command to align the ucsicon and then draw the objects. The left side of the transition makes a 90-degree angle with the bottom.

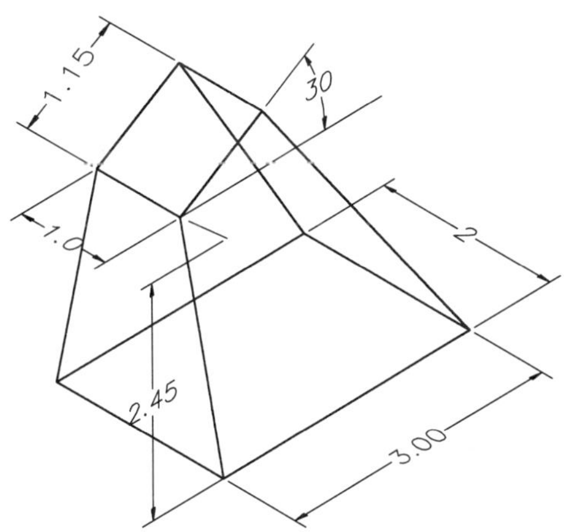

Figure 20-40 Model for Exercise 6

Exercise 7 *General*

Make the drawings shown in Figure 20-41. Assume the missing dimensions for the model. Use the **UCS** command to position the ucsicon and then draw the objects. The center of the top polygon is offset 0.75 units from the center of the bottom polygon

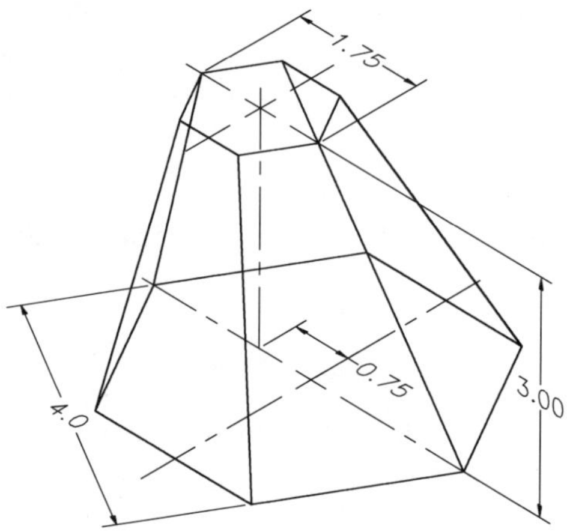

Figure 20-41 *Model for Exercise 7*

Answers to Self-Evaluation Test

1 - F, **2** - T, **3** - F, **4** - T, **5** - world coordinate system, **6** - user, **7** - current, **8** - **UCSICON**, **9** - **Properties**, **10** - first

Chapter 21

Drawing and Viewing 3D Objects

Learning Objectives

After completing this chapter, you will be able to:

- *Understand the need of changing the viewpoint.*
- *Understand various 3D coordinate systems.*
- *Setting thickness and elevation for the new objects.*
- *Use the **HIDE** command.*
- *Understand regions and how to create them.*
- *Create 3D objects.*
- *Dynamically view 3D objects using the **DVIEW** command.*
- *Create shaded models.*
- *Analyze regions using the **MASSPROP** command.*

CHANGING THE VIEWPOINT TO VIEW THE 3D MODELS

Until now, you have drawn only the 2D entities in the XY plane and in the Plan view. But in the Plan view, it is very difficult to find out whether the object displayed is a 2D entity or a 3D model. For example, see Figure 21-1. In this view it appears as though the object displayed is a rectangle. The reason for this is that by default you view the objects in the Plan view from the direction of the positive Z axis. This is a very confusing situation that has to be avoided. This situation can be avoided if the object is viewed from a direction that also displays the Z axis. In order to do that, you need to change the viewpoint so that the object is displayed in the space with all three axes as shown in Figure 21-2. In this view you can also see the Z axis of the model along with the X and the Y axes and so it is clear that the original object is not a rectangle but is a 3D model. The viewpoint can be changed by the **DDVPOINT** command, the **VPOINT** command or using the **View** toolbar.

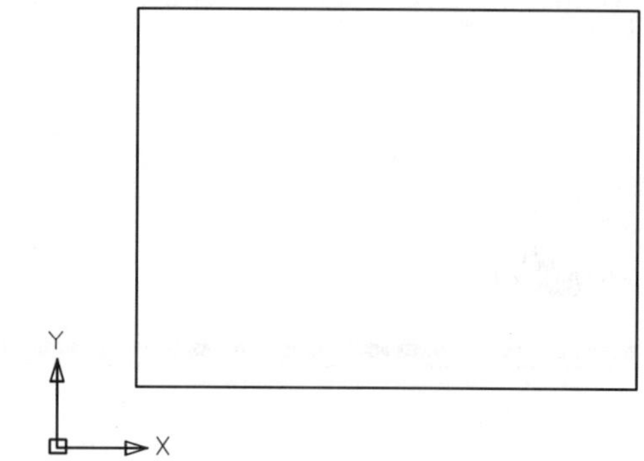

Figure 21-1 *Figure showing a rectangle in the Plan view*

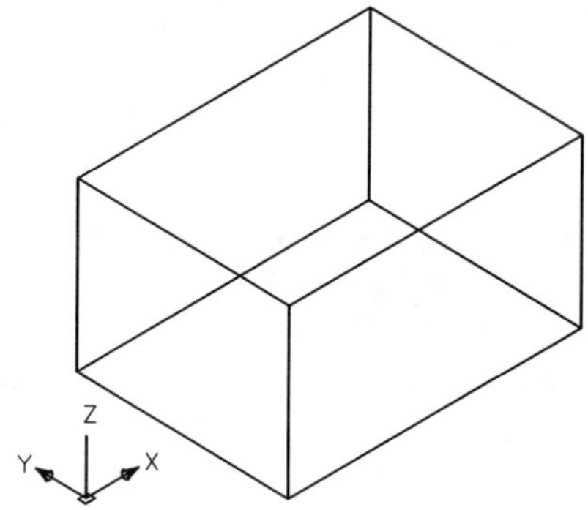

Figure 21-2 *Figure showing the 3D model after changing the viewpoint*

Tip

Bear in mind that changing the viewpoint does not affect the dimensions or the location of the model. When you change the viewpoint, the model is not moved from its original location. Changing the viewpoint only changes the direction from which you are viewing the model.

DDVPOINT Command

Menu:	View > 3D Views > Viewpoint Presets
Command:	DDVPOINT

This command is used to set the viewpoint to view the 3D models. The viewpoint in this command is set with respect to the **angle in the XY plane** and the **angle from the XY plane**. Figure 21-3 shows different view direction parameters.

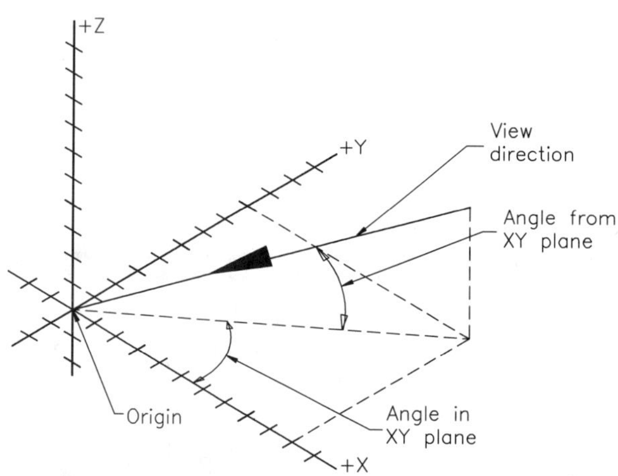

Figure 21-3 *Figure showing view direction parameters*

When you invoke this command, the **Viewpoint Presets** dialog box will be displayed as shown in Figure 21-4.

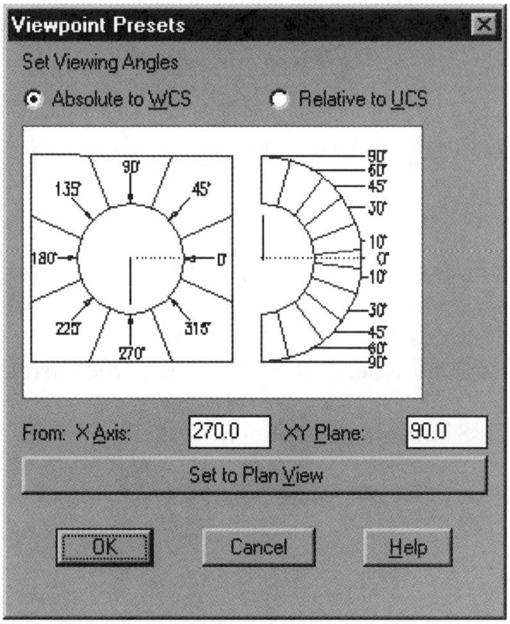

Figure 21-4 ***Viewpoint Presets*** *dialog box*

Viewpoint Presets Dialog Box Options

Absolute to WCS. This radio button is selected to set the viewpoint with respect to the world coordinate system.

Relative to UCS. This radio button is selected to set the viewpoint with respect to the current user coordinate system.

Note
The WCS and the UCS were discussed in Chapter 20, The User Coordinate System.

From: X Axis. This edit box is used to specify the angle in the XY plane from the X axis. You can directly enter the required angle in this edit box or you can select the desired angle from the image tile (Figure 21-5). There are two arms in this image tile. The red arm displays the current viewing angle and the black one displays the new viewing angle. When you select a new angle from the image tile, the new angle is automatically displayed in the **X Axis** edit box. This value can vary between 0 and 359.9.

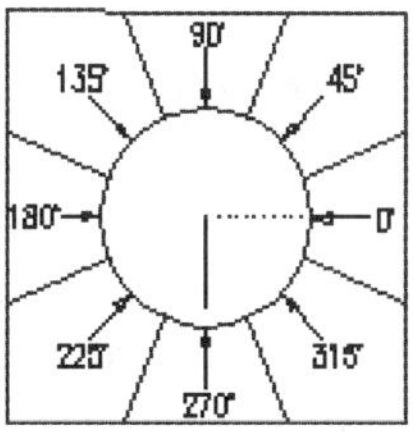

Figure 21-5 *Image tile for selecting the angle in the XY plane*

XY Plane. This edit box is used to specify the angle from the XY plane. You can directly enter the desired angle in this edit box, or can select the angle from the image tile as shown in Figure 21-6. This value can vary from -90 to +90.

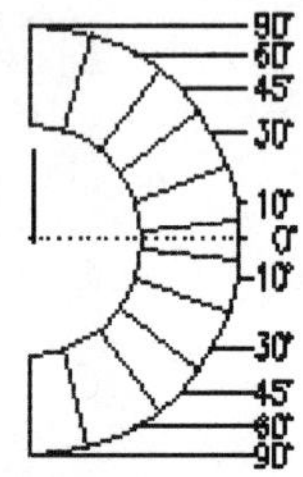

Figure 21-6 *Figure showing the image tile for selecting the angle from the XY plane*

Set to Plan View. This button is chosen to set the viewpoint to the Plan view of the WCS or the UCS. If the **Absolute to WCS** radio button is selected, the viewpoint is set to the Plane view of the WCS and if the **Relative to UCS** radio button is selected, the viewpoint is set to the Plan view of the current UCS.

Figure 21-7 and Figure 21-8 shows a 3D model from different viewing angles.

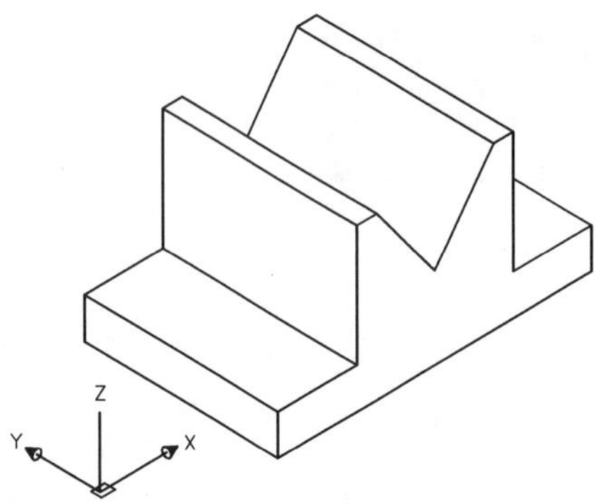

Figure 21-7 *Viewing the 3D model with the angle in the XY plane as 225 and angle from the XY plane as 30*

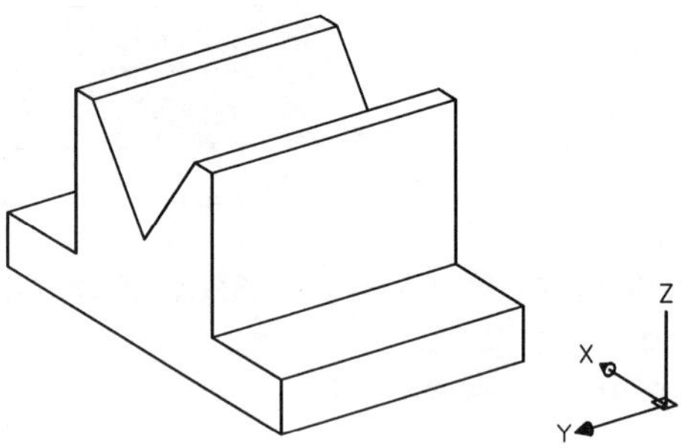

Figure 21-8 *Viewing the 3D model with the angle in the XY plane as 145 and angle from the XY plane as 25*

Tip

If you set a negative angle from the XY plane, then the Z axis will be displayed in the dotted line. See Figure 21-9 and Figure 21-10.

In case of confusion in the direction of the X, Y, and Z axes, use the right-hand rule.

VPOINT Command

Menu:	View > 3D Views > VPOINT

The **VPOINT** command is also used to set the viewpoint for viewing the 3D models. This command allows the user to specify a point in the 3D space that will act as the viewpoint. The prompt sequence that will follow when you invoke this command is:

Current view direction: VIEWDIR=0.0000,0.0000,1.0000
Specify a viewpoint or [Rotate] <display compass and tripod>:

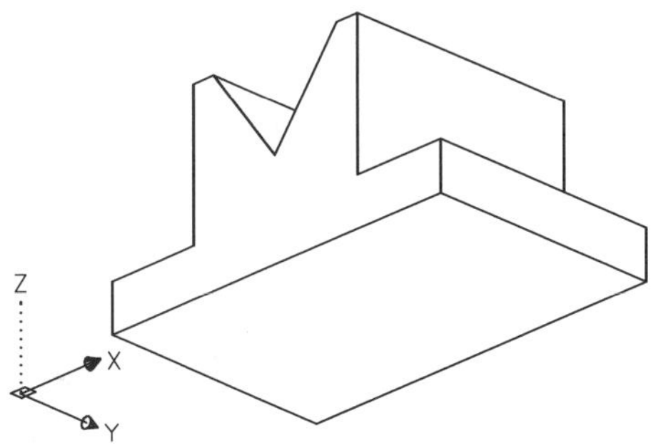

Figure 21-9 *Viewing the 3D model with the angle in the XY plane as 315 and angle from the XY plane as -25*

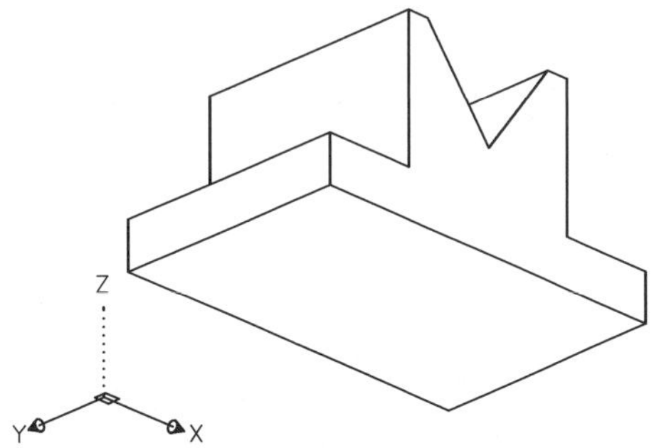

Figure 21-10 *Viewing the 3D model with the angle in the XY plane as 225 and angle from the XY plane as -25*

Specifying a Viewpoint

This is the default option and is used to set a viewpoint by specifying the location of the viewpoint using the X, Y, and Z coordinates of that particular point. AutoCAD LT follows a conventions of the sides of the 3D model for specifying the viewpoint. The convention states that if the UCS is at the World position (default position), then:

1. The side at the positive X axis direction will be taken as the right side of the model.
2. The side at the negative X axis direction will be taken as the left side of the model.
3. The side at the negative Y axis direction will be taken as the front side of the model.
4. The side at the positive Y axis direction will be taken as the back side of the model.
5. The side at the positive Z axis direction will be taken as the top side of the model.
6. The side at the negative Z axis direction will be taken as the bottom side of the model.

Some of the standard viewpoint coordinates and the view they display are shown below:

Vpoint Value	View Displayed
1,0,0	Right side view
-1,0,0	Left side view
0,1,0	Back view
0,-1,0	Front view
0,0,1	Top view
0,0,-1	Bottom view
1,1,1	Right, Back, Top view
-1,1,1	Left, Back, Top view
1,-1,1	Right, Front, Top view
-1,-1,1	Left, Front, Top view
1,1,-1	Right, Back, Bottom view
-1,1,-1	Left, Back, Bottom view
1,-1,-1	Right, Front, Bottom view
-1,-1,-1	Left, Front, Bottom view

Figures 21-11 through 21-14 show the 3D model from different viewpoints.

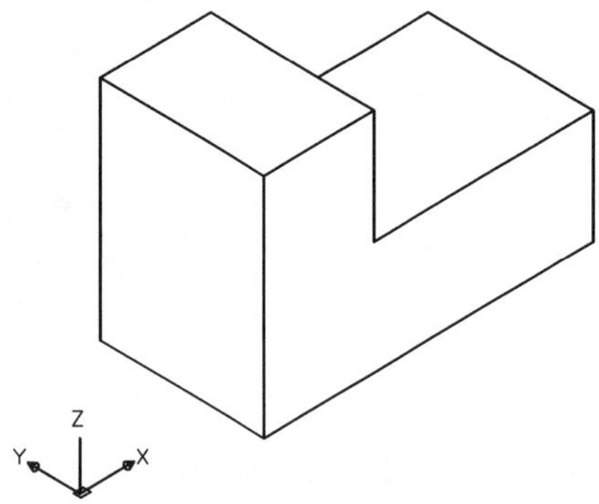

Figure 21-11 *Viewing the model with the viewpoint -1,-1,1*

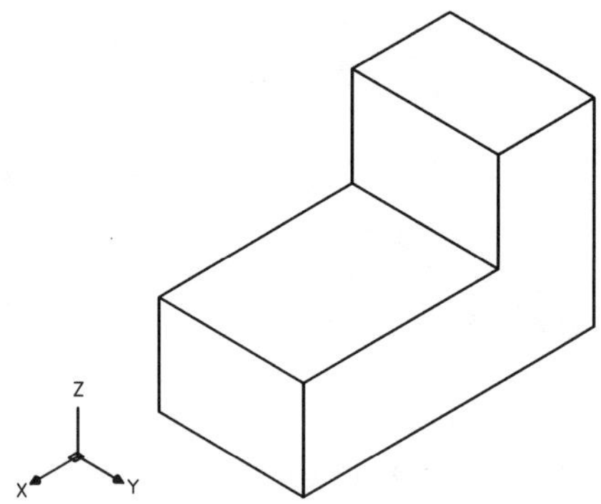

Figure 21-12 *Viewing the model with the viewpoint 1,1,1*

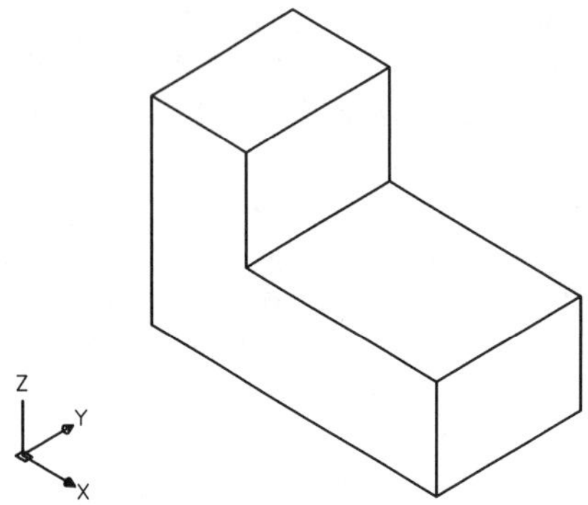

Figure 21-13 *Viewing the model with the viewpoint 1,-1,1*

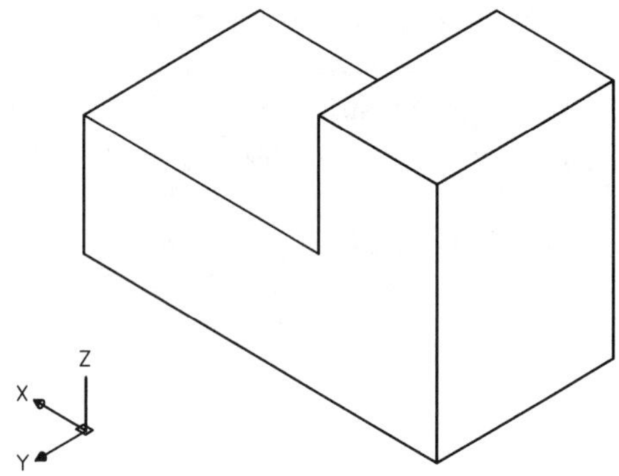

Figure 21-14 *Viewing the model with the viewpoint -1,1,1*

Compass and Tripod

When you press ENTER at the **Specify a viewpoint or [Rotate] <display compass and tripod>** prompt, a compass and an axis tripod is displayed as shown in Figure 21-15. You can directly set the viewpoint using this compass. A crosshair appears on the compass and you can pick any point on this compass to specify the viewpoint. The compass consists of two circles, a smaller circle and a bigger circle. Both of these circles are divided into four quadrants: first, second, third, and fourth. The resultant view will depend upon the quadrant and the circle on which you pick the point. In the first quadrant, both the X and the Y axes are positive; in the second quadrant, the X axis is negative and the Y axis is positive; in third quadrant both the X and Y axes are negative; and in the fourth quadrant, the X axis is positive and the Y axis is negative. Now, if you pick the point inside the inner circle, it will be in the positive Z axis direction and if you pick the point outside the inner circle and inside the outer circle, it will be in the negative Z axis direction. Therefore, if the previously mentioned statements are added, it is concluded that:

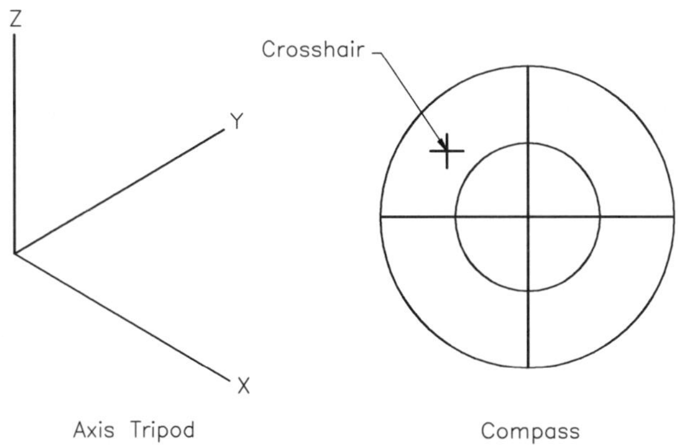

Figure 21-15 Figure showing the Axis tripod and the compass

1. If you pick a point inside the smaller circle in the first quadrant, the resultant view will be the Right, Back, Top view. If you pick a point out side the smaller circle and inside the bigger circle in this quadrant, the resultant view will be the Right, Back, Bottom view (Figure 21-16).

2. If you pick a point inside the smaller circle in the second quadrant, the resultant view will be the Left, Back, Top view. If you pick a point outside the smaller circle and inside the bigger circle in this quadrant, the resultant view will be the Left, Back, Bottom view (Figure 21-16).

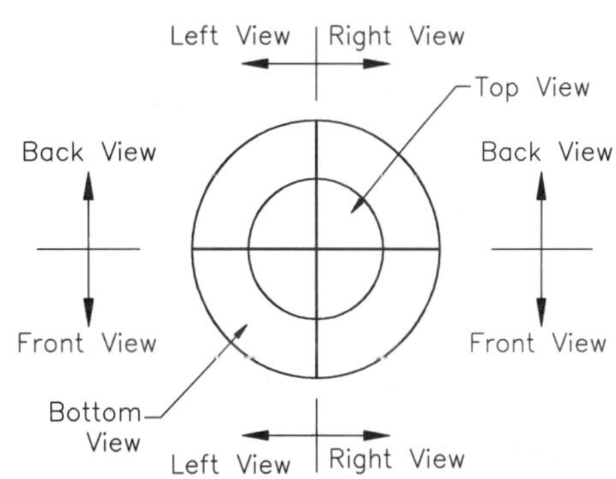

Figure 21-16 *Figures showing the directions for the compass*

3. If you pick a point inside the smaller circle in the third quadrant, the resultant view will be the Left, Front, Top view. If you pick a point outside the smaller circle and inside the bigger circle in this quadrant, the resultant view will be the Left, Front, Bottom view (Figure 21-16).

4. If you pick a point inside the smaller circle in the fourth quadrant, the resultant view will be the Right, Front, Top view. If you pick a point outside the smaller circle and inside the bigger circle in this quadrant, the resultant view will be the Right, Front, Bottom view (Figure 21-16).

Rotate

This option is similar to setting the viewpoint in the **Viewpoint Presets** dialog box displayed by invoking the **DDVPOINT** command. When you select this option, you will be prompted to specify the angle in the XY plane and the angle from the XY plane. The angle in the XY plane will be taken from the X axis.

View Toolbar

Apart from the **DDVPOINT** and the **VPOINT** commands, you can also set the viewpoint using the **View** toolbar (Figure 21-17).

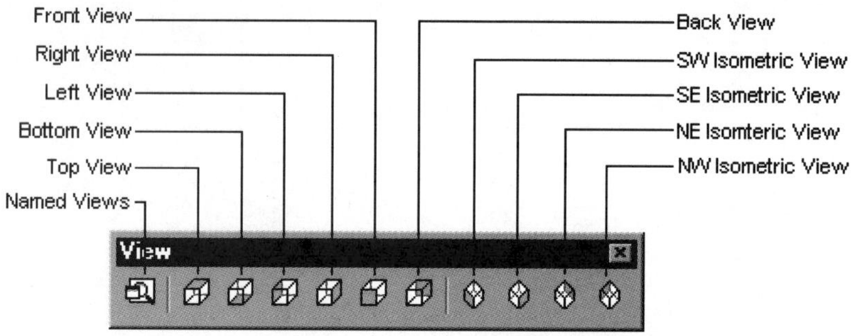

***Figure 21-17** View toolbar*

This toolbar consists of six preset orthographic views: Top View, Bottom View, Left View, Right View, Front View, and Back View. Apart from these orthographic views, this toolbar also has four preset isometric views: Southwest Isometric View, Southeast Isometric View, Northeast Isometric View, and Northwest Isometric View.

Tip

*When you choose any of the preset orthographic views from the **View** toolbar, the UCS is also aligned to that view.*

3D COORDINATE SYSTEMS

Similar to the 2D coordinate systems, there are two main types of coordinate systems. They are as follows.

Absolute Coordinate System

This type of coordinate system is similar to the 2D absolute coordinate system in which the coordinates of the point are calculated from the origin (0,0). The only difference is that here along with the X and Y coordinates, the Z coordinates are also included. For example, if you want to draw a line in 3D space starting from the origin to a point say 10,6,6, the procedure to be followed is:

Command: **LINE**
Specify first point: **0,0,0**
Specify next point or [Undo]: **10,6,6**
Specify next point or [Undo]: Enter

Figure 21-18 shows a figure drawn using the absolute coordinate systems.

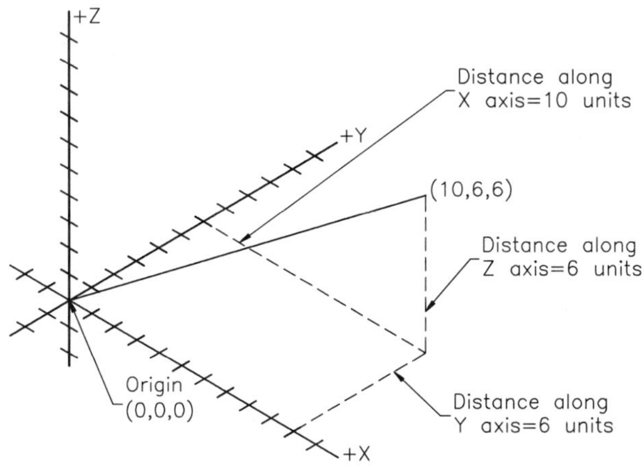

***Figure 21-18** Drawing a line from origin to 10,6,6*

Example 1

In this example you will draw the 3D wireframe model shown in Figure 21-19.

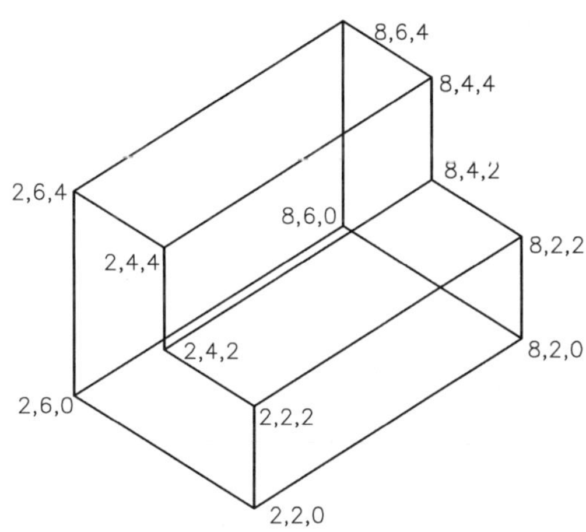

***Figure 21-19** 3D model for Example 1*

1. Choose the **Line** button from the **Draw** toolbar. The prompt sequence is as follows:

   Specify first point: **2,2,0**
   Specify next point or [Undo]: **8,2,0**
   Specify next point or [Undo]: **8,6,0**
   Specify next point or [Close/Undo]: **2,6,0**
   Specify next point or [Close/Undo]: **C**

2. Choose **3D Views > Viewpoint Presets** from the **View** menu to display the **Viewpoint Presets** dialog box (Figure 21-20).

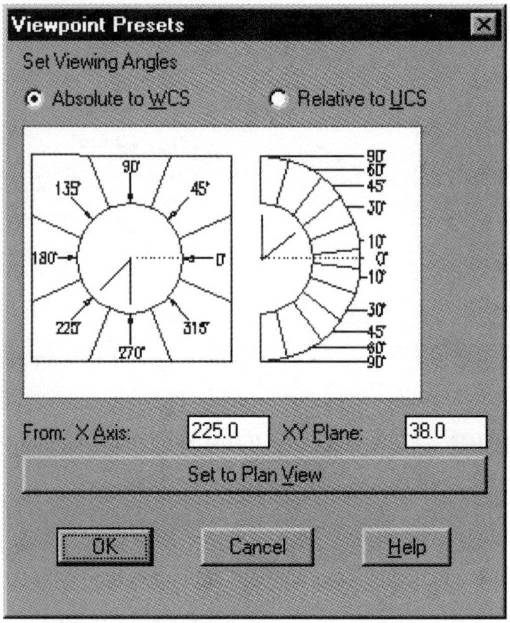

Figure 21-20 ***Viewpoint Presets*** *dialog box*

3. Set the value of the **X Axis** edit box to **225**.

4. Set the value of the **XY Plane** edit box to **38**. Choose **OK**.

5. Choose the **Line** button from the **Draw** toolbar. The prompt sequence is as follows:

   Specify first point: **2,2,0**
   Specify next point or [Undo]: **2,2,2**
   Specify next point or [Undo]: **2,4,2**
   Specify next point or [Close/Undo]: **8,4,2**
   Specify next point or [Close/Undo]: **8,2,2**
   Specify next point or [Close/Undo]: **2,2,2**
   Specify next point or [Close/Undo]: Enter

6. Choose the **Line** button from the **Draw** toolbar. The prompt sequence is as follows:

   Specify first point: **2,4,2**
   Specify next point or [Undo]: **2,4,4**
   Specify next point or [Undo]: **2,6,4**
   Specify next point or [Close/Undo]: **2,6,0**
   Specify next point or [Close/Undo]: Enter

7. Choose **3D Views > Viewpoint Presets** from the **View** menu to display the **Viewpoint Presets** dialog box.

8. Set the value of the **X Axis** edit box to **335**.

9. Set the value of the **XY Plane** edit box to **35**. Choose **OK**.

10. Choose the **Line** button from the **Draw** toolbar. The prompt sequence is as follows:

    Specify first point: **8,2,0**
    Specify next point or [Undo]: **8,2,2**
    Specify next point or [Undo]: Enter

11. Choose the **Line** button from the **Draw** toolbar. The prompt sequence is as follows:

    Specify first point: **8,4,2**
    Specify next point or [Undo]: **8,4,4**
    Specify next point or [Undo]: **8,6,4**
    Specify next point or [Close/Undo]: **8,6,0**
    Specify next point or [Close/Undo]: Enter

12. Choose the **Line** button from the **Draw** toolbar. The prompt sequence is as follows:

    Specify first point: **8,6,4**
    Specify next point or [Undo]: **2,6,4**
    Specify next point or [Undo]: Enter

13. Choose the **Line** button from the **Draw** toolbar. The prompt sequence is as follows:

    Specify first point: **8,4,4**
    Specify next point or [Undo]: **2,4,4**
    Specify next point or [Undo]: Enter

14. Enter **VPOINT** at the Command prompt. The prompt sequence is as follows:

    Current view direction: VIEWDIR=0.7424,-0.3462,0.5736
    Specify a viewpoint or [Rotate] <display compass and tripod>: **-1,-1,1**

15. The final model should look similar to the one shown in Figure 21-21.

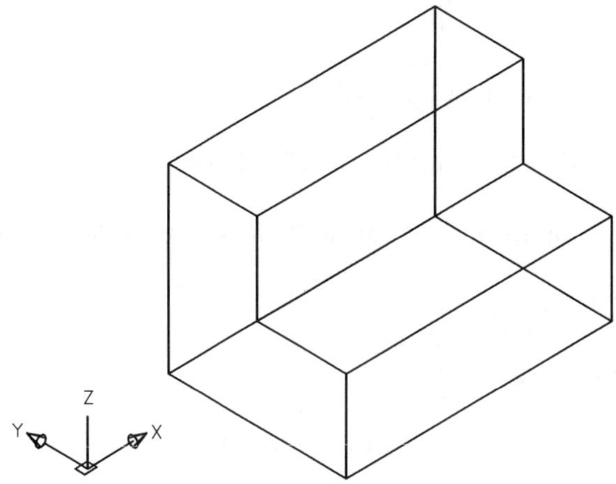

***Figure 21-21** Final 3D model for Example 1*

Relative Coordinate System

In the 3D, you are provided with three types of relative coordinate systems. They are as follows.

Relative Rectangular Coordinate System

This coordinate system is similar to the relative rectangular coordinate system, except that in 3D you have to enter the Z coordinate also along with the X and the Y coordinates. The syntax of the relative rectangular system for the 3D is **@X coordinate,Y coordinate,Z coordinate**.

Relative Cylindrical Coordinate System

In this type of coordinate system you are allowed to locate the point by specifying the distance of the point from the reference point, the angle in the XY plane, and its distance from the XY plane. The syntax of the relative cylindrical coordinate system is **@Distance from the reference point in the XY plane<Angle in the XY plane from the X axis, Distance from the XY plane along the Z axis**. Figure 21-22 shows various components of the relative cylindrical coordinate system.

Relative Spherical Coordinate System

In this type of coordinate system you are allowed to locate the point by specifying the distance of the point from the reference point, the angle in the XY plane, and the angle from the XY plane. The syntax of the relative spherical coordinate system is **@Length of the line from the reference point<Angle in the XY plane from the X axis<Angle from the XY plane**. Figure 21-23 shows various components involved in the relative spherical coordinate system.

Tip

The major difference between the relative cylindrical and relative spherical coordinate system is that in the relative cylindrical coordinate system, the distance you specify is the distance from the reference point in the XY plane, whereas in the relative spherical coordinate system, the distance you specify is the total length of the line in the 3D space.

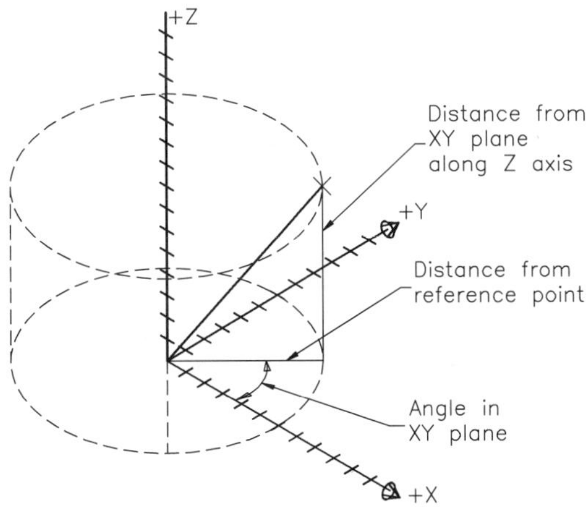

***Figure 21-22** Figure showing various components of the relative cylindrical coordinate system*

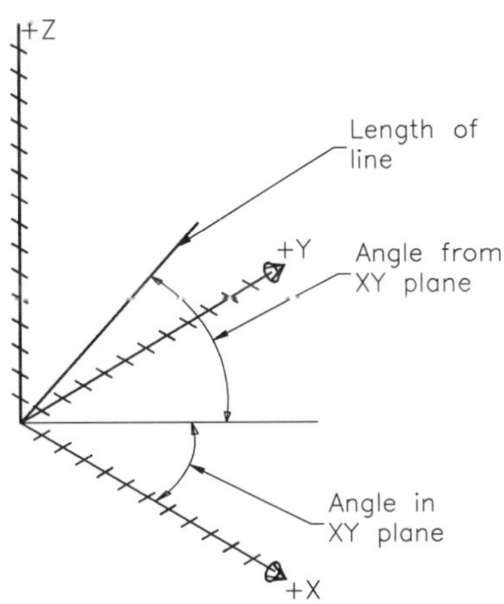

***Figure 21-23** Figure showing various components of the relative spherical coordinate system*

Example 2

In this example you will draw the 3D wireframe model shown in Figure 21-24. The dimensions of the model are given in Figure 21-25 and Figure 21-26.

1. Choose the **SW Isometric View** button from the **View** toolbar to shift to the southwest isometric view.

2. As the start point of the model is not given, therefore, you can start the model from any

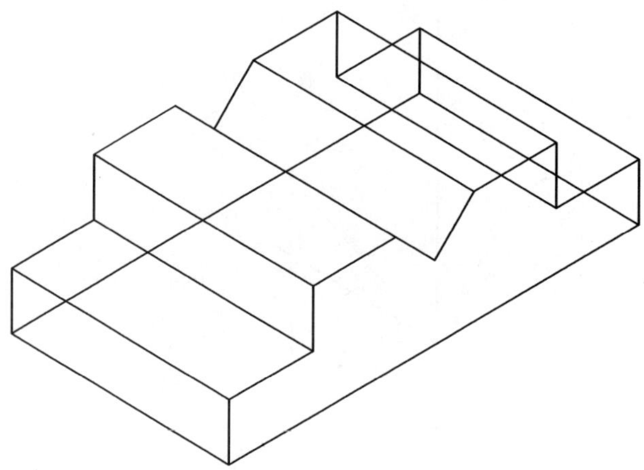

Figure 21-24 Wireframe model for Example 2

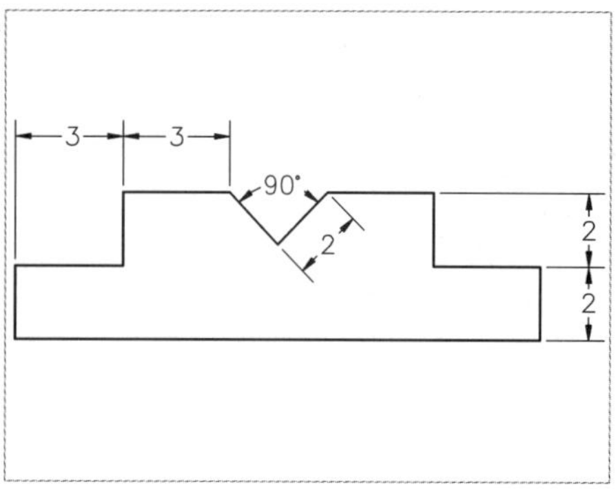

Figure 21-25 Front view of the model

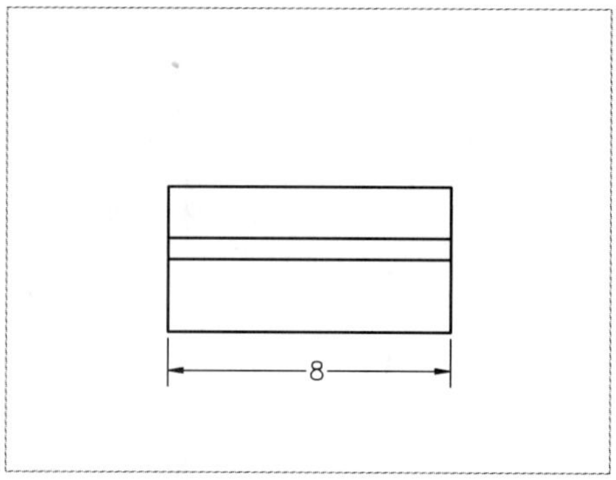

Figure 21-26 Side view of the model

point. Choose the **Line** button from the **Draw** toolbar to invoke the **LINE** command. The prompt sequence is as follows:

Specify first point: **4,2,0**
Specify next point or [Undo]: **@0,0,2**
Specify next point or [Undo]: **@3,0,0**
Specify next point or [Close/Undo]: **@0,0,2**
Specify next point or [Close/Undo]: **@3,0,0**
Specify next point or [Close/Undo]: **@2<0<315**
Specify next point or [Close/Undo]: **@2<0<45**
Specify next point or [Close/Undo]: **@3,0,0**
Specify next point or [Close/Undo]: **@0,0,-2**
Specify next point or [Close/Undo]: **@3,0,0**
Specify next point or [Close/Undo]: **@0,0,-2**
Specify next point or [Close/Undo]: **C**

3. Choose the **Line** button from the **Draw** toolbar. The prompt sequence is as follows:

Specify first point: **4,2,0**
Specify next point or [Undo]: **@0,8,0**
Specify next point or [Undo]: **@0,0,2**
Specify next point or [Close/Undo]: **@3,0,0**
Specify next point or [Close/Undo]: **@0,0,2**
Specify next point or [Close/Undo]: **@3,0,0**
Specify next point or [Close/Undo]: **@2<0<315**
Specify next point or [Close/Undo]: **@2<0<45**
Specify next point or [Close/Undo]: **@3,0,0**
Specify next point or [Close/Undo]: **@0,0,-2**
Specify next point or [Close/Undo]: **@3,0,0**
Specify next point or [Close/Undo]: **@0,0,-2**
Specify next point or [Close/Undo]: **@0,-8,0**
Specify next point or [Close/Undo]: Enter

4. Complete the model by joining the remaining edges using the **LINE** command.

5. The final 3D model should look similar to the one shown in Figure 21-27.

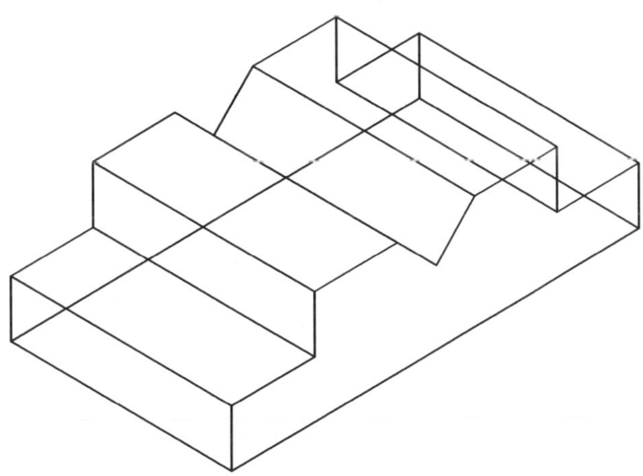

Figure 21-27 *Final 3D model for Example 2*

Exercise 1 *General*

In this exercise you will draw the 3D wireframe model shown in Figure 21-28. The dimensions for the model are given in the figure itself. You can start the model from any point.

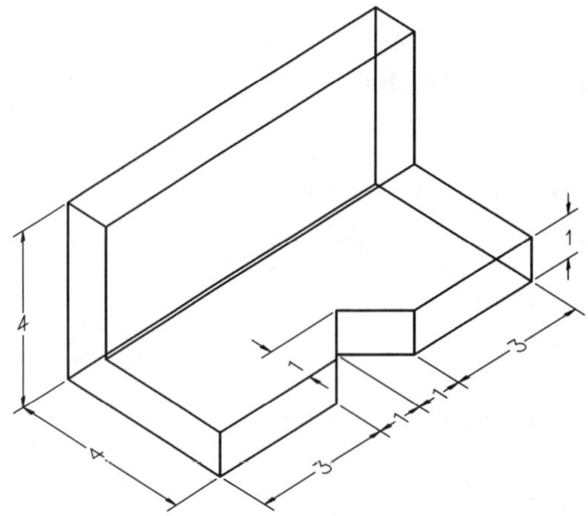

Figure 21-28 Wireframe model for Exercise 1

TRIM, EXTEND, AND FILLET COMMANDS IN 3D

As discussed earlier, you can use various options of these commands in 3D space. You can also use the **PROJMODE** and **EDGEMODE** variables. The **PROJMODE** variable sets the projection mode for trimming and extending. A value of 0 implies **True 3D mode**, that is, no projection. In this case the objects must intersect in 3D space to trim or extend them. It is similar to using the **None** option of the **Project** option of the **EXTEND** and **TRIM** commands. A value of 1 projects onto the XY plane of the current UCS and is similar to using the **UCS** option of the **Project** option of the **TRIM** or **EXTEND** commands. A value of 2 projects onto the current view plane and is like using the **View** option of the **Project** option of the **TRIM** or **EXTEND** commands. The value of the **EDGEMODE** system variable controls how the cutting or trimming boundaries are decided. A value of 0 considers the selected edge with no extension. You can also use the **No Extend** option of the **Edge** option of the **TRIM** or **EXTEND** commands. A value of 1 considers an imaginary extension of the selected edge. This is similar to using the **Extend** option of the **Edge** option of the **TRIM** or **EXTEND** commands.

You can fillet coplanar objects whose extrusion directions are not parallel to the z axis of the current UCS by using the **FILLET** command. The fillet arc exists on the same plane and has the same extrusion direction as the coplanar objects. If the coplanar objects to be filleted exist in opposite directions, the fillet arc will be on the same plane but will be inclined towards the positive Z axis.

SETTING THICKNESS AND ELEVATION FOR THE NEW OBJECTS

You can create the objects with a preset elevation or thickness using the **ELEV** command. This command is discussed next.

ELEV Command

Command: ELEV

This command is a transparent command and is used to set the elevation and thickness for new objects. The prompt sequence that will follow when you invoke this command is:

Specify new default elevation <0.0000>:
Specify new default thickness <0.0000>:

Elevation

This option is used to specify the elevation value for new objects. Setting the elevation is nothing but moving the working plane from its default position. By default, the working plane is on the world XY plane. You can move this working plane using the **Elevation** option of the **ELEV** command. However, remember that the working plane can be moved only along the Z axis (Figure 21-29). The default elevation value is 0.0 and you can set any positive or negative value for elevation. All the objects that will be drawn hereafter will be with the specified elevation. The **Elevation** option sets the elevation only for the new objects and the existing objects are not modified using this option.

Thickness

This option is used to specify the thickness values for the new objects. It is another method of creating the surface models. Specifying the thickness creates the extruded surfaces. The thickness is always taken along the Z axis direction (Figure 21-30). The 3D faces will be automatically applied on the vertical faces of the objects drawn with thickness. The **Thickness** option sets the thickness only for the new objects and the existing objects are not modified using this option.

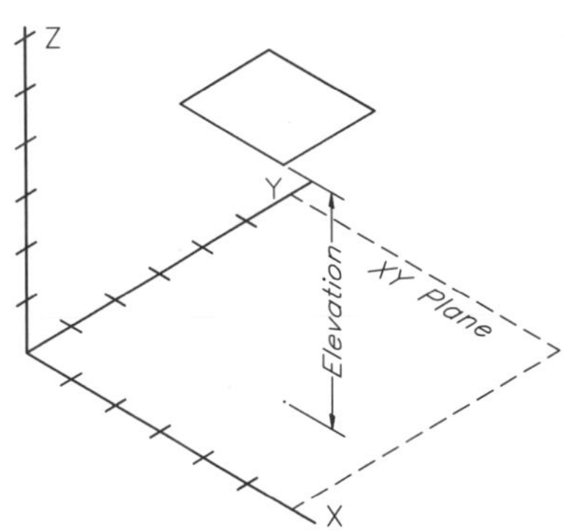

***Figure 21-29** Object drawn with elevation*

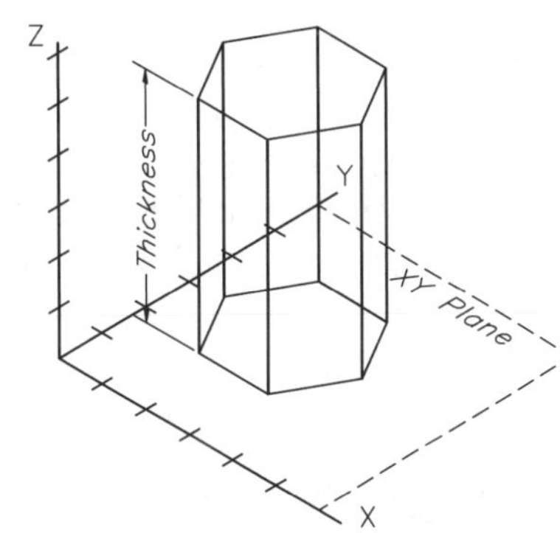

***Figure 21-30** Object drawn with thickness*

Figure 21-31 shows a point, line, ellipse, circle, and a polygon drawn with a thickness and Figure 21-32 shows text with and without thickness.

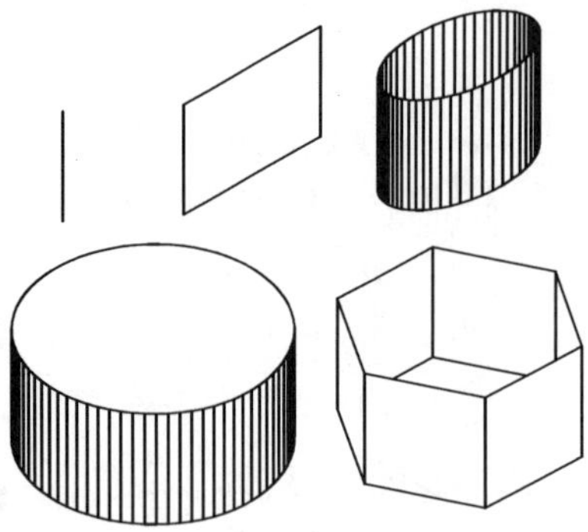

Figure 21-31 *A point, line, ellipse, circle, and a polygon drawn with a thickness of 5 units*

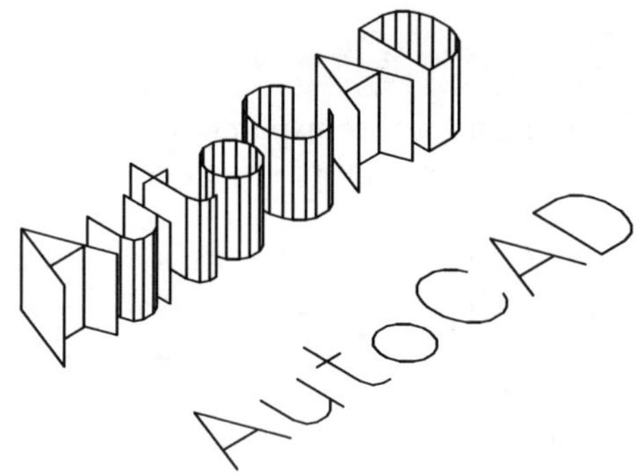

Figure 21-32 *Text with and without thickness*

Tip

The elevation value will be reset to 0.0 when you change the viewpoint using the preset orthographic views in the ***View*** *toolbar.*

The rectangles drawn using the ***RECTANG*** *command do not consider the thickness value set using the* ***ELEV*** *command. To draw a rectangle with thickness, use the* ***Thickness*** *option of the* ***RECTANG*** *command.*

To draw an ellipse with thickness, first set the value of the ***PELLIPSE*** *system variable to* ***1****.*

To write a text with thickness, first write simple text and then change its thickness using any of the commands that modify the properties.

SUPPRESSING THE HIDDEN EDGES

In the 3D objects drawn with some thickness, some of the lines lie behind other lines or objects. These lines are called hidden lines or hidden edges. Sometimes, you may want to suppress the hidden edges in the model to enhance the clarity of the objects. You can suppress these hidden edges using the **HIDE** command discussed next.

HIDE Command

Menu:	View > Hide
Command:	HIDE

This command is used to suppress the hidden lines in the 3D model. When you invoke this command, all the objects on the screen are regenerated and the 3D models are displayed with the hidden edges suppressed. The hidden lines are again included in the drawing when the next regeneration takes place. Note that the circles, solids and wide polylines are considered to be solids if they are drawn with thickness. In this case the bottom and top faces are also considered. Figure 21-33 shows the objects with thickness displaying the hidden edges and Figure 21-34 shows the objects with thickness and the hidden edges suppressed.

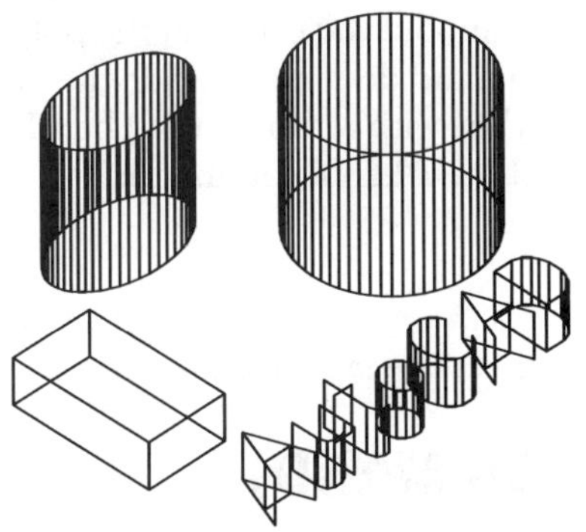

Figure 21-33 Objects before suppressing the hidden edges

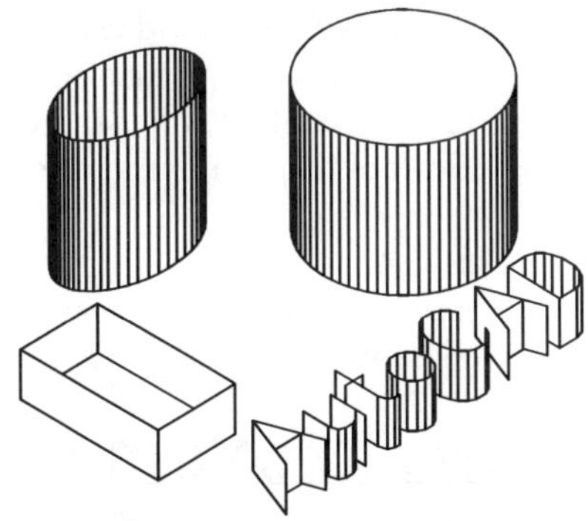

Figure 21-34 Objects after suppressing the hidden edges

You can define the color and linetype of the hidden lines using the **OBSCUREDCOLOR** and the **OBSCUREDLTYPE** variables. If you specify any value for these variables, the hidden lines will not be suppressed when you invoke the **HIDE** command. Instead, these lines will be displayed using the color and linetype specified in their respective variables.

Tip

*The **HIDE** command considers Circles, Solids, Traces, Polylines with width, and Regions as opaque surfaces that will hide objects.*

Exercise 2 *General*

Draw the objects shown in Figure 21-35 and Figure 21-36 using the **Elevation** and **Thickncss** options of the **ELEV** command. Hide the hidden lines and then view the objects using different viewpoints.

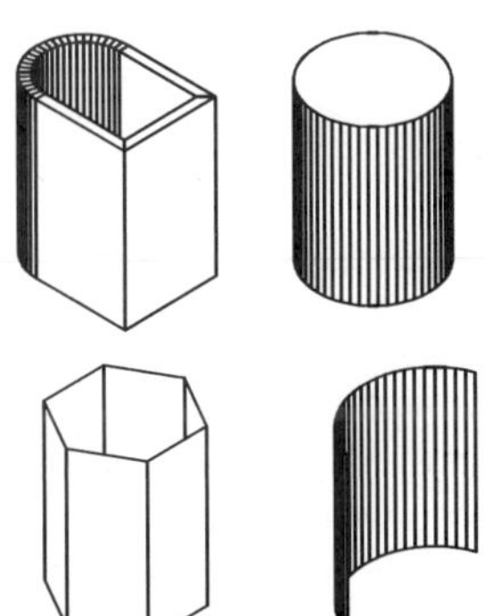

Figure 21-35 Drawing for Exercise 2

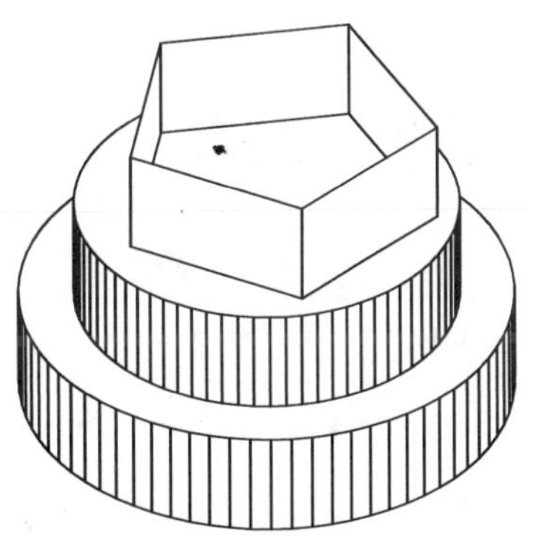

Figure 21-36 Drawing for Exercise 2

DRAWING 3D POLYLINES (3DPOLY COMMAND)

Menu:	Draw > 3D Polyline
Command:	3DPOLY

This command is used to draw straight polylines in a 2D plane or in 3D space. This command is similar to the **PLINE** command except that using this command you can draw the polylines in the plane other then the XY plane also. However, this command does not provide the **Width** or the **Arc** option and therefore, you can not create a 3D polyline with width or arc. The **Close** and the **Undo** options of this command are similar to those of the **PLINE** command.

Tip

You can fit a spline curve about the 3D polyline using the ***PEDIT*** *command. Use the* ***Spline curve*** *option to fit the spline curve and the* ***Decurve*** *option to remove the spline curve.*

CREATING REGIONS (REGION COMMAND)

Toolbar:	Draw > Region
Menu:	Draw > Region
Command:	REGION

This command is used to create regions from the selected loops or closed entities. Regions are the 2D entities with properties of the 3D solids. You can apply the Boolean operation on the regions and you can also calculate the mass properties of the regions. Bear in mind that the 2D entity you want to convert into a region should be a closed loop. Once you have created regions, the original object is deleted automatically. However, if the value of the **DELOBJ** system variable is set to **0**, the original object is retained. The valid selection set for creating the regions are closed polylines, lines, arcs, splines, circles, or ellipses. The current color, layer, linetype, and lineweight will be applied to the regions.

CREATING COMPLEX REGIONS BY APPLYING THE BOOLEAN OPERATIONS

You can create complex regions by applying the Boolean operations on them. The various Boolean operations that can be performed are union, subtract, and intersect. The commands used to apply these Boolean operations are discussed next.

UNION Command

Menu:	Modify > Regions > Union
Command:	UNION

This command is used to apply the union Boolean operations on the selected set of regions. You can create a composite region by combining them using this command. You can combine any number of regions. When you invoke this command, you will be prompted to select the regions to be added (Figure 21-37). Figure 21-38 shows the solid model created by uniting the solids shown in Figure 21-37.

SUBTRACT Command

Menu:	Modify > Regions > Subtract
Command:	SUBTRACT

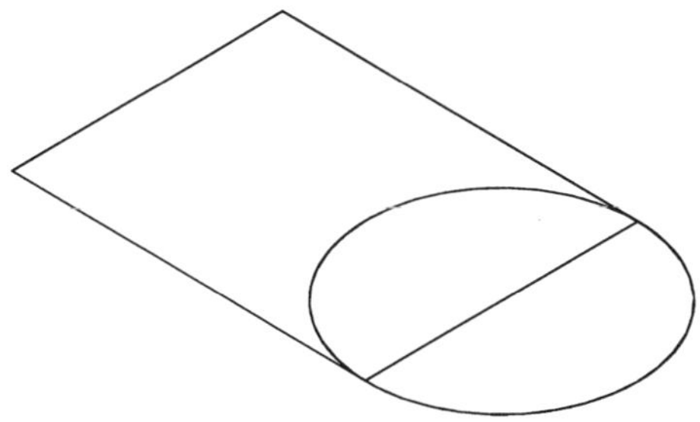

Figure 21-37 *Two different regions before union*

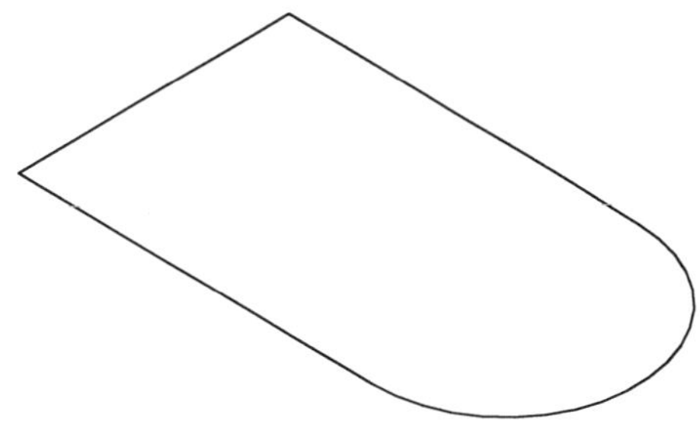

Figure 21-38 *Single region after union*

This command is used to create a composite region by removing the material common to the selected set of regions. When you invoke this command, you will be prompted to select the set of regions to subtract from. Once you have selected the set of regions to subtract from, you will be prompted to select the regions to subtract. The material common to the first selection set and the second selection set is removed from the first selection set. The resultant object will be a single composite region. See Figure 21-39 and Figure 21-40.

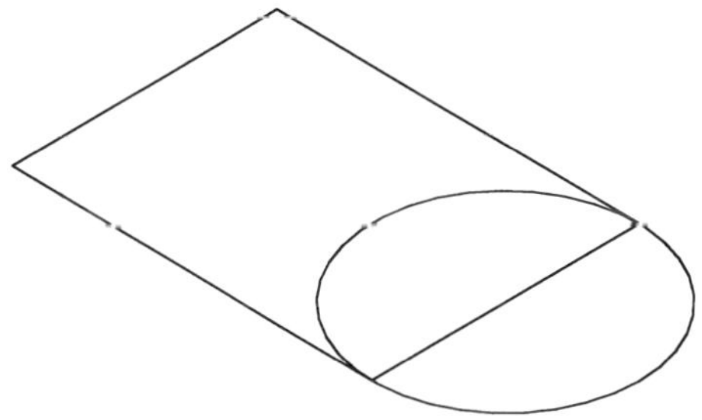

Figure 21-39 *Two different regions before subtracting*

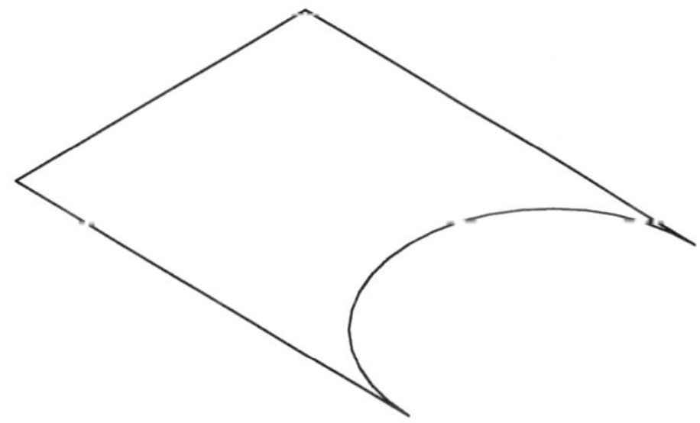

Figure 21-40 *Single region after subtracting*

INTERSECT Command

Menu:	Modify > Regions > Intersect
Command:	INTERSECT

This command is used to create a composite region by retaining the material common to the selected set of regions. When you invoke this command, you will be prompted to select the regions to intersect. The material common to all the selected regions will be retained to create a new composite solid (Figure 21-41 and Figure 21-42).

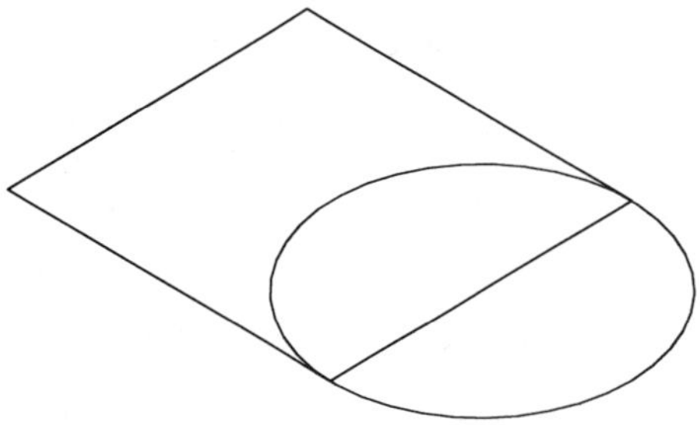

Figure 21-41 Two different regions before intersecting

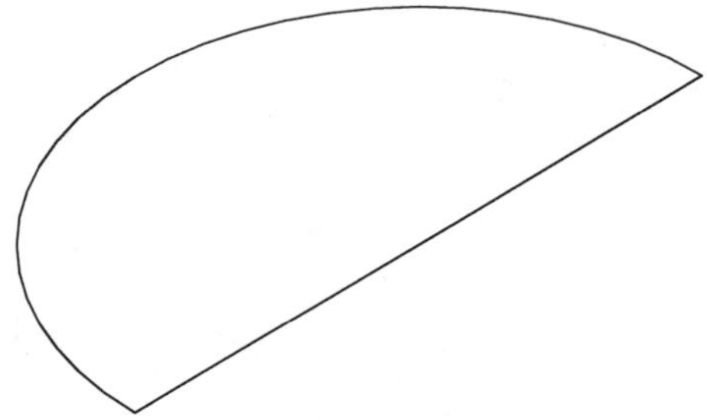

Figure 21-42 Single region after intersecting

Exercise 3 *General*

Draw the objects shown in Figure 21-43 by creating different regions and then bringing them together. Assume the dimensions.

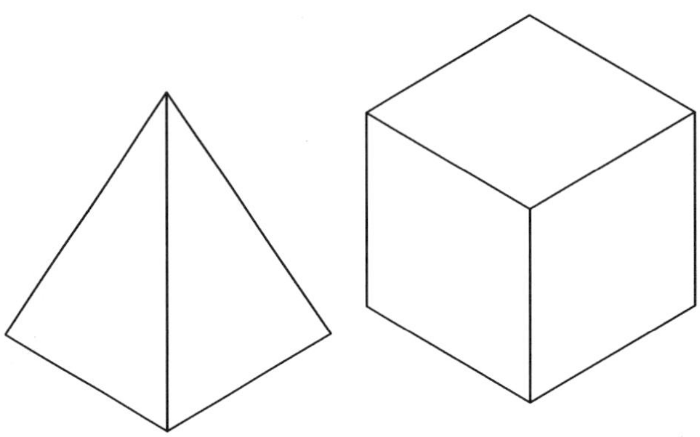

Figure 21-43 Drawing for Exercise 3

DYNAMIC VIEWING OF 3D OBJECTS (DVIEW COMMAND)

With the **DVIEW** command, you can create a parallel projection or a perspective view of objects on the screen. Basically, **DVIEW** is an improvement over the **VPOINT** command in that the **DVIEW** command allows you to visually maneuver around 3D objects to obtain different views. As just mentioned, the **DVIEW** command offers you a choice between parallel viewing and perspective viewing. The difference between the two is that in parallel view, parallel lines in the 3D object remain parallel, whereas in perspective view, parallel lines meet at a vanishing point. These definitions suggest that with the **VPOINT** command, only parallel viewing is possible. When the Perspective view is current, **ZOOM**, **PAN**, **'ZOOM**, **'PAN**, **DSVIEWER** commands, and scroll bars are not available. The **DVIEW** command uses the camera and target concept to visualize an object from any desired position in 3D space. The position from which you want to view the object (the viewer's eye) is the camera, and the focus point is the target. The line formed between these two points is the line of sight, also known as the viewing direction. To get different viewing directions, you can move the camera or target or both. Once you have the required viewing direction, you can change the distance between the camera and the target by moving the camera along the line of sight. The field of view can be changed by attaching a wide-angle lens or a telephoto lens to the camera. You can pan or twist (rotate) the image. The hidden lines in the object being viewed can be suppressed. You can clip those portions of model that you do not want to see. All these options of the **DVIEW** command demonstrate how useful this command is. The **'ZOOM**, **DSVIEWER**, **PAN** commands and Scroll bars are not available in the **DVIEW** command.

Command: **DVIEW** Enter
Select objects or <use DVIEWBLOCK>: *Select the objects you want to view dynamically or press ENTER to use the DVIEWBLOCK.*
Enter option
[CAmera/TArget/Distance/POints/PAn/Zoom/TWist/CLip/Hide/Off/Undo]: *Specify a point or enter an option.*

The point you specify is the start point for dragging. As you move the cursor, the viewing direction changes. The Command prompt is:

Enter direction and magnitude angles: *Enter angles between 0 to 360 degrees or specify on screen.*

Direction angle determines the front of the view and the magnitude angle determines the how far the view is. The different **DVIEW** command options are discussed next.

CAmera Option

With the **CAmera** option, you can rotate the camera about the target point. To invoke the CAmera option, enter CA at the **Enter option [CAmera/TArget/Distance/POints/PAn/Zoom/TWist/CLip/Hide/Off/Undo]** prompt. Once you have invoked this option, the drawing is still, but you can maneuver the camera up and down (above or below the target) or you can move it left or right (clockwise or counterclockwise around the target). Remember that when you are moving the camera, the target is stationary, and that when you are moving the target, the camera is stationary. The next prompt AutoCAD LT displays is:

Specify camera location, or enter angle from XY plane, or [Toggle (angle in)] <current>:

By default, this prompt asks you for the angle from the XY plane of the camera to the current UCS. This is nothing but the vertical (up or down) movement of the camera about the target. There are two ways to specify the angle.

1. You can specify the angle of rotation by moving the graphics cursor in the graphics area until you attain the desired rotation and then choosing the pick button of your pointing device. You will notice that as you move the cursor in the graphics area, the angle of rotation is continuously displayed in the status bar. Also, as you move the cursor, the object also moves dynamically. If you give a horizontal movement to the camera at the

Specify camera location, or enter angle from XY plane, or [Toggle (angle in)] <current> prompt, notice that the angle value displayed in the status bar does not change. This is because at the **Specify camera location, or enter angle from XY plane, or [Toggle (angle in)] <current>** prompt, you are required to specify the vertical movement.

2. The other way to specify the angle of rotation is by entering the required angle of rotation value at the prompt. An angle of 90 degrees (default value) from the XY plane makes the camera look at the object straight down from the top side of the object, providing you with the top view (plan view). In this case, the line of sight is perpendicular to the XY plane of the current UCS. An angle of negative 90 degrees makes the camera look at the object straight up from the bottom side of the object. As you move the camera toward the top side of the object, the angle value increases. If you do not want to specify a vertical movement, press ENTER.

Once you have specified the angle from the XY plane, the next prompt displayed is:

Specify camera location, or enter angle in XY plane from X axis, or [Toggle (angle from)] <current>:

This prompt asks you for the angle of the camera in the XY plane from the X axis. This is nothing but the horizontal (right or left) movement of the camera. You can enter the angle value in the range of -180 to 180 degrees. Moving the camera toward the right side of the object (counterclockwise) can be achieved by increasing the angle value; moving the camera toward the left side of the object (clockwise) can be achieved by decreasing the angle value. You can also toggle between these two prompts using the **Toggle (angle from)** or **Toggle (angle in)** option in the two prompts. The use of the **toggle** option can be illustrated as follows.

There are two ways to enter only the angle in the XY plane from the X axis. The first is to give a null response to the **Specify camera location, or enter angle from XY plane, or [Toggle (angle in)] <default>** prompt. The other is by using the **Toggle (angle in)** option. Hence, to switch from the **Specify camera location, or enter angle from XY plane, or [Toggle (angle in)] <default>** prompt to the **Specify camera location, or enter angle in XY plane from X axis, or [Toggle (angle from)] <current>** prompt, you will have the following prompt sequence:

Specify camera location, or enter angle from XY plane, or [Toggle (angle in)] <default>: **T** Enter
Specify camera location, or enter angle in XY plane from X axis, or [Toggle (angle from)] <current>:

In the same manner, if you want to switch from the **Specify camera location, or enter angle in XY plane from X axis, or [Toggle (angle from)] <current>** prompt to the **Specify camera location, or enter angle from XY plane, or [Toggle (angle in)] <default>** prompt, the following will be the prompt sequence:

Specify camera location, or enter angle in XY plane from X axis, or [Toggle (angle from)] <current>: **T** Enter

Specify camera location, or enter angle from XY plane, or [Toggle (angle in)] <default>:

You have gone through the basic concept of the **CAmera** option of the **DVIEW** command. Now, you will work out some examples to apply the camera concept.

Tip

*The **DVIEWBLOCK** is automatically displayed when you use **DVIEW** without selecting objects. You can define or redefine your own substitute **DVIEWBLOCK**, just like any other block using the **BLOCK** command. Make sure to define it to a 1X1X1 unit size so that it will be properly scaled in the **DVIEW** command.*

Example 3

In Figure 21-44, you have four sections. The house shown can be obtained by pressing ENTER at the **Select objects or <use DVIEWBLOCK>** prompt. The drawing of the house with a window, an open door, and a chimney obtained on the screen is a block named **DVIEWBLOCK**.

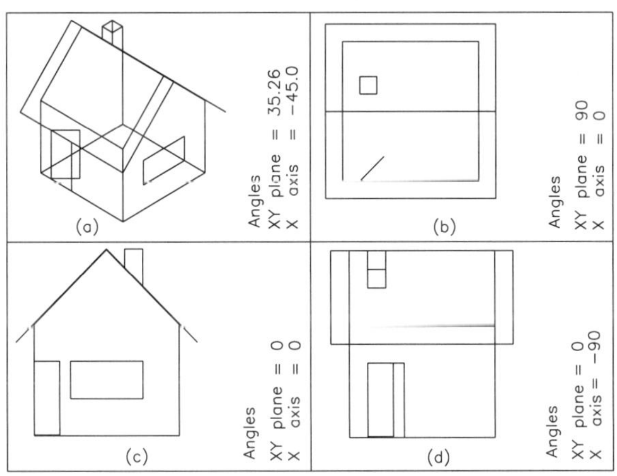

***Figure 21-44** **CAmera** option of the **DVIEW** command*

As you use various options of the **DVIEW** command, the block is updated to reflect the changes. Once you come out of the **DVIEW** command, the entire drawing is regenerated, and the view obtained depends on the view you have selected using the various options of the **DVIEW** command. In this chapter, you will use this block to demonstrate the effect of various options of the **DVIEW** command. However, you can make a custom block of your own. The block should be of unit size. Set the lower left corner of the block as the origin point. The different views of the house block you see in Figure 21-44 can be obtained with the **CAmera** option of the **DVIEW** command. First, use the **DVIEW** command to get the image of the house on the screen.

Command: **DVIEW** [Enter]
Select objects or <use DVIEWBLOCK>: [Enter]

Now, you have the image of the house on the screen. For Figure 21-44(a) (3D view), the following is the prompt sequence:

Enter option
[CAmera/TArget/Distance/POints/PAn/Zoom/TWist/CLip/Hide/Off/Undo]: **CA** Enter
Specify camera location, or enter angle from XY plane, or [Toggle (angle in)] <current>: **35.26** Enter
Specify camera location, or enter angle in XY plane from X axis, or [Toggle (angle from)] <current>: **-45** Enter
Enter option
[CAmera/TArget/Distance/POints/PAn/Zoom/TWist/CLip/Hide/Off/Undo]: Enter
Regenerating model.

Once you exit the **DVIEW** command, the figure of the house is removed from the screen. For Figure 21-44(b) (top view), the following is the prompt sequence:

Command: **DVIEW** Enter
Select objects or <use DVIEWBLOCK>: Enter
Enter option
[CAmera/TArget/Distance/POints/PAn/Zoom/TWist/CLip/Hide/Off/Undo]: **CA** Enter
Specify camera location, or enter angle from XY plane, or [Toggle (angle in)] <current>: **90** Enter
Specify camera location, or enter angle in XY plane from X axis, or [Toggle (angle from)] <current>: **0** Enter
Enter option
[CAmera/TArget/Distance/POints/PAn/Zoom/TWist/CLip/Hide/Off/Undo]: Enter
Regenerating model.

For Figure 21-44(c) (right side view), the following is the prompt sequence:

Command: **DVIEW** Enter
Select objects or <use DVIEWBLOCK>: Enter
Enter option
[CAmera/TArget/Distance/POints/PAn/Zoom/TWist/CLip/Hide/Off/Undo]: **CA**
Specify camera location, or enter angle from XY plane, or [Toggle (angle in)] <current>: **0.00** Enter
Specify camera location, or enter angle in XY plane from X axis, or [Toggle (angle from)] <current>: **0.00** Enter
Enter option
[CAmera/TArget/Distance/POints/PAn/Zoom/TWist/CLip/Hide/Off/Undo]: Enter
Regenerating model.

For Figure 21-44(d) (front view), the following is the prompt sequence:

Command: **DVIEW** Enter
Select objects or <use DVIEWBLOCK>: Enter
Enter option
[CAmera/TArget/Distance/POints/PAn/Zoom/TWist/CLip/Hide/Off/Undo]: **CA** Enter

Specify camera location, or enter angle from XY plane, or [Toggle (angle in)] <current>: **0.00** Enter
Specify camera location, or enter angle in XY plane from X axis, or [Toggle (angle from)] <current>: **-90.00** Enter
Enter option
[CAmera/TArget/Distance/POints/PAn/Zoom/TWist/CLip/Hide/Off/Undo]: Enter
Regenerating model.

TArget Option

With the **TArget** option you can rotate the target point with respect to the camera. To invoke the **TArget** option, enter TA at the **Enter option [CAmera/ TArget/ Distance/ POints/ PAn/ Zoom/ TWist/ CLip/ Hide/ Off/ Undo]** prompt. Once you have invoked this option, the drawing is still, but you can maneuver the target point up or down or left or right about the camera. When you move the target, the camera is stationary. The prompt sequence for the **TArget** option is:

Command: **DVIEW** Enter
Select objects or <use DVIEWBLOCK>: Enter
Enter option
[CAmera/TArget/Distance/POints/PAn/Zoom/TWist/CLip/Hide/Off/Undo]: **TA** Enter
Specify camera location, or enter angle from XY plane, or [Toggle (angle in)] <current>: *Specify the angle about which the target will lie above or below the camera.*
Specify camera location, or enter angle in XY plane from X axis, or [Toggle (angle from)] <current>: *Specify the angle about which the target will lie left or right of the camera.*
Enter option
[CAmera/TArget/Distance/POints/PAn/Zoom/TWist/CLip/Hide/Off/Undo]: Enter

The prompt sequence does not reveal the difference between the **CAmera** option and the **TArget** option. The difference lies in the actual angle of view. For example, if you specify an angle of 90 degrees in the **Specify camera location, or enter angle from XY plane, or [Toggle (angle in)] <current>** prompt of the **CAmera** option, your viewing direction is from the top of the object toward the bottom; if you specify the same angle in the same prompt for the **TArget** option, your viewing direction is from the bottom of the object toward the top.

Example 4

The different views of the house block in Figure 21-45 can be obtained with the **TArget** option of the **DVIEW** command.

First, use the **DVIEW** command to display the image of the house on the screen.

Command: **DVIEW** Enter
Select objects or <use DVIEWBLOCK>: Enter

For Figure 21-45(a) (3D view), the following is the prompt sequence:

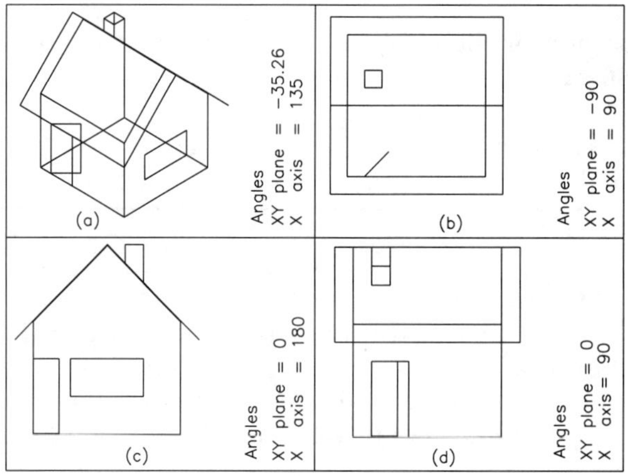

***Figure 21-45** **TArget** option of the **DVIEW** command*

Enter option
[CAmera/TArget/Distance/POints/PAn/Zoom/TWist/CLip/Hide/Off/Undo]: **TA** Enter
Specify camera location, or enter angle from XY plane, or [Toggle (angle in)] <current>: **-35.26** Enter
Specify camera location, or enter angle in XY plane from X axis, or [Toggle (angle from)] <current>: **135** Enter
Enter option
[CAmera/TArget/Distance/POints/PAn/Zoom/TWist/CLip/Hide/Off/Undo]: Enter
Regenerating model.

For Figure 21-45(b) (top view), the following is the prompt sequence:

Command: **DVIEW** Enter
Select objects or <use DVIEWBLOCK>: Enter
Enter option
[CAmera/TArget/Distance/POints/PAn/Zoom/TWist/CLip/Hide/Off/Undo]: **TA** Enter
Specify camera location, or enter angle from XY plane, or [Toggle (angle in)] <current>: **-90.00** Enter
Specify camera location, or enter angle in XY plane from X axis, or [Toggle (angle from)] <current>: **90.00** Enter
Enter option
[CAmera/TArget/Distance/POints/PAn/Zoom/TWist/CLip/Hide/Off/Undo]: Enter
Regenerating model

For Figure 21-45(c) (right side view), the following is the prompt sequence:

Command: **DVIEW** Enter
Select objects or <use DVIEWBLOCK>: Enter
Enter option

[CAmera/TArget/Distance/POints/PAn/Zoom/TWist/CLip/Hide/Off/Undo]: **TA** Enter
Specify camera location, or enter angle from XY plane, or [Toggle (angle in)] <current>: **0.00** Enter
Specify camera location, or enter angle in XY plane from X axis, or [Toggle (angle from)] <current>: **180.00** Enter
Enter option
[CAmera/TArget/Distance/POints/PAn/Zoom/TWist/CLip/Hide/Off/Undo]: Enter
Regenerating model

For Figure 21-45(d) (front view), the following is the prompt sequence:

Command: **DVIEW** Enter
Select objects or <use DVIEWBLOCK>: Enter
Enter option
[CAmera/TArget/Distance/POints/PAn/Zoom/TWist/CLip/Hide/Off/Undo]: **TA** Enter
Specify camera location, or enter angle in XY plane from X axis, or [Toggle (angle from)] <default>: **0.00** Enter
Specify camera location, or enter angle in XY plane from X axis, or [Toggle (angle from)] <default>: **90.00** Enter
Enter option
[CAmera/TArget/Distance/POints/PAn/Zoom/TWist/CLip/Hide/Off/Undo]: Enter
Regenerating model.

Distance Option

As mentioned before, the line obtained on joining the camera position and the target position is known as the line of sight. The **Distance** option can be used to move the camera toward or away from the target along the line of sight. Invoking the **Distance** option enables perspective viewing. Since we have not used the **Distance** option until now, all the previous views were in parallel projection. In perspective display, the objects nearer to the camera appear bigger than objects that are farther away from the camera position. In other words, in perspective views the parallel lines meet at a vanishing point. Another noticeable difference is that the regular coordinate system icon is replaced by the perspective icon. This icon acts as a reminder that perspective viewing is enabled. The **Distance** option can be invoked by entering **D** at the **Enter option [CAmera/TArget/Distance/POints/PAn/Zoom/TWist/CLip/Hide/Off/Undo]** prompt. Once you invoke the **Distance** option, AutoCAD LT prompts you to specify the new distance between the camera and the target.

Specify new camera-target distance <1.0000>: *Enter the desired distance between the camera and the target.*

On the top side of the screen, a slider bar appears. It is marked from 0x to 16x. The current distance is represented by the 1x mark. This is verified by the fact that the slider bar moves right or left with respect to the 1x mark on the slider bar. As you move the slider bar toward the right, the distance between the camera and the target increases; as you move the slider bar toward the left, the distance between the camera and the target decreases. For example, when you move the slider bar to the 16x mark, the distance between the camera and the target increases 16 times, or you can say that the camera moved away from the target on the line of sight sixteen times the previous distance. The distance between the camera and the target is dynamically displayed in the status line. If you cannot display the entire object on the screen by moving the slider bar to the 16x mark, enter a larger distance at the keyboard. To revert to parallel viewing, invoke the **Off** option. If you want to magnify the drawing without turning the Perspective viewing on, you can use the **Zoom** option.

Example 5

Let Figure 21-46(a) be the default figure, in which the distance between camera and target is 4 units, which corresponds to the 1x mark on the slider bar. You can change the distance between camera and target and get different views as follows:

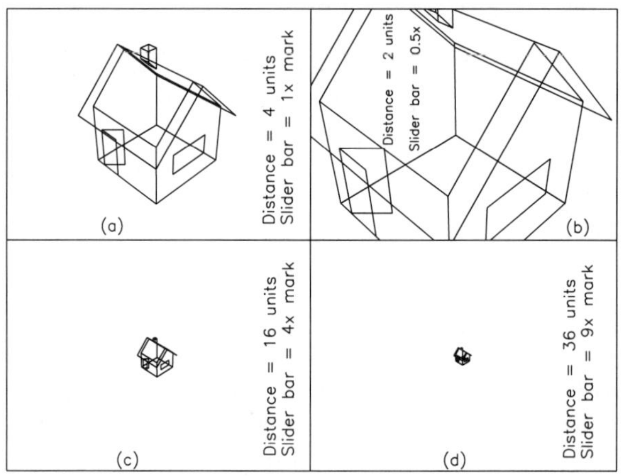

***Figure 21-46 Distance** option of the **DVIEW** command*

For Figure 21-46(b):

Command: **DVIEW** Enter
Select objects or <use DVIEWBLOCK>: Enter
Enter option
[CAmera/TArget/Distance/POints/PAn/Zoom/TWist/CLip/Hide/Off/Undo]: **D** Enter
Specify new camera-target distance <4.0000>: **2.0** Enter

For Figure 21-46(c):

Command: **DVIEW** Enter
Select objects or <use DVIEWBLOCK>: Enter
Enter option
[CAmera/TArget/Distance/POints/PAn/Zoom/TWist/CLip/Hide/Off/Undo]: **D** Enter
Specify new camera-target distance <4.0000>: **16.0** Enter *or move the pointer to the 4x mark on slider bar.*

For Figure 21-46(d):

Command: **DVIEW** [Enter]
Select objects or <use DVIEWBLOCK>: [Enter]
Enter option
[CAmera/TArget/Distance/POints/PAn/Zoom/TWist/CLip/Hide/Off/Undo]: **D** [Enter]
Specify new camera-target distance <4.0000>: **36.0** [Enter] *or move the pointer to the 9x mark on slider bar.*

POints Option

With the **POints** option, you can specify the camera and target positions (points) in X, Y, Z coordinates. You can specify the X, Y, Z coordinates of the point by any method used to specify points, including object snap and .X, .Y, .Z point filters. The X, Y, Z coordinate values are with respect to the current UCS. If you use the object snap to specify the points, you must type the name of the object snap. This option can be invoked by entering PO at the **Enter option [CAmera/TArget/Distance/POints/PAn/Zoom/TWist/CLip/Hide/Off/Undo]:** prompt.

Command: **DVIEW** [Enter]
Select objects or <use DVIEWBLOCK>: [Enter]
Enter option
[CAmera/TArget/Distance/POints/PAn/Zoom/TWist/CLip/Hide/Off/Undo]: **PO** [Enter]

The target point needs to be specified first. A rubber-band line is drawn from the current target position to the drawing crosshairs. This is the line of sight.

Specify target point <current>: *Specify the location of the target or press ENTER.*

Once you have specified the target point, you are prompted to specify the camera point. A rubber-band line is drawn between the target point and the drawing crosshairs. This helps you to place the camera relative to the target.

Enter camera point <current>: *Specify the location of the camera press ENTER.*

Establishment of the new target point and camera point should be carried in parallel projection. If you specify these two points while the perspective projection is active, the perspective projection is temporarily turned off, until you specify the camera and target points. Once this is done, the object is again displayed in perspective. If the viewing direction is changed by the new target location and camera location, the preview image is regenerated to show the change. In Figure 21-47, the target is located at the lower corner of the house, and the camera is located at the corner of the chimney. In Figure 21-48, the camera is located at the lower corner of the house, and the target is located at the corner of the chimney. Both these points are marked by cross marks.

PAn Option

The **PAn** option of the **DVIEW** command resembles the **PAN** command. This option lets you shift the entire drawing with respect to the graphics display area. Just as with the **PAN** command,

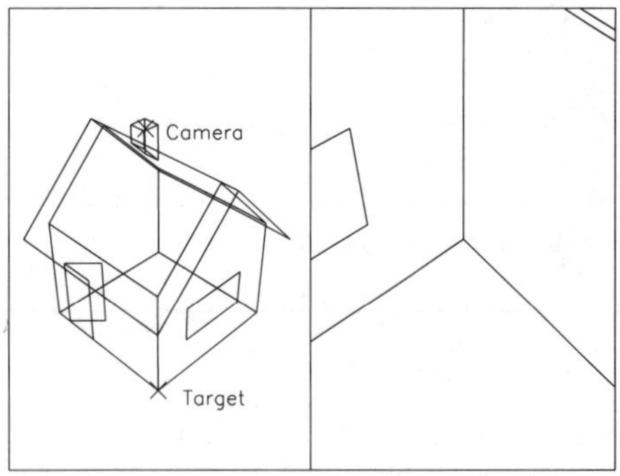

Figure 21-47 *The* ***Point*** *option of the* ***DVIEW*** *command*

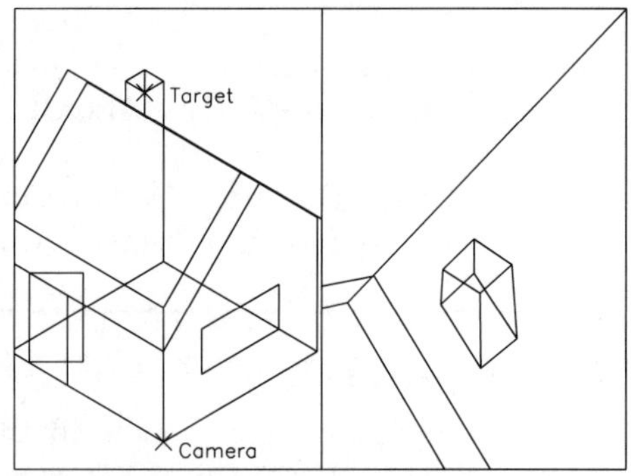

Figure 21-48 *The* ***Point*** *option of the* ***DVIEW*** *command*

you have to specify the pan distance and direction by specifying two points. You must use a pointing device to specify the two points if perspective viewing is active. The prompt sequence for this option is:

Command: **DVIEW** Enter
Select objects or <use DVIEWBLOCK>: Enter
Enter option
[CAmera/TArget/Distance/POints/PAn/Zoom/TWist/CLip/Hide/Off/Undo]: **PA** Enter
Specify displacement base point: *Specify the first point.*
Specify second point: *Specify the second point.*

Zoom Option

The **Zoom** option of the **DVIEW** command resembles the **ZOOM** command. With the help of this option, you can enlarge or reduce the drawing. This option can be invoked by entering Z at the following prompt:

Enter option
[CAmera/TArget/Distance/POints/PAn/Zoom/TWist/CLip/Hide/Off/Undo]: **Z** Enter
Specify lens length <50.000mm>: *Specify the new lens length.*

Just as with the **Distance** option, in the **Zoom** option a slider bar marked from 0x to 16x is displayed on the top side of the screen. The default position on the slider bar is 1x. Two ways to zoom can now be specified. The first is when perspective is enabled. In this case, zooming is defined in terms of lens length. The 1x mark (default position) corresponds to a 50.000 mm lens length. As you move the slider bar toward the right, the lens length increases, and as you move the slider bar toward the left, the lens length decreases. For example, when you move the slider bar to the 16x mark, the lens length increases sixteen times, which is 16 x 50.000 mm = 800.000 mm. You can simulate the telephoto effect by increasing the lens length; by reducing the lens length, you can simulate the wide-angle effect. The lens length is dynamically displayed in the status bar. If perspective is not enabled, zooming is defined in terms of the zoom scale factor. In this case, the **Zoom** option resembles the **ZOOM Center** command, and the center point lies at the center of the current viewport. The 1x mark (default position) corresponds to a scale factor of 1. As you move the slider bar toward the right, the scale factor increases; as you move the slider bar toward the left, the scale factor decreases. For example, when you move the slider bar to the 16x mark, the scale factor increases 16 x 1 = 16 times. The scale factor is dynamically displayed on the status line.

Command: **DVIEW** [Enter]
Select objects or <use DVIEWBLOCK>: *Select the objects.*
Enter option
[CAmera/TArget/Distance/POints/PAn/Zoom/TWist/CLip/Hide/Off/Undo]: **Z** [Enter]
Specify zoom scale factor <1>: *Specify the scale factor.*

Example 6

In this example you will see the effect of the **Zoom** option in perspective projection. In Figure 21-49(a), the lens length is set to 25 mm. This can be realized in the following manner:

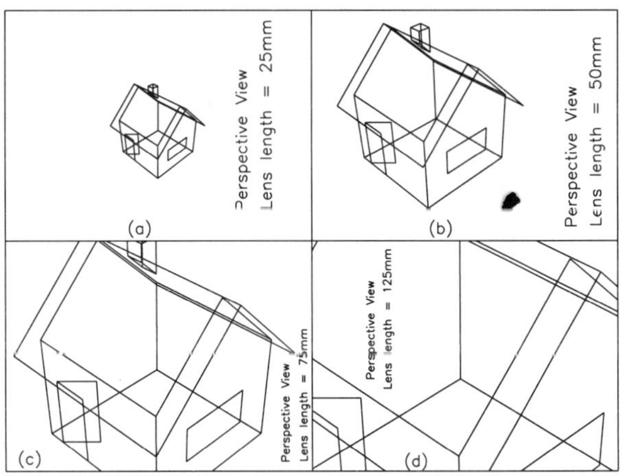

***Figure 21-49 Zoom** option of the **DVIEW** command*

Command: **DVIEW** [Enter]
Select objects or <use DVIEWBLOCK>: [Enter]
Enter option
[CAmera/TArget/Distance/POints/PAn/Zoom/TWist/CLip/Hide/Off/Undo]: **Z** [Enter]
Specify lens length <50.000mm>: **25** [Enter]

Similarly, for Figure 21-49(b), (c), and (d), you can set lens lengths to 50 mm, 75 mm, and 125 mm, respectively.

TWist Option

The **TWist** option allows you to rotate (twist) the view around the line of sight. You can also say that the object on the screen is rotated around the center point of the screen because the display is always adjusted so that the target point is at the center of the screen. If you use a pointing device to specify the angle, the angle value is dynamically displayed on the status line. A rubber-band line is drawn from the center (target point) to the drawing crosshairs, and as you move the crosshairs with a pointing device, the object on the screen is rotated around the line of sight. You can also enter the angle of twist from the keyboard. The angle of twist is measured in a counterclockwise direction starting from the right side.

Command: **DVIEW** Enter
Select objects or <use DVIEWBLOCK>: Enter
Enter option
[CAmera/TArget/Distance/POints/PAn/Zoom/TWist/CLip/Hide/Off/Undo]: **TW** Enter
Specify view twist angle <0.00>: *Specify the angle of rotation (twist).*

In Figure 21-50, different twist angles have been specified. For (a) the twist angle is 0, for (b) it is 338 degrees, for (c) it is 37 degrees, and for (d) it is 360 degrees.

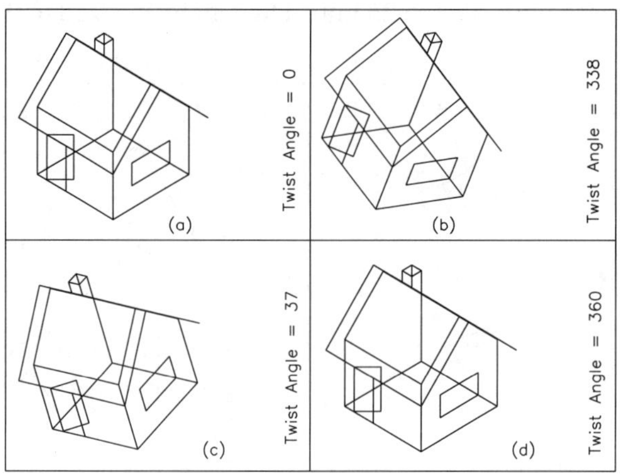

Figure 21-50 ***TWist*** *option of the* ***DVIEW*** *command*

CLip Option

The **CLip** option can be used to clip sections of the drawing (Figure 21-51). AutoCAD LT uses two invisible clipping planes to realize clipping. These clipping walls can be positioned anywhere on the screen and are perpendicular to the line of sight (line between target and camera). Once you position the clipping planes, AutoCAD LT conceals all the lines that are in front of the front clipping plane or behind the back clipping plane. The **CLip** option can be used in both parallel and perspective projections. When perspective is enabled, the front clipping plane is automatically enabled.

Note

If you have specified a positive distance, the clipping plane is placed between the target and the camera. If the distance you have specified is negative, the clipping plane is placed beyond the target.

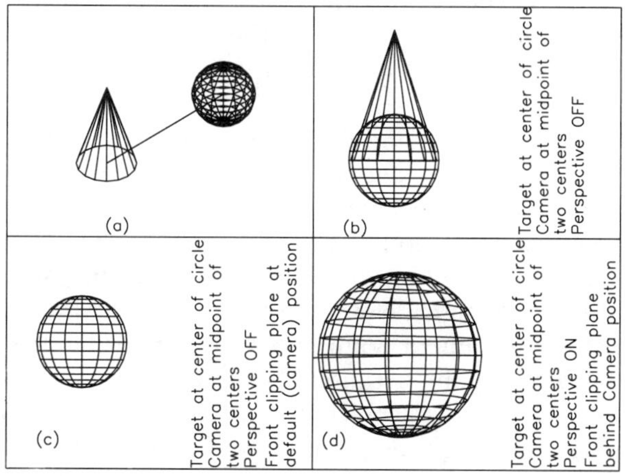

Figure 21-51 *Using the* ***CLip*** *option to clip the view*

The prompt sequence for the **CLip** option is:

Enter option
[CAmera/TArget/Distance/POints/PAn/Zoom/TWist/CLip/Hide/Off/Undo]: **CL** [Enter]
Enter clipping option [Back/Front/Off]<Off>: **B** [Enter]
Specify distance from target or [ON/OFF] <current>:

Once you have specified which clipping plane you want to set (the back clipping plane in our case), a slider bar appears on the screen. As you move the pointer toward the right side of the slider bar, the negative distance between the target and the clipping plane increases. As you move toward the left side of the slider bar, the positive distance between the target and the clipping plane increases; therefore, a greater portion of the drawing is clipped. The rightmost mark on the slider bar corresponds to a distance equal to the distance between the target and the farthermost point on the object you want to clip. After specifying the distance between one of the clipping planes (back clipping plane in this prompt sequence), you need to invoke the **CLip** option again and to specify the position of the front clipping plane in terms of the distance between the front clipping plane and the target. As you move the slider bar toward the right, the negative distance between the target and the front clipping plane increases. As the negative distance increases, a greater portion of the front side of the drawing is clipped. The rightmost mark on the slider bar corresponds to a distance equal to the distance between the target and the back clipping plane.

Enter option
[CAmera/TArget/Distance/POints/PAn/Zoom/TWist/CLip/Hide/Off/Undo]: **CL** [Enter]
Enter clipping option [Back/Front/Off] <Off>: **F** [Enter]
Specify distance from target or [set to Eye(camera)/ON/OFF] <1.0000>: *Specify the distance from target.*

To illustrate the concepts behind clipping, let us display two objects (a sphere and a cone). Let the distance between the two be 10 units. Draw a line between the center of the sphere and the center of the cone. Use the **POints** option to position the target at the center of the sphere and the camera at the midpoint of the line joining the two center points. In this way, you have defined the line between the two center points as the line of sight.

Now that you have defined the target and the camera, you will see the sphere and the cone overlapping because both of them are aligned along the line of sight. You may wonder how you are able to see the cone, since it lies behind the camera position that is analogous to the eye. This is because in parallel projection, the camera is not analogous to the eye, but the line of sight is; therefore, you see everything that lies in the field of vision (determined by the **Zoom** option) about the line of sight. Now, when you use the **CLip** option and the **Front** suboption to positioned at the camera point. Now you cannot see the objects (cone in our case) that are behind the camera point because the default position of the front clipping plane is at the camera position; therefore, any object in front of the front clipping plane is clipped. Also, if you have placed the front clipping plane behind the camera position and the perspective projection is on, the camera position is used as the front clipping plane position.

If you change the front clipping plane position, and then at some stage you want to go back to the default front clipping plane position, you can use the **Eye** option. In perspective, without invoking the **CLip** option, the front clipping plane is placed at the camera position; therefore, you are able to see only what lies in front of the camera.

Example 7

In Figure 21-52, the front clipping plane has been established at a distance of 0.50 units. The distance between the camera and the target is 3.50 units. After invoking the **CLip** option, the prompt sequence is:

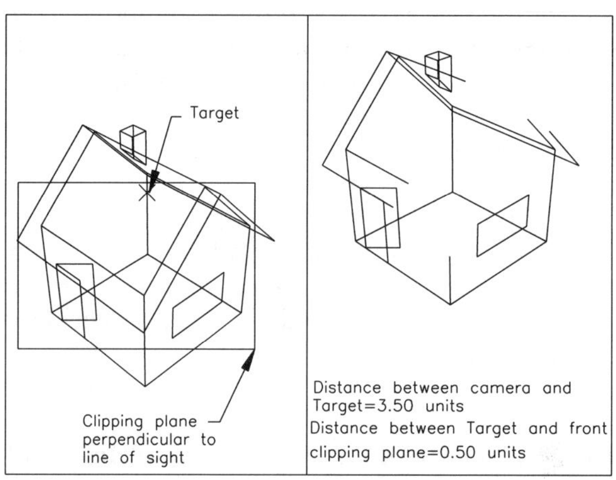

***Figure 21-52 Front** clipping with the **DVIEW** command*

Enter clipping option [Back/Front/Off]<Off>: **F** [Enter]
Specify distance from target or [set to Eye(Camera)/On/OFF] <3.5000>: **0.50** [Enter]

In Figure 21-52, the back clipping plane has been established at a distance of 0.50 units. The distance between the camera and the target is 3.50 units. After invoking the **CLip** option, the prompt sequence is:

Enter clipping option [Back/Front/Off]<Off>: **B** Enter
Specify distance from target or [ON/OFF] <1.0000>: **0.50** Enter

You might have noticed that if you combine the clipped shape in Figure 21-52 and the clipped shape in Figure 21-53, you get the original shape of the house. The reason is that in the first clipped figure, whatever lies between the camera and the front clipping plane is clipped, whereas in the second clipped figure, whatever lies behind the back clipping plane is clipped. Since the distances between the clipping planes and the target, and between camera and target are identical in both figures, then if you combine the clipped figures, you will get the original shape.

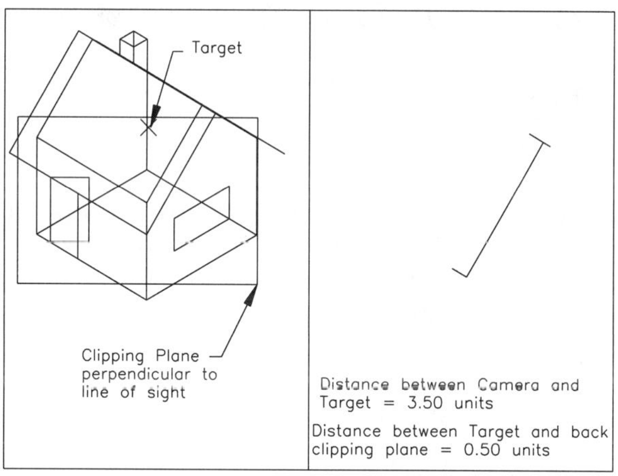

***Figure 21-52 Back** clipping with the **DVIEW** command*

Hide Option

In most 3D drawings, some of the lines lie behind other lines or objects. Sometimes, you may not want these hidden lines to show up on the screen, so you can use the **Hide** option (Figure 21-53). This option is similar to the **HIDE** command. To invoke **Hide** option, enter H at the **Enter option [CAmera/TArget/Distance/POints/PAn/Zoom/TWist/CLip/Hide/Off/Undo]** prompt.

Off Option

The **Off** option turns the perspective projection off. The prompt sequence is:

Enter option
[CAmera/TArget/Distance/POints/PAn/Zoom/TWist/CLip/Hide/Off/Undo]: **O** Enter

When the perspective projection is turned off, you will notice that the perspective icon is

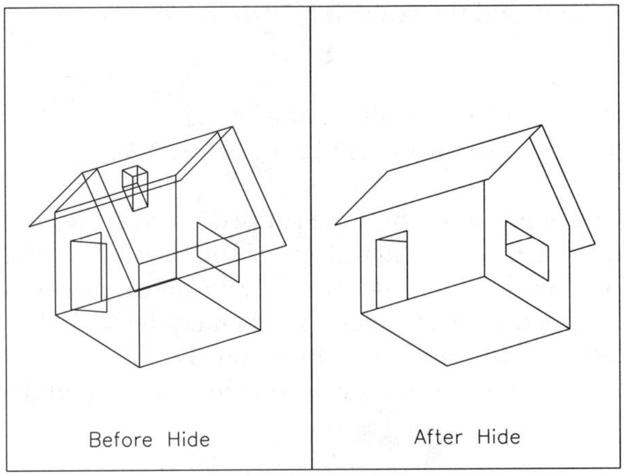

***Figure 21-53 Hide** option of the **DVIEW** command*

replaced by the regular UCS icon.

Undo Option

The **Undo** option is similar to the **UNDO** command. The **Undo** option nullifies the result of the previous **DVIEW** operation. Just as in the case of the **UNDO** command, you can use this option a number of times to undo the results of multiple **DVIEW** operations. To invoke this option, enter U at the following prompt:

Enter option
[CAmera/TArget/Distance/POints/PAn/Zoom/TWist/CLip/Hide/Off/ Undo]: **U** Enter

CREATING SHADED IMAGES

In a 3D object the **HIDE** command can be used to hide the hidden line in the object. By hiding the lines, you can get a better idea about the shape of the object. To get a more realistic image of the object, you can use the **SHADE** and **SHADEMODE** commands.

SHADE Command

Command: SHADE

As you use the **SHADE** command, the hidden lines of the object are removed and the shaded image of the object is displayed. The color of the shaded image depends upon the color of the object. The single light source used here to shade the object is assumed to be placed just over the user's shoulder. Any subsequent changes made in the drawing are not displayed until the **SHADE** command is used again. You have to use the **REGEN** or **ZOOM** command to regenerate the drawing and display the 3D wireframe model.

SHADEMODE Command

Command:	SHADEMODE

The **SHADEMODE** command performs a hide and then creates a flat shaded picture in the current viewport. The shading uses a single light source that is assumed to be located just behind the user over the shoulder. The shaded image can be edited and the **UNDO** command cannot be used to undo shading. Use the **REGEN** or **ZOOM** command to regenerate the drawing and display the 3D wireframe of the object. This command can also be invoked by choosing **Shade** from the **View** menu. The command prompts are:

Command: **SHADEMODE** [Enter]
Enter option [2D wireframe/Hidden]<current>: *Enter an option.*

Setting the Shading Method

You can use the **SHADEDGE** system variable to set the shading method. If **SHADEDGE** is 0, AutoCAD LT creates a shaded image with no edges highlighted. If **SHADEDGE** is 1, AutoCAD LT creates a shaded image with edges highlighted in the background color. To see the effect of these two shading methods, you need 256 color display.

If **SHADEDGE** is 2, AutoCAD LT paints the surfaces of the object in the background color and displays the visible edges in the object's color (Figure 21-54). If **SHADEDGE** is 3 (default setting), AutoCAD LT paints the faces in the object's color and displays the edges in the background color (Figure 21-55).

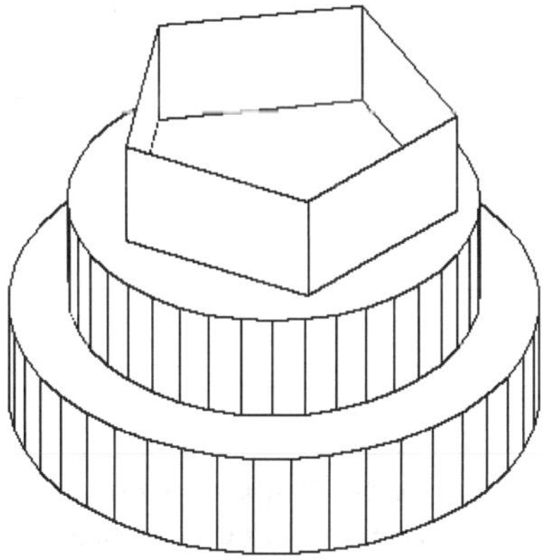

Figure 21-54 *Shading with **SHADEDGE** = 2, object white, and background white*

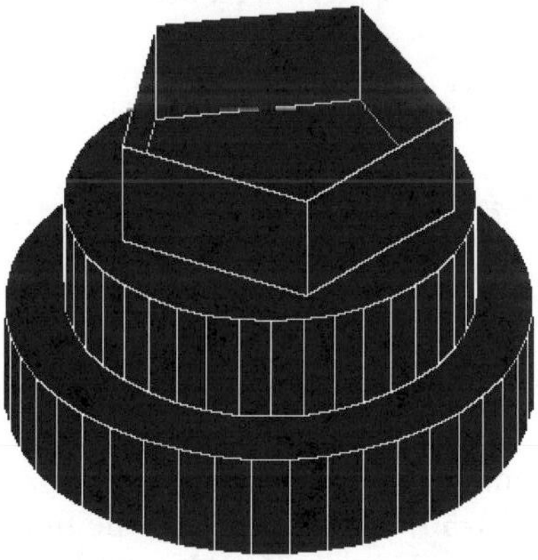

Figure 21-55 *Shading with **SHADEDGE** = 3, object black, and background white*

Setting Diffuse Reflection

The system variable **SHADEDIF** is used to control the amount of light diffused that AutoCAD LT uses to calculate the shade for each surface. The default value of **SHADEDIF** is 70. It can range from 0 to 100.

ANALYZING THE REGIONS (MASSPROP COMMAND)

Toolbar:	Inquiry > Region/Mass Properties
Menu:	Tools > Inquiry > Region/Mass Properties
Command:	MASSPROP

The **MASSPROP** command can be used to analyze a region. This command will automatically calculate the mass properties of the region. When you enter this command, AutoCAD LT will list the properties of the region.

Command: **MASSPROP** Enter
Select objects: *Select the region.*

For Coplanar and Noncoplanar Regions. The information displayed for coplanar regions is similar to the following:

———————— REGIONS ————————

Area: 13.6480
Perimeter: 21.2475
Boundary box: X: 3.3610 — 8.5063
Y: 4.1330 — 6.4829
Centroid: X: 5.9337
Y: 5.3127
Moments of inertia: X: 390.8867
Y: 512.2000
Product of inertia: XY: 430.2535
Radii of gyration: X: 5.3516
Y: 6.1260
Principal moments and X-Y directions about centroid
I: 5.6626 along [1.0000 0.0000]
J: 31.6536 along [0.0000 1.0000]

Write analysis to a file ?<N>:

If you enter Y (Yes) at this last prompt, the **Create Mass and Area Properties File** dialog box (Figure 21-56) is displayed. All the file names of the .MPR type are listed. You can enter the name of the file in the **File Name** edit box. The file is automatically given the .MPR extension.

The various terms displayed on the screen as a result of invoking the **MASSPROP** command are explained next.

Area. It is the enclosed area of the region.

Perimeter. Total length of inside and outside loops of region.

Bounding Box. For regions that are coplanar with the XY plane of the current UCS, the

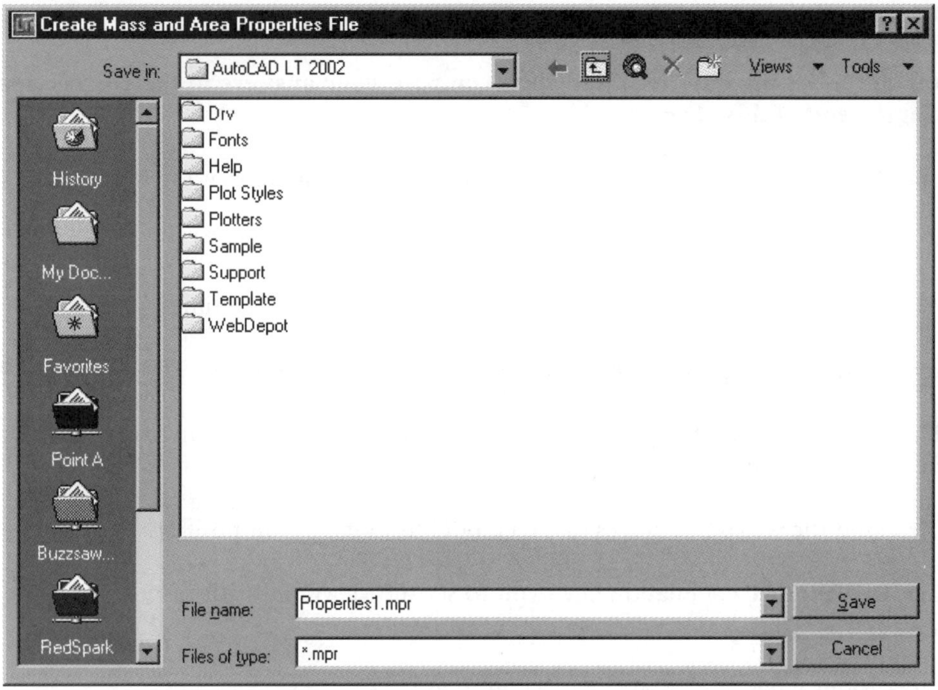

***Figure 21-56 Create Mass and Area Properties File** dialog box*

bounding box is defined by the diagonally opposite corners of a rectangle that encloses the region. For regions that are not coplanar with the XY plane of the current UCS, the bounding box is defined by the diagonally opposite corners of a 3D box that encloses the region.

Centroid. This provides the coordinates of the center of area for the selected region.

Moments of Inertia. This property provides the mass moments of inertia of a region about the two axes. The equation used to calculate this value is:

area_moments_of_inertia = area_of_interest * (square of radius)

Products of Inertia. The value obtained with this property helps to determine the force resulting in the motion of the object. The equation used to calculate this value is:

product _of_inertia YX,XZ = mass * dist centroid_to_YZ * dist centroid_to_XZ

Radii of Gyration. The equation used to calculate this value is:

gyration_radii = (moments_of_inertia/body_mass)1/2

Principal Moments and X-Y-Z Directions about Centroid. This property provides you with the highest, lowest, and middle value for the moment of inertia about an axis passing through the centroid of the object.

Self-Evaluation Test

Answer the following questions and then compare your answers to the correct answers given at the end of this chapter.

1. The **DDVPOINT** command is used to change the viewpoint to view the solid model. (T/F)

2. The 1x mark (default position) corresponds to the scale factor of 1. (T/F)

3. When you move the slider bar to 16x mark, the scale factor is 8 times. (T/F)

4. You can draw a 3D polyline in space using the **PLINE** command. (T/F)

5. The _________ and the _________ commands can be used to change the viewpoint for viewing the models in the 3D space.

6. Changing the viewpoint moves the 3D model from its default position. (T/F)

7. The **DDVPOINT** command allows you to set the viewpoint with respect to _________ and _________.

8. The various types of 3D coordinate systems are _________.

9. The line formed between the target and the camera is known as _________, or _________.

10. The _________ option allows you to shift from parallel viewing to perspective viewing.

Review Questions

Answer the following questions.

1. You can perform Boolean operations on the regions. (T/F)

2. The **ELEV** command is a transparent command. (T/F)

3. You can directly write a text with thickness. (T/F)

4. Negative value for the elevation shifts the workplane in X axis direction. (T/F)

5. Which command is used to create a three-dimensional polyline?

(a) **POLYLINE** (b) **3DPOLY**
(c) **3DPOLYLINE** (c) **POLY3D**

6. Which command is used for setting the elevation and thickness for new objects?

(a) **ELEVATION** (b) **THICKNESS**
(c) **ELEV** (d) **THICK**

7. Which command is used to suppress the hidden edges in the 3D model?

(a) **SUPPRESS** (b) **HIDE**
(c) **3DHIDE** (d) **EDGE**

8. When you open a new drawing, you are by default in which view?

(a) SE Isometric View (b) SW Isometric View
(c) Plan View (d) Bottom View

9. Which system variable has to be modified so that the ellipse should be drawn with thickness?

(a) **ELLIPSEP** (b) **ELLIPSER**
(c) **RELLIPSE** (d) **PELLIPSE**

10. The __________ command is used to suppress the display of the edges that lie behind other object.

11. To draw an ellipse with thickness you have to set the value of the __________ system variable to __________.

12. The __________ system variable is used to define the color of the hidden lines.

13. The __________ system variable is used to define the linetype of the hidden lines

14. The mass properties of a region can be written to a file with an extension of __________.

15. You can define your own block that can be used in the **DVIEW** command. This block can be defined using the __________ command.

Exercises

Exercises 4 through 6 *General*

In these exercises you will create the models shown in Figures 21-57 through 21-59. Assume the missing dimensions for the models.

Hints:
Draw the wireframe models and then convert them into regions using the **REGION** command. Hide them to get the display similar to the one shown in the figures.

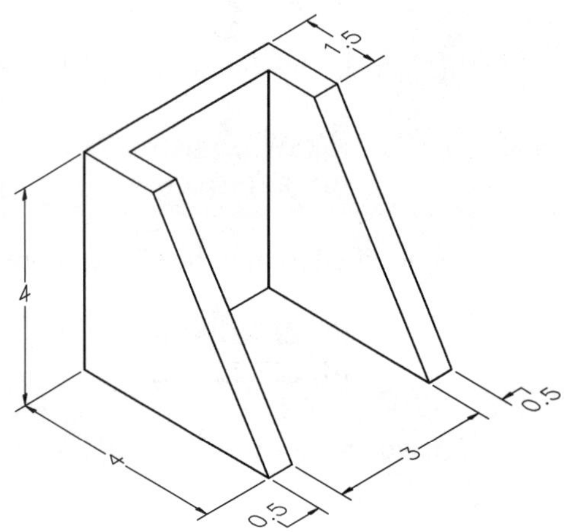

Figure 21-57 *Model for Exercise 4*

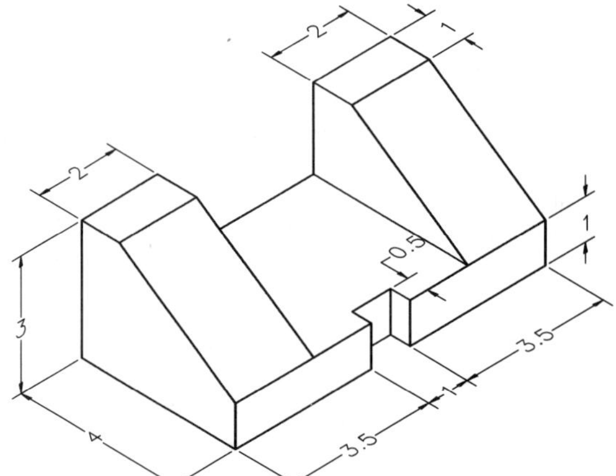

Figure 21-58 *Model for Exercise 5*

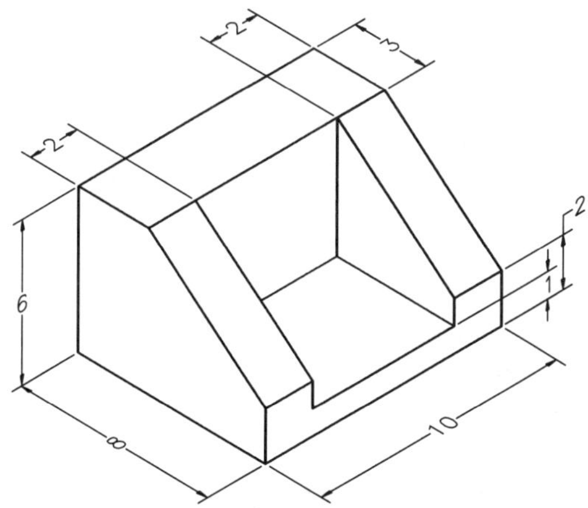

Figure 21-59 *Model for Exercise 6*

Problem Solving Exercise 1 *General*

In this you will create the computer shown in Figure 21-60. You can create the wireframe model and then convert it into regions to get the display similar to the one shown in figure. For you convenience, another image of computer is also shown in Figure 21-61. This figure shows all the lines including the hidden lines that are behind the visible surface. Take the dimensions of the computer you are working on.

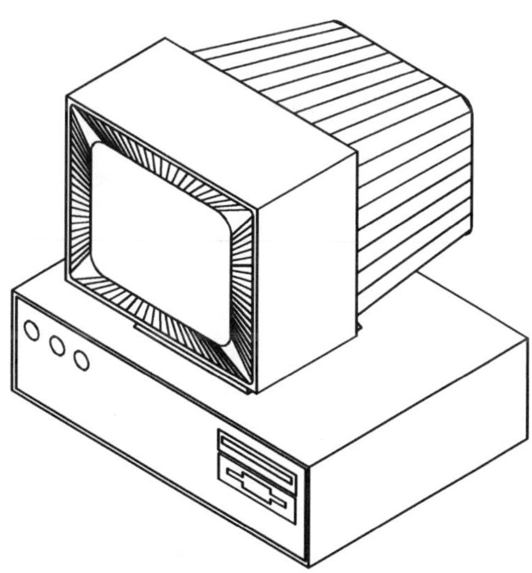

Figure 21-60 *Model for Problem Solving Exercise 1*

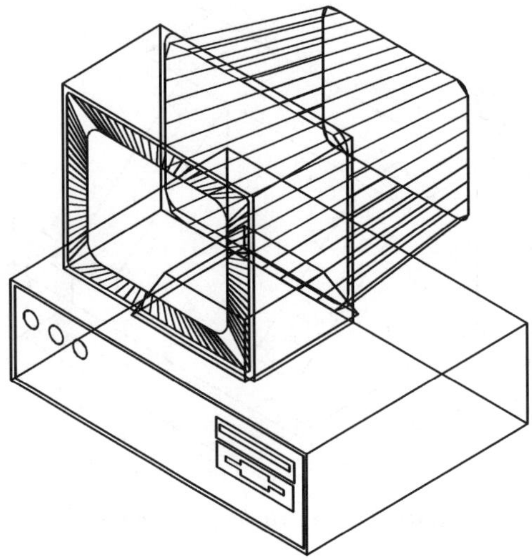

Figure 21-61 *Model for Problem Solving Exercise 1*

Answers to Self-Evaluation Test

1 - T, **2** - T, **3** - F, **4** - F, **5** - **DDVPOINT**, **VPOINT**, **6** - F, **7** - Angle in XY plane from X axis, Angle from XY plane, **8** - Absolute, Relative rectangular, Relative cylindrical, and Relative spherical, **9** - line of sight, view direction, **10** - **Distance**

AutoCAD LT Part III (Customizing)

Author's Web Sites

For Faculty: Please contact the author at **stickoo@calumet.purdue.edu** or **tickoo@cadcim.com** to access the Web site that contains the following:

1. **PowerPoint presentations, programs, and drawings used in this textbook.**
2. **Syllabus, chapter objectives and hints, and questions with answers for every chapter.**

For Students: You can download drawing-exercises, tutorials, programs, and special topics by accessing author's web site at **www.cadcim.com** or **www.calumet.purdue.edu/public/mets/tickoo/index.htm**.

Chapter 22

Template Drawings

Learning Objectives

After completing this chapter, you will be able to:

- *Create template drawings.*
- *Load template drawings using dialog boxes and the command line.*
- *Do an initial drawing setup.*
- *Customize drawings with layers and dimensioning specifications.*
- *Customize drawings with layouts, viewports, and paper space.*

CREATING TEMPLATE DRAWINGS

One way to customize AutoCAD LT is to create template drawings that contain initial drawing setup information and, if desired, visible objects and text. When the user starts a new drawing, the settings associated with the template drawing are automatically loaded. If you start a new drawing from scratch, AutoCAD LT loads default setup values. For example, the default limits are (0.0,0.0), (12.0,9.0) and the default layer is 0 with white color and continuous linetype. Generally, these default parameters need to be reset before generating a drawing on the computer using AutoCAD LT. A considerable amount of time is required to set up the layers, colors, linetypes, limits, snaps, units, text height, dimensioning variables, and other parameters. Sometimes, border lines and a title block may also be needed.

In production drawings, most of the drawing setup values remain the same. For example, the company title block, border, layers, linetypes, dimension variables, text height, LTSCALE, and other drawing setup values do not change. You will save considerable time if you save these values and reload them when starting a new drawing. You can do this by making template drawings, which can contain the initial drawing setup information, set according to company specifications. They can also contain a border, title block, tolerance table, block definitions, floating viewports in the paper space, and perhaps some notes and instructions that are common to all drawings.

STANDARD TEMPLATE DRAWINGS

The AutoCAD LT software package comes with standard template drawings like aclt.dwt, acltiso.dwt, ansi a.dwt, din a.dwt, iso a4.dwt, jisa3.dwt. The ansi, din, and iso template drawings are based on the drawing standards developed by ANSI (American National Standards Institute), DIN (German), and ISO (International Organization for Standardization). When you start a new drawing and you are using traditional startup dialog, AutoCAD LT displays the **Create New Drawing** dialog box on the screen. To load the template drawing, select the **Use a Template** button and AutoCAD LT will display the list of standard template drawings. From this list you can select any template drawing according to your requirements. If you want to start a drawing with default settings, select the **Start from Scratch** button in the **Create New Drawing** dialog box. The following are some of the system variables, with the default values that are assigned to a new drawing:

System Variable Name	Default Value
CHAMFERA	0.5000
CHAMFERB	0.5000
COLOR	Bylayer
DIMALT	Off
DIMALTD	2
DIMALTF	25.4
DIMPOST	None
DIMASO	On
DIMASZ	0.18
FILLETRAD	0.5000
GRID	0.5000
GRIDMODE	0
ISOPLANE	Left
LIMMIN	0.0000,0.0000
LIMMAX	12.0000,9.0000
LTSCALE	1.0
MIRRTEXT	1 (Text mirrored like other objects)
TILEMODE	1 (On)

Example 1

Create a template drawing using **Advanced Setup** of the **Create Drawings tab** with the following specifications. The name of the template drawing is **PROTO1**.

Units	Engineering with precision 0'-0.00"
Angle	Decimal degrees with precision 0.
Angle Direction	Counterclockwise
Area	144' x 96'

Step 1

Choose **New** from the **File** menu to display the **AutoCAD LT Today** window as shown in Figure 22-1. In this dialog box choose the **Create Drawings** tab and then choose the **Wizards** option. Now, select the **Advanced Setup** option.

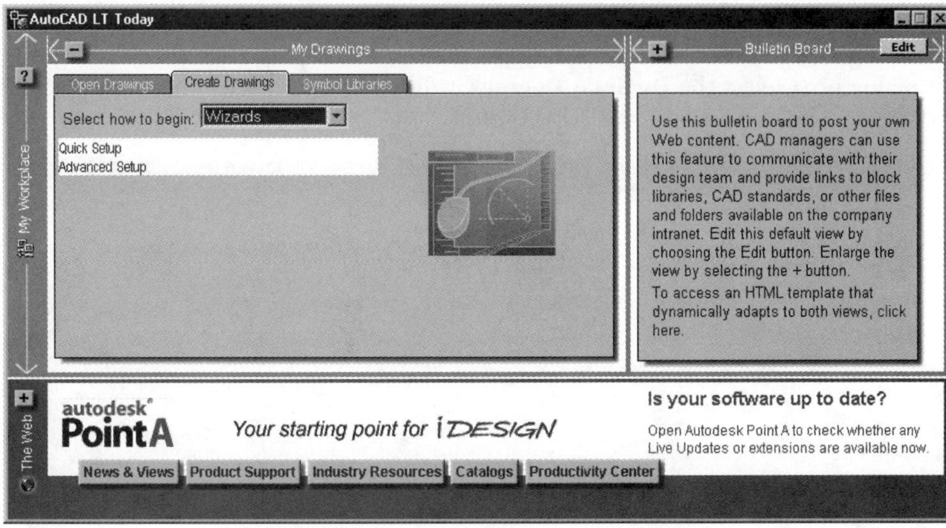

Figure 22-1 AutoCAD LT Today window

Step 2

When you select the **Advanced Setup** option, the **Units** page of the **Advanced Setup** dialog box is displayed as shown in Figure 22-2. Select the **Engineering** radio button. Select **0'-0.00"** precision from the **Precision** drop-down list and then choose the **Next** button.

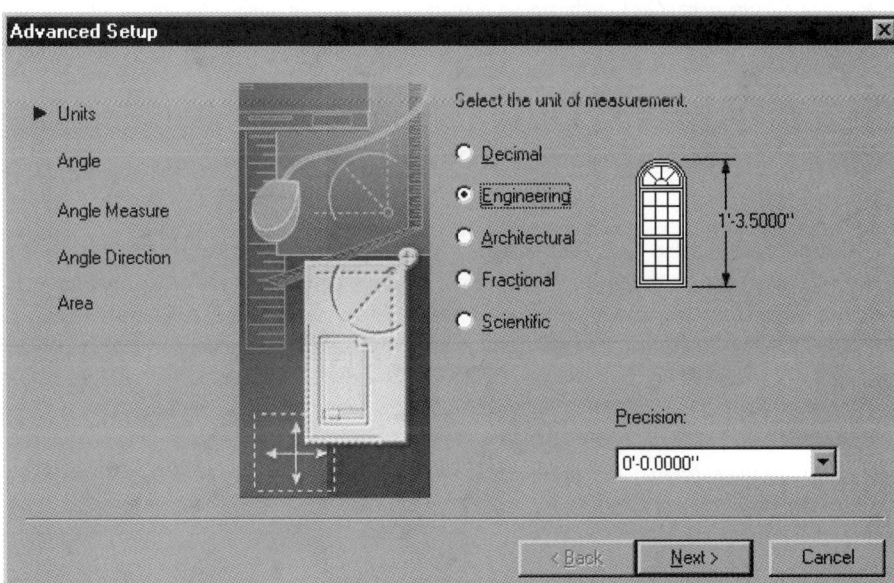

*Figure 22-2 **Advanced Setup** dialog box to define **Units***

Step 3

Now, the **Angle** page of the **Advanced Setup** dialog box is displayed as shown in Figure 22-3. In this dialog box, select the **Decimal Degrees** radio button and select **0** from the **Precision** drop-down list and then choose the **Next** button.

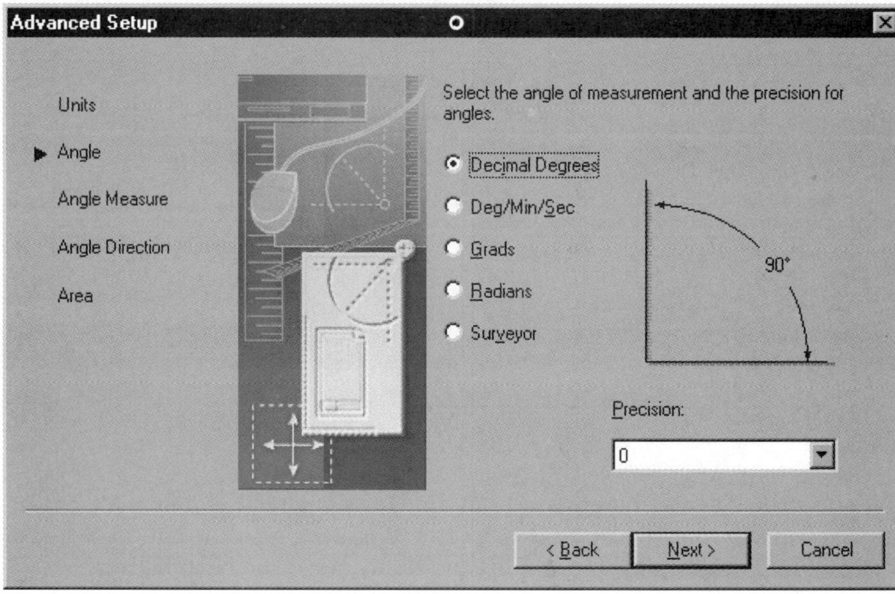

Figure 22-3 ***Advanced Setup*** *dialog box to define* ***Angle***

Step 4

The **Angle Measure** page of the **Advanced Setup** dialog box is displayed. Select the **East** radio button. Choose the **Next** button to display the **Angle Direction** page.

Step 5

Select the **Counterclockwise** radio button and then choose the **Next** button. The **Area** page is displayed. Specify the area as 144' and 96' by entering the value of width and length as **144'** and **96'** in the **Width** and **Length** edit boxes and then choose the **Finish** button. Use the **All** option of the **ZOOM** command to display the new limits on the screen. Save the template drawing as **PROTO1.DWT.**

Note

If you want to customize only units and area, you can use the ***Quick Setup*** *in the* ***Create New Drawing*** *dialog box.*

Example 2

Create a template drawing with the following specifications. The name of the template drawing is PROTO2.

Limits	18.0,12.0
Snap	0.25
Grid	0.50
Text height	0.125
Units	2 digits to the right of decimal point
	Decimal degrees
	2 digits to the right of decimal point
	0 angle along positive X axis (east)
	Angle positive if measured counterclockwise

Step 1

Start AutoCAD LT and select the **Start from Scratch** button in the **Create Drawings** tab of **AutoCAD LT Today** window. You can also invoke the **AutoCAD LT Today** window (Figure 22-4) by selecting **New** in the **File** menu or entering **NEW** at the AutoCAD LT Command prompt. Once you are in the **Drawing Window**, use the following AutoCAD LT commands to set up the values.

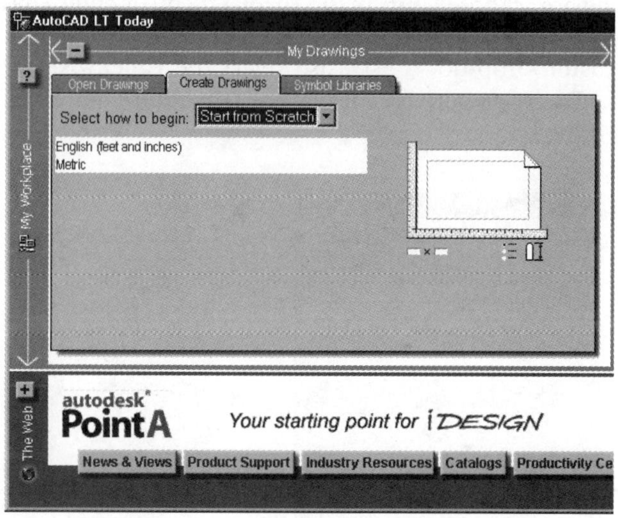

***Figure 22-4 AutoCAD LT Today** window*

Step 2

Setting limits, snap, grid, and text size

The **LIMITS** command can be invoked by choosing **Drawing Limits** from the **Format** menu or by entering **LIMITS** at the Command prompt.

Command: **LIMITS**
Specify lower left corner or [ON/OFF] <0.00,0.00>: **0,0**
Specify upper right corner <12.0,9.0>: **18.0,12.0**

After you set the limits, use the **ZOOM** command with the **All** option to display the new limits on the screen.

To set the **SNAP** and **GRID**, right-click the **Snap** or **Grid** button in the status bar to display the shortcut menu. Choose the **Settings** in the shortcut menu to display the **Drafting Settings** dialog box. You can also choose the **Object Snap Settings** button from the **Object Snap** toolbar to display the **Drafting Settings** dialog box. Choose the **SNAP** and **GRID** tab. Enter **0.25** and **0.25** in the **Snap X spacing** and **Snap Y spacing** edit boxes respectively. Enter **0.5** and **0.5** in the **Grid X spacing** and **Grid Y spacing** edit boxes respectively. Then choose **OK**. You can also use **SNAP** and **GRID** commands to set these values.

Size of the text can be changed by entering **TEXTSIZE** at the Command prompt.

Command: **TEXTSIZE**
Enter new value for TEXTSIZE <0.2000>: **0.125**

Step 3
Setting units
You can use the **Drawing Units** dialog box (Figure 22-5) to set the units. To invoke the **Drawing Units** dialog box, choose **Units** in the **Format** menu or enter **UNITS** at the Command prompt. In the **Angle** area choose **Decimal Degrees** from the **Type** drop-down list and choose **0.00** from the **Precision** drop-down list. Also select the **Clockwise** radio button from the **Angle** area. Select the **Direction** button to display the **Direction Control** dialog box (Figure 22-6) and select the **East** radio button.

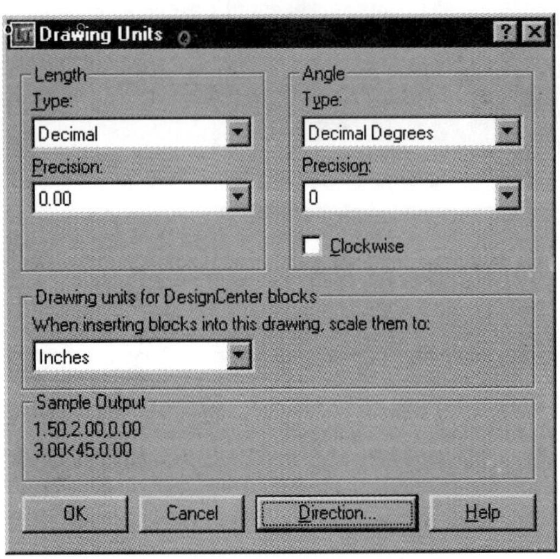

Figure 22-5 ***Drawing Units*** *dialog box*

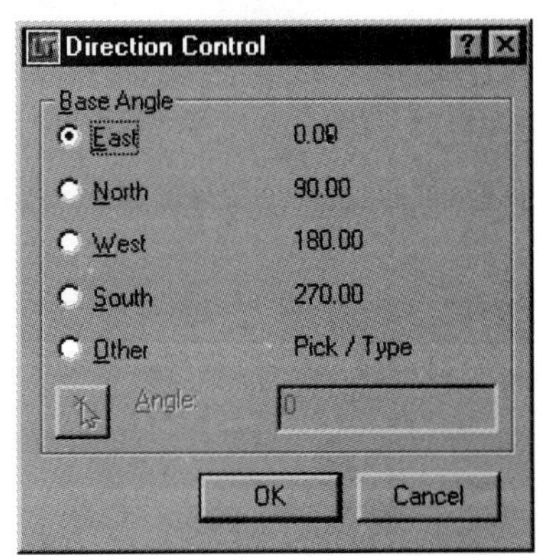

Figure 22-6 ***Direction Control*** *dialog box*

Step 4
Now, save the drawing as **PROTO1.DWT** using AutoCAD LT's **SAVEAS** command. You must select template (*DWT) from the list box in the dialog box. This drawing is now saved as PROTO1.DWT on the default drive. You can also save this drawing on a floppy diskette in drives A or B.

LOADING A TEMPLATE DRAWING

You can use the template drawing any time you want to start a new drawing. To use the preset values of the template drawing, start AutoCAD LT or select the **NEW** button from the **Standard Toolbar**. AutoCAD LT displays the **AutoCAD LT Today** window. You can also start a new drawing by selecting the **New** option from the **File** menu. To load the template drawing, select the **Template** option in the **Create Drawing** tab of **AutoCAD LT Today** window, as shown in Figure 22-7. AutoCAD LT will display the list of template drawings. From this list, select **PROTO1** template drawing. AutoCAD LT will start a new drawing that will have the same setup as that of template drawing, **PROTO1**.

You can have several template drawings, each with a different setup. For example, **PROTOB** for a 18" by 12" drawing, **PROTOC** for a 24" by 18" drawing. Each template drawing can be created according to user-defined specifications. You can then load any of these template drawings as discussed previously.

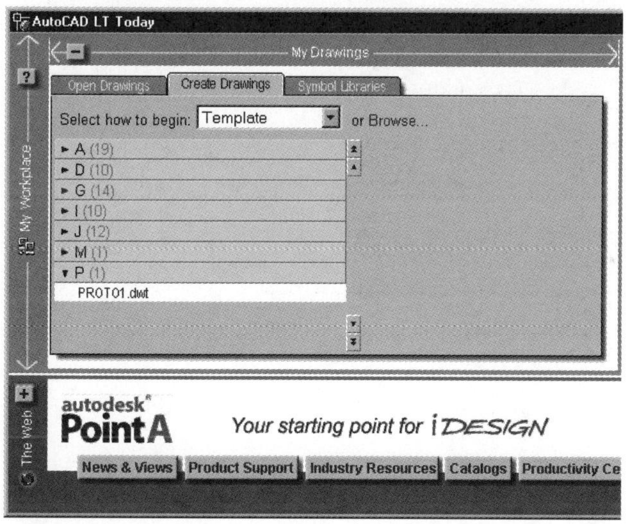

Figure 22-7 ***AutoCAD LT Today*** *window with the list of template drawings*

Note

*You can also create a template drawing by entering **NEW** at the Command prompt and keeping the **FILEDIA** system variable **0**.*

CUSTOMIZING DRAWINGS WITH LAYERS AND DIMENSIONING SPECIFICATIONS

Most production drawings need multiple layers for different groups of objects. In addition to layers, it is a good practice to assign different colors to different layers to control the line width at the time of plotting. You can generate a template drawing that contains the desired number of layers with linetypes and colors according to your company specifications. You can then use this template drawing to make a new drawing. The next example illustrates the procedure used for customizing a drawing with layers, linetypes, and colors.

Example 3

Create a template drawing (**PROTO3**) that has a border and the company's title block, as shown in Figure 22-8. In addition to this, you want the following initial drawing setup:

Limits	48.0,36.0
Text height	0.25
PLINE width	0.02
Ltscale	4.0

DIMENSIONS

Overall dimension scale factor 4.0
Dimension text above the extension line
Dimension text aligned with dimension line

LAYERS

Layer Names	Line Type	Color
0	Continuous	White
OBJ	Continuous	Red
CEN	Center	Yellow
HID	Hidden	Blue
DIM	Continuous	Green
BOR	Continuous	Magenta

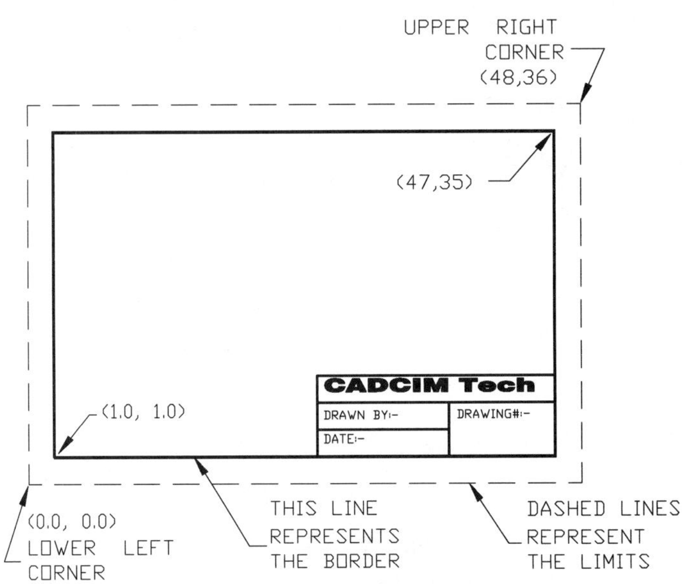

Figure 22-8 *Template drawing for Example 3*

Step 1

Setting limits, text size, polyline width, polyline and linetype scaling

Start a new drawing with default parameters. You can do this by selecting the **Start from Scratch** option in the **Create Drawings** tab of the **AutoCAD LT Today** window. Once you are in the drawing editor, use the AutoCAD LT commands to set up the values as given for this example. Also, draw a border and a title block as shown in Figure 22-8. In this figure the hidden lines indicate the drawing limits. The border lines are 1.0 units inside the drawing limits. For the border lines, use a polyline of width 0.02 units. Use the following procedure to produce the prototype drawing for Example 3:

The **LIMITS** command can be invoked by choosing **Drawing Limits** from the **Format** menu or by entering **LIMITS** at the Command prompt.

Command: **LIMITS**
Specify lower left corner or [ON/OFF] <0.00,0.00>: **0,0**
Specify upper right corner <12.0,9.0>: **48.0,36.0**

Size of the text can be changed by entering **TEXTSIZE** at the Command prompt.

Command: **TEXTSIZE**
Enter new value for TEXTSIZE <0.2000>: **0.25**

Polyline width can be changed by entering **PLINEWID** at the Command prompt.

Command: **PLINEWID**
Enter new value for PLINEWID <0.0000>: **0.02**

To draw the border, use the **PLINE** command. You can invoke the **PLINE** command by choosing the **Polyline** button from the **Draw** toolbar or by entering **PLINE** at the Command prompt.

Command: **PLINE**
Specify start point: **1.0,1.0**
Current line-width is **0.02**
Specify next point or [Arc/Close/Halfwidth/Length/Undo/Width]: **47,1**
Specify next point or [Arc/Close/Halfwidth/Length/Undo/Width]: **47,35**
Specify next point or [Arc/Close/Halfwidth/Length/Undo/Width]: **1,35**
Specify next point or [Arc/Close/Halfwidth/Length/Undo/Width]: **C**

Linetype scale can be changed by entering **LTSCALE** at the Command prompt.

Command: **LTSCALE**
Enter new linetype scale factor<Current>: **4.0**

Step 2

Setting dimensioning parameters

You can use the **Dimension Style Manager** dialog box (Figure 22-9) to set the dimension variables.

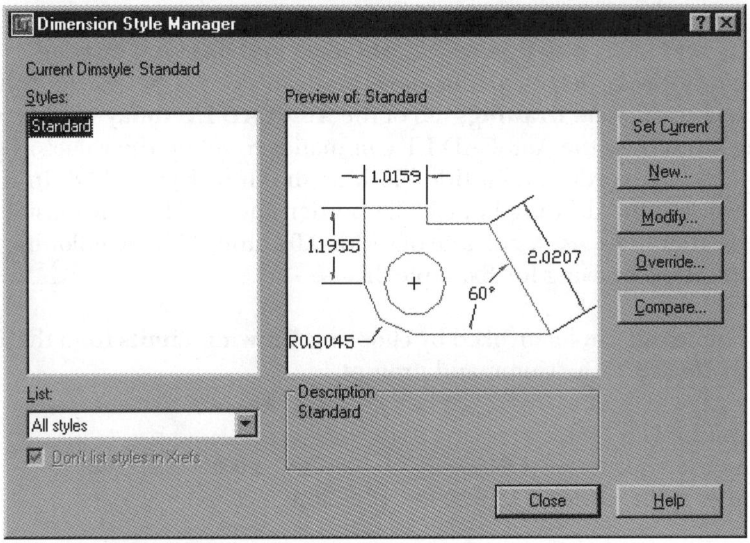

Figure 22-9 ***Dimension Style Manager*** *dialog box*

Choose the **Dimension Style** button from the **Dimension** toolbar or choose **Style** from the **Dimension** menu to invoke the **Dimension Style Manager** dialog box. You can also invoke this dialog box by entering **DIMSTYLE** at the Command prompt. Choose the **New** button from the **Dimension Style Manager** dialog box. The **Create New Dimension Style** dialog box is displayed as shown in the Figure 22-10. Specify new style name as **MYDIM1** in the **New Style Name** edit box and then choose the **Continue** button. The **New Dimension Style** dialog box will be displayed. In the **New Dimension Style** dialog box choose the **Modify** button to display the **New Dimension Style:MYDIM1** dialog box (Figure 22-11).

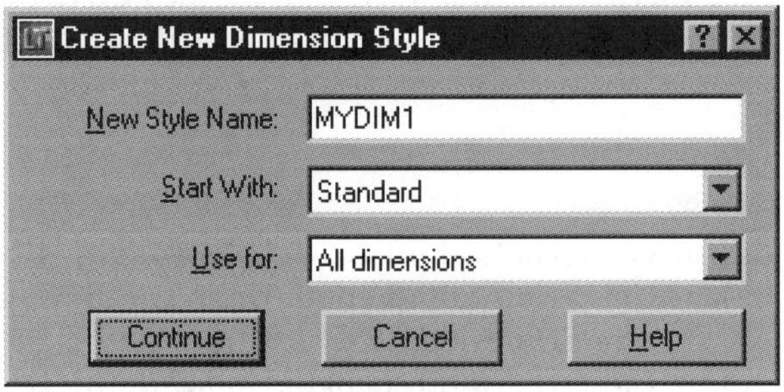

Figure 22-10 ***Create New Dimension Style*** *dialog box*

Overall dimension scale factor

To specify dimension scale factor, choose the **Fit** tab of the **Modify Dimension Style** dialog box. Set the value in the **Use overall scale of** as **4** in the **Scale for Dimension Features** area (Figure 22-11).

Dimension text over the dimension line

Choose the **Text** tab of the **Modify Dimension Style** dialog box. Select the **Above** option from the **Vertical** drop-down list in the **Text Placement** area.

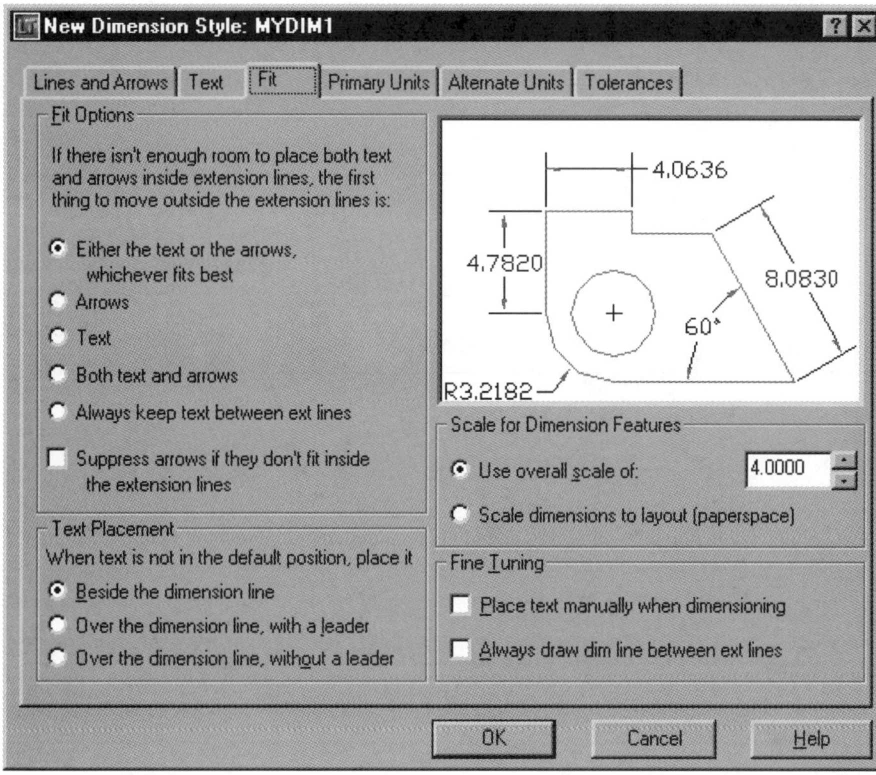

Figure 22-11 New Dimension Style: MYDIM1 dialog box

Dimension text aligned with dimension line

In the **Text Alignment** area **of Text tab** choose the **Aligned with the dimension line** radio button and then choose **OK**. In the **Dimension Style Manager** dialog box, choose the **Close** button to exit.

Step 3

Setting layers

Choose the **Layers** button from the **Object Properties** toolbar or choose **Layer** from the **Format** menu to invoke **Layer Properties Manager** dialog box as shown in Figure 22-12. You can also invoke the **Layer Properties Manager** dialog box by entering **LAYER** at the Command prompt. Choose the **New** button in the **Layer Properties Manager** dialog box and rename **Layer1** as **OBJ**. Select the color swatch of the **OBJ** layer to display the **Select Color** dialog box. Choose the **Red** color and choose **OK,** so the red color will be assigned to the OBJ layer. Again choose the **New** button in the **Layer Properties Manager** dialog box and rename the **Layer1** as **CEN**. Choose the linetype swatch to display the **Select Linetype** dialog box. If the different linetypes are not already loaded, choose the **Load** button to display the **Load or Reload Linetypes** dialog box. Choose the **CENTER** linetype from the **Available Linetypes** area and choose **OK**. Again the **Select Linetype** dialog box will reappear. Choose the **CENTER** linetype from the **Loaded Linetypes** area and choose **OK**. Choose the color swatch to display the **Select Color** dialog box. Choose the **Yellow** color and choose **OK,** so the color yellow and linetype center will be assigned to the layer Cen. Similarly different linetypes and different colors can be set for different layers mentioned in the example.

You can also use the **-LAYER** command to set the layers and linetypes from the Command prompt.

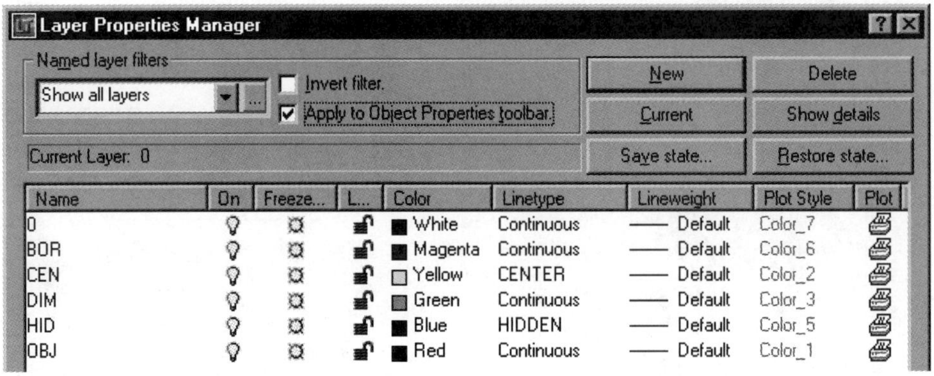

*Figure 22-12 **Layer Properties Manager** dialog box*

Step 4

Next, add the title block and the text as shown in Figure 22-8. After completing the drawing, save it as PROTO3.DWT. You have created a template drawing (PROTO3) that contains all of the information given in Example 3.

CUSTOMIZING DRAWINGS WITH LAYOUT

The Layout (paper space) provides a convenient way to plot multiple views of a 3D drawing or multiple views of a regular 2D drawing. It takes quite some time to set up the viewports in model space with different vpoints and scale factors. You can create prototype drawings that contain predefined viewport settings, with vpoint and other desired information. Now if you create a new drawing, or insert a drawing, the views are automatically generated. The following example illustrates the procedure for generating a prototype drawing with paper space and model space viewports.

Example 4

Create a template drawing as shown in Figure 22-13 with four views in Layout3 (paper space) that display front, top, side, and 3D views of the object. The plot size is 10.5 by 8 inches. The plot scale is 0.5 or 1/2" = 1". The paper space viewports should have the following vpoint setting:

Viewports	Vpoint	View
Top right	1,-1,1	3D view
Top left	0,0,1	Top view
Lower right	1,0,0	Right side view
Lower left	0,-1,0	Front view

Step 1

Start AutoCAD LT and create a new drawing. Use the following commands to set up various parameters.

The first step is to create a layout. To create a layout you can use the **LAYOUT** command. You can also right-click on **Model** or any **Layout** tab to display the shortcut menu. From the shortcut menu select **New layout**.

Command: **LAYOUT**
Enter layout option [Copy/Delete/New/Template/Rename/SAveas/Set/?] <set>: **N**
Enter new Layout name <Layout3>: **Layout3**

Step 2
The next step is to select the new layout (Layout 3) tab. When you select this tab, AutoCAD LT displays the **Page Setup-Layout3** dialog box. Select the **Plot Device** tab and then select the printer or plotter that you want to use. In this example HP LaserJet 4000 is used.

Step 3
Next, select the **Layout Settings** tab and select the paper size that is supported by the selected plotting device. In this example the paper size is 8.5" x 11". Select the **OK** button to accept the settings and exit the dialog box. AutoCAD LT displays the new layout (Layout3) on the screen with default viewport. Use the **ERASE** command to erase this viewport.

Step 4
The next step is to set up a layer (VIEW) for viewports and assign it a color (green). To invoke the **Layer Properties Manager** dialog box, choose the **Layers** button from the **Object Properties** toolbar or choose **Layer** from the **Format** menu. You can also invoke the **Layer Properties Manager** dialog box by entering **LAYER** at the Command prompt. Now the **Layer Properties Manager** dialog box is displayed. Choose the **New** button and name the **Layer1** as **VIEW**. Select the color swatch of the **VIEW** layer to display the **Select Color** dialog box. Select the color **Green** and Choose the **OK** button. This color will be assigned to **View** layer. Also, make the **VIEW** layer current and then select the **OK** button to exit.

Layers can also be set at the Command prompt by entering **-LAYER** at the Command prompt.

Step 5
To create four viewports, use the **MVIEW** command, as shown in Figure 22-13. To invoke the **MVIEW** command, choose **Viewports > 4 Viewport** from the **View** menu or you can directly enter **MVIEW** command at the Command prompt. Then switch to model space to zoom the display to half the size.

Command: **MVIEW**
Specify corner of viewport or
[ON/OFF/Fit/Hideplot/Lock/Restore/2/3/4] <Fit>: **4**
Specify first corner or [Fit] <Fit>: **0.25,0.25**
Specify opposite corner: **10.25,7.75**

Choose **Paper** button in the status bar to activate the model space or enter **MSPACE** at the Command prompt.

Command: **MSPACE** (or **MS**)

Make the first viewport active by selecting a point in the viewport and then use the **ZOOM** command to specify the paper space scale factor to 0.5. The **ZOOM** command can be invoked by choosing **Zoom** > **Scale** from the **View** menu or by entering **ZOOM** at the Command prompt.

Command: **ZOOM**
Specify corner of window, enter a scale factor (nX or nXP), or
[All/Center/Dynamic/Extents/Previous/Scale/Window] <real time>: **0.5XP**

Now, make the next viewport active and specify the scale factor. Do the same for the remaining viewports.

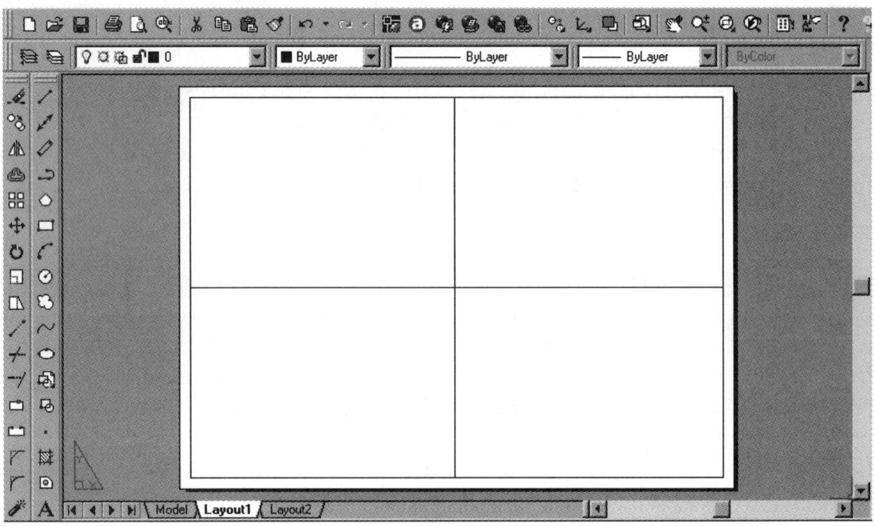

***Figure 22-13** Paper space with four viewports*

Step 6

The next step is to change the vpoints of different paper space viewports by using the **VPOINT** command. The **VPOINT** command can be invoked by choosing **3D Views** > **VPOINT** from the **View** menu or by entering **VPOINT** at the Command prompt. The vpoint values for different viewports are shown in Example 5. To set the viewpoint for the lower left viewport, the Command prompt sequence is as follows:

Command: **VPOINT**
Current view direction: VIEWDIR=0.0000,0.0000,1.0000
Specify a view point or [Rotate] <display compass and tripod>: **0,-1,0**

Similarly use the **VPOINT** command to set the vpoint of other viewports.

Step 7

Use the **Model** button in the status bar to change to paper space and then set a new layer

PBORDER with yellow color. Make the PBORDER layer current, draw a border, and if needed a title block using the **PLINE** command. You can also change to paper space by entering **PSPACE** at the Command prompt.

The **PLINE** command can be invoked by choosing the **Polyline** button from the **Draw** toolbar or by choosing **Polyline** from the **Draw** menu. The **PLINE** Command can also be invoked by entering **PLINE** at the Command prompt.

Command: **PLINE**
Specify start point: **0,0**
Current line-width is 0.0000
Specify next point or [Arc/Close/Halfwidth/Length/Undo/Width]: **0,8.0**
Specify next point or [Arc/Close/Halfwidth/Length/Undo/Width]: **10.5,8.0**
Specify next point or [Arc/Close/Halfwidth/Length/Undo/Width]: **10.5,0**
Specify next point or [Arc/Close/Halfwidth/Length/Undo/Width]: **C**

Step 8

The last step is to select the **Model** tab (or change the **TILEMODE** to 1) and save the prototype drawing. To test the layout that you just created, make the 3D drawing as shown in Figure 22-17 or make any 3D object. Now, if you switch to **Layout 3** tab, you will find four different views of the object (Figure 22-14). If the object views do not appear in the viewports, use the **PAN** commands to position the views in the viewports. You can freeze the VIEW layer so that the viewports do not appear on the drawing. Now you can plot this drawing from The Layout3 with a plot scale factor of 1:1 and the size of the plot will be exactly as specified.

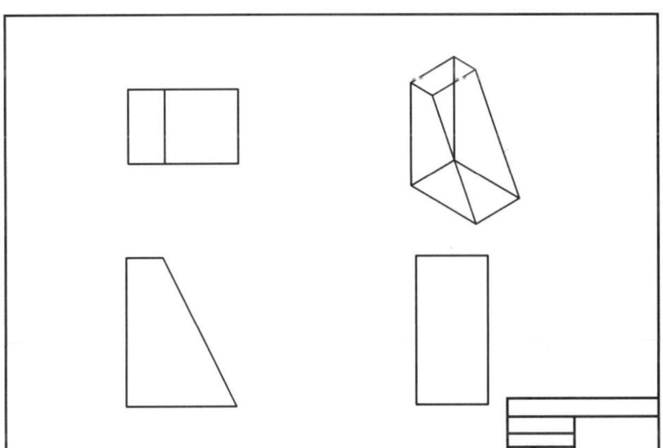

Figure 22-14 *Four views of the 3D object in paper space*

CUSTOMIZING DRAWINGS WITH VIEWPORTS

In certain applications you might need multiple model space viewport configurations to display different views of an object. This involves setting up the desired viewports and then changing the viewpoint for different viewports. You can create a prototype drawing that contains a required number of viewports and the viewpoint information. Now, if you insert a 3D object in one of the viewports of the prototype drawing, you will automatically get different views of the object without setting viewports or viewpoints. The following example illustrates the procedure for creating a prototype drawing with a standard number (four) of viewports and viewpoints.

Example 5

Create a prototype drawing with four viewports, as shown in Figure 22-15. The viewports should have the following viewpoints (vpoints):

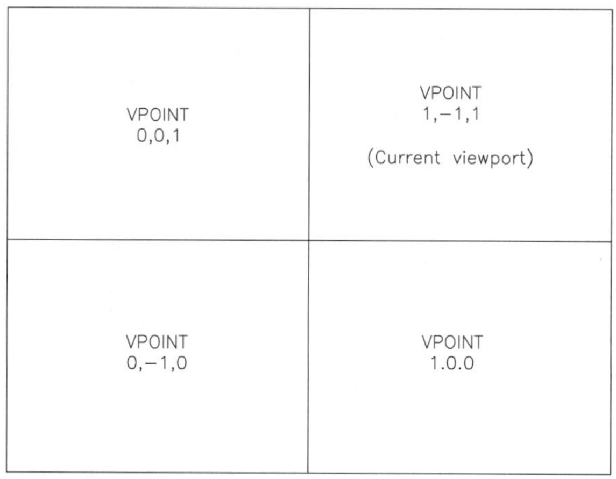

Figure 22-15 *Viewports with different viewpoints*

Viewports	Vpoint	View
Top right	1,-1,1	3D view
Top left	0,0,1	Top view
Lower right	1,0,0	Right side view
Lower left	0,-1,0	Front view

Step 1

Start AutoCAD LT and create a new drawing from scratch. Use the following commands to set the viewports and vpoints.

Step 2

Setting viewports

Viewports and corresponding viewpoints can be set with the **VPORTS** command. You can also choose the **Display Viewports Dialog** button from the **Viewports** toolbar or choose **Viewports > New Viewports** from the **View** menu to display the **Viewports** dialog box as shown in Figure 22-16. Choose **Four:Equal** viewports from the **Standard Viewports** area. In the **Preview** area four equal viewports are displayed. Select **3D** from the **Setup** drop-down list. The four viewports with the different viewpoints will be displayed in the **Preview** area as Top, Front, Right, and SE Isometric respectively. **Top** represents the viewpoints as (0,0,1), **Front** represents the viewpoints as (0,-1,0), **Right** represents the viewpoints as (1,0,0) and **SE Isometric** represents the viewpoints as (1,-1,1) respectively. Choose the **OK** button. Save the drawing as

PROTO5.DWT.

Viewports and viewpoints can also be set by entering **-VPORTS** and **VPOINT** at the Command prompt respectively.

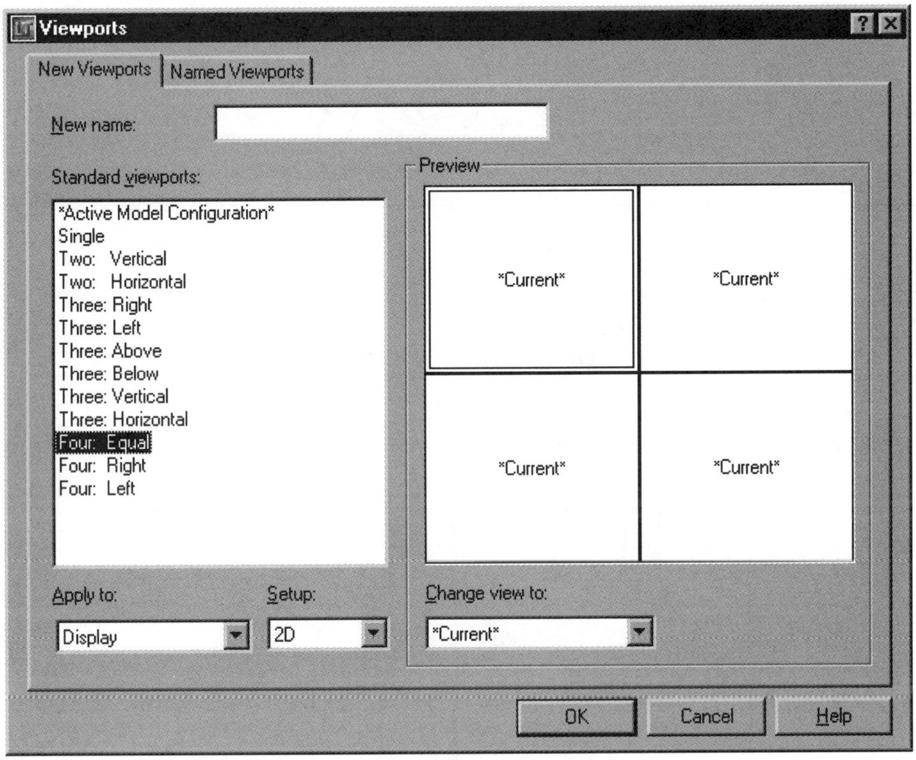

***Figure 22-16 Viewports** dialog box*

Step 3
Now, start a new drawing and draw the 3D tapered block as shown in Figure 22-17.

Step 4
Again, start a new drawing, TEST, using the prototype drawing PROTO5. Make the top right viewport current and insert or create a drawing shown in Figure 22-17. Four different views will be automatically displayed on the screen as shown in Figure 22-18.

CUSTOMIZING DRAWINGS ACCORDING TO PLOT SIZE AND DRAWING SCALE

For controlling the plot area, it is recommended to use layouts. You can make the drawing of any size and then use the layout to specify the sheet size and then draw the border and title block. However you can also plot a drawing in the model space and set up the system variables so that the plotted drawing is to your specifications. You can generate a template drawing

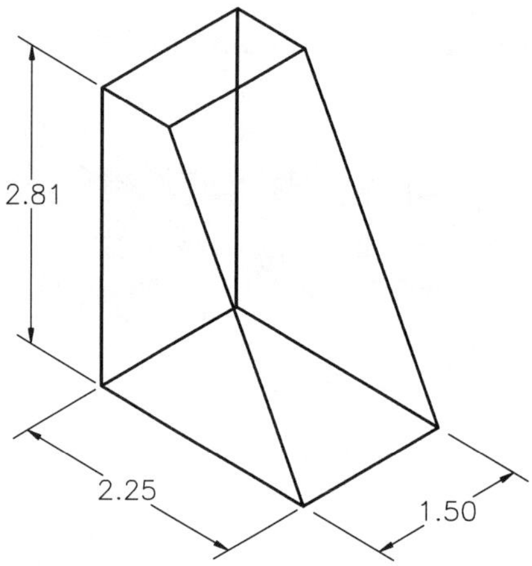

***Figure 22-17** 3D tapered block*

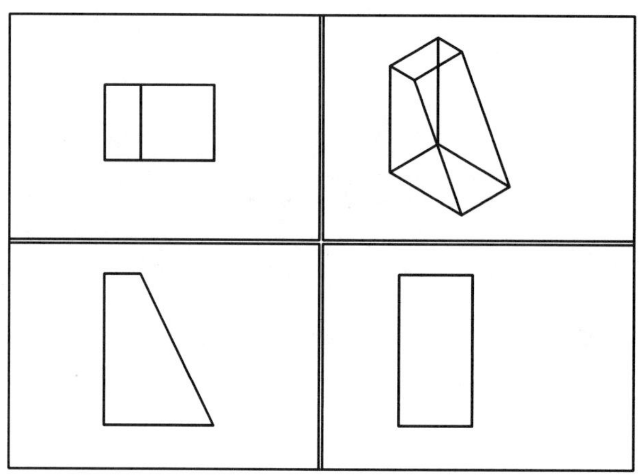

***Figure 22-18** Different views of 3D tapered block*

according to plot size and scale. For example, if the scale is 1/16" = 1' and the drawing is to be plotted on a 36" by 24" area, you can calculate drawing parameters like limits, **DIMSCALE**, and **LTSCALE** and save them in a template drawing. This will save considerable time in the initial drawing setup and provide uniformity in the drawings. The next example explains the procedure involved in customizing a drawing according to a certain plot size and scale. (Note that you can also use the paper space to specify the paper size and scale.)

Example 6

Create a template drawing (**PROTO6**) with the following specifications:

Plotted sheet size 36" by 24" (Figure 22-19)

Scale	1/8" = 1.0'
Snap	3'
Grid	6'
Text height	1/4" on plotted drawing
Linetype scale	Calculate
Dimscale factor	Calculate
Units	Architectural Precision, 16-denominator of smallest fraction Angle in degrees/minutes/seconds Precision, 0d00' Direction control, base angle, east Angle positive if measured counterclockwise
Border	Border should be 1" inside the edges of the plotted drawing sheet, using PLINE 1/32" wide when plotted (Figure 22-19)

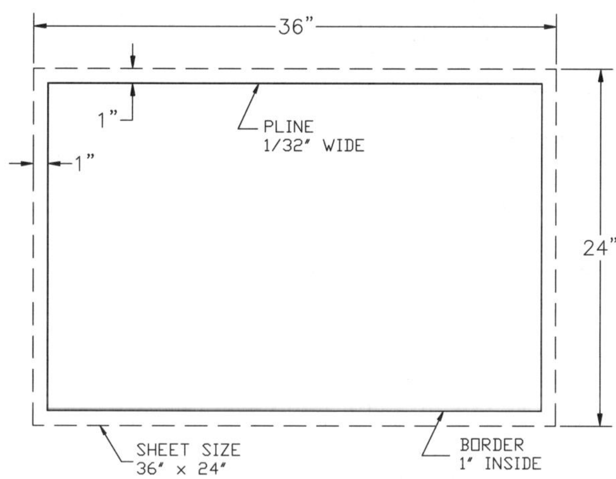

Figure 22-19 *Border of template drawing*

Step 1

Calculating limits, text height, linetype scale, dimension scale and polyline width

In this example, you need to calculate some values before you set the parameters. For example, the limits of the drawing depend on the plotted size of the drawing and the scale of the drawing. Similarly, **LTSCALE** and **DIMSCALE** depend on the plot scale of the drawing. The following calculations explain the procedure for finding the values of limits, ltscale, dimscale, and text height.

Limits

Given:
Sheet size 36" x 24"
Scale 1/8" = 1'
or 1" = 8'
Calculate:
XLimit

YLimit
Since sheet size is 36" x 24" and scale is 1/8"=1'
Therefore, XLimit = 36 x 8' = 288'
YLimit = 24 x 8' = 192'

Text height

Given:
Text height when plotted = 1/4"
Scale 1/8" = 1'
Calculate:
Text height
Since scale is 1/8" = 1'
or 1/8" = 12"
or 1" = 96"
Therefore, scale factor = 96
Text height = 1/4" x 96
= 24" = 2'

Linetype scale and dimension scale

Known:
Since scale is 1/8" = 1'
or 1/8" = 12"
or 1" = 96"

Calculate:
Ltscale and Dimscale
Since scale factor = 96
Therefore, LTSCALE = Scale factor = 96
Similarly, DIMSCALE = 96
(All dimension variables, like DIMTXT and DIMASZ, will be multiplied by 96.)

Polyline width

Given:
Scale is 1/8" = 1'
Calculate:
PLINE width
Since scale is 1/8" = 1'
or 1" = 8'
or 1" = 96"

Therefore,
PLINE width = 1/32 x 96
= 3"

After calculating the parameters, use the following AutoCAD LT commands to set up the drawing, then save the drawing as **PROTO6.DWT.**

Step 2
Setting units

Start a new drawing and choose **Units** from the **Format** menu or enter **UNITS** at the Command prompt to display the **Drawing Units** dialog box. Choose **Architectural** from the **Type** drop-down list in the **Length** area. Choose **0'-01/16"** from the **Precision** drop-down list. Make sure the **Clockwise** radio button in the **Angle** area is not checked. Select **Deg/Min/Sec** from the **Type** drop-down list and select **0d00** from the **Precision** drop-down list in the **Angle** area. Now choose the **Direction** button to display the **Directional Control** dialog box. Choose the **East** radio button if it is not selected in the **Base Angle** area and then choose **OK**.

Step 3
Setting limits, snap and grid, textsize, linetype scale, dimension scale, dimension style and pline

To set the **LIMITS**, select **Drawing Limits** from the **Format** menu or enter **LIMITS** at the Command prompt.

Command: **LIMITS**
Specify lower left corner or [ON/OFF] <0'-0",0'-0">:**0,0**
Specify upper right corner <1'-0",0'-9">: **288',192'**

To set the **SNAP** and **GRID**, right-click on the **Snap** or **Grid** button in the status bar to invoke the shortcut menu. In the shortcut menu choose the **Settings** to display the **Drafting Settings** dialog box. You can also choose the **Object Snap Settings** button from the **Object Snap** toolbar to display the **Drafting Settings** dialog box. In the dialog box choose the **Snap and Grid** tab. Enter **3'** and **3'** in the **Snap X spacing** and **Snap Y spacing** edit boxes respectively. Enter **6'** and **6'** in the **Grid X spacing** and **Grid Y spacing** edit boxes respectively. Then choose **OK**.

You can also set these values by entering **SNAP** and **GRID** at the Command prompt.

The size of the text can be changed by entering **TEXTSIZE** at the Command prompt.

Command: **TEXTSIZE**
Enter new value for TEXTSIZE <current>: **2'**

To set the **LTSCALE**, choose the **Linetype** from the **Format** menu or enter **LINETYPE** at the Command prompt to invoke the **Linetype Manager** dialog box. Choose the **Show details** button. Specify the **Global scale factor** as **96** in the **Global scale factor** edit box.

You can also change the scale of the linetype by entering **LTSCALE** at the Command prompt.

To set the **DIMSTYLE**, choose the **Dimension Style** button from the **Dimension** toolbar or choose **Style** from the **Dimension** menu to invoke the **Dimension Style Manager** dialog box. Choose the **New** button from the **Dimension Style Manager** dialog box to invoke the **Create New Dimension** Style dialog box. Specify the new style name as **MYDIM2** in the **New Style Name** edit box and then choose the **Continue** button. The **New Dimension Style** dialog box will be displayed. Choose the **Modify** button to display the **New Dimension Style: MYDIM2** dialog box. Now choose the **Fit** tab. Set the value in the **Use overall scale of** as **96** in the **Scale**

for Dimension Features area. Now choose the **OK** button to again display the **Dimension Style Manager** dialog box. Choose **Close** to exit the dialog box.

You can invoke **PLINE** command by choosing the **Polyline** button from the **Draw** toolbar or enter **PLINE** at the Command prompt.

Command: **PLINE**
Specify start point: **8',8'**
Current line-width is **0.0000**
Specify next point or [Arc/Close/Halfwidth/Length/Undo/Width]: **W**
Specify starting width<0.00>: **3**
Specify ending width<0'-3">: **ENTER**
Specify next point or [Arc/Close/Halfwidth/Length/Undo/Width]: **280',8'**
Specify next point or [Arc/Close/Halfwidth/Length/Undo/Width]: **280',184'**
Specify next point or [Arc/Close/Halfwidth/Length/Undo/Width]: **8',184'**
Specify next point or [Arc/Close/Halfwidth/Length/Undo/Width]: **C**

Now save the drawing as **PROTO6.DWT**.

Self-Evaluation Test

Answer the following questions and then compare your answers to the correct answers given at the end of this chapter.

1. The template drawings are stored in __________ format.

2. To use a template file, select the __________ option in the **Create Drawings** tab of **AutoCAD LT Today** window.

3. To start a drawing with default setup, select the __________ option in the **Create Drawings** tab of **AutoCAD LT Today** window.

4. If plot size is 36" x 24", and the scale is 1/2" = 1', then XLimit = __________ and YLimit = __________.

5. You can use AutoCAD LT's __________ command to set up a viewport in paper space.

Review Questions

Answer the following questions:

1. The default value of **DIMSCALE** is __________.

2. The default value for **DIMTXT** is __________.

3. The default value for **SNAP** is _________.

4. Architectural units can be selected by using AutoCAD LT's _________ command.

5. Name three standard template drawings that come with AutoCAD LT software _________ , _________ , and _________.

6. If the plot size is 24" x 18", and the scale is 1 = 20, the XLimit = _________ and YLimit = _________.

7. If the plot size is 200 x 150 and limits are (0.00,0.00) and (600.00,450.00), the **LTSCALE** factor = _________.

8. _________ provides a convenient way to plot multiple views of a 3D drawing or multiple views of a regular 2D drawing.

9. You can use AutoCAD LT's _________ command to change to paper space.

10. You can use AutoCAD LT's _________ command to change to model space.

11. The values that can be assigned to **TILEMODE** are _________ and _________.

12. In the model space, if you want to reduce the display size by half, the scale factor you enter in the **ZOOM**-Scale command is _________.

Exercises

Exercise 1 *General*

Create a template drawing (**PROTOE1**) with the following specifications:

Units	Architectural with precision 0'-0 1/16
Angle	Decimal Degrees with precision 0.
Base angle	East.
Angle direction	Counterclockwise
Limits	48' x 36'

Exercise 2 *General*

Create a template drawing (**PROTOE2**) with the following specifications:

Limits	36.0, 24.0
Snap	0.5
Grid	1.0
Text height	0.25
Units	Decimal

Precision 0.00
Decimal degrees
Precision 0
Base angle, East
Angle positive if measured counterclockwise

Exercise 3 *General*

You want to set up a prototype drawing with the following specifications (the name of the drawing is **PROTOE3**).

Limits	36.0,24,0
Border	35.0,23.0
Grid	1.0
Snap	0.5
Text height	0.15
Units	Decimal (up to 2 places)
Ltscale	1
Current layer	Object

Layer Name	Linetype	Color
0	Continuous	White
Object	Continuous	Red
Hidden	Hidden	Yellow
Center	Center	Green
Dim	Continuous	Blue
Border	Continuous	Magenta
Notes	Continuous	White

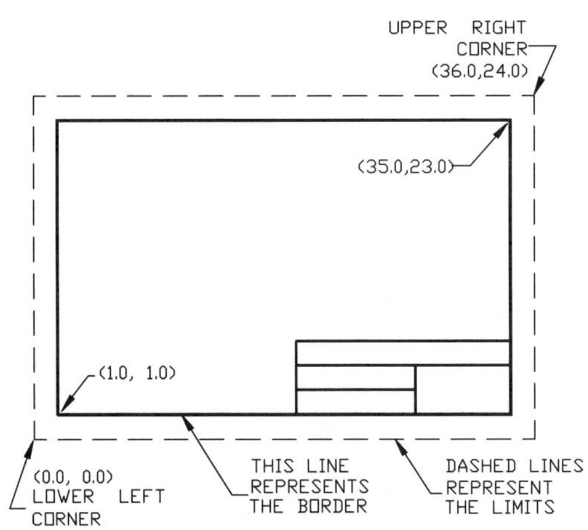

Answers to Self-Evaluation Test

1 - .dwt, **2** - **Template**, **3** - **Start from Scratch**, **4** - 72" x 48", **5** - **MVIEW**

Chapter 23

Script Files and Slide Shows

Learning Objectives

After completing this chapter, you will be able to:

- *Write script files and use the **SCRIPT** command to run script files.*
- *Use the **RSCRIPT** and **DELAY** commands in script files.*
- *Invoke script files when loading AutoCAD LT.*
- *Create a slide show.*
- *Preload slides when running a slide show.*

WHAT ARE SCRIPT FILES?

AutoCAD LT has provided a facility called **script files** that allows you to combine different AutoCAD LT commands and execute them in a predetermined sequence. The commands can be written as a text file using any text editor like Notepad. These files, generally known as script files, have extension **.SCR** (example: **PLOT1.SCR**). A script file is executed with the AutoCAD LT **SCRIPT** command.

Script files can be used to generate a slide show, do the initial drawing setup, or plot a drawing to a predefined specification. They can also be used to automate certain command sequences that are used frequently in generating, editing, or viewing a drawing. Scripts cannot access dialog boxes or menus. When commands that open the file and plot dialog boxes are issued from a script file, AutoCAD LT runs the command line version of the command instead of opening the dialog box.

Example 1

Write a script file that will perform the following initial setup for a drawing (file name **SCRIPT1.SCR**). It is assumed that the drawing will be plotted on 12" x 9" size paper (Scale factor for plotting = 4).

Ortho	On	Zoom	All
Grid	2.0	Text height	0.125
Snap	0.5	Ltscale	4.0
Limits	0,0 48.0,36.0	Dimscale	4.0

Step 1: Understanding commands and prompt entries

Before writing a script file, you need to know the AutoCAD LT commands and the entries required in response to the Command prompts. To find out the sequence of the prompt entries, you can type the command at the keyboard and then respond to different prompts. The following is a list of AutoCAD LT commands and prompt entries for Example 1.

Command: **ORTHO**
Enter mode [ON/OFF] <OFF>: **ON**

Command: **GRID**
Specify grid spacing(X) or [ON/OFF/Snap/Aspect] <1.0>: **2.0**

Command: **SNAP**
Specify snap spacing or [ON/OFF/Aspect/Rotate/Style/Type] <1.0>: **0.5**

Command: **LIMITS**
Reset Model space limits:
Specify lower left corner or [ON/OFF] <0.0,0.0>: **0,0**
Specify upper right corner <12.0,9.0>: **48.0,36.0**

Command: **ZOOM**
Specify corner of window, enter a scale factor (nX or nXP), or
[All/Center/Dynamic/Extents/Previous/Scale/Window] <real time>: **A**

Command: **TEXTSIZE**
Enter new value for TEXTSIZE <0.02>: **0.125**

Command: **LTSCALE**
Enter new linetype scale factor <1.0000>: **4.0**

Command: **DIMSCALE**
Enter new value for DIMSCALE <1.0000>: **4.0**

Step 2: Writing the script file

Once you know the AutoCAD LT commands and the required prompt entries, you can write the script file using any text editor like **Notepad**.

You can invoke the Notepad editor by choosing **Programs > Accessories > Notepad** from the **Start** menu. When you choose this, the Notepad will be opened with a blank file. You can write the script file in this blank file and then save it with the desired extension. The following file is a listing of the script file for Example 1:

```
ORTHO
ON
GRID
2.0
SNAP
0.5
LIMITS
0,0
48.0,36.0
ZOOM
ALL
TEXTSIZE
0.125
LTSCALE
4.0
DIMSCALE 4.0
```

Notice that the commands and the prompt entries in this file are in the same sequence as mentioned before. You can also combine several statements in one line, as shown in the following list:

```
;This is my first script file, SCRIPT1.SCR
ORTHO ON
GRID 2.0
SNAP 0.5
LIMITS 0,0 48.0,36.0 ZOOM ALL
TEXTSIZE 0.125
LTSCALE 4.0
DIMSCALE 4.0
```


Save the script file as **SCRIPT1.SCR** on A or C drive and exit the text editor. Notice the space between the commands and the prompt entries. For example, between **ORTHO** command and **ON** there is a space. Similarly, there is a space between **GRID** and **2.0**.

Note

In the script file, a space is used to terminate a command or a prompt entry. Therefore, spaces are very important in these files. Make sure there are no extra spaces, unless they are required to press ENTER more than once.

After you change the limits, it is a good practice to use the ***ZOOM*** *command with the* ***All*** *option to display the new limits on the screen.*

AutoCAD LT ignores and does not process any lines that begin with a semicolon (;). This allows you to put comments in the file.

SCRIPT COMMAND

The AutoCAD LT **SCRIPT** command allows you to run a script file while you are in the drawing editor. Choose the **Run Script** button from the **Tools** toolbar to invoke the **Select Script File** dialog box as shown in the Figure 23-1. You can also invoke the **Select Script File** dialog box by entering **SCRIPT** at the Command prompt. You can enter the name of the script file or you can accept the default file name. The default script file name is the same as the drawing name. If you want to enter a new file name, type the name of the script file **without** the file extension (**.SCR**). (The file extension is assumed and need not be included with the file name.)

Step 3: Running the script file

To run the script file of Example 1, invoke the **SCRIPT** command, select the file **SCRIPT1**, and then choose the **Open** button in the **Select Script File** dialog box (Figure 23-1). You will see the changes taking place on the screen as the script file commands are executed.

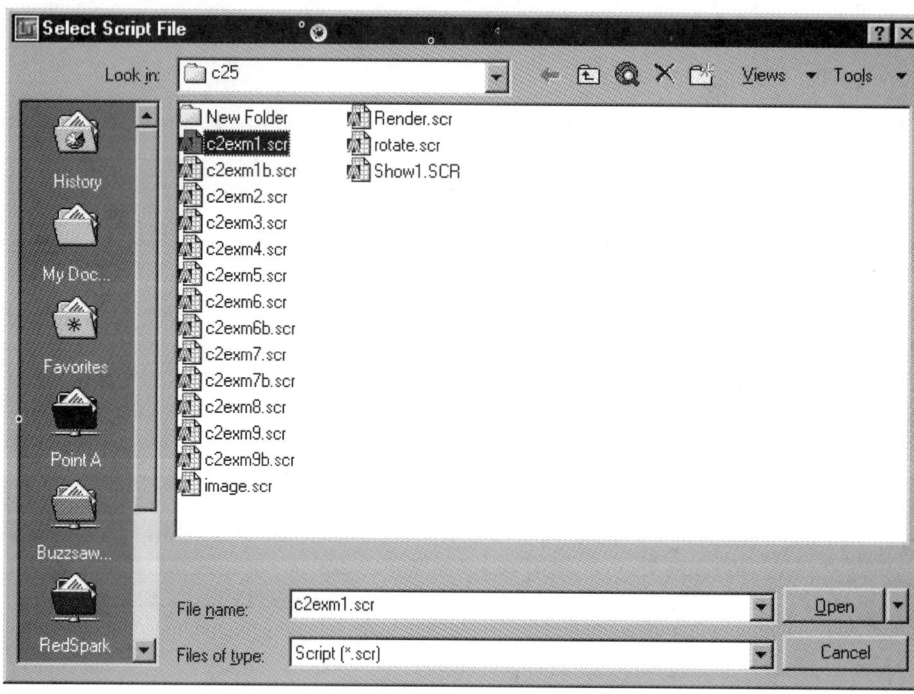

***Figure 23-1 Select Script File** dialog box*

You can also enter the name of the script file at the Command prompt by setting **FILEDIA**=0. The format of the **SCRIPT** command is:

Command: **FILEDIA**
Enter new value for FILEDIA <1>: **0**

Command: **SCRIPT**
Enter script file name <current>: *Script file name.*

Example 2

Write a script file that will set up the following layers with the given colors and linetypes (file name **SCRIPT2.SCR**).

Layer Names	Color	Linetype	Line Weight
Object	Red	Continuous	default
Center	Yellow	Center	default
Hidden	Blue	Hidden	default
Dimension	Green	Continuous	default
Border	Magenta	Continuous	default
Hatch	Cyan	Continuous	0.05

Step 1: Understanding commands and prompt entries

As mentioned earlier, you need to know the AutoCAD LT commands and the required prompt entries before writing a script file. For Example 2, you need the following commands to create the layers with the given colors and linetypes:

Command: **-LAYER**
Enter an option
[?/Make/Set/New/ON/OFF/Color/Ltype/LWeight/Plot/Freeze/Thaw/LOck/Unlock]: **N**
Enter name list for new layer(s): **OBJECT,CENTER,HIDDEN,DIM,BORDER, HATCH**

Enter an option
[?/Make/Set/New/ON/OFF/Color/Ltype/LWeight/Plot/Freeze/Thaw/LOck/Unlock]: **L**
Enter loaded linetype name or [?] <Continuous>: **CENTER**
Enter name list of layer(s) for linetype "CENTER" <0>: **CENTER**

Enter an option
[?/Make/Set/New/ON/OFF/Color/Ltype/LWeight/Plot/Freeze/Thaw/LOck/Unlock]: **L**
Enter loaded linetype name or [?] <Continuous>: **HIDDEN**
Enter name list of layer(s) for linetype "HIDDEN" <0>: **HIDDEN**

Enter an option
[?/Make/Set/New/ON/OFF/Color/Ltype/LWeight/Plot/Freeze/Thaw/LOck/Unlock]: **C**
Enter color name or number (1-255): **RED**
Enter name list of layer(s) for color 1 (red) <0>:**OBJECT**

Enter an option
[?/Make/Set/New/ON/OFF/Color/Ltype/LWeight/Plot/Freeze/Thaw/LOck/Unlock]: **C**
Enter color name or number (1-255): **YELLOW**
Enter name list of layer(s) for color 2 (yellow) <0>: **CENTER**

Enter an option
[?/Make/Set/New/ON/OFF/Color/Ltype/LWeight/Plot/Freeze/Thaw/LOck/Unlock]: **C**
Enter color name or number (1-255): **BLUE**
Enter name list of layer(s) for color 5 (blue)<0>: **HIDDEN**

Enter an option
[?/Make/Set/New/ON/OFF/Color/Ltype/LWeight/Plot/Freeze/Thaw/LOck/Unlock]: **C**
Enter color name or number (1-255): **GREEN**
Enter name list of layer(s) for color 3 (green)<0>: **DIM**

Enter an option
[?/Make/Set/New/ON/OFF/Color/Ltype/LWeight/Plot/Freeze/Thaw/LOck/Unlock]: **C**
Enter color name or number (1-255): **MAGENTA**
Enter name list of layer(s) for color 6 (magenta)<0>: **BORDER**

Enter an option
[?/Make/Set/New/ON/OFF/Color/Ltype/LWeight/Plot/Freeze/Thaw/LOck/Unlock]: **C**
Enter color name or number (1-255): **CYAN**
Enter name list of layer(s) for color 4 (cyan)<0>: **HATCH**

Enter an option
[?/Make/Set/New/ON/OFF/Color/Ltype/LWeight/Plot/Freeze/Thaw/LOck/Unlock]: **LW**
Enter lineweight (0.0mm - 2.11mm): 0.05
Enter name list of layers(s) for lineweight 0.05mm <0>: **HATCH**
[?/Make/Set/New/ON/OFF/Color/Ltype/LWeight/Plot/Freeze/Thaw/LOck/Unlock]: Enter

Step 2: Writing the script file

The following file is a listing of the script file that creates different layers and assigns the given colors and linetypes to these layers:

```
;This script file will create new layers and
;assign different colors and linetypes to layers
LAYER
NEW
OBJECT,CENTER,HIDDEN,DIM,BORDER,HATCH
L
CENTER
CENTER
L
HIDDEN
HIDDEN
C
RED
OBJECT
C
YELLOW
CENTER
```

```
C
BLUE
HIDDEN
C
GREEN
DIM
C
MAGENTA
BORDER
C
CYAN
HATCH
```

(This is a blank line to terminate the **LAYER** command. End of script file.)

Save the script file as **SCRIPT2.SCR**.

Step 3: Running the script file

To run the script file of Example 2, choose the **Run Script** button from the **Tools** menu or enter **SCRIPT** at the Command prompt to invoke the **Select Script File** dialog box. Choose the **SCRIPT2.SCR** and then choose **Open**. You can also enter the **SCRIPT** command and the name of the script file at the Command prompt by setting **FILEDIA**=0.

Example 3

Write a script file that will rotate the circle and the line, as shown in Figure 23-2, around the lower endpoint of the line through 45-degree increments. The script file should be able to produce a continuous rotation of the given objects with a delay of two seconds after every 45-degree rotation (file name **SCRIPT3.SCR**).

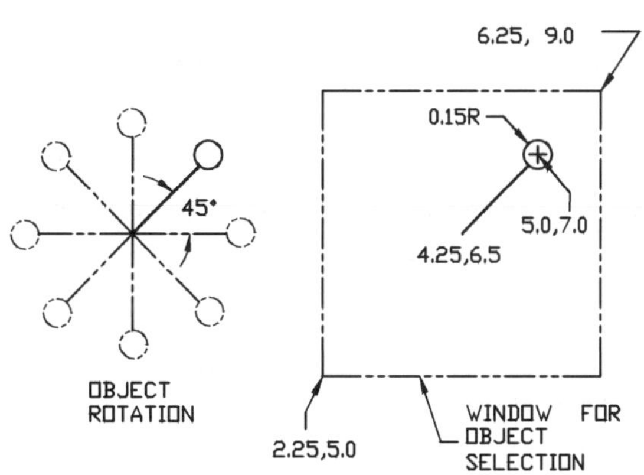

Figure 23-2 Line and circle rotated through 45-degree increments

Step 1: Understanding commands and prompt entries

Before writing the script file, enter the required commands and the prompt entries at the keyboard. Write down the exact sequence of the entries in which they have been entered to perform the given operations. The following is a listing of the AutoCAD LT command sequence needed to rotate the circle and the line around the lower endpoint of the line:

Command: **ROTATE**
Current positive angle in UCS: ANGDIR=counterclockwise ANGBASE=0
Select objects: **W** *(Window option to select object)*
Specify first corner: **2.25, 5.0**
Specify opposite corner: **6.25, 9.0**
Select objects:**<RETURN>**
Specify base point: **4.25,6.5**
Specify rotation angle or [Reference]: **45**

Step 2: Writing the script file

Once the AutoCAD LT commands, command options, and their sequences are known, you can write a script file. As mentioned earlier, you can use any text editor to write a script file. The following file is a listing of the script file that will create the required rotation of the circle and line of Example 3. The line numbers and *(Blank line for Return)* are not a part of the file. They are shown here for reference only.

```
1 ROTATE
2 W
3 2.25,5.0
4 6.25,9.0
5           (Blank line for Return.)
6 4.25,6.5
7 45
```

Line 1
ROTATE
In this line, **ROTATE** is an AutoCAD LT command that rotates the objects.

Line 2
W
In this line, W is the Window option for selecting the objects that need to be edited.

Line 3
2.25,5.0
In this line, 2.25 defines the X coordinate and 5.0 defines the Y coordinate of the lower left corner of the object selection window.

Line 4
6.25,9.0
In this line, 6.25 defines the X coordinate and 9.0 defines the Y coordinate of the upper right corner of the object selection window.

Line 5
Line 5 is a blank line that terminates the object selection process.

Line 6
4.25,6.5
In this line, 4.25 defines the X coordinate and 6.5 defines the Y coordinate of the base point for rotation.

Line 7
45
In this line, 45 is the incremental angle for rotation.

Note
One of the limitations of the script files is that all the information has to be contained within the file. These files do not let you enter information. For instance, in Example 3, if you want to use the Window option to select the objects, the Window option (W) and the two points that define this window must be contained within the script file. The same is true for the base point and all other information that goes in a script file.

Step 3: Running the script file
Choose the **Run Script** button from the **Tools** menu or enter **SCRIPT** at the Command prompt to invoke the **Select Script File** dialog box. Choose the **SCRIPT3.SCR** and then choose **Open**. You can see the changes taking place on the screen. You can also open a script file by entering **SCRIPT** at the Command and setting **FILEDIA**=0.

RSCRIPT COMMAND

The AutoCAD LT **RSCRIPT** command allows the user to execute the script file indefinitely until canceled. It is a very desirable feature when the user wants to run the same file continuously. For example, in the case of a slide show for a product demonstration, the **RSCRIPT** command can be used to run the script file again and again until it is terminated by pressing the ESC (Escape) key from the keyboard. Similarly, in Example 3, the rotation command needs to be repeated indefinitely to create a continuous rotation of the objects. This can be accomplished by adding **RSCRIPT** at the end of the file, as in the following file:

ROTATE
W
2.25,5.0
6.25,9.0
(Blank line for Return.)
4.25,6.5
45
RSCRIPT

The **RSCRIPT** command on line 8 will repeat the commands from line 1 to line 7, and thus set the script file in an indefinite loop. The script file can be stopped by pressing the ESC or the BACKSPACE key.

Note

You cannot provide conditional statements in a script file to terminate the file when a particular condition is satisfied unless you use the AutoLISP functions in the script file.

DELAY COMMAND

In the script files, some of the operations happen very quickly and make it difficult to see the operations taking place on the screen. It might be necessary to intentionally introduce a pause between certain operations in a script file. For example, in a slide show for a product demonstration, there must be a time delay between different slides so that the audience has enough time to see them. This is accomplished by using the AutoCAD LT **DELAY** command, which introduces a delay before the next command is executed. The general format of the **DELAY** command is:

Command: DELAY Time

Where **Command** ------ AutoCAD LT command prompt
DELAY ---------- **DELAY** command
Time ------------ Time in milliseconds

The **DELAY** command is to be followed by the delay time in milliseconds. For example, a delay of 2,000 milliseconds means that AutoCAD LT will pause for approximately two seconds before executing the next command. It is approximately two seconds because computer processing speeds vary. The maximum time delay you can enter is 32,767 milliseconds (about 33 seconds). In Example 3, a two-second delay can be introduced by inserting a **DELAY** command line between line 7 and line 8, as in the following file listing:

ROTATE
W
2.25,5.0
6.25,9.0
(Blank line for Return.)
4.25,6.5
45
DELAY 2000
RSCRIPT

The first seven lines of this file rotate the objects through a 45-degree angle. Before the **RSCRIPT** command on line 8 is executed, there is a delay of 2,000 milliseconds (about two seconds). The **RSCRIPT** command will repeat the script file that rotates the objects through another 45-degree angle. Thus, a slide show is created with a time delay of two seconds after every 45-degree increment.

RESUME COMMAND

If you cancel a script file and then want to continue it, you can do so by using the AutoCAD LT **RESUME** command.

Command: **RESUME**

The **RESUME** command can also be used if the script file has encountered an error that causes it to be suspended. The **RESUME** command will skip the command that caused the error and continue with the rest of the script file. If the error occurred when the command was in progress, use a leading apostrophe with the **RESUME** command (**'RESUME**) to invoke the **RESUME** command in transparent mode.

Command: **'RESUME**

COMMAND LINE SWITCHES

The command line switches can be used as arguments to the aclt.exe file that launches AutoCAD LT. You can also use the **Options** dialog box to set the environment or by adding a set of environment variables in the autoexec.bat file. The command line switches and environment variables override the values set in the **Options** dialog box for the current session only. These switches do not alter the system registry. The following is the list of the command line switches:

Switch	Function
/c	Controls where AutoCAD LT stores and searches for the hardware configuration file. The default file is aclt 2002.cfg.
/s	Specifies which directories to search for support files if they are not in the current directory
/b	Designates a script to run after AutoCAD LT starts
/t	Specifies a template to use when creating a new drawing
/nologo	Starts AutoCAD LT without first displaying the logo screen
/v	Designates a particular view of the drawing to be displayed upon start-up of AutoCAD LT
/r	Reconfigures AutoCAD LT with the default device configuration settings
/p	Specifies the profile to use on start-up

INVOKING A SCRIPT FILE WHEN LOADING AUTOCAD LT

The script files can also be run when loading AutoCAD LT, without getting into the drawing editor. The format of the command for running a script file when loading AutoCAD LT is:

"Drive**Program Files\AutoCAD LT 2002\aclt.exe" [existing-drawing] [/t template] [/v view] /b Script-file**

In the following example, AutoCAD LT will open the existing drawing (Mydwg1) and then run the script file (Setup) through the **Run** dialog box as shown in Figure 23-3.

Example

"C:\Program Files\AutoCAD LT 2002\aclt.exe" Mydwg1 /b Setup

Where **Auto CAD LT 2002** -----AutoCAD LT2002 subdirectory containing AutoCAD LT system files

aclt.exe ---------- AutoCAD LT's command to start AutoCAD LT

MyDwg1 -------- Existing drawing file name

Setup ------------ Name of the script file

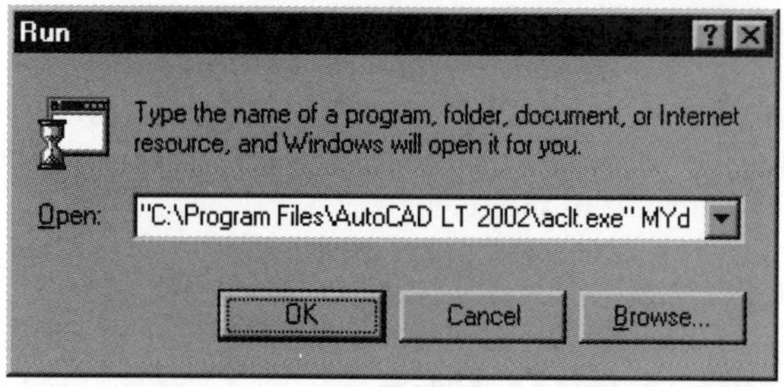

Figure 23-3 *Invoking script file when loading AutoCAD LT using the* ***Run*** *dialog box*

In the following example, AutoCAD LT will start a new drawing with the default name (Drawing), using the template file temp1, and then run the script file (Setup).

Example
"C:\Program Files\AutoCAD LT 2002\aclt.exe" /t temp1 /b Setup
Where **temp1** ------------ Existing template file name
Setup ------------- Name of the script file

or

"C:\ProgramFiles\AutoCAD LT 2002\aclt.exe"/t temp1 "C:\MyFolder"/b Setup

Where C**\Program Files\AutoCAD LT 2002\aclt.exe** Path name for aclt.exe
C:\MyFolder --- Path name for the Setup script file

In the following example, AutoCAD LT will start a new drawing with the default name (Drawing), and then run the script file (Setup).

Example
"C:\Program Files\AutoCAD LT 2002\aclt.exe" /b Setup
Where **Setup** ------------- Name of the script file

Here, it is assumed that the AutoCAD LT system files are loaded in the AutoCAD LT 2002 directory.

Note

For invoking a script file when loading AutoCAD LT, the drawing file or the template file specified in the command must exist in the search path. You cannot start a new drawing with a given name. You can also use any template drawing file that is found in the template directory to run a script file through the ***Run*** *dialog box.*

You should avoid abbreviations to prevent any confusion. For example, a C can be used as a close option when you are drawing lines. It can also be used as a command alias for drawing a circle. If you use both of these in a script file, it might be confusing.

Example 4

Write a script file that can be invoked when loading AutoCAD LT and create a drawing with the following setup (filename SCRIPT4.SCR):

Grid	3.0
Snap	0.5
Limits	0,0
	36.0,24.0
Zoom	All
Text height	0.25
Ltscale	3.0
Dimscale	3.0

Layers

Name	Color	Linetype
Obj	Red	Continuous
Cen	Yellow	Center
Hid	Blue	Hidden
Dim	Green	Continuous

Step 1: Writing the script file

First, write a script file and save the file under the name **SCRIPT4.SCR**. The following file is a listing of this script file that does the initial setup for a drawing:

```
GRID 3.0
SNAP 0.5
LIMITS 0,0 36.0,24.0 ZOOM ALL
TEXTSIZE 0.25
LTSCALE 3
DIMSCALE 3.0
LAYER NEW
OBJ,CEN,HID,DIM
L CENTER CEN
L HIDDEN HID
C RED OBJ
C YELLOW CEN
C BLUE HID
C GREEN DIM
```

(Blank line for ENTER.)

Step 2: Loading the script file through the run dialog box

After you have written and saved the file, quit the drawing editor. To run the script file, SCRIPT4, select **Start > Run**, and then enter the following command line:

"C:\Program Files\AutoCAD LT 2002\aclt.exe" /t EX4 /b SCRIPT4

Where **aclt.exe** ---------- ACLT to load AutoCAD LT
EX4 -------------- Drawing filename
SCRIPT4 ------- Name of the script file

Here it is assumed that the template file (EX4) and the script file (SCRIPT4) is on C drive. When you enter this line, AutoCAD LT is loaded and the file EX4.DWT is opened. The script file, SCRIPT4, is then automatically loaded and the commands defined in the file are executed.

In the following example, AutoCAD LT will start a new drawing with the default name (Drawing), and then run the script file (SCRIPT4) (Figure 23-4).

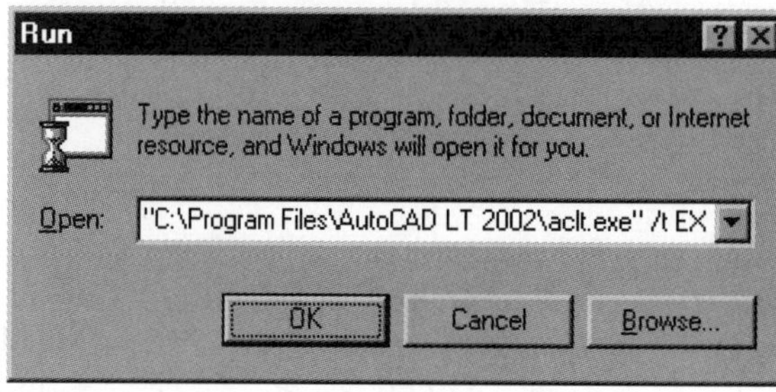

Figure 23-4 *Invoking script file when loading AutoCAD LT using the* ***Run*** *dialog box*

Example
C:\AutoCAD LT 2002\ACAD /b SCRIPT4
Where **SCRIPT4** ------- Name of the script file

Here, it is assumed that the AutoCAD LT system files are loaded in the AutoCAD LT 2002 directory.

Example 5

Write a script file that will plot a 36" by 24" to maximum plot size, using your system printer/plotter. Use the Window option to select the drawing to be plotted.

Step 1: Understanding commands and prompt entries

Before writing a script file to plot a drawing, find out the plotter specifications that must be entered in the script file to obtain the desired output. To determine the prompt entries and their sequences to set up the plotter specifications, enter the AutoCAD LT **-PLOT** command at the keyboard. Note the entries you make and their sequence (the entries for your printer or plotter will probably be different). The following is a listing of the plotter specifications with the new entries:

Command: **-PLOT**
Detailed plot configuration? [Yes/No] <No>: **Yes**
Enter a layout name or [?] <Model>: Enter
Enter an output device name or [?] <HP LaserJet 4000 Series PCL 6>: Enter
Enter paper size or [?] <Letter (8 1/2 x 11 in)>: Enter

Enter paper units [Inches/Millimeters] <Inches>: **I**
Enter drawing orientation [Portrait/Landscape] <Landscape>: ***L***
Plot upside down? [Yes/No] <No>: **N**
Enter plot area [Display/Extents/Limits/View/Window] <Display>: **W**
Enter lower left corner of window <0.000000,0.000000>: **0,0**
Enter upper right corner of window <0.000000,0.000000>: **36,24**
Enter plot scale (Plotted Inches=Drawing Units) or [Fit] <Fit>: **F**
Enter plot offset (x,y) or [Center] <0.00,0.00>: **0,0**
Plot with plot styles? [Yes/No] <Yes>: **Yes**
Enter plot style table name or [?] (enter . for none) <>: .
Plot with lineweights? [Yes/No] <Yes>: Y
Scale lineweights with plot scale? [Yes/No] <No>: N
Plot paper space first? [Yes/No] <No>: N
Remove hidden lines? [Yes/No] <No>: N
Write the plot to a file [Yes/No] <N>: N
Save changes to layout [Yes/No]? <N>N
Proceed with plot [Yes/No] <Y>: Y

Step 2: Writing the script file

Now you can write the script file by entering the responses to these prompts in the file. The following file is a listing of the script file that will plot a 36" by 24" drawing on 9" by 6" paper after making the necessary changes in the plot specifications. The comments on the right are not a part of the file.

```
Plot
y
                (Blank line for ENTER, selects default layout.)
                (Blank line for ENTER, selects default printer.)
                (Blank line for ENTER, selects the default paper size.)
I
L
N
w
0,0
36,24
F
0,0
Y
.               (Enter . for none)
Y
N
N
N
N
N
Y
```

Chapter 23

Saving and running the script file for this example is the same as that which has been described for previous examples. You can use a blank line to accept the default value for a prompt. A blank line in the script file will cause a Return. However, you must not accept the default plot specifications because the file might have been altered by another user or by another script file. Therefore, always enter the actual values in the file so that when you run a script file, it does not take the default values.

Exercise 1 *General*

Write a script file that will plot a 288' by 192' drawing on a 36" x 24" sheet of paper. The drawing scale is 1/8" = 1'. (The filename is SCRIPT9.SCR. In this exercise assume that AutoCAD LT is configured for the HPGL plotter and the plotter description is HPGL-Plotter.)

Example 6

Write a script file to animate a clock with continuous rotation of the second hand (longer needle) through 5 degrees and the minutes hand (shorter needle) through 2 degrees clockwise around the center of the clock (Figure 23-5). The specifications follow.

Figure 23-5 *Drawing for Example 6*

Specification for the rim made of donut

Color of Donut	Blue
Inside diameter of Donut	8.0
Outside diameter of Donut	8.4
Center point of Donut	5,5

Specification for the digit mark made of polyline

Color of the digit mark	Green
Start point of Pline	5,8.5
Initial width of Pline	0.5
Final width of Pline	0.5

Height of Pline	0.5

Specification for second hand (long needle) made of polyline

Color of the second hand	Red
Start point of Pline	5,5
Initial width of Pline	0.5
Final width of Pline	0.0
Length of Pline	3.5
Rotation of the second hand	5 degree clockwise

Specification for minute hand (shorter needle) made of polyline

Color of the minute hand	Cyan
Start point of Pline	5,5
Initial width of Pline	0.35
Final width of Pline	0.0
Length of Pline	3.0
Rotation of the minute hand	2 degree clockwise

Step 1: Understanding the commands and prompt entries for creation of the clock

For this example you can create two script files and the files need to be linked. The first script file will demonstrate the creation of the clock on the screen. The next script file will demonstrate the rotation of the needles of the clock.

First write a script file that deals with the creation of the clock as follows and save the file under the name **CLOCK.SCR**.

Command: **COLOR**
Enter default object color<BYLAYER>: **Blue**
Command: **DONUT**
Specify inside diameter of donut<0.5>: **8.0**
Specify outside diameter of donut<0.5>: **8.4**
Specify center of donut or <exit>: **5,5**
Specify center of donut or <exit>: Enter
Command: **COLOR**
Enter default object color<Blue>: **Green**
Command: **PLINE**
Specify start point: **5,8.5**
Specify next point or [Arc/Close/Halfwidth/Length/Undo/Width]: **Width**
Specify starting width<0.00>: **0.25**
Specify starting width<0.25>: **0.25**
Specify next point or [Arc/Close/Halfwidth/Length/Undo/Width]: **@0.25<270**
Specify next point or [Arc/Close/Halfwidth/Length/Undo/Width]: Enter
Command: **-ARRAY**
Select objects: **Last**
Select objects: Enter
Enter the type of array[Rectangular/Polar]<R>: **Polar**
Specify center point of array: **5,5**

Enter the number of items in the array: **12**
Specify the angle to fill(+= ccw, -=cw)<360>: **360**
Rotate arrayed objects? [Yes/No]<Y>: **Y**
Command: **COLOR**
Enter default object color<Green>: **RED**
Command: **PLINE**
Specify start point: **5,5**
Specify next point or [Arc/Close/Halfwidth/Length/Undo/Width]: **Width**
Specify starting width<0.5>: **0.5**
Specify ending width<0.5>: **0**
Specify next point or [Arc/Close/Halfwidth/Length/Undo/Width]: **@3.5<0**
Specify next point or [Arc/Close/Halfwidth/Length/Undo/Width]: Enter
Command: **COLOR**
Enter default object color<Red>: **Cyan**
Command: **PLINE**
Specify start point: **5,5**
Specify next point or [Arc/Close/Halfwidth/Length/Undo/Width]: **Width**
Specify starting width<0.5>: **0.35**
Specify ending width<0.35>: **0**
Specify next point or [Arc/Close/Halfwidth/Length/Undo/Width]: **@3<90**
Specify next point or [Arc/Close/Halfwidth/Length/Undo/Width]: Enter
Command: **SCRIPT**
ROTATE.SCR

Now you can write the script file by entering the responses to these prompts in the file **CLOCK.SCR**. Listing of the script file is as follows.

Color
Blue
Donut
8.0
8.4
5,5
Blank line for ENTER
Color
Green
Pline
5,8.5
W
0.25
0.25
@0.25<270
Blank line for ENTER
-Array
L
Blank line for ENTER
P

```
5,5
12
360
Y
Color
Red
Pline
5,5
W
0.5
0
@3.5<0
                              Blank line for ENTER
Color
Cyan
Pline
5,5
w
0.35
0
@3<90
                              Blank line for ENTER
Script
ROTATE.scr                    (Name of the script file that will cause rotation)
```


Step 2: Understanding the commands and its sequences for rotation of the needles

The last line in the previous script file is ROTATE.SCR. This is the name of the script file that will rotate the clock hands. Before writing the script file, enter ROTATE command and respond to Command prompts that will cause the desired rotation. The following is a listing of the AutoCAD LT command sequence needed to rotate the objects.

Command: **ROTATE**
Select objects: **L**
Select objects: [Enter]
Specify base point: 5,5
Specify rotation angle or [Reference]: -2
Command: **ROTATE**
Select objects: **C**
Specify first corner:3,3
Specify other corner: 7,7
Select objects: **Remove**
Remove objects: **L**
Remove objects: [Enter]
Specify base point: 5,5
Specify rotation angle or [Reference]: -5

Now you can write the script file by entering the responses to these prompts in the file

ROTATE.SCR. The following is the listing of the script file that will rotate the clock hands.

```
Rotate
L
                                  Blank line for ENTER
5,5
-2
Rotate
c
3,3
7,7
R
L
                                  Blank line for ENTER
5,5
-5
Rscript
```

Save the previous script file as **ROTATE.SCR**. Now run the script file **CLOCK.SCR**. Since this file is linked with **ROTATE.SCR**, it will automatically run **ROTATE.SCR** after running CLOCK.SCR. Also note that ROTATE.SCR file must be saved to a directory in the AutoCAD LT support file search path, or the last line of the CLOCK.SCR must include a fully resolved path to ROTATE.SCR, or AutoCAD LT would not find it.

WHAT IS A SLIDE SHOW?

AutoCAD LT provides a facility using script files to combine the slides in a text file and display them in a predetermined sequence. In this way, you can generate a slide show for a slide presentation. You can also introduce a time delay in the display so that the viewer has enough time to view a slide.

A drawing or parts of a drawing can also be displayed by using the AutoCAD LT display commands. For example, you can use **ZOOM**, **PAN**, or other commands to display the details you want to show. If the drawing is very complicated, it takes quite some time to display the desired information, and it may not be possible to get the desired views in the right sequence. However, with slide shows you can arrange the slides in any order and present them in a definite sequence. In addition to saving time, this will also help to minimize the distraction that might be caused by constantly changing the drawing display. Also, some drawings are confidential in nature and you may not want to display some portions or views of them. You can send a slide show to a client without losing control of the drawings and the information that is contained in them.

WHAT ARE SLIDES?

A **slide** is the snapshot of a screen display; it is like taking a picture of a display with a camera. The slides do not contain any vector information like AutoCAD LT drawings, which means that the entities do not have any information associated with them. For example, the slides do not retain any information about the layers, colors, linetypes, start point, or endpoint of a line or viewpoint. Therefore, slides cannot be edited like drawings. If you want to make any changes in the slide, you need to edit the drawing and then make a new slide from the edited drawing.

MSLIDE COMMAND

Slides are created by using the AutoCAD LT **MSLIDE** command at the Command prompt. If **FILEDIA** is set to 1, the **MSLIDE** command displays the **Create Slide File** dialog box (Figure 23-6) on the screen. You can enter the slide file name in this dialog box. If **FILEDIA** is set to 0, the command will prompt you to enter the slide file name.

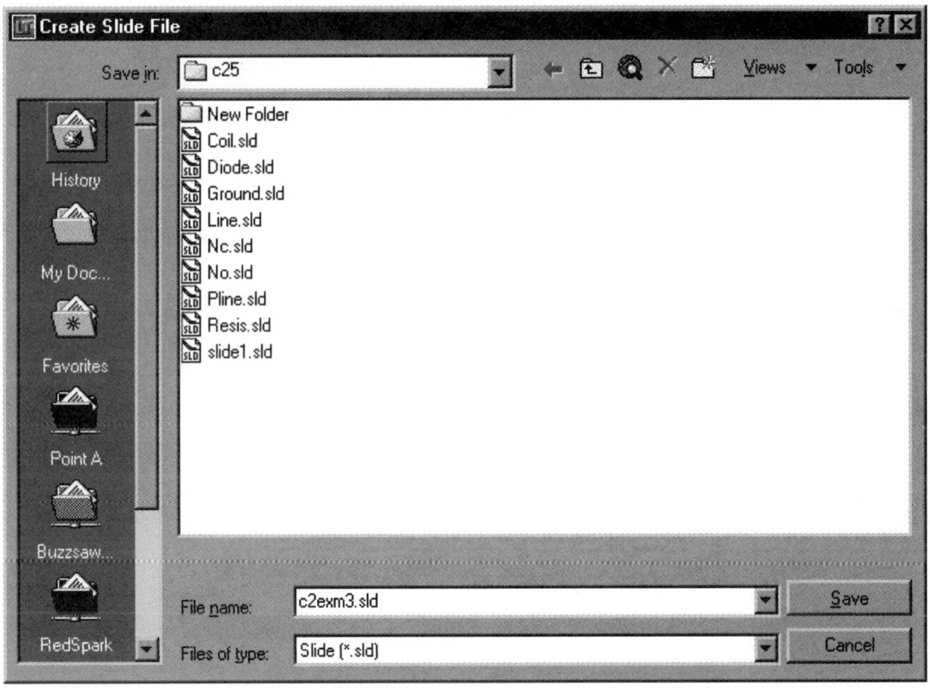

***Figure 23-6 Create Slide File** dialog box*

Command: **MSLIDE**
Enter name of slide file to create <Default>: *Slide file name.*

Example
Command: **MSLIDE**
Slide File: <Drawing1> **SLIDE1**
Where **Drawing1** ------- Default slide file name
SLIDE1 --------- Slide file name

In the preceding example, AutoCAD LT will save the slide file as **SLIDE1.SLD**.

Note

*In model space, you can use the **MSLIDE** command to make a slide of the existing display in the current viewport.*

If you are in the paper space viewport, you can make a slide of the display in the paper space that includes any floating viewports.

*When the viewports are not active, the **MSLIDE** command will make a slide of the current screen display.*

VSLIDE COMMAND

To view a slide, use the **VSLIDE** command at the Command prompt. The **Select Slide File** dialog box is displayed as shown in the Figure 23-7. Choose the file you want to view and then choose **OK**. The corresponding slide will be displayed on the screen. If the FILEDIA is 0, the slides that you want to view can be directly entered at the Command prompt.

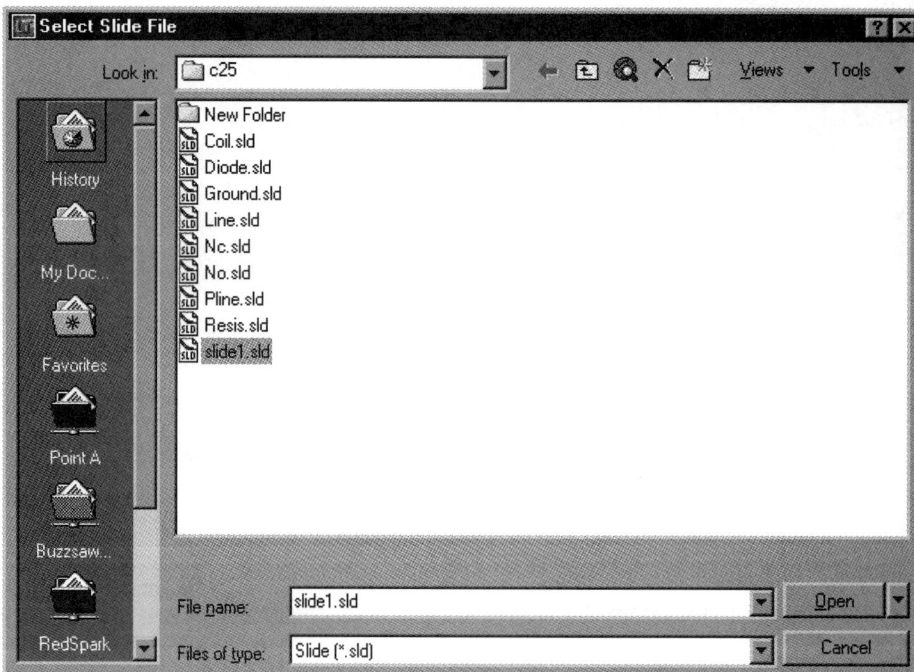

***Figure 23-7 Select Slide Files** dialog box*

Command: **VSLIDE**
Enter name of slide file to create<Default>: *Name*.

Example
Command: **VSLIDE**
Slide file <Drawing1>: SLIDE1
Where **Drawing1** ------- Default slide file name
SLIDE1 --------- Name of slide file

Note

*After viewing a slide, you can use the AutoCAD LT **REDRAW** command, role the wheel on a wheel mouse, or pan with a wheel mouse to remove the slide display and return to the existing drawing on the screen.*

*Any command that is automatically followed by a redraw will also display the existing drawing. For example, AutoCAD LT's **GRID**, **ZOOM ALL**, and **REGEN** commands will automatically return to the existing drawing on the screen.*

You can view the slides on high-resolution or low-resolution monitors. Depending on the resolution of the monitor, AutoCAD LT automatically adjusts the image. However, if you are using a high-resolution monitor, it is better to make the slides on the same monitor to take full advantage of that monitor.

Example 7

Write a script file that will create a slide show of the following slide files, with a time delay of 15 seconds after every slide (Figure 23-8).

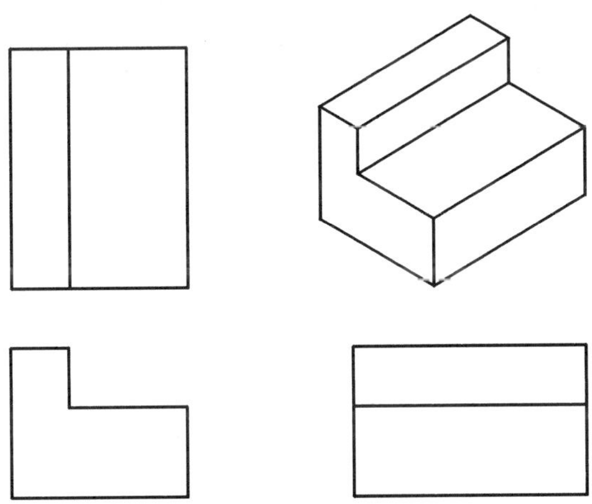

Figure 23-8 *Slides for slide show*

SLIDE1, SLIDE2, SLIDE3, SLIDE4

Step 1: Creating the slides

The first step in a slide show is to create the slides using the **MSLIDE** command. The **MSLIDE** command will invoke the **Create Slide File** dialog box. Enter the name of the slide as **SLIDE1** and choose the **Save** button to exit the dialog box. Similarly other slides can be created and correspondingly saved. Figure 23-8 shows the drawings that have been saved as slide files **SLIDE1, SLIDE2, SLIDE3,** and **SLIDE4**. The slides must be saved to a directory in AutoCAD LT's search path or the script would not find them.

Step 2: Writing the script file

The second step is to find out the sequence in which you want these slides to be displayed, with the necessary time delay, if any, between slides. Then you can use any text editor like the Notepad for writing the script file. Make sure that you save the file in the **.SCR** format. If the file is not saved in this format, you cannot access this file as a script file.

The following file is a listing of the script file that will create a slide show of the slides in Figure 23-8. The name of the script file is **SLDSHOW1**.

```
VSLIDE SLIDE1
DELAY 15000
VSLIDE SLIDE2
DELAY 15000
VSLIDE SLIDE3
DELAY 15000
VSLIDE SLIDE4
DELAY 15000
```

Step 3: Running the script file

To run this slide show, choose the **Run Script** from the **Tools** menu or enter **SCRIPT** at the Command comma Command prompt to invoke the **Select Script File** dialog box. Choose **SLDSHOW1** and choose **Open**. You can see the changes taking place on the screen.

PRELOADING SLIDES

In the script file of Example 7, VSLIDE SLIDE1 in line 1 loads the slide file, **SLIDE1**, and displays it on screen. After a pause of 15,000 milliseconds, it starts loading the second slide file, **SLIDE2**. Depending on the computer and the disk access time, you will notice that it takes some time to load the second slide file; the same is true for the other slides. To avoid the delay in loading the slide files, AutoCAD LT has provided a facility to preload a slide while viewing the previous slide. This is accomplished by placing an asterisk (*) in front of the slide file name.

```
VSLIDE SLIDE1        (View slide, SLIDE1.)
VSLIDE *SLIDE2       (Preload slide, SLIDE2.)
DELAY 15000          (Delay of 15 seconds.)
VSLIDE               (Display slide, SLIDE2.)
VSLIDE *SLIDE3       (Preload slide, SLIDE3.)
DELAY 15000          (Delay of 15 seconds.)
VSLIDE               (Display slide, SLIDE3.)
VSLIDE *SLIDE4
DELAY 15000
VSLIDE
DELAY 15000
RSCRIPT              (Restart the script file.)
```

Example 8

Write a script file to generate a continuous slide show of the following slide files, with a time delay of two seconds between slides SLD1, SLD2, SLD3

The slide files are located in different subdirectories, as shown in Figure 23-9.

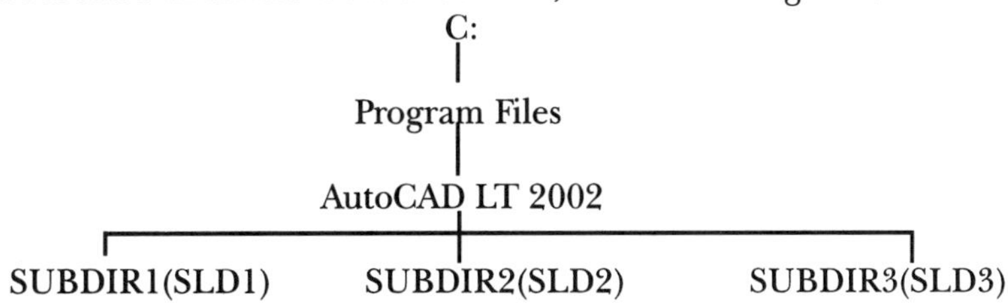

Figure 23-9 *Subdirectories of the C drive*

Where **C:** -------------------- *Root directory.*
Program Files ------------- *Root directory.*
AutoCAD LT 2002 ------ *Subdirectory where the AutoCAD LT files are loaded.*
SUBDIR1 ----------------- *Drawing subdirectory.*
SUBDIR2 ----------------- *Drawing subdirectory.*
SUBDIR3 ----------------- *Drawing subdirectory.*
SLD1 -------------------- *Slide file in SUBDIR1 subdirectory.*
SLD2 -------------------- *Slide file in SUBDIR2 subdirectory.*
SLD3 -------------------- *Slide file in SUBDIR3 subdirectory.*

The following file is the listing of the script files that will generate a slide show for the slides in Example 8:

```
VSLIDE "C:/Program Files/AutoCAD LT 2002/SUBDIR1/SLD1.SLD"
DELAY 2000
VSLIDE "C:/Program Files/AutoCAD LT 2002/SUBDIR2/SLD2.SLD"
DELAY 2000
VSLIDE "C:/Program Files/AutoCAD LT 2002/SUBDIR3/SLD3.SLD"
DELAY 2000
RSCRIPT
```

Line 1
VSLIDE "C:/Program Files/AutoCAD LT 2002/**SUBDIR1/SLD1.SLD"**
In this line, the AutoCAD LT command **VSLIDE** loads the slide file **SLD1**. The path name is mentioned along with the command **VSLIDE.** If the path name directory contains spaces then the path name must be enclosed in quotes.

Line 2
DELAY 2000
This line uses the AutoCAD LT **DELAY** command to create a pause of approximately 2 seconds before the next slide is loaded.

Line 3
VSLIDE "C:/Program Files/**AutoCAD LT 2002/SUBDIR2/SLD2.SLD"**
In this line, the AutoCAD LT command **VSLIDE** loads the slide file **SLD2**, located in the subdirectory **SUBDIR2**. If the slide file is located in a different subdirectory, you need to define the path with the slide file.

Line 5
VSLIDE "C:/Program Files/**AutoCAD LT 2002/SUBDIR3/SLD3.SLD"**
In this line, the **VSLIDE** command loads the slide file SLD3, located in the subdirectory SUBDIR3.

Line 7
RSCRIPT
In this line, the **RSCRIPT** command executes the script file again and displays the slides on the screen. This process continues indefinitely until the script file is canceled by pressing the ESC key or the BACKSPACE key.

SLIDE LIBRARIES

AutoCAD LT provides a utility, SLIDELIB, which constructs a library of the slide files. The format of the SLIDELIB utility command is:

SLIDELIB (Library filename) < (Slide list filename)

Example
SLIDELIB SLDLIB <SLDLIST
Where **SLIDELIB** ------ AutoCAD LT's SLIDELIB utility
SLDLIB --------- Slide library filename
SLDLIST ------- List of slide filenames

The SLIDELIB utility is supplied with the AutoCAD LT software package. You can find this utility (SLIDELIB.EXE) in the support subdirectory. The slide file list is a list of the slide file names that you want in a slide show. It is a text file that can be written by using any text editor like the Notepad. The slide files in the slide file list should not contain any file extension (.SLD). However, if you want to add a file extension it should be .SLD.

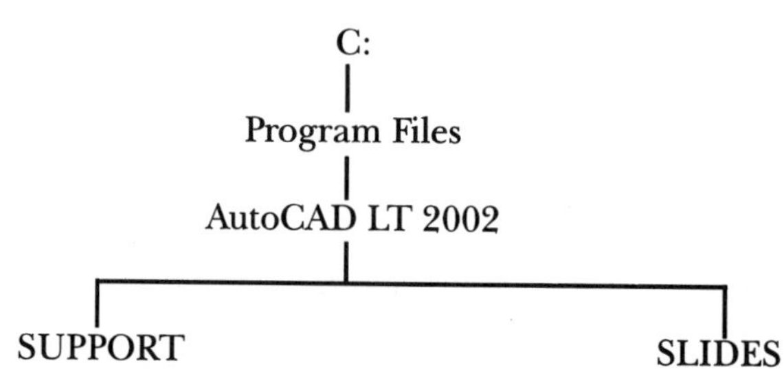

The slide file list can be also created by using the following command, if you have DOS version 5.0 or above. You can take help of make directory (md) or change directory (cd) while making or changing directories.

C:\AutoCAD LT 2002\SLIDES>**DIR *.SLD/B>SLDLIST**

In this example assume that the name of the slide file list is **SLDLIST** and all slide files are in the SLIDES subdirectory. To use this command to create a slide file list, all slide files must be in the same directory.

When you use the SLIDELIB utility, it reads the slide file names from the file that is specified in the slide list and the file is then written to the file specified by the library. In the Example 9, the SLIDELIB utility reads the slide file names from the file SLDLIST and writes them to the library file SLDLIB:

C:\>**SLIDELIB SLDLIB <SLDLIST**

Note

You cannot edit a slide library file. If you want to change anything, you have to create a new list of the slide files and then use the SLIDELIB utility to create a new slide library.

*If you edit a slide while the slide is displayed on the screen, the slide is not edited. Instead, the current drawing that is behind the slide will be edited. Therefore, do not use any editing commands while you are viewing a slide. Use the **VSLIDE** and **DELAY** commands only when viewing a slide.*

The path name is not saved in the slide library; therefore, if you have more than one slide with the same name, even though they are in different subdirectories, only one slide will be saved in the slide library.

Example 9

Use AutoCAD LT's SLIDELIB utility to generate a continuous slide show of the following slide files with a time delay of 2.5 seconds between the slides. (The file names are: SLDLIST for slide list file, SLDSHOW1 for slide library, SHOW1 for script file.)

front, top, rside, 3dview, isoview

The slide files are located in different subdirectories as shown in Figure 23-10.

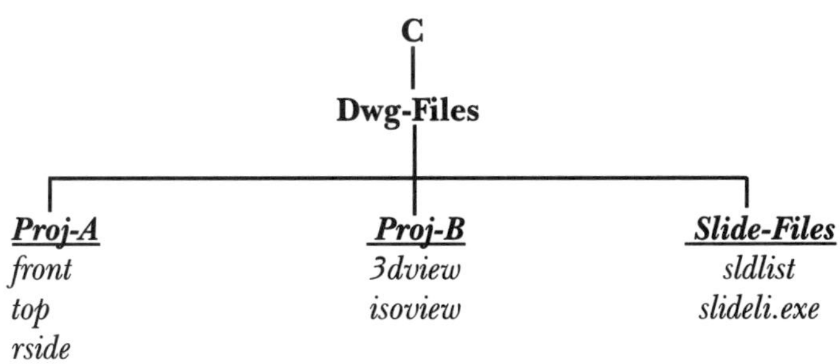

Figure 23-10 *Drawing subdirectories of C drive*

Where **C** ------------------ *(C drive.)*
Dwg-Files ------- *(Subdirectory where drawing files are located)*
Proj-A ------------ *(Drawing subdirectory.)*
Proj-B ------------ *(Drawing subdirectory.)*
Slide-Files ------ *(Directory where Slidelib.exe and sldlist are copied)*

Step 1

The first step is to create a list of the slide file names with the drive and the directory information. Assume that you are in the **Slide-Files** subdirectory. You can use a text editor like the Notepad to create a list of the slide files that you want to include in the slide show. These files do not need a file extension. However, if you choose to give them a file extension, it should be .SLD. The following file is a listing of the file SLDLIST for Example 9:

```
c:\Dwg-Files\Proj-A\front
c:\Dwg-Files\Proj-A\top
c:\Dwg-Files\Proj-A\rside
c:\Dwg-Files\Proj-B\3dview
c:\Dwg-Files\Proj-B\isoview
```

Step 2

The second step is to use AutoCAD LT's SLIDELIB utility program to create the slide library. The name of the slide library is assumed to be **sldshow1** for this example. Before creating the slide library, copy the slide list file (SLDLIST) and the SLIDELIB utility from the support directory to the Slide-Files directory. This way all the needed files are in one directory. Choose **Start > Programs > MS-DOS Prompt** to display the **MS-DOS Prompt** window. See Figure 23-11.

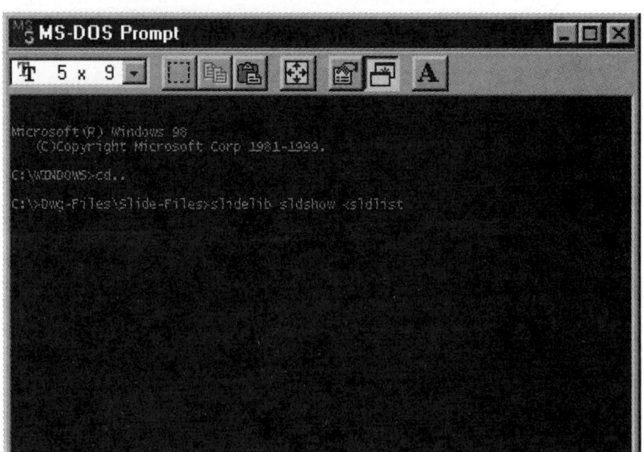

***Figure 23-11 MS-DOS Prompt** window*

Now enter the following to run the SLIDELIB utility to create the slide library. Here it is assumed that Slide-Files directory is current.

C:\Dwg-Files\Slide-Files>SLIDELIB sldshow1 <sldlist

Where **SLIDELIB** ------ AutoCAD LT's SLIDELIB utility
sldshow1 -------- Slide library
sldlist ------------ Slide file list

Step 3

Now you can write a script file for the slide show that will use the slides in the slide library. The name of the script file for this example is assumed to be SHOW1.

```
VSLIDE sldshow1(front)
DELAY 2500
VSLIDE sldshow1(top)
DELAY 2500
VSLIDE sldshow1(rside)
DELAY 2500
VSLIDE sldshow1(3dview)
DELAY 2500
VSLIDE sldshow1(isoview)
DELAY 2500
RSCRIPT
```

Step 4

Invoke the **Select Script File** (Figure 23-12) dialog box by choosing the **Run Script** button from the **Tools** menu or enter the **SCRIPT** command at the Command prompt. You can also enter the **SCRIPT** command from the Command prompt after setting the system variable FILEDIA to 0.

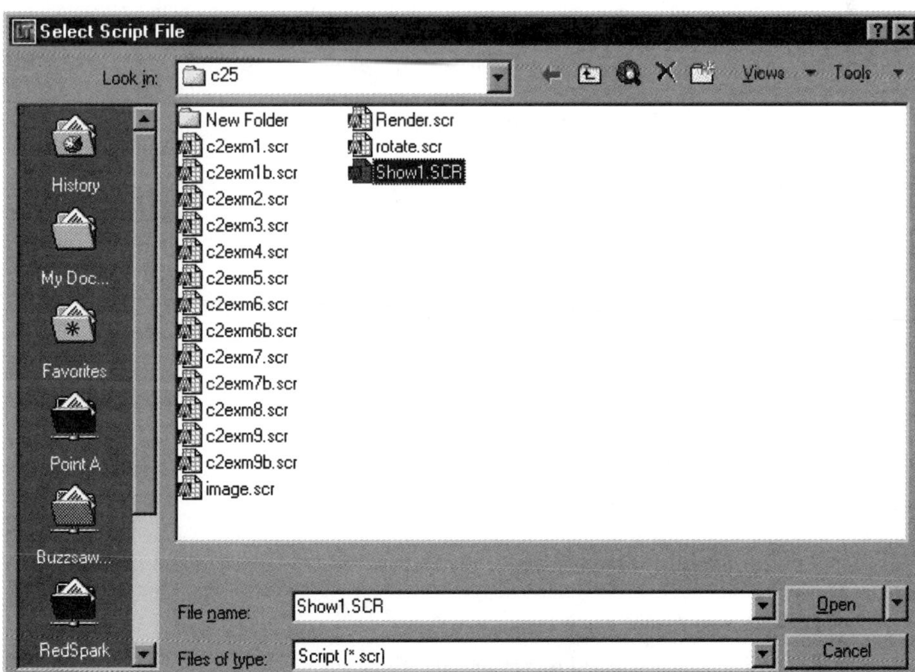

***Figure 23-12 Select Script File** dialog box*

Command: **SCRIPT**
Enter script file name<default>: **SHOW1**

Self-Evaluation Test

Answer the following questions and then compare your answers to the correct answers given at the end of this chapter.

SCRIPT FILES

1. AutoCAD LT has provided a facility of __________ that allows you to combine different AutoCAD LT commands and execute them in a predetermined sequence.

2. Before writing a script file, you need to know the AutoCAD LT __________ and the __________ required in response to the Command prompts.

3. The AutoCAD LT __________ command is used to run a script file.

4. In a script file, the __________ is used to terminate a command or a prompt entry.

5. The **DELAY** command is to be followed by __________ in milliseconds.

SLIDE SHOWS

6. Slides do not contain any __________ information, which means that the entities do not have any information associated with them.

7. Slides __________ edited like a drawing.

8. Slides can be created using the AutoCAD LT __________ command.

9. To view a slide, use the AutoCAD LT __________ command.

10. AutoCAD LT provides a utility that constructs a library of the slide files. This is done with AutoCAD LT's utility program called __________.

Review Questions

Answer the following questions.

SCRIPT FILES

1. The __________ files can be used to generate a slide show, do the initial drawing setup, or plot a drawing to a predefined specification.

2. In a script file, you can __________ several statements in one line.

3. When you run a script file, the default script file name is the same as the __________ name.

4. When you run a script file, type the name of the script file without the file __________.

5. One of the limitations of script files is that all the information has to be contained __________ the file.

6. The AutoCAD LT __________ command allows you to reexecute a script file indefinitely until the command is canceled.

7. You cannot provide a __________ statement in a script file to terminate the file when a particular condition is satisfied.

8. The AutoCAD LT __________ command introduces a delay before the next command is executed.

9. If the script file was canceled and you want to continue the script file, you can do so by using the AutoCAD LT __________ command.

SLIDE SHOWS

10. AutoCAD LT provides a facility through __________ files to combine the slides in a text file and display them in a predetermined sequence.

11. A __________ can also be introduced in the script file so that the viewer has enough time to view a slide.

12. Slides are the __________ of a screen display.

13. In model space, you can use the **MSLIDE** command to make a slide of the __________ display in the __________ viewport.

14. If you are in paper space, you can make a slide of the display in paper space that __________ any floating viewports.

15. If you want to make any change in the slide, you need to __________________ the drawing, then make a new slide from the edited drawing.

16. If the slide is in the slide library and you want to view it, the slide library name has to be __________ with the slide file name.

17. You cannot __________ a slide library file. If you want to change anything, you have to create a new list of the slide files and then use the __________ utility to create a new slide library.

18. The path name __________ be saved in the slide library. Therefore, if you have more than one slide with the same name, although with different subdirectories, only one slide will be saved in the slide library.

Exercises

SCRIPT FILES

Exercise 2 — *General*

Write a script file that will do the following initial setup for a drawing:

Grid	2.0
Snap	0.5
Limits	0,0
	18.0,12.0
Zoom	All
Text height	0.25
Ltscale	2.0

Overall dimension scale factor is 2
Aligned dimension text with the dimension line
Dimension text above the dimension line
Size of the center mark is 0.75

Exercise 3 — *General*

Write a script file that will set up the following layers with the given colors and linetypes (filename SCRIPTE3.SCR):

Contour	Red	Continuous
SPipes	Yellow	Center
WPipes	Blue	Hidden
Power	Green	Continuous
Manholes	Magenta	Continuous
Trees	Cyan	Continuous

Exercise 4 — *General*

Write a script file that will do the following initial setup for a new drawing:

Limits	0,0 24,18
Grid	1.0
Snap	0.25
Ortho	On
Snap	On
Zoom	All

Pline width	0.02	
PLine	0,0 24,0 24,18 0,18 0,0	
Ltscale	1.5	
Units	Decimal units	
	Precision 0.00	
	Decimal degrees	
	Precision 0	
	Base angle East (0.00)	
	Angle measured counterclockwise	

Layers

Name	Color	Linetype
Obj	Red	Continuous
Cen	Yellow	Center
Hid	Blue	Hidden
Dim	Green	Continuous

Exercise 5 *General*

Write a script file that will plot a given drawing according to the following specifications. (Use the plotter for which your system is configured and adjust the values accordingly.)

Plot, using the Window option
Window size (0,0 24,18)
Do not write the plot to file
Size in inch units
Plot origin (0.0,0.0)
Maximum plot size (8.5,11 or the smallest size available on your printer/plotter)
90 degree plot rotation
No removal of hidden lines
Plotting scale (Fit)

Exercise 6 *General*

Write a script file that will continuously rotate a line in 10 degrees increments around its midpoint (Figure 23-13). The time delay between increments is one second.

Exercise 7 *General*

Write a script file that will continuously rotate the following arrangements (Figure 23-14). One set of two circles and one line should rotate clockwise while the other set of two circles and one line should rotate counterclockwise. Assume the rotation to be 5 degrees around the intersection of the lines for both sets of arrangements. Specifications are as follows.

Start point of the horizontal line	2,4
Endpoint of the horizontal line	8,4
Center point of circle at the start point of horizontal line	2,4
Diameter of the circle	1.0

Center point of circle at the end point of horizontal line	8,4
Diameter of circle	1.0
Start point of the vertical line	5,1
Endpoint of the vertical line	5,7
Center point of circle at the start point of the vertical line	5,1
Diameter of the circle	1.0
Center point of circle at the end point of the vertical line	5,7
Diameter of the circle	1.0

(Select one set of two circles and one line and create one group. Similarly select another set of two circles and one line and create another group. Rotate one group clockwise and another group counterclockwise.)

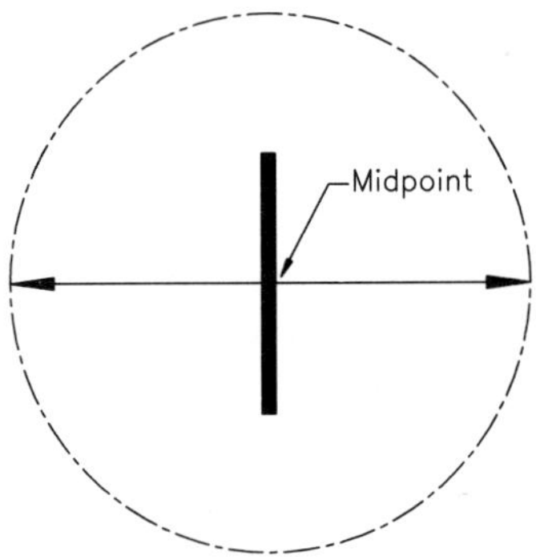

Figure 23-13 *Drawing for Exercise 6*

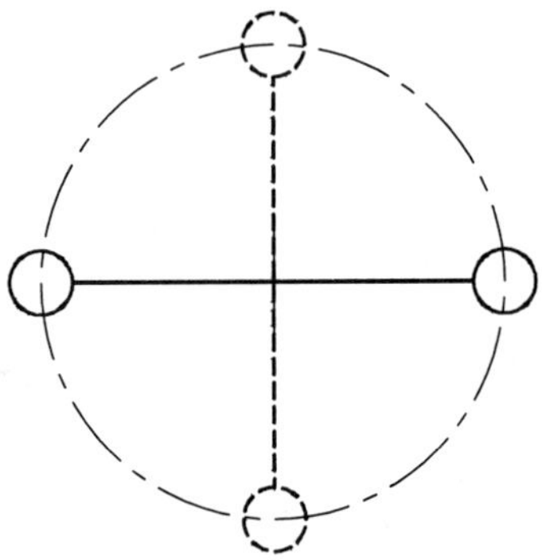

Figure 23-14 *Drawing for Exercise 7*

SLIDE SHOWS

Exercise 8 — *General*

Make the slides shown in Figure 23-15 and write a script file for a continuous slide show. Provide a time delay of 5 seconds after every slide. (You do not have to use the slides shown in Figure 23-15; you can use any slides of your choice.)

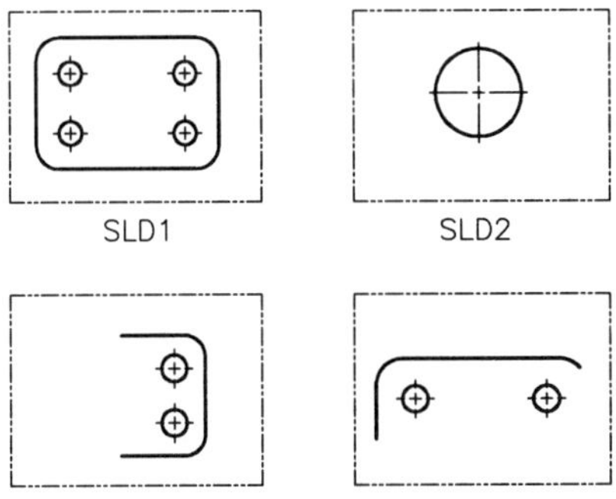

Figure 23-15 *Slides for slide show*

Exercise 9 *General*

From Exercise 8, list the slides in a file SLDLIST2 and create a slide library file SLDLIB2. Then write a script file SHOW2 using the slide library with a time delay of 5 seconds after every slide.

Answers to Self-Evaluation Test

1 - **SCRIPT** files, **2** - Commands, options, **3** - **SCRIPT**, **4** - blank space, **5** - time, **6** - vector, **7** - cannot be, **8** - **MSLIDE**, **9** - **VSLIDE**, **10** - SLIDELIB

Chapter 24

Creating Linetypes and Hatch Patterns

Learning Objectives

After completing this chapter, you will be able to:
Create Linetypes
- *Write linetype definitions.*
- *Create different linetypes.*
- *Create linetype files.*
- *Determine **LTSCALE** for plotting the drawing to given specifications.*
- *Define alternate linetypes and modify existing linetypes.*
- *Create string and shape complex linetypes.*

Create Hatch Patterns
- *Understand hatch pattern definition.*
- *Create new hatch patterns.*
- *Determine the effect of angle and scale factor on hatch.*
- *Create hatch patterns with multiple descriptors.*
- *Save hatch patterns in a separate file.*
- *Define custom hatch pattern file.*

STANDARD LINETYPES

The AutoCAD LT software package comes with a library of standard linetypes that has thirty- eight different standard linetypes and seven complex linetypes, including ISO linetypes. These linetypes are saved in the **ACLT.LIN** file. You can modify existing linetypes or create new ones.

LINETYPE DEFINITION

All linetype definitions consist of two parts: **header line** and **pattern line**.

Header Line

The **header line** consists of an asterisk (*) followed by the name of the linetype and the linetype description. The name and the linetype description should be separated by a comma. If there is no description, the comma that separates the linetype name and the description is not required.

The format of the header line is:

*** Linetype Name, Description**

Example

***HIDDENS,__ __ __ __ __ __ __**

Where * ------------------ Asterisk sign
HIDDENS ------ Linetype name
, ------------------ Comma
__ __ __ __ ------ Linetype description

All linetype definitions require a linetype name. When you want to load a linetype or assign a linetype to an object, AutoCAD LT recognizes the linetype by the name you have assigned to the linetype definition. The names of the linetype definition should be selected to help the user recognize the linetype by its name. For example, the linetype name LINEFCX does not give the user any idea about the type of line. However, a linetype name like DASHDOT gives a better idea about the type of line that a user can expect.

The linetype description is a textual representation of the line. This representation can be generated by using dashes, dots, and spaces at the keyboard. The graphic is used by AutoCAD LT when you want to display the linetypes on the screen by using the AutoCAD LT's **LINETYPE** command with the ? option or by using the dialog box. The linetype description cannot exceed forty-seven characters.

Pattern Line

The **pattern line** contains the definition of the line pattern. The definition of the line pattern consists of the alignment field specification and the linetype specification. The alignment field specification and the linetype specification are separated by a comma.

The format of the pattern line is:

Alignment Field Specification, Linetype Specification

Example

A,.75,-.25,.75

Where **A** ------------------ Alignment field specification
, -------------------- Comma
.75,-.25,.75 ----- Linetype specification

The letter used for alignment field specification is A. This is the only alignment field supported by AutoCAD LT; therefore, the pattern line will always start with the letter A. The linetype specification defines the configuration of the dash-dot pattern to generate a line. The maximum number for dash length specification in the linetype is 12, provided the linetype pattern definition fits on one eighty character line.

ELEMENTS OF LINETYPE SPECIFICATION

All linetypes are created by combining the basic elements in a desired configuration. There are three basic elements that can be used to define a linetype specification.

Dash (Pen down)
Dot (Pen down, 0 length)
Space (Pen up)

Example

______ . ______ . ______ . ______

Where . -------------------- Dot (pen down with 0 length)
Blank space --------------- Space (pen up)
______ -------------------- Dash (pen down with specified length)

The dashes are generated by defining a positive number. For example, .5 will generate a dash 0.5 units long. Similarly, spaces are generated by defining a negative number. For example, -.2 will generate a space 0.2 units long. The dot is generated by defining a 0 length.

Example
A,.5,-.2,0,-.2,.5
Where **0** ------------------- Dot (zero length)
-.2 ----------------- Length of space (pen up)
.5 ------------------ Length of dash (pen down)

CREATING LINETYPES

Before creating a linetype, you need to decide the type of line you want to generate. Draw the line on a piece of paper and measure the length of each element that constitutes the line. You need to define only one segment of the line, because the pattern is repeated when you draw a line. Linetypes can be created or modified by any one of the following methods:

Using a text editor like Notepad
Adding a new linetype in the ACLT.LIN file
Using the AutoCAD LT's LINETYPE command

The following example, Example 1, explains how to create a new linetype by using a text editor, adding a linetype to the **ACLT.LIN** file, and using the AutoCAD LT's **LINETYPE** command.

Example 1

Create linetype DASH3DOT (Figure 24-1) with the following specifications:

Length of the first dash 0.5
Blank space 0.125
Dot
Blank space 0.125
Dot
Blank space 0.125
Dot
Blank space 0.125

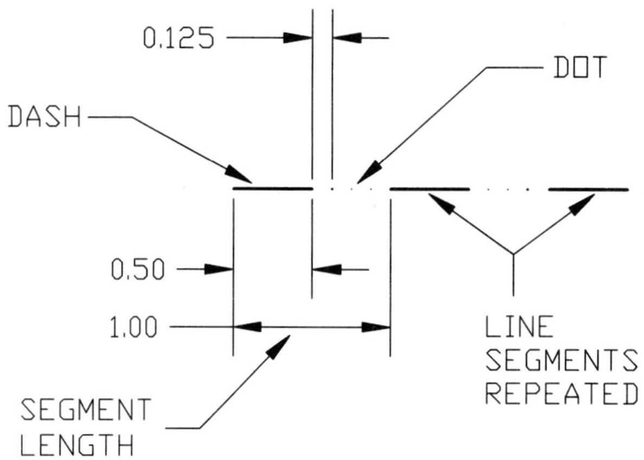

Figure 24-1 *Linetype specifications of DASH3DOT*

Using a Text Editor

Step 1: Writing definition of linetype

You can start a new linetype file and then add the line definitions to this file. Use any text editor like Notepad to start a new file (NEWLT.LIN) and then add the following two lines to the file to define the DASH3DOT linetype. The name and the description must be separated by a comma (,). The description is optional. If you decide not to give one, omit the comma after the linetype name DASH3DOT.

```
*DASH3DOT,___ . . . ___ . . . ___
A,.5,-.125,0,-.125,0,-.125,0,-.125
```

Save it as NEWLT.LIN in AutoCAD LT's Support directory.

Step 2: Loading the linetype

To load this linetype, choose **Linetype** from the **Format** menu to display the **Linetype Manager** dialog box. Choose the **Load** button in the **Linetype Manager** dialog box to display the **Load or Reload Linetypes** dialog box. Choose the **File** button in the **Load or Reload Linetypes** dialog box to display the **Select Linetype File** dialog box as shown in Figure 24-2. Choose the **NEWLT.LIN** file in the **Select Linetype File** dialog box and then choose **Open**. Again the **Load or Reload Linetypes** dialog box is displayed. Choose the **DASH3DOT** linetype in the **Available Linetypes** area and then choose **OK**. Again, the **Linetype Manager** dialog box is displayed. Choose the **DASH3DOT** linetype and then choose the **Current** button to make the selected linetype current. Then choose **OK**.

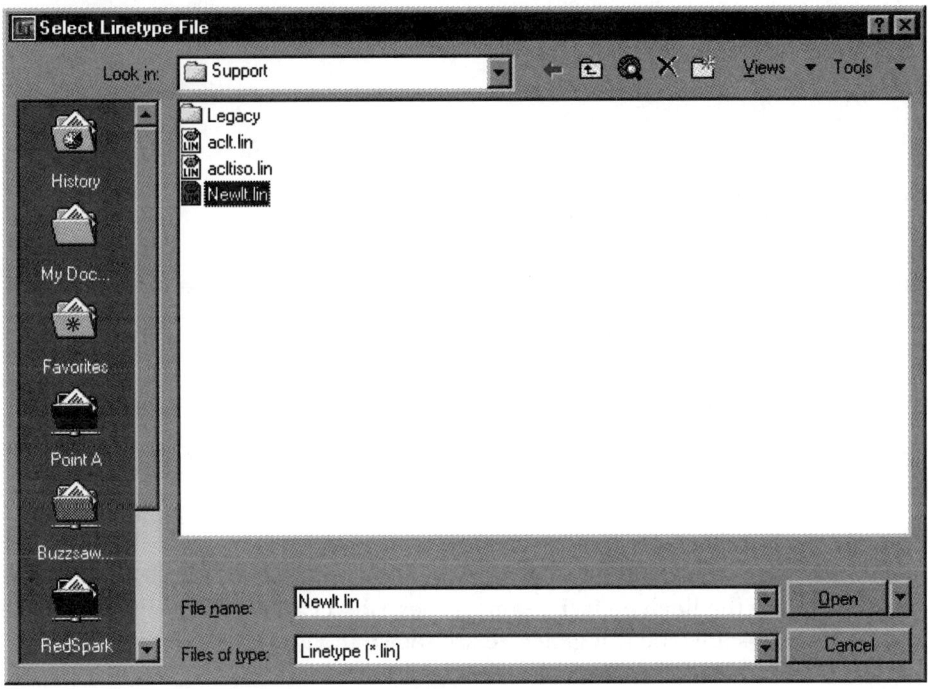

*Figure 24-2 **Select Linetype File** dialog box*

Adding a New Linetype in the ACLT.LIN File

Step 1: Adding a new linetype in the ACLT.LIN

You can also use a text editor (like Notepad) to create a new linetype. Using the text editor, load the file and insert the lines that define the new linetype. The following file is a partial listing of the **ACLT.LIN** file after adding a new linetype to the file:

```
*BORDER,__ __ . __ __ . __ __ . __ __ . __ __ . __ __ .
A,.5,-.25,.5,-.25,0,-.25
*BORDER2,__.__.__.__.__.__.__.__.__.__.__.__.__.
A,.25,-.125,.25,-.125,0,-.125
*BORDERX2,____ ____ . ____ ____ . ____ ____ .
```

```
A,1.0,-.5,1.0,-.5,0,-.5
*CENTERX2,________ __ ________ __ ________ __ __
A,2.5,-.5,.5,-.5
*DASHDOT,__ . __ . __ . __ . __ . __ . __ . __ .
A,.5,-.25,0,-.25
*DOTX2,. . . . . . . . . . . . . . .
A,.25,-.125
*HIDDEN2,_ _ _ _ _ _ _ _ _ _ _ _ _ _ _ _ _ _ _ _ _
A,.125,-.0625
*HIDDENX2,____ ____ ____ ____ ____ ____ ____ ____ ____
A,.5,-.25
*PHANTOM,______ __ __ ______ __ __ ______ __ __
A,1.25,-.25,.25,-.25,.25,-.25
*PHANTOMX2,___________ ____ ____ ___________
|
|
*GAS_LINE,Gas line ----GAS----GAS----GAS----GAS----GAS----GAS--
A,.5,-.2,["GAS",STANDARD,S=.1,R=0.0,X=-0.1,Y=-.05],-.25
*ZIGZAG,Zig zag /\/\/\/\/\/\/\/\/\/\/\/\/\/\/\/\/\/\/\/\/\/\/\/\/\/\/\/\/\/\/\/\/\/\/\/\/
A,.0001,-.2,[ZIG,ltypeshp.shx,x=-.2,s=.2],-.4,[ZIG,ltypeshp.shx,r=180,x=.2,s=.2],-.2
*DASH3DOT,____ . . . ____ . . . ____
A,.5,-.125,0,-.125,0,-.125,0,-.125
```

The last two lines of this file define the new linetype, DASH3DOT. The first line contains the name DASH3DOT and the description of the line (___ . . .___). The second line contains the alignment and the pattern definition.

Step 2: Loading the linetype

Save the file and then load the linetype using the AutoCAD LT **LINETYPE** command. The procedure of loading the linetype is the same as described earlier in this example. The lines and polylines that this linetype will generate are shown in Figure 24-3.

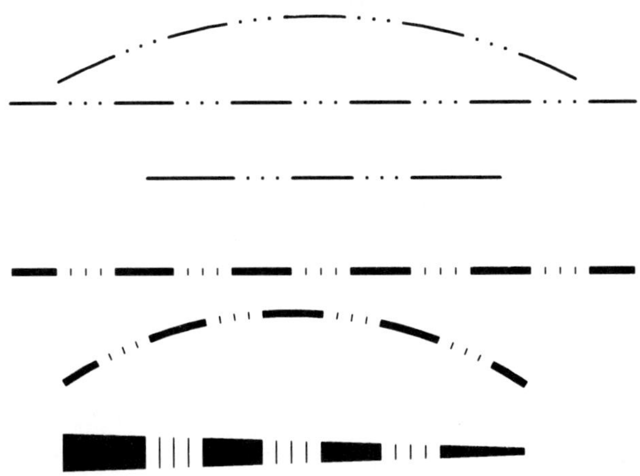

Figure 24-3 Lines created by linetype DASH3DOT

Note

If you change the LTSCALE factor, all lines in the drawing are affected by the new ratio.

Using the AutoCAD LT Linetype Command

Step 1: Creating a linetype

To create a linetype using the AutoCAD LT **LINETYPE** command, first make sure that you are in the drawing editor. Then enter the **-LINETYPE** command and select the Create option to create a linetype.

Command: -**LINETYPE**
Enter an option [?/Create/Load/Set]: **C**

Enter the name of the linetype and the name of the library file in which you want to store the definition of the new linetype.

Enter name of linetype to create: **DASH3DOT**

If **FILEDIA**=1, the **Create or Append Linetype File** dialog box (Figure 24-4) will appear on the screen. If **FILEDIA**=0, AutoCAD LT will prompt you to enter the name of the file.

Enter linetype file name for new linetype definition <default>: **Aclt**

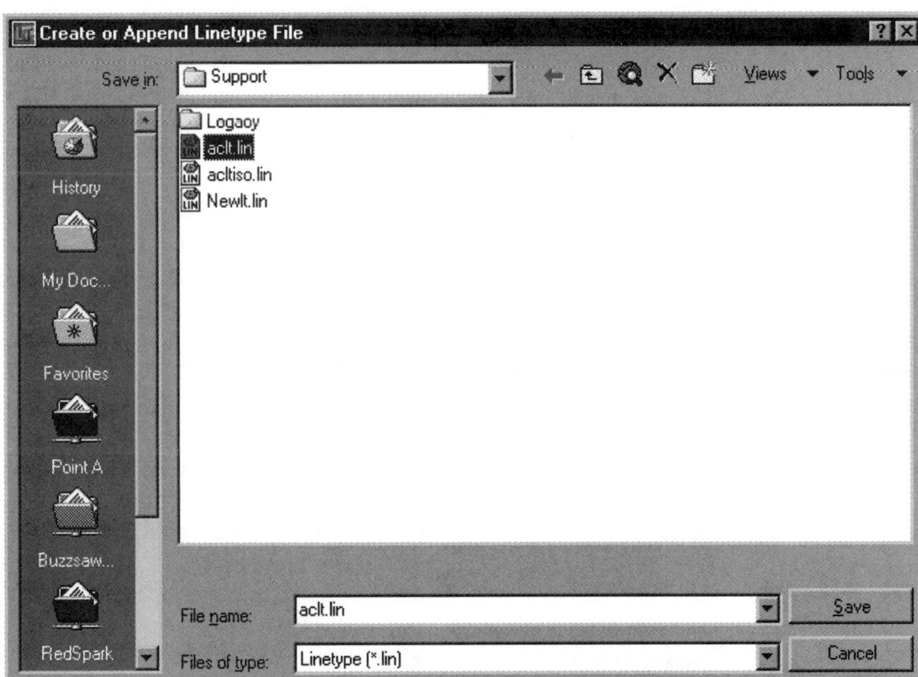

***Figure 24-4 Create or Append Linetype File** dialog box*

If the linetype already exists, the following message will be displayed on the screen:

Wait, checking if linetype already defined...
"Linetype" already exists in this file. Current definition is:
alignment, dash-1, dash-2, _____.
Overwrite?<N>

If you want to redefine the existing line style, enter Y; otherwise, type N or press RETURN to choose the default value of N. You can then repeat the process with a different name of the linetype. After entering the name of the linetype and the library file name, AutoCAD LT will prompt you to enter the descriptive text and the pattern of the line.

Descriptive text: ***DASH3DOT,____ . . . ____ . . . ____**
Enter linetype pattern (on next line):
A,.5,-.125,0,-.125,0,-.125,0,-.125

Descriptive Text

***DASH3DOT,____ . . . ____ . . . ____**

For the descriptive text, you have to type an asterisk (*) followed by the name of the linetype. For Example 1, the name of the linetype is DASH3DOT. The name *DASH3DOT can be followed by the description of the linetype; the length of this description cannot exceed 47 characters. In this example, the description is dashes and dots ____ . . . ____. It could be any text or alphanumeric string. The description is displayed on the screen when you list the linetypes.

Pattern

A,.5,-.125,0,-.125,0,-.125,0,-.125

The line pattern should start with an alignment definition. Currently, AutoCAD LT supports only one type of alignment—A. Therefore, it is automatically displayed on the screen when you select the **LINETYPE** command with the Create option. After entering **A** for pattern alignment, you must define the pen position. A positive number (.5 or 0.5) indicates a "pen-down" position, and a negative number (-.25 or -0.25) indicates a "pen-up" position. The length of the dash or the space is designated by the magnitude of the number. For example, 0.5 will draw a dash 0.5 units long, and -0.25 will leave a blank space of 0.25 units. A dash length of 0 will draw a dot (.). Here are the pattern definition elements for Example 1:

.5	**pen down**	**0.5 units long dash**
-.125	**pen up**	**.125 units blank space**
0	**pen down**	**dot**
-.125	**pen up**	**.125 units blank space**
0	**pen down**	**dot**
-.125	**pen up**	**.125 units blank space**
0	**pen down**	**dot**
-.125	**pen up**	**.125 units blank space**

After you enter the pattern definition, the linetype (DASH3DOT) is automatically saved in the **ACLT.LIN** file

Step 2: Loading the linetype

You can use the **LINETYPE** command to load the linetype or choose **Linetype** in the **Format** menu. The linetype (DASH3DOT) can also be loaded using the AutoCAD LT's **-LINETYPE** command and selecting the **Load** option.

ALIGNMENT SPECIFICATION

The alignment specifies the pattern alignment at the start and the end of the line, circle, or arc. In other words, the line would always start and end with the dash (___). The alignment definition "A" requires that the first element be a dash or dot (pen down), followed by a negative (pen up) segment. The minimum number of dash segments for alignment A is two. If there is not enough space for the line, AutoCAD LT will draw a continuous line.

For example, in the linetype DASH3DOT of Example 1, the length of each line segment is 1.0 (.5 + .125 + .125 + .125 + .125 = 1.0). If the length of the line drawn is less than 1.00, the line will be drawn as a continuous line (Figure 24-5). If the length of the line is 1.00 or greater, the line will be drawn according to DASH3DOT linetype. AutoCAD LT automatically adjusts the length of the dashes and the line will always start and end with a dash. The length of the starting and ending dashes will be at least half the length of the dash as specified in the file. If the length of the dash as specified in the file is 0.5, the length of the starting and ending dashes will be at least 0.25. To fit a line that starts and ends with a dash, the length of these dashes can also increase as shown in Figure 24-5.

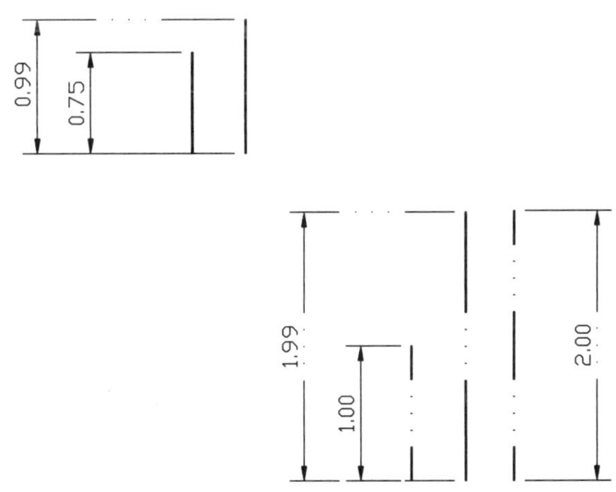

Figure 24-5 *Alignment of linetype DASH3DOT*

LTSCALE COMMAND

As we mentioned previously, the length of each line segment in the DASH3DOT linetype is 1.0 (.5 + .125 + .125 + .125 + .125 = 1.0). If you draw a line that is less than 1.0 unit long,

AutoCAD LT will draw a single dash that looks like a continuous line (Figure 24-6). This problem can be rectified by changing the linetype scale factor variable **LTSCALE** to a smaller value. This can be accomplished by using AutoCAD LT's **LTSCALE** command:

Command: **LTSCALE**
Enter new linetype scale factor <default>: *New value.*

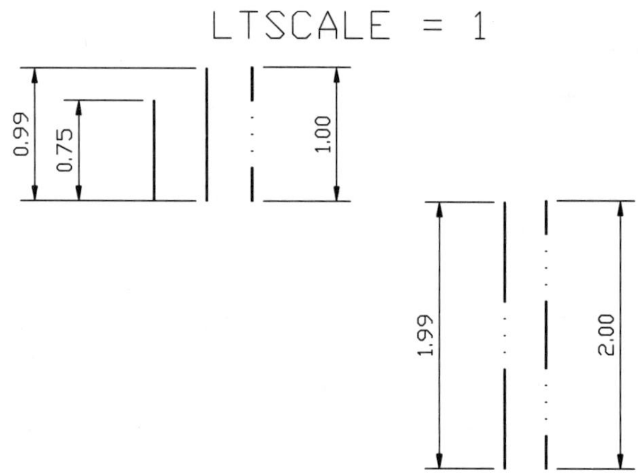

Figure 24-6 *Alignment when Ltscale = 1*

The default value of the **LTSCALE** variable is 1.0. If the LTSCALE is changed to 0.75, the length of each segment is reduced by 0.75 (1.0 x 0.75 = 0.75). Then, if you draw a line 0.75 units or longer, it will be drawn according to the definition of DASH3DOT (___ . . . ___) (Figure 24-7 and Figure 24-8).

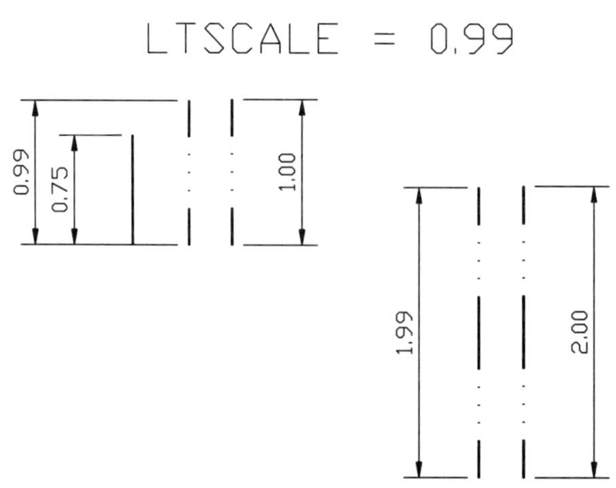

Figure 24-7 *Alignment when Ltscale = 0.99*

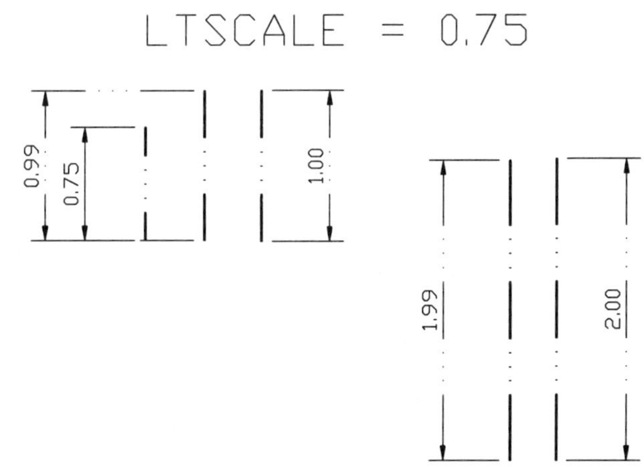

Figure 24-8 Alignment when Ltscale = 0.75

The appearance of the lines is also affected by the limits of the drawing. Most of the AutoCAD LT linetypes work fine for a drawing that has the limits 12,9. Figure 24-9 shows a line of linetype DASH3DOT that is four units long and the limits of the drawing are 12,9. If you increase the limits to 48,36 the lines will appear as continuous lines. If you want the line to appear the same as before **on the screen**, the LTSCALE should be changed. Since the limits of the drawing have increased four times, the LTSCALE should also increase by the same amount. If you change the scale factor to four, the line segments will also increase by a factor of four. As shown in Figure 24-9, the length of the starting and the ending dash has increased to one unit.

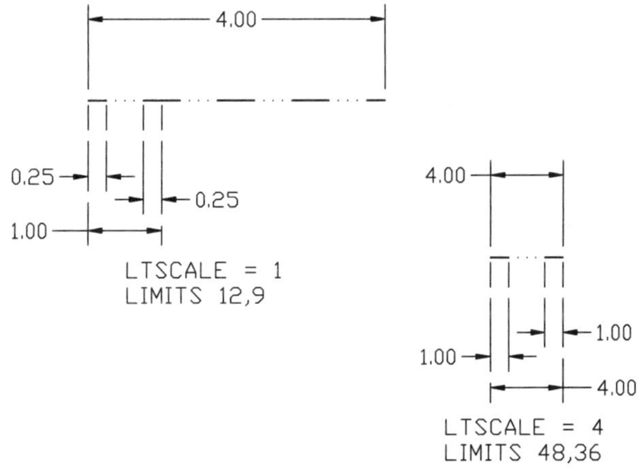

Figure 24-9 Linetype DASH3DOT before and after changing the LTSCALE factor

In general, the approximate LTSCALE factor for **screen display** can be obtained by dividing the X-limit of the drawing by the default X-limit (12.00). However, **it is recommended that the linetype scale must be set according to plot scale** discussed in the next section.

LTSCALE factor for SCREEN DISPLAY = X-limits of the drawing/12.00

Example

Drawing limits are 48,36
LTSCALE factor for screen display= 48/12 = 4

Drawing sheet size is 36,24 and scale is 1/4" = 1'
LTSCALE factor for screen display = 12 x 4 x (36 / 12) = 144

LTSCALE FACTOR FOR PLOTTING

The LTSCALE factor for plotting depends on the size of the sheet you are using to plot the drawing. For example, if the limits are 48 by 36, the drawing scale is 1:1, and you want to plot the drawing on a 48" by 36" size sheet, the LTSCALE factor is 1. If you check the specification of a hidden line in the ACLT.LIN file, the length of each dash is 0.25. Therefore, when you plot a drawing with 1:1 scale, the length of each dash in a hidden line is 0.25.

However, if the drawing scale is 1/8" = 1' and you want to plot the drawing on a 48" by 36" paper, the LTSCALE factor must be 96 (8 x 12 = 96). The length of each dash in the hidden line will increase by a factor of 96 because the LTSCALE factor is 96. Therefore, the length of each dash will be 24 units (0.25 x 96 = 24). At the time of plotting, the scale factor for plotting must be 1:96 to plot the 384' by 288' drawing on a 48" by 36" size paper. Each dash of the hidden line that was 24" long on the drawing will be 0.25 (24/96 = 0.25) inch long when plotted. Similarly, if the desired text size on the paper is 1/8", the text height in the drawing must be 12" (1/8 x 96 = 12").

Ltscale Factor for PLOTTING = Drawing Scale

Sometimes your plotter may not be able to plot a 48" by 36" drawing or you might like to decrease the size of the plot so that the drawing fits within a specified area. To get the correct dash lengths for hidden, center, or other lines, you must adjust the LTSCALE factor. For example, if you want to plot the previously mentioned drawing in a 45" by 34" area, the correction factor is:

Correction factor = 48/45
= 1.0666
New LTSCALE factor = LTSCALE factor x Correction factor
= 96 x 1.0666
= 102.4

New Ltscale Factor for PLOTTING = Drawing Scale x Correction Factor

Note

If you change the LTSCALE factor, all lines in the drawing are affected by the new ratio.

ALTERNATE LINETYPES

One of the problems with the LTSCALE factor is that it affects all the lines in the drawing. As shown in Figure 24-10(a), the length of each segment in all DASH3DOT type lines is approximately equal, no matter how long the lines. You might want to have a small segment length if the lines are small and a longer segment length if the lines are long. You can accomplish this by using CELTSCALE (discussed later in this chapter) or by defining an alternate linetype with a different segment length. For example, you can define a linetype DASH3DOT and DASH3DOTX with different line pattern specifications.

```
*DASH3DOT,____ . . . ____ . . . ____ . . . ____
A,0.5,-.125,0,-.125,0,-.125,0,-.125
*DASH3DOTX,________ . . . __________
A,1.0,-.25,0,-.25,0,-.25,0,-.25
```

In DASH3DOT linetype the segment length is one unit, whereas in DASH3DOTX linetype the segment length is two units. You can have several alternate linetypes to produce the lines with different segment lengths. Figure 24-10(b) shows the lines generated by DASH3DOT and DASH3DOTX.

Note
Although you might have used different linetypes with different segment lengths, the lines will be affected equally when you change the LTSCALE factor. For example, if the LTSCALE factor is 0.5, the segment length of DASH3DOT line will be 0.5 and the segment length of DASH3DTX will be 1.0 unit.

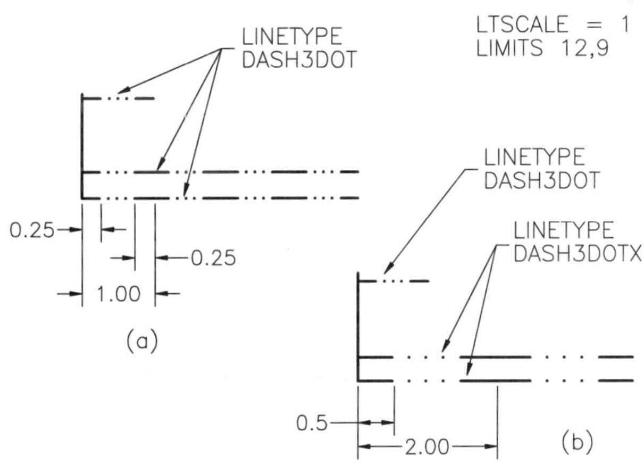

Figure 24-10 *Linetypes generated by DASH3DOT and DASH3DOTX*

MODIFYING LINETYPES

You can also modify the linetypes that are defined in the ACLT.LIN file. You must save a copy of the original ACLT.LIN file before making any changes to it. You need a text editor, such as Notepad, to modify the linetype. You can also use the EDIT function of DOS to edit the files. For example, if you want to change the dash length of the border linetype from 0.5 to 0.75, load the file, then edit the pattern line of the border linetype. The following file is a partial listing of the ACLT.LIN file after changing the border and centerx2 linetypes.

```
;; AutoCAD LT Linetype Definition file, Version 2.0
;; Copyright 1991, 1992, 1993, 1994, 1996 by Autodesk, Inc.
;;
*BORDER,__ __ . __ __ . __ __ . __ __ . __ __ . __ __ . __ __ . __ __ .
A,.75,-.25,.75,-.25,0,-.25
*BORDER2,Border (.5x) __.__.__.__.__.__.__.__.__.__.__.
A,.25,-.125,.25,-.125,0,-.125
*BORDERX2,Border (2x) ____ ____ . ____ ____ . __
A,1.0,-.5,1.0,-.5,0,-.5

*CENTER,Center ____ _ ____ _ ____ _ ____ _ ____ _ ____
A,1.25,-.25,.25,-.25
*CENTER2,Center (.5x) ___ _ ___ _ ___ _ ___ _ ___ _ ___
A,.75,-.125,.125,-.125
*CENTERX2,Center (2x) ________ __ ________ __ _____
A,3.5,-.5,.5,-.5

*DASHDOT,Dash dot __ . __ . __ . __ . __ . __ . __ . __
A,.5,-.25,0,-.25
*DASHDOT2,Dash dot (.5x) _._._._._._._._._._._._._._.
A,.25,-.125,0,-.125
*DASHDOTX2,Dash dot (2x) ____ . ____ . ____ . __
A,1.0,-.5,0,-.5

*DASHED,Dashed __ __ __ __ __ __ __ __ __ __ __ __ __ _
A,.5,-.25
*DASHED2,Dashed (.5x) _ _ _ _ _ _ _ _ _ _ _ _ _ _ _ _ _
A,.25,-.125
*DASHEDX2,Dashed (2x) ____ ____ ____ ____ ____ ___
A,1.0,-.5

*DIVIDE,Divide ____ . . ____ . . ____ . . ____ . . ____
A,.5,-.25,0,-.25,0,-.25
*DIVIDE2,Divide (.5x) __..__..__..__..__..__..__..__.._
A,.25,-.125,0,-.125,0,-.125
*DIVIDEX2,Divide (2x) ________ . . ________ . . _
A,1.0,-.5,0,-.5,0,-.5

*DOT,Dot . . . . . . . . . . . . . . . . . . . . . . . .
A,0,-.25
*DOT2,Dot (.5x) ........................................
A,0,-.125
*DOTX2,Dot (2x) . . . . . . . . . . . . . .
A,0,-.5

*HIDDEN,Hidden __ __ __ __ __ __ __ __ __ __ __ __ __ __
A,.25,-.125
*HIDDEN2,Hidden (.5x) _ _ _ _ _ _ _ _ _ _ _ _ _ _ _ _ _
```

```
A,.125,-.0625
*HIDDENX2,Hidden (2x) ____ ____ ____ ____ ____ ____ ____
A,.5,-.25

*PHANTOM,Phantom ______ __ __ ______ __ __ ______
A,1.25,-.25,.25,-.25,.25,-.25
*PHANTOM2,Phantom (.5x) ___ _ _ ___ _ _ ___ _ _ ___ _ _
A,.625,-.125,.125,-.125,.125,-.125
*PHANTOMX2,Phantom (2x) ____________    ____    ____   _
A,2.5,-.5,.5,-.5,.5,-.5
;;  ISO 128 (ISO/DIS 12011) linetypes
;;  The size of the line segments for each defined ISO line, is
;;  defined for a usage with a pen width of 1 mm. To use them with
;;  the other ISO predefined pen widths, the line has to be scaled
;;  with the appropriate value (e.g, pen width 0,5 mm -> ltscale 0.5).
;;
*ACLT_ISO02W100,ISO dash __ __ __ __ __ __ __ __ __ __ __ __ __
A,12,-3
*ACLT_ISO04W100,ISO long-dash dot ____ . ____ . ____ . ____ . _
A,24,-3,.5,-3

*ACLT_ISO14W100,ISO dash triple-dot __ . . . __ . . . __ . . . _
A,12,-3,.5,-3,.5,-3,.5,-3
*ACLT_ISO15W100,ISO double-dash triple-dot __ __ . . . __ __ . .
A,12,-3,12,-3,.5,-3,.5,-3,.5,-3
;;  Complex linetypes

;;  Complex linetypes have been added to this file.
;;  These linetypes were defined in LTYPESHP.LIN in
;;  Release 13, and are incorporated in ACLT.LIN in Release 14

;;  These linetype definitions use LTYPESHP.SHX.
;;
*FENCELINE1,Fenceline circle ----0-----0-----0-----0-----0-----0--
A,.25,-.1,[CIRC1,ltypeshp.shx,x=-.1,s=.1],-.1,1
*FENCELINE2,Fenceline square ----[]-----[]-----[]-----[]-----[]---
A,.25,-.1,[BOX,ltypeshp.shx,x=-.1,s=.1],-.1,1
*TRACKS,Tracks -|-|-|-|-|-|-|-|-|-|-|-|-|-|-|-|-|-|-|-|-|-|-|-|-|-|-|-|-
A,.15,[TRACK1,ltypeshp.shx,s=.25],.15
*BATTING,Batting SSSSSSSSSSSSSSSSSSSSSSSSSSSSSSSSSSSSSSSSSSSS
A,.0001,-.1,[BAT,ltypeshp.shx,x=-.1,s=.1],-.2,[BAT,ltypeshp.shx,r=180,x=.1,s=.1],-.1
*HOT_WATER_SUPPLY,Hot water supply ---- HW ---- HW ---- HW ----
A,.5,-.2,["HW",STANDARD,S=.1,R=0.0,X=-0.1,Y=-.05],-.2
*GAS_LINE,Gas line ----GAS----GAS----GAS----GAS----GAS----GAS--
A,.5,-.2,["GAS",STANDARD,S=.1,R=0.0,X=-0.1,Y=-.05],-.25
*ZIGZAG,Zig zag /\/\/\/\/\/\/\/\/\/\/\/\/\/\/\/\/\/\/\/\/\/\/\/\/\/\/\
A,.0001,-.2,[ZIG,ltypeshp.shx,x=-.2,s=.2],-.4,[ZIG,ltypeshp.shx,r=180,x=.2,s=.2],-.2
```

Example 2

Create a new file, **NEWLINET.LIN**, and define a linetype, VARDASH, with the following specifications:

Length of first dash 1.0
Blank space 0.25
Length of second dash 0.75
Blank space 0.25
Length of third dash 0.5
Blank space 0.25
Dot
Blank space 0.25
Length of next dash 0.5
Blank space 0.25
Length of next dash 0.75

Step 1: Writing definition of linetype

Use a text editor and insert the following lines that define the new linetype **VARDASH**.

***VARDASH,———— —— — . — —— ————**
A,1,-.25,.75,-.25,.5,-.25,0,-.25,.5,-.25,.75,-.25

Step 2: Loading the linetype

You can use the **LINETYPE** command to load the linetype or choose Linetype in the **Format** menu. The type of lines that this linetype will generate are shown in Figure 24-11.

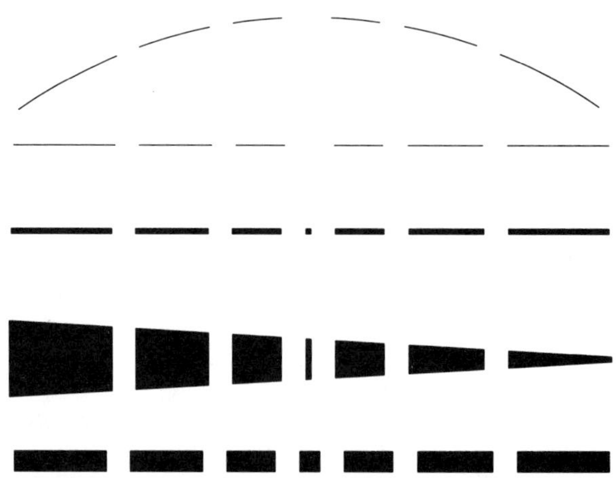

Figure 24-11 *Lines generated by linetype VARDASH*

CURRENT LINETYPE SCALING (CELTSCALE)

Like **LTSCALE**, the **CELTSCALE** system variable controls the linetype scaling. The difference is that **CELTSCALE** determines the current linetype scaling. For example, if you set the

CELTSCALE to 0.5, all lines drawn after setting the new value for **CELTSCALE** will have the linetype scaling factor of 0.5. The value is retained in the **CELTSCALE** system variable. The first line (a) in Figure 24-12 is drawn with the **CELTSCALE** factor of 1 and the second line (b) is drawn with the **CELTSCALE** factor of 0.5. The length of the dashes is reduced by a factor of 0.5 when the **CELTSCALE** is 0.5.

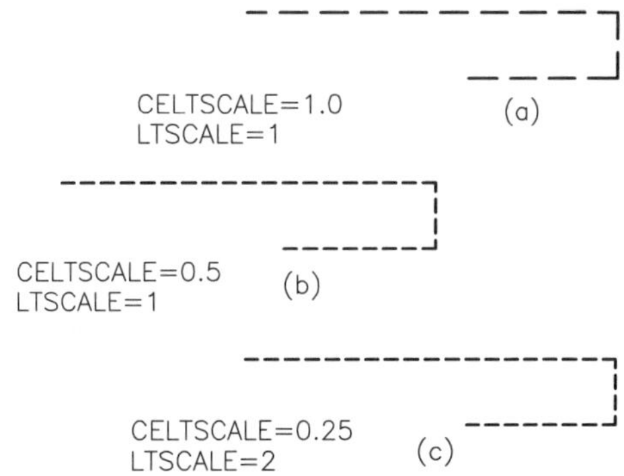

***Figure 24-12** Using **CELTSCALE** to control current linetype scaling*

The **LTSCALE** system variable controls the global scale factor. For example, if **LTSCALE** is set to 2, all lines in the drawing will be affected by a factor of 2. The net scale factor is equal to the product of **CELTSCALE** and **LTSCALE**. Figure 24-12(c) shows a line that is drawn with **LTSCALE** of 2 and **CELTSCALE** of 0.25. The net scale factor is = **LTSCALE** x **CELTSCALE** = 2 x 0.25 = 0.5.

Note

*You can change the current linetype scale factor of a line by using the **Properties** dialog box that can be invoked by choosing the **Properties** tool in the **Standard** toolbar. You can also use the **CHANGE** command and then select the **ltScale** option.*

COMPLEX LINETYPES

AutoCAD LT has provided a facility to create complex linetypes. The complex linetypes has a text string inserted in the line. The facility of creating complex linetypes increases the functionality of lines. For example, if you want to draw a line around a building that indicates the fence line, you can do it by defining a complex linetype that will automatically give you the desired line with the text string (Fence). Similarly, if you want to draw a line that indicates that it is a pipeline carrying the petroleum product, you can define a linetype with a text string inserted in the linetype that says that it is a petroleum line.

Creating a String Complex Linetype

When writing the definition of a string complex linetype, the actual text and its attributes must be included in the linetype definition. The format of the string complex linetype is:

["String", Text Style, Text Height, Rotation, X-Offset, Y-Offset]

String. It is the actual text that you want to insert along the line. The text string must be enclosed in quotation marks (" ").

Text Style. This is the name of the text style file that you want to use for generating the text string. The text style must be predefined.

Text Height. This is the actual height of the text, if the text height defined in the text style is 0. Otherwise, it acts as a scale factor for the text height specified in the text style. In Figure 24-13, the height of the text is 0.1 units.

Rotation. The rotation can be specified as an absolute or relative angle. In absolute rotation the angle is always measured with respect to the positive X axis, no matter what AutoCAD LT's direction setting. The absolute angle is represented by letter "a".

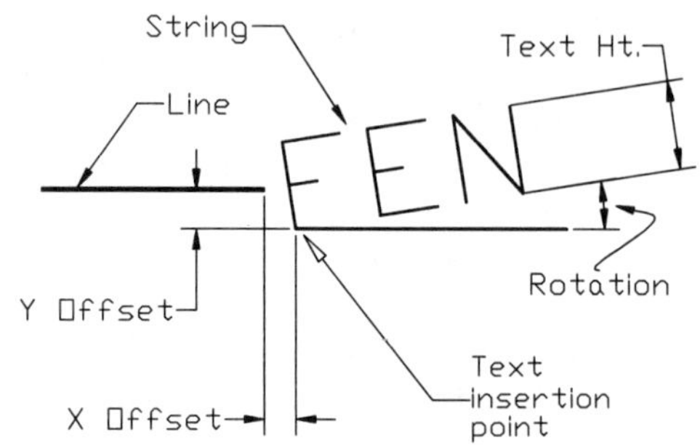

Figure 24-13 The attributes of a string complex linetype

In relative rotation the angle is always measured with respect to orientation of dashes in the linetype. The relative angle is represented by the letter "r". The angle can be specified in radians (r), grads (g), or degrees (d). The default is degrees.

X-Offset. This is the distance of the lower left corner of the text string from the endpoint of the line segment measured along the line. If the line is horizontal, then the X-Offset distance is measured along the X axis. In Figure 24-13, the X-Offset distance is 0.05.

Y-Offset. This is the distance of the lower left corner of the text string from the endpoint of the line segment measured perpendicular to the line. If the line is horizontal, then the Y-Offset distance is measured along the Y axis. In Figure 24-13, the Y-Offset distance is -0.05. The distance is negative because the start point of the text string is 0.05 units below the endpoint of the first line segment.

Example 3

In the following example, you will write the definition of a string complex linetype that consists of the text string "Fence" and line segments. The length of each line segment is 0.75. The height of the text string is 0.1 units, and the space between the end of the text string and the following line segment is 0.05 (Figure 24-14).

Step 1: Determining the line specifications

Before writing the definition of a new linetype, it is important to determine the line specification. One of the ways this can be done is to actually draw the lines and the text the way you want them to appear in the drawing. Once you have drawn the line and the text to your satisfaction, measure the distances needed to define the string complex linetype. In this example, the values are given as follows:

Text string=	Fence
Text style=	Standard
Text height=	0.1
Text rotation=	0
X-Offset=	0.05

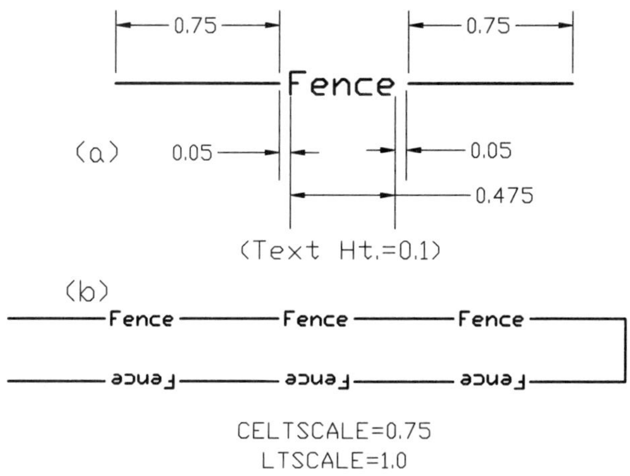

Figure 24-14 The attributes of a string complex linetype and line specifications for Example 3

Y-Offset=	-0.05
Length of the first line segment=	0.75
Distance between the line segments=	0.575

Step 2: Writing the definition of string complex linetype

Use a text editor to write the definition of the string complex linetype. You can add the definition to the AutoCAD LT **ACLT.LIN** file or create a separate file. The extension of the file must be **.LIN**. The following file is the listing of the **FENCE.LIN** file for Example 3. The name of the linetype is NEWFence1.

```
*NEWFence1,New fence boundary line
A,0.75,["Fence",Standard,S=0.1,A=0,X=0.05,Y=-0.05],-0.575
or
A,0.75,-0.05,["Fence",Standard,S=0.1,A=0,X=0,Y=-0.05],-0.525
```

Step 3: Loading the linetype

You can use the **LINETYPE** command to load the linetype or choose Linetype in the **Format** menu. Draw a line or any object to check if the line is drawn to the given specifications as shown in the Figure 24-15. Notice that the text is always drawn along the X axis. Also, when you draw a line at an angle, polyline, circle, or spline, the text string does not align with the object (Figure 24-15).

Step 4: Aligning the text with the line

In the NEWFence linetype definition, the specified angle is 0 degrees (Absolute angle A = 0). Therefore, when you use the NEWFence linetype to draw a line, circle, polyline, or spline, the text string (Fence) will be at 0 degrees. If you want the text string (Fence) to align with the polyline (Figure 24-16), spline, or circle, specify the angle as relative angle (R = 0) in the

Chapter 24

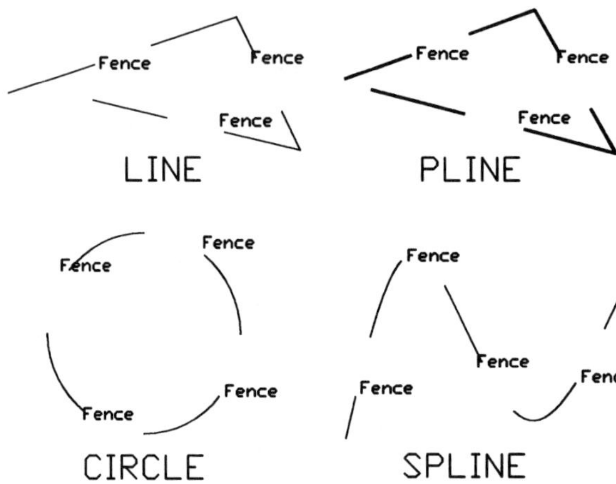

***Figure 24-15** Using string complex linetype with angle A=0*

NEWFence linetype definition. The following is the linetype definition for NEWFence linetype with relative angle R = 0:

```
*NEWFence2,New fence boundary line
A,0.75,["Fence",Standard,S=0.1,R=0,X=0.05,Y=-0.05],-0.575
```

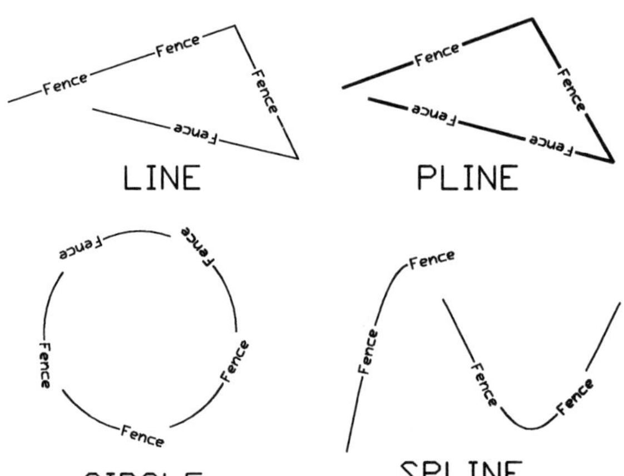

***Figure 24-16** Using a string complex linetype with angle R=0*

Step 5: Aligning the midpoint of text with the line

In Figure 24-16, you might notice that the text string is not properly aligned with the circumference of the circle. This is because AutoCAD LT draws the text string in a direction that is tangent to the circle at the text insertion point. To resolve this problem, you must define the middle point of the text string as the insertion point. Also, the line specifications should be measured accordingly. Figure 24-17 gives the measurements of the NEWFence linetype with the middle point of the text as the insertion point.

The following is the linetype definition for NEWFence linetype:

```
*NEWFence3,New fence boundary line
A,0.75,-0.287,["FENCE",Standard,S=0.1,X=-0.237,Y=-0.05],-0.287
```

Note

If no angle is defined in the line definition, it defaults to angle R = 0. Also, the text does not automatically insert to its midpoint like the regular text with MID justification.

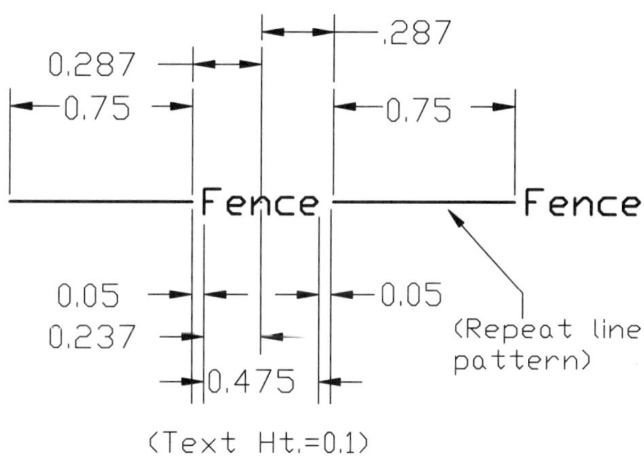

Figure 24-17 *Specifications of a string complex linetype with the middle point of the text string as the text insertion point*

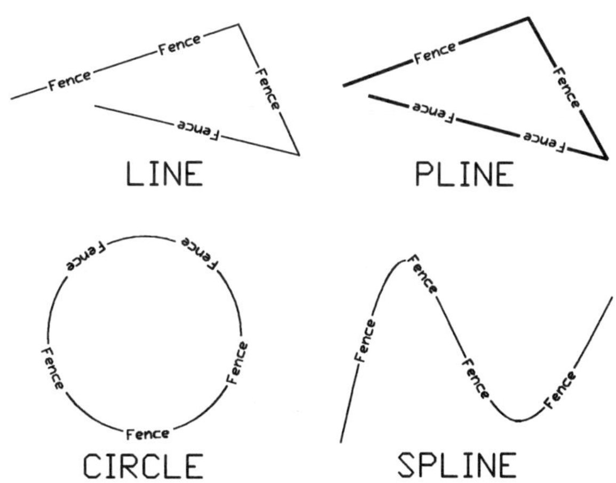

Figure 24-18 *Using a string complex linetype with the middle point of the text string as the text insertion point*

HATCH PATTERN DEFINITION

The AutoCAD LT software comes with a hatch pattern library file, **ACLT.PAT**, that contains sixty-seven hatch patterns. These hatch patterns are sufficient for general drafting work. However, if you need a different hatch pattern, AutoCAD LT lets you create your own. There is no limit to the number of hatch patterns you can define.

The hatch patterns you define can be added to the hatch pattern library file, **ACLT.PAT**. You can also create a new hatch pattern library file, provided the file contains only one hatch pattern definition, and the name of the hatch is the same as the name of the file. The hatch pattern definition consists of the following two parts: **header line and hatch descriptors**.

Header Line

The **header line** consists of an asterisk (*) followed by the name of the hatch pattern. The hatch name is the name used in the hatch command to hatch an area. After the name, you can give the hatch description, which is separated from the hatch name by a comma (,). The general format of the header line is:

***HATCH Name [, Hatch Description]**

Where ***** ---------------------------- Asterisk
HATCH Name ----------- Name of hatch pattern
Hatch Description ------ Description of hatch pattern

The description can be any text that describes the hatch pattern. It can also be omitted, in which case, a comma should not follow the hatch pattern name.

Example

***DASH45, Dashed lines at 45 degrees**

Where **DASH45** --------- Hatch name
Dashed lines at 45 degrees ------ Hatch description

Hatch Descriptors

The **hatch descriptors** consist of one or more lines that contain the definition of the hatch lines. The general format of the hatch descriptor is:

Angle, X-origin, Y-origin, D1, D2 [,Dash Length.....]

Where **Angle** ------------ Angle of hatch lines
X-origin --------- X coordinate of hatch line
Y-origin --------- Y coordinate of hatch line
D1 ---------------- Displacement of second line (Delta-X)
D2 ---------------- Distance between hatch lines (Delta-Y)
Length ----------- Length of dashes and spaces (Pattern line definition)

Example

45,0,0,0,0.5,0.5,-0.125,0,-0.125

Where **45** ----------------- Angle of hatch line

0 ------------------- X-Origin
0 ------------------- Y-Origin
0 ------------------- Delta-X
0.5 ---------------- Delta-Y
0.5 ---------------- Dash (pen down)
-0.125 ------------ Space (pen up)
0 ------------------- Dot (pen down)
-0.125 ------------ Space (pen up)
0.5,-0.125,0,-0.125 Pattern line definition

Hatch Angle

X-origin and Y-origin. The hatch angle is the angle that the hatch lines make with the positive X axis. The angle is positive if measured counterclockwise (Figure 24-19), and negative if the angle is measured clockwise. When you draw a hatch pattern, the first hatch line starts from the point defined by X-origin and Y-origin. The remaining lines are generated by offsetting the first hatch line by a distance specified by delta-X and delta-Y. In Figure 24-20(a), the first hatch line starts from the point with the coordinates X = 0 and Y = 0. In Figure 24-20(b), the first line of hatch starts from a point with the coordinates X = 0 and Y = 0.25.

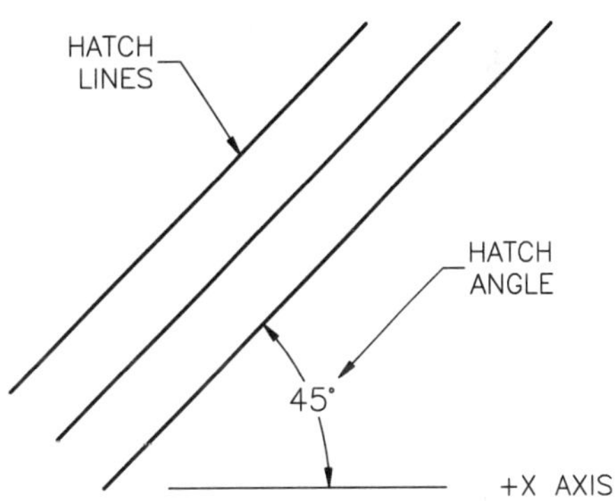

Figure 24-19 Hatch angle

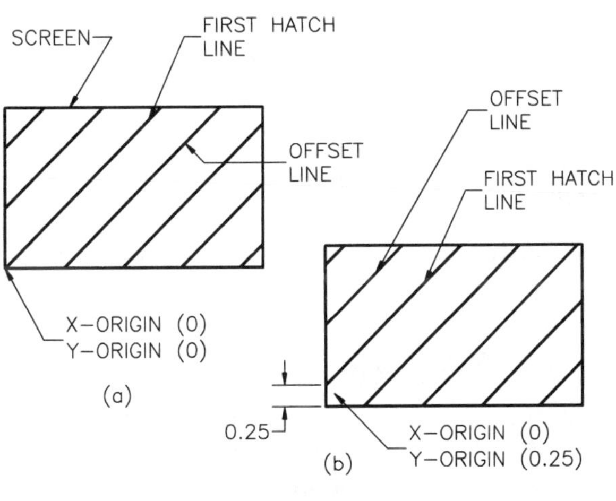

Figure 24-20 X-origin and Y-origin of hatch lines

Delta-X and Delta-Y. Delta-X is the displacement of the offset line in the direction in which the hatch lines are generated. For example, if the lines are drawn at a 0-degree angle and delta-X = 0.5, the offset line will be displaced by a distance delta-X (0.5) along the 0-angle direction. Similarly, if the hatch lines are drawn at a 45-degree angle, the offset line will be displaced by a distance delta-X (0.5) along a 45-degree direction (Figure 24-21). Delta-Y is the displacement of the offset lines measured perpendicular to the hatch lines. For example, if delta-Y = 1.0, the space between any two hatch lines will be 1.0 (Figure 24-21).

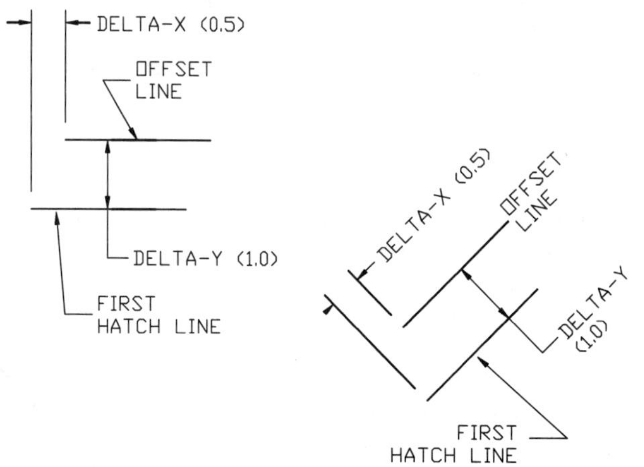

***Figure 24-21** Delta-X and delta-Y of hatch lines*

HOW HATCH WORKS

When you hatch an area, AutoCAD LT generates an infinite number of hatch lines of infinite length. The first hatch line always passes through the point specified by the X-origin and Y-origin. The remaining lines are generated by offsetting the first hatch line in both directions. The offset distance is determined by delta-X and delta-Y as shown in Figure 24-21. All selected entities that form the boundary of the hatch area are then checked for intersection with these lines. Any hatch lines found within the defined hatch boundaries are turned on, and the hatch lines outside the hatch boundary are turned off, as shown in Figure 24-22. Since the hatch lines are generated by offsetting, the hatch lines in different areas of the drawing are automatically aligned relative to the drawing's snap origin. Figure 24-22(a) shows the hatch lines as computed by AutoCAD LT. These lines are not drawn on the screen; they are shown here for illustration only. Figure 24-22(b) shows the hatch lines generated in the circle that was defined as the hatch boundary.

SIMPLE HATCH PATTERN

It is good practice to develop the hatch pattern specification before writing a hatch pattern definition. For simple hatch patterns it may not be that important, but for more complicated hatch patterns you should know the detailed specifications. Example 4 illustrates the procedure for developing a simple hatch pattern.

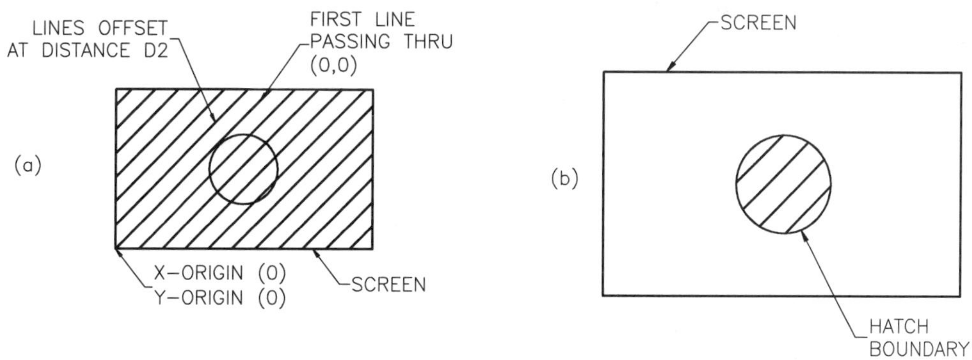

Figure 24-22 *Hatch lines outside the hatch boundary are turned off*

Example 4

Write a hatch pattern definition for the hatch pattern shown in Figure 24-23, with the following specifications:

Name of the hatch pattern =	HATCH1
X-Origin =	0
Y-Origin =	0
Distance between hatch lines =	0.5
Displacement of hatch lines =	0
Hatch line pattern =	Continuous

Step 1: Creating the hatch pattern file

This hatch pattern definition can be added to the existing **ACLT.PAT** hatch file. You can use any text editor (like Notepad) to write the file. Load the **ACLT.PAT** file that is located in **AutoCAD LT 2002\SUPPORT** directory and insert the following two lines at the end of the file.

*HATCH1,Hatch Pattern for Example 4
45,0,0,0,.5

Where **45** ----------------- Hatch angle
0 ------------------- X-origin
0 ------------------- Y-origin
0 ------------------- Displacement of second hatch line
.5 ------------------ Distance between hatch lines

The first field of hatch descriptors contains the angle of the hatch lines. That angle is 45 degrees with respect to the positive X axis. The second and third fields describe the X and Y coordinates of the first hatch line origin. The first line of the hatch pattern will pass through this point. If the values of the X-origin and Y-origin were 0.5 and 1.0, respectively, then the first line would pass through the point with the X coordinate of 0.5 and the Y coordinate of 1.0, with respect to the drawing origin 0,0. The remaining lines are generated by offsetting the first line, as shown in Figure 24-23.

Step 2: Loading the hatch pattern

Choose the **Hatch** button from the **Draw** toolbar or choose **Hatch** from the **Draw** menu to display the **Boundary Hatch** dialog box. Make sure Predefined is selected in the **Type:** edit box. Select the hatch pattern name from the drop-down list or choose the **[...]** button adjacent to the **Pattern** drop-down list to display the **Hatch Pattern Palette** dialog box. Select the hatch pattern file displayed there. Then choose **OK** to display the **Boundary Hatch** dialog box again. Change the **Scale** and **Angle**, if needed, and then hatch an a circle to test the hatch pattern.

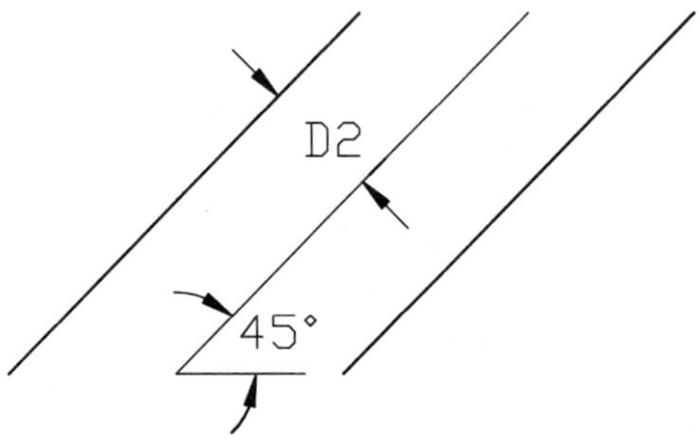

***Figure 24-23** Hatch pattern angle and offset distance*

The **Boundary Hatch** dialog box can also be invoked by entering **BHATCH** at the Command prompt. Hatching can also be achieved by entering **-HATCH** at the Command prompt.

EFFECT OF ANGLE AND SCALE FACTOR ON HATCH

When you hatch an area, you can alter the angle and displacement of hatch lines you have specified in the hatch pattern definition to get a desired hatch spacing. You can do this by entering an appropriate value for angle and scale factor in the AutoCAD LT's **HATCH** command.

To understand how the angle and the displacement can be changed, hatch an area with the hatch pattern HATCH1 in Example 4. You will notice that the hatch lines have been generated according to the definition of hatch pattern HATCH1. Notice the effect of hatch angle and scale factor on the hatch. Figure 24-24(a) shows a hatch with a 0-degree angle and a scale factor of 1.0. If the angle is 0, the hatch will be generated with the same angle as defined in the hatch pattern definition (45 degrees in Example 4). Similarly, if the scale factor is 1.0, the distance between the hatch lines will be the same as defined in the hatch pattern definition. Figure 24-24(b) shows a hatch that is generated when the hatch scale factor is 0.5. If you measure the distance between the successive hatch lines, it will be 0.5 x 0.5 = 0.25. Figures 24-24(c) and (d) show the hatch when the angle is 45 degrees and the scale factors are 1.0 and 0.5, respectively.

Scale and Angle can also be set by entering **-HATCH** at the Command prompt.

HATCH PATTERN WITH DASHES AND DOTS

The lines you can use in a hatch pattern definition are not restricted to continuous lines. You can define any line pattern to generate a hatch pattern. The lines can be a combination of dashes, dots, and spaces in any configuration. However, the maximum number of dashes you can specify in the line pattern definition of a hatch pattern is six. Example 5 uses a dash-dot line to create a hatch pattern.

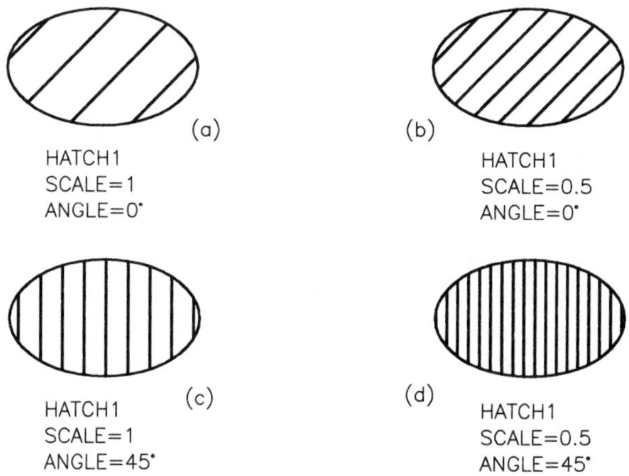

Figure 24-24 *Effect of angle and scale factor on hatch*

Example 5

Write a hatch pattern definition for the hatch pattern shown in Figure 24-25, with the following specifications. Define a new path say C:\Program Files\Hatch1 and save the hatch pattern in that path.

Name of the hatch pattern	HATCH2
Hatch angle =	0
X-origin =	0
Y-origin =	0
Displacement of lines (D1) =	0.25
Distance between lines (D2) =	0.25
Length of each dash =	0.5
Space between dashes and dots =	0.125
Space between dots =	0.125

Step 1: Writing the definition of a hatch pattern

You can use the text editor (Notepad) to edit the **ACLT.PAT** file. The general format of the header line and the hatch descriptors is:

***HATCH NAME, Hatch Description**
Angle, X-Origin, Y-Origin, D1, D2 [,Dash Length.....]

Substitute the values from Example 5 in the corresponding fields of the header line and field descriptor:

*HATCH2,Hatch with dashes and dots
0,0,0,0.25,0.25,0.5,-0.125,0,-0.125,0,-0.125
Where **0** ------------------- Angle

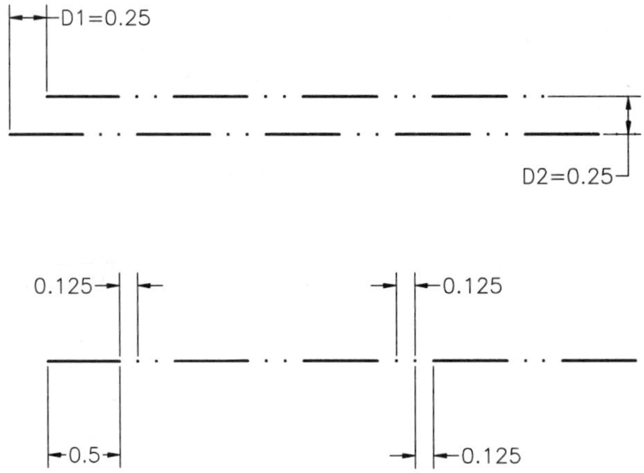

Figure 24-25 Hatch lines made of dashes and dots

0 ------------------ X-origin
0 ------------------ Y-origin
0.25 -------------- Delta-X
0.25 -------------- Delta-Y
0.5 --------------- Length of dash
-0.125 ----------- Space (pen up)
0 ------------------ Dot (pen down)
-0.125 ----------- Space (pen up)
0 ------------------ Dot
-0.125 ----------- Space

Specifying a New Path for Hatch Pattern Files

When you enter a hatch pattern name for hatching, AutoCAD LT looks for that file name in the Support directory or the directory paths specified in AutoCAD LT. You can specify a new path and directory to store your hatch files.

Create a new folder **Hatch1** in C drive under the Program Files. Save the ACLT.PAT file with hatch pattern **HATCH2** definition in the same subdirectory, Hatch1. Right-click in the drawing area to activate the shortcut menu. Choose **Options** from the shortcut menu to display the **Options** dialog box as shown in Figure 24-26. The **Options** dialog box can also be invoked by choosing **Options** from the **Tools** menu or by directly entering **OPTIONS** at the Command prompt. Choose the **Files** tab in the **Options** dialog box to display the **Search paths, file names and file locations** area. Click on the **plus** sign of the **Support File Search Path** to display the different subdirectories of the **Support File Search Path**. Now choose the **Add** button to display the space to add a new subdirectory. Enter the location of the new subdirectory, **C:\Program Files\Hatch1** or click on the **Browse** button to specify the path. Choose the **Apply** button and then choose **OK** to exit the dialog box. Now you have created a subdirectory and you have specified the search path for the hatch files.

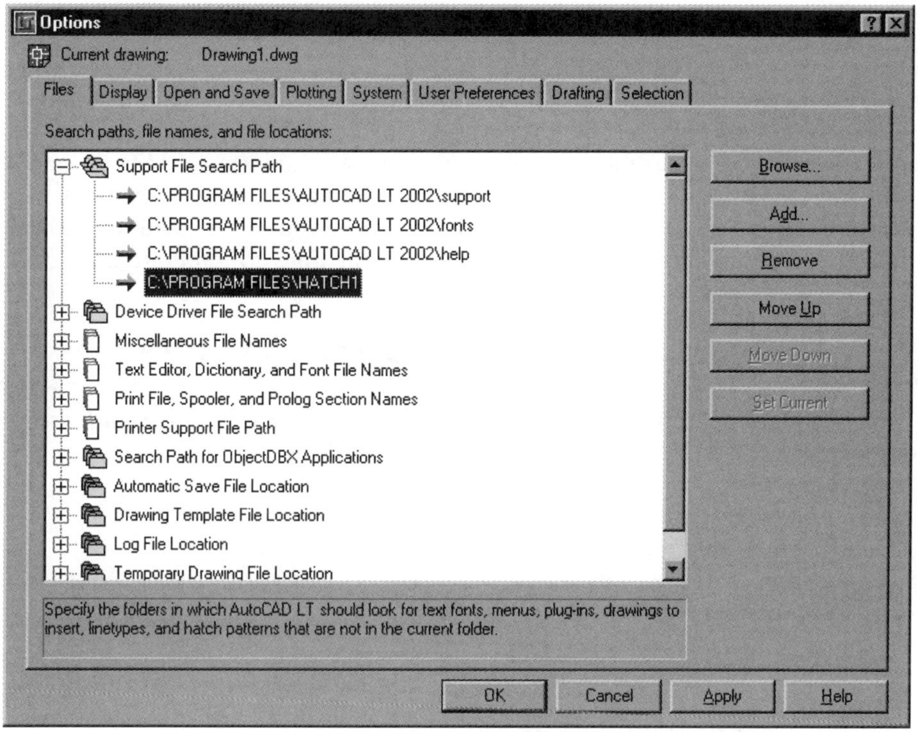

***Figure 24-26 Options** dialog box*

Follow the procedure as described in Example 4 to activate the hatch pattern. The hatch that this hatch definition (Hatch2.pat) will generate is shown in Figure 24-27. Figure 24-27(a) shows the hatch with a 0-degree angle and a scale factor of 1.0. Figure 24-27(b) shows the hatch with a 45-degree angle and a scale factor of 0.5.

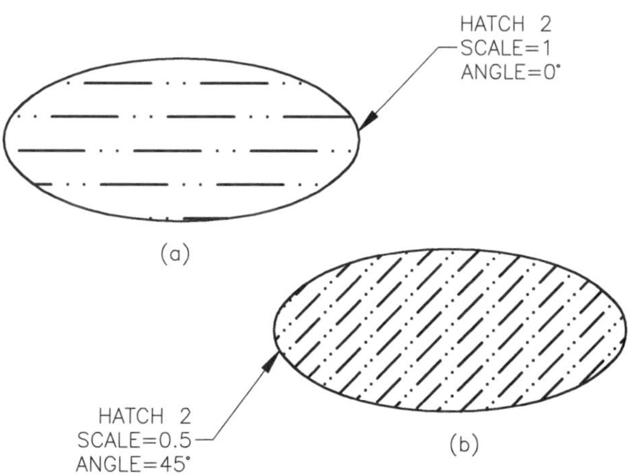

***Figure 24-27** Hatch pattern at different angles and scales*

Chapter 24

HATCH WITH MULTIPLE DESCRIPTORS

Some hatch patterns require multiple lines to generate a shape. For example, if you want to create a hatch pattern of a brick wall, you need a hatch pattern that has four hatch descriptors to generate a rectangular shape. You can have any number of hatch descriptor lines in a hatch pattern definition. It is up to the user to combine them in any conceivable order. However, there are some shapes you cannot generate. A shape that has a nonlinear element, like an arc, cannot be generated by hatch pattern definition. However, you can simulate an arc by defining short line segments because you can use only straight lines to generate a hatch pattern. Example 6 uses three lines to define a triangular hatch pattern.

Example 6

Write a hatch pattern definition for the hatch pattern shown in Figure 24-28, with the following specifications:

Name of the hatch pattern = HATCH3
Vertical height of the triangle = 0.5
Horizontal length of the triangle = 0.5
Vertical distance between the triangles = 0.5
Horizontal distance between the triangles = 0.5

Each triangle in this hatch pattern consists of the following three elements: a vertical line, a horizontal line, and a line inclined at 45 degrees.

Step 1: Defining specifications for vertical line
For the vertical line, the specifications are (Figure 24-29):

Hatch angle = 90 degrees
X-origin = 0
Y-origin = 0
Delta-X (D1) = 0
Delta-Y (D2) = 1.0
Dash length = 0.5
Space = 0.5

Substitute the values from the vertical line specification in various fields of the hatch descriptor to get the following line:

90,0,0,0,1,.5,-.5

Where **90** ---------------- Hatch angle
0 ------------------ X-origin
0 ------------------ Y-origin
0 ------------------ Delta-X
1 ------------------ Delta-Y
.5 ----------------- Dash (pen down)
-.5 ---------------- Space (pen up)

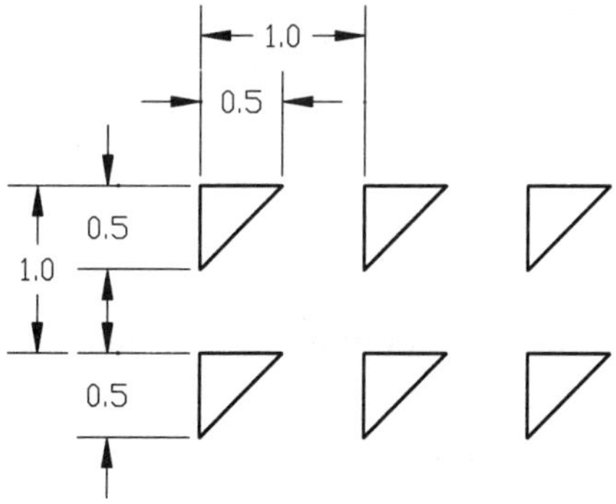

Figure 24-28 Triangle hatch pattern

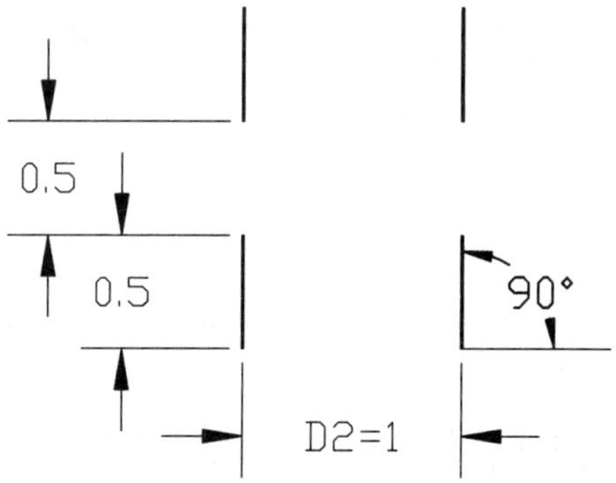

Figure 24-29 Vertical line

Step 2: Defining specifications of horizontal line

For the horizontal line (Figure 24-30), the specifications are:

Hatch angle =	0 degrees
X-origin =	0
Y-origin =	0.5
Delta-X (D1) =	0
Delta-Y (D2) =	1.0
Dash length =	0.5
Space =	0.5

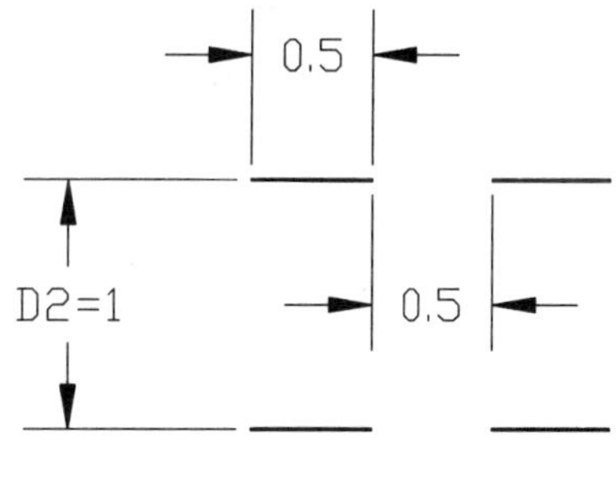

Figure 24-30 Horizontal line

The only difference between the vertical line and the horizontal line is the angle. For the horizontal line, the angle is 0 degrees, whereas for the vertical line, the angle is 90-degrees. Substitute the values from the vertical line specification to obtain the following line:

0,0,0.5,0,1,.5,-.5

Where **0** ------------------- Hatch angle
0 ------------------- X-origin
0.5 ---------------- Y-origin
0 ------------------- Delta-X
1 ------------------- Delta-Y
.5 ------------------ Dash (pen down)
-.5 ----------------- Space (pen up)

Step 3: Defining specifications of the inclined line

This line is at an angle; therefore, you need to calculate the distances delta-X (D1) and delta-Y (D2), the length of the dashed line, and the length of the blank space. Figure 24-31 shows the calculations to find these values.

Hatch angle =	45 degrees
X-Origin =	0
Y-Origin =	0
Delta-X (D1) =	0.7071
Delta-Y (D2) =	0.7071
Dash length =	0.7071
Space =	0.7071

After substituting the values in the general format of the hatch descriptor, you will obtain the following line:

45,0,0,.7071,.7071,.7071,-.7071

Where **45** ----------------- Hatch angle
0 ------------------- X-origin
0 ------------------- Y-origin
.7071 ------------- Delta-X
.7071 ------------- Delta-Y
.7071 ------------- Dash (pen down)
-.7071 ------------ Space (pen up)

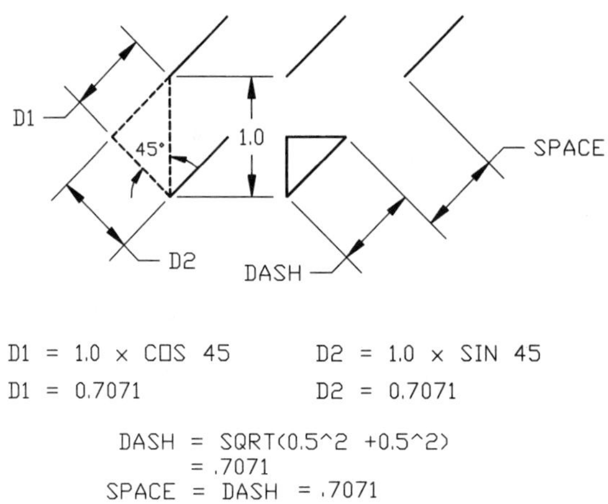

Figure 24-31 *Line inclined at 45 degrees*

Step 4: Loading the hatch pattern

Now you can combine these three lines and insert them at the end of the **ACLT.PAT** file or you can enter the values in a separate hatch file and save it. You can also use the text editor like Notepad to edit the file and insert the lines.

The following file is a partial listing of the **ACLT.PAT** file, after adding the hatch pattern definitions from Examples 4, 5, and 6.

*SOLID, Solid fill
45, 0,0, 0,.125

```
*angle,Angle steel
0, 0,0, 0,.275, .2,-.075
90, 0,0, 0,.275, .2,-.075
*ansi31,ANSI Iron, Brick, Stone masonry
45, 0,0, 0,.125
*ansi32,ANSI Steel
45, 0,0, 0,.375
45, .176776695,0, 0,.375
*ansi33,ANSI Bronze, Brass, Copper
45, 0,0, 0,.25
45, .176776695,0, 0,.25, .125,-.0625
*ansi34,ANSI Plastic, Rubber
45, 0,0, 0,.75
45, .176776695,0, 0,.75
45, .353553391,0, 0,.75
45, .530330086,0, 0,.75
*ansi35,ANSI Fire brick, Refractory material
45, 0,0, 0,.25
45, .176776695,0, 0,.25, .3125,-.0625,0,-.0625
*ansi36,ANSI Marble, Slate, Glass
45, 0,0, .21875,.125, .3125,-.0625,0,-.0625
|
|
|
*swamp,Swampy area
0, 0,0, .5,.866025403, .125,-.875
90, .0625,0, .866025403,.5, .0625,-1.669550806
90, .078125,0, .866025403,.5, .05,-1.682050806
90, .046875,0, .866025403,.5, .05,-1.682050806
0, 0,0, .125,.125, .125,-.125
90, .125,0, .125,.125, .125,-.125
|
|
|
```

***HATCH1,Hatch at 45 Degree Angle**
45,0,0,0,.5
***HATCH2,Hatch with Dashes & Dots:**
0,0,0,.25,.25,0.5,-.125,0,-.125,0,-.125
***HATCH3,Triangle Hatch:**
90,0,0,0,1,.5,-.5
0,0,0.5,0,1,.5,-.5
45,0,0,.7071,.7071,.7071,-.7071

Load the hatch pattern as described in Example 6 (Hatch3.pat) and test the hatch. Figure 24-32 shows the hatch pattern that will be generated by this hatch pattern (HATCH3). In Figure 24-32(a) the hatch pattern is at a 0-degree angle and the scale factor is 0.5. In Figure 24-32(b) the hatch pattern is at a -45-degree angle and the scale factor is 0.5.

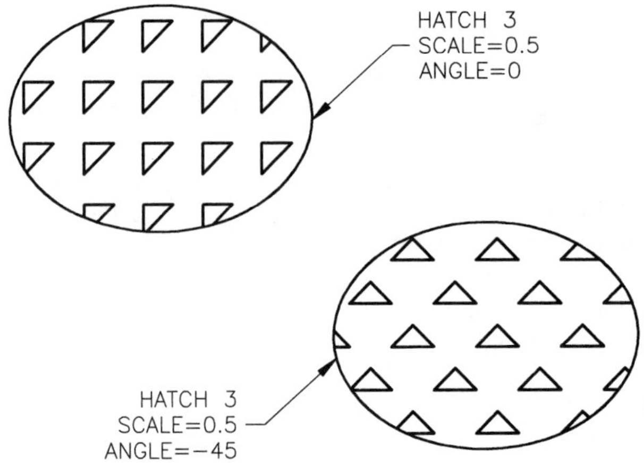

Figure 24-32 *Hatch generated by HATCH3 pattern*

SAVING HATCH PATTERNS IN A SEPARATE FILE

When you load a certain hatch pattern, AutoCAD LT looks for that definition in the ACLT.PAT file; therefore, the hatch pattern definitions must be in that file. However, you can add the new pattern definition to a different file and then copy that file to ACLT.PAT. Be sure to make a copy of the original ACLT.PAT file so that you can copy that file back when needed. Assume the name of the file that contains your custom hatch pattern definitions is CUSTOMH.PAT.

1. Copy ACLT.PAT file to ACLTORG.PAT
2. Copy CUSTOMH.PAT to ACLT.PAT

If you want to use the original hatch pattern file, copy the ACLTORG.PAT file to ACLT.PAT

CUSTOM HATCH PATTERN FILE

As mentioned earlier, you can add the new hatch pattern definitions to the **ACLT.PAT** file. There is no limit to the number of hatch pattern definitions you can add to this file. However, if you have only one hatch pattern definition, you can define a separate file. It has the following three requirements:

1. The name of the file has to be the same as the hatch pattern name.
2. The file can contain only one hatch pattern definition.
3. The hatch pattern name—and, therefore, the hatch file name—should be unique.
4. If you want to save the hatch pattern on the A drive, then the drive letter (A:) should precede the hatch name. For example, if the hatch name is HATCH3, the header line will be ***A:HATCH3, Triangle Hatch:** and the file name **HATCH3.PAT**.

```
*HATCH3,Triangle Hatch:
90,0,0,0,1,.5,-.5
```

0,0,0.5,0,1,.5,-.5
45,0,0,.7071,.7071,.7071,-.7071

Note

The hatch lines can be edited after exploding the hatch with the AutoCAD LT ***EXPLODE*** *command. After exploding, each hatch line becomes a separate object.*

It is good practice not to explode a hatch because it increases the size of the drawing database. For example, if a hatch consists of 100 lines, save it as a single object. However, after you explode the hatch, every line becomes a separate object and you have 99 additional objects in the drawing.

Keep the hatch lines in a separate layer to facilitate editing of the hatch lines.

Assign a unique color to hatch lines so that you can control the width of the hatch lines at the time of plotting.

Tip

1. *The file or the subdirectory in which hatch patterns have been saved must have been defined in the* ***Support File Search Path*** *in the* ***File*** *tab of the* ***Options*** *dialog box.*

2. *The hatch patterns that you create automatically get added to AutoCAD LT's slide library as an integral part of AutoCAD LT 2002 and the hatch patterns are displayed in the* ***Preview*** *area in the* ***Hatch Pattern Palette*** *dialog box under the* ***Boundary Hatch*** *dialog box. Therefore, there is no need to create a slide library.*

Self-Evaluation Test

Answer the following questions and then compare your answers to the correct answers given at the end of this chapter.

CREATING LINETYPES

1. The AutoCAD LT __________ command can be used to change the linetype scale factor.

2. The linetype description should not be more than __________ characters long.

3. A positive number denotes a pen __________ segment.

4. The segment length __________ generates a dot.

5. A negative number denotes a pen __________ segment.

6. The option__________ of linetype Command is used to generate a new linetype.

7. The description in the case of header line is __________ . (optional/necessary)

8. The standard linetypes are stored in the file_________ .

9. The_________ determines the current linetype scaling.

CREATING HATCH PATTERNS

10. The header line consists of an asterisk, the pattern name, and _________.

11. The **ACLT.PAT** file contains _________ number of hatch pattern definitions.

12. The standard hatch patterns are stored in the file _________ .

13. The first hatch line passes through a point whose coordinates are specified by _________ and _________.

Review Questions

CREATING LINETYPES

1. The AutoCAD LT _________ command can be used to create a new linetype.

2. The AutoCAD LT _________ command can be used to load a linetype.

3. In AutoCAD LT, the linetypes are saved in the _________ file.

4. AutoCAD LT supports only _________ alignment field specification.

5. A line pattern definition always starts with _________.

6. A header line definition always starts with _________.

CREATING HATCH PATTERNS

7. The perpendicular distance between the hatch lines in a hatch pattern definition is specified by _________.

8. The displacement of the second hatch line in a hatch pattern definition is specified by _________.

9. The maximum number of dash lengths that can be specified in the line pattern definition of a hatch pattern is _________.

10. The hatch lines in different areas of the drawing will automatically _________ since the hatch lines are generated by offsetting.

11. The hatch angle as defined in the hatch pattern definition can be changed further when you use the AutoCAD LT _________ command.

12. When you load a hatch pattern, AutoCAD LT looks for that hatch pattern in the _________ file.

13. The hatch lines can be edited after _________ the hatch by using the AutoCAD LT _________ command.

Exercises

CREATING LINETYPES

Exercise 1 *General*

Using the AutoCAD LT **LINETYPE** command, create a new linetype "DASH3DASH" with the following specifications:

Length of the first dash 0.75
Blank space 0.125
Dash length 0.25
Blank space 0.125
Dash length 0.25
Blank space 0.125
Dash length 0.25
Blank space 0.125

Exercise 2 *General*

Use a text editor to create a new file, **NEWLT2.LIN**, and a new linetype, DASH2DASH, with the following specifications:

Length of the first dash 0.5
Blank space 0.1
Dash length 0.2
Blank space 0.1
Dash length 0.2
Blank space 0.1

Exercise 3 *General*

a. Write the definition of a string complex linetype (hot water line) as shown in Figure 24-33(a). To determine the length of the HW text string, you should first draw the text (HW) using any text command and then measure its length.

b. Write the definition of a string complex linetype (gas line) as shown in Figure 24-33(b). Determine the length of the text string as mentioned in part a, above.

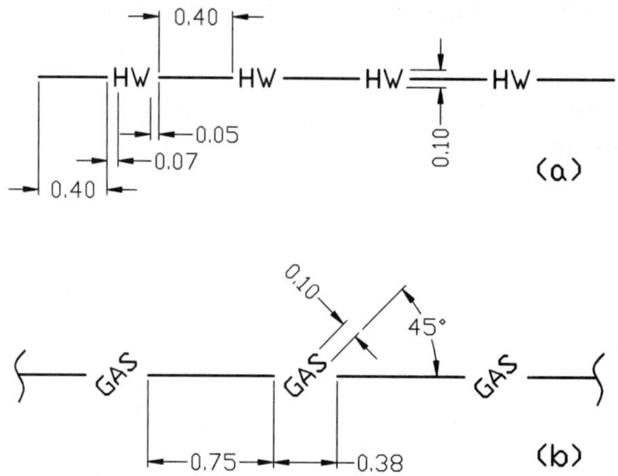

Figure 24-33 Specifications for string a complex linetype

Creating Hatch Patterns

Exercise 4 *General*

Determine the hatch pattern specifications and write a hatch pattern definition for the hatch pattern in Figure 24-34.

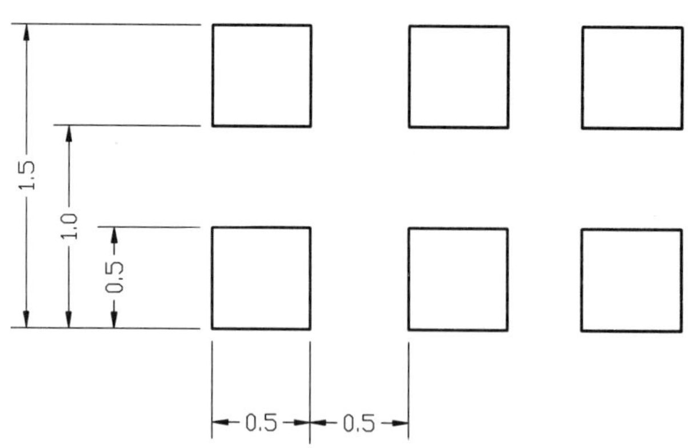

Figure 24-34 Drawing for Exercise 4

Exercise 5 General

Determine the hatch pattern specifications and write a hatch pattern definition for the hatch pattern in Figure 24-35.

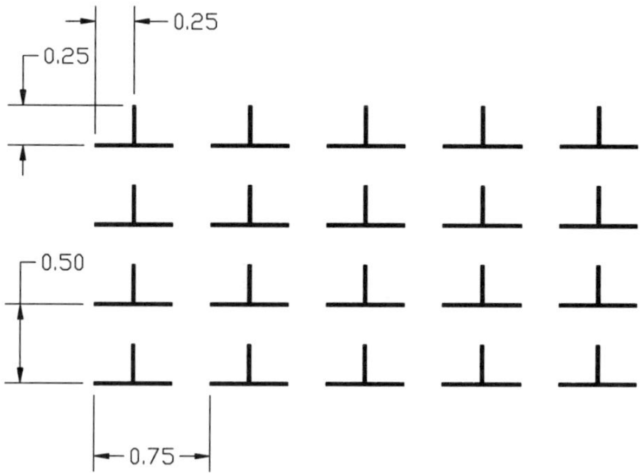

Figure 24-35 *Hatch pattern for Exercise 5*

Answers to Self-Evaluation Test

1 - **LTSCALE**, **2** - 47, **3** - down, **4** - zero, **5** - up, **6** - **Create**, **7** - optional, **8** - **ACLT.LIN**, **9** - **CELTSCALE**, **10** - Pattern Description, **11** - 67, **12** - **ACLT.PAT**, **13** - X-origin, Y-origin

Chapter 25

Pull-down, Shortcut, and Partial Menus and Customizing Toolbars

Learning Objectives

After completing this chapter, you will be able to:
- *Write pull-down menus.*
- *Load menus.*
- *Write cascading submenus in pull-down menus.*
- *Write cursor menus.*
- *Swap pull-down menus.*
- *Write partial menus.*
- *Define accelerator keys.*
- *Write toolbar definitions.*
- *Write menus to access online help.*
- *Customize toolbars.*

AUTOCAD LT MENU

The AutoCAD LT menu provides a powerful tool to customize AutoCAD LT. The AutoCAD LT software package comes with a standard menu file named **ACLT.MNU**. When you start AutoCAD LT, the menu file **ACLT.MNU** is automatically loaded. The AutoCAD LT menu file contains AutoCAD LT commands, separated under different headings for easy identification. For example, all draw commands are under **Draw** and all editing commands are under the **Modify** menu. The headings are named and arranged to make it easier for you to locate and access the commands. However, there are some commands that you may never use. Also, some users might like to regroup and rearrange the commands so that it is easier to access those most frequently used.

AutoCAD LT lets the user eliminate rarely used commands from the menu file and define new ones. This is made possible by editing the existing **ACLT.MNU** file or writing a new menu file. There is no limit to the number of files you can write. You can have a separate menu file for each application. For example, you can have separate menu files for mechanical, electrical, and architectural drawings. You can load these menu files any time by using the AutoCAD LT **MENU** command. The menu files are text files with the extension **.MNU**. These files can be written by using any text editor like Wordpad or Notepad. The menu file can be divided into ten sections, each section identified by a section label. AutoCAD LT uses the following labels to identify different sections of the AutoCAD LT menu file:

*****SCREEN**	
*****TABLET(n)**	**n** is from 1 to 4
*****IMAGE**	
*****POP(n)**	**n** is from 1 to 499 (For Shortcut menus n=0 and 500 to 999)
*****BUTTONS(n)**	**n** is from 1 to 4
*****AUX(n)**	**n** is from 1 to 4
*****MENUGROUP**	
*****TOOLBARS**	
*****HELPSTRING**	
*****ACCELERATORS**	

The tablet menu can have up to four different sections. The **POP** menu (pull-down and cursor menu) can have up to 499 sections, and auxiliary and buttons menus can have up to four sections.

Tablet Menus

***TABLET1
***TABLET2
***TABLET3
***TABLET4

BUTTONS Menus
***BUTTONS1
***BUTTONS2
***BUTTONS3
***BUTTONS4

Auxiliary Menus
***AUX1
***AUX2
***AUX3
***AUX4

Pull-Down and Cursor Menus
***POP0
***POP1
***POP2
***POP3
***POP4
***POP5
***POP6
***POP7
***POP8
|
***POP498
***POP499

STANDARD MENUS

The menu is a part of the AutoCAD LT standard menu file, **ACLT.MNU**. The **ACLT.MNU** file is automatically loaded when you start AutoCAD LT, provided the standard configuration of AutoCAD LT has not been changed. The menus can be selected by moving the crosshairs to the top of the screen, into the menu bar area. If you move the pointing device sideways, different menu bar titles are highlighted and you can select the desired item by pressing the pick button on your pointing device. Once the item is selected, the corresponding menu is displayed directly under the title (Figure 25-1). The menu can have 499 sections, named as POP1, POP2, POP3, . . ., POP499.

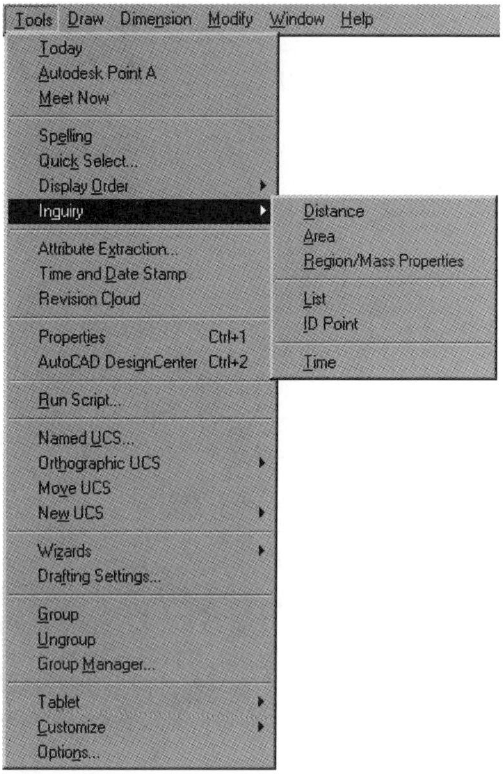

Figure 25-1 Pull-down and cascading menus

WRITING A MENU

Before you write a menu, you need to design a menu and to arrange the commands the way you want them to appear on the screen. To design a menu, you should select and arrange the commands in a way that provides easy access to the most frequently used commands. A careful design will save a lot of time in the long run. Therefore, consider several possible designs with different command combinations, and then select the one best suited for the job. Suggestions from other CAD operators can prove very valuable.

The second important thing in developing a menu is to know the exact sequence of the commands and the prompts associated with each command. To better determine the prompt entries required in a command, you should enter all the commands and the prompt entries at the keyboard. The following is a description of some of the commands and the prompt entries required for Example 1.

<u>**LINE Command**</u>

Command: **LINE**

Notice the command and input sequence:

LINE
<RETURN>

CIRCLE (C,R) Command

Command: **CIRCLE**
Specify center point for circle or [3P/2P/Ttr (tan tan radius)]: *Specify center point.*
Specify radius of circle or [Diameter]: *Enter radius.*

Notice the command and input sequence:

CIRCLE
<RETURN>
Center point
<RETURN>
Radius
<RETURN>

CIRCLE (C,D) Command

Command: **CIRCLE**
Specify center point for circle or [3P/2P/Ttr (tan tan radius)]: *Specify center point.*
Specify radius of circle or [Diameter]: **D**
Specify diameter of circle: *Enter diameter.*

Notice the command and input sequence:

CIRCLE
<RETURN>
Center Point
<RETURN>
D
<RETURN>
Diameter
<RETURN>

CIRCLE (2P) Command

Command: **CIRCLE**
Specify center point for circle or [3P/2P/Ttr (tan tan radius)]: **2P**
Specify first end point of circle's diameter: *Specify first point.*
Specify second end point of circle's diameter: *Specify second point.*

Notice the command and input sequence:

CIRCLE
<RETURN>
2P
<RETURN>
Select first point on diameter

<RETURN>
Select second point on diameter
<RETURN>

ERASE Command

Command: **ERASE**

Notice the command and input sequence:

ERASE
<RETURN>

MOVE Command

Command: **MOVE**

Notice the command and prompt entry sequence:

MOVE
<RETURN>

The difference between the Center-Radius and Center-Diameter options of the **CIRCLE** command is that in the first one the RADIUS is the default, whereas in the second one you need to enter D to use the diameter option. This difference, although minor, is very important when writing a menu file. Similarly, the 2P (two-point) option of the **CIRCLE** command is different from the other two options. Therefore, it is important to know both the correct sequence of the AutoCAD LT commands and the entries made in response to the prompts associated with those commands.

You can use any text editor (like Notepad) to write the file. You can also use the **EDIT** program of the MS-DOS to write the menu file. To use the **EDIT** program of MS-DOS, choose **Programs > MS-DOS Prompt** from the **Start** menu. In the Command prompt, enter **EDIT** to open the text editor for writing the menu file. Note that by default, the file will be saved in the default directory where the MS-DOS prompt was when you opened the MS-DOS window. To save the file in the desired directory, change the directory while saving the file. To understand the process of developing a pull-down menu, consider the following example.

Note
If you are not conversant with the MS-DOS commands, it is advised that you use the Notepad for writing the menu files.

Example 1

Write a pull-down menu for the following AutoCAD LT commands:

LINE **ERASE** **REDRAW** **SAVE**

PLINE
CIRCLE C,R
CIRCLE C,D
CIRCLE 2P
CIRCLE 3P
MOVE
COPY
STRETCH
EXTEND
OFFSET
REGEN
ZOOM ALL
ZOOM WIN
ZOOM PRE
QUIT
PLOT

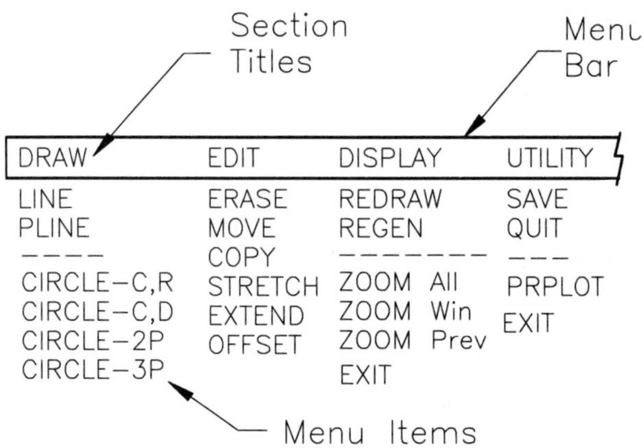

Figure 25-2 *Design of menu*

Step 1: Designing the menu

The first step in writing any menu is to design the menu so that the commands are arranged in the desired configuration. Figure 25-2 shows one of the possible designs of this menu. This menu has four different groups of commands; therefore, it will have four sections: POP1, POP2, POP3, and POP4, and each section will have a section label. The following file is a listing of the pull-down menu file for Example 1. **The line numbers are not a part of the file; they are shown here for reference only.**

Step 2: Writing the menu file

```
***POP1
[DRAW]
[LINE]*^C^CLINE
[PLINE]^C^CPLINE
[—]
[CIR-C,R]^C^CCIRCLE
[CIR-C,D]^C^CCIRCLE \D
[CIR-2P]^C^CCIRCLE 2P
[CIR-3P]^C^CCIRCLE 3P
***POP2
[EDIT]
[ERASE]*^C^CERASE
[MOVE]^C^CMOVE
[COPY]^C^CCOPY
[STRETCH]^C^CSTRETCH;C
[EXTEND]^C^CEXTEND
```

```
17 [OFFSET]^C^COFFSET
18 ***POP3
19 [DISPLAY]
20 [REDRAW]'REDRAW
21 [REGEN]^C^CREGEN
22 [—]
23 [ZOOM-All]^C^C'ZOOM A
24 [ZOOM-Window]'ZOOM W
25 [ZOOM-Prev]'ZOOM PREV
26 [~Exit]^C
27 ***POP4
28 [UTILITY]
29 [SAVE]^C^CSAVE;
30 [QUIT]^C^CQUIT
31 [——]
32 [PLOT]^C^CPLOT
33 [EXIT]^C^CEXIT
```

Explanation

Line 1

*****POP1**

POP1 is the section label for the first pull-down menu. All section labels in the AutoCAD LT menu begin with three asterisks (***), followed by the section label name, such as POP1.

Line 2

[DRAW]

In this menu item **DRAW** is the menu bar title displayed when the cursor is moved in the menu bar area. The title names should be chosen so you can identify the type of commands you expect in that particular pull-down menu. In this example, all the draw commands are under the title **DRAW** (Figure 25-3), all edit commands are under **EDIT**, and so on for other groups of items. The menu bar title can be of any length depending on the display resolution, display device, and the font size of the text. However, it is recommended to keep them short to accommodate other menu items. If a display device at a certain resolution provides a maximum of eighty characters; to have sixteen sections in a single row in the menu bar, the length of each title should not exceed five characters. If the length of the menu bar items exceeds eighty characters, AutoCAD LT will wrap the items that cannot be accommodated in eighty character space, and display them in the next line. This will result in two menu item lines in the menu bar.

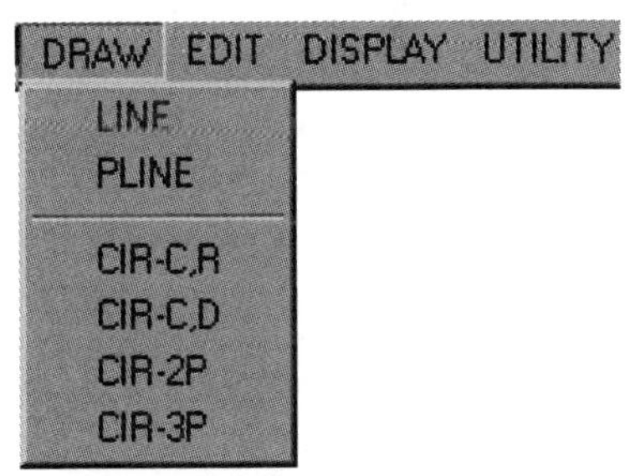

***Figure 25-3** Draw pull-down menu*

If the first line in a menu section is blank, the title of that section is not displayed in the menu bar area. Since the menu bar title is not displayed, you **cannot** access that menu. This allows you to turn off the menu section. For example, if you replace [**DRAW**] with a blank line, the **DRAW** section (POP1) of the menu will be disabled; the second section (POP2) will be displayed in its place.

Example

***POP1	Section label
	Blank line (turns off POP1)
[LINE:]^CLINE	Menu item
[PLINE:]^CPLINE	
[CIRCLE:]^CCIRCLE	

The menu bar titles are left-justified. If the first title is not displayed, the rest of the menu titles will be shifted to the left. In Example 1, if the **DRAW** title is not displayed in the menu bar area, then the **EDIT**, **DISPLAY**, and **FILE** sections of the menu will move to the left.

Line 3

***^C^CLINE**

In this menu item, the command definition starts with an asterisk (*). This feature allows the command to be repeated automatically until it is canceled by pressing ESCAPE, entering CTRL C, or by selecting another menu command. ^C^C cancels the existing command twice; **LINE** is an AutoCAD LT command that generates lines.

***^C^CLINE**

Where ***** ------------------- Repeats the menu item (command)
^C^C ---------- Cancels existing command twice
LINE ------------- AutoCAD LT's **LINE** command

Line 5

[--]

To separate two groups of commands in any section, you can use a menu item that consists of two or more hyphens (--). This line automatically expands to fill the **entire width** of the menu. You can use a blank line in a menu. If any section of a menu (**POP section) has a blank line, it is ignored.

Line 24

[ZOOM-Window]'ZOOM W

In this menu item the single quote (') preceding the **ZOOM** command makes the **ZOOM** Window command transparent. When a command is transparent, the existing command is not canceled. After the **ZOOM** Window command (Figure 25-4), AutoCAD LT will automatically resume the current operation.

[ZOOM-Window]'ZOOM W

Where **W** ----------------- Window option
ZOOM ----------- AutoCAD LT **ZOOM** command
' -------------------- Single quote makes **ZOOM** transparent

Line 26

[~Exit]^C

This menu item is specially meant for canceling the menu. Since this menu item has a tilde (~), the menu item is not available (displayed grayed out), and if you select this item it will not cancel the menu. You can use this feature to disable a menu item or to indicate that the item is not a valid selection. If there is an instruction associated with the item, the instruction will not be executed when you select the item. For example, [~OSNAPS]^C^C$S=OSNAPS will not load the OSNAPS submenu on the screen.

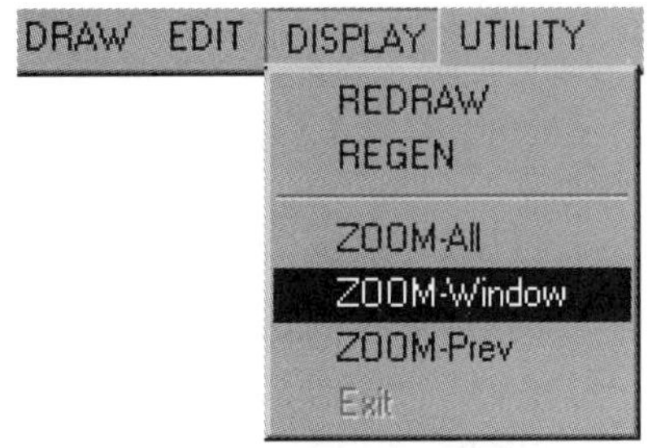

Figure 25-4 Display menu

Line 29
[SAVE]^C^CSAVE;
In this menu item, the semicolon (;) that follows the **SAVE** command enters RETURN. The semicolon is not required; the command will also work without a semicolon.

[SAVE]^C^CSAVE;
Where **SAVE** ------------ AutoCAD LT **SAVE** command
; ------------------ Semicolon enters RETURN

Line 31
[----]
This menu item has four hyphens. The line will extend to fill the width of the menu. If there is only one hyphen [-], AutoCAD LT gives a syntax error when you load the menu.

Line 33
[Exit]^C
In this menu item, **^C** command definition has been used to cancel the menu. This item provides you with one more option for canceling the menu. This is especially useful for new AutoCAD LT users who are not familiar with all AutoCAD LT features. The menu can also be canceled by any of the following actions:

1. Selecting a point.
2. Selecting an item in the screen menu area.
3. Selecting or typing another command.
4. Pressing ESC at the keyboard.
5. Selecting any menu title in the menu bar.

Note

For all menus, the menu items are displayed directly beneath the menu title and are left-justified. If any menu [for example, the rightmost menu (POP16)] does not have enough space to display the entire menu item, the menu will expand to the left to accommodate the entire length of the longest menu item.

You can use // (two forward slashes) for comment lines. AutoCAD LT ignores the lines that start with //.

The last line in the menu file must be terminated by ENTER or else it will be ignored.

From this example it is clear that every statement in the menu is based on the AutoCAD LT commands and the information that is needed to complete these commands. This forms the basis for creating a menu file and should be given consideration. The following is a summary of the AutoCAD LT commands used in Example 1 and their equivalents in the menu file.

AutoCAD LT Commands	Menu File
Command: **LINE**	**[LINE]^C^CLINE**
Command: **CIRCLE** Specify center point for circle or [3P/2P/Ttr (tan tan radius)]: Specify radius of circle or [Diameter]:	**CIR-C,R]^C^CCIRCLE**
Command: **CIRCLE** Specify center point for circle or [3P/2P/Ttr (tan tan radius)]: Specify radius of circle or [Diameter]: D Specify diameter of circle:	**[CIR-C,D]^C^CCIRCLE;\D**
Command: **CIRCLE** Specify center point for circle or [3P/2P/Ttr (tan tan radius)]: **2P** Specify first end point of circle's diameter: Specify second end point of circle's diameter:	**[CIR- 2P]^C^CCIRCLE;2P**
Command: **ERASE**	**[ERASE]^C^CERASE**
Command: **MOVE**	**[MOVE]^C^CMOVE**

LOADING MENUS

AutoCAD LT automatically loads the **ACLT.MNC** file when you get into the AutoCAD LT drawing editor, unless the ACLT.MNU file has been modified or unless a different menu file has been loaded. However, you can also load a different menu file by using the AutoCAD LT **MENULOAD** command. When you enter the **MENULOAD** command, AutoCAD LT displays the **Menu Customization** dialog box. Choose the **Browse** button to display the **Select Menu File** dialog box (Figure 25-5) on the screen. Select the menu file that you want to load and then choose the **Open** button.

You can also load the menu file from the command line, keeping the **FILEDIA** as **0**.

Command: **FILEDIA**
Enter new value for FILEDIA <1>: **0**
Command: **MENULOAD**

Enter name of menu file to load: **PDM1**

Where **PDM1** ------------ Name of menu file

After you enter the **MENULOAD** command, AutoCAD LT will prompt for the file name. Enter the name of the menu file without the file extension **(.MNU)**, since AutoCAD LT assumes the extension .MNU. AutoCAD LT will automatically compile the menu file into **MNC** and **MNR** files. When you load a menu file in windows, AutoCAD LT creates the following files:

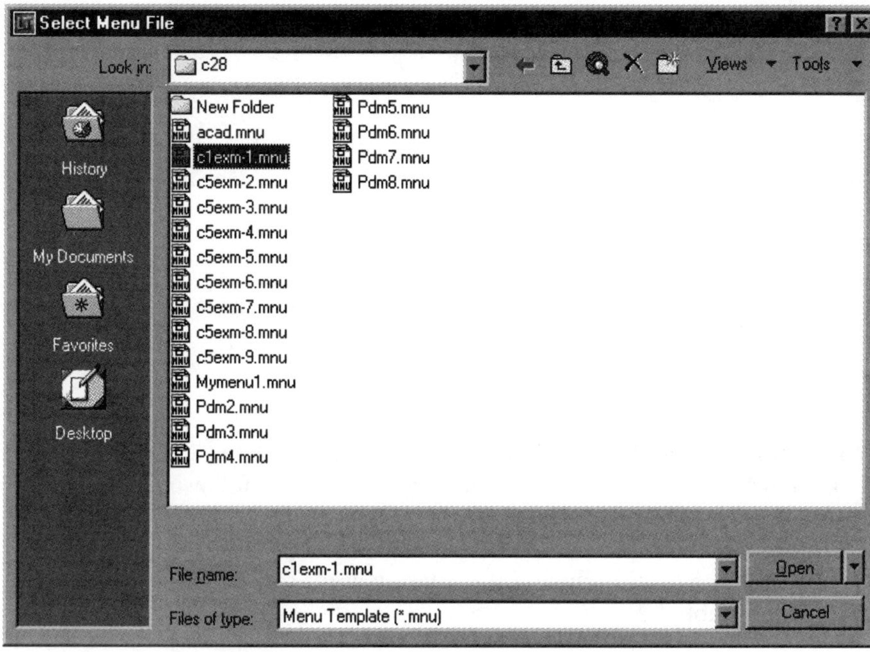

***Figure 25-5 Select Menu File** dialog box*

.mnc and **.mnr** files

When you load a menu file (.mnu), AutoCAD LT compiles the menu file and creates **.mnc** and **.mnr** files. The .mnc file is a compiled menu file. The **.mnr** file contains the bitmaps used by the menu.

.mns file

When you load the menu file, AutoCAD LT also creates an **.mns** file. This is an ASCII file that is the same as the **.mnu** file when you initially load the menu file. Each time you make a change in the contents of the file, AutoCAD LT changes the **.mns** file.

Note

*After you load the new menu, you cannot use the screen menu, buttons menu, or digitizer because the original menu, **ACLT.MNU**, is not present and the new menu does not contain these menu areas.*

*To activate the original menu again, enter **MENULOAD** and then choose the **Browse** button to display the **Select Menu File** dialog box. Select the **ACLT.MNU** file from the **Support** directory and then choose the **Open** button. You can also enter the values at the Command prompt if the **FILEDIA** system variable is 0.*

If you need to use input from a keyboard or a pointing device, use the backslash (\). The system will pause for you to enter data.

There should be no space after the backslash (\).

The menu items, menu labels, and command definition can be uppercase, lowercase, or mixed. You can introduce spaces between the menu items to improve the readability of the menu file. The blank lines are ignored and are not displayed on the screen.

If there are more menu items in the pull-down menu than the number of spaces available, the excess items are not displayed on the screen. For example, if the display device and the screen resolution limits the number of items to 21, items in excess of 21 will not be displayed on the screen and are therefore inaccessible.

If you are using a high-resolution graphics board, you can increase the number of lines that can be displayed on the screen. On some devices this is 80 lines.

RESTRICTIONS

The menus are easy to use and provide a quick access to frequently used AutoCAD LT commands. However, the menu bar and the menus are disabled during the following commands:

TEXT Command
After you assign the text height and the rotation angle to a **TEXT** command, the menu is automatically disabled.

SKETCH Command
The menus are disabled after you set the record increment in the **SKETCH** command.

Exercise 1 *General*

Write a menu for the following AutoCAD LT commands. The menu design is shown in Figure 25-6.

DRAW	EDIT	DISP/TEXT	UTILITY
LINE	FILLET0	TEXT,C	SAVE
PLINE	FILLET	TEXT,L	QUIT
ELLIPSE	CHAMFER	TEXT,R	END
POLYGON	STRETCH	ZOOM WIN	DIR
DONUT	EXTEND	ZOOM PRE	PLOT
	OFFSET		

CASCADING SUBMENUS IN MENUS

The number of items in a menu or cursor menu can be very large, and sometimes they cannot all be accommodated on one screen. For example, consider a display device that can display a maximum of twenty-one menu items. If the pull-down menu or the cursor menu has more items than can be displayed, the excess menu items are not displayed on the screen and cannot be accessed. You can overcome this problem by using cascading menus that let you define smaller groups of items within a menu section. When an item is selected, it loads the cascading menu and displays the items, defined in the cascading menu, on the screen.

The cascading feature of AutoCAD LT allows pull-down and cursor menus to be displayed in

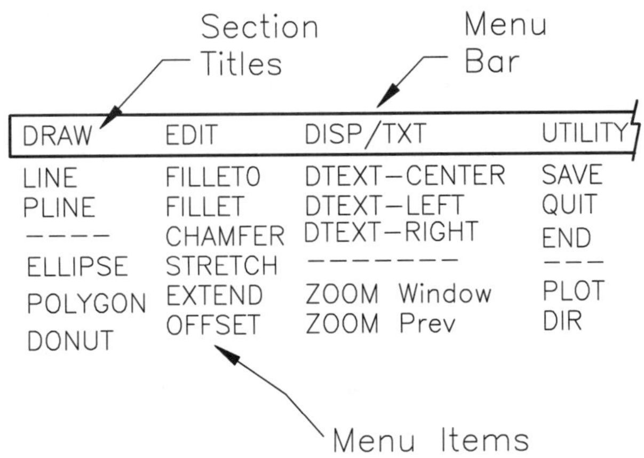

***Figure 25-6** Menu display for Exercise 1*

a hierarchical order that makes it easier to select submenus. To use the cascading feature in pull-down and cursor menus, AutoCAD LT has provided some special characters. For example, -> defines a cascaded submenu and <- designates the last item in the menu. The following table lists some of the characters that can be used with the pull-down or cursor menus.

Character	Character Description
--	The item label consisting of two hyphens automatically expands to fill the entire width of the menu. Example: [--]
+	Used to continue the menu item to the next line. This character has to be the last character of the menu item. Example: [Triang:]^C^CLine;1,1;+3,1;2,2;
->	This label character defines a cascaded submenu; it must precede the name of the submenu. Example: [**->Draw**]
<-	This label character designates the last item of the cascaded pull-down or cursor menu. The character must precede the label item. Example: [**<-CIRCLE 3P**]^**C**^**CCIRCLE;3P**
<-<-...	This label character designates the last item of the pull-down or cursor menu and also terminates the parent menu. The character must precede the label item. Example: [**<-<-Center Mark**]^**C**^**C_dim;_center**
$(	This label character can be used with the pull-down and cursor

	menus to evaluate a DIESEL expression. The character must precede the label item. Example: **$(if,$(getvar,orthomode),Ortho)**
~	This item indicates the label item is not available (displayed grayed out); the character must precede the item. Example: [~Application not available]
!.	When used as a prefix, it displays the item with a check mark.
&	When placed directly before a character, the character is displayed underscored. For example, [W&Block] is displayed as WBlock. It also specifies the character as a menu accelerator key in the pull-down or Shortcut menu.
/c	When placed directly before an item with a character, the character is displayed underscored. For example, [/BWBlock] is displayed as WBlock. It also specifies that the item has a menu accelerator key in the pull-down or Shortcut menu.
\t	The label text to the right of \t is displayed to the right side of the menu.

Consider a display device with a certain resolution and font size that provides space for a maximum of eighty characters. Therefore, if there are ten menus, the length of each menu title should average eight characters. If the combined length of all menu bar titles exceeds eighty characters, AutoCAD LT automatically wraps the excess menu items and displays them on the next line in the menu bar. The following is a list of some additional features of the menu.

1. The section labels of the menus are ***POP1 through ***POP16. The menu bar titles are displayed in the menu bar.

2. The menus can be accessed by selecting the menu title from the menu bar at the top of the screen.

3. A maximum of 999 menu items can be defined in the menu. This includes the items that are defined in the submenus. The menu items in excess of 999 are ignored.

4. The number of menu items that can be displayed depends on the display device you are using. If the cursor or the menu contains more items than can be accommodated on the screen, the excess items are truncated. For example, if your system can display only thirty-five menu items, the menu items in excess of thirty-five are automatically truncated.

Example 2

Write a pull-down menu for the commands shown in Figure 25-7. The menu must use the AutoCAD LT cascading feature.

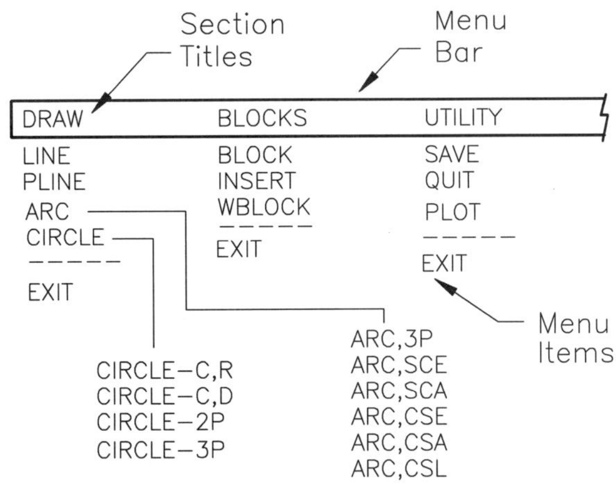

Figure 25-7 *Menu structure for Example 2*

Step 1: Writing the menu file

The following file is a listing of the menu for Example 2. **The line numbers are not a part of the menu; they are shown here for reference only.**

```
***POP1
[DRAW]
[LINE]^C^CLINE
[PLINE]^C^CPLINE
[->ARC]
  [ARC]^C^CARC
  [ARC,3P]^C^CARC
  [ARC,SCE]^C^CARC;\C
  [ARC,SCA]^C^CARC;\C;\A
  [ARC,CSE]^C^CARC;C
  [ARC,CSA]^C^CARC;C;\\A
  [<-ARC,CSL]^C^CARC;C;\\L
[->CIRCLE]
  [CIRCLE C,R]^C^CCIRCLE
  [CIRCLE C,D]^C^CCIRCLE;\D
  [CIRCLE 2P]^C^CCIRCLE;2P
  [<-CIRCLE 3P]^C^CCIRCLE;3P
[—]
[Exit]^C
***POP
[BLOCKS]
[BLOCK]^C^CBLOCK
[INSERT]*^C^CINSERT
[WBLOCK]^C^CWBLOCK
[—]
```

Chapter 25

```
26 [Exit]^C
27 ***POP3
28 [UTILITY]
29 [SAVE]^C^CSAVE
30 [QUIT]^C^CQUIT
31 [PLOT]^C^CPLOT
32 [—]
33 [Exit]^C
```

Explanation

Line 5

[->ARC]

In this menu item, **ARC** is the menu item label that is preceded by the special label characters **->**. These special characters indicate that the menu item has a submenu. The menu items that follow it (Lines 6-12) are the submenu items (Figure 25-8).

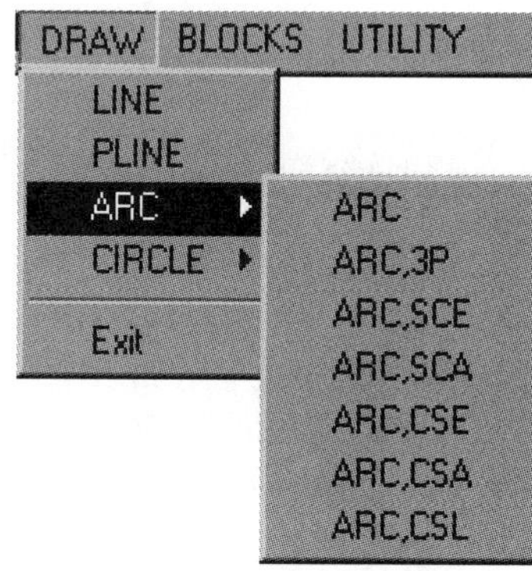

Figure 25-8 *Draw menu*

Line 12

[<-ARC,CSL]^C^CARC;C;\\L

In this line, the menu item label **ARC,CSL** is preceded by another special label characters, **<-**, which indicates the end of the submenu. The item that contains these characters must be the last menu item of the submenu.

Lines 13 and 17

[->CIRCLE]

[<-CIRCLE 3P]^C^CCIRCLE;3P

The special characters -> in front of **CIRCLE** indicate that the menu item has a submenu; the characters <- in front of **CIRCLE 3P** indicate that this item is the last menu item in the submenu. When you select the menu item **CIRCLE** from the menu, it will automatically display the submenu on the side (Figure 25-9).

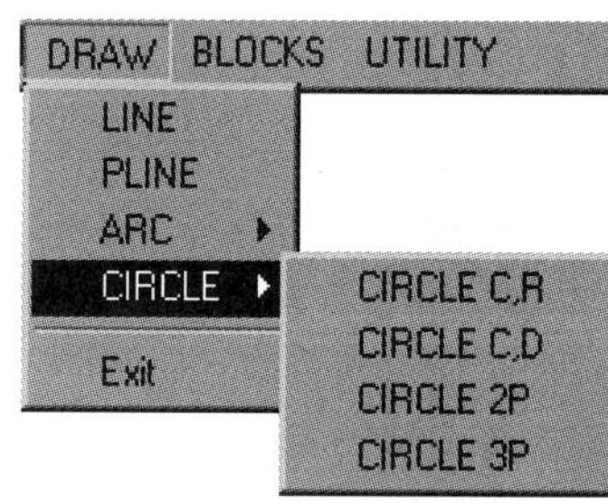

Figure 25-9 *Draw menu with CIRCLE cascading menu*

Step 2: Loading the menu file

Save the menu file with the extension ***.MNU**. Load the menu file using the **MENULOAD** command as described in Example 1.

Example 3

Write a menu that has the cascading submenus for the commands shown in Figure 25-10.

Writing the menu file

The following file is a listing of the menu for Example 3. **The line numbers are not a part of the menu; they are shown here for reference only.**

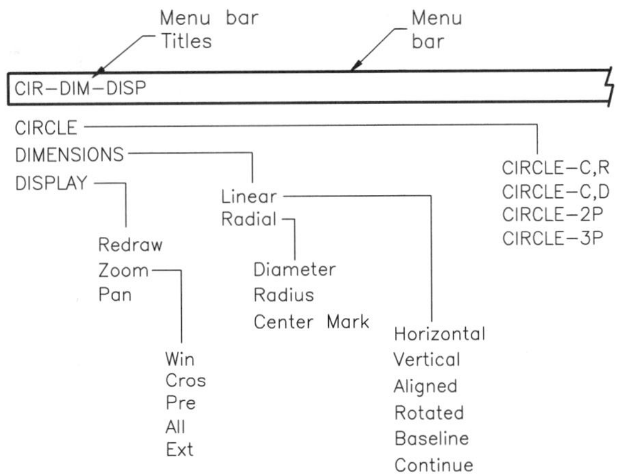

Figure 25-10 *Menu structure for Example 3*

```
***POP1
[DRAW]
[->CIRCLE]
  [CIRCLE C,R]^C^C_CIRCLE
  [CIRCLE C,D]^C^C_CIRCLE;\_D
  [CIRCLE 2P]^C^C_CIRCLE;_2P
  [<-CIRCLE 3P]^C^C_CIRCLE;_3P
[->Dimensions]
    [->Linear]
    [Horizontal]^C^C_dimlinear
   [Vertical]^C^C_dimlinear
   [Aligned]^C^C_dimaligned
  [Rotated]^C^C_dimrotated
  [Baseline]^C^C_dimbaseline
  [<-Continue]^C^C_dimcontinue
   [->Radial]
  [Diameter]^C^C_dimdiameter
   [Radius]^C^C_dimradius
   [<-<-Center Mark]^C^C_dimcenter
[->DISPLAY]
    [REDRAW]^C^CREDRAW
    [->ZOOM]
    [...Win]^C^C_ZOOM;_W
    [...Cros]^C^C_ZOOM;_C
    [...Pre]^C^C_ZOOM;_P
    [...All]^C^C_ZOOM;_A
    [<-...Ext]^C^C_ZOOM;_E
  [<-PAN]^C^C_Pan
```

Chapter 25

Explanation
Lines 8 and 9
[->Dimensions]
[->Linear]
The special label characters **->** in front of the menu item **Dimensions** indicate that it has a submenu, and the characters -> in front of **Linear** indicate that there is another submenu. The second submenu **Linear** is within the first submenu **Dimensions** (Figure 25-11). The menu items on lines 10 to 15 are defined in the **Linear** submenu, and the menu items **Linear** and **Radial** are defined in the submenu **Dimensions**.

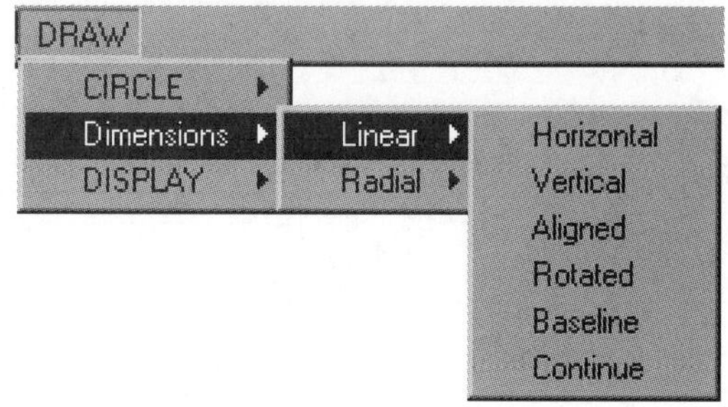

***Figure 25-11** Draw menu with cascading menus*

Line 16
[->Radial]
This menu item defines another submenu; the menu items on line numbers 17, 18, and 19 are part of this submenu.

Line 19
[<-<-Center Mark]^C^C_dim;_center
In this menu item the special label characters **<-<-** terminate the **Radial** and **Dimensions** (parent submenu) submenus.

Lines 27 and 28
[<-...Ext]^C^C_ZOOM;_E
[<-PAN]^C^C_Pan
The special characters **<-** in front of the menu item **...Ext** terminate the **ZOOM** submenu (Figure 25-12); the special character in front of the menu item **PAN** terminates the **DISPLAY** submenu.

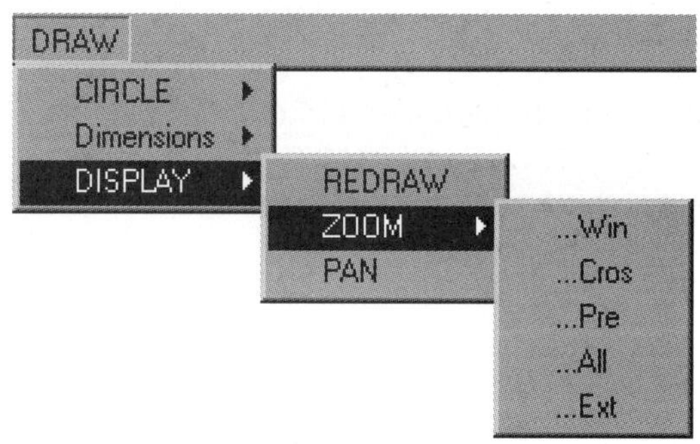

***Figure 25-12** Draw pull-down menu with cascading menus*

SHORTCUT AND CONTEXT MENUS

The Shortcut menus are similar to the pull-down menus, except the Shortcut menu can contain only 499 menu items compared with 999 items in the pull-down menu. The section label of the Shortcut menu can be ***POP0 and ***POP500 to ***POP 999. The Shortcut menus are displayed near or at the cursor location. Therefore, they can be used to provide convenient and quick access to some of the frequently used commands. The Shortcut menus that are in the upper range are also referred to as Context menus. The following is a list of some of the features of the Shortcut menu.

1. The section label of the Shortcut menu are ***POP0 and ***POP500 to ***POP999. The menu bar title defined under this section label is not displayed in the menu bar.

2. On most systems, the menu bar title is not displayed at the top of the Shortcut menu. However, for compatibility reasons you should give a dummy menu bar title.

3. The POP0 menu can be accessed through the **$P0=*** menu command. The shortcut menus POP500 through POP999 must be referenced by their alias names. The reserved alias names for AutoCAD LT use are GRIPS, CMDEFAULT, CMEDIT, and CMCOMMAND. For example, to reference POP500 for grips, use **GRIPS command line under POP500. This command can be issued by a menu item in another menu, such as the button menu, or the auxiliary menu.

4. A maximum of 499 menu items can be defined in the Shortcut menu. This includes the items that are defined in the Shortcut submenus. The menu items in excess of 499 are ignored.

5. The number of menu items that can be displayed on the screen depends on the system you are using. If the Shortcut or pull-down menu contains more items than your screen can accommodate, the excess items are truncated. For example, if your system displays twenty-one menu items, the menu items in excess of twenty-one are automatically truncated.

6. The system variable **SHORTCUTMENU** controls the availability of Default, Edit, and Command mode shortcut menus. If the value is 0, it restores R14 legacy behavior and disables the Default, Edit, and Command mode shortcut menus. The default value of this variable is 11.

Example 4

Write a Shortcut menu for the following AutoCAD LT commands using cascading submenus. The menu should be compatible with foreign language versions of AutoCAD LT. Use the third button of the **BUTTONS** menu to display the cursor menu.

Osnaps	Draw	DISPLAY
Center	Line	REDRAW
Endpoint	PLINE	ZOOM
Intersection	CIR C,R	...Win
Midpoint	CIR 2P	...Cen
Nearest	ARC SCE	...Prev
Perpendicular	ARC CSE	...All
Quadrant		...Ext
Tangent		PAN
None		

Writing the menu file

The following file is a listing of the menu file for Example 4. **The line numbers are not a part of the file; they are for reference only.**

```
1 ***AUX1
2 ;
3 $P0=*
4 ***POP0
5 [Osnaps]
6 [Center]_Center
7 [End point]_Endp
8 [Intersection]_Int
9 [Midpoint]_Mid
10 [Nearest]_Nea
11 [Perpendicular]_Per
12 [Quadrant]_Qua
13 [Tangent]_Tan
14 [None]_Non
15 [—]
16 [->Draw]
17  [Line]^C^C_Line
18  [PLINE]^C^C_Pline
19  [CIR C,R]^C^C_Circle
20  [CIR 2P]^C^C_Circle;_2P
21  [ARC SCE]^C^C_ARC;\C
22  [<-ARC CSE]^C^C_Arc;C
23 [—]
24 [->DISPLAY]
25   [REDRAW]^C^_REDRAW
26   [->ZOOM]
27     [...Win]^C^C_ZOOM;_W
28     [...Cen]^C^C_ZOOM;_C
29     [...Prev]^C^C_ZOOM;_P
30     [...All]^C^C_ZOOM;_A
31     [<-...Ext]^C^C_ZOOM;_E
32   [<-PAN]^C^C_Pan
33 ***POP1
34 [SHORTCUTMENU]
35 [SHORTCUTMENU=0]^C^CSHORTCUTMENU;0
36 [SHORTCUTMENU=1]^C^CSHORTCUTMENU;1
```

Explanation

Line 1

*****AUX1**

AUX1 is the section label for the first auxiliary menu; *** designates the menu section. The menu items that follow it, until the second section label, are a part of this buttons menu.

Lines 2 and 3

;

$P0=*

The semicolon (;) is assigned to the second button of the pointing device (the first button of the pointing device is the pick button); the special command **$P0=*** is assigned to the third button of the pointing device.

Lines 4 and 5
*****POP0**
[Osnaps]
The menu label **POP0** is the menu section label for the Shortcut menu; Osnaps is the menu bar title. The menu bar title is not displayed, but is required. Otherwise, the first item will be interpreted as a title and will be disabled.

Line 6
[Center]_Center
In this menu item, **_Center** is the center object Snap mode. The menu files can be used with foreign language versions of AutoCAD LT, if AutoCAD LT commands and the command options are preceded by the underscore (_) character.

After loading the menu, if you press the third button of your pointing device, the Shortcut menu (Figure 25-13) will be displayed at the cursor (screen crosshairs) location. If the cursor is close to the edges of the screen, the Shortcut menu will be displayed at a location that is closest to the cursor position. When you select a submenu, the items contained in the submenu will be displayed, even if the Shortcut menu is touching the edges of the screen display area.

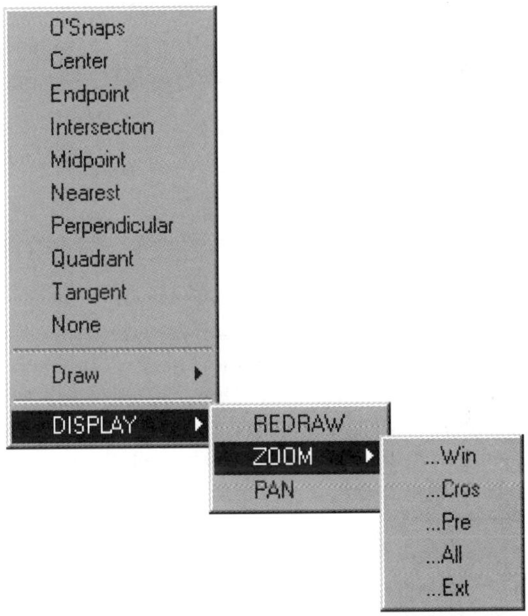

Figure 25-13 *Shortcut menu for Example 4*

Lines 33 and 34
*****POP1**
[Draw]
*****POP1** defines the first pull-down menu. If no POPn sections are defined or the status line is turned off, the Shortcut menu is automatically disabled.

Exercise 2 *General*

Write a menu for the following AutoCAD LT commands. Use a cascading menu for the **LINE** command options in the menu. (The layout of the menu is shown in Figure 25-14.)

LINE	**ZOOM All**	**TIME**
Continue	**ZOOM Win**	**LIST**
Close	**ZOOM Pre**	**DISTANCE**
Undo	**PAN**	**AREA**

CIRCLE
ELLIPSE

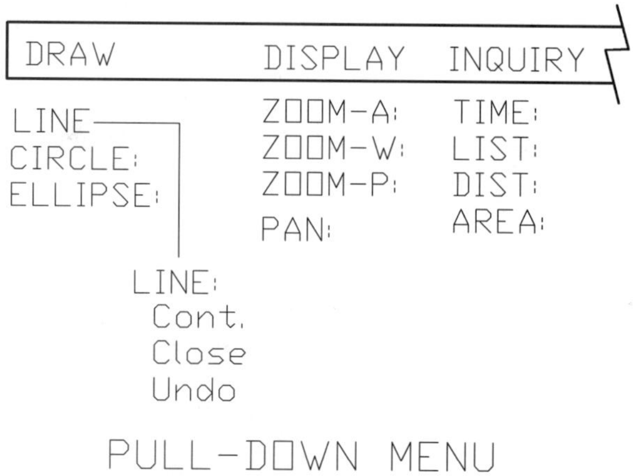

Figure 25-14 Design of menu for Exercise 2

SUBMENUS

The number of items in a pull-down menu or Shortcut menu can be very large and sometimes they cannot all be accommodated on one screen. For example, consider a display device on which the maximum number of items that can be displayed is twenty-one items, the menu excess items are not displayed on the screen and cannot be accessed. You can overcome this problem by using submenus that let you define smaller groups of items within a menu section. When a submenu is selected, it loads the submenu items and displays them on the screen.

The menus that use AutoCAD LT's cascading feature are the most efficient and easy to write. The submenus follow a logical pattern that are easy to load and use without causing any confusion. It is strongly recommended to use the cascading menus whenever you need to write the pull-down or the Shortcut menus. However, AutoCAD LT provides the option to swap the submenus in the menus. These menus can sometimes cause distraction because the original menu is completely replaced by the submenu when swapping the menus.

Submenu Definition

A submenu definition consists of two asterisk signs (**) followed by the name of the submenu. A menu can have any number of submenus and every submenu should have a unique name. The items that follow a submenu, up to the next section label or submenu label, belong to that submenu. Following is the format of a submenu label:

****Name**

Where ****** ---------------- Two asterisk signs (**) designate a submenu
Name ----------- Name of the submenu

Note

The submenu name can be up to thirty-one characters long.

The submenu name can consist of letters, digits, and the special characters like: $ (dollar), - (hyphen), and _ (underscore).

The submenu name cannot have any embedded blanks.

The submenu names should be unique in a menu file.

Submenu Reference

The submenu reference is used to reference or load a submenu. It consists of a "$" sign followed by a letter that specifies the menu section. The letter that specifies a menu section is Pn, where n designates the number of the menu section. The menu section is followed by "=" sign and the name of the submenu that the user wants to activate. The submenu name should be without "**". Following is the format of a submenu reference:

$Section=Submenu

Where **$** ------------------- "$" sign
Section ---------- Menu section specifier
= ------------------ "=" sign
Submenu ------- Name of submenu

Example
$P1=P1A

Where **$P1** --------------- P1-Specifies pull-down menu section 1
P1A --------------- Name of submenu

Displaying a Submenu

When you load a submenu in a menu, the submenu items are not automatically displayed on the screen. For example, when you load a submenu P1A that has DRAW-ARC as the first item, the current title of POP1 will be replaced by the DRAW-ARC. But the items that are defined under DRAW-ARC are not displayed on the screen. To force the display of the new items on the screen, AutoCAD LT uses a special command $Pn=*.

$Pn=*

Where **P** ------------------ P for menu
Pn ----------------- Menu section number (1 to 10)
***** ------------------- Asterisk sign (*)

LOADING MENUS

From the menu, you can load any menu that is defined in the screen or image tile menu sections by using the appropriate load commands. It may not be needed in most of the applications, but if you want to, you can load the menus that are defined in other menu sections.

Loading an Image Tile Menu

You can also load an image tile menu from the menu by using the following load command:

$I=IMAGE1 $I=*

Where **$I=Image1** ----- Load the submenu IMAGE1
$I=* ------------- Display the dialog box

This menu item consists of two load commands. The first load command $I=IMAGE1 loads the image tile submenu IMAGE1 that has been defined in the image tile menu section of the file. The second load command $I=* displays the new dialog box on the screen.

Example 5

Write a pull-down menu for the following AutoCAD LT commands. Use submenus for the **ARC** and **CIRCLE** commands.

LINE
PLINE
ARC
  ARC 3P
  ARC SCE
  ARC SCA
  ARC CSE
  ARC CSA
  ARC CSL
CIRCLE
  CIRCLE C,R
  CIRCLE C,D
  CIRCLE 2P

BLOCK
INSERT
WBLOCK

QUIT
SAVE
PLOT

Step 1: Designing the menu

The layout shown in Figure 25-15 is one of the possible designs for this menu. The **ARC** and **CIRCLE** commands are in separate groups that will be defined as submenus in the menu file.

Step 2: Writing the menu file

The following file is a listing of the menu for Example 5. The line numbers are not a part of the menu file. They are given here for reference only.

```
***POP1
**P1A
[DRAW]
[LINE]^C^CLINE
[PLINE]^C^CPLINE
[—]
[ARC]^C^C$P1=P1B $P1=*
[CIRCLE]^C^C$P1=P1C $P1=*
[—]
```

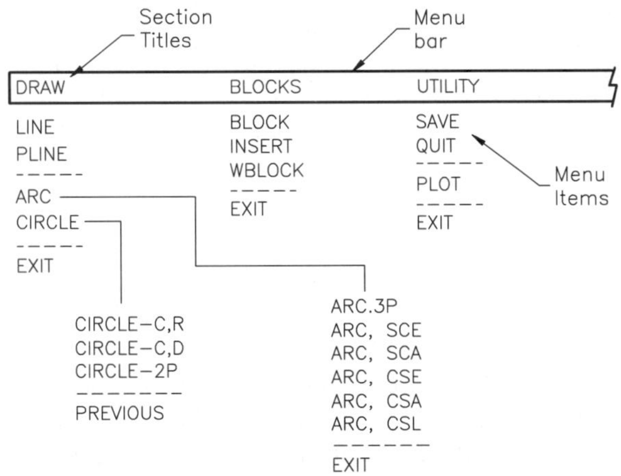

Figure 25-15 *Design of menu for Example 5*

```
[Exit]^C

**P1B
[ARC]
[ARC,3P]^C^CARC
[ARC,SCE]^C^CARC \C
[ARC,SCA]^C^CARC \C \A
[ARC,CSE]^C^CARC C
[ARC,CSA]^C^CARC C \\A
[ARC,CSL]^C^CARC C \\L
[—]
[Exit]^C

**P1C
[CIRCLE]
[CIRCLE C,R]^C^CCIRCLE
[CIRCLE C,D]^C^CCIRCLE \D
[CIRCLE 2P]^C^CCIRCLE 2P
[—]
[PREVIOUS]$P1=P1A $P1=*

***POP2
[BLOCKS]
[BLOCK]^C^CBLOCK
[INSERT]*^C^CINSERT
[WBLOCK]^C^CWBLOCK
[—]
[Exit]$P1=P1A $P1=*
```

```
39 ***POP3
40 [UTILITY]
41 [SAVE]^C^CSAVE
42 [QUIT]^C^CQUIT
43 [~—]
44 [PLOT]^C^CPLOT
45 [~—]
46 [Exit]^C
47
```

Explanation

Line 2

****P1A**

**P1A defines the submenu P1A. All the submenus have two asterisk signs (**) followed by the name of the submenu. The submenu can have any valid name. In this example, P1A has been chosen because it is easy to identify the location of the submenu. P indicates that it is a menu, 1 indicates that it is in the first menu (POP1), and A indicates that it is the first submenu in that section.

Line 6

[--]

The two hyphens enclosed in the brackets will automatically expand to fill the entire width of the menu. This menu item cannot be used to define a command. If it does contain a command definition, the command is ignored. For example, if the menu item is [--]^C^CLINE, the command ^C^CLINE will be ignored.

Line 7

[ARC]^C^C$P1=P1B $P1=*

In this menu item, $P1=P1B loads the submenu P1B and assigns it to the first menu section (POP1), but the new menu is not displayed on the screen (see Figure 25-16). $P1=* forces the display of the new menu on the screen.

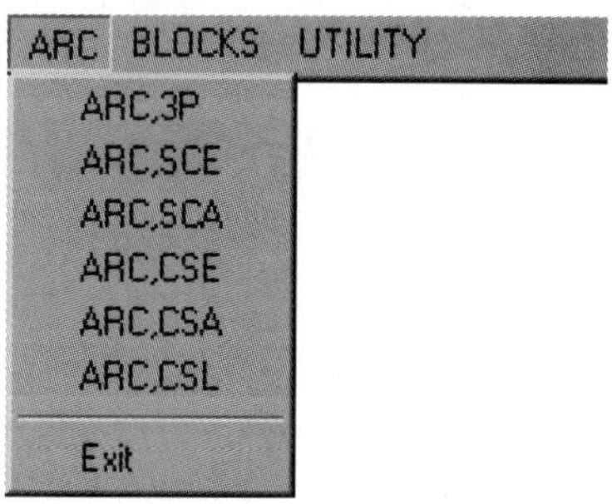

***Figure 25-16** Arc pull-down menu replaces Draw menu*

For example, if you select **CIRCLE** from the first menu (POP1), the menu bar title **DRAW** will be replaced by **CIRCLE**, but the new menu is not displayed on the screen. Now, if you select **CIRCLE** from the menu bar, the command defined under the **CIRCLE** submenu will be displayed in the menu. To force the display of the menu that is currently assigned to POP1, you can use AutoCAD LT's special command **$P1=***. If you select **CIRCLE** from the first menu (POP1), the **CIRCLE** submenu will be loaded and automatically displayed on the screen.

Line 21

[EXIT]^C

When you select this menu item the current menu will be canceled. It will not return to the previous submenu (**DRAW**). If you check the menu bar, it will display **ARC** as the section title, not **DRAW**. Therefore, it is not a good practice to cancel a submenu. It is better to return to the first menu before canceling it or define a command that automatically loads the previous menu and then cancels the menu.

Example

[EXIT]$P1=P1A $P1=* ^C^C

Line 29
[PREVIOUS]$P1=P1A $P1=*
In this menu item $P1=P1A loads the submenu P1A, which happens to be the previous menu in this case. You can also use $P1= to load the previous menu. $P1=* forces the display of the submenu P1A.

Note
When you swap the pull-down menus, the menus will get incremented. For example, if a menu file loads POP1 through POP8 and then a partial menu inserts a new menu at POP5, the existing pull-down menus will be incremented by one. Thus, $P7=P7A $P= will take the new P7A menu and use it to replace what used to be POP6, but has now been pushed over to POP7.*

Step 3: Loading the menu file
Save the menu file and load it using the **MENULOAD** command as described in earlier examples.

PARTIAL MENUS

AutoCAD LT has provided a facility that allows users to write their own menus and then load them in the menu bar. For example, in Windows you can write partial menus, toolbars, and definitions for accelerator keys. After you write the menu, AutoCAD LT lets you load the menu and use it with the standard menu. For example, you could load a partial menu and use it like a menu. You can also unload the menus that you do not want to use. These features make it convenient to use the menus that have been developed by AutoCAD LT users and developers.

Menu Section Labels

The following is a list of the additional menu section labels.

Section label	**Description**
***MENUGROUP	Menu file group name
***TOOLBARS	Toolbar definition
***HELPSTRING	Online help
***ACCELERATORS	Accelerator key definitions

Writing Partial Menus

The following example illustrates the procedure for writing a partial menu.

Example 6

In this example you will write a partial menu for Windows. The menu file has two menus, POP1 (MyDraw) and POP2 (MyEdit), as shown in Figure 25-17.

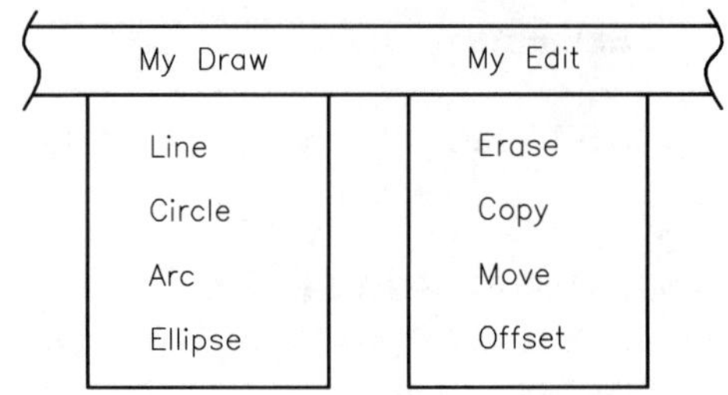

***Figure 25-17** Pull-down menus for Example 6*

Step 1: Writing the menu file

Use a text editor to write the following menu file. The name of the file is assumed to be **MYMENU1.MNU**. The following is the listing of the menu file for this example:

```
1  ***MENUGROUP=Menu1
2  ***POP1
3  [/MMyDraw]
4  [/LLine]^C^CLine
5  [/CCircle]^C^CCircle
6  [/AArc]^C^CArc
7  [/EEllipse]^C^CEllipse
8  ***POP2
9  [/EMyEdit]
10 [/EErase]^C^CErase
11 [/CCopy]^C^CCopy
12 [/VMove]^C^CMove
13 [/OOffset]^C^COffset
```

Explanation

Line 1

*****MENUGROUP=Menu1**

MENUGROUP is the section label and the Menu1 is the name tag for the menu group. The MENUGROUP label must precede all menu section definitions. The name of the MENUGROUP (Menu1) can be up to thirty-two characters long (alphanumeric), excluding spaces and punctuation marks. There is only one MENUGROUP in a menu file. All section labels must be preceded by *** (***MENUGROUP).

Line 2

*****POP1**

POP1 is the menu section label. The items on line numbers 3 through 7 belong to this section. Similarly, the items on line numbers 9 through 13 belong to the menu section **POP2**.

Line 3

[/MMyDraw]

/M defines the mnemonic key you can use to activate the menu item. For example, /M will display an underline under the letter M in the text string that follows it. If you enter the letter M, AutoCAD LT will execute the command defined in that menu item. MyDraw is the menu item label. The text string inside the brackets [ ], except /M, has no function. It is used for displaying the function name so that the user can recognize the command that will be executed by selecting that item.

Line 4

[/LLine]^C^CLine

In this line, the /L defines the mnemonic key, and the Line that is inside the brackets is the menu item label. ^C^C cancels the command twice, and the Line is the AutoCAD LT **LINE** command. The part of the menu item statement that is outside the brackets is executed when you select an item from the menu. When you select Line 4, AutoCAD LT will execute the **LINE** command.

Step 2: Loading the menu file

Save the file as **MYMENU1.MNU**. Choose the **Customize Menu** button from the **Tools** menu or enter the **MENULOAD** command at the Command prompt to invoke the **Menu Customization** dialog box (Figure 25-18). To load the menu file, enter the name of the menu file, **MYMENU1.MNU**, in the **File Name** edit box. You can also use the **Browse** option to invoke the **Select Menu File** dialog box. Select the name of the file, and then use the **OK** button to return to the **Menu Customization** dialog box. To load the selected menu file, choose the **LOAD** button. The name of the menu group (**MENU1**) will be displayed in the **Menu Groups** list box.

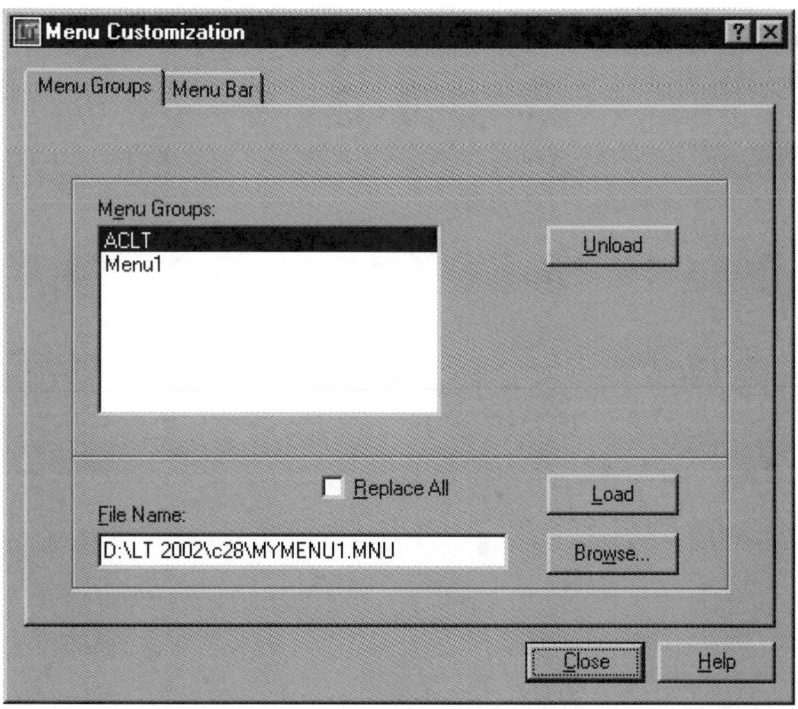

***Figure 25-18 Menu Customization** dialog box (**Menu Groups** tab)*

You can also load the menu file from the command line, as follows:

Command: **FILEDIA**
Enter new value for FILEDIA <1>: **0** *(Disables the file dialog boxes.)*
Command: **MENULOAD**
Enter name of menu file to load: **MYMENU1.MNU**

You can also use the AutoLISP functions to set the **FILEDIA** system variable to 0 and then load the partial menu.

Command: **(SETVAR "FILEDIA" 0)**
Command: **(Command "MENULOAD" "MYMENU1")**

Step 3: Inserting menus in the Menu Bar

In the **Menu Customization** dialog box, select the **Menu Bar** tab to display the menu bar options (Figure 25-19). In the **Menu Groups** list box select Menu1; the menus defined in the menu group (Menu1) will be displayed in the **Menus** list box. In the **Menus** list box select the menu (MyDraw) that you want to insert in the menu bar. In the **Menu Bar** list box select the position where you want to insert the new menu. For example, if you want to insert the new menu in front of Format, select the **Format** menu in the **Menu Bar** list box. Choose the **Insert** button to insert the selected menu (MyDraw) in the menu bar. The menu (MyDraw1) is displayed in the menu bar located at the top of your screen.

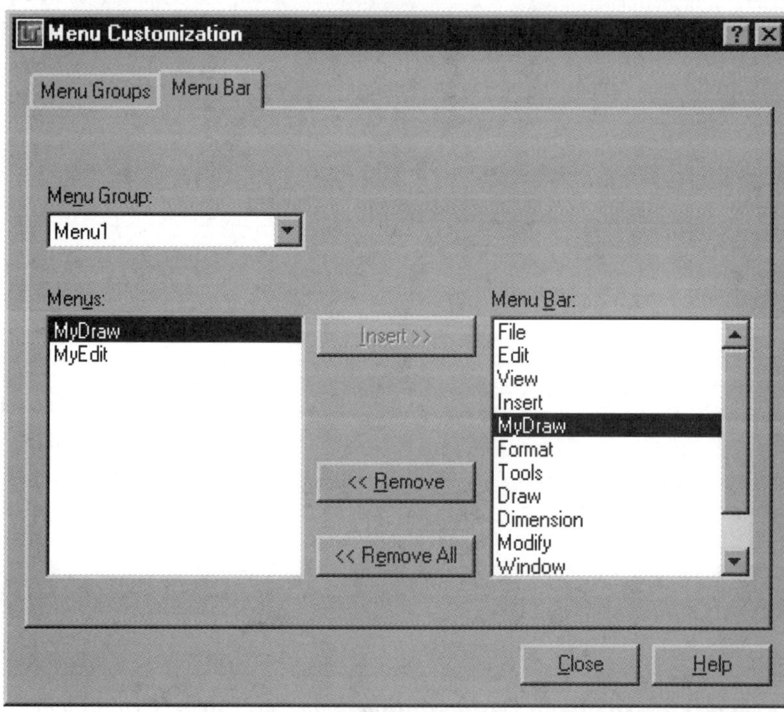

***Figure 25-19 Menu Customization** dialog box (**Menu Bar** tab)*

After you enter these commands, AutoCAD LT will display the menu titles in the menu bar, as shown in Figure 25-20. If you select MyDraw, the corresponding menu as defined in the menu file will be displayed on the screen. Similarly, selecting MyEdit will display the corresponding edit menu.

Figure 25-20 MENUCMD command places the menu titles in the menu bar

Step 4: Unloading the menu

If you want to unload the menu (MyDraw), select the **Menu Bar** tab of the **Menu Customization** dialog box. In the **Menu Bar** list box select the menu item that you want to unload. Choose the **Remove** button to remove the selected item.

Step 5: Unloading the menugroup

You can also unload the menu groups (Menu1) using the **Menu Customization** dialog box. Choose the **Customize > Menus** from the **Tools** menu or enter **MENULOAD** or **MENUUNLOAD** at the Command prompt to invoke the **Menu Customization** dialog box. AutoCAD LT will display the names of the menu files in the **Menu Groups** list box under the **Menu Groups** tab. Select **Menu1** menu, and then choose the **Unload** button. AutoCAD LT will unload the selected menu group. Choose the **Close** button to exit the dialog box.

You can also unload the menu file from the command line, as follows:

Command: **FILEDIA**
Enter new value for FILEDIA <1>: **0**
Command: **MENUUNLOAD**
Enter the name of a MENUGROUP to unload: **MENU1.MNU**

ACCELERATOR KEYS

AutoCAD LT for Windows also supports user-defined accelerator keys. For example, if you enter **C** at the Command prompt, AutoCAD LT draws a circle because it is a command alias for a circle as defined in the ACLT.PGP file. You cannot use the **C** key to enter the **COPY** command. To use the **C** key for entering the **COPY** command, you can define the accelerator keys. You can combine the SHIFT key with C in the menu file so that when you hold down the SHIFT key and then press the C key, AutoCAD LT will execute the **COPY** command. The following example illustrates the use of accelerator keys.

Example 7

In this example you will add the following accelerator keys to the partial menu of Example 6.

CONTROL+"E"	to draw an ellipse (**ELLIPSE** command)
SHIFT+"C"	to copy (**COPY** command)
[CONTROL"Q"]	to quit (**QUIT** command)

Step 1: Writing the accelerators in a file

The following file is the listing of the partial menu file that uses the accelerator keys of Example 7.

```
***MENUGROUP=Menu1
***POP1
**Alias
[/MMyDraw]
[/LLine]^C^CLine
[/CCircle]^C^CCircle
[/AArc]^C^CArc
ID_Ellipse [/EEllipse]^C^CEllipse

***POP2
[/EMyEdit]
[/EErase]^C^CErase
ID_Copy [/CCopy]^C^CCopy
[/OOffset]^C^COffset
[/VMove]^C^CMov

***ACCELERATORS
ID_Ellipse [CONTROL+"E"]
ID_Copy [SHIFT+"C"]
[CONTROL"Q"]^C^CQuit
```

Explanation

This menu file defines three accelerator keys. The **ID_Copy [SHIFT+"C"]** accelerator key consists of two parts. The ID_Copy is the name tag, which must be the same as used earlier in the menu item definition. The SHIFT "C" is the label that contains the modifier (SHIFT) and the keyname (C). The keyname or the string, such as "ESCAPE," must be enclosed in quotation marks.

The accelerator keys can be defined in two ways. One way is to give the name tag followed by the label containing the modifier. The modifier is followed by a single character or a special virtual key enclosed in quotation marks [CONTROL+"E"] or ["ESCAPE"]. You can also use the plus sign (+) to concatenate the modifiers [SHIFT + CONTROL + "L"]. The other way of defining an accelerator key is to give the modifier and the key string, followed by a command sequence [CONTROL "Q"]^C^CQuit.

Step 2: Loading the accelerators

Save the file with the extension **.MNU**. Load the file using the **MENU** command. After you load the file, SHIFT+C will enter the **COPY** command and CTRL+E will draw an ellipse. Similarly, Ctrl+Q will cancel the existing command and enter the **QUIT** command. If it does not work, you may have to unload the partial menu from Example 6 because it uses the same menu group name.

SPECIAL VIRTUAL KEYS

The following are the special virtual keys. These keys must be enclosed in quotation marks when used in the menu file.

String	Description	String	Description
"F1"	F1 key	"NUMBERPAD0"	0 key
"F2"	F2 key	"NUMBERPAD1"	1 key
"F3"	F3 key	"NUMBERPAD2"	2 key
"F4"	F4 key	"NUMBERPAD3"	3 key
"F5"	F5 key	"NUMBERPAD4"	4 key
"F6"	F6 key	"NUMBERPAD5"	5 key
"F7"	F7 key	"NUMBERPAD6"	6 key
"F8"	F8 key	"NUMBERPAD7"	7 key
"F9"	F9 key	"NUMBERPAD8"	8 key
"F10"	F10 key	"NUMBERPAD9"	9 key
"F11"	F11 key	"UP"	UP-ARROW key
"F12"	F12 key	"DOWN"	DOWN-ARROW key
"HOME"	HOME key	"LEFT"	LEFT-ARROW key
"END"	END key	"RIGHT"	RIGHT-ARROW key
"INSERT"	INS key	"ESCAPE"	ESC key
"DELETE"	DEL key		

Valid Modifiers

The following are the valid modifiers:

String	Description
CONTROL	The CTRL key on the keyboard
SHIFT	The SHIFT key (left or right)
COMMAND	The Apple key on Macintosh keyboards
META	The meta key on UNIX keyboards

TOOLBARS

The contents of the toolbar and its default layout can be specified in the Toolbar section (***TOOLBARS) of the menu file. Each toolbar must be defined in a separate submenu.

Toolbar Definition

The following is the general format of the toolbar definition:

```
***TOOLBARS
**MYTOOLS1
TAG1 [Toolbar ("tbarname", orient, visible, xval, yval, rows)]
TAG2 [Button ("btnname", id_small, id_large)]macro
TAG3 [Flyout ("flyname", id_small, id_large, icon, alias)]macro
TAG4 [control (element)]
[—]
```

*****TOOLBARS** is the section label of the toolbar, and **MYTOOLS1** is the name of the submenu that contains the definition of a toolbar. Each toolbar can have five distinct items that control different elements of the toolbar: TAG1, TAG2, TAG3, TAG4, and separator ([—]). The first line of the toolbar (TAG1) defines the characteristics of the toolbar. In this line, **Toolbar** is the keyword, and it is followed by a series of options enclosed in parentheses. The following describes the available options.

tbarname This is a text string that names the toolbar. The tbarname text string must consist of alphanumeric characters with no punctuation other than a dash (-) or an underscore (_).

orient This determines the orientation of the toolbar. The acceptable values are Floating, Top, Bottom, Left, and Right. These values are not case-sensitive.

visible This determines the visibility of the toolbar. The acceptable values are Show and Hide. These values are not case-sensitive.

xval This is a numeric value that specifies the X ordinate in pixels. The X ordinate is measured from the left edge of the screen to the left side of the toolbar.

yval This is a numeric value that specifies the Y ordinate in pixels. The Y ordinate is measured from the top edge of the screen to the top of the toolbar.

rows This is a numeric value that specifies the number of rows.

The second line of the toolbar (TAG2) defines the button. In this line the **Button** is the key word and it is followed by a series of options enclosed in parentheses. The following is the description of the available options.

btnname This is a text string that names the button. The text string must consist of alphanumeric characters with no punctuation other than a dash (-) or an underscore (_). This text string is displayed as ToolTip when you place the cursor over the button.

id_small This is a text string that names the ID string of the small image resource (16 by 16 bitmap). The text string must consist of alphanumeric characters with no punctuation other than a dash (-) or an underscore (_). The id_small text string can also specify a user-defined bitmap (Example: ICON_16_CIRCLE). The **bit map images must exist or you just get a smiley face**.

id_big This is a text string that names the ID string of the large image resource (32 by 32 bitmap). The text string must consist of alphanumeric characters with no punctuation other than a dash (-) or an underscore (_). The id_big text string can also specify a user-defined bitmap

(Example: ICON_32_CIRCLE).

macro The second line (TAG2), which defines a button, is followed by a command string (macro). For example, the macro can consist of ^C^CLine. It follows the same syntax as that of any standard menu item definition.

The third line of the toolbar (TAG3) defines the flyout control. In this line the **Flyout** is the key word, and it is followed by a series of options enclosed in parentheses. The following describes the available options.

flyname This is a text string that names the flyout. The text string must consist of alphanumeric characters with no punctuation other than a dash (—) or an underscore (_). This text string is displayed as ToolTip when you place the cursor over the flyout button.

id_small This is a text string that names the ID string of the small image resource (16 by 16 bitmap). The text string must consist of alphanumeric characters with no punctuation other than a dash (—) or an underscore (_). The id_small text string can also specify a user-defined bitmap.

id_big This is a text string that names the ID string of the large image resource (32 by 32 bitmap). The text string must consist of alphanumeric characters with no punctuation other than a dash (—) or an underscore (_). The id_big text string can also specify a user-defined bitmap.

icon This is a Boolean key word that determines whether the button displays its own icon or the last icon selected. The acceptable values are **ownicon** and **othericon**. These values are not case-sensitive.

alias The alias specifies the name of the toolbar submenu that is defined with the standard ****aliasname** syntax.

macro The third line (TAG3), which defines a flyout control, is followed by a command string (macro). For example, the macro can consist of ^C^CCircle. It follows the same syntax as that of any standard menu item definition.

The fourth line of the toolbar (TAG4) defines a special control element. In this line the Control is the key word, and it is followed by the type of control element enclosed in parentheses. The following describes the available control element types.

element This parameter can have one of the following three values:
Layer: This specifies the layer control element.
Linetype: This specifies the linetype control element.
Color: This specifies the color control element.

The fifth line ([--]) defines a separator.

Example 8

In this example you will write a menu file for a toolbar for the **LINE**, **PLINE**, **CIRCLE**, **ELLIPSE**, and **ARC** commands. The name of the toolbar is MyDraw1 (Figure 25-21).

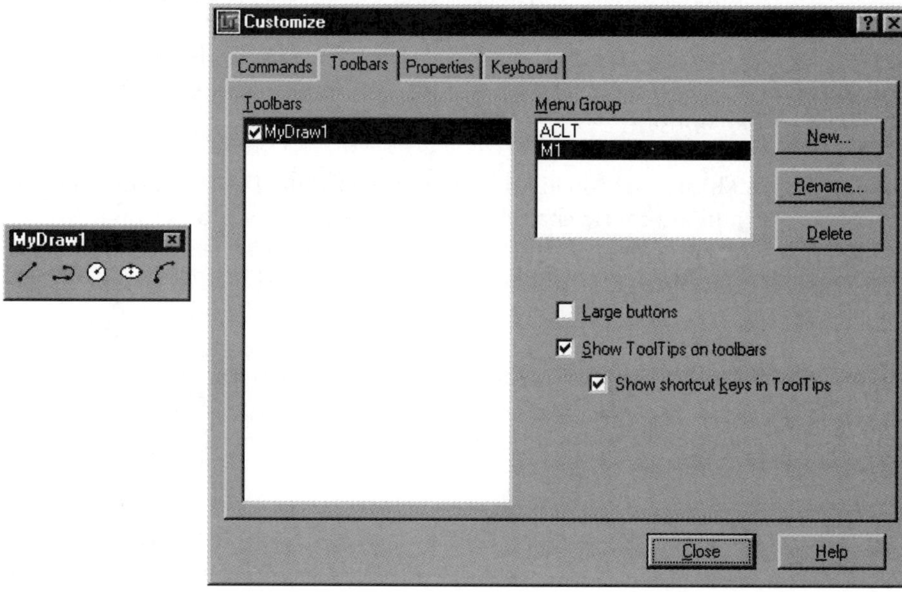

***Figure 25-21** MyDraw1 toolbar for Example 8 and **Toolbars** dialog box*

Step 1: Writing the toolbars

Use any text editor to list the toolbars. The following is the listing of the menu file containing toolbars. In this file listing, ID specifies the name tag.

```
MENUGROUP=M1
***TOOLBARS
**TB_MyDraw1
ID_MyDraw1[_Toolbar("MyDraw1", _Floating, _Hide, 10, 200, 1)]
ID_Line [_Button("Line", ICON_16_LINE, ICON_32_LINE)]^C^C_line
ID_Pline [_Button("Pline", ICON_16_PLine, ICON_32_PLine)]^C^C_PLine
ID_Circle[_Button("Circle", ICON_16_CirRAD, ICON_32_CirRAD)]^C^C_Circle
ID_ELLIPSE[_Button("Ellipse",ICON_16_EllCEN, ICON_32_EllCEN)]^C^C_ELLIPSE
ID_Arc[_Button("Arc 3Point", ICON_16_Arc3Pt, ICON_32_Arc3Pt)]^C^C_Arc
```

Step 2: Loading the menu file containing toolbars

Save the file with the extension **.MNU**. Use the **MENULOAD** command to load the **MyDraw1** menu group as discussed earlier while loading the partial menus. To display the new toolbar (**MyDraw1**) on the screen, select **Toolbars** from the **View** menu and then select **MyDraw1** in the **Menu Groups** list box. Turn the **MyDraw1** toolbar on in the **Toolbars** list box. MyDraw1 toolbar is displayed on the screen.

You can also load the new toolbar from the command line. After using the **MENULOAD** command to load the MyDraw1 menu group, use the **-TOOLBAR** command to display the MyDraw1 toolbar.

Command: **-TOOLBAR**
Enter toolbar name or [ALL]: **MYDRAW1**
Enter an option [Show/Hide/Left/Right/Top/Bottom/Float] <Show>: **S**

Example 9

In this example you will write a menu file for a toolbar with a flyout. The name of the toolbar is MyDraw2 (Figure 25-22), and it contains two buttons, Circle and Arc. When you select the Circle button, it should display a flyout with radius, diameter, 2P, and 3P buttons (Figure 25-23). Similarly, when you select the Arc button, it should display the 3Pts, SCE, and SCA buttons.

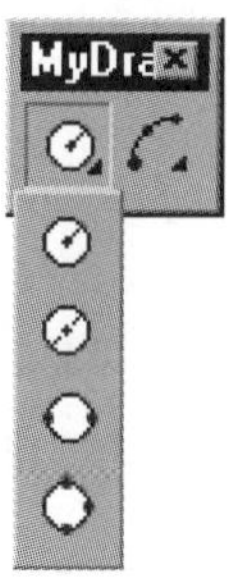

Figure 25-22 *Toolbar for Example 9*

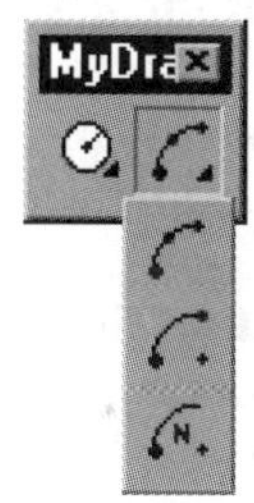

Figure 25-23 *Toolbar with flyout*

Step 1: Writing the toolbars menu file

Use any text editor to write the menu file for the toolbars. The following is the listing of the menu file.

```
***Menugroup=M2
***TOOLBARS
**TB_MyDraw2
ID_MyDraw2[_Toolbar("MyDraw2", _Floating, _Show, 10, 100, 1)]
ID_TbCircle[_Flyout("Circle", ICON_16_Circle, ICON_32_Circle, _OtherIcon,
M2.TB_Circle)]
ID_TbArc[_Flyout("Arc", ICON_16_Arc, ICON_32_Arc, _OtherIcon, M2.TB_Arc)]

**TB_Circle
ID_TbCircle[_Toolbar("Circle", _Floating, _Hide, 10, 150, 1)]
ID_CirRAD[_Button("Circle C,R", ICON_16_CirRAD, ICON_32_CirRAD)]^C^C_Circle
ID_CirDIA[_Button("Circle C,D", ICON_16_CirDIA, ICON_32_CirDIA)]^C^C_Circle;\D
ID_Cir2Pt[_Button("Circle 2Pts", ICON_16_Cir2Pt, ICON_32_Cir2Pt)]^C^C_Circle;2P
ID_Cir3Pt[_Button("Circle 3Pts", ICON_16_Cir3Pt, ICON_32_Cir3Pt)]^C^C_Circle;3P
```

```
**TB_Arc
ID_TbArc[_Toolbar("Arc", _Floating, _Hide, 10, 150, 1)]
ID_Arc3PT[_Button("Arc,3Pts", ICON_16_Arc3PT, ICON_32_Arc3PT)]^C^C_Arc
ID_ArcSCE[_Button("Arc,SCE", ICON_16_ArcSCE, ICON_32_ArcSCE)]^C^C_Arc;\C
ID_ArcSCA[_Button("Arc,SCA", ICON_16_ArcSCA, ICON_32_ArcSCA)]^C^C_Arc;\C;\A
```

Explanation

ID_TbCircle[_Flyout("Circle", ICON_16_Circle, ICON_32_Circle, _OtherIcon, M2.TB_Circle)]

In this line M2 is the MENUGROUP name (***MENUGROUP=M2) and TB_Circle is the name of the toolbar submenu. **M2.TB_Circle** will load the submenu TB_Circle that has been defined in the M2 menu group. If M2 is missing, AutoCAD LT will not display the flyout when you select the Circle button.

ID_CirDIA[_Button("Circle C,D", ICON_16_CirDIA, ICON_32_CirDIA)]^C^C_Circle;\D

CirDIA is a user-defined bitmap that displays the Circle-diameter button. If you use any other name, AutoCAD LT will not display the desired button.

Step 2: Loading the menu file containing the toolbars
To load the toolbar, use the **MENULOAD** command to load the menu file. The **MyDraw1** toolbar will be displayed on the screen.

MENU-SPECIFIC HELP

AutoCAD LT for Windows allows access to online help. For example, if you want to define a helpstring for the **CIRCLE** and **ARC** commands, the syntax is as follows:

***HELPSTRINGS
ID_Copy *[This command will copy the selected object.]*
ID_Ellipse *[(This command will draw an ellipse.]*

The ***HELPSTRING** is the section label for the helpstring menu section. The lines defined in this section start with a name tag (ID_Copy) and are followed by the label enclosed in square brackets. The Helpstring tag names are limited to twelve characters. When a menu item is highlighted, AutoCAD LT searches for the name tag for the corresponding entry in the ***HELPSTRINGS section. If there is a match, the text string contained within the label is displayed in the status line.

CUSTOMIZING THE TOOLBARS

AutoCAD LT has provided several toolbars that should be sufficient for general use. However, sometimes you may need to customize the toolbars so that the commands that you use frequently are grouped in one toolbar. This saves time in selecting commands. It also saves the drawing space because you do not need to have several toolbars on the screen. The following example explains the process involved in creating and editing the toolbars.

Example 10

In this example you will create a new toolbar (My Toolbar1) that has Line, Polyline, Circle (Center, Radius option), Arc (Center, Start, End option), Spline, and Paragraph Text (**MTEXT**) commands. You will also change the image and tooltip of the Line button and perform other operations like deleting toolbars and buttons and copying buttons between toolbars.

Step1

Select the **Toolbars** in the **View** menu. The **Customize** dialog box will appear on the screen with **Toolbars** tab active.

Step 2

Choose the **New** button in the **Toolbars** tab. AutoCAD LT will display the **New Toolbar** dialog box at the top of the existing (**Customize**) dialog box.

Step 3

Enter the name of the toolbar in the **Toolbar name** edit box (Example My Toolbar1) and then choose the **OK** button to exit the dialog box. The name of the new toolbar (My Toolbar1) is displayed in the **Customize** dialog box (Figure 25-24).

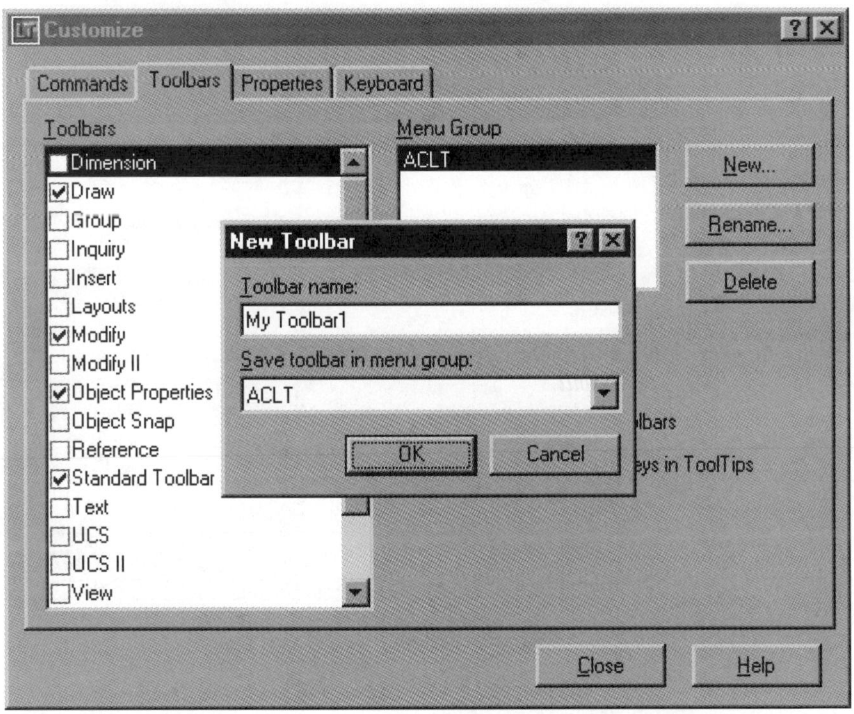

*Figure 25-24 **Customize and New Toolbar** dialog box*

Step 4

Select the new toolbar so that it is highlighted, if it is not already selected, by selecting the

Chapter 25

check box that is located just to the left of the toolbar name (My Toolbar1). The new toolbar (My Toolbar1) appears on the screen.

Step 5

Choose the **Commands** tab in the **Customize** dialog box. The **Categories** and **Commands** list boxes are displayed in the dialog box.

Step 6

In the **Categories** list box select the Draw item. The draw commands are displayed in the Commands area of the **Customize** dialog box.

Step 7

Choose and drag the Line command and position it in the My Toolbar1 toolbar. Repeat the same for Polyline, Circle (Center, Radius option), Arc (Center, Start, End option), Spline, and Text (Paragraph text) buttons.

Step 8

Now, select the Dimension item in the Category list box and then select and drag some of the dimensioning commands to My Toolbar1. See Figure 25-25.

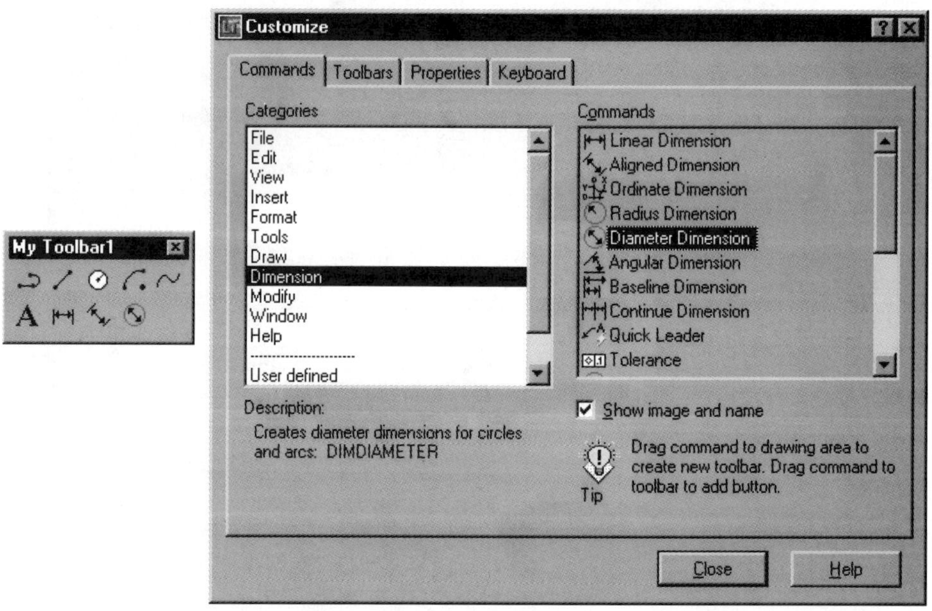

***Figure 25-25** MyToolbar1 toolbar and **Customize Toolbars** dialog box*

Step 9

Similarly, you can open any command category and add commands to the new toolbar.

Step 10

When you are done defining the commands for the new toolbar, choose the **Close** button in the **Customize Toolbars** dialog box to return to the AutoCAD LT screen.

Step 11
Test the buttons in the new toolbar (My Toolbar1).

Creating a New Image and Tooltip for a Button

Step 1
Make sure the button with the image you want to edit is displayed on the screen. In this example we want to edit the image of the **Line** button of My Toolbar1

Step 2
Right-click on any toolbar and select **Customize**; the **Customize** dialog box appears on the screen. Click on the **Properties** tab. Initially, this tab will not provide any option. Select the **Line** button in the **My Toolbar1** toolbar. The **Properties** tab will be converted into the **Button Properties** tab and the button properties are displayed with the image of the existing **Line** button (Figure 25-26).

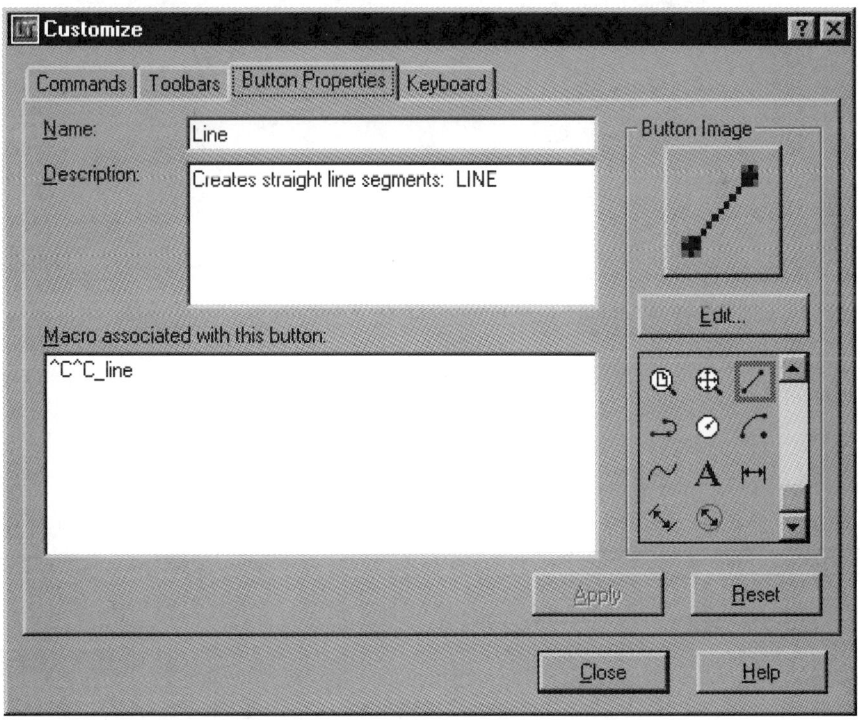

Figure 25-26 ***Button Properties*** *tab of the* ***Customize*** *dialog box*

Step 3
To edit the shape of the image, choose the **Edit** button. The **Button Editor** dialog box will be displayed. Select the Grid to display the grid lines.

You can edit the shape by using different tools in the **Button Editor**. You can draw a Line by choosing the **Line** button and specifying two points. You can draw a circle or an ellipse by using the **Circle** button. The **Erase** button can be used to erase the image.

Step 4

In this example, change the color of Line image. To accomplish this, erase the existing line, and then select the color and draw a line. Also, create the shape L in the lower right corner (Figure 25-27).

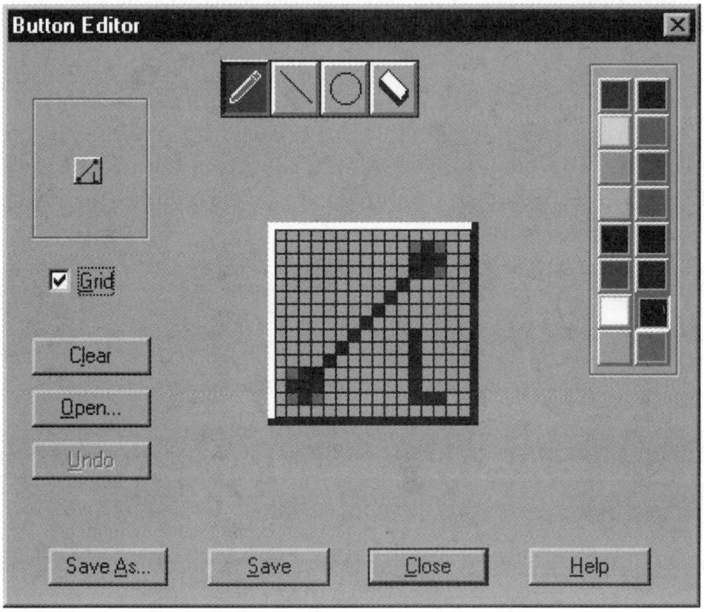

Figure 25-27 ***Button Editor*** *dialog box*

Step 5

Choose the **SaveAs** button and save the image as **MyLine** in the Tutorial directory. Choose the **Close** button to exit the **Button Editor**. Using Save instead of SaveAs will redefine the button image for all existing toolbars

Step 6

In the **Button Properties** tab of the **Customize** dialog box, enter **MyLine** in the **Name** edit box. This changes the tooltip of this button. Now, choose the **Apply** button to apply the changes to the image. Close the dialog boxes to return to the AutoCAD LT screen. Notice the change in the button image and tooltip (Figure 25-28).

Figure 25-28 *MyToolbar1 Toolbar*

Deleting a Button from the Toolbar

Step 1

Right-click on any toolbar to display the shortcut menu. Select **Customize** from the shortcut menu to display the **Customize** dialog box.

Step 2
Click and drag out the **Spline** button from the **My Toolbar1** toolbar. You will be prompted to confirm the deletion of the selected button from the toolbar. If you choose **OK**, the button will be deleted from the toolbar. Repeat the procedure to delete other buttons, if needed. Close the dialog boxes to return to the screen.

Deleting a Toolbar

Select **Toolbars** from the **View** menu. In the **Toolbars** list box select the toolbar you want to delete and then choose the **Delete** button. The toolbar you selected is deleted.

Copying a Tool Button

Step 1
In this example you will copy the **Ordinate Dimension** button from the **Dimension** toolbar to **My Toolbar1** toolbar. Choose **Toolbars** from the **View** menu or right-click on any button in any toolbar to display the shortcut menu and then select **Customize** to display the **Customize** dialog box. In the **Customize** dialog box select the **Toolbars** tab and then select the **Dimension** toolbar from the **Toolbars** list box. The **Dimension** toolbar is displayed.

Step 2
Hold the CTRL key down and then drag the **Ordinate Dimension** button from the **Dimension** toolbar to **My Toolbar1**. The **Ordinate Dimension** button is copied to **My Toolbar1** toolbar.

Note
If you do not hold the CTRL key down and drag the button from one toolbar to the other, the button will be moved from one toolbar to the other instead of getting copied.

Any changes made in the toolbars are saved in the ACLT.MNS and ACLT.MNR files. The following is the partial listing of ACLT.MNS file:

```
**MYTOOLBAR1
ID_MyToolbar1_0 [_Toolbar("MyToolbar1", _Floating, _Show, 512, 177, 1)]
ID_Line_0     [_Button("MyLine", "ICON.bmp", "ICON_24_LINE")]^C^C_line
ID_CircleCenterRadius_0 [_Button("Circle Center Radius", "ICON_16_CIRRAD",
"ICON_24_CIRRAD")]^C^C_circle
ID_Polyline_0 [_Button("Polyline", "ICON_16_PLINE", "ICON_24_PLINE")]^C^C_pline
ID_ArcCenterStartEnd_0 [_Button("Arc Center Start End", "ICON_16_ARCCSE",
"ICON_24_ARCCSE")]^C^C_arc _c
```

Creating Custom Toolbars with Flyout Icons

In this example you will create a custom toolbar with flyout buttons.

Step 1
Right-click on any toolbar and choose Customize from the shortcut menu to display the **Customize** dialog box on the screen. In the **Customize** dialog box, select the **Toolbars** tab and

choose the **New** button to display the **New Toolbar** dialog box. Enter the name of the new toolbar (for example, My Toolbar2) and then choose the **OK** button to exit the box.

Step 2

In the Customize dialog box select the Commands tab and then select the Flyouts from the Categories list box. The names of the flyouts are displayed in the Commands list box. Select and drag the **Draw** flyout button and drop it in the **My Toolbar2** toolbar. The flyout created this way contains all draw commands.

Step 3

In the **Customize** dialog box select the **Commands** tab and then select the **User defined** from the **Categories** list box. The names of the commands are displayed in the **Commands** list box. Select and drag the **User Defined Flyout** button and drop it in the **My Toolbar2** toolbar. Now select the user defined button in **My Toolbar2**; the AutoCAD LT message box appears on the screen. Click on the OK button to return to **Customize** toolbar. In the **Flyout Properties** tab select **Inquiry** and then choose the **Apply** button. The **Inquiry** toolbar gets associated with the custom flyout. Similarly, add some more flyouts to the **My Toolbar2** toolbar (Figure 25-29).

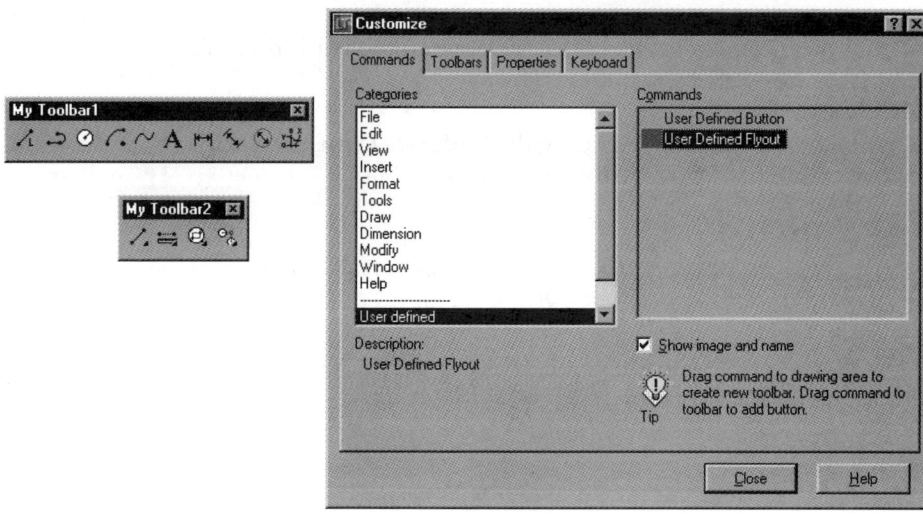

Figure 25-29 *My Toolbar1 toolbar, My Toolbar2 toolbar and* ***Customize*** *dialog box*

Assigning Keyboard Shortcuts to Commands

Step 1

Select the Toolbars from the **View** menu or right-click on any button in any toolbar to display the shortcut menu and then select **Customize** to display the **Customize** dialog box. In the **Customize** dialog box select the **Keyboard** tab. Now, select the **Draw Menu** from the **Categories** drop-down list as shown in Figure 25-30. The draw commands are displayed in the **Commands** list box.

Step 2

In the **Press new shortcut key** edit box, specify a key combination to be used as the keyboard shortcut for the selected menu or toolbar item. To specify a value, simultaneously press CTRL and a letter on the keyboard. You can also simultaneously press CTRL+SHIFT and a letter. In this example, hold the CTRL and SHIFT keys down and press the L key from the keyboard. You can also click on the **Show All** button to display all the shortcut keys.

Step 3

One the combination of the shortcut keys are displayed in the **Press new shortcut key** edit box, choose the **Assign** button. This will assign the shortcut key to the selected command. Choose the Close button to close the Customize dialog box. Now, if you hold the CTRL and the SHIFT key down and press the L key from the keyboard, the LINE command is invoked.

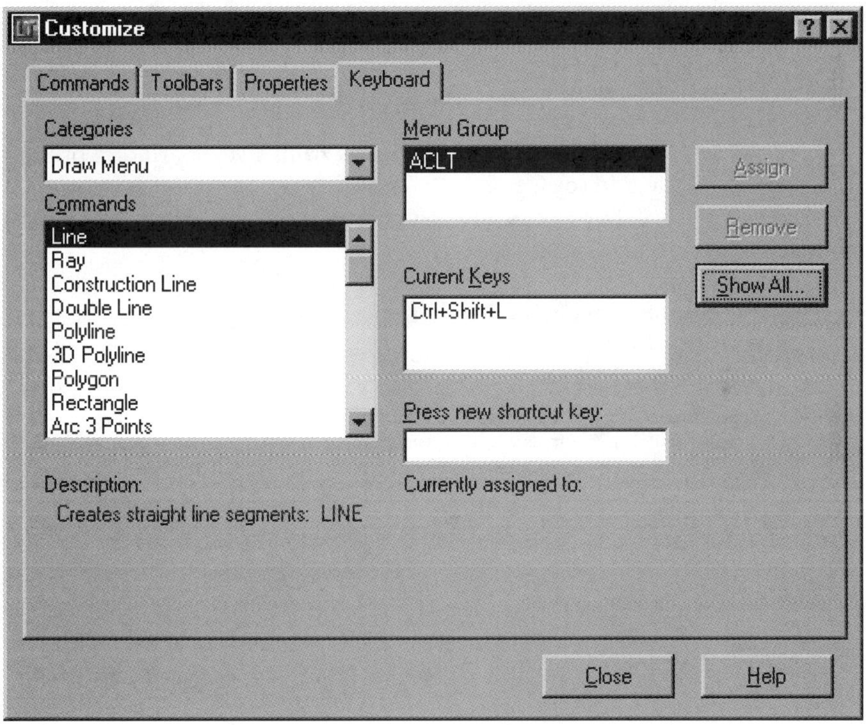

Figure 25-30 *Customize dialog box with Keyboard tab*

Note

You cannot reassign shortcut keys that are internally assigned to Windows, for example, F10, CTRL+F4, CTRL+F6, or CTRL+ALT+DEL. If you use an invalid key combination, AutoCAD LT does not show the combination in the ***Press new shortcut key*** *edit box. Try using another key combination.*

If the keyboard shortcut you specify is already assigned to another AutoCAD LT command, the "Currently assigned to" message is displayed. However, if you still choose the ***Assign*** *button, this combination will be assigned to the command currently selected. This combination of shortcut keys will not work for the previous command.*

Self-Evaluation Test

Answer the following questions and then compare your answers to the correct answers at the end of this chapter.

1. The name of the standard menu file is ________________ .
2. Menu files can be loaded by using the __________ command.
3. Each section in the menu file is identified by a __________.
4. The pop menu can have up to __________.
5. The tablet menus can have up to __________ different sections.
6. __________ can be given if you need to use input from a keyboard or a pointing device.
7. The sign __________ defines a cascaded submenu while __________ designates the last item in the menu.
8. All section labels in the AutoCAD LT menu start with __________.
9. __________ designates the last item of the pull-down or cursor menu.
10. __________ when used as a prefix, displays the item with a check mark.
11. The system variable __________ controls the availability of the default, edit and command mode shortcut menus.
12. A submenu definition consists of __________.
13. The Customize dialog box can be invoked by either __________ or __________.

Review Questions

Answer the following questions.

1. A pull-down menu can have ________________ sections.
2. The length of the section title should not exceed ________________ characters.
3. The section titles in a pull-down menu are __________ justified.
4. In a pull-down menu, a line consisting of two hyphens ([--]) ________________ automatically to fill the ________________ of the menu.

5. If the menu item begins with a tilde (~), the items will be ________________ .

6. Every cascading menu in the menu file should have a ________________ name.

7. The cascading menu name can be ________________ characters long.

8. The cascading menu names should not have any ________________ blanks.

9. In Windows you can write partial menus, toolbars, and accelerator key definitions. (T/F)

10. A menu file can contain only one MENUGROUP. (T/F)

11. You can load the partial menu file by using the AutoCAD LT ___________ command.

Exercises

Exercise 3 *General*

Write a pull-down menu for the following AutoCAD LT commands. (The layout of the menu is shown in Figure 25-31.)

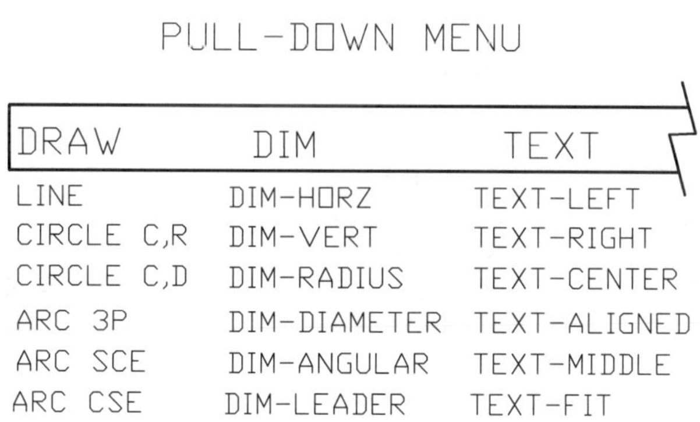

Figure 25-31 *Layout of menu*

LINE
CIRCLE C,R
CIRCLE C,D
ARC 3P
ARC SCE
ARC CSE

DIMLINEAR
DIMALIGNED
DIMRADIUS
DIMDIAMETER
DIMANGULAR
QLEADER

TEXT LEFT
TEXT RIGHT
TEXT CENTER
TEXT ALIGNED
TEXT MIDDLE
TEXT FIT

Exercise 4 *General*

Write a pull-down menu for the following AutoCAD LT commands.

LINE
PLINE
CIRCLE C,R
CIRCLE C,D
ELLIPSE AXIS ENDPOINT
ELLIPSE CENTER
BLOCK
WBLOCK
INSERT
BLOCK LIST
ATTDEF
ATTEDIT

Exercise 5 *General*

Write a partial menu for Windows. The menu file should have two menus, POP1 (MyArc) and POP2 (MyDraw). The MyArc menu should contain all Arc options and must be displayed at the sixth position. Similarly, the MyDraw menu should contain **LINE, CIRCLE, PLINE, TEXT,** and **MTEXT** commands and should occupy the ninth position.

Exercise 6 *General*

Write a menu file for a toolbar with a flyout. The name of the toolbar is MyDrawX1, and it contains two buttons, **Draw** and **Modify**. When you choose the **Draw** button, it should display a flyout with all draw buttons (**Draw** commands). Similarly, when you choose the **Modify** button, it should display a flyout with all the modify buttons (**Modify** commands).

Exercise 7 *General*

Write a menu for the following AutoCAD LT commands.

LAYER NEW
LAYER MAKE
LAYER SET
LAYER LIST
LAYER ON
LAYER OFF
SNAP 0.25
SNAP 0.5
GRID 1.0
GRID 10.0
APERTURE 5
PICKBOX 5
UCS WORLD
UCS PREVIOUS
VPORTS 2
VPORTS 4
VPORTS SING.

Answers to the Self-Evaluation Test

1 - **ACLT.MNU**, **2** - **MENU**, **3** - Section label, **4** - 499 sections, **5** - 4 sections, **6** - backslash, **7** - ->, <-, **8** - ***, **9** - <-<-, **10** - !, **11** - **SHORTCUTMENU**, **12** - **, **13** - choosing **Customize** from the **Tools** menu, right-click on any toolbar and then choose **Customize** from the shortcut menu.

Chapter 26

AutoCAD LT on the Internet

Learning Objectives

After completing this chapter, you will be able to:

- *Launch a Web browser from AutoCAD LT.*
- *Understand the importance of the uniform resource locator (URL).*
- *Use the **AutoCAD LT Today** window for connecting to Autodesk Point A.*
- *Open and save drawings to and from the Internet.*
- *Place hyperlinks in a drawing.*
- *Learn about the Whip plug-in and the DWF file format.*
- *View DWF files and convert drawings to a DWF file format.*

INTRODUCTION

The Internet has become the most important way to exchange information in the world. AutoCAD LT allows you to interact with the Internet in several ways. It can open and save files that are located on the Internet; it can launch a Web browser (AutoCAD LT 2002 includes a simple Web browser); and it can create Drawing Web Format (DWF) files for viewing as drawings on Web pages. Before you can use the Internet features in AutoCAD LT 2002, some components of Microsoft's Internet Explorer must be present in your computer. If Internet Explorer version 4.0 (or later) is already installed on your computer, then you have these required components. If not, then the components are installed automatically during AutoCAD LT 2002's setup when you select: (1) the **Full Install**; or (2) the **Internet Tools** option during the **Custom** installation.

This chapter introduces you to the following Web-related commands:

Browser

Launches a Web browser from within AutoCAD LT.

Hyperlink

Attaches and removes a uniform resource locator (URL) to an object or an area in the drawing.

Hyperlinkfwd

Move to the next hyperlink (an undocumented command).

Hyperlinkback

Moves back to the previous hyperlink (an undocumented command).

Hyperlinkstop

Stops the hyperlink access action (an undocumented command).

Pasteashyperlink

Attaches a URL to an object in the drawing from text stored in the Windows Clipboard (an undocumented command).

Hyperlinkbase

A system variable for setting the path used for relative hyperlinks in the drawing.

In addition, all of AutoCAD LT 2002's file-related dialog boxes are "Web enabled."

CHANGES FROM AUTOCAD LT RELEASE 97

If you used the Internet features in AutoCAD LT Release 97, then you should be aware of changes in AutoCAD LT 2002.

Attached URLs from LT 97

In AutoCAD LT R14, attached URLs were not active until the drawing was converted to a DWF file; in AutoCAD LT 2002, URLs are active in the drawing.

If you attached URLs to drawings in LT 97 or in LT 98, they are converted to AutoCAD LT 2002-style hyperlinks the first time you save the drawings in the AutoCAD LT 2002 DWG format.

When you use the **SAVEAS** command to save an AutoCAD LT 2002 drawing in LT 97 format, hyperlinks are converted back to LT 97-style URLs. Hyperlink descriptions are stored as proxy data and not displayed in LT 97 (this ensures that hyperlink descriptions are available again when the drawing is brought back to AutoCAD LT 2002 format.)

WHIP! V 3.1 LIMITATIONS

AutoCAD LT 2002 can create DWF files compatible with the WHIP v3.1 plug-in. It is recommended, however, that you download WHIP v4.0 from the Autodesk Web site. Whip v3.1 does not display the following AutoCAD LT 2002-specific features:

- Hyperlink descriptions are not saved in v3.1 DWF format.
- All objects, except polylines, are displayed at minimum lineweight.
- Linetypes defined by a plot style table are displayed as continuous lines.

CHANGED INTERNET COMMANDS

Many of AutoCAD LT's previous release Internet-related commands have been discontinued and replaced. The following summary describes the status of all LT 97 Internet commands:

BROWSER (launches a Web browser from within AutoCAD LT): continues to work in AutoCAD LT 2002; the default URL is now http://www.autodesk.com.

ATTACHURL (attaches a URL to an object or an area in the drawing): continues to work in AutoCAD LT 2002, but has been superceded by the **-HYPERLINK** command's **Insert** option.

SELECTURL (selects all objects with attached URLs): continues to work in AutoCAD LT 2002, but has been superceded by the **QSelect** dialog box's **Hyperlink** option.

LISTURL (lists URLs embedded in the drawing): removed from AutoCAD LT 2002. As a replacement, use the **Properties** dialog box's **Hyperlink** option.

DETACHURL (removes the URL from an object): continues to work in AutoCAD LT 2002, but has been superceded by the **-HYPERLINK** command's **Remove** option.

DWFOUT (exports the drawing and embedded URLs as a DWF file): continues to work in AutoCAD LT 2002, but has been superceded by the **Plot** dialog box's **ePlot** option.

DWFOUTD (exports the drawing in DWF format without a dialog box): removed from AutoCAD LT 2002. As a replacement, use the **-PLOT** command's **ePlot** option.

INETCFG (configures AutoCAD LT for Internet access): removed from AutoCAD LT 2002; it has been replaced by the **Internet** applet of the Windows Control Panel (choose the **Connection** tab).

INSERTURL (inserts a block from the Internet into the drawing): automatically executes the **INSERT** command and displays the **Insert** dialog box. You may type a URL for the block name.

OPENURL (opens a drawing from the Internet): automatically executes the **OPEN** command and displays the **Select File** dialog box. You may type a URL for the file name.

SAVEURL (saves the drawing to the Internet): automatically executes the **SAVE** command and displays the **Save Drawing As** dialog box. You may type a URL for the file name.

You are probably already familiar with the best-known uses for the Internet: e-mail (electronic mail) and the Web (short for "World Wide Web"). E-mail lets users exchange messages and data at very low cost. The Web brings together text, graphics, audio, and movies in an easy-to-use format. Other uses of the Internet include file transfer protocol (FTP) for effortless binary file transfer, Gopher (presents data in a structured, subdirectory-like format), and USENET, a collection of more than 10,000 news groups.

AutoCAD LT allows you to interact with the Internet in several ways. AutoCAD LT is able to launch a Web browser from within AutoCAD LT with the **BROWSER** command. Hyperlinks can be inserted in drawings with the **HYPERLINK** command, which lets you link the drawing with other documents on your computer and the Internet. With the **PLOT** command's **ePlot** option (short for "electronic plot), AutoCAD LT creates DWF (short for "drawing Web format") files for viewing drawings in 2D format on Web pages. AutoCAD LT can open, insert, and save drawings to and from the Internet via the **OPEN**, **INSERT**, and **SAVEAS** commands.

UNDERSTANDING URLs

The **uniform resource locator**, known as the URL, is the file-naming system of the Internet. The URL system allows you to find any resource (a file) on the Internet. Example resources include a text file, a Web page, a program file, an audio or movie clip—in short, anything you might also find on your own computer. The primary difference is that these resources are located on somebody else's computer. A typical URL looks like the following examples:

ExampleURL	Meaning
http://www.autodesk.com	Autodesk Primary Web Site
news://adesknews.autodesk.com	Autodesk News Server
ftp://ftp.autodesk.com	Autodesk FTP Server
http://www.autodeskpress.com	Autodesk Press Web Site
http://users.uniserve.com/~ralphg	Editor Ralph Grabowski's Web site

Note that the **http://** prefix is not required. Most of today's Web browsers automatically add in the *routing* prefix, which saves you a few keystrokes.

URLs can access several different kinds of resources such as Web sites, e-mail, and news groups, but they always take on the same general format as follows:

scheme://netloc

The scheme accesses the specific resource on the Internet including these:

Scheme	Meaning
file://	File is located on your computer's hard drive or local network
ftp://	File Transfer Protocol (used for downloading files)
http://	Hyper Text Transfer Protocol (the basis of Web sites)
mailto://	Electronic mail (e-mail)
news://	Usenet news (news groups)
telnet://	Telnet protocol
gopher://	Gopher protocol

The **://** characters indicate a network address. Autodesk recommends the following format for specifying URL-style file names with AutoCAD LT:

<table>
<tr><th>Resource</th><th>URL Format</th></tr>
<tr><td>Web Site</td><td>http://servername/pathname/filename</td></tr>
<tr><td>FTP Site</td><td>ftp://servername/pathname/filename</td></tr>
<tr><td>Local File</td><td>file:///drive:/pathname/filename</td></tr>
<tr><td>or</td><td>drive:\pathname\filename</td></tr>
<tr><td>or</td><td>file:///drive | /pathname/filename</td></tr>
<tr><td>or</td><td>file://\\localPC\pathname\filename</td></tr>
<tr><td>or</td><td>file:////localPC/pathname/filename</td></tr>
<tr><td>Network File</td><td>file://localhost/drive:/pathname/filename</td></tr>
<tr><td>or</td><td>\\localhost\drive:\pathname\filename</td></tr>
<tr><td>or</td><td>file://localhost/drive | /pathname/filename</td></tr>
</table>

The terminology can be confusing. The following definitions will help to clarify these terms.

Term	Meaning
servername	The name or location of a computer on the Internet, for example: *www.autodesk.com*
pathname	The same as a subdirectory or folder name
drive	The driver letter, such as C: or D:
localpc	A file located on your computer
localhost	The name of the network host computer

If you are not sure of the name of the network host computer, use Windows Explorer to check the Network Neighborhood for the network names of computers.

Launching a Web Browser

The **BROWSER** command lets you start a Web browser from within AutoCAD LT. Commonly used Web browsers include Netscape's Navigator, Microsoft's Internet Explorer, and Operasoft's Opera.

By default, the **BROWSER** command uses whatever brand of Web browser program is registered in your computer's Windows operating system. AutoCAD LT prompts you for the uniform resource locator (URL), such as *http://www.autodeskpress.com*. The **BROWSER** command can be used in scripts, toolbars, or menu macros, and AutoLISP routines to automatically access the Internet.

Command: **BROWSER**
Enter Web location (URL) <http://www.autodesk.com/acltuser>: *Enter a URL.*

The default URL is an HTML file added to your computer during AutoCAD LT's installation. After you type the URL and press *, AutoCAD LT launches the Web browser and contacts the Web site. Figure 26-1 shows the popular Internet Explorer with the Autodesk Web site.

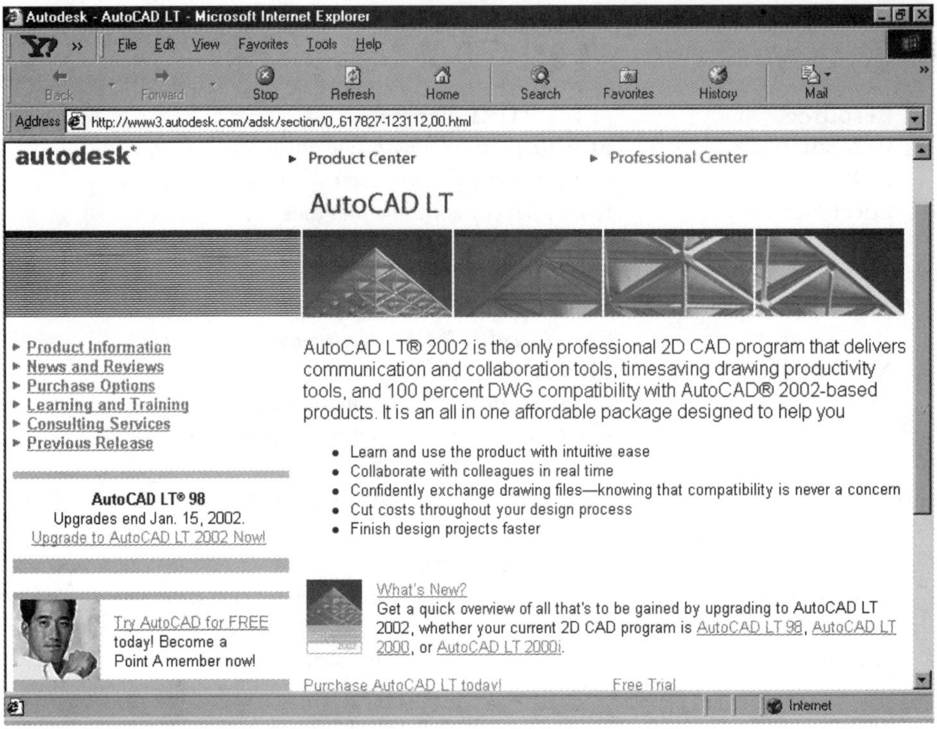

Figure 26-1 Internet Explorer displaying the Autodesk Web Site

Changing the Default Web Site

To change the default Web page that your browser starts with from within AutoCAD LT, change the setting in the **INETLOCATION** system variable. The variable stores the URL used by the **BROWSER** command and the **Browse the Web** dialog box. Make the change, as follows:

Command: **INETLOCATION**
Enter new value for INETLOCATION <"http://www.autodesk.com/acltuser">: *Type URL.*

DRAWINGS ON THE INTERNET

When a drawing is stored on the Internet, you access it from within AutoCAD LT 2002 using the standard **OPEN**, **INSERT**, and **SAVE** commands. (In AutoCAD LT 97, these commands were known as **OPENURL**, **INSERTURL**, and **SAVEURL**.) Instead of specifying the file's location with the usual drive-subdirectory-file name format such as c:\aclt2002\filename.dwg, use the URL format. (Recall that the URL is the universal file-naming system used by the Internet to access any file located on any computer hooked up to the Internet.)

Opening Drawings from the Internet

The drawings from the Internet can be easily opened using the usual **Select File** dialog box. This dialog box can be displayed by invoking the **OPEN** command. Since the file is on the Web, therefore, choose the **Search the Web** button from the **Select File** dialog box (Figure 26-2).

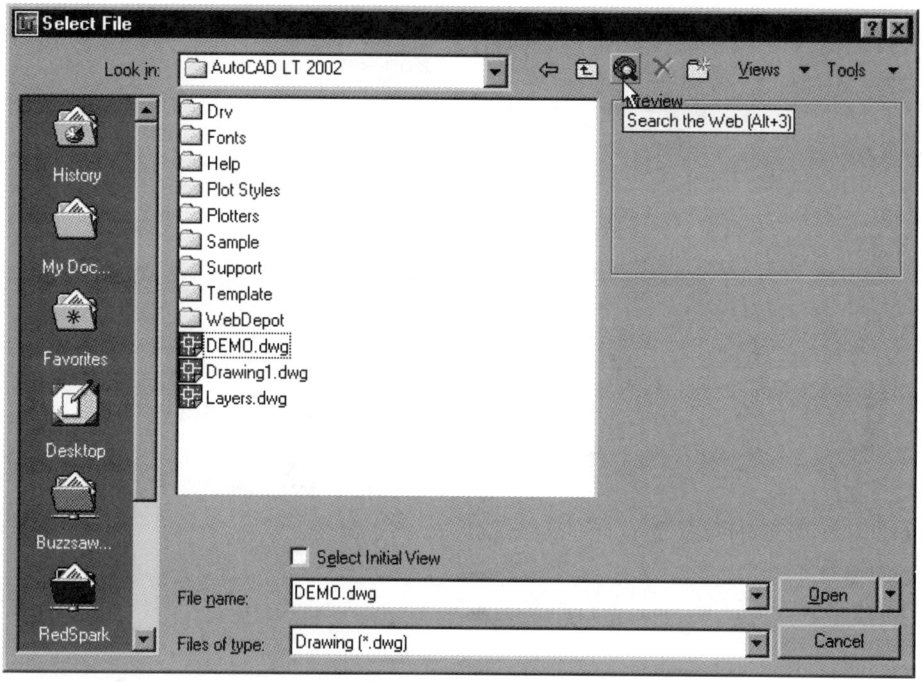

*Figure 26-2 Choosing the **Search the Web** button from the **Select File** dialog box*

When you choose the **Search the Web** button, AutoCAD LT opens the **Browse the Web** dialog box (Figure 26-3). This dialog box is a simplified version of the Microsoft brand of Web browser. The purpose of this dialog box is to allow you to browse files at a Web site.

By default, the **Browse the Web** dialog box displays the contents of the URL stored in the **INETLOCATION** system variable. You can easily change this to another folder or Web site, by entering the required URL in the **Look in** edit box.

Along the top, the dialog box has six buttons:

Back. Go back to the previous URL.

Forward. Go forward to the next URL.

Stop. Halt displaying the Web page (useful if the connection is slow or the page is very large.)

Refresh. Redisplay the current Web page.

Home. Return to the location specified by the **INETLOCATION** system variable.

Favorites. List stored URLs (hyperlinks) or bookmarks. If you have previously used Internet Explorer, you will find all your favorites listed here. Favorites are stored in the **\Windows\Favorites** folder on your computer.

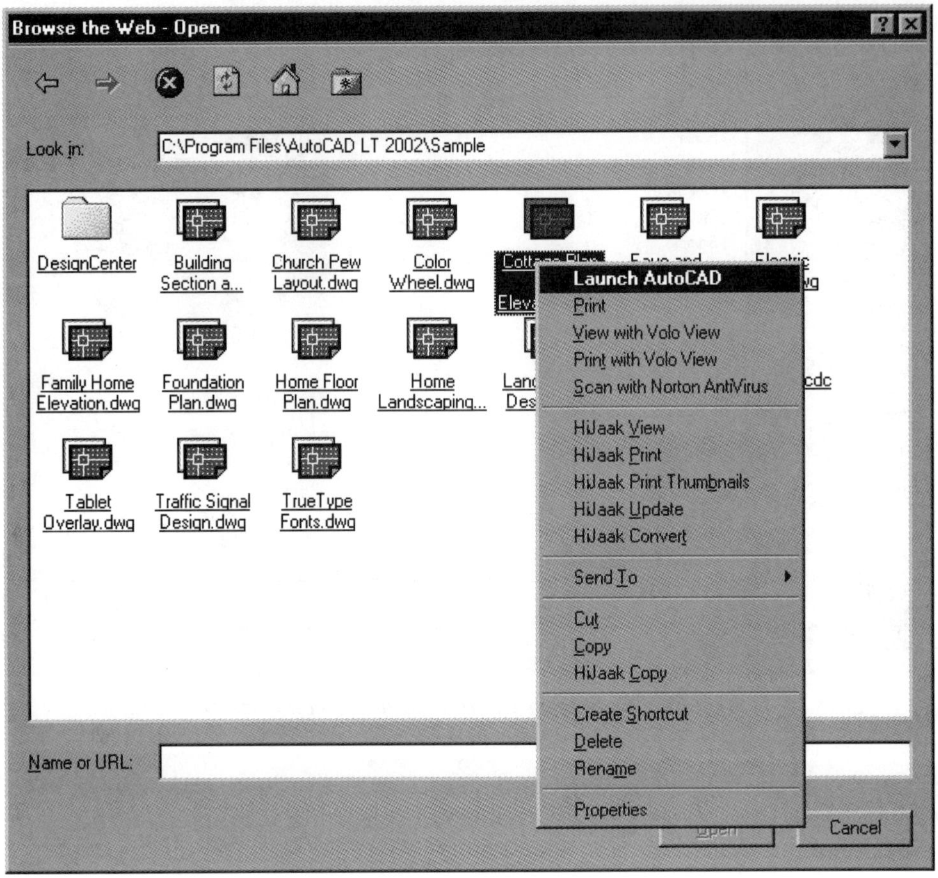

Figure 26-3 ***Browse the Web*** *dialog box*

The **Look in** field allows you to type the URL. Alternatively, click the down arrow to select a previous destination. If you have stored Web site addresses in the Favorites folder, then select a URL from that list.

You can either double-click a file name in the window, or type a URL in the **Name or URL** edit box. The following table gives templates for typing the URL to open a drawing file:

Drawing Location	Template URL
Web or HTTP Site	*http://servername/pathname/filename.dwg*
	http://practicewrench.autodeskpress.com/wrench.dwg
FTP Site	*ftp://servername/pathname/filename.dwg*
	ftp://ftp.autodesk.com
Local File	*drive:\pathname\filename.dwg*
	c:\program files\autocad lt 2002\sample\tablet2002.dwg
Network File	*\\localhost\drive:\pathname\filename.dwg*
	\\upstairs\d:\install\sample.dwg

When you open a drawing over the Internet, it will probably take much longer than opening a file found on your computer. During the file transfer, AutoCAD LT displays a dialog box to report the progress (Figure 26-4). If your computer uses a 28.8 Kbps modem, you should allow about 5 to 10 minutes per megabyte of drawing file size. If your computer has access to a faster T1 connection to the Internet, you should expect a transfer speed of about 1 minute per megabyte.

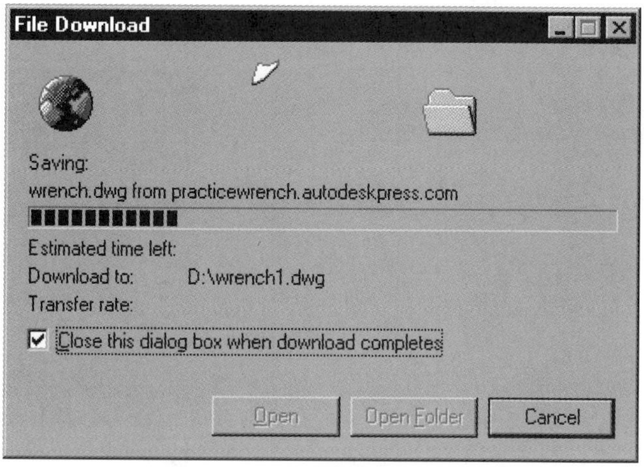

***Figure 26-4 File Download** dialog box*

It may be helpful to understand that the **OPEN** command does not copy the file from the Internet location directly into AutoCAD LT. Instead, it copies the file from the Internet to your computer's designated **Temporary** subdirectory such as C:\Windows\Temp (and then loads the drawing from the hard drive into AutoCAD LT). This is known as *caching*. It helps to speed up the processing of the drawing, since the drawing file is now located on your computer's fast hard drive, instead of the relatively slow Internet.

Note that the **Locate** and **Find File** buttons (in the **Select File** dialog box) do not work for locating files on the Internet.

Example 1 — *General*

The Autodesk Press Web site has an area that allows you to practice using the Internet with AutoCAD LT. In this example, you open a drawing file located at that Web site.

1. Start AutoCAD LT.

2. Ensure that you have a live connection to the Internet. If you normally access the Internet via a telephone (modem) connection, dial your Internet service provider now.

3. Choose the **Open** button from the **Standard Toolbar**. The **Select File** dialog box will be displayed.

4. Choose the **Search the Web** button to display the **Browse the Web** dialog box.

5. In the **Look in** field, type:

   practicewrench.autodeskpress.com

   Press ENTER. After a few seconds, the **Browse the Web** dialog box displays the Web site (Figure 26-5).

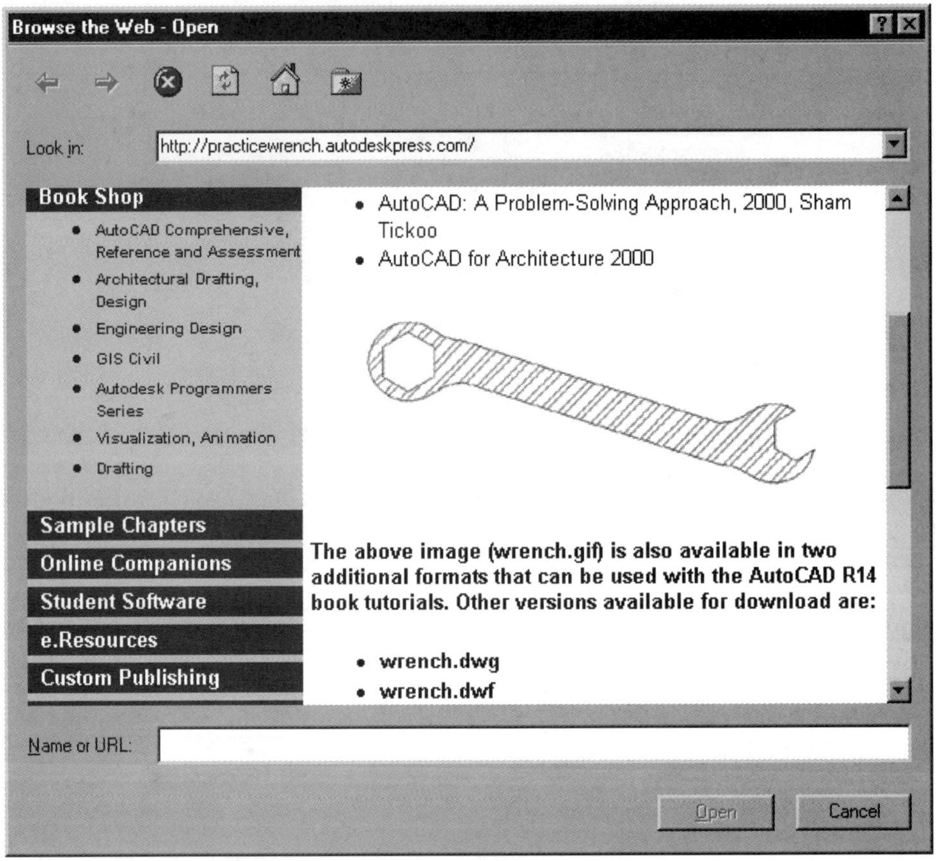

Figure 26-5 *Practice Wrench Web site*

6. In the **Name or URL** edit box, enter the name **Wrench.dwg** and press ENTER. AutoCAD LT begins transferring the file. Depending on the speed of your Internet connection, this will take between a couple of seconds and a half-minute. Once the file download is complete, the drawing of the wrench will be displayed in AutoCAD LT (Figure 26-6).

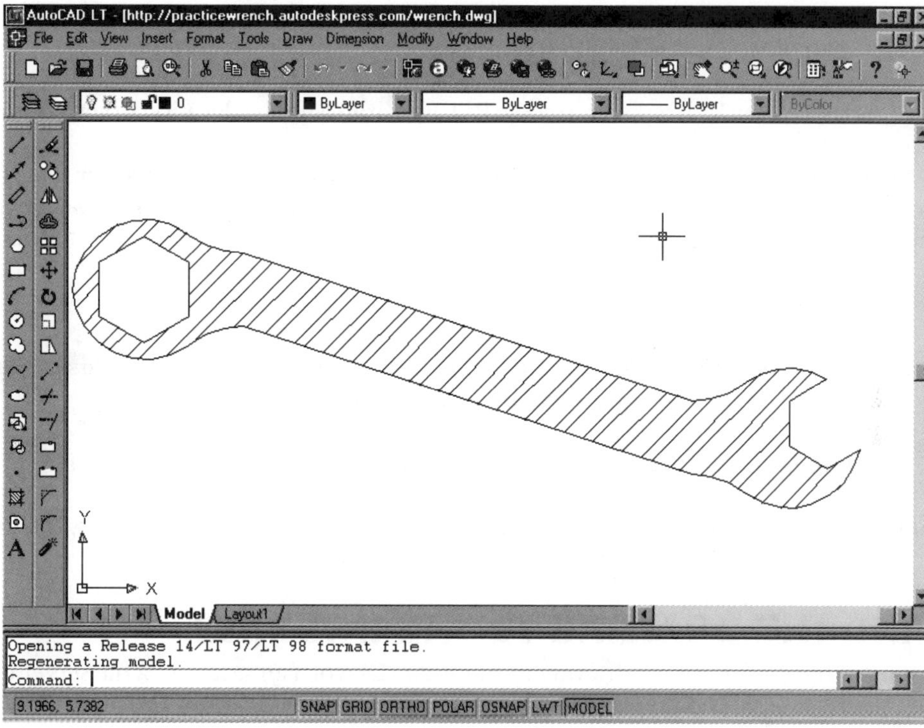

Figure 26-6 *Wrench drawing in AutoCAD LT*

INSERTING A BLOCK FROM THE INTERNET

When a block (symbol) is stored on the Internet, you can access it from within AutoCAD LT using the **INSERT** command. When the **Insert** dialog box appears, choose the **Browse** button to display the **Select Drawing File** dialog box. This is identical to the dialog box discussed previously.

After you choose the select the file, AutoCAD LT downloads the file and continues with the **INSERT** command's familiar prompts.

The process is identical for accessing external reference (xref) and raster image files. Other files that AutoCAD LT can access over the Internet include the Windows metafiles that are saved in the **.WMF** format. This option is available in the **Insert** menu on the menu bar.

ACCESSING OTHER FILES ON THE INTERNET

Most other file-related dialog boxes allow you to access files from the Internet or an intranet. This allows your firm or agency to have a central location that stores drawing standards. When you need to use a linetype or hatch pattern, for example, you access the LIN or PAT file over the Internet. More than likely, you would have the location of these files stored in the Favorites list. Some examples include:

Linetypes. Choose **Linetype** from the **Format** menu. In the **Linetype Manager** dialog box, choose the **Load**, **File**, and **Look in Favorites** buttons.

Hatch Patterns. Use the Web browser to copy .PAT files from a remote location to your computer.

Layer Name. Choose the **Layers** button to display the **Layer Properties Manager** dialog box. In this dialog box choose the **Restore State** button to display the **Layer State Manager** dialog box. In the **Layer Manager** dialog box, choose the **Import** and **Look in Favorites** buttons.

Scripts. Choose **Run Scripts** from the **Tools** menu.

Menus. Choose **Customize > Menus** from the **Tools** menu. In the **Menu Customization** dialog box, choose the **Browse** and **Look in Favorites** buttons.

Images. Choose **Display Image** > **View** from the **Tools** menu.

You cannot access text files, text fonts (SHX and TTF), color settings, lineweights, dimension styles, plot styles, OLE objects, or named UCSs over the Internet.

SAVING THE DRAWING TO THE INTERNET

When you are finished editing a drawing in AutoCAD LT, you can save it to a file server on the Internet with the **SAVE** command. If you inserted the drawing from the Internet (using **INSERT**) into the default Drawing.Dwg drawing, AutoCAD LT insists you first save the drawing to your computer's hard drive.

When a drawing of the same name already exists at that URL, AutoCAD LT warns you, just as it does when you use the **SAVEAS** command. Recall from the **OPEN** command that AutoCAD LT uses your computer system's **Temporary** subdirectory, therefore the reference to it in the dialog box.

AUTOCAD LT TODAY WINDOW

The **AutoCAD LT Today** window is one of the major steps toward making designing in AutoCAD LT more Internet-friendly. This window is one of the most important tools that will help you access various Internet options included in the recent releases of AutoCAD LT. The **AutoCAD LT Today** window is divided into the following three areas as shown in Figure 26-7.

My Drawings Area

This area provides you with various options of opening an existing drawing or creating a new drawing. The options under this area are discussed in earlier chapters.

Bulletin Board Area

This area of the **AutoCAD LT Today** window is used to post the user-defined material. You can create your own template in the **.HTM** or the **.HTML** format and post it on this area. You can also use the example templates provided to post the material on this area. To use the example template, connect to the Internet and then choose **click here** from the last line of the

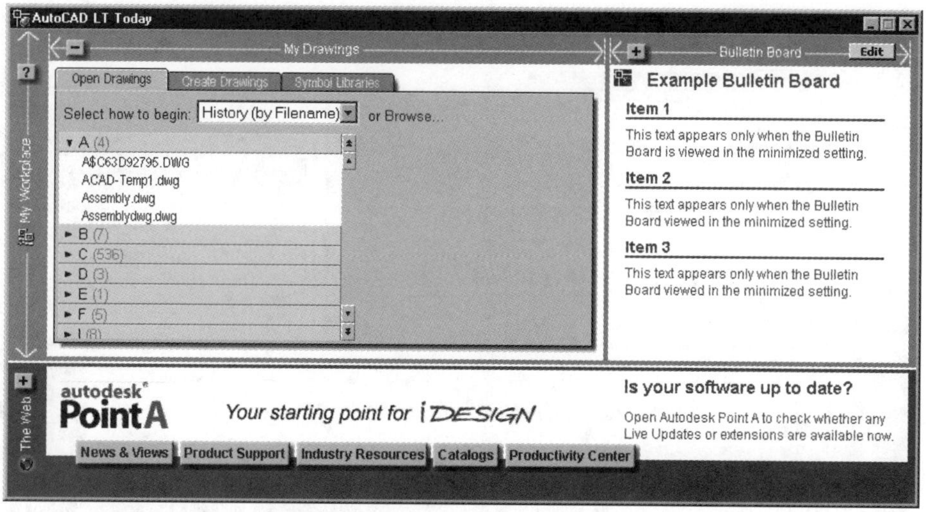

*Figure 26-7 **AutoCAD LT Today** window displaying different areas*

default **Bulletin Board** area. The Microsoft Internet Explorer will be displayed with the example template. Select the text in this template and copy in a new file. Save this new file with the **.HTM** or **.HTML** format. Now choose the **Edit** button provided on the top of the **Bulletin Board** area. In the dialog box choose the **Browse** button to select the path in which the **.HTM** or the **.HTML** file was saved. After selecting the path, choose the **Save Path** button. The **Bulletin Board** area will not display the result of the html file that you created.

Tip

*You can display the default **Bulletin Area** by choosing the **Edit** button and then clearing the path in the dialog box. After clearing the path, choose the **Save Path** button.*

The Web Area

This area utilizes the different Internet options of AutoCAD LT 2002. You can access the **autodesk Point A** Web site by clicking on it. If you are connected to the Internet, the **AutoCAD LT Today** window will be expanded and the **Autodesk Point A** Web site will be displayed. You can also find out about the various updates and extensions of the software using this area. If you have a user name and password provided by Autodesk, you can directly enter this area of the **AutoCAD LT Today** window. **The Web** area of the **AutoCAD LT Today** window allows you to access the following links in addition to the **Autodesk Point A** Web site.

News & Views

If you are connected to the Internet and choose this button, the **AutoCAD LT Today** window expands and displays the latest news and views related to the software, as shown in Figure 26-8.

Project Center

When you choose this button, you are connected to **Autodesk Point A**'s **Project Center**. This area provides you with a wide range of projects that can be easily used.

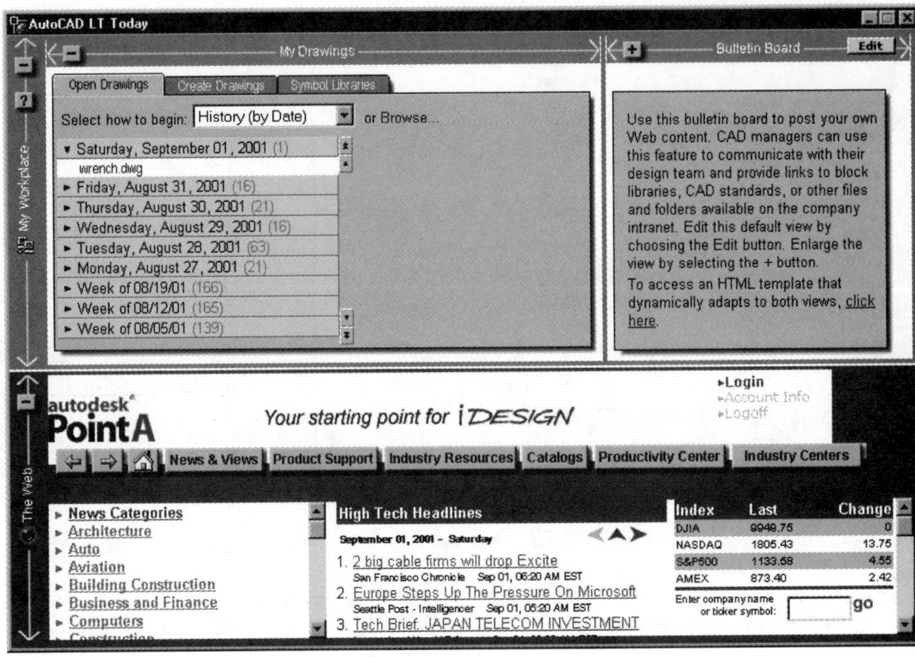

Figure 26-8 Accessing ***News & Views*** *page*

Industry Resources

You can access the **Autodesk Point A**'s **Industry Resources** using this button. You can also access the links related to units convertors, online CAD magazines, informations, articles, and so on using this button.

Catalogs

Using this button you can access the **Catalogs** page of the **Autodesk Point A** Web site. This page can be used to download the symbols or search the symbols that can be used in the drawings. You can also order for purchasing the symbol sets using this page. This area provides you with number of symbols related to the architectural, landscaping, space planning, welding, pipe fitting, and so on, fields. Click on the thumbnail of any of the symbols to view the larger picture of the symbol.

Productivity Center

This button links you to the **CAD Productivity Center** page of the **Autodesk Point A** Web site. The **CAD Productivity Center** can be used to select the links related to the product discussion group, design distribution extension, CAD software center, customizing channel, and so on. You can also locate an Autodesk Training Center or an Autodesk reseller by selecting the links from this page.

Industry Center

This button will be available only when you are connected to the Internet. You can choose this button to access the pages related to the AEC, GIS, Landscaping, Manufacturing or the General

Design industries. When you select any of these industries, the symbols, softwares, support, and solutions related to that industry are displayed. Figure 26-9 shows the links related to a manufacturing industry.

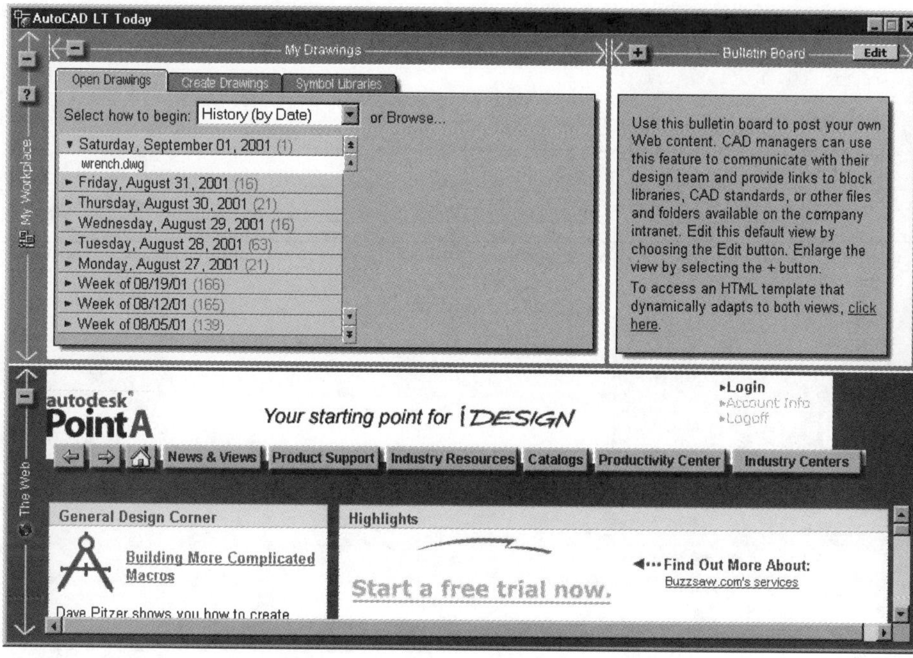

*Figure 26-9 Accessing **Industry Center's General Design** page*

Tip

*You can also access the **Autodesk Point A** Web site by choosing **Autodesk Point A** from the **Tools** menu.*

USING HYPERLINKS WITH AUTOCAD LT

AutoCAD LT 2002 allows you to employ URLs in two ways:

1. Directly within an AutoCAD LT drawing;
2. Indirectly in DWF files displayed by a Web browser.

(URLs are also known as *hyperlinks*), the term that will be used throughout this book. Hyperlinks are created, edited, and removed with the **HYPERLINK** command. You can also use the Command line with the help of the **-HYPERLINKS** command.

HYPERLINK Command

Toolbar:	Standard Toolbar > Insert Hyperlink
Menu:	Insert > Hyperlink
Command:	HYPERLINK

When you invoke this command, you will be prompted to select the objects to which the hyperlinks will be attached. Once the objects are selected, the **Insert Hyperlink** dialog box will be displayed, as shown in Figure 26-10.

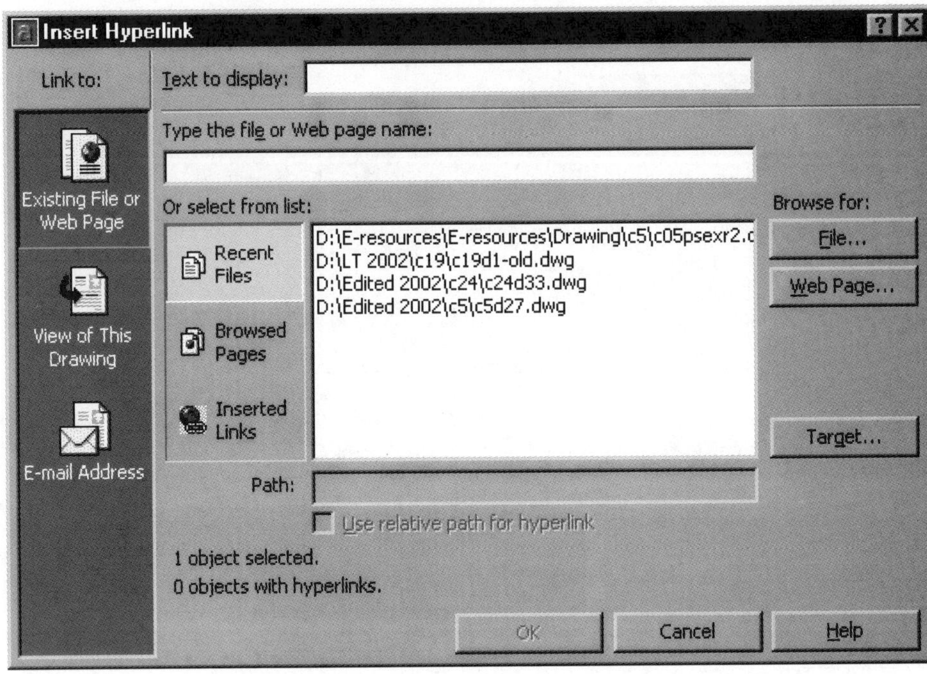

Figure 26-10 *The* ***Insert Hyperlink*** *dialog box*

If you select an object that has a hyperlink already attached to it, a **Remove Link** button will also be included in the dialog box. This button allows you to remove the hyperlink from the object. The same dialog box can also be used to edit the hyperlinks.

If you use the Command line for inserting the hyperlinks (**-HYPERLINK** command), you are allowed to create hyperlink areas. The **hyperlink area** is a rectangular area that can be thought of as a 2D hyperlink (the dialog box-based **HYPERLINK** command does not create hyperlink areas). When you select the **Area** option, the rectangle is placed automatically on layer URLLAYER and colored red (Figure 26-11).

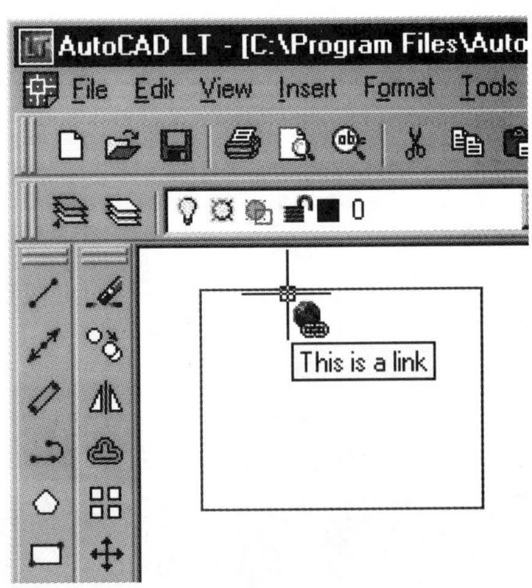

Figure 26-11 *A rectangular hyperlink area*

In the following sections, you learn how to apply and use hyperlinks in an AutoCAD LT drawing and in a Web browser via the dialog box-based **HYPERLINK** command.

Hyperlinks Inside AutoCAD LT

As mentioned earlier, AutoCAD LT allows you to add a hyperlink to any object in the drawing. An object is permitted just a single hyperlink. On the other hand, a single hyperlink may be applied to a selection set of objects.

You can determine whether or not an object has a hyperlink by passing the cursor over it. The cursor displays the "linked Earth" icon, as well as a tooltip describing the link. See Figure 26-12. If, for some reason, you do not want to see the hyperlink cursor, you can turn it off. From the **Tools** menu, choose **Options**, then choose the **User Preferences** tab. The **Display hyperlink cursor and shortcut menu** check box under the **Hyperlink** area toggles the display of the hyperlink cursor, as well as the **Hyperlink** tooltip option on the shortcut menu.

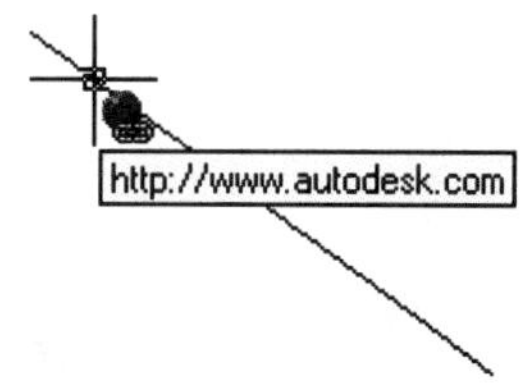

Figure 26-12 *The cursor reveals a hyperlink*

Attaching Hyperlinks

The following example explains attaching the hyperlinks.

Example 2 *General*

In this example, you are given the drawing. To this drawing, you will add hyperlinks. These hyperlinks are another AutoCAD LT drawing. Hyperlinks must be attached to objects. For this reason, place some text in the drawing, then attach the hyperlinks to the text.

1. Start AutoCAD LT 2002.

2. Open the **Family Home Elevation.dwg** file found in the AutoCAD LT 2002 **\Sample** folder.

3. Save this file with the name **Example2.dwg**.

4. Using the **Text** command, write the text as shown in Figure 26-13.

5. Choose the **Insert Hyperlink** button from the **Standard Toolbar**. You will be prompted to select the objects. Select the text **Foundation Plan** to display the **Insert Hyperlink** dialog box.

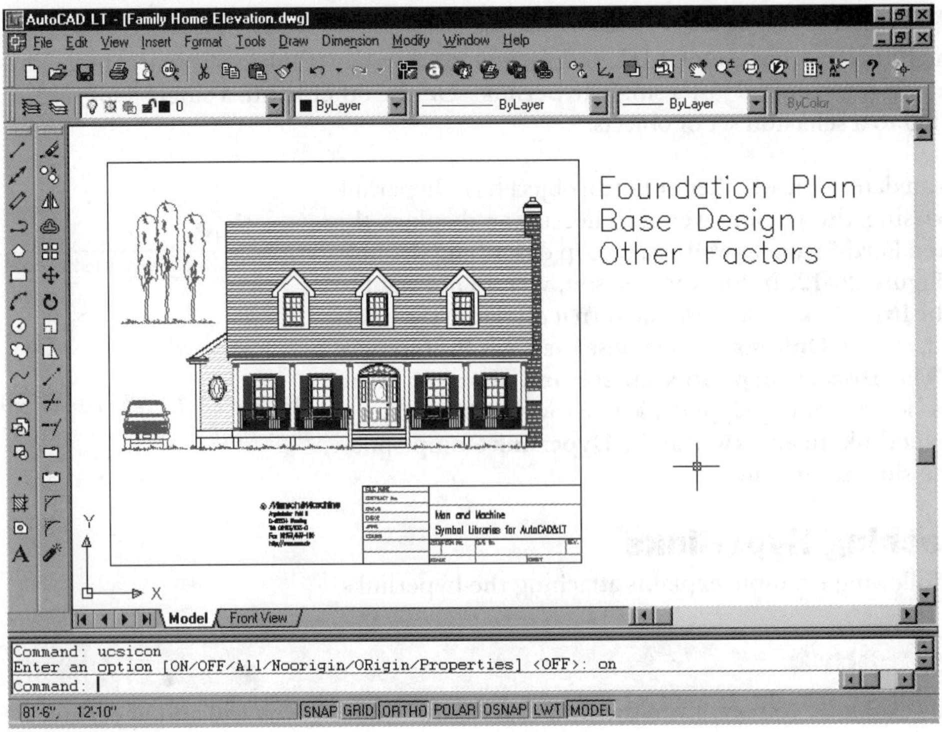

Figure 26-13 *Text placed in drawing*

6. Choose the **File** button to display the **Browse the Web – Select Hyperlink** dialog box (Figure 26-14).

7. Select AutoCAD LT 2002's **Sample** folder and select the **Foundation Plan.dwg** file. Choose **Open**. AutoCAD LT does not open the drawing; rather, it copies the file's name to the **Insert Hyperlinks** dialog box. The name of the drawing you selected will be displayed in the **Type the file or web page name** edit box. The same name and path will also be displayed in the **Text to display** edit box. Remove the path in this edit box and retain only the file name.

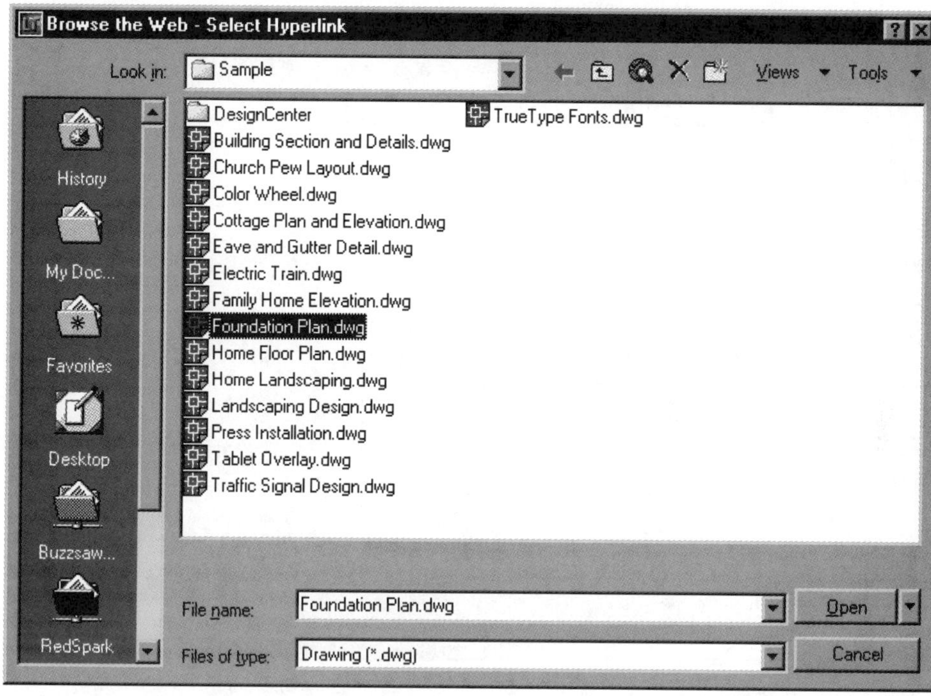

***Figure 26-14 Browse the Web - Select Hyperlink** dialog box*

8. Choose the **OK** button to close the dialog box. Move the cursor over the Site Plan text. Notice the display of the "linked Earth" icon; a moment later, the tooltip displays "City base map.dwg"(Figure 26-15).

***Figure 26-15** The Hyperlink cursor and tooltip*

9. Connect to the Internet and then connect the URL **www.autodesk.com/acltuser** to **Base Design** using the **Web Page** button of the **Insert Hyperlink** dialog box.

10. You can directly open the file attached as the hyperlink to the object using the linked object. This can be done by selecting the object to which the file is linked and then right-clicking to display the shortcut menu. In the shortcut menu choose **Hyperlink > Open (file name)**. In this case the file name is **Foundation Plan.dwg**. Therefore, right-click and select **Hyperlink** > **Open "Foundation Plan.dwg"** from the shortcut menu (Figure 26-16).

11. You can also see both drawings together. This is done by choosing **Tile Vertically** from the **Window** menu (Figure 26-17).

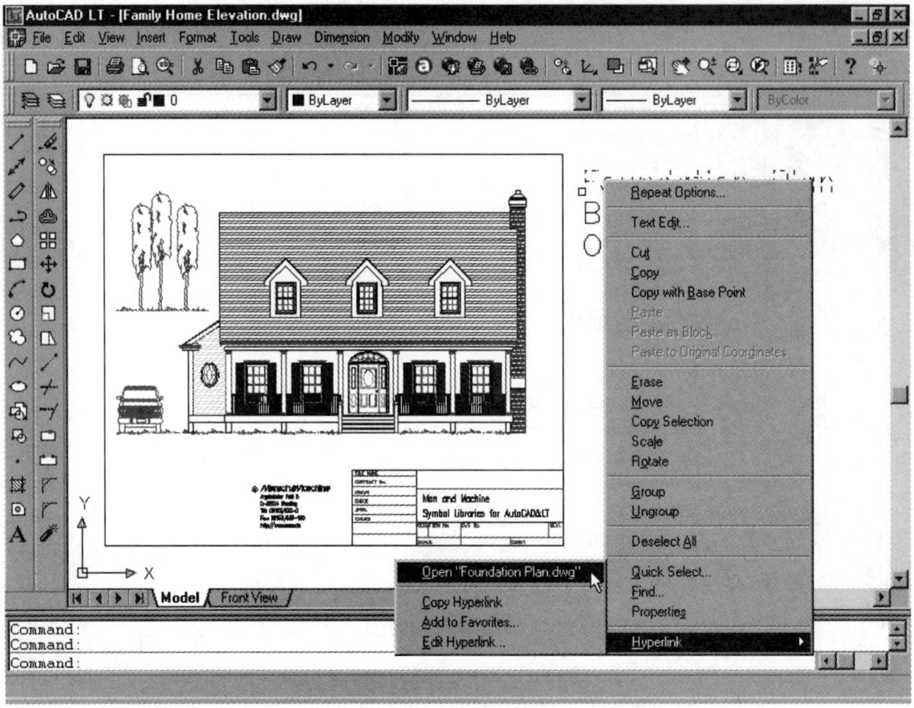

Figure 26-16 *Opening a hyperlink*

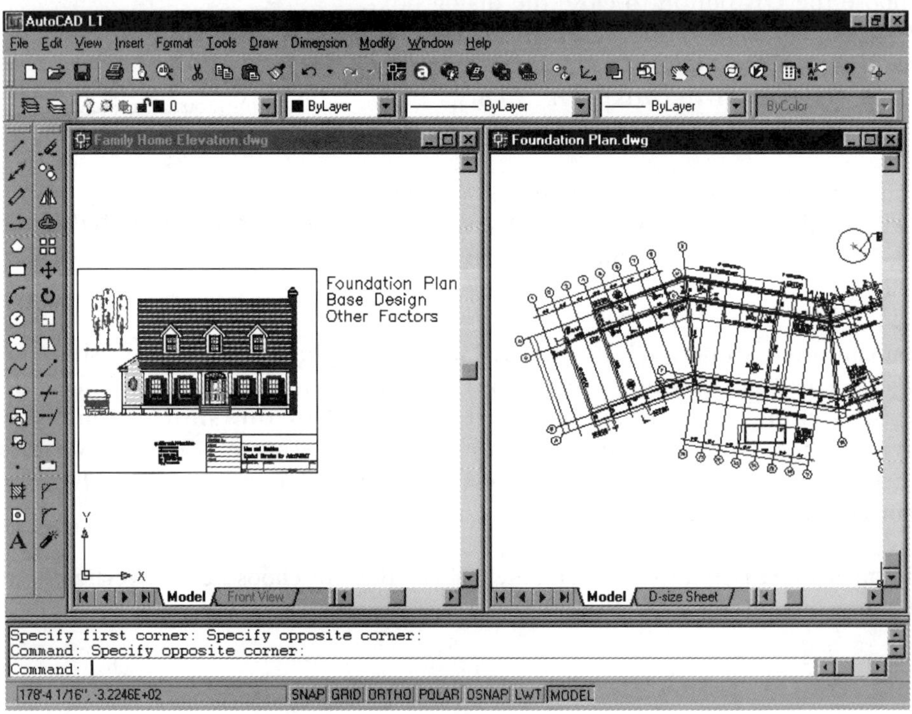

Figure 26-17 *Viewing both the drawings*

12. Select and then right-click on the "**Lighting Specs**" hyperlink. Select **Hyperlink > Open**. Notice that Windows starts your Web browser and opens the URL www.autodesk.com/acltuser (Figure 26-18). Make sure that you are connected to the Internet before opening this URL. If you are not connected to the Internet, this URL will not open.

13. Save this drawing with the name **Example2.dwg**.

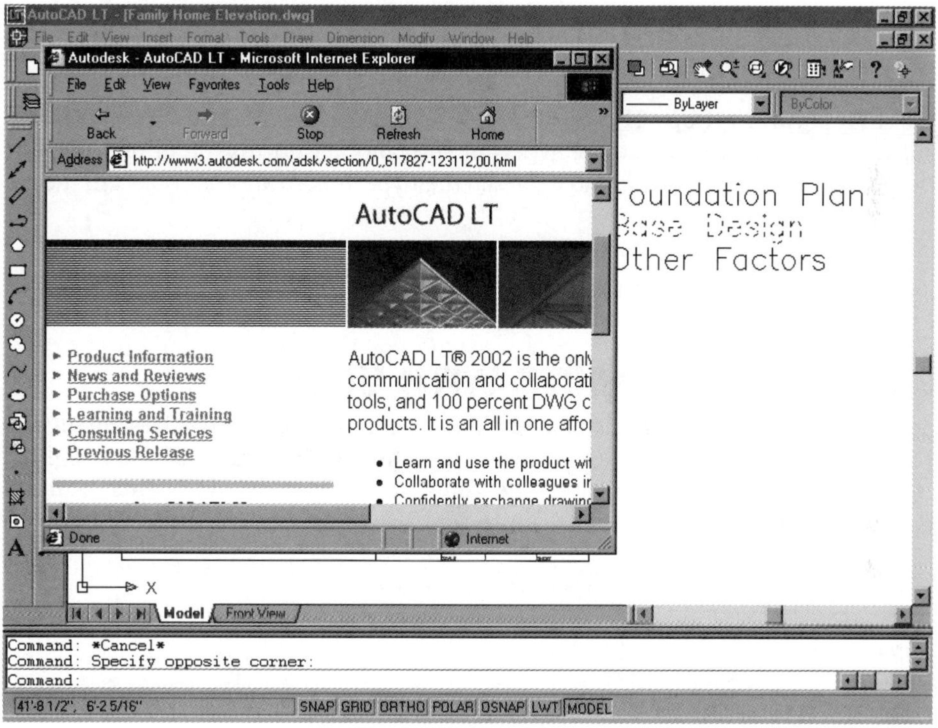

Figure 26-18 Opening the hyperlinked URL

PASTING AS HYPERLINKS

AutoCAD LT 2002 has a shortcut method for creating hyperlinks in the drawing. The undocumented **PASTEASHYPERLINK** command pastes any text in the Windows Clipboard as a hyperlink to any object in the drawing. Here is how it works:

1. In a word processor, select some text and copy it to the Clipboard (via CTRL+C or choose **Edit > Copy**). The text can be a URL (such as http://www.autodeskpress.com) or any other text.

2. Switch to AutoCAD LT and select **Edit > Paste As Hyperlink** from the menu bar. Note that this command does not work (is not available) if anything else is on the Clipboard, such as a picture.

3. Select one or more objects to attach the hyperlink.

4. Move the cursor over the object and note the hyperlink cursor and tooltip. The tooltip displays the same text that you copied from the document.

If the text you copy to the Clipboard is very long, AutoCAD LT displays only portions of it in the tooltip, using ellipses (...) to shorten the text. You cannot select text in the AutoCAD LT drawing to paste it as a hyperlink.

However, you can copy the hyperlink from one object to another. Select the object, right-click, and select **Hyperlink** > **Copy Hyperlink** from the shortcut menu. The hyperlink is copied to the Clipboard. You can now paste the hyperlink into another document, or use AutoCAD LT's **Edit** > **Paste as Hyperlink** command to attach the hyperlink to another object in the drawing.

Tip

*The **MATCHPROP** command does not copy the hyperlinks from the source objects to the destination objects.*

HIGHLIGHTING OBJECTS WITH URLS

Although you can see the rectangle of area URLs, the hyperlinks themselves are invisible. For this reason, AutoCAD LT provides you with the **QSELECT** command, that highlights all objects that match specifications. From the **Tools** menu choose **Quick Select**. AutoCAD LT displays the **Quick Select** dialog box (Figure 26-19).

Enter these specifications in the respective edit boxes:

Apply to: **Entire drawing**
Object type: **Multiple**
Properties: **Hyperlink**
Operator: *** Wildcard Match**
Value: *****

Choose **OK** and AutoCAD LT highlights all objects that have a hyperlink attached to them. Depending on your computer's display system, the highlighting shows up as dashed lines or as another color.

EDITING HYPERLINKS

Once you know where the objects with hyperlinks are located, you can use the **HYPERLINK** command to edit the hyperlinks and related data. Select the hyperlinked object and invoke the **HYPERLINK** command. When the **Edit Hyperlink** dialog box appears (it looks identical to the **Insert Hyperlink** dialog box), make the changes and choose **OK**.

REMOVING HYPERLINKS FROM OBJECTS

To remove a URL from an object, use the **HYPERLINK** command on the object. When the **Edit Hyperlink** dialog box appears, choose the **Remove Hyperlink** button.

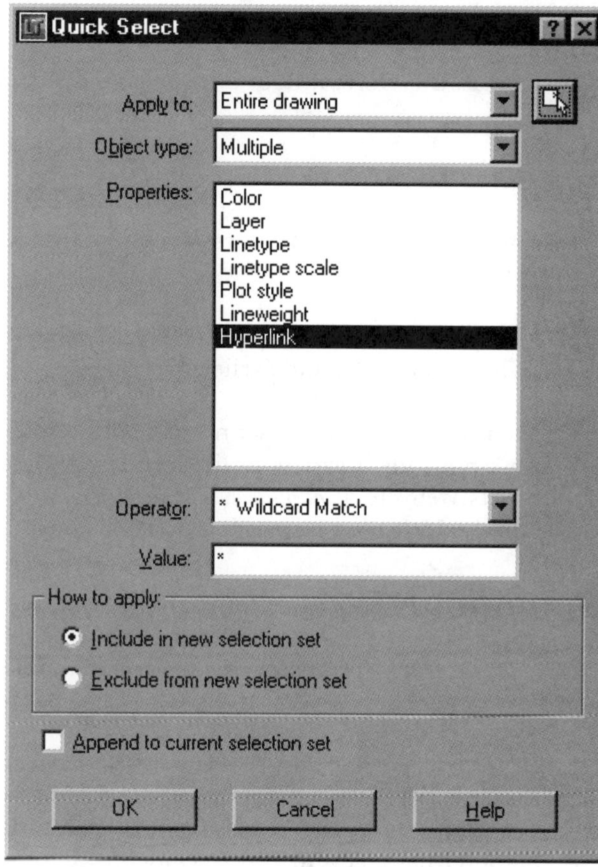

***Figure 26-19** The **Quick Select** dialog box*

To remove a rectangular area hyperlink, you can simply use the **ERASE** command; select the rectangle and AutoCAD LT erases the rectangle. (Unlike in LT 97, AutoCAD LT no longer purges the URLLAYER layer.)

DRAWING WEB FORMAT

To display AutoCAD LT drawings on the Internet, Autodesk invented a file format called drawing Web format (DWF). The DWF file has several benefits and some drawbacks over DWG files. The DWF file is compressed as much as eight times smaller than the original DWG drawing file. Therefore, it takes less time to transmit these files over the Internet, particularly with the relatively slow telephone modem connections. The DWF format is more secure, since the original drawing is not being displayed; another user cannot tamper with the original DWG file.

However, the DWF format has some drawbacks:

- You must go through the extra step of translating a file from DWG to DWF format.

- DWF files cannot display shaded drawings.
- DWF is a flat 2D-file format; therefore, it does not preserve 3D data, although you can export a 3D view.
- AutoCAD LT itself cannot display DWF files.
- DWF files cannot be converted back to DWG format without using file translation software from a third-party vendor.
- Earlier versions of DWF did not handle paperspace objects (version 2.x and earlier), or linewidths, and lineweights (version 3.x and earlier).

To view a DWF file on the Internet, your Web browser needs a *plug-in* software extension called **WHIP!** that lets a Web browser handle a variety of file formats. Autodesk makes this DWF plug-in freely available from its Web site at *http://www.autodesk.com/whip*. It is a good idea to regularly check for updates to the DWF plug-in, which is updated about twice a year.

Other DWF Viewing Options

Volo View Express is a stand-alone viewer that views and prints DWG, DWF, and DXF files. The **Volo View Express** is automatically installed on your computer when you install AutoCAD LT 2002.

In addition to **WHIP** with a Web browser and the **Volo View Express**, you can also use **CADViewer Light** for viewing the DWF files. It is designed to be a DWF viewer that works on all operating systems and computer hardware because it is written in Java. As long as your Windows, Mac, or Unix computer has access to Java (included with most Web browsers), you can view DWF files.

Creating a DWF File

To create a DWF file from AutoCAD LT 2002, follow these steps:

1. Open the file that has to be converted into a DWF file. Choose the **Plot** button from the **Standard Toolbar**. The **Plot** dialog box is displayed as shown in Figure 26-20.
2. Select **DWF ePlot (optimized for plotting).pc3** from the **Name** drop-down list in the **Plotter Configuration** area.
3. Choose the **Properties** button to display the **Plotter Configuration Editor** dialog box (Figure 26-21).
4. Select **Custom Properties** in the tree view. The **Custom Properties** button will appear in the lower half of the dialog box.

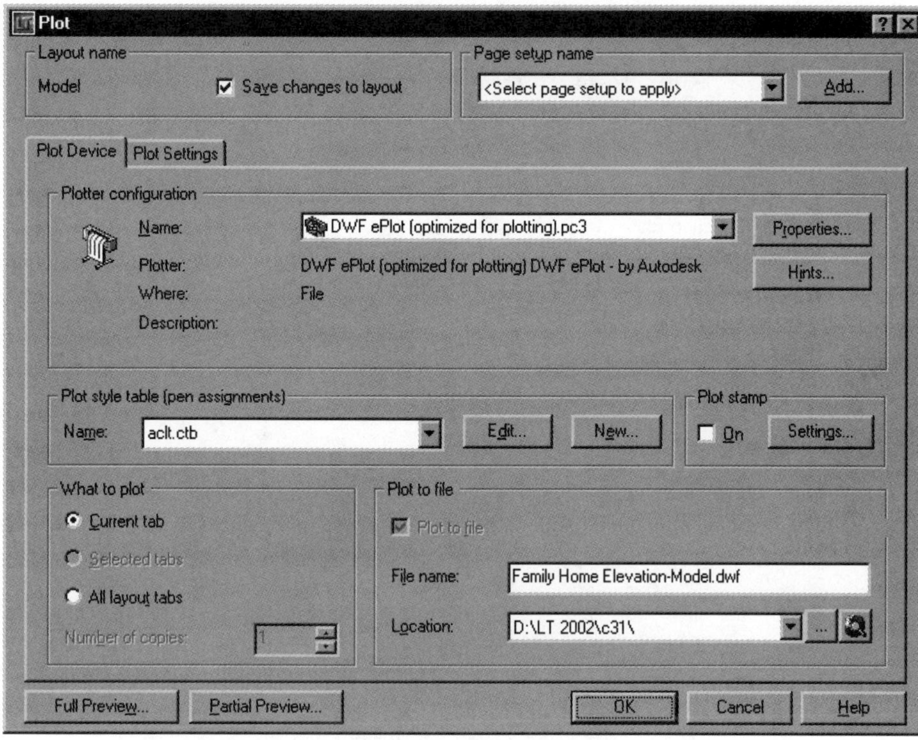

Figure 26-20 ***Plot*** *dialog box*

5. Choose the **Custom Properties** button to display the **DWF Properties** dialog box (Figure 26-22). The options provided under this dialog box are discussed next.

Resolution

Unlike AutoCAD LT DWG files that are based on real numbers, DWF files are saved using integer numbers. The higher the resolution, the better the quality of the DWF file. However, the size of the files also increases with the resolution.

Format

Compression further reduces the size of the DWF file. You should always use compression, unless you know that another application cannot decompress the DWF file. Compressed binary format is seven times smaller than ASCII format. Again, that means the compressed DWF file transmits over the Internet seven times faster than an ASCII DWF file.

Background color shown in viewer

While white is probably the best background color, you may choose any of AutoCAD LT's 255 colors.

Include Layer information

This option includes layers, which allows you to toggle layers off and on when the drawing is viewed by the Web browser.

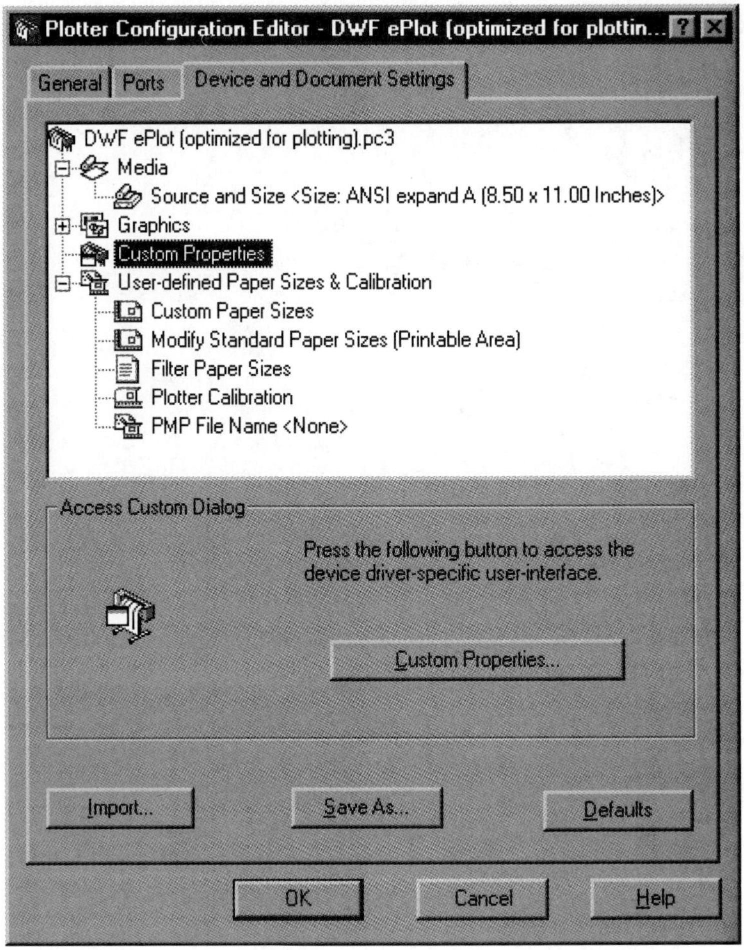

*Figure 26-21 **Plotter Configuration Editor** dialog box*

Include Scale and measurement information

This allows you to use the **Location** option in the Web browser's Whip plug-in to show scaled coordinate data.

Show Paper boundaries

Includes a rectangular boundary at the drawing's extents as displayed in the layouts.

Convert .DWG hyperlink extensions to .DWF

Includes hyperlinks in the DWF file.

6. In most cases, you would turn on all options.

7. Choose **OK** to exit the dialog boxes back to the **Plot** dialog box.

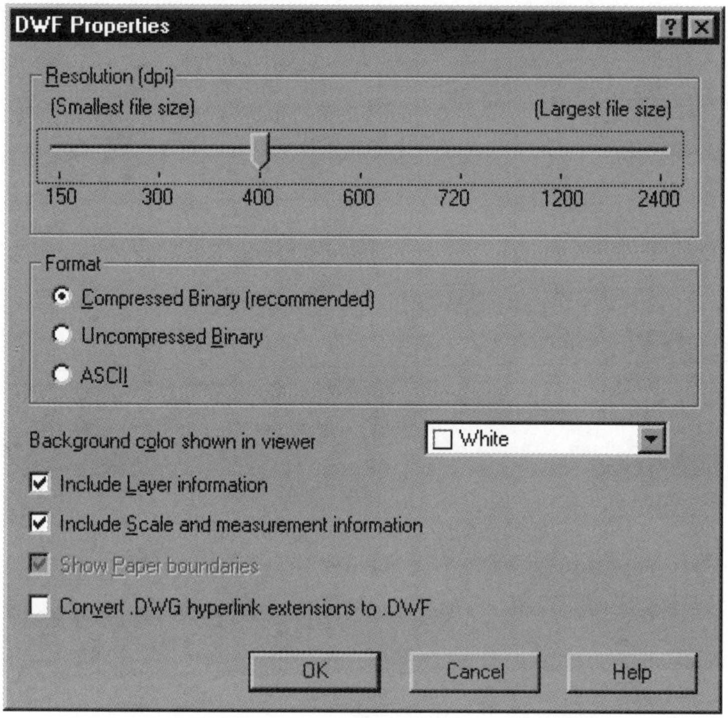

***Figure 26-22 DWF Properties** dialog box*

8. Accept the DWF file name listed in the **File name** text box in the **Plot to file** area, or type a new name. If necessary, change the location that the file will be stored in. Note the two buttons: one has an ellipsis (...), which displays the **Browse for Folder** dialog box. The second button brings up AutoCAD LT's internal Web browser, a simplified version of the Microsoft-branded product.

9. In the **Plot Settings** tab, use the required options to create the DWF file.

10. Choose **OK** to save the drawing in DWF format.

Viewing the DWF Files

One of the major enhancements in AutoCAD LT 2002 is the **Volo View Express** that is automatically installed on your computer when you install AutoCAD LT 2002. This plug-in can be used to view the DWF files in addition to the other Web browsers. You can also use a Web browser with a special plug-in that allows the browser to correctly interpret the file for viewing the DWF files. (Remember: you cannot view a DWF file with AutoCAD LT.) Apart from viewing the DWF files, you can also view the DWG and DXF files using the **Volo View Express**.

Autodesk updates the DWF plug-in approximately twice a year. Each update includes some new features. In summary, all versions of the DWF plug-in perform the following functions:

- Views DWF files created by AutoCAD LT within a browser.
- Right-clicking the DWF image displays a shortcut menu with commands.
- Real-time pan and zoom lets you change the view of the DWF file as quickly as a drawing file in AutoCAD LT.
- Embedded hyperlinks let you display other documents and files.
- File compression means that a DWF file appears in your Web browser much faster than the equivalent DWG drawing file would.
- Print the DWF file alone or along with the entire Web page.
- Works with Netscape Navigator or Microsoft Internet Explorer. A separate plug-in is required, depending upon which of the two browsers you use.
- Allows you to "drag and drop" a DWG file from a Web site into AutoCAD LT as a new drawing or as a block.
- Views a named view stored in the DWF file.
- Can specify a view using x,y- coordinates.
- Toggles layers off and on.

You can use various buttons provided in the **Volo View Express** for manipulating the view of the DWF file. However, keep in mind that the original objects of the DWF file can not be modified. You can also right-click in the display screen to display a shortcut menu. The shortcut menu provides you with the options such as **Pan**, **Zoom**, **Layers**, and **Named Views** (Figure 26-23). To select an option, choose it from the shortcut menu.

Pan is the default option. Press the left mouse button and move the mouse. The cursor changes to an open hand, signaling that you can pan the view around the drawing. This is exactly the same as real-time panning in AutoCAD LT.

The **Zoom** option displays more options for zooming the view. The **Previous**, **Next**, **Window** and the **All** options are the same as those in AutoCAD LT. The **In/Out** option is the **Realtime** option in AutoCAD LT that displays a magnifying glass for zooming in and out of the current display.

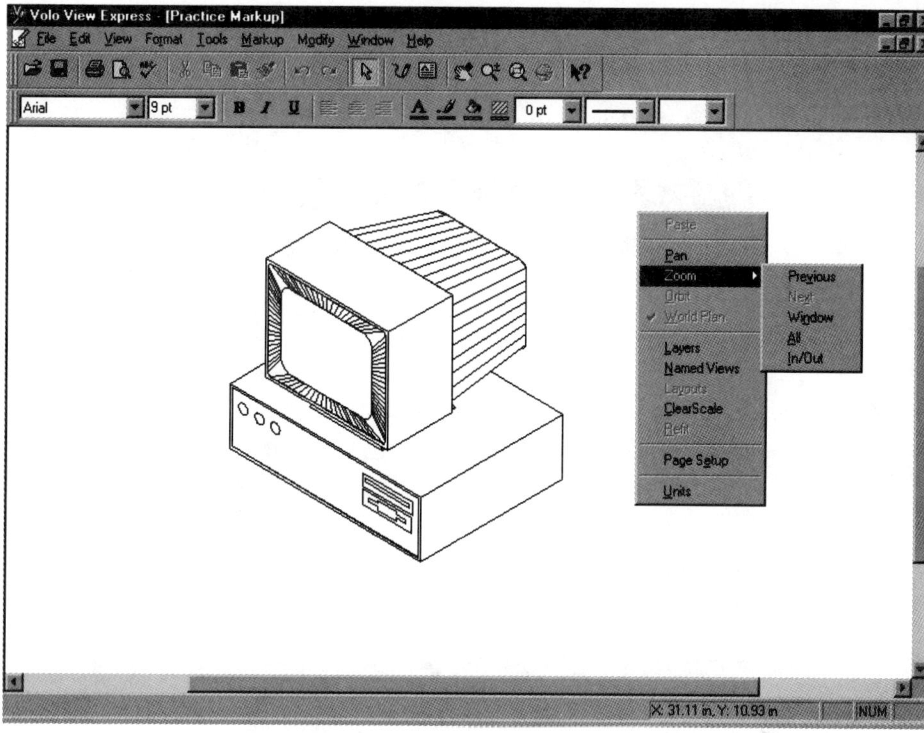

Figure 26-23 Volo View Express' shortcut menu

Layers displays a nonmodal dialog box (Figure 26-24) that lists all layers in the drawing. (A nonmodal dialog box remains on the screen; unlike AutoCAD LT's modal dialog boxes, you do not need to dismiss a nonmodal dialog box to continue working). Select a layer name to toggle its visibility between on (yellow light bulb icon) and off (blue light bulb icon).

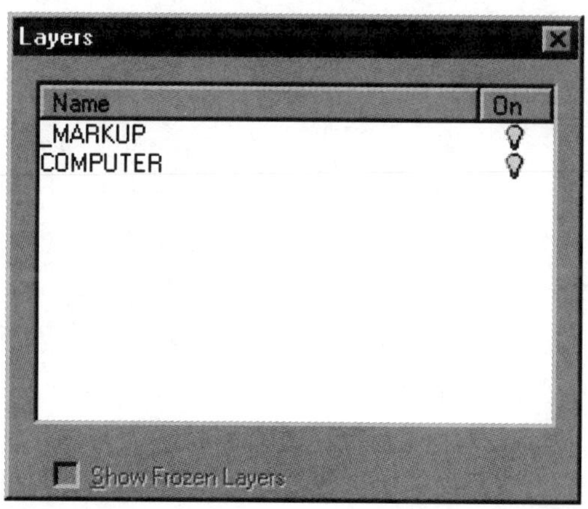

Figure 26-24 Volo View Express' **Layers** dialog box

Named Views works only when the original DWG drawing file contained named views created with the **VIEW** command. Selecting **Named Views** displays a nonmodal dialog box that allows you to select a named view (Figure 26-25). Select a named view to see it; click the small **x** in the upper right corner to dismiss the dialog box.

***Figure 26-25** Volo View Express' **Named Views** dialog box*

Page Setup displays the **Page Setup** dialog box (Figure 26-26) for setting up the pages to plot the DWF file in the **Volo View Express**.

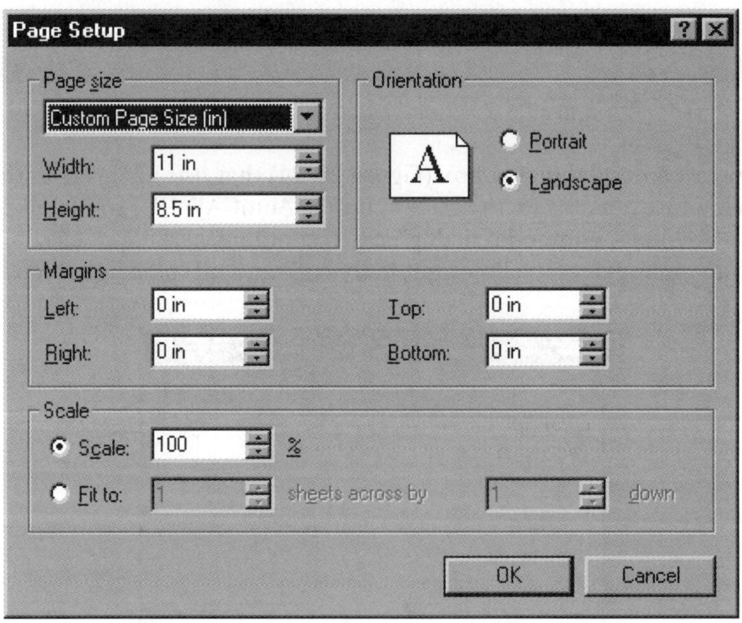

***Figure 26-26 Page Setup** dialog box in Volo View Express*

Units displays the **Print Settings** dialog box for using the units of drawings and text fonts and lineweights. See Figure 26-27.

Apart from all these functions, you can also add text to the DWF file, create an entity using the option similar to the **SKETCH** command in AutoCAD, change the color of the text, and so

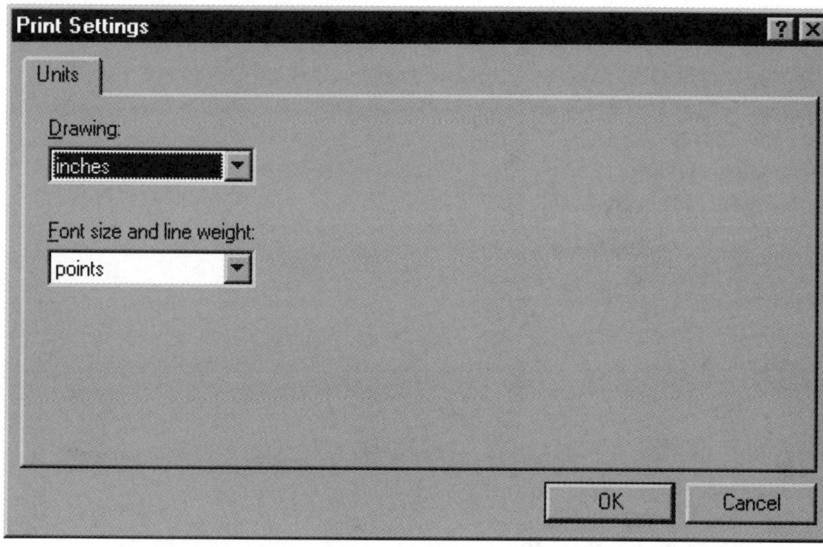

Figure 26-27 ***Print Settings*** *dialog box in Volo View Express*

on. All this can be done using the buttons that are provided in **Volo View Express**.

DRAG AND DROP FROM BROWSER TO DRAWING

If you are using WHIP! for viewing the DWF file, you can perform several "drag and drop" functions. Drag and drop is when you use the mouse to drag an object from one application to another. Hold down the CTRL key to drag a DWF file from the browser into AutoCAD LT. Recall that AutoCAD LT cannot translate a DWF file into DWG drawing format. For this reason, this form of drag and drop only works **when the originating DWG file exists in the same subdirectory as the DWF file**. Also, be sure that the originating DWG file is not be in use. AutoCAD LT executes the **-INSERT** command, as follows:

Enter block name or [?]: D:\LT 2002\c32\c32d1.dwg
Specify insertion point or [Scale/X/Y/Z/Rotate/PScale/PX/PY/PZ/PRotate]: 0,0
Enter X scale factor, specify opposite corner, or [Corner/XYZ] <1>: Enter
Enter Y scale factor <use X scale factor>: Enter
Specify rotation angle <0>: Enter

To see the inserted drawing, you may need to use the **Extents** option of the **ZOOM** command.

Another drag and drop function is to drag a DWF file from the Windows Explorer (or File Manager) into the Web browser. This causes the Web browser to load the DWF plug-in, then display the DWF file.

Finally, you can also drag and drop a DWF file from Windows Explorer (in Windows 95/98) or File Manager (in Windows NT) into AutoCAD LT. This causes AutoCAD LT to launch another program to view the DWF file. This does not work if you do not have other software on your computer system capable of viewing DWF files.

EMBEDDING A DWF FILE

To let others view your DWF file over the Internet, you need to embed the DWF file in a Web page. Here are the steps to embed a DWF file in a Web page.

1. The hypertext markup language (HTML) code is the most basic method of placing a DWF file in your Web page:

   <embed src="filename.dwf">

   The **<embed>** tag embeds an object in a Web page. The **src** option is short for "source." Replace **filename.dwf** with the URL of the DWF file. Remember to keep the quotation marks in place.

2. HTML normally displays an image as large as possible. To control the size of the DWF file, add the **Width** and **Height** options:

   <embed width=800 height=600 src="filename.dwf">

   The **Width** and **Height** values are measured in pixels. Replace 800 and 600 with any appropriate numbers such as 100 and 75 for a "thumbnail" image, or 300 and 200 for a small image.

3. To speed up a Web page's display speed, some users turn off the display of images. For this reason, it is useful to include a description, which is displayed in place of the image:

   <embed width=800 height=600 name=description src="filename.dwf">

   The **Name** option displays a textual description of the image when the browser does not load images. You might replace description with the DWF filename.

4. When the original drawing contains named views created by the **VIEW** command, these are transferred to the DWF file. Specify the initial view for the DWF file:

   <embed width=800 height=600 name=description namedview="viewname"
   src="filename.dwf">

   The **namedview** option specifies the name of the view to display upon loading. Replace **viewname** with the name of a valid view name. When the drawing contains named views, the user can right-click on the DWF image to get a list of all named views.

   As an alternative, you can specify the 2D coordinates of the initial view:

   <embed width=800 height=600 name=description view="0,0 9,12"
   src="filename.dwf">

   The **View** option specifies the x,y-coordinates of the lower left and upper right corners of the initial view. Replace 0,0 9,12 with other coordinates. Since DWF is 2D only, you cannot specify a 3D viewpoint. You can use **View** or **Namedview**, but not both.

5. Before a DWF image can be displayed, the Web browser must have the DWF plug-in called "Whip!". For users of Netscape Communicator, you must include a description of where to get the Whip plug-in when the Web browser is lacking it.

```
<embed pluginspage=http://www.autodesk.com/products/AutoCAD LT/whip/whip.htm
width=800 height=600 name=description view="0,0 9,12" src="filename.dwf">
```

The **pluginspage** option describes the page on the Autodesk Web site where the Whip-DWF plug-in can be downloaded.

The code previously listed works for Netscape Navigator. To provide for users of Internet Explorer, the following HTML code must be added:

```
<object classid ="clsid:B2BE75F3-9197-11CF-ABF4-08000996E931" codebase = "ftp://
ftp.autodesk.com/pub/AutoCAD LT/plugin/whip.cab#version=2,0,0,0" width=800
height=600>
<param name="Filename" value="filename.dwf">
<param name="View" value="0,0 9,12">
<param name="Namedview" value="viewname">
<embed pluginspage=http://www.autodesk.com/products/AutoCAD LT/whip/
whip.htm width=800 height=600 name=description view="0,0 9,12"
src="filename.dwf">
</object>
```

The two **<object>** and three **<param>** tags are ignored by Netscape Navigator; they are required for compatibility with Internet Explorer. The **classid** and **codebase** options tell Explorer where to find the plug-in. Remember that you can use **View** or **Namedview**, but not both.

6. Save the HTML file.

Review Questions

Answer the following questions.

1. Can you launch a Web browser from within AutoCAD LT?

2. What does DWF mean?

3. What is the purpose of DWF files?

4. URL is an acronym for what?

5. Which of the following URLs are valid?

   (a) *www.autodesk.com* (b) *http://www.autodesk.com*
   (c) Both of the above. (d) None of the above.

6. FTP is an acronym for what?

7. What is a "local host"?

8. Are hyperlinks active in an AutoCAD LT 2002 drawing?

9. The purpose of URLs is to let you create _________ between files.

10. When you attach a hyperlink to a block, the hyperlink data is _________ when you scale the block unevenly, stretch the block, or explode it.

11. Can you attach a URL to any object?

12. The ________ is a HTML tag for embedding graphics in a Web page.

13. To see the location of hyperlinks in a drawing, use the _________ command.

14. Rectangular (area) hyperlinks are stored on layer:_________.

15. Compression in the DWF file causes it to take (less, more, the same) time to transmit over the Internet.

Index

Symbols

-Hyperlinks Command 26-15
-Insert Command 26-31
***HELPSTRING 25-38
**aliasname 25-35
-ATTEDIT 15-24
-BHATCH Command 13-18
-BLOCK Command 14-7
-Hyperlink 26-3
-INSERT Command 14-15
-LAYER 22-13
-PURGE Command 14-29
-VIEW Command 7-14
.section label 25-2
1 Point Option 5-37
2 Points Option 5-37
2 Points Select Option 5-37
3 Points option 3-1
3D Coordinate systems 21-8
3D Drawings 22-12
3DPOLY Command 21-19

A

Absolute Coordinate System 1-19, 21-8
ACAD.LIN 24-1, 24-13
ACAD.MNU 25-1, 25-2, 25-10
ACAD.MNU file 25-3
ACAD.PAT 24-22, 24-25
accelerator keys 25-31
Accessing Other Files on the Internet 26-11
Active Assistance 1-42
Add 5-5
Add-A-Plot Style Table Wizard 12-22
Add-a-Plot Style Table wizard 12-6
Add-A-Plotter-Wizard 12-15
Adding an Xref Drawing (Bind option) 16-11
Adding Plotters 12-13
Additional Help Resources 1-44
ADJUSTING GRIP SETTINGS 6-3
Advanced Setup 1-4
Advanced Tab (Boundary hatch dialog box) 13-14
Advantages of Using Blocks 14-2
AERIAL VIEW 7-14
Aerial View window 7-14
alias 25-35
ALIGNED DIMENSIONING 8-14
alignment definition 24-8
alignment field specification 24-2
ALIGNMENT SPECIFICATION 24-9
ALL 5-5
All Option 2-9, 7-4
ALTERNATE LINETYPES 24-12
Alternate Units 10-20
Alternate units 8-5
Angle Option 5-20, 8-12
ANGULAR DIMENSIONING 8-19
Angular dimensioning 8-9
Angular Dimensioning Based on Three Points 8-21
Angular Dimensioning of Circles 8-20
Applying Plot Styles 12-28
Arc 3-28
ARC Command 3-1
AREA Command 6-26
Area option 26-16
Array 5-31
ARRAY Command 5-29
Array dialog box 5-29, 5-31
Arrowheads 8-3, 10-7
ASCII format 26-25
Assigning Color to Layer 4-9
Assigning Keyboard Shortcuts to Commands 25-44
Assigning Linetype to Layer 4-7
Assigning Plot Style to Layer 4-9
ASSOCIATIVE DIMENSIONS 8-6
Attached URLs from R14 26-2
Attaching Hyperlinks 26-17

AttachUrl 26-3
ATTDEF 15-1, 15-2, 15-6
ATTDEF Command 15-2
ATTDIA 15-8
ATTDISP 15-2, 15-3, 15-17
ATTDISP Command 15-17
ATTEDIT 15-2, 15-18
ATTEDIT Command 15-18
ATTEXT 15-1
ATTEXT Command 15-11
Attribute area 15-3
Attribute Definition dialog box 15-2
Attribute Extraction dialog box 15-11
attribute value 15-1
ATTRIBUTES 15-1
AUto 5-6
AutoCAD LT Today Window 22-5
AutoCAD LT Today Window 22-7
AutoCAD LT Today Window 2-1
AutoCAD LT Today Window 26-12
AUTOCAD DESIGNCENTER 4-22
AutoCAD DesignCenter window 6-18
AutoCAD LT Dialog Boxes 1-13
AutoCAD LT Editing Commands 13-25
AutoCAD LT Help Topics 1-42
AutoCAD LT Screen Components 1-7
AutoCAD LT Timed Save 1-39
AutoCAD LT's Help 1-40
AutoSnap 4-30
AutoTrack Settings 4-45

B

back clipping plane 21-34
BASELINE DIMENSIONING 8-16
Basic Dimensioning 8-1
BASIC DISPLAY OPTIONS 7-1
Basic Plotting 2-21
BHATCH Command 13-3
BHATCH Shortcut Menu 13-20
binary file transfer 26-4
BLOCK 15-6
BLOCK Command 14-4
Block Definition dialog box 14-5
Boolean Operations 21-20
Boundary Hatch dialog box 13-3
Boundary Hatch Options 13-4
BOX 5-6
BREAK Command 5-36
Breaking Up a Block 14-25
Browse for Folder dialog box 26-27
Browse the Web dialog box 26-6
Browser 26-1, 26-3
btnname 25-34
Bulletin Board Area 26-12
Bylayer 22-2

C

caching 26-9
Calculating Parameters for an Elliptical Arc 3-21
calibrate the tablet 17-12
camera and target 21-23
Canceling a Command 1-29
Cartesian Coordinate System 1-18
CECOLOR System Variable 4-20
CELTSCALE 24-16
CELTYPE 4-20
CELWEIGHT 4-20
Center Mark and Centerlines 8-4
Center Option 7-4
Chamfer 3-13
CHAMFER Command 5-18
CHAMFER Command 5-18
CHAMFERA 5-19, 5-21, 22-2
CHAMFERB 5-19, 5-21, 22-2
CHAMFERC 5-21
CHAMFERD 5-21
CHAMMODE 5-21
CHANGE Command 18-10
Changed Internet Commands 26-3
Changing Properties and Points of an Object 18-5
CHANGING PROPERTIES USING GRIPS 6-14
Changing the Default Web Site 26-6
Changing the Display Order 19-8
Changing the Point Size 3-36
Changing the Point Type 3-35
Checking Layers 14-3
CHECKING SPELLING 7-35
chord 3-5
CHPROP Command 18-13
CIRCLE Command 1-31
CIRCLERAD 1-32
circular reference 16-15
circumference 6-27

circumscribed 3-23
CLip Option 21-34
Clipboard 26-22
Clipping Raster Images 19-8
Closing a Drawing 1-40
COLOR 22-2
Combining Geometric Characteristics 8-37
Comma Delimited File (CDF) 15-11
Command definition 25-8
COMPARING AND LISTING DIMENSION STYLES 10-31
Compass and Tripod 21-6
COMPLEX FEATURE CONTROL FRAMES 8-37
Composite Position Tolerancing 8-39
Compressed binary format 26-25
compression 26-25
configure the tablet 17-12
Connection tab 26-3
Context menus 25-18
CONTINUE DIMENSIONING 8-17
Continue Option 3-9
CONTROLLING ATTRIBUTE VISIBILITY 15-17
CONTROLLING ATTRIBUTE VISIBILITY(ATTDISP Command) 15-17
CONTROLLING DIMENSION FORMAT 10-9
Controlling the Display of Image Frames 19-8
Controlling the Display of the Hidden Lines in the 11-11
Controlling the Display of the Objects in the View 11-11
Controlling the Layers in the Viewports using the 11-13
Controlling Viewport through the Layer Properties 11-13
Conventions in AutoCAD LT 20-1
CONVERT Command 3-31
Converting DXF Files into Drawing Files 19-4
Converting Objects into a Block 14-4
Coordinate Entry Priority 4-42
coordinate system icon 21-29
Coordinate Systems 1-18
COPY Command 5-8
COPYBASE Command 5-9
Copying a Tool Icon 25-43
CPolygon 5-4
Create New Dimension Style 22-10
Create New Drawing 2-5
Creating a BMP File 19-5
Creating a Data Interchange File 19-2
Creating a DWF File 26-24
Creating a New Image and Tooltip for an Icon 25-41
creating a new plot style table 12-22
CREATING A SELECTION SET 5-1
Creating a String Complex Linetype 24-17
CREATING AND RESTORING DIMENSION STYLES 10-2
Creating and using a Windows Metafile 19-5
Creating Custom Toolbars with Flyout Icons 25-43
Creating Drawing Files (WBLOCK Command) 14-20
Creating Floating Viewports 11-7
CREATING LINETYPES 24-3
Creating New Layers 4-4
CREATING NURBS 17-6
CREATING PARAGRAPH TEXT 7-23
CREATING SINGLE LINE TEXT 7-16
CREATING TEMPLATE DRAWINGS 22-1
CREATING TEXT 3-41
Creating the PostScript Files 19-10
Creating Tiled Viewports 11-3
CREATING TRUE ASSOCIATIVE DIMENSIONS* 8-27
CREATING VIEWS 7-12
Creation of Backup Files 1-40
CURRENT LINETYPE SCALING (CELTSCALE) 24-16
Current Time 6-32
CUSTOM HATCH PATTERN FILE 24-34
Custom Hatch Patterns 13-8
Custom Pattern 13-9
customize the toolbars 25-38
CUSTOMIZING DRAWINGS ACCORDING TO PLOT SIZE AND DRAWING AREA 22-12
CUSTOMIZING DRAWINGS WITH LAYERS AND DIMENSIONING STYLES 22-7
CUSTOMIZING DRAWINGS WITH VIEWPORTS 22-22
CUSTOMIZING THE TOOLBARS 25-38
CVPORT system variable 11-5

Cycling through Groups 18-5
Cycling through Snaps 4-42

D

Data Exchange in AutoCAD LT 19-1
Datum 8-36
DDEDIT Command 7-30, 15-7
DDMODIFY Command 6-13
DDVPOINT Command 21-2
default setup values 22-1
DEFINING ATTRIBUTES 15-2
Defining the Insertion Base Point (BASE Command) 14-24
definition of the line pattern 24-2
DEFINITION POINTS 8-7
DELAY Command 23-10
Deleting a Toolbar 25-43
Deleting Layers 4-10
Deleting the Icons from a Toolbar 25-42
Deleting Unused Blocks 14-29
DELOBJ system variable 17-7
dependent symbols 16-3
Descriptive Text 24-8
DetachUrl 26-3
DETERMINING TEXT HEIGHT 7-34
DIAMETER DIMENSIONING 8-22
Diameter dimensioning 8-9
DIGITIZING DRAWINGS 17-11
DIM and DIM1 Commands 8-9
DIMALT system variable 10-20, 22-2
DIMALT system variable 10-20
DIMALTD system variable 22-2
DIMALTD system variable 10-21, 10-22
DIMALTF system variable 22-2
DIMALTF system variable 10-21
DIMALTRND system variable 10-21
DIMALTTD system variable 10-25
DIMALTU system variable 10-22
DIMALTU system variable 10-21
DIMALTZ system variable 10-22
DIMALTZ system variable 10-22
DIMAPOST system variable 10-22
DIMASO system variable 8-6, 22-2
DIMASZ system variable 10-8, 22-2
DIMASZ system variable 10-8
DIMAUNIT system variable 10-20
DIMAZIN system variable 10-20
DIMBLK1 system variable 10-7
DIMBLK2 system variable 10-7
DIMCEN system variable 10-8
DIMCENTER Command 8-24
DIMCLRD system variable 8-36
DIMCLRD system variable 10-4
DIMCLRE system variable 10-5
DIMCLRT system variable 8-36, 10-10
DIMCLRT system variable 10-10
DIMDEC system variable 10-17
DIMDLE system variable 10-5
DIMDLI system variable 10-5
DIMDSEP system variable 10-18
Dimension Drawing Commands 8-2
Dimension Editing Commands 8-2
Dimension Line 8-3, 10-4
Dimension Style Commands 8-2
DIMENSION STYLE FAMILIES 10-27
Dimension Style Manager dialog box 10-2
Dimension Styles 10-1
Dimension Text 8-3
Dimension Utility Commands 8-2
DIMENSIONING IN AUTOCAD LT 8-2
Dimensioning System Variables 10-1
Dimensioning the Angle Between Two Nonparallel Lines 8-19
Dimensioning the Angle of an Arc 8-20
DIMEXE system variable 10-6
DIMEXO system variable 10-6
DIMFRAC system variable 10-17
DIMGAP system variable 8-36, 10-11, 10-12, 10-24
DIMLFAC system variable 10-19
DIMLFAC system variable 10-19
DIMLIM system variable 10-24
DIMLINEAR Command 8-8
DIMLINEAR Options 8-11
DIMLUNIT system variable 10-17
DIMLWD system variable 10-4
DIMLWE system variable 10-6
DIMPOST system variable 10-18, 22-2
DIMREASSOCIATE Command 8-28
DIMREASSOCIATE Command* 8-27
DIMRND system variable 10-18
DIMSCALE system variable 22-19, 23-3
DIMSCALE system variable 10-15

DIMSD1 system variable 10-5
DIMSD2 system variable 10-5
DIMSE1 system variable 10-6
DIMSE2 system variable 10-6
DIMSOXD system variable 10-15
DIMSTYLE Command 10-2
DIMSTYLE system variable 9-9
DIMTAD system variable 10-11
DIMTDEC system variable 10-24
DIMTEDIT Command 9-6
DIMTFAC system variable 10-10, 10-24
DIMTGAP system variable 10-11
DIMTIH system variable 10-12
DIMTIX system variable 10-15
DIMTM system variable 10-23, 10-24
DIMTMOVE system variable 10-15
DIMTOFL system variable 10-16
DIMTOH system variable 10-12, 10-13
DIMTOH system variable 10-12
DIMTOL system variable 10-23, 10-24
DIMTOLJ system variable 10-25
DIMTP system variable 10-23, 10-24
DIMTVP system variable 10-26
DIMTXSTY system variable 8-36, 10-10
DIMTXT system variable 8-36, 10-10
DIMTXT system variable 10-10
DIMUNIT system variable 10-17
DIMUPT system variable 10-16
DIMUPT system variable 10-16
Direct Distance Entry 1-26
Direction Control dialog box 2-12, 22-6
Displaying a Submenu 25-23
DIST Command 6-29
Distance Option 21-29
DIVIDE Command 5-39
DONUT Command 3-32
DONUTID system variable 3-33
DONUTOD system variable 3-33
DLINE Command 17-1
Drafting Settings dialog box 4-23
Drag and Drop 26-31
Drag and Drop from Browser to Drawing 26-31
Drawing an Ellipse Using the Axis and Eccentricity 3-17
DRAWING ARCS 3-1
DRAWING DONUTS 3-32
DRAWING ELLIPSES 3-16
Drawing Elliptical Arcs 3-19
DRAWING INFINITE LINES 3-37
Drawing Limits 22-5
Drawing Objects for Blocks 14-2
DRAWING POINTS 3-34
DRAWING POLYLINES 3-25
Drawing Properties dialog box 6-33
DRAWING RECTANGLES 3-12
DRAWING REGULAR POLYGONS 3-22
DRAWING SPECIAL CHARACTERS 7-22
Drawing Units dialog box 2-9, 22-6
Drawings on the Internet 26-6
DSVIEWER Command 7-14
DVIEW Command 21-22
DWF Properties dialog box 26-25
DWF Viewing Options 26-24
DwfOut 26-3
DwfOutD 26-3
DWGPROPS Command 6-33
DXF File Format 19-2
DXF Format Extract File (DXX) 15-12
Dynamic Option 7-5

E

Edge Option 5-22
EDGEMODE system variable 5-25
edit a slide 23-27
Edit Attribute Definition dialog box 15-7
Edit Hyperlink dialog box 26-22
EDITING ATTRIBUTE DEFINITION 15-7
Editing Attributes Using the ATTEDIT Command 15-18
EDITING BLOCK ATTRIBUTES VALUES 15-18
EDITING COMMANDS 5-7
Editing Dimension Text 9-6
EDITING Dimensions 9-1
Editing Dimensions by Trimming and Extending 9-3
Editing Dimensions Using the Properties Window 9-8
Editing Dimensions with GRIPS 9-8
EDITING GRIPPED OBJECTS 6-12
Editing Hatch Boundary Using Grips 13-24
Editing Hatch Patterns 13-21
Editing Hyperlinks 26-22

Editing Multiple Polyline 18-28
Editing Plotter Configuration 12-16
Editing Raster Image Files 19-8
EDITING SPLINES 17-7
EDITING TEXT 7-30
Editing Text Using the Properties Window 7-31
Editing the Dimensions by Stretching 9-1
Editing the Objects with Grips 6-4
Editing the Viewports 11-10
EDITING WITH GRIPS 6-1
EFFECT OF ANGLE AND SCALE FACTOR ON HATCH 24-26
Elapsed Timer 6-32
element 25-35
ELEMENTS OF LINETYPE SPECIFICATION 24-3
ELEV Command 21-17
Elevation 3-15, 3-42
ELLIPSE 3-16
ELLIPSE Command 3-16
email 26-3
Embedding a DWF File 26-32
Embedding Objects into AutoCAD LT 19-20
ePlot option 26-3
ERASE Command 1-28
EXIT Command 1-40
EXPLODE Command 14-26, 18-13, 24-35
Export Data Dialog Box 14-23
EXTEND Command 5-23
Extension Line 10-5
Extension Lines 8-4
Extents Option 7-5
External References 16-1
EXTNAMES system variable 14-4
Eye option 21-36

F

Feature Control Frame 8-36
Fence 5-5
File Format 15-11
FILEDIA 22-7
Fillet 3-14
FILLET Command 5-15
Filleting Objects with Different UCS 5-17
FILLETRAD system variable 22-2
FIND Command 7-37
FINDING AND REPLACING TEXT 7-37
floating pointing area 17-13
Floating Screen Pointing Area 17-13
floating screen pointing area 17-13
floating viewports 23-22
flyname 25-35
Force Line Inside 10-16
FORMATION OF BLOCKS 14-2
Formation of Blocks 14-2
Freeze in Active viewport 11-13
front clipping plane 21-34
FTP (file transfer protocol) 26-4
Full Preview 2-24
Full preview 12-12
FUNCTION AND CONTROL KEYS 4-45
FUNDAMENTAL DIMENSIONING TERMS 8-2

G

GENERATING CENTER MARKS AND CENTERLINES 8-24
GEOMETRIC CHARACTERISTICS AND SYMBOLS 8-34
Geometric Characteristics Symbol 8-36
Geometric Tolerance dialog box 8-35, 8-36
GEOMETRY DIALOG BOX 10-3
GLOBAL AND CURRENT LINETYPE SCALING 4-20
Global Editing of Attributes 15-19
Gopher 26-4
Grads 2-11
GRAPHSCR Command 6-25
GRID 22-2, 22-6, 23-2, 23-3
GRIDMODE 22-2
GRIP SYSTEM VARIABLES 6-13
Grip Types 6-2
GRIPBLOCK system variable 6-4
GRIPCOLOR system variable 6-4
GRIPHOT system variable 6-4
GRIPSIZE system variable 6-4
Group 5-6
Group Identification 18-2
Group Manager 18-1
GROUP MANAGER DIALOG BOX OPTIONS 18-2

H

hatch an area 24-24
hatch angle 24-23
hatch boundaries 24-24
HATCH Command 24-26
hatch description 24-22
hatch descriptors 24-22
hatch name 24-22
hatch pattern 24-23
hatch pattern definition 24-24, 24-26
hatch pattern library file 24-22
hatch pattern specification 24-24
HATCH PATTERN WITH DASHES AND DOTS 24-26
Hatch Patterns 13-2
hatch spacing 24-26
HATCH WITH MULTIPLE DESCRIPTORS 24-30
HATCHEDIT Command 13-21
Hatching 13-1
Hatching around Text, Traces, Attributes, Shapes, 13-21
header line 24-2, 24-22
HIDE Command 21-18
Hide Option 21-37
Highlighting Objects with URLs 26-22
Horizontal Option 8-12
Horizontal Placement of Text 10-27
HOW HATCH WORKS 24-24
HPANG system variable 13-7
HPDOUBLE system variable 13-8, 13-32
HPNAME system variable 13-15
HPSPACE system variable 13-7
Hyper Text Transfer Protocol 26-4
Hyperlink 26-2
hyperlink areas 26-16
HYPERLINK Command 26-15
Hyperlink Command 26-15
Hyperlink option 26-3
HyperlinkBack 26-2
Hyperlinkback 26-2
HyperLinkBase 26-2
Hyperlinkbase 26-2
HyperlinkFwd 26-2
Hyperlinkfwd 26-2
hyperlinks 26-1
Hyperlinks Inside AutoCAD LT 26-17
HyperlinkStop 26-2
Hyperlinkstop 26-2

I

icon 25-35
ID Command 6-30
id_big 25-34, 25-35
id_small 25-34, 25-35
Importing PCP/PC2 Configuration Files 12-18
included angle 3-3
INetCfg 26-3
INetLocation system variable 26-6
Information in a DXF File 19-3
initial drawing setup 23-1
INQUIRY COMMANDS 6-25
inscribed 3-23, 3-43
INSERT Command 14-8, 26-3, 26-11
Insert dialog box 26-11
Insert Hyperlink dialog box 26-16
Insert option 26-3
Inserting a Block from the Internet 26-11
Inserting Blocks 14-8
INSERTING BLOCKS WITH ATTRIBUTES 15-8
INSERTING TEXT FILES IN THE DRAWING 15-28
Inserting The LAYOUTS 11-15
InsertUrl 26-3
INSUNITS system variable 14-22
Internet applet 26-3
INTERSECT Command 21-21
INVOKING A SCRIPT FILE WHEN LOADING AUTOCAD LT 23-11
Invoking Commands in AutoCAD LT 1-10
Island Detection Method 13-17
Island Detection Style 13-15
Isometric Snap/Grid 4-27
ISOPLANE 22-2

J

Joining two adjacent viewports 11-5

L

Last 5-2
Last Updated Time 6-32

Launching A Web Browser 26-5
Layer Manager dialog box 26-12
Layer Properties Manager 22-11
Layer Properties Manager dialog box 4-3
Layers 4-1, 26-29
Layers, Colors, Linetypes, and Lineweights for Blocks 14-17
Layout Settings tab 22-13
LAYOUTWIZARD Command 11-19
Leader 8-4
Leader option 8-33
Learning Assistance 1-43
LENGTHEN Command 5-27
LENGTHEN Command 5-27
lens length 21-32
lightweight polylines 3-31
LIMITS 22-5, 22-9, 22-21, 23-2, 23-3
Limits 8-5, 10-24, 10-34, 22-19
LIMITS Command 2-16
LIMMAX 22-2
LIMMIN 22-2
LINE Command 1-14
line pattern 24-8
line style 24-8
LINEAR DIMENSIONING 8-10
Linear dimensioning 8-9
Lines and Arrows Tab 10-3
LINETYPE 24-9
LINETYPE Command 4-16
LINETYPE DEFINITION 24-2
linetype description 24-2
Linetype Manager dialog box 4-16, 26-12
linetype name 24-2
linetype specification 24-2
Lineweight Settings dialog box 4-16
Linking information into AutoCAD LT 19-17
Linking Objects 19-15
Linking Objects into AutoCAD LT 19-18
LIST Command 6-30
ListUrl 26-3
LOADING A TEMPLATE DRAWING 22-7
Loading an Image Tile Menu 25-24
LOADING HYPERLINKS 6-12
LOADING MENUS 25-23
Loading the hatch pattern 24-26
Location option 26-26
Locking the Display in the Viewports 11-11
LTSCALE 22-2, 22-9, 22-19, 23-2, 23-3, 24-10, 24-17
Ltscale and Dimscale 22-20
LTSCALE Command 24-9
LTSCALE factor 24-11
LTSCALE FACTOR FOR PLOTTING 4-21, 24-12
LTSCALE system variable 11-14
LWEIGHT Command 4-16
LWSCALE 12-21

M

macro 25-35
major arc 3-5
Making a Layer Current 4-5
Making a Viewport Current 11-4
MAKING INQUIRIES ABOUT A DRAWING (INQUIRY Commands) 6-25
Managing Contents Using the AutoCAD DesignCenter 6-18
Managing Raster Images (Image Manager dialog box) 19-6
MANIPULATING THE VISIBILITY OF VIEWPORT LAYERS 11-12
MASSPROP Command 21-40
MATCHING PROPERTIES 6-14
MATCHPROP Command 6-14
MatchProp Command 26-22
Material Condition Modifier 8-36
MEASURE Command 5-38
Menu 1-10
menu bar area 25-3
menu bar titles 25-8
Menu Customization dialog box 26-12
Menu-Specific Help 25-38
MENULOAD Command 25-29, 25-31, 25-36
MENUUNLOAD 25-31
Method Option 5-20
Metric Drawings 1-4
minor arc 3-5
MIRROR Command 5-35
MIRROR Command 5-35
Mirror Mode 6-10
MIRRTEXT system variable 5-36, 22-2
MMC 8-36
mnemonic key 25-29

model space 22-12, 23-22
Model Space and Paper Space Dimensioning 9-11
Model Space And Paper Space/Layouts 11-1
Model Space Viewports (Tiled Viewports) 11-2
Modify Dimension Style 22-10
MODIFYING LINETYPES 24-13
MOVE Command 5-8
Move Command 5-7
Move Mode 6-7
MSLIDE 23-22
MSLIDE Command 23-21
MSPACE 22-14
MSPACE Command 11-10
MTEXT Command 7-23
Mtext Option 8-11
Multiline Text Editor 9-4, 15-28
Multiple 5-7
multiple copies 5-8
MULTIPLE DOCUMENT ENVIRONMENT 2-5
multiple polyarcs 3-16
MVIEW 22-13
MYSTYLE 10-27

N

Name option 26-32
Named Views 26-30
namedview option 26-32
NEED FOR DIMENSIONING 8-1
Nesting of Blocks 14-18
new hatch pattern library file 24-22
Next Automatic Save In Time 6-32
non-modal dialog box 26-29
NURBS 17-7

O

Object Linking and Embedding 19-12
OBJECT PROPERTIES 4-18
Object Properties Toolbar 4-18
Object Selection Methods 1-29
Object Selection Modes 18-38
OBJECT SNAPS 4-29
Objects Properties Window 6-13
OFFSET Command 5-11
OFFSETDIST system variable 5-11
OLEHIDE system variable 19-22
OLESCALE Command 19-20
online help 25-38
OOPS Command 1-29
Open as read-only 2-4
OPEN Command 2-3
Open Command 26-3
OPENING AN EXISTING FILE 2-1
Opening Drawings from the Internet 26-6
OpenUrl 26-3
Optimized Polylines 3-31
optimized polylines 3-31
Options 2-26, 2-30, 24-28
Options dialog box 6-3
ORDINATE DIMENSIONING 8-25
Ordinate dimensioning 8-9
orient 25-34
Origin Offset 10-6
ORTHO 23-2, 23-3
Other Data Exchange Formats 19-5
Other DWF Viewing Options 26-24
Other Editing Commands 19-9
Outside Horizontal 10-12
Overriding the Running Snap 4-41

P

PAGESETUP Command 12-18
Pagesetup Command 11-18
Pan 26-28
PAN Command 7-10
PAn Option 21-31
Panning in Realtime 7-11
Paper space 22-12
PAPER SPACE LINETYPE SCALING (PSLTSCALE SYSTEM VAR 11-14
Paper space viewports (Floating Viewports) 11-7
PAPERUPDATE 12-19
parallel projection 21-22
partial menu 25-27
PARTIAL MENUS 25-27
Partial Plot Preview dialog box 2-24, 12-12
Partial Preview 12-12
PasteAsHyperlink 26-2
Pasteashyperlink 26-2
PasteAsHyperlink Command 26-21
PASTEBLOCK Command 5-10

PASTEORIG Command 5-10
Pasting As Hyperlink 26-21
Pattern Line 24-2
PCINWIZARD Command 12-18
PDMODE system variable 3-35
PDSIZE Command 3-37
PEDIT Command 18-15
PELLIPSE system variable 3-16
PERIMETER 6-29
perimeter 6-27
perspective display 21-29
perspective icon 21-29, 21-37
perspective view 21-22
Pick box 1-28
PICKADD system variable 18-40
PICKAUTO system variable 18-41
PICKDRAG system variable 18-40
PICKFIRST system variable 18-39
Play Animation 1-44
PLINE 22-9, 22-15, 22-22
PLINE Command 3-25
PLINEGEN system variable 18-28
PLINEWID system variable 22-9
PLOT Command 2-21, 12-1, 26-4
Plot dialog box 12-1, 12-2
Plot Offset 12-10
Plot Options area 12-11
Plot Preview 12-11, 12-12
Plot Scale area 12-10
Plot Settings Tab (Plot Dialog Box) 12-7
Plot Style Table Editor 12-6, 12-24
PLOTSTYLE Command 12-29
Plotter Configuration Editor 12-5
Plotter Configuration Editor dialog box 26-24
Plottermanager Command 12-13
PLOTTING DRAWINGS IN AUTOCAD LT 12-1
plug-in 26-24
pluginspage option 26-33
POINT Command 3-34
Point Style dialog box 3-35, 3-36
Polar Coordinate System 1-24
Polar Tracking 1-9
POLYGON Command 3-22
polyline 3-25
Polyline Editing 18-14
Polyline Option 5-17, 5-19
POLYSIDES 3-23
POP1 25-7
Positioning Dimension Text 10-26
Postscript Files 19-9
Prefix 10-20
PRELOADING SLIDES 23-24
Previous 5-2
Previous Option 2-8, 7-7
Primary Units 10-17
Project Option 5-22, 5-24
Projected Tolerance Zone 8-40
PROJMODE system variable 5-25
PROPERTIES Command 4-22, 6-13, 13-23
Properties settings dialog box 6-14
Properties Window 4-19
Properties Window (Dimension) 9-8
Properties Window (Leader) 9-10
Prototype Drawing 22-12, 22-15
PSETUPIN Command 12-21
PSLTSCALE Command 11-14
PSPACE 22-14
PSPACE Command 11-10
pull-down menu 25-22
pull-down menu or cursor menu 25-12
PURGE Command 14-29, 18-36

Q

QSELECT Command 6-16
QSelect Command 26-22
Quick Select dialog box 26-22
Quick Setup 1-4
Quitting AutoCAD LT 1-40

R

radians 2-11
RADIUS DIMENSIONING 8-23
Radius dimensioning 8-9
Radius Option 5-16
Raster Images 19-5
RAY 3-37
Ray Casting 13-17
RAY Command 3-40
Read-Only 2-4
Realtime Pan 2-8
Realtime Zoom 2-8
Realtime Zooming 7-3

RECTANG Command 3-12
RECTANGLE 3-12
Rectangular Array 5-29
REDO Command 18-34
REDRAW Command 7-2
REDRAWALL Command 7-2
Reference option 5-15
REGEN Command 2-7, 2-8, 7-2
REGENALL Command 7-2
Rregular Polygon 3-22
Relative Coordinate System 1-21
Relative Cylindrical Coordinate System 21-12
Relative Spherical Coordinate System 21-12
Remove 5-4
Remove option 26-3
Removing Hyperlinks 26-22
Removing Hyperlinks from Objects 26-22
REMOVING THE DIMENSION ASSOCIATIVITY 8-28
RENAME Command 18-35
Renaming Blocks 14-28
Resolution 26-25
RESUME Command 23-10
RESUME Command in transparent mode 23-11
REVCLOUD Command 17-6
Right-hand rule 20-1
Right-hand thumb rule 20-2
ROTATE Command 5-12
Rotate Mode 6-8
ROTATED DIMENSIONING 8-15
Rotated Option 8-12
rows 25-34
RSCRIPT Command 23-9
RUNNING OBJECT SNAP MODE 4-43

S

Save Command 26-3, 26-12
Save Drawing As 1-37
Save Drawing As dialog box 26-3
Savetime 1-13
SaveUrl 26-3
SAVING HATCH PATTERNS IN A SEPARATE FILE 24-34
Saving the Drawing to the Internet 26-12
Saving Your Work 1-35
Scale 10-9
SCALE Command 5-13, 5-14
Scale Mode 6-9
Scale Option 7-8
Scale to Paper Space 10-16
Scaling Raster Images 19-9
Scheme 26-4
script file name 23-4
Select Drawing File dialog box 26-11
Select File dialog box 26-3
Select First Line Option 5-19
Select First Object Option 5-16
Select Initial View 2-4
Select Object to Extend Option 5-24
Select Object to Trim Option 5-21
SELECTING DIMENSIONING COMMANDS 8-8
Selecting Groups 18-4
Selecting Hatch Boundary 13-10
Selection set 1-29
selection set 5-1
Selective Display of Layers 4-11
SelectUrl 26-3
servername 26-5
Setting Grid 4-23
SETTING ORTHO MODE 4-29
Setting Snap 4-26
Setting the CHAMFER System Variable 5-21
Setting the Chamfering System Variables 5-21
Setting the plot parameters 12-18
Setting the TRIMMODE System Variable 5-18
SETVAR 22-6, 22-9, 22-21, 23-2, 23-3
SHADEDGE system variable 21-39
SHADEDIF system variable 21-39
SHADEMODE Command 21-39
sheet sizes 2-17
Shortcut menu 25-18, 25-22
SHORTCUTMENU 25-19
Sign Convention 1-22, 1-24
SIMPLE HATCH PATTERN 24-24
SIngle 5-7
SLIDE LIBRARIES 23-27
slide library file 23-27
slide presentation 23-20
slide show 23-1
SLIDELIB 23-26, 23-28
slider bar 21-29
SNAP 22-6, 23-2, 23-3
SNAPBASE system variable 13-27

Space Delimited File (SDF) 15-11
Specifying a New Path for Hatch Pattern Files 24-28
Specifying an Alternate Default Font 7-33
Specifying Extension Line Origins 8-13
SPELL Command 7-35
SPLFRAME system variable 18-25
SPLINE Command 17-6
SPLINEDIT Command 17-7
SPLINETYPE 18-26
src option 26-32
STANDARD PULL-DOWN MENUS 25-2
standard template drawings 22-2
Starting a Drawing 1-2
Starting a New Drawing 1-37
Starting AutoCAD LT 1-1
Startup 1-2
STRETCH Command 5-26
Stretch Mode 6-4
STYLE Command 7-33
STYLESMANAGER Command 12-22
Submenu Definition 25-22
Submenu Reference 25-23
SUBMENUS 25-22
submenus 25-26
SUBSTITUTING FONTS 7-31
SUBTRACT Command 21-20
Suffix 10-18
Support File Search Path 24-28
Surveyor's units 2-11, 2-29
Symbol Libraries 1-7
SYSTEM VARIABLES 20-23

T

TABLET 17-11
TABLET Command 17-12
Tablet mode 17-11
Tablet Template 1-13
TArget Option 21-27
tbarname 25-34
Telnet protocol 26-4
template drawing 22-7
template drawings 22-1
template file 15-12
Template URL 26-8
Templates 1-3
Temporary model space 11-9
terminators 8-3
TEXT Command 3-41, 7-16
Text Height 10-10
Text height 22-20
Text Mirroring 5-36
Text Option 8-11
Text Options 15-4
TEXT QUALITY AND TEXT FILL 7-36
Text Style dialog box 7-33
TEXTFILL system variable 7-37
TEXTQLTY system variable 7-37
TEXTSCR system variable 6-25
TEXTSIZE system variable 23-3
The 3 Points Option 3-2
The Center and Radius Option 1-31
The Center of Polygon Option 3-23
The Center, Start, Angle Option 3-9
The Center, Start, End Option 3-9
The Center, Start, Length Option 3-9
The Concept of Blocks 14-1
The Crossing Option 1-30
The Drawing Web Format 26-23
The Edge Option 3-24
THE OVERLAY OPTION 16-13
THE STANDARD TEMPLATE DRAWINGS 22-2
The Start, Center, Angle Option 3-3
The Start, Center, End Option 3-3
The Start, Center, Length Option 3-5
The Start, End, Angle Option 3-6
The Start, End, Direction Option 3-6
The Start, End, Radius Option 3-8
The Tangent Tangent Radius Option 1-33
The Three-Point Option 1-33
The Two-Point Option 1-32
The Undo Option 1-17
The Web Area 26-13
The Window Option 1-29
THE WORLD COORDINATE SYSTEM 20-3, 20-4
Thickness 3-15
TILEMODE system variable 11-3, 22-2, 22-13, 22-15
Tiles 1-13
TIME Command 6-31
Tolerance 10-22
TOLERANCE Command 8-35

Tolerance Value and Tolerance Zone Descriptor 8-36
Tolerances 8-5
Toolbar 25-34
toolbar 25-33
Toolbar Definition 25-33
Total Editing Time 6-32
TRIM Command 5-21
TRIM, EXTEND, AND FILLET COMMANDS IN 3D 21-16
Trim Option 5-20
Trimming and Extending with Text, Region, or Splin 5-25
TRIMMODE 5-18
TRIMMODE system variable 5-20

U

U (Undo) Command 1-29
UCS Command 20-6
UCS Dialog Box 20-19
UCS icon 21-38
UCSAXISANG 20-23
UCSFOLLOW 20-23
UCSFOLLOW SYSTEM VARIABLE 11-7
UCSICON 20-23
UCSICON Command 20-4
UCSMAN Command 20-19
UCSNAME 20-23
UCSORG 20-23
UCSXDIR 20-23
UCSYDIR 20-23
Understanding URLs 26-4
Undo 5-7
UNDO Command 18-30
Undo Option 5-24
Uniform Resource Locator (URL) 26-1
UNION Command 21-20
UNITS Command 2-9
Updating Dimensions 9-7
URL 26-4
URLLAYER layer 26-23
USENET 26-4
User Defined Hatch Patterns 13-8
User Defined Page Setups dialog box 12-3
Using AutoCAD DesignCenter to Insert Blocks 14-16
USING AutoTracking 4-43
USING DIMENSION STYLE OVERRIDES 10-30
Using Dimension Styles 10-30
USING EXTERNALLY REFERENCED DIMENSION STYLES 10-32
USING FEATURE CONTROL FRAMES WITH LEADERS 8-40
Using Hyperlinks with AutoCAD LT 26-15
USING PLOT STYLES 12-21
USING STYLES AND VARIABLES TO CONTROL DIMENSIONS 10-1
Using the DDEDIT Command 15-7
Using the Properties Window 15-7
Using the Toolbar and the Menu 8-8

V

Vertical Option 8-12
View dialog box 7-12
View option 26-32
Viewing DWF Files 26-27
Viewports toolbar 22-16
virtual keys 25-33
visible 25-34
Volo View Express 26-24
VPLAYER Command 11-12
VPOINT 22-14
VPOINT Command 21-4
VPORTS 22-16
VPORTS Command 11-3
VSLIDE Command 23-22

W

Web 26-4
Web browser 26-1
Web format (DWF) files 26-1
WHAT ARE SCRIPT FILES? 23-1
WHAT ARE SLIDES? 23-20
WHAT IS A SLIDE SHOW? 23-20
Whip plug-in 26-1
WHIP! v 3.1 Limitations 26-2
Width 3-14
Width and Height options 26-32
Window Option 2-8, 7-8
Working with OLE Objects 19-20
WORLDUCS 20-23

WPolygon 5-3
WRITING A PULL-DOWN MENU 25-3

X

XATTACH 16-16
XLINE 3-37
XLINE Command 3-38
XLINE Options 3-38
XPLODE Command 14-26
xval 25-34

Y

yval 25-34

Z

Zero Suppression 10-20
ZOOM 23-2, 23-3
Zoom 26-28
ZOOM Command 7-3
ZOOM Command 2-7
Zoom Extents Command 26-31
Zoom In and Out 7-10
Zoom Option 21-32